Hilbert Space
and
Quantum Mechanics

Hilbert Space
and
Quantum Mechanics

Franco Gallone

Università degli Studi di Milano, Italy

World Scientific

NEW JERSEY · LONDON · SINGAPORE · BEIJING · SHANGHAI · HONG KONG · TAIPEI · CHENNAI

Published by

World Scientific Publishing Co. Pte. Ltd.

5 Toh Tuck Link, Singapore 596224

USA office: 27 Warren Street, Suite 401-402, Hackensack, NJ 07601

UK office: 57 Shelton Street, Covent Garden, London WC2H 9HE

Library of Congress Cataloging-in-Publication Data

Gallone, Franco, author.

   Hilbert space and quantum mechanics / Franco Gallone, Università degli Studi di Milano, Italy.

    pages cm

   Includes bibliographical references and index.

   ISBN 978-9814635837 (hardcover : alk. paper)

   1. Hilbert space. 2. Quantum theory--Mathematics. 3. Linear operators. 4. Mathematical physics.

5. Nonrelativistic quantum mechanics. I. Title.

   QC174.17.H55G35 2015

   515'.733--dc23

                         2014040049

British Library Cataloguing-in-Publication Data

A catalogue record for this book is available from the British Library.

In-house Editor: Ng Kah Fee

Printed in Singapore

To Kissy, Lilith, Malcy, Micio,
who taught me how to stay focused

Preface

The subjects of this book are the mathematical foundations of non-relativistic quantum mechanics and the mathematical theory they require. In its mathematical part, this book aims at expounding in a complete and self-contained way the mathematical basis for "mathematical" quantum mechanics, namely the branch of mathematical physics that was constructed by David Hilbert, John von Neumann and other mathematicians, notably George Mackey, in order to systematize quantum mechanics, and which was presented in book form for the first time by von Neumann in 1932 (Neumann, 1932). In von Neumann's approach, the language of quantum mechanics is the theory of linear operators in Hilbert space.

Von Neumann's book was the result of work which had been done previously over several years. Hilbert, who had been consulted on numerous aspects of quantum mechanics since its inception, began in 1926 a systematic study of its mathematical foundations. Hilbert taught the course "Mathematical Methods of Quantum Theory" in the academic year 1926-27, and a summary of Hilbert's lessons was published in the spring of 1927 by Hilbert himself and his assistants Lothar Nordheim and von Neumann (Hilbert *et al.*, 1927). In their view, the mathematical framework suitable for quantum mechanics was the mathematical structure that was defined in an abstract way and called a Hilbert space by von Neumann in 1927. Furthermore, between 1926 and 1932, von Neumann proved a number of theorems about operators in Hilbert space which bore upon quantum mechanics (among them, the spectral theorem for unbounded self-adjoint operators), and so did the mathematicians Marshall Stone and Hermann Weyl, who had a keen interest in quantum mechanics. Thus, the theory of linear operators in Hilbert space was actually born as the mathematical basis for quantum mechanics.

Quantum mechanics and the theory of Hilbert space operators constitute one of those rare examples in which there is complete correspondence between physical and mathematical concepts (another example is Euclidean geometry). Actually, it is one of the most stunning examples of "the unreasonable effectiveness of mathematics in the natural sciences" (E.P. Wigner). Unfortunately, this aspect of quantum mechanics is almost completely overlooked in most quantum mechanics textbooks, where too many subtle points are dealt with by means of mathematical shortcuts which not only can hardly convince a mathematically aware reader but also blot out physical subtleties. The main reason for this is that, in the community of physicists, Dirac's quantum mechanics (Dirac, 1958, 1947, 1935, 1930) is by far more popular than von Neumann's quantum mechanics, perhaps exactly because the former requires almost no mathematics. For instance, the idea that self-adjoint operators have a critical domain is almost completely missing in standard quantum mechanics textbooks; however, the domain of an unbounded self-adjoint operator represents exactly the pure states in which the fundamental statistical quantities (expected result and uncertainty) are defined for the observable represented by that operator. This point gets hopelessly blurred in most quantum mechanics books, which treat unbounded observables — like energy, position, momentum, orbital angular momentum — as if they were represented by self-adjoint operators defined on the entire space, while this is impossible on account of the Hellinger–Toeplitz theorem. Another example is the relation existing between the physical idea of compatibility of two observables and the mathematical idea of commutativity of the operators that represent them; for self-adjoint operators, the right notion of commutativity is subtler than the one usually found in quantum mechanics books and depends on the representations of the operators as projection valued measures; however it is exactly through this subtler notion that the physical essence of compatibility can be really grasped. More than anything else, the real way to understand why quantum observables are represented by self-adjoint operators is through the spectral theorem, since quantum observables arise most naturally as projection valued measures, but this is usually outside the scope of standard quantum mechanics books.

One last word about the mathematical framework for quantum mechanics presented in this book. It is undoubtedly very interesting and useful to treat quantum mechanics in the framework of mathematical structures more general than Hilbert space theory, especially in order to study quantum mechanics of systems with an infinite number of degrees of freedom. However, quantum mechanics in Hilbert space is an enthralling subject in its own right, mainly because it is here that one can see most clearly how the mathematical structure is linked to the physical theory in an almost necessary way.

Most books about fundamental quantum mechanics use results in the theory of Hilbert space operators without proving them, while most books about Hilbert space operators do not treat quantum mechanics; moreover, they often use fairly advanced results from other branches of mathematics assuming the reader to be already familiar with them, but this is seldom true. The aim of this book is not to be a complete treatise about Hilbert space operators, but to give a really self-contained treatment of all the elements of this subject that are necessary for a sound and mathematically accurate exposition of the principles of quantum mechanics; this exposition is the object of the final chapters of the book. The main characteristic of the book is that the mathematical theory is developed only assuming familiarity with elementary analysis. Moreover, all the proofs in the book are carried out in a very detailed way. These features make the book easily accessible to readers with only the mathematical experience offered by undergraduate education in mathematics or in physics, and also ideal for individual study. The principles of quantum mechanics are discussed with complete mathematical accuracy and an effort is always made to trace them back to the experimental reality that lies at their root. The treatment of quantum mechanics is axiomatic, with definitions followed by propositions proved in a mathematical fashion. No previous knowledge of quantum mechanics is required. The level of this book is intermediate between advanced undergraduate and graduate. It is a purely theoretical book, in which no exercises are provided.

After the first chapter, whose function is mainly to fix notation and terminology, the first part of the book (Chapters 2–9) is devoted to an exposition of the elements of real and abstract analysis that are needed later in the study of operators in Hilbert space. The reason for this is to make it really self-contained and avoid proving theorems by means of other fairly advanced theorems outside this book. In particular, the chapter devoted to metric spaces (Chapter 2) contains results which are not completely elementary but are necessary in order to prove (in Chapter 6) the theorem about Borel functions that plays an essential role in proving the spectral theorems (in Chapter 15). The chapters about measure and integration (Chapters 5–9) contain results about extensions of measures which are not to be found in first level books on measure theory but which are essential in order to study commuting self-adjoint operators, and also the Riesz–Markov theorem about positive linear functionals which plays an essential role in proving the spectral theorems. Actually, Chapters 1–2 and 5–9 could by themselves be a short book about measure and integration. Chapters 3 and 4 deal with that part of the theory of linear operators in normed spaces that is used later in the study of Hilbert space operators. Moreover, the Stone–Weierstrass approximation theorem is proved in Chapter 4; this theorem plays an essential role in proving the spectral theorems.

The second part of this book (Chapters 10–18) is its core, and contains a treatment of the theory of linear operators in Hilbert space which is particularly well suited for the discussion of the mathematical foundations of quantum mechanics presented later in the book. It contains the spectral theorems for unitary and for self-adjoint operators, one-parameter unitary groups and Stone's theorem, theorems about commuting operators and invariant subspaces, trace class operators, and also Wigner's theorem and the real line special case of Bargmann's theorem about automorphisms of projective Hilbert spaces.

The theory of Hilbert space operators is the backbone of the third and final part of the book, which consists of two chapters (19 and 20). The first of these is by far the longest chapter in the book and endeavours to present the principles of non-relativistic quantum mechanics in a mathematically accurate way, with also an unstinting effort to present some possible physical reasoning behind the constructs that are considered. Since the predictions provided by quantum mechanics are in general statistical ones, in the first part of this chapter general statistical ideas are introduced and it is examined how these ideas are implemented in classical theories; later in the chapter, the statistical aspects of quantum mechanics are compared and contrasted with the same aspects of classical theories. The final chapter deals with an important example of how quantum observables can arise in connection with symmetry principles; moreover, it presents the Stone–von Neumann uniqueness theorem about canonical commutation relations.

Although the book's length might make it difficult to use it as a textbook for a single course, parts of it can easily be used in that way for various courses. Here are some concrete suggestions:

- Chapters 1, 2, 5, 6, 7, 8, 9 for a one-semester course in Real Analysis or in Measure Theory (intermediate, could be either undergraduate or graduate, mathematics);
- Chapters 3, 4, 10, 11, 12, 13, 14, 15, 16, 17, 18 for a two-semester course in Operators in Hilbert Space (graduate, mathematics and physics);
- Chapters 19, 20 (using without proof a large number of results from the previous chapters) for a one-semester course in Mathematical Foundations of Quantum Mechanics (graduate, mathematics and physics).

To make cross-reference as easy as possible, almost every bit of this book is marked with three numbers, the first for the chapter, the second for the section, and the third for the position within the section. Comments also are marked in this way, and they are called "remarks". As already mentioned, all the proofs in this book are written in minute detail; in them, however, previous results are always quoted simply by means of the three numbers code, without spelling them out. This should enable experts to pursue the logic of a proof without too many diversions, and beginners to receive all the support they might need.

I wish to thank Roberto Palazzi for the great job he did of preparing the LaTeX files for the book, and also for useful mathematical comments.

Franco Gallone

Contents

Chapter 1

Sets, Mappings, Groups

Most readers are likely to have a working familiarity with most of the subjects of this introductory chapter. For them, the main function of this chapter is to fix the notation and the terminology that will be used throughout this book and provide ready reference inside the book.

1.1 Symbols, sets, relations

The reader is assumed to be already familiar with the topics of this section, which is only intended for future reference.

1.1.1 *Sets of numbers*

Symbol	Meaning
$\mathbb{N}$	the set of all positive integers, i.e. $\{1, 2, 3, ...\}$
$\mathbb{Z}$	the set of all integers, i.e. $\{0, \pm 1, \pm 2, ...\}$
$\mathbb{Q}$	the set of all rational numbers, i.e. $\{m/n : m, n \in \mathbb{Z}$ and $n \neq 0\}$
$\mathbb{R}$	the set of all real numbers
$[0, \infty)$	the set of all non-negative real numbers
$(0, \infty)$	the set of all positive real numbers
$\mathbb{C}$	the set of all complex numbers

The complex field is always meant to be $\mathbb{R}^2$ endowed with the two operations:

$$(a_1, a_2) + (b_1, b_2) = (a_1 + b_1, a_2 + b_2),$$
$$(a_1, a_2)(b_1, b_2) = (a_1 b_1 - a_2 b_2, a_1 b_2 + a_2 b_1),$$

and $\mathbb{C}$ denotes the set $\mathbb{R}^2$ when $\mathbb{R}^2$ is endowed in this way.

For a complex number $z := (a_1, a_2)$, we define:

$$\operatorname{Re} z := a_1, \operatorname{Im} z := a_2, \overline{z} := (a_1, -a_2), |z| := \sqrt{a_1^2 + a_2^2}.$$

The subset $\{(a, 0) : a \in \mathbb{R}\}$ of $\mathbb{C}$ is identified with $\mathbb{R}$, identifying $(a, 0)$ with a. With this identification, for a complex number z we have $\overline{z}z = z\overline{z} = |z|^2$, and the absolute value of a real number a coincides with $|a|$. Identifying $a \in \mathbb{R}$ with $(a, 0)$ and defining $i := (0, 1)$, we also have $(a_1, a_2) = a_1 + ia_2$. When for a complex number z we write $0 \le z$ (or $0 < z$, $z \le 0$, $z < 0$), we mean $\operatorname{Im} z = 0$ and $0 \le \operatorname{Re} z$ (or $0 < \operatorname{Re} z$, $\operatorname{Re} z \le 0$, $\operatorname{Re} z < 0$). More generally, outside the chapters devoted to measure and integration, when for a symbol x we write $0 \le x$ or $x \ge 0$ we mean $x \in [0, \infty)$; similarly, by $0 < x$ or $x > 0$ we mean $x \in (0, \infty)$. However, in chapters from 5 to 9 by $0 \le x$ or $x \ge 0$ we mean $x \in [0, \infty]$ and by $0 < x$ or $x > 0$ we mean $x \in (0, \infty]$ (i.e. we allow the case $x = \infty$; cf. 5.1.1).

It is always understood that the square root of a positive real number is taken to be positive.

1.1.2 *Proofs*

A proposition is a statement that is either true or false (but not both). By means of logical connectives and brackets, a new proposition can be defined starting from one or more given propositions. We assume known to the reader the logical connectives: "not", "and", "or" ("A or B" means "A or B or both"), "$\Rightarrow$" (if, then), "$\Leftrightarrow$" (if and only if).

Given two propositions P, Q, the proposition $P \Rightarrow Q$ is logically equivalent to the proposition $(\operatorname{not} Q) \Rightarrow (\operatorname{not} P)$, which is called the contrapositive form of $P \Rightarrow Q$. A proof that $(\operatorname{not} Q) \Rightarrow (\operatorname{not} P)$ is true, is called *proof by contraposition* of $P \Rightarrow Q$. The proposition $P \Rightarrow Q$ is also logically equivalent to the proposition $[P \text{ and } (\operatorname{not} Q)] \Rightarrow [R \text{ and } (\operatorname{not} R)]$, for any proposition R. A proof that there is a proposition R such that $[P \text{ and } (\operatorname{not} Q)] \Rightarrow [R \text{ and } (\operatorname{not} R)]$ is true, is called *proof by contradiction* of $P \Rightarrow Q$.

Suppose that, for each positive integer n, we are given a proposition P_n. From the principle of induction it follows that, if the propositions

(a) P_1,
(b) $P_n \Rightarrow P_{n+1}$ is true for each positive integer n

are true, then the proposition

(c) P_n is true for each positive integer n

is true. A proof that propositions a and b are true is called *proof by induction* of proposition c.

Often, for a proposition P, we will write "P" instead of "P is true" or "P holds". Propositions will be written in a rather informal style, mixing logical symbols and ordinary language.

1.1.3 *Symbols and shorthand*

Symbol	Meaning
$x \in S$	x is an element of the set S
	(x is also said to be a *point in* S or a *point of* S)
$x \notin S$	not ($x \in S$)
$\forall x \in S$	for every element x of the set S
$\exists x \in S$	there exists at least one element x of the set S
$\exists! x \in S$	there exists one and only one element x of the set S
$A := B$	A equals B by definition of A
$B =: A$	A equals B by definition of A
s.t.	such that
iff	if and only if
i.e.	that is to say
cf.	see, recall
e.g.	for example

The symbols "$\forall$", "$\exists$", "$\in$" are often used collectively: instead of writing

$$\text{“}\exists x \in S, \exists y \in S\text{” or “}\forall x \in S, \forall y \in S\text{”}$$

one often writes

$$\text{“}\exists x, y \in S\text{” or “}\forall x, y \in S\text{”}.$$

The expressions "$\forall x \in S$" and "for $x \in S$" are regarded as equivalent.

When $n \in \mathbb{N}$, "for $k \in \{1, ..., n\}$" is often written as "for $k = 1, ..., n$".

In definitions, "if" means "if and only if".

When, for a symbol x, we write "$\exists x \geq 0$", or "$\exists x > 0$", or "$\forall x \geq 0$", or "$\forall x > 0$", we mean "$\exists x \in [0, \infty)$", or "$\exists x \in (0, \infty)$", or "$\forall x \in [0, \infty)$", or "$\forall x \in (0, \infty)$" respectively if we are not in chapters from 5 to 9; in chapters from 5 to 9, we mean "$\exists x \in [0, \infty]$", or "$\exists x \in (0, \infty]$", or "$\forall x \in [0, \infty]$", or "$\forall x \in (0, \infty]$" respectively (cf. 5.1.1).

If $I = \{1, ..., N\}$ or $I := \mathbb{N}$, we will often write "$\sum_{n \in I}$" for "$\sum_{n=1}^{N}$" or for "$\sum_{n=1}^{\infty}$".

1.1.4 *Sets*

The words *family* and *collection* will be used synonymously with *set*, e.g. in order to avoid phrases like "set of sets".

The *empty set* is denoted by $\emptyset$, and the *family of all subsets of a set* X is denoted by $\mathcal{P}(X)$. If X is a set and if, for each $x \in X$, $P(x)$ is a proposition involving x, then

$$\{x \in X : P(x)\}$$

denotes the set of all elements x of X for which $P(x)$ is true.

$\{a, b, c, ...\}$ denotes the set which contains the elements that are listed, and $\{x\}$ denotes the set which contains just x (such a set is called a *singleton set*).

For two subsets S_1, S_2 of a set X, we use the following symbols:

Symbol	Meaning
$S_1 \subset S_2$	$x \in S_1 \Rightarrow x \in S_2$
	(S_1 is said to be a *subset* of S_2 or to be *contained* by S_2)
$S_2 \supset S_1$	$S_1 \subset S_2$
$S_1 \not\subset S_2$	not $(S_1 \subset S_2)$, i.e. $\exists x \in S_1$ s.t. $x \notin S_2$
$S_1 = S_2$	$(S_1 \subset S_2)$ and $(S_2 \subset S_1)$, i.e. $x \in S_1 \Leftrightarrow x \in S_2$
$S_1 \neq S_2$	not $(S_1 = S_2)$, i.e. $(S_1 \not\subset S_2)$ or $(S_2 \not\subset S_1)$

If $\mathcal{F}$ is a family of subsets of a set X, we define the *union* and the *intersection* of $\mathcal{F}$:

$$\cup_{S \in \mathcal{F}} S := \{x \in X : \exists S \in \mathcal{F} \text{ such that } x \in S\},$$
$$\cap_{S \in \mathcal{F}} S := \{x \in X : x \in S \text{ for all } S \in \mathcal{F}\}.$$

For a finite family $\mathcal{F} = \{S_1, S_2, ..., S_n\}$, we write

$$S_1 \cup S_2 \cup \cdots \cup S_n := \cup_{S \in \mathcal{F}} S,$$
$$S_1 \cap S_2 \cap \cdots \cap S_n := \cap_{S \in \mathcal{F}} S.$$

If $\mathcal{F}$ is the empty family, we define

$$\cup_{S \in \mathcal{F}} S := \emptyset \text{ and } \cap_{S \in \mathcal{F}} S := X.$$

A family $\mathcal{F}$ of subsets of a set X is said to be *disjoint*, and its elements are said to be disjoint or pairwise disjoint, if $S \cap S' = \emptyset$ for all $S, S' \in \mathcal{F}$ such that $S \neq S'$. For two subsets S, T of a set X, directly from the definitions we have:

$$S \subset T \Leftrightarrow S = S \cap T \Leftrightarrow T = S \cup T.$$

For a subset T of a set X and a family $\mathcal{F}$ of subsets of X, directly from the definitions we have:

$$T \cap (\cup_{S \in \mathcal{F}} S) = \cup_{S \in \mathcal{F}} (T \cap S),$$
$$T \cup (\cap_{S \in \mathcal{F}} S) = \cap_{S \in \mathcal{F}} (T \cup S),$$
$$T \cap (\cap_{S \in \mathcal{F}} S) = \cap_{S \in \mathcal{F}} (T \cap S),$$
$$T \cup (\cup_{S \in \mathcal{F}} S) = \cup_{S \in \mathcal{F}} (T \cup S).$$

For two subsets S_1, S_2 of a set X we define the *difference* of S_2 and S_1:

$$S_2 - S_1 := \{x \in X : x \in S_2 \text{ and } x \notin S_1\};$$

clearly, $S_1 \cap (S_2 - S_1) = \emptyset$.

For a subset S of a set X, $X - S$ is called the *complement* of S in X; we have

$$S \cap (X - S) = \emptyset, X - (X - S) = S, X = S \cup (X - S).$$

For a family $\mathcal{F}$ of subsets of X we have *De Morgan's laws*
$$X - (\cup_{S \in \mathcal{F}} S) = \cap_{S \in \mathcal{F}} (X - S) \text{ and } X - (\cap_{S \in \mathcal{F}} S) = \cup_{S \in \mathcal{F}} (X - S).$$
For two subsets S_1, S_2 of a set X we have, directly from the definitions,
$$S_1 \subset S_2 \Leftrightarrow X - S_2 \subset X - S_1,$$
$$S_1 \cap S_2 = \emptyset \Leftrightarrow S_1 \subset X - S_2,$$
$$S_2 - S_1 = S_2 \cap (X - S_1);$$
then we also have
$$(S_2 - S_1) \cup S_1 = (S_2 \cap (X - S_1)) \cup S_1 = S_2 \cup S_1,$$
$$S_2 - (S_2 - S_1) = S_2 \cap (X - (S_2 \cap (X - S_1))) = S_2 \cap ((X - S_2) \cup S_1) = S_2 \cap S_1;$$
hence, if $S_1 \subset S_2$ we have
$$(S_2 - S_1) \cup S_1 = S_2,$$
$$S_2 - (S_2 - S_1) = S_1,$$
and this implies
$$X - S_1 = X - (S_2 \cap (X - (S_2 - S_1))) = (X - S_2) \cup (S_2 - S_1).$$
Then, for three subsets S_1, S_2, S_3 of X such that $S_1 \subset S_2 \subset S_3$ we have
$$S_3 - S_1 = S_3 \cap ((X - S_2) \cup (S_2 - S_1)) = (S_3 - S_2) \cup (S_2 - S_1).$$

1.1.5 *Relations*

If X and Y are sets, the *cartesian product* of X and Y, written $X \times Y$, is the set of all ordered pairs (x, y) with $x \in X$ and $y \in Y$.

A *relation* in a non-empty set X is a subset R of $X \times X$. If $(x, y) \in R$, we write xRy and say that x is *related by R to y*. If S is a subset of X, then $R \cap (S \times S)$ is a relation in S which is called the relation *induced* by R in S.

A relation R in a set X is said to be an *equivalence relation* if it has the following three properties:

(er_1) $xRx, \forall x \in X$ (R is *reflexive*);
(er_2) $xRy \Rightarrow yRx$ (R is *symmetric*);
(er_3) (xRy and yRz) $\Rightarrow xRz$ (R is *transitive*).

A symbol often used for an equivalence relation is "$\sim$".

Let X be a set equipped with an equivalence relation $\sim$ and let $x \in X$. The *equivalence class* of x for $\sim$ is the set
$$[x] := \{y \in X : y \sim x\},$$
and any element of $[x]$ is called a *representative* of $[x]$.

The following facts can be easily proved:

(a) $x \in [x], \forall x \in X$; thus, every equivalence class is nonempty and $X = \cup_{x \in X}[x]$;
(b) either $[x] = [y]$ or $[x] \cap [y] = \emptyset$ (but not both), $\forall x, y \in X$;
(c) $[x] = [y] \Leftrightarrow x \sim y$;

we notice that, by assertion b, the contrapositive form of statement c is

$$[x] \cap [y] = \emptyset \Leftrightarrow \text{not } (x \sim y).$$

A *partition* of a set X is a family $\mathcal{F}$ of subsets of X which has the following three properties:

(pa_1) $S \neq \emptyset, \forall S \in \mathcal{F}$;
(pa_2) $(S_1, S_2 \in \mathcal{F}, S_1 \neq S_2) \Rightarrow S_1 \cap S_2 = \emptyset$;
(pa_3) $\cup_{S \in \mathcal{F}} S = X$.

Thus, if X is a non-empty set equipped with an equivalence relation, the family of equivalence classes constitute a partition of X. Conversely, it is straightforward to prove that, if $\mathcal{F}$ is a partition of non-empty a set X, then the set

$$R := \{(x, y) \in X \times X : \exists S \in \mathcal{F} \text{ such that } x \in S \text{ and } y \in S\}$$

is an equivalence relation in X and $\mathcal{F}$ is the family of equivalence classes defined by R.

If $\sim$ is an equivalence relation in a non-empty set X, the family of equivalence classes defined by $\sim$ is called the *quotient set* of X by the relation $\sim$ and is denoted by $X/\sim$.

A relation R in a non-empty set X is said to be a *partial ordering* if it has the following three properties:

(po_1) $xRx, \forall x \in X$ (R is *reflexive*);
(po_2) $(xRy \text{ and } yRx) \Rightarrow x = y$ (R is *antisymmetric*);
(po_3) $(xRy \text{ and } yRz) \Rightarrow xRz$ (R is *transitive*).

A partial ordering is called a *total ordering* if it has the following further property:

(po_4) $(xRy \text{ or } yRx), \forall x, y \in X$.

A symbol often used for a partial ordering is "$\leq$". A *partially ordered set* is a pair $(X, \leq)$, where X is a non-empty set and $\leq$ is a partial ordering in X.

Let $(X, \leq)$ be a partially ordered set, S a non-empty subset of X, and x a point of X; the following terms are used:

x is called an *upper bound* for S if $y \leq x$ for each $y \in S$;

x is called a *lower bound* for S if $x \leq y$ for each $y \in S$;

x is called a *least upper bound* (*l.u.b.*) for S if x is an upper bound for S and if, for every upper bound x' for S, we have $x \leq x'$; if a l.u.b. for S exists, then (as can be readily seen) it is the unique l.u.b. for S and is denoted by $\sup S$; if the l.u.b. of S exists and it is an element of S, we write $\max S := \sup S$;

x is called a *greatest lower bound* (*g.l.b.*) for S if x is a lower bound for S and if, for every lower bound x' for S, we have $x' \leq x$; if a g.l.b. for S exists, then it is the unique g.l.b. for S and is denoted by inf S; if the g.l.b. of S exists and it is an element of S, we write min $S := \inf S$.

In the family $\mathcal{P}(X)$ of all subsets of a set X, a relation R is defined by letting

$$R := \{(S_1, S_2) \in \mathcal{P}(X) \times \mathcal{P}(X) : S_1 \subset S_2\}.$$

For $S_1, S_2 \in \mathcal{P}(X)$, one writes $S_1 R S_2$ directly as $S_1 \subset S_2$. This relation is a partial ordering and, for a non-empty subfamily $\mathcal{F} \subset \mathcal{P}(X)$, both sup $\mathcal{F}$ and inf $\mathcal{F}$ exist and in fact

$$\sup \mathcal{F} = \cup_{S \in \mathcal{F}} S \text{ and } \inf \mathcal{F} = \cap_{S \in \mathcal{F}} S.$$

1.2 Mappings

In this section we give a methodical treatment of the subject, since some of the concepts which are contained in this section might not be utterly familiar to all readers. Indeed, for two sets X and Y, we consider mappings from X to Y which are defined on any subset of X. This foreshadows what will happen in the study of linear operators in Hilbert space, where we use the definitions, notations and results of this section.

1.2.1 Definitions. Let X and Y be non-empty sets. A *mapping* φ from X to Y is a rule which assigns to each element x of a non-empty subset D_φ of X a single element of Y, called the *value* of φ at x and denoted by $\varphi(x)$. The set X is called the *initial set* of φ, and Y the *final set* of φ. The mapping φ is said to be *defined in X* and to *have values in Y*. The set D_φ is called the *domain* of φ, and φ is said to be *defined on D_φ*. To indicate that φ is a mapping from X to Y, we write

$$\varphi : D_\varphi \to Y \text{ with } D_\varphi \subset X,$$

or simply

$$\varphi : D_\varphi \to Y$$

if it is already clear that $D_\varphi \subset X$.

If $D_\varphi = X$, we write

$$\varphi : X \to Y$$

and we say that φ is a mapping *on X*.

The *range* of φ is the subset of Y defined by

$$R_\varphi := \{y \in Y : \exists x \in D_\varphi \text{ s.t. } y = \varphi(x)\}.$$

It should be clear that, while D_φ and R_φ are two sets which are completely determined by the mapping φ (D_φ is indeed part of the definition of φ), the initial set X and the final set Y can be replaced with two different sets X' and Y' as long as $D_\varphi \subset X'$ and $R_\varphi \subset Y'$, without altering φ in any essential way. The choice of what sets to use as initial and final sets is often made on the grounds of particular properties they possess, or in order to have a common playground for several mappings.

Mappings are sometimes given different names. A mapping is also called a *map* or a *function*, and we will use the latter name especially when the final set is $\mathbb{C}$ or $\mathbb{R}^*$ (cf. 5.1.1), or some subset of them. When the final set is $\mathbb{C}$ (or $\mathbb{R}$) we sometimes say that the mapping is a *complex* (or a *real*) *function*. A mapping from a cartesian product of two sets to one of them is occasionally called a *binary operation*. A mapping $\varphi : \mathbb{N} \to X$, where X is a non-empty set, is called a *sequence* in X and is denoted by the symbol $\{x_n\}$, where $x_n := \varphi(n)$. Sometimes, given a non-empty set X and a non-empty set I which for psychological reasons we like to think about as a set of indices, the range of a mapping $\varphi : I \to X$ is denoted by the symbol $\{x_i\}_{i \in I}$, where $x_i := \varphi(i)$, and is referred to as a *family* of elements of X *indexed* by the set I. If a family $\mathcal{F}$ of subsets of a set X is obtained in this way, i.e. if $\mathcal{F} = \{S_i\}_{i \in I}$, the union and the intersection of the elements of $\mathcal{F}$ are usually written as follows: $\cup_{i \in I} S_i$ and $\cap_{i \in I} S_i$. If $I := \{1, ..., n\}$ or $I := \mathbb{N}$, "$\cup_{i \in I}$" and "$\cap_{i \in I}$" are written as "$\cup_{i=1}^n$" and "$\cap_{i=1}^n$" or "$\cup_{i=1}^\infty$" and "$\cap_{i=1}^\infty$" respectively.

We can now formalize better the concept of cartesian product, which we have already introduced for two sets. Let $\{X_1, X_2, ..., X_n\}$ be a finite family of sets. If $X_i \neq \emptyset$ for $i = 1, 2, ..., n$, then the *cartesian product* $X_1 \times X_2 \times \cdots \times X_n$ is defined to be the set of all mappings $\varphi : \{1, 2, ..., n\} \to \cup_{i=1}^n X_i$ so that $\varphi(i) \in X_i$ for $i = 1, 2, ..., n$; if there is i so that $X_i = \emptyset$, then $X_1 \times X_2 \times \cdots \times X_n := \emptyset$. If $X_i \neq \emptyset$ for $i = 1, 2, ..., n$, an element φ of $X_1 \times X_2 \times \cdots \times X_n$ is called an *ordered n-tuple*, or simply an *n-tuple*, and is denoted by the symbol $(x_1, x_2, ..., x_n)$, where $x_i := \varphi(i)$. If $E_i \subset X_i$ for $i = 1, 2, ..., n$, then $E_1 \times E_2 \times \cdots \times E_n$ is a subset of $X_1 \times X_2 \times \cdots \times X_n$, and

$$(X_1 \times X_2 \times \cdots \times X_n) - (E_1 \times E_2 \times \cdots \times E_n)$$
$$= \cup_{i=1}^n (X_1 \times \cdots \times X_{i-1} \times (X_i - E_i) \times X_{i+1} \times \cdots \times X_n);$$

if $F_i \subset X_i$ for $i = 1, 2, ..., n$, then

$$(E_1 \times E_2 \times \cdots \times E_n) \cap (F_1 \times F_2 \times \cdots \times F_n)$$
$$= (E_1 \cap F_1) \times (E_2 \cap F_2) \times \cdots \times (E_n \cap F_n).$$

If X is a set so that $X_i = X$ for $i = 1, 2, ..., n$, then we write

$$X^n := X_1 \times X_2 \times \cdots \times X_n.$$

1.2.2 Remark. Given two non-empty sets X and Y, if we want to define a mapping φ from X to Y by using a rule r that assigns elements of Y to some elements of X, we need to define a subset D_φ of X such that the rule r assigns one and only one element of Y to each element of D_φ. After defining D_φ, a mapping φ is defined by assigning to each element of D_φ *the* element $r(x)$ of Y that we obtain by applying the rule r to x. To indicate a mapping defined in this way, we often write

$$\varphi : D_\varphi \to Y$$
$$x \mapsto \varphi(x) := r(x),$$

or equivalently

$$D_\varphi \ni x \mapsto \varphi(x) := r(x) \in Y.$$

When, for a given non-empty subset S of X, the rule r assigns one and only one element of Y to each element of S and we want to define D_φ by setting $D_\varphi := S$, we often write directly

$$\varphi : S \to Y$$
$$x \mapsto \varphi(x) := r(x),$$

or even (without introducing a symbol to denote the mapping)

$$S \ni x \mapsto r(x) \in Y.$$

1.2.3 Definition. Let φ be a mapping from X to Y (by this, here and in the sequel, we mean that X, Y are non-empty sets and φ is a mapping $\varphi : D_\varphi \to Y$ with $D_\varphi \subset X$). The *graph* of φ is the subset of $X \times Y$ defined by

$$G_\varphi := \{(x, y) \in X \times Y : x \in D_\varphi \text{ and } y = \varphi(x)\}.$$

We remark that, when X and Y are replaced with two different sets X' and Y' such that $D_\varphi \subset X'$ and $R_\varphi \subset Y'$, the graph of φ will remain unaltered (but it will be considered as a subset of $X' \times Y'$).

1.2.4 Proposition. *Let X and Y be non-empty sets. For a non-empty subset G of $X \times Y$ the following conditions are equivalent:*

(a) G is the graph of a mapping from X to Y;
(b) $(x, y_1), (x, y_2) \in G \Rightarrow y_1 = y_2$.

Proof. a $\Rightarrow$ b: Let φ be a mapping from X to Y and let $G = G_\varphi$. Then we have

$$(x, y) \in G \Leftrightarrow (x \in D_\varphi \text{ and } y = \varphi(x)) .$$

Hence

$$(x, y_1), (x, y_2) \in G \Rightarrow y_1 = \varphi(x) = y_2.$$

b $\Rightarrow$ a: Assuming condition b, we define

$$D_\varphi := \{x \in X : \exists y \in Y \text{ such that } (x, y) \in G\},$$

and, $\forall x \in D_\varphi$,

$$r(x) := y \text{ if } y \in Y \text{ and } (x, y) \in G.$$

For $x \in D_\varphi$, $\exists! y \in Y$ such that $y \in Y$ and $(x, y) \in G$, by the very definition of D_φ and by condition b. Thus, we can define the mapping

$$\varphi : D_\varphi \to Y$$
$$x \mapsto \varphi(x) := r(x),$$

and we see at once that $G_\varphi = G$. $\qquad\square$

1.2.5 Definitions. Let φ_1, φ_2 be mappings from X to Y. The mapping φ_2 is called an *extension* of φ_1 (or φ_1 a *restriction* of φ_2), and it is said to *extend* φ_1, if we have:

$$D_{\varphi_1} \subset D_{\varphi_2} \text{ and } \varphi_1(x) = \varphi_2(x), \forall x \in D_{\varphi_1},$$

which can be written equivalently as

$$x \in D_{\varphi_1} \Rightarrow (x \in D_{\varphi_2} \text{ and } \varphi_1(x) = \varphi_2(x)).$$

For this we write $\varphi_1 \subset \varphi_2$ (or $\varphi_2 \supset \varphi_1$).

It is immediately clear that $\varphi_1 \subset \varphi_2$ iff $G_{\varphi_1} \subset G_{\varphi_2}$.

The mappings φ_1, φ_2 are said to be *equal* if

$$D_{\varphi_1} = D_{\varphi_2} \text{ and } \varphi_1(x) = \varphi_2(x), \forall x \in D_{\varphi_1},$$

which is equivalent to

$$[x \in D_{\varphi_1} \Rightarrow (x \in D_{\varphi_2} \text{ and } \varphi_1(x) = \varphi_2(x))] \text{ and } (x \in D_{\varphi_2} \Rightarrow x \in D_{\varphi_1}).$$

Clearly, $\varphi_1 = \varphi_2$ iff ($\varphi_1 \subset \varphi_2$ and $\varphi_2 \subset \varphi_1$) iff $G_{\varphi_1} = G_{\varphi_2}$.

Given a mapping φ from X to Y and a non-empty subset S of D_φ, the *restriction of φ to S* is the mapping φ_S defined by

$$\varphi_S : S \to Y$$
$$x \mapsto \varphi_S(x) := \varphi(x).$$

Obviously, $\varphi_S \subset \varphi$.

1.2.6 Examples. We define a few useful mappings.

(a) Let S be a non-empty subset of a set X. The *identity mapping* of S is the mapping id_S defined as follows:

$$id_S : S \to X$$
$$x \mapsto id_S(x) := x.$$

(b) Let S be a subset of a non-empty set X. The *characteristic function* of S is the mapping χ_S defined as follows:

$$\chi_S : X \to \mathbb{R}$$
$$x \mapsto \chi_S(x) := \begin{cases} 1 & \text{if } x \in S, \\ 0 & \text{if } x \notin S. \end{cases}$$

(c) Let X, Y be two non-empty sets. The *projection mappings* of $X \times Y$ are the two mappings π_X, π_Y defined as follows

$$\pi_X : X \times Y \to X$$
$$(x, y) \mapsto \pi_X(x, y) := x,$$

$$\pi_Y : X \times Y \to Y$$
$$(x, y) \mapsto \pi_Y(x, y) := y.$$

1.2.7 Definitions. Let φ be a mapping from X to Y.

For a subset S of D_φ, the *image of S under φ* is the set $\varphi(S)$ defined by

$$\varphi(S) := \{y \in Y : \exists x \in S \text{ s.t. } y = \varphi(x)\}.$$

To mean the set $\varphi(S)$ one sometimes writes

$$\{\varphi(x) : x \in S\},$$

or also

$$\{r(x) : x \in S\},$$

if r is the rule which defines φ as in 1.2.2 and there is no need to mention the mapping φ.

For a subset T of Y, the *counterimage of T under φ* is the set $\varphi^{-1}(T)$ defined by

$$\varphi^{-1}(T) := \{x \in D_\varphi : \varphi(x) \in T\}.$$

1.2.8 Proposition. *Let φ be a mapping from X to Y. For any family $\mathcal{F}$ of subsets of D_φ we have:*

$$\varphi\left(\cup_{S \in \mathcal{F}} S\right) = \cup_{S \in \mathcal{F}} \varphi(S) \text{ and } \varphi\left(\cap_{S \in \mathcal{F}} S\right) \subset \cap_{S \in \mathcal{F}} \varphi(S);$$

we also have:

$$\varphi(\emptyset) = \emptyset;$$

$$\varphi(D_\varphi) = R_\varphi;$$

$$\varphi(S_1) \subset \varphi(S_2) \text{ if } S_1, S_2 \text{ are subsets of } D_\varphi \text{ such that } S_1 \subset S_2.$$

For any family $\mathcal{G}$ of subsets of Y we have:

$$\varphi^{-1}\left(\cup_{T \in \mathcal{G}} T\right) = \cup_{T \in \mathcal{G}} \varphi^{-1}(T) \text{ and } \varphi^{-1}\left(\cap_{T \in \mathcal{G}} T\right) = \cap_{T \in \mathcal{G}} \varphi^{-1}(T);$$

we also have:

$$\varphi^{-1}(\emptyset) = \emptyset;$$

$$\varphi^{-1}(Y) = \varphi^{-1}(R_\varphi) = D_\varphi;$$

$$\varphi^{-1}(T_1) \subset \varphi^{-1}(T_2) \text{ if } T_1, T_2 \text{ are subsets of } Y \text{ such that } T_1 \subset T_2;$$

$$\varphi^{-1}(Y - T) = D_\varphi - \varphi^{-1}(T) \text{ for any subset } T \text{ of } Y.$$

We also have:

$$\varphi(\varphi^{-1}(T)) \subset T \text{ for each subset } T \text{ of } Y;$$

$$S \subset \varphi^{-1}(\varphi(S)) \text{ for each subset } S \text{ of } D_\varphi$$

(equality need not hold in either case).

Proof. Everything follows at once from the definitions. $\qquad\qquad\qquad\square$

1.2.9 Definitions. A mapping φ from X to Y is said to be:

(a) *injective* (or an *injection*) if $[x_1, x_2 \in D_\varphi$ and $\varphi(x_1) = \varphi(x_2)] \Rightarrow x_1 = x_2$,
   i.e. if $[x_1, x_2 \in D_\varphi$ and $x_1 \neq x_2] \Rightarrow \varphi(x_1) \neq \varphi(x_2)$,
   i.e. if $(x_1, y), (x_2, y) \in G_\varphi \Rightarrow x_1 = x_2$,
   i.e. if $\varphi^{-1}(\{y\})$ contains just one point for each $y \in R_\varphi$;
(b) *surjective* (or a *surjection*) onto Y if $R_\varphi = Y$,
   i.e. if $\forall y \in Y, \exists x \in D_\varphi$ s.t. $y = \varphi(x)$;
(c) *bijective* (or a *bijection*) *from X onto Y (or between X and Y)* if $D_\varphi = X$ and
   φ is both injective and surjective onto Y.

As was pointed out before, the final set Y can be replaced with a different set Y' as long as $R_\varphi \subset Y'$. By choosing R_φ as the final set, any mapping φ can be made surjective.

1.2.10 Definitions. A set X is said to be:

   finite if either $X = \emptyset$ or there are $n \in \mathbb{N}$ and a bijection from $\{1, ..., n\}$ onto X;
   denumerable if there is a bijection from $\mathbb{N}$ onto X;
   countable if X is either finite or denumerable;
   uncountable if X is not countable.

The following facts can be proved (cf. e.g. Shilov, 1973, 2.32, 2.33, 2.34, 2.35, 2.41): every subset of a countable set is countable; the union of a countable family of countable sets is countable; the cartesian product of a finite family of countable sets is countable; the set of all rational numbers is countable; the set of all real numbers is uncountable.

1.2.11 Definition. Let φ be a mapping from X to Y. By definition of R_φ we have

$$\forall y \in R_\varphi, \exists x \in D_\varphi \text{ such that } y = \varphi(x),$$

whereas we have

$$\forall y \in R_\varphi, \exists! x \in D_\varphi \text{ such that } y = \varphi(x)$$

iff φ is injective.

   Therefore, if and only if φ is injective can we define a mapping, which we denote by φ^{-1} and call the *inverse* of φ, by setting

$$D_{\varphi^{-1}} := R_\varphi$$

and using the rule

$$\forall y \in D_{\varphi^{-1}}, r(y) := x \text{ if } x \in D_\varphi \text{ and } y = \varphi(x).$$

Thus, if φ is injective we have the mapping

$$\varphi^{-1} : R_\varphi \to X$$
$$y \mapsto \varphi^{-1}(y) := x \text{ if } x \in D_\varphi \text{ and } y = \varphi(x).$$

We recall (cf. 1.2.7) that, for any mapping φ from X to Y, for a subset T of Y we have defined the set

$$\varphi^{-1}(T) := \{x \in D_\varphi : \varphi(x) \in T\}.$$

For an injective mapping φ we have $\{\varphi^{-1}(y)\} = \varphi^{-1}(\{y\})$ for each $y \in R_\varphi$; moreover, for any subset S of R_φ, $\varphi^{-1}(S)$ is the same thing when interpreted as the image of S under the inverse of φ or as the counterimage of S under φ. One can see immediately that the following facts are true for an injective mapping φ:

(a) $R_{\varphi^{-1}} = D_\varphi$;
(b) φ^{-1} is injective and $(\varphi^{-1})^{-1} = \varphi$;
(c) if V denotes the mapping

$$V : X \times Y \to Y \times X$$
$$(x, y) \mapsto V(x, y) := (y, x),$$

then $G_{\varphi^{-1}} = V(G_\varphi)$ (notice that condition b of 1.2.4 is in effect for $G := V(G_\varphi)$ iff φ is injective).

1.2.12 Definition. Let X, Y, Z be non-empty sets, let φ be a mapping from X to Y, and let ψ be a mapping from Y to Z, i.e. $\varphi : D_\varphi \to Y$ with $D_\varphi \subset X$ and $\psi : D_\psi \to Z$ with $D_\psi \subset Y$. If $\varphi^{-1}(D_\psi) \neq \emptyset$, the *composition* of ψ with φ is the mapping $\psi \circ \varphi$ defined as follows:

$$D_{\psi \circ \varphi} := \{x \in D_\varphi : \varphi(x) \in D_\psi\} = \varphi^{-1}(D_\psi),$$

$$\psi \circ \varphi : D_{\psi \circ \varphi} \to Z$$
$$x \mapsto (\psi \circ \varphi)(x) := \psi(\varphi(x)).$$

If $\psi : \mathbb{N} \to X$ is a sequence in X and $\varphi : \mathbb{N} \to \mathbb{N}$ is a mapping such that $\varphi(n_1) < \varphi(n_2)$ whenever $n_1 < n_2$, then the mapping $\psi \circ \varphi$ is called a *subsequence* of ψ. If ψ is denoted by $\{x_n\}$, then $\psi \circ \varphi$ is denoted by $\{x_{\varphi(k)}\}$, or by $\{x_{n_k}\}$ if φ does not need to be specified.

1.2.13 Proposition.

(A) Let φ be a mapping from X to Y. We have:

(a) $\varphi \circ id_X = id_Y \circ \varphi = \varphi$.

If ψ is a mapping from Y to Z such that $\varphi^{-1}(D_\psi) \neq \emptyset$, we have:

(b) $R_{\psi \circ \varphi} \subset R_\psi$;
(c) $D_{\psi \circ \varphi} \subset D_\varphi$;
(d) $D_{\psi \circ \varphi} = D_\varphi$ iff $R_\varphi \subset D_\psi$;
(e) $D_\psi = Y \Rightarrow D_{\psi \circ \varphi} = D_\varphi$;
(f) $(\psi \circ \varphi)^{-1}(S) = \varphi^{-1}(\psi^{-1}(S))$ for every subset S of Z.

If S is a subset of Y, we have:

(g) $(\chi_S \circ \varphi)(x) = \chi_{\varphi^{-1}(S)}(x), \forall x \in D_\varphi$.

(B) *If X and Y are non-empty sets and G is a non-empty subset of $X \times Y$ which satisfies condition b of 1.2.4, then the restriction $(\pi_X)_G$ of the mapping π_X to G is injective, and G is the graph of the mapping $\pi_Y \circ (\pi_X)_G^{-1}$.*

Proof. A: Everything from assertion a to assertion g follows at once from the definitions.

B: The mapping φ of which G is the graph has been constructed in the proof of b $\Rightarrow$ a in 1.2.4, to which we refer in what follows. Condition b of 1.2.4 means exactly that the mapping $(\pi_X)_G$ is injective. The domain of the mapping $(\pi_X)_G^{-1}$ is the range of $(\pi_X)_G$, which is exactly what D_φ was defined to be, and this is also the domain of $\pi_Y \circ (\pi_X)_G^{-1}$ since $D_{\pi_Y} = X \times Y$ (cf. e). We also have

$$\forall x \in D_\varphi, \pi_Y \circ (\pi_X)_G^{-1}(x) = \pi_Y(x, r(x)) = r(x) = \varphi(x).$$

Thus, $\varphi = \pi_Y \circ (\pi_X)_G^{-1}$. □

1.2.14 Proposition.

(A) *If φ is an injective mapping from X to Y, we have:*

$$\varphi^{-1} \circ \varphi = id_{D_\varphi} \subset id_X \text{ and } \varphi \circ \varphi^{-1} = id_{R_\varphi} \subset id_Y.$$

If φ is a bijection from X onto Y, we have

$$\varphi^{-1} \circ \varphi = id_X \text{ and } \varphi \circ \varphi^{-1} = id_Y.$$

(B) *Let φ be an injective mapping from X to Y, ψ an injective mapping from Y to Z, and suppose that $\varphi^{-1}(D_\psi) \neq \emptyset$. Then the mapping $\psi \circ \varphi$ is injective and $(\psi \circ \varphi)^{-1} = \varphi^{-1} \circ \psi^{-1}$.*

Proof. A: Everything follows at once from the definitions.

B: The mapping $\psi \circ \varphi$ is injective since

$$[x_1, x_2 \in D_{\psi \circ \varphi} \text{ and } (\psi \circ \varphi)(x_1) = (\psi \circ \varphi)(x_2)] \Rightarrow$$
$$[x_1, x_2 \in D_\varphi \text{ and } \varphi(x_1) = \varphi(x_2)] \Rightarrow x_1 = x_2.$$

The equality $D_{(\psi \circ \varphi)^{-1}} = D_{\varphi^{-1} \circ \psi^{-1}}$ is proved by

$$z \in D_{(\psi \circ \varphi)^{-1}} \Leftrightarrow z \in R_{\psi \circ \varphi} \Leftrightarrow$$
$$[\exists x \in D_{\psi \circ \varphi} \text{ s.t. } z = (\psi \circ \varphi)(x)] \Leftrightarrow$$
$$[\exists x \in D_\varphi \text{ s.t. } \varphi(x) \in D_\psi \text{ and } z = \psi(\varphi(x))] \Leftrightarrow$$
$$[\exists y \in R_\varphi \text{ s.t. } y \in D_\psi \text{ and } z = \psi(y)] \Leftrightarrow$$
$$[z \in R_\psi \text{ and } \psi^{-1}(z) \in R_\varphi] \Leftrightarrow$$
$$[z \in D_{\psi^{-1}} \text{ and } \psi^{-1}(z) \in D_{\varphi^{-1}}] \Leftrightarrow$$
$$z \in D_{\varphi^{-1} \circ \psi^{-1}}.$$

Finally, for each $z \in R_{\psi \circ \varphi}$, if $x \in D_{\psi \circ \varphi}$ is such that

$$z = (\psi \circ \varphi)(x) = \psi(\varphi(x))$$

then $\psi^{-1}(z) = \varphi(x)$ and hence

$$\varphi^{-1}(\psi^{-1}(z)) = x = (\psi \circ \varphi)^{-1}(z).$$

$\square$

1.2.15 Theorem. *Let φ, ψ be mappings from X to Y such that ψ is injective and $\varphi \subset \psi$. Then φ is injective and $\varphi^{-1} \subset \psi^{-1}$.*

Proof. We have

$$[x_1, x_2 \in D_\varphi, \varphi(x_1) = \varphi(x_2)] \Rightarrow [x_1, x_2 \in D_\psi, \psi(x_1) = \psi(x_2)] \Rightarrow x_1 = x_2,$$

which proves that φ is injective. We also have by 1.2.14A:

$$y \in D_{\varphi^{-1}} = R_\varphi \Rightarrow y = \varphi(\varphi^{-1}(y)) = \psi(\varphi^{-1}(y)) \Rightarrow$$
$$[y \in R_\psi = D_{\psi^{-1}} \text{ and } \psi^{-1}(y) = \psi^{-1}(\psi(\varphi^{-1}(y))) = (\varphi^{-1}(y))],$$

which proves $\varphi^{-1} \subset \psi^{-1}$. $\square$

1.2.16 Theorem. *Let φ be a mapping from X to Y and ψ a mapping from Y to X.*

(a) *If $\psi \circ \varphi = id_{D_\varphi}$, then φ is injective and $\varphi^{-1} \subset \psi$.*
(b) *If $\psi \circ \varphi = id_{D_\varphi}$ and $\varphi \circ \psi = id_{D_\psi}$, then both φ and ψ are injective, $\varphi^{-1} = \psi$ and $\psi^{-1} = \varphi$.*

Proof. a: Assume $\psi \circ \varphi = id_{D_\varphi}$. We have

$$[x_1, x_2 \in D_\varphi, \varphi(x_1) = \varphi(x_2)] \Rightarrow x_1 = \psi(\varphi(x_1)) = \psi(\varphi(x_2)) = x_2,$$

which proves that φ is injective.

From $D_{\psi \circ \varphi} = D_{id_{D_\varphi}} = D_\varphi$ we have $D_{\varphi^{-1}} = R_\varphi \subset D_\psi$ by 1.2.13d; moreover, for $y \in D_{\varphi^{-1}}$ we have $\varphi^{-1}(y) \in D_\varphi$, and hence by 1.2.14A

$$\varphi^{-1}(y) = (\psi \circ \varphi)(\varphi^{-1}(y)) = \psi(\varphi(\varphi^{-1}(y))) = \psi(y);$$

this proves $\varphi^{-1} \subset \psi$.

b: Assume $\psi \circ \varphi = id_{D_\varphi}$ and $\varphi \circ \psi = id_{D_\psi}$. By part a, we have that φ and ψ are both injective, and also $\varphi^{-1} \subset \psi$ and $\psi^{-1} \subset \varphi$. By 1.2.15, $\psi^{-1} \subset \varphi$ implies $\psi = (\psi^{-1})^{-1} \subset \varphi^{-1}$. Thus we have $\varphi^{-1} = \psi$, which implies $\varphi = (\varphi^{-1})^{-1} = \psi^{-1}$. $\square$

1.2.17 Proposition. *Let φ_1 be a mapping from W to X, φ_2 a mapping from X to Y, φ_3 a mapping from Y to Z. We have $(\varphi_3 \circ \varphi_2) \circ \varphi_1 = \varphi_3 \circ (\varphi_2 \circ \varphi_1)$.*

Proof. We have

$$D_{(\varphi_3 \circ \varphi_2) \circ \varphi_1} := \{w \in D_{\varphi_1} : \varphi_1(w) \in D_{\varphi_3 \circ \varphi_2}\}$$
$$= \{w \in D_{\varphi_1} : \varphi_1(w) \in D_{\varphi_2} \text{ and } \varphi_2(\varphi_1(w)) \in D_{\varphi_3}\}$$
$$= \{w \in D_{\varphi_2 \circ \varphi_1} : (\varphi_2 \circ \varphi_1)(w) \in D_{\varphi_3}\}$$
$$= D_{\varphi_3 \circ (\varphi_2 \circ \varphi_1)},$$

and, for each $w \in D_{(\varphi_3 \circ \varphi_2) \circ \varphi_1}$,

$$((\varphi_3 \circ \varphi_2) \circ \varphi_1)(w) = (\varphi_3 \circ \varphi_2)(\varphi_1(w)) = \varphi_3(\varphi_2(\varphi_1(w)))$$
$$= \varphi_3((\varphi_2 \circ \varphi_1)(w)) = (\varphi_3 \circ (\varphi_2 \circ \varphi_1))(w).$$

□

1.2.18 Proposition. *Let φ be a mapping from X to X, and let ψ be bijection from X onto Y. We have:*

(a) $D_\varphi = \psi^{-1}(D_{\psi \circ \varphi \circ \psi^{-1}})$;
(b) $R_{\psi \circ \varphi \circ \psi^{-1}} = \psi(R_\varphi)$.

Proof. a: We have

$$x \in D_\varphi \Rightarrow \psi^{-1}(\psi(x)) \in D_\varphi \Rightarrow \psi(x) \in D_{\varphi \circ \psi^{-1}} \overset{(*)}{=} D_{\psi \circ \varphi \circ \psi^{-1}} \Rightarrow$$
$$x \in \psi^{-1}(D_{\psi \circ \varphi \circ \psi^{-1}}),$$
$$x \in \psi^{-1}(D_{\psi \circ \varphi \circ \psi^{-1}}) \Rightarrow \psi(x) \in D_{\psi \circ \varphi \circ \psi^{-1}} \overset{(*)}{=} D_{\varphi \circ \psi^{-1}} \Rightarrow$$
$$x \in D_{\varphi \circ \psi^{-1} \circ \psi} = D_{\varphi \circ id_X} = D_\varphi,$$

where $(*)$ is true by 1.2.13e.
  b: We have

$$y \in R_{\psi \circ \varphi \circ \psi^{-1}} \Rightarrow [\exists \tilde{y} \in D_{\psi \circ \varphi \circ \psi^{-1}} \text{ s.t. } y = \psi(\varphi(\psi^{-1}(\tilde{y})))] \Rightarrow y \in \psi(R_\varphi),$$

$$y \in \psi(R_\varphi) \Rightarrow [\exists x \in D_\varphi \text{ s.t. } y = \psi(\varphi(x)) = (\psi \circ \varphi \circ \psi^{-1})(\psi(x))] \Rightarrow y \in R_{\psi \circ \varphi \circ \psi^{-1}}.$$

□

1.2.19 Definitions. We define some operations on functions, which will be used in the book. Let X be a non-empty set. For a function φ from X to $\mathbb{C}$, we define:

$$-\varphi : D_\varphi \to \mathbb{C}$$
$$x \mapsto (-\varphi)(x) := -\varphi(x);$$
$$\operatorname{Re} \varphi : D_\varphi \to \mathbb{C}$$
$$x \mapsto (\operatorname{Re} \varphi)(x) := \operatorname{Re} \varphi(x);$$
$$\operatorname{Im} \varphi : D_\varphi \to \mathbb{C}$$
$$x \mapsto (\operatorname{Im} \varphi)(x) := \operatorname{Im} \varphi(x);$$
$$\overline{\varphi} : D_\varphi \to \mathbb{C}$$
$$x \mapsto \overline{\varphi}(x) := \overline{\varphi(x)};$$
$$|\varphi| : D_\varphi \to \mathbb{C}$$
$$x \mapsto |\varphi|(x) := |\varphi(x)|;$$

$$D_{\frac{1}{\varphi}} := \{x \in D_\varphi : \varphi(x) \neq 0\} \quad \text{and}$$

$$\frac{1}{\varphi} : D_{\frac{1}{\varphi}} \to \mathbb{C}$$

$$x \mapsto (\frac{1}{\varphi})(x) := \frac{1}{\varphi(x)};$$

$$e^\varphi : D_\varphi \to \mathbb{C}$$

$$x \mapsto (e^\varphi)(x) := \exp \varphi(x);$$

for $n \in \mathbb{N}$,

$$\varphi^n : D_\varphi \to \mathbb{C}$$

$$x \mapsto (\varphi^n)(x) := (\varphi(x))^n;$$

for $\alpha \in \mathbb{C}$,

$$\alpha\varphi : D_\varphi \to \mathbb{C}$$

$$x \mapsto (\alpha\varphi)(x) := \alpha\varphi(x).$$

For a function φ from X to $\mathbb{R}$, we define:

$$\varphi^+ : D_\varphi \to \mathbb{R}$$

$$x \mapsto \varphi^+(x) := \max\{\varphi(x), 0\};$$

$$\varphi^- : D_\varphi \to \mathbb{R}$$

$$x \mapsto \varphi^-(x) := -\min\{\varphi(x), 0\}.$$

For two functions φ, ψ from X to $\mathbb{C}$ s.t. $D_\varphi \cap D_\psi \neq \emptyset$, we define:

$$\varphi + \psi : D_\varphi \cap D_\psi \to \mathbb{C}$$

$$x \mapsto (\varphi + \psi)(x) := \varphi(x) + \psi(x);$$

$$\varphi\psi : D_\varphi \cap D_\psi \to \mathbb{C}$$

$$x \mapsto (\varphi\psi)(x) := \varphi(x)\psi(x).$$

Clearly, for a function φ from X to $\mathbb{C}$ we have

$$\varphi = \operatorname{Re}\varphi + i\operatorname{Im}\varphi \text{ and } |\varphi|^2 = \varphi\overline{\varphi},$$

and for a function from X to $\mathbb{R}$ we have

$$\varphi = \varphi^+ - \varphi^- \text{ and } |\varphi| = \varphi^+ + \varphi^-;$$

thus, for a function φ from X to $\mathbb{C}$ we have

$$\varphi = (\operatorname{Re}\varphi)^+ - (\operatorname{Re}\varphi)^- + i(\operatorname{Im}\varphi)^+ - i(\operatorname{Im}\varphi)^-.$$

For $\alpha \in \mathbb{C}$, we define the constant function

$$\alpha_X : X \to \mathbb{C}$$

$$x \mapsto \alpha_X(x) := \alpha;$$

we also write $\varphi + \alpha := \varphi + \alpha_X$ for every function φ from X to $\mathbb{C}$.

1.2.20 Remark. If S is a subset of a non-empty set X, we have (for χ_S, cf. 1.2.6b)

$$\chi_{X-S} = 1_X - \chi_S.$$

If S_1 and S_2 are subsets of X, we have

$$\chi_{S_1} + \chi_{S_2} = \chi_{S_1 \cup S_2} + \chi_{S_1 \cap S_2} \text{ and } \chi_{S_1}\chi_{S_2} = \chi_{S_1 \cap S_2}.$$

If $\{S_n\}_{n \in I}$ is a disjoint family of subsets of X and $I := \{1, ..., N\}$ or $I := \mathbb{N}$, then

$$\forall x \in X, \chi_{\cup_{n \in I} S_n}(x) = \sum_{n \in I} \chi_{S_n}(x),$$

where $\sum_{n \in I}$ stands for $\sum_{n=1}^{N}$ or $\sum_{n=1}^{\infty}$ (note that, for each $x \in X$, there is at most one index n so that $\chi_{S_n}(x) \neq 0$).

If S_1 and S_2 are subsets of two non-empty sets X_1 and X_2 respectively, then

$$\forall (x_1, x_2) \in X_1 \times X_2, \chi_{S_1 \times S_2}(x_1, x_2) = \chi_{S_1}(x_1)\chi_{S_2}(x_2).$$

1.2.21 Definition. Let φ be a function from $\mathbb{R}$ to $\mathbb{C}$, i.e. $\varphi : D_\varphi \to \mathbb{C}$ with $D_\varphi \subset \mathbb{R}$, and let x be a point of D_φ for which $\exists \epsilon > 0$ such that $(x - \epsilon, x + \epsilon) \subset D_\varphi$. We say that φ is *differentiable* at x if both $\operatorname{Re}\varphi$ and $\operatorname{Im}\varphi$ are differentiable at x, i.e. if the derivatives of $\operatorname{Re}\varphi$ and $\operatorname{Im}\varphi$ exist at x and are finite. If φ is differentiable at x, we call *derivative* of φ at x the complex number

$$\varphi'(x) := (\operatorname{Re}\varphi)'(x) + i(\operatorname{Im}\varphi)'(x),$$

where $(\operatorname{Re}\varphi)'(x)$ and $(\operatorname{Im}\varphi)'(x)$ stand for the derivatives of $\operatorname{Re}\varphi$ and $\operatorname{Im}\varphi$ at x respectively.

An analogous definition can be given for one-sided differentiability and derivatives.

If φ is differentiable at all $x \in D_\varphi$, we call *derivative* of φ the function φ' that has D_φ as its domain and is defined by assigning $\varphi'(x)$ to each $x \in D_\varphi$.

We will use freely the following axiom.

1.2.22 Axiom (Axiom of choice). Let X and Y be non-empty sets and Φ a mapping $\Phi : X \to \mathcal{P}(Y)$, where $\mathcal{P}(Y)$ denotes the family of all subsets of Y. If $\Phi(x) \neq \emptyset$ for each $x \in X$, then there exists a mapping $\varphi : X \to Y$ such that $\varphi(x) \in \Phi(x)$ for all $x \in X$.

1.3 Groups

We review in this section the few elementary facts about groups that will be used in the book.

1.3.1 Definitions. A *group* is a pair (G, γ), where G is a non-empty set and γ is a mapping $\gamma : G \times G \to G$ with the following properties, which we write with the shorthand notation $g_1 g_2 := \gamma(g_1, g_2)$:

(gr_1) $g_1(g_2g_3) = (g_1g_2)g_3$, $\forall g_1, g_2, g_3 \in G$,

(gr_2) $\exists e \in G$ s.t. $ge = eg = g$ for all $g \in G$,

(gr_3) $\forall g, \exists g'$ s.t. $gg' = g'g = e$.

The mapping γ is called the *product of the group*.

If $u \in G$ is such that $gu = ug = g$ for all $g \in G$, then we have $e = eu = u$. Thus, condition gr_2 identifies a unique element of G, which is called the *identity of the group*.

If, for $g \in G$, $g'' \in G$ is such that $gg'' = g''g = e$, then we have $g' = g'e = g'(gg'') = (g'g)g'' = eg'' = g''$. Thus, for all $g \in G$, the element g' of condition gr_3 is actually unique. It is called the *inverse* of g and denoted by g^{-1}. We see at once that $(g^{-1})^{-1} = g$.

In view of gr_1, we write $g_1g_2g_3 := g_1(g_2g_3)$.

A group (G, γ) is said to be *abelian* if γ has the further property:

(ag) $g_1g_2 = g_2g_1$, $\forall g_1, g_2 \in G$.

For an abelian group, one usually writes "$g_1 + g_2$" instead of "g_1g_2", "*sum*" instead of "product", "*zero*" instead of "identity", "0" instead of "e", "*opposite*" instead of "inverse", "$-g$" instead of "g^{-1}", "$g_1 - g_2$" instead of "$g_1 + (-g_2)$". For elements of G, one writes $g_1 + g_2 + g_3 := g_1 + (g_2 + g_3)$ and $\sum_{i=n}^{m} g_i := g_n + g_{n+1} + \dots + g_m$ if $n < m$; one also writes $\sum_{i \in I} g_i$ to denote the sum of a finite family $\{g_i\}_{i \in I}$ of elements of G.

One often says "the group G" to mean the pair (G, γ), but on the other hand one often speaks of "elements of the group G" to mean "elements of the set G". Tacit conventions of this sort are used whenever one deals with mathematical structures which are composed of sets together with some mappings (as in the case of metric, linear, normed, inner product spaces, algebras, normed algebras, etc.), or together with some relation (as in the case of a partially ordered set), or together with some class of distinguished subsets (as in the case of a measurable space), and will not be mentioned again later on.

1.3.2 Definition. A *subgroup* of a group (G, γ) is a non-empty subset $\tilde{G}$ of G such that:

(sg_1) $g_1, g_2 \in \tilde{G} \Rightarrow g_1g_2 \in \tilde{G}$,

(sg_2) $g \in \tilde{G} \Rightarrow g^{-1} \in \tilde{G}$.

If $\tilde{G}$ is a subgroup of (G, γ), condition sg_1 makes it possible to use $\tilde{G}$ as the final set of $\gamma_{\tilde{G} \times \tilde{G}}$ (the restriction of γ to $\tilde{G} \times \tilde{G}$). Then $(\tilde{G}, \gamma_{\tilde{G} \times \tilde{G}})$ is a group: indeed for this pair condition gr_1 is obviously satisfied; moreover, for any $g \in \tilde{G}$ we have $e = gg^{-1} \in \tilde{G}$ by conditions sg_1 and sg_2, hence condition gr_2 is satisfied for $(\tilde{G}, \gamma_{\tilde{G} \times \tilde{G}})$ with e still playing the role of the identity; finally, condition gr_3 is satisfied for $(\tilde{G}, \gamma_{\tilde{G} \times \tilde{G}})$ since by condition sg_2 we have $g^{-1} \in \tilde{G}$ for all $g \in \tilde{G}$.

1.3.3 Definitions. Let (G_1, γ_1) and (G_2, γ_2) be two groups. A *homomorphism* from G_1 to G_2 is a mapping $\Phi : G_1 \to G_2$ such that:

(hg) $\gamma_2(\Phi(g), \Phi(g')) = \Phi(\gamma_1(g, g'))$, i.e. $\Phi(g)\Phi(g') = \Phi(gg'), \forall g, g' \in G$.

We see that, denoting by e_1 and e_2 the identities of G_1 and G_2 respectively, we have $\Phi(e_1) = e_2$: for any $g \in G_1, e_2 = \Phi(g)^{-1}\Phi(g) = \Phi(g)^{-1}\Phi(ge_1) = \Phi(g)^{-1}\Phi(g)\Phi(e_1) = e_2\Phi(e_1) = \Phi(e_1)$.

We also see that $\Phi(g^{-1}) = \Phi(g)^{-1}$ for all $g \in G_1$: $\Phi(g^{-1}) = \Phi(g^{-1})e_2 = \Phi(g^{-1})\Phi(g)\Phi(g)^{-1} = \Phi(g^{-1}g)\Phi(g)^{-1} = \Phi(e_1)\Phi(g)^{-1} = e_2\Phi(g)^{-1} = \Phi(g)^{-1}$.

A homomorphism from G_1 to G_2 which is also a bijection from G_1 onto G_2 is called an *isomorphism*. It is immediate to see that, when it is defined, the composition (cf. 1.2.12) of two isomorphisms is an isomorphism, and that the inverse (cf. 1.2.11) of an isomorphism is an isomorphism. If $G_1 = G_2$, an isomorphism is called an *automorphism*.

1.3.4 Remark. It can be easily seen that the family of all automorphisms of any group G is a group if the product of two automorphisms is assumed to be their composition as defined in 1.2.12. The identity of this group is id_G (which is clearly an automorphism of G), and the group inverse of an automorphism is its inverse as defined in 1.2.11.

1.3.5 Proposition. *Let Φ be a homomorphism from a group (G_1, γ_1) to a group (G_2, γ_2). We have:*

(a) R_Φ is a subgroup of G_2;
(b) if G_1 is an abelian group, then R_Φ (with the restriction of γ_2 to $R_\Phi \times R_\Phi$) is also an abelian group.

Proof. a: We have

$$g, g' \in R_\Phi \Rightarrow$$
$$[\exists \tilde{g}, \tilde{g}' \in G_1 \text{ s.t. } g = \Phi(\tilde{g}), g' = \Phi(\tilde{g}'), \text{ hence s.t. } gg' = \Phi(\tilde{g}\tilde{g}')] \Rightarrow$$
$$gg' \in R_\Phi$$

and

$$g \in R_\Phi \Rightarrow$$
$$[\exists \tilde{g} \in G_1 \text{ s.t. } g = \Phi(\tilde{g}), \text{ hence s.t. } g^{-1} = \Phi(\tilde{g})^{-1} = \Phi(\tilde{g}^{-1})] \Rightarrow$$
$$g^{-1} \in R_\Phi.$$

b: Let G_1 be abelian. Then we have

$$\Phi(g_1)\Phi(g_2) = \Phi(g_1 g_2) = \Phi(g_2 g_1) = \Phi(g_2)\Phi(g_1), \forall g_1, g_2 \in G.$$

$\square$

Chapter 2

Metric Spaces

This chapter contains just the facts about metric spaces that will be used later in the book and is not intended for a thorough treatment of this subject.

2.1 Distance, convergent sequences

2.1.1 Definition. A *metric space* is a pair (X, d), where X is a non-empty set and d is a function $d : X \times X \to \mathbb{R}$ such that

(di_1) $d(x, y) = d(y, x)$, $\forall x, y \in X$,
(di_2) $d(x, y) \leq d(x, z) + d(z, y)$, $\forall x, y, z \in X$,
(di_3) $d(x, y) = 0 \Leftrightarrow x = y$.

These conditions imply $0 \leq d(x, y), \forall x, y \in X$:

$$0 = \frac{1}{2} d(x, x) \leq \frac{1}{2} \left(d(x, y) + d(y, x) \right) = d(x, y).$$

The function d is called a *distance on X*. The inequality in di_2 is called the *triangle inequality*.

2.1.2 Proposition. *In a metric space (X, d) we have*

$$|d(x_1, y_1) - d(x_2, y_2)| \leq d(x_1, x_2) + d(y_1, y_2), \quad \forall x_1, y_1, x_2, y_2 \in X.$$

Proof. For $x_1, y_1, x_2, y_2 \in X$ we have

$$d(x_1, y_1) \leq d(x_1, x_2) + d(x_2, y_1) \leq d(x_1, x_2) + d(x_2, y_2) + d(y_2, y_1),$$

hence

$$d(x_1, y_1) - d(x_2, y_2) \leq d(x_1, x_2) + d(y_1, y_2).$$

In the same way we have

$$d(x_2, y_2) - d(x_1, y_1) \leq d(x_2, x_1) + d(y_2, y_1).$$

Thus we have

$$|d(x_1, y_1) - d(x_2, y_2)| \leq d(x_1, x_2) + d(y_1, y_2).$$

$\square$

2.1.3 Definition. Let (X, d) be a metric space and S a non-empty subset of X. The function $d_S := d_{S \times S}$ (cf. 1.2.5) is clearly a distance on S. The metric space (S, d_S) is called a *metric subspace* of (X, d) and is said to be *defined by S*.

2.1.4 Example. Define the function
$$d_\mathbb{R} : \mathbb{R} \times \mathbb{R} \to \mathbb{R},$$
$$(x, y) \mapsto d_\mathbb{R}(x, y) := |x - y|,$$
where $|x|$ is the absolute value of $x \in \mathbb{R}$. Directly from the properties of the absolute value it follows that $d_\mathbb{R}$ is a distance on $\mathbb{R}$. We will always consider $\mathbb{R}$ to be the first element of the metric space $(\mathbb{R}, d_\mathbb{R})$ and every subset of $\mathbb{R}$ to be the first element of the metric subspace of $(\mathbb{R}, d_\mathbb{R})$ it defines.

2.1.5 Definition. Let (X_1, d_1) and (X_2, d_2) be metric spaces. An *isomorphism* (or *isometry*) from (X_1, d_1) onto (X_2, d_2) is a mapping $\Phi : X_1 \to X_2$ which is a surjection onto X_2 and has the following property:
$$d_2(\Phi(x), \Phi(y)) = d_1(x, y), \quad \forall x, y \in X_1.$$
An isomorphism is necessarily injective (hence, it is a bijection from X_1 onto X_2):
$$\Phi(x) = \Phi(y) \Rightarrow d_2(\Phi(x), \Phi(y)) = 0 \Rightarrow d_1(x, y) = 0 \Rightarrow x = y.$$
In is obvious that the inverse of an isomorphism from (X_1, d_1) onto (X_2, d_2) is an isomorphism from (X_2, d_2) onto (X_1, d_1).

2.1.6 Definition. Let (X, d) be a metric space. A sequence $\{x_n\}$ in X is said to be *convergent* if the following condition is satisfied:
$$\exists x \in X \text{ such that } \forall \epsilon > 0, \exists N_\epsilon \in \mathbb{N} \text{ such that } n > N_\epsilon \Rightarrow d(x_n, x) < \epsilon.$$
If this condition is satisfied, the point x is unique: assume $x' \in X$ is such that
$$\forall \epsilon > 0, \exists N'_\epsilon \in \mathbb{N} \text{ such that } n > N'_\epsilon \Rightarrow d(x_n, x') < \epsilon;$$
fix $\epsilon > 0$; then, for $n > \max\{N_\epsilon, N'_\epsilon\}$ we have $d(x, x') \leq d(x, x_n) + d(x_n, x') < 2\epsilon$; since ϵ was arbitrary, this proves that $d(x, x') = 0$, which implies $x = x'$.

If the condition of convergence is satisfied, one says that $\{x_n\}$ *converges to x*, calls x the *limit* of $\{x_n\}$, and writes $\lim_{n \to \infty} x_n = x$, or $x_n \to x$ as $n \to \infty$, or $x_n \xrightarrow[n \to \infty]{} x$, or simply $x_n \to x$.

2.1.7 Remarks.

(a) Let (X, d) be a metric space. For $x \in X$ and a sequence $\{x_n\}$ in X, $x_n \to x$ iff $d(x_n, x) \to 0$ in the metric space $(\mathbb{R}, d_\mathbb{R})$.

(b) Let (X, d) be a metric space, $\{x_n\}$ a convergent sequence in X, and $\varphi : \mathbb{N} \to \mathbb{N}$ a mapping such that $\varphi(n_1) < \varphi(n_2)$ whenever $n_1 < n_2$. Then the subsequence $\{x_{\varphi(k)}\}$ is convergent and $\lim_{k \to \infty} x_{\varphi(k)} = \lim_{n \to \infty} x_n$. Indeed, write $x := \lim_{n \to \infty} x_n$ and for each $\varepsilon > 0$ let $N_\varepsilon \in \mathbb{N}$ be such that
$$n > N_\varepsilon \Rightarrow d(x_n, x) < \varepsilon.$$
Then, for each $\varepsilon > 0$,
$$k > \min \varphi^{-1}([N_\varepsilon, \infty) \cap \mathbb{N}) \Rightarrow \varphi(k) > N_\varepsilon \Rightarrow d(x_{\varphi(k)}, x) < \varepsilon.$$

2.1.8 Definition. Let (X, d) be a metric space. A subset S of X is called a *bounded set* (or it is said to be *bounded*) if the following condition is satisfied:

$$\exists m \in (0, \infty), \exists x \in X \text{ such that } d(y, x) < m, \forall y \in S.$$

2.1.9 Proposition. *Let (X, d) be a metric space and $\{x_n\}$ a sequence in X. If $\{x_n\}$ is convergent then the range of $\{x_n\}$, i.e. the set $\{x_n : n \in \mathbb{N}\}$, is a bounded set.*

Proof. Assume that there exists $x \in X$ such that $x_n \to x$. Then

$$\exists N \in \mathbb{N} \text{ such that } n > N \Rightarrow d(x_n, x) < 1.$$

Put $m := \max\{d(x_1, x), ..., d(x_N, x)\}$. Then $d(x_n, x) < m + 1, \forall n \in \mathbb{N}$. $\qquad\square$

2.1.10 Definition. Let (X, γ, d) be a triple so that (X, γ) is an abelian group and (X, d) is a metric space, let $\{x_n\}$ be a sequence in X, and let $s_n := \sum_{k=1}^{n} x_k$ for all $n \in \mathbb{N}$. The sequence $\{s_n\}$ is called the *series* of the x_n's and is denoted by the symbol $\sum_{n=1}^{\infty} x_n$; thus, one says that $\sum_{n=1}^{\infty} x_n$ is *convergent* when the sequence $\{s_n\}$ is convergent. If the sequence $\{s_n\}$ is convergent then one calls $\lim_{n \to \infty} s_n$ the *sum of the series* and denotes $\lim_{n \to \infty} s_n$ by the same symbol $\sum_{n=1}^{\infty} x_n$ as the series, i.e. one writes $\sum_{n=1}^{\infty} x_n := \lim_{n \to \infty} s_n$.

2.2   Open sets

2.2.1 Definition. Let (X, d) be a metric space. If $x \in X$ and $r \in (0, \infty)$, the *open ball* with *center* x and *radius* r is the set

$$B(x, r) := \{y \in X : d(x, y) < r\}.$$

2.2.2 Definition. Let (X, d) be a metric space. A subset G of X is called an *open set* (or it is said to be *open*) if the following condition is satisfied:

$$\forall x \in G, \exists r \in (0, \infty) \text{ such that } B(x, r) \subset G.$$

The family of all open sets is called the *topology* defined by d and is denoted by $\mathcal{T}_d$.

2.2.3 Proposition. *Let (X, d) be a metric space. For all $x \in X$ and $r \in (0, \infty)$, the open ball $B(x, r)$ is an open set (this justifies its name).*

Proof. Let y be a point in $B(x, r)$. We must produce $\bar{r} \in (0, \infty)$ such that $B(y, \bar{r}) \subset B(x, r)$. Since $d(y, x) < r$, we have $0 < r - d(y, x)$. Defining $\bar{r} := r - d(y, x)$, we have:

$$z \in B(y, \bar{r}) \Rightarrow d(z, y) < r - d(y, x) \Rightarrow$$
$$d(z, x) \leq d(z, y) + d(y, x) < r \Rightarrow z \in B(x, r).$$

$\qquad\square$

2.2.4 Theorem. *Let (X, d) be a metric space. The topology $\mathcal{T}_d$ has the following properties:*

(to_1) $\emptyset \in \mathcal{T}_d, X \in \mathcal{T}_d$;
(to_2) if $\mathcal{F}$ is any family of elements of $\mathcal{T}_d$, then $\cup_{S \in \mathcal{F}} S \in \mathcal{T}_d$;
(to_3) if $\mathcal{F}$ is a finite family of elements of $\mathcal{T}_d$, then $\cap_{S \in \mathcal{F}} S \in \mathcal{T}_d$.

Proof. to_1: To show that $\emptyset$ is open, we must show that each point in $\emptyset$ is the center of an open ball contained in $\emptyset$; but since there are no points in $\emptyset$, this requirement is automatically satisfied. The set X is clearly open, since every open ball centered on any of its points is contained in X.

to_2: Every point x in $\cup_{S \in \mathcal{F}} S$ lies in some S_x of the family $\mathcal{F}$. Since S_x is an open set, some open ball centered on x is contained in S_x and hence in $\cup_{S \in \mathcal{F}} S$.

to_3: If $\cap_{S \in \mathcal{F}} S = \emptyset$, then $\cap_{S \in \mathcal{F}} S$ is an open set. Assume then that $\cap_{S \in \mathcal{F}} S$ is non-empty and write $\mathcal{F} = \{S_1, ..., S_n\}$ for some $n \in \mathbb{N}$; let x be a point in $\cap_{S \in \mathcal{F}} S$; for $k = 1, ..., n$, $\exists r_k > 0$ s.t. $B(x, r_k) \subset S_k$; let r be the smallest number in the set $\{r_1, ..., r_n\}$; r is a positive real number and we have $B(x, r) \subset B(x, r_k)$ for $k = 1, ..., n$; therefore $B(x, r) \subset \cap_{S \in \mathcal{F}} S$. $\qquad\square$

2.2.5 Proposition. *Let (X, d) be a metric space and S a non-empty subset of X. A subset T of S is an open set in the metric subspace (S, d_S) iff $\exists G \in \mathcal{T}_d$ such that $T = G \cap S$.*

Proof. First we note that if $x \in S$ then $B(x, r) \cap S$ is the open ball with center x and radius r in the metric subspace (S, d_S).

If $G \in \mathcal{T}_d$ and $x \in G \cap S$, then $\exists r > 0$ such that $B(x, r) \subset G$, hence such that $B(x, r) \cap S \subset G \cap S$. This shows that $G \cap S$ is an open set in (S, d_S).

Conversely, if T is an open set in (S, d_S), then for each $x \in T$ there is $r_x > 0$ s.t. $B(x, r_x) \cap S \subset T$. Then we have $T = \cup_{x \in T} (B(x, r_x) \cap S) = (\cup_{x \in T} B(x, r_x)) \cap S$, with $\cup_{x \in T} B(x, r_x) \in \mathcal{T}_d$ by 2.2.3 and 2.2.4. $\qquad\square$

2.2.6 Definition. *Let (X, d) be a metric space and S a subset of X. The interior of S is the set:*

$$S^o := \cup_{G \in \mathcal{F}} G, \text{ with } \mathcal{F} := \{G \in \mathcal{P}(X) : G \in \mathcal{T}_d \text{ and } G \subset S\}.$$

2.2.7 Theorem. *Let (X, d) be a metric space and S a subset of X. Then:*

(a) $S^o \in \mathcal{T}_d$, $S^o \subset S$, if $G \in \mathcal{T}_d$ and $G \subset S$ then $G \subset S^o$ (thus, S^o is the largest open set that is contained in S);
(b) $S \in \mathcal{T}_d$ iff $S = S^o$;
(c) if T is a subset of X such that $S \subset T$, then $S^o \subset T^o$;
(d) for every subset T of X, $(S \cap T)^o = S^o \cap T^o$.

Let $\mathcal{L}$ be a family of subsets of X. Then

(e) $\cup_{S \in \mathcal{L}} S^o \subset (\cup_{S \in \mathcal{L}} S)^o$.

Proof. a: Everything follows from 2.2.4 and the definition of S^o.

b: This follows immediately from assertion a.

c: This follows immediately from assertion a.

d: From assertion c we have $(S \cap T)^o \subset S^o$ and $(S \cap T)^o \subset T^o$, and hence we have $(S \cap T)^o \subset S^o \cap T^o$. On the other hand, we have $S^o \cap T^o \in \mathcal{T}_d$ by assertion a and 2.2.4 and $S^o \cap T^o \subset S \cap T$ by assertion a, and hence $S^o \cap T^o \subset (S \cap T)^o$ by assertion a.

e: From assertion a and 2.2.4 we have $\cup_{S \in \mathcal{L}} S^o \in \mathcal{T}_d$ and $\cup_{S \in \mathcal{L}} S^o \subset \cup_{S \in \mathcal{L}} S$. By assertion a, this implies $\cup_{S \in \mathcal{L}} S^o \subset (\cup_{S \in \mathcal{L}} S)^o$. $\qquad\square$

2.3 Closed sets

2.3.1 Definition. Let (X, d) be a metric space. A subset F of X is called a *closed set* (or it is said to be *closed*) if $X - F$ is an open set. The family of all closed sets is denoted by $\mathcal{K}_d$. Since $S = X - (X - S)$ and $X - S \in \mathcal{K}_d$ iff (by definition) $X - (X - S) \in \mathcal{T}_d$, for a subset S of X we have $S \in \mathcal{T}_d$ iff $X - S \in \mathcal{K}_d$.

2.3.2 Theorem. *Let (X, d) be a metric space. The family $\mathcal{K}_d$ of all closed sets has the following properties:*

(cl$_1$) $\emptyset \in \mathcal{K}_d, X \in \mathcal{K}_d$;

(cl$_2$) *if $\mathcal{F}$ is any family of elements of $\mathcal{K}_d$, then $\cap_{S \in \mathcal{F}} S \in \mathcal{K}_d$;*

(cl$_3$) *if $\mathcal{F}$ is a finite family of elements of $\mathcal{K}_d$, then $\cup_{S \in \mathcal{F}} S \in \mathcal{K}_d$.*

Proof. Properties cl_1, cl_2, cl_3 follow from 2.3.1, 2.2.4 and De Morgan's laws (cf. 1.1.4). $\qquad\square$

2.3.3 Proposition. *Let (X, d) be a metric space and S a non-empty subset of X. A subset T of S is a closed set in the metric subspace (S, d_S) iff $\exists F \in \mathcal{K}_d$ such that $T = F \cap S$.*

Proof. For a subset T of S and any subset G of X we have (cf. 1.1.4)

$$S - T = G \cap S \overset{(1)}{\Leftrightarrow} T = S - (S - T) = S - (G \cap S) = S \cap (X - (G \cap S))$$
$$= S \cap ((X - G) \cup (X - S)) = S \cap (X - G).$$

Now, a subset T of S is a closed set in (S, d_S) iff $S - T$ is an open set in (S, d_S) iff (by 2.2.5) $\exists G \in \mathcal{T}_d$ s.t. $S - T = G \cap S$ iff (by 1) $\exists G \in \mathcal{T}_d$ s.t. $T = S \cap (X - G)$ iff (set $F := X - G$ or $G := X - F$) $\exists F \in \mathcal{K}_d$ s.t. $T = S \cap F$. $\qquad\square$

2.3.4 Theorem. *Let (X, d) be a metric space. For a subset S of X the following conditions are equivalent:*

(a) S is a closed set;

(b) $[x \in X, \{x_n\}$ a sequence in $S, x_n \to x] \Rightarrow x \in S$.

Proof. $a \Rightarrow b$: Let $\{x_n\}$ be a sequence in S, let x be a point of X, and assume $x_n \to x$. This implies that $B(x,r) \cap S \neq \emptyset$, i.e. $B(x,r) \not\subset X - S$, for all $r \in (0, \infty)$. Then, by 2.2.2 we have

$$x \in X - S \Rightarrow X - S \notin \mathcal{T}_d,$$

which is equivalent to

$$X - S \in \mathcal{T}_d \Rightarrow x \in S.$$

Since $S \in \mathcal{K}_d \Rightarrow X - S \in \mathcal{T}_d$, this proves $a \Rightarrow b$.

$b \Rightarrow a$: We prove (not a)$\Rightarrow$(not b). Assume $S \notin \mathcal{K}_d$. Then $X - S \notin \mathcal{T}_d$, and we have:

$$\exists x \in X - S \text{ s.t. } B(x, \epsilon) \not\subset X - S \text{ for all } \epsilon > 0,$$

which implies

$$\exists x \in X - S \text{ s.t. } B(x, \frac{1}{n}) \not\subset X - S \text{ for all } n \in \mathbb{N},$$

which means

$$\exists x \in X - S \text{ s.t. } B(x, \frac{1}{n}) \cap S \neq \emptyset \text{ for all } n \in \mathbb{N}.$$

By choosing a point x_n in $B(x, \frac{1}{n}) \cap S$ for each $n \in \mathbb{N}$, we construct a sequence $\{x_n\}$ which is in S and such that $x_n \to x$; since $x \notin S$, this proves that proposition b is not true; hence proposition (not b) is true. $\quad\square$

2.3.5 Remark. Using 2.3.4 one can see at once that the set $\{x\}$ is closed, for every point x of any metric space.

2.3.6 Definition. Let (X, d) be a metric space. If $x \in X$ and $r \in (0, \infty)$, the *closed ball* with *center* x and *radius* r is the set

$$K(x,r) := \{y \in X : d(x,y) \leq r\}.$$

2.3.7 Proposition. *Let (X, d) be a metric space. For all $x \in X$ and $r \in (0, \infty)$, the closed ball $K(x,r)$ is a closed set (this justifies its name).*

Proof. Let a sequence $\{y_n\}$ in $K(x,r)$ and $y \in X$ be such that $y_n \to y$. If $y \notin K(x,r)$ were true then $d(x,y) - r > 0$ would be true, and hence there would exist $k \in \mathbb{N}$ such that

$$d(y_k, y) < d(x,y) - r,$$

hence such that

$$d(y_k, x) \geq d(x,y) - d(y_k, y) > r.$$

This proves by contraposition that $y \in K(x,r)$. In view of 2.3.4, this proves that $K(x,r)$ is a closed set. $\quad\square$

2.3.8 Definition. Let (X, d) be a metric space and S a subset of X. The *closure* of S is the set

$$\overline{S} = \cap_{f \in \mathcal{F}} F, \text{ with } \mathcal{F} := \{F \in \mathcal{P}(X) : F \in \mathcal{K}_d \text{ and } S \subset F\}.$$

2.3.9 Theorem. *Let* (X, d) *be a metric space and* S *a subset of* X. *Then:*

(a) $\overline{S} \in \mathcal{K}_d, S \subset \overline{S}$, *if* $F \in \mathcal{K}_d$ *and* $S \subset F$ *then* $\overline{S} \subset F$ *(thus* $\overline{S}$ *is the smallest closed set that contains* S*);*
(b) $X - \overline{S} = (X - S)^o$;
(c) $S \in \mathcal{K}_d$ *iff* $S = \overline{S}$;
(d) *if* T *is a subset of* X *such that* $S \subset T$, *then* $\overline{S} \subset \overline{T}$;
(e) *for every subset* T *of* X, $\overline{S \cup T} = \overline{S} \cup \overline{T}$.

Let $\mathcal{L}$ *be a family of subsets of* X. *Then*

(f) $\overline{(\cap_{S \in \mathcal{L}} S)} \subset \cap_{S \in \mathcal{L}} \overline{S}$.

Proof. a: Everything follows from 2.3.2 and the definition of $\overline{S}$.

b: Using assertion a and 2.2.7a we have:

$$[X - \overline{S} \in \mathcal{T}_d \text{ and } X - \overline{S} \subset X - S] \Rightarrow X - \overline{S} \subset (X - S)^o;$$

$$[(X - S)^o \in \mathcal{T}_d \text{ and } (X - S)^o \subset X - S] \Rightarrow$$
$$[X - (X - S)^o \in \mathcal{K}_d \text{ and } S \subset X - (X - S)^o] \Rightarrow$$
$$\overline{S} \subset X - (X - S)^o \Rightarrow (X - S)^o \subset X - \overline{S}.$$

c, d, e, f: These properties of the closure follow from the corresponding properties b, c, d, e of the interior listed in 2.2.7, by taking the complement in X of every subset involved and using assertion b. $\quad\square$

2.3.10 Theorem. *Let* (X, d) *be a metric space,* S *a subset of* X, *and* x *a point of* X. *The following conditions are equivalent:*

(a) $x \in \overline{S}$;
(b) $\forall \epsilon > 0, \exists y \in S$ *such that* $d(x, y) < \epsilon$;
(c) *there exists a sequence* $\{x_n\}$ *in* S *such that* $x_n \to x$.

Proof. $a \Rightarrow b$: We prove (not b)$\Rightarrow$ (not a). Assume (not b), i.e.

$$\exists \epsilon > 0 \text{ such that } \epsilon \leq d(x, y) \text{ for each } y \in S,$$

which is equivalent to

$$\exists \epsilon > 0 \text{ such that } S \subset X - B(x, \epsilon);$$

since $X - B(x, \epsilon) \in \mathcal{K}_d$ (cf. 2.2.3 and 2.3.1), we have (cf. 2.3.9a)

$$\exists \epsilon > 0 \text{ such that } \overline{S} \subset X - B(x, \epsilon);$$

since $X - B(x, \epsilon) \subset X - \{x\}$, we have

$$\overline{S} \subset X - \{x\},$$

which leads to $\{x\} \subset X - \overline{S}$, i.e. (not a).

$b \Rightarrow c$: Condition b means:

$$\forall \epsilon > 0, B(x, \epsilon) \cap S \neq \emptyset;$$

by choosing a point x_n in $B(x, \frac{1}{n}) \cap S$ for each $n \in \mathbb{N}$, we construct a sequence $\{x_n\}$ which is in S and such that $x_n \to x$. This proves that condition b implies condition c.

$c \Rightarrow a$: Assume condition c. For each $F \in \mathcal{K}_d$ so that $S \subset F$, we have that the sequence $\{x_n\}$ is in F and therefore (by 2.3.4) $x \in F$. By definition of $\overline{S}$, this proves that $x \in \overline{S}$. $\square$

2.3.11 Definition. Let (X, d) be a metric space. A subset S of X is said to be *dense* in (X, d) (or simply dense in X) if $\overline{S} = X$.

2.3.12 Corollary. *Let (X, d) be a metric space and S a subset of X. The following conditions are equivalent:*

(a) $\overline{S} = X$;
(b) $\forall x \in X$, $\forall \epsilon > 0$, $\exists y \in S$ such that $d(x, y) < \epsilon$;
(c) $\forall x \in X$, there exists a sequence $\{x_n\}$ in S such that $x_n \to x$.

Proof. $a \Rightarrow b$: This follows immediately from $a \Rightarrow b$ in 2.3.10.

$b \Rightarrow c$: This follows immediately from $b \Rightarrow c$ in 2.3.10.

$c \Rightarrow a$: If condition c holds, then by $c \Rightarrow a$ in 2.3.10 we have $X \subset \overline{S}$, which is equivalent to condition a. $\square$

2.3.13 Theorem. *Let (X, d) be a metric space, and let S and T be two subsets of X such that $T \subset S$. The following conditions are equivalent:*

(a) T is dense in the metric subspace (S, d_S);
(b) $S \subset \overline{T}$, where $\overline{T}$ means the closure of T in (X, d).

Proof. By 2.3.8, the closure of T in (X, d) is

$$\overline{T} = \cap_{f \in \mathcal{F}} F, \text{ with } \mathcal{F} := \{F \in \mathcal{P}(X) : F \in \mathcal{K}_d \text{ and } T \subset F\},$$

while the closure of T in (S, d_S) is (cf. 2.3.3) $\cap_{f \in \mathcal{F}} (F \cap S)$.

If condition a is true, then

$$S = \cap_{f \in \mathcal{F}} (F \cap S) \subset \cap_{f \in \mathcal{F}} F = \overline{T}.$$

If condition b is true, then $S \subset \cap_{f \in \mathcal{F}} F$ and hence $S = S \cap (\cap_{f \in \mathcal{F}} F) = \cap_{f \in \mathcal{F}} (S \cap F)$, which means that the closure of T in (S, d_S) equals S. $\square$

2.3.14 Corollary. *Let (X, d) be a metric space, and let S and T be two subsets of X such that $T \subset S$. If T is dense in (S, d_S) and S is dense in (X, d), then T is dense in (X, d).*

Proof. If T is dense in (S, d_S), then we have (by 2.3.13) $S \subset \overline{T}$, which implies (by 2.3.9a) $\overline{S} \subset \overline{T}$. If moreover S is dense in (X, d), then we also have $\overline{S} = X$ and hence $X \subset \overline{T}$, which implies $\overline{T} = X$. ☐

2.3.15 Definition. A metric space (X, d) is said to be *separable* if there exists a countable subset of X which is dense in (X, d).

2.3.16 Remark. The metric space $(\mathbb{R}, d_{\mathbb{R}})$ is separable since the set $\mathbb{Q}$ of all rational numbers is both countable and dense in $\mathbb{R}$.

2.3.17 Proposition. *Let (X, d) be a separable metric space. Then there is a countable family $\mathcal{T}_c$ of open balls such that every open set is a union of elements of $\mathcal{T}_c$.*

Proof. Let S be a countable dense subset of X. Let $\mathcal{T}_c$ be the family

$$\mathcal{T}_c := \{B(s, r) : s \in S, r \in (0, \infty) \text{ and } r \in \mathbb{Q}\}.$$

This family $\mathcal{T}_c$ is countable since $\mathbb{Q}$ is countable and the cartesian product of two countable sets is countable. Let G be an arbitrary non-empty open set. For $x \in G$, let $r > 0$ be such that $B(x, r) \subset G$. Since S is dense in X, there is $s_x \in S$ such that $s_x \in B(x, \frac{r}{3})$ (cf. 2.3.12). Let r_x be a rational number such that $\frac{r}{3} < r_x < \frac{2r}{3}$. We have $x \in B(s_x, r_x)$ since $d(x, s_x) < \frac{r}{3} < r_x$. We also have

$$y \in B(s_x, r_x) \Rightarrow d(y, x) \leq d(y, s_x) + d(s_x, x) < r_x + \frac{r}{3} < \frac{2r}{3} + \frac{r}{3} = r,$$

hence $B(s_x, r_x) \subset B(x, r)$, hence $B(s_x, r_x) \subset G$. Then we have $\cup_{x \in G} B(s_x, r_x) = G$, and we note that the family $\{B(s_x, r_x)\}_{x \in G}$ is contained in $\mathcal{T}_c$. ☐

2.3.18 Theorem (Lindelöf's theorem). *Let (X, d) be a separable metric space. Let G be an open set and $\{G_i\}_{i \in I}$ a family of open sets such that $G = \cup_{i \in I} G_i$. Then there is a countable subset I_c of I such that $G = \cup_{i \in I_c} G_i$.*

Proof. If G is the empty set then the statement is trivial. Assume then G non-empty. Let $\mathcal{T}_c$ be the countable family of open balls of 2.3.17. Let x be a point in G. The point x is in some G_i, and we can find an open ball B in $\mathcal{T}_c$ such that $x \in B \subset G_i$. If we do this for each point x in G, we obtain a family of open balls $\{B_n\}_{n \in J}$ such that $\cup_{n \in J} B_n = G$, and this family is countable since it is a subfamily of $\mathcal{T}_c$. Further, for each open ball in this subfamily, we can select $i \in I$ so that G_i contains the ball. The family I_c of i's which arises in this way is countable, since there exists a surjection of the countable set J onto I_c by construction of I_c. Moreover, $\cup_{i \in I_c} G_i = G$ since $G_i \subset G$ for each $i \in I$, and $\forall n \in J, \exists i \in I_c$ such that $B_n \subset G_i$. ☐

2.3.19 Theorem. *Assume that in a metric space (X, d) there is a countable family $\mathcal{T}_c$ of open sets such that every open set is a union of elements of $\mathcal{T}_c$. Then (X, d) is separable.*

Proof. For each element of $\mathcal{T}_c$ choose a point, and let S be the set of all these points. The set S is countable since by its construction there is a surjection from $\mathcal{T}_c$ onto S. For every $x \in X$ and every $\epsilon > 0$, $B(x, \epsilon)$ contains an element of $\mathcal{T}_c$, hence a point $y \in S$, and we have $d(x, y) < \epsilon$. In view of 2.3.12, this shows that $\overline{S} = X$. $\qquad\square$

2.3.20 Proposition. *Let (X, d) be a separable metric space and S a non-empty subset of X. Then the metric subspace (S, d_S) is a separable metric space.*

Proof. Let $\mathcal{T}_c$ be the countable family of open sets of 2.3.17. Consider the countable family
$$\mathcal{T}_c^S := \{B \cap S : B \in \mathcal{T}_c\},$$
which by 2.2.5 is a family of open sets in(S, d_S). By 2.2.5, each open set in (S, d_S) can be written as $G \cap S$, with G an open set in (X, d), and we have $G = \cup_{n \in I} B_n$ with $\{B_n\}_{n \in I}$ a subfamily of $\mathcal{T}_c$; hence we have $G \cap S = \cup_{n \in I}(B_n \cap S)$. This shows that each open set in (S, d_S) is a union of elements of $\mathcal{T}_c^S$. Then, (S, d_S) is separable by 2.3.19. $\qquad\square$

2.3.21 Proposition. *Let Φ be an isomorphism from a metric space (X_1, d_1) onto a metric space (X_2, d_2). Then:*

(a) for every subset S of X_1, $\Phi(\overline{S}) = \overline{\Phi(S)}$;
(b) $S \in \mathcal{K}_{d_1} \Leftrightarrow \Phi(S) \in \mathcal{K}_{d_2}$;
(c) if (X_1, d_1) is separable then (X_2, d_2) is separable.

Proof. a: For $y \in X_2$ we have
$$y \in \overline{\Phi(S)} \Leftrightarrow$$
[there exists a sequence $\{y_n\}$ in $\Phi(S)$ such that $y_n \to y$] $\Leftrightarrow$
[there exists a sequence $\{x_n\}$ in S such that $\Phi(x_n) \to y$] $\Leftrightarrow$
[there exists a sequence $\{x_n\}$ in S such that $x_n \to \Phi^{-1}(y)$] $\Leftrightarrow$
$$\Phi^{-1}(y) \in \overline{S} \Leftrightarrow$$
$$y \in \Phi(\overline{S}),$$
where 2.3.10 has been used twice, and also the fact that both Φ and Φ^{-1} preserve distances.

b: If $S \in \mathcal{K}_{d_1}$ then $S = \overline{S}$ (cf. 2.3.9c), and hence $\Phi(S) = \Phi(\overline{S}) = \overline{\Phi(S)}$ by result a, and hence $\Phi(S) \in \mathcal{K}_{d_2}$. The converse is also true, since $S = \Phi^{-1}(\Phi(S))$ and Φ^{-1} is an isomorphism from (X_2, d_2) onto (X_1, d_1).

c: If (X_1, d_1) is separable then there exists a countable subset S of X_1 such that $\overline{S} = X_1$. Then $\Phi(S)$ is a countable subset of X_2 and $\overline{\Phi(S)} = \Phi(X_1) = X_2$ by result a, and hence (X_2, d_2) is separable. $\qquad\square$

2.4 Continuous mappings

2.4.1 Definitions. Let (X, d) and $(\tilde{X}, \tilde{d})$ be metric spaces. A mapping φ from X to $\tilde{X}$, i.e. $\varphi : D_\varphi \to \tilde{X}$ with $D_\varphi \subset X$, is said to be *continuous at a point* x in D_φ if the following condition is satisfied:

$$\forall \epsilon > 0, \exists \delta_\epsilon > 0 \text{ such that } [y \in D_\varphi \text{ and } d(x, y) < \delta_\epsilon] \Rightarrow \tilde{d}(\varphi(x), \varphi(y)) < \epsilon.$$

It is clear from the definition that, if S is a non-empty subset of D_φ and x is a point in S at which φ is continuous, then the restriction φ_S (cf. 1.2.5) is continuous at x.

The mapping φ is said to be *continuous* if it is continuous at every point in D_φ, i.e. if the following condition is satisfied:

$$\forall x \in D_\varphi, \forall \epsilon > 0, \exists \delta_{x,\epsilon} > 0 \text{ s.t. } [y \in D_\varphi \text{ and } d(x, y) < \delta_{x,\epsilon}] \Rightarrow \tilde{d}(\varphi(x), \varphi(y)) < \epsilon.$$

The mapping φ is said to be *uniformly continuous* if the following condition is satisfied:

$$\forall \epsilon > 0, \exists \delta_\epsilon > 0 \text{ such that } [x, y \in D_\varphi \text{ and } d(x, y) < \delta_\epsilon] \Rightarrow \tilde{d}(\varphi(x), \varphi(y)) < \epsilon.$$

2.4.2 Theorem. *Let (X, d) and $(\tilde{X}, \tilde{d})$ be metric spaces, let $\varphi : D_\varphi \to \tilde{X}$ be a mapping with $D_\varphi \subset X$, and let x be a point in D_φ. The following conditions are equivalent:*

(a) φ is continuous at x;
(b) $[\{x_n\}$ is a sequence in D_φ and $x_n \to x] \Rightarrow \varphi(x_n) \to \varphi(x)$.

Proof. $a \Rightarrow b$: Assume φ continuous at x, i.e.

$$\forall \epsilon > 0, \exists \delta_\epsilon > 0 \text{ s.t. } [y \in D_\varphi \text{ and } d(x, y) < \delta_\epsilon] \Rightarrow \tilde{d}(\varphi(x), \varphi(y)) < \epsilon.$$

Let $\{x_n\}$ be a sequence in D_φ such that $x_n \to x$, i.e. such that

$$\forall \eta > 0, \exists K_\eta \in \mathbb{N} \text{ s.t. } K_\eta < n \Rightarrow d(x_n, x) < \eta.$$

Setting $N_\epsilon := K_{\delta_\epsilon}$, we have for each $\epsilon > 0$

$$N_\epsilon < n \Rightarrow d(x_n, x) < \delta_\epsilon \Rightarrow \tilde{d}(\varphi(x_n), \varphi(x)) < \epsilon.$$

This proves that condition b is true.

$b \Rightarrow a$: We prove (not a)$\Rightarrow$(not b). We assume that condition a is not true, i.e.

$$\exists \epsilon > 0 \text{ s.t. } \forall \delta > 0, \exists y \in D_\varphi \text{ s.t. } d(x, y) < \delta \text{ and } \epsilon \leq \tilde{d}(\varphi(x), \varphi(y)).$$

We fix $\varepsilon > 0$ with this property. Then, for each $n \in \mathbb{N}$, the set

$$S_n := \left\{ y \in D_\varphi : d(x, y) < \frac{1}{n} \text{ and } \epsilon \leq \tilde{d}(\varphi(x), \varphi(y)) \right\}$$

is non-empty. By choosing a point x_n in S_n for each $n \in \mathbb{N}$, we construct a sequence such that $x_n \to x$ but also such that $\{\varphi(x_n)\}$ does not converge to $\varphi(x)$. Therefore, condition b is not true. $\square$

2.4.3 Theorem. *Let (X, d) and $(\tilde{X}, \tilde{d})$ be metric spaces, and let $\varphi : D_\varphi \to \tilde{X}$ be a mapping with $D_\varphi \subset X$. The following conditions are equivalent:*

(a) φ is continuous;
(b) $\forall F \in \mathcal{K}_{\tilde{d}}, \varphi^{-1}(F)$ is a closed set in the metric subspace $(D_\varphi, d_{D_\varphi})$;
(c) $\forall G \in \mathcal{T}_{\tilde{d}}, \varphi^{-1}(G)$ is an open set in the metric subspace $(D_\varphi, d_{D_\varphi})$.

Proof. $a \Rightarrow b$: Assume that φ is continuous and let F be a closed set in $(\tilde{X}, \tilde{d})$. We must prove that $\varphi^{-1}(F)$ is a closed set in $(D_\varphi, d_{D_\varphi})$. Let a sequence $\{x_n\}$ in $\varphi^{-1}(F)$ and $x \in D_\varphi$ be such that $x_n \to x$; by 2.4.2 we have $\varphi(x_n) \to \varphi(x)$; since $\varphi(x_n) \in F$ for each n and F is closed, by 2.3.4 we have $\varphi(x) \in F$, i.e. $x \in \varphi^{-1}(F)$. By 2.3.4, this proves that $\varphi^{-1}(F)$ is a closed set in $(D_\varphi, d_{D_\varphi})$.

$b \Rightarrow c$: This follows immediately from $D_\varphi - \varphi^{-1}(G) = \varphi^{-1}(\tilde{X} - G)$ (cf. 1.2.8) and 2.3.1 (referred to the metric space $(D_\varphi, d_{D_\varphi})$).

$c \Rightarrow a$: Assume condition c and let x be a point in D_φ. For each $\epsilon > 0$, by 2.2.3 $\varphi^{-1}(B(\varphi(x), \epsilon))$ is an open set in $(D_\varphi, d_{D_\varphi})$; since $x \in \varphi^{-1}(B(\varphi(x), \epsilon))$, there is $\delta_\epsilon > 0$ such that for the open ball $B(x, \delta_\epsilon) \cap D_\varphi$ in $(D_\varphi, d_{D_\varphi})$ we have $B(x, \delta_\epsilon) \cap D_\varphi \subset \varphi^{-1}(B(\varphi(x), \epsilon))$, and this means

$$[y \in D_\varphi, d(x, y) < \delta_\epsilon] \Rightarrow \varphi(y) \in B(\varphi(x), \epsilon), \text{ i.e. } \tilde{d}(\varphi(x), \varphi(y)) < \epsilon.$$

This proves condition a. $\square$

2.4.4 Theorem. *Let (X, d), $(\tilde{X}, \tilde{d})$, $(\tilde{\tilde{X}}, \tilde{\tilde{d}})$ be metric spaces, and let $\varphi : D_\varphi \to \tilde{X}$ and $\psi : D_\psi \to \tilde{\tilde{X}}$ be mappings with $D_\varphi \subset X$ and $D_\psi \subset \tilde{X}$. If φ and ψ are continuous, then $\psi \circ \varphi$ is continuous.*

Proof. Assume φ, ψ continuous and let $\tilde{\tilde{G}} \in \mathcal{T}_{\tilde{\tilde{d}}}$. We have (cf. 1.2.13f)

$$(\psi \circ \varphi)^{-1}(\tilde{\tilde{G}}) = \varphi^{-1}(\psi^{-1}(\tilde{\tilde{G}})).$$

By 2.4.3, $\psi^{-1}(\tilde{\tilde{G}})$ is an open set in $(D_\psi, \tilde{d}_{D_\psi})$; hence by 2.2.5 there is $\tilde{G} \in \mathcal{T}_{\tilde{d}}$ such that $\psi^{-1}(\tilde{\tilde{G}}) = \tilde{G} \cap D_\psi$. Similarly, since $\varphi^{-1}(\tilde{G})$ is an open set in $(D_\varphi, d_{D_\varphi})$, there is $G \in \mathcal{T}_d$ such that $\varphi^{-1}(\tilde{G}) = G \cap D_\varphi$. Therefore we have $(\psi \circ \varphi)^{-1}(\tilde{\tilde{G}}) = \varphi^{-1}(\tilde{G}) \cap \varphi^{-1}(D_\psi) = G \cap D_\varphi \cap \varphi^{-1}(D_\psi) = G \cap \varphi^{-1}(D_\psi) = G \cap D_{\psi \circ \varphi}$. By 2.2.5, this proves that $(\psi \circ \varphi)^{-1}(\tilde{\tilde{G}})$ is an open set in $(D_{\psi \circ \varphi}, d_{D_{\psi \circ \varphi}})$. By 2.4.3, this proves that $\psi \circ \varphi$ is continuous. $\square$

2.4.5 Remark. It is obvious that an isomorphism from one metric space onto another is a continuous mapping.

2.5 Characteristic functions of closed and of open sets

2.5.1 Definition. Let (X, d) be a metric space, x a point of X, and S a non-empty subset of X. The *distance of x from S* is the nonnegative number

$$d(x, S) := \inf\{d(x, y) : y \in S\}.$$

2.5.2 Theorem. *Let (X, d) be a metric space, x a point of X, and S a non-empty subset of X. The following conditions are equivalent:*

(a) $d(x, S) = 0$;
(b) $x \in \overline{S}$.

Proof. We have

$$d(x, S) = 0 \Leftrightarrow [\forall \epsilon > 0, \exists y \in S \text{ s.t. } d(x, y) < \epsilon] \Leftrightarrow x \in \overline{S},$$

where the latter equivalence holds by 2.3.10. ☐

2.5.3 Proposition. *Let (X, d) be a metric space and S a non-empty subset of X. Then*

$$|d(x, S) - d(y, S)| \le d(x, y), \quad \forall x, y \in X.$$

Proof. For $x, y \in X$ we have:

$$d(x, S) = \inf\{d(x, z) : z \in S\} \le \inf\{d(x, y) + d(y, z) : z \in S\}$$
$$= d(x, y) + \inf\{d(y, z) : z \in S\} = d(x, y) + d(y, S).$$

Hence we also have $d(y, S) \le d(y, x) + d(x, S)$. ☐

2.5.4 Lemma. *Let (X, d) be a metric space and S a non-empty subset of X. The function*

$$\delta_S : X \to \mathbb{R}$$
$$x \mapsto \delta_S(x) := d(x, S)$$

is uniformly continuous.

Proof. From 2.5.3 we have:

$$\forall \epsilon > 0, [x, y \in X, d(x, y) < \epsilon] \Rightarrow |\delta_S(x) - \delta_S(y)| < \epsilon.$$

☐

2.5.5 Lemma (Urysohn's lemma). *In a metric space, let F_1 and F_2 be closed sets such that $F_1 \cap F_2 = \emptyset$. Then there exists a continuous function $\varphi : X \to \mathbb{R}$ such that: $\varphi(x) = 1, \forall x \in F_1$; $\varphi(x) = 0, \forall x \in F_2$; $0 \le \varphi(x) \le 1, \forall x \in X$.*

Proof. If $F_1 = \emptyset$ take $\varphi := 0_X$; if $F_2 = \emptyset$ take $\varphi := 1_X$ (cf. 1.2.19). Assuming F_1 and F_2 non-empty, by 2.5.2 we have $d(x, F_1) + d(x, F_2) \ne 0, \forall x \in X$. Thus, we can define the function

$$\varphi : X \to \mathbb{R}$$
$$x \mapsto \varphi(x) := \frac{d(x, F_2)}{d(x, F_1) + d(x, F_2)},$$

which is continuous by 2.5.4 and has the other required properties by 2.5.2. ☐

2.5.6 Corollary. *In a metric space, let F_1, F_2 be closed sets such that $F_1 \cap F_2 = \emptyset$. Then, there exist two open sets G_1, G_2 such that $F_1 \subset G_1$, $F_2 \subset G_2$, $G_1 \cap G_2 = \emptyset$.*

Proof. the sets $G_1 := \varphi^{-1}\left(\left(\frac{3}{4}, \infty\right)\right)$ and $G_2 := \varphi^{-1}\left(\left(-\infty, \frac{1}{2}\right)\right)$, where φ is the function of 2.5.5, have the required properties (cf. 1.2.8 and 2.4.3). $\square$

2.5.7 Corollary. *Let F be a closed set in a metric space (X, d). Then there exists a sequence $\{\varphi_n\}$ such that:*

$\forall n \in \mathbb{N}$, φ_n *is a continuous function* $\varphi_n : X \to [0, 1]$;
$\forall x \in X$, $\varphi_n(x) \to \chi_F(x)$ *as* $n \to \infty$.

Proof. If $F = \emptyset$, let $\varphi_n := 0_X$. Assuming $F \neq \emptyset$, for $n \in \mathbb{N}$ the set $F_n := \delta_F^{-1}([\frac{1}{n}, \infty))$ is closed by 2.5.4 and 2.4.3, and $F \cap F_n = \emptyset$ by 2.5.2. Hence, by 2.5.5 there is a continuous function $\varphi_n : X \to [0, 1]$ such that $\varphi_n(x) = 1, \forall x \in F$, and $\varphi_n(x) = 0, \forall x \in F_n$. By 2.5.2 we also have:

$$\forall x \in X - F, \exists N_x \in \mathbb{N} \text{ such that } x \in F_n \text{ for } n > N_x,$$

$$\text{and hence such that } \varphi_n(x) = 0 \text{ for } n > N_x.$$

This proves that

$$\forall x \in X, \varphi_n(x) \to \chi_F(x) \text{ as } n \to \infty.$$

$\square$

2.5.8 Corollary. *Let G be an open set in a metric space (X, d). Then there exists a sequence $\{\psi_n\}$ such that:*

(a) $\forall n \in \mathbb{N}$, ψ_n is a continuous function $\psi_n : X \to [0, 1]$;
(b) $\forall x \in X$, $\psi_n(x) \to \chi_G(x)$ as $n \to \infty$.

Proof. Let $F := X - G$ in 2.5.7 and define $\psi_n := 1_X - \varphi_n$ (cf. 1.2.19). The required properties for $\{\psi_n\}$ follow from the properties of $\{\varphi_n\}$. In particular

$$\forall x \in X, \psi_n(x) = 1 - \varphi_n(x) \to 1 - \chi_{X-G}(x) = \chi_G(x) \text{ as } n \to \infty.$$

$\square$

2.5.9 Definition. Let (X, d) be a metric space.
  The *support* of a function $\varphi : X \to \mathbb{C}$ is the closed subset of X defined by

$$\operatorname{supp} \varphi := \overline{\{x \in X : \varphi(x) \neq 0\}}.$$

2.5.10 Definitions. Let (X, d) be a metric space. For a closed set F and a function $\varphi : X \to \mathbb{C}$, the notation $F \prec \varphi$ will mean that φ is continuous, that $0 \leq \varphi(x) \leq 1$ for each $x \in X$, and that $\varphi(x) = 1$ for each $x \in F$.
  For an open set G and a function $\varphi : X \to \mathbb{C}$, the notation $\varphi \prec G$ will mean that φ is continuous, that $0 \leq \varphi(x) \leq 1$ for each $x \in X$, and that $\operatorname{supp} \varphi \subset G$.
  For $F \in \mathcal{K}_d$, $G \in \mathcal{T}_d$ and $\varphi : X \to \mathbb{C}$, the notation $F \prec \varphi \prec G$ will be used to indicate that both $F \prec \varphi$ and $\varphi \prec G$ hold. Clearly, $F \prec \varphi \prec G$ holds iff the following conditions hold simultaneously:

(a) $F \subset G$,
(b) φ is continuous,
(c) $\chi_F(x) \leq \varphi(x) \leq \chi_G(x), \forall x \in X$,
(d) $\operatorname{supp} \varphi \subset G$.

Clearly, $\emptyset \prec \varphi \prec G$ reduces to $\varphi \prec G$ and $F \prec \varphi \prec X$ reduces to $F \prec \varphi$.

2.5.11 Theorem. *In a metric space (X, d), let F be a closed set, G an open set, and $F \subset G$. Then there exists a function $\varphi : X \to [0, 1]$ such that $F \prec \varphi \prec G$.*

Proof. Since $X - G$ is closed and $(X - G) \cap F = \emptyset$, from 2.5.6 it follows that

$$\exists G_1, G_2 \in \mathcal{T}_d \text{ such that } X - G \subset G_1, F \subset G_2, G_1 \cap G_2 = \emptyset.$$

Since we also have $F \cap (X - G_2) = \emptyset$, by 2.5.5 there exists a continuous function $\varphi : X \to [0, 1]$ such that $\varphi(x) = 1, \forall x \in F$, and $\varphi(x) = 0, \forall x \in X - G_2$, and hence also such that

$$\varphi(x) \neq 0 \Rightarrow x \in G_2 \Rightarrow x \in X - G_1,$$

which implies $\operatorname{supp} \varphi \subset X - G_1$ since $X - G_1$ is closed (cf. 2.3.9a), and hence $\operatorname{supp} \varphi \subset G$ since $X - G_1 \subset G$. Thus, φ has the required properties. $\quad\square$

2.6 Complete metric spaces

2.6.1 Definition. Let (X, d) be a metric space. A sequence $\{x_n\}$ in X is said to be a *Cauchy sequence* if the following condition is satisfied:

$$\forall \epsilon > 0, \exists N_\epsilon \in \mathbb{N} \text{ such that } N_\epsilon < n, m \Rightarrow d(x_n, x_m) < \epsilon.$$

To denote this condition, one sometimes writes $d(x_n, x_m) \to 0$ as $n, m \to \infty$.

2.6.2 Theorem. *Let (X, d) be a metric space and let $\{x_n\}$ be a sequence in X which is convergent. Then $\{x_n\}$ is a Cauchy sequence.*

Proof. Let x be the limit of $\{x_n\}$. For $\epsilon > 0$ let $N_\epsilon \in \mathbb{N}$ be such that $N_\epsilon < n \Rightarrow d(x_n, x) < \frac{\epsilon}{2}$. Then we have

$$N_\epsilon < n, m \Rightarrow d(x_n, x_m) \leq d(x_n, x) + d(x, x_m) < \epsilon.$$

$\quad\square$

2.6.3 Definition. A metric space (X, d) is said to be *complete* if every Cauchy sequence is convergent, i.e. if, for a sequence $\{x_n\}$ in X, the following implication is true:

$$[\forall \epsilon > 0, \exists N_\epsilon \in \mathbb{N} \text{ such that } N_\epsilon < n, m \Rightarrow d(x_n, x_m) < \epsilon] \Rightarrow$$
$$[\exists x \in X \text{ such that } \forall \epsilon > 0, \exists K_\epsilon \in \mathbb{N} \text{ such that } K_\epsilon < n \Rightarrow d(x_n, x) < \epsilon].$$

2.6.4 Proposition. *Let $(X_1, d_1), (X_2, d_2)$ be two metric spaces such that there exists an isomorphism from (X_1, d_1) onto (X_2, d_2). Then (X_1, d_1) is complete iff (X_2, d_2) is complete.*

Proof. Immediate from the definitions.   □

2.6.5 Example. The metric space $(\mathbb{R}, d_\mathbb{R})$ (cf. 2.1.4) is complete (cf. e.g. Rudin, 1976, 3.11).

2.6.6 Proposition. *Let (X, d) be a metric space and let S be a non-empty subset of X.*

(a) If the metric subspace (S, d_S) is a complete metric space, then S is a closed set in (X, d).

(b) If (X, d) is a complete metric space and S is a closed set in (X, d), then the metric subspace (S, d_S) is a complete metric space.

Proof. a: We prove the contrapositive form of the statement. If S is not a closed set, then $\exists x \in X$ s.t. $x \notin S$ and $x = \lim_{n \to \infty} x_n$, with $\{x_n\}$ a sequence in S (cf. 2.3.4). Then $\{x_n\}$ is a Cauchy sequence in (S, d_S) which is not convergent in (S, d_S). Thus, (S, d_S) is not complete.

b: Assume (X, d) complete and S closed. If $\{x_n\}$ is a Cauchy sequence in (S, d_S), then it is a Cauchy sequence in (X, d). Then, since (X, d) is complete, there exists $x \in X$ so that $x_n \to x$. Since S is closed in (X, d), by 2.3.4 we have $x \in S$. Then we have $x_n \to x$ in (S, d_S). This proves that (S, d_S) is complete.   □

2.6.7 Definition. Let (X, d) be a metric space. A *completion* of (X, d) is a pair $((\hat{X}, \hat{d}), \iota)$, where $(\hat{X}, \hat{d})$ is a complete metric space, ι is a mapping $\iota : X \to \hat{X}$, and the following two conditions hold:

(co_1) $\hat{d}(\iota(x), \iota(y)) = d(x, y), \forall x, y \in X$;
(co_2) R_ι is dense in $(\hat{X}, \hat{d})$, i.e. $\overline{R_\iota} = \hat{X}$.

We point out that, as a result of condition co_1, the mapping ι is necessarily injective:

$$\iota(x) = \iota(y) \Rightarrow \hat{d}(\iota(x), \iota(y)) = 0 \Rightarrow d(x, y) = 0 \Rightarrow x = y.$$

If $((\hat{X}, \hat{d}), \iota)$ is a completion of (X, d), then clearly ι is an isomorphism from (X, d) onto the metric subspace $(R_\iota, \hat{d}_{R_\iota})$ of $(\hat{X}, \hat{d})$.

2.6.8 Proposition. *Let (X, d) be a complete metric space and let S be a subset of X such that $\overline{S} = X$ (hence S is non-empty) and $S \neq X$. Then the metric subspace (S, d_S) is not complete and the pair $((X, d), id_S)$ is one of the completions of (S, d_S).*

Proof. Since S is not closed (cf. 2.3.9c), (S, d_S) is not complete by 2.6.6a. It follows directly from the definitions that $((X, d), id_S)$ is a completion of (S, d_S). □

We shall not use the following theorem, also because we shall need completions of metric spaces, but those completions will be constructed without relying on either the statement or the proof of this theorem. For this reason, we state it without giving its proof, which can be found e.g. in 6.3.11 of (Berberian, 1999).

2.6.9 Theorem. *If (X, d) is any metric space, then there exists a completion $((\hat{X}, \hat{d}), \iota)$ of (X, d).*

If $((\tilde{X}, \tilde{d}), \omega)$ is also a completion of (X, d), then there exists an isomorphism Φ from $(\hat{X}, \hat{d})$ onto $(\tilde{X}, \tilde{d})$ such that $\Phi \circ \iota = \omega$, i.e. such that $\Phi(\iota(x)) = \omega(x), \forall x \in X$.

In order that (X, d) be complete, it is necessary and sufficient that ι be surjective onto $\hat{X}$.

2.7 Product of two metric spaces

2.7.1 Theorem. *Let (X, d) and $(\tilde{X}, \tilde{d})$ be metric spaces. The function*

$$d \times \tilde{d} : (X \times \tilde{X}) \times (X \times \tilde{X}) \to \mathbb{R}$$

$$((x, \tilde{x}), (y, \tilde{y})) \mapsto d \times \tilde{d}((x, \tilde{x}), (y, \tilde{y})) := \sqrt{d(x, y)^2 + \tilde{d}(\tilde{x}, \tilde{y})^2}$$

is a distance on $X \times \tilde{X}$.

Proof. One can see that, for $d \times \tilde{d}$, properties di_1 and di_3 of 2.1.1 follow immediately from the corresponding properties for d and for $\tilde{d}$.

As to property di_2, for all $(x, \tilde{x}), (y, \tilde{y}), (z, \tilde{z}) \in X \times \tilde{X}$ we have

$$d \times \tilde{d}((x, \tilde{x}), (y, \tilde{y})) \leq \sqrt{(d(x, z) + d(z, y))^2 + (\tilde{d}(\tilde{x}, \tilde{z}) + \tilde{d}(\tilde{z}, \tilde{y}))^2}$$
$$\leq \sqrt{d(x, z)^2 + \tilde{d}(\tilde{x}, \tilde{z})^2} + \sqrt{d(z, y)^2 + \tilde{d}(\tilde{z}, \tilde{y})^2}$$
$$= d \times \tilde{d}((x, \tilde{x}), (z, \tilde{z})) + d \times \tilde{d}((z, \tilde{z}), (y, \tilde{y})),$$

where the second inequaliy holds since, more in general, the following inequality holds

$$\sqrt{|a_1 + b_1|^2 + |a_2 + b_2|^2} \leq \sqrt{|a_1|^2 + |a_2|^2} + \sqrt{|b_1|^2 + |b_2|^2},$$

for all $a_1, a_2, b_1, b_2 \in \mathbb{C}$ (this inequality will be proved in 10.3.8c). $\square$

2.7.2 Definition. Let (X, d) and $(\tilde{X}, \tilde{d})$ be metric spaces.

The metric space $(X \times \tilde{X}, d \times \tilde{d})$ is called the *product* of the metric spaces (X, d), $(\tilde{X}, \tilde{d})$, and the distance $d \times \tilde{d}$ is called the *product distance*.

2.7.3 Proposition. *Let (X, d) and $(\tilde{X}, \tilde{d})$ be metric spaces. Then we have:*

(a) *A sequence $\{(x_n, \tilde{x}_n)\}$ in $X \times \tilde{X}$ is convergent (with respect to $d \times \tilde{d}$) iff both sequences $\{x_n\}$ and $\{\tilde{x}_n\}$ are convergent (with respect to d and $\tilde{d}$ respectively); in case of convergence, an element $(x, \tilde{x})$ of $X \times \tilde{X}$ is the limit of $\{(x_n, \tilde{x}_n)\}$ iff x is the limit of $\{x_n\}$ and $\tilde{x}$ is the limit of $\{\tilde{x}_n\}$.*

(b) *A sequence $\{(x_n, \tilde{x}_n)\}$ in $X \times \tilde{X}$ is a Cauchy sequence (with respect to $d \times \tilde{d}$) iff both sequences $\{x_n\}$ and $\{\tilde{x}_n\}$ are Cauchy sequences (with respect to d and $\tilde{d}$ respectively).*

(c) *The metric space $(X \times \tilde{X}, d \times \tilde{d})$ is separable iff both metric spaces (X, d) and $(\tilde{X}, \tilde{d})$ are separable.*

(d) *The metric space $(X \times \tilde{X}, d \times \tilde{d})$ is complete iff both metric spaces (X, d) and $(\tilde{X}, \tilde{d})$ are complete.*

Proof. Statements a and b follow directly from the definitions and statement d follows immediately from statements a and b.

As to statement c, assume first (X, d) and $(\tilde{X}, \tilde{d})$ separable, and let S and $\tilde{S}$ be two countable subsets of X and $\tilde{X}$ which are dense in X and $\tilde{X}$ respectively. Then $S \times \tilde{S}$ is a countable subset of $X \times \tilde{X}$. Moreover, for each $(x, \tilde{x}) \in X \times \tilde{X}$, by 2.3.12 there are sequences $\{x_n\}$ and $\{\tilde{x}_n\}$ in S and $\tilde{S}$ respectively such that $x_n \to x$ and $\tilde{x}_n \to \tilde{x}$. Then, by statement a, $\{(x_n, \tilde{x}_n)\}$ is a sequence in $S \times \tilde{S}$ such that $(x_n, \tilde{x}_n) \to (x, \tilde{x})$. By 2.3.12, this proves that $S \times \tilde{S}$ is dense in $X \times \tilde{X}$, and hence that $X \times \tilde{X}$ is separable.

Assume now $(X \times \tilde{X}, d \times \tilde{d})$ separable, and let T be a countable subset of $X \times \tilde{X}$ which is dense in $X \times \tilde{X}$. Let S be the set of first members of T, i.e. $S := \pi_X(T)$ (cf. 1.2.6c). Then S is countable. Fix now $x \in X$ and let $\tilde{x}$ be any element of $\tilde{X}$. By 2.3.12, there is a sequence $\{(x_n, \tilde{x}_n)\}$ in T such that $(x_n, \tilde{x}_n) \to (x, \tilde{x})$, and hence, by statement a, such that $x_n \to x$. Since $x_n \in S$, in view of 2.3.12 this proves that S is dense in X, and hence that X is separable. For $\tilde{X}$, proceed in a similar way.   □

2.7.4 Examples.

(a) The function

$$d_{\mathbb{C}} : \mathbb{C} \times \mathbb{C} \to \mathbb{R}$$

$$(z_1, z_2) \mapsto d_{\mathbb{C}}(z_1, z_2) := |z_1 - z_2|$$

(where $|z|$ is the modulus of $z \in \mathbb{C}$) is a distance on $\mathbb{C}$. Since $\mathbb{C} = \mathbb{R}^2$, with $z = (\operatorname{Re} z, \operatorname{Im} z)$ and $|z| = \sqrt{(\operatorname{Re} z)^2 + (\operatorname{Im} z)^2}$ for $z \in \mathbb{C}$, we see that in fact $d_{\mathbb{C}} = d_{\mathbb{R}} \times d_{\mathbb{R}}$ (cf. 2.1.4). By 2.3.16, 2.6.5 and 2.7.3c,d, $(\mathbb{C}, d_{\mathbb{C}})$ is a separable and complete metric space. We will always consider $\mathbb{C}$ to be the first element of the metric space $(\mathbb{C}, d_{\mathbb{C}})$, and every subset of $\mathbb{C}$ to be the first element of the metric subspace of $(\mathbb{C}, d_{\mathbb{C}})$ it defines.

Since we identify $\mathbb{R}$ with the subset $\{(a, 0) : a \in \mathbb{R}\}$ of $\mathbb{C}$, we can consider the restriction $(d_{\mathbb{C}})_{\mathbb{R}}$ of $d_{\mathbb{C}}$ to $\mathbb{R} \times \mathbb{R}$ (cf. 2.1.4). Clearly, with that identification we have $(d_{\mathbb{C}})_{\mathbb{R}} = d_{\mathbb{R}}$ and the metric subspace of $(\mathbb{C}, d_{\mathbb{C}})$ defined by $\mathbb{R}$ is identified with $(\mathbb{R}, d_{\mathbb{R}})$.

(b) For each $n \in \mathbb{N}$, the function

$$d_n : \mathbb{R}^n \times \mathbb{R}^n \to \mathbb{R}$$

$$((x_1, ..., x_n), (y_1, ..., y_n)) \mapsto d_n((x_1, ..., x_n), (y_1, ..., y_n))$$

$$:= \sqrt{\sum_{k=1}^{n} (x_k - y_k)^2}$$

is the n-fold product $d_{\mathbb{R}} \times \cdots \times d_{\mathbb{R}}$ (defined associatively). In particular, $d_{\mathbb{R}} = d_1$ and $d_{\mathbb{C}} = d_2$. By 2.3.16, 2.6.5 and 2.7.3c,d, $(\mathbb{R}^n, d_n)$ is a separable and complete metric space. We will always consider $\mathbb{R}^n$ to be the first element of the metric space $(\mathbb{R}^n, d_n)$, and every subset of $\mathbb{R}^n$ to be the first element of the metric subspace of $(\mathbb{R}^n, d_n)$ it defines.

2.7.5 Proposition. *Let (X, d), (Y_1, d_1), (Y_2, d_2) be metric spaces.*
Let $\varphi : D_\varphi \to Y_1 \times Y_2$ be a mapping with $D_\varphi \subset X$, and let x be a point in D_φ. Then the mapping φ is continuous at x (with respect to $d_1 \times d_2$) iff both mappings $\pi_{Y_1} \circ \varphi$ and $\pi_{Y_2} \circ \varphi$ (cf. 1.2.6c) are continuous at x (with respect to d_1 and d_2 respectively).

Proof. By 2.4.2, the mapping φ is continuous at x iff the following condition holds:

$$[\{x_n\} \text{ sequence in } D_\varphi \text{ and } x_n \to x] \Rightarrow \varphi(x_n) \to \varphi(x); \tag{1}$$

since $\varphi(x) = ((\pi_{Y_1} \circ \varphi)(x), (\pi_{Y_2} \circ \varphi)(x))$ for all $x \in D_\varphi$, by 2.7.3a condition 1 is equivalent to the condition

$$[\{x_n\} \text{ sequence in } D_\varphi \text{ and } x_n \to x] \Rightarrow (\pi_{Y_i} \circ \varphi)(x_n) \to (\pi_{Y_i} \circ \varphi)(x) \text{ for } i = 1, 2,$$

and by 2.4.2 this condition holds iff $\pi_{Y_1} \circ \varphi$ and $\pi_{Y_2} \circ \varphi$ are both continuous at x. $\square$

2.7.6 Remark. Let (X, d) be a metric space. By 2.7.4a and 2.7.5, a function $\varphi : D_\varphi \to \mathbb{C}$ with $D_\varphi \subset X$ is continuous at $x \in D_\varphi$ iff $\mathrm{Re}\,\varphi$ and $\mathrm{Im}\,\varphi$ (cf. 1.2.19) are both continuous at x.

If φ is a function from $\mathbb{R}$ to $\mathbb{C}$, i.e. $\varphi : D_\varphi \to \mathbb{C}$ with $D_\varphi \subset \mathbb{R}$, and x_0 is a point in D_φ for which $\exists \epsilon > 0$ such that $(x_0 - \epsilon, x_0 + \epsilon) \subset D_\varphi$, we see that φ is differentiable at x_0 (cf. 1.2.21) iff $\exists z \in \mathbb{C}$ such that the following function is continuous at x_0:

$$D_\varphi \to \mathbb{C}$$

$$x \mapsto \begin{cases} \frac{\varphi(x) - \varphi(x_0)}{x - x_0} & \text{if } x \neq x_0, \\ z & \text{if } x = x_0. \end{cases}$$

If z with this property exists, then it is unique and in fact

$$z = (\mathrm{Re}\,\varphi)'(x_0) + i(\mathrm{Im}\,\varphi)'(x_0) = \varphi'(x_0).$$

An analogous remark can be made for one-sided differentiability.

2.7.7 Proposition. *Let (X_1, d_1), (X_2, d_2), (Y, d) be metric spaces, and let a mapping $\varphi : X_1 \times X_2 \to Y$ be continuous at a point $(\overline{x}_1, \overline{x}_2)$ of $X_1 \times X_2$. Then the mapping*

$$\varphi_{\overline{x}_1} : X_2 \to Y$$
$$x_2 \mapsto \varphi_{\overline{x}_1}(x_2) := \varphi(\overline{x}_1, x_2)$$

is continuous at $\overline{x}_2$, and the mapping

$$\varphi^{\overline{x}_2} : X_1 \to Y$$
$$x_1 \mapsto \varphi^{\overline{x}_2}(x_1) := \varphi(x_1, \overline{x}_2)$$

is continuous at $\overline{x}_1$.

Proof. Let $\{x_{2,n}\}$ be a sequence in X_2 such that $x_{2,n} \to \overline{x}_2$. By 2.7.3a, $\{(\overline{x}_1, x_{2,n})\}$ is then a sequence in $X_1 \times X_2$ such that $(\overline{x}_1, x_{2,n}) \to (\overline{x}_1, \overline{x}_2)$. Since φ is continuous at $(\overline{x}_1, \overline{x}_2)$, by 2.4.2 this implies that

$$\varphi_{\overline{x}_1}(x_{2,n}) = \varphi(\overline{x}_1, \overline{x}_{2,n}) \to \varphi(\overline{x}_1, \overline{x}_2) = \varphi_{\overline{x}_1}(x_2).$$

By 2.4.2, this proves that $\varphi_{\overline{x}_1}$ is continuous at $\overline{x}_2$. For $\varphi^{\overline{x}_2}$ one proceeds in a similar way. $\qquad\square$

2.8   Compactness

2.8.1 Definition. *If S is a subset of a set X and $\mathcal{F}$ is a family of subsets of X such that $S \subset \cup_{T \in \mathcal{F}} T$, then $\mathcal{F}$ is called a cover of S and S is said to be covered by $\mathcal{F}$.*

2.8.2 Theorem. *For a subset S of a metric space (X, d) the following three conditions are equivalent:*

(a) the metric subspace (S, d_S) is complete and, for every $\epsilon > 0$, S can be covered by a finite family of open balls with radius ϵ;

(b) for every sequence in S, there exists a subsequence which converges to a point of S;

(c) for every family $\mathcal{G}$ of open sets which is a cover of S, there exists a finite subfamily $\mathcal{G}_f$ of $\mathcal{G}$ which is a cover of S.

Proof. We shall prove $a \Leftrightarrow b$, (a and b) $\Rightarrow c$, $c \Rightarrow b$.

$a \Rightarrow b$: Assume condition a and let $\{x_n\}$ be a sequence in S. We can construct a sequence $\{B_k\}$ of open balls and a sequence $\{n_k\}$ of positive integers such that:

$$\forall k \in \mathbb{N}, \text{ the radius of } B_k \text{ is } \frac{1}{k}, \quad n_k < n_{k+1}, \quad x_{n_k} \in B_l \text{ for all } l \leq k.$$

We proceed inductively as follows. Since there is a finite family of open balls with radius 1 which is a cover of S, at least one of these balls, which we denote by B_1, must contain x_n for infinitely many $n \in \mathbb{N}$; define then the infinite subset $\mathbb{N}_1$ of $\mathbb{N}$ by

$$\mathbb{N}_1 := \{n \in \mathbb{N} : x_n \in B_1\},$$

and choose $n_1 \in \mathbb{N}_1$. Suppose now that we have defined an open ball B_k with radius $\frac{1}{k}$ and an infinite subset $\mathbb{N}_k$ of $\mathbb{N}$ such that $x_n \in B_k$ for all $n \in \mathbb{N}_k$, and that we have chosen $n_k \in \mathbb{N}_k$. Proceed hence as follows; since there is a finite family of open balls with radius $\frac{1}{k+1}$ which is a cover of S and hence of $S \cap B_k$, at least one of these balls, which we denote by B_{k+1}, must contain x_n for infinitely many $n \in \mathbb{N}_k$; define then the infinite subset $\mathbb{N}_{k+1}$ of $\mathbb{N}_k$ by

$$\mathbb{N}_{k+1} := \{n \in \mathbb{N}_k : x_n \in B_{k+1}\},$$

and choose $n_{k+1} \in \mathbb{N}_{k+1}$ such that $n_k < n_{k+1}$ (this is possible because $\mathbb{N}_{k+1}$ is an infinite set). Since $\mathbb{N}_{k+1} \subset \mathbb{N}_k$ for each $k \in \mathbb{N}$, for k and l such that $l < k$ we have $\mathbb{N}_k \subset \mathbb{N}_l$, and hence $n_k \in \mathbb{N}_l$, and hence $x_{n_k} \in B_l$.

Now, $\{x_{n_k}\}$ is a subsequence of $\{x_n\}$ and it is a Cauchy sequence: if $k > l$ then $x_{n_k}, x_{n_l} \in B_l$ and hence $d(x_{n_k}, x_{n_l}) < \frac{2}{l}$. Since (S, d_S) is complete, there exists $x \in S$ such that $x_{n_k} \to x$ as $k \to \infty$.

$b \Rightarrow a$: We shall prove (not a)$\Rightarrow$(not b). Assume (not a), i.e. $[(S, d_S)$ not complete$]$ or $[\exists \epsilon > 0$ s.t. S cannot be covered by a finite family of open balls with radius $\epsilon]$.

If (S, d_S) is not complete, there is a Cauchy sequence $\{x_n\}$ in S with no limit in S. Then, no subsequence of $\{x_n\}$ can converge to a point of S. Indeed, for $\epsilon > 0$ let $N_\epsilon \in \mathbb{N}$ be so that $d(x_n, x_m) < \epsilon$ whenever $n, m > N_\epsilon$; then, if a subsequence $\{x_{n_k}\}$ and $x \in S$ existed such that $x_{n_k} \to x$ as $k \to \infty$, by choosing $k(\epsilon)$ large enough so that $n_{k(\epsilon)} > N_\epsilon$ and $d(x_{n_{k(\epsilon)}}, x) < \epsilon$ we would have

$$n > N_\epsilon \Rightarrow d(x_n, x) \leq d(x_n, x_{n_{k(\epsilon)}}) + d(x_{n_{k(\epsilon)}}, x) < 2\epsilon,$$

and this would prove that $x_n \to x$.

On the other hand, if $\epsilon > 0$ exists such that S cannot be covered by a finite family of open balls with radius ϵ, we can construct a sequence $\{x_n\}$ in S inductively as follows. Choose $x_1 \in S$; having chosen $x_1, ..., x_n$, notice that $S - \cup_{k=1}^n B(x_k, \epsilon) \neq \emptyset$ (otherwise we should have $S \subset \cup_{k=1}^n B(x_k, \epsilon)$) and choose $x_{n+1} \in S - \cup_{k=1}^n B(x_k, \epsilon)$. Then for $n \neq m$ we have $\epsilon \leq d(x_n, x_m)$ (if e.g. $n > m$, then $x_n \notin B(x_m, \epsilon)$), and no subsequence of $\{x_n\}$ can be convergent (cf. 2.6.2).

(a and b) $\Rightarrow$ c: Assume condition b and let $\mathcal{G}$ be a family of open sets which is a cover of S. We can prove by contradiction that

$$\exists n \in \mathbb{N} \text{ such that}$$

$$\left[\left(x \in X \text{ and } B(x, \frac{1}{n}) \cap S \neq \emptyset \right) \Rightarrow \left(\exists G \in \mathcal{G} \text{ s.t. } B(x, \frac{1}{n}) \subset G \right) \right]. \qquad (*)$$

Indeed, suppose to the contrary that

$$\forall n \in \mathbb{N}, \exists x_n \in X \text{ such that}$$

$$\left[B(x_n, \frac{1}{n}) \cap S \neq \emptyset \text{ and } B(x_n, \frac{1}{n}) \not\subset G \text{ for all } G \in \mathcal{G} \right].$$

Then we can construct a sequence in S by choosing $y_n \in B(x_n, \frac{1}{n}) \cap S$; then there are a subsequence $\{y_{n_k}\}$ of $\{y_n\}$ and $y \in S$ so that $y_{n_k} \to y$ as $k \to \infty$. Now, $G_y \in \mathcal{G}$ exists such that $y \in G_y$; then, $\epsilon > 0$ exists so that $B(y, \epsilon) \subset G_y$; hence, if k is large enough so that $\frac{1}{n_k} < \frac{\epsilon}{3}$ and $d(y_{n_k}, y) < \frac{\epsilon}{3}$, we have

$$z \in B(x_{n_k}, \frac{1}{n_k}) \Rightarrow d(z, y) \leq d(z, x_{n_k}) + d(x_{n_k}, y_{n_k}) + d(y_{n_k}, y) < \epsilon,$$

and hence

$$B(x_{n_k}, \frac{1}{n_k}) \subset B(y, \epsilon) \subset G_y,$$

which contradicts $B(x_{n_k}, \frac{1}{n_k}) \not\subset G$ for all $G \in \mathcal{G}$. Thus, condition $(*)$ is proved.

Assuming now condition a as well, there is a finite subset $\{z_1, ..., z_N\}$ of X so that $S \subset \cup_{k=1}^{N} B(z_k, \frac{1}{n})$, and of course we may suppose $B(z_k, \frac{1}{n}) \cap S \neq \emptyset$ for $k = 1, ..., N$. Since $(*)$ holds, we can choose $G_k \in \mathcal{G}$ such that $B(z_k, \frac{1}{n}) \subset G_k$, and we have $S \subset \cup_{k=1}^{N} G_k$.

$c \Rightarrow b$: We shall prove (not b)$\Rightarrow$(not c). Assume (not b), i.e. that there exists a sequence $\{x_n\}$ in S with no subsequence converging to a point of S.

Then the range of $\{x_n\}$ is an infinite set, for otherwise there would exist a subsequence $\{x_{n_k}\}$ and $x \in S$ so that $x_{n_k} = x$ for all $k \in \mathbb{N}$, whence $x_{n_k} \to x$ as $k \to \infty$.

Moreover, for every $x \in S$ there exists $\epsilon_x > 0$ so that $B(x, \epsilon_x)$ contains only a finite number of elements of the range of $\{x_n\}$. Indeed, suppose to the contrary that there exists $x \in S$ such that the sets

$$\mathbb{N}_k := \{n \in \mathbb{N} : x_n \in B(x, \frac{1}{k})\}$$

is infinite for all $k \in \mathbb{N}$. Then we could proceed inductively as follows. We could choose $n_1 \in \mathbb{N}$ and, after choosing $n_k \in \mathbb{N}_k$, we could choose $n_{k+1} \in \mathbb{N}_{k+1}$ in such a way that $n_k < n_{k+1}$ (because $\mathbb{N}_{k+1}$ is an infinite set). Thus we would obtain a subsequence $\{x_{n_k}\}$ of $\{x_n\}$ such that $x_{n_k} \to x$ as $k \to \infty$. But we assumed above that $\{x_n\}$ had no convergent subsequences.

Then, $\{B(x, \epsilon_x)\}_{x \in S}$ is a family of open sets which is a cover of S, but the union of the elements of any finite subfamily of $\{B(x, \epsilon_x)\}_{x \in S}$ can contain only a finite number of elements of the range of $\{x_n\}$, and hence it cannot contain S since S contains the range of $\{x_n\}$ and the range of $\{x_n\}$ is an infinite set. $\square$

2.8.3 Definition. Let (X, d) be a metric space. A subset of X is said to be *compact* if it has the properties a, b and c of 2.8.2. The metric space (X, d) is said to be compact if X is compact (thus X is compact iff for every sequence in X there exists a subsequence which is convergent).

2.8.4 Proposition. *Let (X, d) be a metric space and S a non-empty subset of X. The following conditions are equivalent:*

(a) S is compact;

(b) the metric subspace (S, d_S) is compact.

Proof. $a \Rightarrow b$: Assume S compact. Then for every sequence in S there exists a subsequence which converges (with respect to d) to a point of S; hence this subsequence is convergent in the metric space (S, d_S).

$b \Rightarrow a$: Assume (S, d_S) compact. Then for every sequence in S there exists a subsequence which is convergent in the metric space (S, d_S); hence this subsequence converges (with respect to d) to a point of S. $\qquad\square$

2.8.5 Proposition. *For a subset S of a metric space (X, d) the following conditions are equivalent:*

(a) S is compact;
(b) the metric subspace (S, d_S) is complete and, for each $n \in \mathbb{N}$, there exists a finite family $\{x_{n,1}, ..., x_{n,N_n}\}$ of points of X such that $S \subset \cup_{k=1}^{N_n} K(x_{n,k}, \frac{1}{n})$.

Proof. $a \Rightarrow b$: Assume S compact. Then 2.8.2a implies that (S, d_S) is complete and that, for each $n \in \mathbb{N}$, there exists a finite family $\{x_{n,1}, ..., x_{n,N_n}\}$ of points of X so that $S \subset \cup_{k=1}^{N_n} B(x_{n,k}, \frac{1}{n})$, and hence also so that $S \subset \cup_{k=1}^{N_n} K(x_{n,k}, \frac{1}{n})$.

$b \Rightarrow a$: Assume condition b. Choose $\epsilon > 0$ and let $n \in \mathbb{N}$ be so that $\frac{1}{n} < \epsilon$; then, for a finite family $\{x_{n,1}, ..., x_{n,N_n}\}$ of elements of X such that $S \subset \cup_{k=1}^{N_n} K(x_{n,k}, \frac{1}{n})$, we also have $S \subset \cup_{k=1}^{N_n} B(x_{n,k}, \epsilon)$. Since (S, d_S) is complete, this proves that S has property a of 2.8.2. $\qquad\square$

2.8.6 Proposition. *Let (X, d) be a metric space. A compact subset of X is closed and bounded.*

Proof. Let S be a compact subset of X. Then (S, d_S) is complete, hence S is closed by 2.6.6a. Moreover, there exists a finite subset $\{x_1, ..., x_n\}$ of X such that $S \subset \cup_{k=1}^{n} B(x_k, 1)$, i.e. such that

$$\forall z \in S, \exists k_z \in \{1, ..., n\} \text{ such that } z \in B(x_{k_z}, 1),$$

whence

$$\forall z \in S, d(z, x_1) \le d(z, x_{k_z}) + d(x_{k_z}, x_1) < 1 + \max\{d(x_l, x_1) : l = 1, ..., n\},$$

and this proves that S is bounded. $\qquad\square$

2.8.7 Theorem (Heine–Borel). *In the metric space $(\mathbb{R}^n, d_n)$ (cf. 2.7.4b), every closed and bounded subset of $\mathbb{R}^n$ is compact.*

Proof. Let S be a closed and bounded subset of $\mathbb{R}^n$. Since $(\mathbb{R}^n, d_n)$ is complete, (S, d_S) is complete by 2.6.6b. It remains to prove that S has the second property of condition 2.8.2a. Since for every bounded subset S of $\mathbb{R}^n$ there exists $R > 0$ such that $S \subset Q_R$, with Q_R the cube

$$Q_R := \{(x_1, ..., x_n) \in \mathbb{R}^n : \max\{|x_1|, ..., |x_n|\} \le R\},$$

it is enough to prove that, for every $\epsilon > 0$, Q_R can be covered by a finite family of open balls of radius ϵ. Given $\epsilon > 0$, choose an integer k such that $k > \frac{R\sqrt{n}}{\epsilon}$ and construct a partition of Q_R made of k^n congruent subcubes, by dividing the interval $[-R, R]$ into k intervals of equal length. Each of these subcubes has side length $\frac{2R}{k}$, hence its diameter is $\frac{2R\sqrt{n}}{k} < 2\epsilon$, so it is contained in the open ball that has the center of the subcube as its own center and radius ϵ. Thus, Q_R can be covered by a family of k^n open balls with radius ϵ. $\qquad\square$

2.8.8 Theorem. *In a metric space (X, d), let F and K be subsets of X such that F is closed, K is compact, and $F \subset K$. Then F is compact.*

Proof. Let $\mathcal{G}$ be a family of open sets which is a cover of F. Then

$$K = F \cup (K - F) \subset (\cup_{G \in \mathcal{G}} G) \cup (X - F).$$

Since $\mathcal{G} \cup \{X - F\}$ is a family of open sets and K is compact, this implies that there exists a finite subfamily $\mathcal{G}_f$ of $\mathcal{G}$ such that

$$K \subset (\cup_{G \in \mathcal{G}_f} G) \cup (X - F)$$

and hence such that

$$F = F \cap K \subset F \cap [(\cup_{G \in \mathcal{G}_f} G) \cup (X - F)] = F \cap (\cup_{G \in \mathcal{G}_f} G) \subset \cup_{G \in \mathcal{G}_f} G.$$

This proves that F is compact. $\qquad\square$

2.8.9 Proposition. *Let (X, d) be a metric space and $\{S_1, ..., S_n\}$ a finite family of compact subsets of X. Then $\cup_{k=1}^{n} S_k$ is a compact subset of X.*

Proof. Let $\mathcal{G}$ be a family of open subsets of X which is a cover of $\cup_{k=1}^{n} S_k$. Then, for each $k \in \{1, ..., n\}$, $\mathcal{G}$ is also a cover of S_k, and hence there exists a finite subfamily $\mathcal{G}_k$ of $\mathcal{G}$ which is a cover of S_k. Hence, $\cup_{k=1}^{n} \mathcal{G}_k$ is a finite subfamily of $\mathcal{G}$ which is a cover of $\cup_{k=1}^{n} S_k$. This proves that $\cup_{k=1}^{n} S_k$ is compact. $\qquad\square$

2.8.10 Proposition. *Let (X, d) and $(\tilde{X}, \tilde{d})$ be metric spaces. If S and $\tilde{S}$ are compact subsets in (X, d) and $(\tilde{X}, \tilde{d})$ respectively, then $S \times \tilde{S}$ is a compact subset in the product metric space $(X \times \tilde{X}, d \times \tilde{d})$.*

Proof. Let S and $\tilde{S}$ be compact subsets in (X, d) and $(\tilde{X}, \tilde{d})$ respectively, and let $\{(x_n, \tilde{x}_n)\}$ be a sequence in $S \times \tilde{S}$. Since S is compact, there is a subsequence $\{x_{n_k}\}$ of the sequence $\{x_n\}$ which converges to a point x of S. Since $\{\tilde{x}_{n_k}\}$ is a sequence in $\tilde{S}$ and $\tilde{S}$ is compact, there is a subsequence $\{\tilde{x}_{n_{k_l}}\}$ of $\{\tilde{x}_{n_k}\}$ which converges to a point $\tilde{x}$ of $\tilde{S}$. Now, the subsequence $\{x_{n_{k_l}}\}$ of $\{x_{n_k}\}$ converges to x (cf. 2.1.7b). Then, by 2.7.3a, $\{(x_{n_{k_l}}, \tilde{x}_{n_{k_l}})\}$ is a subsequence of $\{(x_n, \tilde{x}_n)\}$ which converges (with respect to $d \times \tilde{d}$) to $(x, \tilde{x})$, which is a point of $S \times \tilde{S}$. This proves that $S \times \tilde{S}$ is compact. $\qquad\square$

2.8.11 Theorem. *Let (X, d), $(\tilde{X}, \tilde{d})$ be metric spaces, let $\varphi : D_\varphi \to \tilde{X}$ be a continuous mapping with $D_\varphi \subset X$, and let S be a compact subset of X such that $S \subset D_\varphi$. Then $\varphi(S)$ is a compact subset of $\tilde{X}$.*

Proof. Let $\{\tilde{G}_i\}_{i \in I}$ be a family (which for convenience we denote as an indexed family) of open subsets of $\tilde{X}$ such that $\varphi(S) \subset \cup_{i \in I} \tilde{G}_i$. By 2.4.3 and 2.2.5,

$$\forall i \in I, \exists G_i \in \mathcal{T}_d \text{ such that } \varphi^{-1}(\tilde{G}_i) = G_i \cap D_\varphi.$$

Then we have (cf. 1.2.8)

$$S \subset \varphi^{-1}(\varphi(S)) \subset \varphi^{-1}(\cup_{i \in I} \tilde{G}_i) = \cup_{i \in I} \varphi^{-1}(\tilde{G}_i) \subset \cup_{i \in I} G_i.$$

Since S is compact, this implies that there is a finite subset I_f of I such that

$$S \subset \cup_{i \in I_f} G_i,$$

and hence such that

$$S \subset (\cup_{i \in I_f} G_i) \cap D_\varphi = \cup_{i \in I_f} (G_i \cap D_\varphi),$$

and hence such that (cf. 1.2.8)

$$\varphi(S) \subset \varphi(\cup_{i \in I_f} (G_i \cap D_\varphi)) = \cup_{i \in I_f} \varphi(G_i \cap D_\varphi)$$
$$= \cup_{i \in I_f} \varphi(\varphi^{-1}(\tilde{G}_i)) \subset \cup_{i \in I_f} \tilde{G}_i.$$

This proves that $\varphi(S)$ is compact. $\qquad\square$

2.8.12 Corollary. *Let (X, d), $(\tilde{X}, \tilde{d})$ be metric spaces and let $\varphi : D_\varphi \to \tilde{X}$ be a continuous mapping with $D_\varphi \subset X$. If D_φ is a compact subset of X then R_φ is a compact subset of $\tilde{X}$.*

Proof. Immediate, in view of 2.8.11. $\qquad\square$

2.8.13 Definition. *A function φ from a non-empty set X to $\mathbb{C}$, i.e. $\varphi : D_\varphi \to \mathbb{C}$ with $D_\varphi \subset X$, is said to be* bounded *if it has the following property:*

$$\exists m \in [0, \infty) \text{ such that } |\varphi(x)| \leq m \text{ for all } x \in D_\varphi.$$

2.8.14 Corollary. *Let (X, d) be a metric space and let $\varphi : D_\varphi \to \mathbb{C}$ be a continuous function with $D_\varphi \subset X$. If D_φ is a compact subset of X then φ is bounded.*

Proof. Immediate, in view of 2.8.12 and 2.8.6. $\qquad\square$

2.8.15 Theorem. *Let (X, d), $(\tilde{X}, \tilde{d})$ be metric spaces and let $\varphi : D_\varphi \to \tilde{X}$ be a continuous mapping with $D_\varphi \subset X$. If D_φ is a compact subset of X then φ is uniformly continuous.*

Proof. Assume D_φ compact and let $\epsilon > 0$ be given. Since φ is continuous, the following condition is satisfied:

$$\forall x \in D_\varphi, \exists \delta_{x,\epsilon} > 0 \text{ s.t. } [y \in D_\varphi \text{ and } d(x,y) < \delta_{x,\epsilon}] \Rightarrow \tilde{d}(\varphi(x), \varphi(y)) < \frac{\epsilon}{2}.$$

The family $\{B(x, \frac{\delta_{x,\epsilon}}{2})\}_{x \in D_\varphi}$ is a family of open sets and $D_\varphi \subset \cup_{x \in D_\varphi} B(x, \frac{\delta_{x,\epsilon}}{2})$. Since D_φ is compact, there exists a finite subset $\{x_1, ..., x_n\}$ of D_φ such that $D_\varphi \subset \cup_{i=1}^n B(x_i, \frac{\delta_{x_i,\epsilon}}{2})$. We define

$$\delta_\epsilon := \min\left\{ \frac{\delta_{x_i,\epsilon}}{2} : i = 1, ..., n \right\}$$

and we have $\delta_\epsilon > 0$.

Let now x and y be points in D_φ such that $d(x,y) < \delta_\epsilon$. There is $k \in \{1, ..., n\}$ such that $x \in B(x_k, \frac{\delta_{x_k,\epsilon}}{2})$; then we have $\tilde{d}(\varphi(x), \varphi(x_k)) < \frac{\epsilon}{2}$ since $d(x, x_k) < \delta_{x_k,\epsilon}$; we also have $\tilde{d}(\varphi(x_k), \varphi(y)) < \frac{\epsilon}{2}$ since

$$d(x_k, y) \leq d(x_k, x) + d(x, y) < \frac{\delta_{x_k,\epsilon}}{2} + \delta_\epsilon \leq \delta_{x_k,\epsilon};$$

thus we have

$$\tilde{d}(\varphi(x), \varphi(y)) \leq \tilde{d}(\varphi(x), \varphi(x_k)) + \tilde{d}(\varphi(x_k), \varphi(y)) < \frac{\epsilon}{2} + \frac{\epsilon}{2} = \epsilon.$$

This proves that φ is uniformly continuous. $\qquad\square$

2.8.16 Theorem. *In a compact metric space* (X, d), *suppose* F *is a closed set,* $\{G_1, ..., G_N\}$ *a finite family of open sets, and* $F \subset \cup_{n=1}^N G_n$. *Then there exists a family* $\{\psi_1, ..., \psi_N\}$ *of functions* $\psi_n : X \to [0,1]$ *such that* $\psi_n \prec G_n$ *for all* $n \in \{1, ..., N\}$ *and such that* $\left(\sum_{n=1}^N \psi_n\right)(x) = 1$ *for all* $x \in F$.

Proof. Define $I := \{(x, n) \in F \times \{1, ..., N\} : x \in G_n\}$. Since G_n is open,

$$\forall (x, n) \in I, \exists r_{x,n} > 0 \text{ s.t. } B(x, r_{x,n}) \subset G_n.$$

For $(x, n) \in I$, choose $r_{x,n} > 0$ such that $B(x, r_{x,n}) \subset G_n$. Then we have

$$F \subset \cup_{(x,n) \in I} B(x, \frac{r_{x,n}}{2}), \tag{1}$$

since $F \subset \cup_{n=1}^N G_n$ means that $\forall x \in F, \exists n \in \{1, ..., N\}$ such that $(x, n) \in I$.

Since F is closed, F is compact by 2.8.8. Then 1 implies that there is a finite subset I_f of I such that

$$F \subset \cup_{(x,n) \in I_f} B(x, \frac{r_{x,n}}{2}). \tag{2}$$

For $k \in \{1, ..., N\}$, define $I_k := \{(x, n) \in I_f : n = k\}$ (we point out that I_k could be the empty set for some k) and $H_k := \cup_{(x,n) \in I_k} K(x, \frac{r_{x,k}}{2})$ (if I_k is the empty set then $H_k = \emptyset$). From 2 we have

$$F \subset \cup_{k=1}^N \left(\cup_{(x,k) \in I_k} B(x, \frac{r_{x,k}}{2}) \right) \subset \cup_{k=1}^N H_k. \tag{3}$$

For $k \in \{1, ..., N\}$, H_k is closed (cf. 2.3.2 and 2.3.7) and $H_k \subset G_k$ since $r_{x,k}$ has been chosen in such a way that

$$K(x, \frac{r_{x,k}}{2}) \subset B(x, r_{x,k}) \subset G_k, \forall (x, k) \in I_k;$$

then by 2.5.11 there exists a function $\varphi_k : X \to [0,1]$ such that $H_k \prec \varphi_k \prec G_k$.

Define:

$$\psi_1 := \varphi_1,$$
$$\psi_2 := (1_X - \varphi_1)\varphi_2,$$
$$\vdots$$
$$\psi_N := (1_X - \varphi_1)(1_X - \varphi_2) \cdots (1_X - \varphi_{N-1})\varphi_N.$$

For every $n \in \{1, ..., N\}$, ψ_n is a continuous function and $0 \le \psi_n(x) \le 1$ for all $x \in X$ (since ψ_n is a product of continuous functions with values in $[0,1]$), and also $\operatorname{supp} \psi_n \subset G_n$ (since clearly $\operatorname{supp} \psi_n \subset \operatorname{supp} \varphi_n$). Thus, $\psi_n \prec G_n$. It is easily verified, by induction, that

$$\sum_{n=1}^{N} \psi_n = 1_X - (1_X - \varphi_1)(1_X - \varphi_2) \cdots (1_X - \varphi_N). \tag{4}$$

From 3 it follows that $\forall x \in F, \exists k \in \{1, ..., N\}$ s.t. $x \in H_k$, hence s.t. $\varphi_k(x) = 1$, which in view of 4 implies $(\sum_{n=1}^{N} \psi_n)(x) = 1$. $\qquad \square$

2.9 Connectedness

2.9.1 Definition. A metric space (X, d) is said to be *connected* if there does not exist a pair of non-empty open sets G_1 and G_2 such that

$$G_1 \cap G_2 = \emptyset \text{ and } G_1 \cup G_2 = X.$$

A non-empty subset S of X is said to be connected if the metric subspace (S, d_S) is connected.

2.9.2 Proposition. *For a metric space (X, d), the following conditions are equivalent:*

(a) (X, d) is connected;
(b) the only subsets of X which are both open and closed are $\emptyset$ and X;
(c) there does not exist a pair of non-empty closed sets F_1 and F_2 such that

$$F_1 \cap F_2 = \emptyset \text{ and } F_1 \cup F_2 = X.$$

Proof. (not a)$\Rightarrow$(not b): Let G_1 and G_2 be non-empty open sets such that

$$G_1 \cap G_2 = \emptyset \text{ and } G_1 \cup G_2 = X.$$

Then $G_2 = X - G_1$ and hence G_2 is closed and $G_2 \ne X$ (for otherwise $G_1 = \emptyset$).

(not b)$\Rightarrow$(not c): Let S be a subset of X which is both open and closed and such that $S \neq \emptyset$ and $S \neq X$. Then $S' := X - S$ is non-empty and closed and

$$S \cap S' = \emptyset \text{ and } S \cup S' = X.$$

(not c)$\Rightarrow$(not a): Let F_1 and F_2 be non-empty closed sets such that

$$F_1 \cap F_2 = \emptyset \text{ and } F_1 \cup F_2 = X.$$

Then $G_1 := X - F_1$ and $G_2 := X - F_2$ are non-empty ($G_1 = \emptyset$ would imply $F_2 = \emptyset$ since $F_2 = X - F_1$, and similarly for G_2) open sets and

$$G_1 \cap G_2 = X - (F_1 \cup F_2) = \emptyset \text{ and } G_1 \cup G_2 = X - (F_1 \cap F_2) = X.$$

$\square$

2.9.3 Theorem. *Let S be a non-empty subset of $\mathbb{R}$. Then the following conditions are equivalent:*

(a) S is connected, i.e. the metric space (S, d_S) is connected;
(b) S is either $\mathbb{R}$ or an interval or a singleton set, i.e. S is either $\mathbb{R}$ or a non-empty element of one of the families $\mathcal{I}_n$ defined in 6.1.25 for $n = 1, ..., 8$.

Proof. $a \Rightarrow b$: We shall prove (not b)$\Rightarrow$(not a). Suppose that S is neither $\mathbb{R}$ nor an interval nor a singleton set. Then there exist $x, y, z \in \mathbb{R}$ such that

$$x < y < z, \, x, z \in S, \, y \notin S.$$

Then,

$$S = ((-\infty, y) \cap S) \cup ((y, \infty) \cap S)$$

and $(-\infty, y) \cap S$ and $(y, \infty) \cap S$ are two disjoint non-empty sets which are open in the metric space (S, d_S) (cf. 2.2.5). Therefore S is not connected.

$b \Rightarrow a$: If S is a singleton set then it is obviously connected. Then suppose that S is $\mathbb{R}$ or an interval. We shall prove by contradiction that S is connected. Suppose to the contrary that S is not connected. Then (cf. 2.9.2) there exist two non-empty subsets T_1 and T_2 of S which are closed in the metric space (S, d_S) and such that

$$T_1 \cap T_2 = \emptyset \text{ and } T_1 \cup T_2 = S.$$

Since T_1 and T_2 are non-empty we can choose $x_1 \in T_1$ and $x_2 \in T_2$. Since T_1 and T_2 are disjoint, $x_1 \neq x_2$ and (by altering our notation if necessary) we may assume that $x_1 < x_2$. Since S is $\mathbb{R}$ or an interval, $[x_1, x_2] \subset S$ and each point in $[x_1, x_2]$ is in either T_1 or T_2. Since $[x_1, x_2] \cap T_1 \neq \emptyset$, we can define

$$y := \sup([x_1, x_2] \cap T_1).$$

It is clear that $x_1 \leq y \leq x_2$, so $y \in S$. By definition of the l.u.b. (cf. 1.1.5), for each $n \in \mathbb{N}$ we can choose $z_n \in [x_1, x_2] \cap T_1$ such that $y - \frac{1}{n} < z_n$; thus we have a sequence $\{z_n\}$ in T_1 such that $y - \frac{1}{n} < z_n \leq y$; since T_1 is closed in the metric space (S, d_S) and d_S is a restriction of $d_\mathbb{R}$, this proves that $y \in T_1$ (cf. 2.3.4). Since T_1

and T_2 are disjoint, this implies $y < x_2$. For each $n \in \mathbb{N}$ such that $y + \frac{1}{n} \le x_2$ we have $y + \frac{1}{n} \in [x_1, x_2]$ and hence $y + \frac{1}{n} \in S$, and then either $y + \frac{1}{n} \in T_1$ or $y + \frac{1}{n} \in T_2$; however $y + \frac{1}{n} \in T_1$ would imply $y + \frac{1}{n} \in [x_1, x_2] \cap T_1$ and this would contradict the definition of y; therefore, $y + \frac{1}{n} \in T_2$; thus, the sequence $\{y + \frac{1}{n}\}$ is in T_2 for n large enough; since T_2 is closed in the metric space (S, d_S) and d_S is a restriction of $d_\mathbb{R}$, this proves that $y \in T_2$. Therefore y is in both T_1 and T_2, and hence $T_1 \cap T_2 = \emptyset$ and $T_1 \cap T_2 \ne \emptyset$ are both true. This concludes the proof by contradiction. $\square$

2.9.4 Theorem. *Let (X, d) and $(\tilde{X}, \tilde{d})$ be metric spaces and let $\varphi : X \to \tilde{X}$ be a continuous mapping. If (X, d) is connected then R_φ is a connected subset of $\tilde{X}$.*

Proof. The proof is by contraposition. Suppose that R_φ is not a connected subset of $\tilde{X}$. Then (cf. 2.2.5) there exist two open sets G_1 and G_2 in $(\tilde{X}, \tilde{d})$ so that

$$G_1 \cap R_\varphi \ne \emptyset, G_2 \cap R_\varphi \ne \emptyset, (G_1 \cap R_\varphi) \cap (G_2 \cap R_\varphi) = \emptyset,$$

$$(G_1 \cap R_\varphi) \cup (G_2 \cap R_\varphi) = R_\varphi.$$

Since $\varphi^{-1}(G_1) = \varphi^{-1}(G_1 \cap R_\varphi)$ and $\varphi^{-1}(G_2) = \varphi^{-1}(G_2 \cap R_\varphi)$, $\varphi^{-1}(G_1)$ and $\varphi^{-1}(G_2)$ are non-empty, and they are open sets in (X, d) (cf. 2.4.3). Moreover,

$$\varphi^{-1}(G_1) \cap \varphi^{-1}(G_2) = \varphi^{-1}((G_1 \cap R_\varphi) \cap (G_2 \cap R_\varphi)) = \varphi^{-1}(\emptyset) = \emptyset,$$

$$\varphi^{-1}(G_1) \cup \varphi^{-1}(G_2) = \varphi^{-1}((G_1 \cap R_\varphi) \cup (G_2 \cap R_\varphi)) = \varphi^{-1}(R_\varphi) = X.$$

Therefore, (X, d) is not connected. $\square$

2.9.5 Corollary. *Let (X_1, d_1) and (X_2, d_2) be metric spaces such that there exists an isomorphism from (X_1, d_1) onto (X_2, d_2). Then (X_1, d_1) is connected iff (X_2, d_2) is connected.*

Proof. The statement is a direct consequence of 2.9.4 since an isomorphism from (X_1, d_1) onto (X_2, d_2) is a continuous bijection from X_1 onto X_2, and since its inverse is a continuous surjection from X_2 onto X_1. $\square$

2.9.6 Corollary. *The range of a continuous real function defined on a connected metric space is either $\mathbb{R}$ or an interval or a singleton set.*

Proof. The statement follows immediately from 2.9.4 and 2.9.3 $(a \Rightarrow b)$. $\square$

2.9.7 Definition. *Let (X, d) be a metric space. Two subsets S_1 and S_2 of X are said to be separated from one another if $\overline{S_1} \cap \overline{S_2} = \emptyset$.*

2.9.8 Theorem. *Let (X, d) be a metric space and suppose that there exists a family $\mathcal{F}$ of subsets of X such that:*

(a) each element of $\mathcal{F}$ is connected;
(b) $\bigcup_{S \in \mathcal{F}} S = X$;
(c) no two elements of $\mathcal{F}$ are separated from one another.

Then (X, d) is connected.

Proof. Let T be a subset of X which is both open and closed. We shall show that T is either empty or equal to all of X. In view of 2.9.2, this will have proved the statement.

Each element of $\mathcal{F}$ is connected (cf. a), so for any $S \in \mathcal{F}$ we know (cf. 2.9.2) that $T \cap S$ is either empty or all of S, since $T \cap S$ is both open and closed in the metric subspace (S, d_S) (cf. 2.2.5 and 2.3.3).

If $T \cap S = \emptyset$ for all $S \in \mathcal{F}$, then $T = \bigcup_{S \in \mathcal{F}}(T \cap S) = \emptyset$ (cf. b).

The other possibility is that there exists $S_0 \in \mathcal{F}$ such that $T \cap S_0 \neq \emptyset$. Then $T \cap S_0 = S_0$, i.e. $S_0 \subset T$. If S_0 is the only element of $\mathcal{F}$, this gives $T = X$ (cf. b). If not, let S be an element of $\mathcal{F}$ different from S_0; if $T \cap S = \emptyset$ then $S \subset X - T$, and hence $\overline{S} \subset X - T$ since $X - T$ is closed; therefore, $\overline{S_0} \cap \overline{S} = \emptyset$ (we have $\overline{S_0} \subset T$ since T is closed); however, this is not possible since no two elements of $\mathcal{F}$ are separated from one another (cf. c); thus, we must have $T \cap S \neq \emptyset$ and hence $T \cap S = S$. Hence we have $T \cap S = S$ for all $S \in \mathcal{F}$, and hence $T = \bigcup_{S \in \mathcal{F}}(T \cap S) = \bigcup_{S \in \mathcal{F}} S = X$ (cf. b). $\qquad\square$

2.9.9 Theorem. *Let (X, d) and $(\tilde{X}, \tilde{d})$ be connected metric spaces. Then the product metric space $(X \times \tilde{X}, d \times \tilde{d})$ is connected.*

Proof. For each $x \in X$, the subset $\{x\} \times \tilde{X}$ of $X \times \tilde{X}$ is connected by 2.9.5, since there exists an obvious isomorphism from $(\tilde{X}, \tilde{d})$ onto the metric subspace of $(X \times \tilde{X}, d \times \tilde{d})$ defined by $\{x\} \times \tilde{X}$. Similarly, the subset $X \times \{\tilde{x}\}$ is connected for each $\tilde{x} \in \tilde{X}$. Now, for each $(x, \tilde{x}) \in X \times \tilde{X}$, the subsets $\{x\} \times \tilde{X}$ and $X \times \{\tilde{x}\}$ are not separated from one another ($(x, \tilde{x})$ is in both of them), and hence the subset $(\{x\} \times \tilde{X}) \cup (X \times \{\tilde{x}\})$ is connected by 2.9.8.

We have obviously

$$X \times \tilde{X} = \bigcup_{(x,\tilde{x}) \in X \times \tilde{X}} (\{x\} \times \tilde{X}) \cup (X \times \{\tilde{x}\}).$$

For all $(x, \tilde{x}), (x', \tilde{x}') \in X \times \tilde{X}$, $(\{x\} \times \tilde{X}) \cup (X \times \{\tilde{x}\})$ and $(\{x'\} \times \tilde{X}) \cup (X \times \{\tilde{x}'\})$ are not separated from one another since

$$(x, \tilde{x}') \in (\{x\} \times \tilde{X}) \cap (X \times \{\tilde{x}'\}).$$

Therefore $(X \times \tilde{X}, d \times \tilde{d})$ is connected by 2.9.8. $\qquad\square$

2.9.10 Corollary. *The metric space $(\mathbb{R}^n, d_n)$ is connected for all $n \in \mathbb{N}$.*

Proof. We prove the statement by induction.

The metric space $(\mathbb{R}, d_{\mathbb{R}})$ is connected by 2.9.3 ($b \Rightarrow a$). For each $n \in \mathbb{N}$, if $(\mathbb{R}^n, d_n)$ is connected then the product metric space $(\mathbb{R}^n \times \mathbb{R}, d_n \times d_{\mathbb{R}})$ is connected by 2.9.9, and hence $(\mathbb{R}^{n+1}, d_{n+1})$ is connected by 2.9.5 since there exists an obvious isomorphism from $(\mathbb{R}^n \times \mathbb{R}, d_n \times d_{\mathbb{R}})$ onto $(\mathbb{R}^{n+1}, d_{n+1})$. This concludes the proof by induction. $\qquad\square$

Chapter 3

Linear Operators in Linear Spaces

Our main purpose is to study operators in Hilbert spaces, which are in fact linear operators in linear spaces. Hence the subject of this chapter. Throughout the chapter, $\mathbb{K}$ stands for a field. By 0 and 1 we denote the zero and unit elements of $\mathbb{K}$.

3.1 Linear spaces

3.1.1 Definition. A *linear space over* $\mathbb{K}$ (or, simply, a *linear space*) is a triple (X, σ, μ), where X is a non-empty set, σ is a mapping $\sigma : X \times X \to X$, μ is a mapping $\mu : \mathbb{K} \times X \to X$, and the conditions listed under ls_1 and ls_2 are satisfied.

(ls_1) (X, σ) is an abelian group; i.e., with the shorthand notation $f + g := \sigma(f, g)$, we have:

$$f + (g + h) = (f + g) + h, \, \forall f, g, h \in X,$$
$$\exists 0_X \in X \text{ s.t. } f + 0_X = f, \forall f \in X,$$
$$\forall f \in X, \exists f' \in X \text{ s.t. } f + f' = 0_X,$$
$$f + g = g + f, \, \forall f, g \in X;$$

we recall (cf. 1.3.1) that 0_X is the only element of X s.t. $f + 0_X = f$ for all $f \in X$, that for $f \in X$ there is only one element f' of X s.t. $f + f' = 0_X$ and that it is denoted by $-f$, and that we write $f - g := f + (-g)$.

(ls_2) With the shorthand notation $\alpha f := \mu(\alpha, f)$, we have:

$$\alpha(\beta f) = (\alpha\beta)f, \, \forall \alpha, \beta \in \mathbb{K}, \forall f \in X,$$
$$(\alpha + \beta)f = \alpha f + \beta f, \, \forall \alpha, \beta \in \mathbb{K}, \forall f \in X,$$
$$\alpha(f + g) = \alpha f + \alpha g, \, \forall \alpha \in \mathbb{K}, \forall f, g \in X,$$
$$1f = f, \, \forall f \in X.$$

The elements of $\mathbb{K}$ are called *scalars*, and will be preferably denoted by the small Greek letters $\alpha, \beta, \gamma, \dots$. The elements of X are called *vectors*, and will be preferably denoted by the italics $f, g, h, \dots$. The composition law σ is called *vector sum* and the composition law μ is called *scalar multiplication*. Another name for a linear space is *vector space*.

3.1.2 Proposition. *In a linear space X over $\mathbb{K}$ we have:*

(a) $0f = 0_X$, $\forall f \in X$;
(b) $\alpha 0_X = 0_X$, $\forall \alpha \in \mathbb{K}$;
(c) *if $\alpha \in \mathbb{K}$ and $f \in X$ are such that $\alpha f = 0_X$, then $\alpha \neq 0 \Rightarrow f = 0_X$ (or equivalently $f \neq 0_X \Rightarrow \alpha = 0$);*
(d) $(-1)f = -f$, $\forall f \in X$;
(e) $(-\alpha)f = -(\alpha f)$, $\forall \alpha \in \mathbb{K}, \forall f \in X$ *(hence we will write $-\alpha f := (-\alpha)f$).*

Proof. a: We have $f = 1f = (1+0)f = f + 0f$, and hence $0_X = 0f$.

b: We have $\alpha 0_X = \alpha(0_X + 0_X) = \alpha 0_X + \alpha 0_X$, and hence $\alpha 0_X = 0_X$.

c: If $\alpha \neq 0$, we have $\alpha^{-1}\alpha = 1$; thus, if $\alpha f = 0_X$, we have $f = 1f = (\alpha^{-1}\alpha)f = \alpha^{-1}(\alpha f) = \alpha^{-1}0_X = 0_X$ by result b.

d: We have $f + (-1)f = 1f + (-1)f = (1-1)f = 0f = 0_X$ by result a, and hence $(-1)f = -f$.

e: We have $(-\alpha)f = (-1\alpha)f = (-1)(\alpha f) = -(\alpha f)$ by result d. $\square$

3.1.3 Definition. Let (X, σ, μ) be a linear space over $\mathbb{K}$. A *linear manifold* in X is a non-empty subset M of X which has the following properties:

(lm_1) $f, g \in M \Rightarrow f + g \in M$;
(lm_2) $(\alpha \in \mathbb{K}$ and $f \in M) \Rightarrow \alpha f \in M$.

Condition lm_1 is equivalent to $\sigma(M \times M) \subset M$ and condition lm_2 is equivalent to $\mu(\mathbb{K} \times M) \subset M$. Thus, M can be used as final set of the mappings $\sigma_{M \times M}$ and $\mu_{\mathbb{K} \times M}$ if and only if M is a linear manifold. If M is a linear manifold, it is then immediately clear that $(M, \sigma_{M \times M}, \mu_{\mathbb{K} \times M})$ is a linear space over $\mathbb{K}$, since the conditions listed under ls_1 and ls_2 in 3.1.1 trivially hold with X replaced by M. If M is a linear manifold, the linear space $(M, \sigma_{M \times M}, \mu_{\mathbb{K} \times M})$ will be referred to as the linear space M. If M is a linear manifold, we have $0_X \in M$ and, for the zero element 0_M of the linear space M, we have $0_M = 0_X$ (cf. 1.3.2).

A linear manifold is also called linear subspace or vector subspace, but we reserve the name "subspace" for a closed linear manifold in a Banach space (cf. 4.1.9).

3.1.4 Remarks. The following facts are immediately clear:

(a) In any linear space X there are two trivial linear manifolds: $\{0_X\}$ and X.
(b) If M is a linear manifold in a linear space (X, σ, μ) and N is a non-empty subset of M, then N is a linear manifold in (X, σ, μ) iff N is a linear manifold in $(M, \sigma_{M \times M}, \mu_{\mathbb{K} \times M})$.
(c) Conditions lm_1 and lm_2 of 3.1.3 are equivalent to the one condition:

   (lm) $(\alpha, \beta \in \mathbb{K}$ and $f, g \in M) \Rightarrow \alpha f + \beta g \in M$.

3.1.5 Proposition. *Let $\mathcal{F}$ be a family of linear manifolds in a linear space X. Then $\cap_{M \in \mathcal{F}} M$ is a linear manifold in X.*

Proof. We have

$$(\alpha, \beta \in \mathbb{K} \text{ and } f, g \in \cap_{M \in \mathcal{F}} M) \Rightarrow$$

$$(f, g \in M \text{ and hence } \alpha f + \beta g \in M, \forall M \in \mathcal{F}) \Rightarrow \alpha f + \beta g \in \cap_{M \in \mathcal{F}} M.$$

$\square$

3.1.6 Definition. Let S be a subset of a linear space X. Consider the family

$$\mathcal{F} := \{M \in \mathcal{P}(X) : M \text{ is a linear manifold in } X \text{ and } S \subset M\}$$

and define

$$LS := \cap_{M \in \mathcal{F}} M.$$

We have:

(a) LS is a linear manifold in X (cf. 3.1.5);
(b) $S \subset LS$ (immediate from the definition of LS);
(c) if M is a linear manifold in X and $S \subset M$, then $LS \subset M$ (immediate from the definition of LS).

Thus, LS is the smallest linear manifold in X that contains S. For this reason, LS is called the *linear manifold generated* by S (LS is also called the *linear span* or *linear hull* of S, owing to 3.1.7).

From the definition of LS we also have immediately:

(d) $LS = S$ iff S is a linear manifold;
(e) if T is a subset of X such that $S \subset T$, then $LS \subset LT$.

3.1.7 Proposition. *Let S be a non-empty subset of a linear space X over $\mathbb{K}$. We have*

$$LS = \{\sum_{i=1}^{n} \alpha_i f_i : n \in \mathbb{N}, (\alpha_1, ..., \alpha_n) \in \mathbb{K}^n, (f_1, ..., f_n) \in S^n\}$$

$$:= \{f \in X : \exists n \in \mathbb{N}, \exists(\alpha_1, ..., \alpha_n) \in \mathbb{K}^n, \exists(f_1, ..., f_n) \in S^n \text{ s.t. } f = \sum_{i=1}^{n} \alpha_i f_i\}.$$

Proof. Define $M := \{\sum_{i=1}^{n} \alpha_i f_i : n \in \mathbb{N}, (\alpha_1, ..., \alpha_n) \in \mathbb{K}^n, (f_1, ..., f_n) \in S^n\}$. Obviously, M is a linear manifold in X and $S \subset M$; hence $LS \subset M$. On the other hand, since $S \subset LS$ and LS is a linear manifold in X, we have $\sum_{i=1}^{n} \alpha_i f_i \in LS$, $\forall n \in \mathbb{N}, \forall(\alpha_1, ..., \alpha_n) \in \mathbb{K}^n, \forall(f_1, ..., f_n) \in S^n$; thus $M \subset LS$. $\square$

3.1.8 Definition. If S_1 and S_2 are subsets of a linear space X, we define their sum $S_1 + S_2$ by

$$S_1 + S_2 := \{f_1 + f_2 : f_1 \in S_1 \text{ and } f_2 \in S_2\}.$$

It is immediate to see that, if S_1 and S_2 are linear manifolds in X, then $S_1 + S_2$ is a linear manifold in X.

3.1.9 Definition. Let X and Y be linear spaces over the same field $\mathbb{K}$. It is immediate to see that the set $X \times Y$ becomes a linear space over $\mathbb{K}$ if we define vector sum and scalar multiplication by the rules:

$$(f_1, g_1) + (f_2, g_2) := (f_1 + f_2, g_1 + g_2), \qquad \forall (f_1, g_1), (f_2, g_2) \in X \times Y;$$

$$\alpha(f, g) := (\alpha f, \alpha g), \qquad \forall \alpha \in \mathbb{K}, \forall (f, g) \in X \times Y.$$

This linear space is called the *sum of the linear spaces* X, Y and is denoted by $X + Y$. It is immediate to see that the two subsets of $X \times Y$

$$\hat{X} := \{(f, 0_Y) : f \in X\} \text{ and } \hat{Y} := \{(0_X, g) : g \in X\}$$

are linear manifolds in $X + Y$ and that $X \times Y = \hat{X} + \hat{Y}$, with $\hat{X} + \hat{Y}$ defined as in 3.1.8.

3.1.10 Examples.

(a) Let x denote a point (of any set). Define $X := \{x\}$, and vector sum and scalar multiplication by the rules:

$$\sigma(x, x) := x, \quad \mu(\alpha, x) := x \text{ for all } \alpha \in \mathbb{K}.$$

The triple (X, σ, μ) defined in this way is a trivial linear space, which is called a *zero linear space*. If X is a zero linear space, we have $X = \{0_X\}$.

(b) Define $X := \mathbb{K}$, and vector sum and scalar multiplication by the rules:

$$\sigma(z_1, z_2) := z_1 + z_2, \qquad \forall z_1, z_2 \in \mathbb{K},$$

$$\mu(\alpha, z) := \alpha z, \qquad \forall \alpha, z \in \mathbb{K}$$

(where $z_1 + z_2$ and αz are the sum and the product that are defined in the field $\mathbb{K}$). The triple (X, σ, μ) defined in this way is a linear space over $\mathbb{K}$, which is called the *linear space* $\mathbb{K}$.

(c) Let X be a non-empty set and let $\mathcal{F}(X)$ denote the family of all the functions from X to $\mathbb{C}$ that have the whole of X as their domains, i.e. the family of complex functions on X. Define the mappings

$$\sigma : \mathcal{F}(X) \times \mathcal{F}(X) \to \mathcal{F}(X)$$

$$(\varphi, \psi) \mapsto \sigma(\varphi, \psi) := \varphi + \psi,$$

$$\mu : \mathbb{C} \times \mathcal{F}(X) \to \mathcal{F}(X)$$

$$(\alpha, \varphi) \mapsto \mu(\alpha, \varphi) := \alpha \varphi,$$

where $\varphi + \psi$ and $\alpha \varphi$ are defined as in 1.2.19. It is immediate to check that $(\mathcal{F}(X), \sigma, \mu)$ is a linear space over $\mathbb{C}$ (hence, the symbols $\varphi + \psi$ and $\alpha \varphi$ defined in 1.2.19 are in agreement with the shorthand notations used in 3.1.1), with the function

$$0_X : X \to \mathbb{C}$$

$$x \mapsto 0_X(x) := 0$$

(cf. 1.2.19) as zero element, and the function $-\varphi$ (cf. 1.2.19) as the opposite element of an element φ of $\mathcal{F}(X)$.

(d) Let X be a non-empty set, and let $\mathcal{F}_B(X)$ denote the set of all bounded elements of $\mathcal{F}(X)$:

$$\mathcal{F}_B(X) := \{\varphi \in \mathcal{F}(X) : \exists m_\varphi \in [0, \infty) \text{ such that}$$
$$|\varphi(x)| \le m_\varphi \text{ for all } x \in X\}.$$

It is immediate to check that $\mathcal{F}_B(X)$ is a linear manifold in $\mathcal{F}(X)$.

(e) Let (X, d) be a metric space, and define

$$\mathcal{C}(X) := \{\varphi \in \mathcal{F}(X) : \varphi \text{ is continuous}\}.$$

Since a linear combination (cf. 3.1.12) of continuous functions is a continuous function, $\mathcal{C}(X)$ is a linear manifold in $\mathcal{F}(X)$.

We also define

$$\mathcal{C}_B(X) := \mathcal{C}(X) \cap \mathcal{F}_B(X),$$

which is a linear manifold in $\mathcal{F}(X)$ by 3.1.5, and hence in $\mathcal{F}_B(X)$ by 3.1.4b. If (X, d) is a compact metric space, we have $\mathcal{C}(X) = \mathcal{C}_B(X)$ by 2.8.14.

(f) For $a, b \in \mathbb{R}$ such that $a < b$, the family of functions $\mathcal{C}(a, b) := \mathcal{C}([a, b])$ is a linear manifold in $\mathcal{F}_B([a, b])$ since $[a, b]$ is compact (cf. 2.3.7 and 2.8.7).

By $\mathcal{C}^1(a, b)$ we denote the set of all the elements of $\mathcal{C}(a, b)$ that are differentiable at all point of $[a, b]$ and such that their derivatives (cf. 1.2.21 and 2.7.6) are elements of $\mathcal{C}(a, b)$ (differentiability and derivatives at a and b are one-sided). Since a linear combination of differentiable functions is differentiable and its derivative is the linear combination of the derivatives, $\mathcal{C}^1(a, b)$ is a linear manifold in $\mathcal{C}(a, b)$.

(g) By $\mathcal{C}_c(\mathbb{R})$ we denote the family of all continuous complex functions on $\mathbb{R}$ whose support is compact, i.e. we define

$$\mathcal{C}_c(\mathbb{R}) := \{\varphi \in \mathcal{C}(\mathbb{R}) : \operatorname{supp} \varphi \text{ is compact}\}$$

(for $\operatorname{supp} \varphi$ cf. 2.5.9). From 2.8.6 and 2.8.7, for $\varphi \in \mathcal{C}(\mathbb{R})$ we have

$$\varphi \in \mathcal{C}_c(\mathbb{R}) \Leftrightarrow \exists a_\varphi, b_\varphi \in \mathbb{R} \text{ s.t. } a_\varphi < b_\varphi \text{ and } \operatorname{supp} \varphi \subset [a_\varphi, b_\varphi].$$

(h) By $\mathcal{C}^\infty(\mathbb{R})$ we denote the subset of $\mathcal{F}(\mathbb{R})$ defined by

$$\mathcal{C}^\infty(\mathbb{R}) := \{\varphi \in \mathcal{F}(\mathbb{R}) : \varphi \text{ is infinitely differentiable at all points of } \mathbb{R}\}.$$

Clearly, $\mathcal{C}^\infty(\mathbb{R})$ is a linear manifold in $\mathcal{F}(\mathbb{R})$.

Next, we define the *Schwartz space of functions of rapid decrease*:

$$\mathcal{S}(\mathbb{R}) := \{\varphi \in \mathcal{C}^\infty(\mathbb{R}) : \lim_{x \to \pm\infty} x^k \varphi^{(l)}(x) = 0,$$
$$\forall k = 0, 1, 2, ..., \forall l = 0, 1, 2, ...\},$$

where $\varphi^{(l)}$ denotes the l-th derivative of φ (and $\varphi^{(0)} := \varphi$).

The following properties of $\mathcal{S}(\mathbb{R})$ are easily checked:

(1) $\varphi \in \mathcal{S}(\mathbb{R}) \Rightarrow \varphi^{(l)} \in \mathcal{S}(\mathbb{R}), \forall l \in \mathbb{N}$;

(2) $\varphi \in \mathcal{S}(\mathbb{R}) \Rightarrow \overline{\varphi} \in \mathcal{S}(\mathbb{R})$;

(3) $(\alpha, \beta \in \mathbb{C}$ and $\varphi, \psi \in \mathcal{S}(\mathbb{R})) \Rightarrow \alpha\varphi + \beta\psi \in \mathcal{S}(\mathbb{R})$;

(4) $\varphi \in \mathcal{S}(\mathbb{R}) \Rightarrow p\varphi \in \mathcal{S}(\mathbb{R})$, for every polynomial p with complex coefficients;

(5) $\varphi \in \mathcal{S}(\mathbb{R}) \Rightarrow e^{ip}\varphi \in \mathcal{S}(\mathbb{R})$, for every polynomial p with real coefficients;

(6) $\varphi, \psi \in \mathcal{S}(\mathbb{R}) \Rightarrow \varphi\psi \in \mathcal{S}(\mathbb{R})$;

(7) $\varphi \in \mathcal{S}(\mathbb{R}) \Rightarrow \sup\{|x^k\varphi^{(l)}(x)| : x \in \mathbb{R}\} < \infty, \forall k = 0, 1, 2, ..., \forall l = 0, 1, 2, ...$

Property 3 means that $\mathcal{S}(\mathbb{R})$ is a linear manifold in $\mathcal{C}^\infty(\mathbb{R})$.

We will see now just the few facts about linear independence and linear dimension that will be used in our study of Hilbert spaces. Thus, our treatment of these subjects will be nowhere near complete.

3.1.11 Definition. A non-empty subset S of a linear space X over $\mathbb{K}$ is said to be *linearly independent* if for each $n \in \mathbb{N}$ and each subset $\{f_1, ..., f_n\}$ of S the following condition is satisfied:

$$[(\alpha_1, ..., \alpha_n) \in \mathbb{K}^n, \sum_{i=1}^n \alpha_i f_i = 0_X] \Rightarrow \alpha_i = 0 \text{ for } i = 1, ..., n.$$

The set S is said to be *linearly dependent* if it is not linearly independent.

Clearly, every subset of a linearly independent set is linearly independent as well.

3.1.12 Definition. If $\{f_1, ..., f_n\}$ is a finite non-empty subset of a linear space X over $\mathbb{K}$ and $(\alpha_1, ..., \alpha_n) \in \mathbb{K}^n$, then the vector $f := \sum_{i=1}^n \alpha_i f_i$ is called a *linear combination* of $f_1, ..., f_n$.

3.1.13 Definition. A *linear basis* in a non-zero linear space X is a linearly independent subset B of X such that $X = LB$, i.e. (cf. 3.1.7) such that for every vector f of X there is a finite subset of B of which f is a linear combination.

3.1.14 Proposition. *Let $S = \{f_1, ..., f_n\}$ be a finite non-empty subset of a linear space X over $\mathbb{K}$. If $n = 1$, then S is linearly dependent iff $f_1 = 0_X$. If $n > 1$ and $f_1 \neq 0_X$, then S is linearly dependent iff some one of the elements of S is a linear combination of the element of S which precede it (i.e. there exists $k \in \{2, ..., n\}$ such that f_k is a linear combination of $f_1, ..., f_{k-1}$).*

Proof. The first statement follows from 3.1.2b,c. Assume then $n > 1$ and $f_1 \neq 0_X$.

If there exists $k \in \{2, ..., n\}$ such that $f_k = \sum_{i=1}^{k-1} \alpha_i f_i$, with $\alpha_i \in \mathbb{K}$ for $i = 1, ..., k-1$, then $\sum_{i=1}^{k-1} \alpha_i f_i - 1 f_k = 0_X$, and this proves that S is linearly dependent.

Assume now that S is linearly dependent; this means that there are a non-empty subset $\{f_{i_1}, ..., f_{i_r}\}$ of S and $(\alpha_1, ..., \alpha_r) \in \mathbb{K}^r$ such that $(\alpha_1, ..., \alpha_r) \neq (0, ..., 0)$ and $\sum_{l=1}^r \alpha_l f_{i_l} = 0_X$; hence there is $(\beta_1, ..., \beta_n) \in \mathbb{K}^n$ such that $(\beta_1, ..., \beta_n) \neq (0, ..., 0)$ and $\sum_{i=1}^n \beta_i f_i = 0_X$. If β_k is the last non-zero element in $(\beta_1, ..., \beta_n)$, then $k > 1$ (since $f_1 \neq 0_X$) and $f_k = \sum_{i=1}^{k-1} (-\frac{\beta_i}{\beta_k}) f_i$. $\qquad\square$

3.1.15 Theorem. *Let X be a non-zero linear space and assume that there exists a linear basis in X which is finite and contains n elements. Then every linearly independent subset of X which contains n elements is a linear basis.*

Proof. Let $\{e_1, ..., e_n\}$ be the linear basis in X of which we assume the existence, and let $\{f_1, ..., f_n\}$ be a linearly independent subset of X.

Since f_1 is a linear combination of the e_i's, the set

$$S_1 := \{f_1, e_1, ..., e_n\}$$

is linearly dependent; then, by 3.1.14, there is one of the e_i's, say e_{i_1}, which is a linear combinations of the vectors that precede it in S_1; if we delete e_{i_1}, the remaining set

$$S_1' := \{f_1, e_1, ..., e_{i_1-1}, e_{i_1+1}, ..., e_n\}$$

(we have written as if $i_1 \neq 1$ and $i_1 \neq n$; otherwise we would have written $S_1' := \{f_1, e_2, ..., e_n\}$ or $S_1' := \{f_1, e_1, ..., e_{n-1}\}$) is still (as it was $\{e_1, ..., e_n\}$) such that $X = LS_1'$. Just as before, f_2 is a linear combination of the vectors in S_1', so the set

$$S_2 := \{f_1, f_2, e_1, ..., e_{i_1-1}, e_{i_1+1}, ..., e_n\}$$

is linearly dependent; then, by 3.1.14, some vector in S_2 is a linear combination of the vectors which precede it in S_2, and (since $\{f_1, f_2\}$ is a linearly independent set) this vector must be one of the e_i's which are left in S_2, say e_{i_2}; if we delete e_{i_2} from S_2, the remaining set

$$S_2' := \{f_1, f_2, e_1, ..., e_{i_1-1}, e_{i_1+1}, ..., e_{i_2-1}, e_{i_2+1}, ..., e_n\}$$

(we have written as if $i_1 < i_2$, but it could be well the other way round) is still (as it was S_1') such that $X = LS_2'$. Continuing in this way, in the end we are left with the set

$$S_n' := \{f_1, ..., f_n\}$$

which is such that $X = LS_n'$. This proves that $\{f_1, ..., f_n\}$ is a linear basis. $\quad\square$

3.1.16 Corollary. *Let X be a non-zero linear space and assume that there exists a linear basis in X which is finite and contains n elements. Then every linearly independent subset of X is finite and contains at most n elements.*

Proof. Our proof is by contradiction. Assume that there exists a linearly independent subset S of X which is either infinite or finite with more than n elements. In both cases, there is a subset $\{f_1, ..., f_n, f_{n+1}\}$ of S which contains $n+1$ elements. As any subset of S, $\{f_1, ..., f_n, f_{n+1}\}$ is linearly independent. However, $\{f_1, ..., f_n\}$ is also linearly independent, hence a linear basis by 3.1.15. But then f_{n+1} is a linear combination of $f_1, ..., f_n$, and this contradicts the linear independence of $\{f_1, ..., f_n, f_{n+1}\}$. $\quad\square$

3.1.17 Corollary. *If in a linear space X there exists a finite linear basis, then every other linear basis in X is finite and contains the same number of elements.*

Proof. Assume that B_1 is a linear basis in X which is finite and contains n elements, and let B_2 be another linear basis in X. Since B_2 is a linearly independent subset of X, by 3.1.16 B_2 is finite and contains m elements with $m \leq n$. Then, since B_2 is a finite linear basis with m elements and B_1 is a linearly independent subset of X, by 3.1.16 we have $n \leq m$. Thus, $m = n$. $\square$

3.1.18 Definition. A non-zero linear space X is said to be *finite-dimensional* if there exists a linear basis in X which is finite.

If a non-zero linear space X is finite-dimensional, then by 3.1.17 all its bases are finite and contain the same number of elements, which is said to be the *linear dimension* of X.

The linear dimension of a zero linear space (cf. 3.1.10a) is defined to be zero.

3.1.19 Proposition. *Let $\{f_n\}$ be a sequence in a linear space X, and assume that there exists $n \in \mathbb{N}$ such that $f_n \neq 0_X$. Then there exists an N-tuple $(f_{n_1}, ..., f_{n_N})$ or a subsequence $\{f_{n_k}\}$ such that, letting $I := \{1, ..., N\}$ or $I := \mathbb{N}$, $\{f_{n_k}\}_{k \in I}$ is a linearly independent subset of X and $L\{f_{n_k}\}_{k \in I} = L\{f_n\}_{n \in \mathbb{N}}$.*

Proof. We proceed by induction as follows. The set $S_0 := \{n \in \mathbb{N} : f_n \neq 0_X\}$ is not empty, and we define $n_1 := \min S_0$. For $k \in \mathbb{N}$, if the family $\{n_1, ..., n_k\}$ of elements of $\mathbb{N}$ has already been defined in such a way that the set $\{f_{n_1}, ..., f_{n_k}\}$ is linearly independent and f_n is a linear combination of $f_{n_1}, ..., f_{n_k}$ for every $n \in \mathbb{N}$ such that $n \leq n_k$ (note that the family $\{n_1\}$ meets these requirements trivially), let

$$S_k := \{n \in \mathbb{N} : \{f_{n_1}, ..., f_{n_k}, f_n\} \text{ is linearly independent}\};$$

if $S_k \neq \emptyset$ we define $n_{k+1} := \min S_k$ and we note that the set $\{f_{n_1}, ..., f_{n_k}, f_{n_{k+1}}\}$ is obviously linearly independent and that, for $n \in \mathbb{N}$ such that $n \leq n_{k+1}$, f_n is a linear combination of $f_{n_1}, ..., f_{n_k}, f_{n_{k+1}}$ (if $n \leq n_k$ this is already known, for $n = n_{k+1}$ this is obvious, for $n_k < n < n_{k+1}$ this follows from 3.1.14). If $N \in \mathbb{N}$ is such that $S_{N-1} \neq \emptyset$ and $S_N = \emptyset$, this procedure defines an N-tuple $\{f_{n_1}, ..., f_{n_N}\}$; if $S_k \neq \emptyset$ for all $k \in \mathbb{N}$, this procedure defines a subsequence $\{f_{n_k}\}$. We define $I := \{1, ..., N\}$ in the first case and $I := \mathbb{N}$ in the latter. In either case, the set $\{f_{n_k}\}_{k \in I}$ is clearly linearly independent since the set $\{f_{n_1}, ..., f_{n_k}\}$ is linearly independent for each $k \in I$. Also, if $I = \{1, ..., N\}$ then, for each $n \in \mathbb{N}$, f_n is a linear combination of $f_{n_1}, ..., f_{n_N}$; if $I = \mathbb{N}$ then, for each $n \in \mathbb{N}$, f_n is a linear combination of $f_{n_1}, ..., f_{n_k}$ for $n_k \geq n$. Then, in view of 3.1.7 we have

$$L\{f_n\}_{n \in \mathbb{N}} \subset L\{f_{n_k}\}_{k \in I}.$$

Since $\{f_{n_k}\}_{k \in I} \subset \{f_n\}_{n \in \mathbb{N}}$ implies $L\{f_{n_k}\}_{k \in I} \subset L\{f_n\}_{n \in \mathbb{N}}$ (cf. 3.1.6e), we have

$$L\{f_{n_k}\}_{k \in I} = L\{f_n\}_{n \in \mathbb{N}}.$$

$\square$

3.2 Linear operators

3.2.1 Definition. Let X and Y be linear spaces over the same field $\mathbb{K}$. A *linear operator* (or, simply, an *operator*) from X to Y is a mapping A from X to Y, i.e. $A : D_A \to Y$ with $D_A \subset X$ (the set D_A is called the domain of A, cf. 1.2.1), which has the following properties:

(lo_1) D_A is a linear manifold in X;
(lo_2) $A(f + g) = Af + Ag$, $\forall f, g \in D_A$;
(lo_3) $A(\alpha f) = \alpha Af$, $\forall \alpha \in \mathbb{K}, \forall f \in D_A$.

By tradition, linear operators are denoted by capital letters, and the value $A(f)$ of a linear operator A at $f \in D_A$ is written as Af.

When $X = Y$, a linear operator A from X to X is called a linear operator *in* X, and *on* X if $D_A = X$.

When Y is the linear space $\mathbb{K}$ (cf. 3.1.10b), a linear operator from X to $\mathbb{K}$ is called a *linear functional*.

We point out that conditions lo_2 and lo_3 are consistent only when condition lo_1 is assumed, and that conditions lo_1, lo_2, lo_3 are equivalent to the one condition:

(lo) $(\alpha, \beta \in \mathbb{K}$ and $f, g \in D_A) \Rightarrow (\alpha f + \beta g \in D_A$ and $A(\alpha f + \beta g) = \alpha Af + \beta Ag)$.

We denote by the symbol $\mathcal{O}(X, Y)$ the family of all linear operators from X to Y. If $X = Y$, we write $\mathcal{O}(X) := \mathcal{O}(X, X)$.

For $A \in \mathcal{O}(X, Y)$, the *null space* (also called the *kernel*) of A is the subset of X defined by

$$N_A := \{f \in X : f \in D_A \text{ and } Af = 0_Y\}.$$

3.2.2 Definition. Let X and Y be linear spaces over the same field $\mathbb{K}$, and let $A \in \mathcal{O}(X, Y)$. We have:

(a) R_A is a linear manifold in Y (for the range R_φ of a mapping φ, cf. 1.2.1);
(b) N_A is a linear manifold in X.

Proof. a: By its definition, R_A can never be the empty set. Moreover we have:

$(\alpha, \beta \in \mathbb{K}$ and $g_1, g_2 \in R_A) \Rightarrow$
$[\exists f_1, f_2 \in D_A$ s.t. $g_1 = Af_1$ and $g_2 = Af_2$, and hence s.t.
$\alpha f_1 + \beta f_2 \in D_A$ and $\alpha g_1 + \beta g_2 = \alpha Af_1 + \beta Af_2 = A(\alpha f_1 + \beta f_2)] \Rightarrow$
$\alpha g_1 + \beta g_2 \in R_A$.

b: Since D_A is a linear manifold, $0_X \in D_A$ and we have

$$A0_X = A(00_X) = 0A(0_X) = 0_Y$$

(cf. 3.1.2a). Thus, N_A is not the empty set. Moreover we have:

$$(\alpha, \beta \in \mathbb{K} \text{ and } f_1, f_2 \in N_A) \Rightarrow$$
$$[f_1, f_2 \in D_A, \text{ whence } \alpha f_1 + \beta f_2 \in D_A, \text{ and}$$
$$A f_1 = A f_2 = 0_Y, \text{ whence } A(\alpha f_1 + \beta f_2) = \alpha A f_1 + \beta A f_2 = 0_Y] \Rightarrow$$
$$\alpha f_1 + \beta f_2 \in N_A.$$

$\square$

3.2.3 Definition. Let X and Y be linear spaces over the same field. For $A, B \in \mathcal{O}(X, Y)$, in agreement with 1.2.5 we write $A \subset B$ (or $B \supset A$) if

$$f \in D_A \Rightarrow (f \in D_B \text{ and } Af = Bf),$$

and we have $A = B$ iff

$$A \subset B \text{ and } D_B \subset D_A.$$

For $A \in \mathcal{O}(X, Y)$ and a subset M of D_A, one can see at once that the restriction A_M of A to M (cf. 1.2.5) is a linear operator iff M is a linear manifold.

3.2.4 Definition. Let X, Y, Z be linear spaces over the same field $\mathbb{K}$. For $A \in \mathcal{O}(X, Y)$ and $B \in \mathcal{O}(Y, Z)$, we can define the composition $B \circ A$ of B with A (cf. 1.2.12), which by tradition is written as BA and is called the *product* of A and B. Thus, we have:

$$D_{BA} := \{f \in D_A : Af \in D_B\} = A^{-1}(D_B),$$
$$BA : D_{BA} \to Z$$
$$f \mapsto (BA)f := B(Af),$$

where $A^{-1}(D_B)$ denotes the counterimage of D_B under A. From now on we will write $BAf := (BA)f$, for $f \in D_{BA}$. The mapping BA is a linear operator, since condition *lo* of 3.2.1 holds:

$$(\alpha, \beta \in \mathbb{K} \text{ and } f, g \in D_{AB}) \Rightarrow$$
$$[f, g \in D_A \text{ and } Af, Ag \in D_B, \text{ whence } \alpha f + \beta g \in D_A \text{ and}$$
$$A(\alpha f + \beta g) = \alpha Af + \beta Ag \in D_B] \Rightarrow$$
$$[\alpha f + \beta g \in D_{BA} \text{ and } BA(\alpha f + \beta g) = \alpha BAf + \beta BAg].$$

For $A \in \mathcal{O}(X)$ and $n \in \mathbb{N}$, we write $A^n := AA \cdots (n \text{ times}) \cdots A$.

3.2.5 Definition. Let M be a linear manifold in a linear space X. The mapping id_M (cf. 1.2.6a) is obviously a linear operator, which is denoted by the symbol $\mathbb{1}_M$. Thus, we have

$$\mathbb{1}_M : M \to X$$
$$f \mapsto \mathbb{1}_M f := f.$$

Clearly we have $\mathbb{1}_M \subset \mathbb{1}_X$.

3.2.6 Theorem. *Let X and Y be linear spaces over the same field $\mathbb{K}$, and let $A \in \mathcal{O}(X,Y)$. We have:*

(a) A is injective iff $N_A = \{0_X\}$;
(b) if A is injective, then the mapping A^{-1} is a linear operator, i.e. $A^{-1} \in \mathcal{O}(Y,X)$, and we have $A^{-1}A = \mathbb{1}_{D_A}$ and $AA^{-1} = \mathbb{1}_{R_A}$.

Proof. a: If A is injective then we have (recalling that $0_X \in N_A$):

$$f \in N_A \Rightarrow (f \in D_A \text{ and } Af = 0_Y) \Rightarrow (f, 0_X \in D_A \text{ and } Af = A0_X) \Rightarrow f = 0_X;$$

hence $N_A \subset \{0_X\}$, from which $N_A = \{0_X\}$ follows since $0_X \in N_A$.

If $N_A = \{0_X\}$ then we have:

$$(f_1, f_2 \in D_A \text{ and } Af_1 = Af_2) \Rightarrow (f_1 - f_2 \in D_A \text{ and } A(f_1 - f_2) = 0_Y) \Rightarrow$$
$$f_1 - f_2 \in N_A \Rightarrow f_1 - f_2 = 0_X \Rightarrow f_1 = f_2;$$

hence A is injective.

b: Assume A injective. Since $D_{A^{-1}} = R_A$, condition lo_1 holds for A^{-1} (cf. 3.2.2a). As to conditions lo_2 and lo_3, recalling that $R_{A^{-1}} = D_A$ and that $A^{-1} \circ A = \mathbb{1}_{D_A}$ and $A \circ A^{-1} = \mathbb{1}_{R_A}$ (cf. 1.2.14), we have:

$$(\alpha, \beta \in \mathbb{K} \text{ and } g_1, g_2 \in D_{A^{-1}}) \Rightarrow$$
$$[\alpha g_1 + \beta g_2 \in D_{A^{-1}} \text{ and } A^{-1}(\alpha g_1 + \beta g_2) = A^{-1}(\alpha A \circ A^{-1}(g_1) + \beta A \circ A^{-1}(g_2))$$
$$= A^{-1}(\alpha A(A^{-1}(g_1)) + \beta A(A^{-1}(g_2)))$$
$$= A^{-1} \circ A(\alpha A^{-1}(g_1) + \beta A^{-1}(g_2))$$
$$= \alpha A^{-1}(g_1) + \beta A^{-1}(g_2)].$$

Since A^{-1} has been proved to be a linear operator, from now on we will write $A^{-1}A := A^{-1} \circ A$, $AA^{-1} := A \circ A^{-1}$ and $A^{-1}g := A^{-1}(g)$ for $g \in R_A$. $\square$

3.2.7 Remarks. Let X be a linear space. For $A, B \in \mathcal{O}(X)$, recall that:

(a) if B is injective and $A \subset B$, then A is injective and $A^{-1} \subset B^{-1}$ (cf. 1.2.15);
(b) if $BA = \mathbb{1}_{D_A}$, then A is injective and $A^{-1} \subset B$ (cf. 1.2.16a);
(c) if $BA = \mathbb{1}_{D_A}$ and $AB = \mathbb{1}_{D_B}$, then both A and B are injective, $A^{-1} = B$ and $B^{-1} = A$ (cf. 1.2.16b).

3.2.8 Definitions. Let X and Y be linear spaces over the same field $\mathbb{K}$. For $A, B \in \mathcal{O}(X,Y)$, we define the mapping:

$$A + B : D_A \cap D_B \to Y$$
$$f \mapsto (A+B)f := Af + Bf,$$

which is called the *sum* of A and B. Recalling 3.1.5, it is immediate to see that $A + B \in \mathcal{O}(X,Y)$.

For $\alpha \in \mathbb{K}$ and $A \in \mathcal{O}(X,Y)$, we define the mapping

$$\alpha A : D_A \to Y$$
$$f \mapsto (\alpha A)f := \alpha(Af).$$

From now on we will write $\alpha A f := (\alpha A)f$, for $f \in D_A$. It is immediate to see that $\alpha A \in \mathcal{O}(X,Y)$. We will write $-A := (-1)A$ and $B - A := B + (-A)$ for $B \in \mathcal{O}(X,Y)$.

3.2.9 Definition. Let X and Y be linear spaces over the same field $\mathbb{K}$. We define the mapping

$$\mathbb{O}_{X,Y} : X \to Y$$
$$f \mapsto \mathbb{O}_{X,Y}f := 0_Y.$$

Obviously, $\mathbb{O}_{X,Y} \in \mathcal{O}(X,Y)$ and we have:

$$\alpha \mathbb{O}_{X,Y} = \mathbb{O}_{X,Y}, \forall \alpha \in \mathbb{K}, \text{ and } 0A \subset \mathbb{O}_{X,Y}, \forall A \in \mathcal{O}(X,Y).$$

For $X = Y$, we write $\mathbb{O}_X := \mathbb{O}_{X,X}$.

3.2.10 Proposition. *The three binary operations defined in 3.2.4 and 3.2.8 have the following properties*

(a) *If X and Y are linear spaces over the same field $\mathbb{K}$, then we have:*

   (a_1) $A + (B + C) = (A + B) + C$, $\forall A, B, C \in \mathcal{O}(X,Y)$,
   (a_2) $A + \mathbb{O}_{X,Y} = A$, $\forall A \in \mathcal{O}(X,Y)$,
   (a_3) $A - A \subset \mathbb{O}_{X,Y}$, $\forall A \in \mathcal{O}(X,Y)$,
   (a_3') $A - A = \mathbb{O}_{X,Y}$, $\forall A \in \mathcal{O}(X,Y)$ s.t. $D_A = X$,
   (a_4) $A + B = B + A$, $\forall A, B \in \mathcal{O}(X,Y)$,
   (a_5) $\alpha(\beta A) = (\alpha\beta)A$, $\forall \alpha, \beta \in \mathbb{K}, \forall A \in \mathcal{O}(X,Y)$,
   (a_6) $(\alpha + \beta)A = \alpha A + \beta A$, $\forall \alpha, \beta \in \mathbb{K}, \forall A \in \mathcal{O}(X,Y)$,
   (a_7) $\alpha(A + B) = \alpha A + \alpha B$, $\forall \alpha \in \mathbb{K}, \forall A, B \in \mathcal{O}(X,Y)$,
   (a_8) $1A = A$, $\forall A \in \mathcal{O}(X,Y)$.

(b) *If W, X, Y, Z are four linear spaces over the same field $\mathbb{K}$, then we have:*

   (b_1) $(AB)C = A(BC)$, $\forall A \in \mathcal{O}(Y,Z), \forall B \in \mathcal{O}(X,Y), \forall C \in \mathcal{O}(W,X)$,
   (b_2) $AB + AC \subset A(B + C)$, $\forall A \in \mathcal{O}(Y,Z), \forall B, C \in \mathcal{O}(X,Y)$,
   (b_2') $AB + AC = A(B + C)$, $\forall A \in \mathcal{O}(Y,Z)$ s.t. $D_A = Y, \forall B, C \in \mathcal{O}(X,Y)$,
   (b_3) $AC + BC = (A + B)C$, $\forall A, B \in \mathcal{O}(Y,Z), \forall C \in \mathcal{O}(X,Y)$,
   (b_4) $(\alpha A)B = \alpha(AB) = A(\alpha B)$, $\forall \alpha \in \mathbb{K} - \{0\}, \forall A \in \mathcal{O}(Y,Z), \forall B \in \mathcal{O}(X,Y)$,
   (b_4') $(0A)B = 0(AB) \subset A(0B)$, $\forall A \in \mathcal{O}(Y,Z), \forall B \in \mathcal{O}(X,Y)$,
   (b_5) $\mathbb{1}_Y A = A\mathbb{1}_X = A$, $\forall A \in \mathcal{O}(X,Y)$.

Proof. For all the relations we have to prove it is clear that, at a vector which belongs to the intersection of the domains of the two operators which appear on the two sides of the relation, the value of the operator on the left hand side coincides with the value of the operator on the right hand side. Thus, in order to prove the relations between the operators, we need only prove the same relations between their domains. We will examine only the cases that are not completely obvious.

a_1: We have

$$D_{A+(B+C)} = D_A \cap D_{B+C} = D_A \cap (D_B \cap D_C)$$
$$= (D_A \cap D_B) \cap D_C = D_{A+B} \cap D_C = D_{(A+B)+C}.$$

b_1: Cf. 1.2.17.

b_2: We have

$$f \in D_{AB+AC} \Rightarrow f \in D_{AB} \cap D_{AC} \Rightarrow$$
$$(f \in D_B, Bf \in D_A, f \in D_C, Cf \in D_A) \Rightarrow$$
$$(f \in D_B \cap D_C \text{ and } Bf + Cf \in D_A) \Rightarrow$$
$$(f \in D_{B+C} \text{ and } (B+C)f \in D_A) \Rightarrow f \in D_{A(B+C)}.$$

b_2': If $D_A = Y$, we have (cf. 1.2.13e)

$$D_{AB+AC} = D_{AB} \cap D_{AC} = D_B \cap D_C = D_{B+C} = D_{A(B+C)}.$$

b_3: We have

$$f \in D_{AC+BC} \Leftrightarrow f \in D_{AC} \cap D_{BC} \Leftrightarrow$$
$$(f \in D_C, Cf \in D_A, f \in D_C, Cf \in D_B) \Leftrightarrow$$
$$(f \in D_C \text{ and } Cf \in D_A \cap D_B) \Leftrightarrow$$
$$(f \in D_C \text{ and } Cf \in D_{A+B}) \Leftrightarrow f \in D_{(A+B)C}.$$

b_4: If $\alpha \neq 0$, then $D_{AB} = D_{A(\alpha B)}$: since D_A is a linear manifold, for $f \in D_B$ we have $Bf \in D_A \Rightarrow \alpha Bf \in D_A$, and also $\alpha Bf \in D_A \Rightarrow Bf = \alpha^{-1}(\alpha Bf) \in D_A$.

b_4': We have $(0A)B = 0(AB) = (\mathbb{O}_{X,Y})_{D_{AB}}$ and also $A(0B) = (\mathbb{O}_{X,Y})_{D_B}$ because $0Bf = 0_Y \in D_A$ for all $f \in D_B$. Then, we recall that $D_{AB} \subset D_B$ (cf. 1.2.13c). $\qquad\qquad\qquad\square$

3.2.11 Remark. The family $\mathcal{O}(X,Y)$ with the two binary operations defined in 3.2.8, despite the symbols used to denote them, is not a linear space. In fact, $\mathbb{O}_{X,Y}$ is the only element of $\mathcal{O}(X,Y)$ for which condition a_2 of 3.2.10 can hold (if $\tilde{\mathbb{O}}$ is another operator which satisfies that condition, then we have $\tilde{\mathbb{O}} = \tilde{\mathbb{O}} + \mathbb{O}_{X,Y} = \mathbb{O}_{X,Y} + \tilde{\mathbb{O}} = \mathbb{O}_{X,Y}$), and for an operator $A \in \mathcal{O}(X,Y)$ with $D_A \neq X$ no operator $A' \in \mathcal{O}(X,Y)$ can exist such that $A + A' = \mathbb{O}_{X,Y}$, since $D_{A+A'} \subset D_A$ for all $A' \in \mathcal{O}(X,Y)$. Thus, there is no opposite for any element of $\mathcal{O}(X,Y)$ that is not defined on the whole of X.

We also notice that, in condition b_2 of 3.2.10, we do have $AB + AC \neq A(B+C)$ if for instance $R_B \not\subset D_A$ and $C = -B$. In fact, this implies both $D_{AB+AC} = D_{AB} \neq D_B$ (cf. 1.2.13d) and $D_{A(B+C)} = D_B$ (from $B + C = B - B \subset \mathbb{O}_{X,Y}$ we have $(B+C)f = 0_Y \in D_A$ for all $f \in D_B = D_{B+C}$).

3.2.12 Definition. Let X and Y be linear spaces over the same field, and define

$$\mathcal{O}_E(X,Y) := \{A \in \mathcal{O}(X,Y) : D_A = X\}.$$

Thus, $\mathcal{O}_E(X,Y)$ is the family of all the operators from X to Y that are defined everywhere on X. For $X = Y$ we write $\mathcal{O}_E(X) := \mathcal{O}_E(X,X)$.

3.2.13 Remark. All the relations that appear in 3.2.10 are equalities if $\mathcal{O}(W,X), \mathcal{O}(X,Y), \mathcal{O}(Y,Z)$ are replaced by $\mathcal{O}_E(W,X), \mathcal{O}_E(X,Y), \mathcal{O}_E(Y,Z)$ respectively.

3.2.14 Theorem. *Let X and Y be linear spaces over the same field $\mathbb{K}$, and define the mappings*

$$\sigma : \mathcal{O}_E(X,Y) \times \mathcal{O}_E(X,Y) \to \mathcal{O}_E(X,Y)$$
$$(A,B) \mapsto \sigma(A,B) := A + B,$$

$$\mu : \mathbb{C} \times \mathcal{O}_E(X,Y) \to \mathcal{O}_E(X,Y)$$
$$(\alpha, A) \mapsto \mu(\alpha, A) := \alpha A,$$

with $A + B$ and αA defined as in 3.2.8. Then $(\mathcal{O}_E(X,Y), \sigma, \mu)$ is a linear space over $\mathbb{K}$ (thus, the symbols $A + B$ and αA introduced in 3.2.8 are in agreement with the shorthand notation introduced in 3.1.1). The zero element is the operator $\mathbb{O}_{X,Y}$ defined in 3.2.9 and the opposite element of $A \in \mathcal{O}_E(X,Y)$ is the operator $-A$ defined in 3.2.8.

Proof. Everything follows directly from 3.2.10 and 3.2.13.   □

3.2.15 Proposition. *Let X and Y be linear spaces over the same field.*

(a) *A mapping φ from X to Y is a linear operator iff G_φ is a linear manifold in the linear space $X + Y$ (for G_φ, cf. 1.2.3).*
(b) *A linear manifold G in the linear space $X + Y$ is the graph of a mapping from X to Y iff G has the property:*

$$(0_X, g) \in G \Rightarrow g = 0_Y.$$

Proof. a: We assume first $\varphi := A \in \mathcal{O}(X,Y)$. Then G_A is a linear manifold in $X + Y$ since it meets condition *lm* of 3.1.4c:

$$[\alpha, \beta \in \mathbb{K} \text{ and } (f_1, g_1), (f_2, g_2) \in G_A] \Rightarrow$$
$$[f_1, f_2 \in D_A, g_1 = Af_1, g_2 = Af_2, \text{ whence}$$
$$\alpha f_1 + \beta f_2 \in D_A \text{ and } \alpha g_1 + \beta g_2 = A(\alpha f_1 + \beta f_2)] \Rightarrow$$
$$\alpha(f_1, g_1) + \beta(f_2, g_2) = (\alpha f_1 + \beta f_2, \alpha g_1 + \beta g_2) \in G_A.$$

Recall now that a mapping φ from X to Y is the mapping $\pi_Y \circ (\pi_X)^{-1}_{G_\varphi}$ (cf. 1.2.13B). The mappings π_X and π_Y are immediately seen to be linear operators from the linear space $X + Y$ (cf. 3.1.9) to X and Y respectively. If we assume that G_φ is a linear manifold in $X + Y$, then $(\pi_X)_{G_\varphi}$ is a linear operator (cf. 3.2.3), and hence $\pi_Y \circ (\pi_X)^{-1}_{G_\varphi}$ is a linear operator since it is the composition of two linear operators (cf. 3.2.6b and 3.2.4).

b: If G is the graph of a mapping, then by part a this mapping is a linear operator A, and hence we have:

$$(0_X, g) \in G \Rightarrow g = A0_X = 0_Y.$$

If, for a linear manifold G in $X + Y$, we have $(0_X, g) \in G \Rightarrow g = 0_Y$, then we also have

$$(f, g_1), (f, g_2) \in G \Rightarrow (0_X, g_1 - g_2) = (f, g_1) - (f, g_2) \in G \Rightarrow g_1 - g_2 = 0_Y \Rightarrow g_1 = g_2.$$

Then, by 1.2.4, G is the graph of a mapping from X to Y. $\qquad\square$

3.3 The algebra of linear operators

3.3.1 Definition. An *associative algebra* (or, simply, an *algebra*) over $\mathbb{K}$ is a quadruple (X, σ, μ, π), where (X, σ, μ) is a linear space over $\mathbb{K}$ and π is a mapping $\pi : X \times X \to X$ with the following properties, which we write with the shorthand notation $xy := \pi(x, y)$:

(al_1) $(xy)z = x(yz)$, $\forall x, y, z \in X$,
(al_2) $x(y + z) = xy + xz$, $\forall x, y, z \in X$,
(al_3) $(x + y)z = xz + yz$, $\forall x, y, z \in X$,
(al_4) $(\alpha x)y = \alpha(xy) = x(\alpha y)$, $\forall \alpha \in \mathbb{K}, \forall x, y \in X$.

The algebra X is said to be *with identity* if

(al_5) $\exists \tilde{x} \in X$ s.t. $\tilde{x}x = x\tilde{x} = x$ for all $x \in X$.

If $\tilde{x} \in X$ exists such that $\tilde{x}x = x\tilde{x} = x$ for all $x \in X$, then it is obviously the only element of X with this property, and it is called the *identity* of X and denoted by $\mathbb{1}$.

The algebra X is said to be *abelian* if

(al_6) $xy = yx$, $\forall x, y \in X$

The composition law π is called *product*.

For $x \in X$ and $n \in \mathbb{N}$, we define $x^n := xx \cdots (n \text{ times}) \cdots x$. If the algebra X is with identity, we define $x^0 := \mathbb{1}$ for all $x \in X$.

3.3.2 Definition. A *subalgebra* of an associative algebra (X, σ, μ, π) is a non-empty subset M of X which has the following properties:

(sa_1) M is a linear manifold in (X, σ, μ),
(sa_2) $x, y \in M \Rightarrow xy \in M$.

Condition sa_2 is equivalent to $\pi(M \times M) \subset M$. Thus it clear that, if M is a subalgebra, then $(M, \sigma_{M \times M}, \mu_{\mathbb{K} \times M}, \pi_{M \times M})$ is an associative algebra, which will be referred to as the algebra M. If the algebra X is with identity and $\mathbb{1} \in M$, then obviously M is an algebra with identity. If the algebra X is abelian, then obviously M is abelian.

3.3.3 Remarks. The following facts are immediately clear:

(a) Any associative algebra (X, σ, μ, π) has two trivial subalgebras: $\{0_X\}$ and X.

(b) If M is a subalgebra of an associative algebra (X, σ, μ, π) and N is a non-empty subset of M, then N is a subalgebra of (X, σ, μ, π) iff N is a subalgebra of M.

3.3.4 Proposition. *Let $\mathcal{F}$ be a family of subalgebras of an associative algebra (X, σ, μ, π). Then $\cap_{M \in \mathcal{F}} M$ is a subalgebra of (X, σ, μ, π).*

Proof. By 3.1.5, $\cap_{M \in \mathcal{F}} M$ is a linear manifold in (X, σ, μ). Moreover,

$$x, y \in \cap_{M \in \mathcal{F}} M \Rightarrow (x, y \in M \text{ and hence } xy \in M, \forall M \in \mathcal{F}) \Rightarrow xy \in \cap_{M \in \mathcal{F}} M.$$

$\square$

3.3.5 Definitions. Let $(X_1, \sigma_1, \mu_1, \pi_1)$ and $(X_2, \sigma_2, \mu_2, \pi_2)$ be two associative algebras over the same field $\mathbb{K}$. A *homomorphism* from X_1 to X_2 is a mapping $\Phi : X_1 \to X_2$ which has the following properties:

(ha_1) $\sigma_2(\Phi(x), \Phi(y)) = \Phi(\sigma_1(x, y)), \forall x, y \in X_1,$

(ha_2) $\mu_2(\alpha, \Phi(x)) = \Phi(\mu_1(\alpha, x)), \forall \alpha \in \mathbb{K}, \forall x \in X_1,$

(ha_3) $\pi_2(\Phi(x), \Phi(y)) = \Phi(\pi_1(x, y)), \forall x, y \in X_1.$

A homomorphism from X_1 to X_2 that is also a bijection from X_1 onto X_2 is called an *isomorphism*. If $X_1 = X_2$, an isomorphism is called an *automorphism*.

3.3.6 Proposition. *If Φ is a homomorphism from an associative algebra X_1 to an associative algebra X_2, then R_Φ is a subalgebra of X_2. If X_1 is with identity, then the algebra R_Φ is with identity. If X_1 is abelian, then R_Φ is abelian.*

Proof. Everything follows directly from the definitions. $\square$

3.3.7 Theorem. *The linear space $\mathcal{O}_E(X)$ of the operators defined on a linear space X (cf. 3.2.12 and 3.2.14) becomes an associative algebra with identity if we define*

$$\pi : \mathcal{O}_E(X) \times \mathcal{O}_E(X) \to \mathcal{O}_E(X)$$
$$(A, B) \mapsto \pi(A, B) := AB,$$

with AB defined as in 3.2.4 (thus, the symbol AB introduced in 3.2.4 is in agreement with the shorthand notation introduced in 3.3.1). The identity is the operator $\mathbb{1}_X$ defined in 3.2.5.

Proof. Everything follows directly from 3.2.10 and 3.2.13. $\square$

3.3.8 Examples.

(a) For the linear space $\mathcal{F}(X)$ (cf. 3.1.10c) we define

$$\pi : \mathcal{F}(X) \times \mathcal{F}(X) \to \mathcal{F}(X)$$
$$(\varphi, \psi) \mapsto \pi(\varphi, \psi) := \varphi\psi,$$

with $\varphi\psi$ defined as in 1.2.19. It is immediate to check that $(\mathcal{F}(X), \sigma, \mu, \pi)$ is an abelian associative algebra over $\mathbb{C}$ (thus the symbol $\varphi\psi$ defined in 1.2.19 is in agreement with the shorthand notation used in 3.3.1), with the function

$$1_X : X \to \mathbb{C}$$
$$x \mapsto 1_X(x) := 1,$$

(cf. 1.2.19) as identity.

(b) It is immediate to see that the linear manifold $\mathcal{F}_B(X)$ in $\mathcal{F}(X)$ (cf. 3.1.10d) is a subalgebra of $\mathcal{F}(X)$, with identity since $1_X \in \mathcal{F}_B(X)$.

(c) If (X, d) is a metric space, then the linear manifold $\mathcal{C}(X)$ in $\mathcal{F}(X)$ (cf. 3.1.10e) is a subalgebra of $\mathcal{F}(X)$ since the pointwise product of two continuous functions is a continuous function, and it is with identity since $1_X \in \mathcal{C}(X)$. Moreover, $\mathcal{C}_B(X)$ (cf. 3.1.10e) is a subalgebra with identity of $\mathcal{F}(X)$ since it is the intersection of two subalgebras (cf. 3.3.4), and hence it is a subalgebra of both $\mathcal{F}_B(X)$ and $\mathcal{C}(X)$ (cf. 3.3.3b).

Chapter 4

Linear Operators in Normed Spaces

There are properties of a linear operator in a Hilbert space which depend only on the relation between the operator and the norm which is generated by the inner product of the space. Thus, in this chapter we examine what can be said about linear operators in normed spaces. Throughout the chapter, $\mathbb{K}$ stands for the field $\mathbb{C}$ of complex numbers or the field $\mathbb{R}$ of real numbers.

4.1 Normed spaces

4.1.1 Definition. A *normed space over* $\mathbb{K}$ (or simply a normed space) is a quadruple (X, σ, μ, ν), where (X, σ, μ) is a linear space over $\mathbb{K}$ and ν is a function $\nu : X \to \mathbb{R}$ which, with the shorthand notation $\|f\| := \nu(f)$, has the following properties:

(no_1) $\|f + g\| \leq \|f\| + \|g\|, \ \forall f, g \in X$,
(no_2) $\|\alpha f\| = |\alpha| \|f\|, \ \forall \alpha \in \mathbb{K}, \forall f \in X$,
(no_3) $\|f\| = 0 \Rightarrow f = 0_X$.

The function ν is called a *norm* for the linear space (X, σ, μ).

4.1.2 Proposition. *In a normed space X we have:*

(a) $\|0_X\| = 0$;
(b) $\big| \|f\| - \|g\| \big| \leq \|f - g\|, \ \forall f, g \in X$;
(c) $0 \leq \|f\|, \ \forall f \in X$.

Proof. a: $\|0_X\| = \|00_X\| = |0| \|0_X\| = 0$.
  b: For $f, g \in X$ we have

$$\|f\| = \|(f - g) + g\| \leq \|f - g\| + \|g\|, \text{ hence } \|f\| - \|g\| \leq \|f - g\|,$$

and in the same way we also have $\|g\| - \|f\| \leq \|g - f\|$.
  c: Using results a and b, for $f \in X$ we get:

$$0 \leq \big| \|f\| \big| \leq \|f - 0_X\| = \|f\|.$$

$\square$

4.1.3 Theorem. *In a normed space* (X, σ, μ, ν), *the function*

$$d_\nu : X \times X \to \mathbb{R}$$

$$(f, g) \mapsto d_\nu(f, g) := \|f - g\|$$

is a distance on X.

Proof. For d_ν we have the properties of a distance (cf. 2.1.1):

(di_1) for $f, g \in X$,

$$d_\nu(f, g) = \|f - g\| = \|(-1)(g - f)\|$$
$$= |-1|\|g - f\| = \|g - f\| = d_\nu(g, f);$$

(di_2) for $f, g, h \in X$,

$$d_\nu(f, g) = \|f - g\| = \|(f - h) + (h - g)\|$$
$$\leq \|f - h\| + \|h - g\| = d_\nu(f, h) + d_\nu(h, g);$$

(di_3) for $f, g \in X$,

$$d_\nu(f, g) = 0 \Leftrightarrow \|f - g\| = 0 \Leftrightarrow f - g = 0_X \Leftrightarrow f = g.$$

$\square$

Whenever we speak about metric properties in a normed space, we will be referring to the distance defined in 4.1.3.

4.1.4 Example. Recall that $\mathbb{K}$ is a linear space over $\mathbb{K}$ (cf. 3.1.10b). From the properties of the absolute value in $\mathbb{R}$ or of the modulus in $\mathbb{C}$ it follows immediately that the function

$$\nu : \mathbb{K} \to \mathbb{R}$$

$$z \mapsto \nu(z) := |z|$$

is a norm for the linear space $\mathbb{K}$. We have $d_\nu = d_{\mathbb{K}}$ (cf. 2.1.4 and 2.7.4a).

4.1.5 Definition. Let $\{f_n\}$ be a sequence in a normed space. For $n \in \mathbb{N}$, we define $s_n := \sum_{k=1}^n f_k$, which is called the nth *partial sum*. The sequence $\{s_n\}$ is called the *series* of f_n's and denoted by the symbol $\sum_{n=1}^\infty f_n$. If the sequence $\{s_n\}$ is convergent, then the series $\sum_{n=1}^\infty f_n$ is said to be *convergent* and the limit of $\{s_n\}$ is also denoted by the same symbol $\sum_{n=1}^\infty f_n$, i.e. $\sum_{n=1}^\infty f_n := \lim_{n\to\infty} s_n$, and it is called the *sum of the series*. These definitions are in agreement with the ones given in 2.1.10.

4.1.6 Theorem. *Let* (X, σ, μ, ν) *be a normed space. Then:*

(a) the function ν *is continuous (with respect to* d_ν *and* $d_{\mathbb{R}}$*);*
(b) the mapping σ *is continuous (with respect to* $d_\nu \times d_\nu$ *and* d_ν*);*
(c) the mapping μ *is continuous (with respect to* $d_{\mathbb{K}} \times d_\nu$ *and* d_ν*).*

Proof. Use 2.4.2, 2.7.3a, and the following remarks.

(a) For $f \in X$ and a sequence $\{f_n\}$ in X we have:
$$d_{\mathbb{R}}(\|f_n\|, \|f\|) = |\|f_n\| - \|f\|| \leq \|f_n - f\| = d_\nu(f_n, f).$$

(b) For $(f, g) \in X \times X$ and a sequence $\{(f_n, g_n)\}$ in $X \times X$ we have:
$$d_\nu(f_n + g_n, f + g) = \|(f_n + g_n) - (f + g)\|$$
$$\leq \|f_n - f\| + \|g_n - g\| = d_\nu(f_n, f) + d_\nu(g_n, g).$$

(c) For $(\alpha, f) \in \mathbb{K} \times X$ and a sequence $\{(\alpha_n, f_n)\}$ in $\mathbb{K} \times X$ we have:
$$d_\nu(\alpha_n f_n, \alpha f) = \|\alpha_n f_n - \alpha f\| = \|(\alpha_n - \alpha)f_n + \alpha(f_n - f)\|$$
$$\leq |\alpha_n - \alpha|\|f_n\| + |\alpha|\|f_n - f\| = d_{\mathbb{K}}(\alpha_n, \alpha)\|f_n\| + |\alpha| d_\nu(f_n, f).$$

If $(\alpha_n, f_n) \to (\alpha, f)$, then $d_{\mathbb{K}}(\alpha_n, \alpha) \to 0$ and $d_\nu(f_n, f) \to 0$ by 2.7.3a. Besides, $d_\nu(f_n, f) \to 0$ implies $\|f_n\| \to \|f\|$ by result a, and hence the sequence $\{\|f_n\|\}$ is bounded (cf. 2.1.9).

□

4.1.7 Definition. A normed space which is a complete metric space is called a *Banach space*.

4.1.8 Remarks.

(a) Let (X, σ, μ, ν) be a normed space and let M be a linear manifold in the linear space (X, σ, μ). It is immediate to see that $(M, \sigma_{M \times M}, \mu_{\mathbb{K} \times M}, \nu_M)$ is a normed space. If X is a Banach space, then $(M, \sigma_{M \times M}, \mu_{\mathbb{K} \times M}, \nu_M)$ is a Banach space as well iff M is a closed set. This follows at once from 2.6.6. This partially justifies the definition we give in 4.1.9 (which is, however, completely justified only in the context of Banach spaces).

(b) Let $\{f_n\}$ be a sequence in a Banach space. The series $\sum_{n=1}^{\infty} f_n$ is said to be *absolutely convergent* if the series $\sum_{n=1}^{\infty} \|f_n\|$ is convergent. Suppose that the series $\sum_{n=1}^{\infty} f_n$ is absolutely convergent. Then the series $\sum_{n=1}^{\infty} f_n$ is convergent as well. Indeed, if we define $s_n := \sum_{k=1}^{n} f_k$ and $\sigma_n := \sum_{k=1}^{n} \|f_k\|$ for each $n \in \mathbb{N}$, then the sequence $\{\sigma_n\}$ is a Cauchy sequence (cf. 2.6.2), and hence the sequence $\{s_n\}$ is a Cauchy sequence as well since, for $n < m$,
$$\|s_m - s_n\| \leq \sum_{k=n+1}^{m} \|f_k\| = |\sigma_m - \sigma_n|,$$
and this implies that the sequence $\{s_n\}$ is convergent (cf. 2.6.3). Moreover, if β is a bijection from $\mathbb{N}$ onto $\mathbb{N}$ then the series $\sum_{n=1}^{\infty} f_{\beta(n)}$ is convergent and for the sums we have $\sum_{n=1}^{\infty} f_{\beta(n)} = \sum_{n=1}^{\infty} f_n$. Indeed, for any $\varepsilon > 0$ let $N_\varepsilon \in \mathbb{N}$ be so that $|\sigma_n - \sigma_m| < \varepsilon$ for $n, m > N_\varepsilon$. Then $\sum_{k=N_\varepsilon+2}^{p} \|f_k\| < \varepsilon$ for all $p \geq N_\varepsilon + 2$, and hence $\sum_{k \in I} \|f_k\| < \varepsilon$ if I is a finite set of positive integers such that $k > N_\varepsilon + 1$ for all $k \in I$. Let $M_\varepsilon := \max \beta^{-1}(\{1, ..., N_\varepsilon + 1\})$ and note

that $M_\varepsilon \geq N_\varepsilon + 1$ since $\beta^{-1}(\{1, ..., N_\varepsilon + 1\})$ contains $N_\varepsilon + 1$ distinct positive integers. Then, if we define $s'_n := \sum_{k=1}^n f_{\beta(k)}$ for each $n \in \mathbb{N}$, we have

$$\|s'_n - s_n\| < \varepsilon \text{ for } n > M_\varepsilon$$

since $n > M_\varepsilon$ implies that there is a finite set I of positive indices such that

$$k > N_\varepsilon + 1, \forall k \in I, \text{ and } \|s'_n - s_n\| \leq \sum_{k \in I} \|f_k\|.$$

Now let $L_\varepsilon \in \mathbb{N}$ be so that

$$\|\lim_{m\to\infty} s_m - s_n\| < \varepsilon \text{ for } n > L_\varepsilon.$$

Then

$$\|\lim_{m\to\infty} s_m - s'_n\| \leq \|\lim_{m\to\infty} s_m - s_n\| + \|s_n - s'_n\| < 2\varepsilon \text{ for } n > \max\{M_\varepsilon, L_\varepsilon\}.$$

This proves that the sequence $\{s'_n\}$ is convergent and $\lim_{n\to\infty} s'_n = \lim_{n\to\infty} s_n$.

4.1.9 Definition. A closed linear manifold in a normed space X is called a *subspace* of X. Obviously, $\{0_X\}$ and X are (trivial) subspaces of any normed space X (cf. 2.3.5, 2.3.2cl_1, 3.1.4a).

4.1.10 Proposition. *If $\mathcal{F}$ is a family of subspaces of a normed space X, then $\cap_{M\in\mathcal{F}}M$ is a subspace of X.*

Proof. Use 2.3.2cl_2 and 3.1.5. □

4.1.11 Definition. Let S be a subset of a normed space X. Consider the family

$$\mathcal{F} := \{M \in \mathcal{P}(X) : M \text{ is a subspace of } X \text{ and } S \subset M\}$$

and define

$$VS := \cap_{M\in\mathcal{F}}M.$$

We have:

(a) VS is a subspace of X (cf. 4.1.10);
(b) $S \subset VS$ (immediate from the definition of VS);
(c) if M is a subspace of X and $S \subset M$, then $VS \subset M$ (immediate from the definition of VS).

Thus, VS is the smallest subspace of X that contains S. For this reason, VS is called the *subspace generated* by S.

From the definition of VS we also have immediately:

(d) $VS = S$ iff S is a subspace;
(e) if T is a subset of X such that $S \subset T$, then $VS \subset VT$.

4.1.12 Proposition. *Let M be a linear manifold in a normed space X. Then $\overline{M}$ is a linear manifold in X, and hence a subspace of X.*

Proof. We will prove that $\overline{M}$ is a linear manifold by using 2.3.10 and 3.1.4c. For $\alpha, \beta \in \mathbb{K}$ and $f, g \in \overline{M}$, let $\{f_n\}$ and $\{g_n\}$ be sequences in M such that $f_n \to f$ and $g_n \to g$; then by 4.1.6b,c we have $\alpha f_n + \beta g_n \to \alpha f + \beta g$; since $\alpha f_n + \beta g_n \in M$, we have $\alpha f + \beta g \in \overline{M}$. $\qquad\square$

4.1.13 Proposition. *Let S be a subset of a normed space X. Then $VS = \overline{LS}$.*

Proof. Since $S \subset \overline{LS}$ and $\overline{LS}$ is a subspace by 4.1.12, we have $VS \subset \overline{LS}$. Since $S \subset VS$ and VS is a linear manifold, we have $LS \subset VS$, which implies $\overline{LS} \subset VS$ because VS is a closed set. $\qquad\square$

4.1.14 Proposition. *Let M be a linear manifold in a normed space. Then $VM = \overline{M}$.*

Proof. Since M is a linear manifold, we have $LM = M$. Then we use 4.1.13. $\qquad\square$

4.1.15 Proposition. *Let f be any element of a normed space. Then $V\{f\} = L\{f\} = \{\alpha f : \alpha \in \mathbb{K}\}$.*

Proof. The equality $L\{f\} = \{\alpha f : \alpha \in \mathbb{K}\}$ follows directly from 3.1.7. Then, in view of 4.1.13, $V\{f\} = L\{f\}$ is true if the set $\{\alpha f : \alpha \in \mathbb{K}\}$ is closed. If $f = 0_X$, then $\{\alpha f : \alpha \in \mathbb{K}\} = \{0_X\}$, which is a closed set (cf. 2.3.5). Assume next $f \neq 0_X$. Let $\{g_n\}$ be a sequence in $\{\alpha f : \alpha \in \mathbb{K}\}$ and g an element of X such that $g_n \to g$. If β_n is the element of $\mathbb{K}$ such that $g_n = \beta_n f$, the sequence $\{\beta_n\}$ turns out to be a Cauchy sequence because $\{g_n\}$ is such (cf. 2.6.2) and $f \neq 0_X$. Since $\mathbb{K}$ is a complete metric space, there exists $\beta \in \mathbb{K}$ such that $\beta_n \to \beta$, hence by 4.1.6c such that $g_n \to \beta f$. Therefore, $g = \beta f \in \{\alpha f : \alpha \in \mathbb{K}\}$. On account of 2.3.4, this proves that $\{\alpha f : \alpha \in \mathbb{K}\}$ is a closed set. $\qquad\square$

4.1.16 Definition. Let X and Y be normed spaces over the same field, and denote by ν_X and ν_Y their norms. The function

$$\nu : X \times Y \to \mathbb{R}$$

$$(f, g) \mapsto \nu(f, g) := \sqrt{\nu_X^2(f) + \nu_Y^2(g)}$$

is a norm for the linear space $X + Y$ (cf. 3.1.9); in fact, properties no_1, no_2 and no_3 of 4.1.1 follow immediately for ν from the same properties for ν_X and ν_Y, using also (for no_1) the inequality

$$\forall a_1, a_2, b_1, b_2 \in \mathbb{C}, \sqrt{|a_1 + b_1|^2 + |a_2 + b_2|^2} \leq \sqrt{|a_1|^2 + |a_2|^2} + \sqrt{|b_1|^2 + |b_2|^2},$$

which will be proved in 10.3.8c.

The linear space $X + Y$ with this norm ν is called the *sum of the normed spaces* X and Y.

It can be seen immediately that $d_\nu = d_{\nu_X} \times d_{\nu_Y}$. Hence, from 2.7.3d it follows that the normed space $X + Y$ is a Banach space iff X and Y are both Banach spaces.

4.2   Bounded operators

4.2.1 Definition. Let X and Y be normed spaces over the same field and let $A \in \mathcal{O}(X, Y)$. The linear operator A is said to be *bounded* if it has the following property:

$$\exists m \in [0, \infty) \text{ such that } \|Af\| \leq m\|f\| \text{ for all } f \in D_A.$$

For a linear operator, the importance of the condition of being bounded lies in the fact that a linear operator is bounded iff it is continuous, as is shown by the following theorem.

4.2.2 Theorem. *Let X and Y be normed spaces over the same field. For a linear operator $A \in \mathcal{O}(X, Y)$, the following conditions are equivalent:*

(a) A is bounded, i.e. $\exists m \geq 0$ such that $\|Af\| \leq m\|f\|$ for all $f \in D_A$;
(b) A is uniformly continuous;
(c) A is continuous;
(d) $\exists f_0 \in D_A$ such that A is continuous at f_0.

Proof. $a \Rightarrow b$: Assume condition a and let $\epsilon > 0$. Define $\delta_\epsilon := \frac{\epsilon}{m+1}$. Then we have

$$[f, g \in D_A \text{ and } \|f - g\| < \delta_\epsilon] \Rightarrow$$
$$\|Af - Ag\| = \|A(f - g)\| \leq m\|f - g\| < \frac{m}{m+1}\epsilon < \epsilon.$$

This proves that A is uniformly continuous.

$b \Rightarrow c$: This is obvious.

$c \Rightarrow d$: This is obvious.

$d \Rightarrow a$: Assume condition d. Then (setting $\epsilon := 1$ in the condition of continuity of A at f_0 and then $\delta := \delta_1$, cf. 2.4.1) $\exists \delta > 0$ such that

$$[f \in D_A \text{ and } \|f_0 - f\| < \delta] \Rightarrow \|Af_0 - Af\| < 1.$$

Then we have

$$g \in D_A - \{0_X\} \Rightarrow [\frac{\delta}{2\|g\|} g \in D_A \text{ and } \|\frac{\delta}{2\|g\|} g\| < \delta] \Rightarrow$$
$$[f_0 + \frac{\delta}{2\|g\|} g \in D_A \text{ and } \|f_0 - (f_0 + \frac{\delta}{2\|g\|} g)\| < \delta] \Rightarrow$$
$$\frac{\delta}{2\|g\|} \|Ag\| = \|Af_0 - A(f_0 + \frac{\delta}{2\|g\|} g)\| < 1 \Rightarrow \|Ag\| < \frac{2}{\delta}\|g\|.$$

This proves condition a with $m = \frac{2}{\delta}$. $\square$

4.2.3 Theorem. *Let X and Y be normed spaces over the same field. For a linear operator $A \in \mathcal{O}(X, Y)$, the following conditions are equivalent:*

(a) A is injective and A^{-1} is bounded;
(b) $\exists k > 0$ such that $\|Af\| \geq k\|f\|$ for all $f \in D_A$.

Proof. $a \Rightarrow b$: Assuming condition a, there exists $m \in [0, \infty)$ such that

$$\|A^{-1}g\| \le m\|g\|, \forall g \in D_{A^{-1}},$$

and hence (since $Af \in R_A = D_{A^{-1}}, \forall f \in D_A$) such that

$$\|f\| = \|A^{-1}(Af)\| \le m\|Af\| \le (m+1)\|Af\|, \forall f \in D_A.$$

Then we have condition b with $k := \frac{1}{m+1}$.

$b \Rightarrow a$: Assuming condition b, we have

$$f \in N_A \Rightarrow \|Af\| = 0 \Rightarrow \|f\| = 0 \Rightarrow f = 0_X.$$

Hence A is injective by 3.2.6a. Moreover (since $A^{-1}g \in R_{A^{-1}} = D_A, \forall g \in D_{A^{-1}}$),

$$\|g\| = \|A(A^{-1}g)\| \ge k\|A^{-1}g\|, \forall g \in D_{A^{-1}},$$

and hence

$$\|A^{-1}g\| \le \frac{1}{k}\|g\|, \forall g \in D_{A^{-1}},$$

and this proves that A^{-1} is bounded. $\qquad\square$

4.2.4 Definition. Let X and Y be normed spaces over the same field, and let A be a bounded operator from X to Y. Then the set of non-negative real numbers

$$\mathcal{B}_A := \{m \in [0, \infty) : \|Af\| \le m\|f\| \text{ for all } f \in D_A\}$$

is non-empty. Thus, we can define the non-negative number

$$\|A\| := \inf \mathcal{B}_A,$$

which is called the *norm* of A.

We point out that, notwithstanding the symbol and the name, the "norm" just defined cannot be a true norm for the family of all bounded operators from X to Y, because this family cannot be a linear space since a bounded operator which is not defined on the whole of X has no opposite (cf. 3.2.11). However, the symbol and the name will be partially justified by the theorem in 4.2.11.

4.2.5 Proposition. *Let X and Y be normed spaces over the same field (with $X \ne \{0_X\}$), and let A be a bounded operator from X to Y. We have:*

(a) if $B \in \mathcal{O}(X, Y)$ is such that $B \subset A$, then B is bounded and $\|B\| \le \|A\|$;

(b) $\|Af\| \le \|A\|\|f\|, \forall f \in D_A$;

(c) $\|A\|

$\overset{(c_1)}{=} \sup \left\{ \frac{\|Af\|}{\|f\|} : f \in D_A - \{0_X\} \right\}$

$\overset{(c_2)}{=} \sup\{\|Af\| : f \in D_A \text{ and } \|f\| = 1\}$

$\overset{(c_3)}{=} \sup\{\|Af\| : f \in D_A \text{ and } \|f\| \le 1\}.$

Proof. a: Suppose $B \subset A$. Then any non-negative m that implements the condition of 4.2.1 for A implements the same condition for B as well. Hence B is bounded and $\mathcal{B}_A \subset \mathcal{B}_B$, and this implies $\inf \mathcal{B}_B \leq \inf \mathcal{B}_A$.

b: For $f \in D_A - \{0_X\}$ we have $\frac{\|Af\|}{\|f\|} \leq m$ for every $m \in \mathcal{B}_A$. Hence $\frac{\|Af\|}{\|f\|}$ is a lower bound for $\mathcal{B}_A$, and hence $\frac{\|Af\|}{\|f\|} \leq \inf \mathcal{B}_A = \|A\|$ by the definition of g.l.b. (cf. 1.1.5).

c: Equalities c_2 and c_3 are obvious. We prove equality c_1. From result b we have $\frac{\|Af\|}{\|f\|} \leq \|A\|$ for every $f \in D_A - \{0_X\}$, and hence

$$\sup \left\{ \frac{\|Af\|}{\|f\|} : f \in D_A - \{0_X\} \right\} \leq \|A\|.$$

On the other hand, for every $g \in D_A - \{0_X\}$ we have

$$\frac{\|Ag\|}{\|g\|} \leq \sup \left\{ \frac{\|Af\|}{\|f\|} : f \in D_A - \{0_X\} \right\},$$

hence $\sup \left\{ \frac{\|Af\|}{\|f\|} : f \in D_A - \{0_X\} \right\} \in \mathcal{B}_A$, and hence

$$\|A\| \leq \sup \left\{ \frac{\|Af\|}{\|f\|} : f \in D_A - \{0_X\} \right\}$$

(we have used the definition of l.u.b. twice and the definition of g.l.b. once, cf. 1.1.5). □

4.2.6 Theorem. *Let X and Y be normed spaces over the same field, and let A be a bounded operator from X to Y. Assume further that Y is a Banach space. Then there exists one and only one operator $\tilde{A}$ from X to Y with the following properties:*

(a) $D_{\tilde{A}} = \overline{D}_A$;
(b) $A \subset \tilde{A}$;
(c) $\tilde{A}$ is bounded.

We also have

(d) $\|\tilde{A}\| = \|A\|$.

Proof. For $f \in \overline{D}_A$ there is a sequence $\{f_n\}$ in D_A such that $f_n \to f$ (cf. 2.3.10). Then $\{f_n\}$ is a Cauchy sequence (cf. 2.6.2), and hence the sequence $\{Af_n\}$ is also a Cauchy sequence since $\|Af_n - Af_m\| \leq \|A\| \|f_n - f_m\|$ by 4.2.5b. Since Y is a complete metric space, the sequence $\{Af_n\}$ is convergent. For another sequence $\{f'_n\}$ in D_A such that $f'_n \to f$ we have $\lim_{n\to\infty} Af'_n = \lim_{n\to\infty} Af_n$, since

$$\| \lim_{n\to\infty} Af_n - Af'_k \|$$
$$\leq \| \lim_{n\to\infty} Af_n - Af_k \| + \|Af_k - Af'_k\|$$
$$\leq \| \lim_{n\to\infty} Af_n - Af_k \| + \|A\| \|f_k - f'_k\|$$
$$\leq \| \lim_{n\to\infty} Af_n - Af_k \| + \|A\| \|f_k - f\| + \|A\| \|f - f'_k\| \to 0 \text{ as } k \to \infty.$$

Thus, $\lim_{n\to\infty} A f_n$ depends only on f and not on the choice of the sequence $\{f_n\}$ in D_A, as long as $f_n \to f$. Therefore, we can define the mapping

$$\tilde{A} : \overline{D}_A \to Y$$

$$f \mapsto \tilde{A}f := \lim_{n\to\infty} A f_n \quad \text{if } \{f_n\} \text{ is a sequence in } D_A \text{ such that } f_n \to f.$$

The mapping $\tilde{A}$ is a linear operator. In fact, $\overline{D}_A$ is a linear manifold in X by 4.1.12. Moreover, if $f, g \in \overline{D}_A$ and $\{f_n\}$ and $\{g_n\}$ are sequences in D_A such that $f_n \to f$ and $g_n \to g$, then for all scalars α, β the sequence $\{\alpha f_n + \beta g_n\}$ is in D_A and $\alpha f_n + \beta g_n \to \alpha f + \beta g$ by 4.1.6b,c, and we have

$$\tilde{A}(\alpha f + \beta g) = \lim_{n\to\infty} A(\alpha f_n + \beta g_n) = \lim_{n\to\infty} (\alpha A f_n + \beta A g_n)$$

$$= \alpha \lim_{n\to\infty} A f_n + \beta \lim_{n\to\infty} A g_n = \alpha \tilde{A}f + \beta \tilde{A}g,$$

where use has been made of 4.1.6b,c again.

Now, $\tilde{A}$ has property a by its definition, and property b because, if $f \in D_A$, then we can define $\tilde{A}f$ by using the sequence $\{f_n\}$ with $f_n := f$ for every n. Besides, if $f \in \overline{D}_A$ and $\{f_n\}$ is a sequence in D_A such that $f_n \to f$, then we have

$$\|\tilde{A}f\| = \lim_{n\to\infty} \|A f_n\| \le \|A\| \lim_{n\to\infty} \|f_n\| = \|A\| \|f\|,$$

owing to 4.1.6a (used twice) and 4.2.5b. This proves that $\tilde{A}$ has property c and that $\|A\| \in \mathcal{B}_{\tilde{A}}$, whence $\|\tilde{A}\| \le \|A\|$. Since $A \subset \tilde{A}$, we also have $\|A\| \le \|\tilde{A}\|$ by 4.2.5a. Thus, $\|\tilde{A}\| = \|A\|$.

It remains to prove the uniqueness of $\tilde{A}$. Let $B \in \mathcal{O}(X, Y)$ be such that $D_B = \overline{D}_A$, $A \subset B$, B is bounded. Then, if $f \in \overline{D}_A$ and $\{f_n\}$ is a sequence in D_A such that $f_n \to f$, we have

$$B f = \lim_{n\to\infty} B f_n = \lim_{n\to\infty} A f_n = \tilde{A}f,$$

where we have used the continuity of B granted by 4.2.2. Thus, $B = \tilde{A}$. $\qquad\square$

4.2.7 Proposition. *Let X and Y be normed spaces over the same field, and let A, B be bounded operators from X to Y. Then the operator $A + B$ is bounded, and $\|A + B\| \le \|A\| + \|B\|$.*

Proof. Using 4.2.5b we have

$$\forall f \in D_{A+B}, \|(A + B)f\| = \|Af + Bf\| \le \|Af\| + \|Bf\| \le (\|A\| + \|B\|)\|f\|.$$

This proves that $A+B$ is bounded and that $\|A\| + \|B\| \in \mathcal{B}_{A+B}$, whence $\|A + B\| \le \|A\| + \|B\|$. $\qquad\square$

4.2.8 Proposition. *Let X, Y be normed spaces over the same field $\mathbb{K}$, and let A be a bounded operator from X to Y. Then, for each $\alpha \in \mathbb{K}$, the operator αA is bounded and $\|\alpha A\| = |\alpha| \|A\|$.*

Proof. If $\alpha = 0_{\mathbb{K}}$, then we have

$$\forall f \in D_{\alpha A}, \|(\alpha A)f\| = \|0_Y\| = 0\|f\|.$$

Hence αA is bounded and $\|\alpha A\| = 0 = |\alpha|\|A\|$.

Assume now $\alpha \neq 0_{\mathbb{K}}$. Using 4.2.5b we have

$$\forall f \in D_{\alpha A}, \|(\alpha A)f\| = |\alpha|\|Af\| \leq |\alpha|\|A\|\|f\|,$$

which proves that αA is bounded and that $|\alpha|\|A\| \in \mathcal{B}_{\alpha A}$, whence $\|\alpha A\| \leq |\alpha|\|A\|$. Since αA is bounded we also have

$$\forall f \in D_A, \|Af\| = \|\frac{1}{\alpha}(\alpha A)f\| \leq \frac{1}{|\alpha|}\|\alpha A\|\|f\|,$$

which proves that $\frac{1}{|\alpha|}\|\alpha A\| \in \mathcal{B}_A$, whence $\|A\| \leq \frac{1}{|\alpha|}\|\alpha A\|$, whence $|\alpha|\|A\| \leq \|\alpha A\|$. Thus, $\|\alpha A\| = |\alpha|\|A\|$. $\qquad\square$

4.2.9 Proposition. *Let X, Y, Z be normed spaces over the same field, and suppose that $A \in \mathcal{O}(X, Y)$ and $B \in \mathcal{O}(Y, Z)$ are bounded operators. Then the operator BA is bounded and $\|BA\| \leq \|B\|\|A\|$.*

Proof. Using 4.2.5b we have

$$\forall f \in D_{BA}, \|(BA)f\| = \|B(Af)\| \leq \|B\|\|Af\| \leq \|B\|\|A\|\|f\|.$$

This proves that BA is bounded and that $\|B\|\|A\| \in \mathcal{B}_{BA}$, whence $\|BA\| \leq \|B\|\|A\|$. $\qquad\square$

4.2.10 Definition. Let X and Y be normed spaces over the same field. We define

$$\mathcal{B}(X, Y) := \{A \in \mathcal{O}_E(X, Y) : A \text{ is bounded}\}.$$

For $X = Y$, we write $\mathcal{B}(X) := \mathcal{B}(X, X)$.

4.2.11 Theorem. *Let X and Y be normed spaces over the same field. We have:*

(a) *$\mathcal{B}(X, Y)$ is a linear manifold in the linear space $\mathcal{O}_E(X, Y)$ (cf. 3.2.14) and the function*

$$\nu_{\mathcal{B}} : \mathcal{B}(X, Y) \to \mathbb{R}$$
$$A \mapsto \nu_{\mathcal{B}}(A) := \|A\| := \inf \mathcal{B}_A$$

is a norm for the linear space $\mathcal{B}(X, Y)$; hence, $\mathcal{B}(X, Y)$ is a normed space;

(b) *if Y is a Banach space, then $\mathcal{B}(X, Y)$ is also a Banach space.*

Proof. a: On account of 4.2.7 and 4.2.8, conditions lm_1 and lm_2 of 3.1.3 hold for $\mathcal{B}(X, Y)$, and conditions no_1 and no_2 of 4.1.1 hold for $\nu_{\mathcal{B}}$. By 4.2.5b we also have, for $A \in \mathcal{B}(X, Y)$,

$$\|A\| = 0 \Rightarrow [\|Af\| \leq 0, \forall f \in X] \Rightarrow [Af = 0_Y, \forall f \in X] \Rightarrow A = \mathbb{O}_{X,Y},$$

which proves that condition no_3 of 4.1.1 holds for $\nu_{\mathcal{B}}$.

b: Assume Y to be a Banach space, and let $\{A_n\}$ be a Cauchy sequence in the normed space $\mathcal{B}(X,Y)$, i.e. a sequence so that

$$\forall \epsilon > 0, \exists N_\epsilon \in \mathbb{N} \text{ s.t. } N_\epsilon < n, m \Rightarrow \|A_n - A_m\| < \epsilon.$$

For each $f \in X$, by 4.2.5b we have $\|A_n f - A_m f\| \leq \|A_n - A_m\| \|f\|$, and this shows that $\{A_n f\}$ is a Cauchy sequence in Y and hence a convergent sequence since Y is a complete metric space. Therefore, we can define the mapping

$$A : X \to Y$$

$$f \mapsto Af := \lim_{n \to \infty} A_n f.$$

This mapping is a linear operator, since for all $\alpha, \beta \in \mathbb{K}$, $f, g \in X$ we have

$$A(\alpha f + \beta g) = \lim_{n \to \infty} A_n(\alpha f + \beta g) = \lim_{n \to \infty}(\alpha A_n f + \beta A_n g)$$

$$= \alpha \lim_{n \to \infty} A_n f + \beta \lim_{n \to \infty} A_n g = \alpha A f + \beta A g,$$

where use has been made of 4.1.6b,c. Furthermore, for $\epsilon > 0$ and $n > N_\epsilon$ we have

$$\forall f \in X, \|(A - A_n)f\| = \|Af - A_n f\| = \lim_{k \to \infty} \|A_k f - A_n f\|$$

$$= \lim_{k \to \infty} \|(A_k - A_n)f\| \leq \epsilon \|f\|,$$

where use has been made of 4.1.6 and of the inequality $\|(A_k - A_n)f\| \leq \|A_k - A_n\| \|f\|$. This proves first that $A - A_n \in \mathcal{B}(X,Y)$, whence

$$A = (A - A_n) + A_n \in \mathcal{B}(X,Y)$$

since $\mathcal{B}(X,Y)$ is a linear manifold in $\mathcal{O}_E(X,Y)$, and second that

$$\|A - A_n\| \leq \epsilon.$$

As a consequence, the sequence $\{A_n\}$ is convergent and its limit is A. $\qquad\square$

4.2.12 Remark. Let X and Y be normed spaces over the same field, and let $\{A_n\}$ be a sequence in $\mathcal{B}(X,Y)$.

If $\{A_n\}$ is convergent then, letting $A := \lim_{n \to \infty} A_n$, we have

$$\forall f \in X, A f = \lim_{n \to \infty} A_n f,$$

which is proved by

$$\forall f \in X, \forall n \in \mathbb{N}, \|A_n f - A f\| \leq \|A_n - A\| \|f\|.$$

This implies that, if the series $\sum_{n=1}^{\infty} A_n$ is convergent, then for each $f \in X$ the series $\sum_{n=1}^{\infty}(A_n f)$ is convergent and $\left(\sum_{n=1}^{\infty} A_n\right) f = \sum_{n=1}^{\infty}(A_n f)$.

4.2.13 Theorem (The Banach–Steinhaus theorem). *Let X be a Banach space and Y a normed space over the same field, let $\mathcal{F}$ be a family of elements of $\mathcal{B}(X,Y)$, and assume that*

$$\forall f \in X, \exists m_f \in [0, \infty) \text{ such that } \|Af\| \leq m_f, \forall A \in \mathcal{F}.$$

Then

$$\exists m \in [0, \infty) \text{ such that } \|A\| \leq m, \forall A \in \mathcal{F}.$$

Proof. We divide the proof into two steps.

Step 1: We prove by contraposition that the assumption of the statement implies that the following proposition is true

$$P : \exists g \in X, \exists r \in (0, \infty), \exists n \in \mathbb{N} \text{ so that } \|Ah\| \leq n, \forall h \in B(g, r), \forall A \in \mathcal{F},$$

where $B(g, r)$ is the open ball with center g and radius r (cf. 2.2.1), i.e. $B(g, r) := \{h \in X : \|g - h\| < r\}$.

Thus, we assume that the following proposition is true

$$(\text{not } P) : \forall g \in X, \forall r \in (0, \infty), \forall n \in \mathbb{N}, \exists h \in B(g, r), \exists A \in \mathcal{F} \text{ so that } \|Ah\| > n.$$

We define a sequence $\{(h_n, r_n, A_n)\}$ in $X \times (0, \infty) \times \mathcal{F}$ by induction as follows. We fix $g_0 \in X$; since $(\text{not } P)$ is true,

$$\exists h_1 \in B(g_0, 1), \exists A_1 \in \mathcal{F} \text{ so that } \|A_1 h_1\| > 1,$$

and we choose $h_1 \in B(g_0, 1)$ and $A_1 \in \mathcal{F}$ which satisfy this condition; since the function $X \ni h \mapsto \|A_1 h\| \in \mathbb{R}$ is continuous (it is the composition of two continuous mappings),

$$\exists \delta_1 \in (0, \infty) \text{ so that } \|A_1 h\| > 1, \forall h \in B(h_1, \delta_1),$$

and we choose $\delta_1 \in (0, \infty)$ which satisfies this condition; we define

$$r_1 := \min \left\{ \delta_1, \frac{1}{2} \right\};$$

thus,

$$r_1 \leq \frac{1}{2} \text{ and } \|A_1 h\| > 1 \text{ for all } h \in B(h_1, r_1).$$

Next we suppose that, for a definite $n \in \mathbb{N}$, $(h_n, r_n, A_n) \in X \times (0, \infty) \times \mathcal{F}$ has already been defined so that

$$r_n \leq \frac{1}{2^n} \text{ and } \|A_n h\| > n \text{ for all } h \in B(h_n, r_n);$$

since $(\text{not } P)$ is true,

$$\exists h_{n+1} \in B(h_n, r_n), \exists A_{n+1} \in \mathcal{F} \text{ so that } \|A_{n+1} h_{n+1}\| > n + 1,$$

and we choose $h_{n+1} \in B(h_n, r_n)$ and $A_{n+1} \in \mathcal{F}$ which satisfy this condition; since the function $X \ni h \mapsto \|A_{n+1} h\| \in \mathbb{R}$ is continuous,

$$\exists \delta_{n+1} \in (0, \infty) \text{ so that } \|A_{n+1} h\| > n + 1, \forall h \in B(h_{n+1}, \delta_{n+1}),$$

and we choose $\delta_{n+1} \in (0, \infty)$ which satisfies this condition; since $B(h_n, r_n)$ is open and $h_{n+1} \in B(h_n, r_n)$,

$$\exists \rho_{n+1} \in (0, \infty) \text{ so that } B(h_{n+1}, \rho_{n+1}) \subset B(h_n, r_n),$$

and we choose $\rho_{n+1} \in (0, \infty)$ which satisfies this condition; we define

$$r_{n+1} := \min \left\{ \delta_{n+1}, \frac{1}{2} \rho_{n+1}, \frac{1}{2^{n+1}} \right\};$$

thus,

$$r_{n+1} \leq \frac{1}{2^{n+1}},$$
$$\|A_{n+1}h\| > n+1 \text{ for all } h \in B(h_{n+1}, r_{n+1}),$$
$$\overline{B(h_{n+1}, r_{n+1})} \subset B(h_n, r_n).$$

In this way we have defined a sequence $\{(h_n, r_n, A_n)\}$ in $X \times (0, \infty) \times \mathcal{F}$ which is such that, for each $n \in \mathbb{N}$,

$$r_n \leq \frac{1}{2^n}, \quad \|A_n h\| > n \text{ for all } h \in B(h_n, r_n), \quad \overline{B(h_{n+1}, r_{n+1})} \subset B(h_n, r_n).$$

The sequence $\{h_n\}$ is a Cauchy sequence since, for each $k \in \mathbb{N}$,

$$n, m > k \Rightarrow h_n, h_m \in B(h_k, r_k) \Rightarrow \|h_n - h_m\| < 2r_k \leq \frac{1}{2^{k-1}};$$

hence, the sequence $\{h_n\}$ is convergent since X is a complete metric space. We define $h := \lim_{n \to \infty} h_n$. For each $k \in \mathbb{N}$ we have

$$n > k+1 \Rightarrow h_n \in B(h_{k+1}, r_{k+1}),$$

and hence $h \in \overline{B(h_{k+1}, r_{k+1})}$, and hence $h \in B(h_k, r_k)$, and hence

$$\|A_k h\| \geq k.$$

This proves that if proposition P is not true then the assumption of the statement is not true.

Step 2: We prove that if P is true then the conclusion of the statement is true.

Thus, we assume that proposition P is true and we fix $(g, r, n) \in X \times (0, \infty) \times \mathbb{N}$ which satisfies the condition of proposition P. For each $f \in X$ such that $\|f\| < r$, we have $g + f \in B(g, r)$ and hence

$$\|Af\| = \|A(g+f) - Ag\| \leq \|A(f+g)\| + \|Ag\| \leq 2n, \forall A \in \mathcal{F}.$$

Then, for each $f \in X - \{0_X\}$, since $\left\| \frac{r}{2\|f\|} f \right\| < r$ we have

$$\left\| A\left(\frac{r}{2\|f\|} f \right) \right\| \leq 2n, \text{ i.e. } \|Af\| \leq \frac{4n}{r} \|f\|, \forall A \in \mathcal{F}.$$

This proves that

$$\|A\| \leq \frac{4n}{r}, \forall A \in \mathcal{F},$$

which is the conclusion of the statement for $m := \frac{4n}{r}$. $\qquad\square$

4.3 The normed algebra of bounded operators

4.3.1 Definition. A *normed algebra* over $\mathbb{K}$ is a quintuple $(X, \sigma, \mu, \pi, \nu)$, where (X, σ, μ, π) is an associative algebra over $\mathbb{K}$, (X, σ, μ, ν) is a normed space over $\mathbb{K}$, and the following condition is satisfied:

(na) $\|xy\| \le \|x\|\|y\| \ \forall x, y \in X$.

A normed algebra is said to be *with identity* if (X, σ, μ, π) is with identity and $\|\mathbb{1}\| = 1$.

A normed algebra is called a *Banach algebra* if (X, σ, μ, ν) is a Banach space.

4.3.2 Proposition. *If M is a subalgebra of a normed algebra $(X, \sigma, \mu, \pi, \nu)$ over $\mathbb{K}$, then $(M, \sigma_{M \times M}, \mu_{\mathbb{K} \times M}, \pi_{M \times M}, \nu_M)$ is also a normed algebra over $\mathbb{K}$.*

Proof. Since condition na of 4.3.1 holds obviously when X is replaced by any subset of X, we simply recall that a subalgebra of an associative algebra defines an associative algebra (cf. 3.3.2) and a linear manifold in a normed space defines a normed space (cf. 4.1.8a). $\qquad\square$

4.3.3 Theorem. *Let $(X, \sigma, \mu, \pi, \nu)$ be a normed algebra. Then the mapping π is continuous (with respect to $d_\nu \times d_\nu$ and d_ν).*

Proof. We use 2.4.2 and the following remarks. For $(x, y) \in X \times X$ and a sequence (x_n, y_n) in $X \times X$ we have:

$$d_\nu(x_n y_n, xy) = \|x_n y_n - xy\| = \|(x_n y_n - xy_n) + (xy_n - xy)\|$$
$$\le \|x_n - x\|\|y_n\| + \|x\|\|y_n - y\|.$$

If $(x_n, y_n) \to (x, y)$, then $\|x_n - x\| \to 0$ and $\|y_n - y\| \to 0$ by 2.7.3a. Besides, $\|y_n - y\| \to 0$ implies $\|y_n\| \to \|y\|$ by 4.1.6a, and hence the sequence $\{\|y_n\|\}$ is bounded (cf. 2.1.9). $\qquad\square$

4.3.4 Theorem. *Let X be a Banach algebra with identity. For $x \in X$ so that $\|x\| < 1$, we have:*

(a) the series $\sum_{n=1}^{\infty} x^n$ is convergent;
(b) $(\mathbb{1} + \sum_{n=1}^{\infty} x^n)(\mathbb{1} - x) = (\mathbb{1} - x)(\mathbb{1} + \sum_{n=1}^{\infty} x^n) = \mathbb{1}$.

Proof. a: The series $\sum_{n=1}^{\infty} \|x^n\|$ is convergent because $\|x^n\| \le \|x\|^n$ and $\|x\| < 1$. Then the series $\sum_{n=1}^{\infty} x^n$ is convergent by 4.1.8b.

b: We have

$$x + (\sum_{n=1}^{\infty} x^n)x = x + (\lim_{n \to \infty} \sum_{k=1}^{n} x^k)x = x + \lim_{n \to \infty} \sum_{k=1}^{n} x^{k+1}$$

$$= \lim_{n \to \infty} (x + \sum_{k=1}^{n} x^{k+1}) = \lim_{n \to \infty} \sum_{k=1}^{n+1} x^k = \sum_{n=1}^{\infty} x^n,$$

where we have used first 4.3.3 and then 4.1.6b. Then we have

$$(\mathbb{1} + \sum_{n=1}^{\infty} x^n)(\mathbb{1} - x) = \mathbb{1} + \sum_{n=1}^{\infty} x^n - x - (\sum_{n=1}^{\infty} x^n)x = \mathbb{1}.$$

In a similar way we can prove that $(\mathbb{1} - x)(\mathbb{1} + \sum_{n=1}^{\infty} x^n) = \mathbb{1}$. $\quad\square$

4.3.5 Theorem. *Let X be a normed space. Then $\mathcal{B}(X)$ is a subalgebra of the associative algebra $\mathcal{O}_E(X)$ (cf. 3.3.7). With the norm $\nu_\mathcal{B}$ of 4.2.11, $\mathcal{B}(X)$ is a normed algebra with identity. If X is a Banach space, then $\mathcal{B}(X)$ is a Banach algebra.*

Proof. Condition sa_1 of 3.3.2 has been proved for $\mathcal{B}(X)$ in 4.2.11a, and condition sa_2 follows from 4.2.9. Thus, $\mathcal{B}(X)$ is a subalgebra of $\mathcal{O}_E(X)$, and therefore it is also an associative algebra. By 4.2.11a, $\mathcal{B}(X)$ is also a normed space. On account of 4.2.9, for $\mathcal{B}(X)$ we also have property na of 4.3.1. Thus, $\mathcal{B}(X)$ is a normed algebra, and it is with identity since $\mathbb{1}_X \in \mathcal{B}(X)$ and $\|\mathbb{1}_X\| = 1$ (unless X is a zero space). Indeed we have

$$\|\mathbb{1}_X f\| = \|f\| \leq 1\|f\|, \forall f \in X,$$

which proves that $\mathbb{1}_X \in \mathcal{B}(X)$ and $1 \in \mathcal{B}_{\mathbb{1}_X}$, and hence $\|\mathbb{1}_X\| \leq 1$ (cf. 4.2.4). By 4.2.5b we also have

$$\|f\| = \|\mathbb{1}_X f\| \leq \|\mathbb{1}_X\| \|f\|, \forall f \in X,$$

and this implies $1 \leq \|\mathbb{1}_X\|$ if $\exists f \in X$ s.t. $f \neq 0_X$.

Finally, if X is a Banach space then $\mathcal{B}(X)$ is also a Banach space by 4.2.11b. $\quad\square$

4.3.6 Examples.

(a) For the associative algebra $\mathcal{F}_B(X)$ (cf. 3.3.8b), define

$$\nu : \mathcal{F}_B(X) \to \mathbb{R}$$
$$\varphi \mapsto \nu(\varphi) := \|\varphi\|_\infty := \sup\{|\varphi(x)| : x \in X\}.$$

It is easy to see that ν is a norm for the linear space $\mathcal{F}_B(X)$ and that ν has property na of 4.3.1. Therefore, $\mathcal{F}_B(X)$ is a normed algebra, and it is with identity since $\|1_X\|_\infty = 1$.

Actually, $\mathcal{F}_B(X)$ is a Banach algebra. In fact, let $\{\varphi_n\}$ be a Cauchy sequence in $\mathcal{F}_B(X)$; then $\{\varphi_n(x)\}$ is a Cauchy sequence in $\mathbb{C}$ for every $x \in X$, and hence we can define the function

$$X \ni x \mapsto \varphi(x) := \lim_{n\to\infty} \varphi_n(x) \in \mathbb{C};$$

now, for $\epsilon > 0$, let $N_\epsilon \in \mathbb{N}$ be such that $\|\varphi_n - \varphi_m\| < \epsilon$ for $n, m > N_\epsilon$; then we have, for $n > N_\epsilon$,

$$\forall x \in X, |\varphi_n(x) - \varphi(x)| = \lim_{m\to\infty} |\varphi_n(x) - \varphi_m(x)| \leq \epsilon.$$

This proves first that $\varphi_n - \varphi \in \mathcal{F}_B(X)$, and hence that

$$\varphi = (\varphi - \varphi_n) + \varphi_n \in \mathcal{F}_B(X)$$

since $\mathcal{F}_B(X)$ is a linear manifold in $\mathcal{F}(X)$, and second that

$$\|\varphi_n - \varphi\|_\infty \le \epsilon.$$

As a consequence, the sequence $\{\varphi_n\}$ is convergent and its limit is φ. Thus, $\mathcal{F}_B(X)$ is a Banach space.

(b) Let (X, d) be a metric space. Then $\mathcal{C}_B(X)$ is a subalgebra with identity of $\mathcal{F}_B(X)$ (cf. 3.3.8c) and hence it is an associative algebra with identity. Further, $\mathcal{C}_B(X)$ is a closed subset of the Banach space $\mathcal{F}_B(X)$. Indeed, let $\{\varphi_n\}$ be a sequence in $\mathcal{C}_B(X)$, let $\varphi \in \mathcal{F}_B(X)$, and suppose that $\|\varphi_n - \varphi\|_\infty \to 0$; for each $x \in X$ and each $\epsilon > 0$, let $n_\epsilon \in \mathbb{N}$ be such that $\|\varphi_{n_\epsilon} - \varphi\|_\infty < \frac{\epsilon}{3}$, and let $\delta_{x,\epsilon} > 0$ be such that $|\varphi_{n_\epsilon}(x) - \varphi_{n_\epsilon}(y)| < \frac{\epsilon}{3}$ whenever $d(x, y) < \delta_{x,\epsilon}$; then we have

$$d(x, y) < \delta_{x,\epsilon} \Rightarrow |\varphi(x) - \varphi(y)| \le |\varphi(x) - \varphi_{n_\epsilon}(x)| +$$
$$|\varphi_{n_\epsilon}(x) - \varphi_{n_\epsilon}(y)| + |\varphi_{n_\epsilon}(y) - \varphi(y)| < \epsilon;$$

this shows that $\varphi \in \mathcal{C}(X)$, and hence that $\varphi \in \mathcal{C}_B(X)$. Since $\mathcal{C}_B(X)$ is a closed linear manifold in the Banach space $\mathcal{F}_B(X)$, $\mathcal{C}_B(X)$ is a Banach space (cf. 4.1.8a).

Thus, $\mathcal{C}_B(X)$ is a Banach algebra with identity (cf. 4.3.2).

(c) Let $\mathbb{T}$ be the unit circle in the complex plane ($\mathbb{T}$ is also called the *one-dimensional torus*), i.e. we define

$$\mathbb{T} := \{z \in \mathbb{C} : |z| = 1\}.$$

Since $\mathbb{C} \ni z \mapsto |z| \in \mathbb{R}$ is a continuous function, $\mathbb{T}$ is a closed set in the metric space $(\mathbb{C}, d_{\mathbb{C}})$, and it is obviously also bounded. Hence, from 2.8.7 it follows that $\mathbb{T}$ is compact. Hence $\mathcal{C}(\mathbb{T}) = \mathcal{C}_B(\mathbb{T})$ (cf. 3.1.10e).

Define the subset $\mathcal{P}$ of $\mathcal{F}(\mathbb{T})$ (cf. 3.1.10c) by

$$\mathcal{P} := \{p \in \mathcal{F}(\mathbb{T}) : \exists N \ge 0, \exists (\alpha_0, \alpha_1, \alpha_{-1}, ..., \alpha_N, \alpha_{-N}) \in \mathbb{C}^{2N+1} \text{ s.t.}$$

$$p(z) = \sum_{k=-N}^{N} \alpha_k z^k, \forall z \in \mathbb{T}\}.$$

The elements of $\mathcal{P}$ are called trigonometric polynomials since for $\varphi \in \mathcal{F}(\mathbb{T})$ we have

$$\varphi \in \mathcal{P} \text{ iff } \varphi(e^{it}) = \sum_{k=-M}^{M} (\beta_k \cos kt + \gamma_k \sin kt) \text{ for every } t \in \mathbb{R},$$

where M is a non-negative integer and β_k and γ_k are complex numbers. Clearly, $\mathcal{P} \subset \mathcal{C}(\mathbb{T})$ and $\mathcal{P}$ is a subalgebra of $\mathcal{C}(\mathbb{T})$.

The following theorem can be proved (cf. e.g. Shilov, 1974, 1.52):

Stone's theorem for a complex algebra. *Let (X, d) be a compact metric space, and let $\mathcal{A}$ be a subalgebra of the Banach algebra $\mathcal{C}(X)$ which has the following properties:*

(a) $\varphi \in \mathcal{A} \Rightarrow \overline{\varphi} \in \mathcal{A}$;

(b) $1_X \in \mathcal{A}$;

(c) $(x, y \in X$ and $x \neq y) \Rightarrow \exists \varphi \in \mathcal{A}$ s.t. $\varphi(x) \neq \varphi(y)$.

Then $\overline{\mathcal{A}} = \mathcal{C}(X)$.

Now, $\mathcal{P}$ is a subalgebra of $\mathcal{C}(\mathbb{T})$ which has properties a, b, c of Stone's theorem (as to property c, let $\varphi(z) := z$ for all $z \in \mathbb{T}$). Then $\overline{\mathcal{P}} = \mathcal{C}(\mathbb{T})$. However, instead of giving a proof of Stone's theorem (which would require preliminary results outside the scope of this book), in 4.3.7 we give a direct proof (borrowed from Rudin, 1987, 4.24) of the equality $\overline{\mathcal{P}} = \mathcal{C}(\mathbb{T})$, which will play a crucial role in our proof of the spectral theorem for unitary operators (from which we will deduce the spectral theorem for self-adjoint operators).

4.3.7 Theorem (The Stone–Weierstrass approximation theorem). *In the Banach space $\mathcal{C}(\mathbb{T})$ we have $\overline{\mathcal{P}} = \mathcal{C}(\mathbb{T})$.*

Proof. We will show that, for each $\varphi \in \mathcal{C}(\mathbb{T})$, there exists a sequence $\{p_n\}$ in $\mathcal{P}$ such that $\|p_n - \varphi\|_\infty \to 0$. By 2.3.12, this will prove that $\overline{\mathcal{P}} = \mathcal{C}(\mathbb{T})$.

Suppose we have a sequence $\{q_n\}$ in $\mathcal{P}$ such that each q_n has the following properties:

(a) $0 \leq q_n(z)$, $\forall z \in \mathbb{T}$

(b) $\int_{-\pi}^{\pi} q_n(e^{is}) ds = 1$ (by $\int_{-\pi}^{\pi} \ldots ds$ we denote a Riemann integral, cf. 9.3.2);

(c) for each $\delta \in (0, \pi)$, if $m_n(\delta) := \sup\{q_n(e^{it}) : \delta \leq |t| \leq \pi\}$, then $\lim_{n \to \infty} m_n(\delta) = 0$.

Let $\varphi \in \mathcal{C}(\mathbb{T})$. For $n \in \mathbb{N}$, we define the function

$$p_n : \mathbb{T} \to \mathbb{C}$$

$$z \mapsto p_n(z) := \int_{-\pi}^{\pi} \varphi(e^{i(t-s)}) q_n(e^{is}) ds \quad \text{if } t \in (-\pi, \pi] \text{ is so that } z = e^{it}.$$

For all $t \in (-\pi, \pi]$ we have

$$\int_{-\pi}^{\pi} \varphi(e^{i(t-s)}) q_n(e^{is}) ds \overset{(1)}{=} \int_{-\pi}^{\pi} \varphi(e^{i(t+s)}) q_n(e^{-is}) ds$$

$$\overset{(2)}{=} \int_{-\pi+t}^{\pi+t} \varphi(e^{is}) q_n(e^{i(t-s)}) ds \overset{(3)}{=} \int_{-\pi}^{\pi} \varphi(e^{is}) q_n(e^{i(t-s)}) ds,$$

where in 1 we have made the change of variables $s \mapsto -s$, in 2 we have made the change of variables $s \mapsto s - t$, and 3 is true because the integrand is a periodic function of period 2π.

Since $q_n(e^{it}) = \sum_{k=-N_n}^{N_n} \alpha_{k,n} e^{ikt}$ for every $t \in \mathbb{R}$, we have

$$\int_{-\pi}^{\pi} \varphi(e^{is}) q_n(e^{i(t-s)}) ds = \sum_{k=-N_n}^{N_n} \beta_{k,n} e^{ikt}, \text{ with } \beta_{k,n} := \alpha_{k,n} \int_{-\pi}^{\pi} \varphi(e^{is}) e^{-iks} ds.$$

Thus, $p_n \in \mathcal{P}$.

Let now $\epsilon > 0$ be given. Since the function $[-\pi, \pi] \ni t \mapsto \varphi(e^{it}) \in \mathbb{C}$ is continuous, it is uniformly continuous (cf. 2.8.7 and 2.8.15). Therefore, the function $\mathbb{R} \ni t \mapsto \varphi(e^{it}) \in \mathbb{C}$ is uniformly continuous since it is periodic of period 2π. Thus, $\exists \delta_\epsilon > 0$ s.t. $|\varphi(e^{it}) - \varphi(e^{is})| < \epsilon$ whenever $|t - s| < \delta_\epsilon$. For $z \in \mathbb{T}$ and $t \in (-\pi, \pi]$ so that $z = e^{it}$, by property b we have

$$p_n(z) - \varphi(z) = \int_{-\pi}^{\pi} \left(\varphi(e^{i(t-s)}) - \varphi(e^{it}) \right) q_n(e^{is}) ds$$

and property a implies, assuming $0 < \delta_\epsilon < \pi$,

$$|p_n(z) - \varphi(z)| \leq \int_{-\pi}^{\pi} |\varphi(e^{i(t-s)}) - \varphi(e^{it})| q_n(e^{is}) ds$$

$$= \int_{-\pi}^{-\delta_\epsilon} ...ds + \int_{-\delta_\epsilon}^{\delta_\epsilon} ...ds + \int_{\delta_\epsilon}^{\pi} ...ds;$$

now, we have

$$\int_{-\delta_\epsilon}^{\delta_\epsilon} |\varphi(e^{i(t-s)}) - \varphi(e^{it})| q_n(e^{is}) ds \leq \epsilon \int_{-\pi}^{\pi} q_n(e^{is}) ds = \epsilon$$

since $|(t - s) - t| < \delta_\epsilon$ when $s \in (-\delta_\epsilon, \delta_\epsilon)$ and since q_n has property b, and also

$$\int_{-\pi}^{-\delta_\epsilon} |\varphi(e^{i(t-s)}) - \varphi(e^{it})| q_n(e^{is}) ds + \int_{\delta_\epsilon}^{\pi} |\varphi(e^{i(t-s)}) - \varphi(e^{it})| q_n(e^{is}) ds$$

$$\leq 2\|\varphi\|_\infty m_n(\delta_\epsilon) 2(\pi - \delta_\epsilon) < 4\pi \|\varphi\|_\infty m_n(\delta_\epsilon),$$

where the definition of $m_n(\delta_\epsilon)$ has been used; thus we have

$$|p_n(z) - \varphi(z)| \leq \epsilon + 4\pi \|\varphi\|_\infty m_n(\delta_\epsilon).$$

Since this estimate is independent of z, we have

$$\|p_n - \varphi\|_\infty \leq \epsilon + 4\pi \|\varphi\|_\infty m_n(\delta_\epsilon).$$

Recalling property c, let $N_\epsilon \in \mathbb{N}$ be such that $4\pi \|\varphi\|_\infty m_n(\delta_\epsilon) < \epsilon$ whenever $N_\epsilon < n$; then we have

$$N_\epsilon < n \Rightarrow \|p_n - \varphi\|_\infty < 2\epsilon.$$

This proves that $\|p_n - \varphi\|_\infty \to 0$.

It remains to construct a sequence $\{q_n\}$ in $\mathcal{P}$ with properties a, b, c. Let $q_n(z) := \frac{\gamma_n}{4^n}(2 + z + z^{-1})^n$ for every $z \in \mathbb{T}$, with γ_n so that condition b is satisfied. For $z \in \mathbb{T}$ and $t \in (-\pi, \pi]$ so that $z = e^{it}$, we have

$$q_n(z) = \gamma_n \left(\frac{1 + \cos t}{2} \right)^n, \text{ with } \gamma_n := \left(\int_{-\pi}^{\pi} \left(\frac{1 + \cos t}{2} \right)^n dt \right)^{-1}.$$

Since properties a and b are clear, we only need to show condition c.

Since the function $[-\pi, \pi] \ni t \mapsto \left(\frac{1+\cos t}{2}\right)^n \in \mathbb{R}$ is even, property b shows that

$$1 = 2\gamma_n \int_0^\pi \left(\frac{1+\cos t}{2}\right)^n dt > 2\gamma_n \int_0^\pi \left(\frac{1+\cos t}{2}\right)^n \sin t \, dt = 2\gamma_n \frac{2}{n+1}.$$

Since the function $[0, \pi] \ni t \mapsto 1 + \cos t \in \mathbb{R}$ is decreasing, it follows that

$$q_n(e^{it}) \le q_n(e^{i\delta}) \le \frac{n+1}{4}\left(\frac{1+\cos\delta}{2}\right)^n \quad \text{whenever } 0 < \delta \le |t| \le \pi.$$

This implies condition c, since $1 + \cos\delta < 2$ if $0 < \delta \le \pi$. $\qquad\square$

4.4 Closed operators

4.4.1 Definition. Let X and Y be normed spaces over the same field and let $A \in \mathcal{O}(X, Y)$. The linear operator A is said to be *closed* if its graph G_A is a closed subset of the product of the two metric spaces X, Y (cf. 4.1.3 and 2.7.2), i.e. a subspace of the normed space $X + Y$ (cf. 3.2.15 and 4.1.16). From 2.3.4 and 2.7.3a we have that A is closed iff the following condition is satisfied:

$$[f \in X, g \in Y, \{f_n\} \text{ is a sequence in } D_A, f_n \to f, Af_n \to g] \Rightarrow$$
$$[f \in D_A \text{ and } g = Af].$$

This condition can be written in the equivalent way:

$$[\{f_n\} \text{ is a sequence in } D_A \text{ that is convergent in } X \text{ and}$$
$$\{Af_n\} \text{ is a convergent sequence in } Y] \Rightarrow$$
$$[\lim_{n\to\infty} f_n \in D_A \text{ and } A(\lim_{n\to\infty} f_n) = \lim_{n\to\infty} Af_n].$$

4.4.2 Remark. Let X and Y be normed spaces over the same field. For a linear operator $A \in \mathcal{O}(X, Y)$ we have that A is bounded iff the following condition is satisfied (cf. 4.2.2 and 2.4.2):

$$[f \in D_A, \{f_n\} \text{ is a sequence in } D_A, f_n \to f] \Rightarrow [Af_n \to Af].$$

This condition can be written in the equivalent way:

$$[\{f_n\} \text{ is a sequence in } D_A \text{ that is convergent in } X \text{ and } \lim_{n\to\infty} f_n \in D_A] \Rightarrow$$
$$[\{Af_n\} \text{ is a convergent sequence in } Y \text{ and } A(\lim_{n\to\infty} f_n) = \lim_{n\to\infty} Af_n].$$

Thus, for both a bounded (i.e. continuous on account of 4.2.2) operator and a closed one there are conditions, for a convergent sequence $\{f_n\}$ in their domains, which allow one to "commute the operator with the limit". However, while for a bounded operator A one must assume $\lim_{n\to\infty} f_n \in D_A$ in order to obtain that the sequence $\{Af_n\}$ is convergent, for a closed operator A one must assume that the sequence $\{Af_n\}$ is convergent in order to obtain $\lim_{n\to\infty} f_n \in D_A$.

The interplay between the concepts of bounded operator and closed operator is studied in 4.4.3, 4.4.4, 4.4.6.

4.4.3 Theorem. *Let X and Y be normed spaces over the same field, and suppose $A \in \mathcal{O}(X, Y)$. If A is bounded and D_A is closed, then A is closed.*

Proof. Assuming A bounded and D_A closed, A is closed because the following implications are true:

$$[\{f_n\} \text{ is a sequence in } D_A \text{ that is convergent in } X \text{ and}$$
$$\{Af_n\} \text{ is a convergent sequence in } Y] \overset{(1)}{\Rightarrow}$$
$$[\{f_n\} \text{ is a sequence in } D_A \text{ that is convergent in } X \text{ and } \lim_{n\to\infty} f_n \in D_A] \overset{(2)}{\Rightarrow}$$
$$[\lim_{n\to\infty} f_n \in D_A \text{ and } A(\lim_{n\to\infty} f_n) = \lim_{n\to\infty} Af_n],$$

where 1 is true by 2.3.4 because D_A is closed (notice that the condition "$\{Af_n\}$ is a convergent sequence in Y" plays no role) and 2 is true because A is continuous (cf. 4.2.2). □

4.4.4 Theorem. *Let X be a normed space, Y a Banach space over the same field, and $A \in \mathcal{O}(X, Y)$. If A is bounded and closed, then D_A is closed.*

Proof. Assume A bounded and closed. First, we notice that, if $\{f_n\}$ is a sequence in D_A that is convergent in X, then $\{Af_n\}$ is a Cauchy sequence since $\|Af_n - Af_m\| \le \|A\| \|f_n - f_m\|$, and therefore $\{Af_n\}$ is a convergent sequence in Y since Y is a complete metric space. Then, D_A is closed by 2.3.4 since the following implications are true:

$$[\{f_n\} \text{ is a sequence in } D_A \text{ that is convergent in } X] \Rightarrow$$
$$[\{f_n\} \text{ is a sequence in } D_A \text{ that is convergent in } X \text{ and}$$
$$\{Af_n\} \text{ is a convergent sequence in } Y] \overset{(*)}{\Rightarrow}$$
$$\lim_{n\to\infty} f_n \in D_A,$$

where $(*)$ is true because A is closed. □

We state the following theorem without giving its proof, which can be found e.g. in Chapter 10 of (Royden, 1988), since we shall use neither this theorem nor its corollary. We prove less general versions of this theorem and of its corollary in 12.2.3 and in 13.1.9, for an operator in a Hilbert space.

4.4.5 Theorem (Closed graph theorem). *Let X and Y be Banach spaces over the same field. If $A \in \mathcal{O}_E(X, Y)$ (for $\mathcal{O}_E(X, Y)$, cf. 3.2.12) and A is closed, then A is bounded.*

4.4.6 Corollary. *Let X and Y be Banach spaces over the same field, and suppose $A \in \mathcal{O}(X, Y)$. If D_A is closed and A is closed, then A is bounded.*

Proof. If D_A is closed then D_A is a Banach space (cf. 4.1.8a).

If A is closed then G_A is a closed set in the product metric space $X \times Y$. By 2.3.3, G_A is then a closed set in the metric space $D_A \times Y$ as well, since obviously $G_A = G_A \cap (D_A \times Y)$.

Thus, if D_A is closed and A is closed, then D_A and Y are Banach spaces and A is a closed element of $\mathcal{O}_E(D_A, Y)$. Then, by 4.4.5, A is a bounded element of $\mathcal{O}_E(D_A, Y)$, and hence a bounded element of $\mathcal{O}(D_A, Y)$ as well. $\qquad\square$

Assuming X, Y Banach spaces over the same field and $A \in \mathcal{O}(X, Y)$, from 4.4.3, 4.4.4 and 4.4.6 we see that, if two of the three conditions A bounded, A closed, D_A closed are true, then the remaining one is true as well. This rounds off our examination of the interplay between the concepts of bounded operator and closed operator. However, we shall not use either 4.4.5 or 4.4.6 for general closed operators in general Banach spaces, and this is the reason why we have not provided a proof of the closed graph theorem (which, moreover, would require preliminary results outside the scope of this book). Anyway, as already mentioned, the closed graph theorem and its corollary will be proved in 12.2.3 and in 13.1.9 respectively, for operators in Hilbert spaces.

4.4.7 Proposition. *Let X and Y be normed spaces over the same field, and suppose $A \in \mathcal{O}(X, Y)$ and A injective. Then A is closed iff A^{-1} is closed.*

Proof. If A is injective, then $G_{A^{-1}} = V(G_A)$, where V is defined by

$$X \times Y \ni (f, g) \mapsto V(f, g) := (g, f) \in Y \times X$$

(cf. 1.2.11). Now, from 2.7.3a it is clear that a sequence $\{(f_n, g_n)\}$ in the product metric space $X \times Y$ converges to $(f, g) \in X \times Y$ iff the sequence $\{(g_n, f_n)\}$ converges to (g, f) in the product metric space $Y \times X$. Hence, using 2.3.4 we see that G_A is a closed subset of the product metric space $X \times Y$ iff $G_{A^{-1}}$ is a closed subset of the product metric space $Y \times X$. $\qquad\square$

4.4.8 Proposition. *Let X and Y be normed spaces over the same field, and let $A \in \mathcal{O}(X, Y)$. If A is closed then N_A is a subspace of X.*

Proof. Let $f \in X$ and let $\{f_n\}$ be a sequence in N_A so that $f_n \to f$. Then, $Af_n \to 0_Y$. If A is closed, this implies $f \in D_A$ and $Af = 0_Y$, i.e. $f \in N_A$. In view of 2.3.4, this shows that N_A is closed, and hence a subspace of X (cf. 3.2.2b). $\qquad\square$

4.4.9 Proposition. *Let X and Y be normed spaces over the same field, and suppose $A \in \mathcal{O}(X, Y)$, $B \in \mathcal{B}(X, Y)$, A closed. Then $A + B$ is closed.*

Proof. Let $\{f_n\}$ be a sequence in D_{A+B} and assume that there exists $(f, g) \in X \times Y$ so that $f_n \to f$ and $(A + B)f_n \to g$. Since $B \in \mathcal{B}(X, Y)$, we have $Bf_n \to Bf$ and hence $Af_n \to g - Bf$. Since $f_n \in D_A$ for each $n \in \mathbb{N}$ and A is closed, this implies $f \in D_A$ and $g - Bf = Af$, i.e. $f \in D_{A+B}$ and $g = (A + B)f$. This proves that $A + B$ is closed. $\qquad\square$

4.4.10 Definition. Let X and Y be normed spaces over the same field, and let $A \in \mathcal{O}(X,Y)$. The linear operator A is said to be *closable* if the closure (in the product metric space $X \times Y$ or equivalently in the normed space $X + Y$) $\overline{G_A}$ of its graph is the graph of a mapping. If A is closable, then the mapping which has $\overline{G_A}$ as its graph is a closed linear operator from X to Y (cf. 4.1.12 and 3.2.15a) which is called the *closure* of A and is denoted by $\overline{A}$. Clearly, $A \subset \overline{A}$ (cf. 1.2.5) and $\overline{A}$ is the smallest closed operator that contains A: if B is a closed operator that contains A, then $G_A \subset G_B$, and hence $G_{\overline{A}} = \overline{G_A} \subset G_B$, and hence $\overline{A} \subset B$ (cf. 1.2.5).

If A is closable, $G_{\overline{A}} = \overline{G_A}$ means that (cf. 2.3.10):

$$D_{\overline{A}} := \{f \in X : \text{there exists a sequence } \{f_n\} \text{ in } D_A \text{ s.t.}$$
$$f_n \to f \text{ and } \{Af_n\} \text{ is convergent}\},$$
$$\forall f \in D_{\overline{A}}, \overline{A}f = \lim_{n\to\infty} Af_n \text{ if } \{f_n\} \text{ is a sequence in } D_A$$
$$\text{s.t. } f_n \to f \text{ and } \{Af_n\} \text{ is convergent.}$$

It is obvious that A is closed iff (A is closable and $A = \overline{A}$).

4.4.11 Proposition. *Let X and Y be normed spaces over the same field, and let $A \in \mathcal{O}(X,Y)$. Then:*

(a) A is closable iff the following condition holds
$$(0_X, g) \in \overline{G_A} \Rightarrow g = 0_Y;$$
(b) A is closable iff $\exists B \in \mathcal{O}(X,Y)$ such that B is closed and $A \subset B$.

Proof. a: this follows at once from 4.1.12 and 3.2.15b.

b: If A is closable, put $B := \overline{A}$.

If there is $B \in \mathcal{O}(X,Y)$ such that B is closed and $A \subset B$, then we have $\overline{G_A} \subset G_B$. Hence,

$$(0_X, g) \in \overline{G_A} \Rightarrow (0_X, g) \in G_B \Rightarrow g = 0_Y$$

by 3.2.15b. Then A is closable by result a. $\qquad\square$

4.4.12 Proposition. *Let X be a normed space and Y a Banach space over the same field. Let $A \in \mathcal{O}(X,Y)$, and suppose A bounded. Then A is closable, $D_{\overline{A}} = \overline{D_A}$ and $\overline{A}$ is bounded.*

Proof. Let $\tilde{A}$ be the element of $\mathcal{O}(X,Y)$ such that $D_{\tilde{A}} = \overline{D_A}$, $A \subset \tilde{A}$, $\tilde{A}$ is bounded (cf. 4.2.6).

Since $\tilde{A}$ is bounded and $D_{\tilde{A}}$ is closed, $\tilde{A}$ is closed by 4.4.3. Then, since $A \subset \tilde{A}$, A is closable by 4.4.11b and we also have $\overline{A} \subset \tilde{A}$.

For $f \in D_{\tilde{A}}$, in view of $D_{\tilde{A}} = \overline{D_A}$ and 2.3.10 there is a sequence $\{f_n\}$ in D_A s.t. $f_n \to f$, and hence (since $\tilde{A}$ is continuous) also s.t. $\tilde{A}f_n \to \tilde{A}f$. Since $A \subset \overline{A} \subset \tilde{A}$, the sequence $\{f_n\}$ is also in $D_{\overline{A}}$ and $\overline{A}f_n = \tilde{A}f_n$; since $\overline{A}$ is closed, this implies $f \in D_{\overline{A}}$. Thus, $D_{\tilde{A}} \subset D_{\overline{A}}$ and therefore $\overline{A} = \tilde{A}$. $\qquad\square$

4.4.13 Proposition. *Let X and Y be normed spaces over the same field, let $A \in \mathcal{O}(X,Y)$, and suppose A injective and closable. Then A^{-1} is closable iff $\overline{A}$ is injective. If these conditions are satisfied, then $\overline{A^{-1}} = (\overline{A})^{-1}$.*

Proof. If V is the mapping defined in 1.2.11, we have

$$\overline{G_{A^{-1}}} = \overline{V(G_A)} \overset{(*)}{=} V(\overline{G_A}) = V(G_{\overline{A}}),$$

where $(*)$ holds in view of 2.3.10 and of what was noted in the proof of 4.4.7. Since $V(V(f,g)) = (f,g)$ for each $(f,g) \in X \times Y$, for $f \in X$ we also have

$$(0_Y, f) \in V(G_{\overline{A}}) \text{ iff } (f, 0_Y) \in G_{\overline{A}} \text{ iff } f \in N_{\overline{A}}.$$

Thus, for $f \in X$,

$$(0_Y, f) \in \overline{G_{A^{-1}}} \text{ iff } f \in N_{\overline{A}}.$$

In view of 4.4.11a and of 3.2.6a, this shows that A^{-1} is closable iff $\overline{A}$ is injective. If these conditions are satisfied, then we have (cf. 1.2.11)

$$G_{\overline{A^{-1}}} = \overline{G_{A^{-1}}} = V(G_{\overline{A}}) = G_{(\overline{A})^{-1}},$$

and hence $\overline{A^{-1}} = (\overline{A})^{-1}$. □

4.4.14 Proposition. *Let X and Y be normed spaces over the same field, and suppose $A \in \mathcal{O}(X,Y)$, $B \in \mathcal{B}(X,Y)$, A closable. Then $A + B$ is closable and $\overline{A+B} = \overline{A} + B$.*

Proof. Since $\overline{A} + B$ is closed (cf. 4.4.9) and $A + B \subset \overline{A} + B$, we have that $A + B$ is closable (cf. 4.4.11b) and $G_{\overline{A+B}} = \overline{G_{A+B}} \subset G_{\overline{A}+B}$.

Let now $f \in D_{\overline{A}+B}$. Since $D_{\overline{A}+B} = D_{\overline{A}}$, there exists a sequence $\{f_n\}$ in D_A such that $f_n \to f$ and $Af_n \to \overline{A}f$. Since $B \in \mathcal{B}(X,Y)$, we also have $(A+B)f_n \to (\overline{A}+B)f$. Then we have $(f, (\overline{A}+B)f) \in \overline{G_{A+B}}$ since $(f_n, (A+B)f_n)$ is a sequence in G_{A+B} (cf. 2.3.10). This proves that $G_{\overline{A}+B} \subset \overline{G_{A+B}}$, and hence that $\overline{G_{A+B}} = G_{\overline{A}+B}$, which is equivalent to $\overline{A+B} = \overline{A} + B$. □

4.5 The spectrum of a linear operator

In this section, which contains little more than definitions, X denotes a normed space over $\mathbb{K}$ and A denotes an operator in X, i.e. $A \in \mathcal{O}(X)$.

4.5.1 Definitions. The *resolvent set* of A is the set

$$\rho(A) := \{\lambda \in \mathbb{K} : A - \lambda \mathbb{1}_X \text{ is injective}, (A - \lambda \mathbb{1}_X)^{-1} \text{ is bounded}, \overline{R_{A-\lambda \mathbb{1}_X}} = X\}.$$

The *spectrum* of A is the set

$$\sigma(A) := \mathbb{K} - \rho(A).$$

4.5.2 Proposition. *For $\lambda \in \mathbb{K}$, the following conditions are equivalent:*

(a) [$A - \lambda \mathbb{1}_X$ is not injective] or [$A - \lambda \mathbb{1}_X$ is injective and $(A - \lambda \mathbb{1}_X)^{-1}$ is not bounded];

(b) $\forall \epsilon > 0, \exists f_\epsilon \in D_A$ such that $\|Af_\epsilon - \lambda f_\epsilon\| < \epsilon \|f_\epsilon\|$ (hence, $f_\epsilon \neq 0_X$).

Proof. This statement is the statement of 4.2.3, with A replaced by $A - \lambda \mathbb{1}_X$ (note that $D_{A - \lambda \mathbb{1}_X} = D_A$) and the two conditions in contrapositive form. $\qquad \square$

4.5.3 Definition. The *approximate point spectrum* of A is the set
$$Ap\sigma(A) := \{\lambda \in \mathbb{K} : \text{for } \lambda \text{ the conditions of 4.5.2 are true}\}.$$

4.5.4 Remark. Clearly, $Ap\sigma(A) \subset \sigma(A)$.

4.5.5 Proposition. *If A is bounded, then:*
$$Ap\sigma(A) \subset \{\lambda \in \mathbb{K} : |\lambda| \leq \|A\|\}.$$

Proof. Let $\lambda \in Ap\sigma(A)$ and choose $\epsilon > 0$. Then (cf. 4.1.2b)
$$\exists f_\epsilon \in D_A \text{ s.t. } \|f_\epsilon\| = 1 \text{ and } |\lambda| - \|Af_\epsilon\| \leq \|Af_\epsilon - \lambda f_\epsilon\| < \epsilon,$$
and hence (cf. 4.2.5b) $|\lambda| < \|A\| + \epsilon$. Since ϵ was arbitrary, this proves that $|\lambda| \leq \|A\|$. $\qquad \square$

4.5.6 Proposition. *For $\lambda \in \mathbb{K}$ the following conditions are equivalent:*

(a) $A - \lambda \mathbb{1}_X$ is not injective;

(b) $\exists f \in D_A$ such that $f \neq 0_X$ and $Af = \lambda f$;

(c) $N_{A - \lambda \mathbb{1}_X} \neq \{0_X\}$.

Proof. The equivalence of conditions a and c is the contrapositive form of 3.2.6a, with A replaced by $A - \lambda \mathbb{1}_X$. The equivalence of conditions b and c follows from the definition of $N_{A - \lambda \mathbb{1}_X}$ (since $D_{A - \lambda \mathbb{1}_X} = D_A$). $\qquad \square$

4.5.7 Definitions. The *point spectrum* of A is the set
$$\sigma_p(A) := \{\lambda \in \mathbb{K} : \text{for } \lambda \text{ the conditions of 4.5.6 are true}\}.$$
An element λ of $\sigma_p(A)$ is called an *eigenvalue* of A, and the linear manifold $N_{A - \lambda \mathbb{1}_X}$ (cf. 3.2.2b) is called the corresponding *eigenspace*. A non-null vector of $N_{A - \lambda \mathbb{1}_X}$ is called an *eigenvector* of A corresponding to λ.

4.5.8 Remark. Clearly, $\sigma_p(A) \subset Ap\sigma(A)$.

Later in the book (cf. 12.4.24, 12.4.25, 15.3.4), examples will be provided of operators in Hilbert spaces that have empty spectrum, or that have non-empty spectrum but empty approximate point spectrum, or that have non-empty approximate point spectrum but empty point spectrum, or that have non-empty point spectrum. Further, examples will be provided of operators for which the various kinds of spectra coincide and others for which they do not.

4.5.9 Proposition. *If A is closed then $N_{A - \lambda \mathbb{1}_X}$ is a subspace of X, for all $\lambda \in \mathbb{K}$.*

Proof. Use 4.4.9 and 4.4.8. □

4.5.10 Theorem. *Suppose that X is a non-zero Banach space and $A \in \mathcal{B}(X)$. Then:*

$$\sigma(A) \subset \{\lambda \in \mathbb{K} : |\lambda| \leq \|A\|\}.$$

Proof. Since X is a non-zero Banach space, $\mathcal{B}(X)$ is a Banach algebra with identity (cf. 4.3.5). For $\lambda \in \mathbb{K}$ so that $\|A\| < |\lambda|$, we have $\|\frac{1}{\lambda}A\| < 1$. Then, by 4.3.4, the series $\sum_{n=1}^{\infty} \left(\frac{1}{\lambda}A\right)^n$ is convergent in $\mathcal{B}(X)$ and

$$\left(\mathbb{1}_X + \sum_{n=1}^{\infty} \left(\frac{1}{\lambda}A\right)^n\right)\left(\mathbb{1}_X - \frac{1}{\lambda}A\right) = \left(\mathbb{1}_X - \frac{1}{\lambda}A\right)\left(\mathbb{1}_X + \sum_{n=1}^{\infty} \left(\frac{1}{\lambda}A\right)^n\right) = \mathbb{1}_X,$$

and hence

$$\left[-\frac{1}{\lambda}\left(\mathbb{1}_X + \sum_{n=1}^{\infty} \left(\frac{1}{\lambda}A\right)^n\right)\right](A - \lambda\mathbb{1}_X)$$

$$= (A - \lambda\mathbb{1}_X)\left[-\frac{1}{\lambda}\left(\mathbb{1}_X + \sum_{n=1}^{\infty} \left(\frac{1}{\lambda}A\right)^n\right)\right] = \mathbb{1}_X.$$

By 3.2.7c, this implies that $A - \lambda\mathbb{1}_X$ is injective and that

$$(A - \lambda\mathbb{1}_X)^{-1} = -\frac{1}{\lambda}\left(\mathbb{1}_X + \sum_{n=1}^{\infty} \left(\frac{1}{\lambda}A\right)^n\right),$$

and hence that $(A - \lambda\mathbb{1}_X)^{-1} \in \mathcal{B}(X)$, which means $(A - \lambda\mathbb{1}_X)^{-1}$ bounded and $R_{A-\lambda\mathbb{1}_X} = D_{(A-\lambda\mathbb{1}_X)^{-1}} = X$. Thus, $\lambda \in \rho(A)$.

This proves the implication

$$\|A\| < |\lambda| \Rightarrow \lambda \in \rho(A),$$

which is equivalent to

$$\lambda \in \sigma(A) \Rightarrow |\lambda| \leq \|A\|.$$

□

4.5.11 Proposition. *Suppose that X is a Banach space and that A is closable. Then $\sigma(\overline{A}) = \sigma(A)$.*

Proof. Let $\lambda \in \rho(A)$. Then $A - \lambda\mathbb{1}_X$ is injective. Moreover, $A - \lambda\mathbb{1}_X$ is closable (cf. 4.4.14) and $(A - \lambda\mathbb{1}_X)^{-1}$ is bounded and therefore closable (cf. 4.4.12). Then, from 4.4.13 we have that $\overline{A - \lambda\mathbb{1}_X}$ is injective and $\left(\overline{A - \lambda\mathbb{1}_X}\right)^{-1} = \overline{(A - \lambda\mathbb{1}_X)^{-1}}$. Since $\overline{A - \lambda\mathbb{1}_X} = \overline{A} - \lambda\mathbb{1}_X$ (cf. 4.4.14), we have that $\overline{A} - \lambda\mathbb{1}_X$ is injective and $\left(\overline{A} - \lambda\mathbb{1}_X\right)^{-1} = \overline{(A - \lambda\mathbb{1}_X)^{-1}}$. This equality and 4.4.12 imply that $\left(\overline{A} - \lambda\mathbb{1}_X\right)^{-1}$ is bounded. Finally, $\overline{R_{(A-\lambda\mathbb{1}_X)}} = X$ implies $\overline{R_{(\overline{A}-\lambda\mathbb{1}_X)}} = X$ in view of the inclusion $R_{(A-\lambda\mathbb{1}_X)} \subset R_{(\overline{A}-\lambda\mathbb{1}_X)}$. Thus, $\lambda \in \rho(\overline{A})$.

Conversely, suppose $\lambda \in \rho(\overline{A})$. Then $\overline{A} - \lambda \mathbb{1}_X$ is injective and hence (cf. 3.2.7a) $A - \lambda \mathbb{1}_X$ is injective and $(A - \lambda \mathbb{1}_X)^{-1} \subset (\overline{A} - \lambda \mathbb{1}_X)^{-1}$. Since $(\overline{A} - \lambda \mathbb{1}_X)^{-1}$ is bounded, this implies (cf. 4.2.5a) that $(A - \lambda \mathbb{1}_X)^{-1}$ is bounded. Further, $\overline{D_{(\overline{A} - \lambda \mathbb{1}_X)^{-1}}} = \overline{R_{(\overline{A} - \lambda \mathbb{1}_X)}} = X$ implies $D_{(\overline{A} - \lambda \mathbb{1}_X)^{-1}} = X$ by 4.4.4, since $(\overline{A} - \lambda \mathbb{1}_X)^{-1}$ is bounded and closed (cf. 4.4.9 and 4.4.7), and hence $D_{\overline{(A - \lambda \mathbb{1}_X)^{-1}}} = X$ since $(\overline{A} - \lambda \mathbb{1}_X)^{-1} = \overline{(A - \lambda \mathbb{1}_X)^{-1}}$ (see above). We observe now that, if B is a closable operator, then obviously $D_{\overline{B}} \subset \overline{D_B}$. Then, $\overline{D_{(A - \lambda \mathbb{1}_X)^{-1}}} = X$, i.e. $\overline{R_{(A - \lambda \mathbb{1}_X)}} = X$. Thus, $\lambda \in \rho(A)$.

This proves that $\rho(\overline{A}) = \rho(A)$, and hence that $\sigma(\overline{A}) = \sigma(A)$. $\quad\square$

4.5.12 Proposition. *Suppose that X is a Banach space and that A is closed. Then $R_{A - \lambda \mathbb{1}_X} = X$ for each $\lambda \in \rho(A)$.*

Proof. Since A is closed, $A - \lambda \mathbb{1}_X$ is closed for each $\lambda \in \mathbb{K}$ (cf. 4.4.9). Thus, if $\lambda \in \rho(A)$ then $(A - \lambda \mathbb{1}_X)^{-1}$ is closed (cf. 4.4.7) and bounded. Since X is a Banach space, this implies that $D_{(A - \lambda \mathbb{1}_X)^{-1}}$ is closed (cf. 4.4.4), and hence that $R_{A - \lambda \mathbb{1}_X} = \overline{R_{A - \lambda \mathbb{1}_X}} = X$. $\quad\square$

4.5.13 Remark. Some define the resolvent set and the spectrum of A in a different way than we did in 4.5.1, by letting the resolvent set of A be the set

$$\rho'(A) := \{\lambda \in \mathbb{K} : A - \lambda \mathbb{1}_X \text{ is injective and } (A - \lambda \mathbb{1}_X)^{-1} \in \mathcal{B}(X)\}$$

and letting the spectrum of A be the set

$$\sigma'(A) := \mathbb{K} - \rho'(A).$$

However, these definitions are not very useful for non-closed operators. In fact, if $\rho'(A) \neq \emptyset$ then there is $\lambda \in \mathbb{K}$ so that $(A - \lambda \mathbb{1}_X)^{-1}$ exists and $(A - \lambda \mathbb{1}_X)^{-1} \in \mathcal{B}(X)$, hence $(A - \lambda \mathbb{1}_X)^{-1}$ is closed by 4.4.3, hence $A - \lambda \mathbb{1}_X$ is closed by 4.4.7, hence A is closed by 4.4.9. This proves that $\rho'(A) = \emptyset$, and hence $\sigma'(A) = \mathbb{K}$, if A is not closed. Thus, the spectrum as defined above is always trivially the same for all non-closed operators, even when they are closable. This is not true with our definition of spectrum, as indicated by 4.5.11 (cf. also 12.4.25).

If X is Banach space then the definitions given above are actually equivalent to ours, for closed operators. In fact, 4.5.12 proves that $\rho(A) \subset \rho'(A)$ if A is closed and X is a Banach space, and hence $\rho(A) = \rho'(A)$ since $\rho'(A) \subset \rho(A)$ is obvious.

4.6 Isomorphisms of normed spaces

4.6.1 Definitions. Let $(X_1, \sigma_1, \mu_1, \nu_1)$ and $(X_2, \sigma_2, \mu_2, \nu_2)$ be normed spaces over the same field $\mathbb{K}$. An *isomorphism* from X_1 onto X_2 is a mapping $U : X_1 \to X_2$ such that:

(in_1) U is a bijection from X_1 onto X_2;

(in_2) $\sigma_2(U(f), U(g)) = U(\sigma_1(f, g))$, $\forall f, g \in X_1$;

$\qquad \mu_2(\alpha, U(f)) = U(\mu_1(\alpha, f))$, $\forall \alpha \in \mathbb{K}$, $\forall f \in X_1$;

(in_3) $\nu_2(U(f)) = \nu_1(f)$, $\forall f \in X_1$.

If an isomorphism from X_1 onto X_2 exists, then the two normed spaces X_1 and X_2 are said to be *isomorphic*.

If the two normed spaces X_1 and X_2 are the same, an isomorphism from X_1 onto X_2 is called an *automorphism* of X_1.

4.6.2 Remarks.

(a) In 4.6.1, condition in_1 means that U is an "isomorphism" from the set X_1 onto the set X_2 (it preserves the set theoretical "operations", i.e. union, intersection, complementation), conditions in_1 and in_2 mean that U is an isomorphism from the linear space (X_1, σ_1, μ_1) onto the linear space (X_2, σ_2, μ_2) (actually, condition in_2 says that U is a linear operator), and condition in_3 says that U preserves the norm. We laid down condition in_1 the way we did in order to make it clear from the outset that an isomorphism preserves the three level structure of a normed space. However, we could have only asked in in_1 that U be surjective onto X_2, since U is a linear operator by condition in_2 and $N_U = \{0_{X_1}\}$ holds by condition in_3, and hence U is injective by 3.2.6a.

(b) It is obvious (also in view of 3.2.6b) that, if U is an isomorphism from a normed space X_1 onto a normed space X_2, then the inverse mapping U^{-1} (i.e. the linear operator U^{-1}) is an isomorphism from X_2 onto X_1; and also that, if V is an isomorphism from X_2 onto a third normed space X_3, then the composition $V \circ U$ (i.e. the product VU of the linear operators U and V) is an isomorphism from X_1 onto X_3.

(c) For any normed space X, the identity mapping id_X (i.e. the linear operator $\mathbb{1}_X$, cf. 3.2.5) is obviously an automorphism of X. It is immediate to see, also in view of remark b, that the family of all automorphisms of X is a group, with the product of operators as group product, the identity mapping as group identity, the inverse mapping as group inverse.

(d) If U is an isomorphism from a normed space $(X_1, \sigma_1, \mu_1, \nu_1)$ onto a normed space $(X_2, \sigma_2, \mu_2, \nu_2)$, then

$$d_{\nu_2}(U(f), U(g)) := \|U(f) - U(g)\|_2 = \|U(f - g)\|_2$$
$$= \|f - g\|_1 =: d_{\nu_1}(f, g), \forall f, g \in X_1.$$

Thus, U is an isomorphism from the metric space (X_1, d_{ν_1}) onto the metric space (X_2, d_{ν_2}).

(e) As remarked above, an isomorphism U from a normed space X_1 onto a normed space X_2 is a linear operator. It is obvious that U is a bounded operator, i.e. $U \in \mathcal{B}(X_1, X_2)$. Thus, U is a continuous mapping (cf. 4.2.2; however, this was already clear from remark d). Moreover, $\|U\| = 1$ follows immediately from 4.2.5c.

4.6.3 Proposition. *Let X_1 and X_2 be isomorphic normed spaces over the same field, and let U be an isomorphism from X_1 onto X_2. The mapping*

$$T_U : X_1 + X_1 \to X_2 + X_2$$
$$(f, g) \mapsto T_U(f, g) := (Uf, Ug)$$

is an isomorphism from the normed space $X_1 + X_1$ onto the normed space $X_2 + X_2$ (cf. 4.1.16).

For each linear operator $A \in \mathcal{O}(X_1)$ we have $T_U(G_A) = G_{UAU^{-1}}$.

Proof. It follows immediately from the definitions that T_U is an isomorphism from the normed space $X_1 + X_1$ onto the normed space $X_2 + X_2$.

For $(f, g) \in X_1 + X_1$ we have

$$(f, g) \in G_A \Leftrightarrow$$
$$[f \in D_A \text{ and } g = Af] \Leftrightarrow$$
$$[Uf \in D_{AU^{-1}} = D_{UAU^{-1}} \text{ and } Ug = UAU^{-1}Uf] \Leftrightarrow$$
$$(Uf, Ug) \in G_{UAU^{-1}} \Leftrightarrow$$
$$(f, g) \in T_U^{-1}(G_{UAU^{-1}})$$

$(D_{AU^{-1}} = D_{UAU^{-1}}$ is true because $D_U = X_1$). Thus we have $G_A = T_U^{-1}(G_{UAU^{-1}})$ and hence

$$T_U(G_A) = T_U(T_U^{-1}(G_{UAU^{-1}})) = G_{UAU^{-1}},$$

because the counterimage of a subset of $X_2 + X_2$ under the mapping T_U can be considered the image of that subset under the mapping T_U^{-1}, since the mapping T_U is bijective from $X_1 + X_1$ onto $X_2 + X_2$ (cf. 1.2.11). $\qquad\square$

4.6.4 Proposition. *Let X_1 and X_2 be isomorphic normed spaces over the same field, let $A \in \mathcal{O}(X_1)$ and $B \in \mathcal{O}(X_2)$, and let U be an isomorphism from X_1 onto X_2. The following conditions are equivalent:*

(a) $B = UAU^{-1}$;
(b) $A = U^{-1}BU$;
(c) $BU = UA$;
(d) $AU^{-1} = U^{-1}B$;
(e) $D_A = U^{-1}(D_B)$ and $Bf = UAU^{-1}f$, $\forall f \in D_B$;
(f) $D_B = U(D_A)$ and $Ag = U^{-1}BUg$, $\forall g \in D_A$;
(g) $G_B = T_U(G_A)$ (for T_U, cf. 4.6.3).

If these conditions are satisfied then:

(h) $R_B = U(R_A)$;
(i) $\overline{D_B} = U(\overline{D_A})$ and $\overline{R_B} = U(\overline{R_A})$.

Proof. The equivalence of conditions a, b, c, d follows immediately from the associativity of compositions of mappings (cf. 1.2.17) and from $U^{-1}U = \mathbb{1}_{X_1}$ and $UU^{-1} = \mathbb{1}_{X_2}$.

The equivalence of conditions a and g follows from the equality $T_U(G_A) = G_{UAU^{-1}}$ (cf. 4.6.3) and from the fact that two mappings are equal iff their graphs are equal.

If conditions a and e are proved to be equivalent then it is obvious that conditions b and f are also equivalent, since U^{-1} is an isomorphism from X_2 onto X_1 (cf. 4.6.2b).

The equivalence of conditions a and e is proved as follows.

$a \Rightarrow e$: Assume condition a. Then $D_A = U^{-1}(D_B)$ since $D_A = U^{-1}(D_{UAU^{-1}})$ by 1.2.18a. Moreover it is obvious that $Bf = UAU^{-1}f$ for all $f \in D_B$.

$e \Rightarrow a$: If we assume $D_A = U^{-1}(D_B)$ then we have $U^{-1}(D_B) = U^{-1}(D_{UAU^{-1}})$ since $D_A = U^{-1}(D_{UAU^{-1}})$ by 1.2.18a, and hence

$$D_B = U(U^{-1}(D_B)) = U(U^{-1}(D_{UAU^{-1}})) = D_{UAU^{-1}}$$

because counterimages under U can be interpreted as images under U^{-1} since U is bijective (cf. 1.2.11). If we further assume that $Bf = UAU^{-1}f$ for all $f \in D_B$ then we have $B = UAU^{-1}$.

Now suppose that conditions a, b, c, d, e, f, g are satisfied. Then conditions h and i are proved as follows.

h: Condition a implies $R_B = U(R_A)$ by 1.2.18b.

i: Since $D_B = U(D_A)$ and $R_B = U(R_A)$, 2.3.21a implies that $\overline{D_B} = U(\overline{D_A})$ and $\overline{R_B} = U(\overline{R_A})$ since U is an isomorphism from the metric space X_1 onto the metric space X_2 (cf. 4.6.2d). □

4.6.5 Theorem. *Let X_1 and X_2 be isomorphic normed spaces over the same field $\mathbb{K}$. Let $A \in \mathcal{O}(X_1)$, $B \in \mathcal{O}(X_2)$ and suppose that there exists an isomorphism U from X_1 onto X_2 so that $B = UAU^{-1}$. Then:*

(a) *A is injective iff B is injective; if A and B are injective then $B^{-1} = UA^{-1}U^{-1}$;*
(b) *A is bounded iff B is bounded; if A and B are bounded then $\|B\| = \|A\|$;*
(c) *A is closed iff B is closed;*
(d) *A is closable iff B is closable; if A and B are closable, then $\overline{B} = U\overline{A}U^{-1}$;*
(e) *$N_B = U(N_A)$;*
(f) *$B - \lambda\mathbb{1}_{X_2} = U(A - \lambda\mathbb{1}_{X_1})U^{-1}, \forall\lambda \in \mathbb{K}$;*
(g) *$\sigma(B) = \sigma(A)$;*
(h) *$Ap\sigma(B) = Ap\sigma(A)$;*
(i) *$\sigma_p(B) = \sigma_p(A)$.*

Proof. a: Everything follows from 1.2.17 and 1.2.14B (cf. also the equivalence between conditions a and b in 4.6.4)

b: If A is bounded then

$$\|Bf\|_2 = \|UAU^{-1}f\|_2 = \|AU^{-1}f\|_1 \leq \|A\|\|U^{-1}f\|_1 = \|A\|\|f\|_2, \forall f \in D_B$$

(cf. 4.2.5b). This proves that B is bounded and $\|B\| \leq \|A\|$. Since $A = U^{-1}BU$ (cf. the equivalence between conditions a and b in 4.6.4), by the same token it can be proved that if B is bounded then A is bounded and $\|A\| \leq \|B\|$.

c: Since T_U is an isomorphism from $X_1 + X_1$ onto $X_2 + X_2$ as metric spaces (cf. 4.6.3 and 4.6.2d) and since $G_B = T_U(G_A)$ (cf. the equivalence between conditions a and g in 4.6.4), 2.3.21b implies that G_A is closed iff G_B is closed.

d: Since T_U is an isomorphism from $X_1 + X_1$ onto $X_2 + X_2$ as metric spaces and since $G_B = T_U(G_A)$, 2.3.21a implies that $\overline{G_B} = T_U(\overline{G_A})$. Then, if B is closable we have

$$(0_{X_1}, g) \in \overline{G_A} \Rightarrow (0_{X_1}, Ug) \in \overline{G_B} \Rightarrow Ug = 0_{X_2} \Rightarrow g = 0_{X_1},$$

by 4.4.11a. This proves that A is closable if B is closable, in view of 4.4.11a again. The converse statement can be proved by the same token, since $\overline{G_A} = T_U^{-1}(\overline{G_B})$.

Suppose A and B closable. From 4.6.3 we have $T_U(G_{\overline{A}}) = G_{U\overline{A}U^{-1}}$, and hence $G_{\overline{B}} = \overline{G_B} = T_U(\overline{G_A}) = G_{U\overline{A}U^{-1}}$, and hence $\overline{B} = U\overline{A}U^{-1}$.

e: For $f \in X_1$ we have

$$f \in N_A \Leftrightarrow (f, 0_{X_1}) \in G_A \Leftrightarrow (Uf, 0_{X_2}) \in T_U(G_A) = G_B \Leftrightarrow$$
$$Uf \in N_B \Leftrightarrow f \in U^{-1}(N_B).$$

Thus we have $N_A = U^{-1}(N_B)$ and hence $N_B = U(N_A)$.

f: For each $\lambda \in \mathbb{K}$, we have

$$B - \lambda \mathbb{1}_{X_2} = UAU^{-1} - \lambda UU^{-1} = U(AU^{-1} - \lambda U^{-1}) = U(A - \lambda \mathbb{1}_{X_1})U^{-1},$$

where the second equality holds by $3.2.10b_2'$ since $D_U = X_1$ and the last by $3.2.10b_3$.

g: Let $\lambda \in \mathbb{K}$. We have that

$$A - \lambda \mathbb{1}_{X_1} \text{ is injective iff } B - \lambda \mathbb{1}_{X_2} \text{ is injective,}$$

in view of results f and a. Assuming $A - \lambda \mathbb{1}_{X_1}$ and $B - \lambda \mathbb{1}_{X_2}$ injective, we have that

$$(A - \lambda \mathbb{1}_{X_1})^{-1} \text{ is bounded iff } (B - \lambda \mathbb{1}_{X_2})^{-1} \text{ is bounded,}$$

in view of results f, a, b. Finally, we have

$$\overline{R_{A - \lambda \mathbb{1}_{X_1}}} = X_1 \text{ iff } \overline{R_{B - \lambda \mathbb{1}_{X_2}}} = X_2$$

because result f and 4.6.4i imply that

$$\overline{R_{B - \lambda \mathbb{1}_{X_2}}} = U(\overline{R_{A - \lambda \mathbb{1}_{X_1}}}),$$

and also because $R_U = X_2$ and $R_{U^{-1}} = X_1$.

This proves that $\rho(B) = \rho(A)$. Hence, $\sigma(B) = \sigma(A)$.

h: This result is proved by the first two equivalences in the proof of result g.

i: This result is proved by the first equivalence in the proof of result g.   □

4.6.6 Proposition. *Let X_1 and X_2 be Banach spaces over the same field. Suppose that there exists a linear operator V from X_1 to X_2, i.e. $V \in \mathcal{O}(X_1, X_2)$, such that*

$$\overline{D_V} = X_1, \ \overline{R_V} = X_2, \ \|Vf\| = \|f\| \text{ for all } f \in D_V.$$

Then there exists a unique operator $U \in \mathcal{B}(X_1, X_2)$ such that $V \subset U$. The operator U is an isomorphism from X_1 onto X_2. Moreover, the operator U^{-1} is the unique element of $\mathcal{B}(X_1, X_2)$ such that $V^{-1} \subset U^{-1}$, or equivalently such that

$$U^{-1}(Vf) = f, \forall f \in D_V.$$

Proof. From 4.2.6 we have that there exists a unique operator $U \in \mathcal{B}(X_1, X_2)$ such that $V \subset U$. Clearly, condition in_2 of 4.6.1 holds true for U.

Now we fix $f \in X_1$ and let $\{f_n\}$ be a sequence in D_V such that $f_n \to f$ (cf. 2.3.12). Then,

$$Uf = \lim_{n\to\infty} Vf_n$$

and hence

$$\|Uf\| = \lim_{n\to\infty} \|Vf_n\| = \lim_{n\to\infty} \|f_n\| = \|f\|$$

(cf. 4.1.6a). Since f was an arbitrary element of X_1, this proves condition in_3 of 4.6.1 for U. Moreover we fix $g \in X_2$ and let $\{g_n\}$ be a sequence in D_V such that $Vg_n \to g$. Then $\{g_n\}$ is a Cauchy sequence because

$$\|g_n - g_m\| = \|Vg_n - Vg_m\|, \forall n, m \in \mathbb{N},$$

and hence we have

$$U \lim_{n\to\infty} g_n = \lim_{n\to\infty} Vg_n = g.$$

Since g was an arbitrary element of X_2, this proves that $R_U = X_2$. Thus condition in_1 of 4.6.1 holds true for U since U is injective by conditions in_2 and in_3, in view of 3.2.6a.

Finally, $V \subset U$ implies $V^{-1} \subset U^{-1}$ by 1.2.15; this condition can be written as

$$U^{-1}(Vf) = V^{-1}(Vf) = f, \forall f \in D_V.$$

Since $\overline{R_V} = X_2$ and $U^{-1} \in \mathcal{B}(X_2, X_1)$, U^{-1} is determined uniquely by this condition. $\qquad \square$

Chapter 5

The Extended Real Line

There are many situations in integration theory where one finds it unavoidable to deal with infinity. For instance, one wants to be able to integrate over sets of infinite measure. Moreover, even if one is only interested in real-valued functions, the least upper bound or the sum of a sequence of positive real-valued functions may well be infinite at some points. More generally, there are a number of idiomatic expressions about real functions where the word infinity and the symbol ∞ are used, even when these two things have not been given a definite status.

This brief chapter is devoted to the extended real line, which is a way to organize the various rules according to which infinity is dealt with in real analysis, and in particular in the next chapters about measure and integration theory.

5.1 The extended real line as an ordered set

5.1.1 Definitions. Let ∞ and $-\infty$ be two distinct objects, neither of which is a real number. "Many writers use the symbol $+\infty$ for what we write as ∞, but the sign $+$ is a mere nuisance and so we omit it" (Hewitt and Stromberg, 1965, p.54).

The *extended real line* is the set $\mathbb{R}^*$ defined by

$$\mathbb{R}^* := \mathbb{R} \cup \{-\infty, \infty\}.$$

A total ordering is defined in $\mathbb{R}^*$ by specifying that:

for $a, b \in \mathbb{R}$, $a \leq b$ if $a \leq b$ according to the usual ordering in $\mathbb{R}$;
for $a \in \mathbb{R}^*$, $-\infty \leq a$ and $a \leq \infty$.

We will always regard $\mathbb{R}^*$ as endowed with this total ordering. Note that the relation which is induced in $\mathbb{R}$ by this total ordering coincides obviously with the usual ordering in $\mathbb{R}$.

For $a, b \in \mathbb{R}^*$, $a \leq b$ is also written as $b \geq a$, and we write $a < b$ if $a \leq b$ and $a \neq b$.

For $a, b \in \mathbb{R}^*$, we define the sets:

$$(a, b) := \{x \in \mathbb{R}^* : a < x < b\},$$

$$[a, b) := \{x \in \mathbb{R}^* : a \le x < b\},$$

$$(a, b] := \{x \in \mathbb{R}^* : a < x \le b\},$$

$$[a, b] := \{x \in \mathbb{R}^* : a \le x \le b\}.$$

Note that some of these sets can be empty (e.g. $(a, b) = \emptyset$ if $b \le a$) and that

$$\mathbb{R} = (-\infty, \infty), \quad \mathbb{R}^* = [-\infty, \infty], \quad [0, \infty] = [0, \infty) \cup \{\infty\}.$$

Given a non-empty set X, for two functions $\varphi : X \to \mathbb{R}^*$ and $\psi : X \to \mathbb{R}^*$ we write $\varphi \le \psi$ if $\varphi(x) \le \psi(x)$ for all $x \in X$.

5.1.2 Proposition. *If S is a non-empty subset of $\mathbb{R}^*$ then both $\sup S$ (the l.u.b. for S, cf. 1.1.5) and $\inf S$ (the g.l.b. for S) exist.*

Proof. Let S be a non-empty subset of $\mathbb{R}^*$. Notice that ∞ is an upper bound for S and $-\infty$ is a lower bound for S.

If $S \subset \{-\infty, \infty\}$, the assertions we want to prove are obvious. Assume then $S \cap \mathbb{R} \ne \emptyset$ and examine the existence of $\sup S$. If the proposition

$$[\forall m \in \mathbb{R}, \exists x \in S \text{ s.t. } m < x]$$

is true, then ∞ is the only upper bound for S, and hence $\sup S$ exists and $\sup S = \infty$. If the proposition

$$[\exists m \in \mathbb{R} \text{ s.t. } x \le m \text{ for all } x \in S]$$

is true, then an element of $\mathbb{R}$ exists which is $\sup(S - \{-\infty\})$ according to the usual ordering in $\mathbb{R}$ (cf. e.g. Chapter 1 of (Rudin, 1976) or 1.11 in (Apostol, 1974)), and which is therefore $\sup S$ according to the ordering in $\mathbb{R}^*$. The arguments for $\inf S$ are similar. $\qquad\square$

5.2 The extended real line as a metric space

5.2.1 Theorem. *The function φ defined by*

$$\varphi : \left[-\frac{\pi}{2}, \frac{\pi}{2}\right] \to \mathbb{R}^*$$

$$x \mapsto \varphi(x) := \begin{cases} -\infty & \text{if } x = -\frac{\pi}{2}, \\ \tan x & \text{if } x \in (-\frac{\pi}{2}, \frac{\pi}{2}), \\ \infty & \text{if } x = \frac{\pi}{2}. \end{cases}$$

is obviously a bijection from $\left[-\frac{\pi}{2}, \frac{\pi}{2}\right]$ onto $\mathbb{R}^*$. Define then the function

$$\delta : \mathbb{R}^* \times \mathbb{R}^* \to \mathbb{R}^*$$

$$(a, b) \mapsto \delta(a, b) := d_{\mathbb{R}}(\varphi^{-1}(a), \varphi^{-1}(b)) \quad (= |\varphi^{-1}(a) - \varphi^{-1}(b)|, \text{ cf. 2.1.4}).$$

(a) *The function δ is a distance on $\mathbb{R}^*$ ($\mathbb{R}^*$ will always be regarded as the first element of the metric space $(\mathbb{R}^*, \delta)$).*

(b) *For a sequence $\{a_n\}$ in $\mathbb{R}^*$ we have:*

    (b_1) *$a_n \to \infty$ iff $\forall m \in \mathbb{R}$, $\exists N_m \in \mathbb{N}$ such that $n > N_m \Rightarrow a_n > m$;*

    (b_2) *$a_n \to -\infty$ iff $\forall m \in \mathbb{R}$, $\exists N_m \in \mathbb{N}$ such that $n > N_m \Rightarrow a_n < m$;*

    (b_3) *for $a \in \mathbb{R}$,*
$$a_n \to a \text{ iff } \forall \epsilon > 0, \exists N_\epsilon \in \mathbb{N} \text{ such that } n > N_\epsilon \Rightarrow a - \epsilon < a_n < a + \epsilon.$$

(c) *A sequence $\{a_n\}$ in $\mathbb{R}$ is convergent in the metric subspace $(\mathbb{R}, \delta_{\mathbb{R}})$ (i.e. it is convergent in the metric space $(\mathbb{R}^*, \delta)$ and $\lim_{n \to \infty} a_n \in \mathbb{R}$) iff it is convergent in the metric space $(\mathbb{R}, d_{\mathbb{R}})$, and in case of convergence the two limits are equal.*

(d) *The topology (i.e. the family of all open sets) of the metric subspace $(\mathbb{R}, \delta_{\mathbb{R}})$ is the same as the topology of the metric space $(\mathbb{R}, d_{\mathbb{R}})$.*

(e) *The metric subspace $(\mathbb{R}, \delta_{\mathbb{R}})$ is not complete, and one of its completions is the pair $((\mathbb{R}^*, \delta), id_{\mathbb{R}})$.*

Proof. a: The properties of 2.1.1 for δ follow directly from the same properties for $d_{\mathbb{R}}$.

    b: Let $\{a_n\}$ be a sequence in $\mathbb{R}^*$.

    b_1: We have

$$\delta(a_n, \infty) \xrightarrow[n \to \infty]{} 0 \Leftrightarrow$$

$$d_{\mathbb{R}}(\varphi^{-1}(a_n), \frac{\pi}{2}) \xrightarrow[n \to \infty]{} 0 \Leftrightarrow$$

$$\left\{ \forall \epsilon \in (0, \pi), \exists N_\epsilon \in \mathbb{N} \text{ s.t. } n > N_\epsilon \Rightarrow \right.$$
$$\left. [a_n = \infty \text{ or } (a_n \in \mathbb{R} \text{ and } \arctan a_n > \frac{\pi}{2} - \epsilon)] \right\} \overset{(1)}{\Leftrightarrow}$$
$$(\forall m \in \mathbb{R}, \exists N_m \in \mathbb{N} \text{ s.t. } n > N_m \Rightarrow a_n > m).$$

Indeed:

$\overset{(1)}{\Rightarrow}$: for $m \in \mathbb{R}$, put $\epsilon := \frac{\pi}{2} - \arctan m$; then

$$n > N_m := N_\epsilon \Rightarrow$$
$$[a_n = \infty \text{ or } (a_n \in \mathbb{R} \text{ and } \arctan a_n > \arctan m, \text{ i.e. } a_n > m)];$$

$\overset{(1)}{\Leftarrow}$: for $\epsilon \in (0, \pi)$, put $m := \tan(\frac{\pi}{2} - \epsilon)$; then

$$n > N_\epsilon := N_m \Rightarrow a_n > \tan(\frac{\pi}{2} - \epsilon) \Rightarrow$$
$$[a_n = \infty \text{ or } (a_n \in \mathbb{R} \text{ and } \arctan a_n > \frac{\pi}{2} - \epsilon)].$$

    b_2: The proof is analogous to the one given for b_1.

b_3: For $a \in \mathbb{R}$ we have

$$\delta(a_n, a) \xrightarrow[n\to\infty]{} 0 \overset{(2)}{\Leftrightarrow}$$

$$[\exists k \in \mathbb{N} \text{ s.t. } a_n \in \mathbb{R} \text{ for } n > k \text{ and } d_{\mathbb{R}}(\varphi^{-1}(a_{k+n}), \varphi^{-1}(a)) \xrightarrow[n\to\infty]{} 0] \overset{(3)}{\Leftrightarrow}$$

$$[\exists k \in \mathbb{N} \text{ s.t. } a_n \in \mathbb{R} \text{ for } n > k \text{ and } d_{\mathbb{R}}(a_{k+n}, a) \xrightarrow[n\to\infty]{} 0] \overset{(4)}{\Leftrightarrow}$$

$$(\forall \epsilon > 0, \exists N_\epsilon \in \mathbb{N} \text{ s.t. } n > N_\epsilon \Rightarrow a - \epsilon < a_n < a + \epsilon).$$

Indeed:

$\overset{(2)}{\Rightarrow}$: since $\eta := \min\{\delta(a, -\infty), \delta(a, \infty)\} > 0$, there exists $k \in \mathbb{N}$ such that

$$n > k \Rightarrow \delta(a_n, a) < \eta \Rightarrow a_n \notin \{-\infty, \infty\};$$

moreover, $\delta(a_n, a) \xrightarrow[n\to\infty]{} 0$ implies trivially $\delta(a_{k+n}, a) \xrightarrow[n\to\infty]{} 0$;

$\overset{(2)}{\Leftarrow}$: $\delta(a_{k+n}, a) \xrightarrow[n\to\infty]{} 0$ implies trivially $\delta(a_n, a) \xrightarrow[n\to\infty]{} 0$;

$\overset{(3)}{\Leftrightarrow}$: tan and arctan are continuous functions between $(-\frac{\pi}{2}, \frac{\pi}{2})$ and $\mathbb{R}$;

$\overset{(4)}{\Rightarrow}$: $d_{\mathbb{R}}(a_{k+n}, a) \xrightarrow[n\to\infty]{} 0$ implies that

$$\forall \epsilon > 0, \exists n_\epsilon \in \mathbb{N} \text{ s.t. } n > n_\epsilon \Rightarrow a - \epsilon < a_{k+n} < a + \epsilon;$$

let then $N_\epsilon := k + n_\epsilon$;

$\overset{(4)}{\Leftarrow}$: set for instance $k := N_1$ and notice that $d_{\mathbb{R}}(a_n, a) \xrightarrow[n\to\infty]{} 0$ implies trivially $d_{\mathbb{R}}(a_{k+n}, a) \xrightarrow[n\to\infty]{} 0$.

c: This follows immediately from part b_3.

d: From part c and 2.3.4 it follows that a subset of $\mathbb{R}$ is closed in $(\mathbb{R}, \delta_{\mathbb{R}})$ iff it is closed in $(\mathbb{R}, d_{\mathbb{R}})$. Then, from 2.3.1 it follows that a subset of $\mathbb{R}$ is open in $(\mathbb{R}, \delta_{\mathbb{R}})$ iff it is open in $(\mathbb{R}, d_{\mathbb{R}})$.

e: Let $\{a_n\}$ be a Cauchy sequence in $(\mathbb{R}^*, \delta)$; then $\{\varphi^{-1}(a_n)\}$ is a Cauchy sequence in $(\mathbb{R}, d_{\mathbb{R}})$; since $(\mathbb{R}, d_{\mathbb{R}})$ is complete, this implies that there exists $x \in \mathbb{R}$ such that $d_{\mathbb{R}}(\varphi^{-1}(a_n), x) \to 0$, and $x \in [-\frac{\pi}{2}, \frac{\pi}{2}]$ since $[-\frac{\pi}{2}, \frac{\pi}{2}]$ is closed (cf. 2.3.6 and 2.3.4); then we have $\delta(a_n, \varphi(x)) \to 0$ by the definition of δ. This proves that $(\mathbb{R}^*, \delta_{\mathbb{R}})$ is complete. From parts b_1 and b_2 it follows that $\delta(n, \infty) \to 0$ and $\delta(-n, \infty) \to 0$; by 2.3.12, this proves that $\overline{\mathbb{R}} = \mathbb{R}^*$, where $\overline{\mathbb{R}}$ means the closure of $\mathbb{R}$ in the metric space $(\mathbb{R}, \delta)$. The statement we need to prove now follows from 2.6.8. $\square$

5.2.2 Definition. For a sequence $\{a_n\}$ in $\mathbb{R}^*$ we define, for each $n \in \mathbb{N}$,

$$\sup_{k \geq n} a_k := \sup\{a_k : k \geq n\} \text{ and } \inf_{k \geq n} a_k := \inf\{a_k : k \geq n\}$$

(notice that $\sup\{a_k : k \geq n\}$ and $\inf\{a_k : k \geq n\}$ exist by 5.1.2).

5.2.3 Proposition. *Let $\{a_{m,k}\}_{(m,k)\in\mathbb{N}\times\mathbb{N}}$ be a family of elements of $\mathbb{R}^*$. Then,*

$$\sup_{m\geq 1}\left(\sup_{k\geq 1} a_{m,k}\right) = \sup_{k\geq 1}\left(\sup_{m\geq 1} a_{m,k}\right),$$

$$\inf_{m\geq 1}\left(\inf_{k\geq 1} a_{m,k}\right) = \inf_{k\geq 1}\left(\inf_{m\geq 1} a_{m,k}\right).$$

Proof. We have

$$\forall (n,l)\in\mathbb{N}\times\mathbb{N},\ \sup_{m\geq 1}\left(\sup_{k\geq 1} a_{m,k}\right) \geq \sup_{k\geq 1} a_{n,k} \geq a_{n,l},$$

and hence

$$\forall l\in\mathbb{N},\ \sup_{m\geq 1}\left(\sup_{k\geq 1} a_{m,k}\right) \geq \sup_{m\geq 1} a_{m,l},$$

and hence

$$\sup_{m\geq 1}\left(\sup_{k\geq 1} a_{m,k}\right) \geq \sup_{k\geq 1}\left(\sup_{m\geq 1} a_{m,k}\right).$$

In a similar way we can prove

$$\sup_{k\geq 1}\left(\sup_{m\geq 1} a_{m,k}\right) \geq \sup_{m\geq 1}\left(\sup_{k\geq 1} a_{m,k}\right).$$

The proof of the second equality of the statement is analogous. $\qquad\square$

5.2.4 Proposition. *Let $\{a_n\}$ be a sequence in $\mathbb{R}$.*

If $a_n \leq a_{n+1}$ for each $n\in\mathbb{N}$, then $\{a_n\}$ is convergent in the metric space $(\mathbb{R}, d_\mathbb{R})$ iff there exists $m\in\mathbb{R}$ such that $a_n \leq m$ for each $n\in\mathbb{N}$. In case of convergence we have

$$\lim_{n\to\infty} a_n = \sup_{n\geq 1} a_n.$$

If $a_{n+1} \leq a_n$ for each $n\in\mathbb{N}$, then $\{a_n\}$ is convergent in the metric space $(\mathbb{R}, d_\mathbb{R})$ iff there exists $m\in\mathbb{R}$ such that $m \leq a_n$ for each $n\in\mathbb{N}$. In case of convergence we have

$$\lim_{n\to\infty} a_n = \inf_{n\geq 1} a_n.$$

Proof. Suppose $a_n \leq a_{n+1}$ for each $n\in\mathbb{N}$ (the proof is analogous in the other case). If there exists $m\in\mathbb{R}$ s.t. $a_n \leq m$ for each $n\in\mathbb{N}$, then $s := \sup_{n\geq 1} a_n$ is an element of $\mathbb{R}$ since $a_n \leq s \leq m$ for each $n\in\mathbb{N}$ by the definition of l.u.b.. Then, by the same token,

$$\forall \epsilon > 0, \exists N_\epsilon \in \mathbb{N}\ \text{s.t.}\ s - \epsilon < a_{N_\epsilon}.$$

This implies

$$\forall \epsilon > 0, \exists N_\epsilon \in \mathbb{N}\ \text{s.t.}\ n > N_\epsilon \Rightarrow s - \epsilon < a_n \leq s,$$

which implies $d_{\mathbb{R}}(a_n, s) \xrightarrow[n \to \infty]{} 0$.

Conversely, suppose that there exists $a \in \mathbb{R}$ s.t. $a_n \to a$. Then, by 2.1.9,

$$\exists k \in [0, \infty), \exists x \in \mathbb{R} \text{ s.t. } d_{\mathbb{R}}(a_n, x) < k, \forall n \in \mathbb{N}.$$

Hence we have

$$\forall n \in \mathbb{N}, a_n \leq |a_n| \leq |a_n - x| + |x| < k + |x|.$$

$\square$

5.2.5 Proposition. *Let $\{a_n\}$ be a sequence in $\mathbb{R}^*$.*

If $a_n \leq a_{n+1}$ for each $n \in \mathbb{N}$, then $\{a_n\}$ is convergent in the metric space $(\mathbb{R}^*, \delta)$ and $\lim_{n \to \infty} a_n = \sup_{n \geq 1} a_n$.

If $a_{n+1} \leq a_n$ for each $n \in \mathbb{N}$, then $\{a_n\}$ is convergent in the metric space $(\mathbb{R}^*, \delta)$ and $\lim_{n \to \infty} a_n = \inf_{n \geq 1} a_n$.

Proof. Suppose $a_n \leq a_{n+1}$ for each $n \in \mathbb{N}$ (the proof is analogous in the other case).

If $\exists m \in \mathbb{R}$ s.t. $a_n \leq m$ for each $n \in \mathbb{N}$, then the result follows from 5.2.4 and 5.2.1c.

If $\forall m \in \mathbb{R}, \exists N_m \in \mathbb{N}$ s.t. $m < a_{N_m}$, then

$$\forall m \in \mathbb{R}, \exists N_m \in \mathbb{N} \text{ s.t. } n > N_m \Rightarrow a_n > m,$$

hence $a_n \to \infty$ by 5.2.1b_1; also, ∞ is the only upper bound for $\{a_n : n \in \mathbb{N}\}$, hence $\infty = \sup\{a_n : n \in \mathbb{N}\} = \sup_{n \geq 1} a_n$. $\square$

5.2.6 Proposition. *Let $\{a_n\}$ be a convergent sequence in $\mathbb{R}^*$. For each $n \in \mathbb{N}$ define $b_n := \sup_{k \geq n} a_k$ and $c_n := \inf_{k \geq n} a_k$. Then the sequences $\{b_n\}$ and $\{c_n\}$ are convergent in the metric space $(\mathbb{R}^*, \delta)$ and $\lim_{n \to \infty} a_n = \lim_{n \to \infty} b_n = \inf_{n \geq 1} b_n = \lim_{n \to \infty} c_n = \sup_{n \geq 1} c_n$.*

Proof. Clearly, we have $b_{n+1} \leq b_n$ for each $n \in \mathbb{N}$. Hence 5.2.5 implies that the sequence $\{b_n\}$ is convergent and $\lim_{n \to \infty} b_n = \inf_{n \geq 1} b_n$. There are now three possibilities.

If $\lim_{n \to \infty} a_n = \infty$, then (cf. 5.2.1$b_1$)

$$\forall m \in \mathbb{R}, \exists N_m \in \mathbb{N} \text{ s.t. } n > N_m \Rightarrow a_n > m \Rightarrow b_n > m,$$

and this proves that $\lim_{n \to \infty} b_n = \infty$.

If $\lim_{n \to \infty} a_n = -\infty$, then (cf. 5.2.1$b_2$)

$$\forall m \in \mathbb{R}, \exists N_m \in \mathbb{N} \text{ s.t. } n > N_m \Rightarrow a_n < m - 1,$$

and therefore

$$\forall m \in \mathbb{R}, \exists N_m \in \mathbb{N} \text{ s.t. } n > N_m \Rightarrow b_n \leq m - 1 < m,$$

and this proves that $\lim_{n \to \infty} b_n = -\infty$.

If $\lim_{n\to\infty} a_n = a$ with $a \in \mathbb{R}$, then (cf. 5.2.1b_3)

$$\forall \epsilon > 0, \exists N_\epsilon \in \mathbb{N} \text{ s.t. } n > N_\epsilon \Rightarrow a - \frac{\epsilon}{2} < a_n < a + \frac{\epsilon}{2},$$

and therefore

$$\forall \epsilon > 0, \exists N_\epsilon \in \mathbb{N} \text{ s.t. } n > N_\epsilon \Rightarrow a - \frac{\epsilon}{2} < b_n \leq a + \frac{\epsilon}{2} \Rightarrow a - \epsilon < b_n < a + \epsilon,$$

and this proves that $\lim_{n\to\infty} b_n = a$.

For $\{c_n\}$ we can proceed in a similar way. $\qquad\square$

5.3 Algebraic operations in $\mathbb{R}^*$

5.3.1 Definitions. The algebraic operations of $\mathbb{R}$ can, *at least partially*, be extended to $\mathbb{R}^*$.

(a) The mapping $a \mapsto -a$ in $\mathbb{R}$ is extended to $\mathbb{R}^*$ by defining

$$-(\infty) := -\infty \text{ and } -(-\infty) := \infty.$$

(b) The mapping $a \mapsto |a|$ in $\mathbb{R}$ is extended to $\mathbb{R}^*$ by defining

$$|\infty| := \infty \text{ and } |-\infty| := \infty.$$

(c) The product $(a, b) \mapsto ab$ in $\mathbb{R}$ is extended to $\mathbb{R}^*$ by defining

for $a \in (0, \infty]$, $a\infty := \infty a := \infty$ and $a(-\infty) = (-\infty)a = -\infty$,
for $a \in [-\infty, 0)$, $a\infty := \infty a := -\infty$ and $a(-\infty) = (-\infty)a = \infty$,
$0\infty := \infty 0 := 0(-\infty) := (-\infty)0 := 0$.

Note that, for $a, b, c \in \mathbb{R}^*$,

$$ab = ba \text{ and } (ab)c = a(bc).$$

(d) The sum $(a, b) \mapsto a + b$ in $\mathbb{R}$ is *partially* extended to $\mathbb{R}^*$ by defining

for $a \in (-\infty, \infty]$, $a + \infty := \infty + a := \infty$,
for $a \in [-\infty, \infty)$, $a + (-\infty) = (-\infty) + a = -\infty$,

while the sums $\infty + (-\infty)$ and $(-\infty) + \infty$ are *not* defined.

Note that, for $a, b, c \in \mathbb{R}^*$,

$$a + b = b + a \text{ and } (a + b) + c = a + (b + c)$$

whenever the sides of these equations are defined.

For $a, b \in \mathbb{R}^*$, we write

$$a - b := a + (-b)$$

provided that the sum on the right side is defined. The only differences that are not defined are $\infty - \infty$ and $(-\infty) - (-\infty)$.

5.3.2 Remarks.

(a) Note that, for $a, b \in \mathbb{R}^*$, $a \leq b$ iff $-b \leq -a$. As can be easily seen, this implies that, if for $S \subset \mathbb{R}^*$ we write $-S := \{-a : a \in S\}$, then

$$\sup(-S) = -\inf S \text{ and } \inf(-S) = -\sup S.$$

(b) As can be checked directly, if $a, b \in \mathbb{R}^*$ are such that $a \leq b$ then

$$ca \leq cb, \quad \forall c \in [0, \infty].$$

This implies that, if for $S \subset \mathbb{R}^*$ and $c \in [0, \infty]$ we write $cS := \{ca : a \in S\}$, then

$$\sup cS = c \sup S \text{ and } \inf cS = c \inf S, \quad \forall c \in [0, \infty)$$

(these equalities may be false for $c = \infty$, as is shown by $S := \{-\frac{1}{n} : n \in \mathbb{N}\}$ and by $S := \{\frac{1}{n} : n \in \mathbb{N}\}$). In fact, for $c = 0$ both equalities are trivial and for $c \in (0, \infty)$ we have

$$\forall a \in S, \quad ca \leq c \sup S,$$

and, for $m \in \mathbb{R}^*$,

$$[ca \leq m, \forall a \in S] \Rightarrow [a \leq \frac{1}{c}m, \forall a \in S] \Rightarrow \sup S \leq \frac{1}{c}m \Rightarrow c \sup S \leq m,$$

and this proves that $\sup cS = c \sup S$. For the other equality we can proceed in a similar way.

(c) As can be checked directly, if $a, b \in \mathbb{R}^*$ are such that $a \leq b$ then

$$a + c \leq b + c \text{ for all } c \in \mathbb{R}^* \text{ such that the two sides are defined.}$$

This implies that, if for $S \subset \mathbb{R}^*$ and $c \in \mathbb{R}$ we write $S + c := \{a + c : a \in S\}$, then

$$\sup(S + c) = \sup S + c \text{ and } \inf(S + c) = \inf S + c, \quad \forall c \in \mathbb{R}.$$

In fact, we have

$$\forall a \in S, \quad a + c \leq \sup S + c$$

and, for $m \in \mathbb{R}^*$,

$$[a + c \leq m, \forall a \in S] \Rightarrow [a \leq m - c, \forall a \in S] \Rightarrow \sup S \leq m - c \Rightarrow \sup S + c \leq m,$$

and this proves that $\sup(S + c) = \sup S + c$. For the other equality we can proceed in a similar way.

(d) From remark b it follows that, if $a_1, a_2, b_1, b_2 \in [0, \infty]$ are such that $a_i \leq b_i$ for $i = 1, 2$, then

$$a_1 a_2 \leq b_1 a_2 \leq b_1 b_2.$$

(e) From remark c it follows that, if $a_1, a_2, b_1, b_2 \in \mathbb{R}^*$ are so that both $a_1 + a_2$ and $b_1 + b_2$ are defined and also so that $a_i \leq b_i$ for $i = 1, 2$, then

$$a_1 + a_2 \leq b_1 + b_2.$$

Indeed, if $b_1 + a_2$ is defined, then by remark c we have

$$a_1 + a_2 \leq b_1 + a_2 \leq b_1 + b_2.$$

If $b_1 + a_2$ is not defined, then $a_2 = -\infty$ and $b_1 = \infty$ (because $a_2 = \infty$ and $b_1 = -\infty$ would render $a_1 + a_2$ not defined), and hence

$$a_1 + a_2 = -\infty \leq \infty = b_1 + b_2.$$

5.3.3 Remark. In $[0, \infty]$ both the product and the sum are defined without restraints, and they are commutative and associative. Owing to this, for elements of $[0, \infty]$ we will write $a_1 + a_2 + a_3 := a_1 + (a_2 + a_3)$ and $\sum_{k=1}^{n} a_k := a_1 + ... + a_n$; we will also write $\sum_{i \in I} a_i$ to denote the sum of a finite family $\{a_i\}_{i \in I}$ of elements of $[0, \infty]$.

By a straightforward check we see that

$$a(b + c) = ab + ac, \forall a, b, c \in [0, \infty].$$

5.3.4 Proposition. *Suppose that $\{a_n\}$ and $\{b_n\}$ are sequences in $[0, \infty]$ such that $a_n \leq a_{n+1}$ and $b_n \leq b_{n+1}$ for each $n \in \mathbb{N}$, and let $a \in [0, \infty]$. Then the sequences $\{a_n + b_n\}, \{a_n b_n\}$ and $\{ab_n\}$ are convergent in the metric space $(\mathbb{R}^*, \delta)$ and*

$$\lim_{n \to \infty} (a_n + b_n) = \sup_{n \geq 1}(a_n + b_n) = \sup_{n \geq 1} a_n + \sup_{n \geq 1} b_n = \lim_{n \to \infty} a_n + \lim_{n \to \infty} b_n,$$

$$\lim_{n \to \infty} (a_n b_n) = \sup_{n \geq 1}(a_n b_n) = (\sup_{n \geq 1} a_n)(\sup_{n \geq 1} b_n) = (\lim_{n \to \infty} a_n)(\lim_{n \to \infty} b_n),$$

$$\lim_{n \to \infty} (ab_n) = \sup_{n \geq 1}(ab_n) = a \sup_{n \geq 1} b_n = a \lim_{n \to \infty} b_n.$$

Proof. Recall (cf. 5.2.5) that $\{a_n\}$ and $\{b_n\}$ are convergent and $\lim_{n \to \infty} a_n = \sup_{n \geq 1} a_n$ and $\lim_{n \to \infty} b_n = \sup_{n \geq 1} b_n$.

If both $\lim_{n \to \infty} a_n$ and $\lim_{n \to \infty} b_n$ are elements of $\mathbb{R}$, then what we want to prove follows from 5.2.1c and from the continuity of the sum and the product in $\mathbb{R}$. Thus, in what follows we assume e.g. $\lim_{n \to \infty} a_n = \infty$.

We have $\lim_{n \to \infty} a_n + \lim_{n \to \infty} b_n = \infty$; since

$$a_n + b_n \leq a_{n+1} + b_{n+1} \text{ and } a_n \leq a_n + b_n, \forall n \in \mathbb{N},$$

(cf. 5.3.2e), the sequence $\{a_n + b_n\}$ is convergent and

$$\lim_{n \to \infty} (a_n + b_n) = \sup_{n \geq 1}(a_n + b_n) \geq \sup_{n \geq 1} a_n = \infty,$$

and hence

$$\lim_{n \to \infty} (a_n + b_n) = \infty = \lim_{n \to \infty} a_n + \lim_{n \to \infty} b_n.$$

If $\lim_{n\to\infty} b_n = 0$, then $b_n = 0$ and hence $a_n b_n = 0$ for each $n \in \mathbb{N}$; thus,

$$\lim_{n\to\infty} (a_n b_n) = 0 = (\lim_{n\to\infty} a_n)(\lim_{n\to\infty} b_n).$$

If $\lim_{n\to\infty} b_n = \infty$, then there exists $k \in \mathbb{N}$ such that $b_k \geq 1$; now, $\lim_{n\to\infty} a_n = \infty$ implies (cf. $5.2.1b_1$) that

$$\forall m \in \mathbb{R}, \exists N_m \in \mathbb{N} \text{ s.t. } n > N_m \Rightarrow a_n > m;$$

then (letting $\tilde{N}_m := \max\{N_m, k\}$) we have

$$\forall m \in \mathbb{R}, \exists \tilde{N}_m \in \mathbb{N} \text{ s.t. } n > \tilde{N}_m \Rightarrow a_n b_n \geq a_n b_k \geq a_n > m,$$

which proves that

$$\lim_{n\to\infty} (a_n b_n) = \infty = (\lim_{n\to\infty} a_n)(\lim_{n\to\infty} b_n).$$

Assume finally $\lim_{n\to\infty} b_n \in (0, \infty)$; since

$$a_n b_n \leq a_{n+1} b_{n+1}, \forall n \in \mathbb{N}$$

(cf. 5.3.2d), the sequence $\{a_n b_n\}$ is convergent and $\lim_{n\to\infty}(a_n b_n) = \sup_{n\geq 1}(a_n b_n)$; let $k \in \mathbb{N}$ be s.t. $b_k > 0$; since

$$a_n b_k \leq a_n b_n \text{ for all } n \geq k$$

(cf. 5.3.2d), using 5.3.2b (note that $b_k \in (0, \infty)$) we have

$$\sup_{n\geq 1}(a_n b_n) = \sup_{n\geq k}(a_n b_n) \geq \sup_{n\geq k} a_n b_k = b_k \sup_{n\geq k} a_n = b_k \sup_{n\geq 1} a_n = \infty,$$

and hence

$$\lim_{n\to\infty} (a_n b_n) = \infty = (\lim_{n\to\infty} a_n)(\lim_{n\to\infty} b_n).$$

Finally, the sequence $\{ab_n\}$ is the sequence $\{a_n b_n\}$ if $a_n := a$ for all $n \in \mathbb{N}$. $\square$

5.4 Series in $[0, \infty]$

5.4.1 Definition. Let $\{a_n\}$ be a sequence in $[0, \infty]$. For each $n \in \mathbb{N}$, define $s_n := \sum_{k=1}^{n} a_k$. The sequence $\{s_n\}$ is called the *series* of the a_n's and is denoted by the symbol $\sum_{n=1}^{\infty} a_n$. Since $s_n \leq s_{n+1}$ for each $n \in \mathbb{N}$, 5.2.5 implies that the sequence $\{s_n\}$ is convergent in the metric space $(\mathbb{R}^*, \delta)$ and $\lim_{n\to\infty} s_n = \sup_{n\geq 1} s_n$. Then, $\lim_{n\to\infty} s_n$ is called the *sum of the series* of the a_n's and is denoted by the same symbol $\sum_{n=1}^{\infty} a_n$ as the series, i.e.

$$\sum_{n=1}^{\infty} a_n := \lim_{n\to\infty} s_n = \sup_{n\geq 1} s_n$$

(these definitions are in agreement with the ones given in 2.1.10).

If $a_n \in \mathbb{R}$ for each $n \in \mathbb{N}$, then $\lim_{n\to\infty} s_n$ can be either ∞ or an element of $\mathbb{R}$. If $\lim_{n\to\infty} s_n \in \mathbb{R}$, then $\{s_n\}$ converges to $\sum_{n=1}^{\infty} a_n$ also in the metric space $(\mathbb{R}, d_{\mathbb{R}})$ (cf. 5.2.1c). Clearly, $\lim_{n\to\infty} s_n \in \mathbb{R}$ iff $\sum_{n=1}^{\infty} a_n < \infty$ (we will always use the latter expression).

5.4.2 Remarks.

(a) Let $\{a_n\}$ and $\{b_n\}$ be sequences in $[0, \infty]$, and suppose $a_n \leq b_n$ for each $n \in \mathbb{N}$. Then, by induction applied to 5.3.2e,

$$\forall n \in \mathbb{N}, \sum_{k=1}^{n} a_k \leq \sum_{k=1}^{n} b_k \leq \sup_{n \geq 1} \sum_{k=1}^{n} b_k =: \sum_{n=1}^{\infty} b_n,$$

whence

$$\sum_{n=1}^{\infty} a_n := \sup_{n \geq 1} \sum_{k=1}^{n} a_k \leq \sum_{n=1}^{\infty} b_n.$$

(b) For a sequence $\{a_n\}$ in $[0, \infty]$, letting $s_n := \sum_{k=1}^{n} a_k$ we have

$$\sum_{n=1}^{\infty} a_n < \infty \overset{(1)}{\Leftrightarrow}$$

$$[\exists m \in [0, \infty) \text{ s.t. } s_n \leq m, \forall n \in \mathbb{N}] \overset{(2)}{\Leftrightarrow}$$

$$[a_n \in [0, \infty), \forall n \in \mathbb{N}, \text{ and } \{s_n\} \text{ is convergent in } (\mathbb{R}, d_{\mathbb{R}})] \overset{(3)}{\Leftrightarrow}$$

$$[a_n \in [0, \infty), \forall n \in \mathbb{N}, \text{ and }$$

$$\sum_{n=1}^{\infty} a_n \text{ is convergent in the normed space } \mathbb{R} \text{ (cf. 4.1.4)]},$$

where: 1 holds because $\sum_{n=1}^{\infty} a_n = \sup \{s_n : n \in \mathbb{N}\}$; 2 holds by 5.2.4; 3 holds by the definitions given in 4.1.5.

If $\sum_{n=1}^{\infty} a_n < \infty$ then from 5.2.1c it follows that the value of $\sum_{n=1}^{\infty} a_n$ is the same whether it is defined as in 5.4.1 or as in 4.1.5 in the context of the normed space $\mathbb{R}$.

Note that $\sum_{n=1}^{\infty} a_n = \infty$ iff $[(\exists n \in \mathbb{N} \text{ s.t. } a_n = \infty)$ or $(a_n \in [0, \infty), \forall n \in \mathbb{N},$ and the series $\sum_{n=1}^{\infty} a_n$ is not convergent in the normed space $\mathbb{R})]$.

5.4.3 Proposition. *Let $\{a_n\}$ be a sequence in $[0, \infty]$ and let β be a bijection from $\mathbb{N}$ onto $\mathbb{N}$. Then*

$$\sum_{n=1}^{\infty} a_{\beta(n)} = \sum_{n=1}^{\infty} a_n.$$

For this reason, for a countable family $\{b_n\}_{n \in I}$ of elements of $[0, \infty]$, we will write $\sum_{n \in I} b_n$ to denote the sum or the series of the b_n's, with no need to specify the order.

Proof. For $n \in \mathbb{N}$, let $M_n := \max \{\beta(1), ..., \beta(n)\}$; then we have

$$\forall n \in \mathbb{N}, \sum_{k=1}^{n} a_{\beta(k)} \leq \sum_{l=1}^{M_n} a_l \leq \sum_{n=1}^{\infty} a_n,$$

and this implies that

$$\sum_{n=1}^{\infty} a_{\beta(n)} := \sup\left\{\sum_{k=1}^{n} a_{\beta(k)} : n \in \mathbb{N}\right\} \leq \sum_{n=1}^{\infty} a_n.$$

Replacing a_n with $a_{\beta(n)}$ and β with β^{-1}, this also proves that

$$\sum_{n=1}^{\infty} a_n \leq \sum_{n=1}^{\infty} a_{\beta(n)}.$$

$\square$

5.4.4 Corollary. *Let $\{a_n\}$ be a sequence in $[0, \infty)$ and let β be a bijection from $\mathbb{N}$ onto $\mathbb{N}$. Then the series $\sum_{n=1}^{\infty} a_{\beta(n)}$ is convergent in the normed space $\mathbb{R}$ (cf. 4.1.4 and 4.1.5) iff the series $\sum_{n=1}^{\infty} a_n$ is convergent in the normed space $\mathbb{R}$, and in case of convergence the two sums are equal.*

Proof. Use 5.4.2b and 5.4.3. $\square$

5.4.5 Proposition. *Let $\{a_n\}$ be a sequence in $[0, \infty]$ and $a \in [0, \infty]$. Then*

$$\sum_{n=1}^{\infty} (aa_n) = a \sum_{n=1}^{\infty} a_n.$$

Proof. If $a = \infty$, then the two sides are zero if $a_n = 0$ for each $n \in \mathbb{N}$, otherwise the two sides are ∞.

Assuming now $a \in [0, \infty)$, we have (using 5.3.2b)

$$\sum_{n=1}^{\infty} (aa_n) := \sup\left\{\sum_{k=1}^{n} (aa_k) : n \in \mathbb{N}\right\} = \sup\left\{a \sum_{k=1}^{n} a_k : n \in \mathbb{N}\right\}$$

$$= a \sup\left\{\sum_{k=1}^{n} a_k : n \in \mathbb{N}\right\} = a \sum_{n=1}^{\infty} a_n.$$

$\square$

5.4.6 Proposition. *Let $\{a_n\}$ and $\{b_n\}$ be two sequences in $[0, \infty]$. Then*

$$\sum_{n=1}^{\infty} (a_n + b_n) = \sum_{n=1}^{\infty} a_n + \sum_{n=1}^{\infty} b_n.$$

Proof. We have (using 5.3.2e)

$$\forall n \in \mathbb{N}, \sum_{k=1}^{n} (a_k + b_k) = \sum_{k=1}^{n} a_k + \sum_{k=1}^{n} b_k \leq \sum_{n=1}^{\infty} a_n + \sum_{n=1}^{\infty} b_n,$$

and this implies that

$$\sum_{n=1}^{\infty} (a_n + b_n) \leq \sum_{n=1}^{\infty} a_n + \sum_{n=1}^{\infty} b_n.$$

If $\sum_{n=1}^{\infty}(a_n + b_n) = \infty$, this proves the equality of the statement. Assume next $\sum_{n=1}^{\infty}(a_n + b_n) < \infty$; this implies $\sum_{n=1}^{\infty} a_n < \infty$ and $\sum_{n=1}^{\infty} b_n < \infty$ since e.g.

$$\forall n \in \mathbb{N}, \sum_{k=1}^{n} a_k \leq \sum_{k=1}^{n}(a_k + b_k) \leq \sum_{n=1}^{\infty}(a_n + b_n);$$

thus, all three series of the statement are convergent in the normed space $\mathbb{R}$ (cf. 5.4.2b), where the equality holds by the continuity of the sum. $\qquad\square$

5.4.7 Proposition. *Let $\{a_{n,m}\}$ be a family of elements of $[0, \infty]$ indexed by $\mathbb{N} \times \mathbb{N}$ and let σ be a bijection from $\mathbb{N}$ onto $\mathbb{N} \times \mathbb{N}$. Then*

$$\sum_{n=1}^{\infty} a_{\sigma(n)} = \sum_{n=1}^{\infty} \left(\sum_{m=1}^{\infty} a_{n,m} \right).$$

Thus, the sum of the series $\sum_{n=1}^{\infty} a_{\sigma(n)}$ does not depend on the particular bijection that is used, and we can define

$$\sum_{(n,m)\in\mathbb{N}\times\mathbb{N}} a_{n,m} := \sum_{n=1}^{\infty} a_{\sigma(n)}.$$

The following equalities are true:

$$\sum_{(n,m)\in\mathbb{N}\times\mathbb{N}} a_{n,m} = \sum_{n=1}^{\infty} \left(\sum_{m=1}^{\infty} a_{n,m} \right) = \sum_{m=1}^{\infty} \left(\sum_{n=1}^{\infty} a_{n,m} \right).$$

Proof. For $i = 1, 2$, define the mapping

$$\pi_i : \mathbb{N} \times \mathbb{N} \to \mathbb{N}$$

$$(n_1, n_2) \mapsto \pi_i(n_1, n_2) := n_i.$$

Fix $L \in \mathbb{N}$ and let $L_i := \max \pi_i(\sigma(\{1, ..., L\}))$; we have (using induction applied to 5.3.2e)

$$\sum_{k=1}^{L} a_{\sigma(k)} \leq \sum_{n=1}^{L_1} \left(\sum_{m=1}^{L_2} a_{n,m} \right) \leq \sum_{n=1}^{L_1} \left(\sum_{m=1}^{\infty} a_{n,m} \right)$$

$$\leq \sum_{n=1}^{\infty} \left(\sum_{m=1}^{\infty} a_{n,m} \right).$$

Since L was arbitrary, this implies that

$$\sum_{n=1}^{\infty} a_{\sigma(n)} \leq \sum_{n=1}^{\infty} \left(\sum_{m=1}^{\infty} a_{n,m} \right).$$

Fix now $N \in \mathbb{N}$. Applying induction to 5.4.6 we have

$$\sum_{m=1}^{\infty} \left(\sum_{n=1}^{N} a_{n,m} \right) = \sum_{n=1}^{N} \left(\sum_{m=1}^{\infty} a_{n,m} \right). \tag{$*$}$$

Fix also $M \in \mathbb{N}$ and let $K := \max \sigma^{-1}(\{1, ..., N\} \times \{1, ..., M\})$; we have

$$\sum_{m=1}^{M} \sum_{n=1}^{N} a_{n,m} \leq \sum_{l=1}^{K} a_{\sigma(l)} \leq \sum_{n=1}^{\infty} a_{\sigma(n)}.$$

Since M was arbitrary, this implies that

$$\sum_{m=1}^{\infty} \left(\sum_{n=1}^{N} a_{n,m} \right) \leq \sum_{n=1}^{\infty} a_{\sigma(n)},$$

and hence by $(*)$

$$\sum_{n=1}^{N} \left(\sum_{m=1}^{\infty} a_{n,m} \right) \leq \sum_{n=1}^{\infty} a_{\sigma(n)}.$$

Since N was arbitrary, this implies that

$$\sum_{n=1}^{\infty} \left(\sum_{m=1}^{\infty} a_{n,m} \right) \leq \sum_{n=1}^{\infty} a_{\sigma(n)}.$$

Thus,

$$\sum_{n=1}^{\infty} a_{\sigma(n)} = \sum_{n=1}^{\infty} \left(\sum_{m=1}^{\infty} a_{n,m} \right)$$

and we can define

$$\sum_{(n,m) \in \mathbb{N} \times \mathbb{N}} a_{n,m} := \sum_{n=1}^{\infty} a_{\sigma(n)}.$$

Finally, let $b_{n,m} := a_{m,n}$ and denote by γ the bijection from $\mathbb{N} \times \mathbb{N}$ onto $\mathbb{N} \times \mathbb{N}$ defined by $\gamma(n, m) := (m, n)$. Then $\gamma \circ \sigma$ is a bijection from $\mathbb{N}$ onto $\mathbb{N} \times \mathbb{N}$ and

$$\sum_{n=1}^{\infty} a_{\sigma(n)} = \sum_{n=1}^{\infty} b_{(\gamma \circ \sigma)(n)} = \sum_{n=1}^{\infty} \left(\sum_{m=1}^{\infty} b_{n,m} \right)$$

$$= \sum_{n=1}^{\infty} \left(\sum_{m=1}^{\infty} a_{m,n} \right) = \sum_{m=1}^{\infty} \left(\sum_{n=1}^{\infty} a_{n,m} \right).$$

$\square$

5.4.8 Corollary. *Let $\{a_{n,m}\}_{(n,m) \in \mathbb{N} \times \mathbb{N}}$ be a family of elements of $[0, \infty)$ indexed by $\mathbb{N} \times \mathbb{N}$ and let σ be a bijection from $\mathbb{N}$ onto $\mathbb{N} \times \mathbb{N}$. The series $\sum_{n=1}^{\infty} a_{\sigma(n)}$ is convergent in the normed space $\mathbb{R}$ (cf. 4.1.4 and 4.1.5) iff the following two conditions are both satisfied:*

$\forall n \in \mathbb{N}$, the series $\sum_{m=1}^{\infty} a_{n,m}$ is convergent in the normed space $\mathbb{R}$;
the series $\sum_{n=1}^{\infty} \left(\sum_{m=1}^{\infty} a_{n,m} \right)$ is convergent in the normed space $\mathbb{R}$.

In case of convergence, for the sums of the two series we have the equality

$$\sum_{n=1}^{\infty} a_{\sigma(n)} = \sum_{n=1}^{\infty} \left(\sum_{m=1}^{\infty} a_{n,m} \right).$$

Thus, both the convergence of the series $\sum_{n=1}^{\infty} a_{\sigma(n)}$ and its sum do not depend on the particular bijection σ used.

Proof. Use 5.4.2b and 5.4.7. $\qquad\qquad\square$

5.4.9 Proposition. *Let $\{a_{n,k}\}_{(n,k)\in\mathbb{N}\times\mathbb{N}}$ be a family of elements of $[0,\infty]$ such that $a_{n,k} \le a_{n+1,k}$ for each $(n,k) \in \mathbb{N}\times\mathbb{N}$. Then the sequence $\{\sum_{k=1}^{\infty} a_{n,k}\}$ is convergent in the metric space $(\mathbb{R}^*, \delta)$ and*

$$\lim_{n\to\infty} \sum_{k=1}^{\infty} a_{n,k} = \sup_{n\ge 1} \sum_{k=1}^{\infty} a_{n,k} = \sum_{k=1}^{\infty} (\sup_{n\ge 1} a_{n,k})$$

$$= \sum_{k=1}^{\infty} (\lim_{n\to\infty} a_{n,k}).$$

Proof. For each $n \in \mathbb{N}$, induction applied to 5.3.2e implies that

$$\sum_{k=1}^{N} a_{n,k} \le \sum_{k=1}^{N} a_{n+1,k} \le \sum_{k=1}^{\infty} a_{n+1,k}, \forall N \in \mathbb{N},$$

and this implies that

$$\sum_{k=1}^{\infty} a_{n,k} \le \sum_{k=1}^{\infty} a_{n+1,k};$$

then, by 5.2.5 we have that the sequence $\{\sum_{k=1}^{\infty} a_{n,k}\}$ is convergent and

$$\lim_{n\to\infty} \sum_{k=1}^{\infty} a_{n,k} = \sup_{n\ge 1} \sum_{k=1}^{\infty} a_{n,k}.$$

By 5.2.5 we also have

$$\lim_{n\to\infty} a_{n,k} = \sup_{n\ge 1} a_{n,k}, \forall k \in \mathbb{N}.$$

Thus, we need to prove that

$$\sup_{n\ge 1} \sum_{k=1}^{\infty} a_{n,k} = \sum_{k=1}^{\infty} (\sup_{n\ge 1} a_{n,k}),$$

i.e.

$$\sup_{n\ge 1} \left(\sup_{N\ge 1} \sum_{k=1}^{N} a_{n,k} \right) = \sup_{N\ge 1} \left(\sum_{k=1}^{N} (\sup_{n\ge 1} a_{n,k}) \right);$$

for each $n \in \mathbb{N}$, since $a_{n,k} \leq a_{n+1,k}$ for all $n \in \mathbb{N}$ and for $k = 1, ..., N$, we can apply induction to 5.3.4 to obtain

$$\sum_{k=1}^{N} (\sup_{n \geq 1} a_{n,k}) = \sup_{n \geq 1} \sum_{k=1}^{N} a_{n,k};$$

thus, letting $s_{N,n} := \sum_{k=1}^{N} a_{n,k}$, what we need to prove is

$$\sup_{n \geq 1} \left(\sup_{N \geq 1} s_{N,n} \right) = \sup_{N \geq 1} \left(\sup_{n \geq 1} s_{N,n} \right),$$

but this is true by 5.2.3.   □

5.4.10 Remark. Let $\{f_n\}$ be a sequence in a normed space and suppose that the series $\sum_{n=1}^{\infty} f_n$ is convergent (cf. 2.1.10). Then,

$$\left\| \sum_{n=1}^{\infty} f_n \right\| \leq \sum_{n=1}^{\infty} \|f_n\|.$$

If $\sum_{n=1}^{\infty} \|f_n\| = \infty$, this is obvious. Otherwise, it is proved by

$$\left\| \sum_{n=1}^{N} f_n \right\| \leq \sum_{n=1}^{N} \|f_n\| \leq \sum_{n=1}^{\infty} \|f_n\|, \forall N \in \mathbb{N},$$

which implies that

$$\left\| \sum_{n=1}^{\infty} f_n \right\| = \lim_{n \to \infty} \left\| \sum_{n=1}^{N} f_n \right\| \leq \sum_{n=1}^{\infty} \|f_n\|.$$

Chapter 6

Measurable Sets and Measurable Functions

Although this chapter deals with measurable sets and functions, its contents are purely set-theoretic: in this chapter there is still no measure in view. The reason for the adjective "measurable" lies in the following facts, which will be seen in later chapters: a measure is a function defined on a family of measurable sets and an integral is a concept which is consistent only for measurable functions.

6.1 Semialgebras, algebras, σ-algebras

Throughout this section, X stands for a non-empty set.

6.1.1 Definition. A non-empty collection $\mathcal{S}$ of subsets of X is called a *semialgebra on* X if the following conditions are satisfied:

(sa_1) $\emptyset \in \mathcal{S}$;
(sa_2) if $E, F \in \mathcal{S}$ then $E \cap F \in \mathcal{S}$;
(sa_3) if $E \in \mathcal{S}$ then there is a finite and disjoint family $\{F_i\}_{i \in I}$ of elements of $\mathcal{S}$ such that $X - E = \cup_{i \in I} F_i$.

6.1.2 Proposition. *Let $\mathcal{S}$ be a semialgebra on X, let $n \in \mathbb{N}$, and let $\{E_1, ..., E_n\}$ be a disjoint family of elements of $\mathcal{S}$. Then there exists a finite and disjoint family $\{F_i\}_{i \in I}$ of elements of $\mathcal{S}$ such that*

$$X - \cup_{k=1}^{n} E_k = \cup_{i \in I} F_i.$$

Proof. The proof is by induction. For $n = 1$, the statement follows at once from the definition of semialgebra. Assume then that the statement is true for $n = m$ and consider a disjoint family $\{E_1, ..., E_m, E_{m+1}\}$ of elements of $\mathcal{S}$. Then there exists a finite and disjoint family $\{F_i\}_{i \in I}$ of elements of $\mathcal{S}$ such that

$$X - \cup_{k=1}^{m} E_k = \cup_{i \in I} F_i.$$

Since $E_{m+1} \in \mathcal{S}$, there exists also a finite and disjoint family $\{G_j\}_{j \in J}$ of elements of $\mathcal{S}$ such that

$$X - E_{m+1} = \cup_{j \in J} G_j.$$

Then $\{F_i \cap G_j\}_{(i,j)\in I\times J}$ is a finite and disjoint family of elements of $\mathcal{S}$ and

$$X - \cup_{k=1}^{m+1} E_k = (X - \cup_{k=1}^{m} E_k) \cap (X - E_{m+1}) = \cup_{(i,j)\in I\times J} F_i \cap G_j.$$

$\square$

6.1.3 Proposition. *Let $\mathcal{S}$ be a semialgebra on X, let $E \in \mathcal{S}$, let $n \in \mathbb{N}$, and let $\{E_1, ..., E_n\}$ be a disjoint family of elements of $\mathcal{S}$ such that $E_k \subset E$ for $k = 1, ..., n$. Then there exists a finite and disjoint family $\{G_i\}_{i\in I}$ of elements of $\mathcal{S}$ such that:*

$$E_k \cap G_i = \emptyset, \forall k \in \{1, ..., n\}, \forall i \in I,$$

$$E = (\cup_{k=1}^{n} E_k) \cup (\cup_{i\in I} G_i).$$

Proof. By 6.1.2, there exists a finite and disjoint family $\{F_i\}_{i\in I}$ of elements of $\mathcal{S}$ such that

$$X - \cup_{k=1}^{n} E_k = \cup_{i\in I} F_i.$$

Define $G_i := E \cap F_i$ for each $i \in I$. Then $\{G_i\}_{i\in I}$ is a disjoint family of elements of $\mathcal{S}$ and also:

$$\forall k \in \{1, ..., n\}, \forall i \in I, E_k \cap G_i = \emptyset \text{ since } G_i \subset F_i \subset X - \cup_{l=1}^{n} E_l \subset X - E_k,$$

$$(\cup_{k=1}^{n} E_k) \cup (\cup_{i\in I} G_i) = (\cup_{k=1}^{n} E_k) \cup (E \cap (X - \cup_{k=1}^{n} E_k)) = E \cap X = E.$$

$\square$

6.1.4 Proposition. *Let $\mathcal{S}$ be a semialgebra on X, let $n \in \mathbb{N}$, and let $\{A_1, ..., A_n\}$ be a family of elements of $\mathcal{S}$. Then there exists a finite and disjoint family $\{B_j\}_{j\in J}$ of elements of $\mathcal{S}$ such that*

$$\forall k \in \{1, ..., n\}, \exists J_k \subset J \text{ such that } A_k = \cup_{j\in J_k} B_j.$$

Proof. The proof is by induction. For $n = 1$ the statement is obviously true (assume $J := \{1\}$ and $B_1 := A_1$). Assume then that the statement is true for $n = m$ and consider a family $\{A_1, ..., A_m, A_{m+1}\}$ of elements of $\mathcal{S}$. Then there exists a finite and disjoint family $\{B_i\}_{j\in J}$ of elements of $\mathcal{S}$ such that

$$\forall k \in \{1, ..., m\}, \exists J_k \subset J \text{ such that } A_k = \cup_{j\in J_k} B_j.$$

Define $B_{j,1} := A_{m+1} \cap B_j$ for each $j \in J$. From 6.1.3 (with $E := A_{m+1}$ and $\{E_1, ..., E_n\} := \{B_{j,1}\}_{j\in J}$) it follows that there exists a finite and disjoint family $\{G_i\}_{i\in I}$ of elements of $\mathcal{S}$ such that

$$B_{j,1} \cap G_i = \emptyset, \forall j \in J, \forall i \in I,$$

$$A_{m+1} = (\cup_{j\in J} B_{j,1}) \cup (\cup_{i\in I} G_i).$$

From 6.1.3 (with $E := B_j$ and $\{E_1, ..., E_n\} := \{B_{j,1}\}$) it also follows that, for each $j \in J$, there exists a finite and disjoint family $\{B_{j,r}\}_{r=2,...,k_j}$ of elements of $\mathcal{S}$ such that

$$B_{j,1} \cap B_{j,r} = \emptyset, \forall r \in \{2, ..., k_j\},$$

$$B_j = B_{j,1} \cup (\cup_{r=2}^{k_j} B_{j,r}) = \cup_{r=1}^{k_j} B_{j,r}.$$

Now, we have that $\{G_i\}_{i\in I} \cup \{B_{j,r}\}_{j\in J, r=1,...,k_j}$ is a family of elements of $\mathcal{S}$ which are disjoint since:

if $r \neq r'$, $B_{j,r} \cap B_{j,r'} = \emptyset$ (see above),

if $j \neq j'$, $B_{j,r} \cap B_{j',r'} = \emptyset$ since $B_{j,r} \subset B_j, B_{j',r'} \subset B_{j'}, B_j \cap B_{j'} = \emptyset$,

if $i \neq i'$, $G_i \cap G_{i'} = \emptyset$ (see above),

$B_{j,1} \cap G_i = \emptyset$ (see above),

if $r \neq 1$, $B_{j,r} \cap G_i = \emptyset$ since $G_i \subset A_{m+1}$ and $B_{j,r} \subset (X - B_{j,1}) \cap B_j = ((X - A_{m+1}) \cup (X - B_j)) \cap B_j = (X - A_{m+1}) \cap B_j \subset X - A_{m+1}$.

Moreover we have

$$\forall k \in \{1, ..., m\}, A_k = \cup_{j \in J_k} B_j = \cup_{j \in J_k} \cup_{r=1}^{k_j} B_{j,r}.$$

Thus, $\{G_i\}_{i \in I} \cup \{B_{j,r}\}_{j \in J, r=1,...,k_j}$ is a finite and disjoint family of elements of $\mathcal{S}$ such that every element of $\{A_1, ..., A_m, A_{m+1}\}$ can be obtained as the union of some of its elements. $\square$

6.1.5 Definition. A non-empty collection $\mathcal{A}_0$ of subsets of X is called an *algebra* *on* X if the following conditions are satisfied:

(al_1) if $E, F \in \mathcal{A}_0$ then $E \cup F \in \mathcal{A}_0$;

(al_2) if $E \in \mathcal{A}_0$ then $X - E \in \mathcal{A}_0$.

6.1.6 Proposition. *An algebra $\mathcal{A}_0$ on X has the following properties:*

(al_3) *if $E, F \in \mathcal{A}_0$ then $E \cap F \in \mathcal{A}_0$,*

(al_4) $\emptyset \in \mathcal{A}_0$, $X \in \mathcal{A}_0$,

(al_5) *if $n \in \mathbb{N}$ and $\{E_1, ..., E_n\}$ is a family of elements of $\mathcal{A}_0$, then $\cup_{k=1}^{n} E_k \in \mathcal{A}_0$ and $\cap_{k=1}^{n} E_k \in \mathcal{A}_0$,*

(al_6) *if $E, F \in \mathcal{A}_0$ then $E - F \in \mathcal{A}_0$.*

Proof. al_3: We have

$$E, F \in \mathcal{A}_0 \Rightarrow X - E, X - F \in \mathcal{A}_0 \Rightarrow$$
$$X - (E \cap F) = (X - E) \cup (X - F) \in \mathcal{A}_0 \Rightarrow$$
$$E \cap F = X - (X - (E \cap F)) \in \mathcal{A}_0.$$

al_4: Let $E \in \mathcal{A}_0$. Then $X - E \in \mathcal{A}_0$, and hence

$$\emptyset = E \cap (X - E) \in \mathcal{A}_0, \quad X = E \cup (X - E) \in \mathcal{A}_0.$$

al_5: This follows from al_1 and al_3 by elementary induction.

al_6: We have

$$[E, F \in \mathcal{A}_0] \Rightarrow [E, X - F \in \mathcal{A}_0] \Rightarrow [E - F = E \cap (X - F) \in \mathcal{A}_0].$$

$\square$

6.1.7 Remark. It is clear from al_4, al_3, al_2 that, for an algebra, conditions sa_1, sa_2, sa_3 of 6.1.1 are satisfied. Thus, an algebra is also a semialgebra. For any non-empty set X, the collections $\{\emptyset, X\}$ and $\mathcal{P}(X)$ are algebras on X.

6.1.8 Proposition. *Let $\{E_n\}$ be a sequence of subsets of X, and define*

$$F_1 := E_1 \text{ and } F_n := E_n - \cup_{k=1}^{n-1} E_k \text{ for } n > 1.$$

Then:

(a) $F_k \cap F_l = \emptyset$ if $k \neq l$,
(b) $\cup_{n=1}^{N} F_n = \cup_{n=1}^{N} E_n$, $\forall N \in \mathbb{N}$,
(c) $\cup_{n=1}^{\infty} F_n = \cup_{n=1}^{\infty} E_n$,
(d) *if $\mathcal{A}_0$ is an algebra on X and $E_n \in \mathcal{A}_0$ for all $n \in \mathbb{N}$, then $F_n \in \mathcal{A}_0$ for all $n \in \mathbb{N}$.*

Proof. a: If $k \neq l$, assume e.g. $k < l$; then $F_k \subset E_k$ and

$$F_l \subset E_l - E_k \subset X - E_k,$$

and this implies $F_k \cap F_l \subset E_k \cap (X - E_k) = \emptyset$.

b: From $F_n \subset E_n$ for all $n \in \mathbb{N}$ we obtain $\cup_{n=1}^{N} F_n \subset \cup_{n=1}^{N} E_n$. We also have

$$x \in \cup_{n=1}^{N} E_n \Rightarrow [\exists k \in \{1, ..., N\} \text{ s.t. } x \in E_k \text{ and } x \notin E_i \text{ for } i < k] \Rightarrow$$
$$[\exists k \in \{1, ..., N\} \text{ s.t. } x \in F_k] \Rightarrow x \in \cup_{n=1}^{N} F_n.$$

c: Repeat the proof for b, with $\{1, ..., N\}$ replaced by $\mathbb{N}$.

d: This follows at once from al_5 and al_6. $\qquad\qquad\square$

6.1.9 Proposition. *Let Λ be a family of algebras on X. Then $\cap_{\mathcal{A}_0 \in \Lambda} \mathcal{A}_0$ (this intersection is defined within the framework of $\mathcal{P}(X)$) is an algebra on X.*

Proof. al_1: We have

$$E, F \in \cap_{\mathcal{A}_0 \in \Lambda} \mathcal{A}_0 \Rightarrow (E, F \in \mathcal{A}_0, \forall \mathcal{A}_0 \in \Lambda) \Rightarrow$$
$$(E \cup F \in \mathcal{A}_0, \forall \mathcal{A}_0 \in \Lambda) \Rightarrow E \cup F \in \cap_{\mathcal{A}_0 \in \Lambda} \mathcal{A}_0.$$

al_2: We have

$$E \in \cap_{\mathcal{A}_0 \in \Lambda} \mathcal{A}_0 \Rightarrow (X - E \in \mathcal{A}_0, \forall \mathcal{A}_0 \in \Lambda) \Rightarrow X - E \in \cap_{\mathcal{A}_0 \in \Lambda} \mathcal{A}_0.$$

$\qquad\qquad\square$

6.1.10 Theorem.

(a) *Let $\mathcal{F}$ be a family of subsets of X. Then there exists a unique algebra on X, which is called the algebra on X generated by $\mathcal{F}$ and is denoted by $\mathcal{A}_0(\mathcal{F})$, such that*

(ga_1) $\mathcal{F} \subset \mathcal{A}_0(\mathcal{F})$,
(ga_2) *if $\mathcal{A}_0$ is an algebra on X and $\mathcal{F} \subset \mathcal{A}_0$, then $\mathcal{A}_0(\mathcal{F}) \subset \mathcal{A}_0$.*

If $\mathcal{F}$ is the empty family, then $\mathcal{A}_0(\mathcal{F}) = \{0, X\}$.

(b) *Let $\mathcal{F}_1$ and $\mathcal{F}_2$ be families of subsets of X. If $\mathcal{F}_1 \subset \mathcal{F}_2$ or $\mathcal{F}_1 \subset \mathcal{A}_0(\mathcal{F}_2)$, then $\mathcal{A}_0(\mathcal{F}_1) \subset \mathcal{A}_0(\mathcal{F}_2)$.*

Proof. a: As to existence, define

$$\Lambda := \{\mathcal{A}_0 : \mathcal{A}_0 \text{ is an algebra on } X \text{ and } \mathcal{F} \subset \mathcal{A}_0\}$$

and $\mathcal{A}_0(\mathcal{F}) := \cap_{\mathcal{A}_0 \in \Lambda} \mathcal{A}_0$. By 6.1.9, $\mathcal{A}_0(\mathcal{F})$ is an algebra on X. Properties ga_1 and ga_2 are obvious. As to uniqueness, assume that $\tilde{\mathcal{A}}_0$ is an algebra on X such that

$(\tilde{a}_1)$ $\mathcal{F} \subset \tilde{\mathcal{A}}_0$,
$(\tilde{a}_2)$ if $\mathcal{A}_0$ is an algebra on X and $\mathcal{F} \subset \mathcal{A}_0$, then $\tilde{\mathcal{A}}_0 \subset \mathcal{A}_0$.

Then $\mathcal{A}_0(\mathcal{F}) \subset \tilde{\mathcal{A}}_0$ by $\tilde{a}_1$ and ga_2, and also $\tilde{\mathcal{A}}_0 \subset \mathcal{A}_0(\mathcal{F})$ by ga_1 and $\tilde{a}_2$.

If the family $\mathcal{F}$ is empty then $\mathcal{F}$ is contained in every algebra on X, and the intersection of all the algebras on X is $\{\emptyset, X\}$, because $\{\emptyset, X\}$ is an algebra on X and it is contained in all the algebras on X (cf. al_4).

b: If $\mathcal{F}_1 \subset \mathcal{A}_0(\mathcal{F}_2)$, then $\mathcal{A}_0(\mathcal{F}_1) \subset \mathcal{A}_0(\mathcal{F}_2)$ by property ga_2 of $\mathcal{A}_0(\mathcal{F}_1)$. If $\mathcal{F}_1 \subset \mathcal{F}_2$, then $\mathcal{F}_1 \subset \mathcal{A}_0(\mathcal{F}_2)$ by property ga_1 of $\mathcal{A}_0(\mathcal{F}_2)$. $\qquad\square$

6.1.11 Theorem. *Let $\mathcal{S}$ be a semialgebra on X. Then $\mathcal{A}_0(\mathcal{S})$ is the collection of all the unions of finite and disjoint families of elements of $\mathcal{S}$, i.e. by letting*

$$\mathcal{C} := \{E \in \mathcal{P}(X) : \exists n \in \mathbb{N}, \exists \{E_1, ..., E_n\} \text{ s.t.}$$
$$E_k \in \mathcal{S} \text{ for } k = 1, ..., n, E_k \cap E_l = \emptyset \text{ if } k \neq l, E = \cup_{k=1}^n E_k\}$$

we have $\mathcal{A}_0(\mathcal{S}) = \mathcal{C}$.

Proof. It is obvious that $\mathcal{C} \subset \mathcal{A}_0(\mathcal{S})$, since $\mathcal{S} \subset \mathcal{A}_0(\mathcal{S})$ and $\mathcal{A}_0(\mathcal{S})$ has property al_5. Since it is also obvious that $\mathcal{S} \subset \mathcal{C}$, if we can prove that $\mathcal{C}$ is an algebra on X then we can conlude that $\mathcal{A}_0(\mathcal{S}) \subset \mathcal{C}$ by property ga_2 of $\mathcal{A}_0(\mathcal{S})$.

Now, let E, F be two elements of $\mathcal{C}$ and let $\{E_1, ..., E_n\}$, $\{F_1, ..., F_m\}$ be two disjoint families of elements of $\mathcal{S}$ such that $E = \cup_{k=1}^n E_k$ and $F = \cup_{l=1}^m F_l$. From 6.1.4 it follows that there exists a finite and disjoint family $\{B_j\}_{j \in J}$ of elements of $\mathcal{S}$ such that each element in the family $\{E_1, ..., E_n, F_1, ..., F_m\}$ can be obtained as the union of a subfamily of $\{B_j\}_{j \in J}$; it is then clear that $E \cup F$ too can be obtained in this way, and this implies that $E \cup F \in \mathcal{C}$. Thus, $\mathcal{C}$ has property al_1 of 6.1.5. Moreover, if $\{E_1, ..., E_n\}$ is a disjoint family of elements of $\mathcal{S}$, then from 6.1.2 it follows that $X - \cup_{k=1}^n E_k$ can be obtained as the union of a finite and disjoint family of elements of $\mathcal{S}$. This proves that $\mathcal{C}$ has property al_2 of 6.1.5. $\qquad\square$

6.1.12 Corollary. *Let $\mathcal{S}$ be a semialgebra on X. Then $\mathcal{A}_0(\mathcal{S})$ is the collection of all the unions of finite families of elements of $\mathcal{S}$, i.e. by letting*

$$\mathcal{C}' := \{E \in \mathcal{P}(X) : \exists n \in \mathbb{N}, \exists \{E_1, ..., E_n\} \text{ s.t.}$$
$$E_k \in \mathcal{S} \text{ for } k = 1, ..., n \text{ and } E = \cup_{k=1}^n E_k\}$$

we have $\mathcal{A}_0(\mathcal{S}) = \mathcal{C}'$.

Proof. If $\mathcal{C}$ is as in 6.1.11, then clearly $\mathcal{C} \subset \mathcal{C}'$. On the other hand, since $\mathcal{S} \subset \mathcal{C}$ we also have $\mathcal{C}' \subset \mathcal{C}$ by property al_5 of $\mathcal{C}$. Thus, $\mathcal{C}' = \mathcal{C} = \mathcal{A}_0(\mathcal{S})$. $\qquad\square$

6.1.13 Definition. A collection $\mathcal{A}$ of subsets of X is called a *σ-algebra on X* if $\mathcal{A}$ is an algebra on X and it has the following property:

(σa_1) if $\{E_n\}$ is a sequence in $\mathcal{A}$ then $\cup_{n=1}^{\infty} E_n \in \mathcal{A}$.

The pair $(X, \mathcal{A})$, where X is a non-empty set and $\mathcal{A}$ is a σ-algebra on X, is said to be a *measurable space* and the elements of $\mathcal{A}$ are called the *measurable subsets* of X (the reason for these names is that a measure is a function defined on a σ-algebra).

6.1.14 Proposition. *A σ-algebra $\mathcal{A}$ on X has the following property:*

(σa_2) *if $\{E_n\}$ is a sequence in $\mathcal{A}$ then $\cap_{n=1}^{\infty} E_n \in \mathcal{A}$.*

Proof. If $\{E_n\}$ is a sequence in $\mathcal{A}$ then we have

$$(E_n \in \mathcal{A}, \forall n \in \mathbb{N}) \Rightarrow (X - E_n \in \mathcal{A}, \forall n \in \mathbb{N}) \Rightarrow$$
$$X - \cap_{n=1}^{\infty} E_n = \cup_{n=1}^{\infty}(X - E_n) \in \mathcal{A} \Rightarrow$$
$$\cap_{n=1}^{\infty} E_n = X - (X - \cap_{n=1}^{\infty} E_n) \in \mathcal{A}.$$

$\square$

6.1.15 Remark. For any non-empty set X, the collections $\{\emptyset, X\}$ and $\mathcal{P}(X)$ are σ-algebras on X.

6.1.16 Proposition. *Let Λ be a family of σ-algebras on X. Then $\cap_{\mathcal{A} \in \Lambda} \mathcal{A}$ (this intersection is defined within the framework of $\mathcal{P}(X)$) is a σ-algebra on X.*

Proof. From 6.1.9 it follows that $\cap_{\mathcal{A} \in \Lambda} \mathcal{A}$ is an algebra on X. Moreover, if $\{E_n\}$ is a sequence in $\cap_{\mathcal{A} \in \Lambda} \mathcal{A}$, then we have

$$(E_n \in \mathcal{A}, \forall n \in \mathbb{N}, \forall \mathcal{A} \in \Lambda) \Rightarrow (\cup_{n=1}^{\infty} E_n \in \mathcal{A}, \forall \mathcal{A} \in \Lambda) \Rightarrow \cup_{n=1}^{\infty} E_n \in \cap_{\mathcal{A} \in \Lambda} \mathcal{A}.$$

$\square$

6.1.17 Theorem.

(a) *Let $\mathcal{F}$ be a family of subsets of X. Then there exists a unique σ-algebra on X, which is called the σ-algebra on X generated by $\mathcal{F}$ and is denoted by $\mathcal{A}(\mathcal{F})$, such that*

$(g\sigma_1)$ $\mathcal{F} \subset \mathcal{A}(\mathcal{F})$,
$(g\sigma_2)$ *if $\mathcal{A}$ is a σ-algebra on X and $\mathcal{F} \subset \mathcal{A}$, then $\mathcal{A}(\mathcal{F}) \subset \mathcal{A}$.*

If $\mathcal{F}$ is the empty family, then $\mathcal{A}(\mathcal{F}) = \{\emptyset, X\}$.
(b) *Let $\mathcal{F}_1$ and $\mathcal{F}_2$ be families of subsets of X. If $\mathcal{F}_1 \subset \mathcal{F}_2$ or $\mathcal{F}_1 \subset \mathcal{A}(\mathcal{F}_2)$, then $\mathcal{A}(\mathcal{F}_1) \subset \mathcal{A}(\mathcal{F}_2)$.*

Proof. The proof of the present statement is a slight modification of the proof of 6.1.10. $\square$

6.1.18 Proposition. *Let $\mathcal{F}$ be a family of subsets of X. Then*

$$\mathcal{A}(\mathcal{F}) = \mathcal{A}(\mathcal{A}_0(\mathcal{F})).$$

Proof. Since $\mathcal{F} \subset \mathcal{A}_0(\mathcal{F})$ (cf. ga_1), we have $\mathcal{A}(\mathcal{F}) \subset \mathcal{A}(\mathcal{A}_0(\mathcal{F}))$ (cf. 6.1.17b). Since $\mathcal{A}(\mathcal{F})$ is an algebra on X and $\mathcal{F} \subset \mathcal{A}(\mathcal{F})$ (cf. $g\sigma_1$), we have $\mathcal{A}_0(\mathcal{F}) \subset \mathcal{A}(\mathcal{F})$ (cf. ga_2), whence $\mathcal{A}(\mathcal{A}_0(\mathcal{F})) \subset \mathcal{A}(\mathcal{F})$ (cf. 6.1.17b). $\qquad\square$

6.1.19 Proposition. *Let $\mathcal{A}$ be a σ-algebra on X, let Y be a non-empty subset of X, and define the collection $\mathcal{A}^Y$ of subsets of Y by*

$$\mathcal{A}^Y := \{E \cap Y : E \in \mathcal{A}\}.$$

Then $\mathcal{A}^Y$ is a σ-algebra on Y, which is called the σ-algebra induced on Y by $\mathcal{A}$. The measurable space $(Y, \mathcal{A}^Y)$ is said to be a measurable subspace of $(X, \mathcal{A})$, and it is said to be defined by Y.

We have:

(a) $\mathcal{A}^Y \subset \mathcal{A}$ iff $Y \in \mathcal{A}$;
(b) if Z is a non-empty subset of Y, then $(\mathcal{A}^Y)^Z = \mathcal{A}^Z$.

Proof. First we prove that $\mathcal{A}^Y$ is a σ-algebra on Y.

al_1 and σa_1: Let $\{F_n\}_{n \in I}$ be a family of elements of $\mathcal{A}^Y$, with $I := \{1, 2\}$ or $I := \mathbb{N}$. Then there is a family $\{E_n\}_{n \in I}$ of elements of $\mathcal{A}$ such that $F_n = E_n \cap Y$ for all $n \in I$, and we have

$$\cup_{n \in I} F_n = (\cup_{n \in I} E_n) \cap Y,$$

which proves that $\cup_{n \in I} F_n \in \mathcal{A}^Y$ since $\cup_{n \in I} E_n \in \mathcal{A}$.

al_2: If $F \in \mathcal{A}^Y$, then there exists $E \in \mathcal{A}$ such that $F = E \cap Y$, and we have

$$Y - F = Y \cap (X - F) = Y \cap ((X - E) \cup (X - Y)) = (X - E) \cap Y,$$

which proves that $Y - F \in \mathcal{A}^Y$ since $X - E \in \mathcal{A}$.

Now we prove a and b.

a: By property al_3 of $\mathcal{A}$ we have

$$Y \in \mathcal{A} \Rightarrow (E \cap Y \in \mathcal{A}, \forall E \in \mathcal{A}) \Rightarrow \mathcal{A}^Y \subset \mathcal{A},$$

and by property al_4 of $\mathcal{A}^Y$ we have

$$\mathcal{A}^Y \subset \mathcal{A} \Rightarrow Y \in \mathcal{A}.$$

b: If $Z \subset Y$ then $(E \cap Y) \cap Z = E \cap Z$ for all $E \in \mathcal{A}$. $\qquad\square$

6.1.20 Proposition. *Let $\mathcal{F}$ be a family of subsets of X, let Y be a non-empty subset of X, and define the family $\mathcal{F}^Y$ of subsets of Y by*

$$\mathcal{F}^Y := \{F \cap Y : F \in \mathcal{F}\}.$$

Then we can define two σ-algebras on Y: $(\mathcal{A}(\mathcal{F}))^Y$ and $\mathcal{A}(\mathcal{F}^Y)$ (by this symbol we denote the σ-algebra on Y generated by $\mathcal{F}^Y$). However,

$$(\mathcal{A}(\mathcal{F}))^Y = \mathcal{A}(\mathcal{F}^Y).$$

Proof. From the inclusion $\mathcal{F} \subset \mathcal{A}(\mathcal{F})$ (cf. $g\sigma_1$) we have $\mathcal{F}^Y \subset (\mathcal{A}(\mathcal{F}))^Y$, and hence $\mathcal{A}(\mathcal{F}^Y) \subset (\mathcal{A}(\mathcal{F}))^Y$ by property $g\sigma_2$ for $\mathcal{A}(\mathcal{F}^Y)$ (since $(\mathcal{A}(\mathcal{F}))^Y$ is a σ-algebra on Y by 6.1.19).

To prove the opposite inclusion, we define the collection $\mathcal{A}$ of subsets of X by

$$\mathcal{A} := \left\{ E \in \mathcal{P}(X) : E \cap Y \in \mathcal{A}(\mathcal{F}^Y) \right\}.$$

By the definition of $\mathcal{F}^Y$ and property $g\sigma_1$ for $\mathcal{A}(\mathcal{F}^Y)$ we have

$$F \in \mathcal{F} \Rightarrow F \cap Y \in \mathcal{F}^Y \Rightarrow F \cap Y \in \mathcal{A}(\mathcal{F}^Y),$$

and this shows that $\mathcal{F} \subset \mathcal{A}$. We prove now that $\mathcal{A}$ is a σ-algebra on X.

al_1 and σa_1: Let $\{E_n\}_{n \in I}$ be a family of elements of $\mathcal{A}$, with $I := \{1, 2\}$ or $I := \mathbb{N}$. Then we have $E_n \cap Y \in \mathcal{A}(\mathcal{F}^Y)$ for all $n \in I$, and this implies that

$$(\cup_{n \in I} E_n) \cap Y = \cup_{n \in I}(E_n \cap Y) \in \mathcal{A}(\mathcal{F}^Y),$$

which proves that $\cup_{n \in I} E_n \in \mathcal{A}$.

al_2: If $E \in \mathcal{A}$, then $E \cap Y \in \mathcal{A}(\mathcal{F}^Y)$, and we have

$$(X - E) \cap Y = ((X - E) \cup (X - Y)) \cap Y = (X - (E \cap Y)) \cap Y = Y - (E \cap Y) \in \mathcal{A}(\mathcal{F}^Y),$$

which proves that $X - E \in \mathcal{A}$.

Thus $\mathcal{A}$ is a σ-algebra on X and therefore we have $\mathcal{A}(\mathcal{F}) \subset \mathcal{A}$ (cf. $g\sigma_2$), which means

$$E \cap Y \in \mathcal{A}(\mathcal{F}^Y), \forall E \in \mathcal{A}(\mathcal{F}).$$

From this we have that, for $F \in \mathcal{P}(Y)$, the following implications are true

$$F \in (\mathcal{A}(\mathcal{F}))^Y \Rightarrow [\exists E \in \mathcal{A}(\mathcal{F}) \text{ s.t. } F = E \cap Y] \Rightarrow F \in \mathcal{A}(\mathcal{F}^Y).$$

This proves the inclusion $(\mathcal{A}(\mathcal{F}))^Y \subset \mathcal{A}(\mathcal{F}^Y)$ that we wanted to prove. $\qquad\square$

6.1.21 Corollary. *Let d be a distance on X and let Y be a non-empty subset of X. Then we can define two σ-algebras on Y: $(\mathcal{A}(\mathcal{T}_d))^Y$ and $\mathcal{A}(\mathcal{T}_{d_Y})$ (recall that $\mathcal{T}_d$ denotes the family of all open sets in a metric space whose distance is denoted by d; for the metric subspace (Y, d_Y), cf. 2.1.3). However,*

$$(\mathcal{A}(\mathcal{T}_d))^Y = \mathcal{A}(\mathcal{T}_{d_Y}).$$

Proof. The result of 2.2.5 can be rephrased as

$$\mathcal{T}_{d_Y} = (\mathcal{T}_d)^Y.$$

Use then 6.1.20. $\qquad\square$

6.1.22 Definition. Let d be a distance on X. The σ-algebra $\mathcal{A}(\mathcal{T}_d)$ on X is called the *Borel σ-algebra on X* and is denoted by $\mathcal{A}(d)$, i.e.

$$\mathcal{A}(d) := \mathcal{A}(\mathcal{T}_d).$$

Any element of $\mathcal{A}(d)$ is called a *Borel set*.

If Y is a non-empty subset of X, the σ-algebra $\mathcal{A}(\mathcal{T}_{d_Y})$ on Y is called the *Borel σ-algebra on Y*. From 6.1.21 it follows that

$$\mathcal{A}(d_Y) = \mathcal{A}(\mathcal{T}_{d_Y}) = (\mathcal{A}(\mathcal{T}_d))^Y = (\mathcal{A}(d))^Y.$$

6.1.23 Proposition. *Let d be a distance on X. We have*

$$\mathcal{A}(d) = \mathcal{A}(\mathcal{K}_d)$$

(recall that $\mathcal{K}_d$ denotes the family of all closed sets in a metric space whose distance is denoted by d).

Proof. Since

$$E \in \mathcal{K}_d \Rightarrow X - E \in \mathcal{T}_d \Rightarrow X - E \in \mathcal{A}(d) \Rightarrow E = X - (X - E) \in \mathcal{A}(d),$$

we have $\mathcal{K}_d \subset \mathcal{A}(d)$, whence $\mathcal{A}(\mathcal{K}_d) \subset \mathcal{A}(d)$ (cf. 6.1.17b).

Since

$$E \in \mathcal{T}_d \Rightarrow X - E \in \mathcal{K}_d \Rightarrow X - E \in \mathcal{A}(\mathcal{K}_d) \Rightarrow E = X - (X - E) \in \mathcal{A}(\mathcal{K}_d),$$

we have $\mathcal{T}_d \subset \mathcal{A}(\mathcal{K}_d)$, whence $\mathcal{A}(d) \subset \mathcal{A}(\mathcal{K}_d)$. $\qquad\square$

6.1.24 Proposition. *Consider the measurable spaces $(\mathbb{R}, \mathcal{A}(d_{\mathbb{R}}))$ (cf. 2.1.4), $(\mathbb{R}^*, \mathcal{A}(\delta))$ (cf. 5.2.1), and $(\mathbb{C}, \mathcal{A}(d_{\mathbb{C}}))$ (cf. 2.7.4a). We have on $\mathbb{R}$ the three σ-algebras $\mathcal{A}(d_{\mathbb{R}}), (\mathcal{A}(\delta))^{\mathbb{R}}, (\mathcal{A}(d_{\mathbb{C}}))^{\mathbb{R}}$ (the last symbol is consistent since we identify $\mathbb{R}$ with the subset $\{(a, 0) : a \in \mathbb{R}\}$ of $\mathbb{C}$). However,*

$$\mathcal{A}(d_{\mathbb{R}}) = (\mathcal{A}(\delta))^{\mathbb{R}} = (\mathcal{A}(d_{\mathbb{C}}))^{\mathbb{R}}.$$

Proof. From 6.1.21 we have (as already noted in 6.1.22)

$$(\mathcal{A}(\delta))^{\mathbb{R}} = \mathcal{A}(\delta_{\mathbb{R}}) := \mathcal{A}(\mathcal{T}_{\delta_{\mathbb{R}}}).$$

Besides, from 5.2.1d we have $\mathcal{T}_{\delta_{\mathbb{R}}} = \mathcal{T}_{d_{\mathbb{R}}}$. Therefore,

$$(\mathcal{A}(\delta))^{\mathbb{R}} = \mathcal{A}(\mathcal{T}_{d_{\mathbb{R}}}) =: \mathcal{A}(d_{\mathbb{R}}).$$

From 6.1.21 we also have (as already noted in 6.1.22)

$$(\mathcal{A}(d_{\mathbb{C}}))^{\mathbb{R}} = \mathcal{A}((d_{\mathbb{C}})_{\mathbb{R}}).$$

Now, $(d_{\mathbb{C}})_{\mathbb{R}} = d_{\mathbb{R}}$ (cf. 2.7.4a). Therefore,

$$(\mathcal{A}(d_{\mathbb{C}}))^{\mathbb{R}} = \mathcal{A}(d_{\mathbb{R}}).$$

$\qquad\square$

6.1.25 Proposition. *Define the following families of subsets of $\mathbb{R}$:*

$$\mathcal{I}_1 := \{(a, b) : a, b \in \mathbb{R}\},$$
$$\mathcal{I}_2 := \{[a, b) : a, b \in \mathbb{R}\},$$
$$\mathcal{I}_3 := \{[a, b] : a, b \in \mathbb{R}\},$$
$$\mathcal{I}_4 := \{(a, b] : a, b \in \mathbb{R}\},$$
$$\mathcal{I}_5 := \{(-\infty, a] : a \in \mathbb{R}\},$$
$$\mathcal{I}_6 := \{(-\infty, a) : a \in \mathbb{R}\},$$
$$\mathcal{I}_7 := \{[a, \infty) : a \in \mathbb{R}\},$$
$$\mathcal{I}_8 := \{(a, \infty) : a \in \mathbb{R}\}.$$

Then we have

$$\mathcal{A}(d_{\mathbb{R}}) = \mathcal{A}(\mathcal{I}_n) \ \text{for } n = 1, ..., 8.$$

If we define

$$\mathcal{I}_9 := \mathcal{I}_4 \cup \mathcal{I}_5 \cup \mathcal{I}_8,$$

then $\mathcal{I}_9$ is a semialgebra on $\mathbb{R}$ and

$$\mathcal{A}(d_{\mathbb{R}}) = \mathcal{A}(\mathcal{I}_9).$$

Proof. From 2.3.16 and 2.3.17 it follows that every element of $\mathcal{T}_{d_{\mathbb{R}}}$ is the union of a countable family of open balls. Now, the family of open balls in $(\mathbb{R}, d_{\mathbb{R}})$ is $\mathcal{I}_1$. Thus,

$$\mathcal{T}_{d_{\mathbb{R}}} \subset \mathcal{A}(\mathcal{I}_1)$$

by property σa_1 of $\mathcal{A}(\mathcal{I}_1)$, since $\mathcal{I}_1 \subset \mathcal{A}(\mathcal{I}_1)$ (cf. $g\sigma_1$). Therefore (cf. 6.1.17b),

$$\mathcal{A}(d_{\mathbb{R}}) := \mathcal{A}(\mathcal{T}_{d_{\mathbb{R}}}) \subset \mathcal{A}(\mathcal{I}_1).$$

Next, we notice that:

$\forall a, b \in \mathbb{R}, \ (a, b) = \cup_{n=1}^{\infty}[a + \frac{1}{n}, b)$, hence $\mathcal{I}_1 \subset \mathcal{A}(\mathcal{I}_2)$, whence $\mathcal{A}(\mathcal{I}_1) \subset \mathcal{A}(\mathcal{I}_2)$,

$\forall a, b \in \mathbb{R}, \ [a, b) = \cup_{n=1}^{\infty}[a, b - \frac{1}{n}]$, hence $\mathcal{I}_2 \subset \mathcal{A}(\mathcal{I}_3)$, whence $\mathcal{A}(\mathcal{I}_2) \subset \mathcal{A}(\mathcal{I}_3)$,

$\forall a, b \in \mathbb{R}, \ [a, b] = \cap_{n=1}^{\infty}(a - \frac{1}{n}, b]$, hence $\mathcal{I}_3 \subset \mathcal{A}(\mathcal{I}_4)$, whence $\mathcal{A}(\mathcal{I}_3) \subset \mathcal{A}(\mathcal{I}_4)$,

$\forall a, b \in \mathbb{R}, \ (a, b] = (-\infty, b] \cap (\mathbb{R} - (-\infty, a])$, hence $\mathcal{I}_4 \subset \mathcal{A}(\mathcal{I}_5)$, whence $\mathcal{A}(\mathcal{I}_4) \subset \mathcal{A}(\mathcal{I}_5)$,

$\forall a \in \mathbb{R}, \ (-\infty, a] = \cap_{n=1}^{\infty}(-\infty, a + \frac{1}{n})$, hence $\mathcal{I}_5 \subset \mathcal{A}(\mathcal{I}_6)$, whence $\mathcal{A}(\mathcal{I}_5) \subset \mathcal{A}(\mathcal{I}_6)$,

$\forall a \in \mathbb{R}, \ (-\infty, a) = \mathbb{R} - [a, \infty)$, hence $\mathcal{I}_6 \subset \mathcal{A}(\mathcal{I}_7)$, whence $\mathcal{A}(\mathcal{I}_6) \subset \mathcal{A}(\mathcal{I}_7)$,

$\forall a \in \mathbb{R}, \ [a, \infty) = \cap_{n=1}^{\infty}(a - \frac{1}{n}, \infty)$, hence $\mathcal{I}_7 \subset \mathcal{A}(\mathcal{I}_8)$, whence $\mathcal{A}(\mathcal{I}_7) \subset \mathcal{A}(\mathcal{I}_8)$,

$\forall a \in \mathbb{R}, \ (a, \infty) \in \mathcal{T}_{d_{\mathbb{R}}}$, i.e. $\mathcal{I}_8 \subset \mathcal{T}_{d_{\mathbb{R}}}$, whence $\mathcal{A}(\mathcal{I}_8) \subset \mathcal{A}(d_{\mathbb{R}})$.

This proves that $\mathcal{A}(d_{\mathbb{R}}) = \mathcal{A}(\mathcal{I}_1) = ... = \mathcal{A}(\mathcal{I}_8)$.

As to $\mathcal{I}_9$, recall that $(a, b] := \emptyset$ if $b < a$. Thus $\mathcal{I}_9$ has the property sa_1 of 6.1.1, and it is immediate to check that it has properties sa_2 and sa_3 as well. Moreover, from

$$\mathcal{I}_4 \subset \mathcal{A}(d_{\mathbb{R}}), \quad \mathcal{I}_5 \subset \mathcal{A}(d_{\mathbb{R}}), \quad \mathcal{I}_8 \subset \mathcal{A}(d_{\mathbb{R}}),$$

we have $\mathcal{I}_9 \subset \mathcal{A}(d_{\mathbb{R}})$, whence $\mathcal{A}(\mathcal{I}_9) \subset \mathcal{A}(d_{\mathbb{R}})$. And from $\mathcal{I}_4 \subset \mathcal{I}_9$ we have $\mathcal{A}(d_{\mathbb{R}}) = \mathcal{A}(\mathcal{I}_4) \subset \mathcal{A}(\mathcal{I}_9)$. This proves that $\mathcal{A}(d_{\mathbb{R}}) = \mathcal{A}(\mathcal{I}_9)$. $\square$

6.1.26 Proposition. *Define the following families of subsets of $\mathbb{R}^*$:*

$$\mathcal{I}_1^* := \mathcal{A}(d_{\mathbb{R}}) \cup \{\{-\infty\}, \{\infty\}\},$$
$$\mathcal{I}_2^* := \{(a, \infty] : a \in \mathbb{R}\},$$
$$\mathcal{I}_3^* := \{[a, \infty] : a \in \mathbb{R}\}.$$

Then we have

$$\mathcal{A}(\delta) = \mathcal{A}(\mathcal{I}_1^*) = \mathcal{A}(\mathcal{I}_2^*) = \mathcal{A}(\mathcal{I}_3^*).$$

Note that $\mathcal{A}(\delta), \mathcal{A}(\mathcal{I}_1^*), \mathcal{A}(\mathcal{I}_2^*), \mathcal{A}(\mathcal{I}_3^*)$ denote σ-algebras on $\mathbb{R}^*$, while $\mathcal{A}(d_{\mathbb{R}})$ is a σ-algebra on $\mathbb{R}$.

Proof. Consider the metric subspace of $(\mathbb{R}, d_{\mathbb{R}})$ that is defined by $[-\frac{\pi}{2}, \frac{\pi}{2}]$, i.e. the metric space $([-\frac{\pi}{2}, \frac{\pi}{2}], d_r)$, where d_r denotes the restriction of the distance $d_{\mathbb{R}}$ to $[-\frac{\pi}{2}, \frac{\pi}{2}] \times [-\frac{\pi}{2}, \frac{\pi}{2}]$. By the very definition of δ, the mapping φ of 5.2.1 is an isomorphism from the metric space $([-\frac{\pi}{2}, \frac{\pi}{2}], d_r)$ onto the metric space $(\mathbb{R}^*, \delta)$. Therefore, since $([-\frac{\pi}{2}, \frac{\pi}{2}], d_r)$ is separable by 2.3.16 and 2.3.20, $(\mathbb{R}^*, \delta)$ is separable as well (cf. 2.3.21c). Then, by 2.3.17, every element of $\mathcal{T}_\delta$ is the union of a countable family of open balls in $(\mathbb{R}^*, \delta)$. Since φ is an isomorphism, a subset of $\mathbb{R}^*$ is an open ball in $(\mathbb{R}^*, \delta)$ iff it is the image of an open ball in $([-\frac{\pi}{2}, \frac{\pi}{2}], d_r)$. Now, since the family of the open balls in $(\mathbb{R}, d_{\mathbb{R}})$ is the family $\mathcal{I}_1$ of 6.1.25, the family of the open balls in $([-\frac{\pi}{2}, \frac{\pi}{2}], d_r)$ is the family $\mathcal{I}_1^{[-\frac{\pi}{2}, \frac{\pi}{2}]}$ (cf. the proof of 2.2.5), i.e. the family

$$\left\{(a, b) : -\frac{\pi}{2} \le a < b \le \frac{\pi}{2}\right\} \cup \left\{[-\frac{\pi}{2}, a) : -\frac{\pi}{2} < a \le \frac{\pi}{2}\right\}$$
$$\cup \left\{(a, \frac{\pi}{2}] : -\frac{\pi}{2} \le a < \frac{\pi}{2}\right\} \cup \left\{[-\frac{\pi}{2}, \frac{\pi}{2}]\right\}.$$

Therefore, the family of the open balls in $(\mathbb{R}^*, \delta)$ is the family

$$\mathcal{I}^* := \{(a, b) : -\infty \le a < b \le \infty\} \cup \{[-\infty, a) : -\infty < a \le \infty\}$$
$$\cup \{(a, \infty] : -\infty \le a < \infty\} \cup \mathbb{R}^*.$$

Thus, every element of $\mathcal{T}_\delta$ is the union of a countable family of elements of $\mathcal{I}^*$, and this implies that $\mathcal{T}_\delta \subset \mathcal{A}(\mathcal{I}^*)$, and hence that $\mathcal{A}(\mathcal{T}_\delta) \subset \mathcal{A}(\mathcal{I}^*)$ ($\mathcal{A}(\mathcal{T}_\delta)$ and $\mathcal{A}(\mathcal{I}^*)$ denote σ-algebras on $\mathbb{R}^*$). Since $\mathcal{I}^* \subset \mathcal{T}_\delta$, we also have $\mathcal{A}(\mathcal{I}^*) \subset \mathcal{A}(\mathcal{T}_\delta)$. This proves that $\mathcal{A}(\delta) := \mathcal{A}(\mathcal{T}_\delta) = \mathcal{A}(\mathcal{I}^*)$.

Next, we notice that:

$$(a, b) \in \mathcal{A}(d_{\mathbb{R}}) \subset \mathcal{A}(\mathcal{I}_1^*), \text{ for } -\infty \le a < b \le \infty,$$
$$[-\infty, a) = \{-\infty\} \cup (-\infty, a) \in \mathcal{A}(\mathcal{I}_1^*), \text{ for } -\infty < a \le \infty,$$
$$(a, \infty] = (a, \infty) \cup \{\infty\} \in \mathcal{A}(\mathcal{I}_1^*), \text{ for } -\infty \le a < \infty,$$
$$\mathbb{R}^* = \{-\infty\} \cup \mathbb{R} \cup \{\infty\} \in \mathcal{A}(\mathcal{I}_1^*).$$

Thus, $\mathcal{I}^* \subset \mathcal{A}(\mathcal{I}_1^*)$, whence $\mathcal{A}(\mathcal{I}^*) \subset \mathcal{A}(\mathcal{I}_1^*)$.

Moreover,

$$(a, b] = (a, \infty] \cap (\mathbb{R}^* - (b, \infty]), \text{ for } a, b \in \mathbb{R},$$

and this proves that for the family $\mathcal{I}_4$ of 6.1.25 we have the inclusion $\mathcal{I}_4 \subset \mathcal{A}(\mathcal{I}_2^*)$, whence $\mathcal{A}(\mathcal{I}_4) \subset \mathcal{A}(\mathcal{I}_2^*)$ ($\mathcal{A}(\mathcal{I}_4)$ denotes the σ-algebra on $\mathbb{R}^*$ generated by $\mathcal{I}_4$). Now, $\mathbb{R} \in \mathcal{A}(\mathcal{I}_4)$ since $\mathbb{R} = \cup_{n=1}^\infty (-n, n]$, and hence $\mathcal{A}(\mathcal{I}_4)^{\mathbb{R}} \subset \mathcal{A}(\mathcal{I}_4)$ (cf. 6.1.19a); also, $\mathcal{A}(\mathcal{I}_4)^{\mathbb{R}}$ is the σ-algebra on $\mathbb{R}$ generated by $\mathcal{I}_4^{\mathbb{R}}$ (cf. 6.1.20); but $\mathcal{I}_4^{\mathbb{R}} = \mathcal{I}_4$, and hence $\mathcal{A}(\mathcal{I}_4)^{\mathbb{R}} = \mathcal{A}(d_{\mathbb{R}})$ (cf. 6.1.25). Thus, $\mathcal{A}(d_{\mathbb{R}}) \subset \mathcal{A}(\mathcal{I}_2^*)$. We also have

$$\{-\infty\} = \cap_{n=1}^\infty (\mathbb{R}^* - (-n, \infty]) \text{ and } \{\infty\} = \cap_{n=1}^\infty (n, \infty],$$

which proves that $\{\{-\infty\}, \{\infty\}\} \subset \mathcal{A}(\mathcal{I}_2^*)$. Thus we have $\mathcal{I}_1^* \subset \mathcal{A}(\mathcal{I}_2^*)$, whence $\mathcal{A}(\mathcal{I}_1^*) \subset \mathcal{A}(\mathcal{I}_2^*)$.

Since $\mathcal{I}_2^* \subset \mathcal{I}^*$, we also have $\mathcal{A}(\mathcal{I}_2^*) \subset \mathcal{A}(\mathcal{I}^*)$.

Summing up, we have proved that

$$\mathcal{A}(\mathcal{I}^*) \subset \mathcal{A}(\mathcal{I}_1^*) \subset \mathcal{A}(\mathcal{I}_2^*) \subset \mathcal{A}(\mathcal{I}^*),$$

which shows that

$$\mathcal{A}(\delta) = \mathcal{A}(\mathcal{I}^*) = \mathcal{A}(\mathcal{I}_1^*) = \mathcal{A}(\mathcal{I}_2^*).$$

Finally, we have

$$(a, \infty] = \cup_{n=1}^{\infty} [a + \frac{1}{n}, \infty] \in \mathcal{A}(\mathcal{I}_3^*), \forall a \in \mathbb{R},$$

$$[a, \infty] = \cap_{n=1}^{\infty} (a - \frac{1}{n}, \infty] \in \mathcal{A}(\mathcal{I}_2^*), \forall a \in \mathbb{R},$$

and this proves that $\mathcal{I}_2^* \subset \mathcal{A}(\mathcal{I}_3^*)$ and $\mathcal{I}_3^* \subset \mathcal{A}(\mathcal{I}_2^*)$, and hence that $\mathcal{A}(\mathcal{I}_2^*) = \mathcal{A}(\mathcal{I}_3^*)$.

$$\square$$

6.1.27 Definition. Let $N \in \mathbb{N}$ and, for $k = 1, ..., N$, let X_k be a non-empty set. For $k = 1, ..., N$, we define the mapping

$$\pi_k : X_1 \times \cdots \times X_N \to X_k$$
$$(x_1, ..., x_N) \mapsto \pi_k((x_1, ..., x_N)) := x_k,$$

which we will sometimes denote by π_{X_k} (thus, this definition generalizes the definition given in 1.2.6c). Notice that, for $k = 1, ..., N$ and $E_k \in \mathcal{P}(X_k)$,

$$\pi_k^{-1}(E_k) = X_1 \times \cdots \times X_{k-1} \times E_k \times X_{k+1} \times \cdots \times X_N.$$

6.1.28 Definition. Let $N \in \mathbb{N}$, let $(X_k, \mathcal{A}_k)$ be a measurable space for $k = 1, ..., N$, and let $\mathcal{F}$ be the family of subsets of $X_1 \times \cdots \times X_N$ defined by

$$\mathcal{F} := \left\{ \pi_k^{-1}(E_k) : k \in \{1, ..., N\}, E_k \in \mathcal{A}_k \right\}.$$

The σ-algebra on $X_1 \times \cdots \times X_N$ generated by $\mathcal{F}$ is called the *product σ-algebra* of the $\mathcal{A}_k$'s and is denoted by $\mathcal{A}_1 \otimes \cdots \otimes \mathcal{A}_N$, i.e. we define

$$\mathcal{A}_1 \otimes \cdots \otimes \mathcal{A}_N := \mathcal{A}(\mathcal{F}).$$

6.1.29 Proposition. *Let $N \in \mathbb{N}$ and, for $k = 1, ..., N$, let X_k be a non-empty set and $\mathcal{F}_k$ a family of subsets of X_k. Define the families of subsets of $X_1 \times \cdots \times X_N$:*

$$\mathcal{G} := \left\{ \pi_k^{-1}(S_k) : k \in \{1, ..., N\}, S_k \in \mathcal{F}_k \right\},$$
$$\mathcal{G}' := \{ S_1 \times \cdots \times S_N : S_k \in \mathcal{F}_k \text{ or } S_k = X_k \text{ for } k = 1, ..., N \}.$$

Then

$$\mathcal{A}(\mathcal{F}_1) \otimes \cdots \otimes \mathcal{A}(\mathcal{F}_N) = \mathcal{A}(\mathcal{G}) = \mathcal{A}(\mathcal{G}').$$

Note that $\mathcal{A}(\mathcal{F}_k)$ denotes a σ-algebra on X_k for $k = 1, ..., N$, while $\mathcal{A}(\mathcal{G})$ and $\mathcal{A}(\mathcal{G}')$ denote σ-algebras on $X_1 \times \cdots \times X_N$.

Proof. Define the family of subsets of $X_1 \times \cdots \times X_N$:

$$\mathcal{F} := \left\{ \pi_k^{-1}(E_k) : k \in \{1, ..., N\}, E_k \in \mathcal{A}(\mathcal{F}_k) \right\}.$$

Since $\mathcal{F}_k \subset \mathcal{A}(\mathcal{F}_k)$, we have $\mathcal{G} \subset \mathcal{F}$, and hence $\mathcal{A}(\mathcal{G}) \subset \mathcal{A}(\mathcal{F})$.
For $k = 1, ..., N$, define the family of subsets of X_k

$$\mathcal{E}_k := \left\{ T_k \in \mathcal{P}(X_k) : \pi_k^{-1}(T_k) \in \mathcal{A}(\mathcal{G}) \right\}.$$

Since $\mathcal{G} \subset \mathcal{A}(\mathcal{G})$, we have $\mathcal{F}_k \subset \mathcal{E}_k$. Moreover, $\mathcal{E}_k$ is a σ-algebra on X_k since (cf. 1.2.8)

$$\pi_k^{-1}(\cap_{n \in I} T_{k,n}) = \cap_{n \in I} \pi_k^{-1}(T_{k,n})$$

for every family $\{T_{k,n}\}_{n \in I}$ of subsets of X_k, and

$$\pi_k^{-1}(X_k - T_k) = X_1 \times \cdots \times X_N - \pi_k^{-1}(T_k)$$

for each subset T_k of X_k. Thus, we have $\mathcal{A}(\mathcal{F}_k) \subset \mathcal{E}_k$, which can be written as

$$\pi_k^{-1}(E_k) \in \mathcal{A}(\mathcal{G}), \forall E_k \in \mathcal{A}(\mathcal{F}_k),$$

or equivalently as $\mathcal{F} \subset \mathcal{A}(\mathcal{G})$, which implies $\mathcal{A}(\mathcal{F}) \subset \mathcal{A}(\mathcal{G})$. Therefore,

$$\mathcal{A}(\mathcal{F}_1) \otimes \cdots \otimes \mathcal{A}(\mathcal{F}_N) = \mathcal{A}(\mathcal{F}) = \mathcal{A}(\mathcal{G}).$$

As to the second equality we must prove, we notice that $\mathcal{G} \subset \mathcal{G}'$, and hence $\mathcal{A}(\mathcal{G}) \subset \mathcal{A}(\mathcal{G}')$. We notice also that, for each $(S_1, ..., S_N) \in \mathcal{P}(X_1) \times \cdots \times \mathcal{P}(X_N)$,

$$S_1 \times \cdots \times S_N = \cap_{k \in I} \pi_k^{-1}(S_k) \text{ with } I := \{k \in \{1, ..., N\} : S_k \neq X_k\},$$

and hence $\mathcal{G}' \subset \mathcal{A}(\mathcal{G})$, and hence $\mathcal{A}(\mathcal{G}') \subset \mathcal{A}(\mathcal{G})$. $\qquad \square$

6.1.30 Proposition. *Let $N \in \mathbb{N}$ and, for $k = 1, ..., N$, let $(X_k, \mathcal{A}_k)$ be a measurable space.*

(a) The family of subsets of $X_1 \times \cdots \times X_N$ defined by

$$\mathcal{S}_N := \{E_1 \times \cdots \times E_N : E_k \in \mathcal{A}_k \text{ for } k = 1, ..., N\}$$

is a semialgebra on $X_1 \times \cdots \times X_N$ and $\mathcal{A}_1 \otimes \cdots \otimes \mathcal{A}_N = \mathcal{A}(\mathcal{S}_N)$.
(b) If we identify $X_1 \times \cdots \times X_N$ with $((\cdots (X_1 \times X_2) \times \cdots) \times X_{N-1}) \times X_N$, then

$$\mathcal{A}_1 \otimes \cdots \otimes \mathcal{A}_N = ((\cdots (\mathcal{A}_1 \otimes \mathcal{A}_2) \otimes \cdots) \otimes \mathcal{A}_{N-1}) \otimes \mathcal{A}_N.$$

(c) Let Y_k be a non-empty subset of X_k for each $k = 1, ..., N$. Then

$$(\mathcal{A}_1 \otimes \cdots \otimes \mathcal{A}_N)^{Y_1 \times \cdots \times Y_N} = \mathcal{A}_1^{Y_1} \otimes \cdots \otimes \mathcal{A}_N^{Y_N}.$$

Proof. a: We have

$$\emptyset \times \cdots \times \emptyset = \emptyset;$$

$$\forall E_1 \times \cdots \times E_N, F_1 \times \cdots \times F_N \in \mathcal{S}_N,$$

$$(E_1 \times \cdots \times E_N) \cap (F_1 \times \cdots \times F_N) = (E_1 \cap F_1) \times \cdots \times (E_N \cap F_N);$$

$$\forall E_1 \times \cdots \times E_N \in \mathcal{S}_N,$$

$$(X_1 \times \cdots \times X_N) - (E_1 \times \cdots \times E_N)$$

$$= \cup_{k=1}^{N} (X_1 \times \cdots \times X_{k-1} \times (X_k - E_k) \times X_{k+1} \times \cdots \times X_N).$$

This shows that $\mathcal{S}_N$ is a semialgebra on $X_1 \times \cdots \times X_N$. From 6.1.29, with $\mathcal{G}' := \mathcal{S}_N$, we have $\mathcal{A}_1 \otimes \cdots \otimes \mathcal{A}_N = \mathcal{A}(\mathcal{S}_N)$ since obviously $\mathcal{A}(\mathcal{A}_k) = \mathcal{A}_k$.

b: We will prove $\mathcal{A}_1 \otimes \cdots \otimes \mathcal{A}_N = (\mathcal{A}_1 \otimes \cdots \otimes \mathcal{A}_{N-1}) \otimes \mathcal{A}_N$. The result will then follow by induction. From 6.1.29, with

$$\mathcal{G}' := \{F \times E_N : F \in \mathcal{S}_{N-1}, E_N \in \mathcal{A}_N\},$$

and from part a (with N replaced by $N-1$) we have $(\mathcal{A}_1 \otimes \cdots \otimes \mathcal{A}_{N-1}) \otimes \mathcal{A}_N = \mathcal{A}(\mathcal{G}')$. But if we identify $X_1 \times \cdots \times X_N$ with $((\cdots (X_1 \times X_2) \times \cdots) \times X_{N-1}) \times X_N$, $\mathcal{G}'$ gets identified with $\mathcal{S}_N$ and from part a we have $(\mathcal{A}_1 \otimes \cdots \otimes \mathcal{A}_{N-1}) \otimes \mathcal{A}_N = \mathcal{A}_1 \otimes \cdots \otimes \mathcal{A}_N$.

c: We have

$$(\mathcal{A}_1 \otimes \cdots \otimes \mathcal{A}_N)^{Y_1 \times \cdots \times Y_N} \overset{(1)}{=} \mathcal{A}(\mathcal{S}_N)^{Y_1 \times \cdots \times Y_N}$$
$$\overset{(2)}{=} \mathcal{A}(\mathcal{S}_N^{Y_1 \times \cdots \times Y_N}) \overset{(3)}{=} \mathcal{A}_1^{Y_1} \otimes \cdots \otimes \mathcal{A}_N^{Y_N},$$

where 1 holds by part a, 2 holds by 6.1.20, 3 holds by part a (with $(X_k, \mathcal{A}_k)$ replaced by $(Y_k, \mathcal{A}_k^{Y_k})$) since

$$\mathcal{S}_N^{Y_1 \times \cdots \times Y_N} = \{(E_1 \times \cdots \times E_N) \cap (Y_1 \times \cdots \times Y_N) : E_k \in \mathcal{A}_k \text{ for } k = 1, ..., N\}$$
$$= \{(E_1 \cap Y_1) \times \cdots \times (E_N \cap Y_N) : E_k \in \mathcal{A}_k \text{ for } k = 1, ..., N\}$$
$$= \left\{G_1 \times \cdots \times G_N : G_k \in \mathcal{A}_k^{Y_k} \text{ for } k = 1, ..., N\right\}.$$

$\square$

6.1.31 Proposition. *Let $N \in \mathbb{N}$ and, for $k = 1, ..., N$, let (X_k, d^k) be a metric space. Let (X, d) be the product (defined associatively) of the N metric spaces (X_k, d^k), i.e. $X := X_1 \times \cdots \times X_N$ and*

$$d : X \times X \to \mathbb{R}$$
$$((x_1, ..., x_N), (y_1, ..., y_N)) \mapsto d((x_1, ..., x_N), (y_1, ..., y_N))$$
$$:= \sqrt{\sum_{k=1}^{N} d^k(x_k, y_k)^2}.$$

Then

$$\mathcal{A}(d^1) \otimes \cdots \mathcal{A}(d^N) \subset \mathcal{A}(d).$$

If (X_k, d^k) is separable for each $k \in \{1, ..., N\}$, then

$$\mathcal{A}(d^1) \otimes \cdots \mathcal{A}(d^N) = \mathcal{A}(d).$$

Proof. Define the family of subsets of $X_1 \times \cdots \times X_N$

$$\mathcal{G} := \left\{\pi_k^{-1}(G_k) : k = \{1, ..., N\}, G_k \in \mathcal{T}_{d^k}\right\}.$$

By 6.1.29 we have

$$\mathcal{A}(d^1) \otimes \cdots \mathcal{A}(d^N) = \mathcal{A}(\mathcal{G}).$$

Since π_k is continuous (cf. e.g. 2.4.2), we have (cf. 2.4.3) $\mathcal{G} \subset \mathcal{T}_d$, whence

$$\mathcal{A}(d^1) \otimes \cdots \mathcal{A}(d^N) = \mathcal{A}(\mathcal{G}) \subset \mathcal{A}(\mathcal{T}_d) =: \mathcal{A}(d).$$

Now define the function

$$\rho : X \times X \to \mathbb{R}$$

$$((x_1, ..., x_N), (y_1, ..., y_N)) \mapsto \rho((x_1, ..., x_N), (y_1, ..., y_N))$$

$$:= \max \left\{ d^k(x_k, y_k) : k \in \{1, ..., N\} \right\}.$$

It is easy to see that ρ is a distance on X since properties di_1, di_2, di_3 of 2.1.1 for the function ρ follow immediately from the corresponding properties for the functions d^k. It is also immediate to see that, for a sequence $\{(x_{1,n}, ..., x_{N,n})\}$ in X and for $(x_1, ..., x_N) \in X$,

$$\rho((x_{1,n}, ..., x_{N,n}), (x_1, ..., x_N)) \to 0 \text{ as } n \to \infty \Leftrightarrow$$

$$(d^k(x_{k,n}, x_k) \to 0 \text{ as } n \to \infty, \forall k \in \{1, ..., N\}) \Leftrightarrow$$

$$d((x_{1,n}, ..., x_{N,n}), (x_1, ..., x_N)) \to 0 \text{ as } n \to \infty.$$

By 2.3.4, this implies $\mathcal{K}_\rho = \mathcal{K}_d$, and hence $\mathcal{T}_\rho = \mathcal{T}_d$.

Suppose now that (X_k, d^k) is separable for each $k \in \{1, ..., N\}$. Then, by 2.7.3c, (X, d) is separable. Since $\mathcal{K}_\rho = \mathcal{K}_d$, (X, ρ) is separable as well. By 2.3.17, this implies that every element of $\mathcal{T}_\rho$ is a countable union of open balls in X defined with respect to ρ. But for any such ball $B_\rho((x_1, ..., x_N), r)$ we have (denoting by $B_{d^k}(x_k, r)$ a ball in X_k defined with respect to d^k):

$$B_\rho((x_1, ..., x_N), r) = B_{d^1}(x_1, r) \times \cdots \times B_{d^N}(x_N, r)$$

$$= \cap_{k=1}^N \pi_k^{-1}(B_{d^k}(x_k, r)) \in \mathcal{A}(\mathcal{G}).$$

This proves the inclusion $\mathcal{T}_\rho \subset \mathcal{A}(\mathcal{G})$, i.e. $\mathcal{T}_d \subset \mathcal{A}(\mathcal{G})$, and therefore also the inclusion $\mathcal{A}(d) := \mathcal{A}(\mathcal{T}_d) \subset \mathcal{A}(\mathcal{G})$. In view of the inclusion $\mathcal{A}(\mathcal{G}) \subset \mathcal{A}(d)$ proved above, this shows that

$$\mathcal{A}(d^1) \otimes \cdots \mathcal{A}(d^N) = \mathcal{A}(\mathcal{G}) = \mathcal{A}(d).$$

$\square$

6.1.32 Corollary. *For $n \in \mathbb{N}$,*

$$\mathcal{A}(d_n) = \mathcal{A}(d_\mathbb{R}) \otimes \cdots n \text{ times} \cdots \otimes \mathcal{A}(d_\mathbb{R})$$

(d_n is the distance on $\mathbb{R}^n$ defined in 2.7.4b).

Proof. Use 6.1.31 and 2.3.16.

$\square$

6.1.33 Proposition. *Define the following family of subsets of $\mathbb{C}$:*

$$\mathcal{R} := \{(a, b) \times (c, d) : a, b, c, d \in \mathbb{R}\}.$$

Then we have

$$\mathcal{A}(d_\mathbb{C}) = \mathcal{A}(\mathcal{R}).$$

Proof. Since $d_{\mathbb{C}} = d_2$ (cf. 2.7.4b), by 6.1.32 we have

$$\mathcal{A}(d_{\mathbb{C}}) = \mathcal{A}(d_{\mathbb{R}}) \otimes \mathcal{A}(d_{\mathbb{R}}).$$

From 6.1.25 we also have

$$\mathcal{A}(d_{\mathbb{R}}) = \mathcal{A}(\mathcal{I}_1), \text{ with } \mathcal{I}_1 := \{(a, b) : a, b \in \mathbb{R}\}.$$

Then by 6.1.29 we have $\mathcal{A}(d_{\mathbb{C}}) = \mathcal{A}(\mathcal{G})$ if we define

$$\mathcal{G} := \{(a, b) \times \mathbb{R}, \mathbb{R} \times (c, d) : a, b, c, d \in \mathbb{R}\}.$$

Now, the equality

$$(a, b) \times (c, d) = ((a, b) \times \mathbb{R}) \cap (\mathbb{R} \times (c, d))$$

shows that $\mathcal{R} \subset \mathcal{A}(\mathcal{G})$, and this implies that $\mathcal{A}(\mathcal{R}) \subset \mathcal{A}(\mathcal{G})$. Moreover, the equalities

$$(a, b) \times \mathbb{R} = \cup_{n=1}^{\infty}(a, b) \times (-n, n) \text{ and } \mathbb{R} \times (c, d) = \cup_{n=1}^{\infty}(-n, n) \times (c, d)$$

show that $\mathcal{G} \subset \mathcal{A}(\mathcal{R})$, and this implies that $\mathcal{A}(\mathcal{G}) \subset \mathcal{A}(\mathcal{R})$. Thus, we have $\mathcal{A}(d_{\mathbb{C}}) = \mathcal{A}(\mathcal{G}) = \mathcal{A}(\mathcal{R})$. $\qquad\square$

6.1.34 Definition. A non-empty collection $\mathcal{C}$ of subsets of X is called a *monotone class on X* if the following conditions are satisfied:

(mo_1) if $\{E_n\}$ is a sequence in $\mathcal{C}$ such that $E_n \subset E_{n+1}$ for each $n \in \mathbb{N}$, then $\cup_{n=1}^{\infty} E_n \in \mathcal{C}$;

(mo_2) if $\{E_n\}$ is a sequence in $\mathcal{C}$ such that $E_{n+1} \subset E_n$ for each $n \in \mathbb{N}$, then $\cap_{n=1}^{\infty} E_n \in \mathcal{C}$.

6.1.35 Remarks. Every σ-algebra on X is a monotone class on X (cf. σa_1 in 6.1.13 and σa_2 in 6.1.14).

Proceeding as in 6.1.16, it is easy to see that, if Λ is a family of monotone classes on X, then $\cap_{\mathcal{C} \in \Lambda} \mathcal{C}$ (this intersection is defined within the framework of $\mathcal{P}(X)$) is a monotone class on X. Proceeding as in 6.1.10a it is then easy to see that, if $\mathcal{F}$ is a family of subsets of X, then there exists a unique monotone class on X, which is called the *monotone class on X generated by $\mathcal{F}$* and is denoted by $\mathcal{C}(\mathcal{F})$, such that

(gm_1) $\mathcal{F} \subset \mathcal{C}(\mathcal{F})$,

(gm_2) If $\mathcal{C}$ is a monotone class on X and $\mathcal{F} \subset \mathcal{C}$, then $\mathcal{C}(\mathcal{F}) \subset \mathcal{C}$.

If $\mathcal{F}$ is the empty family, then $\mathcal{C}(\mathcal{F}) = \{\emptyset, X\}$.

6.1.36 Theorem. *If $\mathcal{A}_0$ is an algebra on X, then the monotone class $\mathcal{C}(\mathcal{A}_0)$ generated by $\mathcal{A}_0$ is the same as the σ-algebra $\mathcal{A}(\mathcal{A}_0)$ generated by $\mathcal{A}_0$, i.e. $\mathcal{C}(\mathcal{A}_0) = \mathcal{A}(\mathcal{A}_0)$.*

Proof. Since a σ-algebra is always a monotone class and $\mathcal{A}_0 \subset \mathcal{A}(\mathcal{A}_0)$ (cf. $g\sigma_1$ in 6.1.17), we have $\mathcal{C}(\mathcal{A}_0) \subset \mathcal{A}(\mathcal{A}_0)$ (cf. gm_2). Hence it is sufficient to show that $\mathcal{C}(\mathcal{A}_0)$ is a σ-algebra, because then from $\mathcal{A}_0 \subset \mathcal{C}(\mathcal{A}_0)$ (cf. gm_1) we can derive $\mathcal{A}(\mathcal{A}_0) \subset \mathcal{C}(\mathcal{A}_0)$ (cf. $g\sigma_2$ in 6.1.17). Besides, if we prove that $\mathcal{C}(\mathcal{A}_0)$ is an algebra on

X, then for any sequence $\{E_n\}$ in $\mathcal{C}(\mathcal{A}_0)$ we have $\cup_{n=1}^{N} E_n \in \mathcal{C}(\mathcal{A}_0)$ for each $N \in \mathbb{N}$, and hence $\cup_{n=1}^{\infty} E_n = \cup_{N=1}^{\infty}(\cup_{n=1}^{N} E_n) \in \mathcal{C}(\mathcal{A}_0)$ by property mo_1 of $\mathcal{C}(\mathcal{A}_0)$, and we can conclude that $\mathcal{C}(\mathcal{A}_0)$ is a σ-algebra on X.

For each $E \in \mathcal{C}(\mathcal{A}_0)$ we define a collection $\mathcal{C}(E)$ of subsets of X by

$$\mathcal{C}(E) := \{F \in \mathcal{C}(\mathcal{A}_0) : E - F, F - E, E \cap F \in \mathcal{C}(\mathcal{A}_0)\}.$$

Clearly, $\emptyset \in \mathcal{C}(E)$ (since $\emptyset \in \mathcal{A}_0 \subset \mathcal{C}(\mathcal{A}_0)$) and hence $\mathcal{C}(E)$ is not empty. Moreover, if $\{F_n\}$ is a sequence in $\mathcal{C}(E)$ such that $F_n \subset F_{n+1}$ for each $n \in \mathbb{N}$, then:

$$E - (\cup_{n=1}^{\infty} F_n) = E \cap (\cap_{n=1}^{\infty}(X - F_n)) = \cap_{n=1}^{\infty}(E - F_n) \in \mathcal{C}(\mathcal{A}_0)$$

by property mo_2 of $\mathcal{C}(\mathcal{A}_0)$, since $E - F_{n+1} \subset E - F_n$ for each $n \in \mathbb{N}$;

$$(\cup_{n=1}^{\infty} F_n) - E = \cup_{n=1}^{\infty}(F_n - E) \in \mathcal{C}(\mathcal{A}_0)$$

by property mo_1 of $\mathcal{C}(\mathcal{A}_0)$, since $F_n - E \subset F_{n+1} - E$ for each $n \in \mathbb{N}$;

$$E \cap (\cup_{n=1}^{\infty} F_n) = \cup_{n=1}^{\infty}(E \cap F_n) \in \mathcal{C}(\mathcal{A}_0)$$

by property mo_1 of $\mathcal{C}(\mathcal{A}_0)$, since $E \cap F_n \subset E \cap F_{n+1}$ for each $n \in \mathbb{N}$.

This proves property mo_1 for $\mathcal{C}(E)$. Property mo_2 can be proved for $\mathcal{C}(E)$ in a similar way. Thus, $\mathcal{C}(E)$ is a monotone class.

For each $E \in \mathcal{A}_0$, it is clear (from 6.1.6 and $\mathcal{A}_0 \subset \mathcal{C}(\mathcal{A}_0)$) that $\mathcal{A}_0 \subset \mathcal{C}(E)$, so that $\mathcal{C}(\mathcal{A}_0) \subset \mathcal{C}(E)$ (cf. gm_2). Hence, for each $F \in \mathcal{C}(\mathcal{A}_0)$, we have

$$E \in \mathcal{A}_0 \Rightarrow F \in \mathcal{C}(E) \Rightarrow E \in \mathcal{C}(F),$$

where the second implication follows from the symmetry of the definition of $\mathcal{C}(E)$. This proves that, for each $F \in \mathcal{C}(\mathcal{A}_0)$, $\mathcal{A}_0 \subset \mathcal{C}(F)$ and hence $\mathcal{C}(\mathcal{A}_0) \subset \mathcal{C}(F)$ (cf. gm_2). Thus, if $E, F \in \mathcal{C}(\mathcal{A}_0)$ then $E \in \mathcal{C}(F)$ and hence $F - E$ and $F \cap E$ are elements of $\mathcal{C}(\mathcal{A}_0)$. Since $X \in \mathcal{A}_0 \subset \mathcal{C}(\mathcal{A}_0)$, this implies that $\mathcal{C}(\mathcal{A}_0)$ is an algebra on X:

if $E \in \mathcal{C}(\mathcal{A}_0)$, then $X - E \in \mathcal{C}(\mathcal{A}_0)$;
if $E, F \in \mathcal{C}(\mathcal{A}_0)$, then $X - E, X - F \in \mathcal{C}(\mathcal{A}_0)$, and then
$E \cup F = X - ((X - E) \cap (X - F)) \in \mathcal{C}(\mathcal{A}_0)$.

This completes the proof. $\qquad\square$

6.2 Measurable mappings

6.2.1 Definition. Let $(X_1, \mathcal{A}_1)$ and $(X_2, \mathcal{A}_2)$ be measurable spaces, i.e. let X_1, X_2 be non-empty sets and let $\mathcal{A}_1, \mathcal{A}_2$ be σ-algebras on X_1, X_2 respectively. A mapping $\varphi : X_1 \to X_2$ is said to be *measurable w.r.t. (with respect to)* $\mathcal{A}_1$ *and* $\mathcal{A}_2$ (or, simply, *measurable* when no confusion can occur) if the following condition holds:

$$\varphi^{-1}(E) \in \mathcal{A}_1, \forall E \in \mathcal{A}_2.$$

6.2.2 Remark. Let X_1, X_2 be non-empty sets, and assume that $\varphi : X_1 \to X_2$ is a constant mapping, i.e. that there is $x_2 \in X$ such that $\varphi(x) = x_2$ for all $x \in X_1$. Then φ is measurable w.r.t. any σ-algebras $\mathcal{A}_1$ on X_1 and $\mathcal{A}_2$ on X_2, since the only possible counterimages under φ are $\emptyset$ and X_1, and $\emptyset, X_1 \in \mathcal{A}_1$ for any σ-algebra $\mathcal{A}_1$ on X_1 (cf. al_4 in 6.1.6).

6.2.3 Proposition. *Let $(X_1, \mathcal{A}_1), (X_2, \mathcal{A}_2)$ be measurable spaces, and suppose that $\varphi : X_1 \to X_2$ is a measurable mapping w.r.t. $\mathcal{A}_1$ and $\mathcal{A}_2$. If Y is a non-empty subset of X_1, the restriction φ_Y of φ to Y is measurable w.r.t. $\mathcal{A}_1^Y$ and $\mathcal{A}_2$.*

Proof. Notice that

$$\varphi_Y^{-1}(S) = \varphi^{-1}(S) \cap Y, \forall S \in \mathcal{P}(Y).$$

Thus,

$$\varphi_Y^{-1}(E) \in \mathcal{A}_1^Y, \forall E \in \mathcal{A}_2.$$

$\square$

6.2.4 Proposition. *Let $(X_1, \mathcal{A}_1)$ and $(X_2, \mathcal{A}_2)$ be measurable spaces. For a mapping $\varphi : X_1 \to X_2$, let Y be a subset of X_2 such that $R_\varphi \subset Y$. Then the final set X_2 can be replaced by Y, i.e. we can consider φ as $\varphi : X_1 \to Y$ (cf. 1.2.1). However, the following are equivalent conditions:*

(a) φ is measurable w.r.t. $\mathcal{A}_1$ and $\mathcal{A}_2$;
(b) φ is measurable w.r.t. $\mathcal{A}_1$ and $\mathcal{A}_2^Y$.

Proof. Notice that

$$\varphi^{-1}(S) = \varphi^{-1}(S) \cap X_1 = \varphi^{-1}(S) \cap \varphi^{-1}(Y) = \varphi^{-1}(S \cap Y), \forall S \in \mathcal{P}(Y).$$

Thus,

$$\varphi^{-1}(E) \in \mathcal{A}_1, \forall E \in \mathcal{A}_1,$$

is equivalent to

$$\varphi^{-1}(E) \in \mathcal{A}_1, \forall E \in \mathcal{A}_2^Y.$$

$\square$

6.2.5 Theorem. *Let $(X_1, \mathcal{A}_1)$, $(X_2, \mathcal{A}_2)$, $(X_3, \mathcal{A}_3)$ be measurable spaces, and let $\varphi : X_1 \to X_2$ be a measurable mapping w.r.t. $\mathcal{A}_1$ and $\mathcal{A}_2$ and $\psi : X_2 \to X_3$ a measurable mapping w.r.t. $\mathcal{A}_2$ and $\mathcal{A}_3$. Then $\psi \circ \varphi$ is a measurable mapping w.r.t $\mathcal{A}_1$ and $\mathcal{A}_3$.*

Proof. Use the definition of measurable mapping (cf. 6.2.1) and (cf. 1.2.13f)

$$(\psi \circ \varphi)^{-1}(S) = \varphi^{-1}(\psi^{-1}(S)), \forall S \in \mathcal{P}(X_3).$$

$\square$

6.2.6 Corollary. *Let* $(X_1, \mathcal{A}_1)$, $(X_2, \mathcal{A}_2)$, $(X_3, \mathcal{A}_3)$ *be measurable spaces, let* φ *be a mapping from* X_1 *to* X_2 *(i.e.* $\varphi : D_\varphi \to X_2$ *with* $D_\varphi \subset X_1$*), and let* ψ *be a mapping from* X_2 *to* X_3 *(i.e.* $\psi : D_\psi \to X_3$ *with* $D_\psi \subset X_2$*). Suppose that* φ *is measurable w.r.t.* $\mathcal{A}_1^{D_\varphi}$ *and* $\mathcal{A}_2$ *and that* ψ *is measurable w.r.t.* $\mathcal{A}_2^{D_\psi}$ *and* $\mathcal{A}_3$*. Then* $\psi \circ \varphi$ *is measurable w.r.t.* $\mathcal{A}_1^{D_{\psi \circ \varphi}}$ *and* $\mathcal{A}_3$*.*

Proof. Letting $D := D_{\psi \circ \varphi}$, we have $\psi \circ \varphi = \psi \circ \varphi_D$; in fact

$$R_{\varphi_D} = \varphi_D(D) = \varphi(\varphi^{-1}(D_\psi)) \subset D_\psi,$$

and this implies (cf. 1.2.13d) that $D_{\psi \circ \varphi_D} = D_{\varphi_D} = D_{\psi \circ \varphi}$; moreover,

$$(\psi \circ \varphi_D)(x) = \psi(\varphi(x)) = (\psi \circ \varphi)(x), \forall x \in D_{\psi \circ \varphi}.$$

Next, from 6.2.3 we have that φ_D is measurable w.r.t. $(\mathcal{A}_1^{D_\varphi})^D$ and $\mathcal{A}_2$, hence (cf. 6.1.19b) w.r.t. $\mathcal{A}_1^D$ and $\mathcal{A}_2$. Besides, since $R_{\varphi_D} \subset D_\psi$, from 6.2.4 we have that φ_D can be considered as a mapping $\varphi_D : D \to D_\psi$ which is measurable w.r.t. $\mathcal{A}_1^D$ and $\mathcal{A}_2^{D_\psi}$. Then $\psi \circ \varphi_D$ is measurable w.r.t. $\mathcal{A}_1^D$ and $\mathcal{A}_3$ by 6.2.5. $\qquad \square$

6.2.7 Theorem. *Let* $(X_1, \mathcal{A}_1)$ *be a measurable space,* X_2 *a non-empty set, and* $\mathcal{F}$ *a family of subsets of* X_2*. For a mapping* $\varphi : X_1 \to X_2$ *the following are equivalent conditions:*

(a) φ *is measurable w.r.t* $\mathcal{A}_1$ *and* $\mathcal{A}(\mathcal{F})$*;*
(b) $\varphi^{-1}(F) \in \mathcal{A}_1, \forall F \in \mathcal{F}$*.*

Proof. $a \Rightarrow b$: Recall that $\mathcal{F} \subset \mathcal{A}(\mathcal{F})$.

$b \Rightarrow a$: Assume condition b and define

$$\mathcal{A}_\varphi := \{E \in \mathcal{P}(X_2) : \varphi^{-1}(E) \in \mathcal{A}_1\}.$$

First we prove that $\mathcal{A}_\varphi$ is a σ-algebra on X_2, by showing that it has properties al_1 and al_2 of 6.1.5 and σa_1 of 6.1.13.

al_1 and σa_1: Let $\{E_n\}_{n \in I}$ be a family of elements of $\mathcal{A}_\varphi$, with $I := \{1, 2\}$ or $I := \mathbb{N}$. Then $\varphi^{-1}\left(\bigcup_{n \in I} E_n\right) = \bigcup_{n \in I} \varphi^{-1}(E_n) \in \mathcal{A}_1$, and this proves that $\bigcup_{n \in I} E_n \in \mathcal{A}_\varphi$.

al_2: If $E \in \mathcal{A}_\varphi$ then $\varphi^{-1}(X_2 - E) = X_1 - \varphi^{-1}(E) \in \mathcal{A}_1$, and this proves that $X_2 - E \in \mathcal{A}_\varphi$.

Next we notice that condition b means that $\mathcal{F} \subset \mathcal{A}_\varphi$. Thus $\mathcal{A}(\mathcal{F}) \subset \mathcal{A}_\varphi$, and this means that

$$\varphi^{-1}(F) \in \mathcal{A}_1, \forall F \in \mathcal{A}(\mathcal{F}),$$

which is condition a. $\qquad \square$

6.2.8 Corollary. *Let* (X_1, d^1) *and* (X_2, d^2) *be metric spaces.*

If a mapping $\varphi : X_1 \to X_2$ *is continuous then it is measurable w.r.t.* $\mathcal{A}(d^1)$ *and* $\mathcal{A}(d^2)$*.*

Proof. If φ is continuous, then (cf. 2.4.3)

$$\varphi^{-1}(E) \in \mathcal{T}_{d^1}, \text{ and hence } \varphi^{-1}(E) \in \mathcal{A}(d^1), \forall E \in \mathcal{T}_{d^2}.$$

Therefore, φ is measurable w.r.t. $\mathcal{A}(d^1)$ and $\mathcal{A}(d^2)$ by 6.2.7, since $\mathcal{A}(d^2) = \mathcal{A}(\mathcal{T}_{d^2})$.
$\square$

6.2.9 Proposition. *Let $N \in \mathbb{N}$ and let $(X_k, \mathcal{A}_k)$, $(Y_k, \mathcal{B}_k)$ be measurable spaces for $k = 1, ..., N$. For $k = 1, ..., N$, let $\varphi_k : X_k \to Y_k$ be a measurable mapping w.r.t. $\mathcal{A}_k$ and $\mathcal{B}_k$.*

(a) The mapping

$$\varphi_1 \times \cdots \times \varphi_N : X_1 \times \cdots \times X_N \to Y_1 \times \cdots \times Y_N$$

$$(x_1, ..., x_N) \mapsto (\varphi_1 \times \cdots \times \varphi_N)(x_1, ..., x_N)$$

$$:= (\varphi_1(x_1), ..., \varphi_N(x_N))$$

is measurable w.r.t. $\mathcal{A}_1 \otimes \cdots \otimes \mathcal{A}_N$ and $\mathcal{B}_1 \otimes \cdots \otimes \mathcal{B}_N$.

(b) If $(Z, \mathcal{C})$ is a measurable space and $\rho : Y_1 \times \cdots \times Y_N \to Z$ is a measurable mapping w.r.t. $\mathcal{B}_1 \otimes \cdots \otimes \mathcal{B}_N$ and $\mathcal{C}$, then the mapping

$$\chi : X_1 \times \cdots \times X_N \to Z$$

$$(x_1, ..., x_N) \mapsto \chi(x_1, ..., x_N) := \rho(\varphi_1(x_1), ..., \varphi_N(x_N))$$

is measurable w.r.t. $\mathcal{A}_1 \otimes \cdots \otimes \mathcal{A}_N$ and $\mathcal{C}$.

Proof. a: By 6.1.30a we have $E_1 \times \cdots \times E_N \in \mathcal{A}_1 \otimes \cdots \otimes \mathcal{A}_N$ if $E_k \in \mathcal{A}_k$ for $k = 1, ..., N$. Then by the measurability of the φ_k's we have

$$[F_k \in \mathcal{B}_k, \forall k \in \{1, ..., N\}] \Rightarrow$$

$$(\varphi_1 \times \cdots \times \varphi_N)^{-1}(F_1 \times \cdots \times F_N) = \varphi_1^{-1}(E_1) \times \cdots \times \varphi_N^{-1}(E_N) \in \mathcal{A}_1 \otimes \cdots \otimes \mathcal{A}_N.$$

By 6.1.30a and 6.2.7, this proves the measurability of $\varphi_1 \times \cdots \times \varphi_N$.

b: This follows at once from part a and 6.2.5, since $\chi = \rho \circ (\varphi_1 \times \cdots \times \varphi_N)$. $\square$

6.2.10 Proposition. *Let $(X, \mathcal{A})$ be a measurable space, let $N \in \mathbb{N}$, and let $(Y_k, \mathcal{B}_k)$ be a measurable space for $k = 1, ..., N$. For $k = 1, ..., N$, let $\varphi_k : X \to Y_k$ be a measurable mapping w.r.t. $\mathcal{A}$ and $\mathcal{B}_k$.*

(a) The mapping

$$\varphi : X \to Y_1 \times \cdots \times Y_N$$

$$x \mapsto \varphi(x) := (\varphi_1(x), ..., \varphi_N(x))$$

is measurable w.r.t. $\mathcal{A}$ and $\mathcal{B}_1 \otimes \cdots \otimes \mathcal{B}_N$.

(b) If $(Z, \mathcal{C})$ is a measurable space and $\rho : Y_1 \times \cdots \times Y_N \to Z$ is a measurable mapping w.r.t. $\mathcal{B}_1 \otimes \cdots \otimes \mathcal{B}_N$ and $\mathcal{C}$, then the mapping

$$\chi : X \to Z$$

$$x \mapsto \chi(x) := \rho(\varphi_1(x), ..., \varphi_N(x))$$

is measurable w.r.t. $\mathcal{A}$ and $\mathcal{C}$.

Proof. a: Define the mapping

$$\iota : X \to X \times \cdots \; n \text{ times } \cdots \times X$$

$$x \mapsto \iota(x) := (x, ..., x, ..., x).$$

We have

$$(E_k \in \mathcal{A}, \forall k \in \{1, ..., N\}) \Rightarrow \iota^{-1}(E_1 \times \cdots \times E_N) = \bigcap_{k=1}^{N} E_k \in \mathcal{A}.$$

By 6.1.30a and 6.2.7, this proves that the mapping ι is measurable w.r.t. $\mathcal{A}$ and $\mathcal{A} \otimes \cdots \; n \text{ times } \cdots \otimes \mathcal{A}$. Now, by 6.2.9a the mapping $\varphi_1 \times \cdots \times \varphi_N$ is measurable w.r.t. $\mathcal{A} \otimes \cdots \; n \text{ times } \cdots \otimes \mathcal{A}$ and $\mathcal{B}_1 \otimes \cdots \otimes \mathcal{B}_N$. Then φ is measurable w.r.t. $\mathcal{A}$ and $\mathcal{B}_1 \otimes \cdots \otimes \mathcal{B}_N$ by 6.2.5, since $\varphi = (\varphi_1 \times \cdots \times \varphi_N) \circ \iota$.

b: This follows at once from part a and 6.2.5, since $\chi = \rho \circ \varphi$. $\qquad\square$

6.2.11 Definition. Let $(X, \mathcal{A})$ be a measurable space.

A function $\varphi : X \to \mathbb{R}$ is said to be $\mathcal{A}$-*measurable* if it is measurable w.r.t. $\mathcal{A}$ and $\mathcal{A}(d_{\mathbb{R}})$.

A function $\varphi : X \to \mathbb{R}^*$ is said to be $\mathcal{A}$-*measurable* if it is measurable w.r.t. $\mathcal{A}$ and $\mathcal{A}(\delta)$.

A function $\varphi : X \to \mathbb{C}$ is said to be $\mathcal{A}$-*measurable* if it is measurable w.r.t. $\mathcal{A}$ and $\mathcal{A}(d_{\mathbb{C}})$.

6.2.12 Theorem. *Let $(X, \mathcal{A})$ be a measurable space. For a function $\varphi : X \to \mathbb{C}$, the following conditions are equivalent:*

(a) φ is $\mathcal{A}$-measurable;
(b) $\mathrm{Re}\,\varphi$ and $\mathrm{Im}\,\varphi$ are both $\mathcal{A}$-measurable.

Proof. $a \Rightarrow b$: Assume condition a and notice that the two functions

$$\pi_1 : \mathbb{C} \to \mathbb{R}$$

$$z \mapsto \pi_1(z) := \mathrm{Re}\,z$$

and

$$\pi_2 : \mathbb{C} \to \mathbb{R}$$

$$z \mapsto \pi_2(z) := \mathrm{Im}\,z$$

are continuous, hence they are $\mathcal{A}(d_{\mathbb{C}})$-measurable by 6.2.8. Then $\mathrm{Re}\,\varphi$ and $\mathrm{Im}\,\varphi$ are $\mathcal{A}$-measurable by 6.2.5, since $\mathrm{Re}\,\varphi = \pi_1 \circ \varphi$ and $\mathrm{Im}\,\varphi = \pi_2 \circ \varphi$.

$b \Rightarrow a$: Use 6.2.10a with $N := 2$, $(Y_1, \mathcal{B}_1) := (Y_2, \mathcal{B}_2) := (\mathbb{R}, \mathcal{A}(d_{\mathbb{R}}))$, $\varphi_1 := \mathrm{Re}\,\varphi$, $\varphi_2 := \mathrm{Im}\,\varphi$, and notice that $\mathcal{A}(d_{\mathbb{R}}) \otimes \mathcal{A}(d_{\mathbb{R}}) = \mathcal{A}(d_{\mathbb{C}})$ (cf. 6.1.32 with $n := 2$ and 2.7.4b). $\qquad\square$

6.2.13 Proposition. *Let $(X, \mathcal{A})$ be a measurable space.*

(a) *A function $\varphi : X \to \mathbb{R}$ is $\mathcal{A}$-measurable iff the following condition holds:*

$$\varphi^{-1}(E) \in \mathcal{A}, \forall E \in \mathcal{F},$$

where $\mathcal{F}$ is any of the nine families $\mathcal{I}_n$ (with $n = 1, ..., 9$) of 6.1.25, or else $\mathcal{F} := \mathcal{T}_{d_\mathbb{R}}$, or else $\mathcal{F} = \mathcal{K}_{d_\mathbb{R}}$.

(b) *A function $\varphi : X \to \mathbb{R}^*$ is $\mathcal{A}$-measurable iff the following condition holds:*

$$\varphi^{-1}(E) \in \mathcal{A}, \forall E \in \mathcal{F}^*,$$

where $\mathcal{F}^*$ is any of the three families $\mathcal{I}_n^*$ (with $n = 1, 2, 3$) of 6.1.26, or else $\mathcal{F}^* := \mathcal{T}_\delta$, or else $\mathcal{F}^* = \mathcal{K}_\delta$.

Moreover, φ is $\mathcal{A}$-measurable iff the following condition holds:

$$\varphi^{-1}(\{-\infty\}), \varphi^{-1}(\{\infty\}) \in \mathcal{A} \text{ and } \varphi_{\varphi^{-1}(\mathbb{R})} \text{ is } \mathcal{A}^{\varphi^{-1}(\mathbb{R})}\text{-measurable.}$$

(c) *A function $\varphi : X \to \mathbb{C}$ is $\mathcal{A}$-measurable iff the following condition holds:*

$$\varphi^{-1}(E) \in \mathcal{A}, \forall E \in \mathcal{G},$$

where $\mathcal{G}$ is the family R of 6.1.33, or else $\mathcal{G} := \mathcal{T}_{d_\mathbb{C}}$, or else $\mathcal{G} := \mathcal{K}_{d_\mathbb{C}}$.

Proof. With the exception of the second assertion of part b, everything follows at once from 6.2.7 along with 6.1.25, 6.1.26, 6.1.33, 6.1.22, 6.1.23.

As to the second assertion of part b, we notice that the condition

$$\varphi^{-1}(E) \in \mathcal{A}, \forall E \in \mathcal{I}_1^*$$

is precisely the condition

$$\varphi^{-1}(\{-\infty\}), \varphi^{-1}(\{\infty\}) \in \mathcal{A} \text{ and } \varphi^{-1}(E) \in \mathcal{A} \text{ for all } E \in \mathcal{A}(d_\mathbb{R}).$$

We also notice that the condition

$$\varphi^{-1}(E) \in \mathcal{A}, \forall E \in \mathcal{A}(d_\mathbb{R})$$

can be written as

$$\exists F \in \mathcal{A} \text{ s.t. } \varphi^{-1}_{\varphi^{-1}(\mathbb{R})}(E) = \varphi^{-1}(E) = F = F \cap \varphi^{-1}(\mathbb{R}), \forall E \in \mathcal{A}(d_\mathbb{R})$$

and that this in its turn can be written as

$$\varphi^{-1}_{\varphi^{-1}(\mathbb{R})}(E) \in \mathcal{A}^{\varphi^{-1}(\mathbb{R})}, \forall E \in \mathcal{A}(d_\mathbb{R}).$$

Finally, we notice that the last condition is the condition of $\mathcal{A}^{\varphi^{-1}(\mathbb{R})}$-measurability for the mapping $\varphi_{\varphi^{-1}(\mathbb{R})}$. $\qquad\qquad\square$

6.2.14 Remark. Let $(X, \mathcal{A})$ be a measurable space. From 6.2.4 and 6.1.24 it follows that, for the $\mathcal{A}$-measurability of a function $\varphi : X \to \mathbb{R}$, it is immaterial what choice is made among $\mathbb{R}$, $\mathbb{R}^*$ and $\mathbb{C}$ for the final set of φ. For this reason, in what follows we consider only functions whose final sets are either $\mathbb{R}^*$ or $\mathbb{C}$.

6.2.15 Definition. For a measurable space $(X, \mathcal{A})$, we denote by $\mathcal{M}(X, \mathcal{A})$ the family of all the functions, with X as domain and $\mathbb{C}$ as final set, that are $\mathcal{A}$-measurable, i.e. $\mathcal{M}(X, \mathcal{A})$ is the family of all $\mathcal{A}$-measurable complex functions on X:

$$\mathcal{M}(X, \mathcal{A}) := \{\varphi \in \mathcal{F}(X) : \varphi \text{ is } \mathcal{A}\text{-measurable}\}$$

(for $\mathcal{F}(X)$ cf. 3.1.10c).

6.2.16 Theorem. *Let $(X, \mathcal{A})$ be a measurable space. Then we have*

(lm$_1$) $\varphi + \psi \in \mathcal{M}(X, \mathcal{A})$, $\forall \varphi, \psi \in \mathcal{M}(X, \mathcal{A})$,
(lm$_2$) $\alpha\varphi \in \mathcal{M}(X, \mathcal{A})$, $\forall \alpha \in \mathbb{C}$, $\forall \varphi \in \mathcal{M}(X, \mathcal{A})$,
(sa$_2$) $\varphi\psi \in \mathcal{M}(X, \mathcal{A})$, $\forall \varphi, \psi \in \mathcal{M}(X, \mathcal{A})$.

This means that $\mathcal{M}(X, \mathcal{A})$ is a subalgebra of the abelian associative algebra $\mathcal{F}(X)$ (cf. 3.3.8a). Since $1_X \in \mathcal{M}(X, \mathcal{A})$, $\mathcal{M}(X, \mathcal{A})$ is with identity.

Proof. For lm_1 and sa_2, use 6.2.10b with $N := 2$, $(Y_1, \mathcal{B}_1) = (Y_2, \mathcal{B}_2) = (\mathbb{C}, \mathcal{A}(d_{\mathbb{C}}))$, $\varphi_1 := \varphi$, $\varphi_2 := \psi$ and either

$$\mathbb{C} \times \mathbb{C} \ni (z_1, z_2) \mapsto \rho(z_1, z_2) := z_1 + z_2 \in \mathbb{C}$$

or

$$\mathbb{C} \times \mathbb{C} \ni (z_1, z_2) \mapsto \rho(z_1, z_2) := z_1 z_2 \in \mathbb{C}.$$

Notice in fact that in either case ρ is $\mathcal{A}(d_{\mathbb{C}} \times d_{\mathbb{C}})$-measurable since it is continuous (cf. 6.2.8), and that $\mathcal{A}(d_{\mathbb{C}} \times d_{\mathbb{C}}) = \mathcal{A}(d_{\mathbb{C}}) \otimes \mathcal{A}(d_{\mathbb{C}})$ by 6.1.31.

Condition lm_2 follows from sa_2, since $\alpha\varphi = \alpha_X\varphi$ (for the constant function α_X cf. 1.2.19) and $\alpha_X \in \mathcal{M}(X, \mathcal{A})$ by 6.2.2.

Finally, $1_X \in \mathcal{M}(X, \mathcal{A})$ by 6.2.2. $\qquad\square$

6.2.17 Proposition. *Let $(X, \mathcal{A})$ be a measurable space and $\varphi \in \mathcal{M}(X, \mathcal{A})$. Then:*

$\overline{\varphi} \in \mathcal{M}(X, \mathcal{A})$;
$|\varphi|^n \in \mathcal{M}(X, \mathcal{A})$, $\forall n \in \mathbb{N}$;
the function $\frac{1}{\varphi}$ is $\mathcal{A}^{D_{\frac{1}{\varphi}}}$-measurable

(for the functions $\overline{\varphi}$, $|\varphi|^n$ and $\frac{1}{\varphi}$, cf. 1.2.19).

Proof. Notice that the function

$$\mathbb{C} \ni z \mapsto \psi(z) := \overline{z} \in \mathbb{C}$$

is continuous, and hence it is $\mathcal{A}(d_{\mathbb{C}})$-measurable by 6.2.8. Use then 6.2.5 to obtain the $\mathcal{A}$-measurability of $\overline{\varphi}$, since $\overline{\varphi} = \psi \circ \varphi$.

For $|\varphi|^n$ the proof is analogous, using the function

$$\mathbb{C} \ni z \mapsto \psi(z) := |z|^n \in \mathbb{C}.$$

Finally, $\frac{1}{\varphi}$ is $\mathcal{A}^{D\frac{1}{\varphi}}$-measurable by 6.2.6. Indeed, $\frac{1}{\varphi} = \psi \circ \varphi$ if ψ is the function

$$\mathbb{C} - \{0\} \ni z \mapsto \psi(z) := \frac{1}{z} \in \mathbb{C}.$$

Now ψ is continuous, hence $\mathcal{A}(d_{\mathbb{C}-\{0\}})$-measurable by 6.2.8 (we have denoted by $d_{\mathbb{C}-\{0\}}$ the restriction of $d_{\mathbb{C}}$ to $(\mathbb{C} - \{0\}) \times (\mathbb{C} - \{0\})$). Moreover, $\mathcal{A}(d_{\mathbb{C}-\{0\}}) = (\mathcal{A}(d_{\mathbb{C}}))^{\mathbb{C}-\{0\}} = (\mathcal{A}(d_{\mathbb{C}}))^{D_\psi}$ by 6.1.21 (cf. also 6.1.22). $\qquad\square$

6.2.18 Definitions. Let X be a non-empty set, and let $\{\varphi_n\}$ be a sequence of functions $\varphi_n : X \to \mathbb{R}^*$. For each $n \in \mathbb{N}$ we define the functions (cf. 5.2.2):

$$\sup_{k \geq n} \varphi_k : X \to \mathbb{R}^*$$

$$x \mapsto (\sup_{k \geq n} \varphi_k)(x) := \sup_{k \geq n} \varphi_k(x),$$

$$\inf_{k \geq n} \varphi_k : X \to \mathbb{R}^*$$

$$x \mapsto (\inf_{k \geq n} \varphi_k)(x) := \inf_{k \geq n} \varphi_k(x).$$

If the sequence $\{\varphi_n(x)\}$ is convergent (in the metric space $(\mathbb{R}^*, \delta)$) for all $x \in X$, we define the function

$$\lim_{n \to \infty} \varphi_n : X \to \mathbb{R}^*$$

$$x \mapsto (\lim_{n \to \infty} \varphi_n)(x) := \lim_{n \to \infty} \varphi_n(x),$$

and from 5.2.6 we have $\lim_{n \to \infty} \varphi_n = \inf_{n \geq 1}(\sup_{k \geq n} \varphi_k) = \sup_{n \geq 1}(\inf_{k \geq n} \varphi_k)$.

6.2.19 Proposition. *Let $(X, \mathcal{A})$ be a measurable space and let $\{\varphi_n\}$ be a sequence of $\mathcal{A}$-measurable functions $\varphi_n : X \to \mathbb{R}^*$.*

(a) For each $n \in \mathbb{N}$, the functions $\sup_{k \geq n} \varphi_k$ and $\inf_{k \geq n} \varphi_k$ are $\mathcal{A}$-measurable.
(b) If the sequence $\{\varphi_n(x)\}$ is convergent for all $x \in X$, then the function $\lim_{n \to \infty} \varphi_n$ is measurable.

Proof. a: Recall that, for $S \subset \mathbb{R}^*$, we have

(1) $s \leq \sup S, \forall s \in S$,
(2) for $a \in \mathbb{R}^*$,

$$(s \leq a, \forall s \in S) \Rightarrow \sup S \leq a, \quad \text{or equivalently}$$

$$a < \sup S \Rightarrow (\exists s \in S \text{ s.t. } a < s).$$

Fix now $a \in \mathbb{R}$. For $x \in X$ we have

$$x \in (\sup_{k \geq n} \varphi_k)^{-1}((a, \infty]) \Leftrightarrow a < \sup_{k \geq n} \varphi_k(x) \overset{(3)}{\Leftrightarrow}$$

$$[\exists k \geq n \text{ s.t. } a < \varphi_k(x)] \Leftrightarrow x \in \bigcup_{n=k}^{\infty} \varphi_k^{-1}((a, \infty]),$$

where $\overset{(3)}{\Leftarrow}$ holds by 1 and $\overset{(3)}{\Rightarrow}$ holds by 2. This proves that

$$(\sup_{k\geq n} \varphi_k)^{-1}((a,\infty]) = \bigcup_{n=k}^{\infty} \varphi_k^{-1}((a,\infty]).$$

Since this is true for all $a \in \mathbb{R}$, it proves that $\sup_{k\geq n} \varphi_k$ is $\mathcal{A}$-measurable, in view of 6.2.13b with $\mathcal{F}^* := \mathcal{I}_2^*$.

For $\inf_{k\geq n} \varphi_k$ the proof is analogous.

b: Let the sequence $\{\varphi_n(x)\}$ be convergent for all $x \in X$. Then $\lim_{n\to\infty} \varphi_n = \inf_{n\geq 1}(\sup_{k\geq n} \varphi_k)$. Now, $\sup_{k\geq n} \varphi_k$ is $\mathcal{A}$-measurable for each $n \in \mathbb{N}$, in view of part a. Then, in view of part a once again, $\inf_{n\geq 1}(\sup_{k\geq n} \varphi_k)$ is $\mathcal{A}$-measurable. $\qquad\square$

6.2.20 Corollaries. Let $(X, \mathcal{A})$ be a measurable space.

(a) If the functions $\varphi : X \to \mathbb{R}^*$ and $\psi : X \to \mathbb{R}^*$ are measurable, then the functions

$$\max\{\varphi, \psi\} : X \to \mathbb{R}^*$$
$$x \mapsto (\max\{\varphi, \psi\})(x) := \max\{\varphi(x), \psi(x)\}$$

and

$$\min\{\varphi, \psi\} : X \to \mathbb{R}^*$$
$$x \mapsto (\min\{\varphi, \psi\})(x) := \min\{\varphi(x), \psi(x)\}$$

are $\mathcal{A}$-measurable.

(b) A function $\varphi : X \to \mathbb{R}$ is $\mathcal{A}$-measurable iff both φ^+ and φ^- (cf. 1.2.19) are $\mathcal{A}$-measurable.

(c) Let $\{\varphi_n\}$ be a sequence in $\mathcal{M}(X, \mathcal{A})$ and suppose that the sequence $\{\varphi_n(x)\}$ is convergent (in the metric space $(\mathbb{C}, d_{\mathbb{C}})$) for all $x \in X$. Then the function

$$\lim_{n\to\infty} \varphi_n : X \to \mathbb{C}$$
$$x \mapsto (\lim_{n\to\infty} \varphi_n)(x) := \lim_{n\to\infty} \varphi_n(x)$$

is an element of $\mathcal{M}(X, \mathcal{A})$.

(d) Let $\{\varphi_n\}$ be a sequence in $\mathcal{M}(X, \mathcal{A})$ and suppose that the series $\sum_{n=1}^{\infty} \varphi_n(x)$ is convergent for all $x \in X$. Then the function

$$\sum_{n=1}^{\infty} \varphi_n : X \to \mathbb{C}$$
$$x \mapsto \left(\sum_{n=1}^{\infty} \varphi_n\right)(x) := \sum_{n=1}^{\infty} \varphi_n(x).$$

is an element of $\mathcal{M}(X, \mathcal{A})$.

Proof. a: If we define $\varphi_1 := \varphi$ and $\varphi_n := \psi$ for $n \geq 2$, then we have

$$\max\{\varphi, \psi\} = \sup_{k\geq 1} \varphi_k \text{ and } \min\{\varphi, \psi\} = \inf_{k\geq 1} \varphi_k,$$

and the result follows from 6.2.19a.

b: If φ is $\mathcal{A}$-measurable, then φ^+ and φ^- are $\mathcal{A}$-measurable by corollary a, since the function 0_X is $\mathcal{A}$-measurable (cf. 6.2.2). If φ^+ and φ^- are $\mathcal{A}$-measurable, then φ is $\mathcal{A}$-measurable by 6.2.16 since $\varphi = \varphi^+ - \varphi^-$.

c: By 2.7.3a we have that, for all $x \in X$, the sequences $\{\operatorname{Re}\varphi_n(x)\}$ and $\{\operatorname{Im}\varphi_n(x)\}$ are convergent (in the metric space $(\mathbb{R}, d_{\mathbb{R}})$, and hence in the metric space $(\mathbb{R}^*, \delta)$ as well: cf. 5.2.1c) and

$$\lim_{n\to\infty} \varphi_n(x) = \lim_{n\to\infty} \operatorname{Re}\varphi_n(x) + i \lim_{n\to\infty} \operatorname{Im}\varphi_n(x).$$

Therefore we have

$$\lim_{n\to\infty} \varphi_n = \lim_{n\to\infty} \operatorname{Re}\varphi_n + i \lim_{n\to\infty} \operatorname{Im}\varphi_n.$$

Now, the functions $\operatorname{Re}\varphi_n$ and $\operatorname{Im}\varphi_n$ are $\mathcal{A}$-measurable for each $n \in \mathbb{N}$ (cf. 6.2.12). Thus, 6.2.19b implies that $\lim_{n\to\infty} \operatorname{Re}\varphi_n$ and $\lim_{n\to\infty} \operatorname{Im}\varphi_n$ are $\mathcal{A}$-measurable (we have also used 6.2.14 twice). Then, $\lim_{n\to\infty} \varphi_n$ is $\mathcal{A}$-measurable by 6.2.16 (or by 6.2.10a and the equality $\mathcal{A}(d_{\mathbb{C}}) = \mathcal{A}(d_{\mathbb{R}}) \otimes \mathcal{A}(d_{\mathbb{R}})$).

d: This result follows from 6.2.16 and corollary c, since

$$\sum_{n=1}^{\infty} \varphi_n = \lim_{n\to\infty} \sum_{k=1}^{n} \varphi_k.$$

$\square$

6.2.21 Proposition. *Let $(X, \mathcal{A})$ be a measurable space and S a subset of X. Then the characteristic function χ_S (cf. 1.2.6b) is $\mathcal{A}$-measurable iff $S \in \mathcal{A}$.*

Proof. The only counterimages under χ_S are the sets $\emptyset, S, X - S, X$. $\square$

6.2.22 Definition. Let $(X, \mathcal{A})$ be a measurable space. A function $\psi : X \to \mathbb{C}$ is said to be $\mathcal{A}$-*simple* if there exist $n \in \mathbb{N}$, a family $\{\alpha_1, ..., \alpha_n\}$ of elements of $\mathbb{C}$, and a disjoint family $\{E_1, ..., E_n\}$ of elements of $\mathcal{A}$ so that

$$\psi = \sum_{k=1}^{n} \alpha_k \chi_{E_k}.$$

We denote by $\mathcal{S}(X, \mathcal{A})$ the family of all $\mathcal{A}$-simple functions, i.e. we define

$$\mathcal{S}(X, \mathcal{A}) := \left\{ \psi \in \mathcal{F}(X) : \exists n \in \mathbb{N}, \exists (\alpha_1, ..., \alpha_n) \in \mathbb{C}^n, \exists (E_1, ..., E_n) \in \mathcal{A}^n \text{ s.t.} \right.$$

$$\left. E_i \cap E_j = \emptyset \text{ if } i \neq j \text{ and } \psi = \sum_{k=1}^{n} \alpha_k \chi_{E_k} \right\}.$$

We point out the obvious fact that, for $\psi \in \mathcal{S}(X, \mathcal{A})$, its representation $\psi = \sum_{k=1}^{n} \alpha_k \chi_{E_k}$ as in the definition above is never unique (it would be if we required $\alpha_i \neq \alpha_j$ for $i \neq j$ and $\bigcup_{k=1}^{n} E_k = X$, but we do not).

6.2.23 Proposition. *Let $(X, \mathcal{A})$ be a measurable space. For a function $\psi : X \to \mathbb{C}$, the following conditions are equivalent:*

(a) $\psi \in \mathcal{S}(X, \mathcal{A})$,

(b) $\psi \in \mathcal{M}(X, \mathcal{A})$ *and* ψ *has only finitely many values, i.e.* R_ψ *is a finite set.*

Proof. $a \Rightarrow b$: Let $n \in \mathbb{N}$, $(\alpha_1, ..., \alpha_N) \in \mathbb{C}^n$, $(E_1, ..., E_n) \in \mathcal{A}$ be so that

$$E_i \cap E_j = \emptyset \text{ if } i \neq j \text{ and } \psi = \sum_{k=1}^{n} \alpha_k \chi_{E_k}.$$

Then $\chi_{E_k} \in \mathcal{M}(X, \mathcal{A})$ for $k = 1, ..., n$ (cf. 6.2.21), and hence $\psi \in \mathcal{M}(X, \mathcal{A})$ by 6.2.16. Moreover, ψ has finitely many values since the only possible values of ψ are the numbers $0, \alpha_1, ..., \alpha_n$.

$b \Rightarrow a$: Assume condition b and let $\{\alpha_1, ..., \alpha_n\} := R_\psi$ and $E_k := \psi^{-1}(\{\alpha_k\})$ for $k = 1, ..., n$.

We have $E_k \in \mathcal{A}$ by 6.2.13c (with $\mathcal{G} = \mathcal{K}_{d_{\mathbb{C}}}$) since ψ is $\mathcal{A}$-measurable and $\{\alpha_k\} \in \mathcal{K}_{d_{\mathbb{C}}}$. Moreover,

$$E_i \cap E_j = \psi^{-1}(\{\alpha_i\} \cap \{\alpha_j\}) = \psi^{-1}(\emptyset) = \emptyset \text{ if } i \neq j.$$

Finally, since $X = \bigcup_{k=1}^{n} E_k$ we have

$$\forall x \in X, \exists! i \in \{1, ..., n\} \text{ s.t. } x \in E_i, \text{ hence s.t. } \psi(x) = \alpha_i = \sum_{k=1}^{n} \alpha_k \chi_{E_k}(x),$$

and this proves that $\psi = \sum_{k=1}^{n} \alpha_k \chi_{E_k}$. $\qquad\qquad\square$

6.2.24 Theorem. *Let* $(X, \mathcal{A})$ *be a measurable space. Then we have*

(lm$_1$) $\psi_1 + \psi_2 \in \mathcal{S}(X, \mathcal{A})$, $\forall \psi_1, \psi_2 \in \mathcal{S}(X, \mathcal{A})$,

(lm$_2$) $\alpha\psi \in \mathcal{S}(X, \mathcal{A})$, $\forall \alpha \in \mathbb{C}$, $\forall \psi \in \mathcal{S}(X, \mathcal{A})$,

(sa$_2$) $\psi_1 \psi_2 \in \mathcal{S}(X, \mathcal{A})$, $\forall \psi_1, \psi_2 \in \mathcal{S}(X, \mathcal{A})$.

This means that $\mathcal{S}(X, \mathcal{A})$ *is a subalgebra of the abelian associative algebra* $\mathcal{M}(X, \mathcal{A})$ *(cf. 6.2.16). Since* $1_X \in \mathcal{S}(X, \mathcal{A})$, $\mathcal{S}(X, \mathcal{A})$ *is with identity.*

Proof. In view of 6.2.23 and 6.2.16, we only need to notice that, if $\alpha \in \mathbb{C}$ and $\psi_1, \psi_2, \psi \in \mathcal{S}(X, \mathcal{A})$, then $\psi_1 + \psi_2, \alpha\psi, \psi_1 \psi_2$ have only finitely many values. $\qquad\square$

6.2.25 Definitions. Let $(X, \mathcal{A})$ be a measurable space. We define

$$\mathcal{S}^+(X, \mathcal{A}) := \{\psi \in \mathcal{S}(X, \mathcal{A}) : \psi(x) \in [0, \infty), \forall x \in X\}.$$

We denote by $\mathcal{L}^+(X, \mathcal{A})$ the family of all the functions, with X as domain and $[0, \infty]$ as final set, that are $\mathcal{A}$-measurable, i.e. we define

$$\mathcal{L}^+(X, \mathcal{A}) := \{\varphi : X \to [0, \infty] : \varphi \text{ is } \mathcal{A}\text{-measurable}\}.$$

Obviously, $\mathcal{S}^+(X, \mathcal{A}) \subset \mathcal{L}^+(X, \mathcal{A})$.

6.2.26 Theorem. *Let* $(X, \mathcal{A})$ *be a measurable space and* $\varphi \in \mathcal{L}^+(X, \mathcal{A})$. *Then there exists a sequence* $\{\psi_n\}$ *in* $\mathcal{S}^+(X, \mathcal{A})$ *such that:*

(a) $\psi_n \leq \psi_{n+1} \leq \varphi$, $\forall n \in \mathbb{N}$ *(for the notation $\varphi \leq \psi$, cf. 5.1.1);*
(b) $\psi_n(x) \to \varphi(x)$ *as* $n \to \infty$, $\forall x \in X$;
(c) *if a subset Y of X and $m \in [0, \infty)$ are such that $\varphi(x) \leq m$ for all $x \in Y$, then*
$\sup\{\varphi(x) - \psi_n(x) : x \in Y\} \to 0$ *as* $n \to \infty$.

Proof. For each $n \in \mathbb{N}$ define

$$E_{0,n} := \varphi^{-1}([n, \infty]) \text{ and } E_{k,n} := \varphi^{-1}\left(\left[\frac{k-1}{2^n}, \frac{k}{2^n}\right)\right) \text{ for } k = 1, ..., n2^n.$$

For all $a, b \in \mathbb{R}$, we have $[a, b) \in \mathcal{A}(d_\mathbb{R})$ from 6.1.25, and hence $[a, b) \in \mathcal{A}(\delta)$ from 6.1.26. For each $a \in \mathbb{R}$ we have $[a, \infty] \in \mathcal{A}(\delta)$ from 6.1.26. Thus, we have $E_{k,n} \in \mathcal{A}$ for $k = 0, 1, ..., n2^n$, since φ is $\mathcal{A}$-measurable. Besides, the family $\{E_{k,n}\}_{k=0,1,...,n2^n}$ is a disjoint family because such is the family $\{[n, \infty]\} \cup \left\{\left[\frac{k-1}{2^n}, \frac{k}{2^n}\right)\right\}_{k=1,...,n2^n}$. Thus, the function

$$\psi_n := \sum_{k=1}^{n2^n} \frac{k-1}{2^n} \chi_{E_{k,n}} + n \chi_{E_{0,n}}$$

is an element of $\mathcal{S}^+(X, \mathcal{A})$, and clearly $\psi_n \leq \varphi$.

We will prove that the sequence $\{\psi_n\}$ has properties a, b, c.

a: Fix $n \in \mathbb{N}$, notice that $X = \left(\bigcup_{k=1}^{n2^n} E_{k,n}\right) \cup E_{0,n}$, and consider $x \in X$. Suppose first that there exists $k \in \{1, ..., n2^n\}$ such that $x \in E_{k,n}$. Since

$$\left[\frac{k-1}{2^n}, \frac{k}{2^n}\right) = \left[\frac{(2k-1)-1}{2^{n+1}}, \frac{2k-1}{2^{n+1}}\right) \cup \left[\frac{2k-1}{2^{n+1}}, \frac{2k}{2^{n+1}}\right),$$

then:

either $x \in E_{2k-1,n+1}$, and then $\psi_n(x) = \frac{k-1}{2^n} = \frac{(2k-1)-1}{2^{n+1}} = \psi_{n+1}(x)$,
or $x \in E_{2k,n+1}$, and then $\psi_n(x) = \frac{k-1}{2^n} < \frac{2k-1}{2^{n+1}} = \psi_{n+1}(x)$.

Suppose next that $x \in E_{0,n}$. Then:

either $\varphi(x) < n + 1$, and then $\exists k \in \{1, ..., (n+1)2^{n+1}\}$ s.t. $n \leq \frac{k-1}{2^{n+1}}$ and $x \in E_{k,n+1}$ (recall that $n \leq \varphi(x)$), whence $\psi_n(x) = n \leq \frac{k-1}{2^{n+1}} = \psi_{n+1}(x)$,
or $n + 1 \leq \varphi(x)$, i.e. $x \in E_{0,n+1}$, whence $\psi_n(x) = n < n + 1 = \psi_{n+1}(x)$.

This shows that $\psi_n \leq \psi_{n+1}$.

b: Consider $x \in X$, and suppose first that $\varphi(x) < \infty$. Then for $n > \varphi(x)$ there exists $k \in \{1, ..., n2^n\}$ such that $x \in E_{k,n}$, and hence

$$0 \leq \varphi(x) - \psi_n(x) \leq \frac{1}{2^n},$$

and this shows (cf. $5.1.2b_3$) that $\psi_n(x) \to \varphi(x)$ as $n \to \infty$. Suppose next that $\varphi(x) = \infty$. Then $x \in E_{0,n}$, hence $\psi_n(x) = n$ for all $n \in \mathbb{N}$, and this shows (cf. $5.2.1b_1$) that $\psi_n(x) \to \varphi(x)$ as $n \to \infty$.

c: If $Y \in \mathcal{P}(X)$ and $m \in [0, \infty)$ are such that $\varphi(x) \leq m$ for all $x \in y$, then for $n \geq m$ we have

$$\varphi(x) - \psi_n(x) \leq \frac{1}{2^n}, \forall x \in Y.$$

$\square$

6.2.27 Corollary. *Let $(X, \mathcal{A})$ be a measurable space and $\varphi \in \mathcal{M}(X, \mathcal{A})$. Then there exists a sequence $\{\psi_n\}$ in $\mathcal{S}(X, \mathcal{A})$ such that:*

(a) $|\psi_n| \leq |\varphi|, \forall n \in \mathbb{N}$,
(b) $\psi_n(x) \to \varphi(x)$ *as* $n \to \infty, \forall x \in X$,
(c) if a subset Y of X and $m \in [0, \infty)$ are such that $|\varphi(x)| \leq m$ for all $x \in Y$, then
$\sup\{|\varphi(x) - \psi_n(x)| : x \in Y\} \to 0$ *as* $n \to \infty$.

Proof. The functions

$$\varphi_1 := (\operatorname{Re}\varphi)^+, \varphi_2 := (\operatorname{Re}\varphi)^-, \varphi_3 := (\operatorname{Im}\varphi)^+, \varphi_4 := (\operatorname{Im}\varphi)^-$$

are elements of $\mathcal{L}^+(X, \mathcal{A})$ by 6.2.12 and 6.2.20b. Then 6.2.26 implies that, for $i = 1, ..., 4$, there exists a sequence $\{\psi_n^i\}$ in $\mathcal{S}^+(X, \mathcal{A})$ such that:

$0 \leq \psi_n^i \leq \varphi_i, \forall n \in \mathbb{N}$,
$\psi_n^i(x) \to \varphi_i(x)$ as $n \to \infty, \forall x \in X$,
if $Y \in \mathcal{P}(X)$ and $m \in [0, \infty)$ are such that $|\varphi(x)| \leq m, \forall x \in Y$, then (since $\varphi_i \leq |\varphi|$) $\sup\{\varphi_i(x) - \psi_n^i(x) : x \in Y\} \to 0$ as $n \to \infty$.

For each $n \in \mathbb{N}$ the function $\psi_n := \psi_n^1 - \psi_n^2 + i(\psi_n^3 - \psi_n^4)$ is an element of $\mathcal{S}(X, \mathcal{A})$ by 6.2.24, and we will prove that conditions a, b, c are satisfied for the sequence $\{\psi_n\}$.

a: For each $n \in \mathbb{N}$ we have

$$|\psi_n^1 - \psi_n^2| \leq \varphi_1 + \varphi_2 = |\operatorname{Re}\varphi| \text{ and } |\psi_n^3 - \psi_n^4| \leq \varphi_3 + \varphi_4 = |\operatorname{Im}\varphi|,$$

and this implies that $|\psi_n| \leq |\varphi|$.

b and c: We have

$$|\varphi - \psi_n| \leq \sum_{i=1}^{4} |\varphi_i - \psi_n^i|.$$

This implies in the first place that

$$\psi_n(x) \to \varphi(x) \text{ as } n \to \infty, \forall x \in X,$$

and in the second place that, if $Y \in \mathcal{P}(X)$ and $m \in [0, \infty)$ are such that $|\varphi(x)| \leq m$ for all $x \in Y$, then

$$\sup\{|\varphi(x) - \psi_n(x)| : x \in Y\} \leq \sum_{i=1}^{4} \sup\{\varphi_i(x) - \psi_n^i(x) : x \in Y\} \to 0 \text{ as } n \to \infty.$$

$\square$

6.2.28 Definition. For a measurable space $(X, \mathcal{A})$, we denote by $\mathcal{M}_B(X, \mathcal{A})$ the family of all bounded $\mathcal{A}$-measurable complex functions on X, i.e. we define

$$\mathcal{M}_B(X, \mathcal{A}) := \mathcal{M}(X, \mathcal{A}) \cap \mathcal{F}_B(X)$$

(for $\mathcal{F}_B(X)$, cf. 3.1.10d).

6.2.29 Remarks. Let $(X, \mathcal{A})$ be a measurable space. Since $\mathcal{M}(X, \mathcal{A})$ and $\mathcal{F}_B(X)$ are subalgebras of the associative algebra $\mathcal{F}(X)$ (cf. 6.2.16 and 3.3.8b), $\mathcal{M}_B(X, \mathcal{A})$ is a subalgebra of the associative algebra $\mathcal{F}(X)$ (cf. 3.3.4), and hence of the associative algebras $\mathcal{M}(X, \mathcal{A})$ and $\mathcal{F}_B(X)$ as well (cf. 3.3.3b).

Since $\mathcal{F}_B(X)$ is a normed algebra (cf. 4.3.6a), $\mathcal{M}_B(X, \mathcal{A})$ is a normed algebra as well (cf. 4.3.2).

Since convergence with respect to the $\| \ \|_\infty$ norm (cf. 4.3.6a) implies pointwise convergence, 6.2.20c shows that $\mathcal{M}_B(X, \mathcal{A})$ is a closed subset of $\mathcal{F}_B(X)$. Hence, since $\mathcal{F}_B(X)$ is a Banach space (cf. 4.3.6a), $\mathcal{M}_B(X, \mathcal{A})$ is a Banach space as well (cf. 4.1.8a) and hence a Banach algebra.

Clearly, $\mathcal{S}(X, \mathcal{A}) \subset \mathcal{M}_B(X, \mathcal{A})$. Since $\mathcal{S}(X, \mathcal{A})$ is a subalgebra of the associative algebra $\mathcal{M}(X, \mathcal{A})$ (cf. 6.2.24), $\mathcal{S}(X, \mathcal{A})$ is a subalgebra of the associative algebra $\mathcal{M}_B(X, \mathcal{A})$ as well (cf. 3.3.3b).

Finally, 6.2.27c and 2.3.12 show that $\mathcal{S}(X, \mathcal{A})$ is dense (in the $\| \ \|_\infty$ norm) in $\mathcal{M}_B(X, \mathcal{A})$.

6.2.30 Definitions. Let X be a non-empty set and $\varphi : D_\varphi \to [0, \infty]$ a function with $D_\varphi \subset X$.

If $\psi : D_\psi \to [0, \infty]$ is a function with $D_\psi \subset X$, then we define the functions:

$$\varphi + \psi : D_\varphi \cap D_\psi \to [0, \infty]$$
$$x \mapsto (\varphi + \psi)(x) := \varphi(x) + \psi(x),$$

$$\varphi\psi : D_\varphi \cap D_\psi \to [0, \infty]$$
$$x \mapsto (\varphi\psi)(x) := \varphi(x)\psi(x).$$

Clearly, if $R_\varphi \subset [0, \infty)$ and $R_\psi \subset [0, \infty)$ then these definitions are in agreement with the ones given in 1.2.19.

If $a \in [0, \infty]$, then we define the function

$$a\varphi : D_\varphi \to [0, \infty]$$
$$x \mapsto (a\varphi)(x) := a\varphi(x).$$

Clearly, if $R_\varphi \subset [0, \infty)$ and $a \in [0, \infty)$ then this definition is in agreement with the one given in 1.2.19.

6.2.31 Theorem. *Let $(X, \mathcal{A})$ be a measurable space. If $\varphi_1, \varphi_2 \in \mathcal{L}^+(X, \mathcal{A})$, then $\varphi_1 + \varphi_2 \in \mathcal{L}^+(X, \mathcal{A})$ and $\varphi_1\varphi_2 \in \mathcal{L}^+(X, \mathcal{A})$. If $a \in [0, \infty]$ and $\varphi \in \mathcal{L}^+(X, \mathcal{A})$, then $a\varphi \in \mathcal{L}^+(X, \mathcal{A})$.*

Proof. It is obvious that $R_{\varphi_1 + \varphi_2} \subset [0, \infty]$. By 6.2.26, there are two sequences $\{\psi_n^1\}$ and $\{\psi_n^2\}$ in $\mathcal{S}^+(X, \mathcal{A})$ so that, for $i = 1, 2$,

$$\forall x \in X, \ \psi_n^i(x) \leq \psi_{n+1}^i(x) \text{ and } \varphi_i(x) = \lim_{n \to \infty} \psi_n^i(x).$$

Then by 5.3.4 we have

$$\forall x \in X, (\varphi_1 + \varphi_2)(x) = \lim_{n \to \infty} \psi_n^1(x) + \lim_{n \to \infty} \psi_n^2(x)$$

$$= \lim_{n \to \infty} (\psi_n^1(x) + \psi_n^2(x)) = \lim_{n \to \infty} (\psi_n^1 + \psi_n^2)(x),$$

and $\psi_n^1 + \psi_n^2$ is $\mathcal{A}$-measurable by 6.2.16. This implies, by 6.2.19b, that $\varphi_1 + \varphi_2$ is $\mathcal{A}$-measurable. Thus $\varphi_1 + \varphi_2 \in \mathcal{L}^+(X, \mathcal{A})$.

The proof for $\varphi_1\varphi_2$ is analogous, and for $a\varphi$ we notice that $a\varphi = a_X\varphi$ if a_X denotes the constant function on X with value a, which is an element of $\mathcal{L}^+(X, \mathcal{A})$ by 6.2.2. $\qquad \square$

6.2.32 Corollary. *Let $(X, \mathcal{A})$ be a measurable space, let $\{\varphi_n\}$ be a sequence in $\mathcal{L}^+(X, \mathcal{A})$, and define the function (cf. 5.4.1)*

$$\sum_{n=1}^{\infty} \varphi_n : X \to [0, \infty]$$

$$x \mapsto \left(\sum_{n=1}^{\infty} \varphi_n \right)(x) := \sum_{n=1}^{\infty} \varphi_n(x).$$

Then $\sum_{n=1}^{\infty} \varphi_n \in \mathcal{L}^+(X, \mathcal{A})$.

Proof. Using 5.4.1 and the definition given in 6.2.18, we have

$$\sum_{n=1}^{\infty} \varphi_n = \lim_{n \to \infty} \sum_{k=1}^{n} \varphi_k.$$

The result then follows from 6.2.31 and 6.2.19b. $\qquad \square$

6.3 Borel functions

In this section we prove a result about Borel functions which has hardly anything to do with the theory of measure and integration, but which will play an essential role in our proof of the spectral theorem for unitary operators (from which we will deduce the spectral theorem for self-adjoint operators).

6.3.1 Definition. Let X be a non-empty set, $\{\varphi_n\}$ a sequence in $\mathcal{F}(X)$ (for $\mathcal{F}(X)$, cf. 3.1.10c), and $\varphi \in \mathcal{F}(X)$. We say that φ is the *uniformly bounded pointwise limit*, in short *ubp limit*, of $\{\varphi_n\}$ and we write $\varphi_n \xrightarrow{ubp} \varphi$ if the following two conditions are satisfied:

$\exists m \in [0, \infty)$ such that $|\varphi_n(x)| \le m, \forall x \in X, \forall n \in \mathbb{N}$;
$\varphi_n(x) \to \varphi(x)$ as $n \to \infty, \forall x \in X$.

Clearly, if $\varphi_n \xrightarrow{ubp} \varphi$ then $\varphi \in \mathcal{F}_B(X)$ (for $\mathcal{F}_B(X)$, cf. 3.1.10d).

A family of functions $\mathcal{V} \subset \mathcal{F}(X)$ is said to be *ubp closed* if the following condition is satisfied:

$$[\varphi \in \mathcal{F}(X), \{\varphi_n\} \text{ a sequence in } \mathcal{V}, \varphi_n \xrightarrow{ubp} \varphi] \Rightarrow \varphi \in \mathcal{V}.$$

6.3.2 Lemma. *Let (X, d) be a metric space, define the collection of families of functions*

$$\Gamma := \{\mathcal{V} \subset \mathcal{F}(X) : \mathcal{V} \text{ is ubp closed and } \mathcal{C}_B(X) \subset \mathcal{V}\}$$

(for $\mathcal{C}_B(X)$, cf. 3.1.10e), and then define the family of functions

$$\mathcal{E} := \bigcap_{\mathcal{V} \in \Gamma} \mathcal{V}.$$

The family $\mathcal{E}$ has the following properties:

(a) $\mathcal{E} \in \Gamma$;
(b) $\mathcal{E} \subset \mathcal{F}_B(X)$;
(c) $\varphi + \psi \in \mathcal{E}$ and $\varphi\psi \in \mathcal{E}$, $\forall \varphi, \psi \in \mathcal{E}$,
   $\alpha\varphi \in \mathcal{E}$, $\forall \alpha \in \mathbb{C}$, $\forall \varphi \in \mathcal{E}$;*
(d) $\chi_E \in \mathcal{E}$, $\forall E \in \mathcal{A}(d)$.

Proof. a: This follows immediately from the definitions.

b: If a sequence $\{\varphi_n\}$ in $\mathcal{F}_B(X)$ and $\varphi \in \mathcal{F}(X)$ are such that $\varphi_n \xrightarrow{ubp} \varphi$, then $\varphi \in \mathcal{F}_B(X)$ since φ is a ubp limit. Thus, $\mathcal{F}_B(X)$ is ubp closed. Moreover, $\mathcal{C}_B(X) \subset \mathcal{F}_B(X)$. This shows that $\mathcal{F}_B(X) \subset \Gamma$, and hence that $\mathcal{E} \subset \mathcal{F}_B(X)$.

c: Choose $h \in \mathcal{C}_B(X)$ and define

$$\mathcal{V}_h := \{f \in \mathcal{F}(X) : f + h \in \mathcal{E}\}.$$

We prove that $\mathcal{V}_h \in \Gamma$. If a sequence $\{f_n\}$ in $\mathcal{V}_h$ and $f \in \mathcal{F}(X)$ are such that $f_n \xrightarrow{ubp} f$, then clearly $f_n + h \xrightarrow{ubp} f + h$, and this implies (since $f_n + h \in \mathcal{E}$ and $\mathcal{E}$ is ubp closed, cf. property a) that $f + h \in \mathcal{E}$, i.e. $f \in \mathcal{V}_h$. Thus, $\mathcal{V}_h$ is ubp closed. Moreover, $\mathcal{C}_B(X) \subset \mathcal{V}_h$ since

$$f \in \mathcal{C}_B(X) \Rightarrow f + h \in \mathcal{C}_B(X) \overset{(1)}{\Rightarrow} f + h \in \mathcal{E} \Rightarrow f \in \mathcal{V}_h,$$

where 1 holds because $\mathcal{C}_B(X) \subset \mathcal{E}$ (cf. property a). Therefore, $\mathcal{V}_h \in \Gamma$. This implies that $\mathcal{E} \subset \mathcal{V}_h$, i.e. that

$$\psi + h \in \mathcal{E}, \forall \psi \in \mathcal{E}.$$

Since h was an arbitrary element of $\mathcal{C}_B(X)$, we have proved that

$$h \in C_B(X) \Rightarrow \psi + h \in \mathcal{E}, \forall \psi \in \mathcal{E}. \tag{2}$$

Choose now $\psi \in \mathcal{E}$ and define

$$\mathcal{V}_\psi := \{f \in \mathcal{F}(X) : f + \psi \in \mathcal{E}\}.$$

We prove that $\mathcal{V}_\psi \in \Gamma$. If a sequence $\{f_n\}$ in $\mathcal{V}_\psi$ and $f \in \mathcal{F}(X)$ are such that $f_n \xrightarrow{ubp} f$, then $f_n + \psi \xrightarrow{ubp} f + \psi$ (since $\psi \in \mathcal{F}_B(X)$, cf. property b), and this implies (since $f_n + \psi \in \mathcal{E}$ and $\mathcal{E}$ is ubp closed, cf. property a) that $f + \psi \in \mathcal{E}$, i.e. that $f \in \mathcal{V}_\psi$. Thus, $\mathcal{V}_\psi$ is ubp closed. Moreover, $\mathcal{C}_B(X) \subset \mathcal{V}_\psi$ since

$$h \in \mathcal{C}_B(X) \overset{(3)}{\Rightarrow} h + \psi \in \mathcal{E} \Rightarrow h \in \mathcal{V}_\psi,$$

where 3 follows from 2. Therefore, $\mathcal{V}_\psi \in \Gamma$. This implies that $\mathcal{E} \subset \mathcal{V}_\psi$, i.e. that

$$\varphi + \psi \in \mathcal{E}, \forall \varphi \in \mathcal{E}.$$

Since ψ was an arbitrary element of $\mathcal{E}$, we have indeed proved that

$$\varphi + \psi \in \mathcal{E}, \forall \varphi, \psi \in \mathcal{E}.$$

Proceeding as above, with pointwise sum of functions replaced by pointwise product, we can prove that

$$\varphi\psi \in \mathcal{E}, \forall \varphi, \psi \in \mathcal{E}.$$

Finally, we note that $\alpha_X \in \mathcal{C}_B(X)$ and hence (cf. property a) $\alpha_X \in \mathcal{E}$, for each $\alpha \in \mathbb{C}$ (for the constant function α_X cf. 1.2.19). Then we have

$$\alpha\varphi = (\alpha_X)\varphi \in \mathcal{E}, \forall \alpha \in \mathbb{C}, \forall \varphi \in \mathcal{E}.$$

d: Define the collection of subsets of X

$$\mathcal{A} := \{E \in \mathcal{P}(X) : \chi_E \in \mathcal{E}\}.$$

First, we prove that $\mathcal{A}$ is a σ-algebra, by showing that it has properties al_1 and al_2 of 6.1.5 and σa_1 of 6.1.13.

al_1: If $E, F \in \mathcal{A}$ then $\chi_E, \chi_F \in \mathcal{E}$, and hence (cf. 1.2.20 and property c)

$$\chi_{E \cup F} = \chi_E + \chi_F - \chi_E\chi_F \in \mathcal{E},$$

and this shows that $E \cup F \in \mathcal{A}$.

al_2: If $E \in \mathcal{A}$ then $\chi_E \in \mathcal{E}$; besides, $1_X \in \mathcal{E}$ since $\mathcal{C}_B(X) \subset \mathcal{E}$ (cf. property a); hence (cf. 1.2.20 and property c)

$$\chi_{X-E} = 1_X - \chi_E \in \mathcal{E},$$

and this shows that $X - E \in \mathcal{A}$.

σa_1: If $\{E_n\}$ is a sequence in $\mathcal{A}$, then induction applied to al_1 shows that $\bigcup_{n=1}^N E_n \in \mathcal{A}$, and hence that $\chi_{\bigcup_{n=1}^N E_n} \in \mathcal{E}$ for each $N \in \mathbb{N}$. Besides, directly from the definition of union we have

$$\forall x \in X, \chi_{\bigcup_{n=1}^N E_n}(x) \to \chi_{\bigcup_{n=1}^\infty E_n}(x) \text{ as } N \to \infty,$$

and hence $\chi_{\bigcup_{n=1}^N E_n} \xrightarrow{ubp} \chi_{\bigcup_{n=1}^\infty E_n}$. Since $\mathcal{E}$ is ubp closed (cf. property a), this implies that $\chi_{\bigcup_{n=1}^\infty E_n} \in \mathcal{E}$, and hence that $\bigcup_{n=1}^\infty E_n \in \mathcal{A}$.

Next, we prove that $\mathcal{T}_d \subset \mathcal{A}$. If $G \in \mathcal{T}_d$, then there exists a sequence $\{\psi_n\}$ in $\mathcal{C}_B(X)$ such that $\psi_n \xrightarrow{ubp} \chi_G$ (cf. 2.5.8). Since $\mathcal{C}_B(X) \subset \mathcal{E}$ and $\mathcal{E}$ is ubp closed (cf. property a), this shows that $\chi_G \in \mathcal{E}$, and hence that $G \in \mathcal{A}$.

Thus, we have $\mathcal{A}(d) := \mathcal{A}(\mathcal{T}_d) \subset \mathcal{A}$ (cf. $g\sigma_2$ of 6.1.17, with $\mathcal{F} := \mathcal{T}_d$), and this proves that

$$\chi_E \in \mathcal{E}, \forall E \in \mathcal{A}(d).$$

$\square$

6.3.3 Definition. Let (X, d) be a metric space. A function $\varphi \in \mathcal{F}(X)$ is called a *Borel function* if it is $\mathcal{A}(d)$-measurable, i.e. if it is measurable w.r.t $\mathcal{A}(d)$ and $\mathcal{A}(d_{\mathbb{C}})$. Thus, $\mathcal{M}(X, \mathcal{A}(d))$ is the family of all Borel functions, and $\mathcal{M}_B(X, \mathcal{A}(d))$ is the family of all bounded Borel functions.

6.3.4 Theorem. *Let (X, d) be a metric space. The family $\mathcal{M}_B(X, \mathcal{A}(d))$ of all bounded Borel functions is the smallest family of complex functions on X that is ubp closed and that contains $\mathcal{C}_B(X)$. More explicitly:*

(a) $\mathcal{M}_B(X, \mathcal{A}(d))$ is ubp closed and $\mathcal{C}_B(X) \subset \mathcal{M}_B(X, \mathcal{A}(d))$;
(b) if $\mathcal{V} \subset \mathcal{F}(X)$, $\mathcal{V}$ is ubp closed, and $\mathcal{C}_B(X) \subset \mathcal{V}$, then $\mathcal{M}_B(X, \mathcal{A}(d)) \subset \mathcal{V}$.

Proof. a: Suppose that a sequence $\{\varphi_n\}$ in $\mathcal{M}_B(X, \mathcal{A}(d))$ and $\varphi \in \mathcal{F}(X)$ are such that $\varphi_n \xrightarrow{ubp} \varphi$. Then $\varphi \in \mathcal{M}(X, \mathcal{A}(d))$ by 6.2.20c, and $\varphi \in \mathcal{F}_B(X)$ since φ is a ubp limit. This shows that $\mathcal{M}_B(X, \mathcal{A}(d))$ is ubp closed. Moreover, the inclusion $\mathcal{C}_B(X) \subset \mathcal{M}_B(X, \mathcal{A}(d))$ holds by 6.2.8.

b: We prove this property of $\mathcal{M}_B(X, \mathcal{A}(d))$ by proving the inclusion $\mathcal{M}_B(X, \mathcal{A}(d)) \subset \mathcal{E}$, where $\mathcal{E}$ is the family of functions defined in 6.3.2. Indeed, for every $\varphi \in \mathcal{M}_B(X, \mathcal{A}(d))$, by 6.2.27 there exists a sequence $\{\psi_n\}$ in $\mathcal{S}(X, \mathcal{A}(d))$ such that $\psi_n \xrightarrow{ubp} \varphi$. Now, properties c and d of 6.3.2 imply that $\mathcal{S}(X, \mathcal{A}(d)) \subset \mathcal{E}$. Since $\mathcal{E}$ is ubp closed (cf. 6.3.2a), this shows that $\varphi \in \mathcal{E}$. $\qquad\square$

6.3.5 Remark. The statement of 6.3.4 is clearly equivalent to the equality

$$\mathcal{M}_B(X, \mathcal{A}(d)) = \mathcal{E},$$

where $\mathcal{E}$ is the family of functions defined in 6.3.2.

6.3.6 Remark. By substituting pointwise limits for ubp limits and dropping everywhere any condition of boundedness, the whole reasoning of this section can be rerun to prove that, for any metric space (X, d), the family $\mathcal{M}(X, \mathcal{A}(d))$ of all Borel functions is the smallest subset of $\mathcal{F}(X)$ that is closed with respect to pointwise convergence and that contains $\mathcal{C}(X)$ (for $\mathcal{C}(X)$, cf. 3.1.10e).

Chapter 7

Measures

7.1 Additive functions, premeasures, measures

Throughout this section, X stands for a non-empty set.

7.1.1 Definitions. Let $\mathcal{A}_0$ be an algebra on X. An *additive function on* $\mathcal{A}_0$ is a function $\mu_0 : \mathcal{A}_0 \to [0, \infty]$ with the following properties:

(af_1) $\mu_0(\emptyset) = 0$;
(af_2) for every finite and disjoint family $\{E_1, ..., E_n\}$ of elements of $\mathcal{A}_0$,

$$\mu_0(\cup_{k=1}^n E_k) = \sum_{k=1}^n \mu_0(E_k)$$

(this property of μ_0 is called *additivity*).

If a function $\mu_0 : \mathcal{A}_0 \to [0, \infty]$ is so that there exists $E \in \mathcal{A}_0$ for which $\mu_0(E) < \infty$, then property af_2 implies property af_1: for $E \in \mathcal{A}_0$, property af_2 implies $\mu_0(E) = \mu_0(E) + \mu_0(\emptyset)$, and this implies $\mu_0(\emptyset) = 0$ if $\mu_0(E) < \infty$.

An additive function μ_0 on $\mathcal{A}_0$ is said to be *σ-finite* if there exists a countable family $\{E_n\}_{n \in I}$ of elements of $\mathcal{A}_0$ so that $\mu_0(E_n) < \infty$ for all $n \in I$ and $X = \cup_{n \in I} E_n$.

An additive function μ_0 on $\mathcal{A}_0$ is said to be *finite* if $\mu_0(X) < \infty$. Clearly, a finite additive function is also σ-finite.

7.1.2 Proposition. *Let $\mathcal{A}_0$ be an algebra on X and μ_0 an additive function on $\mathcal{A}_0$. Then:*

(a) if $E, F \in \mathcal{A}_0$, then

$$E \subset F \Rightarrow \mu_0(E) \leq \mu_0(F)$$

(this property of μ_0 is called monotonicity);
(b) if $\{E_1, ..., E_N\}$ is a finite family of elements of $\mathcal{A}_0$, then

$$\mu_0(\cup_{n=1}^N E_n) \leq \sum_{n=1}^N \mu_0(E_n)$$

(this property of μ_0 is called subadditivity).

Proof. a: If $E, F \in \mathcal{A}_0$ are such that $E \subset F$, then $(F - E) \cup E = F$. Since $(F-E) \cap E = \emptyset$, by af_2 we have $\mu_0(F-E) + \mu_0(E) = \mu_0(F)$, whence $\mu_0(E) \leq \mu_0(F)$ (since $0 \leq \mu_0(F - E)$ implies $\mu_0(E) \leq \mu_0(F - E) + \mu_0(E)$, cf. 5.3.2c).

b: If $\{E_1, ..., E_N\}$ is a finite family of elements of $\mathcal{A}_0$, then there exists a disjoint family $\{F_1, ..., F_N\}$ of elements of $\mathcal{A}_0$ such that

$$\cup_{n=1}^{N} F_n = \cup_{n=1}^{N} E_n \text{ and } F_n \subset E_n \text{ for } n = 1, ..., N$$

(cf. 6.1.8, with $E_n := \emptyset$ for $n > N$). Then we have

$$\mu_0(\cup_{n=1}^{N} E_n) = \mu_0(\cup_{n=1}^{N} F_n) = \sum_{n=1}^{N} \mu_0(F_n) \leq \sum_{n=1}^{N} \mu_0(E_n)$$

by af_2, by the monotonicity of μ_0 proved in a, and by induction applied to 5.3.2e. $\square$

7.1.3 Definition. Let $\mathcal{A}_0$ be an algebra on X. An additive function μ_0 on $\mathcal{A}_0$ is said to be a *premeasure* if it has the following property (which is called *σ-additivity*):

(pm) for every sequence $\{E_n\}$ in $\mathcal{A}_0$ such that $\cup_{n=1}^{\infty} E_n \in \mathcal{A}_0$ and $E_i \cap E_j = \emptyset$ if $i \neq j$,

$$\mu_0(\cup_{n=1}^{\infty} E_n) = \sum_{n=1}^{\infty} \mu_0(E_n)$$

7.1.4 Proposition. *Let $\mathcal{A}_0$ be an algebra on X and μ_0 a premeasure on $\mathcal{A}_0$. Then:*

(a) if $\{E_n\}$ is a sequence in $\mathcal{A}_0$ such that $\cup_{n=1}^{\infty} E_n \in \mathcal{A}_0$, then

$$\mu_0(\cup_{n=1}^{\infty} E_n) \leq \sum_{n=1}^{\infty} \mu_0(E_n)$$

(this property of μ_0 is called σ-subadditivity);

(b) if $\{E_n\}$ is a sequence in $\mathcal{A}_0$ such that $\cup_{n=1}^{\infty} E_n \in \mathcal{A}_0$ and $E_n \subset E_{n+1}$ for all $n \in \mathbb{N}$, then

$$\mu_0(\cup_{n=1}^{\infty} E_n) = \lim_{n \to \infty} \mu_0(E_n) = \sup_{n \geq 1} \mu_0(E_n)$$

(this property of μ_0 is called continuity from below);

(c) if $\{E_n\}$ is a sequence in $\mathcal{A}_0$ such that $\cap_{n=1}^{\infty} E_n \in \mathcal{A}_0$ and $E_{n+1} \subset E_n$ for all $n \in \mathbb{N}$, and if there exists $l \in \mathbb{N}$ so that $\mu_0(E_l) < \infty$, then

$$\mu_0(\cap_{n=1}^{\infty} E_n) = \lim_{n \to \infty} \mu_0(E_n) = \inf_{n \geq 1} \mu_0(E_n)$$

(this property of μ_0 is called continuity from above).

Proof. a: If $\{E_n\}$ is a sequence in $\mathcal{A}_0$, then there exists a sequence $\{F_n\}$ in $\mathcal{A}_0$ such that

$$\cup_{n=1}^{\infty} F_n = \cup_{n=1}^{\infty} E_n, F_n \subset E_n \text{ for all } n \in \mathbb{N}, F_k \cap F_l = \emptyset \text{ if } k \neq l$$

(cf. 6.1.8). If $\cup_{n=1}^{\infty} E_n \in \mathcal{A}_0$, then we have

$$\mu_0(\cup_{n=1}^{\infty} E_n) = \mu_0(\cup_{n=1}^{\infty} F_n) = \sum_{n=1}^{\infty} \mu_0(F_n) \leq \sum_{n=1}^{\infty} \mu_0(E_n)$$

by *pm*, by the monotonicity of μ_0, and by 5.4.2a.

b: If $\{E_n\}$ is a sequence in $\mathcal{A}_0$ such that $E_n \subset E_{n+1}$ for all $n \in \mathbb{N}$, then there exists a sequence $\{F_n\}$ in $\mathcal{A}_0$ such that

$$\cup_{n=1}^{\infty} F_n = \cup_{n=1}^{\infty} E_n, \cup_{k=1}^{n} F_k = E_n \text{ for all } n \in \mathbb{N}, F_k \cap F_l = \emptyset \text{ if } k \neq l$$

(cf. 6.1.8). By af_2 we have

$$\forall n \in \mathbb{N}, \sum_{k=1}^{n} \mu_0(F_k) = \mu_0(\cup_{k=1}^{n} F_k) = \mu_0(E_n),$$

and hence, if $\cup_{n=1}^{\infty} E_n \in \mathcal{A}_0$, by *pm* we have

$$\mu_0(\cup_{n=1}^{\infty} E_n) = \mu_0(\cup_{n=1}^{\infty} F_n)$$
$$= \sum_{n=1}^{\infty} \mu_0(F_n) = \lim_{n \to \infty} \sum_{k=1}^{n} \mu_0(F_k) = \lim_{n \to \infty} \mu_0(E_n),$$

or

$$\mu_0(\cup_{n=1}^{\infty} E_n) = \sum_{n=1}^{\infty} \mu_0(F_n) = \sup_{n \geq 1} \sum_{k=1}^{n} \mu_0(F_k) = \sup_{n \geq 1} \mu_0(E_n).$$

c: Let $\{E_n\}$ be a sequence in $\mathcal{A}_0$ such that $\cap_{n=1}^{\infty} E_n \in \mathcal{A}_0$ and $E_{n+1} \subset E_n$ for all $n \in \mathbb{N}$, and suppose that there exists $l \in \mathbb{N}$ so that $\mu_0(E_l) < \infty$. We may assume $l = 1$ since, if it is not already so, we may replace the sequence $\{E_n\}$ with the sequence $\{E_{l+n}\}$ since $\cap_{n=1}^{\infty} E_n = \cap_{n=1}^{\infty} E_{l+n}$ and of course $\lim_{n \to \infty} \mu_0(E_n) = \lim_{n \to \infty} \mu_0(E_{l+n})$.

Letting $F_n := E_1 - E_n$ for all $n \in \mathbb{N}$, we have a sequence $\{F_n\}$ in $\mathcal{A}_0$ such that $F_n \subset F_{n+1}$ for all $n \in \mathbb{N}$, and also such that $\cup_{n=1}^{\infty} F_n \in \mathcal{A}_0$ since

$$\cup_{n=1}^{\infty} F_n = \cup_{n=1}^{\infty} (E_1 \cap (X - E_n)) = E_1 \cap (X - \cap_{n=1}^{\infty} E_n) = E_1 - \cap_{n=1}^{\infty} E_n.$$

Then, by part b, we have

$$\mu_0(\cup_{n=1}^{\infty} F_n) = \lim_{n \to \infty} \mu_0(F_n).$$

Notice that $\lim_{n \to \infty} \mu_0(F_n) < \infty$ since $\cup_{n=1}^{\infty} F_n \subset E_1$ implies, by the monotonicity of μ_0, $\mu_0(\cup_{n=1}^{\infty} F_n) \leq \mu_0(E_1) < \infty$. We also have

$$E_1 = (\cap_{n=1}^{\infty} E_n) \cup (\cup_{n=1}^{\infty} F_n) \text{ and } (\cap_{n=1}^{\infty} E_n) \cap (\cup_{n=1}^{\infty} F_n) = \emptyset,$$

whence, by af_2,

$$\mu_0(E_1) = \mu_0(\cap_{n=1}^{\infty} E_n) + \mu_0(\cup_{n=1}^{\infty} F_n),$$

which implies, since $\mu_0(\cup_{n=1}^{\infty} F_n) < \infty$,

$$\mu_0(\cap_{n=1}^{\infty} E_n) = \mu_0(E_1) - \mu_0(\cup_{n=1}^{\infty} F_n).$$

Thus we have

$$\mu_0(\cap_{n=1}^\infty E_n) = \mu_0(E_1) - \lim_{n\to\infty} \mu_0(F_n).$$

From

$$\forall n \in \mathbb{N}, E_1 = E_n \cup F_n \text{ and } E_n \cap F_n = \emptyset$$

we have

$$\forall n \in \mathbb{N}, \mu_0(E_1) = \mu_0(E_n) + \mu_0(F_n),$$

whence, since $E_n \subset E_1$ implies $\mu_0(E_n) \le \mu_0(E_1) < \infty$,

$$\forall n \in \mathbb{N}, \mu_0(F_n) = \mu_0(E_1) - \mu_0(E_n),$$

and this implies, since all the terms involved are in $\mathbb{R}$ and so is $\lim_{n\to\infty}\mu_0(F_n)$, that the sequence $\{\mu_0(E_n)\}$ is convergent to a limit in $\mathbb{R}$ and that

$$\lim_{n\to\infty}\mu_0(F_n) = \mu_0(E_1) - \lim_{n\to\infty}\mu_0(E_n).$$

From all this we derive (recall that $\mu_0(E_1) < \infty$)

$$\mu_0(\cap_{n=1}^\infty E_n) = \mu_0(E_1) - (\mu_0(E_1) - \lim_{n\to\infty}\mu_0(E_n)) = \lim_{n\to\infty}\mu_0(E_n).$$

Finally, from 7.1.2a and from 5.2.5 we obtain $\lim_{n\to\infty}\mu_0(E_n) = \inf_{n\ge1}\mu_0(E_n)$. $\square$

7.1.5 Theorem (Alexandroff's theorem). *Suppose that we have a distance d on X, an algebra $\mathcal{A}_0$ on X, an additive function μ_0 on $\mathcal{A}_0$, and that the following two conditions are satisfied:*

(a) for each $E \in \mathcal{A}_0$,

$$\mu_0(E) = \sup\{\mu_0(F) : F \in \mathcal{A}_0, \overline{F} \subset E, \overline{F} \text{ is compact}\}$$

(this condition is consistent because $F \in \mathcal{A}_0$ and $\overline{F} \subset E$ imply $\mu_0(F) \le \mu_0(E)$ by the monotonicity of μ_0, and because $\emptyset \in \mathcal{A}_0$, $\overline{\emptyset} = \emptyset \subset E$, $\emptyset$ is compact);
(b) for each $E \in \mathcal{A}_0$ such that $\mu_0(E) < \infty$,

$$\forall \epsilon > 0, \exists G_\epsilon \in \mathcal{A}_0 \text{ s.t. } E \subset G_\epsilon^\circ \text{ and } \mu_0(G_\epsilon) - \mu_0(E) < \epsilon.$$

Then μ_0 is a premeasure.
If $\mu_0(X) < \infty$ then condition a implies condition b.

Proof. First we prove that conditions a and b imply that μ_0 has property *pm* of 7.1.3. Let then $\{E_n\}$ be a sequence in $\mathcal{A}_0$ such that $\cup_{n=1}^\infty E_n \in \mathcal{A}_0$ and $E_i \cap E_j = \emptyset$ if $i \ne j$. The additivity and the monotonicity of μ_0 imply that

$$\forall N \in \mathbb{N}, \sum_{n=1}^N \mu_0(E_n) = \mu_0(\cup_{n=1}^N E_n) \le \mu_0(\cup_{n=1}^\infty E_n),$$

whence

$$\sum_{n=1}^\infty \mu_0(E_n) := \sup_{N\ge1}\sum_{n=1}^N \mu_0(E_n) \le \mu_0(\cup_{n=1}^\infty E_n). \tag{1}$$

If $\sum_{n=1}^{\infty} \mu_0(E_n) = \infty$, 1 implies that

$$\sum_{n=1}^{\infty} \mu_0(E_n) = \mu_0(\cup_{n=1}^{\infty} E_n).$$

Thus our task is now to prove that conditions a and b imply that

$$\mu_0(\cup_{n=1}^{\infty} E_n) \leq \sum_{n=1}^{\infty} \mu_0(E_n),$$

assuming that $\sum_{n=1}^{\infty} \mu_0(E_n) < \infty$ (we will see that, in this part of the proof, no role is played by the condition $E_i \cap E_j = \emptyset$ if $i \neq j$, which however has already played its role in the proof of 1). Assume then $\sum_{n=1}^{\infty} \mu_0(E_n) < \infty$ (this implies that $\mu_0(E_n) < \infty$ for all $n \in \mathbb{N}$), and consider any $F \in \mathcal{A}_0$ such that $\overline{F} \subset \cup_{n=1}^{\infty} E_n$ and such that $\overline{F}$ is compact. Choose $\epsilon > 0$. For each $n \in \mathbb{N}$, condition b for E_n implies that

$$\exists G_{n,\epsilon} \in \mathcal{A}_0 \text{ s.t. } E_n \subset G_{n,\epsilon}^{\circ} \text{ and } \mu_0(G_{n,\epsilon}) - \mu_0(E_n) < \frac{\epsilon}{2^n}.$$

Since $\cup_{n=1}^{\infty} G_{n,\epsilon}^{\circ} \supset \cup_{n=1}^{\infty} E_n \supset \overline{F}$ and $\overline{F}$ is compact, there exists $N \in \mathbb{N}$ so that $\cup_{n=1}^{N} G_{n,\epsilon}^{\circ} \supset \overline{F}$, and hence so that

$$\cup_{n=1}^{N} G_{n,\epsilon} \supset \cup_{n=1}^{N} G_{n,\epsilon}^{\circ} \supset \overline{F} \supset F.$$

Then we have

$$\sum_{n=1}^{\infty} \mu_0(E_n) \geq \sum_{n=1}^{N} \mu_0(E_n) > \sum_{n=1}^{N} (\mu_0(G_{n,\epsilon}) - \frac{\epsilon}{2^n}) > \sum_{n=1}^{N} \mu_0(G_{n,\epsilon}) - \epsilon$$
$$\geq \mu_0(\cup_{n=1}^{N} G_{n,\epsilon}) - \epsilon \geq \mu_0(F) - \epsilon,$$

where the subadditivity and the monotonicity of μ_0 have been used. Since ϵ was arbitrary, this proves that

$$\sum_{n=1}^{\infty} \mu_0(E_n) \geq \mu_0(F).$$

Since F was any element of $\mathcal{A}_0$ such that $\overline{F} \subset \cup_{n=1}^{\infty} E_n$ and such that $\overline{F}$ was compact, condition a for $\cup_{n=1}^{\infty} E_n$ implies that

$$\sum_{n=1}^{\infty} \mu_0(E_n) \geq \mu_0(\cup_{n=1}^{\infty} E_n),$$

which along with 1 proves that

$$\sum_{n=1}^{\infty} \mu_0(E_n) = \mu_0(\cup_{n=1}^{\infty} E_n).$$

Now suppose that $\mu_0(X) < \infty$. Notice that this implies that $\mu_0(H) < \infty$ for all $H \in \mathcal{A}_0$, by the monotonicity of μ_0. We will prove that from this it follows that condition a implies condition b. Assume then condition a, and consider any $E \in \mathcal{A}_0$. Choose $\epsilon > 0$. Since $X - E \in \mathcal{A}_0$, condition a for $X - E$ implies that

$$\exists F_\epsilon \in \mathcal{A}_0 \text{ s.t. } \overline{F}_\epsilon \subset X - E \text{ and } \mu_0(X - E) - \mu_0(F_\epsilon) < \epsilon.$$

Letting $G_\epsilon := X - F_\epsilon$, we have:

$G_\epsilon \in \mathcal{A}_0$;

$E \subset G_\epsilon^\circ$ (because $\overline{F}_\epsilon \subset X - E$ implies $E \subset X - \overline{F}_\epsilon$, and $X - \overline{F}_\epsilon \subset (X - F_\epsilon)^\circ$ follows from $X - \overline{F}_\epsilon \in \mathcal{T}_d$ and $X - \overline{F}_\epsilon \subset X - F_\epsilon$);

$\mu_0(G_\epsilon) - \mu_0(E) = \mu_0(X) - \mu_0(F_\epsilon) - \mu_0(E) = \mu_0(X - E) - \mu_0(F_\epsilon) < \epsilon$, where $\mu_0(G_\epsilon) = \mu_0(X) - \mu_0(F_\epsilon)$ and $\mu_0(X) - \mu_0(E) = \mu_0(X - E)$ are true because μ_0 is finite.

Since ϵ was arbitrary, this proves condition b for E. $\qquad\qquad\square$

7.1.6 Corollary. *Suppose that we have a distance d on X, a semialgebra $\mathcal{S}$ on X, an additive function μ_0 on $\mathcal{A}_0(\mathcal{S})$ (the algebra on X generated by $\mathcal{S}$), and that the following two conditions are satisfied:*

(a) for each $E \in \mathcal{S}$,

$$\mu_0(E) = \sup\left\{\mu_0(F) : F \in \mathcal{S}, \overline{F} \subset E, \overline{F} \text{ is compact}\right\}$$

(this condition is consistent by the monotonicity of μ_0 and because $\emptyset \in \mathcal{S}$);
(b) for each $E \in \mathcal{S}$ such that $\mu_0(E) < \infty$,

$$\forall \epsilon > 0, \exists G_\epsilon \in \mathcal{S} \text{ s.t. } E \subset G_\epsilon^\circ \text{ and } \mu_0(G_\epsilon) - \mu_0(E) < \epsilon.$$

Then μ_0 is a premeasure.
If $\mu_0(X) < \infty$, then condition a is enough to make μ_0 a premeasure.

Proof. We will prove this corollary by proving that condition a of the statement implies condition 7.1.5a and that condition b implies condition 7.1.5b.

Consider then $E \in \mathcal{A}_0(\mathcal{S})$. By 6.1.11, there is a finite and disjoint family $\{E_1, ..., E_n\}$ of elements of $\mathcal{S}$ so that $E = \cup_{k=1}^n E_k$.

Assume condition a and suppose first that $\mu_0(E_0) = \infty$. Then there exists $l \in \{1, ..., n\}$ such that $\mu_0(E_l) = \infty$, for otherwise we would have $\mu_0(E) < \infty$ by the additivity of μ_0. Notice that, since $\mathcal{S} \subset \mathcal{A}_0(\mathcal{S})$ and $E_l \subset E$,

$$\left\{F \in \mathcal{S} : \overline{F} \subset E_l, \overline{F} \text{ is compact}\right\} \subset \left\{F \in \mathcal{A}_0(\mathcal{S}) : \overline{F} \subset E, \overline{F} \text{ is compact}\right\}.$$

Condition a for E_l is

$$\infty = \mu_0(E_l) = \sup\left\{\mu_0(F) : F \in \mathcal{S}, \overline{F} \subset E_l, \overline{F} \text{ is compact}\right\}.$$

Hence we have

$$\sup\left\{\mu_0(F) : F \in \mathcal{A}_0(\mathcal{S}), \overline{F} \subset E, \overline{F} \text{ is compact}\right\} = \infty = \mu_0(E),$$

which is condition 7.1.5a for E.

Assume next condition a and suppose that $\mu_0(E) < \infty$. Then $\mu_0(E_k) < \infty$ for $k = 1, ..., n$ by the monotonicity of μ_0. Choose $\epsilon > 0$. For $k = 1, ..., n$, condition a for E_k implies that

$$\exists F_{k,\epsilon} \in \mathcal{S} \text{ s.t. } \overline{F}_{k,\epsilon} \subset E_k, \overline{F}_{k,\epsilon} \text{ is compact}, \mu_0(E_k) - \mu_0(F_{k,\epsilon}) < \frac{\epsilon}{n}.$$

Letting $F_\epsilon := \cup_{k=1}^n F_{k,\epsilon}$, we have:

$F_\epsilon \in \mathcal{A}_0(\mathcal{S})$ (since $\mathcal{S} \subset \mathcal{A}_0(\mathcal{S})$);

$\overline{F}_\epsilon = \cup_{k=1}^n \overline{F}_{k,\epsilon} \subset \cup_{k=1}^n E_k = E$ (cf. 2.3.9e);

$\overline{F}_\epsilon$ is compact (since $\overline{F}_\epsilon = \cup_{k=1}^n \overline{F}_{k,\epsilon}$, cf. 2.8.9);

$\mu_0(E) - \mu_0(F_\epsilon) = \sum_{k=1}^n \mu_0(E_k) - \sum_{k=1}^n \mu_0(F_{k,\epsilon}) < \epsilon$, where the additivity of μ_0 has been used (note that $F_{k,\epsilon} \subset E_k$ implies $F_{i,\epsilon} \cap F_{j,\epsilon} = \emptyset$ if $i \neq j$).

Since ϵ was arbitrary, this proves condition 7.1.5a for E.

Finally, assume condition b and suppose that $\mu_0(E) < \infty$. Choose $\epsilon > 0$. For $k = 1, ..., n$, since $\mu_0(E_k) < \infty$, condition b for E_k implies that

$$\exists G_{k,\epsilon} \in \mathcal{S} \text{ s.t. } E_k \subset G_{k,\epsilon}^\circ \text{ and } \mu_0(G_{k,\epsilon}) - \mu_0(E_k) < \frac{\epsilon}{n}.$$

Letting $G_\epsilon := \cup_{k=1}^n G_{k,\epsilon}$, we have:

$G_\epsilon \in \mathcal{A}_0(\mathcal{S})$ (since $\mathcal{S} \subset \mathcal{A}_0(\mathcal{S})$);

$E = \cup_{k=1}^n E_k \subset \cup_{k=1}^n G_{k,\epsilon}^\circ \subset (\cup_{k=1}^n G_{k,\epsilon})^\circ = G_\epsilon^\circ$ (cf. 2.2.7e);

$\mu_0(G_\epsilon) - \mu_0(E) \leq \sum_{k=1}^n \mu_0(G_{k,\epsilon}) - \sum_{k=1}^n \mu_0(E_k) < \epsilon$, where the subadditivity of μ_0 has been used.

Since ϵ was arbitrary, this proves condition 7.1.5b for E. $\qquad\square$

7.1.7 Definitions. Let $\mathcal{A}$ be a σ-algebra on X. A premeasure on $\mathcal{A}$ is said to be a *measure*. Thus, a measure on $\mathcal{A}$ is an additive function on $\mathcal{A}$ which has the following property:

(*me*) for every sequence $\{E_n\}$ in $\mathcal{A}$ such that $E_i \cap E_j = \emptyset$ if $i \neq j$,

$$\mu(\cup_{n=1}^\infty E_n) = \sum_{n=1}^\infty \mu(E_n).$$

To prove that a function $\mu : \mathcal{A} \to [0, \infty]$ is a measure, it is enough to prove that it has the property af_1 of 7.1.1 and me, since these two properties imply property af_2 of 7.1.1: for any finite and disjoint family $\{E_1, ..., E_n\}$ of elements of $\mathcal{A}$, consider the sequence $\{E_k\}$ defined by letting $E_k := \emptyset$ for $k > n$.

If μ is a measure on $\mathcal{A}$, the triple $(X, \mathcal{A}, \mu)$ is called a *measure space*, and it is said to be σ-*finite* if μ is σ-finite.

A measure μ on $\mathcal{A}$ is called a *probability measure* if $\mu(X) = 1$.

A measure μ on $\mathcal{A}$ is said to be *complete* if it has the following property

(*cm*) $[F \in \mathcal{P}(X), \exists E \in \mathcal{A} \text{ s.t. } F \subset E \text{ and } \mu(E) = 0] \Rightarrow F \in \mathcal{A}$.

The *null measure* on $\mathcal{A}$ is the function that assigns the value 0 to each element of $\mathcal{A}$.

7.1.8 Proposition. *Let $\mathcal{A}$ be a σ-algebra on X, μ a measure on $\mathcal{A}$, and $a \in [0, \infty]$. Then the function*

$$a\mu : \mathcal{A} \to [0, \infty]$$
$$E \mapsto (a\mu)(E) := a\mu(E)$$

is a measure on $\mathcal{A}$.

Proof. Use 5.4.5.   □

7.1.9 Definition. Let $\mathcal{A}$ be a σ-algebra on X and μ a measure on $\mathcal{A}$. Let E be an element of $\mathcal{A}$ and, for each $x \in E$, let $P(x)$ be a proposition, i.e. a statement about x which is either true or false (but not both). We write "$P(x)$ μ-a.e. on E" or "$P(x)$ is true μ-a.e. on E" or "$P(x)$ is true for μ-a.e. $x \in E$" ("a.e." is read "*almost everywhere*" in the first two cases, and "*almost every*" in the third one) when the following condition is satisfied:

$$\exists F \in \mathcal{A} \text{ such that } \mu(F) = 0 \text{ and } P(x) \text{ is true for all } x \in E - F.$$

7.1.10 Remark. Let $\mathcal{A}, \mu, E, P(x)$ be as in 7.1.9. Note that, if $P(x)$ is true μ-a.e. on E and $F \in \mathcal{A}$ is s.t. $\mu(F) = 0$ and $P(x)$ is true for all $x \in E - F$, then

$$E_f := \{x \in E : P(x) \text{ is false}\} \subset E \cap F \text{ and } \mu(E \cap F) = 0 \text{ (cf. 7.1.2a)}.$$

However, this does not imply $\mu(E_f) = 0$, because E_f need not to be an element of $\mathcal{A}$; if it is, then $\mu(E_f) = 0$ is true by the monotonicity of μ.

It is obvious that, if μ is complete, then

$$P(x) \text{ } \mu\text{-a.e. on } E \Leftrightarrow (E_f \in \mathcal{A} \text{ and } \mu(E_f) = 0).$$

It must be pointed out that sometimes the nature of $P(x)$ makes E_f an element of $\mathcal{A}$. An istance of this is when $E := X$ and, for $\varphi, \psi \in \mathcal{M}(X, \mathcal{A})$, $P(x)$ is "$\varphi(x) = \psi(x)$". Indeed, in this case, $E_f = (\varphi - \psi)^{-1}(\mathbb{C} - \{0\}) \in \mathcal{A}$ by 6.2.16 and 6.2.13c (with $\mathcal{G} := \mathcal{T}_d$). When the nature of $P(x)$ makes E_f an element of $\mathcal{A}$, then it is obvious that

$$P(x) \text{ } \mu\text{-a.e. on } X \Leftrightarrow \mu(E_f) = 0$$

even if μ is not complete.

7.2 Outer measures

Throughout this section, X stands for a non-empty set.

Carathéodory's theorem, along with the construction of an outer measure set forth in 7.2.4, shows how we can construct a measure on a σ-algebra on X starting from almost any non-negative function defined on almost any family of subsets of X.

7.2.1 Definitions. An *outer measure on* X is a function $\mu^* : \mathcal{P}(X) \to [0, \infty]$ which satisfies the following conditions:

(om_1) $\mu^*(\emptyset) = 0$;
(om_2) if $E, F \in \mathcal{P}(X)$ are such that $E \subset F$, then $\mu^*(E) \leq \mu^*(F)$;
(om_3) for every sequence $\{E_n\}$ in $\mathcal{P}(X)$, $\mu^*(\cup_{n=1}^{\infty} E_n) \leq \sum_{n=1}^{\infty} \mu^*(E_n)$.

If μ^* is an outer measure on X and $\{E_1, ..., E_N\}$ is a finite family of subsets of X, define $E_n := \emptyset$ for $n > N$ to obtain, by conditions om_3 and om_1,

$$\mu^*(\cup_{n=1}^N E_n) \doteq \mu^*(\cup_{n=1}^\infty E_n) \le \sum_{n=1}^\infty \mu^*(E_n) = \sum_{n=1}^N \mu^*(E_n).$$

If μ^* is an outer measure on X, a set $E \in \mathcal{P}(X)$ is called μ^*-*measurable* if

$$\mu^*(A) = \mu^*(A \cap E) + \mu^*(A \cap (X - E)), \forall A \in \mathcal{P}(X).$$

7.2.2 Proposition. *Let μ^* be an outer measure on X. A set $E \in \mathcal{P}(X)$ is μ^*- measurable if*

$$\mu^*(A \cap E) + \mu^*(A \cap (X - E)) \le \mu^*(A) \text{ for all } A \in \mathcal{P}(X) \text{ such that } \mu^*(A) < \infty.$$

Proof. For each $E \in \mathcal{P}(X)$, conditions om_3 and om_1 imply that

$$\mu^*(A) \le \mu^*(A \cap E) + \mu^*(A \cap (X - E)), \forall A \in \mathcal{P}(X), \tag{1}$$

since $A = (A \cap E) \cup (A \cap (X - E))$, and 1 implies that

$$\mu^*(A) = \mu^*(A \cap E) + \mu^*(A \cap (X - E)) \text{ for all } A \in \mathcal{P}(X) \text{ such that } \mu^*(A) = \infty. \tag{2}$$

If $E \in \mathcal{P}(X)$ is such that

$$\mu^*(A \cap E) + \mu^*(A \cap (X - E)) \le \mu^*(A) \text{ for all } A \in \mathcal{P}(X) \text{ such that } \mu^*(A) < \infty,$$

then in view of 1 we have

$$\mu^*(A) = \mu^*(A \cap E) + \mu^*(A \cap (X - E)) \text{ for all } A \in \mathcal{P}(X) \text{ such that } \mu^*(A) < \infty,$$

which, together with 2, proves that E is μ^*-measurable. $\qquad\square$

7.2.3 Theorem (Carathéodory's theorem). *Let μ^* be an outer measure on X. Then the collection $\mathcal{M}$ of μ^*-measurable subsets of X is a σ-algebra on X and the restriction $\mu^*_\mathcal{M}$ of μ^* to $\mathcal{M}$ is a complete measure on $\mathcal{M}$.*

Proof. First, we observe that $X - E \in \mathcal{M}$ whenever $E \in \mathcal{M}$, since the definition of μ^*-measurability is symmetric in E and $X - E$ (since $X - (X - E) = E$).

Next, suppose $E, F \in \mathcal{M}$ and let A be an arbitrary subset of X. Since

$$E \cup F = (E \cap F) \cup (E \cap (X - F)) \cup ((X - E) \cap F),$$

by conditions om_3 and om_1 we have

$$\mu^*(A \cap (E \cup F)) \le \mu^*(A \cap E \cap F) + \mu^*(A \cap E \cap (X - F)) + \mu^*(A \cap (X - E) \cap F),$$

and hence

$$\mu^*(A \cap (E \cup F)) + \mu^*(A \cap (X - (E \cup F)))$$
$$\le \mu^*(A \cap E \cap F) + \mu^*(A \cap E \cap (X - F))$$
$$+ \mu^*(A \cap (X - E) \cap F) + \mu^*(A \cap (X - E) \cap (X - F))$$
$$= \mu^*(A \cap E) + \mu^*(A \cap (X - E)) = \mu^*(A),$$

since $E, F \in \mathcal{M}$. In view of 7.2.2, this shows that $E \cup F \in \mathcal{M}$. Thus, $\mathcal{M}$ is an algebra on X.

Suppose now that $E_1 \in \mathcal{M}$, $E_2 \in \mathcal{P}(X)$ and $E_1 \cap E_2 = \emptyset$. Then

$$\mu^*(E_1 \cup E_2) = \mu^*((E_1 \cup E_2) \cap E_1) + \mu^*((E_1 \cup E_2) \cap (X - E_1))$$
$$= \mu^*(E_1) + \mu^*(E_2).$$

This proves that, for every disjoint pair E_1, E_2 of elements of $\mathcal{M}$,

$$\mu^*(E_1 \cup E_2) = \mu^*(E_1) + \mu^*(E_2).$$

Applying induction to this result, we obtain property af_2 of 7.1.1 for $\mu_\mathcal{M}^*$. Since property af_1 is ensured by condition om_1, we can conclude that $\mu_\mathcal{M}^*$ is an additive function on $\mathcal{M}$.

We will now show that $\mathcal{M}$ and μ^* have the following properties: if $\{E_n\}$ is a sequence in $\mathcal{M}$ such that $E_i \cap E_j = \emptyset$ for $i \neq j$, then

$$\cup_{n=1}^\infty E_n \in \mathcal{M} \text{ and } \mu^*(\cup_{n=1}^\infty E_n) = \sum_{n=1}^\infty \mu^*(E_n).$$

In view of 6.1.8, this will prove first that $\mathcal{M}$ has property σa_1 of 6.1.13, and hence that $\mathcal{M}$ is a σ-algebra on X, and second that $\mu_\mathcal{M}^*$ has property pm of 7.1.3, and hence that $\mu_\mathcal{M}^*$ is a measure on $\mathcal{M}$.

Then let $\{E_n\}$ be a sequence in $\mathcal{M}$ such that $E_i \cap E_j = \emptyset$ for $i \neq j$, define the sets $F_n := \cup_{k=1}^n E_k$ for all $n \in \mathbb{N}$ and $F := \cup_{n=1}^\infty E_n$, and let A be an arbitrary subset of X. We will prove by induction that

$$\forall n \in \mathbb{N}, \mu^*(A \cap F_n) = \sum_{k=1}^n \mu^*(A \cap E_k). \tag{1}$$

Obviously, we have

$$\mu^*(A \cap F_1) = \mu^*(A \cap E_1).$$

Assume next that for $m \in \mathbb{N}$ we have

$$\mu^*(A \cap F_m) = \sum_{k=1}^m \mu^*(A \cap E_k);$$

then we have, since $E_{m+1} \in \mathcal{M}$,

$$\mu^*(A \cap F_{m+1}) = \mu^*(A \cap F_{m+1} \cap E_{m+1}) + \mu^*(A \cap F_{m+1} \cap (X - E_{m+1}))$$
$$= \mu^*(A \cap E_{m+1}) + \mu^*(A \cap F_m)$$
$$= \mu^*(A \cap E_{m+1}) + \sum_{k=1}^m \mu^*(A \cap E_k) = \sum_{k=1}^{m+1} \mu^*(A \cap E_k).$$

Thus, 1 is proved. Since $F_n \in \mathcal{M}$, from 1, condition om_2, and 5.3.2c we obtain

$$\forall n \in \mathbb{N}, \mu^*(A) = \mu^*(A \cap F_n) + \mu^*(A \cap (X - F_n))$$
$$\geq \sum_{k=1}^n \mu^*(A \cap E_k) + \mu^*(A \cap (X - F)).$$

From this, using 5.3.2c, om_3, om_1, we obtain

$$\mu^*(A) \geq \sup_{n \geq 1}(\sum_{k=1}^{n} \mu^*(A \cap E_k) + \mu^*(A \cap (X - F)))$$

$$= \sup_{n \geq 1} \sum_{k=1}^{n} \mu^*(A \cap E_k) + \mu^*(A \cap (X - F))$$

$$= \sum_{n=1}^{\infty} \mu^*(A \cap E_n) + \mu^*(A \cap (X - F))$$

$$\geq \mu^*(\cup_{n=1}^{\infty}(A \cap E_n)) + \mu^*(A \cap (X - F))$$

$$= \mu^*(A \cap F) + \mu^*(A \cap (X - F)) \geq \mu^*(A).$$

Thus, all the inequalities in this last calculation are in fact equalities. Since A was an arbitrary subset of X, this proves that $\cup_{n=1}^{\infty} E_n =: F \in \mathcal{M}$. Moreover, letting $A := F$ and using condition om_1, we have

$$\mu^*(\cup_{n=1}^{\infty} E_n) = \mu^*(F) = \sum_{n=1}^{\infty} \mu^*(F \cap E_n) = \sum_{n=1}^{\infty} \mu^*(E_n).$$

Finally, we prove that the measure $\mu_{\mathcal{M}}^*$ is complete. If $F \in \mathcal{P}(X)$ and there exists $E \in \mathcal{M}$ such that $F \subset E$ and $\mu_{\mathcal{M}}(F) = 0$, then $\mu^*(F) = 0$ by condition om_2. By conditions om_1, om_2, om_3, this implies that

$$\forall A \in \mathcal{P}(X), \mu^*(A) \leq \mu^*(A \cap F) + \mu^*(A \cap (X - F))$$

$$= \mu^*(A \cap (X - F)) \leq \mu^*(A),$$

so that $F \in \mathcal{M}$. $\qquad\square$

7.2.4 Proposition. *Let $\mathcal{E}$ be a family of subsets of X so that $\emptyset, X \in \mathcal{E}$ and let $\rho : \mathcal{E} \to [0, \infty]$ be a function so that $\rho(\emptyset) = 0$. For each $E \in \mathcal{P}(X)$, define the subset R_E of $[0, \infty]$ by*

$$R_E := \left\{ \sum_{n=1}^{\infty} \rho(A_n) : \{A_n\} \text{ is a sequence in } \mathcal{E} \text{ s.t. } E \subset \cup_{n=1}^{\infty} A_n \right\}$$

and note that R_E is non-empty (take $A_n := X$ for all $n \in \mathbb{N}$). Then the function

$$\mu^* : \mathcal{P}(X) \to [0, \infty]$$

$$E \mapsto \mu^*(E) := \inf R_E$$

is an outer measure on X.

Proof. Obviously $\mu^*(\emptyset) = 0$ since $A_n := \emptyset$ defines a sequence in $\mathcal{E}$ such that $\emptyset \subset \cup_{n=1}^{\infty} A_n$. Moreover, if $E, F \in \mathcal{P}(X)$ are such that $E \subset F$ then $R_F \subset R_E$ and hence $\mu^*(E) \leq \mu^*(F)$. Thus, conditions om_1 and om_2 are satisfied by μ^*. To prove condition om_3, consider a sequence $\{E_n\}$ in $\mathcal{P}(X)$. If there is $n \in \mathbb{N}$ so that $\mu^*(E_n) = \infty$, then the inequality

$$\mu^*(\cup_{n=1}^{\infty} E_n) \leq \sum_{n=1}^{\infty} \mu^*(E_n)$$

is obvious. Suppose then $\mu^*(E_k) < \infty$ for all $k \in \mathbb{N}$ and choose $\epsilon > 0$. For each $k \in \mathbb{N}$ there exists a sequence $\{A_n^k\}$ in $\mathcal{E}$ such that

$$E_k \subset \cup_{n=1}^\infty A_n^k \text{ and } \sum_{n=1}^\infty \rho(A_n^k) < \mu^*(E_k) + \frac{\epsilon}{2^k}.$$

Then we have

$$\cup_{k=1}^\infty E_k \subset \cup_{(k,n)\in\mathbb{N}\times\mathbb{N}} A_n^k$$

and

$$\sum_{(k,n)\in\mathbb{N}\times\mathbb{N}} \rho(A_n^k) = \sum_{k=1}^\infty (\sum_{n=1}^\infty \rho(A_n^k))$$

$$\leq \sum_{k=1}^\infty (\mu^*(E_k) + \frac{\epsilon}{2^k}) = \sum_{k=1}^\infty \mu^*(E_k) + \epsilon,$$

where 5.4.7, 5.4.2a and 5.4.6 have been used. Thus,

$$\mu^*(\cup_{k=1}^\infty E_k) \leq \sum_{(k,n)\in\mathbb{N}\times\mathbb{N}} \rho(A_n^k) \leq \sum_{k=1}^\infty \mu^*(E_k) + \epsilon.$$

Since ϵ was arbitrary, this proves that

$$\mu^*(\cup_{k=1}^\infty E_k) \leq \sum_{k=1}^\infty \mu^*(E_k).$$

$\square$

7.3   Extension theorems

7.3.1 Theorem. *Let $\mathcal{S}$ be a semialgebra on a non-empty set X.*

(A) Let $\nu : \mathcal{S} \to [0,\infty]$ be a function which satisfies the following conditions:
  (a) $\nu(\emptyset) = 0$;
  (b) for every finite and disjoint family $\{E_1, ..., E_N\}$ of elements of $\mathcal{S}$ such that $\cup_{n=1}^N E_n \in \mathcal{S}$,

$$\nu(\cup_{n=1}^N E_n) = \sum_{n=1}^N \nu(E_n).$$

  Then there exists a unique additive function μ_0 on $\mathcal{A}_0(\mathcal{S})$ (the algebra on X generated by $\mathcal{S}$) which is an extension of ν.
(B) If ν satisfies the further condition

  (c) for every sequence $\{E_n\}$ in $\mathcal{S}$ such that $\cup_{n=1}^\infty E_n \in \mathcal{S}$ and $E_i \cap E_j = \emptyset$ if $i \neq j$,

$$\nu(\cup_{n=1}^\infty E_n) \leq \sum_{n=1}^\infty \nu(E_n),$$

  then μ_0 is a premeasure on $\mathcal{A}_0(\mathcal{S})$.

Proof. A, existence: Let $E \in \mathcal{A}_0(\mathcal{S})$. Then there exists a finite and disjoint family $\{E_1, ..., E_N\}$ of elements of $\mathcal{S}$ such that $E = \cup_{n=1}^N E_n$ (cf. 6.1.11). Suppose now that there exists another finite and disjoint family $\{F_1, ..., F_L\}$ of elements of $\mathcal{S}$ such that $E = \cup_{l=1}^L F_l$. If we define $G_{n,l} := E_n \cap F_l$, we have $G_{n,l} \in \mathcal{S}$ for all $n = 1, ..., N$ and $l = 1, ..., L$, and $G_{n,l} \cap G_{m,k} = \emptyset$ if $(n, l) \neq (m, k)$. By condition b, we also have

$$\text{for } n = 1, ..., N, E_n = \cup_{l=1}^L G_{n,l}, \text{ whence } \nu(E_n) = \sum_{l=1}^L \nu(G_{n,l}),$$

$$\text{for } l = 1, ..., L, F_l = \cup_{n=1}^N G_{n,l}, \text{ whence } \nu(F_l) = \sum_{n=1}^N \nu(G_{n,l}),$$

and hence

$$\sum_{n=1}^N \nu(E_n) = \sum_{n=1}^N \sum_{l=1}^L \nu(G_{n,l}) = \sum_{l=1}^L \sum_{n=1}^N \nu(G_{n,l}) = \sum_{l=1}^L \nu(F_l).$$

This shows that we can define the function

$$\mu_0 : \mathcal{A}_0(\mathcal{S}) \to [0, \infty]$$

$$E \mapsto \mu_0(E) := \sum_{n=1}^N \nu(E_n) \text{ if } \{E_1, ..., E_N\} \text{ is a finite and disjoint family}$$

$$\text{of elements of } \mathcal{S} \text{ s.t. } E = \cup_{n=1}^N E_n.$$

Obviously, μ_0 is an extension of ν.

Since μ_0 is an extension of ν, μ_0 has property af_1 of 7.1.1 because ν satisfies condition a. Let now $\{E_1, E_2\}$ be a disjoint pair of elements of $\mathcal{A}_0(\mathcal{S})$, and let $\{E_1^1, ..., E_{N_1}^1\}$ and $\{E_1^2, ..., E_{N_2}^2\}$ be finite and disjoint families of elements of $\mathcal{S}$ such that $E_i = \cup_{n=1}^{N_i} E_n^i$ for $i = 1, 2$. Then $\{E_n^i\}_{i=1,2;n=1,...,N_i}$ is a finite and disjoint family of elements of $\mathcal{S}$ and $E_1 \cup E_2 = \cup_{i=1,2} \cup_{n=1}^{N_i} E_n^i$, and hence

$$\mu_0(E_1 \cup E_2) = \sum_{i=1,2} \sum_{n=1}^{N_i} \nu(E_n^i) = \sum_{n=1}^{N_1} \nu(E_n^1) + \sum_{n=1}^{N_2} \nu(E_n^2) = \mu_0(E_1) + \mu_0(E_2).$$

Applying induction to this result, we obtain property af_2 of 7.1.1 for μ_0, which is therefore an additive function on $\mathcal{A}_0(\mathcal{S})$.

A, uniqueness: Suppose that $\tilde{\mu}_0$ is an additive function on $\mathcal{A}_0(\mathcal{S})$ which extends ν. For any $E \in \mathcal{A}_0(\mathcal{S})$, let $\{E_1, ..., E_N\}$ be a finite and disjoint family of elements of $\mathcal{S}$ such that $E = \cup_{n=1}^N E_n$. Then we have, by the additivity of $\tilde{\mu}_0$ and the definition of μ_0,

$$\tilde{\mu}_0(E) = \sum_{n=1}^N \tilde{\mu}_0(E_n) = \sum_{n=1}^N \nu(E_n) = \mu_0(E).$$

B: Assume now condition c, and suppose that $\{F_n\}$ is a sequence in $\mathcal{A}_0(\mathcal{S})$ such that $F := \cup_{n=1}^{\infty} F_n \in \mathcal{A}_0(\mathcal{S})$ and $F_i \cap F_j = \emptyset$ if $i \neq j$.

There are finite and disjoint families $\{A_1, ..., A_N\}$ and $\{B_{n,1}, ..., B_{n,N_n}\}$ (for each $n \in \mathbb{N}$) of elements of $\mathcal{S}$ so that (cf. 6.1.11)

$$F = \cup_{k=1}^{N} A_k \text{ and } F_n := \cup_{l=1}^{N_n} B_{n,l} \text{ (for each } n \in \mathbb{N}).$$

Define

$$I := \{(k, n, l) : k = 1, ..., N, n \in \mathbb{N}, l = 1, ..., N_n\},$$
$$C_{k,n,l} := A_k \cap B_{n,l} \text{ for } (k, n, l) \in I.$$

Clearly, $\{C_{k,n,l}\}_{(k,n,l) \in I}$ is a disjoint family of elements of $\mathcal{S}$ and

$$\cup_{k=1}^{N} C_{k,n,l} = B_{n,l}, \quad \cup_{l=1}^{N_n} C_{k,n,l} = A_k \cap F_n, \quad \cup_{n=1}^{\infty}(A_k \cap F_n) = A_k.$$

We have

$$\nu(B_{n,l}) = \sum_{k=1}^{N} \nu(C_{k,n,l})$$

since ν satisfies condition b, and

$$\nu(A_k) \leq \sum_{n=1}^{\infty} \nu(A_k \cap F_n) = \sum_{n=1}^{\infty}(\sum_{l=1}^{N_n} \nu(C_{k,n,l}))$$

since ν satisfies conditions c and b. Then we have

$$\mu_0(F) = \sum_{k=1}^{N} \nu(A_k) \leq \sum_{k=1}^{N}(\sum_{n=1}^{\infty}(\sum_{l=1}^{N_n} \nu(C_{k,n,l}))) = \sum_{n=1}^{\infty}(\sum_{k=1}^{N}(\sum_{l=1}^{N_n} \nu(C_{k,n,l})))$$

$$= \sum_{n=1}^{\infty}(\sum_{l=1}^{N_n} \nu(B_{n,l})) = \sum_{n=1}^{\infty} \mu_0(F_n),$$

where we have used induction applied to 5.3.2e, induction applied to 5.4.6, and 5.3.3.

On the other hand, the additivity and the monotonicity of μ_0 imply that

$$\forall N \in \mathbb{N}, \sum_{n=1}^{N} \mu_0(F_n) = \mu_0(\cup_{n=1}^{N} F_n) \leq \mu_0(F),$$

and hence that

$$\sum_{n=1}^{\infty} \mu_0(F_n) = \sup_{N \geq 1} \sum_{n=1}^{N} \mu_0(F_n) \leq \mu_0(F).$$

Thus, $\mu_0(F) = \sum_{n=1}^{\infty} \mu_0(F_n)$. This proves that, if condition c is satisfied, then μ_0 has the property pm of 7.1.3. $\square$

7.3.2 Theorem (Hahn's theorem). *Let $\mathcal{A}_0$ be an algebra on X and μ_0 a pre-measure on $\mathcal{A}_0$. For each $E \in \mathcal{P}(X)$, define the subset M_E of $[0, \infty]$ by*

$$M_E := \left\{ \sum_{n=1}^{\infty} \mu_0(A_n) \; : \{A_n\} \text{ is a sequence in } \mathcal{A}_0 \text{ such that} \right.$$

$$\left. A_i \cap A_j = \emptyset \text{ if } i \neq j \text{ and } E \subset \cup_{n=1}^{\infty} A_n \right\}$$

and note that M_E is non-empty (take $A_1 := X$ and $A_n := \emptyset$ for $n > 1$). Then the function

$$\mu : \mathcal{A}(\mathcal{A}_0) \to [0, \infty]$$
$$E \mapsto \mu(E) := \inf M_E$$

is a measure on $\mathcal{A}(\mathcal{A}_0)$ (the σ-algebra on X generated by $\mathcal{A}_0$) and μ is an extension of μ_0, i.e.

$$\mu(E) = \mu_0(E), \forall E \in \mathcal{A}_0.$$

If $\tilde{\mu}$ is another measure on $\mathcal{A}(\mathcal{A}_0)$ that is an extension of μ_0, then:

$\tilde{\mu}(E) \leq \mu(E), \forall E \in \mathcal{A}(\mathcal{A}_0)$;
$\tilde{\mu}(E) = \mu(E)$ for each $E \in \mathcal{A}(\mathcal{A}_0)$ such that $\mu(E) < \infty$;
$\tilde{\mu} = \mu$ if μ_0 is σ-finite.

Proof. Define R_E and μ^* as in 7.2.4, with $\mathcal{E} := \mathcal{A}_0$ and $\rho := \mu_0$. Then μ^* is an outer measure on X and from 7.2.3 it follows that the collection $\mathcal{M}$ of μ^*-measurable subsets of X is a σ-algebra on X, and that the restriction $\mu^*_{\mathcal{M}}$ of μ^* to $\mathcal{M}$ is a measure on $\mathcal{M}$. From these facts, all the assertions of the statements can be derived. It will be convenient to divide this derivation into several steps.

Step 1: We prove that $\mu^*(E) = \inf M_E$ for each $E \in \mathcal{P}(X)$.

For each $E \in \mathcal{P}(X)$, on the one hand clearly $M_E \subset R_E$ and this implies that $\inf R_E \leq \inf M_E$. On the other hand, if $\{A_n\}$ is a sequence in $\mathcal{A}_0$ such that $E \subset \cup_{n=1}^{\infty} A_n$, then by 6.1.8 there is a sequence $\{B_n\}$ in $\mathcal{A}_0$ such that

$$B_n \subset A_n \text{ for all } n \in \mathbb{N}, B_i \cap B_j = \emptyset \text{ if } i \neq j \text{ and } E \subset \cup_{n=1}^{\infty} A_n = \cup_{n=1}^{\infty} B_n,$$

and we have, by the monotonicity of μ_0 and 5.4.2a,

$$\sum_{n=1}^{\infty} \mu_0(B_n) \leq \sum_{n=1}^{\infty} \mu_0(A_n).$$

This proves that

$$\forall a \in R_E, \exists a' \in M_E \text{ s.t. } a' \leq a,$$

and hence that

$$\forall a \in R_E, \inf M_E \leq a,$$

and hence that $\inf M_E \leq \inf R_E$. Thus, $\mu^*(E) := \inf R_E = \inf M_E$.

Step 2: We prove that $\mathcal{A}(\mathcal{A}_0) \subset \mathcal{M}$.

Suppose $E \in \mathcal{A}_0$. Let $A \in \mathcal{P}(X)$ be such that $\mu^*(A) < \infty$ and choose $\epsilon > 0$. Since $\mu^*(A) = \inf R_A$, there is a sequence $\{A_n\}$ in $\mathcal{A}_0$ such that

$$A \subset \cup_{n=1}^{\infty} A_n \text{ and } \sum_{n=1}^{\infty} \mu_0(A_n) < \mu^*(A) + \epsilon.$$

By the additivity of μ_0 we also have

$$\mu_0(A_n) = \mu_0(A_n \cap E) + \mu_0(A_n \cap (X - E)), \forall n \in \mathbb{N},$$

which implies

$$\mu^*(A \cap E) + \mu^*(A \cap (X - E))$$
$$\leq \sum_{n=1}^{\infty} \mu_0(A_n \cap E) + \sum_{n=1}^{\infty} \mu_0(A_n \cap (X - E)) = \sum_{n=1}^{\infty} \mu_0(A_n),$$

where the equality holds by 5.4.6 and the inequality holds because $\{A_n \cap E\}$ is a sequence in $\mathcal{A}_0$ such that $A \cap E \subset \cup_{n=1}^{\infty}(A_n \cap E)$ and $\{A_n \cap (X - E)\}$ is a sequence in $\mathcal{A}_0$ such that $A \cap (X - E) \subset \cup_{n=1}^{\infty}(A_n \cap (X - E))$. Thus we have

$$\mu^*(A \cap E) + \mu^*(A \cap (X - E)) < \mu^*(A) + \epsilon.$$

Since ϵ was arbitrary, this proves that

$$\mu^*(A \cap E) + \mu^*(A \cap (X - E)) \leq \mu^*(A).$$

In view of 7.2.2, this proves that $E \in \mathcal{M}$.

Step 3: From steps 1 and 2 it follows that μ is a measure, since it is the restriction of the measure $\mu_{\mathcal{M}}^*$ to $\mathcal{A}(\mathcal{A}_0)$.

Step 4: We prove that $\mu(E) = \mu_0(E)$ for all $E \in \mathcal{A}_0$.

Suppose $E \in \mathcal{A}_0$. If $\{A_n\}$ is any sequence in $\mathcal{A}_0$ such that $A_i \cap A_j = \emptyset$ if $i \neq j$ and $E \subset \cup_{n=1}^{\infty} A_n$, let $B_n := E \cap A_n$ for all $n \in \mathbb{N}$. Then $B_n \in \mathcal{A}_0$ for all $n \in \mathbb{N}$, $B_i \cap B_j = \emptyset$ if $i \neq j$ and $E = \cup_{n=1}^{\infty} B_n$, so by the σ-additivity and the monotonicity of μ_0 and by 5.4.2a we have

$$\mu_0(E) = \sum_{n=1}^{\infty} \mu_0(B_n) \leq \sum_{n=1}^{\infty} \mu_0(A_n).$$

This proves that $\mu_0(E) \leq \mu(E)$. On the other hand, $A_1 := E$ and $A_n := \emptyset$ for $n > 1$ defines a sequence in $\mathcal{A}_0$ such that $A_i \cap A_j = \emptyset$ if $i \neq j$ and $E \subset \cup_{n=1}^{\infty} A_n$, and this proves that $\mu(E) \leq \mu_0(E)$.

Step 5: We prove that, if $\tilde{\mu}$ is a measure on $\mathcal{A}(\mathcal{A}_0)$ which is an extension of μ_0, then the three assertions about $\tilde{\mu}$ in the statement are true.

Let $\tilde{\mu}$ be a measure on $\mathcal{A}(\mathcal{A}_0)$ such that $\tilde{\mu}(F) = \mu_0(F), \forall F \in \mathcal{A}_0$, and suppose $E \in \mathcal{A}(\mathcal{A}_0)$.

For any sequence $\{A_n\}$ in $\mathcal{A}_0$ such that $A_i \cap A_j = \emptyset$ if $i \neq j$ and $E \subset \cup_{n=1}^{\infty} A_n$, by the monotonicity and the σ-additivity of $\tilde{\mu}$ we have

$$\tilde{\mu}(E) \leq \tilde{\mu}(\cup_{n=1}^{\infty} A_n) = \sum_{n=1}^{\infty} \tilde{\mu}(A_n) = \sum_{n=1}^{\infty} \mu_0(A_n).$$

This proves that $\tilde{\mu}(E) \leq \mu(E)$.

Assume next $\mu(E) < \infty$ and choose $\epsilon > 0$. Then we can choose the A_n's above so that

$$\mu(\cup_{n=1}^{\infty} A_n) = \sum_{n=1}^{\infty} \mu(A_n) < \mu(E) + \epsilon$$

and obtain, by the monotonicity of μ and the σ-additivity of μ and $\tilde{\mu}$,

$$\begin{aligned}
\mu(E) \leq \mu(\cup_{n=1}^{\infty} A_n) &= \sum_{n=1}^{\infty} \mu_0(A_n) = \sum_{n=1}^{\infty} \tilde{\mu}(A_n) \\
&= \tilde{\mu}(\cup_{n=1}^{\infty} A_n) = \tilde{\mu}(E) + \tilde{\mu}((\cup_{n=1}^{\infty} A_n) - E) \\
&\leq \tilde{\mu}(E) + \mu((\cup_{n=1}^{\infty} A_n) - E) = \tilde{\mu}(E) + \mu(\cup_{n=1}^{\infty} A_n) - \mu(E) \\
&< \tilde{\mu}(E) + \epsilon.
\end{aligned}$$

Since ϵ was arbitrary, this proves that $\mu(E) \leq \tilde{\mu}(E)$, and hence that $\tilde{\mu}(E) = \mu(E)$.

Finally, assume that μ_0 is σ-finite. This means that there exists a sequence $\{E_n\}$ in $\mathcal{A}_0$ so that $\mu_0(E_n) < \infty$ for all $n \in \mathbb{N}$ and $X = \cup_{n=1}^{\infty} E_n$. By 6.1.8, there exists a sequence $\{F_n\}$ in $\mathcal{A}_0$ such that $F_n \subset E_n$ (and hence $\mu_0(F_n) < \infty$ by the monotonicity of μ_0) for all $n \in \mathbb{N}$, and such that $F_k \cap F_l = \emptyset$ if $k \neq l$ and $X = \cup_{n=1}^{\infty} F_n$. Then we have, by the σ-additivity of μ and $\tilde{\mu}$,

$$\mu(E) = \mu(\cup_{n=1}^{\infty}(E \cap F_n)) = \sum_{n=1}^{\infty} \mu(E \cap F_n) = \sum_{n=1}^{\infty} \tilde{\mu}(E \cap F_n) = \tilde{\mu}(E)$$

since $\mu(E \cap F_n) < \infty$ by the monotonicity of μ. $\qquad\square$

7.3.3 Theorem. *Let (X, d) be a metric space, let $\mathcal{S}$ be a semialgebra on X, and let $\nu : \mathcal{S} \to [0, \infty]$ be a function which satisfies the following conditions:*

(a) $\nu(\emptyset) = 0$;

(b) for each finite and disjoint family $\{E_1, ..., E_N\}$ of elements of $\mathcal{S}$ such that $\cup_{n=1}^{N} E_n \in \mathcal{S}$,

$$\nu(\cup_{n=1}^{N} E_n) = \sum_{n=1}^{N} \nu(E_n);$$

(c) for each $E \in \mathcal{S}$,

$$\nu(E) = \sup\left\{\nu(F) : F \in \mathcal{S}, \overline{F} \subset E, \overline{F} \text{ is compact}\right\}$$

(this condition is consistent because conditions a and b imply, by 7.3.1A, that ν is the restriction of an additive function on $\mathcal{A}_0(\mathcal{S})$ and hence that $\nu(F) \leq \nu(E)$ if $E, F \in \mathcal{S}$ are such that $\overline{F} \subset E$, and also because $\emptyset \in \mathcal{S}, \overline{\emptyset} = \emptyset \subset E$ for each $E \in \mathcal{S}$, and $\overline{\emptyset} = \emptyset$ is compact);

(d) for each $E \in \mathcal{S}$ such that $\nu(E) < \infty$,

$$\forall \epsilon > 0, \exists G_\epsilon \in \mathcal{S} \text{ s.t. } E \subset G_\epsilon^\circ \text{ and } \nu(G_\epsilon) - \nu(E) < \epsilon$$

(if there exists a finite family $\{E_1, ..., E_N\}$ of elements of $\mathcal{S}$ such that $\nu(E_n) < \infty$ for $n = 1, ..., N$ and $X = \cup_{n=1}^{N} E_n$, then condition d is redundant).

Then there exists a measure on $\mathcal{A}(\mathcal{S})$ (the σ-algebra on X generated by $\mathcal{S}$) that is an extension of ν.

If ν satisfies the further condition

(e) there exists a countable family $\{E_n\}_{n\in I}$ of elements of $\mathcal{S}$ such that $\nu(E_n) < \infty$ for all $n \in I$ and $X = \cup_{n\in I}E_n$,

then the measure on $\mathcal{A}(\mathcal{S})$ that extends ν is unique, and it is σ-finite.

Proof. Conditions a and b imply that there exists a unique additive function μ_0 on $\mathcal{A}_0(\mathcal{S})$ that extends ν (cf. 7.3.1A). Then conditions c and d become respectively conditions a and b of 7.1.6 for μ_0, and this implies that μ_0 is a premeasure. If there is a finite family $\{E_1, ..., E_N\}$ of elements of $\mathcal{S}$ such that $\nu(E_n) < \infty$ for $n = 1, ..., N$ and $X = \cup_{n=1}^{N}E_n$, then the additive function μ_0 that exists on $\mathcal{A}_0(\mathcal{S})$ is finite (since it is subadditive), and hence condition c is enough to make μ_0 a premeasure (cf. 7.1.6).

Since μ_0 is a premeasure on $\mathcal{A}_0(\mathcal{S})$, 7.3.2 implies that there is a measure μ on $\mathcal{A}(\mathcal{A}_0(\mathcal{S}))$ that extends μ_0. Since $\mathcal{A}(\mathcal{A}_0(\mathcal{S})) = \mathcal{A}(\mathcal{S})$ (cf. 6.1.18) and μ_0 extends ν, this proves that there exists a measure μ on $\mathcal{A}(\mathcal{S})$ that extends ν.

Finally, assume that ν satisfies also condition e. Then clearly μ and μ_0 are σ-finite. Suppose that $\tilde{\mu}$ is another measure on $\mathcal{A}(\mathcal{S})$ that extends ν. Then the restriction $\tilde{\mu}_{\mathcal{A}_0(\mathcal{S})}$ of $\tilde{\mu}$ to $\mathcal{A}_0(\mathcal{S})$ is an additive function on $\mathcal{A}_0(\mathcal{S})$ that extends ν and we have $\tilde{\mu}_{\mathcal{A}_0(\mathcal{S})} = \mu_0$ by the uniqueness asserted in 7.3.1A. Hence, $\tilde{\mu}$ is an extension of μ_0 and we have $\tilde{\mu} = \mu$ by the uniqueness asserted in 7.3.2 in the event that μ_0 is σ-finite. $\square$

7.4   Finite measures in metric spaces

The content of the first part of this section will be used mainly in the study of the product of two commuting projection valued measures, in Section 13.5. Indeed, for that study, which is a necessary step for the spectral theory of two commuting self-adjoint operators, it is essential to prove first that every finite measure on the Borel σ-algebra $\mathcal{A}(d_{\mathbb{R}})$ on $\mathbb{R}$ is regular. This can be achieved in several ways. The way we adopt here is borrowed from Section 2.7 of (Parthasarathy, 2005), and will allow to prove a more general result about commuting projection valued measures than the one that is required by the spectral theory of two commuting self-adjoint operators.

Lusin's theorem is presented in the second part of this section. It will be used to prove that $C(a, b)$ is isomorphic to a dense linear manifold in the Hilbert space $L^2(a, b)$ (cf. 11.2.1).

Throughout this section, (X, d) stands for a metric space.

7.4.1 Proposition. *Let μ be a finite measure on the Borel σ-algebra $\mathcal{A}(d)$ (cf. 6.1.22). Then, for each $E \in \mathcal{A}(d)$ the following conditions are both satisfied:*

(a) $\mu(E) = \sup\{\mu(F) : F \subset E, F \in \mathcal{K}_d\}$;
(b) $\mu(E) = \inf\{\mu(G) : E \subset G, G \in \mathcal{T}_d\}$.

Proof. We prove first that, for $E \in \mathcal{A}(d)$, conditions a and b together are equivalent to the one condition

(c) $\forall \epsilon > 0, \exists F_\epsilon \in \mathcal{K}_d, \exists G_\epsilon \in \mathcal{T}_d$ s.t. $F_\epsilon \subset E \subset G_\epsilon$ and $\mu(G_\epsilon - F_\epsilon) < \epsilon$.

On the one hand in fact, if conditions a and b are true for $E \in \mathcal{A}(d)$, then (since μ is finite) for $\epsilon > 0$ there exist $F_\epsilon \in \mathcal{K}_d$ and $G_\epsilon \in \mathcal{T}_d$ such that $F_\epsilon \subset E \subset G_\epsilon$ and
$$\mu(E) - \mu(F_\epsilon) < \frac{\epsilon}{2} \text{ and } \mu(G_\epsilon) - \mu(E) < \frac{\epsilon}{2}. \tag{1}$$
Since $G_\epsilon - F_\epsilon = (G_\epsilon - E) \cup (E - F_\epsilon)$ (cf. 1.1.4) and $(G_\epsilon - E) \cap (E - F_\epsilon) = \emptyset$, and since μ is finite, 1 implies that
$$\mu(G_\epsilon - F_\epsilon) = \mu(G_\epsilon - E) + \mu(E - F_\epsilon) = \mu(G_\epsilon) - \mu(E) + \mu(E) - \mu(F_\epsilon) < \epsilon,$$
and this shows that condition c is true for E. On the other hand, if condition c is true for $E \in \mathcal{A}(d)$, then for $\epsilon > 0$ we have
$$\mu(E) - \mu(F_\epsilon) = \mu(E - F_\epsilon) \leq \mu(G_\epsilon - F_\epsilon) < \epsilon \text{ and}$$
$$\mu(G_\epsilon) - \mu(E) = \mu(G_\epsilon - E) \leq \mu(G_\epsilon - F_\epsilon) < \epsilon$$
since $E - F_\epsilon \subset G_\epsilon - F_\epsilon$ and $G_\epsilon - E \subset G_\epsilon - F_\epsilon$ and since μ is finite, and this shows that conditions a and b are true for E.

Let now $\mathcal{B}$ denote the collection of all the subsets of X for which condition c is satisfied. We will prove that $\mathcal{B}$ is a σ-algebra on X and that $\mathcal{K}_d \subset \mathcal{B}$.

In the first place, we prove that condition al_2 of 6.1.5 is satisfied for $\mathcal{B}$. Indeed, let $E \in \mathcal{B}$ and $\epsilon > 0$. Then there exist $F_\epsilon \in \mathcal{K}_d$ and $G_\epsilon \in \mathcal{T}_d$ such that $F_\epsilon \subset E \subset G_\epsilon$ and $\mu(G_\epsilon - F_\epsilon) < \epsilon$. Hence, $X - G_\epsilon \in \mathcal{K}_d$, $X - F_\epsilon \in \mathcal{T}_d$, $X - G_\epsilon \subset X - E \subset X - F_\epsilon$ and $(X - F_\epsilon) - (X - G_\epsilon) = (X - F_\epsilon) \cap G_\epsilon = G_\epsilon - F_\epsilon$, whence $\mu((X - F_\epsilon) - (X - G_\epsilon)) < \epsilon$. This shows that $X - E \in \mathcal{B}$. Note next that $\emptyset \in \mathcal{B}$, since $\emptyset \in \mathcal{K}_d$ and $\emptyset \in \mathcal{T}_d$, and thus condition c is trivially satisfied for $\emptyset$. Therefore, to prove both al_1 of 6.1.5 and σa_1 of 6.1.13 for $\mathcal{B}$ it is enough to prove that $\cup_{n=1}^\infty E_n \in \mathcal{B}$ whenever $\{E_n\}$ is a sequence in $\mathcal{B}$. Let then $\{E_n\}$ be a sequence in $\mathcal{B}$ and $\epsilon > 0$. For each $n \in \mathbb{N}$ there exist $F_{n,\epsilon} \in \mathcal{K}_d$ and $G_{n,\epsilon} \in \mathcal{T}_d$ such that $F_{n,\epsilon} \subset E_n \subset G_{n,\epsilon}$ and $\mu(G_{n,\epsilon} - F_{n,\epsilon}) < \frac{\epsilon}{3^n}$. Letting $S := \cup_{n=1}^\infty F_{n,\epsilon} = \cup_{N=1}^\infty(\cup_{n=1}^N F_{n,\epsilon})$, from 7.1.4b it follows (since μ is finite) that we can choose N_ϵ so large that
$$\mu(S - \cup_{n=1}^{N_\epsilon} F_{n,\epsilon}) = \mu(S) - \mu(\cup_{n=1}^{N_\epsilon} F_{n,\epsilon}) < \frac{\epsilon}{2}.$$
Letting now $F_\epsilon := \cup_{n=1}^{N_\epsilon} F_{n,\epsilon}$ and $G_\epsilon := \cup_{n=1}^\infty G_{n,\epsilon}$, we have $F_\epsilon \in \mathcal{K}_d$, $G_\epsilon \in \mathcal{T}_d$, $F_\epsilon \subset \cup_{n=1}^\infty E_n \subset G_\epsilon$ and
$$\mu(G_\epsilon - F_\epsilon) = \mu(G_\epsilon - S) + \mu(S - F_\epsilon)$$
$$\leq \mu(\cup_{n=1}^\infty(G_{n,\epsilon} - F_{n,\epsilon})) + \mu(S - F_\epsilon)$$
$$\leq \sum_{n=1}^\infty \mu(G_{n,\epsilon} - F_{n,\epsilon}) + \mu(S - F_\epsilon) < \sum_{n=1}^\infty \frac{\epsilon}{3^n} + \frac{\epsilon}{2} = \epsilon,$$

where we have used the following facts: $G_\epsilon - F_\epsilon = (G_\epsilon - S) \cup (S - F_\epsilon)$ (cf. 1.1.4) and $(G_\epsilon - S) \cap (S - F_\epsilon) = \emptyset$; the monotonicity of μ and the calculation

$$G_\epsilon - S = (\cup_{n=1}^\infty G_{n,\epsilon}) \cap (X - \cup_{k=1}^\infty F_{k,\epsilon}) = (\cup_{n=1}^\infty G_{n,\epsilon}) \cap (\cap_{k=1}^\infty (X - F_{k,\epsilon}))$$
$$= \cup_{n=1}^\infty (G_{n,\epsilon} \cap (\cap_{k=1}^\infty (X - F_{k,\epsilon})))$$
$$\subset \cup_{n=1}^\infty (G_{n,\epsilon} \cap (X - F_{n,\epsilon})) = \cup_{n=1}^\infty (G_{n,\epsilon} - F_{n,\epsilon});$$

the σ-subadditivity of μ. This shows that $\cup_{n=1}^\infty E_n \in \mathcal{B}$. Thus, $\mathcal{B}$ is a σ-algebra on X.

We prove now that $\mathcal{K}_d \subset \mathcal{B}$. Indeed, fix $E \in \mathcal{K}_d$. Then, by 2.3.9c and 2.5.2, we have

$$E = \{x \in X : \delta_E(x) = 0\} = \cap_{n=1}^\infty \left\{x \in X : \delta_E(x) < \frac{1}{n}\right\},$$

where δ_E is the non-negative function defined in 2.5.4. For each $n \in \mathbb{N}$ define

$$G_n := \left\{x \in X : \delta_E(x) < \frac{1}{n}\right\} = \delta_E^{-1}((-\infty, \frac{1}{n})).$$

Since δ_E is continuous and $(-\infty, \frac{1}{n})$ is an open subset of $\mathbb{R}$, we have $G_n \in \mathcal{T}_d$ by 2.4.3. Also, since $G_{n+1} \subset G_n$ for all $n \in \mathbb{N}$ and μ is finite, from 7.1.4c it follows that we can choose $n_\epsilon \in \mathbb{N}$ so large that

$$\mu(G_{n_\epsilon} - E) = \mu(G_{n_\epsilon}) - \mu(E) < \epsilon.$$

Letting $F_\epsilon := E$ and $G_\epsilon := G_{n_\epsilon}$, this proves that condition c is satisfied for E, and hence that $E \in \mathcal{B}$.

Thus, $\mathcal{B}$ is a σ-algebra on X and $\mathcal{K}_d \subset \mathcal{B}$. Therefore,

$$\mathcal{A}(d) = \mathcal{A}(\mathcal{K}_d) \subset \mathcal{B}$$

(cf. 6.1.23 and $g\sigma_2$ of 6.1.17a with $\mathcal{F} := \mathcal{K}_d$ and $\mathcal{A} := \mathcal{B}$), and this inclusion proves the assertion of the statement. $\square$

7.4.2 Corollary. *If μ_1 and μ_2 are finite measures on $\mathcal{A}(d)$ and either $\mu_1(F) = \mu_2(F)$ for all $F \in \mathcal{K}_d$ or $\mu_1(G) = \mu_2(G)$ for all $G \in \mathcal{T}_d$, then $\mu_1 = \mu_2$.*

Proof. This result is obtained immediately from 7.4.1, by using either condition a or condition b for all elements of $\mathcal{A}(d)$. $\square$

7.4.3 Definition. A measure μ on the Borel σ-algebra $\mathcal{A}(d)$ is said to be *regular* if, for each $E \in \mathcal{A}(d)$, the following conditions are both satisfied:

(α) $\mu(E) = \sup\{\mu(C) : C \subset E, C \text{ is compact}\}$;
(β) $\mu(E) = \inf\{\mu(G) : E \subset G, G \in \mathcal{T}_d\}$.

Note that condition α is consistent because, if C is compact, then $C \in \mathcal{K}_d$ by 2.8.6.

7.4.4 Definition. A finite measure μ on the Borel σ-algebra $\mathcal{A}(d)$ is said to be *tight* if the following condition is satisfied:

$$\forall \epsilon > 0, \text{ there exists a compact subset } K_\epsilon \text{ of } X \text{ such that } \mu(X - K_\epsilon) < \epsilon.$$

7.4.5 Theorem. *Let μ be a finite and tight measure on the Borel σ-algebra $\mathcal{A}(d)$.* *Then μ is regular.*

Proof. Let E be an arbitrary element of $\mathcal{A}(d)$. Condition β of 7.4.3 coincides with condition b of 7.4.1, and therefore we know that it is satisfied since μ is finite. Let now $\epsilon > 0$. Since μ is finite, from 7.4.1 we know that

$$\exists F_\epsilon \in \mathcal{K}_d \text{ s.t. } F_\epsilon \subset E \text{ and } \mu(E) - \mu(F_\epsilon) < \frac{\epsilon}{2}.$$

Since μ is tight, there is a compact subset K_ϵ of X such that $\mu(X - K_\epsilon) < \frac{\epsilon}{2}$. Define then $C_\epsilon := F_\epsilon \cap K_\epsilon$. We have $C_\epsilon \subset E$. Also, C_ϵ is closed since K_ϵ is closed by 2.8.6, and hence C_ϵ is compact by 2.8.8. Since μ is finite, we also have

$$\mu(E) - \mu(C_\epsilon) = \mu(E) - \mu(F_\epsilon) + \mu(F_\epsilon) - \mu(C_\epsilon) < \frac{\epsilon}{2} + \mu(F_\epsilon - C_\epsilon) \le \frac{\epsilon}{2} + \mu(X - K_\epsilon) < \epsilon,$$

where we have used the monotonicity of μ and the calculation

$$F_\epsilon - C_\epsilon = F_\epsilon \cap ((X - F_\epsilon) \cup (X - K_\epsilon)) = F_\epsilon - K_\epsilon \subset X - K_\epsilon.$$

Since μ is finite, this shows that condition α is satisfied for E. $\qquad\square$

7.4.6 Theorem. *If the metric space (X, d) is complete and separable, then every finite measure on the Borel σ-algebra $\mathcal{A}(d)$ is tight.*

Proof. Let (X, d) be complete and separable, and let μ be a finite measure on $\mathcal{A}(d)$. Choose $\epsilon > 0$. For each $n \in \mathbb{N}$, the family of open balls $\{B(x, \frac{1}{n})\}_{x \in X}$ is so that $X = \cup_{x \in X} B(x, \frac{1}{n})$. Then, by 2.3.18, there is a countable family $\{x_{n,k}\}_{k \in I_n}$ of points of X so that $X = \cup_{k \in I_n} B(x_{n,k}, \frac{1}{n})$, and hence so that $X = \cup_{k \in I_n} K(x_{n,k}, \frac{1}{n})$. We can assume that either $I_n = \{1, ..., N_n\}$ or $I_n = \mathbb{N}$. If $I = \mathbb{N}$, we have $X = \cup_{N=1}^{\infty} (\cup_{k=1}^{N} K(x_{n,k}, \frac{1}{n}))$ and 7.1.4b implies (since μ is finite) that there exists $N_n \in \mathbb{N}$ so large that

$$\mu(X - \cup_{k=1}^{N_n} K(x_{n,k}, \frac{1}{n})) = \mu(X) - \mu(\cup_{k=1}^{N_n} K(x_{n,k}, \frac{1}{n})) < \frac{\epsilon}{2^n}.$$

Thus, for each $n \in \mathbb{N}$, in either case there is a finite family $\{x_{n,1}, ..., x_{n,N_n}\}$ of points so that

$$\mu(X - \cup_{k=1}^{N_n} K(x_{n,k}, \frac{1}{n})) < \frac{\epsilon}{2^n}.$$

Let then $K_\epsilon := \cap_{n=1}^{\infty} (\cup_{k=1}^{N_n} K(x_{n,k}, \frac{1}{n}))$. The set K_ϵ is closed (cf. 2.3.7 and 2.3.2) and hence the metric subspace $(K_\epsilon, d_{K_\epsilon})$ is complete (cf. 2.6.6b). Moreover,

$$\forall n \in \mathbb{N}, K_\epsilon \subset \cup_{k=1}^{N_n} K(x_{n,k}, \frac{1}{n}).$$

Therefore K_ϵ is compact by 2.8.5. Moreover, we have

$$X - K_\epsilon = \cup_{n=1}^{\infty} (X - \cup_{k=1}^{N_n} K(x_{n,k}, \frac{1}{n}))$$

and this implies, by the σ-subadditivity of μ, that

$$\mu(X - K_\epsilon) \le \sum_{n=1}^{\infty} \mu(X - \cup_{k=1}^{N_n} K(x_{n,k}, \frac{1}{n})) < \sum_{n=1}^{\infty} \frac{\epsilon}{2^n} = \epsilon.$$

This shows that μ is tight. $\qquad\square$

7.4.7 Corollary. *If the metric space (X, d) is complete and separable, then every finite measure on the Borel σ-algebra $\mathcal{A}(d)$ is regular.*

Proof. Use 7.4.6 and then 7.4.5. □

7.4.8 Theorem (Lusin's theorem). *Let μ be a finite measure on the Borel σ-algebra $\mathcal{A}(d)$, and let $\varphi \in \mathcal{M}(X, \mathcal{A}(d))$. Then, for each $\varepsilon > 0$ there exists $\tilde{\varphi} \in \mathcal{C}_B(X)$ (for $\mathcal{C}_B(X)$, cf. 3.1.10e) such that*

$$\mu(\{x \in X : \varphi(x) \neq \tilde{\varphi}(x)\}) < \varepsilon$$

and

$$\sup\{|\tilde{\varphi}(x)| : x \in X\} \leq \sup\{|\varphi(x)| : x \in X\}.$$

Proof. First of all we point out that the statement is consistent because

$$\{x \in X : \varphi(x) \neq \tilde{\varphi}(x)\} \in \mathcal{A}(d)$$

(since $\{0\} \in \mathcal{K}_{d_{\mathbb{C}}}$, this follows from 6.2.8, 6.2.16, 6.2.13c).

We divide the proof into five steps.

Step 1: Besides $\varphi \in \mathcal{M}(X, \mathcal{A}(d))$, we assume $0 \leq \varphi(x) < 1$ for all $x \in X$. Then if we define, for each $n \in \mathbb{N}$,

$$E_{k,n} := \varphi^{-1}\left(\left[\frac{k-1}{2^n}, \frac{k}{2^n}\right)\right), \quad \text{for } k = 2, 3, ..., 2^n,$$

and

$$\psi_n := \sum_{k=2}^{2^n} \frac{k-1}{2^n} \chi_{E_{k,n}},$$

we have a sequence $\{\psi_n\}$ in $\mathcal{S}^+(X, \mathcal{A}(d))$ such that $\lim_{n \to \infty} \psi_n(x) = \varphi(x)$ for all $x \in X$ (cf. the proof of 6.2.26).

We have

$$\psi_1 = \frac{1}{2}\chi_{E_{2,1}},$$

which can be written as

$$\psi_1 = \frac{1}{2}\chi_{E_1}$$

if we define

$$E_1 := E_{2,1}.$$

For $n > 1$ we have

$$\psi_n = \frac{1}{2^n}\chi_{E_{2,n}} + \sum_{h=2}^{2^{n-1}} \frac{(2h-1)-1}{2^n}\chi_{E_{2h-1,n}} + \sum_{h=2}^{2^{n-1}} \frac{2h-1}{2^n}\chi_{E_{2h,n}},$$

and also, for $h = 2, ..., 2^{n-1}$,

$$E_{h,n-1} = \varphi^{-1}\left(\left[\frac{h-1}{2^{n-1}}, \frac{h}{2^{n-1}}\right)\right)$$

$$= \varphi^{-1}\left(\left[\frac{(2h-1)-1}{2^n}, \frac{2h-1}{2^n}\right)\right) \cup \varphi^{-1}\left(\left[\frac{2h-1}{2^n}, \frac{2h}{2^n}\right)\right)$$

$$= E_{2h-1,n} \cup E_{2h,n},$$

and hence, since $E_{k,n} \cap E_{k',n} = \emptyset$ if $k \neq k'$,

$$\psi_{n-1} = \sum_{h=2}^{2^{n-1}} \frac{h-1}{2^{n-1}} \chi_{E_{h,n-1}} = \sum_{h=2}^{2^{n-1}} \frac{h-1}{2^{n-1}} \chi_{E_{2h-1,n}} + \sum_{h=2}^{2^{n-1}} \frac{h-1}{2^{n-1}} \chi_{E_{2h,n}}.$$

Thus, for $n > 1$ we have

$$\psi_n - \psi_{n-1} = \frac{1}{2^n} \chi_{E_{2,n}} + \sum_{h=2}^{2^{n-1}} \left(\frac{2h-1}{2^n} - \frac{h-1}{2^{n-1}}\right) \chi_{E_{2h,n}}$$

$$= \frac{1}{2^n}\left(\chi_{E_{2,n}} + \sum_{h=2}^{2^{n-1}} \chi_{E_{2h,n}}\right),$$

which can be written as

$$\psi_n - \psi_{n-1} = \frac{1}{2^n} \chi_{E_n}$$

if we define

$$E_n := E_{2,n} \cup \left(\bigcup_{h=2}^{2^{n-1}} E_{2h,n}\right).$$

Now fix $\varepsilon > 0$. From 7.4.1 and its proof we have that, for each $n \in \mathbb{N}$,

$$\exists F_n \in \mathcal{K}_d, \exists G_n \in \mathcal{T}_d \text{ such that } F_n \subset E_n \subset G_n \text{ and } \mu(G_n - F_n) < \frac{\varepsilon}{2^n},$$

and from 2.5.11 we have that

$$\exists \varphi_n \in C(X) \text{ such that } 0 \leq \varphi_n(x) \leq 1, \forall x \in X, \text{ and } F_n \prec \varphi_n \prec G_N.$$

Then we define the function

$$\psi : X \to \mathbb{R}$$

$$x \mapsto \psi(x) := \sum_{n=1}^{\infty} \frac{1}{2^n} \varphi_n(x).$$

Since

$$0 \leq \psi(x) \leq \sum_{n=1}^{\infty} \frac{1}{2^n} = 1, \forall x \in X,$$

we have $\psi \in \mathcal{F}_B(X)$, and since

$$\left\|\sum_{n=1}^{N} \frac{1}{2^n} \varphi_n - \psi\right\|_{\infty} \leq \sum_{n=N+1}^{\infty} \frac{1}{2^n} \to 0 \text{ as } N \to \infty,$$

we have $\psi \in \mathcal{C}_B(X)$ (cf. 4.3.6b).

We note that

$$\varphi(x) = \sum_{n=1}^{\infty} \frac{1}{2^n} \chi_{E_n}(x), \forall x \in X,$$

because

$$\sum_{n=1}^{N} \frac{1}{2^n} \chi_{E_n}(x) = \psi_1(x) + \sum_{n=2}^{N} (\psi_n(x) - \psi_{n-1}(x))$$

$$= \psi_N(x) \to \varphi(x) \text{ as } N \to \infty, \forall x \in X.$$

Since

$$\varphi_n(x) = \chi_{E_n}(x), \forall x \in F_n \cup (X - G_n), \forall n \in \mathbb{N}$$

(because $\varphi_n(x) = 1 = \chi_{E_n}(x)$ if $x \in F_n$ and $\varphi_n(x) = 0 = \chi_{E_n}(x)$ if $x \in X - G_n$), we have

$$\bigcap_{n=1}^{\infty} (F_n \cup (X - G_n)) \subset \{x \in X : \varphi(x) = \psi(x)\},$$

and hence

$$\{x \in X : \varphi(x) \neq \psi(x)\} \subset \bigcup_{n=1}^{\infty} ((X - F_n) \cap G_n) = \bigcup_{n=1}^{\infty} (G_n - F_n),$$

and hence, by the σ-subadditivity of μ (cf. 7.1.4a)

$$\mu(\{x \in X : \varphi(x) \neq \psi(x)\}) \leq \sum_{n=1}^{\infty} \mu(G_n - F_n) < \sum_{n=1}^{\infty} \frac{\varepsilon}{2^n} = \varepsilon.$$

Step 2: Besides $\varphi \in \mathcal{M}(X, \mathcal{A}(d))$, we assume $0 \leq \varphi(x)$ for all $x \in X$ and $\varphi \in \mathcal{F}_B(X)$. Then the function

$$\varphi_0 := \frac{1}{\|\varphi\|_\infty + 1} \varphi$$

is an element of $\mathcal{M}(X, \mathcal{A}(d))$ such that $0 \leq \varphi_0(x) < 1$ for all $x \in X$, and therefore the result of step 1 implies that if we fix $\varepsilon > 0$ then there exists $\psi_0 \in \mathcal{C}_B(X)$ such that

$$\mu(\{x \in X : \varphi_0(x) \neq \psi_0(x)\}) < \varepsilon.$$

Then the function

$$\psi := (\|\varphi\|_\infty + 1)\psi_0$$

is an element of $\mathcal{C}_B(X)$ such that

$$\mu(\{x \in X : \varphi(x) \neq \psi(x)\}) < \varepsilon.$$

Step 3: Besides $\varphi \in \mathcal{M}(X, \mathcal{A}(d))$, we assume $\varphi \in \mathcal{F}_B(X)$. Then the functions

$$\varphi_1 := (\operatorname{Re} \varphi)^+, \varphi_2 := (\operatorname{Re} \varphi)^-, \varphi_3 := (\operatorname{Im} \varphi)^+, \varphi_4 := (\operatorname{Im} \varphi)^-$$

are elements of $\mathcal{M}(X, \mathcal{A}(d)) \cap \mathcal{F}_B(X)$ such that $0 \leq \varphi_i(x)$ for all $x \in X$, for $i = 1, 2, 3, 4$. Therefore, the result of step 2 implies that if we fix $\varepsilon > 0$ then, for $i = 1, 2, 3, 4$, there exists $\psi_i \in \mathcal{C}_B(X)$ such that

$$\mu(\{x \in X : \varphi_i(x) \neq \psi_i(x)\}) < \frac{\varepsilon}{4}.$$

Then the function

$$\psi := \psi_1 - \psi_2 + i\psi_3 - i\psi_4$$

is an element of $\mathcal{C}_B(X)$ such that

$$\{x \in X : \varphi(x) \neq \psi(x)\} \subset \bigcup_{i=1}^{4} \{x \in X : \varphi_i(x) \neq \psi_i(x)\},$$

and hence, by the subadditivity of μ (cf. 7.1.2b), such that

$$\mu(\{x \in X : \varphi(x) \neq \psi(x)\}) < 4\frac{\varepsilon}{4} = \varepsilon.$$

Step 4: We make no further assumptions about φ besides $\varphi \in \mathcal{M}(X, \mathcal{A}(d))$. For each $n \in \mathbb{N}$ we define

$$A_n := \{x \in X : n < |\varphi(x)|\},$$

and we have $A_n \in \mathcal{A}(d)$ (cf. 6.2.17 and 6.2.13a), $A_{n+1} \subset A_n$, and $\bigcap_{n=1}^{\infty} A_n = \emptyset$. Then, 7.1.4c implies that $\lim_{n \to \infty} \mu(A_n) = 0$. Therefore, if we fix $\varepsilon > 0$ then there exists $k \in \mathbb{N}$ so that

$$\mu(A_k) < \frac{\varepsilon}{2}.$$

Now obviously

$$\chi_{X - A_k} \varphi \in \mathcal{M}(X, \mathcal{A}(d)) \cap \mathcal{F}_B(X)$$

and hence the result of step 3 implies that there exists $\psi \in \mathcal{C}_B(X)$ such that

$$\mu(\{x \in X : (\chi_{X - A_k} \varphi)(x) \neq \psi(x)\}) < \frac{\varepsilon}{2}.$$

Thus, from

$$(X - A_k) \cap \{x \in X : (\chi_{X - A_k} \varphi)(x) = \psi(x)\} \subset \{x \in X : \varphi(x) = \psi(x)\}$$

we have

$$\{x \in X : \varphi(x) \neq \psi(x)\} \subset A_k \cup \{x \in X : (\chi_{X - A_k} \varphi)(x) \neq \psi(x)\}$$

and hence, by the subadditivity of μ

$$\mu(\{x \in X : \varphi(x) \neq \psi(x)\}) < 2\frac{\varepsilon}{2} = \varepsilon.$$

Step 5: We make no further assumptions about φ besides $\varphi \in \mathcal{M}(X, \mathcal{A}(d))$. In step 4 it was proved that if we fix $\varepsilon > 0$ then there exists $\psi \in \mathcal{C}_B(X)$ such that

$$\mu(\{x \in X : \varphi(x) \neq \psi(x)\}) < \varepsilon.$$

Now we prove that there exists $\tilde{\varphi} \in C_B(X)$ as in the statement. If

$$\sup\{|\varphi(x)| : x \in X\} = \infty$$

then it is enough to define $\tilde{\varphi} := \psi$. If

$$M := \sup\{|\varphi(x)| : x \in X\} < \infty,$$

we first define the function

$$\eta : \mathbb{C} \to \mathbb{C}$$

$$z \mapsto \eta(z) := \begin{cases} z & \text{if } |z| \leq M, \\ Mz|z|^{-1} & \text{if } |z| > M, \end{cases}$$

and then we define $\tilde{\varphi} := \eta \circ \psi$. The function η is continuous and hence $\tilde{\varphi} \in C(X)$ (cf. 2.4.4), and it is bounded and hence $\tilde{\varphi} \in C_B(X)$. Furthermore, it is obvious that, for $x \in X$,

$$\varphi(x) = \psi(x) \Rightarrow [\varphi(x) = \psi(x) \text{ and } |\psi(x)| \leq M] \Rightarrow \varphi(x) = \tilde{\varphi}(x).$$

Therefore,

$$\{x \in X : \varphi(x) \neq \tilde{\varphi}(x)\} \subset \{x \in X : \varphi(x) \neq \psi(x)\}$$

and hence, by the monotonicity of μ (cf. 7.1.2a),

$$\mu(\{x \in X : \varphi(x) \neq \tilde{\varphi}(x)\}) < \varepsilon.$$

Finally, it is obvious that

$$|\tilde{\varphi}(x)| \leq M, \forall x \in X,$$

and hence that

$$\sup\{|\tilde{\varphi}(x)| : x \in X\} \leq \sup\{|\varphi(x)| : x \in X\}.$$

$\square$

Chapter 8

Integration

8.1 Integration of positive functions

In this section, $(X, \mathcal{A}, \mu)$ denotes an abstract measure space.

8.1.1 Proposition. *Let $n, m \in \mathbb{N}$, let $\{a_1, ..., a_n\}$ and $\{b_1, ..., b_m\}$ be families of elements of $[0, \infty)$, let $\{E_1, ..., E_n\}$ and $\{F_1, ..., F_m\}$ be disjoint (i.e. $E_i \cap E_j = \emptyset$ and $F_i \cap F_j = \emptyset$ if $i \neq j$) families of elements of $\mathcal{A}$, and suppose that*

$$\sum_{k=1}^{n} a_k \chi_{E_k} = \sum_{l=1}^{m} b_l \chi_{F_l}.$$

Then

$$\sum_{k=1}^{n} a_k \mu(E_k) = \sum_{l=1}^{m} b_l \mu(F_l).$$

Proof. We define

$$a_{n+1} := b_{m+1} := 0, \quad E_{n+1} := X - \bigcup_{k=1}^{n} E_k, \quad F_{m+1} := X - \bigcup_{l=1}^{m} F_l.$$

Then, $E_k \cap F_l \in \mathcal{A}$ for $k = 1, ..., n+1$ and $l = 1, ..., m+1$, $(E_k \cap F_l) \cap (E_{k'} \cap F_{l'}) = \emptyset$ if $(k, l) \neq (k', l')$, and we have:

$$\forall k \in \{1, ..., n+1\}, E_k = \bigcup_{l=1}^{m+1} (E_k \cap F_l), \text{ whence } \mu(E_k) = \sum_{l=1}^{m+1} \mu(E_k \cap F_l),$$

$$\forall l \in \{1, ..., m+1\}, F_l = \bigcup_{k=1}^{n+1} (E_k \cap F_l), \text{ whence } \mu(F_l) = \sum_{k=1}^{n+1} \mu(E_k \cap F_l).$$

We also notice that

$$\sum_{k=1}^{n+1} a_k \chi_{E_k} = \sum_{k=1}^{n} a_k \chi_{E_k} = \sum_{l=1}^{m} b_l \chi_{F_l} = \sum_{l=1}^{m+1} b_l \chi_{F_l}.$$

This implies that, for $k = 1, ..., n+1$ and $l = 1, ..., m+1$, if $\mu(E_k \cap F_l) \neq 0$ and hence $E_k \cap F_l \neq \emptyset$, then $a_k = b_l$.

From these facts and from 5.3.3 we obtain

$$\sum_{k=1}^{n} a_k \mu(E_k) = \sum_{k=1}^{n+1} a_k \mu(E_k) = \sum_{k=1}^{n+1} \sum_{l=1}^{m+1} a_k \mu(E_k \cap F_l)$$

$$= \sum_{l=1}^{m+1} \sum_{k=1}^{n+1} b_l \mu(E_k \cap F_l) = \sum_{l=1}^{m+1} b_l \mu(F_l) = \sum_{l=1}^{m} b_l \mu(F_l).$$

$\square$

8.1.2 Definition. Let $\psi \in \mathcal{S}^+(X, \mathcal{A})$ (for $\mathcal{S}^+(X, \mathcal{A})$, cf. 6.2.25). Then there are $n \in \mathbb{N}$, a family $\{a_1, ..., a_n\}$ of elements of $[0, \infty)$, and a disjoint family $\{E_1, ..., E_n\}$ of elements of $\mathcal{A}$ so that $\psi = \sum_{k=1}^{n} a_k \chi_{E_k}$. We define the *integral (with respect to μ)* of ψ by

$$\int_X \psi d\mu := \sum_{k=1}^{n} a_k \mu(E_k),$$

which is an element of $[0, \infty]$ determined by ψ without ambiguity in view of 8.1.1 (it depends only on ψ, and not on the representation $\psi = \sum_{k=1}^{n} a_k \chi_{E_k}$, which is not unique).

8.1.3 Remarks.

(a) For each $E \in \mathcal{A}$, we have $\chi_E \in \mathcal{S}^+(X, \mathcal{A})$. Thus, immediately from the definition in 8.1.2, we have

$$\int_X \chi_E d\mu = \mu(E).$$

Hence in particular $\int_X 0_X d\mu = \mu(\emptyset) = 0$ (even if $\mu(X) = \infty$) since $0_X = \chi_\emptyset$, and $\int_X 1_X d\mu = \mu(X)$ since $1_X = \chi_X$.

(b) From the definition in 8.1.2 and from 7.1.2a we have that if $\mu(X) = 0$ then $\int_X \psi d\mu = 0$ for all $\psi \in \mathcal{S}^+(X, \mathcal{A})$.

8.1.4 Proposition. *Let* $\psi_1, \psi_2 \in \mathcal{S}^+(X, \mathcal{A})$.

(a) If $a, b \in [0, \infty)$ *then* $a\psi_1 + b\psi_2 \in \mathcal{S}^+(X, \mathcal{A})$ *and*

$$\int_X (a\psi_1 + b\psi_2) d\mu = a \int_X \psi_1 d\mu + b \int_X \psi_2 d\mu.$$

(b) If $\psi_1 \leq \psi_2$ *then* $\int_X \psi_1 d\mu \leq \int_X \psi_2 d\mu$ *(for the notation* $\varphi \leq \psi$, *cf. 5.1.1).*

Proof. Let us write $\psi_1 = \sum_{k=1}^{n} a_k \chi_{E_k}$ and $\psi_2 = \sum_{l=1}^{m} b_l \chi_{F_l}$, with $a_k \in [0, \infty)$ for $k = 1, ..., n$, $b_l \in [0, \infty)$ for $l = 1, ..., m$, $\{E_1, ..., E_n\}$ and $\{F_1, ..., F_m\}$ disjoint families of elements of $\mathcal{A}$.

We define

$$a_{n+1} := b_{m+1} := 0, \quad E_{n+1} := X - \bigcup_{k=1}^{n} E_k, \quad F_{m+1} := X - \bigcup_{l=1}^{m} F_l.$$

We have
$$\psi_1 = \sum_{k=1}^{n+1} a_k \chi_{E_k} \text{ and } \psi_2 = \sum_{l=1}^{m+1} b_l \chi_{F_l}.$$

The family $\{E_k \cap F_l\}_{(k,l)\in I}$, with $I := \{1,...,n+1\} \times \{1,...,m+1\}$, is a disjoint family of elements of $\mathcal{A}$, and from $E_k = \bigcup_{l=1}^{m+1}(E_k \cap F_l)$ and $F_l = \bigcup_{k=1}^{n+1}(E_k \cap F_l)$ we have

$$\chi_{E_k} = \sum_{l=1}^{m+1} \chi_{E_k \cap F_l} \text{ for } k = 1,...,n+1 \text{ and}$$

$$\chi_{F_l} = \sum_{k=1}^{n+1} \chi_{E_k \cap F_l} \text{ for } l = 1,...,m+1,$$

and hence

$$\psi_1 = \sum_{(k,l)\in I} a_k \chi_{E_k \cap F_l} \text{ and } \psi_2 = \sum_{(k,l)\in I} b_l \chi_{E_k \cap F_l}.$$

a: We have, for $a, b \in [0, \infty)$,
$$a\psi_1 + b\psi_2 = \sum_{(k,l)\in I} (aa_k + bb_l)\chi_{E_k \cap F_l}.$$

This shows that $a\psi_1 + b\psi_2 \in \mathcal{S}^+(X, \mathcal{A})$ (this was already clear from 6.2.24). Moreover (cf. 5.3.3)

$$\int_X (a\psi_1 + b\psi_2)d\mu = \sum_{(k,l)\in I} (aa_k + bb_l)\mu(E_k \cap F_l)$$

$$= a\sum_{k=1}^{n+1} a_k \sum_{l=1}^{m+1} \mu(E_k \cap F_l) + b\sum_{l=1}^{m+1} b_l \sum_{k=1}^{n+1} \mu(E_k \cap F_l)$$

$$= a\sum_{k=1}^{n+1} a_k\mu(E_k) + b\sum_{l=1}^{m+1} b_l\mu(F_l) = a\int_X \psi_1 d\mu + b\int_X \psi_2 d\mu.$$

b: Suppose $\psi_1 \le \psi_2$, i.e.
$$\sum_{(k,l)\in I} a_k \chi_{E_k \cap F_l}(x) \le \sum_{(k,l)\in I} b_l \chi_{E_k \cap F_l}(X), \forall x \in X.$$

Then $a_k \le b_l$ whenever $E_k \cap F_l \ne \emptyset$, and hence (cf. 5.3.2b,e)

$$\int_X \psi_1 d\mu = \sum_{(k,l)\in I} a_k\mu(E_k \cap F_l) \le \sum_{(k,l)\in I} b_l\mu(E_k \cap F_l) = \int_X \psi_2 d\mu.$$

$\square$

8.1.5 Proposition. *Let* $\psi \in \mathcal{S}^+(X, \mathcal{A})$. *Then the function*
$$\nu : \mathcal{A} \to [0, \infty]$$

$$E \mapsto \nu(E) := \int_X \chi_E \psi d\mu.$$

is a measure on $\mathcal{A}$.

Proof. We have $\nu(\emptyset) = \int_X 0_X d\mu = 0$ (cf. 8.1.3a). Thus, ν has property af_1 of 7.1.1.

Write now $\psi = \sum_{k=1}^m a_k \chi_{F_k}$, with $a_k \in [0, \infty)$ for $k = 1, ..., m$ and $\{F_1, ..., F_m\}$ a disjoint family of elements of $\mathcal{A}$, and notice that, for every $E \in \mathcal{A}$, the equality $\chi_E \psi = \sum_{k=1}^m a_k \chi_{E \cap F_k}$ shows that $\chi_E \psi \in \mathcal{S}^+(X, \mathcal{A})$ (this was already clear from 6.2.24) and $\int_X \chi_E \psi d\mu = \sum_{k=1}^m a_k \mu(E \cap F_k)$. Then, if $\{E_n\}$ is a sequence in $\mathcal{A}$ such that $E_i \cap E_j = \emptyset$ whenever $i \neq j$, we have (by 5.4.5 and induction applied to 5.4.6)

$$\nu\left(\bigcup_{n=1}^\infty E_n\right) = \int_X \chi_{\bigcup_{n=1}^\infty E_n} \psi d\mu = \sum_{k=1}^m a_k \mu\left(\left(\bigcup_{n=1}^\infty E_n\right) \cap F_k\right)$$

$$= \sum_{k=1}^m a_k \sum_{n=1}^\infty \mu(E_n \cap F_k) = \sum_{n=1}^\infty \sum_{k=1}^m a_k \mu(E_n \cap F_k)$$

$$= \sum_{n=1}^\infty \int_X \chi_{E_n} \psi d\mu = \sum_{n=1}^\infty \nu(E_n).$$

Thus, ν has property *me* of 7.1.7. $\qquad\qquad\square$

8.1.6 Definition. From 8.1.4b it is clear that, for $\varphi \in \mathcal{S}^+(X, \mathcal{A})$,

$$\int_X \varphi d\mu = \sup\left\{\int_X \psi d\mu : \psi \in \mathcal{S}^+(X, \mathcal{A}), \psi \leq \varphi\right\},$$

where $\int_X \varphi d\mu$ and $\int_X \psi d\mu$ are defined as in 8.1.2. It is therefore consistent with the definition given in 8.1.2 to define the *integral* (*with respect to* μ) of any $\varphi \in \mathcal{L}^+(X, \mathcal{A})$ (for $\mathcal{L}^+(X, \mathcal{A})$, cf. 6.2.25) by

$$\int_X \varphi d\mu = \sup\left\{\int_X \psi d\mu : \psi \in \mathcal{S}^+(X, \mathcal{A}), \psi \leq \varphi\right\},$$

which is an element of $[0, \infty]$ (notice that this definition is consistent also because $0_X \in \mathcal{S}^+(X, \mathcal{A})$ and $0_X \leq \varphi$ for each $\varphi \in \mathcal{L}^+(X, \mathcal{A})$, and hence the set for which the l.u.b. is taken is non-empty).

8.1.7 Proposition. *If* $\varphi_1, \varphi_2 \in \mathcal{L}^+(X, \mathcal{A})$ *are so that* $\varphi_1 \leq \varphi_2$, *then*

$$\int_X \varphi_1 d\mu \leq \int_X \varphi_2 d\mu.$$

Proof. If $\varphi_1, \varphi_2 \in \mathcal{L}^+(X, \mathcal{A})$ are so that $\varphi_1 \leq \varphi_2$, then

$$\{\psi \in \mathcal{S}^+(X, \mathcal{A}), \psi \leq \varphi_1\} \subset \{\psi \in \mathcal{S}^+(X, \mathcal{A}), \psi \leq \varphi_2\},$$

and this implies the inequality we want to prove. $\qquad\qquad\square$

8.1.8 Theorem (Monotone convergence theorem). *Let* $\{\varphi_n\}$ *be a sequence in* $\mathcal{L}^+(X, \mathcal{A})$ *and suppose that* $\varphi_n \leq \varphi_{n+1}$ *for all* $n \in \mathbb{N}$. *Then (for* $\lim_{n \to \infty} \varphi_n$ *and* $\sup_{n \geq 1} \varphi_n$, *cf. 6.2.18):*

(a) the sequence $\{\varphi_n(x)\}$ *is convergent (in the metric space* $(\mathbb{R}^*, \delta)$*) for all* $x \in X$, $\lim_{n \to \infty} \varphi_n = \sup_{n \geq 1} \varphi_n$, *and* $\lim_{n \to \infty} \varphi_n \in \mathcal{L}^+(X, \mathcal{A})$;

(b) $\int_X (\lim_{n \to \infty} \varphi_n) d\mu = \lim_{n \to \infty} \int_X \varphi_n d\mu = \sup_{n \geq 1} \int_X \varphi_n d\mu.$

Proof. a: By 5.2.5 we have that, for each $x \in X$, the sequence $\{\varphi_n(x)\}$ is convergent and $\lim_{n \to \infty} \varphi_n(x) = \sup_{n \geq 1} \varphi_n(x)$, and hence also that $\lim_{n \to \infty} \varphi_n(x) \in [0, \infty]$. Thus, $\lim_{n \to \infty} \varphi_n = \sup_{n \geq 1} \varphi_n$, and $\lim_{n \to \infty} \varphi_n \in \mathcal{L}^+(X, \mathcal{A})$ follows from 6.2.19b.

b: From 8.1.7 and 5.2.5 it follows that the sequence $\{\int_X \varphi_n d\mu\}$ is convergent (in the metric space $(\mathbb{R}^*, \delta)$) and that

$$\lim_{n \to \infty} \int_X \varphi_n d\mu = \sup_{n \geq 1} \int_X \varphi_n d\mu.$$

Moreover, $\varphi_k \leq \sup_{n \geq 1} \varphi_n$ for all $k \in \mathbb{N}$, so

$$\sup_{n \geq 1} \int_X \varphi_n d\mu \leq \int_X (\sup_{n \geq 1} \varphi_n) d\mu$$

by 8.1.7. The reverse inequality is obvious if $\sup_{n \geq 1} \int_X \varphi_n d\mu = \infty$. Assume then $\sup_{n \geq 1} \int_X \varphi_n d\mu < \infty$ and fix $\psi \in \mathcal{S}^+(X, \mathcal{A})$ such that $\psi \leq \sup_{n \geq 1} \varphi_n$. Choose $a \in (0, 1)$ and define

$$E_n := \{x \in X : a\psi(x) \leq \varphi_n(x)\} \text{ for all } n \in \mathbb{N}.$$

From 6.2.31 and 6.1.26 we have $E_n \in \mathcal{A}$ for all $n \in \mathbb{N}$. Also, $E_n \subset E_{n+1}$ for all $n \in \mathbb{N}$ and $X = \bigcup_{n=1}^{\infty} E_n$ (to see this, notice that if $\psi(x) = 0$ then $x \in E_1$; if $0 < \psi(x)$, then $a\psi(x) < \psi(x) \leq \sup_{n \geq 1} \varphi_n(x)$ and hence there exists $n \in \mathbb{N}$ so that $a\psi(x) < \varphi_n(x)$). Then, 8.1.4a, 8.1.5 (since $a\psi \in \mathcal{S}^+(X, \mathcal{A})$ by 6.2.24) and 7.1.4b imply that

$$a \int_X \psi d\mu = \int_X a\psi d\mu = \sup_{n \geq 1} \int_X \chi_{E_n} a\psi d\mu.$$

Now 8.1.7 implies that

$$\int_X \chi_{E_n} a\psi d\mu \leq \int_X \varphi_n d\mu.$$

Thus,

$$a \int_X \psi d\mu \leq \sup_{n \geq 1} \int_X \varphi_n d\mu.$$

Since this is true for every $a \in (0, 1)$, we have in particular

$$\left(1 - \frac{1}{k}\right) \int_X \psi d\mu \leq \sup_{n \geq 1} \int_X \varphi_n d\mu, \quad \forall k \in \mathbb{N},$$

whence (notice that $\int_X \psi d\mu < \infty$ since we are assuming $\sup_{n \geq 1} \int_X \varphi_n d\mu < \infty$)

$$\int_X \psi d\mu = \lim_{k \to \infty} \left(1 - \frac{1}{k}\right) \int_X \psi d\mu \leq \sup_{n \geq 1} \int_X \varphi_n d\mu.$$

Since this is true for every $\psi \in \mathcal{S}^+(X, \mathcal{A})$ such that $\psi \leq \sup_{n \geq 1} \varphi_n$, we have

$$\int_X (\sup_{n \geq 1} \varphi_n) d\mu \leq \sup_{n \geq 1} \int_X \varphi_n d\mu.$$

This proves that

$$\int_X (\sup_{n \geq 1} \varphi_n) d\mu = \sup_{n \geq 1} \int_X \varphi_n d\mu,$$

which, in view of part a, can be written as

$$\int_X (\lim_{n \to \infty} \varphi_n) d\mu = \sup_{n \geq 1} \int_X \varphi_n d\mu.$$

$\square$

8.1.9 Proposition. *Let $\varphi_1, \varphi_2 \in \mathcal{L}^+(X, \mathcal{A})$. Then*

$$\varphi_1 + \varphi_2 \in \mathcal{L}^+(X, \mathcal{A}) \ and \ \int_X (\varphi_1 + \varphi_2) d\mu = \int_X \varphi_1 d\mu + \int_X \varphi_2 d\mu.$$

Proof. We have $\varphi_1 + \varphi_2 \in \mathcal{L}^+(X, \mathcal{A})$ from 6.2.31. By 6.2.26, there are two sequences $\{\psi_n^1\}$ and $\{\psi_n^2\}$ in $\mathcal{S}^+(X, \mathcal{A})$ so that, for $i = 1, 2$, $\psi_n^i \leq \psi_{n+1}^i$ and $\lim_{n \to \infty} \psi_n^i = \varphi_i$. By 8.1.4a, $\{\psi_n^1 + \psi_n^2\}$ is a sequence in $\mathcal{S}^+(X, \mathcal{A})$ and

$$\int_X (\psi_n^1 + \psi_n^2) d\mu = \int_X \psi_n^1 d\mu + \int_X \psi_n^2 d\mu.$$

Since

$$\forall x \in X, \ (\psi_n^1 + \psi_n^2)(x) \leq (\psi_{n+1}^1 + \psi_{n+1}^2)(x) \ \text{and}$$
$$\lim_{n \to \infty} (\psi_n^1 + \psi_n^2)(x) = (\varphi_1 + \varphi_2)(x)$$

(cf. 5.3.4), by 8.1.8, 8.1.4a and 5.3.4 we have

$$\int_X (\varphi_1 + \varphi_2) d\mu = \lim_{n \to \infty} \int_X (\psi_n^1 + \psi_n^2) d\mu$$
$$= \lim_{n \to \infty} \int_X \psi_n^1 d\mu + \lim_{n \to \infty} \int_X \psi_n^2 d\mu = \int_X \varphi_1 d\mu + \int_X \varphi_2 d\mu.$$

$\square$

8.1.10 Proposition. *Let $\{\varphi_n\}$ be a sequence in $\mathcal{L}^+(X, \mathcal{A})$. Then (for $\sum_{n=1}^\infty \varphi_n$, cf. 6.2.32)*

$$\sum_{n=1}^\infty \varphi_n \in \mathcal{L}^+(X, \mathcal{A}) \ and \ \int_X \left(\sum_{n=1}^\infty \varphi_n\right) d\mu = \sum_{n=1}^\infty \int_X \varphi_n d\mu.$$

Proof. We have $\sum_{n=1}^{\infty} \varphi_n \in \mathcal{L}^+(X, \mathcal{A})$ from 6.2.32. Applying induction to 8.1.9 we have

$$\sum_{k=1}^{n} \varphi_k \in \mathcal{L}^+(X, \mathcal{A}) \text{ and } \int_X \left(\sum_{k=1}^{n} \varphi_k \right) d\mu = \sum_{k=1}^{n} \int_X \varphi_k d\mu, \forall n \in \mathbb{N}.$$

Then, since $\sum_{k=1}^{n} \varphi_k \leq \sum_{k=1}^{n+1} \varphi_k$ for all $n \in \mathbb{N}$, by 8.1.8 and 5.4.1 we have

$$\int_X \left(\sum_{n=1}^{\infty} \varphi_n \right) d\mu = \lim_{n \to \infty} \int_X \left(\sum_{k=1}^{n} \varphi_k \right) d\mu$$

$$= \lim_{n \to \infty} \sum_{k=1}^{n} \int_X \varphi_k d\mu = \sum_{n=1}^{\infty} \int_X \varphi_n d\mu.$$

$\square$

8.1.11 Proposition. *Let $\varphi_1, \varphi_2 \in \mathcal{L}^+(X, \mathcal{A})$. Then:*

(a) if $\varphi_1(x) \leq \varphi_2(x)$ μ-a.e. on X, then $\int_X \varphi_1 d\mu \leq \int_X \varphi_2 d\mu$;
(b) if $\varphi_1(x) = \varphi_2(x)$ μ-a.e. on X, then $\int_X \varphi_1 d\mu = \int_X \varphi_2 d\mu$.

Proof. a: let $E \in \mathcal{A}$ be so that $\mu(E) = 0$ and $\varphi_1(x) \leq \varphi_2(x)$ for all $x \in X - E$. For $i = 1, 2$ we have:

$$\varphi_i = \chi_E \varphi_i + \chi_{X-E} \varphi_i, \quad \chi_E \varphi_i \in \mathcal{L}^+(X, \mathcal{A}), \quad \int_X \chi_E \varphi_i d\mu = 0.$$

Indeed, $\chi_E \varphi_i \in \mathcal{L}^+(X, \mathcal{A})$ follows from 6.2.31 and, if $\psi \in \mathcal{S}^+(X, \mathcal{A})$ is such that $\psi \leq \chi_E \varphi_i$, then $\psi \leq (\max\{\psi(x) : x \in X\}) \chi_E$ and hence (cf. 8.1.4a,b and 8.1.3a)

$$\int_X \psi d\mu \leq (\max\{\psi(x) : x \in X\}) \int_X \chi_E d\mu = (\max\{\psi(x) : x \in X\}) \mu(E) = 0.$$

Then, by 8.1.9 and 8.1.7,

$$\int_X \varphi_1 d\mu = \int_X \chi_E \varphi_1 d\mu + \int_X \chi_{X-E} \varphi_1 d\mu$$

$$= \int_X \chi_{X-E} \varphi_1 d\mu \leq \int_X \chi_{X-E} \varphi_2 d\mu = \int_X \varphi_2 d\mu.$$

b: This follows immediately from part a. $\square$

8.1.12 Proposition. *Let $\varphi \in \mathcal{L}^+(X, \mathcal{A})$. Then:*

(a) $\int_X \varphi d\mu = 0$ iff $\varphi(x) = 0$ μ-a.e. on X;
(b) if $\int_X \varphi d\mu < \infty$ then $\mu(\varphi^{-1}(\{\infty\})) = 0$, i.e. $\varphi(x) < \infty$ μ-a.e. on X.

Proof. a: If $\varphi(x) = 0$ μ-a.e. on X, then 8.1.11b implies that

$$\int_X \varphi d\mu = \int_X 0_X d\mu = 0.$$

On the other hand, letting $E := \varphi^{-1}((0, \infty])$ and $E_n := \varphi^{-1}\left(\left(\frac{1}{n}, \infty\right]\right)$ for each $n \in \mathbb{N}$, we have (cf. 6.1.26):

$$E \in \mathcal{A}, E_n \in \mathcal{A} \text{ for each } n \in \mathbb{N}, E = \bigcup_{n=1}^{\infty} E_n.$$

Since $\frac{1}{n}\chi_{E_n} \leq \varphi$, by 8.1.7 we have

$$\frac{1}{n}\mu(E_n) = \int_X \frac{1}{n}\chi_{E_n} d\mu \leq \int_X \varphi d\mu, \forall n \in \mathbb{N}.$$

Thus, if $\int_X \varphi d\mu = 0$ then $\mu(E_n) = 0$ for each $n \in \mathbb{N}$, and hence (cf. 7.1.4a) $\mu(E) = 0$. Since

$$\varphi(x) = 0, \forall x \in X - E,$$

this shows that $\int_X \varphi d\mu = 0$ implies $\varphi(x) = 0$ μ-a.e. on X.

b: We have $\varphi^{-1}(\{\infty\}) \in \mathcal{A}$ by 6.1.26. Defining $\psi_n := n\chi_{\varphi^{-1}(\{\infty\})}$ for each $n \in \mathbb{N}$, we have

$$\psi_n \in \mathcal{S}^+(X, \mathcal{A}), \quad \psi_n \leq \varphi, \quad \int_X \psi_n d\mu = n\mu(\varphi^{-1}(\{\infty\})).$$

In view of 8.1.6, this shows that if $\mu(\varphi^{-1}(\{\infty\})) \neq 0$ then $\int_X \varphi d\mu = \infty$. $\qquad \square$

8.1.13 Proposition. *Let $a \in [0, \infty]$ and $\varphi \in \mathcal{L}^+(X, \mathcal{A})$. Then,*

$$a\varphi \in \mathcal{L}^+(X, \mathcal{A}) \text{ and } \int_X a\varphi d\mu = a \int_X \varphi d\mu.$$

Proof. We have $a\varphi \in \mathcal{L}^+(X, \mathcal{A})$ from 6.2.31.

If $a = 0$, then the equality of the statement is obvious.

If $a = \infty$ and $\int_X \varphi d\mu = 0$, then (cf. 8.1.12a) $\varphi(x) = 0$ μ-a.e. on X, and hence $(a\varphi)(x) = 0$ μ-a.e. on X, and hence (cf. 8.1.12a)

$$\int_X a\varphi d\mu = 0 = a \int_X \varphi d\mu.$$

If $a = \infty$ and $\int_X \varphi d\mu > 0$, then there exists $n \in \mathbb{N}$ so that $\mu\left(\varphi^{-1}\left(\left(\frac{1}{n}, \infty\right]\right)\right) \neq 0$, for otherwise we would have $\mu(\varphi^{-1}((0, \infty])) = 0$ (cf. the proof of 8.1.12a) and hence $\varphi(x) = 0$ μ-a.e. on X and hence $\int_X \varphi d\mu = 0$ by 8.1.12a. Since

$$\varphi^{-1}\left(\left(\frac{1}{n}, \infty\right]\right) \subset (a\varphi)^{-1}(\{\infty\}), \forall n \in \mathbb{N},$$

by 7.1.2a we have $\mu((a\varphi)^{-1}(\{\infty\})) \neq 0$, and hence (cf. 8.1.12b)

$$\int_X a\varphi d\mu = \infty = a \int_X \varphi d\mu.$$

Finally, suppose that $a \in (0, \infty)$. If $\psi \in \mathcal{S}^+(X, \mathcal{A})$ is such that $\psi \leq \varphi$, then $a\psi \in \mathcal{S}^+(X, \mathcal{A})$ (cf. 6.2.24) and $a\psi \leq a\varphi$ (cf. 5.3.2b). Then (cf. 8.1.4a,b and 5.3.2b),

$$
\int_X a\varphi d\mu \geq \sup \left\{ \int_X a\psi d\mu : \psi \in \mathcal{S}^+(X, \mathcal{A}), \psi \leq \varphi \right\}
$$

$$
= \sup \left\{ a \int_X \psi d\mu : \psi \in \mathcal{S}^+(X, \mathcal{A}), \psi \leq \varphi \right\}
$$

$$
= a \sup \left\{ \int_X \psi d\mu : \psi \in \mathcal{S}^+(X, \mathcal{A}), \psi \leq \varphi \right\} = a \int_X \varphi d\mu.
$$

Now replace, in the equality just obtained, a with $\frac{1}{a}$ and φ with $a\varphi$ (which is still an element of $\mathcal{L}^+(X, \mathcal{A})$, by 6.2.31). Then,

$$
\int_X \varphi d\mu \geq \frac{1}{a} \int_X a\varphi d\mu,
$$

whence (cf. 5.3.2b)

$$
a \int_X \varphi d\mu \geq \int_X a\varphi d\mu,
$$

and this concludes the proof of the equality of the statement. □

8.1.14 Definitions. We denote by $\mathcal{L}^+(X, \mathcal{A}, \mu)$ the family of functions from X to $[0, \infty]$ that is defined as follows:

$$
\mathcal{L}^+(X, \mathcal{A}, \mu) := \{\varphi : D_\varphi \to [0, \infty] : D_\varphi \in \mathcal{A}, \mu(X - D_\varphi) = 0, \varphi \in \mathcal{L}^+(D_\varphi, \mathcal{A}^{D_\varphi})\}.
$$

Clearly, $\mathcal{L}^+(X, \mathcal{A}) \subset \mathcal{L}^+(X, \mathcal{A}, \mu)$.

If $\varphi \in \mathcal{L}^+(X, \mathcal{A}, \mu)$ and $D_\varphi \neq X$, then φ has any number of extensions which are elements of $\mathcal{L}^+(X, \mathcal{A})$. One of these extensions is for instance the function

$$
\varphi_e : X \to [0, \infty]
$$

$$
x \mapsto \varphi_e(x) := \begin{cases} \varphi(x) & \text{if } x \in D_\varphi, \\ 0 & \text{if } x \notin D_\varphi, \end{cases}
$$

which we call the *standard extension* of φ and which is $\mathcal{A}$-measurable since, for $S \subset [0, \infty]$, $\varphi_e^{-1}(S)$ is either $\varphi^{-1}(S)$ or $\varphi^{-1}(S) \cup (X - D_\varphi)$, and $\mathcal{A}^{D_\varphi} \subset \mathcal{A}$ (cf. 6.1.19a). Now, if φ_1 and φ_2 are two elements of $\mathcal{L}^+(X, \mathcal{A})$ which are extensions of φ then $\varphi_1(x) = \varphi_2(x)$ μ-a.e. on X (since $\varphi_1(x) = \varphi_2(x)$ for all $x \in X - (X - D_\varphi)$, $X - D_\varphi \in \mathcal{A}$, and $\mu(X - D_\varphi) = 0$), and this implies (cf. 8.1.11b)

$$
\int_X \varphi_1 d\mu = \int_X \varphi_2 d\mu.
$$

Thus, we can define the *integral (with respect to μ)* of φ as the integral of any element of $\mathcal{L}^+(X, \mathcal{A})$ that is an extension of φ, for instance as

$$
\int_X \varphi d\mu := \int_X \varphi_e d\mu,
$$

Actually, since the value of $\int_X \varphi d\mu$ is independent from what extension of φ is used in order to define it, this extension need not be specified, unless this is useful for calculations.

As a matter of convenience, for $\varphi \in \mathcal{L}^+(X, \mathcal{A}, \mu)$ we will sometimes write

$$\int_X \varphi(x) d\mu(x) := \int_X \varphi d\mu.$$

8.1.15 Proposition. *For every* $\varphi \in \mathcal{L}^+(X, \mathcal{A}, \mu)$ *we have*

$$\int_X \varphi d\mu = \lim_{n \to \infty} \left[\sum_{k=2}^{n2^n} \frac{k-1}{2^n} \mu \left(\varphi^{-1} \left(\left[\frac{k-1}{2^n}, \frac{k}{2^n} \right) \right) \right) + n\mu(\varphi^{-1}([n, \infty])) \right]$$

$$= \sup_{n \geq 1} \left[\sum_{k=2}^{n2^n} \frac{k-1}{2^n} \mu \left(\varphi^{-1} \left(\left[\frac{k-1}{2^n}, \frac{k}{2^n} \right) \right) \right) + n\mu(\varphi^{-1}([n, \infty])) \right].$$

Proof. We have

$$\int_X \varphi d\mu := \int_X \varphi_e d\mu,$$

where φ_e is the standard extension of φ (cf. 8.1.14).

For each $n \in \mathbb{N}$ we define:

$$E_{0,n} := \varphi_e^{-1}([n, \infty]),$$

$$E_{k,n} := \varphi_e^{-1} \left(\left[\frac{k-1}{2^n}, \frac{k}{2^n} \right) \right) \quad \text{for } k = 1, ..., n2^n,$$

$$\psi_n := \sum_{k=1}^{n2^n} \frac{k-1}{2^n} \chi_{E_{k,n}} + n\chi_{E_{0,n}}.$$

We recall (cf. the proof of 6.2.26) that $\{\psi_n\}$ is a sequence in $\mathcal{S}^+(X, \mathcal{A})$ so that

$$\psi_n \leq \psi_{n+1} \text{ for all } n \in \mathbb{N} \text{ and } \lim_{n \to \infty} \psi_n = \varphi_e.$$

Then, from 8.1.8 it follows that

$$\int_X \varphi_e d\mu = \lim_{n \to \infty} \int_X \psi_n d\mu = \lim_{n \to \infty} \left[\sum_{k=1}^{n2^n} \frac{k-1}{2^n} \mu(E_{k,n}) + n\mu(E_{0,n}) \right]$$

$$= \sup_{n \geq 1} \left[\sum_{k=1}^{n2^n} \frac{k-1}{2^n} \mu(E_{k,n}) + n\mu(E_{0,n}) \right].$$

Now, we notice that, for each $n \in \mathbb{N}$,

$$\frac{k-1}{2^n} = 0 \text{ for } k = 1, E_{k,n} = \varphi^{-1} \left(\left[\frac{k-1}{2^n}, \frac{k}{2^n} \right) \right) \text{ for } k > 1, E_{0,n} = \varphi^{-1}([n, \infty]).$$

Thus, we have the equalities of the statement. $\square$

8.1.16 Remark. Proposition 8.1.15 shows that we could have defined, for every $\varphi \in \mathcal{L}^+(X, \mathcal{A}, \mu)$, the integral of φ with respect to μ by

$$\int_X \varphi d\mu := \sup_{n \geq 1} \left[\sum_{k=2}^{\lceil n2^n \rceil} \frac{k-1}{2^n} \mu \left(\varphi^{-1} \left(\left[\frac{k-1}{2^n}, \frac{k}{2^n} \right) \right) \right) + n\mu(\varphi^{-1}([n, \infty])) \right],$$

without going through 8.1.2, 8.1.6, 8.1.14; this would have been close to Lebesgue's original way of defining his integral (cf. Shilov and Gurevich, 1966, 6.6). This way of defining the integral has the merit of showing at the outset why the integral can be defined for measurable functions only (if φ were not measurable with D_φ measurable, then the sets $\varphi^{-1}\left(\left[\frac{k-1}{2^n}, \frac{k}{2^n}\right)\right)$ and $\varphi^{-1}([n, \infty])$ would not be elements of $\mathcal{A}$ for all n's and k's, and the whole formula would be meaningless since the domain of μ is $\mathcal{A}$). Indeed, the definition in 8.1.6 would not be contradictory if it were given for all functions with X as domain and $[0, \infty]$ as final set, and only later does it become clear why the functions must be measurable (e.g., the proof of additivity given in 8.1.9 requires the measurability of the functions in an essential way).

8.1.17 Theorem. *Let $\varphi, \psi \in \mathcal{L}^+(X, \mathcal{A}, \mu)$. Then:*

(a) $a\varphi + b\psi \in \mathcal{L}^+(X, \mathcal{A}, \mu)$ and $\int_X (a\varphi + b\psi)d\mu = a\int_X \varphi d\mu + b\int_X \psi d\mu$, $\forall a, b \in [0, \infty]$;
(b) if $\varphi(x) \leq \psi(x)$ μ-a.e. on $D_\varphi \cap D_\psi$, then $\int_X \varphi d\mu \leq \int_X \psi d\mu$;
(c) if $\varphi(x) = \psi(x)$ μ-a.e. on $D_\varphi \cap D_\psi$, then $\int_X \varphi d\mu = \int_X \psi d\mu$;
(d) $\varphi\psi \in \mathcal{L}^+(X, \mathcal{A}, \mu)$.

Proof. a: For $a, b \in [0, \infty]$ we have:

$D_{a\varphi + b\psi} = D_\varphi \cap D_\psi \in \mathcal{A}$ (cf. 6.2.30);
$\mu(X - D_\varphi \cap D_\psi) = \mu((X - D_\varphi) \cup (X - D_\psi)) = 0$ (cf. 7.1.2b);
$\varphi_{D_\varphi \cap D_\psi}, \psi_{D_\varphi \cap D_\psi} \in \mathcal{L}^+(D_\varphi \cap D_\psi, \mathcal{A}^{D_\varphi \cap D_\psi})$ (cf. 6.2.3 and 6.1.19b) and hence
$a\varphi + b\psi = a\varphi_{D_\varphi \cap D_\psi} + b\psi_{D_\varphi \cap D_\psi} \in \mathcal{L}^+(D_\varphi \cap D_\psi, \mathcal{A}^{D_\varphi \cap D_\psi})$ (cf. 6.2.31).

This proves that $a\varphi + b\psi \in \mathcal{L}^+(X, \mathcal{A}, \mu)$.

Further, if $\tilde{\varphi}, \tilde{\psi} \in \mathcal{L}^+(X, \mathcal{A})$ are extensions of φ, ψ respectively (cf. 8.1.14), then $a\tilde{\varphi} + b\tilde{\psi} \in \mathcal{L}^+(X, \mathcal{A})$ and $a\tilde{\varphi} + b\tilde{\psi}$ is an extension of $a\varphi + b\psi$. Then,

$$\int_X (a\varphi + b\psi)d\mu = \int_X (a\tilde{\varphi} + b\tilde{\psi})d\mu = a\int_X \tilde{\varphi}d\mu + b\int_X \tilde{\psi}d\mu = a\int_X \varphi d\mu + b\int_X \psi d\mu,$$

where 8.1.9 and 8.1.13 have been used.

b: Let $E \in \mathcal{A}$ be so that

$$\mu(E) = 0 \text{ and } \varphi(x) \leq \psi(x) \text{ for all } x \in D_\varphi \cap D_\psi \cap (X - E).$$

If $\tilde{\varphi}, \tilde{\psi} \in \mathcal{L}^+(X, \mathcal{A})$ are extensions of φ, ψ respectively, then we have

$$\tilde{\varphi}(x) \leq \tilde{\psi}(x) \text{ for all } x \in D_\varphi \cap D_\psi \cap (X - E) = X - ((X - D_\varphi) \cup (X - D_\psi) \cup E).$$

Since $(X - D_\varphi) \cup (X - D_\psi) \cup E \in \mathcal{A}$ and $\mu((X - D_\varphi) \cup (X - D_\psi) \cup E) = 0$ (cf. 7.1.2b), we have

$$\tilde{\varphi}(x) \leq \tilde{\psi}(x) \ \mu\text{-a.e. on } X,$$

and hence, by 8.1.11a,

$$\int_X \varphi d\mu = \int_X \tilde{\varphi} d\mu \leq \int_X \tilde{\psi} = \int_X \psi d\mu.$$

c: This follows immediately from part b.

d: Since $\varphi\psi = \varphi_{D_\varphi \cap D_\psi} \psi_{D_\varphi \cap D_\psi}$ (cf. 6.2.30), the proof is quite similar to the one given in part a for $a\varphi + b\psi$.   □

8.1.18 Theorem. *Let $\varphi \in \mathcal{L}^+(X, \mathcal{A}, \mu)$. Then:*

(a) $\int_X \varphi d\mu = 0$ iff $\varphi(x) = 0$ μ-a.e. on D_φ;
(b) if $\mu(X) = 0$ then $\int_X \varphi = 0$;
(c) if ν is a measure on $\mathcal{A}$ such that $\nu(E) \leq \mu(E)$ for all $E \in \mathcal{A}$, then

$$\varphi \in \mathcal{L}^+(X, \mathcal{A}, \nu) \text{ and } \int_X \varphi d\nu \leq \int_X \varphi d\mu.$$

Proof. a: Let $\tilde{\varphi} \in \mathcal{L}^+(X, \mathcal{A})$ be an extension of φ. Then,

$$\tilde{\varphi}(x) = 0 \ \mu\text{-a.e. on } X \text{ iff } \varphi(x) = 0 \ \mu\text{-a.e. on } D_\varphi.$$

Indeed, the "only if" is obvious and, if $E \in \mathcal{A}$ is so that

$$\mu(E) = 0 \text{ and } \varphi(x) = 0 \text{ for all } x \in D_\varphi \cap (X - E),$$

then

$$\mu((X - D_\varphi) \cup E) = 0 \ (\text{cf. 7.1.2b}) \text{ and } \tilde{\varphi}(x) = 0 \text{ for all } x \in X - ((X - D_\varphi) \cup E).$$

The result now follows from 8.1.12a, since $\int_X \varphi d\mu := \int_X \tilde{\varphi} d\mu$.

b: If $\mu(X) = 0$, then $\varphi(x) = 0$ μ-a.e. on D_φ, and the result follows from part a.

c: If ν is a measure on $\mathcal{A}$ such that $\nu(E) \leq \mu(E)$ for all $E \in \mathcal{A}$, then the inequality $\nu(X - D_\varphi) \leq \mu(X - D_\varphi) = 0$ shows that $\varphi \in \mathcal{L}^+(X, \mathcal{A}, \nu)$. Moreover, by 5.3.2b and induction applied to 5.3.2e, for each $n \in \mathbb{N}$ we have

$$\sum_{k=2}^{n2^n} \frac{k-1}{2^n} \nu\left(\varphi^{-1}\left(\left[\frac{k-1}{2^n}, \frac{k}{2^n}\right)\right)\right) + n\nu(\varphi^{-1}([n, \infty]))$$

$$\leq \sum_{k=2}^{n2^n} \frac{k-1}{2^n} \mu\left(\varphi^{-1}\left(\left[\frac{k-1}{2^n}, \frac{k}{2^n}\right)\right)\right) + n\mu(\varphi^{-1}([n, \infty])),$$

and hence, by 8.1.15,

$$\int_X \varphi d\nu \leq \int_X \varphi d\mu.$$

                  □

8.1.19 Theorem (Monotone convergence theorem (2nd version)). *Let $\{\varphi_n\}$ be a sequence in $\mathcal{L}^+(X, \mathcal{A}, \mu)$ and suppose that*

$$\forall n \in \mathbb{N}, \varphi_n(x) \leq \varphi_{n+1}(x) \ \mu\text{-a.e on } D_{\varphi_n} \cap D_{\varphi_{n+1}}.$$

Then:

(a) there exists $\tilde{\varphi} \in \mathcal{L}^+(X, \mathcal{A})$ such that $\tilde{\varphi}(x) = \lim_{n \to \infty} \varphi_n(x)$ μ-a.e. on $\bigcap_{n=1}^{\infty} D_{\varphi_n}$;

(b) if $\varphi \in \mathcal{L}^+(X, \mathcal{A}, \mu)$ and $\varphi(x) = \lim_{n \to \infty} \varphi_n(x)$ μ-a.e. on $D_\varphi \cap (\bigcap_{n=1}^{\infty} D_{\varphi_n})$, then

$$\int_X \varphi d\mu = \lim_{n \to \infty} \int_X \varphi_n d\mu = \sup_{n \geq 1} \int_X \varphi_n d\mu.$$

Proof. a: For each $n \in \mathbb{N}$, let $E_n \in \mathcal{A}$ be so that

$$\mu(E_n) = 0 \text{ and } \varphi_n(x) \leq \varphi_{n+1}(x) \text{ for all } x \in D_{\varphi_n} \cap D_{\varphi_{n+1}} \cap (X - E_n).$$

Letting $E := \bigcup_{n=1}^{\infty}((X - D_{\varphi_n}) \cup E_n)$, we have $E \in \mathcal{A}$ and $\mu(E) = 0$ (cf. 7.1.4a). Since $X - E \subset D_{\varphi_n}$, for each $n \in \mathbb{N}$ we can define the function

$$\tilde{\varphi}_n : X \to [0, \infty]$$

$$x \mapsto \tilde{\varphi}_n(x) := \begin{cases} \varphi_n(x) & \text{if } x \in X - E, \\ 0 & \text{if } x \in E, \end{cases}$$

which is an element of $\mathcal{L}^+(X, \mathcal{A})$ since, for every $S \subset [0, \infty]$, $\tilde{\varphi}_n^{-1}(S)$ is either $\varphi_n^{-1}(S) \cap (X - E)$ or $(\varphi_n^{-1}(S) \cap (X - E)) \cup E$, and $\mathcal{A}^{D_{\varphi_n}} \subset \mathcal{A}$. We have $\tilde{\varphi}_n \leq \tilde{\varphi}_{n+1}$ since $X - E \subset D_{\varphi_n} \cap D_{\varphi_{n+1}} \cap (X - E_n)$. Hence, by 8.1.8a, the sequence $\{\tilde{\varphi}_n(x)\}$ is convergent for all $x \in X$ and $\lim_{n \to \infty} \tilde{\varphi}_n \in \mathcal{L}^+(X, \mathcal{A})$. Letting $\tilde{\varphi} := \lim_{n \to \infty} \tilde{\varphi}_n$, we also have

$$\tilde{\varphi}(x) := \lim_{n \to \infty} \tilde{\varphi}_n(x) = \lim_{n \to \infty} \varphi_n(x), \forall x \in X - E = \left(\bigcap_{n=1}^{\infty} D_{\varphi_n} \right) \cap (X - E),$$

and hence

$$\tilde{\varphi}(x) = \lim_{n \to \infty} \varphi_n(x) \ \mu\text{-a.e. on } \bigcap_{n=1}^{\infty} D_{\varphi_n}.$$

b: Let $\tilde{\varphi}_n$ and $\tilde{\varphi}$ denote the same functions as in the proof of part a. By 8.1.8b we have

$$\int_X \tilde{\varphi} d\mu = \lim_{n \to \infty} \int_X \tilde{\varphi}_n d\mu = \sup_{n \geq 1} \int_X \tilde{\varphi}_n d\mu. \tag{1}$$

For each $n \in \mathbb{N}$ we have

$$\tilde{\varphi}_n(x) = \varphi_n(x) \text{ for all } x \in X - E = D_{\varphi_n} \cap (X - E),$$

and hence

$$\tilde{\varphi}_n(x) = \varphi_n(x) \ \mu\text{-a.e. on } D_{\varphi_n},$$

and hence, by 8.1.17c,

$$\int_X \tilde{\varphi}_n d\mu = \int_X \varphi_n d\mu. \tag{2}$$

Similarly, if $\varphi \in \mathcal{L}^+(X, \mathcal{A}, \mu)$ and $\varphi(x) = \lim_{n \to \infty} \varphi_n(x)$ μ-a.e. on $D_\varphi \cap (\bigcap_{n=1}^\infty D_{\varphi_n})$, and if $F \in \mathcal{A}$ is so that

$$\mu(F) = 0 \text{ and } \varphi(x) = \lim_{n \to \infty} \varphi_n(x) \text{ for all } x \in D_\varphi \cap \left(\bigcap_{n=1}^\infty D_{\varphi_n} \right) \cap (X - F),$$

then

$$\varphi(x) = \tilde{\varphi}(x) \text{ for all } x \in D_\varphi \cap \left(\bigcap_{n=1}^\infty D_{\varphi_n} \right) \cap (X - F) \cap (X - E) = D_\varphi \cap (X - (F \cup E)).$$

Since $F \cup E \in \mathcal{A}$ and $\mu(F \cup E) = 0$ (cf. 7.1.2b), this shows that

$$\varphi(x) = \tilde{\varphi}(x) \text{ } \mu\text{-a.e. on } D_\varphi,$$

and hence that

$$\int_X \varphi d\mu = \int_X \tilde{\varphi} d\mu. \tag{3}$$

What we want to prove follows from 1, 2, 3. $\qquad\qquad\qquad\qquad\qquad\square$

8.1.20 Lemma (Fatou's lemma). *Let $\{\varphi_n\}$ be a sequence in $\mathcal{L}^+(X, \mathcal{A}, \mu)$ and let $\varphi \in \mathcal{L}^+(X, \mathcal{A}, \mu)$. Suppose that*

$$\varphi(x) = \lim_{n \to \infty} \varphi_n(x) \text{ } \mu\text{-a.e. on } D_\varphi \cap \left(\bigcap_{n=1}^\infty D_{\varphi_n} \right)$$

and that there exists $M \in [0, \infty)$ such that $\int_X \varphi_n d\mu \le M$ for all $n \in \mathbb{N}$. Then,

$$\int_X \varphi d\mu \le M.$$

Proof. Let $E \in \mathcal{A}$ be so that

$$\mu(E) = 0 \text{ and } \varphi(x) = \lim_{n \to \infty} \varphi_n(x) \text{ for all } x \in D_\varphi \cap \left(\bigcap_{n=1}^\infty D_{\varphi_n} \right) \cap (X - E),$$

and define $F := (X - D_\varphi) \cup (\bigcup_{n=1}^\infty (X - D_{\varphi_n})) \cup E$. We have $F \in \mathcal{A}$ and $\mu(F) = 0$ (cf. 7.1.4a). Since $X - F \subset D_\varphi$ and $X - F \subset D_{\varphi_n}$ for each $n \in \mathbb{N}$, we can define the functions

$$\tilde{\varphi} : X \to [0, \infty]$$

$$x \mapsto \tilde{\varphi}(x) := \begin{cases} \varphi(x) & \text{if } x \in X - F, \\ 0 & \text{if } x \in F, \end{cases}$$

and, for each $n \in \mathbb{N}$,

$$\tilde{\varphi}_n : X \to [0, \infty]$$

$$x \mapsto \tilde{\varphi}_n(x) := \begin{cases} \varphi_n(x) & \text{if } x \in X - F, \\ 0 & \text{if } x \in F. \end{cases}$$

Proceeding as in the proof of 8.1.19a, we see that:

$\tilde{\varphi} \in \mathcal{L}^+(X, \mathcal{A})$ and $\tilde{\varphi}(x) = \varphi(x)$ μ-a.e. on D_φ;
$\tilde{\varphi}_n \in \mathcal{L}^+(X, \mathcal{A})$ and $\tilde{\varphi}_n(x) = \varphi_n(x)$ μ-a.e. on D_{φ_n}, $\forall n \in \mathbb{N}$.

Thus, 8.1.17c implies that

$$\int_X \tilde{\varphi} d\mu = \int_X \varphi d\mu \text{ and } \int_X \tilde{\varphi}_n = \int_X \varphi_n d\mu, \quad \forall n \in \mathbb{N}.$$

Since $X - F = D_\varphi \cap (\bigcap_{n=1}^\infty D_{\varphi_n}) \cap (X - E)$, we also have

$$\tilde{\varphi}(x) = \lim_{n \to \infty} \tilde{\varphi}_n(x), \forall x \in X,$$

and hence (cf. 6.2.18) $\tilde{\varphi} = \sup_{n \geq 1}(\inf_{k \geq n} \tilde{\varphi}_k)$. Now, for each $k \in \mathbb{N}$,

$$\inf_{k \geq n} \tilde{\varphi}_k \in \mathcal{L}^+(X, \mathcal{A}) \text{ (cf. 6.2.19a) and } \inf_{k \geq n} \tilde{\varphi}_k \leq \inf_{k \geq n+1} \tilde{\varphi}_k.$$

By 8.1.8, this implies that

$$\int_X \tilde{\varphi} d\mu = \sup_{n \geq 1} \int_X (\inf_{k \geq n} \tilde{\varphi}_k) d\mu.$$

Moreover, for each $n \in \mathbb{N}$ we have $\inf_{k \geq n} \tilde{\varphi}_k \leq \tilde{\varphi}_n$ and hence (cf. 8.1.7)

$$\int_X (\inf_{k \geq n} \tilde{\varphi}_k) d\mu \leq \int_X \tilde{\varphi}_n d\mu = \int_X \varphi_n d\mu \leq M.$$

Then we have

$$\int_X \varphi d\mu = \int_X \tilde{\varphi} d\mu \leq M.$$

$\square$

8.2 Integration of complex functions

In this section, $(X, \mathcal{A}, \mu)$ denotes an abstract measure space.

8.2.1 Definition. We denote by $\mathcal{M}(X, \mathcal{A}, \mu)$ the family of functions from X to $\mathbb{C}$ which is defined as follows:

$$\mathcal{M}(X, \mathcal{A}, \mu) := \{\varphi : D_\varphi \to \mathbb{C} : D_\varphi \in \mathcal{A}, \mu(X - D_\varphi) = 0, \varphi \in \mathcal{M}(D_\varphi, \mathcal{A}^{D_\varphi})\}$$

(for $\mathcal{M}(X, \mathcal{A})$, cf. 6.2.15). The elements of $\mathcal{M}(X, \mathcal{A}, \mu)$ are called *μ-measurable functions*.

8.2.2 Theorem. *We have:*

$\alpha\varphi + \beta\psi \in \mathcal{M}(X, \mathcal{A}, \mu)$, $\forall \alpha, \beta \in \mathbb{C}$, $\forall \varphi, \psi \in \mathcal{M}(X, \mathcal{A}, \mu)$;
$\varphi\psi \in \mathcal{M}(X, \mathcal{A}, \mu)$, $\forall \varphi, \psi \in \mathcal{M}(X, \mathcal{A}, \mu)$.

However, $\mathcal{M}(X, \mathcal{A}, \mu)$ is not an associative algebra nor a linear space, unless the only element E of $\mathcal{A}$ such that $\mu(E) = 0$ is $E = \emptyset$.

Proof. For $\alpha, \beta \in \mathbb{C}$ and $\varphi, \psi \in \mathcal{M}(X, \mathcal{A}, \mu)$ we have:

$D_{\alpha\varphi+\beta\psi} = D_\varphi \cap D_\psi \in \mathcal{A}$ (cf. 1.2.19);

$\mu(X - D_\varphi \cap D_\psi) = \mu((X - D_\varphi) \cup (X - D_\psi)) = 0$ (cf. 7.1.2b);

$\varphi_{D_\varphi \cap D_\psi}, \psi_{D_\varphi \cap D_\psi} \in \mathcal{M}(D_\varphi \cap D_\psi, \mathcal{A}^{D_\varphi \cap D_\psi})$ (cf. 6.2.3 and 6.1.19b) and hence $\alpha\varphi + \beta\psi = \alpha\varphi_{D_\varphi \cap D_\psi} + \beta\psi_{D_\varphi \cap D_\psi} \in \mathcal{M}(D_\varphi \cap D_\psi, \mathcal{A}^{D_\varphi \cap D_\psi})$ (cf. 6.2.16).

This proves that $\alpha\varphi + \beta\psi \in \mathcal{M}(X, \mathcal{A}, \mu)$.

The proof for $\varphi\psi$ is analogous.

If there exists $E \in \mathcal{A}$ so that $E \neq \emptyset$ and $\mu(E) = 0$, then obviously there exists $\varphi \in \mathcal{M}(X, \mathcal{A}, \mu)$ so that $D_\varphi \neq X$ (e.g., a constant function with $X - E$ as its domain), and hence so that it has no opposite (the situation is quite similar to the one discussed in 3.2.11). Therefore, $\mathcal{M}(X, \mathcal{A}, \mu)$ is not a linear space, and hence it cannot be an associative algebra.

On the other hand, if the only element E of $\mathcal{A}$ such that $\mu(E) = 0$ is $E = \emptyset$, then $\mathcal{M}(X, \mathcal{A}, \mu) = \mathcal{M}(X, \mathcal{A})$, which is an associative algebra and hence a linear space as well (cf. 6.2.16). $\square$

8.2.3 Definitions. We denote by $\mathcal{L}^1(X, \mathcal{A}, \mu)$ the family of functions that is defined as follows:

$$\mathcal{L}^1(X, \mathcal{A}, \mu) := \left\{ \varphi \in \mathcal{M}(X, \mathcal{A}, \mu) : \int_X \varphi_i d\mu < \infty \text{ for } i = 1, ..., 4 \right\},$$

with, for $\varphi \in \mathcal{M}(X, \mathcal{A}, \mu)$,

$$\varphi_1 := (\operatorname{Re}\varphi)^+, \varphi_2 := (\operatorname{Re}\varphi)^-, \varphi_3 := (\operatorname{Im}\varphi)^+, \varphi_4 := (\operatorname{Im}\varphi)^-$$

(cf. 1.2.19). Note that, for $i = 1, ..., 4$, the condition $\int_X \varphi_i d\mu < \infty$ is consistent because $\varphi_i \in \mathcal{L}^+(X, \mathcal{A}, \mu)$ (cf. 6.2.12 and 6.2.20b). The elements of $\mathcal{L}^1(X, \mathcal{A}, \mu)$ are called *Lebesgue integrable functions (with respect to μ)* or *μ-integrable functions* or, simply, *integrable functions*.

For all $\varphi \in \mathcal{L}^1(X, \mathcal{A}, \mu)$, we define the *Lebesgue integral* of φ *(with respect to μ)* as the element $\int_X \varphi d\mu$ of $\mathbb{C}$ defined by

$$\int_X \varphi d\mu := \int_X (\operatorname{Re}\varphi)^+ d\mu - \int_X (\operatorname{Re}\varphi)^- d\mu + i \int_X (\operatorname{Im}\varphi)^+ d\mu - i \int_X (\operatorname{Im}\varphi)^- d\mu.$$

This definition is consistent with the one given in 8.1.14 because $\varphi = (\operatorname{Re}\varphi)^+$ and $(\operatorname{Re}\varphi)^- = (\operatorname{Im}\varphi)^+ = (\operatorname{Im}\varphi)^- = 0_X$ for all $\varphi \in \mathcal{L}^1(X, \mathcal{A}, \mu) \cap \mathcal{L}^+(X, \mathcal{A}, \mu)$.

For $\varphi \in \mathcal{M}(X, \mathcal{A}, \mu)$, it follows immediately from the definitions that

$$\varphi \in \mathcal{L}^1(X, \mathcal{A}, \mu) \text{ iff } \operatorname{Re}\varphi, \operatorname{Im}\varphi \in \mathcal{L}^1(X, \mathcal{A}, \mu)$$

and that

$$\text{if } \varphi \in \mathcal{L}^1(X, \mathcal{A}, \mu) \text{ then } \int_X \varphi d\mu = \int_X (\operatorname{Re}\varphi) d\mu + i \int_X (\operatorname{Im}\varphi) d\mu.$$

Thus, for $\varphi \in \mathcal{L}^1(X, \mathcal{A}, \mu)$, $\overline{\varphi} \in \mathcal{L}^1(X, \mathcal{A}, \mu)$ and

$$\int_X \overline{\varphi} d\mu = \int_X (\operatorname{Re}\overline{\varphi}) d\mu + i \int_X (\operatorname{Im}\overline{\varphi}) d\mu = \int_X (\operatorname{Re}\varphi) d\mu - i \int_X (\operatorname{Im}\varphi) d\mu = \overline{\left(\int_X \varphi d\mu\right)}.$$

As a matter of convenience, for $\varphi \in \mathcal{L}^1(X, \mathcal{A}, \mu)$ we will sometimes write

$$\int_X \varphi(x) d\mu(x) := \int_X \varphi d\mu.$$

8.2.4 Proposition. *For $\varphi \in \mathcal{M}(X, \mathcal{A}, \mu)$, the following conditions are equivalent:*

(a) $\varphi \in \mathcal{L}^1(X, \mathcal{A}, \mu)$;
(b) $\int_X |\varphi| d\mu < \infty$.

Proof. First, note that condition b is consistent because $|\varphi| \in \mathcal{L}^+(X, \mathcal{A}, \mu)$ by 6.2.17.

$a \Rightarrow b$: Notice that, by the triangle inequality in $\mathbb{C}$,

$$|\varphi|(x) \leq (\operatorname{Re} \varphi)^+(x) + (\operatorname{Re} \varphi)^-(x) + (\operatorname{Im} \varphi)^+(x) + (\operatorname{Im} \varphi)^-(x), \forall x \in D_\varphi,$$

and use 8.1.17a,b.

$b \Rightarrow a$: Notice that

$$(\operatorname{Re} \varphi)^\pm(x) \leq |\operatorname{Re} \varphi|(x) \leq |\varphi|(x) \text{ and } (\operatorname{Im} \varphi)^\pm(x) \leq |\operatorname{Im} \varphi|(x) \leq |\varphi|(x), \forall x \in D_\varphi,$$

and use 8.1.17b. $\qquad \square$

8.2.5 Corollary. *If $\varphi \in \mathcal{M}(X, \mathcal{A}, \mu)$, $\psi \in \mathcal{L}^1(X, \mathcal{A}, \mu)$ and $|\varphi(x)| \leq |\psi(x)|$ μ-a.e. on $D_\varphi \cap D_\psi$, then $\varphi \in \mathcal{L}^1(X, \mathcal{A}, \mu)$.*

Proof. Use 8.1.17b and 8.2.4. $\qquad \square$

8.2.6 Corollary. *Suppose that*

$$\varphi \in \mathcal{M}(X, \mathcal{A}, \mu), \quad \exists k \in [0, \infty) \text{ s.t. } |\varphi(x)| \leq k \text{ } \mu\text{-a.e. on } D_\varphi, \quad \mu(X) < \infty.$$

Then $\varphi \in \mathcal{L}^1(X, \mathcal{A}, \mu)$.

Proof. The result follows from 8.2.4 and 8.2.5, since $\int_X k d\mu = k\mu(X)$. $\qquad \square$

8.2.7 Theorem. *Suppose that $\varphi \in \mathcal{L}^1(X, \mathcal{A}, \mu)$, $\psi \in \mathcal{M}(X, \mathcal{A}, \mu)$ and*

$$\varphi(x) = \psi(x) \text{ } \mu\text{-a.e. on } D_\varphi \cap D_\psi.$$

Then

$$\psi \in \mathcal{L}^1(X, \mathcal{A}, \mu) \text{ and } \int_X \psi d\mu = \int_X \varphi d\mu.$$

Proof. We have

$$(\operatorname{Re} \varphi)^\pm(x) = (\operatorname{Re} \psi)^\pm(x) \text{ and } (\operatorname{Im} \varphi)^\pm(x) = (\operatorname{Im} \psi)^\pm(x) \text{ } \mu\text{-a.e. on } D_\varphi \cap D_\psi.$$

The result then follows from 8.1.17c. $\qquad \square$

8.2.8 Proposition. *If $\mu(X) = 0$, then $\mathcal{L}^1(X, \mathcal{A}, \mu) = \mathcal{M}(X, \mathcal{A}, \mu)$ and*

$$\int_X \varphi d\mu = 0, \quad \forall \varphi \in \mathcal{M}(X, \mathcal{A}, \mu).$$

Proof. Use 8.1.18b (or else, notice that if $\mu(X) = 0$ then $\varphi(x) = 0$ μ-a.e. for each $\varphi \in \mathcal{M}(X, \mathcal{A}, \mu)$, and use 8.2.7). $\qquad \square$

8.2.9 Theorem. *Let $\alpha, \beta \in \mathbb{C}$ and $\varphi, \psi \in \mathcal{L}^1(X, \mathcal{A}, \mu)$. Then*

$$\alpha\varphi + \beta\psi \in \mathcal{L}^1(X, \mathcal{A}, \mu) \text{ and } \int_X (\alpha\varphi + \beta\psi)d\mu = \alpha \int_X \varphi d\mu + \beta \int_X \psi d\mu.$$

Proof. We already know that $\alpha\varphi + \beta\psi \in \mathcal{M}(X, \mathcal{A}, \mu)$ (cf. 8.2.2). By 8.1.17a,b and 8.2.4 we also have

$$\int_X |\alpha\varphi + \beta\psi|d\mu \leq \int_X (|\alpha||\varphi| + |\beta||\psi|)d\mu = |\alpha| \int_X |\varphi|d\mu + |\beta| \int_X |\psi|d\mu < \infty,$$

which proves (cf. 8.2.4) that $\alpha\varphi + \beta\psi \in \mathcal{L}^1(X, \mathcal{A}, \mu)$.

To prove the second part of the statement, it is clearly sufficient to prove that, if $\varphi, \psi \in \mathcal{L}^1(X, \mathcal{A}, \mu)$, then

$$\int_X (\varphi + \psi)d\mu = \int_X \varphi d\mu + \int_X \psi d\mu, \tag{1}$$

and that, if $\alpha \in \mathbb{C}$ and $\varphi \in \mathcal{L}^1(X, \mathcal{A}, \mu)$, then

$$\int_X \alpha\varphi d\mu = \alpha \int_X \varphi d\mu. \tag{2}$$

To prove 1 for $\varphi, \psi \in \mathcal{L}^1(X, \mathcal{A}, \mu)$, we note that

$$(\text{Re}(\varphi + \psi))^+ - (\text{Re}(\varphi + \psi))^- = \text{Re}(\varphi + \psi) = \text{Re}\,\varphi + \text{Re}\,\psi$$
$$= (\text{Re}\,\varphi)^+ - (\text{Re}\,\varphi)^- + (\text{Re}\,\psi)^+ - (\text{Re}\,\psi)^-,$$

or

$$(\text{Re}(\varphi + \psi))^+ + (\text{Re}\,\varphi)^- + (\text{Re}\,\psi)^- = (\text{Re}\,\varphi)^+ + (\text{Re}\,\psi)^+ + (\text{Re}(\varphi + \psi))^-$$

(all the terms of these equalities are functions with domain $D_\varphi \cap D_\psi$). Hence, by 8.1.17a,

$$\int_X (\text{Re}(\varphi + \psi))^+ d\mu + \int_X (\text{Re}\,\varphi)^- d\mu + \int_X (\text{Re}\,\psi)^- d\mu$$
$$= \int_X (\text{Re}\,\varphi)^+ d\mu + \int_X (\text{Re}\,\psi)^+ d\mu + \int_X (\text{Re}(\varphi + \psi))^- d\mu$$

and, since each of these integrals is finite, we may transpose and obtain

$$\int_X (\text{Re}(\varphi + \psi))^+ d\mu - \int_X (\text{Re}(\varphi + \psi))^- d\mu$$
$$= \int_X (\text{Re}\,\varphi)^+ d\mu - \int_X (\text{Re}\,\varphi)^- d\mu + \int_X (\text{Re}\,\psi)^+ d\mu - \int_X (\text{Re}\,\psi)^- d\mu.$$

In exactly the same way we obtain

$$\int_X (\text{Im}(\varphi + \psi))^+ d\mu - \int_X (\text{Im}(\varphi + \psi))^- d\mu$$
$$= \int_X (\text{Im}\,\varphi)^+ d\mu - \int_X (\text{Im}\,\varphi)^- d\mu + \int_X (\text{Im}\,\psi)^+ d\mu - \int_X (\text{Im}\,\psi)^- d\mu.$$

By multiplying by i the second equation and summing, we obtain 1.

Let now $\varphi \in \mathcal{L}^1(X, \mathcal{A}, \mu)$. If $\alpha \geq 0$, then 2 follows from 8.1.17a and from the equalities

$$(\operatorname{Re}(\alpha\varphi))^{\pm} = \alpha(\operatorname{Re}\varphi)^{\pm} \text{ and } (\operatorname{Im}(\alpha\varphi))^{\pm} = \alpha(\operatorname{Im}\varphi)^{\pm}.$$

If $\alpha = -1$, then 2 follows from the equalities

$$(\operatorname{Re}(-\varphi))^{\pm} = (\operatorname{Re}\varphi)^{\mp} \text{ and } (\operatorname{Im}(-\varphi))^{\pm} = (\operatorname{Im}\varphi)^{\mp}.$$

If $\alpha = i$, then 2 follows from the equalities

$$(\operatorname{Re}(i\varphi))^{\pm} = (\operatorname{Im}\varphi)^{\mp} \text{ and } (\operatorname{Im}(i\varphi))^{\pm} = (\operatorname{Re}\varphi)^{\pm}.$$

Combining these cases with 1, we obtain 2 for any $\alpha \in \mathbb{C}$. $\qquad\square$

8.2.10 Theorem. *If $\varphi \in \mathcal{L}^1(X, \mathcal{A}, \mu)$, then*

$$\left| \int_X \varphi d\mu \right| \leq \int_X |\varphi| d\mu.$$

Proof. If $\int_X \varphi d\mu = 0$, then the inequality is obvious. Now assume $\int_X \varphi d\mu \neq 0$ and let $\alpha := \overline{\left(\int_X \varphi d\mu \right)} \left| \int_X \varphi d\mu \right|^{-1}$. Then

$$\left| \int_X \varphi d\mu \right| = \alpha \int_X \varphi d\mu = \int_X \alpha\varphi d\mu.$$

This shows that $\int_X \alpha\varphi d\mu$ is an element of $\mathbb{R}$. Hence

$$\left| \int_X \varphi d\mu \right| = \operatorname{Re} \int_X \alpha\varphi d\mu = \int_X (\operatorname{Re}(\alpha\varphi))^+ d\mu - \int_X (\operatorname{Re}(\alpha\varphi))^- d\mu$$

$$\leq \int_X (\operatorname{Re}(\alpha\varphi))^+ d\mu + \int_X (\operatorname{Re}(\alpha\varphi))^- d\mu$$

$$= \int_X |\operatorname{Re}(\alpha\varphi)| d\mu \leq \int_X |\alpha\varphi| d\mu = \int_X |\varphi| d\mu,$$

where the second inequality holds by 8.1.17b and the last equality holds since $|\alpha| = 1$. $\qquad\square$

8.2.11 Theorem (Lebesgue's dominated convergence theorem).
Suppose that $\{\varphi_n\}$ is a sequence in $\mathcal{M}(X, \mathcal{A}, \mu)$ such that the following two conditions are satisfied:

the sequence $\{\varphi_n(x)\}$ is convergent (in the metric space $(\mathbb{C}, d_{\mathbb{C}})$) μ-a.e. on $\bigcap_{n=1}^{\infty} D_{\varphi_n}$;
there exists $\psi \in \mathcal{L}^1(X, \mathcal{A}, \mu)$ so that $|\varphi_n(x)| \leq \psi(x)$ μ-a.e. on $D_{\varphi_n} \cap D_\psi$, $\forall n \in \mathbb{N}$ (ψ is said to be a dominating function).

Then:

(a) $\varphi_n \in \mathcal{L}^1(X, \mathcal{A}, \mu)$, $\forall n \in \mathbb{N}$;
(b) there exists $\varphi \in \mathcal{M}(X, \mathcal{A}, \mu)$ s.t. $\varphi(x) = \lim_{n\to\infty} \varphi_n(x)$, $\forall x \in D_\varphi \cap (\bigcap_{n=1}^{\infty} D_{\varphi_n})$;

*(c) if $\varphi \in \mathcal{M}(X, \mathcal{A}, \mu)$ is s.t. $\varphi(x) = \lim_{n \to \infty} \varphi_n(x)$ μ-a.e. on $D_\varphi \cap (\bigcap_{n=1}^{\infty} D_{\varphi_n})$,
then $\varphi \in \mathcal{L}^1(X, \mathcal{A}, \mu)$, $\lim_{n \to \infty} \int_X |\varphi_n - \varphi| d\mu = 0$, $\int_X \varphi d\mu = \lim_{n \to \infty} \int_X \varphi_n d\mu$.*

Proof. a: For each $n \in \mathbb{N}$, we note that $|\varphi_n(x)| \le \psi(x)$ entails $\psi(x) \in [0, \infty)$. Thus we have $|\varphi_n(x)| \le |\psi(x)|$ μ-a.e. on $D_{\varphi_n} \cap D_\psi$, and this implies $\varphi_n \in \mathcal{L}^1(X, \mathcal{A}, \mu)$ by 8.2.5.

b: Let $E \in \mathcal{A}$ be so that

$$\mu(E) = 0 \text{ and } \{\varphi_n(x)\} \text{ is convergent for all } x \in \left(\bigcap_{n=1}^{\infty} D_{\varphi_n} \right) \cap (X - E).$$

Letting $S := (\bigcap_{n=1}^{\infty} D_{\varphi_n}) \cap (X - E)$, we have $S \in \mathcal{A}$. We define the function

$$\varphi : S \to \mathbb{C}$$
$$x \mapsto \varphi(x) := \lim_{n \to \infty} \varphi_n(x).$$

Since $(\varphi_n)_S \in \mathcal{M}(S, \mathcal{A}^S)$ (cf. 6.2.3 and 6.1.19b), we have $\varphi \in \mathcal{M}(S, \mathcal{A}^S)$ by 6.2.20c, and hence $\varphi \in \mathcal{M}(X, \mathcal{A}, \mu)$ since $\mu(X - S) = \mu((\bigcup_{n=1}^{\infty}(X - D_{\varphi_n})) \cup E) = 0$ (cf. 7.1.4a). From the definition of φ we have

$$\varphi(x) = \lim_{n \to \infty} \varphi_n(x), \forall x \in D_\varphi = D_\varphi \cap \left(\bigcap_{n=1}^{\infty} D_{\varphi_n} \right).$$

c: Let $\varphi \in \mathcal{M}(X, \mathcal{A}, \mu)$ and let $F \in \mathcal{A}$ be so that

$$\mu(F) = 0 \text{ and } \varphi(x) = \lim_{n \to \infty} \varphi_n(x) \text{ for all } x \in D_\varphi \cap \left(\bigcap_{n=1}^{\infty} D_{\varphi_n} \right) \cap (X - F).$$

For each $n \in \mathbb{N}$, let $G_n \in \mathcal{A}$ be so that

$$\mu(G_n) = 0 \text{ and } |\varphi_n(x)| \le \psi(x) \text{ for all } x \in D_{\varphi_n} \cap D_\psi \cap (X - G_n).$$

We define

$$H := \left(\bigcup_{n=1}^{\infty}(X - D_{\varphi_n}) \right) \cup (X - D_\psi) \cup (X - D_\varphi) \cup F \cup \left(\bigcup_{n=1}^{\infty} G_n \right).$$

We have $H \in \mathcal{A}$ and $\mu(H) = 0$ (cf. 7.1.4a). Moreover, we note that

$$X - H = \left(\bigcap_{n=1}^{\infty} D_{\varphi_n} \right) \cap D_\psi \cap D_\varphi \cap (X - F) \cap \left(\bigcap_{n=1}^{\infty}(X - G_n) \right).$$

We have

$$|\varphi(x)| \le \psi(x), \forall x \in X - H = D_\varphi \cap D_\psi \cap (X - H),$$

and hence $\varphi \in \mathcal{L}^1(X, \mathcal{A}, \mu)$ by 8.2.5.

Now we define the functions:

$$\tilde{\varphi}_n := (\varphi_n)_{X-H} \text{ for each } n \in \mathbb{N}, \quad \tilde{\psi} := \psi_{X-H}, \quad \tilde{\varphi} := \varphi_{X-H}.$$

These functions are elements of $\mathcal{M}(X - H, \mathcal{A}^{X-H})$ by 6.2.3 and 6.1.19b, and hence elements of $\mathcal{M}(X, \mathcal{A}, \mu)$ since $\mu(H) = 0$. Moreover, $\tilde{\psi} \in \mathcal{L}^1(X, \mathcal{A}, \mu)$ by 8.2.7.

For each $n \in \mathbb{N}$, we define the function

$$\tilde{\psi}_n : X - H \to [0, \infty]$$

$$x \mapsto \tilde{\psi}_n(x) := \sup_{k \geq n} |\tilde{\varphi}_k(x) - \tilde{\varphi}(x)|$$

(in this proof we characterize with a tilde the functions whose domain is $X - H$). By 6.2.16, 6.2.17, 6.2.19a we have $\tilde{\psi}_n \in \mathcal{M}(X - H, \mathcal{A}^{X-H})$ and hence $\tilde{\psi}_n \in \mathcal{M}(X, \mathcal{A}, \mu)$. From

$$|\tilde{\varphi}_k(x)| \leq \tilde{\psi}(x) \text{ for each } k \in \mathbb{N} \text{ and } |\tilde{\varphi}(x)| \leq \tilde{\psi}(x), \quad \forall x \in X - H,$$

we have

$$\tilde{\psi}_n(x) \leq 2\tilde{\psi}(x), \forall x \in X - H,$$

and hence $\tilde{\psi}_n \in \mathcal{L}^1(X, \mathcal{A}, \mu)$ (cf. 8.2.5) and $2\tilde{\psi} - \tilde{\psi}_n \in \mathcal{L}^+(X, \mathcal{A}, \mu)$. We also have $2\tilde{\psi} - \tilde{\psi}_n \leq 2\tilde{\psi} - \tilde{\psi}_{n+1}$ since obviously $\tilde{\psi}_{n+1} \leq \tilde{\psi}_n$. Furthermore, by 5.2.6 we have

$$\lim_{n \to \infty} \tilde{\psi}_n(x) = \lim_{n \to \infty} |\tilde{\varphi}_n(x) - \tilde{\varphi}(x)|$$

$$= \lim_{n \to \infty} |\varphi_n(x) - \varphi(x)| = 0, \forall x \in X - H,$$

and hence

$$2\tilde{\psi}(x) = \lim_{n \to \infty} (2\tilde{\psi}(x) - \tilde{\psi}_n(x)), \forall x \in X - H.$$

Then, by 8.1.19 and 8.2.9 we have

$$\int_X 2\tilde{\psi} d\mu = \lim_{n \to \infty} \int_X (2\tilde{\psi} - \tilde{\psi}_n) d\mu = \lim_{n \to \infty} \left(\int_X 2\tilde{\psi} d\mu - \int_X \tilde{\psi}_n d\mu \right),$$

and hence

$$\lim_{n \to \infty} \int_X \tilde{\psi}_n d\mu = 0. \tag{1}$$

Moreover, for each $n \in \mathbb{N}$ we have

$$|\varphi_n(x) - \varphi(x)| = |\tilde{\varphi}_n(x) - \tilde{\varphi}(x)| \leq \tilde{\psi}_n(x), \forall x \in X - H = D_{\varphi_n - \varphi} \cap D_{\tilde{\psi}_n} \cap (X - H),$$

and hence, by 8.2.9, 8.2.10, 8.1.17b,

$$\left| \int_X \varphi_n d\mu - \int_X \varphi d\mu \right| \leq \int_X |\varphi_n - \varphi| d\mu \leq \int_X \tilde{\psi}_n d\mu. \tag{2}$$

Now, 1 and 2 imply that

$$\lim_{n \to \infty} \int_X |\varphi_n - \varphi| d\mu = 0 \text{ and } \int_X \varphi d\mu = \lim_{n \to \infty} \int_X \varphi_n d\mu.$$

$\square$

8.2.12 Definition. In $\mathcal{M}(X, \mathcal{A}, \mu)$ we define a relation, denoted by $\sim$, as follows:

$$\varphi \sim \psi \text{ if } \varphi(x) = \psi(x) \ \mu\text{-a.e. on } D_\varphi \cap D_\psi.$$

This relation is obviously reflexive and symmetric. It is also transitive. Suppose in fact that $\varphi_1, \varphi_2, \varphi_3 \in \mathcal{M}(X, \mathcal{A}, \mu)$ are so that $\varphi_1 \sim \varphi_2$ and $\varphi_2 \sim \varphi_3$, and let:

$E \in \mathcal{A}$ be so that $\mu(E) = 0$ and $\varphi_1(x) = \varphi_2(x), \forall x \in D_{\varphi_1} \cap D_{\varphi_2} \cap (X - E)$;
$F \in \mathcal{A}$ be so that $\mu(F) = 0$ and $\varphi_2(x) = \varphi_3(x), \forall x \in D_{\varphi_2} \cap D_{\varphi_3} \cap (X - F)$;

letting $G := (X - D_{\varphi_2}) \cup E \cup F$, we have $G \in \mathcal{A}$, $\mu(G) = 0$ (cf. 7.1.2b) and

$$\varphi_1(x) = \varphi_3(x), \forall x \in D_{\varphi_1} \cap D_{\varphi_3} \cap (X - G),$$

and this shows that $\varphi_1 \sim \varphi_3$.

Thus, $\sim$ is an equivalence relation (cf. 1.1.5) and this justifies the symbol used. We denote by $M(X, \mathcal{A}, \mu)$ the quotient set defined by $\sim$, i.e. we define

$$M(X, \mathcal{A}, \mu) := \mathcal{M}(X, \mathcal{A}, \mu)/\sim .$$

In every element of $M(X, \mathcal{A}, \mu)$ there exists a representative which is an element of $\mathcal{M}(X, \mathcal{A})$. Indeed, for each $\varphi \in \mathcal{M}(X, \mathcal{A}, \mu)$, the function

$$\varphi_e : X \to \mathbb{C}$$

$$x \mapsto \varphi_e(x) := \begin{cases} \varphi(x) & \text{if } x \in D_\varphi, \\ 0 & \text{if } x \notin D_\varphi \end{cases}$$

is an element of $\mathcal{M}(X, \mathcal{A})$ (this can be seen as in 8.1.14) and clearly $\varphi_e \sim \varphi$.

8.2.13 Proposition. *The following definitions, of the mappings σ, μ, π, are consistent:*

$$\sigma : M(X, \mathcal{A}, \mu) \times M(X, \mathcal{A}, \mu) \to M(X, \mathcal{A}, \mu)$$
$$([\varphi], [\psi]) \mapsto \sigma([\varphi], [\psi]) := [\varphi] + [\psi] := [\varphi + \psi];$$

$$\mu : \mathbb{C} \times M(X, \mathcal{A}, \mu) \to M(X, \mathcal{A}, \mu)$$
$$(\alpha, [\psi]) \mapsto \mu(\alpha, [\psi]) := \alpha[\varphi] := [\alpha\varphi];$$

$$\pi : M(X, \mathcal{A}, \mu) \times M(X, \mathcal{A}, \mu) \to M(X, \mathcal{A}, \mu)$$
$$([\varphi], [\psi]) \mapsto \pi([\varphi], [\psi]) := [\varphi][\psi] := [\varphi\psi].$$

Then, $(M(X, \mathcal{A}, \mu), \sigma, \mu, \pi)$ is an abelian associative algebra over $\mathbb{C}$, with $[1_X]$ as identity. The zero element is $[0_X]$ and the opposite of $[\varphi] \in M(X, \mathcal{A}, \mu)$ is $[-\varphi]$.

It must be remarked that the second mapping above has been denoted by μ, which is the symbol used for scalar multiplication throughout the book, even though the same symbol μ had already been chosen to denote a measure in the present chapter. However, no confusion should arise since the roles of the two things denoted by μ are utterly different.

Proof. The only thing to prove is that the mappings σ, μ, π can indeed be defined as they were in the statement, while it is immediate to check all the rest.

We already know that, if $\varphi, \psi \in \mathcal{M}(X, \mathcal{A}, \mu)$, then $\varphi + \psi \in \mathcal{M}(X, \mathcal{A}, \mu)$ (cf. 8.2.2). Suppose now that $\varphi, \varphi', \psi, \psi' \in \mathcal{M}(X, \mathcal{A}, \mu)$ are so that $\varphi' \sim \varphi$, $\psi' \sim \psi$, and let:

$E \in \mathcal{A}$ be so that $\mu(E) = 0$ and $\varphi'(x) = \varphi(x)$, $\forall x \in D_{\varphi'} \cap D_\varphi \cap (X - E)$;
$F \in \mathcal{A}$ be so that $\mu(F) = 0$ and $\psi'(x) = \psi(x)$, $\forall x \in D_{\psi'} \cap D_\psi \cap (X - F)$.

Then

$$\varphi'(x) + \psi'(x) = \varphi(x) + \psi(x),$$
$$\forall x \in D_{\varphi'} \cap D_\varphi \cap (X - E) \cap D_{\psi'} \cap D_\psi \cap (X - F)$$
$$= D_{\varphi' + \psi'} \cap D_{\varphi + \psi} \cap (X - (E \cup F)),$$

which proves that $\varphi' + \psi' \sim \varphi + \psi$. This shows that the equivalence class $[\varphi + \psi]$ does not depend on the particular elements φ and ψ (of the classes $[\varphi]$ and $[\psi]$) through which it has been defined. Hence, the rule which assigns $[\varphi + \psi]$ to a pair $([\varphi], [\psi]) \in M(X, \mathcal{A}, \mu) \times M(X, \mathcal{A}, \mu)$ does assign one and only one element of $M(X, \mathcal{A}, \mu)$ to $([\varphi], [\psi])$.

The arguments for μ and for π are analogous. $\quad\square$

8.2.14 Definition. We define the subset $S(X, \mathcal{A}, \mu)$ of $M(X, \mathcal{A}, \mu)$ as follows:

$$S(X, \mathcal{A}, \mu) := \{[\varphi] \in M(X, \mathcal{A}, \mu) : \exists \psi \in \mathcal{S}(X, \mathcal{A}) \text{ so that } \psi \in [\varphi]\}.$$

In view of 6.2.24, $S(X, \mathcal{A}, \mu)$ is a subalgebra of the abelian associative algebra $M(X, \mathcal{A}, \mu)$.

8.2.15 Theorem. *The following definition, of the set $L^1(X, \mathcal{A}, \mu)$, is consistent:*

$$L^1(X, \mathcal{A}, \mu) := \left\{ [\varphi] \in M(X, \mathcal{A}, \mu) : \int_X |\varphi| d\mu < \infty \right\}.$$

Then, $L^1(X, \mathcal{A}, \mu)$ is a linear manifold in the linear space $M(X, \mathcal{A}, \mu)$.
The following definition, of the function ν, is consistent:

$$\nu : L^1(X, \mathcal{A}, \mu) \to \mathbb{R}$$
$$[\varphi] \mapsto \nu([\varphi]) := \|[\varphi]\|_{L^1} := \int_X |\varphi| d\mu.$$

Then, ν is a norm for the linear space $L^1(X, \mathcal{A}, \mu)$.
The following definition, of the function I, is consistent:

$$I : L^1(X, \mathcal{A}, \mu) \to \mathbb{C}$$
$$[\varphi] \mapsto I([\varphi]) := \int_X \varphi d\mu.$$

Then, I is a continuous linear functional.

Proof. To prove that $L^1(X, \mathcal{A}, \mu)$ can indeed be defined as it was in the statement, we must show that the implication

$$[\varphi', \varphi \in \mathcal{M}(X, \mathcal{A}, \mu), \varphi' \sim \varphi, \int_X |\varphi| d\mu < \infty] \Rightarrow \int_X |\varphi'| d\mu < \infty \qquad (*)$$

holds true, because then the condition $\int_X |\varphi| d\mu < \infty$ is actually a condition for the equivalence class $[\varphi]$ even though it is expressed through a particular element of it. Now, $(*)$ is true by 8.1.17c.

Similar arguments, based on 8.1.17c and 8.2.7, show that ν and I can be defined as they were in the statement.

Since, for $\varphi \in \mathcal{M}(X, \mathcal{A}, \mu)$, $\int_X |\varphi| d\mu < \infty$ is equivalent to $\varphi \in \mathcal{L}^1(X, \mathcal{A}, \mu)$ (cf. 8.2.4), 8.2.9 proves that $L^1(X, \mathcal{A}, \mu)$ is a linear manifold in $M(X, \mathcal{A}, \mu)$ and that I is a linear functional.

To prove that ν is a norm, we notice that:

$$\forall [\varphi], [\psi] \in L^1(X, \mathcal{A}, \mu), \|[\varphi] + [\psi]\|_{L^1} = \int_X |\varphi + \psi| d\mu \leq \int_X (|\varphi| + |\psi|) d\mu$$

$$= \int_X |\varphi| d\mu + \int_X |\psi| d\mu = \|[\varphi]\|_{L^1} + \|[\psi]\|_{L^1}$$

$$\forall \alpha \in \mathbb{C}, \forall [\varphi] \in L^1(X, \mathcal{A}, \mu), \|\alpha[\varphi]\|_{L^1} = \int_X |\alpha \varphi| d\mu = |\alpha| \int_X |\varphi| d\mu = \|\alpha\| \|[\varphi]\|_{L^1},$$

$$\|[\varphi]\|_{L^1} = 0 \Rightarrow \int_X |\varphi| d\mu = 0 \Rightarrow \varphi(x) = 0 \ \mu\text{-a.e. on } D_\varphi \Rightarrow [\varphi] = [0_X],$$

where we have used 8.1.17a,b and 8.1.18a.

To prove that I is continuous, we notice that, by 8.2.10,

$$\forall [\varphi] \in L^1(X, \mathcal{A}, \mu), |I([\varphi])| = \left| \int_X \varphi d\mu \right| \leq \int_X |\varphi| d\mu = \|[\varphi]\|_{L^1},$$

and we use 4.2.2. $\qquad \square$

8.2.16 Proposition. *The intersection $S^1(X, \mathcal{A}, \mu) := S(X, \mathcal{A}, \mu) \cap L^1(X, \mathcal{A}, \mu)$ is a dense linear manifold in the normed space $L^1(X, \mathcal{A}, \mu)$.*

Proof. Since $S(X, \mathcal{A}, \mu)$ and $L^1(X, \mathcal{A}, \mu)$ are linear manifolds in the linear space $M(X, \mathcal{A}, \mu)$, the same is true for their intersection (cf. 3.1.5). Then $S^1(X, \mathcal{A}, \mu)$ is a linear manifold in the linear space $L^1(X, \mathcal{A}, \mu)$, too (cf. 3.1.4b).

Let $[\varphi] \in L^1(X, \mathcal{A}, \mu)$ and assume that the representative φ is an element of $\mathcal{M}(X, \mathcal{A})$ (cf. 8.2.12). By 6.2.27 there exists a sequence $\{\psi_n\}$ in $\mathcal{S}(X, \mathcal{A})$ such that

$|\psi_n(x)| \leq |\varphi(x)|, \forall x \in X, \forall n \in \mathbb{N},$
$\varphi(x) = \lim_{n \to \infty} \psi_n(x), \forall x \in X.$

Then 8.2.11 implies that $\psi_n \in \mathcal{L}^1(X, \mathcal{A}, \mu)$ for all $n \in \mathbb{N}$ and that

$$\lim_{n \to \infty} \|[\psi_n] - [\varphi]\| = \lim_{n \to \infty} \int_X |\psi_n - \varphi| d\mu = 0.$$

In view of 2.3.12, this proves that $S^1(X, \mathcal{A}, \mu)$ is dense in $L^1(X, \mathcal{A}, \mu)$. $\qquad \square$

8.3 Integration with respect to measures constructed from other measures

8.3.1 Proposition. *Let $(X, \mathcal{A}, \mu)$ be a measure space, let E be a non-empty element of $\mathcal{A}$, and denote by μ_E the restriction of μ to the σ-algebra $\mathcal{A}^E$ (cf. 6.1.19), i.e. define*

$$\mu_E : \mathcal{A}^E \to [0, \infty]$$
$$F \mapsto \mu_E(F) := \mu(F).$$

(a) μ_E is a measure on $\mathcal{A}^E$.
(b) For $\varphi \in \mathcal{L}^+(X, \mathcal{A}, \mu)$, denote by φ_E the restriction of φ to $D_\varphi \cap E$, i.e. define $\varphi_E := \varphi_{D_\varphi \cap E}$. Then $\varphi_E \in \mathcal{L}^+(E, \mathcal{A}^E, \mu_E)$, $\chi_E \varphi \in \mathcal{L}^+(X, \mathcal{A}, \mu)$ and

$$\int_E \varphi_E \, d\mu_E = \int_X \chi_E \varphi \, d\mu.$$

(c) For $\psi \in \mathcal{M}(X, \mathcal{A}, \mu)$, denote by ψ_E the restriction of ψ to $D_\psi \cap E$, i.e. define $\psi_E := \psi_{D_\psi \cap E}$. Then $\psi_E \in \mathcal{M}(E, \mathcal{A}^E, \mu_E)$, $\chi_E \psi \in \mathcal{M}(X, \mathcal{A}, \mu)$, and:

$$\psi_E \in \mathcal{L}^1(E, \mathcal{A}^E, \mu_E) \text{ iff } \chi_E \psi \in \mathcal{L}^1(X, \mathcal{A}, \mu),$$

$$\text{if } \psi_E \in \mathcal{L}^1(E, \mathcal{A}^E, \mu_E) \text{ then } \int_E \psi_E \, d\mu_E = \int_X \chi_E \psi \, d\mu.$$

Proof. a: This is obvious.

b: Let $\varphi \in \mathcal{L}^+(X, \mathcal{A}, \mu)$. Then $D_\varphi \cap E \in \mathcal{A}^E$ and

$$\mu_E(E - D_\varphi \cap E) = \mu(E \cap (X - D_\varphi)) \leq \mu(X - D_\varphi) = 0.$$

Moreover, φ_E is measurable w.r.t. $\mathcal{A}^{D_\varphi \cap E} = (\mathcal{A}^{D_\varphi})^{D_\varphi \cap E}$ and $\mathcal{A}(\delta)$ since φ is measurable w.r.t. $\mathcal{A}^{D_\varphi}$ and $\mathcal{A}(\delta)$ (cf. 6.2.3 and 6.1.19b). Thus, $\varphi_E \in \mathcal{L}^+(E, \mathcal{A}^E, \mu_E)$. We have $\chi_E \varphi \in \mathcal{L}^+(X, \mathcal{A}, \mu)$ by 8.1.17d.

Notice now that, for each $S \subset (0, \infty]$,

$$\varphi_E^{-1}(S) = \varphi^{-1}(S) \cap E = (\chi_E \varphi)^{-1}(S).$$

Then, by 8.1.15, we have

$$\int_E \varphi_E \, d\mu_E = \sup_{n \geq 1} \left[\sum_{k=2}^{\lceil n2^n \rceil} \frac{k-1}{2^n} \mu_E \left(\varphi_E^{-1} \left(\left[\frac{k-1}{2^n}, \frac{k}{2^n} \right) \right) \right) + n\mu_E(\varphi_E^{-1}([n, \infty])) \right]$$

$$= \sup_{n \geq 1} \left[\sum_{k=2}^{\lceil n2^n \rceil} \frac{k-1}{2^n} \mu \left((\chi_E \varphi)^{-1} \left(\left[\frac{k-1}{2^n}, \frac{k}{2^n} \right) \right) \right) + n\mu((\chi_E \varphi)^{-1}([n, \infty])) \right]$$

$$= \int_X \chi_E \varphi \, d\mu.$$

c: Let $\psi \in \mathcal{M}(X, \mathcal{A}, \mu)$. In the same way as for φ_E in part b, it can be proved that $\psi_E \in \mathcal{M}(E, \mathcal{A}^E, \mu_E)$. Also, $\chi_E \psi \in \mathcal{M}(X, \mathcal{A}, \mu)$ by 8.2.2. Then, the rest of the statement follows from the equalities

$$(\text{Re } \psi_E)^\pm = ((\text{Re } \psi)^\pm)_E, \ (\text{Im } \psi_E)^\pm = ((\text{Im } \psi)^\pm)_E,$$

$$\chi_E(\operatorname{Re}\psi)^{\pm} = (\operatorname{Re}(\chi_E\psi))^{\pm}, \ \chi_E(\operatorname{Im}\psi)^{\pm} = (\operatorname{Im}(\chi_E\psi))^{\pm},$$

from the definitions given in 8.2.3, and from what was proved in part b for every $\varphi \in \mathcal{L}^+(X, \mathcal{A}, \mu)$. $\qquad\qquad\qquad\qquad\qquad\qquad\qquad\qquad\qquad\qquad\square$

8.3.2 Definition. Let $(X, \mathcal{A}, \mu)$ be a measure space.

If E is a non-empty element of $\mathcal{A}$, we write:

for $\varphi \in \mathcal{L}^+(X, \mathcal{A}, \mu)$, $\int_E \varphi d\mu := \int_E \varphi d\mu_E = \int_X \chi_E \varphi d\mu$;
for $\psi \in \mathcal{M}(X, \mathcal{A}, \mu)$ such that $\chi_E\psi \in \mathcal{L}^1(X, \mathcal{A}, \mu)$, $\int_E \psi d\mu := \int_E \psi d\mu_E = \int_X \chi_E \psi d\mu$.

The integrals $\int_E \varphi d\mu$ and $\int_E \psi d\mu$ above are said to be *integrals over* E.

If $E = \emptyset$, we define:

$$\int_E \varphi d\mu := 0, \forall \varphi \in \mathcal{L}^+(X, \mathcal{A}, \mu);$$

$$\int_E \psi d\mu := 0, \forall \psi \in \mathcal{M}(X, \mathcal{A}, \mu).$$

8.3.3 Proposition. *Let $(X, \mathcal{A}, \mu)$ be a measure space and let E be a non-empty element of $\mathcal{A}$ such that $\mu(X - E) = 0$. Then:*

(a) for $\varphi \in \mathcal{L}^+(X, \mathcal{A}, \mu)$ we have

$$\int_E \varphi d\mu = \int_X \varphi d\mu;$$

(b) for $\psi \in \mathcal{M}(X, \mathcal{A}, \mu)$ we have

$$\chi_E\psi \in \mathcal{L}^1(X, \mathcal{A}, \mu) \ \text{iff} \ \psi \in \mathcal{L}^1(X, \mathcal{A}, \mu),$$

if $\psi \in \mathcal{L}^1(X, \mathcal{A}, \mu)$ then $\int_E \psi d\mu = \int_X \psi d\mu.$

Proof. a: We have $\chi_E\varphi \in \mathcal{L}^+(X, \mathcal{A}, \mu)$ (cf. 8.1.17d) and $\chi_E(x)\varphi(x) = \varphi(x)$ μ-a.e. on D_φ. The result then follows from 8.1.17c.

b: We have $\chi_E\psi \in \mathcal{M}(X, \mathcal{A}, \mu)$ (cf. 8.2.2) and $\chi_E(x)\psi(x) = \psi(x)$ μ-a.e. on D_ψ. The result then follows from 8.2.7. $\qquad\qquad\qquad\qquad\qquad\qquad\qquad\square$

8.3.4 Proposition. *Let $(X, \mathcal{A}, \mu)$ be a measure space, let $\rho \in \mathcal{L}^+(X, \mathcal{A}, \mu)$, and define the function*

$$\nu : \mathcal{A} \to [0, \infty]$$

$$E \mapsto \nu(E) := \int_E \rho d\mu.$$

(a) ν is a measure on $\mathcal{A}$ and $\nu(E) = 0$ whenever $E \in \mathcal{A}$ is such that $\mu(E) = 0$.

If $\rho(x) < \infty$ μ-a.e. on D_ρ and the measure μ is σ-finite, then the measure ν is σ-finite as well.

(b) For every $\varphi \in \mathcal{L}^+(X, \mathcal{A}, \mu)$ we have $\varphi \in \mathcal{L}^+(X, \mathcal{A}, \nu)$, $\varphi\rho \in \mathcal{L}^+(X, \mathcal{A}, \mu)$ and

$$\int_X \varphi d\nu = \int_X \varphi\rho d\mu.$$

(c) Assuming $\rho(x) < \infty$ for all $x \in D_\rho$, for every $\psi \in \mathcal{M}(X, \mathcal{A}, \mu)$ we have $\psi\rho \in \mathcal{M}(X, \mathcal{A}, \mu)$ and:

$$\psi \in \mathcal{L}^1(X, \mathcal{A}, \nu) \text{ iff } \psi\rho \in \mathcal{L}^1(X, \mathcal{A}, \mu);$$

if $\psi \in \mathcal{L}^1(X, \mathcal{A}, \nu)$ then $\int_X \psi d\nu = \int_X \psi\rho d\mu.$

Proof. a: Let $\tilde{\rho} \in \mathcal{L}^+(X, \mathcal{A})$ be an extension of ρ (cf. 8.1.14). Then, for each $E \in \mathcal{A}$, $\chi_E \tilde{\rho} \in \mathcal{L}^+(X, \mathcal{A})$ (cf. 6.2.31) and $\chi_E \tilde{\rho}$ is an extension of $\chi_E \rho$, and hence $\nu(E) = \int_X \chi_E \tilde{\rho} d\mu$.

We have:

$\nu(\emptyset) = 0$ (cf. 8.3.2);

for a sequence $\{E_n\}$ in $\mathcal{A}$ such that $E_i \cap E_j = \emptyset$ if $i \neq j$, $\chi_{\bigcup_{n=1}^\infty E_n} \tilde{\rho} = \sum_{n=1}^\infty (\chi_{E_n} \tilde{\rho})$ (with $\sum_{n=1}^\infty (\chi_{E_n} \tilde{\rho})$ defined as in 6.2.32), and hence by 8.1.10

$$\nu\left(\bigcup_{n=1}^\infty E_n\right) = \int_X \chi_{\bigcup_{n=1}^\infty E_n} \tilde{\rho} d\mu = \sum_{n=1}^\infty \int_X \chi_{E_n} \tilde{\rho} d\mu = \sum_{n=1}^\infty \nu(E_n).$$

Thus, ν is a measure on $\mathcal{A}$.

If $E \in \mathcal{A}$ and $\mu(E) = 0$, then $\chi_E(x)\tilde{\rho}(x) = 0$ μ-a.e. on X and hence by 8.1.12a

$$\nu(E) = \int_X \chi_E \tilde{\rho} d\mu = 0.$$

Assume now $\rho(x) < \infty$ μ-a.e. on D_ρ. Then, if $E \in \mathcal{A}$ is so that

$$\mu(E) = 0 \text{ and } \rho(x) < \infty \text{ for all } x \in D_\rho \cap (X - E),$$

we have $\tilde{\rho}^{-1}(\{\infty\}) \subset X - (D_\rho \cap (X - E)) = (X - D_\rho) \cup E$; thus $\tilde{\rho}^{-1}(\{\infty\})$ is an element of $\mathcal{A}$ (cf. 6.1.26) so that (cf. 7.1.2a,b)

$$\mu(\tilde{\rho}^{-1}(\{\infty\})) \leq \mu((X - D_\varphi) \cup E) = 0.$$

Assume also that μ is σ-finite. Then there exists a countable family $\{E_n\}_{n \in I}$ so that

$$E_n \in \mathcal{A} \text{ and } \mu(E_n) < \infty \text{ for all } n \in \mathbb{N}, \text{ and } X = \bigcup_{n \in I} E_n.$$

Define

$$F_0 := \tilde{\rho}^{-1}(\{\infty\}) \text{ and } F_k := \tilde{\rho}^{-1}([k-1, k)) \text{ for } k \in \mathbb{N}.$$

Then, $\{F_0\} \cup \{E_n \cap F_k\}_{(n,k) \in I \times \mathbb{N}}$ is a countable family of elements of $\mathcal{A}$ (cf. 6.1.26 and 6.1.25) and:

$\nu(F_0) = 0$ since $\mu(F_0) = 0$,

$$\forall (n, k) \in I \times \mathbb{N}, \nu(E_n \cap F_k) = \int_X \chi_{E_n \cap F_k} \tilde{\rho} d\mu \leq k \int_X \chi_{E_n \cap F_k} d\mu$$

$$= k\mu(E_n \cap F_k) \leq k\mu(E_n) < \infty$$

(cf. 8.1.7, 8.1.3a, 7.1.2a). Moreover,

$$X = F_0 \cup \left(\bigcup_{k=1}^{\infty} F_k \right) = F_0 \cup \left(\left(\bigcup_{n=1}^{\infty} E_n \right) \cap \left(\bigcup_{k=1}^{\infty} F_k \right) \right)$$

$$= F_0 \cup \left(\bigcup_{(n,k)\in I\times\mathbb{N}} (E_n \cap F_k) \right).$$

This shows that ν is σ-finite.

b: For each $E \in \mathcal{A}$, from 8.1.3a we have

$$\int_X \chi_E d\nu = \nu(E) = \int_X \chi_E \tilde{\rho} d\mu.$$

Hence, for each $\tau \in \mathcal{S}^+(X,\mathcal{A})$, by 8.1.9 and 8.1.13 we have

$$\tau\tilde{\rho} \in \mathcal{L}^+(X,\mathcal{A}) \text{ and } \int_X \tau d\nu = \int_X \tau\tilde{\rho} d\mu.$$

Now let $\varphi \in \mathcal{L}^+(X,\mathcal{A},\mu)$. Then $\varphi\rho \in \mathcal{L}^+(X,\mathcal{A},\mu)$ by 8.1.17d and $\varphi \in \mathcal{L}^+(X,\mathcal{A},\nu)$ since $\mu(X - D_\varphi) = 0$ implies $\nu(X - D_\varphi) = 0$. Let $\tilde{\varphi} \in \mathcal{L}^+(X,\mathcal{A})$ be an extension of φ (cf. 8.1.14). Then $\tilde{\varphi}\tilde{\rho} \in \mathcal{L}^+(X,\mathcal{A})$ (cf. 6.2.31), and $\tilde{\varphi}\tilde{\rho}$ is an extension of $\varphi\rho$. Let $\{\tau_n\}$ be a sequence in $\mathcal{S}^+(X,\mathcal{A})$ so that (cf. 6.2.26)

$$\tau_n \le \tau_{n+1} \text{ for all } n \in \mathbb{N} \text{ and } \tilde{\varphi} = \lim_{n\to\infty} \tau_n$$

(the function $\lim_{n\to\infty} \tau_n$ is defined as in 6.2.18). Then $\{\tau_n\tilde{\rho}\}$ is a sequence in $\mathcal{L}^+(X,\mathcal{A})$ so that (cf. 5.3.2b and 5.3.4)

$$\tau_n\tilde{\rho} \le \tau_{n+1}\tilde{\rho} \text{ for all } n \in \mathbb{N} \text{ and } \tilde{\varphi}\tilde{\rho} = \lim_{n\to\infty} \tau_n\tilde{\rho},$$

and 8.1.8 implies that

$$\int_X \varphi d\nu = \int_X \tilde{\varphi} d\nu = \lim_{n\to\infty} \int \tau_n d\nu = \lim_{n\to\infty} \int \tau_n\tilde{\rho} d\mu = \int_X \tilde{\varphi}\tilde{\rho} d\mu = \int_X \varphi\rho d\mu.$$

c: Suppose $\rho(x) < \infty$ for all $x \in D_\rho$; then $\rho \in \mathcal{M}(X,\mathcal{A},\mu)$. If $\psi \in \mathcal{M}(X,\mathcal{A},\mu)$, then $\psi\rho \in \mathcal{M}(X,\mathcal{A},\mu)$ by 8.2.2 and $\psi \in \mathcal{M}(X,\mathcal{A},\nu)$ since $\mu(X - D_\psi) = 0$ implies $\nu(X - D_\psi) = 0$. The rest of the statement about ψ follows from the definitions given in 8.2.3 and from the results proved in part b. □

8.3.5 Proposition. *Let* $(X,\mathcal{A})$ *be a measurable space.*

(a) *Let* $\{\mu_k\}$ *be a sequence of measures on* $\mathcal{A}$, *let* $\{a_k\}$ *be a sequence in* $[0,\infty]$, *and define the function*

$$\mu : \mathcal{A} \to [0,\infty]$$

$$E \mapsto \mu(E) := \sum_{k=1}^{\infty} a_k\mu_k(E).$$

Then μ *is a measure on* $\mathcal{A}$ *and:*

letting $J := \{k \in \mathbb{N} : a_k > 0\}$,

$$\mathcal{L}^+(X, \mathcal{A}, \mu) = \bigcap_{k \in J} \mathcal{L}^+(X, \mathcal{A}, \mu_k) \text{ and } \mathcal{M}(X, \mathcal{A}, \mu) = \bigcap_{k \in J} \mathcal{M}(X, \mathcal{A}, \mu_k);$$

$\forall \varphi \in \mathcal{L}^+(X, \mathcal{A}, \mu)$, $\int_X \varphi d\mu = \sum_{k=1}^{\infty} a_k \int_X \varphi d\mu_k$;
for $\psi \in \mathcal{M}(X, \mathcal{A}, \mu)$, $\psi \in \mathcal{L}^1(X, \mathcal{A}, \mu)$ *iff* $\sum_{k=1}^{\infty} a_k \int_X |\psi| d\mu_k < \infty$;
$\forall \psi \in \mathcal{L}^1(X, \mathcal{A}, \mu)$, $\int_X \psi d\mu = \sum_{k=1}^{\infty} a_k \int_X \psi d\mu_k$.

(b) *Let* μ, ν *be measures on* $\mathcal{A}$ *and let* $a, b, \in (0, \infty)$. *Then the function*

$$a\mu + b\nu : \mathcal{A} \to [0, \infty]$$

$$E \mapsto (a\mu + b\nu)(E) := a\mu(E) + b\nu(E)$$

is a measure on $\mathcal{A}$ *and:*

$\mathcal{L}^+(X, \mathcal{A}, a\mu + b\nu) = \mathcal{L}^+(X, \mathcal{A}, \mu) \cap \mathcal{L}^+(X, \mathcal{A}, \nu)$ *and* $\mathcal{M}(X, \mathcal{A}, a\mu + b\nu) =$
$\mathcal{M}(X, \mathcal{A}, \mu) \cap \mathcal{M}(X, \mathcal{A}, \nu)$;
for $\varphi \in \mathcal{L}^+(X, \mathcal{A}, a\mu + b\nu)$, $\int_X \varphi d(a\mu + b\nu) = a \int_X \varphi d\mu + b \int_X \psi d\nu$;
$\mathcal{L}^1(X, \mathcal{A}, a\mu + b\nu) = \mathcal{L}^1(X, \mathcal{A}, \mu) \cap \mathcal{L}^1(X, \mathcal{A}, \nu)$;
for $\psi \in \mathcal{L}^1(X, \mathcal{A}, a\mu + b\nu)$, $\int_X \psi d(a\mu + b\nu) = a \int_X \psi d\mu + b \int_X \psi d\nu$.

Proof. a: We have:

$\mu(\emptyset) = \sum_{k=1}^{\infty} a_k \mu_k(\emptyset) = 0$;
for a sequence $\{E_n\}$ in $\mathcal{A}$ such that $E_i \cap E_j = \emptyset$ if $i \neq j$, by 5.4.5 and 5.4.7,

$$\mu\left(\bigcup_{n=1}^{\infty} E_n\right) = \sum_{k=1}^{\infty} a_k \mu_k\left(\bigcup_{n=1}^{\infty} E_n\right) = \sum_{k=1}^{\infty} a_k \sum_{n=1}^{\infty} \mu_k(E_n)$$

$$= \sum_{n=1}^{\infty} \sum_{k=1}^{\infty} a_k \mu_k(E_n) = \sum_{n=1}^{\infty} \mu(E_n).$$

Thus, μ is a measure on $\mathcal{A}$.

Now notice that, for $E \in \mathcal{A}$, $\mu(E) = 0$ iff $\mu_k(E) = 0$ for all $k \in J$. This proves that

$$\mathcal{L}^+(X, \mathcal{A}, \mu) = \bigcap_{k \in J} \mathcal{L}^+(X, \mathcal{A}, \mu_k) \text{ and } \mathcal{M}(X, \mathcal{A}, \mu) = \bigcap_{k \in J} \mathcal{M}(X, \mathcal{A}, \mu_k).$$

For each $E \in \mathcal{A}$ we have

$$\int_X \chi_E d\mu = \mu(E) = \sum_{k=1}^{\infty} a_k \mu_k(E) = \sum_{k=1}^{\infty} a_k \int_X \chi_E d\mu_k.$$

Hence, for $\tau \in \mathcal{S}^+(X, \mathcal{A})$, letting $\tau = \sum_{n=1}^{N} b_n \chi_{E_n}$ with $\{E_1, ..., E_n\}$ a disjoint family of elements of $\mathcal{A}$ and $b_n \in [0, \infty)$ for $n = 1, ..., N$, we have (cf. 5.4.5, 5.4.6, 5.3.3)

$$\int_X \tau d\mu = \sum_{n=1}^{N} b_n \int_X \chi_{E_n} d\mu = \sum_{n=1}^{N} b_n \sum_{k=1}^{\infty} a_k \int_X \chi_{E_n} d\mu_k$$

$$= \sum_{k=1}^{\infty} a_k \sum_{n=1}^{N} b_n \int_X \chi_{E_n} d\mu_k = \sum_{k=1}^{\infty} a_k \int_X \tau d\mu_k.$$

For $\varphi \in \mathcal{L}^+(X, \mathcal{A}, \mu)$, let $\tilde{\varphi} \in \mathcal{L}^+(X, \mathcal{A})$ be an extension of φ (cf. 8.1.14) and let $\{\tau_n\}$ be a sequence in $\mathcal{S}^+(X, \mathcal{A})$ so that (cf. 6.2.26)

$$\tau_n \leq \tau_{n+1} \text{ for all } n \in \mathbb{N} \text{ and } \tilde{\varphi} = \lim_{n \to \infty} \tau_n$$

(the function $\lim_{n \to \infty} \tau_n$ is defined as in 6.2.18). Then we have, by 8.1.7 and 5.3.2b,

$$a_k \int_X \tau_n d\mu_k \leq a_k \int_X \tau_{n+1} d\mu_k, \forall (n, k) \in \mathbb{N} \times \mathbb{N},$$

and hence, by 8.1.8, 5.4.9, 5.3.4,

$$\int_X \varphi d\mu = \lim_{n \to \infty} \int_X \tau_n d\mu = \lim_{n \to \infty} \sum_{k=1}^{\infty} a_k \int \tau_n d\mu_k$$

$$= \sum_{k=1}^{\infty} \left(\lim_{n \to \infty} a_k \int \tau_n d\mu_k \right) = \sum_{k=1}^{\infty} \left(a_k \lim_{n \to \infty} \int \tau_n d\mu_k \right) = \sum_{k=1}^{\infty} a_k \int \varphi d\mu_k.$$

The part of the statement about $\psi \in \mathcal{M}(X, \mathcal{A}, \mu)$ follows easily from what has just been proved for $\varphi \in \mathcal{L}^+(X, \mathcal{A}, \mu)$, from 8.2.4, and from the definitions given in 8.2.3.

b: In part a of the statement, assume $\mu_1 := \mu$, $\mu_2 := \nu$, $a_1 := a$, $a_2 := b$, and, for $k > 2$, a_k any positive number and μ_k the null measure on $\mathcal{A}$. Then everything asserted in part b follows at once from part a.   □

8.3.6 Proposition. *Let $(X, \mathcal{A})$ be a measurable space and let $x_0 \in X$ be so that $\{x_0\} \in \mathcal{A}$. Then the function*

$$\mu_{x_0} : \mathcal{A} \to [0, \infty]$$

$$E \mapsto \mu_{x_0}(E) := \begin{cases} 1 & \text{if } x_0 \in E \\ 0 & \text{if } x_0 \notin E \end{cases}$$

is a measure on $\mathcal{A}$, which is called the Dirac measure in x_0, and:

$\forall \varphi \in \mathcal{L}^+(X, \mathcal{A}, \mu_{x_0})$, $x_0 \in D_\varphi$ and $\int_X \varphi d\mu_{x_0} = \varphi(x_0)$;
$\mathcal{L}^1(X, \mathcal{A}, \mu_{x_0}) = \mathcal{M}(X, \mathcal{A}, \mu_{x_0})$;
$\forall \psi \in \mathcal{M}(X, \mathcal{A}, \mu_{x_0})$, $x_0 \in D_\psi$ and $\int_X \psi d\mu_{x_0} = \psi(x_0)$.

Proof. By a straightforward check, we see that μ_{x_0} has properties af_1 of 7.1.1 and me of 7.1.7.

If $\varphi \in \mathcal{L}^+(X, \mathcal{A}, \mu)$, then $x_0 \in D_\varphi$ (otherwise, $\mu_{x_0}(X - D_\varphi) = 1$) and

$$\varphi(x) = \varphi(x_0) \ \mu\text{-a.e.}$$

(since $\mu_{x_0}(X - \{x_0\}) = 0$). Then (cf. 8.1.17c and 8.1.3a)

$$\int_X \varphi d\mu_{x_0} = \int_X \varphi(x_0) d\mu_{x_0} = \varphi(x_0) \mu_{x_0}(X) = \varphi(x_0).$$

The part of the statement about $\mathcal{M}(X, \mathcal{A}, \mu_{x_0})$ follows from this and from the definitions given in 8.2.3.   □

8.3.7 Proposition. *Let μ be a non-null measure on $\mathcal{A}(d_\mathbb{R})$ and suppose that $\mu(E)$ is either 0 or 1 for each $E \in \mathcal{A}(d_\mathbb{R})$. Then there exists $x_0 \in \mathbb{R}$ so that μ is the Dirac measure in x_0.*

Proof. Since $\mathbb{R} = \bigcup_{n \in \mathbb{Z}}[n, n+1]$, the σ-subadditivity of μ (cf. 7.1.4a) implies that

$$\exists \bar{n} \in \mathbb{Z} \text{ such that } \mu([\bar{n}, \bar{n}+1]) = 1.$$

Then the family

$$X := \{[a, b] : a, b \in \mathbb{R}, \bar{n} \le a \le b \le \bar{n}+1, \mu([a, b]) = 1\}$$

is non-empty because $[\bar{n}, \bar{n}+1] \in X$. The subadditivity of μ (cf. 7.1.2b) implies that

$$\forall [a, b] \in X, \mu\left(\left[a, \frac{a+b}{2}\right]\right) = 0 \Rightarrow \mu\left(\left[\frac{a+b}{2}, b\right]\right) = 1.$$

Thus, we can define the mapping

$$\varphi : X \to X$$

$$[a, b] \mapsto \varphi([a, b]) := \begin{cases} \left[a, \frac{a+b}{2}\right] & \text{if } \mu\left(\left[a, \frac{a+b}{2}\right]\right) = 1 \\ \left[\frac{a+b}{2}, b\right] & \text{if } \mu\left(\left[a, \frac{a+b}{2}\right]\right) = 0. \end{cases}$$

Next, we define a sequence $\{[a_n, b_n]\}$ by letting:

$$[a_1, b_1] := [\bar{n}, \bar{n}+1],$$

$$[a_{n+1}, b_{n+1}] := \varphi([a_n, b_n]) \text{ for each } n \in \mathbb{N}.$$

Clearly,

$$a_n \le a_{n+1} \le \bar{n}+1 \text{ and } \bar{n} \le b_{n+1} \le b_n, \forall n \in \mathbb{N}.$$

This, along with $|b_n - a_n| = \frac{1}{2^{n-1}}$ for each $n \in \mathbb{N}$, implies that

$$\exists x_0 \in \mathbb{R} \text{ such that } x_0 = \lim_{n \to \infty} a_n = \lim_{n \to \infty} b_n,$$

and it is easy to see that

$$\{x_0\} = \bigcap_{n=1}^{\infty} [a_n, b_n].$$

Then, since $\mu([a_n, b_n]) = 1$ for each $n \in \mathbb{N}$, by 7.1.4c we have

$$\mu(\{x_0\}) = \lim_{n \to \infty} \mu([a_n, b_n]) = 1,$$

and hence we also have, by the additivity of μ,

$$\mu(\mathbb{R} - \{x_0\}) = 0.$$

Then, for $E \in \mathcal{A}(d_\mathbb{R})$ we have, by the monotonicity of μ,

$$x_0 \in E \Rightarrow \{x_0\} \subset E \Rightarrow \mu(\{x_0\}) \le \mu(E) \Rightarrow \mu(E) = 1,$$

$$x_0 \notin E \Rightarrow E \subset \mathbb{R} - \{x_0\} \Rightarrow \mu(E) \le \mu(\mathbb{R} - \{x_0\}) \Rightarrow \mu(E) = 0.$$

$\square$

8.3.8 Proposition. *Let $(X, \mathcal{A})$ be a measurable space. Suppose we have a family $\{x_n\}_{n \in I}$, with $I := \{1, ..., N\}$ or $I := \mathbb{N}$, of points of X so that the singleton set $\{x_n\}$ is an element of $\mathcal{A}$ for each $n \in I$, suppose we have a function $I \ni n \mapsto a_n \in [0, \infty]$, and for each $E \in \mathcal{A}$ define the set of indices*

$$I_E := \{n \in I : x_n \in E\}.$$

Then the function

$$\mu : \mathcal{A} \to [0, \infty]$$

$$E \mapsto \mu(E) := \sum_{n \in I_E} a_n$$

(for $\sum_{n \in I_E}$, cf. 5.3.3 and 5.4.3) is a measure on $\mathcal{A}$ and:

$\forall \varphi \in \mathcal{L}^+(X, \mathcal{A}, \mu)$, $\int_X \varphi d\mu = \sum_{n \in I} a_n \varphi(x_n)$;
for $\psi \in \mathcal{M}(X, \mathcal{A}, \mu)$, $\psi \in \mathcal{L}^1(X, \mathcal{A}, \mu)$ *iff* $\sum_{n \in I} a_n |\psi(x_n)| < \infty$;
$\forall \psi \in \mathcal{L}^1(X, \mathcal{A}, \mu)$, $\int_X \psi d\mu = \sum_{n \in I} a_n \psi(x_n)$
(if $I = \mathbb{N}$, $\sum_{n \in I} a_n \psi(x_n) := \sum_{n=1}^{\infty} a_n \psi(x_n)$).

Proof. We notice that

$$\forall E \in \mathcal{A}, \mu(E) = \sum_{n \in I} a_n \mu_{x_n}(E),$$

where μ_{x_n} is the Dirac measure in x_n (cf. 8.3.6). Then we use 8.3.5 and 8.3.6. $\square$

8.3.9 Remark. For the measure μ defined in 8.3.8 we have $\mu(X - \{x_n\}_{n \in I}) = 0$.

Conversely, suppose that we have a measure space $(X, \mathcal{A}, \mu)$ such that there exists a family $\{x_n\}_{n \in I}$, with $I := \{1, ..., N\}$ or $I := \mathbb{N}$, of points of X so that the singleton set $\{x_n\}$ is an element of $\mathcal{A}$ for each $n \in I$ and $\mu(X - \{x_n\}_{n \in I}) = 0$. Then, for each $E \in \mathcal{A}$, we have

$$\mu(E) = \mu(E \cap \{x_n\}_{n \in I}) = \sum_{n \in I_E} \mu(\{x_n\})$$

if we define $I_E := \{n \in I : x_n \in E\}$. Thus, μ turns out to be the measure defined in 8.3.8, with $a_n := \mu(\{x_n\})$ for each $n \in I$.

The measures of this kind, i.e. the ones that can be constructed as in 8.3.8, are said to be *discrete*.

8.3.10 Remarks.

(a) In 8.3.8, let $X := \mathbb{N}$, $\mathcal{A} := \mathcal{P}(\mathbb{N})$ (cf. 6.1.15), $I := \mathbb{N}$, $x_n := n$ and $a_n := 1$ for each $n \in \mathbb{N}$. Then the measure μ is called the *counting measure* on $\mathbb{N}$ (since, for $E \subset \mathbb{N}$, $\mu(E)$ is the number of the points that are contained in E), $\mathcal{L}^+(X, \mathcal{A}, \mu)$ is the family of all sequences in $[0, \infty]$ and $\mathcal{M}(X, \mathcal{A}, \mu)$ is the family of all sequences in $\mathbb{C}$. For a sequence $\varphi := \{y_n\}$ in $[0, \infty]$ we have

$$\int_X \varphi d\mu = \sum_{n=1}^{\infty} y_n,$$

and for a sequence $\psi := \{z_n\}$ in $\mathbb{C}$ we have:

$\psi \in \mathcal{L}^1(X, \mathcal{A}, \mu)$ iff $\sum_{n=1}^{\infty} |z_n| < \infty$;
if $\psi \in \mathcal{L}^1(X, \mathcal{A}, \mu)$ then $\int_X \psi d\mu = \sum_{n=1}^{\infty} z_n$.

Thus, all the results about integrals of Section 8.1 and 8.2 have corollaries which are results about series.

(b) In 8.3.8, let $X := \{1, ..., N\}$, $\mathcal{A} := \mathcal{P}(\{1, ..., N\})$, $I := \{1, ..., N\}$, $x_n := n$ and $a_n := 1$ for each $n \in \{1, ..., N\}$. Then the measure μ is called the *counting measure* on $\{1, ..., N\}$, the equalities $\mathcal{L}^1(X, \mathcal{A}, \mu) = \mathcal{M}(X, \mathcal{A}, \mu) = \mathbb{C}^N$ hold true, and for an N-tuple $\psi := (z_1, ..., z_N) \in \mathbb{C}^N$ we have

$$\int_X \psi d\mu = \sum_{n=1}^{N} z_n.$$

8.3.11 Theorem (Change of variable theorem). *Let* $(X_1, \mathcal{A}_1, \mu_1)$ *be a measure space, let* $(X_2, \mathcal{A}_2)$ *be a measurable space, let* $\pi : D_\pi \to X_2$ *be a mapping from* X_1 *to* X_2 *which is measurable w.r.t.* $\mathcal{A}_1^{D_\pi}$ *and* $\mathcal{A}_2$, *and so that* $D_\pi \in \mathcal{A}_1$ *and* $\mu_1(X_1 - D_\pi) = 0$.

(a) The function

$$\mu_2 : \mathcal{A}_2 \to [0, \infty]$$
$$E \mapsto \mu_2(E) := \mu_1(\pi^{-1}(E))$$

is a measure on $\mathcal{A}_2$.

(b) For $\varphi \in \mathcal{L}^+(X_2, \mathcal{A}_2, \mu_2)$ *we have:*

$\varphi \circ \pi \in \mathcal{L}^+(X_1, \mathcal{A}_1, \mu_1)$,
$\int_{X_2} \varphi d\mu_2 = \int_{X_1} (\varphi \circ \pi) d\mu_1$ *and* $\int_E \varphi d\mu_2 = \int_{\pi^{-1}(E)} (\varphi \circ \pi) d\mu_1$, $\forall E \in \mathcal{A}_2$.

(c) For $\psi \in \mathcal{M}(X_2, \mathcal{A}_2, \mu_2)$ *we have:*

$\psi \circ \pi \in \mathcal{M}(X_1, \mathcal{A}_1, \mu_1)$;
$\psi \in \mathcal{L}^1(X_2, \mathcal{A}_2, \mu_2)$ *iff* $\psi \circ \pi \in \mathcal{L}^1(X_1, \mathcal{A}_1, \mu_1)$;
if $\psi \in \mathcal{L}^1(X_2, \mathcal{A}_2, \mu_2)$ *then* $\int_{X_2} \psi d\mu_2 = \int_{X_1} (\psi \circ \pi) d\mu_1$ *and* $\int_E \psi d\mu_2 = \int_{\pi^{-1}(E)} (\psi \circ \pi) d\mu_1$, $\forall E \in \mathcal{A}_2$.

Proof. a: We have (cf. 1.2.8):

$\mu_2(\emptyset) = \mu_1(\pi^{-1}(\emptyset)) = \mu_1(\emptyset) = 0$;
for a sequence $\{E_n\}$ in $\mathcal{A}_2$ such that $E_i \cap E_j = \emptyset$ if $i \neq j$,

$$\mu_2 \left(\bigcup_{n=1}^{\infty} E_n \right) = \mu_1 \left(\pi^{-1} \left(\bigcup_{n=1}^{\infty} E_n \right) \right) = \mu_1 \left(\bigcup_{n=1}^{\infty} \pi^{-1}(E_n) \right)$$
$$= \sum_{n=1}^{\infty} \mu_1(\pi^{-1}(E_n)) = \sum_{n=1}^{\infty} \mu_2(E_n).$$

Thus, μ_2 is a measure on $\mathcal{A}_2$.

b: Let $\varphi \in \mathcal{L}^+(X_2, \mathcal{A}_2, \mu_2)$. Then:

$$D_{\varphi \circ \pi} = \pi^{-1}(D_\varphi) \in \mathcal{A}_1^{D_\pi} \subset \mathcal{A}_1;$$

$$\mu_1(X_1 - D_{\varphi \circ \pi}) = \mu_1(D_\pi - \pi^{-1}(D_\varphi)) = \mu_1(\pi^{-1}(X_2 - D_\varphi)) = \mu_2(X_2 - D_\varphi) = 0,$$

where the first equality is true because $\mu_1(X_1 - D_\pi) = 0$; further, $\varphi \circ \pi$ is measurable w.r.t. $\mathcal{A}_1^{D_{\varphi \circ \pi}}$ and $\mathcal{A}(\delta)$ since π is measurable w.r.t. $\mathcal{A}_1^{D_\pi}$ and $\mathcal{A}_2$, and φ is measurable w.r.t. $\mathcal{A}_2^{D_\varphi}$ and $\mathcal{A}(\delta)$ (cf. 6.2.6). Thus, $\varphi \circ \pi \in \mathcal{L}^+(X_1, \mathcal{A}_1, \mu_1)$.

Then, by 8.1.15 and 1.2.13Af we have

$$\int_{X_2} \varphi d\mu_2 = \sup_{n \geq 1} \left[\sum_{k=2}^{n2^n} \frac{k-1}{2^n} \mu_2 \left(\varphi^{-1} \left(\left[\frac{k-1}{2^n}, \frac{k}{2^n} \right) \right) \right) + n\mu_2(\varphi^{-1}([n, \infty))) \right]$$

$$= \sup_{n \geq 1} \left[\sum_{k=2}^{n2^n} \frac{k-1}{2^n} \mu_1 \left((\varphi \circ \pi)^{-1} \left(\left[\frac{k-1}{2^n}, \frac{k}{2^n} \right) \right) \right) + n\mu_1((\varphi \circ \pi)^{-1}([n, \infty))) \right]$$

$$= \int_{X_1} (\varphi \circ \pi) d\mu_1.$$

From this we also obtain, for $E \in \mathcal{A}_2$,

$$\int_E \varphi d\mu_2 = \int_{X_2} \chi_E \varphi d\mu_2 = \int_{X_1} (\chi_E \circ \pi)(\varphi \circ \pi) d\mu_1$$

$$= \int_{X_1} \chi_{\pi^{-1}(E)} (\varphi \circ \pi) d\mu_1 = \int_{\pi^{-1}(E)} (\varphi \circ \pi) d\mu_1,$$

since $(\chi_E \circ \pi)(x) = \chi_{\pi^{-1}(E)}(x)$ for each $x \in D_\pi$ (cf. 1.2.13Ag).

c: We have $\psi \circ \pi \in \mathcal{M}(X_1, \mathcal{A}_1, \mu_1)$ for each $\psi \in \mathcal{M}(X_2, \mathcal{A}_2, \mu_2)$ by the first result of part b, in view of 6.2.12 and 6.2.20b. The rest of the statement follows easily from what was proved in part b and from the definitions given in 8.2.3. $\quad\square$

8.4 Integration on product spaces

The subject of this section is integration of functions defined on the cartesian product of two σ-finite measure spaces. Actually, this section could be part of the preceding one, since it deals with integration with respect to measures which are constructed out of previously given measures. However, since the treatment of the subject goes through several steps, it is perhaps better to deal with it in a separate section.

We start with two set-theoretical concepts, which we define in 8.4.1 and examine in 8.4.2.

8.4.1 Definitions. Let X_1 and X_2 be sets.

(a) Given a subset E of $X_1 \times X_2$ and a point x_1 of X_1, the subset of X_2 defined by

$$E_{x_1} := \{x_2 \in X_2 : (x_1, x_2) \in E\}$$

is called the *section of E at x_1*. Similarly, for $x_2 \in X_2$, the subset of X_1 defined by

$$E^{x_2} := \{x_1 \in X_1 : (x_1, x_2) \in E\}$$

is called the *section of E at x_2*.

(b) Assume X_1 and X_2 non-empty. Given a non-empty set Y, a mapping $\varphi : X_1 \times X_2 \to Y$, and a point x_1 of X_1, the mapping

$$\varphi_{x_1} : X_2 \to Y$$
$$x_2 \mapsto \varphi_{x_1}(x_2) := \varphi(x_1, x_2)$$

is called the *section of φ at x_1*. Similarly, for $x_2 \in X_2$, the mapping

$$\varphi^{x_2} : X_1 \to Y$$
$$x_1 \mapsto \varphi^{x_2}(x_1) := \varphi(x_1, x_2)$$

is called the *section of φ at x_2*.

8.4.2 Proposition. *Let $(X_1, \mathcal{A}_1)$ and $(X_2, \mathcal{A}_2)$ be measurable spaces.*

(a) Let $E \in \mathcal{A}_1 \otimes \mathcal{A}_2$ (cf. 6.1.28). Then

$$E_{x_1} \in \mathcal{A}_2, \forall x_1 \in X_1, \text{ and } E^{x_2} \in \mathcal{A}_1, \forall x_2 \in X_2.$$

(b) Let $(Y, \mathcal{B})$ be a measurable space and let a mapping $\varphi : X_1 \times X_2 \to Y$ be measurable w.r.t. $\mathcal{A}_1 \otimes \mathcal{A}_2$ and $\mathcal{B}$. Then:

$$\varphi_{x_1} \text{ is measurable w.r.t. } \mathcal{A}_2 \text{ and } \mathcal{B}, \forall x_1 \in X_1;$$
$$\varphi^{x_2} \text{ is measurable w.r.t. } \mathcal{A}_1 \text{ and } \mathcal{B}, \forall x_2 \in X_2.$$

Proof. a: Let $\mathcal{E}$ and $\mathcal{S}_2$ be the two collections of subsets of $X_1 \times X_2$ defined by

$$\mathcal{E} := \{E \subset X_1 \times X_2 : E_{x_1} \in \mathcal{A}_2, \forall x_1 \in X_1, \text{ and } E^{x_2} \in \mathcal{A}_1, \forall x_2 \in X_2\},$$
$$\mathcal{S}_2 := \{E_1 \times E_2 : E_k \in \mathcal{A}_k \text{ for } k = 1, 2\}.$$

For every $E_1 \subset X_1$ and $E_2 \subset X_2$, we have:

$$(E_1 \times E_2)_{x_1} = E_2 \text{ if } x_1 \in E_1 \text{ or } (E_1 \times E_2)_{x_1} = \emptyset \text{ if } x_1 \notin E_1;$$
$$(E_1 \times E_2)^{x_2} = E_1 \text{ if } x_2 \in E_2 \text{ or } (E_1 \times E_2)^{x_2} = \emptyset \text{ if } x_2 \notin E_2.$$

This shows that $\mathcal{S}_2 \subset \mathcal{E}$. Moreover, for every subset E of $X_1 \times X_2$, every $x_1 \in X_1$, every $x_2 \in X_2$, we have

$$(X_1 \times X_2 - E)_{x_1} = X_2 - E_{x_1} \text{ and } (X_1 \times X_2 - E)^{x_2} = X_1 - E^{x_2},$$

and for every family $\{E_i\}_{i \in I}$ of subsets of $X_1 \times X_2$, every $x_1 \in X_1$, every $x_2 \in X_2$, we have

$$\left(\bigcup_{i \in I} E_i \right)_{x_1} = \bigcup_{i \in I} E_{x_1} \text{ and } \left(\bigcup_{i \in I} E_i \right)^{x_2} = \bigcup_{i \in I} E^{x_2}.$$

This implies that $\mathcal{E}$ is a σ-algebra, and hence that $\mathcal{A}(\mathcal{S}_2) \subset \mathcal{E}$ (cf. $g\sigma_2$ of 6.1.17). Since $\mathcal{A}_1 \otimes \mathcal{A}_2 = \mathcal{A}(\mathcal{S}_2)$ (cf. 6.1.30a), this proves the statement.

b: Let $x_1 \in X_1$. Then, for every subset F of Y,

$$\varphi_{x_1}^{-1}(F) = \{x_2 \in X_2 : \varphi(x_1, x_2) \in F\}$$
$$= \{x_2 \in X_2 : (x_1, x_2) \in \varphi^{-1}(F)\} = (\varphi^{-1}(F))_{x_1}.$$

Thus, if $F \in \mathcal{B}$ then $\varphi^{-1}(F) \in \mathcal{A}_1 \otimes \mathcal{A}_2$, and hence (in view of the result proved in part a) $\varphi_{x_1}^{-1}(F) \in \mathcal{A}_2$. This proves that φ_{x_1} is measurable w.r.t. $\mathcal{A}_2$ and $\mathcal{B}$.

The proof for φ^{x_2}, for every $x_2 \in X_2$, in analogous. $\qquad\square$

We construct now the measure with respect to which functions defined on a product space can be integrated.

8.4.3 Theorem. Let $(X_1, \mathcal{A}_1, \mu_1)$ and $(X_2, \mathcal{A}_2, \mu_2)$ be σ-finite measure spaces. Then there exists a unique measure μ on $\mathcal{A}_1 \otimes \mathcal{A}_2$ such that

$$\forall (E_1, E_2) \in \mathcal{A}_1 \times \mathcal{A}_2, \mu(E_1 \times E_2) = \mu_1(E_1)\mu_2(E_2).$$

The measure μ is σ-finite.

Proof. We recall that the family

$$\mathcal{S}_2 := \{E_1 \times E_2 : E_k \in \mathcal{A}_k \text{ for } k = 1, 2\}$$

is a semialgebra in $X_1 \times X_2$ and $\mathcal{A}_1 \otimes \mathcal{A}_2 = \mathcal{A}(\mathcal{S}_2)$ (cf. 6.1.30a). The function

$$\nu : \mathcal{S}_2 \to [0, \infty]$$
$$E_1 \times E_2 \mapsto \nu(E_1 \times E_2) := \mu_1(E_1)\mu_2(E_2)$$

has obviously property a of 7.3.1. We will prove that it has properties b and c as well. Indeed, suppose that $\{E_{1,n} \times E_{2,n}\}_{n \in I}$ is a disjoint family of elements of $\mathcal{S}_2$ such that $I = \{1, ..., N\}$ or $I = \mathbb{N}$, and such that $\bigcup_{n \in I}(E_{1,n} \times E_{2,n}) \in \mathcal{S}_2$, i.e. $\bigcup_{n \in I}(E_{1,n} \times E_{2,n}) = E_1 \times E_2$ with $E_1 \in \mathcal{A}_1$ and $E_2 \in \mathcal{A}_2$. Then, for each $(x_1, x_2) \in X_1 \times X_2$,

$$\chi_{E_1}(x_1)\chi_{E_2}(x_2) = \chi_{E_1 \times E_2}(x_1, x_2)$$
$$= \sum_{n \in I} \chi_{E_{1,n} \times E_{2,n}}(x_1, x_2) = \sum_{n \in I} \chi_{E_{1,n}}(x_1)\chi_{E_{2,n}}(x_2)$$

(cf. 1.2.20), where $\sum_{n \in I}$ stands for $\sum_{n=1}^{N}$ or $\sum_{n=1}^{\infty}$. Using this along with 8.1.9 or 8.1.10, for each $x_2 \in X_2$ we obtain

$$\mu_1(E_1)\chi_{E_2}(x_2) = \int_{X_1} \chi_{E_1}(x_1)\chi_{E_2}(x_2)d\mu_1(x_1)$$
$$= \sum_{n \in I} \int_{X_1} \chi_{E_{1,n}}(x_1)\chi_{E_{2,n}}(x_2)d\mu_1(x_1) = \sum_{n \in I} \mu_1(E_{1,n})\chi_{E_{2,n}}(x_2),$$

and hence, using 8.1.13 as well,

$$\nu\left(\bigcup_{n\in I}(E_{1,n}\times E_{2,n})\right)=\mu_1(E_1)\mu_2(E_2)=\int_{X_2}\mu_1(E_1)\chi_{E_2}(x_2)d\mu_2(x_2)$$

$$=\sum_{n\in I}\mu_1(E_{1,n})\int_{X_2}\chi_{E_{2,n}}(x_2)d\mu_2(x_2)$$

$$=\sum_{n\in I}\mu_1(E_{1,n})\mu_2(E_{2,n})=\sum_{n\in I}\nu(E_{1,n}\times E_{2,n}).$$

Thus, ν has properties a, b, c of 7.3.1, and hence there exists a unique additive function μ_0 on the algebra $\mathcal{A}_0(\mathcal{S}_2)$ which is an extension of ν, and μ_0 is a premeasure. Then, by 7.3.2 there exists a measure μ on $\mathcal{A}(\mathcal{A}_0(\mathcal{S}_2))$ which is an extension of μ_0. Since $\mathcal{A}(\mathcal{S}_2)=\mathcal{A}(\mathcal{A}_0(\mathcal{S}_2))$ (cf. 6.1.18), this proves that there exists a measure μ on $\mathcal{A}_1\otimes\mathcal{A}_2$ which is an extension of ν.

If $\tilde{\mu}$ is another measure which is an extension of ν, then the restriction of $\tilde{\mu}$ to $\mathcal{A}_0(\mathcal{S}_2)$ must coincide with μ_0 on account of the uniqueness asserted in 7.3.1A, and then $\tilde{\mu}$ must coincide with μ on account of the uniqueness asserted in 7.3.2, since μ_0 is σ-finite. Indeed, for $k=1,2$ there exists a countable family $\{F_{k,n}\}_{n\in I_k}$ of elements of $\mathcal{A}_k$ so that $\mu_k(F_{k,n})<\infty$ for all $n\in I_k$ and $X_k=\bigcup_{n\in I_k}F_{k,n}$. Then $\mu_0(F_{1,n}\times F_{2,m})=\mu_1(F_{1,n})\mu_2(F_{2,m})<\infty$ for all $(n,m)\in I_1\times I_2$ and $X_1\times X_2=\bigcup_{(n,m)\in I_1\times I_2}(F_{1,n}\times F_{2,m})$, so μ_0 is σ-finite. Obviously, this also proves that μ is σ-finite. $\qquad\square$

8.4.4 Definitions. Let $(X_1,\mathcal{A}_1,\mu_1)$ and $(X_2,\mathcal{A}_2,\mu_2)$ be σ-finite measure spaces.

The measure μ whose existence and uniqueness was proved in 8.4.3 is called the *product* of μ_1 and μ_2 and is denoted by $\mu_1\otimes\mu_2$.

The measure space $(X_1\times X_2,\mathcal{A}_1\otimes\mathcal{A}_2,\mu_1\otimes\mu_2)$ is called the *product measure space* of $(X_1,\mathcal{A}_1,\mu_1)$ and $(X_2,\mathcal{A}_2,\mu_2)$.

8.4.5 Proposition. *Let $N\in\mathbb{N}$ be so that $N>2$ and let $(X_k,\mathcal{A}_k,\mu_k)$ be a σ-finite measure space for $k=1,...,N$. We define*

$$\mu_1\otimes\cdots\otimes\mu_N:=((\cdots((\mu_1\otimes\mu_2)\otimes\mu_3)\otimes\cdots)\otimes\mu_{N-1})\otimes\mu_N.$$

We identify $((\cdots((X_1\times X_2)\times X_3)\times\cdots)\times X_{N-1})\times X_N$ with $X_1\times X_2\times\cdots\times X_N$. Then, $\mu_1\otimes\cdots\otimes\mu_N$ is a σ-finite measure on the σ-algebra $\mathcal{A}_1\otimes\mathcal{A}_2\otimes\cdots\otimes\mathcal{A}_N$, and it is the only measure on $\mathcal{A}_1\otimes\mathcal{A}_2\otimes\cdots\otimes\mathcal{A}_N$ such that

$$(\mu_1\otimes\cdots\otimes\mu_N)(E_1\times E_2\times\cdots\times E_N)=\mu_1(E_1)\mu_2(E_2)\cdots\mu_N(E_N),$$
$$\forall(E_1,E_2,...,E_N)\in\mathcal{A}_1\times\mathcal{A}_2\times\cdots\mathcal{A}_N.$$

If $1<i_1<i_2<...i_r<N$, we identify $X_1\times X_2\times\cdots\times X_N$ with

$$(\cdots(((X_1\times\cdots X_{i_1})\times(X_{i_1+1}\times\cdots\times X_{i_2}))\times(X_{i_2+1}\times\cdots\times X_{i_3}))\times\cdots)$$
$$\times(X_{i_r+1}\times\cdots\times X_N).$$

Then

$$\mu_1 \otimes \cdots \otimes \mu_N$$
$$= (\cdots(((\mu_1 \otimes \cdots \otimes \mu_{i_1}) \otimes (\mu_{i_1+1} \otimes \cdots \otimes \mu_{i_2})) \otimes (\mu_{i_2+1} \otimes \cdots \otimes \mu_{i_3})) \otimes \cdots)$$
$$\otimes (\mu_{i_r+1} \otimes \cdots \otimes \mu_N).$$

Proof. From 8.4.3 and 8.4.4 we have that $\mu_1 \otimes \cdots \otimes \mu_N$ is a σ-finite measure on the σ-algebra $((\cdots((\mathcal{A}_1 \otimes \mathcal{A}_2) \otimes \mathcal{A}_3) \otimes \cdots) \otimes \mathcal{A}_{N-1}) \otimes \mathcal{A}_N$ and that

$$(\mu_1 \otimes \cdots \otimes \mu_N)(((\cdots((E_1 \times E_2) \times E_3) \times \cdots) \times E_{N-1}) \times E_N)$$
$$= \mu_1(E_1)\mu_2(E_2)\cdots\mu_N(E_N).$$

Since we identify $((\cdots((X_1 \times X_2) \times X_3) \times \cdots) \times X_{N-1}) \times X_N$ with $X_1 \times X_2 \times \cdots \times X_N$, we have

$$((\cdots((\mathcal{A}_1 \otimes \mathcal{A}_2) \otimes \mathcal{A}_3) \otimes \cdots) \otimes \mathcal{A}_{N-1}) \otimes \mathcal{A}_N = \mathcal{A}_1 \otimes \mathcal{A}_2 \otimes \cdots \otimes \mathcal{A}_N$$

(cf. 6.1.30b) and

$$((\cdots((E_1 \times E_2) \times E_3) \times \cdots) \times E_{N-1}) \times E_N = E_1 \times E_2 \times \cdots \times E_N,$$
$$\forall(E_1, E_2, ..., E_N) \in \mathcal{A}_1 \times \mathcal{A}_2 \times \cdots \times \mathcal{A}_N.$$

Suppose now that $\tilde{\mu}$ is a measure on $\mathcal{A}_1 \otimes \mathcal{A}_2 \otimes \cdots \otimes \mathcal{A}_N$ such that

$$\tilde{\mu}(E_1 \times E_2 \times \cdots \times E_N) = \mu_1(E_1)\mu_2(E_2)\cdots\mu_N(E_N),$$
$$\forall(E_1, E_2, ..., E_N) \in \mathcal{A}_1 \times \mathcal{A}_2 \times \cdots \times \mathcal{A}_N.$$

Then the restrictions of the measures $\tilde{\mu}$ and $\mu_1 \otimes \mu_2 \otimes \cdots \otimes \mu_N$ to $\mathcal{S}_N$ (defined as in 6.1.30a) coincide, and hence their restrictions to $\mathcal{A}_0(\mathcal{S}_N)$ must coincide as well by the uniqueness asserted in 7.3.1A, and hence $\tilde{\mu}$ and $\mu_1 \otimes \mu_2 \otimes \cdots \otimes \mu_N$ must coincide altogether by the uniqueness asserted in 7.3.2, in view of the equality $\mathcal{A}(\mathcal{A}_0(\mathcal{S}_N)) = \mathcal{A}(\mathcal{S}_N) = \mathcal{A}_1 \otimes \mathcal{A}_2 \otimes \cdots \mathcal{A}_N$ (cf. 6.1.30a) and of the fact that the restriction of $\mu_1 \otimes \mu_2 \otimes \cdots \otimes \mu_N$ to $\mathcal{A}_0(\mathcal{S}_N)$ is σ-finite (this can be seen by an argument similar to the one which led to the σ-finiteness of μ_0 in the proof of 8.4.3).

Finally, we notice that, since we identify $X_1 \times X_2 \times \cdots \times X_N$ with

$$(\cdots(((X_1 \times \cdots X_{i_1}) \times (X_{i_1+1} \times \cdots \times X_{i_2})) \times (X_{i_2+1} \times \cdots \times X_{i_3})) \times \cdots)$$
$$\times (X_{i_r+1} \times \cdots \times X_N),$$

we have, for every $(E_1, E_2, ..., E_N) \in \mathcal{A}_1 \times \mathcal{A}_2 \times \cdots \mathcal{A}_N$,

$$(\cdots(((\mu_1 \otimes \cdots \otimes \mu_{i_1}) \otimes (\mu_{i_1+1} \otimes \cdots \otimes \mu_{i_2})) \otimes (\mu_{i_2+1} \otimes \cdots \otimes \mu_{i_3})) \otimes \cdots)$$
$$\otimes (\mu_{i_r+1} \otimes \cdots \otimes \mu_N)(E_1, E_2, ..., E_N) = \mu_1(E_1)\mu_2(E_2)\cdots\mu_N(E_N).$$

and we use the uniqueness result proved above. $\qquad\qquad\square$

8.4.6 Proposition. *Let $N \in \mathbb{N}$ be so that $N > 1$ and, for $k = 1, ..., N$, let $(X_k, \mathcal{A}_k, \mu_k)$ be a σ-finite measure space. For $k = 1, ..., N$, let Y_k be a non-empty element of $\mathcal{A}_k$. Then (cf. 8.3.1 for μ_E)*

$$(\mu_1 \otimes \cdots \otimes \mu_N)_{Y_1 \times \cdots \times Y_N} = (\mu_1)_{Y_1} \otimes \cdots \otimes (\mu_N)_{Y_N}.$$

Proof. We recall that the measure $(\mu_1 \otimes \cdots \otimes \mu_N)_{Y_1 \times \cdots \times Y_N}$ is defined on the σ-algebra $(\mathcal{A}_1 \otimes \cdots \otimes \mathcal{A}_N)^{Y_1 \times \cdots \times Y_N}$ and that the measure $(\mu_1)_{Y_1} \otimes \cdots \otimes (\mu_N)_{Y_N}$ is defined on the σ-algebra $\mathcal{A}_1^{Y_1} \otimes \cdots \otimes \mathcal{A}_N^{Y_N}$. Moreover, from 6.1.30c and its proof we know that

$$(\mathcal{A}_1 \otimes \cdots \otimes \mathcal{A}_N)^{Y_1 \times \cdots \times Y_N} = \mathcal{A}_1^{Y_1} \otimes \cdots \otimes \mathcal{A}_N^{Y_N} = \mathcal{A}(\mathcal{S}_N^{Y_1 \times \cdots \times Y_N}),$$

with

$$\mathcal{S}_N^{Y_1 \times \cdots \times Y_N} = \{G_1 \times \cdots \times G_N : G_k \in \mathcal{A}_k^{Y_k} \text{ for } k = 1, ..., N\}.$$

We also know that $\mathcal{S}_N^{Y_1 \times \cdots \times Y_N}$ is a semialgebra on $Y_1 \times \cdots \times Y_N$ (cf. 6.1.30a, with $(X_k, \mathcal{A}_k)$ replaced by $(Y_k, \mathcal{A}_k^{Y_k})$). For each $G_1 \times \cdots \times G_N \in \mathcal{S}_N^{Y_1 \times \cdots \times Y_N}$ we have

$$\begin{aligned}
(\mu_1 \otimes \cdots \otimes \mu_N)_{Y_1 \times \cdots \times Y_N}(G_1 \times \cdots \times G_N) &= (\mu_1 \otimes \cdots \otimes \mu_N)(G_1 \times \cdots \times G_N) \\
&= \mu_1(G_1) \cdots \mu_N(G_N) = (\mu_1)_{Y_1}(G_1) \cdots (\mu_N)_{Y_N}(G_N) \\
&= ((\mu_1)_{Y_1} \otimes \cdots \otimes (\mu_N)_{Y_N})(G_1 \times \cdots \times G_N).
\end{aligned}$$

Since the restrictions of $(\mu_1 \otimes \cdots \otimes \mu_N)_{Y_1 \times \cdots \times Y_N}$ and $(\mu_1)_{Y_1} \otimes \cdots \otimes (\mu_N)_{Y_N}$ to $\mathcal{S}_N^{Y_1 \times \cdots \times Y_N}$ coincide and since $\mathcal{S}_N^{Y_1 \times \cdots \times Y_N}$ is a semialgebra on $Y_1 \times \cdots \times Y_N$, an argument similar to the one used at the end of 8.4.5 leads to the equality between the measures $(\mu_1 \otimes \cdots \otimes \mu_N)_{Y_1 \times \cdots \times Y_N}$ and $(\mu_1)_{Y_1} \otimes \cdots \otimes (\mu_N)_{Y_N}$. $\qquad\square$

Finally, we come to the theorems that govern integration on product spaces. Fubini's theorem follows from Tonelli's theorem, which follows from the lemma in 8.4.7. In the remainder of this section we make the tacit assumption that all the measures we consider are non-null (in all the statements, if one of the measures was null then either the statement would hold trivially or it would be of no interest).

8.4.7 Lemma. *Let* $(X_1, \mathcal{A}_1, \mu_1)$ *and* $(X_2, \mathcal{A}_2, \mu_2)$ *be* σ*-finite measure spaces. For each* $E \in \mathcal{A}_1 \otimes \mathcal{A}_2$, *the functions*

$$\psi_1^E : X_1 \to [0, \infty]$$
$$x_1 \mapsto \psi_1^E(x_1) := \mu_2(E_{x_1})$$

and

$$\psi_2^E : X_2 \to [0, \infty]$$
$$x_2 \mapsto \psi_2^E(x_2) := \mu_1(E^{x_2})$$

are defined consistently, are elements of $\mathcal{L}^+(X_1, \mathcal{A}_1)$ *and* $\mathcal{L}^+(X_2, \mathcal{A}_2)$ *respectively, and*

$$(\mu_1 \otimes \mu_2)(E) = \int_{X_1} \psi_1^E d\mu_1 = \int_{X_2} \psi_2^E d\mu_2,$$

which can also be written as

$$(\mu_1 \otimes \mu_2)(E) = \int_{X_1} \mu_2(E_{x_1}) d\mu_1(x_1) = \int_{X_2} \mu_1(E^{x_2}) d\mu_2(x_2).$$

Proof. It follows from 8.4.2a that the definitions of the functions ψ_1^E and ψ_2^E are consistent, for each $E \in \mathcal{A}_1 \otimes \mathcal{A}_2$.

First, suppose that $(\mu_1 \otimes \mu_2)(X_1 \times X_2) < \infty$. Since $(\mu_1 \otimes \mu_2)(X_1 \times X_2) = \mu_1(X_1)\mu_2(X_2)$, this is equivalent to $\mu_1(X_1) < \infty$ and $\mu_2(X_2) < \infty$.

Define

$$\mathcal{C} := \{E \in \mathcal{A}_1 \otimes \mathcal{A}_2 : \ \psi_k^E \in \mathcal{L}^+(X_k, \mathcal{A}_k) \text{ and}$$

$$(\mu_1 \otimes \mu_2)(E) = \int_{X_k} \psi_k^E d\mu_k, \text{ for } k = 1, 2\}.$$

If $\{E_1, .., E_N\}$ is a finite and disjoint family of elements of $\mathcal{C}$ and $E := \bigcup_{n=1}^N E_n$, then $E \in \mathcal{C}$. In fact, for each $x_1 \in X_1$, $\{(E_1)_{x_1}, ..., (E_N)_{x_1}\}$ is a disjoint family of elements of $\mathcal{A}_2$ and $E_{x_1} := \bigcup_{n=1}^N (E_n)_{x_1}$, and therefore

$$\psi_1^E(x_1) = \mu_2(E_{x_1}) = \sum_{n=1}^N \mu_2((E_n)_{x_1}) = \sum_{n=1}^N \psi_1^{E_n}(x_1).$$

Thus, $\psi_1^E = \sum_{n=1}^N \psi_1^{E_n}$ and hence (cf. 8.1.9) $\psi_1^E \in \mathcal{L}^+(X_1, \mathcal{A}_1)$ and

$$\int_{X_1} \psi_1^E d\mu_1 = \sum_{n=1}^N \int_{X_1} \psi_1^{E_n} d\mu_1 = \sum_{n=1}^N (\mu_1 \otimes \mu_2)(E_n) = (\mu_1 \otimes \mu_2)(E).$$

And the analogous facts are true for ψ_2^E.

The family

$$\mathcal{S}_2 := \{E_1 \times E_2 : E_k \in \mathcal{A}_k \text{ for } k = 1, 2\}$$

is a semialgebra on $X_1 \times X_2$ and $\mathcal{A}_1 \otimes \mathcal{A}_2 = \mathcal{A}(\mathcal{S}_2)$ (cf. 6.1.30a). We also have $\mathcal{S}_2 \subset \mathcal{C}$. Indeed, it is easy to see that, for $E := E_1 \times E_2 \in \mathcal{S}$, $\mu_2(E_{x_1}) = \mu_2(E_2)\chi_{E_1}(x_1)$. Thus, $\psi_1^E = \mu(E_2)\chi_{E_1}$ and hence (cf. 8.1.13) $\psi_1^E \in \mathcal{L}^+(X_1, \mathcal{A}_1)$ and

$$\int_{X_1} \psi_1^E d\mu_1 = \mu_2(E_2) \int_{X_1} \chi_{E_1} d\mu_1 = \mu_2(E_2)\mu_1(E_1) = (\mu_1 \otimes \mu_2)(E).$$

And the analogous facts are true for ψ_2^E.

From what was proved above and from 6.1.11, it follows that $\mathcal{A}_0(\mathcal{S}_2) \subset \mathcal{C}$.

Now we will show that $\mathcal{C}$ is a monotone class. Let us assume that $\{E_n\}$ is a sequence in $\mathcal{C}$ such that either $E_n \subset E_{n+1}$ or $E_{n+1} \subset E_n$ for each $n \in \mathbb{N}$. This implies that, for every $x_1 \in X_1$, either $(E_n)_{x_1} \subset (E_{n+1})_{x_1}$ of $(E_{n+1})_{x_1} \subset (E_n)_{x_1}$ for each $n \in \mathbb{N}$. If we define E as either $E := \bigcup_{n=1}^\infty E_n$ or $E := \bigcap_{n=1}^\infty E_n$, then we have either $E_{x_1} = \bigcup_{n=1}^\infty (E_n)_{x_1}$ or $E_{x_1} = \bigcap_{n=1}^\infty (E_n)_{x_1}$ for every $x_1 \in X_1$. Since $(\mu_1 \otimes \mu_2)(X_1 \times X_2) < \infty$ and $\mu_2(X_2) < \infty$, by either 7.1.4b or 7.1.4c we have

$$(\mu_1 \otimes \mu_2)(E) = \lim_{n \to \infty} (\mu_1 \otimes \mu_2)(E_n)$$

and

$$\psi_1^E(x_1) = \mu_2(E_{x_1}) = \lim_{n \to \infty} \mu_2((E_n)_{x_1}) = \lim_{n \to \infty} \psi_1^{E_n}(x_1), \forall x_1 \in X_1.$$

This shows that $\psi_1^E \in \mathcal{L}^+(X_1, \mathcal{A}_1)$ (cf. 6.2.19b). Moreover, we notice that ψ_1^E, $\psi_1^{E_n}$ for each $n \in \mathbb{N}$, and the constant function

$$\psi_1 : X_1 \to [0, \infty]$$
$$x_1 \mapsto \psi_1(x_1) := \mu_2(X_2),$$

which are elements of $\mathcal{L}^+(X_1, \mathcal{A}_1)$ (cf. also 6.2.2), are in fact elements of $\mathcal{M}(X_1, \mathcal{A}_1)$ since $\mu_2(X_2) < \infty$ (cf. also 7.1.2a). Further, $\psi_1 \in \mathcal{L}^1(X_1, \mathcal{A}_1, \mu_1)$ since $\mu_1(X_1) < \infty$ (cf. 8.2.6). Then, by 8.2.11 with ψ_1 as dominating function, we have

$$(\mu_1 \otimes \mu_2)(E) = \lim_{n \to \infty} (\mu_1 \otimes \mu_2)(E_n) = \lim_{n \to \infty} \int_{X_1} \psi_1^{E_n} d\mu_1 = \int_{X_1} \psi_1^E d\mu_1.$$

And the analogous facts are true for ψ_2^E.

Thus, $\mathcal{A}_0(\mathcal{S}_2) \subset \mathcal{C}$ and $\mathcal{C}$ is a monotone class. Then, $\mathcal{C}(\mathcal{A}_0(\mathcal{S}_2)) \subset \mathcal{C}$ (cf. gm_2 in 6.1.35), where $\mathcal{C}(\mathcal{A}_0(\mathcal{S}_2))$ is the monotone class generated by $\mathcal{A}_0(\mathcal{S}_2)$. But $\mathcal{C}(\mathcal{A}_0(\mathcal{S}_2)) = \mathcal{A}(\mathcal{A}_0(\mathcal{S}_2))$ (cf. 6.1.36) and $\mathcal{A}(\mathcal{A}_0(\mathcal{S}_2)) = \mathcal{A}(\mathcal{S}_2)$ (cf. 6.1.18). This proves that $\mathcal{A}_1 \otimes \mathcal{A}_2 \subset \mathcal{C}$, and therefore that the statement is true, if $(\mu_1 \otimes \mu_2)(X_1 \times X_2) < \infty$.

Suppose now $(\mu_1 \otimes \mu_2)(X_1 \times X_2) = \infty$. Then μ_1 and μ_2 cannot be both finite. Since μ_1 and μ_2 are σ-finite, for $k = 1, 2$ there is a family $\{F_{k,n}\}_{n \in \mathbb{N}}$ of elements of $\mathcal{A}_k$ so that $\mu_k(F_{k,n}) < \infty$ for all $n \in \mathbb{N}$ and $X_k = \bigcup_{n=1}^{\infty} F_{k,n}$ (if μ_k happens to be finite, take $F_{k,n} := X_k$ for all $n \in \mathbb{N}$); then, for all $N \in \mathbb{N}$, we define $X_{k,N} := \bigcup_{\hat{n}=1}^{N} F_{k,n}$ and we have $X_{k,N} \in \mathcal{A}_k$ and $\mu_k(X_{k,N}) < \infty$ (cf. 7.1.2b); we write $\mathcal{A}_{k,N} := \mathcal{A}_k^{X_{k,N}}$ and $\mu_{k,N} := (\mu_k)_{X_{k,N}}$ and we consider the measure space $(X_{k,N}, \mathcal{A}_{k,N}, \mu_{k,N})$ (cf. 8.3.1); notice that $\mu_{k,N}$ is a finite measure. Notice that $X_k = \bigcup_{N=1}^{\infty} X_{k,N}$ for $k = 1, 2$, and hence $X_1 \times X_2 = \bigcup_{N=1}^{\infty}(X_{1,N} \times X_{2,N})$.

Fix $E \in \mathcal{A}_1 \otimes \mathcal{A}_2$. For each $N \in \mathbb{N}$, define $E_N := E \cap (X_{1,N} \times X_{2,N})$. We have $E_N \in \mathcal{A}_{1,N} \otimes \mathcal{A}_{2,N}$ (cf. 6.1.30c). Since $\mu_{1,N}$ and $\mu_{2,N}$ are finite measures, what was proved before implies that the function

$$\tilde{\psi}_1^{E_N} : X_{1,N} \to [0, \infty]$$
$$x_1 \mapsto \tilde{\psi}_1^{E_N}(x_1) := \mu_{2,N}((E_N)_{x_1})$$

is an element of $\mathcal{L}^+(X_{1,N}, \mathcal{A}_{1,N})$ and that

$$(\mu_{1,N} \otimes \mu_{2,N})(E_N) = \int_{X_{1,N}} \tilde{\psi}_1^{E_N} d\mu_{1,N}.$$

We notice now that

$$\psi_1^{E_N}(x_1) = \begin{cases} \tilde{\psi}_1^{E_N}(x_1) & \text{if } x_1 \in X_{1,N} \text{ (since } \mu_{2,N}((E_N)_{x_1}) = \mu_2((E_N)_{x_1})), \\ 0 & \text{if } x_1 \in X - X_{1,N} \text{ (since } E_N \subset X_{1,N} \times X_{2,N}). \end{cases}$$

Clearly, $(\psi_1^{E_N})_{X_{1,N}} = \tilde{\psi}_1^{E_N}$ and $\psi_1^{E_N} = \chi_{X_{1,N}} \psi_1^{E_N}$. Moreover, $\psi_1^{E_N} \in \mathcal{L}^+(X_1, \mathcal{A}_1)$ since, for $F \in \mathcal{A}(d_{\mathbb{R}})$,

$$(\psi_1^{E_N})^{-1}(F) = \begin{cases} (\tilde{\psi}_1^{E_N})^{-1}(F) & \text{if } 0 \notin F, \\ (\tilde{\psi}_1^{E_N})^{-1}(F) \cup (X_1 - X_{1,N}) & \text{if } 0 \in F, \end{cases}$$

and $(\psi_1^{E_N})^{-1}(F) \in \mathcal{A}_{1,N} \subset \mathcal{A}_1$ (cf. 6.1.19a). Then we have

$$(\mu_1 \otimes \mu_2)(E_N) = (\mu_1 \otimes \mu_2)_{X_{1,N} \times X_{2,N}}(E_N) \overset{(1)}{=} (\mu_{1,N} \otimes \mu_{2,N})(E_N)$$

$$= \int_{X_{1,N}} (\psi_1^{E_N})_{X_{1,N}} d\mu_{1,N} \overset{(2)}{=} \int_{X_1} \chi_{X_{1,N}} \psi_1^{E_N} d\mu_1 = \int_{X_1} \psi_1^{E_N} d\mu_1,$$

where 1 holds by 8.4.6 and 2 holds by 8.3.1b.

Notice that $E_N \subset E_{N+1}$ for all $N \in \mathbb{N}$ and $E = \bigcup_{N=1}^{\infty} E_N$. This implies that, for each $x_1 \in X_1$, $(E_N)_{x_1} \subset (E_{N+1})_{x_1}$ for all $N \in \mathbb{N}$ and $E_{x_1} = \bigcup_{N=1}^{\infty} (E_N)_{x_1}$. Then, by 7.1.4b, we have

$$(\mu_1 \otimes \mu_2)(E) = \lim_{n \to \infty} (\mu_1 \otimes \mu_2)(E_N)$$

and

$$\forall x_1 \in X_1, \; \lim_{N \to \infty} \psi_1^{E_N}(x_1) = \lim_{N \to \infty} \mu_2((E_N)_{x_1}) = \mu_2(E_{x_1}) = \psi_1^E(x_1).$$

By 7.1.2a we also have, for all $N \in \mathbb{N}$,

$$\forall x_1 \in X_1, \psi_1^{E_N}(x_1) = \mu_2((E_N)_{x_1}) \le \mu_2((E_{N+1})_{x_1}) = \psi_1^{E_{N+1}}(x_1).$$

Then, by 8.1.8, we have $\psi_1^E \in \mathcal{L}^+(X_1, \mathcal{A}_1)$ and

$$(\mu_1 \otimes \mu_2)(E) = \lim_{N \to \infty} \int_{X_1} \psi_1^{E_N} d\mu_1 = \int_{X_1} \psi_1^E d\mu_1.$$

In a similar way we can prove that $\psi_2^E \in \mathcal{L}^+(X_2, \mathcal{A}_2)$ and $(\mu_1 \otimes \mu_2)(E) = \int_{X_2} \psi_2^E d\mu_2$. $\qquad \square$

8.4.8 Theorem (Tonelli's theorem). *Let* $(X_1, \mathcal{A}_1, \mu_1)$ *and* $(X_2, \mathcal{A}_2, \mu_2)$ *be σ-finite measure spaces. For each* $\varphi \in \mathcal{L}^+(X_1 \times X_2, \mathcal{A}_1 \otimes \mathcal{A}_2)$, *the functions*

$$\psi_1^{\varphi} : X_1 \to [0, \infty]$$

$$x_1 \mapsto \psi_1^{\varphi}(x_1) := \int_{X_2} \varphi_{x_1} d\mu_2$$

and

$$\psi_2^{\varphi} : X_2 \to [0, \infty]$$

$$x_2 \mapsto \psi_2^{\varphi}(x_2) := \int_{X_1} \varphi^{x_2} d\mu_1$$

are defined consistently, are elements of $\mathcal{L}^+(X_1, \mathcal{A}_1)$ *and* $\mathcal{L}^+(X_2, \mathcal{A}_2)$ *respectively, and*

$$\int_{X_1 \times X_2} \varphi \, d(\mu_1 \otimes \mu_2) = \int_{X_1} \psi_1^{\varphi} d\mu_1 = \int_{X_2} \psi_2^{\varphi} d\mu_2,$$

which can also be written as

$$\int_{X_1 \times X_2} \varphi \, d(\mu_1 \otimes \mu_2) = \int_{X_1} \left(\int_{X_2} \varphi(x_1, x_2) d\mu_2(x_2) \right) d\mu_1(x_1)$$

$$= \int_{X_2} \left(\int_{X_1} \varphi(x_1, x_2) d\mu_1(x_1) \right) d\mu_2(x_2)$$

(the second equality is often referred to by saying that "the order of integration may be reversed").

Proof. It follows from 8.4.2b that the definitions of the functions ψ_1^φ and ψ_2^φ are consistent, for each $\varphi \in \mathcal{L}^+(X_1 \times X_2, \mathcal{A}_1 \otimes \mathcal{A}_2)$.

Suppose $E \in \mathcal{A}_1 \otimes \mathcal{A}_2$ and $\varphi := \chi_E$. Then

$$\forall x_2 \in X_2, \varphi_{x_1}(x_2) = \chi_E(x_1, x_2) = \chi_{E_{x_1}}(x_2)$$

and hence, if we define ψ_1^E as in 8.4.7,

$$\forall x_1 \in X_1, \psi_1^\varphi(x_1) = \int_{X_2} \chi_{E_{x_1}} d\mu_2 = \mu_2(E_{x_1}) = \psi_1^E(x_1).$$

Thus, in view of 8.4.7 we have $\psi_1^\varphi \in \mathcal{L}^+(X_1, \mathcal{A}_1)$ and

$$\int_{X_1 \times X_2} \varphi d(\mu_1 \otimes \mu_2) = \int_{X_1 \times X_2} \chi_E d(\mu_1 \otimes \mu_2) = (\mu_1 \otimes \mu_2)(E) = \int_{X_1} \psi_1^\varphi d\mu_1.$$

And similarly for ψ_2^φ. Thus, the conclusions of the statement are true for $\varphi = \chi_E$ with $E \in \mathcal{A}_1 \otimes \mathcal{A}_2$.

For $a, b \in [0, \infty)$ and $\varphi, \tilde{\varphi} \in \mathcal{L}^+(X_1 \times X_2, \mathcal{A}_1 \otimes \mathcal{A}_2)$, we have

$$a\varphi + b\tilde{\varphi} \in \mathcal{L}^+(X_1 \times X_2, \mathcal{A}_1 \otimes \mathcal{A}_2)$$

(cf. 6.2.31) and also, for each $x_1 \in X_1$,

$$\forall x_2 \in X_2, (a\varphi + b\tilde{\varphi})_{x_1}(x_2) = a\varphi_{x_1}(x_2) + b\tilde{\varphi}_{x_1}(x_2),$$

and hence $\psi_1^{a\varphi + b\tilde{\varphi}} = a\psi_1^\varphi + b\psi_1^{\tilde{\varphi}}$, and similarly for $\psi_2^{a\varphi + b\tilde{\varphi}}$. From this and from what was proved above for a characteristic function, by linearity (cf. 8.1.9 and 8.1.13) we have that the conclusions of the statement are true for all the elements of $\mathcal{S}^+(X_1 \times X_2, \mathcal{A}_1 \otimes \mathcal{A}_2)$.

Suppose now $\varphi \in \mathcal{L}^+(X_1 \times X_2, \mathcal{A}_1 \otimes \mathcal{A}_2)$. Then there is a sequence $\{\varphi_n\}$ in $\mathcal{S}^+(X_1 \times X_2, \mathcal{A}_1 \otimes \mathcal{A}_2)$ such that

$$\varphi_n \leq \varphi_{n+1}, \forall n \in \mathbb{N}, \text{ and } \varphi_n(x_1, x_2) \xrightarrow[n \to \infty]{} \varphi(x_1, x_2), \forall (x_1, x_2) \in X_1 \times X_2$$

(cf. 6.2.26). Now, for each $x_1 \in X_1$, $\varphi_{x_1} \in \mathcal{L}^+(X_2, \mathcal{A}_2)$, $\{(\varphi_n)_{x_1}\}$ is a sequence in $\mathcal{L}^+(X_2, \mathcal{A}_2)$ (cf. 8.4.2b), $(\varphi_n)_{x_1} \leq (\varphi_{n+1})_{x_1}$, and $(\varphi_n)_{x_1}(x_2) \xrightarrow[n \to \infty]{} \varphi_{x_1}(x_2)$ for all $x_2 \in X_2$. By 8.1.7 and 8.1.8, this implies that

$$\forall x_1 \in X_1, \psi_1^{\varphi_n}(x_1) \leq \psi_1^{\varphi_{n+1}}(x_1) \text{ and } \psi_1^{\varphi_n}(x_1) \xrightarrow[n \to \infty]{} \psi_1^\varphi(x_1).$$

Since the conclusions of the statement are true for φ_n for all $n \in \mathbb{N}$, this implies, by 8.1.8 used twice, that $\psi_1^\varphi \in \mathcal{L}^+(X_2, \mathcal{A}_2)$ and that

$$\int_{X_1 \times X_2} \varphi d(\mu_1 \otimes \mu_2) = \lim_{n \to \infty} \int_{X_1 \times X_2} \varphi_n d(\mu_1 \otimes \mu_2)$$

$$= \lim_{n \to \infty} \int_{X_1} \psi_1^{\varphi_n} d\mu_1 = \int_{X_1} \psi_1^\varphi d\mu_1.$$

And similarly for ψ_2^φ. $\qquad\qquad\qquad\qquad\qquad\qquad\qquad\qquad\qquad\qquad\quad$ $\square$

8.4.9 Corollary. *Let $(X_1, \mathcal{A}_1, \mu_1)$ and $(X_2, \mathcal{A}_2, \mu_2)$ be σ-finite measure spaces and suppose that, for $\varphi \in \mathcal{M}(X_1 \times X_2, \mathcal{A}_1 \otimes \mathcal{A}_2)$, there exist $\varphi_1 \in \mathcal{L}^1(X_1, \mathcal{A}_1, \mu_1)$ and $\varphi_2 \in \mathcal{L}^1(X_2, \mathcal{A}_2, \mu_2)$ so that*

$$|\varphi(x_1, x_2)| = |\varphi_1(x_1)||\varphi_2(x_2)|, \forall (x_1, x_2) \in X_1 \times X_2.$$

Then $\varphi \in \mathcal{L}^1(X_1 \times X_2, \mathcal{A}_1 \otimes \mathcal{A}_2, \mu_1 \otimes \mu_2)$.

Proof. Letting $I_k := \int_{X_k} |\varphi_k| d\mu_k$, we have $I_k < \infty$ for $k = 1, 2$ (cf. 8.2.4). Since $|\varphi| \in \mathcal{L}^+(X_1 \times X_2, \mathcal{A}_1 \otimes \mathcal{A}_2)$ (cf. 6.2.17), from 8.4.8 (with φ replaced by $|\varphi|$) we have

$$\int_{X_1 \times X_2} |\varphi| d(\mu_1 \otimes \mu_2) = \int_{X_1} \psi_1^{|\varphi|} d\mu_1.$$

Now, $\psi_1^{|\varphi|} = I_2 |\varphi_1|$ and hence

$$\int_{X_1 \times X_2} |\varphi| d(\mu_1 \otimes \mu_2) = I_2 I_1 < \infty,$$

which shows (cf. 8.2.4) that $\varphi \in \mathcal{L}^1(X_1 \times X_2, \mathcal{A}_1 \otimes \mathcal{A}_2, \mu_1 \otimes \mu_2)$. $\square$

8.4.10 Theorem (Fubini's theorem). *Let $(X_1, \mathcal{A}_1, \mu_1)$ and $(X_2, \mathcal{A}_2, \mu_2)$ be σ-finite measure spaces, and let $\varphi \in \mathcal{L}^1(X_1 \times X_2, \mathcal{A}_1 \otimes \mathcal{A}_2, \mu_1 \otimes \mu_2)$ be such that $D_\varphi = X_1 \times X_2$. Then:*

(a) $\varphi_{x_1} \in \mathcal{L}^1(X_2, \mathcal{A}_2, \mu_2)$ for μ_1-a.e. $x_1 \in X_1$ and $\varphi^{x_2} \in \mathcal{L}^1(X_1, \mathcal{A}_1, \mu_1)$ for μ_2-a.e. $x_2 \in X_2$;

(b) the function

$$\rho_1^\varphi : D_1 \to \mathbb{C}$$

$$x_1 \mapsto \rho_1^\varphi(x_1) := \int_{X_2} \varphi_{x_1} d\mu_2,$$

with $D_1 := \{x_1 \in X_1 : \varphi_{x_1} \in \mathcal{L}^1(X_2, \mathcal{A}_2, \mu_2)\}$, is an element of $\mathcal{L}^1(X_1, \mathcal{A}_1, \mu_1)$, and the function

$$\rho_2^\varphi : D_2 \to \mathbb{C}$$

$$x_2 \mapsto \rho_2^\varphi(x_2) := \int_{X_1} \varphi^{x_2} d\mu_1$$

with $D_2 := \{x_2 \in X_2 : \varphi^{x_2} \in \mathcal{L}^1(X_1, \mathcal{A}_1, \mu_1)\}$, is an element of $\mathcal{L}^1(X_2, \mathcal{A}_2, \mu_2)$;

(c)

$$\int_{X_1 \times X_2} \varphi d(\mu_1 \otimes \mu_2) = \int_{X_1} \rho_1^\varphi d\mu_1 = \int_{X_2} \rho_2^\varphi d\mu_2,$$

which can also be written as

$$\int_{X_1 \times X_2} \varphi d(\mu_1 \otimes \mu_2) = \int_{X_1} \left(\int_{X_2} \varphi(x_1, x_2) d\mu_2(x_2) \right) d\mu_1(x_1)$$

$$= \int_{X_2} \left(\int_{X_1} \varphi(x_1, x_2) d\mu_1(x_1) \right) d\mu_2(x_2),$$

with the understanding that the expressions $\int_{X_2} \varphi(x_1, x_2) d\mu_2(x_2)$ and $\int_{X_1} \varphi(x_1, x_2) d\mu_1(x_1)$ are to be considered only for $x_1 \in D_1$ and for $x_2 \in D_2$, and hence only for μ_1-a.e. $x_1 \in X_1$ and for μ_2-a.e. $x_2 \in X_2$ respectively (the second equality is often referred to by saying that "the order of integration may be reversed").

Proof. We will prove the conclusions of the statement for φ_{x_1} and for ρ_1^φ. The proof for φ^{x_2} and ρ_2^φ is similar.

a: From 8.4.2b we have $\varphi_{x_1} \in \mathcal{M}(X_2, \mathcal{A}_2)$ for each $x_1 \in X_1$. Moreover we have $|\varphi| \in \mathcal{L}^+(X_1 \times X_2, \mathcal{A}_1 \otimes \mathcal{A}_2)$ (cf. 6.2.17) and $\int_{X_1 \times X_2} |\varphi| d(\mu_1 \otimes \mu_2) < \infty$ (cf. 8.2.4). Then from 8.4.8 (with φ replaced by $|\varphi|$) we have $\psi_1^{|\varphi|} \in \mathcal{L}^+(X_1, \mathcal{A}_1)$ and

$$\int_{X_1} \psi_1^{|\varphi|} d\mu_1 = \int_{X_1 \times X_2} |\varphi| d(\mu_1 \otimes \mu_2) < \infty,$$

and hence (since obviously $|\varphi_{x_1}| = |\varphi|_{x_1}$) that the the set

$$D_1^\infty := (\psi_1^{|\varphi|})^{-1}(\{\infty\}) = \left\{ x_1 \in X_1 : \int_{X_2} |\varphi_{x_1}| d\mu_2 = \infty \right\}$$

is an element of $\mathcal{A}_1$ (cf. 6.1.26) and $\mu_1(D_1^\infty) = 0$ (cf. 8.1.12b). This proves that $\varphi_{x_1} \in \mathcal{L}^1(X_2, \mathcal{A}_2, \mu_2)$ for μ_1-a.e. $x_1 \in X_1$ (cf. 8.2.4).

b: We define

$$\varphi_1 := (\operatorname{Re} \varphi)^+, \varphi_2 := (\operatorname{Re} \varphi)^-, \varphi_3 := (\operatorname{Im} \varphi)^+, \varphi_4 := (\operatorname{Im} \varphi)^-$$

and we notice that $\varphi_{x_1} = (\varphi_1)_{x_1} - (\varphi_2)_{x_1} + i(\varphi_3)_{x_1} - i(\varphi_4)_{x_1}$ for each $x_1 \in X_1$ (cf. 1.2.19).

Fix now $i \in \{1, 2, 3, 4\}$. We have $\varphi_i \in \mathcal{L}^+(X_1 \times X_2, \mathcal{A}_1 \otimes \mathcal{A}_2)$ (cf. 6.2.12 and 6.2.20b) and $\varphi_i \leq |\varphi|$, and hence $(\varphi_i)_{x_1} \in \mathcal{L}^+(X_2, \mathcal{A}_2)$ (cf. 8.4.2b) and $(\varphi_i)_{x_1} \leq |\varphi_{x_1}|$ for each $x_1 \in X_1$. We define $\psi_i := (\psi_1^{\varphi_i})_{D_1}$, with $\psi_1^{\varphi_i}$ as in 8.4.8 (with φ replaced by φ_i). From 8.4.8 we have $\psi_1^{\varphi_i} \in \mathcal{L}^+(X_1, \mathcal{A}_1)$, and hence $\psi_i \in \mathcal{L}^+(D_1, \mathcal{A}_1^{D_1})$ (cf. 6.2.3). As a matter of fact, $\psi_i \in \mathcal{M}(D_1, \mathcal{A}_1^{D_1})$ because $D_1 = X_1 - D_1^\infty$ (cf. 8.2.4) and hence

$$\forall x_1 \in D_1, \psi_i(x_1) = \psi_1^{\varphi_i}(x_1) = \int_{X_2} (\varphi_i)_{x_1} d\mu_2 \leq \int_{X_2} |\varphi_{x_1}| d\mu_2 < \infty.$$

Thus, $\psi_i \in \mathcal{M}(X_1, \mathcal{A}_1, \mu_1)$ since $D_1 \in \mathcal{A}_1$ and $\mu_1(X_1 - D_1) = \mu_1(D_1^\infty) = 0$ (cf. 8.2.1). Moreover, $\psi_i \in \mathcal{L}^+(X_1, \mathcal{A}_1, \mu_1)$ and from 8.1.14, 8.4.8, 8.1.11a we have

$$\int_{X_1} \psi_i d\mu_1 = \int_{X_1} \psi_1^{\varphi_i} d\mu_1 = \int_{X_1 \times X_2} \varphi_i d(\mu_1 \otimes \mu_2) \leq \int_{X_1 \times X_2} |\varphi| d(\mu_1 \otimes \mu_2) < \infty,$$

and this shows (cf. 8.2.4) that $\psi_i \in \mathcal{L}^1(X_1, \mathcal{A}_1, \mu_1)$.

Finally, for each $x_1 \in D_1$, we have

$$\rho_1^\varphi(x_1) = \int_{X_2} \varphi_{x_1} d\mu_2$$

$$= \int_{X_2} (\varphi_1)_{x_1} d\mu_2 - \int_{x_2} (\varphi_2)_{x_1} d\mu_2 + i \int_{x_2} (\varphi_3)_{x_1} d\mu_2 - i \int_{x_2} (\varphi_4)_{x_1} d\mu_2$$

$$= \psi_1(x_1) - \psi_2(x_1) + i\psi_3(x_1) - i\psi_4(x_1).$$

This shows that $\rho_1^\varphi = \psi_1 - \psi_2 + i\psi_3 - i\psi_4$, and hence that $\rho_1^\varphi \in \mathcal{L}^1(X_1, \mathcal{A}_1, \mu_1)$ (cf. 8.2.9).

c: From 8.2.9 we have

$$
\int_{X_1 \times X_2} \varphi d(\mu_1 \otimes \mu_2) = \int_{X_1 \times X_2} \varphi_1 d(\mu_1 \otimes \mu_2) - \int_{X_1 \times X_2} \varphi_2 d(\mu_1 \otimes \mu_2)
$$

$$
+ i \int_{X_1 \times X_2} \varphi_3 d(\mu_1 \otimes \mu_2) - i \int_{X_1 \times X_2} \varphi_4 d(\mu_1 \otimes \mu_2)
$$

$$
= \int_{X_1} \psi_1 d\mu_1 - \int_{X_1} \psi_2 d\mu_1
$$

$$
+ i \int_{X_1} \psi_3 d\mu_1 - i \int_{X_1} \psi_4 d\mu_1 = \int_{X_1} \rho_1^\varphi d\mu_1.
$$

$\square$

8.4.11 Remarks.

(a) Let $N \in \mathbb{N}$ be so that $N > 2$ and let $(X_k, \mathcal{A}_k, \mu_k)$ be a σ-finite measure space for $k = 1, ..., N$. If $1 < i_1 < i_2 < ... < i_r < N$, then (cf. 8.4.5)

$$
\mu_1 \otimes \cdots \otimes \mu_N
$$
$$
= (\cdots (((\mu_1 \otimes \cdots \otimes \mu_{i_1}) \otimes (\mu_{i_1+1} \otimes \cdots \otimes \mu_{i_2})) \otimes
$$
$$
(\mu_{i_2+1} \otimes \cdots \otimes \mu_{i_3})) \otimes \cdots) \otimes (\mu_{i_r+1} \otimes \cdots \otimes \mu_N).
$$

Thus, if $\varphi \in \mathcal{L}^+(X_1 \times \cdots X_N, \mathcal{A}_1 \otimes \cdots \otimes \mathcal{A}_N)$, from 8.4.8 we have

$$
\int_{X_1 \times \cdots \times X_N} \varphi d(\mu_1 \otimes \cdots \otimes \mu_N)
$$

$$
= \int_{X_{i_r+1} \times \cdots \times X_N} \left(\cdots \left(\int_{X_{i_1+1} \times \cdots \times X_{i_2}} \left(\int_{X_1 \times \cdots X_{i_1}} \varphi(x_1, ..., x_N) \right. \right. \right.
$$
$$
\left. d(\mu_1 \otimes \cdots \otimes \mu_{i_1})(x_1, ..., x_{i_1}) \right) d(\mu_{i_1+1} \otimes \cdots \otimes \mu_{i_2})(x_{i_1+1}, ..., x_{i_2}) \left. \right) \cdots \right)
$$
$$
d(\mu_{i_r+1} \otimes \cdots \otimes \mu_N)(x_{i_r+1}, ..., x_N).
$$

If $\varphi \in \mathcal{L}^1(X_1 \times \cdots X_N, \mathcal{A}_1 \otimes \cdots \otimes \mathcal{A}_N, \mu_1 \otimes \cdots \otimes \mu_N)$ and $D_\varphi = X_1 \times \cdots \times X_N$, the same is true by 8.4.10.

(b) Let $(X, \mathcal{A}, \nu)$ be a σ-finite measure space and put $\nu_N := \nu \otimes \cdots N$ times $\cdots \otimes \nu$ for $N > 1$ and $\nu_1 := \nu$. From the result found in remark a it follows that, by repeated use of 8.4.8 or of 8.4.10, we can reverse the order of integration in any two variables when integrating with respect to ν_N for $N > 2$. Indeed, for

$$
\varphi \in \mathcal{L}^+(X^N, \mathcal{A} \otimes \cdots N \text{ times } \cdots \otimes \mathcal{A}),
$$

or for

$$
\varphi \in \mathcal{L}^1(X^N, \mathcal{A} \otimes \cdots N \text{ times } \cdots \otimes \mathcal{A}, \nu_N) \text{ such that } D_\varphi = X^N,
$$

we have (supposing $j < N$ and $k < j - 1$; if $j = N$ or $k = j - 1$, it is obvious how to simplify the calculation below)

$$\int_{X^N} \varphi(x_1, ..., x_N) d\nu_N(x_1, ..., x_N)$$

$$= \int_{X^{N-j}} \left(\int_X \left(\int_{X^{j-1-k}} \left(\int_X \left(\int_{X^{k-1}} \varphi(x_1, ..., x_N) d\nu_{k-1}(x_1, ..., x_{k-1}) \right) \right. \right. \right.$$
$$\left. \left. \left. d\nu(x_k) \right) d\nu_{j-1-k}(x_{k+1}, ..., x_{j-1}) \right) d\nu(x_j) \right) d\nu_{N-j}(x_{j+1}, ..., x_N)$$

$$= \int_{X^{N-j}} \left(\int_{X^{j-1-k}} \left(\int_X \left(\int_X \left(\int_{X^{k-1}} \varphi(x_1, ..., x_N) d\nu_{k-1}(x_1, ..., x_{k-1}) \right) \right. \right. \right.$$
$$\left. \left. \left. d\nu(x_k) \right) d\nu(x_j) \right) d\nu_{j-1-k}(x_{k+1}, ..., x_{j-1}) \right) d\nu_{N-j}(x_{j+1}, ..., x_N)$$

$$= \int_{X^{N-j}} \left(\int_{X^{j-1-k}} \left(\int_X \left(\int_X \left(\int_{X^{k-1}} \varphi(x_1, ..., x_N) d\nu_{k-1}(x_1, ..., x_{k-1}) \right) \right. \right. \right.$$
$$\left. \left. \left. d\nu(x_j) \right) d\nu(x_k) \right) d\nu_{j-1-k}(x_{k+1}, ..., x_{j-1}) \right) d\nu_{N-j}(x_{j+1}, ..., x_N)$$

$$= \int_{X^{N-j}} \left(\int_X \left(\int_{X^{j-1-k}} \left(\int_X \left(\int_{X^{k-1}} \varphi(x_1, ..., x_N) d\nu_{k-1}(x_1, ..., x_{k-1}) \right) \right. \right. \right.$$
$$\left. \left. \left. d\nu(x_j) \right) d\nu_{j-1-k}(x_{k+1}, ..., x_{j-1}) \right) d\nu(x_k) \right) d\nu_{N-j}(x_{j+1}, ..., x_N)$$

$$= \int_{X^N} \varphi(x_1, .., x_N) d\nu_N(x_1, ..., x_{k-1}, x_j, x_{k+1}, ..., x_{j-1}, x_k, x_{j+1}, ..., x_N)$$

$$= \int_{X^N} \varphi(x_1, ..., x_{k-1}, x_j, x_{k+1}, ..., x_{j-1}, x_k, x_{j+1}, ..., x_N) d\nu_N(x_1, ..., x_N),$$

where the last equality holds because the names we give to variables are immaterial (while their positions are essential). Thus, the two functions φ and

$$(x_1, ..., x_N) \mapsto \varphi(x_1, ..., x_{k-1}, x_j, x_{k+1}, ..., x_{j-1}, x_k, x_{j+1}, ..., x_N)$$

have the same integrals with respect to ν_N.

8.4.12 Remark. In 8.4.8 and in 8.4.10 we assumed $D_\varphi = X_1 \times X_2$ for the functions φ in the statements. However, both Tonelli's theorem and Fubini's theorem can be generalized to the case of functions defined only $\mu_1 \otimes \mu_2$-a.e. We examine here the case of Tonelli's theorem. For Fubini's theorem the treatment would be analogous.

Let $(X_1, \mathcal{A}_1, \mu_1)$ and $(X_2, \mathcal{A}_2, \mu_2)$ be σ-finite measure spaces, and suppose that for a function φ we have $\varphi \in \mathcal{L}^+(X_1 \times X_2, \mathcal{A}_1 \otimes \mathcal{A}_2, \mu_1 \otimes \mu_2)$. This entails (cf. 8.1.14) $D_\varphi \in \mathcal{A}_1 \otimes \mathcal{A}_2$, $(\mu_1 \otimes \mu_2)(X_1 \times X_2 - D_\varphi) = 0$, $\varphi \in \mathcal{L}^+(D_\varphi, (\mathcal{A}_1 \otimes \mathcal{A}_2)^{D_\varphi})$. For each $x_1 \in \pi_{X_1}(D_\varphi)$ (cf. 1.2.6c), we define

$$\varphi_{x_1} : (D_\varphi)_{x_1} \to [0, \infty]$$
$$x_2 \mapsto \varphi_{x_1}(x_2) := \varphi(x_1, x_2)$$

(the condition $x_1 \in \pi_{X_1}(D_\varphi)$ implies $(D_\varphi)_{x_1} \neq \emptyset$). We have $(D_\varphi)_{x_1} \in \mathcal{A}_2$ (cf. 8.4.2a). We also have, for every subset F of $\mathbb{R}^*$,

$$\varphi_{x_1}^{-1}(F) = \{x_2 \in (D_\varphi)_{x_1} : \varphi(x_1, x_2) \in F\}$$
$$= \{x_2 \in X_2 : (x_1, x_2) \in D_\varphi \text{ and } \varphi(x_1, x_2) \in F\}$$
$$= \{x_2 \in X_2 : (x_1, x_2) \in \varphi^{-1}(F)\} = (\varphi^{-1}(F))_{x_1};$$

thus, if $F \in \mathcal{A}(\delta)$, then $\varphi^{-1}(F) = E \cap D_\varphi$ with $E \in \mathcal{A}_1 \otimes \mathcal{A}_2$, and hence

$$\varphi_{x_1}^{-1}(F) = (E \cap D_\varphi)_{x_1} = E_{x_1} \cap (D_\varphi)_{x_1} \in (\mathcal{A}_2)^{(D_\varphi)_{x_1}}$$

since $E_x \in \mathcal{A}_2$ (cf. 8.4.2a); this implies that $\varphi_{x_1} \in \mathcal{L}^+((D_\varphi)_{x_1}, \mathcal{A}_2^{(D_\varphi)_{x_1}})$. Moreover,

$$\int_{X_1} \mu_2((X_1 \times X_2 - D_\varphi)_{x_1}) d\mu_1(x_1) = (\mu_1 \otimes \mu_2)(X_1 \times X_2 - D_\varphi) = 0$$

(cf. 8.4.7) implies (cf. 8.1.12a) that $\mu_2((X_1 \times X_2 - D_\varphi)_{x_1}) = 0$ μ_1-a.e. on X_1; since $(X_1 \times X_2 - D_\varphi)_{x_1} = X_2 - (D_\varphi)_{x_1}$, this implies that

$$\varphi_{x_1} \in \mathcal{L}^+(X_2, \mathcal{A}_2, \mu_2) \text{ } \mu_1\text{-a.e. on } X_1.$$

Let then $E_1 \in \mathcal{A}_1$ be such that $\mu_1(E_1) = 0$ and $\varphi_{x_1} \in \mathcal{L}^+(X_2, \mathcal{A}_2, \mu_2)$ for each $x_1 \in X_1 - E_1$.

Now, if $\tilde{\varphi} \in \mathcal{L}^+(X_1 \times X_2, \mathcal{A}_1 \otimes \mathcal{A}_2)$ is an extension of φ, we have (cf. 8.1.14)

$$\int_{X_1 \times X_2} \varphi \, d(\mu_1 \otimes \mu_2) = \int_{X_1 \times X_2} \tilde{\varphi} \, d(\mu_1 \otimes \mu_2).$$

Moreover, for each $x_1 \in \pi_{X_1}(D_\varphi)$, $\tilde{\varphi}_{x_1}$ is an element of $\mathcal{L}^+(X_2, \mathcal{A}_2)$ (cf. 8.4.2b) which is obviously an extension of φ_{x_1} for each $x_1 \in X - E_1$, and hence we have (cf. 8.1.14),

$$\int_{X_2} \varphi_{x_1} d\mu_2 = \int_{X_2} \tilde{\varphi}_{x_1} d\mu_2, \forall x \in X_1 - E_1.$$

Thus, the function

$$X_1 - E_1 \ni x_i \mapsto \int_{X_2} \varphi_{x_1} d\mu_2 \in [0, \infty]$$

is the restriction of the function $\psi_1^{\tilde{\varphi}}$ (cf. 8.4.8 with φ replaced by $\tilde{\varphi}$) to $X_1 - E_1$, and hence (cf. 6.2.3) it is an element of $\mathcal{L}^+(X_1, \mathcal{A}_1, \mu_1)$ and (cf. 8.1.14)

$$\int_{X_1} \left(\int_{X_2} \varphi_{x_1}(x_2) d\mu_2(x_2) \right) d\mu_1(x_1) = \int_{X_1} \psi_1^{\tilde{\varphi}} d\mu_1.$$

Then, 8.4.8 (with φ replaced by $\tilde{\varphi}$) implies that

$$\int_{X_1 \times X_2} \varphi \, d(\mu_1 \otimes \mu_2) = \int_{X_1 \times X_2} \tilde{\varphi} \, d(\mu_1 \otimes \mu_2) = \int_{X_1} \psi_1^{\tilde{\varphi}_1} d\mu_1$$

$$= \int_{X_1} \left(\int_{X_2} \varphi(x_1, x_2) d\mu_2(x_2) \right) d\mu_1(x_1).$$

In a similar way it can be proved that

$$\int_{X_1 \times X_2} \varphi \, d(\mu_1 \otimes \mu_2) = \int_{X_2} \left(\int_{X_1} \varphi(x_1, x_2) d\mu_1(x_1) \right) d\mu_2(x_2).$$

8.4.13 Remark. Having generalized 8.4.8 in 8.4.12, we can generalize 8.4.9 in a similar way: let $(X_1, \mathcal{A}_1, \mu_1)$ and $(X_2, \mathcal{A}_2, \mu_2)$ be σ-finite measure spaces and suppose that, for $\varphi \in \mathcal{M}(X_1 \times X_2, \mathcal{A}_1 \otimes \mathcal{A}_2, \mu_1 \otimes \mu_2)$, there exist $\varphi_1 \in \mathcal{L}^1(X_1, \mathcal{A}_1, \mu_1)$ and $\varphi_2 \in \mathcal{L}^1(X_2, \mathcal{A}_2, \mu_2)$ so that

$$|\varphi(x_1, x_2)| = |\varphi_1(x_1)||\varphi_2(x_2)| \quad \mu_1 \otimes \mu_2\text{-a.e. on } X_1 \times X_2.$$

then $\varphi \in \mathcal{L}^1(X_1 \times X_2, \mathcal{A}_1 \otimes \mathcal{A}_2, \mu_1 \otimes \mu_2)$.

From Fubini's theorem we can derive the results about double series that we present in 8.4.14. These results, it must be said, can be obtained by more elementary means (cf. e.g. Apostol, 1974, th. 8.42).

8.4.14 Proposition. *Suppose φ is a function $\varphi : \mathbb{N} \times \mathbb{N} \to \mathbb{C}$, and let $\sigma : \mathbb{N} \to \mathbb{N} \times \mathbb{N}$ be a bijection from $\mathbb{N}$ onto $\mathbb{N} \times \mathbb{N}$.*

(a) The following are equivalent conditions:

$$\sum_{n=1}^{\infty} \left(\sum_{s=1}^{\infty} |\varphi(n, s)| \right) < \infty \quad \left(\text{and hence } \sum_{s=1}^{\infty} |\varphi(n, s)| < \infty \text{ for each } n \in \mathbb{N}\right);$$

$$\sum_{s=1}^{\infty} \left(\sum_{n=1}^{\infty} |\varphi(n, s)| \right) < \infty \quad \left(\text{and hence } \sum_{n=1}^{\infty} |\varphi(n, s)| < \infty \text{ for each } s \in \mathbb{N}\right);$$

$$\sum_{k=1}^{\infty} |\varphi(\sigma(k))| < \infty.$$

(b) If the conditions in part a are satisfied, then all the series written below are convergent and

$$\sum_{n=1}^{\infty} \left(\sum_{s=1}^{\infty} \varphi(n, s) \right) = \sum_{s=1}^{\infty} \left(\sum_{n=1}^{\infty} \varphi(n, s) \right) = \sum_{k=1}^{\infty} \varphi(\sigma(k)).$$

Proof. a: Use 5.4.7 (and 5.4.2b).

b: Let μ be the counting measure on $\mathbb{N}$ (cf. 8.3.10a). Then

$$\tilde{\mu} : \mathcal{P}(\mathbb{N} \times \mathbb{N}) \to [0, \infty]$$
$$E \mapsto \tilde{\mu}(E) := \mu(\sigma^{-1}(E))$$

is a measure on $\mathcal{P}(\mathbb{N} \times \mathbb{N})$ (cf. 8.3.11a). We note now that $\mathcal{P}(\mathbb{N} \times \mathbb{N}) = \mathcal{P}(\mathbb{N}) \otimes \mathcal{P}(\mathbb{N})$ and that, for all $E_1, E_2 \in \mathcal{P}(\mathbb{N})$,

$$\tilde{\mu}(E_1 \times E_2) = \text{ number of points in } E_1 \times E_2$$
$$= (\text{number of points in } E_1)(\text{number of points in } E_2) = \mu(E_1)\mu(E_2).$$

(by "number of points in a set" we mean here ∞ if the set is not finite). Thus, $\tilde{\mu} = \mu \otimes \mu$ (cf. 8.4.3).

Assume now that the conditions in part a are satisfied. Then all the series in part b are absolutely convergent and hence convergent (also because $|\sum_{s=1}^{\infty} \varphi(n, s)| \leq$

$\sum_{s=1}^{\infty} |\varphi(n,s)|$ and $|\sum_{n=1}^{\infty} \varphi(n,s)| \le \sum_{n=1}^{\infty} |\varphi(n,s)|$), and from 8.3.10a we have the equalities

$$\sum_{n=1}^{\infty} \left(\sum_{n=1}^{\infty} \varphi(n,s) \right) = \int_{\mathbb{N}} \left(\int_{\mathbb{N}} \varphi(n,s) d\mu(s) \right) d\mu(n),$$

$$\sum_{s=1}^{\infty} \left(\sum_{n=1}^{\infty} \varphi(n,s) \right) = \int_{\mathbb{N}} \left(\int_{\mathbb{N}} \varphi(n,s) d\mu(n) \right) d\mu(s),$$

$$\sum_{k=1}^{\infty} \varphi(\sigma(k)) = \int_{\mathbb{N}} (\varphi \circ \sigma) d\mu.$$

Now, the equalities we need to prove follow from 8.4.10 since from 8.3.11c we have

$$\int_{\mathbb{N}} (\varphi \circ \sigma) d\mu = \int_{\mathbb{N} \times \mathbb{N}} \varphi d\tilde{\mu} = \int_{\mathbb{N} \times \mathbb{N}} \varphi d(\mu \otimes \mu).$$

$\square$

8.4.15 Corollary. *Let* $I := \{1, ..., N\}$ *or* $I := \mathbb{N}$, *and for each* $n \in I$ *let* $I_n := \{(n,1), ..., (n, N_n)\}$ *or* $I_n := \{(n,s) : s \in \mathbb{N}\}$. *Let* $J := \bigcup_{n \in I} I_n$ *and let* $\{\alpha_{(n,s)}\}_{(n,s) \in J}$ *be a family of elements of* $\mathbb{C}$.

(a) The following are equivalent conditions:

$$\sum_{(n,s) \in I_n} |\alpha_{(n,s)}| < \infty \text{ for each } n \in I \text{ and } \sum_{n \in I} \left(\sum_{(n,s) \in I_n} |\alpha_{(n,s)}| \right) < \infty;$$

$$\sum_{(n,s) \in J} |\alpha_{(n,s)}| < \infty.$$

(b) If the conditions in part a are satisfied, then the series that may appear below are convergent and

$$\sum_{n \in I} \left(\sum_{(n,s) \in I_n} \alpha_{(n,s)} \right) = \sum_{(n,s) \in J} \alpha_{(n,s)}.$$

Note that, for the various sums or series of this statement to be defined properly, an ordering must be assumed in the various index sets. However, the orderings we use in part a are immaterial in view of 5.4.3, and the orderings we use in part b are immaterial in view of 4.1.8b because, if the conditions in part a are satisfied, then all the series that may appear in part b are absolutely convergent.

Proof. If $I = \mathbb{N}$ and $I_n = \{(n,s) : s \in \mathbb{N}\}$ for each $n \in \mathbb{N}$, this follows immediately from 8.4.14. Otherwise, define $I' := \mathbb{N}$, $I'_n := \{(n,s) : s \in \mathbb{N}\}$ for all $n \in \mathbb{N}$, and

$$\alpha'_{(n,s)} := \begin{cases} \alpha_{(n,s)} & \text{if } (n,s) \in J, \\ 0 & \text{if } (n,s) \notin J, \end{cases} \quad \text{for all } (n,s) \in \mathbb{N} \times \mathbb{N},$$

and then use 8.4.14. $\square$

8.5 The Riesz–Markov theorem

In this section we prove the Riesz–Markov theorem for compact metric spaces, which will play an essential role in our proof of the spectral theorem for unitary operators (from which we will deduce the spectral theorem for self-adjoint operators).

8.5.1 Definition. Let M be a linear manifold in the linear space $\mathcal{F}(X)$ (cf. 3.1.10c) and let L be a linear functional with $D_L = M$. The linear functional L is said to be *positive* if $0 \le L\varphi$ whenever $\varphi \in M$ is such that $0_X \le \varphi$.

Notice that, if L is positive and $\varphi, \psi \in M$ are such that $\varphi \le \psi$, then $L\varphi \le L\psi$ since $L\psi = L\varphi + L(\psi - \varphi)$ and $0_X \le \psi - \varphi$.

8.5.2 Remark. Let (X, d) be a compact metric space and μ a finite measure on the Borel σ-algebra $\mathcal{A}(d)$. For $\mathcal{C}(X)$ (cf. 3.1.10e) we have $\mathcal{C}(X) \subset \mathcal{L}^1(X, \mathcal{A}(d), \mu)$ by 6.2.8, 2.8.14 and 8.2.6. Thus, we can define the mapping

$$L_\mu : \mathcal{C}(X) \to \mathbb{C}$$

$$\varphi \mapsto L_\mu \varphi := \int_X \varphi d\mu,$$

which is a positive linear functional on $\mathcal{C}(X)$ (cf. 8.2.9).

The Riesz–Markov theorem proves that every positive linear functional on $\mathcal{C}(X)$ can be obtained in this way.

8.5.3 Theorem (The Riesz–Markov theorem). *Let (X, d) be a compact metric space and let L be a positive linear functional on $\mathcal{C}(X)$. Then there exists a unique finite measure μ on the Borel σ-algebra $\mathcal{A}(d)$ so that*

$$L\varphi = \int_X \varphi d\mu, \forall \varphi \in \mathcal{C}(X).$$

Proof. *Existence:* For each $G \in \mathcal{T}_d$, the family $\{\varphi \in \mathcal{C}(X) : \varphi \prec G\}$ is not empty (it contains the function 0_X; for $\varphi \prec G$, cf. 2.5.10). Thus, we can define the function

$$\nu : \mathcal{T}_d \to [0, \infty]$$
$$G \mapsto \nu(G) := \sup\{L\varphi : \varphi \in \mathcal{C}(X) \text{ and } \varphi \prec G\}. \tag{1}$$

If $G_1, G_2 \in \mathcal{T}_d$ are so that $G_1 \subset G_2$, then

$$\{\varphi \in \mathcal{C}(X) : \varphi \prec G_1\} \subset \{\varphi \in \mathcal{C}(X) : \varphi \prec G_2\},$$

and this implies that

$$\text{if } G_1, G_2 \in \mathcal{T}_d \text{ are so that } G_1 \subset G_2, \text{ then } \nu(G_1) \le \nu(G_2). \tag{2}$$

We recall that, for $\varphi \in \mathcal{C}(X)$ and $G \in \mathcal{T}_d$, $\varphi \prec G$ implies $\varphi \le 1_X$; thus, since $1_X \prec X$ and L is positive, we have

$$\nu(X) = L1_X. \tag{3}$$

We define the function

$$\mu^* : \mathcal{P}(X) \to [0, \infty]$$
$$E \mapsto \mu^*(E) := \inf\{\nu(G) : G \in \mathcal{T}_d \text{ and } E \subset G\}. \tag{4}$$

From 2 we have

$$\forall G \in \mathcal{T}_d, \mu^*(G) = \nu(G). \tag{5}$$

If $E_1, E_2 \in \mathcal{P}(X)$ are so that $E_1 \subset E_2$, then

$$\{G \in \mathcal{T}_d : E_2 \subset G\} \subset \{G \in \mathcal{T}_d : E_1 \subset G\},$$

and this implies that

$$\text{if } E_1, E_2 \in \mathcal{P}(X) \text{ are so that } E_1 \subset E_2, \text{ then } \mu^*(E_1) \leq \mu^*(E_2). \tag{6}$$

We notice that 6, 5 and 3 imply that

$$\forall E \in \mathcal{P}(X), \mu^*(E) \leq \mu^*(X) = \nu(X) < \infty. \tag{7}$$

We want to prove that μ^* is an outer measure on X.

The only $\varphi \in \mathcal{C}(X)$ such that $\varphi \prec \emptyset$ is $\varphi = 0_X$. Thus, $\nu(\emptyset) = 0$ since L is linear, and hence $\mu^*(\emptyset) = 0$ by 5.

Since 6 has already been proved, it remains to show that

$$\text{for every sequence } \{E_n\} \text{ in } \mathcal{P}(X), \mu^* \left(\bigcup_{n=1}^{\infty} E_n \right) \leq \sum_{n=1}^{\infty} \mu^*(E_n). \tag{8}$$

Consider first $G_1, G_2 \in \mathcal{T}_d$ and let $\varphi \in \mathcal{C}(X)$ be such that $\varphi \prec G_1 \cup G_2$. Since $\operatorname{supp} \varphi$ is closed and $\operatorname{supp} \varphi \subset G_1 \cup G_2$, there are $\psi_1, \psi_2 \in \mathcal{C}(X)$ such that $\psi_1 \prec G_1$, $\psi_2 \prec G_2$ and $\psi_1(x) + \psi_2(x) = 1$ for all $x \in \operatorname{supp} \varphi$ (cf. 2.8.16). Hence, $\psi_1\varphi \prec G_1$, $\psi_2\varphi \prec G_2$, and $\varphi = \psi_1\varphi + \psi_2\varphi$. So, by the linearity of L and by 1 we have

$$L\varphi = L(\psi_1\varphi) + L(\psi_2\varphi) \leq \nu(G_1) + \nu(G_2).$$

Since this is true for every $\varphi \in \mathcal{C}(X)$ such that $\varphi \prec G_1 \cup G_2$, by 1 we have

$$\nu(G_1 \cup G_2) \leq \nu(G_1) + \nu(G_2). \tag{9}$$

Now, let $\{E_n\}$ be a sequence in $\mathcal{P}(X)$ and choose $\varepsilon > 0$. By 4 and 7, for every $n \in \mathbb{N}$ there exists $G_n \in \mathcal{T}_d$ so that $E_n \subset G_n$ and $\nu(G_n) < \mu^*(E_n) + \varepsilon 2^{-n}$. Let now $\varphi \in \mathcal{C}(X)$ be such that $\varphi \prec \bigcup_{n=1}^{\infty} G_n$. Since $\operatorname{supp} \varphi$ is closed and X is compact, by 2.8.8 there exists $N \in \mathbb{N}$ so that $\varphi \prec \cup_{n=1}^{N} G_n$. Then, by 1, by induction applied to 9, and by 5.4.6, we have

$$L\varphi \leq \nu \left(\bigcup_{n=1}^{N} G_n \right) \leq \sum_{n=1}^{N} \nu(G_n) < \sum_{n=1}^{\infty} \mu^*(E_n) + \varepsilon.$$

Since this is true for every $\varphi \in \mathcal{C}(X)$ such that $\varphi \prec \bigcup_{n=1}^{\infty} G_n$, by 1 we have

$$\nu \left(\bigcup_{n=1}^{\infty} G_n \right) \leq \sum_{n=1}^{\infty} \mu^*(E_n) + \varepsilon.$$

Since $\bigcup_{n=1}^{\infty} E_n \subset \bigcup_{n=1}^{\infty} G_n$, in view of 4 this implies

$$\mu^* \left(\bigcup_{n=1}^{\infty} E_n \right) \leq \sum_{n=1}^{\infty} \mu^*(E_n) + \varepsilon.$$

Since ε was arbitrary, 8 is proved.

Since μ^* is an outer measure on X, from Carathéodory's theorem (cf. 7.2.3) it follows that the restriction of μ^* to the σ-algebra $\mathcal{M}$ of μ^*-measurable subsets of X is a measure, which is finite by 7.

We want to prove that $\mathcal{T}_d \subset \mathcal{M}$. In view of 7.2.2 and of 7, we need to show that, for each $G \in \mathcal{T}_d$ and each $A \in \mathcal{P}(X)$,

$$\mu^*(A \cap G) + \mu^*(A - G) \leq \mu^*(A). \tag{10}$$

Assuming $G \in \mathcal{T}_d$, suppose first $A \in \mathcal{T}_d$. Then $A \cap G \in \mathcal{T}_d$, so given $\varepsilon > 0$ we can find $\varphi \in \mathcal{C}(X)$ such that $\varphi \prec A \cap G$ and $\nu(A \cap G) - \varepsilon < L\varphi$ (cf. 1 and 7). Moreover, $A - \operatorname{supp}\varphi$ is open, so we can find $\psi \in \mathcal{C}(X)$ such that $\psi \prec A - \operatorname{supp}\varphi$ and $\nu(A - \operatorname{supp}\varphi) - \varepsilon < L\psi$. But then $A - G \subset A - \operatorname{supp}\varphi$ and $\varphi + \psi \prec A$, so (cf. 6, 5, 1)

$$\mu^*(A \cap G) + \mu^*(A - G) - 2\varepsilon \leq \mu^*(A \cap G) + \mu^*(A - \operatorname{supp}\varphi) - 2\varepsilon$$
$$= \nu(A \cap G) + \nu(A - \operatorname{supp}\varphi) - 2\varepsilon$$
$$< L\varphi + L\psi = L(\varphi + \psi) \leq \nu(A) = \mu^*(A).$$

Since ε was arbitrary, 10 is proved for $A \in \mathcal{T}_d$. For the general case $A \in \mathcal{P}(X)$, given $\varepsilon > 0$ we can find $U \in \mathcal{T}_d$ so that $A \subset U$ and $\nu(U) < \mu^*(A) + \varepsilon$ (cf. 4 and 7), and this implies (cf. 6, 8, 5) that

$$\mu^*(A \cap G) + \mu^*(A - G) \leq \mu^*(U \cap G) + \mu^*(U - G) \leq \mu^*(U) = \nu(U) < \mu^*(A) + \varepsilon.$$

Since ε was arbitrary, 10 is proved for the general case $A \in \mathcal{P}(X)$.

Having thus proved that $\mathcal{T}_d \subset \mathcal{M}$, we have $\mathcal{A}(d) \subset \mathcal{M}$. Then the restriction of μ^* to $\mathcal{A}(d)$ is a finite measure on $\mathcal{A}(d)$ (since the restriction of μ^* to $\mathcal{M}$ was a finite measure on $\mathcal{M}$), which we denote by μ.

To conclude the proof of existence, it remains to prove that

$$L\varphi = \int_X \varphi d\mu, \forall \varphi \in \mathcal{C}(X).$$

For $\varphi \in \mathcal{C}(X)$, we have $L\varphi = L(\operatorname{Re}\varphi) + iL(\operatorname{Im}\varphi)$ by the linearity of L. Then, it is enough to carry out the proof for $\varphi \in \mathcal{C}(X)$ such that $\varphi(x) \in \mathbb{R}$ for all $x \in \mathbb{R}$. Moreover, it is enough to prove that

$$L\varphi \leq \int_X \varphi d\mu \text{ for all real } \varphi \in \mathcal{C}(X). \tag{11}$$

For once 11 is established, the linearity of L shows that

$$-L\varphi = L(-\varphi) \leq \int_X (-\varphi) d\mu = -\int_X \varphi d\mu \text{ for all real } \varphi \in \mathcal{C}(X),$$

which, together with 11, shows that equality holds in 11.

Let φ be a real element of $\mathcal{C}(X)$. From 2.8.14 it follows that there are $a, b \in \mathbb{R}$ so that $a < b$ and $R_\varphi \subset [a, b]$. Choose $\varepsilon > 0$ and choose $y_0, y_1, ..., y_n \in \mathbb{R}$ so that

$$y_0 < a < y_1 < ... < y_n = b \text{ and } y_i - y_{i-1} < \varepsilon \text{ for } i = 1, ..., n.$$

Put $E_i := \varphi^{-1}((y_{i-1}, y_i])$ for $i = 1, ..., n$. Since φ is continuous, φ is $\mathcal{A}(d)$-measurable (cf. 6.2.8), and the sets E_i are therefore disjoint elements of $\mathcal{A}(d)$ whose union is X. By 4 and 7, for $i = 1, ..., n$ there exists $G_i \in \mathcal{T}_d$ so that

$$E_i \subset G_i \text{ and } \mu(G_i) < \mu(E_i) + \frac{\varepsilon}{n}. \tag{12}$$

For $i = 1, ..., n$, by letting $\tilde{G}_i := G_i \cap \varphi^{-1}((-\infty, y_i + \varepsilon))$ we have $E_i \subset \tilde{G}_i$ and $\tilde{G}_i \in \mathcal{T}_d$ since φ is continuous (cf. 2.4.3). Since $\bigcup_{i=1}^n \tilde{G}_i = X$, by 2.8.16 there exists a family $\{\psi_1, ..., \psi_n\}$ so that $\psi_i \in \mathcal{C}(X)$ and $\psi_i \prec \tilde{G}_i$ for $i = 1, ..., n$, and so that $\sum_{i=1}^n \psi_i = 1_X$. Hence, by 3 and the linearity of L we have

$$\mu(X) = \sum_{i=1}^n L\psi_i. \tag{13}$$

We also have

$$\varphi = \sum_{i=1}^n \psi_i \varphi. \tag{14}$$

Then we have

$$L\varphi \overset{(15)}{=} \sum_{n=1}^n L(\psi_i \varphi) \overset{(16)}{\leq} \sum_{i=1}^n (y_i + \varepsilon) L\psi_i$$

$$= \sum_{i=1}^n (|a| + y_i + \varepsilon) L\psi_i - |a| \sum_{i=1}^n L\psi_i$$

$$\overset{(17)}{\leq} \sum_{i=1}^n (|a| + y_i + \varepsilon) \mu(\tilde{G}_i) - |a| \mu(X)$$

$$\overset{(18)}{\leq} \sum_{i=1}^n (|a| + y_i + \varepsilon) \left(\mu(E_i) + \frac{\varepsilon}{n} \right) - |a| \mu(X)$$

$$\overset{(19)}{\leq} \sum_{i=1}^n (y_i + \varepsilon) \mu(E_i) + \frac{\varepsilon}{n} \sum_{i=1}^n (|a| + y_i + \varepsilon)$$

$$\overset{(20)}{\leq} \sum_{i=1}^n (y_i - \varepsilon) \mu(E_i) + 2\varepsilon \mu(X) + \varepsilon |a| + \varepsilon b + \varepsilon^2$$

$$\overset{(21)}{\leq} \int_X \varphi d\mu + \varepsilon (2\mu(X) + |a| + b + \varepsilon),$$

where: 15 holds by 14 and the linearity of L; 16 holds by the linearity and the positivity of L, since $\psi_i \varphi \leq (y_i + \varepsilon) \psi_i$ by the definition of $\tilde{G}_i$; 17 holds by 1 and by 13; 18 holds by 12 since $\tilde{G}_i \subset G_i$; 19 holds because $\sum_{i=1}^n \mu(E_i) = \mu\left(\bigcup_{i=1}^n E_i\right) = \mu(X)$;

20 holds because $\sum_{i=1}^{n} \mu(E_i) = \mu(X)$ and because $y_i \le b$ for $i = 1, ..., n$; 21 holds because

$$\sum_{i=1}^{n}(y_i - \varepsilon)\chi_{E_i} \le \sum_{i=1}^{n} y_{i-1}\chi_{E_i} \le \varphi,$$

$$\sum_{i=1}^{n}(y_i - \varepsilon)\mu(E_i) = \int_X \sum_{i=1}^{n}(y_i - \varepsilon)\chi_{E_i} d\mu.$$

Since ε was arbitrary and $\mu(X) < \infty$, 11 is established and the proof of existence is complete.

Uniqueness: Let $\tilde{\mu}$ be a finite measure on $\mathcal{A}(d)$ such that

$$L\varphi = \int_X \varphi d\tilde{\mu}, \forall \varphi \in \mathcal{C}(X).$$

For every closed set F, by 2.5.7 there exists a sequence $\{\varphi_n\}$ in $\mathcal{C}(X)$ so that

$$\forall x \in X, 0 \le \varphi_n(x) \le 1 \text{ and } \varphi_n(x) \to \chi_F(x) \text{ as } n \to \infty.$$

Then, by Lebesgue's dominated convergence theorem (cf. 8.2.11, with 1_X as dominating function),

$$\tilde{\mu}(F) = \int_X \chi_F d\tilde{\mu} = \lim_{n \to \infty} \int_X \varphi_n d\tilde{\mu}$$

$$= \lim_{n \to \infty} L\varphi_n = \lim_{n \to \infty} \int_X \varphi_n d\mu = \int_X \chi_F d\mu = \mu(F).$$

In view of 7.4.2, this proves that $\tilde{\mu} = \mu$. $\qquad\qquad\square$

Chapter 9

Lebesgue Measure

In this chapter we study the Lebesgue measure on $\mathbb{R}^n$, which according to our definition is a measure on the Borel σ-algebra $\mathcal{A}(d_n)$. We warn the reader that many books call Lebesgue measure a measure which is in fact an extension of our Lebesgue measure.

9.1 Lebesgue–Stieltjes and Lebesgue measures

9.1.1 Theorem. *Suppose we have a function $F : \mathbb{R} \to \mathbb{R}$ which is monotone increasing and right continuous, i.e. such that:*

(a) if $x', x'' \in \mathbb{R}$ are so that $x' < x''$ then $F(x') \leq F(x'')$;
(b) for each $x \in \mathbb{R}$, if $\{\delta_n\}$ is a sequence in $[0, \infty)$ so that $\delta_n \xrightarrow[n \to \infty]{} 0$ then

$$F(x + \delta_n) \xrightarrow[n \to \infty]{} F(x).$$

Then there exists a unique measure μ_F on the Borel σ-algebra $\mathcal{A}(d_\mathbb{R})$ on $\mathbb{R}$ (cf. 6.1.22 and 2.1.4) such that

$$\mu_F((a, b]) = F(b) - F(a), \text{ for all } a, b \in \mathbb{R} \text{ so that } a < b.$$

The measure μ_F is σ-finite and is called the Lebesgue–Stieltjes measure associated to F.

Proof. Recall that $\mathcal{I}_9$ denotes a semialgebra on $\mathbb{R}$ such that $\mathcal{A}(d_\mathbb{R}) = \mathcal{A}(\mathcal{I}_9)$ (cf. 6.1.25), and define the function

$$\nu : \mathcal{I}_9 \to [0, \infty]$$

$$E \mapsto \nu(E) := \begin{cases} 0 & \text{if } E = \emptyset, \\ F(b) - F(a) & \text{if } E = (a, b] \text{ with } a, b \in \mathbb{R} \text{ s.t. } a < b, \\ F(b) - \lim_{n \to \infty} F(-n) & \text{if } E = (-\infty, b] \text{ with } b \in \mathbb{R}, \\ \lim_{n \to \infty} F(n) - F(a) & \text{if } E = (a, \infty) \text{ with } a \in \mathbb{R}. \end{cases}$$

(notice that $\lim_{n \to \infty} F(-n)$ and $\lim_{n \to \infty} F(n)$ do exist by 5.2.5). We will show that ν satisfies conditions a, b, c, d, e of 7.3.3.

a: This conditions holds by the definition of ν.

b: Let $E \in \mathcal{S}$ be the union of a finite and disjoint family $\{E_1, ..., E_N\}$ of elements of $\mathcal{I}_9$. In view of condition a, we may assume $E \neq \emptyset$ and $E_n \neq \emptyset$ for $n = 1, ..., N$. If $E = (a, b]$ (with $a, b \in \mathbb{R}$ so that $a < b$), then we must have $E_n = (a_n, b_n]$ (with $a_n, b_n \in \mathbb{R}$ so that $a_n < b_n$) for $n = 1, ..., N$ and, after perhaps relabelling the index n, we must also have $a = a_1 < b_1 = a_2 < b_2 = a_3 < \cdots < b_N = b$, and hence

$$\nu((a, b]) = F(b) - F(a) = \sum_{n=1}^{N}(F(b_n) - F(a_n)) = \sum_{n=1}^{N} \nu((a_n, b_n]).$$

If either $E = (-\infty, b]$ (with $b \in \mathbb{R}$) or $E = (a, \infty)$ (with $a \in \mathbb{R}$), then we proceed as above, either with a replaced by $-\infty$ and $F(a)$ by $\lim_{n \to \infty} F(-n)$ or with b replaced by ∞ and $F(b)$ by $\lim_{n \to \infty} F(n)$.

c: If $E = \emptyset$, this condition holds trivially.

If $E = (a, b]$ (with $a, b \in \mathbb{R}$ so that $a < b$): for $n \in \mathbb{N}$ large enough, $\left(a + \frac{1}{n}, b\right] \in \mathcal{S}$, $\overline{\left(a + \frac{1}{n}, b\right]} = \left[a + \frac{1}{n}, b\right] \subset (a, b]$, $\overline{\left(a + \frac{1}{n}, b\right]}$ is compact, and also (since F is right continuous)

$$\forall \varepsilon > 0, \exists n_\varepsilon \in \mathbb{N} \text{ s.t. } \nu((a, b]) - \nu\left(\left(a + \frac{1}{n_\varepsilon}, b\right]\right) = F\left(a + \frac{1}{n_\varepsilon}\right) - F(a) < \varepsilon.$$

If $E = (-\infty, b]$ (with $b \in \mathbb{R}$): $\forall n \in \mathbb{N}$, $(-n, b] \in \mathcal{S}$, $\overline{(-n, b]} \subset (-\infty, b]$, $\overline{(-n, b]}$ is compact; if $\lim_{n \to \infty} F(-n) \in \mathbb{R}$, then (from the definition of limit)

$$\forall \varepsilon > 0, \exists n_\varepsilon \in \mathbb{N} \text{ s.t. } \nu((-\infty, b]) - \nu((-n_\varepsilon, b]) = F(-n_\varepsilon) - \lim_{n \to \infty} F(-n) < \varepsilon;$$

if $\lim_{n \to \infty} F(-n) = -\infty$, then $\nu((-\infty, b]) = \infty$ and (cf. 5.3.2c, 5.3.2a, 5.2.5)

$$\sup_{n \geq 1} \nu((-n, b]) = \sup_{n \geq 1}(F(b) - F(-n)) = F(b) - \inf_{n \geq 1} F(-n)$$
$$= F(b) - \lim_{n \to \infty} F(-n) = F(b) + \infty = \infty.$$

If $E = (a, \infty)$ (with $a \in \mathbb{R}$): $\forall n \in \mathbb{N}$, $\left(a + \frac{1}{n}, n\right] \in \mathcal{S}$, $\overline{\left(a + \frac{1}{n}, n\right]} \subset (a, \infty)$, $\overline{\left(a + \frac{1}{n}, n\right]}$ is compact; if $\lim_{n \to \infty} F(n) \in \mathbb{R}$, then (from the definition of limit and the right continuity of F)

$$\forall \varepsilon > 0, \exists n_\varepsilon \in \mathbb{N} \text{ s.t.}$$
$$\nu((a, \infty)) - \nu\left(\left(a + \frac{1}{n_\varepsilon}, n_\varepsilon\right]\right) = \lim_{n \to \infty} F(n) - F(n_\varepsilon) + F\left(a + \frac{1}{n_\varepsilon}\right) - F(a) < \varepsilon;$$

if $\lim_{n \to \infty} F(n) = \infty$, then $\nu((a, \infty)) = \infty$ and

$$\sup_{n \geq 1} \nu\left(\left(a + \frac{1}{n}, n\right]\right) = \sup_{n \geq 1}\left(F(n) - F\left(a + \frac{1}{n}\right)\right)$$
$$\geq \sup_{n \geq 1}(F(n) - F(a + 1)) = \sup_{n \geq 1} F(n) - F(a + 1)$$
$$= \lim_{n \to \infty} F(n) - F(a + 1) = \infty - F(a + 1) = \infty.$$

d: If $E = \emptyset$, this condition holds trivially.

If $E = (a, b]$ (with $a, b \in \mathbb{R}$ so that $a < b$): $\forall n \in \mathbb{N}$, $\left(a, b + \frac{1}{n}\right] \in \mathcal{S}$, $(a, b] \subset \left(a, b + \frac{1}{n}\right) = \left(a, b + \frac{1}{n}\right]^{\circ}$, and also (since F is right continuous)

$$\forall \varepsilon > 0, \exists n_{\varepsilon} \in \mathbb{N} \text{ s.t. } \nu\left(\left(a, b + \frac{1}{n_{\varepsilon}}\right]\right) - \nu((a, b]) = F\left(b + \frac{1}{n_{\varepsilon}}\right) - F(b) < \varepsilon.$$

If $E = (-\infty, b]$ (with $b \in \mathbb{R}$) and $\nu((-\infty, b]) < \infty$, i.e. $\lim_{n \to \infty} F(-n) \in \mathbb{R}$: $\forall n \in \mathbb{N}$, $\left(-\infty, b + \frac{1}{n}\right] \in \mathcal{S}$, $(-\infty, b] \subset \left(-\infty, b + \frac{1}{n}\right) = \left(-\infty, b + \frac{1}{n}\right]^{\circ}$, and also (since F is right continuous)

$$\forall \varepsilon > 0, \exists n_{\varepsilon} \in \mathbb{N} \text{ s.t. } \nu\left(\left(-\infty, b + \frac{1}{n_{\varepsilon}}\right]\right) - \nu((-\infty, b]) = F\left(b + \frac{1}{n_{\varepsilon}}\right) - F(b) < \varepsilon.$$

If $E = (a, \infty)$ (with $a \in \mathbb{R}$) and $\nu((a, \infty)) < \infty$, simply notice that $(a, \infty)^{\circ} = (a, \infty)$.

e: We have $\mathbb{R} = \bigcup_{n \in \mathbb{Z}}(n, n + 1]$ and $\nu((n, n + 1]) = F(n + 1) - F(n) < \infty$ for all $n \in \mathbb{Z}$.

Since ν satisfies conditions a, b, c, d, e and $\mathcal{A}(d_{\mathbb{R}}) = \mathcal{A}(\mathcal{I}_9)$, 7.3.3 implies that there exists a unique measure μ_F which is an extension of ν, and that μ_F is σ-finite. Since μ_F extends ν, we have

$$\forall a, b \in \mathbb{R} \text{ so that } a < b, \mu_F((a, b]) = \nu((a, b]) = F(b) - F(a).$$

Suppose now that μ is a measure on $\mathcal{A}(d_{\mathbb{R}})$ such that

$$\forall a, b \in \mathbb{R} \text{ so that } a < b, \mu((a, b]) = F(b) - F(a).$$

Then we have, by 7.1.4b:

$$\forall b \in \mathbb{R}, \mu((-\infty, b]) = \mu\left(\bigcup_{n=1}^{\infty}(-n, b]\right) = \lim_{n \to \infty} \mu((-n, b]) = F(b) - \lim_{n \to \infty} F(-n);$$

$$\forall a \in \mathbb{R}, \mu((a, \infty)) = \mu\left(\bigcup_{n=1}^{\infty}(a, n]\right) = \lim_{n \to \infty} \mu((a, n]) = \lim_{n \to \infty} F(n) - F(a).$$

Thus, μ is an extension of ν, and hence $\mu = \mu_F$ since μ_F is the only measure on $\mathcal{A}(d_{\mathbb{R}})$ that is an extension of ν. $\qquad\square$

9.1.2 Definition. We call *Lebesgue measure on* $\mathbb{R}$ and denote by m the Lebesgue–Stieltjes measure μ_{ξ} associated to the function

$$\xi : \mathbb{R} \to \mathbb{R}$$
$$x \mapsto \xi(x) := x.$$

9.1.3 Theorem. *For the Lebesgue measure m on $\mathbb{R}$ we have*

$\forall a, b \in \mathbb{R}$ *so that* $a < b$, $m((a, b]) = m((a, b)) = m([a, b)) = m([a, b]) = b - a$;
$\forall a \in \mathbb{R}$, $m(\{a\}) = 0$ *and* $m((-\infty, a)) = m((a, \infty)) = m((-\infty, a]) = m([a, \infty)) = \infty$.

The Lebesgue measure m is the only measure on $\mathcal{A}(d_{\mathbb{R}})$ with the property:
$$\forall a, b \in \mathbb{R} \text{ so that } a < b, m((a, b)) = b - a.$$
(or the same proposition with (a, b) replaced by $[a, b)$ or by $[a, b]$).

Proof. We know that m is the only measure on $\mathcal{A}(d_{\mathbb{R}})$ such that
$$\forall a, b \in \mathbb{R} \text{ so that } a < b, m((a, b]) = b - a.$$
Then we have, by 7.1.4b and 7.1.4c, for all $a, b \in \mathbb{R}$ so that $a < b$:

$$m((a, b)) = m\left(\bigcup_{n=1}^{\infty}\left(a, b - \frac{1}{n}\right]\right)$$

$$= \lim_{n\to\infty} m\left(\left(a, b - \frac{1}{n}\right]\right) = \lim_{n\to\infty}\left(b - \frac{1}{n} - a\right) = b - a;$$

$$m([a, b)) = m\left(\bigcap_{n=1}^{\infty}\left(a - \frac{1}{n}, b\right)\right)$$

$$= \lim_{n\to\infty} m\left(\left(a - \frac{1}{n}, b\right)\right) = \lim_{n\to\infty}\left(b - a + \frac{1}{n}\right) = b - a;$$

$$m([a, b]) = m\left(\bigcap_{n=1}^{\infty}\left[a, b + \frac{1}{n}\right)\right)$$

$$= \lim_{n\to\infty} m\left(\left[a, b + \frac{1}{n}\right)\right) = \lim_{n\to\infty}\left(b + \frac{1}{n} - a\right) = b - a.$$

Similarly we have, for all $a \in \mathbb{R}$:

$$m(\{a\}) = m\left(\bigcap_{n=1}^{\infty}\left(a - \frac{1}{n}, a\right]\right) = \lim_{n\to\infty} m\left(\left(a - \frac{1}{n}, a\right]\right) = \lim_{n\to\infty}\frac{1}{n} = 0;$$

$$m((-\infty, a)) = m\left(\bigcup_{n=1}^{\infty}(-n, a)\right) = \lim_{n\to\infty} m((-n, a)) = \lim_{n\to\infty}(a + n) = \infty;$$

$$m((a, \infty)) = m\left(\bigcup_{n=1}^{\infty}(a, n)\right) = \lim_{n\to\infty} m((a, n)) = \lim_{n\to\infty}(n - a) = \infty;$$

$$m((-\infty, a]) = m([a, \infty)) = \infty \text{ since } (-\infty, a) \subset (-\infty, a] \text{ and } (a, \infty) \subset [a, \infty)$$
(cf. 7.1.2a).

Suppose now that $\tilde{m}$ is a measure on $\mathcal{A}(d_{\mathbb{R}})$ such that
$$\forall a, b \in \mathbb{R} \text{ so that } a < b, \tilde{m}((a, b)) = b - a.$$
Then

$$\forall a, b \in \mathbb{R} \text{ so that } a < b, \tilde{m}((a, b]) = \tilde{m}\left(\bigcap_{n=1}^{\infty}\left(a, b + \frac{1}{n}\right)\right)$$

$$= \lim_{n\to\infty} \tilde{m}\left(\left(a, b + \frac{1}{n}\right)\right)$$

$$= \lim_{n\to\infty}\left(b + \frac{1}{n} - a\right) = b - a,$$

and hence $\tilde{m} = m$ by the uniqueness property of m quoted above. The proofs for (a, b) replaced by $[a, b)$ or by $[a, b]$ are analogous. $\qquad\square$

9.1.4 Definition. For $n \in \mathbb{N}$ so that $n > 1$, we call *Lebesgue measure on* $\mathbb{R}^n$ and denote by m_n the measure $m \otimes \cdots n$ times $\cdots \otimes m$.

Since $\mathcal{A}(d_n) = \mathcal{A}(d_\mathbb{R}) \otimes \cdots n$ times $\cdots \otimes \mathcal{A}(d_\mathbb{R})$ (cf. 6.1.32), m_n is a measure on $\mathcal{A}(d_n)$. The measure m_n is σ-finite (cf. 8.4.5).

9.1.5 Lemma. *For $n \in \mathbb{N}$, a subset I of $\mathbb{R}^n$ which is defined by*

$$I := (a_1, b_1] \times \cdots (a_n, b_n], \quad \text{with } a_k, b_k \in \mathbb{R} \text{ so that } a_k < b_k \text{ for } k = 1, ..., n,$$

is called a half-open interval. Every open subset of $\mathbb{R}^n$ is the union of a countable and disjoint family of half-open intervals.

Proof. For each $r \in \mathbb{N}$, let $\mathcal{H}_r$ be the family of the hyperplanes

$$\left\{ (x_1, ..., x_n) \in \mathbb{R}^n : x_k = \frac{i}{2^r} \right\}, \quad \text{with } k = 1, ..., n \text{ and } i \in \mathbb{Z}.$$

The family $\mathcal{H}_r$ defines in an obvious way a partition of $\mathbb{R}^n$ into a countable and disjoint family of half-open intervals.

Fix $G \in \mathcal{T}_{d_n}$. We define by induction a family of half-open intervals as follows. Let $\{I_1^p\}_{p \in J_1}$ (where J_1 is a countable set of indices) be the family of the half-open intervals defined by $\mathcal{H}_1$ that are contained in G. For $r > 1$, let $\{I_r^p\}_{p \in J_r}$ be the family of the half-open intervals defined by $\mathcal{H}_r$ that are contained in G but are not contained in (and hence are disjoint from) any interval I_s^q with $q \in J_s$ and $s < r$. Clearly, $\{I_r^p : r \in \mathbb{N}, p \in J_r\}$ is a countable and disjoint family, and

$$\bigcup_{r=1}^{\infty} \left(\bigcup_{p \in J_r} I_r^p \right) \subset G.$$

The function

$$\rho : \mathbb{R}^n \times \mathbb{R}^n \to \mathbb{R}$$

$$((x_1, ..., x_n), (y_1, ..., y_n)) \mapsto \rho((x_1, ..., x_n), (y_1, ..., y_n))$$

$$:= \max\{|x_k - y_k| : k = 1, ..., n\}$$

is a distance on $\mathbb{R}^n$ and $\mathcal{T}_\rho = \mathcal{T}_{d_n}$ (cf. the proof of 6.1.31). For each $(x_1, ..., x_n) \in G$, since $G \in \mathcal{T}_\rho$ there exists $\varepsilon > 0$ so that, for $(y_1, ..., y_n) \in \mathbb{R}^n$,

$$[|x_k - y_k| < \varepsilon \text{ for } k = 1, ..., n] \Rightarrow (y_1, ..., y_n) \in G;$$

then, if $r \in \mathbb{N}$ is so that $\frac{1}{2^r} < \varepsilon$, the half-open interval defined by $\mathcal{H}_r$ that contains $(x_1, ..., x_n)$ must be contained in G, and therefore either this half-open interval is contained in an interval I_s^q with $q \in J_s$ and $s < r$ or this half-open interval itself is an interval I_r^p with $p \in J_r$. This shows that each point of G is contained in an element of the family $\{I_r^p : r \in \mathbb{N}, p \in J_r\}$, and hence that

$$G \subset \bigcup_{r=1}^{\infty} \left(\bigcup_{p \in J_r} I_r^p \right).$$

$\square$

9.1.6 Theorem. *For $n \in \mathbb{N}$, the Lebesgue measure m_n is the only measure on $\mathcal{A}(d_n)$ with the property: for all $(a_1, ..., a_n), (b_1, ..., b_n) \in \mathbb{R}^n$ so that $a_k < b_k$ for $k = 1, ..., n$,*

$$m_n((a_1, b_1] \times \cdots (a_n, b_n]) = (b_1 - a_1) \cdots (b_n - a_n).$$

The measure m_n is also the only measure on $\mathcal{A}(d_n)$ with the property: for all $(a_1, ..., a_n), (b_1, ..., b_n) \in \mathbb{R}^n$ so that $a_k < b_k$ for $k = 1, ..., n$,

$$m_n((a_1, b_1) \times \cdots (a_n, b_n)) = (b_1 - a_1) \cdots (b_n - a_n).$$

Proof. Clearly, m_n has the two properties of the statement since $m((a, b]) = m((a, b)) = b - a$ for all $a, b \in \mathbb{R}$ so that $a < b$.

Let now μ be a measure on $\mathcal{A}(d_n)$ such that, for all $(a_1, ..., a_n), (b_1, ..., b_n) \in \mathbb{R}^n$ so that $a_k < b_k$ for $k = 1, ..., n$,

$$\mu((a_1, b_1] \times \cdots (a_n, b_n]) = (b_1 - a_1) \cdots (b_n - a_n).$$

Then, 9.1.5 and the σ-additivity of μ and of m_n imply that

$$\mu(G) = m_n(G), \forall G \in \mathcal{T}_{d_n}.$$

For $N \in \mathbb{N}$, let Q_N be the open cube

$$Q_N := \{(x_1, ..., x_n) \in \mathbb{R}^n : -N < x_k < N \text{ for } k = 1, ..., N\}.$$

Then μ_{Q_N} and $(m_n)_{Q_N}$ are finite measures (this is proved for example by the inclusion $Q_n \subset (-N, N] \times \cdots n \text{ times } \cdots \times (-N, N]$ and by the monotonicity of μ and m_n) on the σ-algebra $\mathcal{A}(d_n)^{Q_N} = \mathcal{A}((d_n)_{Q_N})$ (cf. 6.1.21) and

$$\mu_{Q_N}(G \cap Q_N) = \mu(G \cap Q_N) = m_n(G \cap Q_N) = (m_n)_{Q_N}(G \cap Q_N), \forall G \in \mathcal{T}_{d_n},$$

and hence (cf. 2.2.5)

$$\mu_{Q_N}(G) = (m_n)_{Q_N}(G), \forall G \in \mathcal{T}_{(d_n)_{Q_N}}.$$

Then $\mu_{Q_N} = (m_n)_{Q_N}$ by 7.4.2. Thus, for each $E \in \mathcal{A}(d_n)$, from $E = \bigcup_{N=1}^{\infty}(E \cap Q_N)$ and 7.1.4b we have

$$\mu(E) = \lim_{N \to \infty} \mu_{Q_N}(E \cap Q_N) = \lim_{n \to \infty}(m_n)_{Q_N}(E \cap Q_N) = m_n(E),$$

and this proves that $\mu = m_n$.

If $\tilde{\mu}$ is a measure on $\mathcal{A}(d_n)$ such that, for all $(a_1, ..., a_n), (b_1, ..., b_n) \in \mathbb{R}^n$ so that $a_k < b_k$ for $k = 1, ..., n$,

$$\tilde{\mu}((a_1, b_1) \times \cdots (a_n, b_n)) = (b_1 - a_1) \cdots (b_n - a_n),$$

then, for all $(a_1, ..., a_n), (b_1, ..., b_n) \in \mathbb{R}^n$ so that $a_k < b_k$ for $k = 1, ..., n$, by 7.1.4c we have

$$\tilde{\mu}((a_1, b_1] \times \cdots \times (a_n, b_n]) = \tilde{\mu}\left(\bigcap_{k=1}^{\infty} \left(\left(a_1, b_1 + \frac{1}{k}\right) \times \cdots \times \left(a_n, b_n + \frac{1}{k}\right) \right) \right)$$

$$= \lim_{k \to \infty} \tilde{\mu}\left(\left(a_1, b_1 + \frac{1}{k}\right) \times \cdots \times \left(a_n, b_n + \frac{1}{k}\right) \right)$$

$$= \lim_{k \to \infty} \left(b_1 + \frac{1}{k} - a_1\right) \times \cdots \times \left(b_n + \frac{1}{k} - a_n\right)$$

$$= (b_1 - a_1) \cdots (b_n - a_n).$$

In view of what was proved before, this entails $\mu = m_n$. $\qquad\square$

9.2 Invariance properties of Lebesgue measure

The first result of this section is that Lebesgue measure is invariant under translation.

9.2.1 Theorem. *Let $n \in \mathbb{N}$ and $(c_1, ..., c_n) \in \mathbb{R}^n$. Then:*

(a) for each $E \in \mathcal{A}(d_n)$, if we define

$$E + (c_1, ..., c_n) := \{(x_1 + c_1, .., x_n + c_n) : (x_1, ..., x_n) \in E\}$$

then

$$E + (c_1, ..., c_n) \in \mathcal{A}(d_n) \text{ and } m_n(E + (c_1, ..., c_n)) = m_n(E);$$

(b) for each $\varphi \in \mathcal{L}^+(\mathbb{R}^n, \mathcal{A}(d_n), m_n)$ (or $\varphi \in \mathcal{L}^1(\mathbb{R}^n, \mathcal{A}(d_n), m_n)$), if we define

$$\varphi_c : D_\varphi + (c_1, ..., c_n) \to [0, \infty] \quad (or \ \mathbb{C})$$

$$x \mapsto \varphi_c(x_1, ..., x_n) := \varphi(x_1 - c_1, ..., x_n - c_n)$$

then $\varphi_c \in \mathcal{L}^+(\mathbb{R}^n, \mathcal{A}(d_n), m_n)$ (or $\varphi_c \in \mathcal{L}^1(\mathbb{R}^n, \mathcal{A}(d_n), m_n)$) and

$$\int_{\mathbb{R}^n} \varphi_c \, dm_n = \int_{\mathbb{R}^n} \varphi \, dm_n.$$

Proof. a: The function (which is called *translation* by $(-c_1, ..., -c_n)$)

$$\tau_c : \mathbb{R}^n \to \mathbb{R}^n$$

$$(x_1, ..., x_n) \mapsto \tau_c(x_1, ..., x_n) := (x_1 - c_1, ..., x_n - c_n)$$

is continuous, and hence (cf. 6.2.8) it is measurable. Then,

$$\forall E \in \mathcal{A}(d_\mathbb{R}), E + (c_1, ..., c_n) = \tau_c^{-1}(E) \in \mathcal{A}(d_n).$$

The function

$$\mu_c : \mathcal{A}(d_n) \to [0, \infty]$$

$$E \mapsto \mu_c(E) := m(\tau_c^{-1}(E))$$

is a measure on $\mathcal{A}(d_n)$ (cf. 8.3.11 with $\mu_1 := m_n$ and $\pi := \tau_c$) and we have, for all $(a_1, ..., a_n), (b_1, ..., b_n) \in \mathbb{R}^n$ so that $a_k < b_k$ for $k = 1, ..., n$,

$$\mu_c((a_1, b_1] \times \cdots \times (a_n, b_n]) = m_n((a_1 + c_1, b_1 + c_1] \times \cdots \times (a_n + c_n, b_n + c_n])$$
$$= (b_1 - a_1) \cdots (b_n - a_n).$$

Then, by the uniqueness asserted in 9.1.6, we have $\mu_c = \mu_n$, i.e.

$$\forall E \in \mathcal{A}(d_n), m_n(E + (c_1, ..., c_n)) = \mu_c(E) = m_n(E).$$

b: Since $\varphi_c = \varphi \circ \tau_c$ and $m_n(E) = m_n(\tau_c^{-1}(E))$ for each $E \in \mathcal{A}(d_\mathbb{R})$, the assertions of the statement follow from 8.3.11. $\qquad \square$

We now investigate the behaviour of Lebesgue measure under linear transformations. We need a lemma, which as a matter of fact already says how things are in the one-dimensional case.

9.2.2 Lemma. *Let $c \in \mathbb{R} - \{0\}$. Then:*

(a) for each $E \in \mathcal{A}(d_{\mathbb{R}})$, if we define

$$cE := \{cx : x \in E\}$$

then

$$cE \in \mathcal{A}(d_{\mathbb{R}}) \text{ and } m(cE) = |c|m(E);$$

(b) for each $\varphi \in \mathcal{L}^+(\mathbb{R}, \mathcal{A}(d_{\mathbb{R}}), m)$ (or $\varphi \in \mathcal{L}^1(\mathbb{R}, \mathcal{A}(d_{\mathbb{R}}), m)$), if we define

$$\varphi^c : cD_\varphi \to [0, \infty] \quad (or \ \mathbb{C})$$

$$x \mapsto \varphi^c(x) := \varphi\left(\frac{x}{c}\right)$$

then $\varphi^c \in \mathcal{L}^+(\mathbb{R}, \mathcal{A}(d_{\mathbb{R}}), m)$ (or $\varphi^c \in \mathcal{L}^1(\mathbb{R}, \mathcal{A}(d_{\mathbb{R}}), m)$) and

$$\int_{\mathbb{R}} \varphi^c dm = |c| \int_{\mathbb{R}} \varphi dm.$$

Proof. a: The function (which is called *dilatation* by $\frac{1}{c}$)

$$\delta_c : \mathbb{R} \to \mathbb{R}$$

$$x \mapsto \delta_c(x) := \frac{x}{c}$$

is continuous, and hence (cf. 6.2.8) it is $\mathcal{A}(d_{\mathbb{R}})$-measurable. Thus,

$$\forall E \in \mathcal{A}(d_{\mathbb{R}}), cE = \delta_c^{-1}(E) \in \mathcal{A}(d_{\mathbb{R}}).$$

The function

$$\mu^c : \mathcal{A}(d_{\mathbb{R}}) \to [0, \infty]$$

$$E \mapsto \mu^c(E) := m(\delta_c^{-1}(E))$$

is a measure on $\mathcal{A}(d_{\mathbb{R}})$ (cf. 8.3.11 with $\mu_1 := m$ and $\pi := \delta_c$) and for the measure $\frac{1}{|c|}\mu^c$ (cf. 7.1.8) we have, for all $a, b \in \mathbb{R}$ so that $a < b$,

$$\frac{1}{|c|}\mu^c((a, b)) = \frac{1}{|c|}|c|(b - a) = b - a,$$

since $m((ca, cb)) = c(b - a)$ if $c > 0$ and $m((cb, ca)) = (-c)(b - a)$ if $c < 0$. Then, by the uniqueness asserted in 9.1.3 we have $\frac{1}{|c|}\mu^c = m$, i.e.

$$\forall E \in \mathcal{A}(d_{\mathbb{R}}), m(cE) = \mu^c(E) = |c|m(E).$$

b: Since $\varphi^c = \varphi \circ \delta_c$ and $(|c|m)(E) = m(\delta_c^{-1}(E))$ for each $E \in \mathcal{A}(d_{\mathbb{R}})$, the assertions of the statement follow from 8.3.11 (and from 8.3.5b with $a := |c|$, $\mu := m$, ν the null measure). $\qquad\qquad\square$

9.2.3 Remark. We denote by $\mathrm{GL}(n, \mathbb{R})$ the family of all injective linear operators on the linear space $\mathbb{R}^n$. If $A, B \in \mathrm{GL}(n, \mathbb{R})$ then $AB \in \mathrm{GL}(n, \mathbb{R})$ (cf. 1.2.14B and 3.2.4), and if $A \in \mathrm{GL}(n, \mathbb{R})$ then $A^{-1} \in \mathrm{GL}(n, \mathbb{R})$ (cf. 3.2.6b). Our next result requires a few facts which are known from linear algebra (cf. e.g. Munkres, 1991, Chapter 1). Every $A \in \mathrm{GL}(n, \mathbb{R})$ determines a unique matrix $[A_{ik}]$ so that

$$A(x_1, ..., x_n) = (x_1', ..., x_n') \text{ with } x_i' = \sum_{k=1}^{n} A_{ik} x_k$$

$$\text{for } i = 1, ..., n, \forall (x_1, ..., x_n) \in \mathbb{R}^n;$$

from this it is clear that A is a continuous mapping; also, denoting by $\det A$ the determinant of the matrix $[A_{ik}]$, we have:

$$\det(AB) = \det A \det B, \forall A, B \in \mathrm{GL}(n, \mathbb{R});$$

$$\det A \neq 0 \text{ and } \det A^{-1} = (\det A)^{-1}, \forall A \in \mathrm{GL}(n, \mathbb{R}).$$

If $n = 1$, then $Ax = \alpha x$ with $\alpha \in \mathbb{R}$ and $\det A = \alpha$. Every $A \in \mathrm{GL}(n, \mathbb{R})$ can be obtained as the product of finitely many elements of $\mathrm{GL}(n, \mathbb{R})$ of the following three types: for all $(x_1, ..., x_n) \in \mathbb{R}^n$,

- $A_1(x_1, ..., x_n) := (x_1, ..., x_{i-1}, cx_i, x_{i+1}, ..., x_n)$, with $c \in \mathbb{R} - \{0\}$,
- $A_2(x_1, ..., x_n) := (x_1, ..., x_{i-1}, x_i + cx_k, x_{i+1}, ..., x_n)$, with $c \in \mathbb{R}$ and $i \neq k$,
- $A_3(x_1, ..., x_{i-1}, x_i, x_{i+1}, ..., x_{k-1}, x_k, x_{k+1}, ...x_n)$
  $:= (x_1, ..., x_{i-1}, x_k, x_{i+1}, ..., x_{k-1}, x_i, x_{k+1}, ..., x_n).$

It is easy to see that $\det A_1 = c$, $\det A_2 = 1$, $\det A_3 = -1$.

9.2.4 Theorem. *Let $n \in \mathbb{N}$ and $A \in \mathrm{GL}(n, \mathbb{R})$. Then:*

(a) for each $E \in \mathcal{A}(d_n)$, if we define

$$A(E) := \{A(x_1, ..., x_n) : (x_1, ..., x_n) \in E\}$$

then

$$A(E) \in \mathcal{A}(d_n) \text{ and } m_n(A(E)) = |\det A| m_n(E);$$

(b) for each $\varphi \in \mathcal{L}^+(\mathbb{R}^n, \mathcal{A}(d_n), m_n)$ (or $\varphi \in \mathcal{L}^1(\mathbb{R}^n, \mathcal{A}(d_n), m_n)$) we have

$$\varphi \circ A^{-1} \in \mathcal{L}^+(\mathbb{R}^n, \mathcal{A}(d_n), m_n) \quad (\text{or } \varphi \circ A^{-1} \in \mathcal{L}^1(\mathbb{R}^n, \mathcal{A}(d_n), m_n))$$

and

$$\int_{\mathbb{R}^n} (\varphi \circ A^{-1}) dm_n = |\det A| \int_{\mathbb{R}^n} \varphi \, dm_n.$$

Proof. a: For each $\psi \in \mathcal{L}^+(\mathbb{R}^n, \mathcal{A}(d_n))$ and each $T \in \mathrm{GL}(n, \mathbb{R})$, the continuity of T implies that $\psi \circ T \in \mathcal{L}^+(\mathbb{R}^n, \mathcal{A}(d_n))$ (cf. 6.2.8 and 6.2.5).
 For each $\psi \in \mathcal{L}^+(\mathbb{R}^n, \mathcal{A}(d_n))$ we have:

for $c \in \mathbb{R} - \{0\}$, assuming $i < n$ (otherwise, it is easy to simplify the calculation below),

$$|c| \int_{\mathbb{R}^n} \psi(x_1, ..., x_{i-1}, cx_i, x_{i+1}, ..., x_n) dm_n(x_1, ..., x_n)$$

$$= |c| \int_{\mathbb{R}^{n-i}} \left(\int_{\mathbb{R}} \left(\int_{\mathbb{R}^{i-1}} \psi(x_1, ..., x_{i-1}, cx_i, x_{i+1}, ..., x_n) dm_{i-1}(x_1, ..., x_{i-1}) \right) dm(x_i) \right) dm_{n-i}(x_{i+1}, ..., x_n)$$

$$= |c| \int_{\mathbb{R}^{n-i}} \left(\int_{\mathbb{R}^{i-1}} \left(\int_{\mathbb{R}} \psi(x_1, ..., x_{i-1}, cx_i, x_{i+1}, ..., x_n) dm(x_i) \right) dm_{i-1}(x_1, ..., x_{i-1}) \right) dm_{n-i}(x_{i+1}, ..., x_n)$$

$$= \int_{\mathbb{R}^{n-i}} \left(\int_{\mathbb{R}^{i-1}} \left(\int_{\mathbb{R}} \psi(x_1, ..., x_n) dm(x_i) \right) dm_{i-1}(x_1, ..., x_{i-1}) \right) dm_{n-i}(x_{i+1}, ..., x_n)$$

$$= \int_{\mathbb{R}^n} \psi(x_1, ..., x_n) dm_n(x_1, ..., x_n),$$

where 8.4.11a, 8.4.8 and 9.2.2 (with c replaced by $\frac{1}{c}$) have been used;

for $c \in \mathbb{R}$ and $i, k = 1, ..., n$ so that $i \neq k$, assuming $i < n$ (otherwise, it is easy to simplify the calculation below),

$$\int_{\mathbb{R}^n} \psi(x_1, ..., x_{i-1}, x_i + cx_k, x_{i+1}, ..., x_n) dm_n(x_1, ..., x_n)$$

$$= \int_{\mathbb{R}^{n-i}} \left(\int_{\mathbb{R}} \left(\int_{\mathbb{R}^{i-1}} \psi(x_1, ..., x_{i-1}, x_i + cx_k, x_{i+1}, ..., x_n) dm_{i-1}(x_1, ..., x_{i-1}) \right) dm(x_i) \right) dm_{n-i}(x_{i+1}, ..., x_n)$$

$$= \int_{\mathbb{R}^{n-i}} \left(\int_{\mathbb{R}^{i-1}} \left(\int_{\mathbb{R}} \psi(x_1, ..., x_{i-1}, x_i + cx_k, x_{i+1}, ..., x_n) dm(x_i) \right) dm_{i-1}(x_1, ..., x_{i-1}) \right) dm_{n-i}(x_{i+1}, ..., x_n)$$

$$= \int_{\mathbb{R}^{n-i}} \left(\int_{\mathbb{R}^{i-1}} \left(\int_{\mathbb{R}} \psi(x_1, ..., x_n) dm(x_i) \right) dm_{i-1}(x_1, ..., x_{i-1}) \right) dm_{n-i}(x_{i+1}, ..., x_n)$$

$$= \int_{\mathbb{R}^n} \psi(x_1, ..., x_n) dm_n(x_1, ..., x_n),$$

where 8.4.11a, 8.4.8 and 9.2.1 (with $n = 1$) have been used;

for $i, k = 1, ..., n$ so that $i < k$,

$$\int_{\mathbb{R}^n} \psi(x_1, ..., x_{i-1}, x_k, x_{i+1}, ..., x_{k-1}, x_i, x_{k+1}, ..., x_n) dm_n(x_1, ..., x_n)$$

$$= \int_{\mathbb{R}^n} \psi(x_1, ..., x_n) dm_n(x_1, ..., x_n)$$

(this was proved in 8.4.11b).

Thus, for each $\psi \in \mathcal{L}^+(\mathbb{R}^n, \mathcal{A}(d_n))$ and for the three elements A_1, A_2, A_3 of $\mathrm{GL}(n, \mathbb{R})$ introduced in 9.2.3, we have

$$|\det A_k| \int_{\mathbb{R}^n} (\psi \circ A_k) dm_n = \int_{\mathbb{R}^n} \psi dm_n.$$

We notice now that, if T_1 and T_2 are elements of $\mathrm{GL}(n, \mathbb{R})$ so that

$$|\det T_k| \int_{\mathbb{R}^n} (\psi \circ T_k) dm_n = \int_{\mathbb{R}^n} \psi dm_n, \forall \psi \in \mathcal{L}^+(\mathbb{R}^n, \mathcal{A}(d_n)), \text{ for } k = 1, 2,$$

then

$$|\det(T_1 T_2)| \int_{\mathbb{R}^n} (\psi \circ (T_1 T_2)) dm_n = |\det T_1||\det T_2| \int_{\mathbb{R}^n} ((\psi \circ T_1) \circ T_2) dm_n$$

$$= |\det T_1| \int_{\mathbb{R}^n} (\psi \circ T_1) dm_n$$

$$= \int_{\mathbb{R}^n} \psi dm_n, \forall \psi \in \mathcal{L}^+(\mathbb{R}^n, \mathcal{A}(d_n)).$$

On account of the result of linear algebra mentioned in 9.2.3, this proves that, for each $\psi \in \mathcal{L}^+(\mathbb{R}^n, \mathcal{A}(d_n))$ and each $T \in \mathrm{GL}(n, \mathbb{R})$,

$$|\det T| \int_{\mathbb{R}^n} (\psi \circ T) dm_n = \int_{\mathbb{R}^n} \psi dm_n.$$

Let now $E \in \mathcal{A}(d_n)$. Then $A(E) = (A^{-1})^{-1}(E)$. Since A^{-1} is continuous, it is measurable (cf. 6.2.8) and hence $A(E) \in \mathcal{A}(d_n)$. Moreover,

$$(x_1, ..., x_n) \in A(E) \Leftrightarrow A^{-1}(x_1, ..., x_n) \in E$$

proves that $\chi_{A(E)} = \chi_E \circ A^{-1}$ and hence (using the result obtained above) that

$$m_n(A(E)) = \int_{\mathbb{R}^n} \chi_{A(E)} dm_n = \int_{\mathbb{R}^n} (\chi_E \circ A^{-1}) dm_n$$

$$= \frac{1}{|\det A^{-1}|} \int_{\mathbb{R}^n} \chi_E dm_n = |\det A| m_n(E).$$

b: If in 8.3.11 we assume $\mu_1 := m_n$ and $\pi := A^{-1}$, we have $\mu_2 = |\det A| m_n$ by the result of part a. Then, the assertion of the statement follow from 8.3.11 (and from 8.3.5b with $a := |\det A|$, $\mu := m_n$, ν the null measure). $\qquad \square$

9.3 The Lebesgue integral as an extension of the Riemann integral

In this section we see how the Riemann integral can be subsumed in the Lebesgue integral, when this is defined with respect to the Lebesgue measure on a bounded interval.

9.3.1 Definition. Let $a, b \in \mathbb{R}$ be so that $a < b$. We call *Lebesgue measure on* $[a, b]$ the restriction $m_{[a,b]}$ of m to the σ-algebra $\mathcal{A}_{[a,b]} := (\mathcal{A}(d_\mathbb{R}))^{[a,b]}$ (cf. 8.3.1).

9.3.2 Definitions. Let $a, b \in \mathbb{R}$ be so that $a < b$. By a partition of $[a, b]$ we mean here a family $P := \{t_0, t_1, ..., t_N\}$, where $N \in \mathbb{N}$, $t_n \in [a, b]$ for $n = 1, ..., N$ and $a = t_0 < t_1 < ... < t_n < t_{n+1} < ... < t_N = b$. We denote by $\mathcal{P}$ the family of all partitions of $[a, b]$. Let $\varphi : [a, b] \to \mathbb{R}$ be a bounded function. For each partition $P := \{t_0, t_1, ..., t_N\}$ we define

$$s_P(\varphi) := \sum_{i=1}^{N} m_i(\varphi)(t_i - t_{i-1}) \text{ and } S_P(\varphi) := \sum_{n=1}^{N} M_i(\varphi)(t_i - t_{i-1}),$$

where

$$m_i(\varphi) := \inf\{\varphi(t) : t \in [t_{i-1}, t_i]\} \text{ and } M_i(\varphi) := \sup\{\varphi(t) : t \in [t_{i-1}, t_i]\}$$

(notice that $m_i(\varphi)$ and $M_i(\varphi)$ are elements of $\mathbb{R}$ since φ is bounded). The function φ is said to be *Riemann-integrable* if

$$\sup\{s_P(\varphi) : P \in \mathcal{P}\} = \inf\{S_P(\varphi) : P \in \mathcal{P}\}.$$

If φ is Riemann-integrable, then the *Riemann integral* of φ is defined by

$$\int_a^b \varphi(x)dx := \sup\{s_P(\varphi) : P \in \mathcal{P}\}$$

(notice that $\int_a^b \varphi(x)dx$ is an element of $\mathbb{R}$ since φ is bounded).

A bounded function $\varphi : [a, b] \to \mathbb{C}$ is said to be *Riemann integrable* if $\operatorname{Re}\varphi$ and $\operatorname{Im}\varphi$ (which are bounded functions) are Riemann integrable. If φ is Riemann integrable, then the *Riemann integral* of φ is defined by

$$\int_a^b \varphi(x)dx := \int_a^b (\operatorname{Re}\varphi)(x)dx + i \int_a^b (\operatorname{Im}\varphi)(x)dx.$$

9.3.3 Theorem. *Let $a, b \in \mathbb{R}$ be such that $a < b$, and let $\varphi : [a, b] \to \mathbb{C}$ be a bounded function which is $\mathcal{A}_{[a,b]}$-measurable. Then $\varphi \in \mathcal{L}^1([a, b], \mathcal{A}_{[a,b]}, m_{[a,b]})$. If φ is Riemann-integrable, then*

$$\int_{[a,b]} \varphi \, dm_{[a,b]} = \int_a^b \varphi(x)dx.$$

Proof. Since φ is bounded and $m_{[a,b]}([a, b]) = b - a < \infty$, $\varphi \in \mathcal{L}^1([a, b], \mathcal{A}_{[a,b]}, m_{[a,b]})$ by 8.2.6. Then $\operatorname{Re}\varphi, \operatorname{Im}\varphi \in \mathcal{L}^1([a, b], \mathcal{A}_{[a,b]}, m_{[a,b]})$ (cf. 8.2.3).

Suppose now that φ is Riemann integrable, denote by $\tilde{\varphi}$ either $\operatorname{Re}\varphi$ or $\operatorname{Im}\varphi$, and let $I := \int_a^b \tilde{\varphi}(x)dx$. Then, using the symbols introduced in 9.3.2,

$$\forall n \in \mathbb{N}, \exists P_n \in \mathcal{P} \text{ s.t. } I - \frac{1}{n} < s_{P_n}(\tilde{\varphi}).$$

Thus, $I = \lim_{n \to \infty} s_{P_n}(\tilde{\varphi})$. Define now a sequence $\{P_n'\}$ in $\mathcal{P}$ by letting

$$P_1' := P_1 \text{ and } P_{n+1}' := P_{n+1} \cup P_n' \text{ for } n \in \mathbb{N},$$

where $P_{n+1} \cup P_n'$ denotes the partition of $[a, b]$ that is obtained by reordering the union of the families P_{n+1} and P_n'. For each $n \in \mathbb{N}$, write

$$\{t_0^n, t_1^n, ..., t_{N_n}^n\} := P_n' \text{ and } m_i^n(\tilde{\varphi}) := \inf\{\tilde{\varphi}(t) : t \in [t_{i-1}^n, t_i^n]\} \text{ for } i = 1, ..., N_n,$$

and define $\psi_n := \sum_{n=1}^{N_n} m_i^n(\tilde{\varphi})\chi_{(t_{i-1}^n, t_i^n]}$. Then, for each $n \in \mathbb{N}$:

(a) $\psi_n \leq \psi_{n+1}$ since P'_{n+1} is a refinement of P'_n, and $\psi_n \leq \tilde{\varphi}$;

(b) $s_{P_n}(\tilde{\varphi}) \leq s_{P'_n}(\tilde{\varphi})$ since P'_n is a refinement of P_n;

(c) ψ_n is obviously $\mathcal{A}_{[a,b]}$-measurable, $\psi_n \in \mathcal{L}^1([a,b], \mathcal{A}_{[a,b]}, m_{[a,b]})$ since ψ_n is bounded, and $\int_{[a,b]} \psi_n dm_{[a,b]} = s_{P'_n}(\tilde{\varphi})$.

From (a) and from 5.2.4 it follows that we can define the function

$$\psi : [a,b] \to \mathbb{R}$$
$$x \mapsto \psi(x) := \lim_{n \to \infty} \psi_n(x)$$

and that $\psi \leq \tilde{\varphi}$. By 6.2.20c, ψ is $\mathcal{A}_{[a,b]}$-measurable. From (b) it follows that

$$s_{P'_n}(\tilde{\varphi}) \to I \text{ as } n \to \infty,$$

and this and (c) imply that

$$I = \lim_{n \to \infty} \int_{[a,b]} \psi_n dm_{[a,b]}.$$

Then we have $\psi \in \mathcal{L}^1([a,b], \mathcal{A}_{[a,b]}, m_{[a,b]})$ and

$$\int_{[a,b]} \psi dm_{[a,b]} = \lim_{n \to \infty} \int_{[a,b]} \psi_n dm_{[a,b]} = I$$

by 8.2.11 (with dominating function any constant function which majorizes $|\varphi|$).

We can show in a similar way that there exists $\chi \in \mathcal{L}^1([a,b], \mathcal{A}_{[a,b]}, m_{[a,b]})$ such that

$$\tilde{\varphi} \leq \chi \text{ and } \int_{[a,b]} \chi dm_{[a,b]} = I.$$

Then we have

$$\int_{[a,b]} (\chi - \psi) dm_{[a,b]} = \int_{[a,b]} \chi dm_{[a,b]} - \int_{[a,b]} \psi dm_{[a,b]} = 0.$$

Since $0 \leq \chi - \psi$, by 8.1.18a we have $\psi(x) = \chi(x)$ $m_{[a,b]}$-a.e. on $[a,b]$, and hence $\psi(x) = \tilde{\varphi}(x)$ $m_{[a,b]}$-a.e on $[a,b]$ since $\psi \leq \tilde{\varphi} \leq \chi$. From this we obtain, by 8.2.7,

$$\int_{[a,b]} \tilde{\varphi} dm_{[a,b]} = \int_{[a,b]} \psi dm_{[a,b]} = I = \int_a^b \tilde{\varphi}(x) dx.$$

Thus we have (cf. 8.2.3)

$$\int_{[a,b]} \varphi dm_{[a,b]} = \int_{[a,b]} \text{Re}\,\varphi dm_{[a,b]} + i \int_{[a,b]} \text{Im}\,\varphi dm_{[a,b]}$$
$$= \int_a^b (\text{Re}\,\varphi)(x) dx + i \int_a^b (\text{Im}\,\varphi)(x) dx = \int_a^b \varphi(x) dx.$$

$\square$

Chapter 10

Hilbert Spaces

In this chapter we study inner product spaces, and Hilbert spaces in particular, which we only consider over the complex field $\mathbb{C}$. While linear operators in Hilbert spaces are studied in later chapters in connection with the concept of adjoint operator, we present here what more can be said about the concepts previously introduced for linear operators when the linear spaces in which they are defined are actually inner product or Hilbert spaces.

10.1 Inner product spaces

10.1.1 Definition. Let (X, σ, μ) be a linear space over $\mathbb{C}$. A *sesquilinear form in* X is a function ψ from $X \times X$ to $\mathbb{C}$, i.e. $\psi : D_\psi \to \mathbb{C}$ with $D_\psi \subset X \times X$, which has the following properties:

(sf_1) there exists a linear manifold M_ψ in X so that $D_\psi = M_\psi \times M_\psi$;
(sf_2) $\psi(f, \alpha g_1 + \beta g_2) = \alpha\psi(f, g_1) + \beta\psi(f, g_2), \forall \alpha, \beta \in \mathbb{C}, \forall f, g_1, g_2 \in M_\psi$;
(sf_3) $\psi(\alpha f_1 + \beta f_2, g) = \overline{\alpha}\psi(f_1, g) + \overline{\beta}\psi(f_2, g), \forall \alpha, \beta \in \mathbb{C}, \forall f_1, f_2, g \in M_\psi$

($\overline{\alpha}$ denotes the complex conjugate of a complex number α). We point out that conditions sf_2 and sf_3 are consistent only when condition sf_1 is assumed.

A sesquilinear form ψ is said to be *on* X if $M_\psi = X$.

10.1.2 Proposition. *Let X be a linear space over $\mathbb{C}$ and ψ a sesquilinear form in X. Then:*

(a) $\psi(f, g) = \sum_{n=1}^{4} \frac{1}{4i^n}\psi(f + i^n g, f + i^n g), \forall f, g \in M_\psi$
 (i denotes the complex number $(0, 1)$); this is called the polarization identity;
(b) $\psi(f, 0_X) = \psi(0_X, f) = 0, \forall f \in X.$

Proof. a: Conditions sf_2 and sf_3 imply that, for all $f, g \in M_\psi$,

$$\psi(f + i^n g, f + i^n g) = \psi(f, f) + i^n\psi(f, g) + (-i)^n\psi(g, f) + \psi(g, g).$$

247

Then, it is enough to note that

$$\sum_{n=1}^{4} \frac{1}{4i^n} i^n = 1 \text{ and } \sum_{n=1}^{4} \frac{1}{4i^n} = \sum_{n=1}^{4} \frac{1}{4i^n}(-i)^n = 0.$$

b: For every $f \in M_\psi$ we have

$$\psi(f, 0_X) = \psi(f, 0f) = 0\psi(f, f) = 0,$$

and similarly for $\psi(0_X, f) = 0$.   □

10.1.3 Definition. An *inner product space* is a quadruple (X, σ, μ, ϕ), where (X, σ, μ) is a linear space over $\mathbb{C}$ and ϕ is a function $\phi : X \times X \to \mathbb{C}$ which, with the shorthand notation $(f|g) := \phi(f, g)$, has the following properties:

(ip_1) $(f|\alpha g_1 + \beta g_2) = \alpha (f|g_1) + \beta (f|g_2), \forall \alpha, \beta \in \mathbb{C}, \forall f, g_1, g_2 \in X$;
(ip_2) $(f|g) = \overline{(g|f)}; \forall f, g \in X$;
(ip_3) $0 \le (f|f), \forall f \in X$;
(ip_4) $(f|f) = 0 \Rightarrow f = 0_X$.

The function ϕ is called an *inner product* for the linear space (X, σ, μ). An inner product is also called a *scalar product*.

10.1.4 Remarks.

(a) It is immediately clear that, in every inner product space X, conditions ip_1 and ip_2 imply the following condition:

(ip_5) $(\alpha f_1 + \beta f_2|g) = \overline{\alpha} (f_1|g) + \overline{\beta} (f_2|g), \forall \alpha, \beta \in \mathbb{C}, \forall f_1, f_2, g \in X$.

Thus, an inner product for X is a sesquilinear form on X.

(b) The reader should be aware that some define an inner product with condition ip_1 replaced by condition

(ip_1') $(f|\alpha g_1 + \beta g_2) = \overline{\alpha} (f|g_1) + \overline{\beta} (f|g_2), \forall \alpha, \beta \in \mathbb{C}, \forall f, g_1, g_2 \in X$.

Then, condition ip_5 gets replaced by condition

(ip_5') $(\alpha f_1 + \beta f_2|g) = \alpha (f_1|g) + \beta (f_2|g), \forall \alpha, \beta \in \mathbb{C}, \forall f_1, f_2, g \in X$.

Of course, the two definitions of an inner product are fully equivalent. However, care must be taken not to mix formulae obtained on the basis of different definitions.

10.1.5 Examples.

(a) Let ℓ_f denote the family of all the sequences in $\mathbb{C}$ that have just a finite number of non-zero elements, i.e.

$$\ell_f := \{\{x_n\} \in \mathcal{F}(\mathbb{N}) : \exists N_{\{x_n\}} \in \mathbb{N} \text{ such that } n > N_{\{x_n\}} \Rightarrow x_n = 0\}.$$

Obviously, ℓ_f is a linear manifold in the linear space $\mathcal{F}(\mathbb{N})$ (cf. 3.1.10c), and therefore it is a linear space over $\mathbb{C}$ (cf. 3.1.3). It is immediately clear that the function

$$\phi : \ell_f \times \ell_f \to \mathbb{C}$$

$$(\{x_n\}, \{y_n\}) \mapsto \phi(\{x_n\}, \{y_n\}) := \sum_{n=1}^{\infty} \overline{x_n} y_n$$

(note that the series $\sum_{n=1}^{\infty} \overline{x_n} y_n$ is actually a finite sum) is an inner product for the linear space ℓ_f.

(b) For $a, b \in \mathbb{R}$ such that $a < b$, let $\mathcal{C}(a, b)$ be the linear space over $\mathbb{C}$ defined by the linear manifold $\mathcal{C}(a, b)$ in $\mathcal{F}([a, b])$ introduced in 3.1.10f.

For all $\varphi, \psi \in \mathcal{C}(a, b)$ we have $\overline{\varphi}\psi \in \mathcal{L}^1([a, b], (\mathcal{A}(d_{\mathbb{R}}))^{[a,b]}, m_{[a,b]})$, where $m_{[a,b]}$ is the Lebesgue measure on $[a, b]$ (cf. 6.2.8, 6.2.17, 6.2.16, 2.8.7, 2.8.14, 8.2.6). Thus, we can define the function

$$\phi : \mathcal{C}(a, b) \times \mathcal{C}(a, b) \to \mathbb{C}$$

$$(\varphi, \psi) \mapsto \phi(\varphi, \psi) := \int_{[a,b]} \overline{\varphi}\psi \, dm_{[a,b]}$$

and it is immediately clear that this function has properties ip_1, ip_2, ip_3 of 10.1.3. As to property ip_4, we note first that if $\varphi \in \mathcal{C}(a, b)$ is such that $\varphi(x) = 0$ m-a.e. on $[a, b]$ then $\varphi(x) = 0$ for all $x \in [a, b]$. Indeed, as can be easily seen, if for $\varphi \in \mathcal{C}(a, b)$ there exists $x_0 \in (a, b)$ so that $\varphi(x_0) \neq 0$, then there exists $\delta > 0$ so that $(x_0 - \delta, x_0 + \delta) \subset [a, b]$ and $\varphi(x) \neq 0$ for all $x \in (x_0 - \delta, x_0 + \delta)$, and hence it cannot be that $\varphi(x) = 0$ m-a.e. on $[a, b]$, since $m((x_0 - \delta, x_0 + \delta)) = 2\delta > 0$. Now, for $\varphi \in \mathcal{C}(a, b)$, $(\varphi|\varphi) = 0$ implies $\varphi(x) = 0$ m-a.e. on $[a, b]$ by 8.1.12a, and hence $\varphi = 0_{\mathcal{C}(a,b)}$. This shows that ϕ has property ip_4.

It is worth remarking that this example can be formulated without recourse to Lebesgue integration, but using Riemann integration instead. In fact, 9.3.3 implies that

$$\phi(\varphi, \psi) = \int_a^b \overline{\varphi}(x)\psi(x)dx, \forall \varphi, \psi \in \mathcal{C}(a, b),$$

since $\overline{\varphi}\psi \in \mathcal{C}(a, b)$ for all $\varphi, \psi \in \mathcal{C}(a, b)$ and the elements of $\mathcal{C}(a, b)$ are Riemann-integrable. Moreover, the argument presented above to prove property ip_4 of ϕ can be replaced with an argument suited to the definition of ϕ by means of Riemann integrals.

(c) The linear space $\mathcal{S}(\mathbb{R})$ (cf. 3.1.10h) is an inner product space, with the inner product defined by

$$\phi : \mathcal{S}(\mathbb{R}) \times \mathcal{S}(\mathbb{R}) \to \mathbb{C}$$

$$(\varphi, \psi) \mapsto \phi(\varphi, \psi) := \int_{\mathbb{R}} \overline{\varphi}\psi \, dm$$

(m denotes the Lebesgue measure on $\mathbb{R}$).

Indeed, once it is proved that $\overline{\varphi}\psi \in \mathcal{L}^1(\mathbb{R}, \mathcal{A}(d_\mathbb{R}), m)$ for all $\varphi, \psi \in \mathcal{S}(\mathbb{R})$, it will be immediate to see that conditions ip_1, ip_2, ip_3 of 10.1.3 are fulfilled, and condition ip_4 will be proved in the same way as in example b, by noting first that if $\varphi \in \mathcal{C}(\mathbb{R})$ is such that $\varphi(x) = 0$ m-a.e. on $\mathbb{R}$ then $\varphi(x) = 0$ for all $x \in \mathbb{R}$. Since $\overline{\varphi}\psi \in \mathcal{S}(\mathbb{R})$ for all $\varphi\psi \in \mathcal{S}(\mathbb{R})$ (this follows from 2 and 6 of 3.1.10h), it remains to prove the inclusion $\mathcal{S}(\mathbb{R}) \subset \mathcal{L}^1(\mathbb{R}, \mathcal{A}(d_\mathbb{R}), m)$.

As a preliminary step, we note that

$$\int_\mathbb{R} \frac{1}{1+x^2} dm(x) < \infty. \tag{$*$}$$

Indeed, since

$$\chi_{[-n,n]} \leq \chi_{[-n-1,n+1]} \text{ and}$$

$$\int_\mathbb{R} \chi_{[-n,n]}(x) \frac{1}{1+x^2} dm(x) = \int_{-n}^n \frac{1}{1+x^2} dx = 2\arctan n, \forall n \in \mathbb{N}$$

(cf. 9.3.3), and since

$$\frac{1}{1+x^2} = \lim_{n\to\infty} \chi_{[-n,n]}(x) \frac{1}{1+x^2}, \forall x \in \mathbb{R},$$

8.1.8 implies that

$$\int_\mathbb{R} \frac{1}{1+x^2} dm(x) = \lim_{n\to\infty} 2\arctan n = \pi.$$

Now let $\varphi \in \mathcal{S}(\mathbb{R})$. We have $\varphi \in \mathcal{M}(\mathbb{R}, \mathcal{A}(d_\mathbb{R}), m)$ by 6.2.8. Furthermore,

$$\int_\mathbb{R} |\varphi| dm = \int_\mathbb{R} \frac{1}{1+x^2}(|\varphi(x)| + x^2|\varphi(x)|)dm(x)$$

$$\leq (\sup\{|\varphi(x)| : x \in \mathbb{R}\} + \sup\{x^2|\varphi(x)| : x \in \mathbb{R}\}) \int_\mathbb{R} \frac{1}{1+x^2} dm(x) < \infty,$$

where the first inequality holds by 8.1.7 and the second inequality holds by $(*)$ because

$$\sup\{|\varphi(x)| : x \in \mathbb{R}\} < \infty \text{ and } \sup\{x^2|\varphi(x)| : x \in \mathbb{R}\} < \infty$$

(if a function defined on $\mathbb{R}$ is continuous and has finite limits as $x \to \pm\infty$, then it can be easily proved to be bounded; in any case, cf. 7 in 3.1.10h). In view of 8.2.4, this proves that $\varphi \in \mathcal{L}^1(\mathbb{R}, \mathcal{A}(d_\mathbb{R}), m)$.

We point out that this example can be presented using Riemann integration instead of Lebesgue integration. In fact, for every function $\varphi \in \mathcal{S}(\mathbb{R})$,

$$|\chi_{[-n,n]}(x)\varphi(x)| \leq |\varphi(x)|, \forall x \in \mathbb{R}, \forall n \in \mathbb{N},$$

and

$$\lim_{n\to\infty} \chi_{[-n,n]}(x)\varphi(x) = \varphi(x), \forall x \in \mathbb{R},$$

imply

$$\int_\mathbb{R} \varphi dm = \lim_{n\to\infty} \int_\mathbb{R} \chi_{[-n,n]}\varphi dm = \lim_{n\to\infty} \int_{[-n,n]} \varphi dm$$

by 8.2.11, and hence

$$\int_{\mathbb{R}} \varphi \, dm = \lim_{n \to \infty} \int_{-n}^{n} \varphi(x) \, dx$$

by 9.3.3. Thus,

$$\phi(\varphi, \psi) = \lim_{n \to \infty} \int_{-n}^{n} \overline{\varphi}(x) \psi(x) \, dx, \forall \varphi, \psi \in \mathcal{S}(\mathbb{R}).$$

10.1.6 Remark. Let (X, σ, μ, ϕ) be an inner product space and M a linear manifold in the linear space (X, σ, μ). It is immediate to see that $(M, \sigma_{M \times M}, \mu_{\mathbb{C} \times M}, \phi_{M \times M})$ is an inner product space, since $(M, \sigma_{M \times M}, \mu_{\mathbb{C} \times M})$ is a linear space over $\mathbb{C}$ (cf. 3.1.3) and conditions ip_1, ip_2, ip_3, ip_4 of 10.1.3 hold trivially if X is replaced by M.

10.1.7 Proposition. *Let f, g be two elements of an inner product space X. Then:*

(a) $|(f|g)| \leq \sqrt{(f|f)}\sqrt{(g|g)}$
 (by $\sqrt{(f|f)}$ we mean the non-negative square root of $(f|f)$, which is non-negative by ip_3); this is called the Schwarz inequality;
(b) *we have $|(f|g)| = \sqrt{(f|f)}\sqrt{(g|g)}$ iff the set $\{f, g\}$ is linearly dependent.*

Proof. As a preliminary step, we note that if $f \neq 0_X$ then $(f|f) \neq 0$, by property ip_4, and that, by properties ip_1 and ip_5,

$$\left(g - \frac{(f|g)}{(f|f)} f \Big| g - \frac{(f|g)}{(f|f)} f \right) = (g|g) - \frac{|(f|g)|^2}{(f|f)}. \tag{$*$}$$

 a: If $f = 0_X$ we have (cf. 10.1.2b)

$$|(f|g)| = 0 = 0\sqrt{(g|g)} = \sqrt{(f|f)}\sqrt{(g|g)}.$$

If $f \neq 0_X$, from $(*)$ we have, by properties ip_3,

$$0 \leq (g|g) - \frac{|(f|g)|^2}{(f|f)},$$

and hence

$$|(f|g)| \leq \sqrt{(f|f)}\sqrt{(g|g)}.$$

 b: If the set $\{f, g\}$ is linearly dependent, then there exist $\alpha, \beta \in \mathbb{C}$ so that $(\alpha, \beta) \neq (0, 0)$ and $\alpha f + \beta g = 0_X$; assuming for instance $\alpha \neq 0$, we have $f = -\frac{\beta}{\alpha} g$ and hence

$$|(f|g)| = \left| -\frac{\beta}{\alpha} \right| |(g|g)| = \left| -\frac{\beta}{\alpha} \right| (g|g) = \sqrt{\left(-\frac{\beta}{\alpha} g \Big| -\frac{\beta}{\alpha} g \right)} \sqrt{(g|g)} = \sqrt{(f|f)}\sqrt{(g|g)}.$$

If $|(f|g)| = \sqrt{(f|f)}\sqrt{(g|g)}$ and $f \neq 0_X$ ($f = 0_X$ would make the set $\{f, g\}$ linearly dependent in any case), from $(*)$ we have

$$\left(g - \frac{(f|g)}{(f|f)} f \Big| g - \frac{(f|g)}{(f|f)} f \right) = 0,$$

and hence

$$g - \frac{(f|g)}{(f|f)} f = 0_X,$$

which shows that the set $\{f, g\}$ is linearly dependent. $\qquad\square$

10.1.8 Theorem. *Let (X, σ, μ, ϕ) be an inner product space. The function*

$$\nu_\phi : X \to \mathbb{R}$$

$$f \mapsto \nu_\phi(f) := \|f\|_\phi := \sqrt{(f|f)}$$

is a norm for the linear space (X, σ, μ).

Proof. For ν_ϕ, condition no_2 of 4.1.1 follows from properties ip_1 and ip_5 of an inner product:

$$\|\alpha f\|_\phi = \sqrt{(\alpha f|\alpha f)} = |\alpha|\sqrt{(f|f)} = |\alpha|\|f\|_\phi, \forall \alpha \in \mathbb{C}, \forall f \in X.$$

Condition no_3 of 4.1.1 is actually the same as condition ip_4 of 10.1.3. It remains to verify condition no_1 of 4.1.1. Now, for all $f, g \in X$,

$$(f + g|f + g) = (f|f) + (g|g) + 2\,\mathrm{Re}\,(f|g),$$

and, in view of 10.1.7a,

$$\mathrm{Re}\,(f|g) \le |(f|g)| \le \sqrt{(f|f)}\sqrt{(g|g)}.$$

Thus we have

$$\|f + g\|_\phi^2 \le \|f\|_\phi^2 + \|g\|_\phi^2 + 2\|f\|_\phi\|g\|_\phi.$$

which implies

$$\|f + g\|_\phi \le \|f\|_\phi + \|g\|_\phi.$$

$\square$

10.1.9 Remark. Whenever we consider an inner product space as a normed space, we will refer to the norm defined in 10.1.8. Unless confusion can arise, we will drop the index ϕ in ν_ϕ and in $\|\ \|_\phi$. Thus, in an inner product space X, the Schwarz inequality is written simply as follows:

$$|(f|g)| \le \|f\|\|g\|, \forall f, g \in X.$$

10.1.10 Proposition. *Let X be an inner product space. Then:*

(a) for every linear operator $A \in \mathcal{O}(X)$,

$$(f|Ag) = \sum_{n=1}^{4} \frac{1}{4i^n}\,(f + i^n g|A(f + i^n g)), \forall f, g \in D_A,$$

$$(Af|g) = \sum_{n=1}^{4} \frac{1}{4i^n}\,(A(f + i^n g)|f + i^n g), \forall f, g \in D_A;$$

(b) $(f|g) = \sum_{n=1}^{4} \frac{1}{4i^n}\|f + i^n g\|^2, \forall f, g \in X.$

Proof. a: We note that the functions
$$D_A \times D_A \ni (f,g) \mapsto (f|Ag) \in \mathbb{C}$$
and
$$D_A \times D_A \ni (f,g) \mapsto (Af|g) \in \mathbb{C}$$
are sesquilinear forms in the linear space (X, σ, μ), and use 10.1.2a.

b: Set $A := \mathbb{1}_X$ in part a of the statement. $\qquad\square$

10.1.11 Proposition. *Let X_1 and X_2 be inner product spaces. For a linear operator $A \in \mathcal{O}(X_1, X_2)$, the following conditions are equivalent:*

(a) $(Af|Ag)_2 = (f|g)_1$, $\forall f, g \in D_A$;
(b) $\|Af\|_2 = \|f\|_1$, $\forall f \in D_A$

(we have indexed by 1 and 2 the inner products and the norms in X_1 and X_2 respectively, as we will do whenever a similar situation arises).

Proof. $a \Rightarrow b$: This follows immediately from the definition of ν_ϕ in 10.1.8.

$b \Rightarrow a$: If condition b holds true then, by 10.1.10b, we have for all $f, g \in D_A$

$$(Af|Ag)_2 = \sum_{n=1}^{4} \frac{1}{4i^n} \|Af + i^n Ag\|_2^2$$
$$= \sum_{n=1}^{4} \frac{1}{4i^n} \|A(f + i^n g)\|_2^2 = \sum_{n=1}^{4} \frac{1}{4i^n} \|f + i^n g\|_1^2 = (f|g)_1 .$$
$\qquad\square$

10.1.12 Proposition. *In an inner product space X we have*
$$\|f + g\|^2 + \|f - g\|^2 = 2\|f\|^2 + 2\|g\|^2, \forall f, g \in X;$$
this is called the parallelogram law.

Proof. A straightforward computation, starting from the definition of ν_ϕ (cf. 10.1.8). $\qquad\square$

10.1.13 Remarks.

(a) We saw in 10.1.12 that if a norm is derived from an inner product as in 10.1.8 then it satisfies the parallelogram law.

The converse is also true, namely if a norm ν for a linear space (X, σ, μ) over $\mathbb{C}$ is such that
$$\nu(f + g)^2 + \nu(f - g)^2 = 2\nu(f)^2 + 2\nu(g)^2, \forall f, g \in X, \qquad (*)$$
then there exists a unique inner product ϕ for (X, σ, μ) so that $\nu = \nu_\phi$. The idea of the proof is as follows. If an inner product ϕ does exist so that $\nu = \nu_\phi$, then
$$\phi(f,g) = \sum_{n=1}^{4} \frac{1}{4i^n} \nu(f + i^n g)^2, \forall f, g \in X,$$

must be true, in view of 10.1.10b. Thus, one is led to define the function

$$\phi : X \times X \to \mathbb{C}$$

$$(f,g) \mapsto \phi(f,g) := \sum_{n=1}^{4} \frac{1}{4i^n} \nu(f + i^n g)^2,$$

and to check that this function has properties ip_1, ip_2, ip_3, ip_4 of 10.1.3; in this check, properties no_1, no_2, no_3 of 4.1.1 and condition $(*)$ are used (cf. Weidmann, 1980, p.10–11). After that, one notes that, for every $f \in X$,

$$\nu_\phi(f)^2 = \phi(f,f) = \sum_{n=1}^{4} \frac{1}{4i^n} \nu(f + i^n f)^2$$

$$= \frac{1}{4} \left(\frac{1}{i} 2 + \frac{1}{-1} 0 + \frac{1}{-i} 2 + \frac{1}{1} 4 \right) \nu(f)^2 = \nu(f)^2.$$

We do not give the details of the aforementioned checks because we shall not use this result.

(b) There are norms which do not satisfy the parallelogram law and which therefore cannot be derived from any inner product. Such is e.g. the norm defined in 4.3.6a.

10.1.14 Proposition. *Let X and Y be inner product spaces and $A \in \mathcal{O}(X,Y)$. We suppose $D_A \neq \{0_X\}$ and $Y \neq \{0_Y\}$, and we set*

$$k := \sup \left\{ \frac{|(f|Ag)|}{\|f\|\|g\|} : f \in Y - \{0_Y\}, g \in D_A - \{0_X\} \right\}$$

(we have denoted by the same symbol the norms in X and in Y). Then,

$$k = \sup\{|(f|Ag)| : f \in Y, g \in D_A, \|f\| = \|g\| = 1\}.$$

The operator A is bounded iff $k < \infty$. If A is bounded then $\|A\| = k$.

Proof. The equality between the two least upper bounds of the statement is obvious.

Now assume A bounded. Then, by 10.1.7a and 4.2.5b,

$$|(f|Ag)| \leq \|f\|\|Ag\| \leq \|A\|\|f\|\|g\|, \forall f \in Y, \forall g \in D_A;$$

thus, $k \leq \|A\|$ and therefore $k < \infty$.

Conversely, assume $k < \infty$. Then,

$$\|Ah\| = \left(\frac{(Ah|Ah)}{\|Ah\|\|h\|} \right) \|h\| \leq k\|h\|, \forall h \in D_A - N_A,$$

and this implies

$$\|Af\| \leq k\|f\|, \forall f \in D_A;$$

thus, A is bounded, $k \in \mathcal{B}_A$ and therefore $\|A\| \leq k$ (cf. 4.2.4).

The statement follows from the two arguments above.   □

10.1.15 Remark. Let (X, σ, μ, ϕ) be an inner product space. From 10.1.8 and 4.1.3 we have that the function

$$d_\phi := d_{\nu_\phi} : X \times X \to \mathbb{R}$$
$$(f, g) \mapsto d_\phi(f, g) := \nu_\phi(f - g) = \sqrt{\phi(f - g, f - g)}$$

is a distance on X. Whenever we use metric concepts in an inner product space, we will refer to this distance. For instance, if we say that the inner product space X is complete or separable we mean that the metric space (X, d_ϕ) is such.

If M is a linear manifold in X, it is immediately clear that we obtain the same metric space by first defining the inner product space $(M, \sigma_{M \times M}, \mu_{\mathbb{C} \times M}, \phi_{M \times M})$ (cf. 10.1.6) and then the metric space $(M, d_{\phi_{M \times M}})$, or by first defining the metric space (X, d_ϕ) and then the metric subspace $(M, (d_\phi)_M)$ (cf. 2.1.3). Thus, there can be no ambiguity when we refer to M as a metric space.

10.1.16 Theorem. *Let (X, σ, μ, ϕ) be an inner product space. Then:*

(a) the mapping σ is continuous (with respect to $d_\phi \times d_\phi$ and d_ϕ);
(b) the mapping μ is continuous (with respect to $d_\mathbb{C} \times d_\phi$ and d_ϕ);
(c) the inner product ϕ is continuous (with respect to $d_\phi \times d_\phi$ and $d_\mathbb{C}$).

Proof. a, b: These follow from 4.1.6b,c.

c: This follows from the continuity of σ and μ, from the continuity of ν_ϕ (with respect to d_ϕ and $d_\mathbb{R}$, cf. 4.1.6a), from the continuity of sum and product in $\mathbb{C}$, and from 10.1.10b. $\qquad\square$

10.1.17 Definitions. Let $(X_1, \sigma_1, \mu_1, \phi_1)$ and $(X_2, \sigma_2, \mu_2, \phi_2)$ be inner product spaces. An *isomorphism* from X_1 onto X_2 is a mapping $U : X_1 \to X_2$ such that:

(is_1) U is a bijection from X_1 onto X_2;
(is_2) $\sigma_2(U(f), U(g)) = U(\sigma_1(f, g))$, $\forall f, g \in X_1$;
$\qquad \mu_2(\alpha, U(f)) = U(\mu_1(\alpha, f))$, $\forall \alpha \in \mathbb{C}$, $\forall f \in X_1$;
(is_3) $\phi_2(U(f), U(g)) = \phi_1(f, g)$, $\forall f, g \in X_1$.

If an isomorphism from X_1 onto X_2 exists, then the two inner product spaces X_1 and X_2 are said to be *isomorphic*.

If the two inner product spaces X_1 and X_2 are the same, an isomorphism from X_1 onto X_2 is called an *automorphism* of X_1.

10.1.18 Remark. In 10.1.17, condition is_1 means that U is an "isomorphism" from the set X_1 onto the set X_2 (U preserves the set theoretical "operations", i.e. union, intersection, complementation), conditions is_1 and is_2 mean that U is an isomorphism from the linear space (X_1, σ_1, μ_1) onto the linear space (X_2, σ_2, μ_2) (condition is_2 says that U is a linear operator), and condition is_3 says that U preserves the inner product. We formulated condition is_1 the way we did in order to make it clear from the definition that an isomorphism preserves the three levels of the structure of an inner product space. However, 10.1.19 proves that the injectivity part of condition is_1 and condition is_2 altogether are in fact redundant.

10.1.19 Theorem. *Let X_1 and X_2 be inner product spaces and U a mapping from X_1 to X_2 such that D_U is a linear manifold in X_1 and*

$$(U(f)|U(g))_2 = (f|g)_1, \forall f, g, \in D_U.$$

Then U is an injective linear operator.

Proof. For all $f, g \in D_U$ we have

$$\|U(f + g) - U(f) - U(g)\|_2^2$$
$$= \|U(f + g)\|_2^2 + \|U(f) + U(g)\|_2^2$$
$$-2 \operatorname{Re}(U(f + g)|U(f))_2 - 2 \operatorname{Re}(U(f + g)|U(g))_2.$$

We also have, by the condition assumed for U:

$$\|U(f + g)\|_2^2 = (U(f + g)|U(f + g))_2 = (f + g|f + g)_1 = \|f + g\|_1^2;$$
$$\|U(f) + U(g)\|_2^2 = \|U(f)\|_2^2 + \|U(g)\|_2^2 + 2 \operatorname{Re}(U(f)|U(g))_2$$
$$= \|f\|_1^2 + \|g\|_1^2 + 2 \operatorname{Re}(f|g)_1 = \|f + g\|_1^2;$$
$$\operatorname{Re}(U(f + g)|U(f))_2 + \operatorname{Re}(U(f + g)|U(g))_2$$
$$= \operatorname{Re}(f + g|f)_1 + \operatorname{Re}(f + g|g)_1 = \operatorname{Re}((f + g|f)_1 + (f + g|g)_1)$$
$$= \operatorname{Re}(f + g|f + g)_1 = (f + g|f + g)_1 = \|f + g\|_1^2.$$

This shows that

$$\|U(f + g) - U(f) - U(g)\|_2 = 0,$$

and therefore that

$$U(f + g) = U(f) + U(g).$$

For all $\alpha \in \mathbb{C}$ and $f \in D_U$ we have

$$\|U(\alpha f) - \alpha U(f)\|_2^2 = \|U(\alpha f)\|_2^2 + |\alpha|^2 \|U(f)\|_2^2 - 2 \operatorname{Re}(U(\alpha f)|\alpha U(f))_2.$$

We also have:

$$\|U(\alpha f)\|_2^2 = \|\alpha f\|_1^2 = |\alpha|^2 \|f\|_1^2;$$
$$\|U(f)\|_2^2 = \|f\|_1^2;$$
$$\operatorname{Re}(U(\alpha f)|\alpha U(f))_2 = \operatorname{Re}(\alpha (U(\alpha f)|U(f))_2)$$
$$= \operatorname{Re}(\alpha (\alpha f|f)_1) = \operatorname{Re}|\alpha|^2 (f|f)_1 = |\alpha|^2 \|f\|_1^2.$$

This shows that

$$\|U(\alpha f) - \alpha U(f)\|_2 = 0,$$

and therefore that

$$U(\alpha f) = \alpha U(f).$$

This proves that U is a linear operator.

Finally, we have

$$\|Uf\|_2 = \|f\|_1, \forall f \in D_U,$$

and hence

$$f \in N_U \Rightarrow \|f\|_1 = \|Uf\|_2 = 0 \Rightarrow f = 0_{X_1}.$$

By 3.2.6a, this proves that U is injective. $\square$

10.1.20 Theorem. *Let X_1 and X_2 be inner product spaces and U a mapping from X_1 to X_2. The following conditions are equivalent:*

(a) U is an isomorphism from X_1 onto X_2;

(b) $D_U = X_1$, $R_U = X_2$, U is a linear operator, and

$$\|Uf\|_2 = \|f\|_1, \forall f \in X_1;$$

(c) $D_U = X_1$, $R_U = X_2$, and

$$(U(f)|U(g))_2 = (f|g)_1, \forall f, g, \in X_1.$$

Proof. $a \Rightarrow b$: This is obvious.

$b \Rightarrow c$: This follows from 10.1.11.

$c \Rightarrow a$: Assuming condition c, U is an injective linear operator by 10.1.19. Then U has all the properties required by the definition of an isomorphism. $\square$

10.1.21 Remark. Let $(X_1, \sigma_1, \mu_1, \phi_1)$ and $(X_2, \sigma_2, \mu_2, \phi_2)$ be inner product spaces and U a mapping from X_1 to X_2. The equivalence of conditions a and b in 10.1.20 proves that U is an isomorphism from the inner product space X_1 onto the inner product space X_2 iff U is an isomorphism from the normed space $(X_1, \sigma_1, \mu_1, \nu_{\phi_1})$ onto the normed space $(X_2, \sigma_2, \mu_2, \nu_{\phi_2})$ (cf. 4.6.1 and 4.6.2a). Therefore, all the remarks made in 4.6.2 about isomorphisms of normed spaces apply also to isomorphisms of inner product spaces. In particular, if two inner product spaces X_1 and X_2 are isomorphic then so are the metric spaces (X_1, d_{ϕ_1}) and (X_2, d_{ϕ_2}), and hence X_1 is complete iff X_2 is complete (cf. 2.6.4) and X_1 is separable iff X_2 is separable (cf. 2.3.21c).

10.2 Orthogonality in inner product spaces

Throughout this section, X stands for an abstract inner product space.

10.2.1 Definitions. An element f of X is said to be *orthogonal to* an element g of X if $(f|g) = 0$. Property ip_2 of an inner product implies that if f is orthogonal to g then g is orthogonal to f.

A subset S of X is said to be *orthogonal* if $(f|g) = 0$ whenever f and g are different elements of S.

10.2.2 Proposition. *Let S be an orthogonal set of non-zero elements of X. Then S is linearly independent.*

Proof. Suppose that $\{f_1, ..., f_n\}$ is a subset of S and $(\alpha_1, ..., \alpha_n) \in \mathbb{C}^n$ is so that $\sum_{i=1}^{n} \alpha_i f_i = 0_X$. For $k = 1, ..., n$ we have

$$0 = \left(f_k \Big| \sum_{i=1}^{n} \alpha_i f_i \right) = \sum_{i=1}^{n} \alpha_i \, (f_k | f_i) = \alpha_k \| f_k \|^2;$$

since $f_k \neq 0_X$, this implies $\alpha_k = 0$. $\qquad\square$

10.2.3 Proposition. *Let $\{f_1, ..., f_n\}$ be an orthogonal subset of X. Then,*

$$\left\| \sum_{i=1}^{n} f_i \right\|^2 = \sum_{i=1}^{n} \| f_i \|^2.$$

This is called the Pythagorean theorem.

Proof. We have

$$\left\| \sum_{i=1}^{n} f_i \right\|^2 = \left(\sum_{i=1}^{n} f_i \Big| \sum_{k=1}^{n} f_k \right)$$

$$= \sum_{i=1}^{n} (f_i | f_i) + \sum_{i=1}^{n} \sum_{k \neq i}^{n} (f_i | f_k) = \sum_{i=1}^{n} \| f_i \|^2.$$

$\qquad\square$

10.2.4 Definition. An orthogonal subset S of X is called an *orthonormal system* (briefly, *o.n.s.*) if $\|u\| = 1$ for all $u \in S$. Thus, an indexed family $\{u_i\}_{i \in I}$ of elements of X is an o.n.s. iff $(u_i | u_k) = \delta_{i,k}$ for all $i, k \in I$ ($\delta_{i,k}$ denotes the Kronecker delta, i.e. $\delta_{i,i} = 1$ for all $i \in I$ and $\delta_{i,k} = 0$ for $i \neq k$). The reason for the "normal" part of the name "orthonormal system" is that an element f of X is said to be *normalized* if $\|f\| = 1$.

10.2.5 Examples.

(a) For each $k \in \mathbb{N}$, let δ_k be the element of the inner product space ℓ_f (cf. 10.1.5a) defined by

$$\delta_k := \{\delta_{k,n}\},$$

i.e. δ_k is the sequence whose elements are all zero but the k-th, which is one. The family $\{\delta_k\}_{k \in \mathbb{N}}$ is an o.n.s. in ℓ_f, since it is obvious that $(\delta_k | \delta_l) = \delta_{k,l}$ for all $k, l \in \mathbb{N}$.

(b) We define a family $\{u_n\}_{n \in \mathbb{Z}}$ of elements of the inner product space $\mathcal{C}(0, 2\pi)$ (cf. 10.1.5b) by

$$u_n(x) := \frac{1}{\sqrt{2\pi}} e^{inx}, \forall x \in [0, 2\pi], \forall n \in \mathbb{Z}.$$

The family $\{u_n\}_{n\in\mathbb{Z}}$ is an o.n.s. in $\mathcal{C}(0, 2\pi)$ since $(u_n|u_n) = 1$ is obvious and, for $n \neq m$,

$$(u_m|u_n) = \frac{1}{2\pi} \int_0^{2\pi} e^{i(n-m)x} dx$$

$$= \frac{1}{2\pi} \int_0^{2\pi} \cos(n-m)x dx + i\frac{1}{2\pi} \int_0^{2\pi} \sin(n-m)x dx = 0.$$

For each $n \in \mathbb{N}$ we define the elements v_n and w_n of $\mathcal{C}(0, 2\pi)$ by

$$v_n(x) := \frac{1}{\sqrt{\pi}} \cos nx \text{ and } w_n(x) := \frac{1}{\sqrt{\pi}} \sin nx, \forall x \in [0, 2\pi].$$

Since

$$v_n = \frac{1}{\sqrt{2}}(u_n + u_{-n}) \text{ and } w_n = \frac{1}{\sqrt{2}i}(u_n - u_{-n}), \forall n \in \mathbb{N},$$

a straightforward computation shows that the family $\{u_0\}\cup\{v_n\}_{n\in\mathbb{N}}\cup\{w_n\}_{n\in\mathbb{N}}$ is an o.n.s., and 3.1.7 implies that

$$L(\{u_0\} \cup \{v_n\}_{n\in\mathbb{N}} \cup \{w_n\}_{n\in\mathbb{N}}) \subset L\{u_n\}_{n\in\mathbb{Z}}.$$

However, since

$$u_n = \frac{1}{\sqrt{2}}(v_n + iw_n) \text{ and } u_{-n} = \frac{1}{\sqrt{2}}(v_n - iw_n), \forall n \in \mathbb{N},$$

3.1.7 implies also that

$$L\{u_n\}_{n\in\mathbb{Z}} \subset L(\{u_0\} \cup \{v_n\}_{n\in\mathbb{N}} \cup \{w_n\}_{n\in\mathbb{N}}).$$

Thus,

$$L\{u_n\}_{n\in\mathbb{Z}} = L(\{u_0\} \cup \{v_n\}_{n\in\mathbb{N}} \cup \{w_n\}_{n\in\mathbb{N}}).$$

10.2.6 Theorem (Gram–Schmidt orthonormalization). *Let $\{f_n\}_{n\in I}$ be a countable and linearly independent subset of X, and suppose $I := \{1, ..., N\}$ or $I := \mathbb{N}$. Then:*

(a) a family $\{u_n\}_{n\in I}$ can be consistently defined by induction as

$$u_1 := \|f_1\|^{-1}f_1,$$

$$u_n := \left\| f_n - \sum_{k=1}^{n-1} (u_k|f_n) u_k \right\|^{-1} \left(f_n - \sum_{k=1}^{n-1} (u_k|f_n) u_k \right)$$

for $n \in I$ such that $n > 1$,

i.e. $\|f_1\| \neq 0$ and $\left\| f_n - \sum_{k=1}^{n-1} (u_k|f_n) u_k \right\| \neq 0$;

(b) $\{u_n\}_{n\in I}$ is an o.n.s. in X;
(c) $L\{u_1, ..., u_n\} = L\{f_1, ..., f_n\}, \forall n \in I$;
(d) $L\{u_n\}_{n\in I} = L\{f_n\}_{n\in I}$;

(e) if $\{g_n\}_{n\in I}$ is an orthogonal set such that $g_n \neq 0_X$ and g_n is a linear combination
  of $f_1, ..., f_n$ for each $n \in I$, then for each $n \in I$ there exists $\alpha_n \in \mathbb{C}$ so that
  $g_n = \alpha_n u_n$.

Proof. a, b, c, d: We define, for each $n \in I$, the proposition

$$P_n := \begin{cases} \text{the family } \{u_1, ..., u_n\} \text{ can be consistently defined,} \\ \text{the family } \{u_1, ..., u_n\} \text{ is an o.n.s. in } X, \\ L\{u_1, ..., u_n\} = L\{f_1, ..., f_n\}, \end{cases}$$

and we prove by induction that proposition P_n is true for each $n \in I$ (if $I :=$ $\{1, ..., N\}$, to comply with the definition of a proof by induction given in 1.1.2 we can define P_n to be a trivially true proposition for all $n > N$).

Proposition P_1 is true because $\|f_1\| \neq 0$ (in fact, $f_1 \neq 0_X$ since $\{f_n\}_{n\in I}$ is a linearly independent set), because $\|u_1\| = 1$ holds trivially, because

$$L\{u_1\} = \{\alpha u_1 : \alpha \in \mathbb{C}\} \text{ and } L\{f_1\} = \{\alpha f_1 : \alpha \in \mathbb{C}\}$$

(cf. 3.1.7), and because $\{\alpha u_1 : \alpha \in \mathbb{C}\} = \{\alpha f_1 : \alpha \in \mathbb{C}\}$ is obvious.

Assume now that, for $n \in I$ such that $n > 1$, proposition P_{n-1} is true. Then, $\left\| f_n - \sum_{k=1}^{n-1} (u_k | f_n) u_k \right\| \neq 0$ (in fact, if $f_n - \sum_{k=1}^{n-1} (u_k | f_n) u_k = 0_X$ were true then by 3.1.7 and by proposition P_{n-1} we should have $f_n \in L\{u_1, ..., u_{n-1}\} = L\{f_1, ..., f_{n-1}\}$ and hence $\{f_n\}_{n\in I}$ would not be a linearly independent set); this and proposition P_{n-1} imply that the family $\{u_1, ..., u_n\}$ can be consistently defined. Furthermore, $\|u_n\| = 1$ holds trivially and, for $l = 1, ..., n-1$,

$$(u_l | u_n) = \left\| f_n - \sum_{k=1}^{n-1} (u_k | f_n) u_k \right\|^{-1} \left((u_l | f_n) - \sum_{k=1}^{n-1} (u_k | f_n)(u_l | u_k) \right)$$

$$= \left\| f_n - \sum_{k=1}^{n-1} (u_k | f_n) u_k \right\|^{-1} \left((u_l | f_n) - \sum_{k=1}^{n-1} (u_k | f_n) \delta_{l,k} \right) = 0;$$

these facts and proposition P_{n-1} imply that $\{u_1, ..., u_n\}$ is an o.n.s. in X. Finally, the definition of u_n, 3.1.7 and proposition P_{n-1} imply that

$$L\{u_1, ..., u_{n-1}, u_n\} \subset L\{u_1, ..., u_{n-1}, f_n\} \subset L\{f_1, ..., f_{n-1}, f_n\},$$

and also that

$$L\{f_1, ..., f_{n-1}, f_n\} \subset L\{f_1, ..., f_{n-1}, u_1, ..., u_{n-1}, u_n\} \subset L\{u_1, ..., u_{n-1}, u_n\}.$$

Thus, proposition P_n is true.

This proves by induction that proposition P_n is true for each $n \in I$. Then parts a, b and c are obviously true, and part d is true since 3.1.7 implies that

$$L\{u_n\}_{n\in I} = \bigcup_{n\in I} L\{u_1, ..., u_n\} \text{ and } L\{f_n\}_{n\in I} = \bigcup_{n\in I} L\{f_1, ..., f_n\}.$$

e: let $\{g_n\}_{n \in I}$ be as in the statement. Then, for each $n \in I$, we have

$$g_k \in L\{f_1, ..., f_k\} = L\{u_1, ..., u_k\} \subset L\{u_1, ..., u_n\} \text{ for } 1 \leq k \leq n;$$

since $\{u_1, ..., u_n\}$ is obviously a linear basis in the linear space $L\{u_1, ..., u_n\}$ (cf. 10.2.2), and since $\{g_1, ..., g_n\}$ is a linearly independent set (cf. 10.2.2), 3.1.15 implies that $\{g_1, ..., g_n\}$ is a linear basis in the linear space $L\{u_1, ..., u_n\}$; thus, we have

$$L\{g_1, ..., g_n\} = L\{u_1, ..., u_n\}.$$

Then, for each $k \in I$ there exists $(\alpha_{k,1}, ..., \alpha_{k,k}) \in \mathbb{C}^k$ so that

$$g_k = \sum_{i=1}^{k} \alpha_{k,i} u_i$$

and for each $l \in I$ there exists $(\beta_{l,1}, ..., \beta_{l,l}) \in \mathbb{C}^l$ so that

$$u_l = \sum_{j=1}^{l} \beta_{l,j} g_j.$$

We note that, for $k \in I$ and $l \leq k$,

$$(u_l | g_k) = \left(u_l \Big| \sum_{i=1}^{k} \alpha_{k,i} u_i \right) = \sum_{i=1}^{k} \alpha_{k,i} (u_l | u_i) = \sum_{i=1}^{k} \alpha_{k,i} \delta_{l,i} = \alpha_{k,l}$$

and also

$$(u_l | g_k) = \left(\sum_{j=1}^{l} \beta_{l,j} g_j \Big| g_k \right) = \sum_{j=1}^{l} \overline{\beta_{l,j}} (g_j | g_k) = 0 \text{ if } l < k.$$

This proves that, for each $k \in I$, $\alpha_{k,l} = 0$ if $l < k$ and hence that $g_k = \alpha_{k,k} u_k$. $\quad\square$

10.2.7 Theorem. *Let $I := \{0\} \cup \mathbb{N}$. For each $n \in I$, let f_n be the function*

$$f_n : \mathbb{R} \to \mathbb{C}$$

$$x \mapsto f_n(x) := x^n e^{-\frac{x^2}{2}}.$$

The family $\{f_n\}_{n \in I}$ is a linearly independent subset of the inner product space $\mathcal{S}(\mathbb{R})$ (cf. 3.1.10h and 10.1.5c).

For each $n \in I$, let g_n be the function

$$g_n : \mathbb{R} \to \mathbb{C}$$

$$x \mapsto g_n(x) := e^{\frac{x^2}{2}} \frac{d^n e^{-x^2}}{dx^n}.$$

For each $n \in I$, there exists a polynomial H_n of degree n and with real coefficients which contains only even (respectively, odd) powers of x if n is even (respectively, odd), and which is so that

$$g_n(x) = H_n(x) e^{-\frac{x^2}{2}}, \forall x \in \mathbb{R}.$$

For each $n \in I$, it is obvious that g_n is a non-zero element of $\mathcal{S}(\mathbb{R})$; then, let $c_n := \|g_n\|^{-1}$ and

$$h_n := c_n g_n.$$

The family $\{h_n\}_{n \in I}$ is an o.n.s. in $\mathcal{S}(\mathbb{R})$ and

$$L\{h_n\}_{n \in I} = L\{f_n\}_{n \in I}.$$

The function h_n is called the nth Hermite function and the polynomial H_n is called the Hermite polynomial of degree n.

Proof. Since the function $x \mapsto e^{-\frac{x^2}{2}}$ is obviously an element of $\mathcal{S}(\mathbb{R})$, we have $f_n \in \mathcal{S}(\mathbb{R})$ for each $n \in I$ (cf. 4 in 3.1.10h). The family $\{f_n\}_{n \in I}$ is a linearly independent subset of $\mathcal{S}(\mathbb{R})$ because for every polynomial $\sum_{k=1}^{n} \alpha_k x^k$ we have

$$\left(\sum_{k=1}^{n} \alpha_k x^k = 0, \forall x \in \mathbb{R} \right) \Rightarrow (\alpha_k = 0 \text{ for } k = 1, ..., n).$$

For each $n \in I$, we define a function H_n on $\mathbb{R}$ by

$$H_n(x) := g_n(x) e^{\frac{x^2}{2}} = e^{x^2} \frac{d^n e^{-x^2}}{dx^n}, \forall x \in \mathbb{R}.$$

We prove by induction that H_n is a polynomial with the stated properties, for each $n \in I$. Since

$$H_0(x) = 1 \text{ and } H_1(x) = -2x, \forall x \in \mathbb{R},$$

H_0 and H_1 have the required properties. Now suppose that, for $n \in I$, H_n is a polynomial of degree n and with real coefficients, i.e. that $(\alpha_1, ..., \alpha_n) \in \mathbb{R}^n$ exists so that $\alpha_n \neq 0$ and

$$H_n(x) = \sum_{k=1}^{n} \alpha_k x^k, \forall x \in \mathbb{R};$$

then

$$\frac{d^n e^{-x^2}}{dx^n} = H_n(x) e^{-x^2} = \sum_{k=1}^{n} \alpha_k x^k e^{-x^2}, \forall x \in \mathbb{R},$$

and hence

$$H_{n+1}(x) = g_{n+1}(x) e^{\frac{x^2}{2}} = e^{x^2} \frac{d^{n+1} e^{-x^2}}{dx^{n+1}}$$

$$= e^{x^2} \frac{d}{dx} \left(\sum_{k=1}^{n} \alpha_k x^k e^{-x^2} \right) = \sum_{k=1}^{n} \alpha_k (k x^{k-1} - 2x^{k+1}), \forall x \in \mathbb{R};$$

thus, H_{n+1} is a polynomial of degree $n + 1$ and with real coefficients. Moreover, if H_n contains only even (respectively, odd) powers of x then H_{n+1} contains only odd (respectively, even) powers of n. This proves that H_n has the required properties for each $n \in I$.

For each $n \in I$, since

$$g_n(x) = H_n(x)e^{-\frac{x^2}{2}}, \forall x \in \mathbb{R},$$

it is obvious that g_n is a non-zero element of $\mathcal{S}(\mathbb{R})$ (cf. 4 in 3.1.10h) and that g_n is a linear combination of $f_1, ..., f_n$.

Now we prove that $\{g_n\}_{n \in I}$ is an orthogonal set. In order to see this, we need to prove beforehand that, for all $m, n \in I$ such that $m < n$, we have

$$\int_{\mathbb{R}} x^m \frac{d^n e^{-x^2}}{dx^n} dm(x) = 0. \tag{*}$$

First, we note that the function

$$x \mapsto \varphi_{m,n}(x) := x^m \frac{d^n e^{-x^2}}{dx^n}$$

is an element of $\mathcal{S}(\mathbb{R})$ for all $m, n \in I$, since the function $x \mapsto e^{-x^2}$ is obviously so (cf. 1 and 4 in 3.1.0h). Therefore, $\varphi_{m,n} \in \mathcal{L}^1(\mathbb{R}, \mathcal{A}(d_{\mathbb{R}}), m)$ (cf. 10.1.5c). We proved in 10.1.5c that for every $\varphi \in \mathcal{S}(\mathbb{R})$ we have

$$\int_{\mathbb{R}} \varphi dm = \lim_{k \to \infty} \int_{-k}^{k} \varphi(x) dx.$$

Thus, for all $m, n \in I$ such that $m < n$, we have

$$\int_{\mathbb{R}} x^m \frac{d^n e^{-x^2}}{dx^n} dm(x) = \lim_{k \to \infty} \int_{-k}^{k} x^m \frac{d^n e^{-x^2}}{dx^n} dx$$

$$\overset{(1)}{=} \lim_{k \to \infty} (\varphi_{m,n-1}(k) - \varphi_{m,n-1}(-k)) - \lim_{k \to \infty} \int_{-k}^{k} m x^{m-1} \frac{d^{n-1} e^{-x^2}}{dx^{n-1}} dx$$

$$\overset{(2)}{=} -m \lim_{k \to \infty} \int_{-k}^{k} x^{m-1} \frac{d^{n-1} e^{-x^2}}{dx^{n-1}} dx = \cdots$$

$$= (-1)^m m! \lim_{k \to \infty} \int_{-k}^{k} \frac{d^{n-m} e^{-x^2}}{dx^{n-m}} dx$$

$$= (-1)^m m! (\varphi_{0,n-m-1}(k) - \varphi_{0,n-m-1}(-k)) \overset{(3)}{=} 0,$$

where: 1 follows from integration by parts for the Riemann integral, which can be used since the functions $x \mapsto x^m$ and $x \mapsto \frac{d^{n-1} e^{-x^2}}{dx^{n-1}}$ are differentiable and their derivatives are continuous; 2 holds because $\varphi_{m,n-1} \in \mathcal{S}(\mathbb{R})$; 3 holds because $\varphi_{0,n-m-1} \in \mathcal{S}(\mathbb{R})$. Having thus proved $(*)$, let $m, n \in I$ be such that $m < n$; we have

$$(g_m|g_n) = \int_{\mathbb{R}} H_m(x) e^{-\frac{x^2}{2}} g_n(x) dm(x) = \int_{\mathbb{R}} H_m(x) \frac{d^n e^{-x^2}}{dx^n} dm(x) = 0$$

since H_m is a polynomial which contains powers of x of degree less than n.

Thus, the families $\{f_n\}_{n \in I}$ and $\{g_n\}_{n \in I}$ are subsets of the inner product space $\mathcal{S}(\mathbb{R})$ which satisfy the conditions of 10.2.6. Therefore, for each $n \in I$ there exists $\alpha_n \in \mathbb{C}$ so that $g_n = \alpha_n u_n$, where $\{u_n\}_{n \in I}$ is an o.n.s. in $\mathcal{S}(\mathbb{R})$ such that

$$L\{u_n\}_{n \in I} = L\{f_n\}_{n \in I}.$$

Since $\{g_n\}_{n \in I}$ is an orthogonal set, it is obvious that $\{h_n\}_{n \in I}$ is an o.n.s. Moreover, since $h_n = c_n \alpha_n u_n$ and $c_n \alpha_n \neq 0$ for each $n \in I$, from 3.1.7 it follows that

$$L\{h_n\}_{n \in I} = L\{u_n\}_{n \in I}.$$

Hence we have

$$L\{h_n\}_{n \in I} = L\{f_n\}_{n \in I}.$$

$\square$

10.2.8 Theorem.

(a) Let $\{u_1, ..., u_n\}$ be a finite o.n.s. in X. Then

$$\sum_{k=1}^{n} |(u_k|f)|^2 \leq \|f\|^2, \forall f \in X.$$

(b) Let $\{u_i\}_{i \in I}$ be any o.n.s. in X and, for every $f \in X$, define

$$I_f := \{i \in I : (u_i|f) \neq 0\}.$$

Then the set I_f is countable and

$$\sum_{i \in I_f} |(u_i|f)|^2 \leq \|f\|^2, \forall f \in X.$$

Note that the total ordering in I_f that is necessary for the definition of the sum or the series $\sum_{i \in I_f} |(u_i|f)|^2$ need not be specified in view of 5.4.3. This inequality is called Bessel's inequality. For any $f, g \in X$ such that the set $I_f \cap I_g$ is denumerable, the series

$$\sum_{i \in I_f \cap I_g} (f|u_i)(u_i|g)$$

is absolutely convergent in the Banach space $\mathbb{C}$ (cf. 4.1.4 and 2.7.4a). Hence, the total ordering of $I_f \cap I_g$ that is necessary for the definition of the series $\sum_{i \in I_f \cap I_g} (f|u_i)(u_i|g)$ need not be specified (cf. 4.1.8b).

Proof. a: We have

$$0 \leq \left(f - \sum_{k=1}^{n} (u_k|f) u_k \bigg| f - \sum_{l=1}^{n} (u_l|f) u_l \right)$$

$$= (f|f) - \sum_{k=1}^{n} \overline{(u_k|f)}(u_k|f) - \sum_{l=1}^{n} (u_l|f)(f|u_l) + \sum_{k=1}^{n} \sum_{l=1}^{n} \overline{(u_k|f)}(u_l|f) \delta_{k,l}$$

$$= \|f\|^2 - \sum_{k=1}^{n} |(u_k|f)|^2.$$

b: Suppose $f \in X$ and define, for each $n \in \mathbb{N}$,

$$I_{f,n} := \{i \in I : \frac{1}{n} \leq |(u_i|f)|\}.$$

The result obtained in part a shows that the number of the elements of $I_{f,n}$ can not exceed $n^2\|f\|^2$; thus, $I_{f,n}$ is a finite set for each $n \in \mathbb{N}$, and this implies that I_f is a countable set since $I_f = \bigcup_{n\in\mathbb{N}} I_{f,n}$ (cf. 1.2.10). If I_f is finite then the inequality of the statement follows from the result obtained in part a. If I_f is denumerable, let $\{i_k\}_{k\in\mathbb{N}} := I_f$ be an ordering in I_f; then (cf. 5.4.1)

$$\sum_{i\in I_f} |(u_i|f)|^2 = \sup_{n\geq 1} \sum_{k=1}^{n} |(u_{i_k}|f)|^2,$$

and the inequality of the statement follows once again from the result obtained in part a.

For any $\alpha, \beta \in \mathbb{C}$, the inequality $|\alpha\beta| \leq \frac{1}{2}(|\alpha|^2+|\beta|^2)$ follows from $0 \leq (|\alpha|-|\beta|)^2$; then, for all $f, g \in X$ we have (whatever ordering is chosen in $I_f \cap I_g$)

$$\sum_{i\in I_f\cap I_g} |(f|u_i)(u_i|g)| \leq \sum_{i\in I_f\cap I_g} \frac{1}{2}(|(f|u_i)|^2 + |(u_i|g)|^2)$$

$$\leq \frac{1}{2} \sum_{i\in I_f\cap I_g} |(f|u_i)|^2 + \frac{1}{2} \sum_{i\in I_f\cap I_g} |(u_i|g)|^2$$

$$\leq \frac{1}{2}\|f\|^2 + \frac{1}{2}\|g\|^2,$$

where 5.4.2a, 5.4.5 and 5.4.6 have been used if $I_f \cap I_g$ is denumerable. This proves that if $I_f \cap I_g$ is denumerable then the series $\sum_{i\in I_f\cap I_g} (f|u_i)(u_i|g)$ is absolutely convergent. □

10.2.9 Definition. For every subset S of X, we define

$$S^\perp := \{f \in X : (f|g) = 0, \forall g \in S\},$$

which is called the *orthogonal complement of S in X*.

10.2.10 Proposition. *The following statements hold true:*

(a) $\{0_X\}^\perp = X$ *and* $X^\perp = \{0_X\}$;

(b) *if S_1 and S_2 are subsets of X such that $S_1 \subset S_2$, then $S_2^\perp \subset S_1^\perp$;*

(c) *if $\{S_i\}_{i\in I}$ is a family of subsets of X, then $\left(\bigcup_{i\in I} S_i\right)^\perp = \bigcap_{i\in I} S_i^\perp$;*

(d) *for every subset S of X, $S \subset S^{\perp\perp} := (S^\perp)^\perp$;*

(e) *for every subset S of X, $S^\perp = S^{\perp\perp\perp} := (S^{\perp\perp})^\perp$;*

(f) *for every subset S of X, $S \cap S^\perp = \{0_X\}$ if $0_X \in S$ and $S \cap S^\perp = \emptyset$ if $0_X \notin S$.*

Proof. a: Use 10.1.2b and property ip_4 of an inner product.

b and c: These are obvious.

d: We have

$$g \in S \Rightarrow [(g|f) = \overline{(f|g)} = 0, \forall f \in S^\perp] \Rightarrow g \in (S^\perp)^\perp.$$

e: From part d we have $S \subset S^{\perp\perp}$, and hence by part b we have $(S^{\perp\perp})^\perp \subset S^\perp$. On the other hand, if we substitute $S^\perp$ for S in part d we have $S^\perp \subset (S^\perp)^{\perp\perp}$.

f: We have
$$f \in S \cap S^\perp \Rightarrow (f|f) = 0 \Rightarrow f = 0_X. \tag{*}$$
If $0_X \in S$ then $0_X \in S \cap S^\perp$ since $0_X \in S^\perp$ follows from 10.1.2b, and hence $(*)$ proves that $S \cap S^\perp = \{0_X\}$. If $0_X \notin S$ then $(*)$ proves that $S \cap S^\perp = \emptyset$. $\qquad\square$

10.2.11 Proposition. *For every subset S of X, we have*
$$S^\perp = (\overline{S})^\perp = (LS)^\perp = (VS)^\perp.$$

Proof. $S^\perp = (\overline{S})^\perp$: From $S \subset \overline{S}$ we have $(\overline{S})^\perp \subset S^\perp$ (cf. 10.2.10b). Now suppose $f \in S^\perp$ and $g \in \overline{S}$; by 2.3.10, there exists a sequence $\{g_n\}$ in S such that $g_n \to g$; then, by 10.1.16c, 2.4.2, 2.7.3a, we have
$$(f|g) = \lim_{n \to \infty} (f|g_n) = 0;$$
this proves that $f \in (\overline{S})^\perp$.

$\quad$ $S^\perp = (LS)^\perp$: From $S \subset LS$ we have $(LS)^\perp \subset S^\perp$ (cf. 10.2.10b). Now suppose $f \in S^\perp$ and $g \in LS$; by 3.1.7, there exist $n \in \mathbb{N}$, $(\alpha_1, ..., \alpha_n) \in \mathbb{C}^n$, $(g_1, ..., g_n) \in S^n$ so that $g = \sum_{i=1}^n \alpha_i g_i$; then we have
$$(f|g) = \sum_{i=1}^n \alpha_i (f|g_i) = 0;$$
this proves that $f \in (LS)^\perp$.

$\quad$ $(LS)^\perp = (VS)^\perp$: Since $VS = \overline{LS}$ (cf. 4.1.13), this follows from $S^\perp = \overline{S}^\perp$ with S replaced by LS. $\qquad\square$

10.2.12 Proposition. *Let A and B be linear operators in X, i.e. $A, B \in \mathcal{O}(X)$, and suppose that $\overline{D_A} = X$, $D_A \subset D_B$, and*
$$(Au|u) = (Bu|u), \forall u \in D_A \cap \tilde{X},$$
with $\tilde{X} := \{u \in X : \|u\| = 1\}$. Then $A \subset B$.

Proof. We have
$$(Af|f) = \|f\|^2 \left(A \frac{1}{\|f\|} f \Big| \frac{1}{\|f\|} f \right)$$
$$= \|f\|^2 \left(B \frac{1}{\|f\|} f \Big| \frac{1}{\|f\|} f \right) = (Bf|f), \forall f \in D_A - \{0_{\mathcal{H}}\},$$
and hence
$$(Af|f) = (Bf|f), \forall f \in D_A.$$
Then, by 10.1.10a we have
$$(Af|g) = (Bf|g), \forall f, g \in D_A,$$
and hence
$$Af - Bf \in D_A^\perp, \forall f \in D_A.$$
Since $D_A^\perp = (\overline{D_A})^\perp = X^\perp = \{0_X\}$ (cf. 10.2.11 and 10.2.10a), we have
$$Af = Bf, \forall f \in D_A.$$

$\qquad\square$

10.2.13 Theorem. *For every subset S of X, $S^\perp$ is a subspace of X.*

Proof. For all $\alpha, \beta \in \mathbb{C}$ and $f, g \in S^\perp$, we have

$$(\alpha f + \beta g | h) = \overline{\alpha}\,(f|h) + \overline{\beta}\,(g|h) = 0, \forall h \in S,$$

and hence $\alpha f + \beta g \in S^\perp$. This proves that $S^\perp$ is a linear manifold in X (cf. 3.1.4c). Now, let $f \in X$ and let $\{f_n\}$ be a sequence in $S^\perp$ so that $f_n \to f$; then, by 10.1.16c we have

$$(f|g) = \lim_{n \to \infty} (f_n|g) = 0, \forall g \in S,$$

and hence $f \in S^\perp$. This proves that $S^\perp$ is a closed subset of X (cf. 2.3.4). $\qquad\square$

10.2.14 Definition. A subset S_1 of X is said to be *orthogonal to* a subset S_2 of X if $S_1 \subset S_2^\perp$. If S_1 is orthogonal to S_2 then S_2 is orthogonal to S_1 since

$$S_1 \subset S_2^\perp \Rightarrow S_2 \subset S_2^{\perp\perp} \subset S_1^\perp,$$

where 10.2.10d and 10.2.10b have been used.

10.2.15 Proposition. *Let S_1 and S_2 be two subsets of X such that $S_1 \subset S_2^\perp$ and $X = S_1 + S_2$. Then $S_1 = S_2^\perp$ and $S_2 = S_1^\perp$.*

Proof. We prove that $S_1 = S_2^\perp$ by proving that $S_2^\perp \subset S_1$. For each $f \in S_2^\perp$, there exists a pair $(f_1, f_2) \in S_1 \times S_2$ so that $f = f_1 + f_2$ and hence so that $f - f_1 = f_2$; now, $f - f_1 \in S_2^\perp$ since $f_1 \in S_1 \subset S_2^\perp$ and $S_2^\perp$ is a linear manifold (cf. 10.2.13), while $f_2 \in S_2$; thus, $f - f_1 = 0_X$ (cf. 10.2.10f), and hence $f = f_1 \in S_1$.

Since $S_1 \subset S_2^\perp$ implies $S_2 \subset S_1^\perp$ (cf. 10.2.14), by the same reasoning we can prove that $S_2 = S_1^\perp$. $\qquad\square$

10.2.16 Proposition. *Let X_1 and X_2 be isomorphic inner product spaces, let U be an isomorphism from X_1 onto X_2, and let S be a subset of X_1. Then,*

$$U(S^\perp) = (U(S))^\perp.$$

Proof. For $f \in X_2$ we have

$$
\begin{aligned}
&f \in U(S^\perp) \Leftrightarrow \\
&U^{-1}f \in S^\perp \Leftrightarrow \\
&[(f|Ug) = (UU^{-1}f|Ug) = \left(U^{-1}f|g\right) = 0, \forall g \in S] \Leftrightarrow \\
&f \in (U(S))^\perp.
\end{aligned}
$$

$\qquad\square$

10.3 Completions, direct sums, unitary and antiunitary operators in Hilbert spaces

10.3.1 Definition. A *Hilbert space* is an inner product space (X, σ, μ, ϕ) such that the metric space (X, d_ϕ) is complete (equivalently, such that the normed space $(X, \sigma, \mu, \nu_\phi)$ is a Banach space).

The symbol $\mathcal{H}$ denotes an abstract Hilbert space throughout the book.

10.3.2 Theorem. *Let M be a linear manifold in a Hilbert space $(\mathcal{H}, \sigma, \mu, \phi)$. Then the inner product space $(M, \sigma_{M \times M}, \mu_{\mathbb{C} \times M}, \phi_{M \times M})$ (cf. 10.1.6) is a Hilbert space iff M is a closed set in the metric space (X, d_ϕ). This fully explains why in 4.1.9 a closed linear manifold was called a subspace.*

Proof. The statement follows from 2.6.6. $\qquad\qquad\qquad\qquad\qquad\qquad\qquad\quad\square$

10.3.3 Theorem. *If two inner product spaces are isomorphic and one of them is a Hilbert space, then the other one is also a Hilbert space.*

Proof. The statement follows from 2.6.4 (cf. 10.1.21). $\qquad\qquad\qquad\qquad\qquad\quad\square$

10.3.4 Definition. Let (X, σ, μ, ϕ) be an inner product space. A *completion* of (X, σ, μ, ϕ) is a pair $((\hat{X}, \hat{\sigma}, \hat{\mu}, \hat{\phi}), \iota)$, where $(\hat{X}, \hat{\sigma}, \hat{\mu}, \hat{\phi})$ is a Hilbert space, ι is a mapping $\iota : X \to \hat{X}$, and the following two conditions hold:

(co_1) $(\iota(f)|\iota(g)) = (f|g)$, $\forall f, g \in X$;
(co_2) R_ι is dense in $\hat{X}$, i.e. $\overline{R_\iota} = \hat{X}$.

10.3.5 Remark. Let $((\hat{X}, \hat{\sigma}, \hat{\mu}, \hat{\phi}), \iota)$ be a completion of an inner product space (X, σ, μ, ϕ). Then ι is a linear operator (cf. 10.1.19) and therefore R_ι can be considered as an inner product space (cf. 3.2.2a and 10.1.6). Moreover, ι is injective (cf. 10.1.19). Thus, condition co_1 in 10.3.4 is equivalent to the condition that R_ι be a linear manifold in $\hat{X}$ and ι be an isomorphism from the inner product space (X, σ, μ, ϕ) onto the inner product space $(R_\iota, \hat{\sigma}_{R_\iota \times R_\iota}, \hat{\mu}_{\mathbb{C} \times R_\iota}, \hat{\phi}_{R_\iota \times R_\iota})$ (cf. 10.1.6 and 10.1.20).

Since ι is a linear operator, it follows directly from the definitions that $((\hat{X}, d_{\hat{\phi}}), \iota)$ is a completion of the metric space (X, d_ϕ) (cf. 2.6.7).

We shall not use the following theorem, also because the completions of inner product spaces that we need will be constructed without using either the statement or the proof of this theorem. For this reason we state it without giving its proof, which can be found e.g. in 4.11 of (Weidmann, 1980).

10.3.6 Theorem. *For every inner product space (X, σ, μ, ϕ), there exists a completion $((\hat{X}, \hat{\sigma}, \hat{\mu}, \hat{\phi}), \iota)$ of (X, σ, μ, ϕ).*

If $((\tilde{X}, \tilde{\sigma}, \tilde{\mu}, \tilde{\phi}), \omega)$ is also a completion of (X, σ, μ, ϕ), then there exists an iso-morphism U from $(\hat{X}, \hat{\sigma}, \hat{\mu}, \hat{\phi})$ onto $(\tilde{X}, \tilde{\sigma}, \tilde{\mu}, \tilde{\phi})$ such that $U \circ \iota = \omega$, i.e. such that $U(\iota(f)) = \omega(f), \forall f \in X$.

10.3.7 Theorem. *Let $\{(\mathcal{H}_n, \sigma_n, \mu_n, \phi_n)\}_{n \in I}$ be a countable family of Hilbert spaces, and suppose $I := \{1, ..., N\}$ or $I := \mathbb{N}$. If $I = \{1, ..., N\}$, we define*

$$\sum_{n \in I}^{\oplus} \mathcal{H}_n := \mathcal{H}_1 \times \cdots \times \mathcal{H}_N,$$

and we denote an element of $\sum_{n \in I}^{\oplus} \mathcal{H}_n$ by $\{f_n\}$ (thus, this symbols stands here also for an N-tuple). If $I = \mathbb{N}$, we denote by $\prod_{n \in \mathbb{N}} \mathcal{H}_n$ the family of all the sequences $\{f_n\}$ in $\bigcup_{n \in \mathbb{N}} \mathcal{H}_n$ that are such that $f_n \in \mathcal{H}_n$ for all $n \in \mathbb{N}$, and we define

$$\sum_{n \in I}^{\oplus} \mathcal{H}_n := \left\{ \{f_n\} \in \prod_{n \in \mathbb{N}} \mathcal{H}_n : \sum_{n=1}^{\infty} \|f_n\|_n^2 < \infty \right\}$$

(we have denoted by $\| \ \|_n$ the norm of $\mathcal{H}_n$).

The following definitions, of the mappings σ, μ, ϕ, are consistent:

$$\sigma : \sum_{n \in I}^{\oplus} \mathcal{H}_n \times \sum_{n \in I}^{\oplus} \mathcal{H}_n \to \sum_{n \in I}^{\oplus} \mathcal{H}_n$$

$$(\{f_n\}, \{g_n\}) \mapsto \sigma(\{f_n\}, \{g_n\}) := \{\sigma_n(f_n, g_n)\},$$

$$\mu : \mathbb{C} \times \sum_{n \in I}^{\oplus} \mathcal{H}_n \to \sum_{n \in I}^{\oplus} \mathcal{H}_n$$

$$(\alpha, \{f_n\}) \mapsto \mu(\alpha, \{f_n\}) := \{\mu_n(\alpha, f_n)\},$$

$$\phi : \sum_{n \in I}^{\oplus} \mathcal{H}_n \times \sum_{n \in I}^{\oplus} \mathcal{H}_n \to \mathbb{C}$$

$$(\{f_n\}, \{g_n\}) \mapsto \phi(\{f_n\}, \{g_n\}) := \sum_{n \in I} \phi_n(f_n, g_n)$$

($\sum_{n \in I}$ stands for either $\sum_{n=1}^{N}$ or $\sum_{n=1}^{\infty}$). The quadruple $\left(\sum_{n \in I}^{\oplus} \mathcal{H}_n, \sigma, \mu, \phi \right)$ is a Hilbert space, which is called the direct sum of the family $\{\mathcal{H}_n\}_{n \in I}$. The symbol $\sum_{n \in I}^{\oplus} \mathcal{H}_n$ is written as $\sum_{n=1}^{N \oplus} \mathcal{H}_n$ or as $\mathcal{H}_1 \oplus \cdots \oplus \mathcal{H}_N$ if $I = \{1, ..., N\}$, and as $\sum_{n=1}^{\infty \oplus} \mathcal{H}_n$ if $I = \mathbb{N}$.

Proof. We expound the proof for $I = \mathbb{N}$, from which the proof for $I = \{1, ..., N\}$ can be obtained by obvious simplifications.

To prove that the definition of σ is consistent, we note first that the inequality

$$|\alpha\beta| \le \frac{1}{2}(|\alpha|^2 + |\beta|^2) \text{ (i.e. } 0 \le (|\alpha| - |\beta|)^2), \forall \alpha, \beta \in \mathbb{C} \tag{1}$$

implies the inequality

$$(|\alpha| + |\beta|)^2 \le 2(|\alpha|^2 + |\beta|^2), \forall \alpha, \beta \in \mathbb{C}. \tag{2}$$

Then, for $\{f_n\}, \{g_n\} \in \sum_{n=1}^{\infty \oplus} \mathcal{H}_n$ we have

$$\sum_{n=1}^{\infty} \|f_n + g_n\|_n^2 \le \sum_{n=1}^{\infty} (\|f_n\|_n + \|g_n\|_n)^2$$

$$\le 2 \sum_{n=1}^{\infty} (\|f_n\|_n^2 + \|g_n\|_n^2) = 2 \sum_{n=1}^{\infty} \|f_n\|_n^2 + 2 \sum_{n=1}^{\infty} \|g_n\|_n^2 < \infty$$

(where 5.4.2a, 5.4.5, 5.4.6 and inequality 2 have been used), which proves that $\{\sigma_n(f_n, g_n)\} \in \sum_{n=1}^{\infty \oplus} \mathcal{H}_n$.

As to the definition of μ, for $\alpha \in \mathbb{C}$ and $\{f_n\} \in \sum_{n=1}^{\infty \oplus} \mathcal{H}_n$ we have

$$\sum_{n=1}^{\infty} \|\alpha f_n\|_n^2 = \sum_{n=1}^{\infty} |\alpha|^2 \|f_n\|_n^2 = |\alpha|^2 \sum_{n=1}^{\infty} \|f_n\|_n^2 < \infty$$

(where 5.4.5 has been used), which proves that $\{\mu_n(\alpha, f_n)\} \in \sum_{n=1}^{\infty \oplus} \mathcal{H}_n$.

As to the definition of ϕ, for $\{f_n\}, \{g_n\} \in \sum_{n=1}^{\infty \oplus} \mathcal{H}_n$ we have

$$\sum_{n=1}^{\infty} |(f_n|g_n)_n| \le \sum_{n=1}^{\infty} \|f_n\|_n \|g_n\|_n$$

$$\le \sum_{n=1}^{\infty} \frac{1}{2}(\|f_n\|^2 + \|g_n\|_n^2) = \frac{1}{2} \sum_{n=1}^{\infty} \|f_n\|_n^2 + \frac{1}{2} \sum_{n=1}^{\infty} \|g_n\|_n^2 < \infty$$

(where 10.1.7a, 5.4.2a, 5.4.5, 5.4.6 and inequality 1 have been used), which proves that the series $\sum_{n=1}^{\infty} \phi_n(f_n, g_n)$ is absolutely convergent and hence convergent.

Then, it is easy to see that $\left(\sum_{n\in I}^{\oplus} \mathcal{H}_n, \sigma, \mu, \phi\right)$ is an inner product space. Properties ls_1 and ls_2 of 3.1.1 follow directly from the definitions of σ and μ (the zero vector of $\sum_{n=1}^{\infty \oplus} \mathcal{H}_n$ is the sequence $\{0_{\mathcal{H}_n}\}$, and the opposite of $\{f_n\} \in \sum_{n=1}^{\infty \oplus} \mathcal{H}_n$ is the sequence $\{-f_n\}$), and properties ip_1, ip_2, ip_3, ip_4 of 10.1.3 follow from the definition of ϕ and from the continuity of sum and product in $\mathbb{C}$ (for ip_1) or the continuity of complex conjugation (for ip_2).

Finally, we prove that the metric space $\left(\sum_{n=1}^{\infty \oplus} \mathcal{H}_n, d_\phi\right)$ is complete. Let $\{\varphi_k\}$ be a Cauchy sequence in $\sum_{n=1}^{\infty \oplus} \mathcal{H}_n$. This means that $\varphi_k := \{f_{k,n}\} \in \prod_{n\in\mathbb{N}} \mathcal{H}_n$ and $\sum_{n=1}^{\infty} \|f_{k,n}\|_n^2 < \infty$ for each $k \in \mathbb{N}$, and that $\forall \varepsilon > 0, \exists N_\varepsilon \in \mathbb{N}$ so that

$$N_\varepsilon < k, l \Rightarrow \left(\sum_{n=1}^{\infty} \|f_{k,n} - f_{l,n}\|_n^2\right)^{\frac{1}{2}} = d_\phi(\varphi_k, \varphi_l) < \varepsilon. \tag{3}$$

This implies that, for each $n \in \mathbb{N}$,

$$N_\varepsilon < k, l \Rightarrow d_{\phi_n}(f_{k,n}, f_{l,n}) = \|f_{k,n} - f_{l,n}\| < \varepsilon.$$

Thus, for each $n \in \mathbb{N}$, $\{f_{k,n}\}$ (where k is the index within the sequence) is a Cauchy sequence in $\mathcal{H}_n$. Therefore (since $\mathcal{H}_n$ is a complete metric space) there exists $f_n \in \mathcal{H}_n$ so that $f_n = \lim_{k \to \infty} f_{k,n}$. Moreover, 3 implies that, for each $p \in \mathbb{N}$,

$$N_\varepsilon < k, l \Rightarrow \sum_{n=1}^{p} \|f_{k,n} - f_{l,n}\|_n^2 \leq \varepsilon^2,$$

and therefore (in view of the continuity of σ_n and ν_{ϕ_n}) also that, for each $p \in \mathbb{N}$,

$$N_\varepsilon < k \Rightarrow \sum_{n=1}^{p} \|f_{k,n} - f_n\|_n^2 = \lim_{l \to \infty} \sum_{n=1}^{p} \|f_{k,n} - f_{l,n}\|_n^2 \leq \varepsilon^2,$$

and therefore also that

$$N_\varepsilon < k \Rightarrow \sum_{n=1}^{\infty} \|f_{k,n} - f_n\|_n^2 \leq \varepsilon^2. \tag{4}$$

Now, if we fix $k > N_\varepsilon$, 4 implies that the sequence $\psi_k := \{f_{k,n} - f_n\} \in \prod_{n \in \mathbb{N}} \mathcal{H}_n$ is an element of $\sum_{n=1}^{\infty \oplus} \mathcal{H}_n$, and hence that the sequence $\varphi := \{f_n\} \in \prod_{n \in \mathbb{N}} \mathcal{H}_n$ is an element of $\sum_{n=1}^{\infty \oplus} \mathcal{H}_n$ as well since $\varphi_k - \psi_k = \varphi$. Then, 4 can be written as

$$N_\varepsilon < k \Rightarrow d_\phi(\varphi_k, \varphi) \leq \varepsilon$$

and this shows that the sequence $\{\varphi_k\}$ is convergent. $\qquad\square$

10.3.8 Examples.

(a) We define an inner product ϕ for a zero linear space (cf. 3.1.10a) by letting $\phi(0_X, 0_X) := 0$. This trivial inner product space is obviously a Hilbert space, which is called a *zero Hilbert space*. It is obvious that two zero Hilbert spaces are isomorphic and that an inner product space which is isomorphic to a zero Hilbert space is also a zero Hilbert space.

(b) The function

$$\phi : \mathbb{C} \times \mathbb{C} \to \mathbb{C}$$
$$(x_1, x_2) \mapsto \phi(x_1, x_2) := \overline{x_1} x_2$$

is an inner product for the linear space $\mathbb{C}$ (cf. 3.1.10b) and $d_\phi = d_\mathbb{C}$ (cf. 2.7.4a), as can be immediately seen. Since $(\mathbb{C}, d_\mathbb{C})$ is a complete metric space, the inner product space $\mathbb{C}$ is a Hilbert space.

(c) Let $N \in \mathbb{N}$ and let $\mathcal{H}_n := \mathbb{C}$ for $n = 1, ..., N$. The Hilbert space $\sum_{n=1}^{N \oplus} \mathcal{H}_n$ (cf. 10.3.7) is then denoted by $\mathbb{C}^N$ (this is consistent with the definition $\mathbb{C}^N := \mathbb{C} \times \cdots N$ times $\cdots \times \mathbb{C}$ given in 1.2.1). Explicitly, the mappings σ, μ, ϕ are defined by

$$\sigma((x_1, ..., x_N), (y_1, ..., y_N)) := (x_1 + y_1, ..., x_N + y_N),$$
$$\forall (x_1, ..., x_N), (y_1, ..., y_N) \in \mathbb{C}^N,$$
$$\mu(\alpha, (x_1, ..., x_N)) := (\alpha x_1, ..., \alpha x_N), \forall \alpha \in \mathbb{C}, \forall (x_1, ..., x_N) \in \mathbb{C}^N,$$
$$\phi((x_1, ..., x_N), (y_1, ..., y_N)) := \sum_{n=1}^{N} \overline{x_n} y_n, \forall (x_1, ..., x_N), (y_1, ..., y_N) \in \mathbb{C}^N.$$

The norm ν_ϕ turns out to be

$$\nu_\phi((x_1,...,x_N)) = \sqrt{\sum_{n=1}^{N}|x_n|^2}, \forall(x_1,...,x_N) \in \mathbb{C}^N.$$

Thus, altogether independently from the concepts of product of metric spaces and of the sum of normed spaces (where the equality we are about to state was used), condition no_1 of 4.1.1 turns out to be, for the norm ν_ϕ,

$$\sqrt{\sum_{n=1}^{N}|x_n+y_n|^2} \leq \sqrt{\sum_{n=1}^{N}|x_n|^2} + \sqrt{\sum_{n=1}^{N}|y_n|^2},$$

$$\forall(x_1,...,x_N),(y_1,...,y_N) \in \mathbb{C}^N,$$

which is known as the *triangle inequality* in $\mathbb{C}^N$.

Moreover 10.1.7a turns out to be

$$\left|\sum_{n=1}^{N}\overline{x_n}y_n\right| \leq \sqrt{\sum_{n=1}^{N}|x_n|^2}\sqrt{\sum_{n=1}^{N}|y_n|^2},$$

$$\forall(x_1,...,x_N),(y_1,...,y_N) \in \mathbb{C}^N,$$

which is therefore the Schwarz inequality in $\mathbb{C}^N$.

Finally, we note that the distance d_ϕ coincides with $d_\mathbb{C} \times \cdots$ N times $\cdots \times d_\mathbb{C} = d_{2N}$ (cf. 2.7.4b).

(d) Let $\mathcal{H}_n := \mathbb{C}$ for all $n \in \mathbb{N}$. The Hilbert space $\sum_{n=1}^{\infty\oplus}\mathcal{H}_n$ (cf. 10.3.7) is then denoted by ℓ^2. Explicitly, the set ℓ^2 is defined by

$$\ell^2 := \left\{\{x_n\} \in \mathcal{F}(\mathbb{N}) : \sum_{n=1}^{\infty}|x_n|^2 < \infty\right\}$$

and the mappings σ, μ, ϕ are defined by

$$\sigma(\{x_n\},\{y_n\}) := \{x_n+y_n\}, \forall\{x_n\},\{y_n\} \in \ell^2,$$

$$\mu(\alpha,\{x_n\}) := \{\alpha x_n\}, \forall\alpha \in \mathbb{C}, \forall\{x_n\} \in \ell^2,$$

$$\phi(\{x_n\},\{y_n\}) := \sum_{n=1}^{\infty}\overline{x_n}y_n, \forall\{x_n\},\{y_n\} \in \ell^2.$$

The norm ν_ϕ turns out to be

$$\nu_\phi(\{x_n\}) = \sqrt{\sum_{n=1}^{\infty}|x_n|^2}, \forall\{x_n\} \in \ell^2.$$

Thus, condition no_1 of 4.1.1 turns out to be, for the norm ν_ϕ,

$$\sqrt{\sum_{n=1}^{\infty}|x_n+y_n|^2} \leq \sqrt{\sum_{n=1}^{\infty}|x_n|^2} + \sqrt{\sum_{n=1}^{\infty}|y_n|^2}, \forall\{x_n\},\{y_n\} \in \ell^2,$$

which is known as the *triangle inequality* in ℓ^2.

Moreover 10.1.7a turns out to be

$$\left| \sum_{n=1}^{\infty} \overline{x_n} y_n \right| \leq \sqrt{\sum_{n=1}^{\infty} |x_n|^2} \sqrt{\sum_{n=1}^{\infty} |y_n|^2}, \forall \{x_n\}, \{y_n\} \in \ell^2,$$

which is therefore the Schwarz inequality in ℓ^2.

We notice that ℓ_f (cf. 10.1.5a) is a linear manifold in ℓ^2, and that its structure of an inner product space is exactly the one it inherits from ℓ^2 in the way explained in 10.1.6. Moreover, we have $\overline{\ell_f} = \ell^2$. In fact, let $\xi := \{x_n\} \in \ell^2$; for each $k \in \mathbb{N}$, define $\xi_k := \{x_{k,n}\}$ by $x_{k,n} := x_n$ if $n \leq k$ and $x_{k,n} := 0$ if $k < n$; then, $\xi_k \in \ell_f$ and

$$d(\xi, \xi_k) = \sqrt{\sum_{n=1}^{\infty} |x_n - x_{k,n}|^2} = \sqrt{\sum_{n=k+1}^{\infty} |x_n|^2} \to 0 \text{ as } k \to \infty$$

since $\sum_{n=1}^{\infty} |x_n|^2 < \infty$. Thus, if we define $\iota := id_{\ell_f}$, i.e. (cf. 1.2.6)

$$\iota : \ell_f \to \ell^2$$
$$\{x_n\} \mapsto \iota(\{x_n\}) := \{x_n\},$$

then the pair (ℓ^2, ι) is a completion of the inner product space ℓ_f. We point out that, in view of 2.6.8, this shows that ℓ_f is not a Hilbert space, since obviously $\ell_f \neq \ell^2$ (for instance, the sequence $\{\frac{1}{n}\}$ is an element of ℓ^2 and not an element of ℓ_f).

10.3.9 Definition. An isomorphism from a Hilbert space onto another is called a *unitary operator*. The family (which can be empty) of all unitary operators from a Hilbert space $\mathcal{H}_1$ onto a Hilbert space $\mathcal{H}_2$ is denoted by the symbol $\mathcal{U}(\mathcal{H}_1, \mathcal{H}_2)$. For a Hilbert space $\mathcal{H}$ we write $\mathcal{U}(\mathcal{H}) := \mathcal{U}(\mathcal{H}, \mathcal{H})$ and an element of $\mathcal{U}(\mathcal{H})$ is called a *unitary operator in* $\mathcal{H}$.

10.3.10 Remark. For a Hilbert space $\mathcal{H}$, the family $\mathcal{U}(\mathcal{H})$ is a group with product of operators as group product, the operator $\mathbb{1}_{\mathcal{H}}$ as group identity, the operator U^{-1} as group inverse of U for every $U \in \mathcal{U}(\mathcal{H})$ (cf. 10.1.21 and 4.6.2c).

10.3.11 Definition. Let $\mathcal{H}_1$ and $\mathcal{H}_2$ be isomorphic Hilbert spaces. Two linear operators $A \in \mathcal{O}(\mathcal{H}_1)$ and $B \in \mathcal{O}(\mathcal{H}_2)$ are said to be *unitarily equivalent* if there exists $U \in \mathcal{U}(\mathcal{H}_1, \mathcal{H}_2)$ so that $B = UAU^{-1}$.

10.3.12 Remarks.

(a) For a Hilbert space $\mathcal{H}$, the set

$$R := \{(A, B) \in \mathcal{O}(\mathcal{H}) \times \mathcal{O}(\mathcal{H}) : \exists U \in \mathcal{U}(\mathcal{H}) \text{ such that } B = UAU^{-1}\}$$

defines a relation in $\mathcal{O}(\mathcal{H})$ which can be easily seen to be an equivalence relation. This justifies the term "unitarily equivalent" used in 10.3.11. The fact that R

is an equivalence relation is linked to the group structure of $\mathcal{U}(\mathcal{H})$ (cf. 10.3.10). In fact, reflexivity holds because $\mathbb{1}_{\mathcal{H}}$ is a unitary operator, symmetry because the inverse of a unitary operator is a unitary operator, transitivity because the product of two unitary operators is a unitary operator, and these are exactly the facts that form the basis for the group structure of $\mathcal{U}(\mathcal{H})$.

(b) Let $\mathcal{H}_1$ and $\mathcal{H}_2$ be isomorphic Hilbert spaces. For $A \in \mathcal{O}(\mathcal{H}_1)$, $B \in \mathcal{O}(\mathcal{H}_2)$, $U \in \mathcal{U}(\mathcal{H}_1, \mathcal{H}_2)$, suppose that $B = UAU^{-1}$. Then, in view of 10.1.21, all the conditions and all the propositions listed in 4.6.4 and in 4.6.5 hold true (as to condition 4.6.4g, the operator T_U defined in 4.6.3 is now a unitary operator from the Hilbert space $\mathcal{H}_1 \oplus \mathcal{H}_1$ onto the Hilbert space $\mathcal{H}_2 \oplus \mathcal{H}_2$). Thus, A and B have the same abstract properties related to set theory, metric space theory, normed space theory. We shall see that this is true also for the properties related to inner product.

10.3.13 Definition. Let X and Y be linear spaces over $\mathbb{C}$. An *antilinear operator* from X to Y is a mapping A from X to Y, i.e. $A : D_A \to Y$ with $D_A \subset X$, which has the following properties:

(ao_1) D_A is a linear manifold in X;
(ao_2) $A(f + g) = Af + Ag, \forall f, g \in D_A$;
(ao_3) $A(\alpha f) = \overline{\alpha} A f, \forall \alpha \in \mathbb{C}, \forall f \in D_A$.

10.3.14 Remark. All the definitions and the symbols introduced in Section 3.2 for linear operators can be extended to the family of all linear or antilinear operators, and it is easy to see that all the results proved in Section 3.2 hold for this wider family, with only the following exceptions: $3.2.10b_4$ must be supplemented with

$$(\overline{\alpha} A)B = \overline{\alpha}(AB) = A(\alpha B), \forall \alpha \in \mathbb{C} - \{0\}, \text{ for every antilinear } A$$

$$\text{and every linear or antilinear } B;$$

3.2.15 is not true for an antilinear operator.

The product of two antilinear operators is a linear operator and the product of a linear operator and an antilinear one (in either order) is an antilinear operator; for the sum of two operators to give a linear or an antilinear operator, the two operators must be both linear or both antilinear.

Moreover, if X and Y are normed spaces over $\mathbb{C}$, all the definitions, the symbols and the results set out about linear operators in Section 4.2 can be extended to the family of all linear or antilinear operators (in the extended version of 4.2.7, both the operators A and B must be either linear or antilinear).

10.3.15 Definition. An *antiunitary operator* from a Hilbert space $\mathcal{H}_1$ onto a Hilbert space $\mathcal{H}_2$ is a mapping $V : \mathcal{H}_1 \to \mathcal{H}_2$ such that:

(au_1) V is a bijection from $\mathcal{H}_1$ onto $\mathcal{H}_2$;
(au_2) V is an antilinear operator;
(au_3) $(Vf|Vg)_2 = (g|f)_1, \forall f, g \in \mathcal{H}_1$.

The family (which can be empty) of all antiunitary operators from $\mathcal{H}_1$ onto $\mathcal{H}_2$ is denoted by the symbol $\mathcal{A}(\mathcal{H}_1, \mathcal{H}_2)$. We also write

$$\mathcal{U}\mathcal{A}(\mathcal{H}_1, \mathcal{H}_2) := \mathcal{U}(\mathcal{H}_1, \mathcal{H}_2) \cup \mathcal{A}(\mathcal{H}_1, \mathcal{H}_2).$$

For a Hilbert space $\mathcal{H}$, we write

$$\mathcal{A}(\mathcal{H}) := \mathcal{A}(\mathcal{H}, \mathcal{H}) \text{ and } \mathcal{U}\mathcal{A}(\mathcal{H}) := \mathcal{U}\mathcal{A}(\mathcal{H}, \mathcal{H}).$$

An element of $\mathcal{A}(\mathcal{H})$ is called an *antiunitary operator in* $\mathcal{H}$.

10.3.16 Remarks.

(a) The reason why we take antiunitary operators into consideration is that they play an essential role in Wigner's theorem (cf. Section 10.9).

(b) If $\mathcal{H}_1$ and $\mathcal{H}_2$ are Hilbert spaces and $V \in \mathcal{A}(\mathcal{H}_1, \mathcal{H}_2)$, it is immediate to see that $V^{-1} \in \mathcal{A}(\mathcal{H}_2, \mathcal{H}_1)$.

(c) If $\mathcal{H}_1, \mathcal{H}_2, \mathcal{H}_3$ are Hilbert spaces, $U \in \mathcal{U}\mathcal{A}(\mathcal{H}_1, \mathcal{H}_2)$, and $V \in \mathcal{U}\mathcal{A}(\mathcal{H}_2, \mathcal{H}_3)$, it is immediate to see that $VU \in \mathcal{U}\mathcal{A}(\mathcal{H}_1, \mathcal{H}_3)$, and that $VU \in \mathcal{U}(\mathcal{H}_1, \mathcal{H}_3)$ iff U and V are both unitary or both antiunitary.

(d) For every Hilbert space $\mathcal{H}$, 10.3.10 and remarks b and c above imply that the family $\mathcal{U}\mathcal{A}(\mathcal{H})$ is a group with product of operators as group product, the operator $\mathbb{1}_{\mathcal{H}}$ as group identity, the operator T^{-1} as group inverse of T for every $T \in \mathcal{U}\mathcal{A}(\mathcal{H})$.

(e) If $\mathcal{H}_1$ and $\mathcal{H}_2$ are Hilbert spaces and $V \in \mathcal{A}(\mathcal{H}_1, \mathcal{H}_2)$, it is immediate to see that

$$\|Vf - Vg\|_2 = \|f - g\|_1, \forall f, g \in \mathcal{H}_1.$$

Thus, V is an isomorphism from the metric space $\mathcal{H}_1$ onto the metric space $\mathcal{H}_2$.

(f) The result of 10.2.16 holds true also for an antiunitary operator (if X_1 and X_2 in that proposition are Hilbert spaces and U is an antiunitary operator, the proof remains essentially the same).

10.3.17 Theorem. *Let $\mathcal{H}_1$ and $\mathcal{H}_2$ be Hilbert spaces and V a mapping from $\mathcal{H}_1$ to $\mathcal{H}_2$. The following conditions are equivalent:*

(a) $V \in \mathcal{A}(\mathcal{H}_1, \mathcal{H}_2)$;
(b) $D_V = \mathcal{H}_1$, $R_V = \mathcal{H}_2$, V is an antilinear operator, and

$$\|Vf\|_2 = \|f\|_1, \forall f \in \mathcal{H}_1;$$

(c) $D_V = \mathcal{H}_1$, $R_V = \mathcal{H}_2$, and

$$(V(f)|V(g))_2 = (g|f)_1, \forall f, g \in \mathcal{H}_1.$$

Proof. The proof is an obvious modification of the proof of 10.1.20, and it follows from obvious modifications of 10.1.11 and 10.1.19 and their proofs. $\qquad\square$

10.3.18 Definition. Let $\mathcal{H}_1$ and $\mathcal{H}_2$ be Hilbert spaces such that the family $\mathcal{A}(\mathcal{H}_1, \mathcal{H}_2)$ is not empty. Two linear operators $A \in \mathcal{O}(\mathcal{H}_1)$ and $B \in \mathcal{O}(\mathcal{H}_2)$ are said to be *antiunitarily equivalent* if there exists $V \in \mathcal{A}(\mathcal{H}_1, \mathcal{H}_2)$ so that $B = VAV^{-1}$.

10.3.19 Remark. Let $\mathcal{H}_1$ and $\mathcal{H}_2$ be Hilbert spaces such that the family $\mathcal{UA}(\mathcal{H}_1, \mathcal{H}_2)$ is not empty. For $A \in \mathcal{O}(\mathcal{H}_1)$, $B \in \mathcal{O}(\mathcal{H}_2)$, $U \in \mathcal{A}(\mathcal{H}_1, \mathcal{H}_2)$, suppose that $B = UAU^{-1}$. Then it is easy to check that all the conditions listed in 4.6.4 still hold true. Indeed, conditions from a to h depend on U being a bijection from $\mathcal{H}_1$ onto $\mathcal{H}_2$, and condition i depends on U being an isomorphism of metric spaces (the mapping T_U defined in 4.6.3 is now an antiunitary operator from the Hilbert space $\mathcal{H}_1 \oplus \mathcal{H}_1$ onto the Hilbert space $\mathcal{H}_2 \oplus \mathcal{H}_2$; note that conditions e, f, g, h, i are still consistent because the image of a linear manifold under an antiunitary operator is a linear manifold, as can be easily seen). Furthermore, it is easy to check that propositions from a to e in 4.6.5 still hold true, while propositions from f to i get replaced by:

(f') $B - \bar{\lambda}\mathbb{1}_{\mathcal{H}_2} = U(A - \lambda\mathbb{1}_{\mathcal{H}_1})U^{-1}, \forall \lambda \in \mathbb{C}$;
(g') $\sigma(B) = \overline{\sigma(A)}$;
(h') $Ap\sigma(B) = \overline{Ap\sigma(A)}$;
(i') $\sigma_p(B) = \overline{\sigma_p(A)}$

(the bar means here complex conjugation, not closure).

10.3.20 Definition. Let $\mathcal{H}_1$ and $\mathcal{H}_2$ be Hilbert spaces such that the family $\mathcal{UA}(\mathcal{H}_1, \mathcal{H}_2)$ is not empty. Two linear operators $A \in \mathcal{O}(\mathcal{H}_1)$ and $B \in \mathcal{O}(\mathcal{H}_2)$ are said to be *unitarily-antiunitarily equivalent* if there exists $T \in \mathcal{UA}(\mathcal{H}_1, \mathcal{H}_2)$ so that $B = TAT^{-1}$.

10.3.21 Remark. For a Hilbert space $\mathcal{H}$, it is easy to check that the relation in $\mathcal{O}(\mathcal{H})$ of unitary-antiunitary equivalence is indeed an equivalence relation, in analogy with what we saw in 10.3.12a. This in linked to the group structure of $\mathcal{UA}(\mathcal{H})$ (cf. 10.3.16d).

10.4 Orthogonality in Hilbert spaces

The orthogonal decomposition theorem (also known as the projection theorem) is the cornerstone of the spectral theory in Hilbert spaces since the definition of a projection operator relies on this theorem, and projection operators are the building blocks of the spectral decomposition of unitary and self-adjoint operators. The first part of this section is devoted to this theorem and its corollaries. The second part deals with series of mutually orthogonal vectors.

10.4.1 Theorem (The orthogonal decomposition theorem). *Let M be a subspace of a Hilbert space $\mathcal{H}$. Then,*

$$\forall f \in \mathcal{H}, \exists!(f_1, f_2) \in M \times M^{\perp} \text{ so that } f = f_1 + f_2.$$

The pair (f_1, f_2) is called the orthogonal decomposition of f with respect to M.

Proof. Let $f \in \mathcal{H}$ and $d := \inf\{\|f - g\| : g \in M\}$. For every $n \in \mathbb{N}$ there exists $g \in M$ so that $\|f - g\| < d + \frac{1}{n}$ (if this were not true, we should have $d + \frac{1}{n} \leq \|f - g\|$ for all $g \in M$, contrary to d being the greater lower bound for $\{\|f - g\| : g \in M\}$), and for each $n \in \mathbb{N}$ we choose $g_n \in M$ so that $\|f - g_n\| < d + \frac{1}{n}$. Since obviously $d \leq \|f - g_n\|$, we have a sequence $\{g_n\}$ in M which is so that $\|f - g_n\| \to d$ as $n \to \infty$.

For all $n, m \in \mathbb{N}$, the equalities (cf. 10.1.12)

$$\|g_n - g_m\|^2 = \|(g_n - f) - (g_m - f)\|^2 = 2\|g_n - f\|^2 + 2\|g_m - f\|^2 - \|g_n + g_m - 2f\|^2,$$

together with $d \leq \|\frac{1}{2}(g_n + g_m) - f\|$ (since $\frac{1}{2}(g_n + g_m) \in M$), imply that

$$\|g_n - g_m\|^2 < 2\left(d + \frac{1}{n}\right)^2 + 2\left(d + \frac{1}{m}\right)^2 - 4d^2,$$

and this shows that $\{g_n\}$ is a Cauchy sequence. Since $\mathcal{H}$ is a complete metric space, there exists $g_0 \in \mathcal{H}$ so that $g_n \to g_0$ as $n \to \infty$, and $g_0 \in M$ because M is a closed subset of $\mathcal{H}$ (cf. 2.3.4). By the continuity of the sum and of the norm in $\mathcal{H}$ (cf. 4.1.6) we have $\|f - g_0\| = d$.

Now we prove that $f - g_0 \in M^\perp$. For every $h \in M$ we have

$$\|f - g_0\|^2 = d^2 \leq \|f - (g_0 - \alpha h)\|^2 = \|(f - g_0) + \alpha h\|^2, \forall \alpha \in \mathbb{C}$$

(since $g_0 - \alpha h \in M$), and hence

$$0 \leq |\alpha|^2 \|h\|^2 + 2 \operatorname{Re} \alpha (f - g_0|h), \forall \alpha \in \mathbb{C}.$$

From this, for every $h \in M - \{0_\mathcal{H}\}$, by putting $\alpha := -\|h\|^{-2} (h|f - g_0)$ we have

$$0 \leq \|h\|^{-2}(|(h|f - g_0)|^2 - 2|(h|f - g_0)|^2),$$

which obviously implies $(h|f - g_0) = 0$.

Thus, by letting $f_1 := g_0$ and $f_2 := f - g_0$, we have proved that

$$\exists (f_1, f_2) \in M \times M^\perp \text{ so that } f = f_1 + f_2.$$

To prove uniqueness, suppose that $(f_1', f_2') \in M \times M^\perp$ is so that $f = f_1' + f_2'$. Then,

$$f_1 - f_1' = f_2' - f_2.$$

Since $f_1 - f_1' \in M$ and $f_2' - f_2 \in M^\perp$ (cf. 10.2.13), we have $f_1 - f_1' = f_2' - f_2 = 0_\mathcal{H}$ by 10.2.10f, and hence $(f_1', f_2') = (f_1, f_2)$. $\qquad \square$

10.4.2 Remarks.

(a) The existence part of the statement of 10.4.1 can be rephrased as follows (cf. 3.1.8): if M is a subspace of a Hilbert space $\mathcal{H}$, then $\mathcal{H} = M + M^\perp$.

(b) In 10.4.1, the condition that the linear manifold M be closed is essential. This is proved by the following counterexample. In the Hilbert space ℓ^2 (cf. 10.3.8d), ℓ_f is a linear manifold and it is not a closed set since $\overline{\ell_f} = \ell^2$ and $\ell_f \neq \ell^2$ (cf. 2.3.9c). Now, for any Hilbert space $\mathcal{H}$ and any linear manifold M in $\mathcal{H}$ such that $\overline{M} = \mathcal{H}$ and $M \neq \mathcal{H}$, we have $M^\perp = \mathcal{H}^\perp = \{0_\mathcal{H}\}$ (cf. 10.2.11 and 10.2.10a) and hence $\mathcal{H} \neq M + M^\perp$.

(c) In 10.4.1, the condition that the inner product space $\mathcal{H}$ be complete is essential. This is proved by the following counterexample. For $a, b \in \mathbb{R}$, let $c \in (a, b)$ and define the subset $M(a, c)$ and $M(c, b)$ of the inner product space $\mathcal{C}(a, b)$ (cf. 10.1.5.b) by letting:

$$M(a, c) := \{\varphi \in \mathcal{C}(a, b) : \varphi(x) = 0, \forall x \in (c, b)\};$$
$$M(c, b) := \{\varphi \in \mathcal{C}(a, b) : \varphi(x) = 0, \forall x \in (a, c)\}.$$

Obviously, $M(c, b) \subset M(a, c)^{\perp}$. We will prove by contraposition the inclusion $M(a, c)^{\perp} \subset M(c, b)$, i.e. that

$$[\varphi \in \mathcal{C}(a, b), \exists x_0 \in (a, c) \text{ s.t. } \varphi(x_0) \neq 0] \Rightarrow$$
$$[\exists \psi \in M(a, c) \text{ s.t. } (\varphi|\psi) \neq 0].$$

Assume that $\varphi \in \mathcal{C}(a, b)$ and $x_0 \in (a, c)$ are so that $\varphi(x_0) \neq 0$, and suppose e.g. that $\operatorname{Re} \varphi(x_0) > 0$ (the argument would be analogue if $\operatorname{Re} \varphi(x_0) < 0$, or $\operatorname{Im} \varphi(x_0) > 0$, or $\operatorname{Im} \varphi(x_0) < 0$); since $\operatorname{Re} \varphi$ is a continuous function (cf. 2.7.6), there exists $\varepsilon > 0$ so that $\operatorname{Re} \varphi(x) > 0$ for all $x \in (x_0 - \varepsilon, x_0 + \varepsilon) \cap [a, b]$, and we can choose ε so that $(x_0 - \varepsilon, x_0 + \varepsilon) \subset (a, c)$; then the function

$\psi : [a, b] \to \mathbb{C}$

$$x \mapsto \psi(x) := \begin{cases} 0 & \text{if } x \notin (x_0 - \varepsilon, x_0 + \varepsilon), \\ x - (x_0 - \varepsilon) & \text{if } x \in (x_0 - \varepsilon, x_0), \\ -x + (x_0 + \varepsilon) & \text{if } x \in [x_0, x_0 + \varepsilon) \end{cases}$$

is a continuous function such that $\psi(x) = 0$ for all $x \notin (x_0 - \varepsilon, x_0 + \varepsilon)$ and $\psi(x) > 0$ for all $x \in (x_0 - \varepsilon, x_0 + \varepsilon)$. Now, $\psi \in M(a, c)$ and

$$\operatorname{Re}(\varphi|\psi) = \int_{[a,b]} (\operatorname{Re} \varphi)\psi \, dm > 0$$

since $(\operatorname{Re} \varphi)\psi \in \mathcal{L}^{+}([a, b], (\mathcal{A}(d_{\mathbb{R}}))^{[a,b]})$ and therefore $\int_{[a,b]}(\operatorname{Re} \varphi)\psi \, dm = 0$ would imply $\operatorname{Re} \varphi(x)\psi(x) = 0$ m-a.e. on $[a, b]$ and hence (cf. 10.1.5b) $\operatorname{Re} \varphi(x)\psi(x) = 0$ for all $x \in [a, b]$, which is not true. This proves by contraposition that $M(a, c)^{\perp} \subset M(c, b)$. Hence, $M(c, b) = M(a, c)^{\perp}$. The equality $M(a, c) = M(c, b)^{\perp}$ can be proved in a similar way. Thus, $M(a, c)$ is a subspace of $\mathcal{C}(a, b)$ (cf. 10.2.13) but

$$\mathcal{C}(a, b) \neq M(a, c) + M(c, b) = M(a, c) + M(a, c)^{\perp},$$

since $\varphi(c) = 0$ for all $\varphi \in M(a, c) + M(c, b)$. In view of 10.4.1, this proves that $\mathcal{C}(a, b)$ is not a Hilbert space.

Summing up, we have proved that the inner product space $\mathcal{C}(a, b)$ is not a Hilbert space, and that a subspace M of $\mathcal{C}(a, b)$ exists such that $\mathcal{C}(a, b) \neq M + M^{\perp}$.

10.4.3 Corollary. *Let M and N be subspaces of a Hilbert space and suppose that $M \subset N$. Then,*

$$\forall f \in N, \exists! (f_1, f_2) \in M \times (M^{\perp} \cap N) \text{ so that } f = f_1 + f_2.$$

Proof. Since N is a subspace, it can be regarded as a Hilbert space on its own (cf. 10.3.2) and M can be considered a subspace of N (cf. 3.1.4b and 2.3.3). Since the orthogonal complement of M in the Hilbert space N is obviously $M^\perp \cap N$, the equality of the statement follows from 10.4.1 with N substituted for $\mathcal{H}$ and $M^\perp \cap N$ for $M^\perp$. □

10.4.4 Corollaries. Let $\mathcal{H}$ be a Hilbert space. Then:

(a) A subset S of $\mathcal{H}$ is a subspace of $\mathcal{H}$ iff $S = S^{\perp\perp}$.
(b) For every subset S of $\mathcal{H}$, $VS = S^{\perp\perp}$.
(c) For every linear manifold M in $\mathcal{H}$, $\overline{M} = M^{\perp\perp}$.
(d) For a linear manifold M in $\mathcal{H}$, $\overline{M} = \mathcal{H}$ iff $M^\perp = \{0_\mathcal{H}\}$.

Proof. a: Let S be a subset of $\mathcal{H}$. If $S = S^{\perp\perp}$ then S is a subspace of $\mathcal{H}$ by 10.2.13. If S is a subspace, then $\mathcal{H} = S + S^\perp$ by 10.4.2a and this implies $S = S^{\perp\perp}$ by 10.2.15.

b: Let S be a subset of $\mathcal{H}$. From corollary a we have $VS = (VS)^{\perp\perp}$ and from 10.2.11 we have $(VS)^\perp = S^\perp$ and hence $(VS)^{\perp\perp} = S^{\perp\perp}$.

c: For a linear manifold M in $\mathcal{H}$ we have $\overline{M} = VM$ (cf. 4.1.14), and hence $\overline{M} = M^{\perp\perp}$ by corollary b.

d: For any subset S of any inner product space we have

$$S^\perp = \{0_X\} \Leftrightarrow S^{\perp\perp} = X,$$

in view of 10.2.10a,e. Thus, for a linear manifold M in $\mathcal{H}$ we have

$$M^\perp = \{0_\mathcal{H}\} \Leftrightarrow \overline{M} = \mathcal{H}$$

by corollary c. □

10.4.5 Remark. In all the corollaries of 10.4.1 proved in 10.4.4, the condition that the inner product space $\mathcal{H}$ be complete is essential. We prove this by a counterexample, which shows that if $\mathcal{H}$ were not complete then the statement of corollary 10.4.4d would not be true. Since each corollary listed in 10.4.4 implies the following one (see the proof of 10.4.4), this actually shows that if $\mathcal{H}$ were not complete then no corollary listed in 10.4.4 would be true.

In the inner product space ℓ_f (cf. 10.1.5a), which is not a Hilbert space (cf. 10.3.8d), let

$$M := \left\{ \{x_n\} \in \ell_f : \sum_{n=1}^\infty \frac{1}{n} x_n = 0 \right\}.$$

It is obvious that M is a linear manifold.

Suppose that a sequence $\{\xi_k\}$ in M and $\xi \in \ell_f$ are given so that $\xi = \lim_{k\to\infty} \xi_k$; setting $\{y_n\} := \xi$ and $\{x_{k,n}\} := \xi_k$, by the Schwarz inequality in ℓ^2 (cf. 10.3.8d) we have, for every $k \in \mathbb{N}$,

$$\left| \sum_{n=1}^{\infty} \frac{1}{n} y_n \right| = \left| \sum_{n=1}^{\infty} \frac{1}{n} y_n - \sum_{n=1}^{\infty} \frac{1}{n} x_{k,n} \right| = \left| \sum_{n=1}^{\infty} \frac{1}{n} (y_n - x_{k,n}) \right|$$

$$\leq \sqrt{\sum_{n=1}^{\infty} \frac{1}{n^2}} \sqrt{\sum_{n=1}^{\infty} |y_k - x_{k,n}|^2} = \sqrt{\sum_{n=1}^{\infty} \frac{1}{n^2}} \|\xi - \xi_k\|;$$

this shows that $\sum_{n=1}^{\infty} \frac{1}{n} y_n = 0$ and hence that $\xi \in M$; thus, M is a closed subset of ℓ_f (cf. 2.3.4), and hence $\overline{M} = M$ (cf. 2.3.9c).

Now, for each $k \in \mathbb{N}$ let η_k be the element of M defined by $\eta_k := \{w_{k,n}\}$ with

$$w_{k,n} = \begin{cases} \frac{1}{k} & \text{if } n = 1, \\ -\delta_{k,n} & \text{if } n \neq 1 \end{cases};$$

then we have, for $\zeta := \{z_n\} \in \ell_f$,

$$0 = (\eta_k | \zeta) = \frac{1}{k} z_1 - z_k \Rightarrow z_k = \frac{1}{k} z_1;$$

therefore, if $\zeta \in M^{\perp}$ then $z_k = \frac{1}{k} z_1$ for all $k \in \mathbb{N}$, and hence (since $\zeta \in \ell_f$) $z_1 = 0$, and hence $z_k = 0$ for all $k \in \mathbb{N}$. This proves that $M^{\perp} = \{0_{\ell_f}\}$. However, $\overline{M} = M \neq \ell_f$.

10.4.6 Corollary. *Let A be a linear operator in a Hilbert space. Then, the spectrum of A is a closed subset of $\mathbb{C}$.*

Proof. We denote by $\mathcal{H}$ the Hilbert space in which A is defined. We prove the statement by proving that the resolvent set of A is open (cf. 4.5.1 and 2.3.1). If $\rho(A) = \emptyset$ then there is nothing to prove. If $\mathcal{H}$ is the zero Hilbert space (cf. 10.3.8a), then it is immediate to see that $\rho(A) = \mathbb{C}$. In what follows we assume that $\rho(A) \neq \emptyset$ and that $\mathcal{H}$ is a non-zero Hilbert space.

Let λ be an arbitrary element of $\rho(A)$. Then the operator $A - \lambda \mathbb{1}_{\mathcal{H}}$ is injective, the operator $(A - \lambda \mathbb{1}_{\mathcal{H}})^{-1}$ is bounded, and $\|(A - \lambda \mathbb{1}_{\mathcal{H}})^{-1}\| \neq 0$ (indeed, if we had $\|(A - \lambda \mathbb{1}_{\mathcal{H}})^{-1}\| = 0$ then we should have $R_{A-\lambda \mathbb{1}_{\mathcal{H}}} = D_{(A-\lambda \mathbb{1}_{\mathcal{H}})^{-1}} = \{0_{\mathcal{H}}\}$ since $(A - \lambda \mathbb{1}_{\mathcal{H}})^{-1}$ is injective, but this would be contradictory to $\overline{R_{A-\lambda \mathbb{1}_{\mathcal{H}}}} = \mathcal{H}$). We will prove that $\rho(A)$ is open by proving that

$$\left[\mu \in \mathbb{C}, |\mu - \lambda| < \frac{1}{\|(A - \lambda \mathbb{1}_{\mathcal{H}})^{-1}\|} \right] \Rightarrow \mu \in \rho(A)$$

(cf. 2.2.2). Let then $\mu \in \mathbb{C}$ be such that $|\mu - \lambda| < \frac{1}{\|(A-\lambda \mathbb{1}_{\mathcal{H}})^{-1}\|}$. We prove that $\mu \in \rho(A)$ in two steps. In step I we prove that the operator $A - \mu \mathbb{1}_{\mathcal{H}}$ is injective and the operator $(A - \mu \mathbb{1}_{\mathcal{H}})^{-1}$ is bounded, and in step II we prove that $\overline{R_{A-\mu \mathbb{1}_{\mathcal{H}}}} = \mathcal{H}$.

Step I: For every $f \in D_A$ we have

$$\|f\| = \|(A - \lambda \mathbb{1}_{\mathcal{H}})^{-1}(A - \lambda \mathbb{1}_{\mathcal{H}})f\| \leq \|(A - \lambda \mathbb{1}_{\mathcal{H}})^{-1}\| \|(A - \lambda \mathbb{1}_{\mathcal{H}})f\|$$

(cf. 4.2.5b), and also

$$\|(A - \lambda \mathbb{1}_{\mathcal{H}})f\| - |\mu - \lambda| \|f\| \leq \|(A - \lambda \mathbb{1}_{\mathcal{H}})f - (\mu - \lambda)f\| = \|(A - \mu \mathbb{1}_{\mathcal{H}})f\|$$

(cf. 4.1.2b), and hence

$$\|f\| - |\mu - \lambda|\|(A - \lambda \mathbb{1}_{\mathcal{H}})^{-1}\|\|f\| \leq \|(A - \lambda \mathbb{1}_{\mathcal{H}})^{-1}\|(\|(A - \lambda \mathbb{1}_{\mathcal{H}})f\| - |\mu - \lambda|\|f\|)$$
$$\leq \|(A - \lambda \mathbb{1}_{\mathcal{H}})^{-1}\|\|(A - \mu \mathbb{1}_{\mathcal{H}})f\|,$$

and hence

$$\frac{1 - |\mu - \lambda|\|(A - \lambda \mathbb{1}_{\mathcal{H}})^{-1}\|}{\|(A - \lambda \mathbb{1}_{\mathcal{H}})^{-1}\|}\|f\| \leq \|(A - \mu \mathbb{1}_{\mathcal{H}})f\|.$$

Since $0 < 1 - |\mu - \lambda|\|(A - \lambda \mathbb{1}_{\mathcal{H}})^{-1}\|$, we have by 4.2.3 that the operator $A - \mu \mathbb{1}_{\mathcal{H}}$ is injective and the operator $(A - \mu \mathbb{1}_{\mathcal{H}})^{-1}$ is bounded.

Step II: We prove by contraposition that $\overline{R_{A - \mu \mathbb{1}_{\mathcal{H}}}} = \mathcal{H}$. Assume to the contrary that $\overline{R_{A - \mu \mathbb{1}_{\mathcal{H}}}} \neq \mathcal{H}$. Since $R_{A - \mu \mathbb{1}_{\mathcal{H}}}$ is a linear manifold in $\mathcal{H}$ (cf. 3.2.2a), this implies by 10.4.4d that there exists $f \in \mathcal{H}$ such that $f \neq 0_{\mathcal{H}}$ and $f \in (R_{A - \mu \mathbb{1}_{\mathcal{H}}})^{\perp}$. Since $\overline{R_{A - \lambda \mathbb{1}_{\mathcal{H}}}} = \mathcal{H}$, there exists a sequence $\{f_n\}$ in $D_{(A - \lambda \mathbb{1}_{\mathcal{H}})^{-1}} = R_{A - \lambda \mathbb{1}_{\mathcal{H}}}$ such that $f_n \to f$ (cf. 2.3.12). For each $n \in \mathbb{N}$, we define $g_n := (A - \lambda \mathbb{1}_{\mathcal{H}})^{-1} f_n$; we have $g_n \in R_{(A - \lambda \mathbb{1}_{\mathcal{H}})^{-1}} = D_{A - \lambda \mathbb{1}_{\mathcal{H}}} = D_A$, and hence $g_n \in D_{A - \mu \mathbb{1}_{\mathcal{H}}}$; since $(f|(A - \mu \mathbb{1}_{\mathcal{H}})g_n) = 0$, we have

$$\|f\|^2 \leq \|(A - \mu \mathbb{1}_{\mathcal{H}})g_n\|^2 + \|f\|^2 = \|(A - \mu \mathbb{1}_{\mathcal{H}})g_n - f\|^2$$

(cf. 10.2.3); we also have

$$\|(A - \mu \mathbb{1}_{\mathcal{H}})g_n - (A - \lambda \mathbb{1}_{\mathcal{H}})g_n\| = |\mu - \lambda|\|g_n\| \leq |\mu - \lambda|\|(A - \lambda \mathbb{1}_{\mathcal{H}})^{-1}\|\|f_n\|;$$

thus, we have

$$\|f\| \leq \|(A - \mu \mathbb{1}_{\mathcal{H}})g_n - f\|$$
$$\leq \|(A - \mu \mathbb{1}_{\mathcal{H}})g_n - (A - \lambda \mathbb{1}_{\mathcal{H}})g_n\| + \|(A - \lambda \mathbb{1}_{\mathcal{H}})g_n - f\|$$
$$\leq |\mu - \lambda|\|(A - \lambda \mathbb{1}_{\mathcal{H}})^{-1}\|\|f_n\| + \|f_n - f\|.$$

By the continuity of the norm (cf. 4.1.6a), this implies

$$\|f\| \leq \lim_{n \to \infty} (|\mu - \lambda|\|(A - \lambda \mathbb{1}_{\mathcal{H}})^{-1}\|\|f_n\| + \|f_n - f\|) = |\mu - \lambda|\|(A - \lambda \mathbb{1}_{\mathcal{H}})^{-1}\|\|f\|,$$

and this implies, since $\|f\| \neq 0$,

$$1 \leq |\mu - \lambda|\|(A - \lambda \mathbb{1}_{\mathcal{H}})^{-1}\|,$$

which is contrary to the hypothesis that was assumed for μ. $\qquad\square$

The results we prove in 10.4.8, which are sometimes known as the Riesz–Fisher theorem, are corollaries of the next theorem, which is an extension of 10.2.3.

10.4.7 Theorem. *Let $\{f_n\}$ be a sequence in a inner product space X and suppose that $(f_n|f_m) = 0$ if $n \neq m$. Then:*

(a) *if the series $\sum_{n=1}^{\infty} f_n$ is convergent then the series $\sum_{n=1}^{\infty} \|f_n\|^2$ is convergent in $\mathbb{R}$, i.e. $\sum_{n=1}^{\infty} \|f_n\|^2 < \infty$, and $\sum_{n=1}^{\infty} \|f_n\|^2 = \|\sum_{n=1}^{\infty} f_n\|^2$;*

(b) if X is a Hilbert space and $\sum_{n=1}^{\infty} \|f_n\|^2 < \infty$ then the series $\sum_{n=1}^{\infty} f_n$ is convergent.

Proof. For each $n \in \mathbb{N}$, let $s_n := \sum_{k=1}^{n} f_k$ and $\sigma_n := \sum_{k=1}^{n} \|f_k\|^2$. We recall that the series $\sum_{n=1}^{\infty} f_n$ is said to be convergent if the sequence $\{s_n\}$ is convergent, and that we write $\sum_{n=1}^{\infty} f_n := \lim_{n\to\infty} s_n$ when $\{s_n\}$ is convergent (cf. 2.1.10). Similarly, the series $\sum_{n=1}^{\infty} \|f_n\|^2$ is said to be convergent in $\mathbb{R}$ if the sequence $\{\sigma_n\}$ is convergent in $\mathbb{R}$, and we write $\sum_{n=1}^{\infty} \|f_n\|^2 := \lim_{n\to\infty} \sigma_n$ when $\{\sigma_n\}$ is convergent (cf. 5.4.1).

a: Assume that the series $\sum_{n=1}^{\infty} f_n$ is convergent. Then the continuity of the norm (cf. 4.1.6a) implies that the sequence $\{\|s_n\|^2\}$ is convergent in $\mathbb{R}$ and $\lim_{n\to\infty} \|s_n\|^2 = \|\lim_{n\to\infty} s_n\|^2$ (cf. 2.4.2). Since $\|s_n\|^2 = \sigma_n$ by 10.2.3, this means that the series $\sum_{n=1}^{\infty} \|f_n\|^2$ is convergent in $\mathbb{R}$, i.e. $\sum_{n=1}^{\infty} \|f_n\|^2 < \infty$ (cf. 5.4.1), and $\sum_{n=1}^{\infty} \|f_n\|^2 = \|\sum_{n=1}^{\infty} f_n\|^2$.

b: Assume $\sum_{n=1}^{\infty} \|f_n\|^2 < \infty$, i.e. that the sequence $\{\sigma_n\}$ is convergent in $\mathbb{R}$. Then $\{\sigma_n\}$ is a Cauchy sequence (cf. 2.6.2). Since $|\sigma_m - \sigma_n| = \sum_{k=n+1}^{m} \|f_n\|^2 = \|\sum_{k=n+1}^{m} f_n\|^2 = \|s_m - s_n\|^2$ for all $m, n \in \mathbb{N}$ such that $n < m$ (cf. 10.2.3), this implies that $\{s_n\}$ is a Cauchy sequence as well, and hence a convergent sequence if X is a complete metric space. $\qquad \square$

10.4.8 Corollaries. Let $\{u_n\}_{n\in\mathbb{N}}$ be an o.n.s. in an inner product space X, and let $\{\alpha_n\}$ be a sequence in $\mathbb{C}$. Then:

(a) if the series $\sum_{n=1}^{\infty} \alpha_n u_n$ is convergent then $\sum_{n=1}^{\infty} |\alpha_n|^2 < \infty$ and $\sum_{n=1}^{\infty} |\alpha_n|^2 = \|\sum_{n=1}^{\infty} \alpha_n u_n\|^2$;

(b) if X is a Hilbert space and $\sum_{n=1}^{\infty} |\alpha_n|^2 < \infty$ then the series $\sum_{n=1}^{\infty} \alpha_n u_n$ is convergent;

(c) if the series $\sum_{n=1}^{\infty} \alpha_n u_n$ is convergent then $\alpha_k = (u_k | \sum_{n=1}^{\infty} \alpha_n u_n)$ for all $k \in \mathbb{N}$.

Proof. Letting $f_n := \alpha_n u_n$ for all $n \in \mathbb{N}$, statements a and b follow immediately from 10.4.7.

c: If the series $\sum_{n=1}^{\infty} \alpha_n u_n$ is convergent, then the continuity of the inner product (cf. 10.1.16c) implies that

$$\left(u_k \,\Big|\, \sum_{n=1}^{\infty} \alpha_n u_n \right) = \left(u_k \,\Big|\, \lim_{n\to\infty} \sum_{l=1}^{n} \alpha_l u_l \right)$$

$$= \lim_{n\to\infty} \sum_{l=1}^{n} \alpha_l \delta_{k,l} = \alpha_k, \forall k \in \mathbb{N}.$$

$\qquad \square$

10.4.9 Proposition. *Let $\{f_n\}$ be a sequence in a Hilbert space such that $(f_n|f_m) = 0$ if $n \neq m$, and let β be a bijection from $\mathbb{N}$ onto $\mathbb{N}$. Then the series $\sum_{n=1}^{\infty} f_{\beta(n)}$ is convergent iff the series $\sum_{n=1}^{\infty} f_n$ is convergent. If these series are convergent then their sums are the same, i.e. $\sum_{n=1}^{\infty} f_{\beta(n)} = \sum_{n=1}^{\infty} f_n$.*

Proof. By 10.4.7, the series $\sum_{n=1}^{\infty} f_{\beta(n)}$ is convergent iff $\sum_{n=1}^{\infty} \|f_{\beta(n)}\|^2 < \infty$ and the series $\sum_{n=1}^{\infty} f_n$ is convergent iff $\sum_{n=1}^{\infty} \|f_n\|^2 < \infty$. Now, $\sum_{n=1}^{\infty} \|f_{\beta(n)}\|^2 = \sum_{n=1}^{\infty} \|f_n\|^2$ by 5.4.3.

Suppose that the series $\sum_{n=1}^{\infty} f_{\beta(n)}$ and $\sum_{n=1}^{\infty} f_n$ are convergent. Then, by the continuity of the inner product,

$$\left(f_k \Big| \sum_{n=1}^{\infty} f_{\beta(n)} - \sum_{n=1}^{\infty} f_n\right) = \sum_{n=1}^{\infty} \left(f_k | f_{\beta(n)}\right) - \sum_{n=1}^{\infty} \left(f_k | f_n\right)$$

$$= (f_k | f_k) - (f_k | f_k) = 0, \forall k \in \mathbb{N}.$$

Hence,

$$\sum_{n=1}^{\infty} f_{\beta(n)} - \sum_{n=1}^{\infty} f_n \in \{f_n\}_{n \in \mathbb{N}}^{\perp} = (V\{f_n\}_{n \in \mathbb{N}})^{\perp}$$

by 10.2.11. We also have

$$\sum_{n=1}^{\infty} f_{\beta(n)} - \sum_{n=1}^{\infty} f_n = \lim_{n \to \infty} \sum_{k=1}^{n} f_{\beta(k)} - \lim_{n \to \infty} \sum_{k=1}^{n} f_k \in V\{f_n\}_{n \in \mathbb{N}}$$

by 4.1.13, 2.3.10, and 3.1.7. Then, $\sum_{n=1}^{\infty} f_{\beta(n)} = \sum_{n=1}^{\infty} f_n$ by 10.2.10f. $\qquad \square$

10.4.10 Proposition. *Let $\{f_{n,s}\}_{(n,s) \in \mathbb{N} \times \mathbb{N}}$ be a family of vectors of a Hilbert space such that $(f_{n,s} | f_{m,t}) = 0$ if $(n,s) \neq (m,t)$. Then, the following conditions are equivalent:*

(a) the series $\sum_{(n,s) \in \mathbb{N} \times \mathbb{N}} f_{n,s}$ is convergent (there is no need to specify what ordering in $\mathbb{N} \times \mathbb{N}$ is used in order to define this series, in view of 10.4.9);

(b) the series $\sum_{s=1}^{\infty} f_{n,s}$ is convergent for all $n \in \mathbb{N}$ and the series $\sum_{n=1}^{\infty} \left(\sum_{s=1}^{\infty} f_{n,s}\right)$ is convergent.

If the series in conditions a and b are convergent, then for their sums we have $\sum_{(n,s) \in \mathbb{N} \times \mathbb{N}} f_{n,s} = \sum_{n=1}^{\infty} \left(\sum_{s=1}^{\infty} f_{n,s}\right)$.

Proof. By 10.4.7, condition a is true iff $\sum_{(n,s) \in \mathbb{N} \times \mathbb{N}} \|f_{n,s}\|^2 < \infty$ (cf. 5.4.7 for the symbol $\sum_{(n,s) \in \mathbb{N} \times \mathbb{N}} \|f_{n,s}\|^2$). Further, condition b is true iff $\sum_{s=1}^{\infty} \|f_{n,s}\|^2 < \infty$ for all $n \in \mathbb{N}$ and $\sum_{n=1}^{\infty} \left\|\sum_{s=1}^{\infty} f_{n,s}\right\|^2 < \infty$, since $\left(\sum_{s=1}^{\infty} f_{n,s} \Big| \sum_{t=1}^{\infty} f_{m,t}\right) = \sum_{s=1}^{\infty} \sum_{t=1}^{\infty} (f_{n,s} | f_{m,t}) = 0$ if $n \neq m$; also, if $\sum_{s=1}^{\infty} \|f_{n,s}\|^2 < \infty$ then $\left\|\sum_{s=1}^{\infty} f_{n,s}\right\|^2 = \sum_{s=1}^{\infty} \|f_{n,s}\|^2$ by 10.4.7; hence, condition b is true iff $\sum_{n=1}^{\infty} \left(\sum_{s=1}^{\infty} \|f_{n,s}\|^2\right) < \infty$. Then, conditions a and b are equivalent by 5.4.7.

Suppose that the series of conditions a and b are convergent. Then, by the same procedure as in 10.4.9, we see that

$$\left(f_{m,t} \Big| \sum_{(n,s) \in \mathbb{N} \times \mathbb{N}} f_{n,s} - \sum_{n=1}^{\infty} \left(\sum_{s=1}^{\infty} f_{n,s}\right)\right) = (f_{m,t} | f_{m,t}) - (f_{m,t} | f_{m,t})$$

$$= 0, \forall (m,t) \in \mathbb{N} \times \mathbb{N},$$

and hence that $\sum_{(n,s) \in \mathbb{N} \times \mathbb{N}} f_{n,s} = \sum_{n=1}^{\infty} \left(\sum_{s=1}^{\infty} f_{n,s}\right)$. $\qquad \square$

10.4.11 Remark. By using 10.3.6, it is easy to see that the statements of 10.4.9 and 10.4.10 hold true even if the inner product space of the statements, which we denote here by X, is not a Hilbert space. Simply, let $(\hat{X}, \iota)$ be a completion of X, substitute the vectors f_n or $f_{n,s}$ with $\iota(f_n)$ or $\iota(f_{n,s})$, and note that the series $\sum_{n=1}^{\infty} f_n$ (for instance) is convergent (in the metric space X) iff the series $\sum_{n=1}^{\infty} \iota(f_n)$ is convergent (in the metric space $\hat{X}$) and the sum $\sum_{n=1}^{\infty} \iota(f_n)$ is an element of R_ι.

10.5　The Riesz–Fréchet theorem

We present here the Riesz–Fréchet theorem and a result about bounded sesquilinear forms which follows from it. The Riesz–Fréchet theorem is actually a corollary of the orthogonal decomposition theorem, since its proof relies on 10.4.4a. The Riesz–Fréchet theorem is also known as the Riesz representation theorem, but we prefer to call it Riesz–Fréchet theorem in order to distinguish it from several other "Riesz representation theorems". For the same reason, we called Riesz–Markov theorem the theorem in 8.5.3, which is often named after the first author only.

10.5.1 Proposition. *Let h be a vector of an inner product space X and M a linear manifold in X. Then the function*

$$F_h : M \to \mathbb{C}$$

$$f \mapsto F_h f := (h|f)$$

is a continuous linear functional (for the definition of a linear functional, cf. 3.2.1), $\|F_h\| \le \|h\|$ and $\|F_h\| = \|h\|$ if $h \in M$.

Proof. The function F_h is a linear operator by property ip_1 of an inner product. Moreover, by 10.1.9 we have

$$|F_h f| \le \|h\| \|f\|, \forall f \in M,$$

and this shows that the linear operator F_h is bounded, and hence continuous (cf. 4.2.2), and that $\|F_h\| \le \|h\|$ (cf. 4.2.4). If $h = 0_X$ then $\|F_h\| = 0$. If $h \ne 0_X$ and $h \in M$ then $|F_h h| = \|h\| \|h\|$ shows that $\|F_h\| \ge \|h\|$ and hence that $\|F_h\| = \|h\|$. □

10.5.2 Theorem (The Riesz–Fréchet theorem). *Let $\mathcal{H}$ be a Hilbert space and F a continuous linear functional on $\mathcal{H}$, i.e. $F \in \mathcal{B}(\mathcal{H}, \mathbb{C})$. Then,*

$$\exists! h \in \mathcal{H} \text{ such that } F = F_h, \text{ i.e. } Ff = (h|f), \forall f \in \mathcal{H}.$$

Proof. *Existence:* If $N_F = \mathcal{H}$, take $h := 0_\mathcal{H}$. Now suppose $N_F \ne \mathcal{H}$. By 4.2.2, 4.4.3, 4.4.8, N_F is a subspace of $\mathcal{H}$. Then, $N_F^\perp \ne \{0_\mathcal{H}\}$ by 10.4.4a ($N_F^\perp = \{0_\mathcal{H}\}$ would imply $N_F = N_F^{\perp\perp} = \mathcal{H}$). Fix a non-zero element g of $N_F^\perp$, and define $h := \frac{\overline{Fg}}{\|g\|^2} g$. Since $Fg \ne 0$ because $N_F \cap N_F^\perp = \{0_\mathcal{H}\}$ (cf. 10.2.10f), we can write

$$f = \left(f - \frac{Ff}{Fg} g\right) + \frac{Ff}{Fg} g, \quad \forall f \in \mathcal{H},$$

and we have

$$F\left(f - \frac{Ff}{Fg}g\right) = 0, \text{ i.e. } f - \frac{Ff}{Fg}g \in N_F, \quad \forall f \in \mathcal{H},$$

and hence, as $h \in N_F^\perp$,

$$(h|f) = \left(h\Big|\frac{Ff}{Fg}g\right) = \frac{Fg}{\|g\|^2}\frac{Ff}{Fg}(g|g) = Ff, \quad \forall f \in \mathcal{H}.$$

Uniqueness: Suppose that h, h' are so that

$$Ff = (h|f) = (h'|f), \forall f \in \mathcal{H}.$$

Then $h' - h \in \mathcal{H}^\perp$ and hence $h' - h = 0_\mathcal{H}$ (cf. 10.2.10a), i.e. $h' = h$. $\qquad\square$

10.5.3 Remarks.

(a) The plan for the proof of 10.5.2 is prompted by the following considerations. If the theorem is true, then $N_F = \{h\}^\perp$ and hence

$$N_F^\perp = \{h\}^{\perp\perp} = V\{h\} = \{\alpha h : \alpha \in \mathbb{C}\}$$

(cf. 10.4.4b and 4.1.15). Thus, if we assume that the theorem is true and that $N_F \neq \mathcal{H}$ and hence $h \neq 0_\mathcal{H}$, for any non-zero element g of $N_F^\perp$ there exists $\alpha \in \mathbb{C}$ such that $\alpha \neq 0$ and $g = \alpha h$, and hence also such that $Fg = (h|g) = \overline{\alpha}^{-1}(g|g)$, which implies $\alpha^{-1} = \frac{\overline{F(g)}}{\|g\|^2}$. Therefore, if the theorem is true and $N_F \neq \mathcal{H}$, we must have $h = \frac{\overline{F(g)}}{\|g\|^2}g$ for any non-zero element g of $N_F^\perp$.

(b) In 10.5.2, the condition that the inner product space $\mathcal{H}$ be complete is essential. This is readily seen as follows. Let $\mathcal{H}$ be a Hilbert space and M a linear manifold in $\mathcal{H}$ such that $M \neq \mathcal{H}$ and $\overline{M} = \mathcal{H}$ (such are e.g. ℓ^2 and ℓ_f, cf. 10.3.8d). Then, M can be regarded as an inner product space (cf. 10.1.6), which is not complete by 2.6.8. Let $g \in \mathcal{H}$ be such that $g \notin M$. Then the function

$$M \ni f \mapsto Ff := (g|f) \in \mathbb{C}$$

is a continuous linear functional defined on M (cf. 10.5.1). However, there exists no $h \in M$ so that

$$Ff = (h|f), \forall f \in M,$$

since this would imply $h - g \in M^\perp$, and hence $h - g = 0_\mathcal{H}$ (cf. 10.4.4d), and hence $g \in M$.

10.5.4 Definition. A sesquilinear form ψ in a inner product space X is said to be *bounded* if it has the following property:

$$\exists m \in [0, \infty) \text{ such that } |\psi(f, g)| \leq m\|f\|\|g\|, \forall f, g \in D_\psi.$$

10.5.5 Proposition. *Let A be a bounded linear operator in a inner product space X. Then the function*

$$\psi : D_A \times D_A \to \mathbb{C}$$
$$(f, g) \mapsto \psi(f, g) := (Af|g)$$

is a bounded sesquilinear form in X.

  The same is true if ψ is defined by $\psi(f, g) := (f|Ag)$, $\forall (f, g) \in D_A \times D_A$.

Proof. For both the definitions of ψ given in the statement, the function ψ is a sesquilinear form since A is a linear operator and an inner product is a sesquilinear form. Moreover, for both definitions,

$$|\psi(f, g)| \leq \|A\| \|f\| \|g\|, \forall (f, g) \in D_A \times D_A,$$

by 10.1.7a and 4.2.5b. $\square$

10.5.6 Theorem. *Let $\mathcal{H}$ be a Hilbert space and ψ a bounded sesquilinear form on $\mathcal{H}$. Then*

$$\exists! A \in \mathcal{O}_E(\mathcal{H}) \text{ such that } \psi(f, g) = (Af|g), \forall f, g \in \mathcal{H}$$

(cf. 3.2.12 for the definition of $\mathcal{O}_E(\mathcal{H})$), and

$$\exists! B \in \mathcal{O}_E(\mathcal{H}) \text{ such that } \psi(f, g) = (f|Bg), \forall f, g \in \mathcal{H}.$$

The linear operators A and B are bounded, i.e. $A, B \in \mathcal{B}(\mathcal{H})$.

Proof. Existence: Let $m \geq 0$ be such that $|\psi(f, g)| \leq m\|f\| \|g\|$, $\forall f, g \in \mathcal{H}$. For each $f \in \mathcal{H}$, define the function

$$F_f : \mathcal{H} \to \mathbb{C}$$
$$g \mapsto F_f(g) := \psi(f, g),$$

which is a linear functional in view of property sf_2 of ψ (cf. 10.1.1), and is continuous since (cf. 4.2.2)

$$|F_f g| \leq (m\|f\|)\|g\|, \forall g \in \mathcal{H};$$

hence, by 10.5.2,

$$\exists! h_f \in \mathcal{H} \text{ such that } \psi(f, g) = F_f g = (h_f|g), \forall g \in \mathcal{H}.$$

Then, we can define the mapping

$$A : \mathcal{H} \to \mathcal{H}$$
$$f \mapsto Af := h_f \text{ if } h_f \in \mathcal{H} \text{ is such that } (h_f|g) = \psi(f, g), \forall g \in \mathcal{H},$$

which is obviously such that $\psi(f, g) = (Af|g)$, $\forall f, g \in \mathcal{H}$.

  The mapping A is a linear operator since, for all $\alpha, \beta \in \mathbb{C}$ and $f_1, f_2 \in \mathcal{H}$,

$$(\alpha A f_1 + \beta A f_2|g) = \overline{\alpha}(Af_1|g) + \overline{\beta}(Af_2|g)$$
$$= \overline{\alpha}\psi(f_1, g) + \overline{\beta}\psi(f_2, g)$$
$$= \psi(\alpha f_1 + \beta f_2, g) = (A(\alpha f_1 + \beta f_2)|g), \forall g \in \mathcal{H},$$

in view of property sf_3 of ψ, and hence $\alpha Af_1 + \beta Af_2 = A(\alpha f_1 + \beta f_2)$. Moreover,

$$|(Af|g)| = |\psi(f,g)| \le m\|f\|\|g\|, \forall f, g \in \mathcal{H},$$

proves that A is bounded, in view of 10.1.14.

Finally, we note that the function

$$\tilde{\psi} : \mathcal{H} \times \mathcal{H} \to \mathbb{C}$$

$$(f, g) \mapsto \tilde{\psi}(f, g) := \overline{\psi(g, f)}$$

is obviously a bounded sesquilinear form on $\mathcal{H}$. Therefore, what was proved above implies that there exists $B \in \mathcal{B}(\mathcal{H})$ such that

$$\tilde{\psi}(f, g) = (Bf|g), \forall f, g \in \mathcal{H},$$

and hence such that

$$\psi(g, f) = (g|Bf), \forall f, g \in \mathcal{H}.$$

Uniqueness: If $A, A' \in \mathcal{O}_E(\mathcal{H})$ are such that

$$\psi(f, g) = (Af|g) = (A'f|g), \forall f, g \in \mathcal{H},$$

then $A = A'$ by 10.2.12. And similarly for the uniqueness of B. $\qquad\square$

10.6 Complete orthonormal systems

Throughout this section, $\mathcal{H}$ denotes an abstract Hilbert space.

10.6.1 Proposition. *Let $\{u_i\}_{i \in I}$ be any o.n.s. in $\mathcal{H}$. The family of indices $I_f := \{i \in I : (u_i|f) \ne 0\}$ is countable, for every $f \in \mathcal{H}$. If I_f is denumerable then the series $\sum_{i \in I_f} (u_i|f)\, u_i$ is convergent and its sum is the same whatever ordering is chosen in I_f for the definition of this series. Thus, we can define*

$$\sum_{i \in I} (u_i|f)\, u_i := \sum_{i \in I_f} (u_i|f)\, u_i, \forall f \in \mathcal{H},$$

where some ordering in I_f is understood. We have

$$\sum_{i \in I} (u_i|f)\, u_i \in V\{u_i\}_{i \in I} \text{ and } f - \sum_{i \in I} (u_i|f)\, u_i \in (V\{u_i\}_{i \in I})^\perp, \forall f \in \mathcal{H}.$$

Proof. For every $f \in \mathcal{H}$, it was proved in 10.2.8b that I_f was countable. Now, suppose that I_f is denumerable. Since

$$((u_i|f)\, u_i|\, (u_k|f)\, u_k) = \overline{(u_i|f)}\, (u_k|f)\, (u_i|u_k) = 0 \text{ if } i \ne k,$$

10.4.9 proves that the choice of the ordering in I_f, which is necessary for the definition of the series $\sum_{i \in I_f} (u_i|f)\, u_i$, is immaterial both for the convergence of the series and, in case of convergence, for its sum. Moreover, it was proved in 10.2.8b that, for whatever ordering in I_f,

$$\sum_{i \in I_f} |(u_i|f)|^2 \le \|f\|^2,$$

and hence 10.4.8b implies that the series $\sum_{i \in I_f} (u_i|f) u_i$ is convergent. From 4.1.13, 2.3.10, 3.1.7 we have

$$\sum_{i \in I_f} (u_i|f) u_i \in V\{u_i\}_{i \in I}.$$

For each $k \in I$ we also have, using the continuity of inner product if I_f is denumerable,

$$\left(u_k \Big| f - \sum_{i \in I_f} (u_i|f) u_i \right) = (u_k|f) - \sum_{i \in I_f} (u_i|f) \delta_{k,i}$$

$$= \begin{cases} 0 & \text{if } k \notin I_f, \\ (u_k|f) - (u_k|f) & \text{if } k \in I_f. \end{cases}$$

In view of 10.2.11, this proves that

$$f - \sum_{i \in I_f} (u_i|f) u_i \in (\{u_i\}_{i \in I})^\perp = (V\{u_i\}_{i \in I})^\perp.$$

$\square$

10.6.2 Proposition. *Let $\{u_i\}_{i \in I}$ be an o.n.s. in $\mathcal{H}$. For $f \in \mathcal{H}$, if I_f is denumerable then the series $\sum_{i \in I_f} |(u_i|f)|^2$ is convergent and its sum is the same whatever ordering is chosen in I_f in order to define this series. Thus, we can define*

$$\sum_{i \in I} |(u_i|f)|^2 := \sum_{i \in I_f} |(u_i|f)|^2, \forall f \in \mathcal{H},$$

where some ordering in I_f is understood. We have

$$\sum_{i \in I} |(u_i|f)|^2 = \left\| \sum_{i \in I} (u_i|f) u_i \right\|^2, \forall f \in \mathcal{H}.$$

For $f, g \in \mathcal{H}$, if $I_f \cap I_g$ is denumerable then the series $\sum_{i \in I_f \cap I_g} (f|u_i)(u_i|g)$ is convergent and its sum is the same whatever ordering is chosen in $I_f \cap I_g$ for the definition of this series. Thus, we can define

$$\sum_{i \in I} (f|u_i)(u_i|g) := \sum_{i \in I_f \cap I_g} (f|u_i)(u_i|g), \forall f, g \in \mathcal{H},$$

where some ordering in $I_f \cap I_g$ is understood. We have

$$\sum_{i \in I} (f|u_i)(u_i|g) = \left(\sum_{i \in I} (u_i|f) u_i \Big| \sum_{k \in I} (u_k|g) u_k \right), \forall f, g \in \mathcal{H}.$$

Proof. For the part of the statement concerning converge of series and independence of sums from the orderings, cf. 10.2.8b.

For $f, g \in \mathcal{H}$, suppose that $I_f \cup I_g$ is denumerable and let $\{i_n\}_{n \in \mathbb{N}} := I_f \cup I_g$. Then the continuity of inner product (cf. 10.1.6c and 2.4.2) implies that

$$\lim_{N \to \infty} \left(\sum_{n=1}^{N} (u_{i_n}|f) \, u_{i_n} \Big| \sum_{k=1}^{N} (u_{i_k}|g) \, u_{i_k} \right) = \left(\sum_{n=1}^{\infty} (u_{i_n}|f) \, u_{i_n} \Big| \sum_{k=1}^{\infty} (u_{i_k}|g) \, u_{i_k} \right)$$

$$= \left(\sum_{i \in I_f} (u_i|f) \, u_i \Big| \sum_{k \in I_g} (u_k|g) \, u_k \right).$$

Also, by properties ip_1, ip_2, ip_5 of inner product,

$$\lim_{N \to \infty} \left(\sum_{n=1}^{N} (u_{i_n}|f) \, u_{i_n} \Big| \sum_{k=1}^{N} (u_{i_k}|g) \, u_{i_k} \right) = \lim_{N \to \infty} \sum_{n=1}^{N} \sum_{k=1}^{N} \overline{(u_{i_n}|f)} \, (u_{i_k}|g) \, \delta_{i,k}$$

$$= \lim_{N \to \infty} \sum_{n=1}^{N} (f|u_{i_n}) \, (u_{i_n}|g) = \sum_{n=1}^{\infty} (f|u_{i_n}) \, (u_{i_n}|g) = \sum_{i \in I_f \cap I_g} (f|u_i) \, (u_i|g).$$

Thus,

$$\sum_{i \in I_f \cap I_g} (f|u_i) \, (u_i|g) = \left(\sum_{i \in I_f} (u_i|f) \, u_i \Big| \sum_{k \in I_g} (u_k|g) \, u_k \right).$$

If $I_f \cup I_g$ is finite then this equality follows solely from properties ip_1, ip_2, ip_5 of an inner product.

By letting $g := f$ in the equality above, for every $f \in \mathcal{H}$ we have

$$\sum_{i \in I_f} |(u_i|f)|^2 = \left\| \sum_{i \in I_f} (u_i|f) \, u_i \right\|^2.$$

$\square$

10.6.3 Definitions. An o.n.s. $\{u_i\}_{i \in I}$ in $\mathcal{H}$ is said to be *complete in a subspace* M of $\mathcal{H}$ if $V\{u_i\}_{i \in I} = M$.

An o.n.s. in $\mathcal{H}$ which is complete in $\mathcal{H}$ is called a *complete orthonormal system* (briefly, *c.o.n.s.*) in $\mathcal{H}$.

10.6.4 Theorem. *Let M be a subspace of $\mathcal{H}$ and $\{u_i\}_{i \in I}$ an o.n.s. in $\mathcal{H}$ such that $u_i \in M$ for all $i \in I$. Then, the following conditions are equivalent:*

(a) $\{u_i\}_{i \in I}$ *is complete in M;*
(b) $f = \sum_{i \in I} (u_i|f) \, u_i, \, \forall f \in M$;
(c) $(f|g) = \sum_{i \in I} (f|u_i) \, (u_i|g), \, \forall f, g \in M$;
(d) $\|f\|^2 = \sum_{i \in I} |(u_i|f)|^2, \, \forall f \in M$;
(e) $[f \in M \text{ and } (u_i|f) = 0, \forall i \in I] \Rightarrow f = 0_{\mathcal{H}}.$

If $M = \mathcal{H}$, the equality in condition b is called Fourier expansion and the equalities in conditions c and d are called Parseval's identities.

Proof. $a \Rightarrow b$: For every $f \in M$, 4.1.13, 2.3.10, and 3.1.7 imply that

$$f - \sum_{i \in I} (u_i|f) \, u_i \in M,$$

and hence, in view of 10.6.1,

$$f - \sum_{i \in I} (u_i|f) \, u_i \in M \cap (V\{u_i\}_{i \in I})^{\perp}.$$

Therefore, if condition a is true then

$$f - \sum_{i \in I} (u_i|f) \, u_i \in M \cap M^{\perp} = \{0_{\mathcal{H}}\}, \forall f \in M$$

(cf. 10.2.10f). This proves that condition a implies condition b.

$b \Rightarrow c$: This is obvious, since in 10.6.2 we saw that

$$\sum_{i \in I} (f|u_i)\,(u_i|g) = \left(\sum_{i \in I} (u_i|f)\, u_i \middle| \sum_{i \in I} (u_k|g)\, u_k \right), \forall f, g \in \mathcal{H}.$$

$c \Rightarrow d$: Set $g := f$ in condition c.

$d \Rightarrow e$: We prove this by contraposition. Assume that $f \in M$ exists such that $f \neq 0_{\mathcal{H}}$ and $(u_i|f) = 0$ for all $i \in I$. Then,

$$\|f\|^2 \neq 0 = \sum_{i \in I} |(u_i|f)|^2.$$

$e \Rightarrow a$: We prove this by contraposition. First note that $\{u_i\}_{i \in I} \subset M$ implies $V\{u_i\}_{i \in I} \subset M$ and hence, in view of 10.4.3,

$$M = V\{u_i\}_{i \in I} + ((V\{u_i\}_{i \in I})^{\perp} \cap M).$$

Therefore, if $V\{u_i\}_{i \in I} \neq M$ then $(\{u_i\}_{i \in I})^{\perp} \cap M \neq \{0_{\mathcal{H}}\}$ and hence

$$\exists f \in M \text{ such that } f \in (V\{u_i\}_{i \in I})^{\perp} = (\{u_i\}_{i \in I})^{\perp} \text{ and } f \neq 0_{\mathcal{H}},$$

where 10.2.11 has been used. This proves that if condition a is not true then condition e is not true. $\qquad \square$

10.6.5 Remarks.

(a) The equivalence of conditions a and e in 10.6.4 can be rephrased as follows: an o.n.s. $\{u_i\}_{i \in I}$ in $\mathcal{H}$ is complete in a subspace M of $\mathcal{H}$ iff $\{u_i\}_{i \in I} \subset M$ and $(\{u_i\}_{i \in I})^{\perp} \cap M = \{0_{\mathcal{H}}\}$; in particular, $\{u_i\}_{i \in I}$ is a c.o.n.s. in $\mathcal{H}$ iff $(\{u_i\}_{i \in I})^{\perp} = \{0_{\mathcal{H}}\}$.

(b) Suppose that $\{u_i\}_{i \in I}$ is a c.o.n.s. in $\mathcal{H}$ and M is a linear manifold in $\mathcal{H}$ such that $\{u_i\}_{i \in I} \subset M$. Then M is dense in $\mathcal{H}$. Indeed,

$$\{u_i\}_{i \in I} \subset M \Rightarrow L\{u_i\}_{i \in I} \subset M \Rightarrow$$
$$\mathcal{H} = V\{u_i\}_{i \in I} = \overline{L\{u_i\}_{i \in I}} \subset \overline{M} \Rightarrow \overline{M} = \mathcal{H}$$

(cf. 3.1.6c, 4.1.13, 2.3.9d).

(c) Let M be a subspace of $\mathcal{H}$. A subset $\{u_i\}_{i\in I}$ of M is clearly an o.n.s. in the framework of the Hilbert space $(\mathcal{H}, \sigma, \mu, \phi)$ iff it is an o.n.s. in the framework of the Hilbert space $(M, \sigma_{M\times M}, \mu_{\mathbb{C}\times M}, \phi_{M\times M})$ (cf. 10.3.2). Moreover, conditions from b to e in 10.6.4 are clearly the same whether they are interpreted in the framework of $(\mathcal{H}, \sigma, \mu, \phi)$ or in the framework of $(M, \sigma_{M\times M}, \mu_{\mathbb{C}\times M}, \phi_{M\times M})$. Therefore, an o.n.s. in $\mathcal{H}$ is complete in the subspace M iff it is a c.o.n.s. in the Hilbert space $(M, \sigma_{M\times M}, \mu_{\mathbb{C}\times M}, \phi_{M\times M})$ (note that, in condition $V\{u_i\}_{i\in I} = M$ of 10.6.3, $V\{u_i\}_{i\in I}$ stands for the intersection of all the subspaces of $\mathcal{H}$, not of all the subspaces of M, that contain $\{u_i\}_{i\in I}$).

10.6.6 Corollaries.

(a) Assume that, for $N \in \mathbb{N}$, an o.n.s. $\{u_1, ..., u_N\}$ exists in $\mathcal{H}$. Then,

$$V\{u_1, ..., u_N\} = \left\{ \sum_{n=1}^{N} \alpha_n u_n : (\alpha_1, ..., \alpha_N) \in \mathbb{C}^N \right\}.$$

(b) Assume that a denumerable o.n.s. $\{u_n\}_{n\in\mathbb{N}}$ exists in $\mathcal{H}$. Then,

$$V\{u_n\}_{n\in\mathbb{N}} = \left\{ \sum_{n=1}^{\infty} \alpha_n u_n : \{\alpha_n\} \in \ell^2 \right\}.$$

Proof. a: The o.n.s. $\{u_1, ..., u_N\}$ is obviously complete in the subspace $V\{u_1, ..., u_N\}$. Then 10.6.4 proves that

$$f = \sum_{n=1}^{N} (u_n|f)\, u_n, \forall f \in V\{u_1, ..., u_n\}.$$

Hence,

$$V\{u_1, ..., u_N\} \subset \left\{ \sum_{n=1}^{N} \alpha_n u_n : (\alpha_1, ..., \alpha_N) \in \mathbb{C}^N \right\}.$$

Since the opposite inclusion is obvious, we have the equality of the statement.

b: First, we note that the condition $\{\alpha_n\} \in \ell^2$ is necessary and sufficient for the series $\sum_{n=1}^{\infty} \alpha_n u_n$ to converge (cf. 10.4.8a,b). Then, since the o.n.s. $\{u_n\}_{n\in\mathbb{N}}$ is obviously complete in the subspace $V\{u_n\}_{n\in\mathbb{N}}$, 10.6.4 proves that

$$f = \sum_{n=1}^{\infty} (u_n|f)\, u_n, \forall f \in V\{u_n\}_{n\in\mathbb{N}};$$

moreover, $\{(u_n|f)\} \in \ell^2$ for every $f \in \mathcal{H}$ (cf. 10.2.8b). Thus we have

$$V\{u_n\}_{n\in\mathbb{N}} \subset \left\{ \sum_{n=1}^{\infty} \alpha_n u_n : \{\alpha_n\} \in \ell^2 \right\}.$$

Since the opposite inclusion follows from 4.1.13, 2.3.10, 3.1.7, we have the equality of the statement. $\qquad\square$

10.6.7 Examples.

(a) Let $N \in \mathbb{N}$ and, for $1 \leq k \leq N$, define the element e_k of $\mathbb{C}^N$ by

$$e_k := (\delta_{k,1}, ..., \delta_{k,n}, ..., \delta_{k,N}).$$

The family $\{e_1, ..., e_N\}$ is an o.n.s. in the Hilbert space $\mathbb{C}^N$ (cf. 10.3.8c) since it is obvious that $(e_k|e_l) = \delta_{k,l}$ for $k, l = 1, ..., N$, and it is complete by 10.6.4 because

$$x_k = (e_k|\xi), \forall \xi := (x_1, ..., x_N) \in \mathbb{C}^N, \forall k \in \{1, ..., N\}$$

proves that

$$[\xi \in \mathbb{C}^N \text{ and } (e_k|\xi) = 0, \forall k \in \{1, ..., N\}] \Rightarrow \xi = 0_{\mathbb{C}^N}.$$

(b) The family $\{\delta_k\}_{k\in\mathbb{N}}$, which is an o.n.s. in ℓ_f (cf. 10.2.5a), is obviously an o.n.s. in the Hilbert space ℓ^2 (cf. 10.3.8d) as well, and it is complete by 10.6.4 because

$$x_k = (\delta_k|\xi), \forall \xi := \{x_n\} \in \ell^2, \forall k \in \mathbb{N}$$

proves that

$$[\xi \in \ell^2 \text{ and } (\delta_k|\xi) = 0, \forall k \in \mathbb{N}] \Rightarrow \xi = 0_{\ell^2}.$$

10.6.8 Proposition. *Let $\mathcal{H}_1$ and $\mathcal{H}_2$ be Hilbert spaces such that the family $\mathcal{UA}(\mathcal{H}_1, \mathcal{H}_2)$ is not empty, and let $U \in \mathcal{UA}(\mathcal{H}_1, \mathcal{H}_2)$. Then:*

(a) for every o.n.s. $\{u_i\}_{i\in I}$ in $\mathcal{H}_1$, $\{Uu_i\}_{i\in I}$ is an o.n.s. in $\mathcal{H}_2$;
(b) for every c.o.n.s. $\{u_i\}_{i\in I}$ in $\mathcal{H}_1$, $\{Uu_i\}_{i\in I}$ is a c.o.n.s. in $\mathcal{H}_2$.

Proof. a: For every o.n.s. $\{u_i\}_{i\in I}$ in $\mathcal{H}_1$ we have:

$$(Uu_i|Uu_k)_2 = (u_i|u_k)_1 = \delta_{i,k} \text{ if } U \in \mathcal{U}(\mathcal{H}_1, \mathcal{H}_2);$$
$$(Uu_i|Uu_k)_2 = (u_k|u_i)_1 = \delta_{i,k} \text{ if } U \in \mathcal{A}(\mathcal{H}_1, \mathcal{H}_2).$$

b: For any family $\{u_i\}_{i\in I}$ of vectors of $\mathcal{H}_1$, suppose that $f \in \mathcal{H}_2$ is such that

$$(Uu_i|f)_2 = 0, \forall i \in I.$$

Since $(Uu_i|f)_2 = (u_i|U^{-1}f)_1$ if $U \in \mathcal{U}(\mathcal{H}_1, \mathcal{H}_2)$ and $(Uu_i|f)_2 = (U^{-1}f|u_i)_1$ if $U \in \mathcal{A}(\mathcal{H}_1, \mathcal{H}_2)$, in either case we have

$$(u_i|U^{-1}f)_1 = 0, \forall i \in I.$$

If $\{u_i\}_{i\in I}$ is a c.o.n.s. in $\mathcal{H}_1$, this implies $U^{-1}f = 0_{\mathcal{H}_1}$ (cf. 10.6.4) and hence $f = 0_{\mathcal{H}_2}$ since U is a linear or antilinear operator. In view of statement a and of 10.6.4, this proves that $\{Uu_i\}_{i\in I}$ is a c.o.n.s. in $\mathcal{H}_2$ if $\{u_i\}_{i\in I}$ is a c.o.n.s. in $\mathcal{H}_1$. $\square$

10.6.9 Proposition. *Let $\mathcal{H}_1$ and $\mathcal{H}_2$ be Hilbert spaces such that a c.o.n.s. $\{u_i\}_{i \in I}$ exists in $\mathcal{H}_1$ and a c.o.n.s. $\{v_i\}_{i \in I}$ exists in $\mathcal{H}_2$ which are indexed by the same set I of indices. For every $f \in \mathcal{H}_1$, the set $I_f := \{i \in I : (u_i|f)_1 \neq 0\}$ is countable. If I_f is denumerable then the series $\sum_{i \in I_f} (u_i|f)_1 v_i$ and $\sum_{i \in I_f} (f|u_i)_1 v_i$ are convergent and their sums are independent from the orderings chosen in I_f for their definitions. The mapping*

$$U : \mathcal{H}_1 \to \mathcal{H}_2$$
$$f \mapsto Uf := \sum_{i \in I} (u_i|f)_1 v_i := \sum_{i \in I_f} (u_i|f)_1 v_i$$

is an element of $\mathcal{U}(\mathcal{H}_1, \mathcal{H}_2)$, while the mapping

$$V : \mathcal{H}_1 \to \mathcal{H}_2$$
$$f \mapsto Vf := \sum_{i \in I} (f|u_i)_1 v_i := \sum_{i \in I_f} (f|u_i)_1 v_i$$

is an element of $\mathcal{A}(\mathcal{H}_1, \mathcal{H}_2)$. For the inverse operators we have

$$U^{-1} g = \sum_{i \in I} (v_i|g)_2 u_i := \sum_{i \in I_g} (v_i|g)_2 u_i, \forall g \in \mathcal{H}_2,$$
$$V^{-1} g = \sum_{i \in I} (g|v_i)_2 u_i := \sum_{i \in I_g} (g|v_i)_2 u_i, \forall g \in \mathcal{H}_2,$$

where $I_g := \{i \in I : (v_i|g)_2 \neq 0\}$.

Proof. We set out the proof for U, from which the proof for V can be obtained by obvious modifications.

For each $f \in \mathcal{H}_1$, it was proved in 10.2.8b that I_f was countable and $\sum_{i \in I_f} |(u_i|f)_1|^2 < \infty$; then, if I_f is denumerable, 10.4.9 implies that the convergence of the series $\sum_{i \in I_f} (u_i|f)_1 v_i$ and its sum do not depend on the ordering chosen in I_f, and the series is convergent by 10.4.8b.

For each $g \in \mathcal{H}_2$, the same arguments as above prove that I_g is countable and that if I_g is denumerable then the series $\sum_{i \in I_g} (v_i|g)_2 u_i$ is convergent and its sum is independent from the ordering chosen in I_g, and we can define the vector f of $\mathcal{H}_1$ by

$$f := \sum_{i \in I_g} (v_i|g)_2 u_i;$$

we have

$$(u_i|f)_1 = \begin{cases} (v_i|g)_2 & \text{if } i \in I_g, \\ 0 & \text{if } i \notin I_g \end{cases}$$

(cf. 10.4.8c); thus, $I_f = I_g$ and

$$Uf = \sum_{i \in I_f} (u_i|f)_1 v_i = \sum_{i \in I_g} (v_i|g)_2 v_i = g$$

since $\{v_i\}_{i \in I}$ is a c.o.n.s. in $\mathcal{H}_2$ (cf. 10.6.4b). This proves that $R_U = \mathcal{H}_2$. Moreover, for all $f, h \in \mathcal{H}_1$,

$$(Uf|Uh)_2 = \sum_{i \in I_f} \sum_{k \in I_h} \overline{(u_i|f)_1} \, (u_k|h)_1 \, (v_i|v_k)_2$$

$$= \sum_{i \in I_f \cap I_h} \overline{(u_i|f)_1} \, (u_i|h)_1 = (f|h)_1$$

since $\{u_i\}_{i \in I}$ is a c.o.n.s. in $\mathcal{H}_1$ (cf. 10.6.4c). In view of 10.1.20, this proves that $U \in \mathcal{U}(\mathcal{H}_1, \mathcal{H}_2)$ (in the proof for V, 10.1.20 must be replaced by 10.3.17).

Since U is an isomorphism, it is injective and the proof of surjectivity given above for U proves also the part of the statement concerning U^{-1}.   □

10.6.10 Remark. Suppose that a c.o.n.s. $\{u_i\}_{i \in I}$ exists in a Hilbert space $\mathcal{H}$. Then the mapping

$$V : \mathcal{H} \to \mathcal{H}$$

$$f \mapsto Vf := \sum_{i \in I} (f|u_i) \, u_i$$

is an element of $\mathcal{A}(\mathcal{H})$ (cf. 10.6.9) and $V^2 = \mathbb{1}_{\mathcal{H}}$, as can be easily seen. Thus, every antiunitary operator in $\mathcal{H}$ is the product of a unitary operator multiplied by V. In fact, for $A \in \mathcal{A}(\mathcal{H})$, $A = (AV)V$ and $AV \in \mathcal{U}(\mathcal{H})$ (cf. 10.3.16c).

10.7   Separable Hilbert spaces

It can be proved that there exists a c.o.n.s. in any non-zero Hilbert space, if the axiom of choice is assumed, in its equivalent form called Zorn's lemma (cf. e.g. Weidmann, 1980, th. 3.10). However, it is possible to prove that there exists a c.o.n.s. in every separable non-zero Hilbert space without using the axiom of choice. In this section, we give the proof of the existence of a c.o.n.s. in this reduced form only, because in our opinion the idea of a c.o.n.s. is really useful in separable Hilbert spaces only (mainly because, as we see below, a c.o.n.s. is countable iff the Hilbert space is separable). The importance of a theorem which proves the existence of a c.o.n.s. is that it justifies all the procedures in which complete orthonormal systems are used.

10.7.1 Theorem. *Suppose that a Hilbert space $\mathcal{H}$ is separable and non-zero. Then a countable c.o.n.s. exists in $\mathcal{H}$.*

Proof. Since $\mathcal{H}$ is separable, there exists a countable subset S of $\mathcal{H}$ so that $\overline{S} = \mathcal{H}$. It is easy to see that S must be denumerable since $\mathcal{H}$ is non-zero, and we can write $\{f_n\}_{n \in \mathbb{N}} := S$. Now, 3.1.19 implies that there exists a countable subset $\{f_{n_k}\}_{k \in I}$ of $\{f_n\}_{n \in \mathbb{N}}$ which is linearly independent and such that

$$L\{f_{n_k}\}_{k \in I} = L\{f_n\}_{n \in \mathbb{N}}.$$

Since $\{f_{n_k}\}_{k\in I}$ is a linearly independent subset of $\mathcal{H}$, 10.2.6 implies that there exists an o.n.s. $\{u_n\}_{n\in I}$ in $\mathcal{H}$ such that

$$L\{u_n\}_{n\in I} = L\{f_{n_k}\}_{k\in I},$$

and hence such that

$$L\{u_n\}_{n\in I} = LS.$$

Then we have

$$V\{u_n\}_{n\in I} = \overline{L\{u_n\}_{n\in I}} = \overline{LS} \supset \overline{S} = \mathcal{H}$$

(cf. 4.1.13, 3.1.6b, 2.3.9d), and hence $V\{u_n\}_{n\in I} = \mathcal{H}$. $\qquad\square$

10.7.2 Corollary. *Suppose that a Hilbert space $\mathcal{H}$ is separable, that M is a subspace of $\mathcal{H}$, and that $M \neq \{0_{\mathcal{H}}\}$. Then there exists a countable o.n.s. in $\mathcal{H}$ which is complete in the subspace M.*

Proof. Since M is a subspace, it can be regarded as a Hilbert space on its own (cf. 10.1.6 and 10.3.2), and it is not a zero Hilbert space since $M \neq \{0_{\mathcal{H}}\}$. Moreover, M is separable (cf. 2.3.20 and 10.1.15). Then, 10.7.1 proves that there exists a countable c.o.n.s. in the Hilbert space M, and hence a countable o.n.s. in $\mathcal{H}$ which is complete in the subspace M (cf. 10.6.5c). $\qquad\square$

10.7.3 Corollary. *Suppose that a Hilbert space $\mathcal{H}$ is separable, and let $\{u_i\}_{i\in I}$ be an o.n.s. in $\mathcal{H}$. Then there exists a c.o.n.s. in $\mathcal{H}$ which contains $\{u_i\}_{i\in I}$.*

Proof. If $(\{u_i\}_{i\in I})^{\perp} = \{0_{\mathcal{H}}\}$ then $\{u_i\}_{i\in I}$ is a c.o.n.s. in $\mathcal{H}$ (cf. 10.6.5a). Now assume $(\{u_i\}_{i\in I})^{\perp} \neq \{0_{\mathcal{H}}\}$. Then, 10.2.13 and 10.7.2 imply that there exists an o.n.s. $\{v_j\}_{j\in J}$ in $\mathcal{H}$ such that $V\{v_j\}_{j\in J} = (\{u_i\}_{i\in I})^{\perp}$. It is obvious that the family $\{u_i\}_{i\in I} \cup \{v_j\}_{j\in J}$ is an o.n.s. in $\mathcal{H}$. Moreover,

$$(\{v_j\}_{j\in J})^{\perp} = (V\{v_j\}_{j\in J})^{\perp} = ((\{u_i\}_{i\in I})^{\perp})^{\perp}$$

(cf. 10.2.11), and hence

$$(\{u_i\}_{i\in I} \cup \{v_j\}_{j\in J})^{\perp} = (\{u_i\}_{i\in I})^{\perp} \cap (\{v_j\}_{j\in J})^{\perp}$$
$$= (\{u_i\}_{i\in I})^{\perp} \cap ((\{u_i\}_{i\in I})^{\perp})^{\perp} = \{0_{\mathcal{H}}\}$$

(cf. 10.2.10c,f). This proves that $\{u_i\}_{i\in I} \cup \{v_j\}_{j\in J}$ is a c.o.n.s. in $\mathcal{H}$ (cf. 10.6.5a). $\qquad\square$

10.7.4 Remark. In the proof of the orthogonal decomposition theorem that was given in 10.4.1, the axiom of choice (cf. 1.2.22) was used in the construction of the sequence $\{g_n\}$ in M which was such that $\|f - g_n\| \to d$. Now, corollary 10.7.2 makes it possible to prove the orthogonal decomposition theorem without resorting to the axiom of choice, if the Hilbert space is separable. Indeed, if the Hilbert space $\mathcal{H}$ is separable and M is a non-zero subspace of $\mathcal{H}$, 10.7.2 proves that there exists an o.n.s. $\{u_i\}_{i \in I}$ in $\mathcal{H}$ which is complete in M (this o.n.s. is countable, but this has no relevance here). Since 10.6.1 proves that

$$\sum_{i \in I} (u_i | f) \, u_i \in V\{u_i\}_{i \in I} \text{ and } f - \sum_{i \in I} (u_i | f) \, u_i \in (V\{u_i\}_{i \in I})^{\perp}, \forall f \in \mathcal{H},$$

then for each $f \in \mathcal{H}$ we actually have a pair $(f_1, f_2) \in M \times M^{\perp}$ such that $f = f_1 + f_2$ if we define

$$f_1 := \sum_{i \in I} (u_i | f) \, u_i \text{ and } f_2 := f - \sum_{i \in I} (u_i | f) \, u_i$$

(the uniqueness of the pair can then be proved as in 10.4.1).

The next two theorems round off our exposition of the relation between separability of a Hilbert space and countability of orthonormal systems. Theorem 10.7.5 is the converse of theorem 10.7.1.

10.7.5 Theorem. *If there exists a countable c.o.n.s. in a Hilbert space $\mathcal{H}$ then $\mathcal{H}$ is separable.*

Proof. Assume that there exists a countable c.o.n.s. $\{u_n\}_{n \in I}$ in $\mathcal{H}$. We set out the proof of the separability of $\mathcal{H}$ for I denumerable, from which the proof for I finite can be derived easily. Let then $I := \mathbb{N}$, and fix $f \in \mathcal{H}$ and $\varepsilon > 0$. Since $V\{u_n\}_{n \in \mathbb{N}} = \mathcal{H}$, 4.1.13 and 2.3.12 imply that

$$\exists f_\varepsilon \in L\{u_n\}_{n \in \mathbb{N}} \text{ such that } \|f - f_\varepsilon\| < \frac{\varepsilon}{2},$$

and 3.1.7 implies that

$$\exists N_\varepsilon \in \mathbb{N}, \exists (\alpha_1^\varepsilon, ..., \alpha_{N_\varepsilon}^\varepsilon) \in \mathbb{C}^{N_\varepsilon} \text{ such that } f_\varepsilon = \sum_{n=1}^{N_\varepsilon} \alpha_n^\varepsilon u_n.$$

Since $\mathbb{Q}$ is dense in $\mathbb{R}$ (cf. 2.3.16), there exist $(a_1^\varepsilon, ..., a_{N_\varepsilon}^\varepsilon), (b_1^\varepsilon, ..., b_{N_\varepsilon}^\varepsilon) \in \mathbb{Q}^{N_\varepsilon}$ such that

$$|\operatorname{Re} \alpha_n^\varepsilon - a_n^\varepsilon| < \frac{\varepsilon}{4N_\varepsilon} \text{ and } |\operatorname{Im} \alpha_n^\varepsilon - b_n^\varepsilon| < \frac{\varepsilon}{4N_\varepsilon}, \forall n \in \{1, ..., N_\varepsilon\},$$

and hence such that

$$\left\| f - \sum_{n=1}^{N_\varepsilon} (a_n^\varepsilon + ib_n^\varepsilon) u_n \right\| \leq \|f - f_\varepsilon\| + \left\| f_\varepsilon - \sum_{n=1}^{N_\varepsilon} (a_n^\varepsilon + ib_n^\varepsilon) u_n \right\|$$

$$< \frac{\varepsilon}{2} + \sum_{n=1}^{N_\varepsilon} |\alpha_n^\varepsilon - (a_n^\varepsilon + ib_n^\varepsilon)|$$

$$\leq \frac{\varepsilon}{2} + \sum_{n=1}^{N_\varepsilon} |\operatorname{Re} \alpha_n^\varepsilon - a_n^\varepsilon| + \sum_{n=1}^{N_\varepsilon} |\operatorname{Im} \alpha_n^\varepsilon - b_n^\varepsilon| < \varepsilon.$$

This proves that the subset $\mathcal{H}_0$ of $\mathcal{H}$ defined by

$$\mathcal{H}_0 := \left\{ \sum_{n=1}^{N} (a_n + ib_n) u_n : N \in \mathbb{N}, (a_1, ..., a_N), (b_1, ..., b_N) \in \mathbb{Q}^N \right\}$$

is dense in $\mathcal{H}$. Now, the mapping

$$\bigcup_{N=1}^{\infty} (\mathbb{Q}^N \times \mathbb{Q}^N) \to \mathcal{H}_0$$

$$((a_1, ..., a_N), (b_1, ..., b_N)) \mapsto \sum_{n=1}^{N} (a_n + ib_n) u_n$$

is easily seen to be a bijection, and this proves that $\mathcal{H}_0$ is countable in view of the facts that are mentioned in 1.2.10. Thus, $\mathcal{H}$ is separable. $\qquad\square$

10.7.6 Remark. From 10.7.5 and 10.6.7 we have that the Hilbert spaces $\mathbb{C}^N$ and ℓ^2 (cf. 10.3.8c,d) are separable.

10.7.7 Theorem. *If a Hilbert space $\mathcal{H}$ is separable then every o.n.s. in $\mathcal{H}$ is countable.*

Proof. If $\mathcal{H}$ is a zero Hilbert space then there is nothing to prove. Now assume that $\mathcal{H}$ is a non-zero separable Hilbert space. Then there exists a countable subset S of $\mathcal{H}$ which is dense in $\mathcal{H}$. Let $\{u_i\}_{i \in I}$ be an o.n.s. in $\mathcal{H}$ and define

$$S_i := \left\{ f \in S : \|u_i - f\| < \frac{1}{\sqrt{2}} \right\}, \forall i \in I.$$

In view of 2.3.12, $S_i \neq \emptyset$ for each $i \in I$. Then, by the axiom of choice (cf. 1.2.22) there exists a mapping $\varphi : I \to S$ such that $\varphi(i) \in S_i$ for each $i \in I$. Moreover,

$$S_i \cap S_k = \emptyset \text{ if } i \neq k;$$

in fact, if $f \in S$ existed such that $f \in S_i \cap S_k$, we would have

$$\|u_i - u_k\| \leq \|u_i - f\| + \|f - u_k\| < \frac{2}{\sqrt{2}} = \sqrt{2},$$

while we have, if $i \neq k$,

$$\|u_i - u_k\| = \sqrt{(u_i - u_k | u_i - u_k)} = \sqrt{(u_i | u_i) + (u_k | u_k)} = \sqrt{2}.$$

Then, the mapping φ is injective and hence φ is a bijection from I onto R_φ, which is a countable set since it is a subset of S (cf. 1.2.10). Thus, I is countable and so is $\{u_i\}_{i \in I}$. $\qquad\square$

10.7.8 Proposition. *Suppose that a set X is not finite. Then there exists an injection $i : \mathbb{N} \to X$.*

Proof. We must produce a sequence $\{x_n\}$ in X such that $x_n \neq x_k$ if $n \neq k$. We define x_n for all $n \in \mathbb{N}$ by induction as follows. Since X is not finite, $X \neq \emptyset$; choose $x_1 \in X$. Then $X \neq \{x_1\}$ since $\{x_1\}$ is finite, so $X - \{x_1\} \neq \emptyset$; choose $x_2 \in X - \{x_1\}$. Assuming $x_1, ..., x_n$ already defined, $X \neq \{x_1, ..., x_n\}$ and hence $X - \{x_1, ..., x_n\} \neq \emptyset$; choose $x_{n+1} \in X - \{x_1, ..., x_n\}$. $\qquad\square$

10.7.9 Proposition. *Let $(\mathcal{H}, \sigma, \mu, \phi)$ be a Hilbert space. Then:*

(a) *If an o.n.s. in the Hilbert space $\mathcal{H}$ is a linear basis in the linear space $(\mathcal{H}, \sigma, \mu)$, then it is a finite set.*

(b) *If a c.o.n.s. in the Hilbert space $\mathcal{H}$ is finite, then it is a linear basis in the linear space $(\mathcal{H}, \sigma, \mu)$.*

Proof. a: The proof is by contradiction. Assume that $\{u_i\}_{i \in I}$ is an o.n.s. in the Hilbert space $\mathcal{H}$, and that it is not finite. Then there exists an injection $i : \mathbb{N} \to I$ (cf. 10.7.8) and we can define the vector $f := \sum_{n=1}^{\infty} \frac{1}{n} u_{i(n)}$ (cf. 10.4.8b). Next, assume that $\{u_i\}_{i \in I}$ is a linear basis in the linear space (X, σ, μ). Then there exist a finite subfamily $\{u_{i_1}, ..., u_{i_n}\}$ of $\{u_i\}_{i \in I}$ and $(\alpha_1, ..., \alpha_n) \in \mathbb{C}^n$ so that

$$f = \sum_{k=1}^{n} \alpha_k u_{i_k},$$

and hence so that

$$(u_i | f) = 0, \forall i \in I - \{i_1, ..., i_n\}.$$

But this is in contradiction to

$$\left(u_{i(n)} | f\right) = \frac{1}{n}, \forall n \in \mathbb{N},$$

which holds true by 10.4.8c.

b: If a c.o.n.s. in the Hilbert space $\mathcal{H}$ is finite, then 10.2.2 and 10.6.4b prove that it is a linear basis in the linear space (X, σ, μ). $\qquad\square$

10.7.10 Theorem. *Let $\mathcal{H}$ be a separable and non-zero Hilbert space. Then a countable c.o.n.s. S exists in $\mathcal{H}$ and:*

(a) *if S is finite then every other c.o.n.s. in $\mathcal{H}$ is finite and contains the same number of vectors as S;*

(b) *if S is denumerable then every other c.o.n.s. in $\mathcal{H}$ is denumerable.*

Proof. We already know that a countable c.o.n.s. exists in $\mathcal{H}$ (cf. 10.7.1).

a: If S is finite, then S is a linear basis by 10.7.9b. Then, every other c.o.n.s. S' in $\mathcal{H}$ must be finite by 3.1.16, since S' is a linearly independent subset of $\mathcal{H}$ by 10.2.2. But then S' is a linear basis by 10.7.9b, and hence S' must contain the same number of vectors as S by 3.1.17.

b: If S is denumerable then any other c.o.n.s. S' in $\mathcal{H}$ cannot be finite, for otherwise S would be finite by part a of the statement. Hence S' must be denumerable since it is countable by 10.7.7. $\qquad\square$

10.7.11 Definitions. We say that the *orthogonal dimension* of a separable and non-zero Hilbert space $\mathcal{H}$ is *finite and equal to N* if a c.o.n.s. in $\mathcal{H}$ is finite and contains N vectors (10.7.10a proves that if this is true for a c.o.n.s. in $\mathcal{H}$ then it is true for every other c.o.n.s. in $\mathcal{H}$), and we say that the orthogonal dimension of $\mathcal{H}$ is *denumerable* if a c.o.n.s. in $\mathcal{H}$ is denumerable (10.7.10b proves that if this is true for a c.o.n.s. in $\mathcal{H}$ then it is true for every other c.o.n.s. in $\mathcal{H}$).

The orthogonal dimension of a zero Hilbert space (cf. 10.3.8a) is defined to be zero.

If M is a subspace of a separable Hilbert space then M can be considered a separable Hilbert space (cf. the proof of 10.7.2), and the orthogonal dimension of this Hilbert space is said to be the *orthogonal dimension of the subspace M*.

10.7.12 Remarks.

(a) From 10.6.7a we have that the orthogonal dimension of the Hilbert space $\mathbb{C}^N$ is finite and equal to N, and from 10.6.7b we have that the orthogonal dimension of the Hilbert space ℓ^2 is denumerable.

(b) If M is a subspace of a separable Hilbert space whose orthogonal dimension is finite, then the orthogonal dimension of M is finite as well. This follows immediately from 10.7.2 and 10.7.3.

10.7.13 Theorem. *Suppose that the orthogonal dimension of a separable Hilbert space $(\mathcal{H}, \sigma, \mu, \phi)$ is denumerable. Then no countable linear basis exists in the linear space $(\mathcal{H}, \sigma, \mu)$.*

Proof. The proof is by contraposition. Assume that a countable linear basis B exists in the linear space $(\mathcal{H}, \sigma, \mu)$. Then, 10.2.6 implies that an o.n.s. S exists in $\mathcal{H}$ such that

$$LS = LB = \mathcal{H}.$$

Since S is a linearly independent subset of the linear space $(\mathcal{H}, \sigma, \mu)$ (cf. 10.2.2), this implies that S is a linear basis in the linear space $(\mathcal{H}, \sigma, \mu)$, and hence (cf. 10.7.9a) that S is a finite set, and hence that the orthogonal dimension of the Hilbert space $\mathcal{H}$ is finite since $LS = \mathcal{H}$ implies $VS = \mathcal{H}$, i.e. that S is a c.o.n.s. in $\mathcal{H}$. $\qquad\square$

10.7.14 Theorem. *Let $\mathcal{H}_1$ and $\mathcal{H}_2$ be Hilbert spaces, and suppose that $\mathcal{H}_1$ is separable. Then the following conditions are equivalent:*

(a) $\mathcal{H}_2$ is separable and the orthogonal dimensions of $\mathcal{H}_1$ and $\mathcal{H}_2$ are equal;
(b) $\mathcal{U}(\mathcal{H}_1, \mathcal{H}_2)$ is not empty (i.e. $\mathcal{H}_1$ and $\mathcal{H}_2$ are isomorphic);
(c) $\mathcal{A}(\mathcal{H}_1, \mathcal{H}_2)$ is not empty.

If $\mathcal{H}_1$ is not a zero space and if these conditions are satisfied, then a mapping $T : \mathcal{H}_1 \to \mathcal{H}_2$ is a unitary (or antiunitary) operator iff there exist a c.o.n.s. $\{u_n\}_{n \in I}$ in $\mathcal{H}_1$ and a c.o.n.s. $\{v_n\}_{n \in I}$ in $\mathcal{H}_2$, with $I := \{1, ..., N\}$ or $I := \mathbb{N}$, so that

$$Tf = \sum_{n \in I} (u_n | f)_1 \, v_n \quad \left(or \ Tf = \sum_{n \in I} (f | u_n)_1 \, v_n\right), \quad \forall f \in \mathcal{H}_1,$$

where $\sum_{n \in I}$ stands for $\sum_{n=1}^{N}$ or $\sum_{n=1}^{\infty}$.

Proof. The first half of the statement is trivial if $\mathcal{H}_1$ is a zero space. Then we assume that $\mathcal{H}_1$ is not a zero space.

The implications "$a \Rightarrow b$" and "$a \Rightarrow c$", as well as the "if" part of the second half of the statement are proved by 10.6.9.

The implications "$b \Rightarrow a$" and "$c \Rightarrow a$" are proved by 10.6.8b.

As to the "only if" part of the second half of the statement, let $T \in \mathcal{UA}(\mathcal{H}_1, \mathcal{H}_2)$ and let $\{u_n\}_{n \in I}$ be a c.o.n.s. in $\mathcal{H}_1$. We may assume $I := \{1, ..., N\}$ or $I := \mathbb{N}$ by 10.7.7. If we define $v_n := T u_n$ for all $n \in I$, then $\{T u_n\}_{n \in I}$ is a c.o.n.s. in $\mathcal{H}_2$ by 10.6.8b and we have, in view of 10.6.4b,

$$Tf = T \sum_{n \in I} (u_n|f)_1 \, u_n = \sum_{n \in I} (u_n|f)_1 \, T u_n = \sum_{n \in I} (u_n|f)_1 \, v_n, \forall f \in \mathcal{H}_1,$$

if $T \in \mathcal{U}(\mathcal{H}_1, \mathcal{H}_2)$ or

$$Tf = T \sum_{n \in I} (u_n|f)_1 \, u_n = \sum_{n \in I} \overline{(u_n|f)_1} T u_n = \sum_{n \in I} (f|u_n)_1 \, v_n, \forall f \in \mathcal{H}_1,$$

if $T \in \mathcal{A}(\mathcal{H}_1, \mathcal{H}_2)$, since T is a linear or antilinear operator and (if $I = \mathbb{N}$) since T is a continuous mapping in either case (cf. 10.1.21 and 4.6.2.d, or 10.3.16e, and 2.4.5). $\qquad \square$

10.7.15 Remarks.

(a) From 10.7.14 and 10.7.12 we have that any non-zero separable Hilbert space is isomorphic either to $\mathbb{C}^N$ for a suitable $N \in \mathbb{N}$ or to ℓ^2.

(b) If the orthogonal dimension of a separable Hilbert space $\mathcal{H}$ is finite and equal to N, for every c.o.n.s. $\{u_1, ..., u_N\}$ in $\mathcal{H}$ the mapping

$$U : \mathcal{H} \to \mathbb{C}^N$$
$$f \mapsto Uf := ((u_1|f), ..., (u_N|f))$$

is a unitary operator from $\mathcal{H}$ onto $\mathbb{C}^N$. This follows immediately from 10.6.9 with $\{u_i\}_{i \in I} := \{u_1, ..., u_N\}$ and $\{v_i\}_{i \in I} := \{e_1, ..., e_N\}$ (cf. 10.6.7a).

(c) If the orthogonal dimension of a separable Hilbert space $\mathcal{H}$ is denumerable, for every c.o.n.s. $\{u_n\}_{n \in \mathbb{N}}$ in $\mathcal{H}$ the mapping

$$U : \mathcal{H} \to \ell^2$$
$$f \mapsto Uf := \{(u_n|f)\}$$

is a unitary operator from $\mathcal{H}$ onto ℓ^2. This follows immediately from 10.6.9 with $\{u_i\}_{i \in I} := \{u_n\}_{n \in \mathbb{N}}$ and $\{v_i\}_{i \in I} := \{\delta_n\}_{n \in \mathbb{N}}$ (cf. 10.6.7b). In fact, for each $f \in \mathcal{H}$, the sequence $\xi := \{(u_n|f)\}$ is an element of ℓ^2 by 10.2.8b and $\sum_{n=1}^{\infty} |(u_n|f)|^2 < \infty$ implies that

$$\left\| \xi - \sum_{n=1}^{N} (u_n|f) \, \delta_n \right\|^2 = \sum_{n=N+1}^{\infty} |(u_n|f)|^2 \to 0 \text{ as } N \to \infty.$$

(d) The reason why, notwithstanding remark a, other separable Hilbert spaces are worth studying besides $\mathbb{C}^N$ and ℓ^2 is that there are problems which can be formulated in separable Hilbert spaces and which, although in their abstract form they are the same in all isomorphic Hilbert spaces, are actually easier to solve or even to phrase in certain Hilbert spaces than in others.

10.8 The finite-dimensional case

In the finite-dimensional case there is no distinction between inner product spaces and Hilbert spaces, nor between linear manifolds and subspaces. Moreover, all linear operators are bounded, and hence continuous. Furthermore, all propositions in a finite-dimensional Hilbert space can be expressed in the language of matrices.

10.8.1 Proposition. *Let (X, σ, μ, ϕ) be an inner product space and let M be a linear manifold in X. The following conditions are equivalent:*

(a) *M is finite-dimensional, i.e. the linear space $(M, \sigma_{M \times M}, \mu_{\mathbb{C} \times M})$ is finite-dimensional;*

(b) *the inner product space $(M, \sigma_{M \times M}, \mu_{\mathbb{C} \times M}, \phi_{M \times M})$ is a separable Hilbert space and its orthogonal dimension is finite.*

If these conditions are satisfied, then the following conditions also hold true:

(c) *M is a subspace of X;*

(d) *the linear dimension of the linear space $(M, \sigma_{M \times M}, \mu_{\mathbb{C} \times M})$ and the orthogonal dimension of the separable Hilbert space $(M, \sigma_{M \times M}, \mu_{\mathbb{C} \times M}, \phi_{M \times M})$ are equal.*

Proof. If $M = \{0_X\}$ then all the conditions of the statement are true. Therefore, suppose $M \neq \{0_X\}$.

$a \Rightarrow [b, c, d]$: Assume condition a, let N be the linear dimension of $(M, \sigma_{M \times M}, \mu_{\mathbb{C} \times M})$, and let $\{f_1, ..., f_N\}$ be a linear basis in M. Then by 10.2.6 there exists an o.n.s. $\{u_1, ..., u_N\}$ in X such that

$$L\{u_1, ..., u_N\} = L\{f_1, ..., f_N\} = M.$$

Now suppose that $\{g_n\}$ is a Cauchy sequence in M. Then,

$$\forall n \in \mathbb{N}, \exists \{\alpha_1^{(n)}, ... \alpha_N^{(n)}\} \in \mathbb{C}^N \text{ such that } g_n = \sum_{k=1}^{N} \alpha_k^{(n)} u_k,$$

and, by 10.2.3,

$$\sum_{k=1}^{N} |\alpha_k^{(n)} - \alpha_k^{(m)}|^2 = \|g_n - g_m\|^2 \to 0 \text{ as } n, m \to \infty.$$

Thus, for all $k \in \{1, ..., N\}$, the sequence $\{\alpha_k^{(n)}\}$ is a Cauchy sequence in $\mathbb{C}$ and therefore there exists $\alpha_k \in \mathbb{C}$ such that $\alpha_k = \lim_{n \to \infty} \alpha_k^{(n)}$. By the continuity of vector sum and of scalar multiplication, this implies that

$$\sum_{k=1}^{N} \alpha_k u_k = \lim_{n \to \infty} g_n.$$

This proves that the metric space $(M, d_{\phi_{M \times M}})$ is complete and consequently that $(M, \sigma_{M \times M}, \mu_{\mathbb{C} \times M}, \phi_{M \times M})$ is a Hilbert space. From 2.6.6a it follows that M is a closed subset of X, and hence a subspace of X. Further, suppose that $f \in M$ is such that $(u_k|f) = 0$ for all $k \in \{1, ..., N\}$. Then there exists $(\beta_1, ..., \beta_N) \in \mathbb{C}^N$ such that $f = \sum_{k=1}^{N} \beta_k u_k$ and also such that

$$\beta_k = \sum_{i=1}^{N} \beta_i \delta_{k,i} = \left(u_k \Big| \sum_{i=1}^{N} \beta_i u_i \right) = (u_k|f) = 0, \forall k \in \{1, ..., N\}.$$

This implies that $f = 0_M$. By 10.6.4, this proves that $\{u_1, ..., u_N\}$ is a c.o.n.s. in the Hilbert space $(M, \sigma_{M \times M}, \mu_{\mathbb{C} \times M}, \phi_{M \times M})$. Therefore, the Hilbert space $(M, \sigma_{M \times M}, \mu_{\mathbb{C} \times M}, \phi_{M \times M})$ is separable (cf. 10.7.5) and its orthogonal dimension is N.

$b \Rightarrow a$: This is proved by 10.7.9b. $\qquad\qquad\qquad\qquad\qquad\qquad\square$

10.8.2 Proposition. *Suppose that an inner product space X is finite-dimensional as a linear space. Then X is a separable Hilbert space, its orthogonal dimension is finite, and every linear manifold in X is a subspace of X.*

Proof. From 10.8.1 we obtain immediately that X is a separable Hilbert space and that its orthogonal dimension is finite.

Let M be a linear manifold in X. If $M = \{0_X\}$ then M is a subspace. Now suppose $M \neq \{0_X\}$. If we had proved in Section 3.1 that every linear manifold in a finite-dimensional linear space was finite-dimensional then we would have that M is a subspace of X by 10.8.1. However we did not prove that result and therefore we must take a different tack. From 2.3.20 we have that M is separable. Hence, there exists a countable subset S of M such that $M \subset \overline{S}$ (cf. 2.3.13). Proceeding as in the proof of 10.7.1 we see that there exists an o.n.s. $\{u_n\}_{n \in I}$ in X such that

$$L\{u_n\}_{n \in I} = LS,$$

and hence such that

$$M \subset \overline{S} \subset \overline{LS} = \overline{L\{u_n\}_{n \in I}}.$$

Now, $\{u_n\}_{n \in I}$ is a linearly independent subset of X (cf. 10.2.2) and hence it must be a finite set (cf. 3.1.16). Then, $L\{u_n\}_{n \in I}$ is a finite-dimensional linear manifold in X and hence it is a subspace of X by 10.8.1. Thus,

$$\overline{L\{u_n\}_{n \in I}} = L\{u_n\}_{n \in I}$$

and hence

$$M \subset L\{u_n\}_{n \in I}.$$

Since M is a linear manifold, we have

$$L\{u_n\}_{n \in I} = LS \subset M$$

and hence $M = L\{u_n\}_{n \in I}$. This proves that M is a subspace of X. $\qquad\square$

10.8.3 Proposition.

(A) Let A be a linear operator from an inner product space X to a normed space Y and suppose that the linear manifold D_A is finite-dimensional. Then A is bounded, and hence continuous.

(B) We say that a Hilbert space is finite-dimensional if it is separable and its orthogonal dimension is finite; in view of 10.7.9b and 10.8.2, this is equivalent to its being finite dimensional as a linear space.

Every linear operator in a finite-dimensional Hilbert space is bounded.

Proof. A: From 10.8.1 we have that D_A is a separable Hilbert space and that its orthogonal dimension is finite. Then, let $\{u_1, ..., u_N\}$ be a c.o.n.s. in the Hilbert space D_A, and define

$$K := \max\{\|Au_n\| : n \in \{1, ..., N\}\}.$$

We have, for all $f \in D_A$,

$$\|Af\| = \left\|A\sum_{n=1}^{N} (u_n|f)\, u_n\right\| \leq \sum_{n=1}^{N} |(u_n|f)|\, \|Au_n\|$$

$$\leq K \sum_{n=1}^{N} |(u_n|f)| \leq K\sqrt{N} \sqrt{\sum_{n=1}^{N} |(u_n|f)|^2} = K\sqrt{N}\|f\|,$$

where 10.6.4b,d and the Schwarz inequality in $\mathbb{C}^N$ (cf. 10.3.8c) have been used. This proves that A is bounded, and hence continuous (cf. 4.2.2).

B: Let A be a linear operator in a finite-dimensional Hilbert space. The domain D_A is a linear manifold in $\mathcal{H}$, and hence it is a subspace of $\mathcal{H}$ (cf. 10.8.2). Therefore, the orthogonal dimension of D_A is finite (cf. 10.7.12b). Then, proceeding as in the proof of statement A, we see that the operator A is bounded. $\qquad\square$

10.8.4 Remark. For $N \in \mathbb{N}$, we denote by $\mathbb{M}(N)$ the family of all N by N matrices with complex entries. We assume that the reader is familiar with the following facts:

the family $\mathbb{M}(N)$ can be given the structure of an associative algebra over $\mathbb{C}$ (cf. 3.3.1);

the associative algebra $\mathcal{O}_E(\mathbb{C}^N)$ (cf. 3.3.7) can be represented faithfully by $\mathbb{M}(N)$, i.e. an isomorphism (cf. 3.3.5) can be defined from the associative algebra $\mathcal{O}_E(\mathbb{C}^N)$ onto the associative algebra $\mathbb{M}(N)$, in such a way that if a linear operator $A \in \mathcal{O}_E(\mathbb{C}^N)$ is represented by the matrix $[\alpha_{ik}] \in \mathbb{M}(N)$ then

$$A(x_1, ..., x_i, ..., x_N) = \left(\sum_{k=1}^{N} \alpha_{1k}x_k, ..., \sum_{k=1}^{N} \alpha_{ik}x_k, ..., \sum_{k=1}^{N} \alpha_{Nk}x_k\right), \forall(x_1, ..., x_N) \in \mathbb{C}^N.$$

(if vectors of $\mathbb{C}^N$ are represented as column matrices, then A transforms a vector by the row-by-column multiplication of the matrix $[\alpha_{ik}]$ by the column matrix that represents the vector).

Now, let $\mathcal{H}$ be a finite-dimensional Hilbert space, let N be its dimension, and let $\{u_1, ..., u_N\}$ be a c.o.n.s. in $\mathcal{H}$. The mapping U of 10.7.15b is an isomorphism from $\mathcal{H}$ onto $\mathbb{C}^N$ and it is easy to see that the mapping

$$\mathcal{O}_E(\mathcal{H}) \ni A \mapsto UAU^{-1} \in \mathcal{O}_E(\mathbb{C}^N)$$

is an isomorphism from the associative algebra $\mathcal{O}_E(\mathcal{H})$ (cf. 3.3.7) onto the associative algebra $\mathcal{O}_E(\mathbb{C}^N)$. The composition of this isomorphism with the one from $\mathcal{O}_E(\mathbb{C}^N)$ onto $\mathbb{M}(N)$ mentioned above is obviously an isomorphism Φ_U from $\mathcal{O}_E(\mathcal{H})$ onto $\mathbb{M}(N)$. Now, for $A \in \mathcal{O}_E(\mathcal{H})$, $\Phi_U(A)$ is the element $[\alpha_{ik}]$ of $\mathbb{M}(N)$ such that, for all $(x_1, ..., x_N) \in \mathbb{C}^N$,

$$UAU^{-1}(x_1, ..., x_i, ..., x_N) = \left(\sum_{k=1}^{N} \alpha_{1k} x_k, ..., \sum_{k=1}^{N} \alpha_{ik} x_k, ..., \sum_{k=1}^{N} \alpha_{Nk} x_k \right);$$

we also have

$$UAU^{-1}(x_1, ..., x_i, ..., x_N) = UA \sum_{k=1}^{N} x_k u_k = U \sum_{k=1}^{N} x_k A u_k$$

$$= \left(\left(u_1 | \sum_{k=1}^{N} x_k A u_k \right), ..., \left(u_i | \sum_{k=1}^{N} x_k A u_k \right), ..., \left(u_N | \sum_{k=1}^{N} x_k A u_k \right) \right)$$

$$= \left(\sum_{k=1}^{N} (u_1 | A u_k) x_k, ..., \sum_{k=1}^{N} (u_i | A u_k) x_k, ..., \sum_{k=1}^{N} (u_N | A u_k) x_k \right);$$

this proves that $\Phi_U(A) = [(u_i | A u_k)]$. We underline the fact that the isomorphism Φ_U depends in a crucial way on the c.o.n.s. $\{u_1, ..., u_N\}$ in $\mathcal{H}$ that was chosen in order to define the isomorphism U.

10.9   Projective Hilbert spaces and Wigner's theorem

In the mathematical formalism of quantum mechanics, symmetries are represented by what we call automorphisms of a projective Hilbert space, and so is reversible and conservative time evolution. Wigner's theorem proves that every such automorphism can be implemented by a unitary or an antiunitary operator. The theorem was presented by Eugene P. Wigner in the book he wrote in 1931. Yet, " in Wigner's book the theorem is not proved in full detail. The construction of the mapping U, however, is clearly indicated, so that is not difficult to close the gaps in the proof" (Bargmann, 1964). And indeed those gaps were closed by Valentine Bargmann in 1964 (loc. cit.). A major difference between Wigner's proof and Bargmann's is that Wigner relies on the existence of a c.o.n.s. in the Hilbert space under consideration, while Bargmann does not. Hence, Bargmann's approach is better suited to our exposition since we proved the existence of a c.o.n.s. only for a separable Hilbert space. In this section we follow Bargmann's arguments faithfully, with the only difference that we find it more convenient to arrange his reasoning in two theorems instead of one.

10.9.1 Definitions. Let $\mathcal{H}$ be a Hilbert space. In $\mathcal{H}$ we define a relation, denoted by $\div$, as follows:

$$f \div g \text{ if } \exists z \in \mathbb{T} \text{ such that } f = zg,$$

where $\mathbb{T} := \{z \in \mathbb{C} : |z| = 1\}$. This relation is obviously an equivalence relation, and we denote by $\mathcal{H}_q$ the quotient set $\mathcal{H}/\div$. Obviously, if $f \neq 0_{\mathcal{H}}$ then the equivalence class $[f]$ is an infinite set, while the equivalence class $[0_{\mathcal{H}}]$ contains just the vector $0_{\mathcal{H}}$.

It is obvious that the following definitions of mappings are consistent:

$$\tau : \mathcal{H}_q \times \mathcal{H}_q \to [0, \infty)$$
$$([f], [g]) \mapsto \tau([f], [g]) := |(f|g)|;$$
$$\mathcal{H}_q \to [0, \infty)$$
$$[f] \mapsto \|[f]\| := \|f\|;$$
$$[0, \infty) \times \mathcal{H}_q \to \mathcal{H}_q$$
$$(a, [f]) \mapsto a[f] := [af].$$

We note that

$$\tau([f], [f]) = \|f\|^2 = \|[f]\|^2, \forall f \in \mathcal{H}.$$

We also note that, for $\alpha \in \mathbb{C}$ and $f \in \mathcal{H}$, if $\alpha = |\alpha|e^{i\theta}$ with $\theta \in \mathbb{R}$ then $\alpha f = e^{i\theta}|\alpha|f$, and hence $\alpha f \div |\alpha|f$; thus,

$$[\alpha f] = [|\alpha|f] = |\alpha|[f], \text{ or } \alpha f \in |\alpha|[f], \forall \alpha \in \mathbb{C}, \forall f \in \mathcal{H},$$

and also

$$\||\alpha|[f]\| = \|[\alpha f]\| = \|\alpha f\| = |\alpha|\|f\| = |\alpha|\|[f]\|, \forall \alpha \in \mathbb{C}, \forall f \in \mathcal{H}.$$

When another Hilbert space $\mathcal{H}'$ is discussed in this section, the same symbols as above will be used, with τ replaced by τ' for the first mapping above; moreover, the same symbols will be used for the norms and the inner products in $\mathcal{H}$ and in $\mathcal{H}'$.

10.9.2 Proposition. *Let $\mathcal{H}$ and $\mathcal{H}'$ be Hilbert spaces and suppose that there exists a linear or an antilinear operator $U : \mathcal{H} \to \mathcal{H}'$ which fulfils either one of the following conditions:*

$$(Uf|Ug) = (f|g), \forall f, g \in \mathcal{H}, \text{ if } U \text{ is linear;}$$
$$(Uf|Ug) = (g|f), \forall f, g \in \mathcal{H}, \text{ if } U \text{ is antilinear.}$$

Then the mapping

$$\Phi_U : \mathcal{H}_q \to \mathcal{H}'_q$$
$$[f] \mapsto \Phi_U([f]) := [Uf]$$

is defined consistently and it has the following properties:

$$\Phi_U(a[f]) = a\Phi_U([f]), \forall a \in [0,\infty), \forall f \in \mathcal{H};$$
$$\tau'(\Phi_U([f]), \Phi_U([g])) = \tau([f],[g]), \forall f,g \in \mathcal{H}.$$

For each $z \in \mathbb{T}$, the operator zU has all the properties assumed for U above and $\Phi_{zU} = \Phi_U$.

Proof. The definition of Φ_U is consistent because, for $f,g \in \mathcal{H}$,

$$f \div g \Rightarrow$$
$$[\exists z \in \mathbb{T} \text{ s.t. } f = zg \text{ and hence s.t. either } Uf = zUg \text{ or } Uf = \bar{z}Ug] \Rightarrow$$
$$Uf \div Ug,$$

and hence the equivalence class $[Uf]$ does not depend on the choice of the representative f in the equivalence class $[f]$.

For all $a \in [0,\infty)$ and $f \in \mathcal{H}$, we have

$$\Phi_U(a[f]) = \Phi_U([af]) = [U(af)] = [aUf] = a[Uf] = a\Phi_U([f]).$$

For all $f,g \in \mathcal{H}$, we have

$$\tau'(\Phi_U([f]), \Phi_U([g])) = \tau'([Uf],[Ug]) = |(Uf|Ug)| = |(f|g)| = \tau([f],[g]).$$

Finally, for each $z \in \mathbb{T}$, it is immediate to check that zU has all the properties assumed for U in the statement and that $\Phi_{zU} = \Phi_U$. $\square$

10.9.3 Theorem. *Let $\mathcal{H}$ and $\mathcal{H}'$ be Hilbert spaces and suppose that there exists a mapping $\Phi : \mathcal{H}_q \to \mathcal{H}'_q$ which has the following properties:*

$$\Phi(a[f]) = a\Phi([f]), \forall a \in [0,\infty), \forall f \in \mathcal{H};$$
$$\tau'(\Phi([f]), \Phi([g])) = \tau([f],[g]), \forall f,g \in \mathcal{H}.$$

Then the following properties are true:

(A) The mapping Φ is injective.

(B) There exists a linear or an antilinear operator $U : \mathcal{H} \to \mathcal{H}'$ which is so that

$$\Phi_U = \Phi, \text{ i.e. } [Uf] = \Phi([f]) \text{ for all } f \in \mathcal{H},$$

and which fulfils either one of the following conditions:

$$(Uf|Ug) = (f|g), \forall f,g \in \mathcal{H}, \text{ if } U \text{ is linear;}$$
$$(Uf|Ug) = (g|f), \forall f,g \in \mathcal{H}, \text{ if } U \text{ is antilinear.}$$

(C) If the mapping Φ is surjective onto $\mathcal{H}'_q$ then the operator U is surjective onto $\mathcal{H}'$ and hence it is a unitary or an antiunitary operator, i.e. $U \in \mathcal{UA}(\mathcal{H}, \mathcal{H}')$.

(D) If $\mathcal{H}$ is not one-dimensional as a linear space, and if a mapping $V : \mathcal{H} \to \mathcal{H}'$ is such that

$$V(f + g) = Vf + Vg, \forall f,g \in \mathcal{H},$$

and

$$[Vf] = \Phi([f]), \forall f \in \mathcal{H},$$

then there exists $z \in \mathbb{T}$ so that $Vf = zUf$ for all $f \in \mathcal{H}$.

Proof. A: We have

$$\|\Phi([f])\|^2 = \tau'(\Phi([f]), \Phi([f])) = \tau([f], [f]) = \|[f]\|^2, \forall f \in \mathcal{H}. \tag{1}$$

This implies that, for $f \in \mathcal{H}$,

$$\Phi([f]) = [0_{\mathcal{H}'}] \text{ iff } f = 0_{\mathcal{H}}. \tag{2}$$

Moreover, if $f, g \in \mathcal{H} - \{0_{\mathcal{H}}\}$ are so that $\Phi([f]) = \Phi([g])$, then 1 implies that

$$\tau'(\Phi([f]), \Phi([g])) = \|[f]\|^2 = \|[g]\|^2,$$

and hence

$$\|f\|\|g\| = \|[f]\|\|[g]\| = \tau'(\Phi([f]), \Phi([g])) = \tau([f], [g]) = |(f|g)|;$$

by 10.1.7b and 3.1.14, this implies that

$$\exists z \in \mathbb{T} \text{ such that } f = zg, \text{ i.e. } [f] = [g].$$

This proves that the mapping Φ is injective.

B: We prove in what follows the existence of U with the required properties. If $\mathcal{H}$ is a zero Hilbert space, this is true trivially. In the other cases, the proof is by construction.

First we assume that, as a linear space, $\mathcal{H}$ is one-dimensional, and we choose an element $u \in \mathcal{H}$ such that $\|u\| = 1$. Then,

$$\forall f \in \mathcal{H}, \exists! \alpha \in \mathbb{C} \text{ so that } f = \alpha u.$$

Therefore, if we choose $u' \in \Phi([u])$, we can define the mappings

$$U_1 : \mathcal{H} \to \mathcal{H}'$$
$$\alpha u \mapsto U_1(\alpha u) := \alpha u'$$

and

$$U_2 : \mathcal{H} \to \mathcal{H}'$$
$$\alpha u \mapsto U_2(\alpha u) := \bar{\alpha} u'.$$

The mapping U_1 is linear since

$$U_1(\alpha(\beta u) + \gamma(\delta u)) = (\alpha\beta + \gamma\delta)u' = \alpha(\beta u') + \gamma(\delta u')$$
$$= \alpha U_1(\beta u) + \gamma U_1(\delta u), \forall \alpha, \beta, \gamma, \delta \in \mathbb{C};$$

we also have

$$[U_1(\alpha u)] = [\alpha u'] = |\alpha|[u'] = |\alpha|\Phi([u]) = \Phi(|\alpha|[u]) = \Phi([\alpha u]), \forall \alpha \in \mathbb{C},$$

and

$$(U_1(\alpha u)|U_1(\beta u)) = (\alpha u'|\beta u') = \bar{\alpha}\beta = (\alpha u|\beta u), \forall \alpha, \beta \in \mathbb{C},$$

which is true since 1 implies that

$$\|u'\| = \|\Phi([u])\| = \|[u]\| = \|u\| = 1.$$

Similarly, the mapping U_2 is antilinear since

$$U_2(\alpha(\beta u) + \gamma(\delta u)) = (\overline{\alpha}\overline{\beta} + \overline{\gamma}\overline{\delta})u' = \overline{\alpha}(\overline{\beta}u') + \overline{\gamma}(\overline{\delta}u')$$
$$= \overline{\alpha}U_2(\beta u) + \overline{\gamma}U_2(\delta u), \forall \alpha, \beta, \gamma, \delta \in \mathbb{C};$$

we also have

$$[U_2(\alpha u)] = [\overline{\alpha}u'] = |\alpha|[u'] = |\alpha|\Phi([u]) = \Phi(|\alpha|[u]) = \Phi([\alpha u]), \forall \alpha \in \mathbb{C},$$

and

$$(U_2(\alpha u)|U_2(\beta u)) = (\overline{\alpha}u'|\overline{\beta}u') = \alpha\overline{\beta} = (\beta u|\alpha u), \forall \alpha, \beta \in \mathbb{C}.$$

Thus, both the mappings U_1 and U_2 have the properties required for U in the statement, and the proof for the one-dimensional case is concluded.

In what follows we assume that $\mathcal{H}$, as a linear space, is neither zero-dimensional nor one-dimensional. During the construction of U we shall need the results that we collect in the following preliminary remarks.

Remark 1: Suppose that $f \in \mathcal{H}$ is so that Uf has been defined and $[Uf] = \Phi([f])$. If $g \in \mathcal{H}$ is also so that Ug has been defined and $[Ug] = \Phi([g])$, then

$$|(Uf|Ug)| = \tau'([Uf], [Ug]) = \tau'(\Phi([f]), \Phi([g])) = \tau([f], [g]) = |(f|g)|. \tag{3}$$

Hence, for $g := f$,

$$\|Uf\| = \|f\|. \tag{4}$$

Now suppose $f \neq 0_{\mathcal{H}}$ and also that, for all $\alpha \in \mathbb{C}$, $U(\alpha f)$ has been defined and $[U(\alpha f)] = \Phi([\alpha f])$; then

$$U(\alpha f) \in \Phi([\alpha f]) = |\alpha|\Phi([f]) = |\alpha|[Uf] = [\alpha Uf], \forall \alpha \in \mathbb{C},$$

and hence

$$\forall \alpha \in \mathbb{C}, \exists z \in \mathbb{T} \text{ so that } U(\alpha f) = z\alpha Uf,$$

and hence

$$\forall \alpha \in \mathbb{C}, \exists! \alpha_f \in \mathbb{C} \text{ so that } |\alpha_f| = |\alpha| \text{ and } U(\alpha f) = \alpha_f Uf, \tag{5}$$

where uniqueness holds since $Uf \neq 0_{\mathcal{H}'}$ (cf. 4). This defines the function

$$\mathbb{C} \ni \alpha \mapsto \chi_f(\alpha) := \alpha_f \in \mathbb{C}.$$

Obviously, $\chi_f(1) = 1$.

Remark 2: Let $\{u_i\}_{i \in I}$ be a finite o.n.s. in $\mathcal{H}$. If we choose $u_i' \in \Phi([u_i])$ for each $i \in I$, then $\{u_i'\}_{i \in I}$ is an o.n.s. in $\mathcal{H}'$ since

$$|(u_i'|u_j')| = \tau'(\Phi([u_i]), \Phi([u_j])) = \tau([u_i], [u_j]) = |(u_i|u_j)|, \forall i, j \in I.$$

If $f = \sum_{i \in I} \alpha_i u_i$ with $\alpha_i \in \mathbb{C}$ for all $i \in I$, then for each $f' \in \Phi([f])$ we have $f' = \sum_{i \in I} \alpha_i' u_i'$ with $\alpha_i' \in \mathbb{C}$ such that $|\alpha_i'| = |\alpha_i|$ for all $i \in I$. In fact, 1 implies $\|f'\| = \|f\|$; moreover, we have $\alpha_i = (u_i|f)$ for all $i \in I$; thus, if we define $\alpha_i' := (u_i'|f')$ for all $i \in I$ then we have

$$|\alpha_i'| = |(u_i'|f')| = \tau'(\Phi([u_i]), \Phi([f])) = \tau([u_i], [f]) = |(u_i|f)| = |\alpha_i|, \forall i \in I,$$

and also

$$\|f'\|^2 = \left\| \left(f' - \sum_{i \in I} \alpha_i' u_i' \right) + \sum_{i \in I} \alpha_i' u_i' \right\|^2$$

$$= \left\| f' - \sum_{i \in I} \alpha_i' u_i' \right\|^2 + \left\| \sum_{i \in I} \alpha_i' u_i' \right\|^2 = \left\| f' - \sum_{i \in I} \alpha_i' u_i' \right\|^2 + \sum_{i \in I} |\alpha_i'|^2$$

(cf. 10.6.1 and 10.2.3), and hence

$$\left\| f' - \sum_{i \in I} \alpha_i' u_i' \right\|^2 = \|f\|^2 - \sum_{i \in I} |\alpha_i|^2 = 0.$$

These remarks attended to, now we proceed to the construction of U, which we divide into five steps for the sake of clarity.

Step 1: As in the one-dimensional case, we choose an element u of $\mathcal{H}$ such that $\|u\| = 1$, and we choose $u' \in \Phi([u])$. The arbitrariness of the selection of u' after u has been selected is the one arbitrary point in the construction of U, and therein lies the non-uniqueness of U which is clear from the last part of the statement in 10.9.2. In fact, as we will see, U is constructed in such a way that $Uu = u'$; if another element u'' of $\Phi([u])$ was chosen at the outset and the same procedure that leads to U was followed, we should obtain a mapping V such that $Vu = u''$, and hence such that $Vu = zUu$ if $z \in \mathbb{T}$ is so that $u'' = zu'$.

Step 2: In this step we define Ug and $U(u + g)$ for all $g \in \{u\}^\perp$.

First we prove that

$$\forall g \in \{u\}^\perp - \{0_\mathcal{H}\}, \exists! g' \in \Phi([g]) \text{ such that } u' + g' \in \Phi([u + g]). \tag{6}$$

Let $g \in \{u\}^\perp - \{0_\mathcal{H}\}$ and set $f := u + g$; we have

$$f = u + \|g\| \left(\frac{1}{\|g\|} g \right)$$

and we note that $\left\{ u, \frac{1}{\|g\|} g \right\}$ is an o.n.s. in $\mathcal{H}$; hence, if we choose $u_g' \in \Phi\left(\left[\frac{1}{\|g\|} g \right] \right)$ and $f' \in \Phi([f])$, from remark 2 we have that $\{u', u_g'\}$ is an o.n.s. in $\mathcal{H}'$ and

$$\exists (\alpha_0', \alpha_1') \in \mathbb{C}^2 \text{ such that } |\alpha_0'| = 1, |\alpha_1'| = \|g\|, f' = \alpha_0' u' + \alpha_1' u_g'$$

(we will see presently that the particular choices made for u_g' and f' are immaterial for the construction of U); we note that

$$u' + \frac{\alpha_1'}{\alpha_0'} u_g' = (\alpha_0')^{-1} f' \in \Phi([f]) = \Phi([u + g])$$

and

$$\frac{\alpha_1'}{\alpha_0'} u_g' \in \left| \frac{\alpha_1'}{\alpha_0'} \right| [u_g'] = \|g\| \Phi\left(\left[\frac{1}{\|g\|} g \right] \right) = \Phi\left(\|g\| \left[\frac{1}{\|g\|} g \right] \right) = \Phi([g]);$$

this proves that

$$\exists g' \in \Phi([g]) \text{ such that } u' + g' \in \Phi([u + g]).$$

Now suppose that $g'' \in \Phi([g])$ is such that $u' + g'' \in \Phi([u + g])$. Then there exists $z \in \mathbb{T}$ so that

$$u' + g'' = z(u' + g'), \text{ i.e. } (1 - z)u' = zg' - g'';$$

since

$$|(u'|g')| = |(u'|g'')| = \tau'(\Phi([u]), \Phi([g])) = \tau([u], [g]) = |(u|g)| = 0,$$

we have

$$(1 - z)u' = 0_{\mathcal{H}'} \text{ and } zg' = g'',$$

and hence $z = 1$ and $g'' = g'$. Thus, 6 is proved.

Moreover, we note that

$$\exists! g' \in \Phi([0_{\mathcal{H}}]) \text{ such that } u' + g' \in \Phi([u + 0_{\mathcal{H}}]);$$

indeed, the vector $g' := 0_{\mathcal{H}'}$ satisfies the above condition trivially since $\Phi([0_{\mathcal{H}}]) = [0_{\mathcal{H}'}]$ (cf. 2), and it is the only one that does so since the equivalence class $[0_{\mathcal{H}'}]$ contains just the vector $0_{\mathcal{H}'}$.

Thus, we have proved that

$$\forall g \in \{u\}^\perp, \exists! g' \in \Phi([g]) \text{ such that } u' + g' \in \Phi([u + g]).$$

Therefore we can define, for each $g \in \{u\}^\perp$,

$$Ug := g' \text{ and } U(u + g) := u' + g'$$

if g' is the unique element of $\Phi([g])$ such that $u' + g' \in \Phi([u + g])$. Since $g' = 0_{\mathcal{H}'}$ if $g = 0_{\mathcal{H}}$ (see above), this entails

$$U0_{\mathcal{H}} = 0_{\mathcal{H}'} \text{ and } Uu = u',$$

and hence

$$U(u + g) = Uu + Ug, \forall g \in \{u\}^\perp. \tag{7}$$

We point out that

$$[Ug] = \Phi([g]), \forall g \in \{u\}^\perp, \tag{8}$$

and

$$[U(u + g)] = \Phi([u + g]), \forall g \in \{u\}^\perp. \tag{9}$$

Step 3: In this step we prove that, for each $v \in \{u\}^\perp$ such that $\|v\| = 1$,

$$\text{either } \chi_v(\alpha) = \alpha \text{ for all } \alpha \in \mathbb{C} \text{ or } \chi_v(\alpha) = \overline{\alpha} \text{ for all } \alpha \in \mathbb{C}. \tag{10}$$

Let g and h be elements of $\{u\}^\perp$. From 7, 9, 3 we have

$$|(Uu + Ug|Uu + Uh)|^2 = |(u + g|u + h)|^2;$$

from 8, 9, 3 we have

$$(Uu|Ug) = (Uu|Uh) = 0;$$

then, since $\|Uu\| = 1$ (cf. 4), we obtain

$$|1 + (Ug|Uh)|^2 = |1 + (g|h)|^2,$$

which can be written as

$$1 + |(Ug|Uh)|^2 + 2\operatorname{Re}(Ug|Uh) = 1 + |(g|h)|^2 + 2\operatorname{Re}(g|h),$$

and this, in view of 3, is equivalent to

$$\operatorname{Re}(Ug|Uh) = \operatorname{Re}(g|h). \tag{11}$$

If $\operatorname{Im}(g|h) = 0$, 3 and 11 imply that $\operatorname{Im}(Ug|Uh) = 0$; therefore,

$$\text{if } (g|h) \in \mathbb{R} \text{ then } (Ug|Uh) \in \mathbb{R} \text{ and hence } (Ug|Uh) = (g|h). \tag{12}$$

Now let $v \in \{u\}^\perp$ be such that $\|v\| = 1$. Since $\|Uv\| = 1$ (cf. 8 and 4), from 5 and 11 we obtain, for all $\alpha, \beta \in \mathbb{C}$,

$$\begin{aligned}
\operatorname{Re}\left(\overline{\chi_v(\beta)}\chi_v(\alpha)\right) &= \operatorname{Re}(\chi_v(\beta)Uv|\chi_v(\alpha)Uv) \\
&= \operatorname{Re}(U(\beta v)|U(\alpha v)) = \operatorname{Re}(\beta v|\alpha v) = \operatorname{Re}(\overline{\beta}\alpha)
\end{aligned} \tag{13}$$

In this, set $\beta := 1$. Since $\chi_v(1) = 1$, we obtain, for all $\alpha \in \mathbb{C}$,

$$\operatorname{Re}\chi_v(\alpha) = \operatorname{Re}\alpha. \tag{14}$$

In this, set $\alpha := i$. We obtain $\operatorname{Re}\chi_v(i) = 0$, which together with $|\chi_v(i)| = 1$ (cf. 5) implies $\chi_v(i) = \eta i$ with either $\eta = 1$ or $\eta = -1$, and hence $i = \eta\chi_v(i)$. This and 13 imply that, for all $\alpha \in \mathbb{C}$,

$$\operatorname{Im}\chi_v(\alpha) = \operatorname{Re}(-i\chi_v(\alpha)) = \operatorname{Re}(\eta\overline{\chi_v(i)}\chi_v(\alpha)) = \eta\operatorname{Re}(-i\alpha) = \eta\operatorname{Im}\alpha. \tag{15}$$

Combining 14 and 15 we obtain 10.

We point out that 10 implies the following relations:

$$\chi_v(\alpha + \beta) = \chi_v(\alpha) + \chi_v(\beta), \forall \alpha, \beta \in \mathbb{C}; \tag{16}$$

$$\chi_v(\alpha\beta) = \chi_v(\alpha)\chi_v(\beta), \forall \alpha, \beta \in \mathbb{C}; \tag{17}$$

$$\overline{\chi_v(\alpha)} = \chi_v(\overline{\alpha}), \forall \alpha \in \mathbb{C}. \tag{18}$$

Step 4: In this step we prove that, for all $g, h \in \{u\}^\perp$,

$$U(g + h) = Ug + Uh, \tag{19}$$

$$U(\alpha h) = \chi(\alpha)Uh, \forall \alpha \in \mathbb{C}, \tag{20}$$

$$(Ug|Uh) = \chi((g|h)), \tag{21}$$

where the function $\chi : \mathbb{C} \to \mathbb{C}$ is defined either by $\chi(\alpha) := \alpha$ for all $\alpha \in \mathbb{C}$ or by $\chi(\alpha) := \overline{\alpha}$ for all $\alpha \in \mathbb{C}$.

First we suppose that the linear manifold $\{u\}^\perp$ is one-dimensional.

Let v be an element of $\{u\}^\perp$ such that $\|v\| = 1$. Then $\{u\}^\perp = \{\alpha v : \alpha \in \mathbb{C}\}$. For $g, h \in \{u\}^\perp$, by 8, 4, 5, 16, 17, 18 we have, if $\beta, \gamma \in \mathbb{C}$ are so that $g = \beta v$ and $h = \gamma v$:

$$U(g + h) = U(\beta v + \gamma v) = U((\beta + \gamma)v) = \chi_v(\beta + \gamma)Uv$$
$$= (\chi_v(\beta) + \chi_v(\gamma))Uv = \chi_v(\beta)Uv + \chi_v(\gamma)Uv$$
$$= U(\beta v) + U(\gamma v) = U(g) + U(h);$$

$$U(\alpha h) = U(\alpha \gamma v) = \chi_v(\alpha \gamma)Uv = \chi_v(\alpha)\chi_v(\gamma)Uv$$
$$= \chi_v(\alpha)U(\gamma v) = \chi_v(\alpha)U(h), \forall \alpha \in \mathbb{C};$$

$$(Ug|Uh) = (U(\beta v)|U(\gamma v)) = (\chi_v(\beta)Uv|\chi_v(\gamma)Uv)$$
$$= \overline{\chi_v(\beta)}\chi_v(\gamma) = \chi_v(\overline{\beta}\gamma) = \chi_v((\beta v|\gamma v)) = \chi_v((g|h)).$$

This proves that 19, 20, 21 are true for all $g, h \in \{u\}^\perp$, with $\chi := \chi_v$. Also, recall 10.

Next we suppose that the linear manifold $\{u\}^\perp$ is not one-dimensional.

Let $\{u_1, u_2\}$ be any o.n.s. in $\{u\}^\perp$ (the linear space $\{u\}^\perp$ is not a zero space by 10.4.2a and 10.2.11 since the linear space $\mathcal{H}$ is not one-dimensional, and it is not one-dimensional; then there exist two linearly independent vectors in $\{u\}^\perp$, and hence there exists an o.n.s. in $\{u\}^\perp$ which contains two elements, by 10.2.6 with $I := \{1, 2\}$). For $i = 1, 2$, $Uu_i \in \Phi([u_i])$ (cf. 8). Hence, remark 2 implies that $\{Uu_1, Uu_2\}$ is an o.n.s. in $\mathcal{H}'$ and that, for each $(\alpha_1, \alpha_2) \in \mathbb{C}^2$,

$$U(\alpha_1 u_1 + \alpha_2 u_2) = \alpha_1' Uu_1 + \alpha_2' Uu_2,$$

with $\alpha_i' \in \mathbb{C}$ so that $|\alpha_i'| = |\alpha_i|$ for $i = 1, 2$, since $U(\alpha_1 u_1 + \alpha_2 u_2) \in \Phi([\alpha_1 u_1 + \alpha_2 u_2])$ (cf. 8). Now we prove that

$$\alpha_i' = \chi_{u_i}(\alpha_i) \text{ for } i = 1, 2. \tag{22}$$

This is trivial if $\alpha_i = 0$ (cf. 10). Then, let $i = 1$ or $i = 2$ be so that $\alpha_i \neq 0$, and set $\beta_i := \overline{\alpha_i}^{-1}$. We have

$$1 = (\beta_i u_i | \alpha_i u_i) = (\beta_i u_i | \alpha_1 u_1 + \alpha_2 u_2).$$

Since the inner products above are real, from 12, 8, 5 we have

$$(\beta_i u_i | \alpha_i u_i) = (U(\beta_i u_i)|U(\alpha_i u_i)) = (\chi_{u_i}(\beta_i)Uu_i|\chi_{u_i}(\alpha_i)Uu_i) = \overline{\chi_{u_i}(\beta_i)}\chi_{u_i}(\alpha_i),$$

and also

$$(\beta_i u_i | \alpha_1 u_1 + \alpha_2 u_2) = (U(\beta_i u_i)|U(\alpha_1 u_1 + \alpha_2 u_2))$$
$$= (\chi_{u_i}(\beta_i)Uu_i|\alpha_1' Uu_1 + \alpha_2' Uu_2) = \overline{\chi_{u_i}(\beta_i)}\alpha_i'.$$

This implies

$$\overline{\chi_{u_i}(\beta_i)}\chi_{u_i}(\alpha_i) = \overline{\chi_{u_i}(\beta_i)}\alpha_i',$$

and hence (since $\chi_{u_i}(\beta_i) \neq 0$ as $\beta_i \neq 0$)

$$\alpha_i' = \chi_{u_i}(\alpha_i).$$

Thus, 22 is proved. Since 22 is true, we can prove that

$$\chi_{u_1}(\alpha) = \chi_{u_2}(\alpha), \forall \alpha \in \mathbb{C}. \tag{23}$$

Indeed, from 22 we have

$$U(u_1 + u_2) = \chi_{u_1}(1)Uu_1 + \chi_{u_2}(1)Uu_2 = Uu_1 + Uu_2,$$

and hence

$$\begin{aligned}
U(\alpha(u_1 + u_2)) &= \chi_{u_1+u_2}(\alpha)U(u_1 + u_2) \\
&= \chi_{u_1+u_2}(\alpha)Uu_1 + \chi_{u_1+u_2}(\alpha)Uu_2, \forall \alpha \in \mathbb{C},
\end{aligned} \tag{24}$$

(cf. 8 and 5); from 22 we also have

$$U(\alpha(u_1 + u_2)) = U(\alpha u_1 + \alpha u_2) = \chi_{u_1}(\alpha)Uu_1 + \chi_{u_2}(\alpha)Uu_2, \forall \alpha \in \mathbb{C}; \tag{25}$$

since $\{Uu_1, Uu_2\}$ is an o.n.s., 24 and 25 imply that

$$\chi_{u_1}(\alpha) = \chi_{u_1+u_2}(\alpha) \text{ and } \chi_{u_2}(\alpha) = \chi_{u_1+u_2}(\alpha), \forall \alpha \in \mathbb{C},$$

and this proves 23. Then, for $g, h \in L\{u_1, u_2\}$, from 22 and 23 (and 16, 17, 18 as well) we have, if $(\beta_1, \beta_2), (\gamma_1, \gamma_2) \in \mathbb{C}^2$ are so that $g = \beta_1 u_1 + \beta_2 u_2$ and $h = \gamma_1 u_1 + \gamma_2 u_2$:

$$\begin{aligned}
U(g + h) &= U((\beta_1 + \gamma_1)u_1 + (\beta_2 + \gamma_2)u_2) \\
&= \chi_{u_1}(\beta_1 + \gamma_1)Uu_1 + \chi_{u_1}(\beta_2 + \gamma_2)Uu_2 \\
&= \chi_{u_1}(\beta_1)Uu_1 + \chi_{u_1}(\gamma_1)Uu_1 + \chi_{u_1}(\beta_2)Uu_2 + \chi_{u_1}(\gamma_2)Uu_2 \\
&= Ug + Uh;
\end{aligned} \tag{26}$$

$$\begin{aligned}
U(\alpha h) &= U(\alpha\gamma_1 u_1 + \alpha\gamma_2 u_2) = \chi_{u_1}(\alpha\gamma_1)Uu_1 + \chi_{u_1}(\alpha\gamma_2)Uu_2 \\
&= \chi_{u_1}(\alpha)(\chi_{u_1}(\gamma_1)Uu_1 + \chi_{u_1}(\gamma_2)Uu_2) = \chi_{u_1}(\alpha)Uh, \forall \alpha \in \mathbb{C};
\end{aligned} \tag{27}$$

$$\begin{aligned}
(Ug|Uh) &= (\chi_{u_1}(\beta_1)Uu_1 + \chi_{u_1}(\beta_2)Uu_2|\chi_{u_1}(\gamma_1)Uu_1 + \chi_{u_1}(\gamma_2)U_2) \\
&= \overline{\chi_{u_1}(\beta_1)}\chi_{u_1}(\gamma_1) + \overline{\chi_{u_1}(\beta_2)}\chi_{u_1}(\gamma_2) = \chi_{u_1}(\overline{\beta}_1\gamma_1 + \overline{\beta}_2\gamma_2) \\
&= \chi_{u_1}((\beta_1 u_1 + \beta_2 u_2|\gamma_1 u_1 + \gamma_2 u_2)) = \chi_{u_1}((g|h)).
\end{aligned} \tag{28}$$

Now we are ready to prove 19, 20, 21 for all $g, h \in \{u\}^\perp$. Fix $g_0 \in \{u\}^\perp - \{0_\mathcal{H}\}$ and define $u_1 := \frac{1}{\|g_0\|}g_0$. For each $h \in \{u\}^\perp$ there exists $u_2 \in \{u\}^\perp$ so that $\{u_1, u_2\}$ is an o.n.s. and $h \in L\{u_1, u_2\}$: u_2 is obtained as in 10.2.6 with $I := \{1, 2\}$, $f_1 := g_0$, and either $f_2 := h$ if h is not a multiple of g_0 or f_2 any element of $\{u\}^\perp$ which is not a multiple of g_0 if h is a multiple of g_0. Then, from 26 we obtain

$$U(g_0 + h) = Ug_0 + Uh, \forall h \in \{u\}^\perp.$$

Since g_0 is an arbitrary element of $\{u\}^\perp - \{0_\mathcal{H}\}$ and since 19 holds trivially for all $h \in \{u\}^\perp$ if $g = 0_\mathcal{H}$ (recall that $U0_\mathcal{H} = 0_{\mathcal{H}'}$), this proves 19 for all $g, h \in \{u\}^\perp$.

Furthermore, from 27 we obtain 20 for all $h \in \{u\}^\perp$, with $\chi := \chi_{u_1}$. Finally, from 28 we obtain

$$(Ug_0|Uh) = \chi_{u_1}((g_0|h)), \forall h \in \{u\}^\perp.$$

If g is any element of $\{u\}^\perp - \{0_\mathcal{H}\}$ different from g_0, proceeding as above we obtain

$$(Ug|Uh) = \chi_{u_{1,g}}((g|h)), \forall h \in \{u\}^\perp,$$

with $u_{1,g} := \frac{1}{\|g\|}g$. However, 27 proves that $\chi_h = \chi_{u_1}$ for all $h \in \{u\}^\perp - \{0_\mathcal{H}\}$. Thus we have

$$(Ug|Uh) = \chi_{u_1}((g|h)), \forall g \in \{u\}^\perp - \{0_\mathcal{H}\}, \forall h \in \{u\}^\perp.$$

Since 21 holds trivially for all $h \in \{u\}^\perp$ if $g = 0_\mathcal{H}$, this proves 21 for all $g, h \in \{u\}^\perp$, with $\chi := \chi_{u_1}$. Finally, recall 10.

Step 5: In this step we define Uf for all $f \in \mathcal{H}$ and conclude the proof of part B of the statement.

From 10.4.1 we have

$$\forall f \in \mathcal{H}, \exists!(f_1, f_2) \in V\{u\} \times (V\{u\})^\perp \text{ so that } f = f_1 + f_2.$$

Since $V\{u\} = \{\alpha u : \alpha \in \mathbb{C}\}$ (cf. 4.1.15) and $(V\{u\})^\perp = \{u\}^\perp$ (cf. 10.2.11), we have

$$\forall f \in \mathcal{H}, \exists!(\alpha, g) \in \mathbb{C} \times \{u\}^\perp \text{ so that } f = \alpha u + g.$$

Therefore, we can define Uf for all $f \in \mathcal{H}$ by letting

$$U(\alpha u + g) := \chi(\alpha)Uu + Ug, \forall \alpha \in \mathbb{C}, \forall g \in \{u\}^\perp,$$

where χ is the function of step 4. For $\alpha = 0$ or $\alpha = 1$, this definition coincides with the ones already given in step 2 since $\chi(0) = 0$ and $\chi(1) = 1$ (cf. 7).

We have already noted that $[Ug] = \Phi([g])$ for all $g \in \{u\}^\perp$ (cf. 8). Moreover, for every $\alpha \in \mathbb{C} - \{0\}$ and every $g \in \{u\}^\perp$ we have

$$[U(\alpha u + g)] = [\chi(\alpha)(Uu + U(\alpha^{-1}g))] = |\chi(\alpha)|[U(u + \alpha^{-1}g)]$$
$$= |\alpha|\Phi([u + \alpha^{-1}g]) = \Phi([\alpha u + g]),$$

where 20, 7, 9 have been used. Thus,

$$[Uf] = \Phi([f]), \forall f \in \mathcal{H}.$$

Finally, for $f_1, f_2 \in \mathcal{H}$ we have, if $\alpha_1, \alpha_2 \in \mathbb{C}$ and $g_1, g_2 \in \{u\}^\perp$ are so that $f_1 = \alpha_1 u + g_1$ and $f_2 = \alpha_2 u + g_2$,

$$\begin{aligned} U(f_1 + f_2) &= U((\alpha_1 + \alpha_2)u + (g_1 + g_2)) \\ &= \chi(\alpha_1 + \alpha_2)Uu + U(g_1 + g_2) \\ &= \chi(\alpha_1)Uu + Ug_1 + \chi(\alpha_2)Uu + Ug_2 \\ &= U(\alpha_1 u + g_1) + U(\alpha_2 u + g_2) = Uf_1 + Uf_2, \end{aligned}$$

where 16 and 19 have been used; we also have

$$\begin{aligned} U(\alpha f_1) &= U(\alpha\alpha_1 u + \alpha g_1) = \chi(\alpha\alpha_1)Uu + U(\alpha g_1) \\ &= \chi(\alpha)(\chi(\alpha_1)Uu + Ug_1) = \chi(\alpha)Uf_1, \forall \alpha \in \mathbb{C}, \end{aligned}$$

where 17 and 20 have been used; finally, we have

$$(Uf_1|Uf_2) = (\chi(\alpha_1)Uu + Ug_1|\chi(\alpha_2)Uu + Ug_2)$$
$$= \overline{\chi(\alpha_1)}\chi(\alpha_2) + (Ug_1|Ug_2) = \chi(\overline{\alpha}_1\alpha_2) + \chi((g_1|g_2))$$
$$= \chi(\overline{\alpha}_1\alpha_2 + (g_1|g_2)) = \chi((\alpha_1 u + g_1|\alpha_2 u + g_2)) = \chi((f_1|f_2)),$$

where 3, 18, 17, 21, 16 have been used. Thus, U has all the required properties.

C: Suppose that the mapping Φ is surjective onto $\mathcal{H}_q'$. Then,

$$\forall f' \in \mathcal{H}', \exists f \in \mathcal{H} \text{ so that } [f'] = \Phi([f]) = [Uf],$$

and hence

$$\forall f' \in \mathcal{H}', \exists f \in \mathcal{H}, \exists z \in \mathbb{T} \text{ so that } f' = zUf,$$

and hence

$$\forall f' \in \mathcal{H}', \exists f \in \mathcal{H}, \exists z \in \mathbb{T} \text{ so that } f' = U(zf) \text{ if } U \text{ is linear or}$$
$$f' = U(\overline{z}f) \text{ if } U \text{ is antilinear.}$$

Thus, the mapping U is surjective onto $\mathcal{H}'$.

D: If $\mathcal{H}$ is a zero space then the statement is trivial. Now suppose that $\mathcal{H}$ is neither a zero space nor a one-dimensional linear space. Then for every vector $f \in \mathcal{H}$ there exists a vector in $\mathcal{H}$ which is not a multiple of f.

Preliminarily, we note that if $f, g \in \mathcal{H}$ are linearly independent then so are Uf, Ug. In fact, for any $f, g \in \mathcal{H}$ we have

$$|(Uf|Ug)| = |(f|g)|, \|Uf\| = \|f\|, \|Ug\| = \|g\|,$$

and hence 10.1.7b proves that f and g are linearly dependent iff so are Uf and Ug.

Now let V be as in the statement. Then

$$\forall f \in \mathcal{H}, \exists z_f \in \mathbb{T} \text{ so that } Vf = z_f Uf.$$

We point out that z_f is uniquely fixed by this condition if $f \neq 0_\mathcal{H}$, since $Uf = 0_{\mathcal{H}'}$ iff $f = 0_\mathcal{H}$. Also, this condition implies that $V0_\mathcal{H} = 0_{\mathcal{H}'}$.

For $f, g \in \mathcal{H}$ we have

$$V(f + g) = Vf + Vg, U(f + g) = Uf + Ug, V(f + g) = z_{f+g}U(f + g),$$

and hence

$$z_f Uf + z_g Ug = z_{f+g}Uf + z_{f+g}Ug.$$

If f and g are linearly independent then so are Uf and Ug and hence

$$z_f = z_{f+g} \text{ and } z_g = z_{f+g}, \text{ and hence } z_f = z_g.$$

Now fix $g_0 \in \mathcal{H} - \{0_\mathcal{H}\}$ and set $z_0 := z_{g_0}$. If $f \in \mathcal{H}$ is such that f and g_0 are linearly independent, then $z_f = z_0$. If $f \in \mathcal{H} - \{0_\mathcal{H}\}$ is such that f and g_0 are linearly dependent then there exists $\alpha \in \mathbb{C} - \{0\}$ so that $f = \alpha g_0$ (cf. 3.1.14); if we choose $h \in \mathcal{H}$ which is not a multiple of g_0, then g_0 and h are linearly independent and so are f and h, and hence $z_0 = z_h$ and $z_f = z_h$, and hence $z_f = z_0$. This proves that

$$Vf = z_0 Uf, \forall f \in \mathcal{H} - \{0\},$$

and hence that

$$Vf = z_0 Uf, \forall f \in \mathcal{H}.$$

$\square$

10.9.4 Definitions. Let $\mathcal{H}$ be a non-zero Hilbert space. We define the subset $\tilde{\mathcal{H}}$ of $\mathcal{H}$ by letting

$$\tilde{\mathcal{H}} := \{u \in \mathcal{H} : \|u\| = 1\}$$

(this notation is consistent with the one introduced in 10.2.12). The relation $\dot{\div}$ defined in 10.9.1 induces an equivalence relation in $\tilde{\mathcal{H}}$, and we denote by $\hat{\mathcal{H}}$ the quotient set of $\tilde{\mathcal{H}}$ by this equivalence relation. Obviously,

$$\hat{\mathcal{H}} = \{[u] \in \mathcal{H}_q : \|u\| = 1\}.$$

An element of $\hat{\mathcal{H}}$ is called a *ray*.

We denote by the same symbol τ the restriction to $\hat{\mathcal{H}} \times \hat{\mathcal{H}}$ of the function τ defined in 10.9.1. The pair $(\hat{\mathcal{H}}, \tau)$ is called a *projective Hilbert space*.

Two projective Hilbert spaces $(\hat{\mathcal{H}}, \tau)$ an $(\hat{\mathcal{H}}', \tau')$ (where $\mathcal{H}$ and $\mathcal{H}'$ are non-zero Hilbert spaces, not necessarily different from each other) are said to be *isomorphic* if there exists a mapping $\omega : \hat{\mathcal{H}} \to \hat{\mathcal{H}}'$ which is bijective from $\hat{\mathcal{H}}$ onto $\hat{\mathcal{H}}'$ and such that

$$\tau'(\omega([u]), \omega([v])) = \tau([u], [v]), \forall u, v \in \tilde{\mathcal{H}}.$$

If such a mapping exists, it is called an *isomorphism* from $\hat{\mathcal{H}}$ onto $\hat{\mathcal{H}}'$. If $\mathcal{H} = \mathcal{H}'$, an isomorphism is called an *automorphism*. It is obvious that the family Aut $\hat{\mathcal{H}}$ of all automorphisms of $\hat{\mathcal{H}}$ is a group with respect to the usual composition of mappings.

10.9.5 Proposition. *Let $\mathcal{H}$ and $\mathcal{H}'$ be non-zero Hilbert spaces and suppose that the family $\mathcal{UA}(\mathcal{H}, \mathcal{H}')$ is not empty. For every $U \in \mathcal{UA}(\mathcal{H}, \mathcal{H}')$, the mapping*

$$\omega_U : \hat{\mathcal{H}} \to \hat{\mathcal{H}}'$$
$$[u] \mapsto \omega_U([u]) := [Uu]$$

is defined consistently and it is an isomorphism from $\hat{\mathcal{H}}$ onto $\hat{\mathcal{H}}'$. For all $z \in \mathbb{T}$, $zU \in \mathcal{UA}(\mathcal{H}, \mathcal{H}')$ and $\omega_{zU} = \omega_U$.

Proof. The mapping ω_U is the restriction to $\hat{\mathcal{H}}$ of the mapping Φ_U defined in 10.9.2. Moreover, $Uu \in \tilde{\mathcal{H}}'$ for all $u \in \tilde{\mathcal{H}}$. Hence, the definition of ω_U is consistent and

$$\tau'(\omega_U([u]), \omega_U([v])) = \tau([u], [v]), \forall u, v \in \tilde{\mathcal{H}}.$$

Furthermore, ω_U is injective because so is Φ_U (cf. 10.9.3A). Finally, for each $u' \in \tilde{\mathcal{H}}'$ we have

$$U^{-1}u' \in \tilde{\mathcal{H}} \text{ and } \omega_U([U^{-1}u']) = [u'],$$

and this proves that ω_U is surjective onto $\hat{\mathcal{H}}'$. Thus, ω_U is an isomorphism from $\hat{\mathcal{H}}$ onto $\hat{\mathcal{H}}'$.

The last part of the statement is obvious (besides, it follows immediately from the last part of the statement in 10.9.2). $\qquad\square$

10.9.6 Theorem (Wigner's theorem). *Let $\mathcal{H}$ and $\mathcal{H}'$ be non-zero Hilbert spaces and suppose that there exists a mapping $\omega : \hat{\mathcal{H}} \to \hat{\mathcal{H}}'$ which is surjective from $\hat{\mathcal{H}}$ onto $\hat{\mathcal{H}}'$ and such that*

$$\tau'(\omega([u]), \omega([v])) = \tau([u], [v]), \forall u, v \in \tilde{\mathcal{H}}.$$

Then ω is an isomorphism from $\hat{\mathcal{H}}$ onto $\hat{\mathcal{H}}'$ and there exists $U \in \mathcal{UA}(\mathcal{H}, \mathcal{H}')$ such that

$$\omega_U = \omega, \ i.e. \ [Uu] = \omega([u]) \text{ for all } u \in \tilde{\mathcal{H}}.$$

The operator U is said to be an implementation of ω.

If $\mathcal{H}$ and $\mathcal{H}'$ are not one-dimensional as linear spaces and if $V \in \mathcal{UA}(\mathcal{H}, \mathcal{H}')$ is such that $\omega_V = \omega$, then there exists $z \in \mathbb{T}$ so that $V = zU$. Thus, an implementation of ω is unique up to a multiplicative factor in $\mathbb{T}$.

Proof. We define the mapping

$$\Phi_\omega : \mathcal{H}_q \to \mathcal{H}'_q$$

$$[f] \mapsto \Phi_\omega([f]) := \begin{cases} [0_{\mathcal{H}'}] & \text{if } [f] = [0_{\mathcal{H}}], \\ \|f\|\omega\left(\left[\frac{1}{\|f\|}f\right]\right) & \text{if } [f] \neq [0_{\mathcal{H}}]. \end{cases}$$

This definition is consistent because if f and g are non-zero elements of $\mathcal{H}$ and $f \div g$ then

$$\|f\| = \|g\| \text{ and } \frac{1}{\|f\|}f \div \frac{1}{\|g\|}g,$$

and therefore

$$\|f\|\omega\left(\left[\frac{1}{\|f\|}f\right]\right) = \|g\|\omega\left(\left[\frac{1}{\|g\|}g\right]\right).$$

Thus, the element $\|f\|\omega\left(\left[\frac{1}{\|f\|}f\right]\right)$ of $\mathcal{H}'_q$ does not depend on the choice of the representative f in the equivalence class $[f]$. It is obvious that the mapping Φ_ω is an extension of the mapping ω. Moreover, we have:

$$\Phi_\omega(a[f]) = \Phi_\omega([af]) = a\|f\|\omega\left(\left[\frac{1}{a\|f\|}af\right]\right)$$

$$= a\Phi_\omega([f]), \forall a \in (0, \infty), \forall f \in \mathcal{H} - \{0_{\mathcal{H}}\};$$

$$\Phi_\omega(a[f]) = [0_{\mathcal{H}'}] = a\Phi_\omega([f]) \text{ if } a = 0 \text{ or } f = 0_{\mathcal{H}};$$

$$\tau'(\Phi_\omega([f]), \Phi_\omega([g])) = \tau'\left(\|f\|\omega\left(\left[\frac{1}{\|f\|}f\right]\right), \|g\|\omega\left(\left[\frac{1}{\|g\|}g\right]\right)\right)$$

$$= \|f\|\|g\|\tau'\left(\omega\left(\left[\frac{1}{\|f\|}f\right]\right), \omega\left(\left[\frac{1}{\|g\|}g\right]\right)\right)$$

$$= \|f\|\|g\|\tau\left(\left[\frac{1}{\|f\|}f\right], \left[\frac{1}{\|g\|}g\right]\right)$$

$$= \tau\left(\|f\|\left[\frac{1}{\|f\|}f\right], \|g\|\left[\frac{1}{\|g\|}g\right]\right)$$

$$= \tau([f], [g]), \forall f, g \in \mathcal{H} - \{0_{\mathcal{H}}\};$$

$$\tau'(\Phi_\omega([f]), \Phi_\omega([g])) = 0 = \tau([f], [g]) \text{ if } f = 0_{\mathcal{H}} \text{ or } g = 0_{\mathcal{H}}.$$

Thus, Φ_ω has all the properties that were assumed for Φ in 10.9.3. Then, Φ_ω is injective by 10.9.3A and so is ω. Thus, ω is an isomorphism from $\hat{\mathcal{H}}$ onto $\hat{\mathcal{H}}'$. Further, if $f' \in \mathcal{H}' - \{0_{\mathcal{H}'}\}$ then $\frac{1}{\|f'\|}f' \in \tilde{\mathcal{H}}'$ and hence (since ω is surjective onto $\hat{\mathcal{H}}'$) there exists $u \in \tilde{\mathcal{H}}$ so that

$$\omega([u]) = \left[\frac{1}{\|f'\|}f'\right],$$

and hence so that

$$\Phi_\omega([\|f'\|u]) = \|f'\|\Phi_\omega([u]) = \|f'\|\omega([u]) = \|f'\|\left[\frac{1}{\|f'\|}f'\right] = [f'].$$

This proves that the mapping Φ_ω is surjective onto $\mathcal{H}'_q$. Then, by 10.9.3B,C there exists $U \in \mathcal{UA}(\mathcal{H}, \mathcal{H}')$ such that

$$[Uf] = \Phi_\omega([f]), \forall f \in \mathcal{H},$$

and hence such that

$$[Uu] = \Phi_\omega([u]) = \omega([u]), \forall u \in \tilde{\mathcal{H}}.$$

If $V \in \mathcal{UA}(\mathcal{H}, \mathcal{H}')$ is such that $\omega_V = \omega$, then we have:

$$[Vf] = \|f\|\left[V\left(\frac{1}{\|f\|}f\right)\right] = \|f\|\omega_V\left(\left[\frac{1}{\|f\|}f\right]\right)$$

$$= \|f\|\omega\left(\left[\frac{1}{\|f\|}f\right]\right) = \Phi_\omega([f]), \forall f \in \mathcal{H} - \{0_{\mathcal{H}}\};$$

$$[V0_{\mathcal{H}}] = [0_{\mathcal{H}'}] = \Phi_\omega([0_{\mathcal{H}}]).$$

Therefore, if $\mathcal{H}$ and $\mathcal{H}'$ are not one-dimensional linear spaces, 10.9.3D implies that there exists $z \in \mathbb{T}$ so that $V = zU$. $\square$

10.9.7 Remarks.

(a) If $\mathcal{H}$ and $\mathcal{H}'$ are non-zero Hilbert spaces and ω is an isomorphism from $\hat{\mathcal{H}}$ onto $\hat{\mathcal{H}}'$, we say that an element U of $\mathcal{UA}(\mathcal{H}, \mathcal{H}')$ *implements* ω if $\omega_U = \omega$.

Suppose that $\mathcal{H}$ and $\mathcal{H}'$ are non-zero Hilbert spaces and that the projective spaces $(\hat{\mathcal{H}}, \tau)$, $(\hat{\mathcal{H}}', \tau')$ are isomorphic. Further, suppose that $\mathcal{H}$ and $\mathcal{H}'$ are not one-dimensional as linear spaces. If U is an element of $\mathcal{UA}(\mathcal{H}, \mathcal{H}')$ which implements an isomorphism ω from $\hat{\mathcal{H}}$ onto $\hat{\mathcal{H}}'$, then it is clear from 10.9.5 and 10.9.6 that another element V of $\mathcal{UA}(\mathcal{H}, \mathcal{H}')$ implements ω iff there exists $z \in \mathbb{T}$ so that $V = zU$. Then, the operators in $\mathcal{UA}(\mathcal{H}, \mathcal{H}')$ which implement a given isomorphism are either all unitary or all antiunitary.

(b) Let $\mathcal{H}$ be a non-zero Hilbert space. It is obvious that the mapping

$$\mathcal{UA}(\mathcal{H}) \ni U \mapsto \omega_U \in \operatorname{Aut}\hat{\mathcal{H}}$$

is a homomorphism from the group $\mathcal{UA}(\mathcal{H})$ (cf. 10.3.16d) to the group $\operatorname{Aut}\hat{\mathcal{H}}$. Wigner's theorem proves that this homomorphism is surjective onto $\operatorname{Aut}\hat{\mathcal{H}}$.

Chapter 11

L^2 Hilbert Spaces

This chapter deals with actualizations of the concept of abstract Hilbert space which are used in most applications of Hilbert space theory.

11.1 $L^2(X, \mathcal{A}, \mu)$

In this section, $(X, \mathcal{A}, \mu)$ denotes an abstract measure space.

11.1.1 Definition. We denote by $\mathcal{L}^2(X, \mathcal{A}, \mu)$ the subset of $\mathcal{M}(X, \mathcal{A}, \mu)$ which is defined as follows:

$$\mathcal{L}^2(X, \mathcal{A}, \mu) := \left\{ \varphi \in \mathcal{M}(X, \mathcal{A}, \mu) : \int_X |\varphi|^2 d\mu < \infty \right\}$$

(this definition is consistent because $\varphi \in \mathcal{M}(X, \mathcal{A}, \mu)$ implies $|\varphi|^2 \in \mathcal{L}^+(X, \mathcal{A}, \mu)$, in view of 6.2.17).

Clearly, for $\varphi \in \mathcal{M}(X, \mathcal{A}, \mu)$, $\varphi \in \mathcal{L}^2(X, \mathcal{A}, \mu)$ iff $|\varphi|^2 \in \mathcal{L}^1(X, \mathcal{A}, \mu)$.

The elements of $\mathcal{L}^2(X, \mathcal{A}, \mu)$ are called *square integrable functions*.

11.1.2 Proposition. *Let* $\varphi, \psi \in \mathcal{L}^2(X, \mathcal{A}, \mu)$. *Then:*

(a) $\alpha\varphi + \beta\psi \in \mathcal{L}^2(X, \mathcal{A}, \mu)$, $\forall \alpha, \beta \in \mathbb{C}$;
(b) $\varphi\psi \in \mathcal{L}^1(X, \mathcal{A}, \mu)$.

Proof. a: For any $\alpha, \beta \in \mathbb{C}$, we have $\alpha\varphi + \beta\psi \in \mathcal{M}(X, \mathcal{A}, \mu)$ by 8.2.2. We also have

$$|\alpha\varphi(x) + \beta\psi(x)|^2 \le 2(|\alpha|^2 |\varphi(x)|^2 + |\beta|^2 |\psi(x)|^2), \forall x \in D_\varphi \cap D_\psi$$

(cf. inequality 2 in the proof of 10.3.7), and this implies $|\alpha\varphi + \beta\psi|^2 \in \mathcal{L}^1(X, \mathcal{A}, \mu)$ by 8.2.9 and 8.2.5, and hence $\alpha\varphi + \beta\psi \in \mathcal{L}^2(X, \mathcal{A}, \mu)$.

b: We have $\varphi\psi \in \mathcal{M}(X, \mathcal{A}, \mu)$ by 8.2.2. We also have

$$|\varphi(x)\psi(x)| \le \frac{1}{2}(|\varphi(x)|^2 + |\psi(x)|^2), \forall x \in D_\varphi \cap D_\psi$$

(cf. inequality 1 in the proof of 10.3.7), and this implies $\varphi\psi \in \mathcal{L}^1(X, \mathcal{A}, \mu)$ by 8.2.9 and 8.2.5. $\qquad\square$

11.1.3 Corollary. *If the measure μ is finite, then $\mathcal{L}^2(X, \mathcal{A}, \mu) \subset \mathcal{L}^1(X, \mathcal{A}, \mu)$.*

Proof. If $\mu(X) < \infty$ the obviously $1_X \in \mathcal{L}^2(X, \mathcal{A}, \mu)$ and we can assume $\psi := 1_X$ in 11.1.2b. $\qquad\square$

11.1.4 Theorem. *A subset of $M(X, \mathcal{A}, \mu)$, which we denote by $L^2(X, \mathcal{A}, \mu)$, is consistently defined as follows:*

$$L^2(X, \mathcal{A}, \mu) := \left\{ [\varphi] \in M(X, \mathcal{A}, \mu) : \int_X |\varphi|^2 d\mu < \infty \right\}.$$

Clearly, for $[\varphi] \in M(X, \mathcal{A}, \mu)$, $[\varphi] \in L^2(X, \mathcal{A}, \mu)$ iff $\varphi \in \mathcal{L}^2(X, \mathcal{A}, \mu)$.
$L^2(X, \mathcal{A}, \mu)$ is a linear manifold in the linear space $M(X, \mathcal{A}, \mu)$.
A function ϕ is consistently defined by

$$\phi : L^2(X, \mathcal{A}, \mu) \times L^2(X, \mathcal{A}, \mu) \to \mathbb{C}$$

$$([\varphi], [\psi]) \mapsto \phi([\varphi], [\psi]) := \int_X \overline{\varphi}\psi d\mu$$

and ϕ is an inner product for the linear space $L^2(X, \mathcal{A}, \mu)$.

Proof. To prove the consistency of the definition of the set $L^2(X, \mathcal{A}, \mu)$, we need to check the implication

$$\left[\varphi', \varphi \in \mathcal{M}(X, \mathcal{A}, \mu), \varphi' \sim \varphi, \int_X |\varphi|^2 d\mu < \infty \right] \Rightarrow \int_X |\varphi'|^2 d\mu < \infty,$$

because if this implication is true then the condition $\int_X |\varphi|^2 d\mu < \infty$ is actually a condition for the equivalence class $[\varphi]$ even though it is expressed through a particular representative of it. Now, the above implication follows from 8.1.17c. Then, 11.1.2a proves that $L^2(X, \mathcal{A}, \mu)$ is a linear manifold in $M(X, \mathcal{A}, \mu)$.

To prove the consistency of the definition of the function ϕ, first we note that 11.1.2b implies that $\overline{\varphi}\psi \in \mathcal{L}^1(X, \mathcal{A}, \mu)$ for all $\varphi, \psi \in \mathcal{L}^2(X, \mathcal{A}, \mu)$. Next, if $\varphi', \varphi, \psi', \psi \in \mathcal{L}^2(X, \mathcal{A}, \mu)$ are so that $\varphi' \sim \varphi$ and $\psi' \sim \psi$, then

$$\overline{\varphi'(x)}\psi'(x) = \overline{\varphi(x)}\psi(x) \quad \mu\text{-a.e. on } D_{\varphi\psi} \cap D_{\varphi'\psi'}$$

(the proof is similar to the one given in 8.2.13 for $\varphi' + \psi' \sim \varphi + \psi$) and hence

$$\int_X \overline{\varphi'}\psi' d\mu = \int_X \overline{\varphi}\psi\mu$$

by 8.2.7. Thus, the number $\int_X \overline{\varphi}\psi d\mu$ depends only on the equivalence classes $[\varphi]$ and $[\psi]$, and not on the particular representatives φ and ψ through which it is obtained.

As to the properties listed in 10.1.3 which ϕ must have in order to be an inner product, ip_1 follows from 8.2.9, ip_2 from

$$\int_X \overline{\varphi}\psi d\mu = \overline{\left(\int_X \overline{\psi}\varphi d\mu \right)}, \forall \varphi, \psi \in \mathcal{L}^2(X, \mathcal{A}, \mu)$$

(cf. 8.2.3), ip_3 is obvious. Finally, for $\varphi \in \mathcal{L}^2(X, \mathcal{A}, \mu)$ we have

$$\int_X |\varphi|^2 d\mu = ([\varphi]|[\varphi]) = 0 \Rightarrow \varphi(x) = 0 \ \mu\text{-a.e. on } D_\varphi \Rightarrow [\varphi] = 0_{L^2(X, \mathcal{A}, \mu)}$$

(cf. 8.1.18a), and this shows that ϕ has property ip_4. $\qquad\square$

11.1.5 Remark. Since in every element of $M(X, \mathcal{A}, \mu)$ there exists a representative which is an element of $\mathcal{M}(X, \mathcal{A})$ (cf. 8.2.12), it is easy to see that the inner product space constructed in 11.1.4 is trivially isomorphic to the one that is constructed as follows (we will still use the symbols $\mathcal{L}^2(X, \mathcal{A}, \mu)$ and $L^2(X, \mathcal{A}, \mu)$, although they have now a different meaning than before). One defines a subset of $\mathcal{M}(X, \mathcal{A})$ by

$$\mathcal{L}^2(X, \mathcal{A}, \mu) := \left\{ \varphi \in \mathcal{M}(X, \mathcal{A}) : \int_X |\varphi|^2 d\mu < \infty \right\},$$

then defines an equivalence relation in $\mathcal{L}^2(X, \mathcal{A}, \mu)$ by

$$\varphi \sim \psi \text{ if } \varphi(x) = \psi(x) \text{ } \mu\text{-a.e. on } X,$$

and then writes $L^2(X, \mathcal{A}, \mu) := \mathcal{L}^2(X, \mathcal{A}, \mu)/\sim$ for the quotient set. Then one defines vector sum, scalar multiplication and inner product by

$$[\varphi] + [\psi] := [\varphi + \psi], \forall [\varphi], [\psi] \in L^2(X, \mathcal{A}, \mu),$$
$$\alpha[\varphi] := [\alpha\varphi], \forall \alpha \in \mathbb{C}, \forall [\varphi] \in L^2(X, \mathcal{A}, \mu),$$
$$([\varphi] | [\psi]) := \int_X \overline{\varphi}\psi d\mu, \forall [\varphi], [\psi] \in L^2(X, \mathcal{A}, \mu).$$

This way of defining the inner product space $L^2(X, \mathcal{A}, \mu)$ is more frequent than ours.

In a similar way one can give an alternative but equivalent definition of the normed space $L^1(X, \mathcal{A}, \mu)$ (cf. 8.2.15).

11.1.6 Theorem. *The normed space $L^1(X, \mathcal{A}, \mu)$ is a Banach space and the inner product space $L^2(X, \mathcal{A}, \mu)$ is a Hilbert space.*

Proof. In what follows, p stands for either 1 or 2. We need to prove that the metric space $(L^p(X, \mathcal{A}, \mu), d)$ is complete, where d is the distance on $L^p(X, \mathcal{A}, \mu)$ defined by

$$d([\varphi], [\psi]) := \|[\varphi] - [\psi]\| = \left(\int_X |\varphi - \psi|^p d\mu \right)^{\frac{1}{p}}, \forall [\varphi], [\psi] \in L^p(X, \mathcal{A}, \mu).$$

Then, let $\{[\varphi_n]\}$ be a Cauchy sequence in $L^p(X, \mathcal{A}, \mu)$, and for each $\varepsilon > 0$ let $N_\varepsilon \in \mathbb{N}$ be so that

$$n, m > N_\varepsilon \Rightarrow \|[\varphi_n] - [\varphi_m]\| < \varepsilon.$$

We define a subsequence $\{[\varphi_{n_k}]\}$ of $\{[\varphi_n]\}$ by induction as follows:

we choose n_1 such that $n_1 > N_{\frac{1}{2}}$;
for $k > 1$, assuming that we have already chosen $n_2, ..., n_k$ so that $n_i > N_{\frac{1}{2^i}}$ and $n_i > n_{i-1}$ for $i = 2, ..., k$, we choose n_{k+1} such that $n_{k+1} > N_{\frac{1}{2^{k+1}}}$ and $n_{k+1} > n_k$.

It is expedient to choose, for each $n \in \mathbb{N}$, the representative φ_n so that $D_{\varphi_n} = X$ (cf. 8.2.12). We define

$$\psi_m := \sum_{k=1}^{m} |\varphi_{n_{k+1}} - \varphi_{n_k}|, \forall m \in \mathbb{N},$$

and we have $\psi_m \in \mathcal{L}^p(X, \mathcal{A}, \mu)$ (cf. 8.2.9 or 11.1.2a). We also define

$$\psi_\infty := \sum_{k=1}^{\infty} |\varphi_{n_{k+1}} - \varphi_{n_k}|$$

($\psi_\infty(x)$ may be ∞ for some $x \in X$), and we have $\psi_\infty \in \mathcal{L}^+(X, \mathcal{A})$ (cf. 6.2.32). Since

$$\left(\int_X |\psi_m|^p d\mu \right)^{\frac{1}{p}} = \|[\psi_m]\| \leq \sum_{k=1}^{m} \|[\varphi_{n_{k+1}}] - [\varphi_{n_k}]\| < \sum_{k=1}^{m} \frac{1}{2^k} < 1, \forall m \in \mathbb{N},$$

by 8.1.20 we have

$$\int_X |\psi_\infty|^p d\mu \leq 1.$$

In view of 8.1.12b, this implies that if we define $E := \psi_\infty^{-1}(\{\infty\})$ then

$$E \in \mathcal{A} \text{ and } \mu(E) = 0.$$

The series $\sum_{k=1}^{\infty} (\varphi_{n_{k+1}}(x) - \varphi_{n_k}(x))$ is convergent for all $x \in X - E$ (cf. 4.1.8b) and hence we can define the function

$$\varphi : X - E \to \mathbb{C}$$

$$x \mapsto \varphi(x) := \varphi_{n_1}(x) + \sum_{k=1}^{\infty} (\varphi_{n_{k+1}}(x) - \varphi_{n_k}(x)),$$

which is an element of $\mathcal{M}(X, \mathcal{A}, \mu)$ (cf. 6.2.3 and 6.2.20d). Since

$$\varphi_{n_1} + \sum_{k=1}^{i-1} (\varphi_{n_{k+1}} - \varphi_{n_k}) = \varphi_{n_i}, \forall i \in \mathbb{N},$$

we see that

$$\varphi(x) = \lim_{i \to \infty} \varphi_{n_i}(x), \forall x \in X - E.$$

Now fix $\varepsilon > 0$ and let $n > N_\varepsilon$. We have

$$|\varphi(x) - \varphi_n(x)|^p = \lim_{i \to \infty} |\varphi_{n_i}(x) - \varphi_n(x)|^p, \forall x \in X - E.$$

We also have

$$\int_X |\varphi_{n_i} - \varphi_n|^p d\mu < \varepsilon^p \text{ for all } i \in \mathbb{N} \text{ s.t. } n_i > N_\varepsilon.$$

Then, by 8.1.20 we have

$$\int_X |\varphi - \varphi_n|^p d\mu \leq \varepsilon^p.$$

This proves first that $\varphi - \varphi_n \in \mathcal{L}^p(X, \mathcal{A}, \mu)$ and hence that $\varphi \in \mathcal{L}^p(X, \mathcal{A}, \mu)$ since $\varphi = (\varphi - \varphi_n) + \varphi_n$ (cf. 8.2.9 or 11.1.2a), and thereafter also that

$$\|[\varphi] - [\varphi_n]\| \leq \varepsilon.$$

Thus, we have proved that there exists $[\varphi] \in L^p(X, \mathcal{A}, \mu)$ such that $\lim_{n \to \infty} [\varphi_n] = [\varphi]$. $\qquad \square$

11.1.7 Remark. From the proof of 11.1.6 we obtain the following result: if $\{[\varphi_n]\}$ is a convergent sequence in $L^1(X, \mathcal{A}, \mu)$ or in $L^2(X, \mathcal{A}, \mu)$ and $[\varphi] := \lim_{n \to \infty} [\varphi_n]$, then there exists a subsequence $\{[\varphi_{n_k}]\}$ so that

$$\varphi(x) = \lim_{k \to \infty} \varphi_{n_k}(x) \ \mu\text{-a.e. on } X.$$

11.1.8 Remark. For $X := \{1, ..., N\}$ or $X := \mathbb{N}$, if $\mathcal{A} = \mathcal{P}(X)$ and μ is the counting measure on X (cf. 8.3.10) then $L^2(X, \mathcal{A}, \mu)$ is the Hilbert space $\mathbb{C}^N$ or the Hilbert space ℓ^2, respectively (cf. 10.3.8c,d).

11.1.9 Proposition. *The intersection* $S^2(X, \mathcal{A}, \mu) := S(X, \mathcal{A}, \mu) \cap L^2(X, \mathcal{A}, \mu)$ *is a dense linear manifold in the Hilbert space* $L^2(X, \mathcal{A}, \mu)$ *(for* $S(X, \mathcal{A}, \mu)$*, cf. 8.2.14).*

Proof. Since $S(X, \mathcal{A}, \mu)$ and $L^2(X, \mathcal{A}, \mu)$ are linear manifolds in the linear space $M(X, \mathcal{A}, \mu)$, the same holds true for their intersection (cf. 3.1.5). Then $S^2(X, \mathcal{A}, \mu)$ is a linear manifold in the linear space $L^2(X, \mathcal{A}, \mu)$ as well (cf. 3.1.4b).

Let $[\varphi] \in L^2(X, \mathcal{A}, \mu)$ and assume that the representative φ is an element of $M(X, \mathcal{A})$ (cf. 8.2.12). Then, by 6.2.27 there exists a sequence $\{\psi_n\}$ in $S(X, \mathcal{A})$ such that

$$|\psi_n(x)| \leq |\varphi(x)|, \forall x \in X, \forall n \in \mathbb{N},$$
$$\lim_{n \to \infty} \psi_n(x) = \varphi(x), \forall x \in X,$$

and hence such that

$$|\psi_n(x) - \varphi(x)|^2 \leq 4|\varphi(x)|^2, \forall x \in X,$$
$$\lim_{n \to \infty} |\psi_n(x) - \varphi(x)|^2 = 0, \forall x \in X.$$

Then $\psi_n \in \mathcal{L}^2(X, \mathcal{A}, \mu)$ by 8.2.5, and by 8.2.11 we have

$$\lim_{n \to \infty} \|[\psi_n] - [\varphi]\| = \lim_{n \to \infty} \left(\int_X |\psi_n - \varphi|^2 d\mu \right)^{\frac{1}{2}} = 0.$$

In view of 2.3.12, this proves that $S^2(X, \mathcal{A}, \mu)$ is dense in $L^2(X, \mathcal{A}, \mu)$. $\square$

11.1.10 Proposition. *Let* $(X_1, \mathcal{A}_1, \mu_1)$ *and* $(X_2, \mathcal{A}_2, \mu_2)$ *be measure spaces, and suppose that* $\pi : D_\pi \to X_2$ *is an injective mapping from* X_1 *to* X_2 *such that the following conditions are true:*

$D_\pi \in \mathcal{A}_1, R_\pi \in \mathcal{A}_2$;

π *is measurable w.r.t.* $\mathcal{A}_1^{D_\pi}$ *and* $\mathcal{A}_2, \pi^{-1}$ *is measurable w.r.t.* $\mathcal{A}_2^{R_\pi}$ *and* $\mathcal{A}_1$;

$\mu_1(X_1 - D_\pi) = 0$;

$\mu_2(E) = \mu_1(\pi^{-1}(E)), \forall E \in \mathcal{A}_2$.

Then, for every $\varphi \in \mathcal{L}^2(X_2, \mathcal{A}_2, \mu_2)$ *the function* $\varphi \circ \pi$ *is an element of* $\mathcal{L}^2(X_1, \mathcal{A}_1, \mu_1)$. *Moreover, for* $\varphi, \psi \in \mathcal{L}^2(X_2, \mathcal{A}_2, \mu_2)$, *if* $\varphi \sim \psi$ *then* $\varphi \circ \pi \sim \psi \circ \pi$. *Therefore, the definition of the mapping*

$$U : L^2(X_2, \mathcal{A}_2, \mu_2) \to L^2(X_1, \mathcal{A}_1, \mu_1)$$
$$[\varphi] \mapsto U[\varphi] := [\varphi \circ \pi]$$

is consistent.

The definition of the mapping

$$V : L^2(X_1, \mathcal{A}_1, \mu_1) \to L^2(X_2, \mathcal{A}_2, \mu_2)$$
$$[\psi] \mapsto V[\psi] := [\psi \circ \pi^{-1}]$$

is also consistent.

The mappings U and V are unitary operators and $V = U^{-1}$.

Proof. In view of 8.3.11c, if $\varphi \in \mathcal{L}^2(X_2, \mathcal{A}_2, \mu_2)$ then $\varphi \in \mathcal{M}(X_2, \mathcal{A}_2, \mu_2)$ and hence $\varphi \circ \pi \in \mathcal{M}(X_1, \mathcal{A}_1, \mu_1)$; also, $|\varphi|^2 \in \mathcal{L}^1(X_2, \mathcal{A}_2, \mu_2)$ and hence $|\varphi \circ \pi|^2 \in \mathcal{L}^1(X_1, \mathcal{A}_1, \mu)$ and hence $\varphi \circ \pi \in \mathcal{L}^2(X_1, \mathcal{A}_1, \mu)$. Next, suppose that $\varphi, \psi \in \mathcal{L}^2(X_2, \mathcal{A}_2, \mu_2)$ are such that $\varphi \sim \psi$; then there exists $E \in \mathcal{A}_2$ such that $\mu_2(E) = 0$ and

$$\varphi(x) = \psi(x), \forall x \in D_\varphi \cap D_\psi \cap (X_2 - E);$$

and this implies that

$$(\varphi \circ \pi)(y) = (\psi \circ \pi)(y), \forall y \in \pi^{-1}(D_\varphi \cap D_\psi \cap (X_2 - E));$$

since

$$\pi^{-1}(D_\varphi \cap D_\psi \cap (X_2 - E)) = \pi^{-1}(D_\varphi) \cap \pi^{-1}(D_\psi) \cap \pi^{-1}(X_2 - E)$$
$$= D_{\varphi \circ \pi} \cap D_{\psi \circ \pi} \cap (D_\pi - \pi^{-1}(E))$$
$$= D_{\varphi \circ \pi} \cap D_{\psi \circ \pi} \cap (X_1 - \pi^{-1}(E))$$

(the last equality holds because e.g. $D_{\varphi \circ \pi} \subset D_\pi$) and $\mu_1(\pi^{-1}(E)) = \mu_2(E) = 0$, this proves that $\varphi \circ \pi \sim \psi \circ \pi$. Therefore, the definition of the mapping U is consistent.

Now, define $\tau := \pi^{-1}$. Obviously $D_\tau = R_\pi$. Also,

$$\mu_2(X_2 - D_\tau) = \mu_1(\pi^{-1}(X_2 - R_\pi)) = \mu_1(\emptyset) = 0.$$

Moreover, for every $F \in \mathcal{A}_1$ we have $F = ((X_1 - D_\pi) \cap F) \cup (D_\pi \cap F)$ and hence, since $\mu_1((X_1 - D_\pi) \cap F) \le \mu_1(X_1 - D_\pi) = 0$,

$$\mu_1(F) = \mu_1(D_\pi \cap F) = \mu_1(\pi^{-1}(\pi(D_\pi \cap F))) = \mu_2(\pi(D_\pi \cap F)) = \mu_2(\tau^{-1}(F))$$

(the second and the fourth equalities are true because π is injective). Thus, the assumptions are actually completely symmetrical with respect to the indices 1 and 2, and therefore the definition of the mapping V is also consistent.

For every $\varphi \in \mathcal{L}^2(X_2, \mathcal{A}_2, \mu_2)$ we have $\varphi \circ \pi \circ \pi^{-1} = \varphi \circ id_{R_\pi} = \varphi_{D_\varphi \cap R_\pi}$, and hence $\varphi \sim \varphi \circ \pi \circ \pi^{-1}$ since

$$\varphi(x) = \varphi_{D_\varphi \cap R_\pi}(x), \forall x \in D_\varphi \cap R_\pi = D_\varphi \cap (D_\varphi \cap R_\pi) \cap (X_2 - (X_2 - R_\pi))$$

and $\mu_2(X_2 - R_\pi) = 0$. Thus we have

$$V(U[\varphi]) = [\varphi \circ \pi \circ \pi^{-1}] = [\varphi], \forall [\varphi] \in L^2(X_2, \mathcal{A}_2, \mu_2),$$

and symmetrically

$$U(V[\psi]) = [\psi], \forall [\psi] \in L^2(X_1, \mathcal{A}_1, \mu_1).$$

By 1.2.16b, this proves that U is injective and $U^{-1} = V$, and therefore also that $R_U = L^2(X_1, \mathcal{A}_1, \mu_1)$. Moreover, for all $[\varphi], [\psi] \in L^2(X_2, \mathcal{A}_2, \mu_2)$ we have

$$\int_{X_1} \overline{(\varphi \circ \pi)}(\psi \circ \pi) d\mu_1 = \int_{X_2} \overline{\varphi}\psi d\mu_2$$

by 8.3.11c. Thus, U is a unitary operator in view of 10.1.20. Symmetrically, V is a unitary operator as well (this can be deduced also on the basis of 4.6.2b and 10.1.21). $\qquad\square$

11.1.11 Remark. It is immediate to see that the mapping

$$V : L^2(X, \mathcal{A}, \mu) \to L^2(X, \mathcal{A}, \mu)$$
$$[\varphi] \mapsto V[\varphi] := [\overline{\varphi}]$$

is defined consistently, also in view of 6.2.17, and that V is an antiunitary operator as a result of the basic properties of the Lebesgue integral. Clearly, $V^2 = \mathbb{1}_{L^2(X,\mathcal{A},\mu)}$ and hence every antiunitary operator A in $L^2(X, \mathcal{A}, \mu)$ is the product of a unitary operator multiplied by V, since $A = (AV)V$ and AV is a unitary operator (cf. 10.3.16c).

11.2 $L^2(a, b)$

In this section, a and b are two real numbers such that $a < b$.

We write $\mathcal{M}(a, b) := \mathcal{M}(X, \mathcal{A}, \mu)$, $\mathcal{L}^1(a, b) := \mathcal{L}^1(X, \mathcal{A}, \mu)$, $\mathcal{L}^2(a, b) := \mathcal{L}^2(X, \mathcal{A}, \mu)$, $L^2(a, b) := L^2(X, \mathcal{A}, \mu)$ if $X := [a, b]$, $\mathcal{A} := (\mathcal{A}(d_{\mathbb{R}}))^{[a,b]}$, $\mu := m_{[a,b]}$, where $m_{[a,b]}$ is the Lebesgue measure on $[a, b]$ (cf. 9.3.1). Moreover, we denote $(\mathcal{A}(d_{\mathbb{R}}))^{[a,b]}$ by the symbol $\mathcal{A}_{[a,b]}$.

11.2.1 Theorem. *The inclusion $\mathcal{C}(a, b) \subset \mathcal{L}^2(a, b)$ holds true. If the mapping ι is defined by*

$$\iota : \mathcal{C}(a, b) \to L^2(a, b)$$
$$\varphi \mapsto \iota(\varphi) := [\varphi],$$

then the pair $(L^2(a, b), \iota)$ is a completion of the inner product space $\mathcal{C}(a, b)$ (for which, cf. 10.1.5b).

Proof. For every $\varphi \in \mathcal{C}(a, b)$, $\varphi \in \mathcal{M}(a, b)$ (cf. 6.2.8). Moreover, the function $|\varphi|^2$ is bounded since it is an element of $\mathcal{C}(a, b)$ (cf. 3.1.10f), and therefore $|\varphi|^2 \in \mathcal{L}^1(a, b)$ (cf. 8.2.6), and hence $\varphi \in \mathcal{L}^2(a, b)$.

If the inner products in $\mathcal{C}(a, b)$ and in $L^2(a, b)$ are denoted by ϕ and by $\hat{\phi}$ respectively, directly from their definitions we have

$$\hat{\phi}(\iota(\varphi), \iota(\psi)) = \int_{[a,b]} \overline{\varphi}\psi \, dm_{[a,b]} = \phi(\varphi, \psi), \forall \varphi, \psi \in \mathcal{C}(a, b).$$

To complete the proof of the statement it remains to prove that R_ι is dense in $L^2(a, b)$, which we do below by proving that

$$\forall [\varphi] \in L^2(a, b), \forall \varepsilon > 0, \exists \psi \in \mathcal{C}(a, b) \text{ such that } \|[\varphi] - [\psi]\| < \varepsilon$$

(cf. 2.3.12). We fix $[\varphi] \in L^2(a, b)$ and $\varepsilon > 0$, and for convenience we assume $D_\varphi = [a, b]$ for the representative φ (cf. 8.2.12). For each $n \in \mathbb{N}$ we define the function

$$\eta_n : \mathbb{C} \to \mathbb{C}$$

$$z \mapsto \eta_n(z) := \begin{cases} z & \text{if } |z| \le n, \\ nz|z|^{-1} & \text{if } |z| > n, \end{cases}$$

and then the function $\varphi_n := \eta_n \circ \varphi$. Since the function η_n is continuous, the function φ_n is an element of $\mathcal{M}(a, b)$ (cf. 6.2.5 and 6.2.8), and it is an element of $\mathcal{L}^2(a, b)$ since $|\varphi_n(x)| \le |\varphi(x)|$ for all $x \in [a, b]$ (cf. 8.2.5). Since

$$|\varphi(x) - \varphi_n(x)|^2 \le 4|\varphi(x)|^2, \forall x \in X, \forall n \in \mathbb{N},$$

and

$$\lim_{n \to \infty} \varphi_n(x) = \varphi(x) \text{ and hence } \lim_{n \to \infty} |\varphi(x) - \varphi_n(x)|^2 = 0, \forall x \in X,$$

by 8.2.11 we have

$$\lim_{n \to \infty} \int_{[a,b]} |\varphi - \varphi_n|^2 dm_{[a,b]} = 0.$$

Thus, there exists $k \in \mathbb{N}$ such that

$$\int_{[a,b]} |\varphi - \varphi_k|^2 dm_{[a,b]} < \frac{\varepsilon^2}{4}.$$

Moreover, by 7.4.8 there exists $\psi \in \mathcal{C}(a, b)$ such that

$$m_{[a,b]}(\{x \in [a, b] : \varphi_k(x) \ne \psi(x)\}) < \frac{\varepsilon^2}{16k^2}$$

and

$$\sup\{|\psi(x)| : x \in [a, b]\} \le \sup\{|\varphi_k(x)| : x \in [a, b]\} \le k.$$

Thus, if we write $E := \{x \in [a, b] : \varphi_k(x) \ne \psi(x)\}$ we have

$$\int_{[a,b]} |\varphi_k - \psi|^2 dm_{[a,b]} = \int_{[a,b]} \chi_E |\varphi_k - \psi|^2 dm_{[a,b]}$$

$$\le \int_{[a,b]} \chi_E (|\varphi_k| + |\psi|)^2 dm_{[a,b]}$$

$$\le 4k^2 \int_{[a,b]} \chi_E dm_{[a,b]} = 4k^2 m_{[a,b]}(E) < \frac{\varepsilon^2}{4},$$

and hence

$$\|[\varphi] - [\psi]\| \le \|[\varphi] - [\varphi_k]\| + \|[\varphi_k] - [\psi]\| < \varepsilon.$$

$\square$

11.2.2 Remarks.

(a) Since $\overline{\iota(\mathcal{C}(a,b))} = L^2(a,b)$ and $\iota(\mathcal{C}(a,b)) \neq L^2(a,b)$, $\iota(\mathcal{C}(a,b))$ is not a closed subset of $L^2(a,b)$ (cf. 2.3.9c). Hence, the inner product space $\iota(\mathcal{C}(a,b))$ (cf. 10.3.5) is not a Hilbert space (cf. 2.6.6a). Since the inner product spaces $\iota(\mathcal{C}(a,b))$ and $\mathcal{C}(a,b)$ are isomorphic (cf. 10.3.5), this furnishes another proof (besides the one given in 10.4.2c) that the inner product space $\mathcal{C}(a,b)$ is not a Hilbert space (cf. 10.1.21).

(b) The mapping ι of 11.2.1 is a linear operator and it is injective (cf. 10.3.5). Thus, if $\varphi, \psi \in \mathcal{C}(a,b)$ are such that $[\varphi] = [\psi]$ then $\varphi = \psi$. Namely, if an element of $L^2(a,b)$ contains a continuous function then this function is the only continuous one it contains.

(c) The mapping ι of 11.2.1 is not surjective onto $L^2(a,b)$. To see this, let $x_0 \in (a,b)$ and consider the element $[\chi_{[a,x_0]}]$ of $L^2(a,b)$. If $\varphi \in [\chi_{[a,x_0]}]$ and $D_\varphi = [a,b]$ then for each $n \in \mathbb{N}$ large enough there exists $x'_n \in \left[x_0 - \frac{1}{n}, x_0 \right]$ such that $\varphi(x'_n) = 1$ (otherwise we should have $\varphi(x) \neq \chi_{[a,x_0]}(x)$ for all $x \in \left[x_0 - \frac{1}{n}, x_0 \right]$, and this would be in contradiction with $\varphi \sim \chi_{[a,x_0]}$), and similarly there exists $x''_n \in \left(x_0, x_0 + \frac{1}{n} \right]$ such that $\varphi(x''_n) = 0$. Hence, there exist two sequences $\{x'_n\}$ and $\{x''_n\}$ in $[a,b]$ such that $\lim_{n\to\infty} x'_n = \lim_{n\to\infty} x''_n = x_0$, $\lim_{n\to\infty} \varphi(x'_n) = 1$, $\lim_{n\to\infty} \varphi(x''_n) = 0$. By 2.4.2, this proves that φ is not continuous at x_0, and hence that $\varphi \notin \mathcal{C}(a,b)$. Another proof that ι is not surjective is by contraposition as follows: if ι were surjective onto $L^2(a,b)$ then ι would be an isomorphism from $\mathcal{C}(a,b)$ onto $L^2(a,b)$ and hence $\mathcal{C}(a,b)$ would be a Hilbert space (cf. 10.1.21), which is not true.

11.2.3 Proposition. *Let π be the function*

$$\pi : [a,b] \to [0, 2\pi]$$

$$x \mapsto \pi(x) := 2\pi \frac{x-a}{b-a}.$$

The definitions of the mappings

$$U : L^2(0, 2\pi) \to L^2(a,b)$$

$$[\varphi] \mapsto U[\varphi] := \left(\frac{2\pi}{b-a} \right)^{\frac{1}{2}} [\varphi \circ \pi]$$

and

$$V : L^2(a,b) \to L^2(0, 2\pi)$$

$$[\psi] \mapsto V[\psi] := \left(\frac{b-a}{2\pi} \right)^{\frac{1}{2}} [\psi \circ \pi^{-1}]$$

are consistent.

  The mappings U and V are unitary operators and $V = U^{-1}$.

Proof. The proof is based on 11.1.10. Let $X_1 := [a,b]$, $\mathcal{A}_1 := \mathcal{A}_{[a,b]}$, $\mu_1 := m_{[a,b]}$, $X_2 := [0, 2\pi]$, $\mathcal{A}_2 := \mathcal{A}_{[0,2\pi]}$, $\mu_2 := \frac{b-a}{2\pi} m_{[0,2\pi]}$ (for which, cf. 8.3.5b with $\mu := m_{[0,2\pi]}$ and ν the null measure on $\mathcal{A}_{[0,2\pi]}$). In view of 9.2.1a and 9.2.2a we have, for every $E \in \mathcal{A}_{[0,2\pi]}$,

$$\frac{b-a}{2\pi} m_{[0,2\pi]}(E) = \frac{b-a}{2\pi} m(E) = m\left(\frac{b-a}{2\pi} E + a\right)$$

$$= m_{[a,b]}\left(\frac{b-a}{2\pi} E + a\right) = m_{[a,b]}(\pi^{-1}(E)).$$

Then, 11.1.10 proves that the mapping

$$W : L^2\left([0,2\pi], \mathcal{A}_{[0,2\pi]}, \frac{b-a}{2\pi} m_{[0,2\pi]}\right) \to L^2(a,b)$$

$$[\varphi] \mapsto W[\varphi] := [\varphi \circ \pi]$$

is a unitary operator. Since $\mathcal{M}\left([0,2\pi], \mathcal{A}_{[0,2\pi]}, \frac{b-a}{2\pi} m_{[0,2\pi]}\right) = \mathcal{M}(0, 2\pi)$ and $\mathcal{L}^1\left([0,2\pi], \mathcal{A}_{[0,2\pi]}, \frac{b-a}{2\pi} m_{[0,2\pi]}\right) = \mathcal{L}^1(0, 2\pi)$ (cf. 8.3.5b), and since trivially, for $E \in \mathcal{A}_{[0,2\pi]}$, $\frac{b-a}{2\pi} m_{[0,2\pi]}(E) = 0$ iff $m_{[0,2\pi]}(E) = 0$, we have $L^2\left([0,2\pi], \mathcal{A}_{[0,2\pi]}, \frac{b-a}{2\pi} m_{[0,2\pi]}\right) = L^2(0, 2\pi)$. Thus, the mapping

$$T : L^2(0, 2\pi) \to L^2\left([0,2\pi], \mathcal{A}_{[0,2\pi]}, \frac{b-a}{2\pi} m_{[0,2\pi]}\right)$$

$$[\varphi] \mapsto T[\varphi] := \left(\frac{2\pi}{b-a}\right)^{\frac{1}{2}} [\varphi]$$

is defined consistently and $R_T = L^2\left([0,2\pi], \mathcal{A}_{[0,2\pi]}, \frac{b-a}{2\pi} m_{[0,2\pi]}\right)$. In view of 8.3.5b, we also have

$$\int_{[0,2\pi]} \left(\frac{2\pi}{b-a}\right)^{\frac{1}{2}} \overline{\varphi} \left(\frac{2\pi}{b-a}\right)^{\frac{1}{2}} \psi \, d\left(\frac{b-a}{2\pi} m_{[0,2\pi]}\right) = \int_{[0,2\pi]} \overline{\varphi} \psi \, dm_{[0,2\pi]}.$$

Thus, T is a unitary operator in view of 10.1.20, and hence U is a unitary operator since $U = W \circ T$ (cf. 10.1.21 and 4.6.2b).

Now, it is obvious that

$$V(U[\varphi]) = [\varphi], \forall [\varphi] \in L^2(0, 2\pi).$$

Then, 1.2.16a implies $U^{-1} \subset V$, and hence $U^{-1} = V$. This also proves that V is a unitary operator (cf. 10.1.21 and 4.6.2b). $\qquad\square$

11.2.4 Theorem. *For each $n \in \mathbb{Z}$ let u_n be the element of $C(a,b)$ defined by*

$$u_n(x) := \left(\frac{1}{b-a}\right)^{\frac{1}{2}} \exp\left(i2\pi n \frac{x-a}{b-a}\right), \forall x \in [a,b],$$

and for each $n \in \mathbb{N}$ let v_n and w_n be the elements of $C(a,b)$ defined by

$$v_n(x) := \left(\frac{2}{b-a}\right)^{\frac{1}{2}} \cos\left(2\pi n \frac{x-a}{b-a}\right), \forall x \in [a,b],$$

$$w_n(x) := \left(\frac{2}{b-a}\right)^{\frac{1}{2}} \sin\left(2\pi n \frac{x-a}{b-a}\right), \forall x \in [a, b].$$

Both the families $\{[u_n]\}_{n\in\mathbb{Z}}$ *and* $\{[u_n]\} \cup \{[v_n]\}_{n\in\mathbb{N}} \cup \{[w_n]\}_{n\in\mathbb{N}}$ *are complete orthonormal systems in the Hilbert space* $L^2(a, b)$.

Proof. First we consider the special case $a := 0$ and $b := 2\pi$. We already know that both the families $\{u_n\}_{n\in\mathbb{Z}}$ and $\{u_0\} \cup \{v_n\}_{n\in\mathbb{N}} \cup \{w_n\}_{n\in\mathbb{N}}$ are orthonormal systems in the inner product space $\mathcal{C}(0, 2\pi)$ (cf. 10.2.5b). Hence so are in $L^2(0, 2\pi)$ the two families of the statement, owing to property co_1 of 10.3.4 possessed by the mapping ι of 11.2.1. To prove that these orthonormal systems are complete in $L^2(0, 2\pi)$, we first prove that $L\{u_n\}_{n\in\mathbb{Z}}$ is dense in R_ι. Then, fix $\varphi \in \mathcal{C}(0, 2\pi)$ and $\varepsilon > 0$. For $n \in \mathbb{N}$ large enough, let χ_n be the element of $\mathcal{C}(0, 2\pi)$ defined by

$$\chi_n(x) := \begin{cases} nx & \text{if } 0 \le x < \frac{1}{n}, \\ 1 & \text{if } \frac{1}{n} \le x \le 2\pi - \frac{1}{n}, \\ n(2\pi - x) & \text{if } 2\pi - \frac{1}{n} < x \le 2\pi, \end{cases}$$

and let $\varphi_n := \chi_n \varphi$. Clearly, for $n \in \mathbb{N}$ large enough, $\varphi_n \in \mathcal{C}(0, 2\pi)$ and

$$|\varphi_n(x)| \le |\varphi(x)| \text{ and hence } |\varphi(x) - \varphi_n(x)|^2 \le 4|\varphi(x)|^2, \forall x \in [0, 2\pi].$$

Moreover,

$$\lim_{n\to\infty} \varphi_n(x) = \varphi(x) \text{ and hence } \lim_{n\to\infty} |\varphi(x) - \varphi_n(x)|^2 = 0, \forall x \in (0, 2\pi).$$

By 8.2.11, this implies that

$$\lim_{n\to\infty} \|[\varphi] - [\varphi_n]\| = \lim_{n\to\infty} \left(\int_{[a,b]} |\varphi - \varphi_n|^2 dm_{[a,b]}\right)^{\frac{1}{2}} = 0.$$

Then, fix $k \in \mathbb{N}$ so that $\|[\varphi] - [\varphi_k]\| < \frac{\varepsilon}{2}$. Since $\varphi_k(0) = 0 = \varphi_k(2\pi)$, φ_k can be identified in an obvious way with an element of $\mathcal{C}(\mathbb{T})$ and conversely any trigonometric polynomial can be identified with an element of $L\{u_n\}_{n\in\mathbb{Z}}$ by 3.1.7 (for $\mathcal{C}(\mathbb{T})$ and the trigonometric polynomials, cf. 4.3.6c). Then, 4.3.7 implies that there exists $\psi \in L\{u_n\}_{n\in\mathbb{Z}}$ such that

$$\sup\{|\varphi_k(x) - \psi(x)| : x \in [0, 2\pi)\} < \frac{\varepsilon}{2(b-a)^{\frac{1}{2}}}$$

(cf. 2.3.12), and hence such that

$$\|[\varphi_k] - [\psi]\| = \left(\int_{[a,b]} |\varphi_k - \psi|^2 dm\right)^{\frac{1}{2}} \le \left((b-a)\frac{\varepsilon^2}{4(b-a)}\right)^{\frac{1}{2}} = \frac{\varepsilon}{2},$$

and hence such that

$$\|[\varphi] - [\psi]\| \le \|[\varphi] - [\varphi_k]\| + \|[\varphi_k] - [\psi]\| < \varepsilon.$$

Now, $[\psi] \in L\{[u_n]\}_{n\in\mathbb{Z}}$ by 3.1.7 since ι is a linear operator. By 2.3.12, this proves that $L\{[u_n]\}_{n\in\mathbb{Z}}$ is dense in R_ι. Since R_ι is dense in $L^2(0, 2\pi)$ (cf. 11.2.1), $L\{[u_n]\}_{n\in\mathbb{Z}}$ is dense in $L^2(0, 2\pi)$ by 2.3.14. Thus we have

$$V\{[u_n]\}_{n\in\mathbb{Z}} = \overline{L\{[u_n]\}_{n\in\mathbb{Z}}} = L^2(0, 2\pi)$$

(cf. 4.1.13), and also

$$V(\{[u_0]\} \cup \{[v_n]\}_{n\in\mathbb{N}} \cup \{[w_n]\}_{n\in\mathbb{N}}) = L^2(0, 2\pi)$$

since

$$L\{u_n\}_{n\in\mathbb{Z}} = L(\{u_0\} \cup \{v_n\}_{n\in\mathbb{N}} \cup \{w_n\}_{n\in\mathbb{N}})$$

(cf. 10.2.5b) implies

$$L\{[u_n]\}_{n\in\mathbb{Z}} = L(\{[u_0]\} \cup \{[v_n]\}_{n\in\mathbb{N}} \cup \{[w_n]\}_{n\in\mathbb{N}})$$

in view of 3.1.7 and of the linearity of ι. This proves that the two orthonormal systems of the statement are complete in $L^2(a, b)$ if $a := 0$ and $b := 2\pi$.

For any $a, b \in \mathbb{R}$ such that $a < b$, the two families of vectors of the statement can be obtained from the same families for $a := 0$ and $b := 2\pi$ by means of the unitary operator U of 11.2.3. In view of 10.6.8b, this proves that they are complete orthonormal systems in $L^2(a, b)$. $\qquad\square$

11.2.5 Remark. In view of 10.7.5, 11.2.4 proves that the Hilbert space $L^2(a, b)$ is separable.

11.2.6 Proposition. *Let u be the element of $\mathcal{C}(a, b)$ defined by*

$$u(x) := \left(\frac{1}{b-a}\right)^{\frac{1}{2}}, \forall x \in [a, b],$$

and, for each $n \in \mathbb{N}$, let c_n and s_n be the elements of $\mathcal{C}(a, b)$ defined by

$$c_n(x) := \left(\frac{2}{b-a}\right)^{\frac{1}{2}} \cos\left(\pi n \frac{x-a}{b-a}\right), \forall x \in [a, b],$$

$$s_n(x) := \left(\frac{2}{b-a}\right)^{\frac{1}{2}} \sin\left(\pi n \frac{x-a}{b-a}\right), \forall x \in [a, b].$$

Both the families $\{[u]\} \cup \{[c_n]\}_{n\in\mathbb{N}}$ and $\{[s_n]\}_{n\in\mathbb{N}}$ are complete orthonormal systems in the Hilbert space $L^2(a, b)$.

Proof. We prove the statement for the family $\{[s_n]\}_{n\in\mathbb{N}}$.

First we examine the special case $a := 0$ and $b := 2\pi$, for which we denote s_n by σ_n. We have

$$\sigma_n(x) := \left(\frac{1}{\pi}\right)^{\frac{1}{2}} \sin\frac{nx}{2}, \forall x \in [0, 2\pi], \forall n \in \mathbb{N}.$$

We consider 8.3.11 with

$$(X_1, \mathcal{A}_1, \mu_1) := ([-2\pi, 0], \mathcal{A}_{[-2\pi, 0]}, m_{[-2\pi, 0]}),$$
$$(X_2, \mathcal{A}_2) := ([0, 2\pi], \mathcal{A}_{[0, 2\pi]}),$$
$$\pi : [-2\pi, 0] \to [0, 2\pi]$$
$$x \mapsto \pi(x) := -x.$$

By 9.2.2a (with $c := -1$) we have

$$\mu_2(E) := m_{[-2\pi, 0]}((-1)E) = m((-1)E) = m(E) = m_{[0, 2\pi]}(E), \forall E \in \mathcal{A}_{[0, 2\pi]}.$$

Then we have

$$\int_{[0, 2\pi]} \psi dm_{[0, 2\pi]} = \int_{[-2\pi, 0]} (\psi \circ \pi) dm_{[-2\pi, 0]}, \forall \psi \in \mathcal{L}^1(-2\pi, 2\pi)$$

(cf. 8.3.11c and note that if $\psi \in \mathcal{L}^1(-2\pi, 2\pi)$ then $\psi_{[0, 2\pi]} \in \mathcal{L}^1(0, 2\pi)$; in the integrals we simply denote by ψ the restrictions of ψ that are actually used). Hence, if ψ is an even function, i.e. if $D_\psi = [-2\pi, 2\pi]$ and $(\psi \circ \pi)(x) = \psi(-x) = \psi(x)$ for all $x \in [-2\pi, 0]$, then

$$\int_{[-2\pi, 2\pi]} \psi dm_{[-2\pi, 2\pi]} = \int_{[-2\pi, 2\pi]} \chi_{[-2\pi, 0)} \psi dm_{[-2\pi, 2\pi]} + \int_{[-2\pi, 2\pi]} \chi_{[0, 2\pi]} \psi dm_{[-2\pi, 2\pi]}$$

$$= \int_{[-2\pi, 0]} \psi dm_{[-2\pi, 0]} + \int_{[0, 2\pi]} \psi dm_{[0, 2\pi]} = 2 \int_{[0, 2\pi]} \psi dm_{[0, 2\pi]}.$$

If ψ is an odd function, i.e. if $D_\psi = [-2\pi, 2\pi]$ and $(\psi \circ \pi)(x) = \psi(-x) = -\psi(x)$ for all $x \in [-2\pi, 0]$, then

$$\int_{[-2\pi, 2\pi]} \psi dm_{[-2\pi, 2\pi]} = \int_{[-2\pi, 0]} \psi dm_{[-2\pi, 0]} + \int_{[0, 2\pi]} \psi dm_{[0, 2\pi]}$$

$$= -\int_{[0, 2\pi]} \psi dm_{[0, 2\pi]} + \int_{[0, 2\pi]} \psi dm_{[0, 2\pi]} = 0.$$

Now we consider the c.o.n.s. $\{[u_0]\} \cup \{[v_n]\}_{n \in \mathbb{N}} \cup \{[w_n]\}_{n \in \mathbb{N}}$ of 11.2.4, in the special case $a := -2\pi$ and $b := 2\pi$. We note that, for each $n \in \mathbb{N}$,

$$w_n(x) = (-1)^n \left(\frac{1}{2\pi}\right)^{\frac{1}{2}} \sin \frac{nx}{2}, \forall x \in [-2\pi, 2\pi],$$

and hence

$$\sigma_n(x) = (-1)^n \sqrt{2} w_n(x), \forall x \in [0, 2\pi].$$

For all $n, k \in \mathbb{N}$ we have

$$\int_{[0, 2\pi]} \overline{\sigma_n} \sigma_k dm_{[0, 2\pi]} = (-1)^n (-1)^k 2 \int_{[0, 2\pi]} \overline{w_n} w_k dm_{[0, 2\pi]}$$

$$= (-1)^n (-1)^k \int_{[-2\pi, 2\pi]} \overline{w_n} w_k dm_{[-2\pi, 2\pi]}$$

$$= (-1)^n (-1)^k \delta_{n,k} = \delta_{n,k},$$

where the second equality holds because the function $\overline{w_n}w_k$ is even and the third holds because $\{[w_n]\}_{n\in\mathbb{N}}$ is an o.n.s. in $L^2(-2\pi, 2\pi)$. Thus, $\{[\sigma_n]\}_{n\in\mathbb{N}}$ is an o.n.s. in $L^2(0, 2\pi)$. For each $[\varphi] \in L^2(0, 2\pi)$, assuming for convenience that for the representative φ we have $D_\varphi = [0, 2\pi]$ (cf. 8.2.12), we define the function

$$\tilde{\varphi} : [-2\pi, 2\pi] \to \mathbb{C}$$

$$x \mapsto \tilde{\varphi}(x) := \begin{cases} -\varphi(-x) & \text{if } x \in [-2\pi, 0), \\ \varphi(x) & \text{if } x \in [0, 2\pi]. \end{cases}$$

From 8.3.11c we have that $\varphi \circ \pi \in \mathcal{M}(-2\pi, 0)$ and $|\varphi \circ \pi|^2 \in \mathcal{L}^1(-2\pi, 0)$; it is then easy to see that $\tilde{\varphi} \in \mathcal{L}^2(-2\pi, 2\pi)$. Then we have

$$2\int_{[0,2\pi]} |\varphi|^2 dm_{[0,2\pi]} = \int_{[-2\pi,2\pi]} |\tilde{\varphi}|^2 dm_{[-2\pi,2\pi]}$$

$$= \left|\int_{[-2\pi,2\pi]} \overline{u_0}\tilde{\varphi}\, dm_{[-2\pi,2\pi]}\right|^2 + \sum_{n=1}^\infty \left|\int_{[-2\pi,2\pi]} \overline{v_n}\tilde{\varphi}\, dm_{[-2\pi,2\pi]}\right|^2$$

$$+ \sum_{n=1}^\infty \left|\int_{[-2\pi,2\pi]} \overline{w_n}\tilde{\varphi}\, dm_{[-2\pi,2\pi]}\right|^2$$

$$= \sum_{n=1}^\infty \left|2\int_{[0,2\pi]} \overline{w_n}\tilde{\varphi}\, dm_{[0,2\pi]}\right|^2,$$

where the first equality holds because the function $|\tilde{\varphi}|^2$ is even, the second holds by 10.6.4d, the third holds because the functions $\overline{u_0}\tilde{\varphi}$ and $\overline{v_n}\tilde{\varphi}$ are odd and the functions $\overline{w_n}\tilde{\varphi}$ are even. Thus we have

$$\int_{[0,2\pi]} |\varphi|^2 dm_{[0,2\pi]} = \sum_{n=1}^\infty \left|\sqrt{2}\int_{[0,2\pi]} \overline{w_n}\tilde{\varphi}\, dm_{[0,2\pi]}\right|^2 = \sum_{n=1}^\infty \left|\int_{[0,2\pi]} \overline{\sigma_n}\tilde{\varphi}\, dm_{[0,2\pi]}\right|^2.$$

This proves that condition 10.6.4d (with $M := L^2(0, 2\pi)$) holds true for the o.n.s. $\{\sigma_n\}_{n\in\mathbb{N}}$, which is therefore a c.o.n.s. in $L^2(0, 2\pi)$.

Now, if U is the unitary operator of 11.2.3 then $\{U[\sigma_n]\}_{n\in\mathbb{N}}$ is a c.o.n.s. in $L^2(a, b)$ by 10.6.8b, and we note that

$$U[\sigma_n] = [s_n], \forall n \in \mathbb{N}.$$

This completes the proof for the family $\{[s_n]\}_{n\in\mathbb{N}}$.

The proof that the statement is true for the family $\{[u]\}\cup\{[c_n]\}_{n\in\mathbb{N}}$ is analogous, with $\tilde{\varphi}$ defined by

$$\tilde{\varphi}(x) := \begin{cases} \varphi(-x) & \text{if } x \in [-2\pi, 0), \\ \varphi(x) & \text{if } x \in [0, 2\pi]. \end{cases}$$

so that the functions $\overline{u_0}\tilde{\varphi}$ and $\overline{v_n}\tilde{\varphi}$ are even and the functions $\overline{w_n}\tilde{\varphi}$ are odd. □

11.3 $L^2(\mathbb{R})$

We write $\mathcal{M}(\mathbb{R}) := \mathcal{M}(X, \mathcal{A}, \mu)$, $\mathcal{L}^1(\mathbb{R}) := \mathcal{L}^1(X, \mathcal{A}, \mu)$, $\mathcal{L}^2(\mathbb{R}) := \mathcal{L}^2(X, \mathcal{A}, \mu)$, $L^2(\mathbb{R}) := L^2(X, \mathcal{A}, \mu)$ if $X := \mathbb{R}$, $\mathcal{A} := \mathcal{A}(d_\mathbb{R})$, $\mu = m$, where m is the Lebesgue measure on $\mathbb{R}$.

11.3.1 Proposition. *The inclusion $\mathcal{S}(\mathbb{R}) \subset \mathcal{L}^2(\mathbb{R})$ holds true (for $\mathcal{S}(\mathbb{R})$, cf. 3.1.10h and 10.1.5c). The mapping*

$$\iota : \mathcal{S}(\mathbb{R}) \to L^2(\mathbb{R})$$
$$\varphi \mapsto \iota(\varphi) := [\varphi]$$

is a linear operator and

$$(\iota(\varphi)|\iota(\psi))_{L^2(\mathbb{R})} = (\varphi|\psi)_{\mathcal{S}(\mathbb{R})}, \forall \varphi, \psi \in \mathcal{S}(\mathbb{R}).$$

Proof. Every $\varphi \in \mathcal{S}(\mathbb{R})$ is a continuous function, and hence $\varphi \in \mathcal{M}(\mathbb{R})$ by 6.2.8; moreover, $|\varphi|^2 = \overline{\varphi}\varphi$ shows that $|\varphi|^2 \in \mathcal{S}(\mathbb{R})$ (cf. 2 and 6 in 3.1.10h), and hence that $|\varphi|^2 \in \mathcal{L}^1(\mathbb{R})$ (cf. 10.1.5c). This proves the inclusion $\mathcal{S}(\mathbb{R}) \subset \mathcal{L}^2(\mathbb{R})$.

The fact that the mapping ι is a linear operator follows at once from the definitions of vector sum and scalar multiplication in $\mathcal{S}(\mathbb{R})$ and in $M(X, \mathcal{A}, \mu)$, and the fact that ι preserves the inner product follows at once from the definitions of the inner products in $\mathcal{S}(\mathbb{R})$ and in $L^2(X, \mathcal{A}, \mu)$. $\qquad\square$

11.3.2 Definitions. We denote by ξ the function defined as follows

$$\xi : \mathbb{R} \to \mathbb{C}$$
$$x \mapsto \xi(x) := x.$$

More generally, if the rule that defines a function from $\mathbb{R}$ to $\mathbb{C}$ in an expression containing $x \in \mathbb{R}$ and if the domain of the function is the set of all $x \in \mathbb{R}$ for which this expression is consistent as the definition of an element of $\mathbb{C}$, then we denote the function by the same expression with x replaced by ξ. Thus, for example, $\sin \xi$ is the function defined by

$$\mathbb{R} \ni x \mapsto \sin x \in \mathbb{C}$$

and $\frac{1}{1-\xi}$ is the function defined by

$$\mathbb{R} - \{1\} \ni x \mapsto \frac{1}{1-x} \in \mathbb{C}.$$

11.3.3 Theorem. *The family $\{[h_n]\}_{n \in I}$, where $I := \{0\} \cup \mathbb{N}$ and $\{h_n\}_{n \in I}$ is the Hermite o.n.s. in $\mathcal{S}(\mathbb{R})$ (cf. 10.2.7), is a c.o.n.s. in the Hilbert space $L^2(\mathbb{R})$.*

Proof. It is obvious that the family $\{[h_n]\}_{n \in I}$ is an o.n.s. in $L^2(\mathbb{R})$, since $\{h_n\}_{n \in I}$ is an o.n.s. in $\mathcal{S}(\mathbb{R})$ and the mapping ι of 11.3.1 preserves the inner product. To prove that $\{[h_n]\}_{n \in I}$ is complete in $L^2(\mathbb{R})$ we proceed as follows. We prove below that

$$(\{[f_n]\}_{n \in I})^\perp = \{0_{L^2(\mathbb{R})}\},$$

where $\{f_n\}_{n \in I}$ is as in 10.2.7, i.e. $f_n = \xi^n e^{-\frac{\xi^2}{2}}$ for all $n \in I$. From the equality

$$L\{h_n\}_{n \in I} = L\{f_n\}_{n \in I}$$

in $\mathcal{S}(\mathbb{R})$ (cf. 10.2.7), we obtain the equality

$$L\{[h_n]\}_{n \in I} = L\{[f_n]\}_{n \in I}$$

in $L^2(\mathbb{R})$, by 3.1.7 and the linearity of the mapping ι of 11.3.1. Then we have

$$(\{[h_n]\}_{n \in I})^\perp = (L\{[h_n]\}_{n \in I})^\perp = (L\{[f_n]\}_{n \in I})^\perp = (\{[f_n]\}_{n \in I})^\perp = \{0_{L^2(\mathbb{R})}\}$$

(cf. 10.2.11), and this proves that $\{[h_n]\}_{n \in I}$ is a c.o.n.s. in $L^2(\mathbb{R})$ (cf. 10.6.5a).

Now we prove that $(\{[f_n]\}_{n \in I})^\perp = \{0_{L^2(\mathbb{R})}\}$. Then, let $[\varphi] \in L^2(\mathbb{R})$ be such that

$$([\varphi]|[f_n]) = 0, \forall n \in I.$$

We assume for convenience that for the representative φ we have $D_\varphi = \mathbb{R}$ (cf. 8.2.12), and we write $\varphi_1 := \operatorname{Re}\varphi$ and $\varphi_2 := \operatorname{Im}\varphi$. In what follows, fix $i = 1$ or $i = 2$. We have $\varphi_i \in \mathcal{L}^2(\mathbb{R})$ (cf. 6.2.12 and 8.1.17b) and

$$([\varphi_i]|[f_n]) = 0, \forall n \in I,$$

because f_n is a real function. For any $a \in \mathbb{R}$, the function $e^{ia\xi} e^{-\frac{\xi^2}{2}}$ is an element of $\mathcal{S}(\mathbb{R})$ (cf. 5 in 3.1.10h) and hence of $\mathcal{L}^2(\mathbb{R})$ (cf. 11.3.1). Then $\varphi_i e^{ia\xi} e^{-\frac{\xi^2}{2}} \in \mathcal{L}^1(\mathbb{R})$ (cf. 11.1.2b) and

$$
\begin{aligned}
\int_{\mathbb{R}} \varphi_i e^{ia\xi} e^{-\frac{\xi^2}{2}} \, dm &= \int_{\mathbb{R}} \varphi_i(x) \sum_{n=0}^{\infty} \frac{(iax)^n}{n!} e^{-\frac{x^2}{2}} \, dm(x) \\
&= \int_{\mathbb{R}} \lim_{N \to \infty} \varphi_i(x) \sum_{n=0}^{N} \frac{(iax)^n}{n!} e^{-\frac{x^2}{2}} \, dm(x) \\
&= \lim_{N \to \infty} \int_{\mathbb{R}} \sum_{n=0}^{N} \frac{(ia)^n}{n!} \varphi_i(x) x^n e^{-\frac{x^2}{2}} \, dm(x) \\
&= \lim_{N \to \infty} \sum_{n=0}^{N} \frac{(ia)^n}{n!} ([\varphi_i]|[f_n]) = 0,
\end{aligned}
\tag{1}
$$

where the third equality holds true by 8.2.11, with the function $|\varphi_i| e^{|a\xi|} e^{-\frac{\xi^2}{2}}$ as dominating function. Indeed,

$$
\left| \varphi_i(x) \sum_{n=0}^{N} \frac{(iax)^n}{n!} e^{-\frac{x^2}{2}} \right| \leq |\varphi_i(x)| \sum_{n=0}^{\infty} \frac{|ax|^n}{n!} e^{-\frac{x^2}{2}}
$$

$$
= |\varphi_i(x)| e^{|ax|} e^{-\frac{x^2}{2}}, \forall x \in \mathbb{R}, \forall N \in \mathbb{N};
$$

moreover, $e^{|a\xi|} e^{-\frac{\xi^2}{2}} \in \mathcal{L}^2(\mathbb{R})$ (it can be seen that $e^{2|a\xi|} e^{-\xi^2} \in \mathcal{L}^1(\mathbb{R})$ in the same way as for an element of $\mathcal{S}(\mathbb{R})$ in 10.1.5c) and therefore $|\varphi_i| e^{|a\xi|} e^{-\frac{\xi^2}{2}} \in \mathcal{L}^1(\mathbb{R})$ (cf.

11.1.2b). For each $l \in \mathbb{N}$, let the function $\psi_l : \mathbb{R} \to \mathbb{C}$ be the periodic function with period $2l$ that is defined as follows for $x \in [-l, l)$:

$$\psi_l(x) := \begin{cases} 1 & \text{if } \varphi_i(x) > 0, \\ 0 & \text{if } \varphi_i(x) = 0, \\ -1 & \text{if } \varphi_i(x) < 0. \end{cases}$$

We have $\psi_l \in \mathcal{M}(\mathbb{R})$ since, if we denote by $\varphi_{i,0}$ the restriction of φ_i to $[-l, l)$, then the three sets $\varphi_{i,0}^{-1}((0, \infty))$, $\varphi_{i,0}^{-1}(\{0\})$, $\varphi_{i,0}^{-1}((-\infty, 0))$ are elements of $\mathcal{A}(d_{\mathbb{R}})$ (cf. 6.2.3, 6.2.13a, 6.1.19a) and the same is true for their translations (cf. 9.2.1a). We also have

$$\lim_{l \to \infty} \varphi_i(x) \psi_l(x) e^{-\frac{x^2}{2}} = |\varphi_i(x)| e^{-\frac{x^2}{2}}, \forall x \in \mathbb{R},$$

since $\varphi_i(x) \psi_l(x) = |\varphi_i(x)|$ for all $x \in [-l, l)$, and

$$|\varphi_i(x) \psi_l(x) e^{-\frac{x^2}{2}}| \leq |\varphi_i(x) e^{-\frac{x^2}{2}}|, \forall x \in \mathbb{R},$$

since $|\psi_l(x)| \leq 1$ for all $x \in \mathbb{R}$. Hence, by 8.2.11 with the function $|\varphi_i| e^{-\frac{\xi^2}{2}}$ (which is an element of $\mathcal{L}^1(\mathbb{R})$ by 11.1.2b) as dominating function, we have

$$\lim_{l \to \infty} \int_{\mathbb{R}} \varphi_i \psi_l e^{-\frac{\xi^2}{2}} dm = \int_{\mathbb{R}} |\varphi_i| e^{-\frac{\xi^2}{2}} dm.$$

Therefore, if we fix $\varepsilon > 0$, there exists $r \in \mathbb{N}$ such that

$$\left| \int_{\mathbb{R}} |\varphi_i| e^{-\frac{\xi^2}{2}} dm - \int_{\mathbb{R}} \varphi_i \psi_r e^{-\frac{\xi^2}{2}} dm \right| < \varepsilon. \tag{2}$$

Now, the restriction of ψ_r to the interval $[0, 2r)$ is obviously an element of $\mathcal{L}^2(0, 2r)$, and therefore there is a function $q := \sum_{k=-m}^{m} \alpha_k e^{i\frac{\pi}{r}k\xi}$ such that

$$\int_{[0,2r)} |\psi_r - q|^2 dm = \int_{[0,2r]} |\psi_r - q|^2 dm < \varepsilon^2 \left(2 \sum_{n=0}^{\infty} e^{-4r^2 n^2} \right)^{-1}$$

(cf. 10.6.4b in the Hilbert space $L^2(0, 2r)$, with the c.o.n.s. $\{[u_k]\}_{k \in \mathbb{Z}}$ of 11.2.4). Now, $(\psi_r - q)e^{-\frac{\xi^2}{2}}$ is an element of $\mathcal{L}^2(\mathbb{R})$ since both ψ_r and q are bounded functions, and

$$\begin{aligned}
\|[(\psi_r - q)e^{-\frac{\xi^2}{2}}]\|^2 &= \int_{\mathbb{R}} |\psi_r - q|^2 e^{-\xi^2} dm \\
&= \sum_{n=0}^{\infty} \int_{[2rn, 2r(n+1))} |\psi_r - q|^2 e^{-\xi^2} dm \\
&\quad + \sum_{n=0}^{\infty} \int_{[-2r(n+1), -2rn)} |\psi_r - q|^2 e^{-\xi^2} dm \\
&\leq \sum_{n=0}^{\infty} e^{-4r^2 n^2} \int_{[2rn, 2r(n+1))} |\psi_r - q|^2 dm \\
&\quad + \sum_{n=0}^{\infty} e^{-4r^2 n^2} \int_{[-2r(n+1), -2rn)} |\psi_r - q|^2 dm \\
&= 2 \sum_{n=0}^{\infty} e^{-4r^2 n^2} \int_{[0,2r)} |\psi_r - q|^2 dm < \varepsilon^2,
\end{aligned} \tag{3}$$

where the second equality holds by 8.3.4a and the last equality holds because ψ_r and q are periodic functions with period $2r$. Moreover, from 1 we have

$$\int_{\mathbb{R}} \varphi_i q e^{-\frac{\xi^2}{2}} dm = \sum_{k=-m}^{m} \alpha_k \int_{\mathbb{R}} \varphi_i e^{i\frac{\pi}{r}k\xi} e^{-\frac{\xi^2}{2}} dm = 0. \tag{4}$$

From 2, 3, 4 and 10.1.7a we have

$$\int_{\mathbb{R}} |\varphi_i| e^{-\frac{\xi^2}{2}} dm = \left| \int_{\mathbb{R}} |\varphi_i| e^{-\frac{\xi^2}{2}} dm - \int_{\mathbb{R}} \varphi_i q e^{-\frac{\xi^2}{2}} dm \right|$$

$$\leq \left| \int_{\mathbb{R}} |\varphi_i| e^{-\frac{\xi^2}{2}} dm - \int_{\mathbb{R}} \varphi_i \psi_r e^{-\frac{\xi^2}{2}} dm \right|$$

$$+ \left| \int_{\mathbb{R}} \varphi_i \psi_r e^{-\frac{\xi^2}{2}} dm - \int_{\mathbb{R}} \varphi_i q e^{-\frac{\xi^2}{2}} dm \right|$$

$$< \varepsilon + \left| \left([\varphi_i] \mid \left[(\psi_r - q) e^{-\frac{\xi^2}{2}} \right] \right) \right|$$

$$\leq \varepsilon + \|[\varphi_i]\| \|[(\psi_r - q) e^{-\frac{\xi^2}{2}}]\| \leq (1 + \|[\varphi_i]\|)\varepsilon.$$

Since ε was fixed arbitrarily, this implies that

$$\int_{\mathbb{R}} |\varphi_i| e^{-\frac{\xi^2}{2}} dm = 0,$$

and hence, by 8.1.12a,

$$\varphi_i(x) = 0 \text{ } m\text{-a.e. on } \mathbb{R}.$$

Since this holds for both $i = 1$ and $i = 2$, this implies that

$$\varphi(x) = 0 \text{ } m\text{-a.e. on } \mathbb{R},$$

and hence $[\varphi] = 0_{L^2(\mathbb{R})}$. This proves that $(\{[f_n]\}_{n\in I})^{\perp} = \{0_{L^2(\mathbb{R})}\}$. $\qquad\square$

11.3.4 Remark. In view of 10.7.5, 11.3.3 proves that the Hilbert space $L^2(\mathbb{R})$ is separable.

11.3.5 Theorem. *The pair $(L^2(\mathbb{R}), \iota)$, with ι defined as in 11.3.1, is a completion of the inner product space $\mathcal{S}(\mathbb{R})$.*

Proof. We already know that ι fulfils condition co_1 of 10.3.4 (cf. 11.3.1). Condition co_2 follows from 11.3.3 and 10.6.5b. $\qquad\square$

11.3.6 Remarks.

(a) By reasoning as in 11.2.2a, we can see that the inner product space $\mathcal{S}(\mathbb{R})$ (cf. 10.1.5c) is not a Hilbert space.

(b) The mapping ι of 11.3.1 is injective (cf. 10.1.19). This means that if an element of $L^2(\mathbb{R})$ contains an element φ of $\mathcal{S}(\mathbb{R})$, then φ is the only element of $\mathcal{S}(\mathbb{R})$ it contains. As a rule, when we denote by $[\varphi]$ an element of R_ι, the representative φ by which we mark the equivalence class is meant to be the element of $\mathcal{S}(\mathbb{R})$ that is contained by the class.

(c) The mapping ι of 11.3.1 is not surjective. In fact, by reasoning as in 11.2.2c we see that if ι were surjective then $\mathcal{S}(\mathbb{R})$ would be a Hilbert space. Actually, there are elements of $L^2(\mathbb{R})$ which do not contain any continuous function. For instance, by reasoning as in 11.2.2c, we can see that for any finite interval $[a,b]$ the characteristic function $\chi_{[a,b]}$ is an element of $\mathcal{L}^2(\mathbb{R})$ such that $[\chi_{[a,b]}]$ does not contain any continuous function.

(d) If an element of $L^2(\mathbb{R})$ contains an element φ of $C(\mathbb{R})$ (cf. 3.1.10e), then φ is the only element of $C(\mathbb{R})$ it contains (this generalizes what was seen for $\mathcal{S}(\mathbb{R})$ in remark b). In fact suppose that, for $\varphi, \psi \in \mathcal{L}^2(\mathbb{R}) \cap C(\mathbb{R})$, there exists $x_0 \in \mathbb{R}$ so that $|\varphi(x_0) - \psi(x_0)| = k \neq 0$; then there exists $\delta > 0$ such that

$$[x \in \mathbb{R} \text{ and } |x - x_0| < \delta] \Rightarrow [\varphi(x) - \psi(x)] > \frac{1}{2}k,$$

and hence such that

$$\int_{\mathbb{R}} |\varphi - \psi|^2 dm \geq \frac{1}{2}k^2\delta,$$

and this implies $[\varphi] \neq [\psi]$.

11.4 The Fourier transform on $L^2(\mathbb{R})$

The Fourier transform is an important topic of functional analysis, and it can be defined in a variety of contexts, as for instance the Hilbert space $L^2(\mathbb{R}^n)$, the Banach space $L^1(\mathbb{R}^n)$, the spaces of distributions and of generalized functions. In this section we study the Fourier transform as an operator on $L^2(\mathbb{R})$ (from a structural point of view it would be the same to study it in the framework of $L^2(\mathbb{R}^n)$, we limit ourselves to $n = 1$ essentially in order to keep the notation as simple as possible). The Fourier transform on $L^2(\mathbb{R})$ plays a major role in fundamental non-relativistic quantum mechanics as it implements the unitary equivalence of the operators that represent the observables position and momentum of a non-relativistic quantum particle (cf. 20.1.7b and 20.3.6c).

In this section we denote elements of $\mathcal{L}^1(\mathbb{R})$ or $\mathcal{L}^2(\mathbb{R})$ by the letters $f, g, ...$, while we denote elements of $C^\infty(\mathbb{R})$ (for the definition of $C^\infty(\mathbb{R})$, cf. 3.1.10h) by the letters $\varphi, \psi, ...$.

11.4.1 Definitions. Suppose $f \in \mathcal{L}^1(\mathbb{R})$. The *Fourier transform* of f is the function

$$\hat{f} : \mathbb{R} \to \mathbb{C}$$

$$x \mapsto \hat{f}(x) := (2\pi)^{-\frac{1}{2}} \int_{\mathbb{R}} e^{-ixy} f(y) dm(y),$$

and the *inverse Fourier transform* of f is the function

$$\check{f} : \mathbb{R} \to \mathbb{C}$$

$$x \mapsto \check{f}(x) := (2\pi)^{-\frac{1}{2}} \int_{\mathbb{R}} e^{ixy} f(y) dm(y).$$

These definitions are consistent in view of 8.2.4.

11.4.2 Proposition. *Let $f \in \mathcal{L}^1(\mathbb{R})$ be such that $\xi^l f \in \mathcal{L}^1(\mathbb{R})$ for all $l \in \mathbb{N}$ (ξ is the function defined in 11.3.2).*

Then $\hat{f} \in C^\infty(\mathbb{R})$ and

$$(\hat{f})^{(l)} = (-i)^l (\xi^l f)\hat{\ }, \forall l \in \mathbb{N}$$

($(\hat{f})^{(l)}$ denotes the l-th derivative of $\hat{f}$).

Similarly, $\check{f} \in C^\infty(\mathbb{R})$ and

$$(\check{f})^{(l)} = i^l (\xi^l f)\check{\ }, \forall l \in \mathbb{N}.$$

Proof. Let $\{t_n\}$ be a sequence in $\mathbb{R} - \{0\}$ such that $t_n \to 0$. For all $x \in \mathbb{R}$, we have

$$\lim_{n\to\infty} \frac{1}{t_n}(\hat{f}(x+t_n) - \hat{f}(x)) = \lim_{n\to\infty} (2\pi)^{-1} \int_\mathbb{R} \frac{1}{t_n}(e^{-it_n y} - 1)e^{-ixy} f(y) dm(y)$$

$$= (2\pi)^{-1} \int_\mathbb{R} (-iy)e^{-ixy} f(y) dm(y) = -i(\xi f)\hat{\ }(x)$$

by 8.2.11 (with $|\xi f|$ as dominating function), in view of the equation

$$\lim_{n\to\infty} \frac{1}{t_n}(e^{-it_n y} - 1) = -iy, \forall y \in \mathbb{R},$$

and of the inequality

$$|e^{i\alpha} - 1| \le |\alpha|, \forall \alpha \in \mathbb{R};$$

this proves that

$$\hat{f} \text{ is differentiable at } x \text{ and } (\hat{f})^{(1)}(x) = -i(\xi f)\hat{\ }(x), \forall x \in \mathbb{R}.$$

In the same way we can prove that, for each $l \in \mathbb{N}$, if

$$\hat{f} \text{ is } l \text{ times differentiable at all points of } \mathbb{R} \text{ and } (\hat{f})^{(l)} = (-i)^l (\xi^l f)\hat{\ }$$

then

$$\hat{f} \text{ is } l+1 \text{ times differentiable at all points of } \mathbb{R} \text{ and } (\hat{f})^{(l+1)} = (-i)^{l+1}(\xi^{l+1} f)\hat{\ }.$$

This proves by induction the part of the statement about $\hat{f}$. The proof for $\check{f}$ is analogous. $\qquad \square$

11.4.3 Remark. We recall (cf. 10.1.5c) that we have, for all $\varphi \in \mathcal{S}(\mathbb{R})$:

(a) $\varphi \in \mathcal{L}^1(\mathbb{R})$;

(b) $\int_\mathbb{R} \varphi dm = \lim_{n\to\infty} \int_{-n}^n \varphi(x) dx$
   (the integrals on the right hand side of this equation are Riemann integrals).

11.4.4 Proposition. *Let $\varphi \in \mathcal{S}(\mathbb{R})$. Then*

$$(\varphi^{(k)})\hat{\ } = (i\xi)^k \hat{\varphi} \quad and \quad (\varphi^{(k)})\check{\ } = (-i\xi)^k \check{\varphi}, \forall k \in \mathbb{N}.$$

Proof. Preliminarily we note that, for all $k \in \mathbb{N}$, $\varphi^{(k)} \in \mathcal{S}(\mathbb{R})$ (cf. 3.1.10h-1) and hence $\varphi^{(k)} \in \mathcal{L}^1(\mathbb{R})$ (cf. 11.4.3a). Thus, the statement is consistent.

For all $x \in \mathbb{R}$, we have

$$(\varphi^{(1)})\hat{\,}(x) \overset{(1)}{=} (2\pi)^{-\frac{1}{2}} \lim_{n\to\infty} \int_{-n}^{n} e^{-ixy} \varphi^{(1)}(y) dy$$

$$\overset{(2)}{=} (2\pi)^{-\frac{1}{2}} \lim_{n\to\infty} \left(e^{-ixn} \varphi(n) - e^{ixn} \varphi(-n) + ix \int_{-n}^{n} e^{-ixy} \varphi(y) dy \right)$$

$$\overset{(3)}{=} ix(2\pi)^{-\frac{1}{2}} \lim_{n\to\infty} \int_{-n}^{n} e^{-ixy} \varphi(y) dy \overset{(4)}{=} ix\hat{\varphi}(x),$$

where 1 and 4 hold true by 11.4.3b, 2 is integration by parts for the Riemann integrals, 3 holds true because $\lim_{n\to\infty} \varphi(\pm n) = 0$. This proves that

$$(\varphi^{(1)})\hat{\,} = i\xi\hat{\varphi}.$$

In the same way we can prove that, for each $k \in \mathbb{N}$, if

$$(\varphi^{(k)})\hat{\,} = (i\xi)^k \hat{\varphi}$$

then

$$(\varphi^{(k+1)})\hat{\,} = (i\xi)^{k+1} \hat{\varphi}.$$

This proves by induction the part of the statement about $\hat{\varphi}$. The proof for $\check{\varphi}$ is analogous. $\qquad \square$

11.4.5 Proposition. *Let* $\varphi \in \mathcal{S}(\mathbb{R})$. *Then*

$$\xi^k(\hat{\varphi})^{(l)} = (-i)^{l+k}((\xi^l\varphi)^{(k)})\hat{\,} \quad and \quad \xi^k(\check{\varphi})^{(l)} = i^{l+k}((\xi^l\varphi)^{(k)})\check{\,}, \ \forall k, l \in \mathbb{N}.$$

Proof. Preliminarily we note that $(\xi^l\varphi)^{(k)} \in \mathcal{S}(\mathbb{R})$ (cf. 3.1.10h-1,4) and hence $(\xi^l\varphi)^{(k)} \in \mathcal{L}^1(\mathbb{R})$ (cf. 11.4.3a), for all $k, l \in \mathbb{N}$. Moreover, $\hat{\varphi}$ and $\check{\varphi}$ are elements of $C^\infty(\mathbb{R})$ since $\xi^l\varphi \in \mathcal{L}^1(\mathbb{R})$, for all $l \in \mathbb{N}$ (cf. 11.4.2). Thus, the statement is consistent.

We fix $k, l \in \mathbb{N}$. From 11.4.2 we have

$$\xi^k(\hat{\varphi})^{(l)} = (-i)^l \xi^k (\xi^l\varphi)\hat{\,}. \tag{1}$$

Since $\xi^l\varphi \in \mathcal{S}(\mathbb{R})$, we can write the first equality in 11.4.4 with φ replaced by $\xi^l\varphi$, to obtain

$$((\xi^l\varphi)^{(k)})\hat{\,} = (i\xi)^k(\xi^l\varphi)\hat{\,}. \tag{2}$$

Now, 1 and 2 yield the first equality of the statement. The proof of the second equality is analogous. $\qquad \square$

11.4.6 Theorem. *Let* $\varphi \in \mathcal{S}(\mathbb{R})$. *Then*

$$\hat{\varphi} \in \mathcal{S}(\mathbb{R}) \quad and \quad \check{\varphi} \in \mathcal{S}(\mathbb{R}).$$

Proof. As already noted in the proof of 11.4.5, $\hat{\varphi}$ and $\check{\varphi}$ are elements of $\mathcal{C}^{\infty}(\mathbb{R})$. Moreover we have, for all $k = 0, 1, 2, \ldots$ and all $l = 0, 1, 2, \ldots$,

$$\sup\{|x^{k+1}(\hat{\varphi})^{(l)}(x)| : x \in \mathbb{R}\} \overset{(1)}{=} \sup\{|((\xi^l\varphi)^{(k+1)})^{\hat{}}(x)| : x \in \mathbb{R}\}$$

$$\overset{(2)}{\leq} (2\pi)^{-1} \int_{\mathbb{R}} |(\xi^l\varphi)^{(k+1)}| dm \overset{(3)}{<} \infty,$$

where: 1 holds true by 11.4.5 if $l \in \mathbb{N}$ or by 11.4.4 if $l = 0$; 2 follows from 8.2.10; 3 holds true in view of 11.4.3a since $(\xi^l\varphi)^{(k+1)} \in \mathcal{S}(\mathbb{R})$ (cf. 3.1.10h-1,4). Then,

$$\lim_{x \to \pm\infty} x^k(\hat{\varphi})^{(l)}(x) = \lim_{x \to \pm\infty} \frac{1}{x} x^{k+1}(\hat{\varphi})^{(l)}(x) = 0.$$

The proof for $\check{\varphi}$ is analogous. $\square$

11.4.7 Remark. The function

$$\mathbb{R} \ni x \mapsto e^{-x^2} \in \mathbb{R}$$

is obviously an element of $\mathcal{S}(\mathbb{R})$, and hence an element of $\mathcal{L}^1(\mathbb{R})$ (cf. 11.4.3a). We take it that the reader already knows the equation

$$\int_{\mathbb{R}} e^{-x^2} dm(x) = \sqrt{\pi}$$

(cf. e.g. Rudin, 1976, 8.21 or exercise 10, p.290). Then we also have

$$\int_{\mathbb{R}} e^{-c^{-2}x^2} dm(x) = c\sqrt{\pi}, \forall c \in (0, \infty)$$

(cf. 9.2.2b).

11.4.8 Lemma. *For each $a \in (0, \infty)$, let the function γ_a be defined by*

$$\gamma_a : \mathbb{R} \to \mathbb{C}$$

$$x \mapsto \gamma_a(x) := e^{-\frac{1}{2}ax^2}.$$

Then, $\gamma_a \in \mathcal{S}(\mathbb{R})$ and

$$\hat{\gamma}_a(x) = \check{\gamma}_a(x) = a^{-\frac{1}{2}} e^{-\frac{1}{2}a^{-1}x^2}, \forall x \in \mathbb{R}.$$

Proof. We fix $a \in (0, \infty)$. It is obvious that $\gamma_a \in \mathcal{S}(\mathbb{R})$. Then $\hat{\gamma}_a \in \mathcal{C}^{\infty}(\mathbb{R})$ and

$$(\hat{\gamma}_a)^{(1)} = -i(\xi\gamma_a)^{\hat{}} \tag{1}$$

(cf. 11.4.2). Moreover, from

$$\gamma_a'(x) = -ax\gamma_a(x), \forall x \in \mathbb{R},$$

we obtain

$$(\xi\gamma_a)^{\hat{}} = -a^{-1}(\gamma_a^{(1)})^{\hat{}} = -a^{-1}i\xi\hat{\gamma}_a, \tag{2}$$

in view of 11.4.4. From 1 and 2, we see that

$$(\hat{\gamma}_a)^{(1)} + a^{-1}\xi\hat{\gamma}_a = 0.$$

Moreover,

$$\hat{\gamma}_a(0) = (2\pi)^{-\frac{1}{2}} \int_{\mathbb{R}} e^{-\frac{1}{2}ax^2} dm(x) = (2\pi)^{-\frac{1}{2}} (2a^{-1})^{\frac{1}{2}} \sqrt{\pi} = a^{-\frac{1}{2}}$$

(cf. 11.4.7). Now, there exists a unique element φ of $\mathcal{C}^{\infty}(\mathbb{R})$ such that

$$\varphi'(x) + a^{-1}x\varphi(x) = 0, \forall x \in \mathbb{R}, \text{ and } \varphi(0) = a^{-\frac{1}{2}},$$

and φ is defined by

$$\varphi(x) := a^{-\frac{1}{2}} e^{-\frac{1}{2}a^{-1}x^2}.$$

Thus,

$$\hat{\gamma}_a(x) = a^{-\frac{1}{2}} e^{-\frac{1}{2}a^{-1}x^2}, \forall x \in \mathbb{R}.$$

As to $\check{\gamma}_a$, we note that

$$\overline{(\bar{f})}\,\hat{} = \check{f}, \forall f \in \mathcal{L}^1(\mathbb{R}),$$

since complex conjugation commutes with integration (cf. 8.2.3). $\qquad\qquad\square$

11.4.9 Theorem. *Let* $\varphi \in \mathcal{S}(\mathbb{R})$. *Then*

$$(\hat{\varphi})\check{} = \varphi \text{ and } (\check{\varphi})\hat{} = \varphi.$$

Proof. The first equation of the statement is proved by the following equalities, where x is a fixed but arbitrary element of $\mathbb{R}$:

$$(\hat{\varphi})\check{}(x) = (2\pi)^{-\frac{1}{2}} \int_{\mathbb{R}} e^{ixt} \hat{\varphi}(t) dm(t)$$

$$\overset{(1)}{=} (2\pi)^{-\frac{1}{2}} \lim_{n \to \infty} \int_{\mathbb{R}} e^{-\frac{1}{2}n^{-2}t^2} e^{ixt} \left(\int_{\mathbb{R}} e^{-ity} \varphi(y) dm(y) \right) dm(t)$$

$$\overset{(2)}{=} (2\pi)^{-\frac{1}{2}} \lim_{n \to \infty} \int_{\mathbb{R}} \varphi(y) \left(\int_{\mathbb{R}} e^{-i(y-x)t} e^{-\frac{1}{2}n^{-2}t^2} dm(t) \right) dm(y)$$

$$\overset{(3)}{=} (2\pi)^{-\frac{1}{2}} \lim_{n \to \infty} \int_{\mathbb{R}} \varphi(y) n e^{-\frac{1}{2}n^2(y-x)^2} dm(y)$$

$$\overset{(4)}{=} (2\pi)^{-\frac{1}{2}} \lim_{n \to \infty} \int_{\mathbb{R}} \varphi \left(\frac{1}{n}s + x \right) e^{-\frac{1}{2}s^2} dm(s)$$

$$\overset{(5)}{=} (2\pi)^{-\frac{1}{2}} \int_{\mathbb{R}} \varphi(x) e^{-\frac{1}{2}s^2} dm(s)$$

$$\overset{(6)}{=} (2\pi)^{-\frac{1}{2}} \varphi(x) (2\pi)^{\frac{1}{2}} = \varphi(x).$$

The explanations of the above equalities are as follows:

1 holds true in view of 8.2.11 with dominating function $|\hat{\varphi}|$, which is an element of $\mathcal{L}^1(\mathbb{R})$ in view of 11.4.6 and 11.4.3a;

2 holds true in view of 8.4.9 and 8.4.10c, because

$$\left| e^{-\frac{1}{2}n^{-2}t^2} e^{ixt} e^{-ity} \varphi(y) \right| = e^{-\frac{1}{2}n^{-2}t^2} |\varphi(y)|, \forall(y,t) \in \mathbb{R}^2,$$

and because γ_{n-2} and φ are elements of $\mathcal{S}(\mathbb{R})$ and hence of $\mathcal{L}^1(\mathbb{R})$ (cf. 11.4.3a);

    3 follows from 11.4.8;

    4 follows from the change of variable

$$s := n(y - x),$$

in view of 9.2.1 and 9.2.2;

    5 holds in view of 8.2.11, since φ is continuous and hence

$$\lim_{n\to\infty} \varphi\left(\frac{1}{n}s + x\right) e^{-\frac{1}{2}s^2} = \varphi(x)e^{-\frac{1}{2}s^2}, \forall s \in \mathbb{R},$$

and since

$$\left|\varphi\left(\frac{1}{n}s + x\right) e^{-\frac{1}{2}s^2}\right| \le \sup\{|\varphi(y)| : y \in \mathbb{R}\}e^{-\frac{1}{2}s^2}, \forall s \in \mathbb{R}$$

(also, recall 3.1.10h-7);

    6 follows from 11.4.7.

The proof of the second equation of the statement is analogous. $\qquad\square$

11.4.10 Theorem. *Let $\varphi, \psi \in \mathcal{S}(\mathbb{R})$. Then*

$$\int_{\mathbb{R}} \overline{\varphi}\psi\, dm = \int_{\mathbb{R}} \overline{\hat{\varphi}}\psi\, dm = \int_{\mathbb{R}} \overline{\varphi}\hat{\psi}\, dm.$$

Proof. Preliminarily we note that $\overline{\varphi}\psi$, $\overline{\hat{\varphi}}\hat{\psi}$, $\overline{\varphi}\hat{\psi}$ are elements of $\mathcal{S}(\mathbb{R})$ (in view of 3.1.10h-2,6 and 11.4.6) and hence of $\mathcal{L}^1(\mathbb{R})$ (in view of 11.4.3a).

The first equation of the statement is proved by the following equalities:

$$\int_{\mathbb{R}} \overline{\varphi}\psi \overset{(1)}{=} \int_{\mathbb{R}} \overline{\varphi}(\hat{\psi})\check{}\, dm$$

$$= \int_{\mathbb{R}} \overline{\varphi}(x)\left((2\pi)^{-\frac{1}{2}}\int_{\mathbb{R}} e^{ixy}\hat{\psi}(y)dm(y)\right)dm(x)$$

$$\overset{(2)}{=} \int_{\mathbb{R}} \hat{\psi}(y)\left((2\pi)^{-\frac{1}{2}}\int_{\mathbb{R}} e^{ixy}\overline{\varphi}(x)dm(x)\right)dm(y)$$

$$\overset{(3)}{=} \int_{\mathbb{R}} \hat{\psi}(y)\overline{\left((2\pi)^{-\frac{1}{2}}\int_{\mathbb{R}} e^{-ixy}\varphi(x)dm(x)\right)}dm(y)$$

$$= \int_{\mathbb{R}} \hat{\psi}\overline{\hat{\varphi}}\, dm.$$

The explanations of the above equalities are as follows:

    1 holds true by 11.4.9;

    2 holds true in view of 8.4.9 and 8.4.10c, because

$$|\overline{\varphi}(x)e^{ixy}\hat{\psi}(y)| = |\varphi(x)||\hat{\psi}(y)|, \forall (x, y) \in \mathbb{R}^2,$$

and because $\varphi, \hat{\psi} \in \mathcal{L}^1(\mathbb{R})$ (cf. 11.4.6 and 11.4.3a);

    3 holds true because complex conjugation commutes with integration (cf. 8.2.3).

The proof of the second equation of the statement is analogous. $\qquad\square$

11.4.11 Remark. We define the mappings

$$\hat{F} : \mathcal{S}(\mathbb{R}) \to \mathcal{S}(\mathbb{R})$$
$$\varphi \mapsto \hat{F}\varphi := \hat{\varphi}$$

and

$$\check{F} : \mathcal{S}(\mathbb{R}) \to \mathcal{S}(\mathbb{R})$$
$$\varphi \mapsto \check{F}\varphi := \check{\varphi}.$$

These definitions are consistent in view of 11.4.6. The mappings $\hat{F}$ and $\check{F}$ are obviously linear operators on the linear space $\mathcal{S}(\mathbb{R})$ (cf. 3.1.10h). The statement of 11.4.9 can be written as

$$\check{F}\hat{F} = \hat{F}\check{F} = \mathbb{1}_{\mathcal{S}(\mathbb{R})}.$$

In view of 1.2.16b, this implies that both $\hat{F}$ and $\check{F}$ are injective and that

$$\check{F} = \hat{F}^{-1} \text{ and } \hat{F} = \check{F}^{-1}.$$

Since $R_{\hat{F}} = D_{\hat{F}^{-1}}$ and $R_{\check{F}} = D_{\check{F}^{-1}}$, both $\hat{F}$ and $\check{F}$ are surjective. By means of the inner product for $\mathcal{S}(\mathbb{R})$ (cf. 10.1.5c), the statement of 11.4.10 can be written as

$$\left(\hat{F}\varphi|\hat{F}\psi\right) = \left(\check{F}\varphi|\check{F}\psi\right) = (\varphi|\psi), \forall \varphi, \psi \in \mathcal{S}(\mathbb{R}).$$

Therefore, $\hat{F}$ and $\check{F}$ are automorphisms of the inner product space $\mathcal{S}(\mathbb{R})$ (cf. 10.1.17).

11.4.12 Theorem. *There exists a unique operator $F \in \mathcal{B}(L^2(\mathbb{R}))$ such that*

$$F[\varphi] = [\hat{\varphi}], \forall \varphi \in \mathcal{S}(\mathbb{R}).$$

The operator F is a unitary operator in $L^2(\mathbb{R})$. The operator F^{-1} is the unique element of $\mathcal{B}(L^2(\mathbb{R}))$ such that

$$F^{-1}[\varphi] = [\check{\varphi}], \forall \varphi \in \mathcal{S}(\mathbb{R}),$$

or equivalently such that

$$F^{-1}[\hat{\varphi}] = [\varphi], \forall \varphi \in \mathcal{S}(\mathbb{R}).$$

The operator F is called the Fourier transform on $L^2(\mathbb{R})$.

Proof. We recall that the pair $(L^2(\mathbb{R}), \iota)$, with ι defined as in 11.3.1, is a completion of the inner product space $\mathcal{S}(\mathbb{R})$ (cf. 11.3.5). We define the mapping

$$F_0 : R_\iota \to L^2(\mathbb{R})$$
$$[\varphi] \mapsto F_0[\varphi] := [\hat{\varphi}]$$

(cf. 11.3.6b). Clearly, F_0 is a linear operator in $L^2(\mathbb{R})$. Moreover, from 11.4.9 we have

$$F_0[\check{\varphi}] = [\varphi], \forall \varphi \in \mathcal{S}(\mathbb{R}), \tag{1}$$

and this proves that $R_{F_0} = R_\iota$. Thus we have

$$\overline{D_{F_0}} = \overline{R_{F_0}} = L^2(\mathbb{R}).$$

Furthermore, from 11.4.10 we have

$$\|F_0[\varphi]\|^2 = \int_\mathbb{R} |\hat\varphi|^2 dm = \int_\mathbb{R} |\varphi|^2 dm = \|[\varphi]\|^2, \forall\varphi \in \mathcal{S}(\mathbb{R}).$$

Then, in view of 4.6.6, there exists a unique operator $F \in \mathcal{B}(L^2(\mathbb{R}))$ such that $F_0 \subset F$, i.e. such that

$$F[\varphi] = [\hat\varphi], \forall\varphi \in \mathcal{S}(\mathbb{R}),$$

and F is a unitary operator in $L^2(\mathbb{R})$.

We already know that $D_{F_0^{-1}} = R_\iota$, and 1 implies that

$$F_0^{-1}[\varphi] = [\check\varphi], \forall\varphi \in \mathcal{S}(\mathbb{R}).$$

Then, in view of 4.6.6, F^{-1} is the unique element of $\mathcal{B}(L^2(\mathbb{R}))$ such that

$$F^{-1}[\varphi] = F_0^{-1}[\varphi] = [\check\varphi], \forall\varphi \in \mathcal{S}(\mathbb{R}),$$

or equivalently such that

$$F^{-1}[\hat\varphi] = F^{-1}(F_0[\varphi]) = [\varphi], \forall\varphi \in \mathcal{S}(\mathbb{R}).$$

$\square$

11.4.13 Remark. The Fourier transform F is an operator on $L^2(\mathbb{R})$. Thus, $F[f]$ is a vector in $L^2(\mathbb{R})$ for each $[f] \in L^2(\mathbb{R})$. However we do not have a formula which permits us to construct the vector $F[f]$ directly from $[f]$, unless $f \in \mathcal{S}(\mathbb{R})$. Indeed, for each $[f] \in L^2(\mathbb{R})$, from the continuity of F we have that

$$F[f] = \lim_{n\to\infty} [\hat\varphi_n]$$

if $\{\varphi_n\}$ is a sequence in $\mathcal{S}(\mathbb{R})$ such that $[f] = \lim_{n\to\infty}[\varphi_n]$ (such a sequence exists in view of 11.3.5 and 2.3.12; these limits are with respect to the distance on $L^2(\mathbb{R})$), but the construction of a sequence $\{\varphi_n\}$ in $\mathcal{S}(\mathbb{R})$ such that $[f] = \lim_{n\to\infty}[\varphi_n]$ is not a straightforward procedure.

Now, we point out that the formula

$$(2\pi)^{-\frac{1}{2}} \int_\mathbb{R} e^{-ixy} f(y) dm(y),$$

for $x \in \mathbb{R}$ and $f \in \mathcal{L}^2(\mathbb{R})$, has no meaning unless $f \in \mathcal{L}^1(\mathbb{R})$ (note that there exist elements of $\mathcal{L}^2(\mathbb{R})$ which are not in $\mathcal{L}^1(\mathbb{R})$, as for instance the function $g : \mathbb{R} \to \mathbb{C}$ defined by $g(x) := 0$ for all $x \in [-1,1]$ and $g(x) := x^{-1}$ for all $x \in \mathbb{R} - [-1,1]$). However, for $f \in \mathcal{L}^2(\mathbb{R}) \cap \mathcal{L}^1(\mathbb{R})$ we can actually define both $F[f]$, which is an element of $L^2(\mathbb{R})$, and the function $\hat f$ as in 11.4.1, and we can wonder whether the following guess is right:

$$\hat f \in \mathcal{L}^2(\mathbb{R}) \text{ and } F[f] = [\hat f].$$

In 11.4.22 we prove that this guess is correct. Then, the "Fourier transform of $[f]$" is an unambiguous expression when both $F[f]$ and $[\hat f]$ exist, that is to say when $f \in \mathcal{L}^2(\mathbb{R}) \cap \mathcal{L}^1(\mathbb{R})$. In order to prove 11.4.22 we need a few preliminary results, which have corollaries of interest on their own.

11.4.14 Proposition. *Let* $\psi \in \mathcal{S}(\mathbb{R})$ *and* $f \in \mathcal{L}^1(\mathbb{R})$. *Then:*

(a) the function

$$\varphi : \mathbb{R} \to \mathbb{C}$$

$$x \mapsto \varphi(x) := \int_{\mathbb{R}} \psi(x - y)f(y)dm(y)$$

is defined consistently and

$$\varphi(x) = \int_{\mathbb{R}} \psi(s)f(x - s)dm(s), \forall x \in \mathbb{R};$$

(b) $\varphi \in \mathcal{C}^{\infty}(\mathbb{R})$ *and*

$$\varphi^{(n)}(x) = \int_{\mathbb{R}} \psi^{(n)}(x - y)f(y)dm(y), \forall x \in \mathbb{R}, \forall n \in \mathbb{N}.$$

Proof. a: For each $x \in \mathbb{R}$, we have

$$|\psi(x - y)f(y)| \leq \sup\{|\psi(t)| : t \in \mathbb{R}\}|f(y)|, \forall y \in \mathbb{R}.$$

This proves that the definition of the function φ is consistent, in view of 3.1.10h-7 and 8.2.5.

The equality in the second part of statement a follows from the change of variable

$$s := x - y,$$

in view of 9.2.1 and 9.2.2.

b: Preliminarily we note that the right hand side of the equation we want to prove is defined consistently in view of statement a, since $\psi^{(n)} \in \mathcal{S}(\mathbb{R})$ for all $n \in \mathbb{N}$ (cf. 3.1.10h-1).

For simplicity, we assume $D_f = \mathbb{R}$ (if this was not true, we could replace f with its extension f_e defined in 8.2.12 and everything in the statement would remain unchanged, in view of 8.2.7). Now let x be a fixed element of $\mathbb{R}$. For each sequence $\{t_n\}$ in $\mathbb{R} - \{0\}$ such that $t_n \to 0$ we have

$$\lim_{n \to \infty} \frac{1}{t_n}(\psi(x + t_n - y) - \psi(x - y)) = \psi'(x - y), \forall y \in \mathbb{R};$$

moreover, the mean value theorem implies that

$$\forall y \in \mathbb{R}, \forall n \in \mathbb{N}, \exists s_{y,n} \in \mathbb{R} \text{ s.t. } \psi(x + t_n - y) - \psi(x - y) = \psi'(s_{y,n})t_n,$$

and hence that

$$\left| \frac{1}{t_n}(\psi(x + t_n - y) - \psi(x - y)) \right| \leq \sup\{|\psi'(s)| : s \in \mathbb{R}\}, \forall y \in \mathbb{R}, \forall n \in \mathbb{N}.$$

Then, by 8.2.11 (with $\sup\{|\psi'(s)| : s \in \mathbb{R}\}|f|$ as dominating function, cf. 3.1.10h-7) we have

$$\lim_{n \to \infty} \frac{1}{t_n}(\varphi(x + t_n) - \varphi(x)) = \int_{\mathbb{R}} \psi'(x - y)f(y)dm(y).$$

This proves that

$$\varphi \text{ is differentiable at } x \text{ and } \varphi'(x) = \int_{\mathbb{R}} \psi'(x - y)f(y)dm(y).$$

In the same way we can prove that, for each $n \in \mathbb{N}$, if

$$\varphi \text{ is } n \text{ times differentiable at } x \text{ and } \varphi^{(n)}(x) = \int_{\mathbb{R}} \psi^{(n)}(x - y)f(y)dm(y)$$

then

$$\varphi \text{ is } n + 1 \text{ times differentiable at } x \text{ and } \varphi^{(n+1)}(x) = \int_{\mathbb{R}} \psi^{(n+1)}(x - y)f(y)dm(y).$$

Since x was an arbitrary element of $\mathbb{R}$, this proves statement b by induction.   □

11.4.15 Definitions. For each $t \in \mathbb{R}$ and each $f \in \mathcal{L}^2(\mathbb{R})$, we define the functions

$$f^t : D_f \to \mathbb{C}$$
$$x \mapsto f^t(x) := e^{itx}f(x)$$

and

$$f_{-t} : D_f - t \to \mathbb{C}$$
$$x \mapsto f_{-t}(x) := f(x + t)$$

(the definition of f_{-t} is consistent with the definition of φ_c given in 9.2.1b, while the definition of f^t has nothing to do with the definition of φ^c given in 9.2.2). It is obvious that $f^t \in \mathcal{L}^2(\mathbb{R})$, while $f_{-t} \in \mathcal{L}^2(\mathbb{R})$ follows from 9.2.1b.

It is obvious that, for $f, g \in \mathcal{L}^2(\mathbb{R})$,

$$f \sim g \Rightarrow f^t \sim g^t,$$

while the implication

$$f \sim g \Rightarrow f_{-t} \sim g_{-t}$$

follows from 9.2.1a.

In view of the remarks above, for each $t \in \mathbb{R}$ we can define the mappings

$$U_t : L^2(\mathbb{R}) \to L^2(\mathbb{R})$$
$$[f] \mapsto U_t[f] := [f^t]$$

and

$$V_t : L^2(\mathbb{R}) \to L^2(\mathbb{R})$$
$$[f] \mapsto V_t[f] := [f_{-t}].$$

It is obvious that U_t and V_t are linear operators. Moreover,

$$\|U_t[f]\| = \|[f]\|, \forall[f] \in L^2(\mathbb{R}),$$

is obvious, while

$$\|V_t[f]\| = \|[f]\|, \forall[f] \in L^2(\mathbb{R}),$$

follows from 9.2.1b. Thus, U_t and V_t are elements of $\mathcal{B}(L^2(\mathbb{R}))$.

11.4.16 Proposition. *For all $t \in \mathbb{R}$,*

$$V_t = F^{-1}U_t F.$$

For all $[f] \in L^2(\mathbb{R})$, the mappings

$$\mathbb{R} \ni t \mapsto U_t[f] \in L^2(\mathbb{R}) \quad and \quad \mathbb{R} \ni t \mapsto V_t[f] \in L^2(\mathbb{R})$$

are continuous.

Proof. For each $\varphi \in \mathcal{S}(\mathbb{R})$, we have

$$((\hat{\varphi})^t)^{\check{}}(x) = (2\pi)^{-1} \int_{\mathbb{R}} e^{ixy} e^{ity} \hat{\varphi}(y) dm(y) = (\hat{\varphi})^{\check{}}(x + t)$$

$$= \varphi(x + t) = \varphi_{-t}(x), \forall x \in \mathbb{R}, \forall t \in \mathbb{R}$$

(cf. 11.4.9; note that $\psi^t \in \mathcal{S}(\mathbb{R})$ for all $\psi \in \mathcal{S}(\mathbb{R})$, cf. 3.1.10h-5). This proves that

$$F^{-1}U_t F[\varphi] = V_t[\varphi], \forall \varphi \in \mathcal{S}(\mathbb{R}), \forall t \in \mathbb{R}.$$

By 4.2.6 this proves that

$$F^{-1}U_t F = V_t, \forall t \in \mathbb{R},$$

since $F^{-1}U_t F$ and V_t are elements of $\mathcal{B}(L^2(\mathbb{R}))$ and R_ι is dense in $L^2(\mathbb{R})$.

Now we fix $[f] \in L^2(\mathbb{R})$ and $t_0 \in \mathbb{R}$. For each sequence $\{t_n\}$ in $\mathbb{R}$ such that $t_n \to t_0$, we have

$$\lim_{n \to \infty} |e^{it_n x} - e^{it_0 x}|^2 |f(x)|^2 = 0, \forall x \in D_f,$$

and

$$|(e^{it_n x} - e^{it_0 x})f(x)|^2 \le 4|f(x)|^2, \forall x \in D_f.$$

Then, by 8.2.11 (with $4|f|^2$ as dominating function) we have

$$\lim_{n \to \infty} \int_{\mathbb{R}} |e^{it_n x} f(x) - e^{it_0 x} f(x)|^2 dm(x) = 0,$$

or

$$\lim_{n \to \infty} \|U_{t_n}[f] - U_{t_0}[f]\| = 0.$$

Since $[f]$ was an arbitrary element of $L^2(\mathbb{R})$ and t_0 an arbitrary element of $\mathbb{R}$, this proves that the mapping

$$\mathbb{R} \ni t \mapsto U_t[f] \in L^2(\mathbb{R})$$

is continuous for all $[f] \in L^2(\mathbb{R})$. Then the mapping

$$\mathbb{R} \ni t \mapsto U_t F[f] \in L^2(\mathbb{R})$$

is continuous for all $[f] \in L^2(\mathbb{R})$, and hence so is the mapping

$$\mathbb{R} \ni t \mapsto V_t[f] = F^{-1}U_t F[f] \in L^2(\mathbb{R})$$

since the mapping F^{-1} is continuous. $\qquad\square$

11.4.17 Definitions. We define

$$\mathcal{C}_c^\infty(\mathbb{R}) := \mathcal{C}_c(\mathbb{R}) \cap \mathcal{C}^\infty(\mathbb{R})$$

(for $\mathcal{C}_c(\mathbb{R})$ and $\mathcal{C}^\infty(\mathbb{R})$, cf. 3.1.10g,h). Obviously, $\mathcal{C}_c^\infty(\mathbb{R}) \subset \mathcal{S}(\mathbb{R})$ and $\mathcal{C}_c^\infty(\mathbb{R})$ is a linear manifold in the linear space $\mathcal{F}(\mathbb{R})$.

For $a, b \in \mathbb{R}$ such that $a < b$, we define

$$\mathcal{C}_0^\infty(a, b) := \{\varphi \in \mathcal{C}(a, b) : \varphi \text{ is infinitely differentiable at all points of } (a, b)$$
$$\text{and } \operatorname{supp} \varphi \subset (a, b)\}$$

(for $\operatorname{supp} \varphi$, cf. 2.5.9). Obviously, $\mathcal{C}_0^\infty(a, b)$ is a linear manifold in the linear space $\mathcal{C}(a, b)$.

11.4.18 Lemma. *Let $f \in L^2(\mathbb{R})$, let $a, b \in \mathbb{R}$ be such that $a < b$, and suppose that*

$$f(x) = 0 \text{ m-a.e. on } D_f - [a, b].$$

Then, for each $\varepsilon_1 > 0$ and each $\varepsilon_2 > 0$, there exists $\varphi \in \mathcal{C}_c^\infty(\mathbb{R})$ so that

$$\|[f] - [\varphi]\| \leq \varepsilon_1$$

and

$$\operatorname{supp} \varphi \subset [a - \varepsilon_2, b + \varepsilon_2].$$

Proof. For each $n \in \mathbb{N}$, we define the interval $I_n := \left(-\frac{1}{n}, \frac{1}{n}\right)$ and the function

$$\psi_n : \mathbb{R} \to \mathbb{C}$$

$$x \mapsto \psi_n(x) := \begin{cases} k_n \exp(x^2 - n^{-2})^{-1} & \text{if } x \in I_n, \\ 0 & \text{if } x \notin I_n, \end{cases}$$

where $k_n \in \mathbb{R}$ is so that $\int_{\mathbb{R}} \psi_n dm = 1$. It is easy to see that $\psi_n \in \mathcal{C}^\infty(\mathbb{R})$. Hence $\psi_n \in \mathcal{S}(\mathbb{R})$.

Now, we fix $\varepsilon_1 > 0$. Since the mapping

$$\mathbb{R} \ni t \mapsto V_t[f] \in L^2(\mathbb{R})$$

is continuous (cf. 11.4.16), there exists $\delta > 0$ so that

$$|t| < \delta \Rightarrow \|[f] - [f_t]\| < \varepsilon_1.$$

Moreover, we fix $\varepsilon_2 > 0$ and then $n \in \mathbb{N}$ such that $n^{-1} < \min\{\delta, \varepsilon_2\}$.

We note that $f \in L^1(\mathbb{R})$ (this follows from 11.1.2b with $\varphi := \chi_{[a,b]}$ and $\psi := f$) and define the function

$$\varphi : \mathbb{R} \to \mathbb{C}$$

$$x \mapsto \varphi(x) := \int_{\mathbb{R}} \psi_n(s) f(x - s) dm(s),$$

which is an element of $\mathcal{C}^\infty(\mathbb{R})$ (cf. 11.4.14). Moreover, if $x \notin [a - \varepsilon_2, b + \varepsilon_2]$ then

$$s \in I_n \Rightarrow |s| < \varepsilon_2 \Rightarrow x - s \notin [a, b],$$

and hence

$$\psi_n(s)f(x-s) = 0 \text{ for } m\text{-a.e. } s \in D_f + x,$$

and hence

$$\varphi(x) = 0.$$

This proves that supp $\varphi \subset [a - \varepsilon_2, b + \varepsilon_2]$, and hence also that $\varphi \in \mathcal{C}_c^\infty(\mathbb{R})$.

Now we want to prove that

$$\|[f] - [\varphi]\| \le \varepsilon_1. \tag{1}$$

In view of the Schwarz inequality (cf. 10.1.9) we have

$$\|[f] - [\varphi]\| = \sup\{|\,([h]|[f] - [\varphi])\,| : [h] \in L^2(\mathbb{R}) \text{ s.t. } \|[h]\| = 1\}. \tag{2}$$

We fix $[h] \in L^2(\mathbb{R})$ such that $\|[h]\| = 1$. We have

$$([h]|[\varphi]) = \int_\mathbb{R} \overline{h}(x) \left(\int_\mathbb{R} \psi_n(s)f(x-s)dm(s) \right) dm(x).$$

We note that the function

$$\mathbb{R} \ni x \mapsto \int_\mathbb{R} \psi_n(x)|f(x-s)|dm(s) \in [0, \infty)$$

is an element of $\mathcal{C}_c^\infty(\mathbb{R})$ (by the same argument as above, with f replaced by $|f|$) and hence of $L^2(\mathbb{R})$, and hence

$$\int_\mathbb{R} |\overline{h}(x)| \left(\int_\mathbb{R} \psi_n(s)|f(x-s)|dm(s) \right) dm(x) < \infty$$

(cf. 11.1.2b). Then, by Tonelli's theorem (cf. 8.4.8) the function

$$\mathbb{R}^2 \ni (x, s) \mapsto \overline{h}(x)\psi_n(s)f(x-s) \in \mathbb{C}$$

is an element of $\mathcal{L}^1(\mathbb{R}^2, \mathcal{A}(d_\mathbb{R}) \otimes \mathcal{A}(d_\mathbb{R}), m \otimes m)$, and hence by Fubini's theorem (cf. 8.4.10c) we have

$$\int_\mathbb{R} \overline{h}(x) \left(\int_\mathbb{R} \psi_n(s)f(x-s)dm(s) \right) dm(x)$$

$$= \int_\mathbb{R} \psi_n(s) \left(\int_\mathbb{R} \overline{h}(x)f(x-s)dm(x) \right) dm(s)$$

$$= \int_\mathbb{R} \psi_n(s) \,([h]|[f_s])\, dm(s) = \int_{I_n} \psi_n(s) \,([h]|[f_s])\, dm(s).$$

Moreover we have

$$([h]|[f]) = \int_{I_n} \psi_n(s) \,([h]|[f])\, dm(s)$$

since $\int_{I_n} \psi_n dm = 1$. Therefore we have

$$([h]|[f] - [\varphi]) = \int_{I_n} \psi_n(s) \,([h]|[f] - [f_s])\, dm(s),$$

and hence

$$| \, ([h]|[f] - [\varphi]) \, | \overset{(3)}{\leq} \int_{I_n} \psi_n(s)| \, ([h]|[f] - [f_s]) \, | dm(s)$$

$$\overset{(4)}{\leq} \int_{I_n} \psi_n(s) \|[f] - [f_s]\| dm(s) \overset{(5)}{\leq} \varepsilon_1 \int_{I_n} \psi_n dm = \varepsilon_1,$$

where 3 holds by 8.2.10, 4 by the Schwarz inequality, and 5 because

$$s \in I_n \Rightarrow s < \delta \Rightarrow \|[f] - [f_s]\| < \varepsilon_1.$$

Since $[h]$ was an arbitrary normalized element of $L^2(\mathbb{R})$, this proves that

$$| \, ([h]|[f] - [\varphi]) \, | \leq \varepsilon_1, \forall [h] \in L^2(\mathbb{R}) \text{ s.t. } \|[h]\| = 1,$$

and hence (in view of 2) it proves 1. $\qquad \square$

11.4.19 Corollary. *The family $\iota(C_c^\infty(\mathbb{R}))$ (with ι defined as in 11.3.1) is dense in $L^2(\mathbb{R})$.*

Proof. We fix $f \in L^2(\mathbb{R})$ and $\varepsilon > 0$. We define $f_n := \chi_{[-n,n]} f$ for all $n \in \mathbb{N}$. By 8.2.11 (with $|f|^2$ as dominating function) we have

$$\lim_{n \to \infty} \int_{\mathbb{R}} |f - f_n|^2 dm = 0.$$

Therefore, there exists $k \in \mathbb{N}$ such that

$$\|[f] - [f_k]\| < \frac{\varepsilon}{2}.$$

Moreover, by 11.4.18 there exists $\varphi \in C_c^\infty(\mathbb{R})$ such that

$$\|[f_k] - [\varphi]\| < \frac{\varepsilon}{2}.$$

Thus, there exists $\varphi \in C_c^\infty(\mathbb{R})$ such that

$$\|[f] - [\varphi]\| < \varepsilon.$$

Since f and ε were arbitrary, this proves that

$$\forall [f] \in L^2(\mathbb{R}), \forall \varepsilon > 0, \exists \varphi \in C_c^\infty(\mathbb{R}) \text{ s.t. } \|[f] - [\varphi]\| < \varepsilon,$$

and hence (cf. 2.3.12) that $\iota(C_c^\infty(\mathbb{R}))$ is dense in $L^2(\mathbb{R})$. $\qquad \square$

11.4.20 Corollary. *Let $f \in L^2(\mathbb{R})$, let $a, b \in \mathbb{R}$ be such that $a < b$, and suppose that*

$$f(x) = 0 \ m\text{-a.e. on } D_f - [a, b].$$

Then, for each $\varepsilon > 0$ there exists $\varphi \in C_c^\infty(\mathbb{R})$ so that

$$\|[f] - [\varphi]\| < \varepsilon$$

and

$$\mathrm{supp} \, \varphi \subset (a, b).$$

Proof. For each $n \in \mathbb{N}$ we define $f_n := \chi_{[a+\frac{1}{n},b-\frac{1}{n}]}f$. By 8.2.11 (with $|f|^2$ as dominating function) we have

$$\lim_{n\to\infty} \int_\mathbb{R} |f - f_n|^2 dm = 0.$$

Therefore, there exists $k \in \mathbb{N}$ such that

$$\|[f] - [f_k]\| < \frac{\varepsilon}{2}.$$

Now let $\varepsilon_2 > 0$ be such that $\varepsilon_2 < \frac{1}{k}$. In view of 11.4.18, there exists $\varphi \in C_c^\infty(\mathbb{R})$ such that

$$\|[f_k] - [\varphi]\| \le \varepsilon_1 := \frac{\varepsilon}{2},$$

and hence

$$\|[f] - [\varphi]\| < \varepsilon,$$

and also such that

$$\operatorname{supp}\varphi \subset \left[a + \frac{1}{k} - \varepsilon_2, b - \frac{1}{k} + \varepsilon_2\right] \subset (a,b).$$

$\square$

11.4.21 Corollary. *Let* $a, b \in \mathbb{R}$ *be such that* $a < b$. *The family* $\iota(C_0^\infty(a,b))$ *(with* ι *defined as in 11.2.1) is dense in* $L^2(a,b)$.

Proof. This follows immediately from 11.4.20, since each element of $\mathcal{L}^2(a,b)$ is extended trivially by an element of $\mathcal{L}^2(\mathbb{R})$ which satisfies the condition of 11.4.20, and each element φ of $C_c^\infty(\mathbb{R})$ such that $\operatorname{supp}\varphi \subset (a,b)$ becomes an element of $C_0^\infty(a,b)$ when it is restricted to $[a,b]$. $\square$

11.4.22 Theorem. *Let* $f \in \mathcal{L}^2(\mathbb{R}) \cap \mathcal{L}^1(\mathbb{R})$. *Then:*

$$\hat{f} \in \mathcal{L}^2(\mathbb{R}) \text{ and } F[f] = [\hat{f}];$$
$$\check{f} \in \mathcal{L}^2(\mathbb{R}) \text{ and } F^{-1}[f] = [\check{f}].$$

Proof. We define $f_n := \chi_{[-n,n]}f$ for all $n \in \mathbb{N}$. For each $n \in \mathbb{N}$, let $\varphi_n \in C_c^\infty(\mathbb{R})$ be such that

$$\|[f_n] - [\varphi_n]\| < \frac{1}{n} \text{ and } \operatorname{supp}\varphi_n \subset (-n,n)$$

(φ_n with these properties exists by 11.4.20).

From the condition $f \in \mathcal{L}^2(\mathbb{R})$ we obtain

$$\|[f] - [\varphi_n]\| \le \|[f] - [f_n]\| + \|[f_n] - [\varphi_n]\| \xrightarrow[n\to\infty]{} 0, \tag{1}$$

since

$$\lim_{n\to\infty} \int_\mathbb{R} |f - f_n|^2 dm = 0$$

by 8.2.11 (with $|f|^2$ as dominating function).

By the Schwarz inequality (for the vectors $[\chi_{[-n,n]}]$ and $[|f_n - \varphi_n|]$ in $L^2(\mathbb{R})$) we have, for each $n \in \mathbb{N}$,

$$\int_{[-n,n]} |f - \varphi_n| dm = \int_{\mathbb{R}} \chi_{[-n,n]} |f_n - \varphi_n| dm$$

$$\leq \left(\int_{\mathbb{R}} \chi_{[-n,n]} dm \right)^{\frac{1}{2}} \left(\int_{\mathbb{R}} |f_n - \varphi_n|^2 dm \right)^{\frac{1}{2}} < (2n)^{\frac{1}{2}} \frac{1}{n} = \left(\frac{2}{n} \right)^{\frac{1}{2}};$$

moreover, by 8.2.11 (with $|f|$ as dominating function) we have

$$\int_{\mathbb{R}-[-n,n]} |f| dm = \int_{\mathbb{R}} |f - f_n| \xrightarrow[n \to \infty]{} 0;$$

therefore we have

$$\int_{\mathbb{R}} |f - \varphi_n| dm = \int_{[-n,n]} |f - \varphi_n| dm + \int_{\mathbb{R}-[-n,n]} |f| dm \xrightarrow[n \to \infty]{} 0. \qquad (2)$$

In obtaining this, the condition $f \in \mathcal{L}^1(\mathbb{R})$ was essential.

Now we prove the statement for $\hat{f}$. The proof for $\check{f}$ would be similar. We set $[h] := F[f]$. Then from 1 we have

$$[h] = \lim_{n \to \infty} F[\varphi_n] = \lim_{n \to \infty} [\hat{\varphi}_n],$$

by the continuity of F and since $\varphi_n \in \mathcal{S}(\mathbb{R})$ for all $n \in \mathbb{N}$. In view of 11.1.7, this implies that there exists a subsequence $\{\hat{\varphi}_{n_k}\}$ of the sequence $\{\hat{\varphi}_n\}$ so that

$$h(x) = \lim_{k \to \infty} \hat{\varphi}_{n_k}(x) \text{ } m\text{-a.e. on } \mathbb{R}.$$

Moreover, from 2 we have

$$|\hat{f}(x) - \hat{\varphi}_n(x)| \leq (2\pi)^{-1} \int_{\mathbb{R}} |f - \varphi_n| dm \xrightarrow[n \to \infty]{} 0, \forall x \in \mathbb{R}$$

(the inequality holds by 8.2.10). Therefore we have

$$h(x) = \hat{f}(x) \text{ } m\text{-a.e. on } \mathbb{R}.$$

Since $h \in \mathcal{L}^2(\mathbb{R})$, this proves that $\hat{f} \in \mathcal{L}^2(\mathbb{R})$ (cf. 8.1.17c) and that $[h] = [\hat{f}]$, i.e. $F[f] = [\hat{f}]$. $\qquad \square$

11.4.23 Remark. For all $[f] \in L^2(\mathbb{R})$, on the basis of 11.4.22 we can find a formula which yields $F[f]$ more directly than the mere definition of F does. Indeed, let $[f] \in L^2(\mathbb{R})$, let $\{a_n\}$ and $\{b_n\}$ be sequences in $\mathbb{R}$ such that

$$a_n < b_n \text{ for all } n \in \mathbb{N}, \quad \lim a_n = -\infty, \quad \lim b_n = \infty,$$

and define

$$f_n := \chi_{[a_n,b_n]} f, \forall n \in \mathbb{N}.$$

For all $n \in \mathbb{N}$, $f_n \in \mathcal{L}^2(\mathbb{R})$ is obvious and $f_n \in \mathcal{L}^1(\mathbb{R})$ follows from 11.1.2b. Moreover,

$$\lim_{n \to \infty} \|[f] - [f_n]\| = 0$$

follows from 8.2.11 (with $|f|^2$ as dominating function). Then, in view of the continuity of F and of 11.4.22, we have

$$F[f] = \lim_{n \to \infty} F[f_n] = \lim_{n \to \infty} [\hat{f}_n],$$

with

$$\hat{f}_n(x) = (2\pi)^{-\frac{1}{2}} \int_{[a_n, b_n]} e^{-ixy} f(y) dm(y), \forall x \in \mathbb{R}, \forall n \in \mathbb{N}.$$

The sequences $\{a_n\}$ and $\{b_n\}$ can be chosen in order to make the computation of the limit above as easy as possible.

Chapter 12

Adjoint Operators

In this chapter we study the idea of adjoint operator, which is in a sense the main tool for dealing with linear operators in Hilbert space. Throughout the chapter, $\mathcal{H}$ denotes an abstract Hilbert space. We recall that $\mathcal{O}(\mathcal{H})$ denotes the family of all linear operators in $\mathcal{H}$ (cf. 3.2.1).

12.1 Basic properties of adjoint operators

Adjoint operators can be defined and investigated in a geometrical way connected with the concept of graph of a linear operator. However, we prefer to resort to this approach only when it really makes things easier, and we give the definition of adjoint in a more direct way, which makes the reason behind the definition immediately clear and also leads directly to most results.

12.1.1 Definition. For any $A \in \mathcal{O}(\mathcal{H})$, we define a subset D_A^* of $\mathcal{H}$ by letting

$$D_A^* := \{g \in \mathcal{H} : \exists g^* \in \mathcal{H} \text{ so that } (Af|g) = (f|g^*), \forall f \in D_A\}.$$

We have $D_A^* \neq \emptyset$ since $0_\mathcal{H} \in D_A^*$ (for $g := 0_\mathcal{H}$ take $g^* := 0_\mathcal{H}$).

The proposition

$$\forall g \in D_A^*, \exists! g^* \text{ so that } (Af|g) = (f|g^*), \forall f \in D_A,$$

is true iff $\overline{D_A} = \mathcal{H}$. Indeed, while the existence of g^* is obvious in any case, for its uniqueness we can argue as follows. Let $g \in D_A^*$ and $g^* \in \mathcal{H}$ be such that

$$(Af|g) = (f|g^*), \forall f \in D_A.$$

Then, for $g' \in \mathcal{H}$,

$$[(Af|g) = (f|g'), \forall f \in D_A] \Leftrightarrow g' - g^* \in D_A^\perp.$$

Therefore, if $\overline{D_A} = \mathcal{H}$ then $D_A^\perp = \{0_\mathcal{H}\}$ (cf. 10.4.4d) and hence

$$[(Af|g) = (f|g'), \forall f \in D_A] \Rightarrow g' = g^*.$$

On the other hand, if $\overline{D_A} \neq \mathcal{H}$ then there exists $h \in D_A^\perp$ such that $h \neq 0_\mathcal{H}$ (cf. 10.4.4d) and $g' := g^* + h$ is such that

$$g' \neq g^* \text{ and } (Af|g) = (f|g'), \forall f \in D_A.$$

Thus, if and only if $\overline{D_A} = \mathcal{H}$ can we define a mapping $A^\dagger$ from $\mathcal{H}$ to $\mathcal{H}$ by letting $D_{A^\dagger} := D_A^*$ and

$$A^\dagger : D_{A^\dagger} \to \mathcal{H}$$
$$g \mapsto A^\dagger g := g^* \text{ if } g^* \in \mathcal{H} \text{ and } (Af|g) = (f|g^*), \forall f \in D_A.$$

If $\overline{D_A} = \mathcal{H}$ then the mapping $A^\dagger$ is called the *adjoint* of A and the operator A is said to be *adjointable*.

12.1.2 Theorem. *For every adjointable operator A in $\mathcal{H}$, the mapping $A^\dagger$ is a linear operator.*

Proof. Let $A \in \mathcal{O}(\mathcal{H})$ be such that $\overline{D_A} = \mathcal{H}$. For any $g_1, g_2 \in D_A^*$, there exist $g_1^*, g_2^* \in \mathcal{H}$ so that

$$(Af|g_1) = (f|g_1^*) \text{ and } (Af|g_2) = (f|g_2^*), \forall f \in D_A,$$

and hence so that, for all $\alpha, \beta \in \mathbb{C}$,

$$(Af|\alpha g_1 + \beta g_2) = (f|\alpha g_1^* + \beta g_2^*), \forall f \in D_A;$$

therefore, $\alpha g_1 + \beta g_2 \in D_A^*$ and $A^\dagger(\alpha g_1 + \beta g_2) = \alpha g_1^* + \beta g_2^* = \alpha A^\dagger g_1 + \beta A^\dagger g_2$. This proves that condition *lo* of 3.2.1 holds for $A^\dagger$.   □

12.1.3 Proposition. *Let A be an adjointable operator in $\mathcal{H}$. Then:*

(A) $(Af|g) = (f|A^\dagger g), \forall f \in D_A, \forall g \in D_{A^\dagger}.$
(B) For a mapping $\psi : D_\psi \to \mathcal{H}$ with $D_\psi \subset \mathcal{H}$, the following conditions are equivalent:

    (a) $(Af|g) = (f|\psi(g)), \forall f \in D_A, \forall g \in D_\psi;$
    (b) $\psi \subset A^\dagger.$

Proof. A: This follows directly from the definition of $A^\dagger$.

B: In view of part A, it is obvious that condition b implies condition a. On the other hand, if condition a holds true then we obtain, directly from the definitions of D_A^* and of $A^\dagger$,

$$g \in D_\psi \Rightarrow [(Af|g) = (f|\psi(g)), \forall f \in D_A] \Rightarrow [g \in D_A^* \text{ and } A^\dagger g = \psi(g)],$$

i.e. $\psi \subset A^\dagger$ (cf. 1.2.5).   □

12.1.4 Proposition. *Let A be an adjointable operator in $\mathcal{H}$. If $B \in \mathcal{O}(\mathcal{H})$ is such that $A \subset B$, then B is adjointable and $B^\dagger \subset A^\dagger$.*

Proof. If $B \in \mathcal{O}(\mathcal{H})$ is such that $A \subset B$, then $D_A \subset D_B$ and hence $\overline{D_B} = \mathcal{H}$ (cf. 2.3.9d). Moreover, from 12.1.3A (written with the operator A replaced by the operator B) we obtain

$$(Af|g) = (Bf|g) = (f|B^\dagger g), \forall f \in D_A, \forall g \in D_{B^\dagger},$$

which implies $B^\dagger \subset A^\dagger$ by 12.1.3B.   □

To prove results 12.1.6 and 12.1.8, it is expedient to go over to the geometric characterization of the adjoint that is obtained in 12.1.5.

12.1.5 Theorem. *We define the mapping*

$$W : \mathcal{H} \oplus \mathcal{H} \to \mathcal{H} \oplus \mathcal{H}$$
$$(f_1, f_2) \mapsto W(f_1, f_2) := (f_2, -f_1)$$

(for the Hilbert space $\mathcal{H} \oplus \mathcal{H}$, cf. 10.3.7). Then:

(a) *the mapping W is a unitary operator;*
(b) *for $A \in \mathcal{O}(\mathcal{H})$, $\overline{D_A} = \mathcal{H}$ iff $(W(G_A))^\perp$ is the graph of a mapping from $\mathcal{H}$ to $\mathcal{H}$ (the orthogonal complement $(W(G_A))^\perp$ is defined with respect to the Hilbert space $\mathcal{H} \oplus \mathcal{H}$);*
(c) *for an adjointable operator A in $\mathcal{H}$, $G_{A^\dagger} = (W(G_A))^\perp$.*

Proof. a: It is obvious that the mapping W satisfies the conditions of 10.1.20c.

b: For $A \in \mathcal{O}(\mathcal{H})$, $(W(G_A))^\perp$ is a linear manifold in the Hilbert space $\mathcal{H} \oplus \mathcal{H}$ (cf. 10.2.13). Hence, $(W(G_A))^\perp$ is the graph of a mapping from $\mathcal{H}$ to $\mathcal{H}$ iff the following condition is true (cf. 3.2.15b):

$$(0_\mathcal{H}, g) \in (W(G_A))^\perp \Rightarrow g = 0_\mathcal{H}.$$

Now, for $g \in \mathcal{H}$,

$$(0_\mathcal{H}, g) \in (W(G_A))^\perp \Leftrightarrow [(0_\mathcal{H}|Af) - (g|f) = 0, \forall f \in D_A] \Leftrightarrow g \in D_A^\perp,$$

and the condition

$$g \in D_A^\perp \Rightarrow g = 0_\mathcal{H}$$

is equivalent to $D_A^\perp = \{0_\mathcal{H}\}$, which is equivalent to $\overline{D_A} = \mathcal{H}$ by 10.4.4d.

c: Let A be an adjointable operator in $\mathcal{H}$. Then, for $(g, g^*) \in \mathcal{H}$,

$$(g, g^*) \in G_{A^\dagger} \Leftrightarrow [g \in D_A^* \text{ and } g^* = A^\dagger g]$$
$$\Leftrightarrow [(Af|g) = (f|g^*), \forall f \in D_A]$$
$$\Leftrightarrow [(W(f, Af)|(g, g^*))_{\mathcal{H} \oplus \mathcal{H}} = 0, \forall f \in D_A]$$
$$\Leftrightarrow (g, g^*) \in (W(G_A))^\perp.$$

$\square$

12.1.6 Theorem. *Let A be an adjointable operator in $\mathcal{H}$. Then:*

(a) *The operator $A^\dagger$ is closed.*
(b) *The operator A is closable iff $\overline{D_{A^\dagger}} = \mathcal{H}$. If these conditions hold true, then*

$$\overline{A} = A^{\dagger\dagger} \text{ and } A^\dagger = (\overline{A})^\dagger$$

(note that $\overline{A}$ is adjointable since $A \subset \overline{A}$), and hence

$$A \subset A^{\dagger\dagger} \text{ and } A^\dagger = A^{\dagger\dagger\dagger},$$

where we have written $A^{\dagger\dagger} := (A^\dagger)^\dagger$ and $A^{\dagger\dagger\dagger} := ((A^\dagger)^\dagger)^\dagger$.

(c) The operator A is closed iff $[\overline{D_{A^\dagger}} = \mathcal{H}$ and $A = A^{\dagger\dagger}]$.

Proof. a: This follows from 12.1.5c since $(W(G_A))^\perp$ is a subspace of $\mathcal{H} \oplus \mathcal{H}$ (cf. 10.2.13).

b: We have

$$(W(G_{A^\dagger}))^\perp \overset{(1)}{=} (W((W(G_A))^\perp))^\perp \overset{(2)}{=} ((W^2(G_A))^\perp)^\perp \overset{(3)}{=} (G_A^\perp)^\perp \overset{(4)}{=} \overline{G_A},$$

where: 1 holds by 12.1.5c; 2 holds by 10.2.16; 3 holds because $W^2 = -\mathbb{1}_{\mathcal{H} \oplus \mathcal{H}}$ and G_A is a linear manifold in $\mathcal{H} \oplus \mathcal{H}$ (cf. 3.2.15a); 4 holds by 10.4.4c. Now, A is closable iff $\overline{G_A}$ is the graph of a mapping, and $(W(G_{A^\dagger}))^\perp$ is the graph of a mapping iff $\overline{D_{A^\dagger}} = \mathcal{H}$ (cf. 12.1.5b). Thus, A is closable iff $\overline{D_{A^\dagger}} = \mathcal{H}$.

If A is closable and $\overline{D_{A^\dagger}} = \mathcal{H}$, then (cf. 12.1.5c)

$$G_{A^{\dagger\dagger}} = (W(G_{A^\dagger}))^\perp = \overline{G_A} = G_{\overline{A}},$$

and hence $A^{\dagger\dagger} = \overline{A}$ because two mappings are equal if their graphs are equal. Moreover,

$$G_{(\overline{A})^\dagger} = (W(G_{\overline{A}}))^\perp = (W(\overline{G_A}))^\perp \overset{(5)}{=} (\overline{W(G_A)})^\perp \overset{(6)}{=} (W(G_A))^\perp = G_{A^\dagger},$$

where: 5 holds by 10.1.21, 4.6.2d, 2.3.21a; 6 holds by 10.2.11. This proves that $(\overline{A})^\dagger = A^\dagger$.

c: This follows immediately from result b, since A is closed iff $[A$ is closable and $A = \overline{A}]$. $\qquad \square$

12.1.7 Proposition. *For every adjointable operator A in $\mathcal{H}$,*

$$N_{A^\dagger} = R_A^\perp.$$

Proof. Let $A \in \mathcal{O}(\mathcal{H})$ be such that $\overline{D_A} = \mathcal{H}$. Then, for $g \in \mathcal{H}$,

$$g \in N_{A^\dagger} \Leftrightarrow [g \in D_A^* \text{ and } A^\dagger g = 0_{\mathcal{H}}]$$
$$\Leftrightarrow [(Af|g) = 0 = (f|0_{\mathcal{H}}), \forall f \in D_A]$$
$$\Leftrightarrow g \in R_A^\perp.$$

$\qquad \square$

12.1.8 Theorem. *Let $A \in \mathcal{O}(\mathcal{H})$ be such that $\overline{D_A} = \mathcal{H}$ and $N_A = \{0_{\mathcal{H}}\}$ (thus, the operators $A^\dagger$ and A^{-1} are defined). Then, $\overline{D_{A^{-1}}} = \mathcal{H}$ iff $N_{A^\dagger} = \{0_{\mathcal{H}}\}$ (thus, the operator $(A^{-1})^\dagger$ is defined iff the operator $(A^\dagger)^{-1}$ is defined). If these conditions hold true, then*

$$(A^{-1})^\dagger = (A^\dagger)^{-1}.$$

Proof. The parenthetical remarks of the statement are true by 12.1.1 and by 3.2.6a.

We have

$$\overline{R_A} = \mathcal{H} \Leftrightarrow R_A^\perp = \{0_{\mathcal{H}}\}$$

by 10.4.4d, and hence

$$\overline{D_{A^{-1}}} = \mathcal{H} \Leftrightarrow N_{A^\dagger} = \{0_\mathcal{H}\}$$

because $D_{A^{-1}} = R_A$ and $N_{A^\dagger} = R_A^\perp$ (cf. 12.1.7).

Now, suppose $\overline{D_{A^{-1}}} = \mathcal{H}$ and $N_{A^\dagger} = \{0_\mathcal{H}\}$, and define the mapping

$$V : \mathcal{H} \oplus \mathcal{H} \to \mathcal{H} \oplus \mathcal{H}$$

$$(f_1, f_2) \mapsto V(f_1, f_2) := (f_2, f_1)$$

which is a unitary operator since it obviously satisfies the conditions of 10.1.20c. Then,

$$G_{(A^{-1})^\dagger} = (W(G_{A^{-1}}))^\perp = (W(V(G_A)))^\perp$$
$$= (V(W(G_A)))^\perp = V((W(G_A))^\perp) = V(G_{A^\dagger}) = G_{(A^\dagger)^{-1}},$$

where 12.1.5c and 1.2.11c have been used twice, as well as 10.2.16 and the equation

$$W(V(G_A)) = V(W(G_A)),$$

which is true because $WV = -VW$ and G_A is a linear manifold in $\mathcal{H} \oplus \mathcal{H}$. This proves that $(A^{-1})^\dagger = (A^\dagger)^{-1}$. $\qquad\square$

12.1.9 Example. In what follows, we provide an example of an operator which is not closable.

Let $\mathcal{H}$ be a separable Hilbert space which is not finite-dimensional and let $\{u_n\}_{n\in\mathbb{N}}$ be a c.o.n.s. in $\mathcal{H}$. Let $\{x_n\}$ be a sequence in $\mathbb{C}$ and let u be a non-zero element of $\mathcal{H}$. We define a mapping A by letting

$$D_A := \left\{ f \in \mathcal{H} : \sum_{n=1}^\infty |x_n (u_n|f)| < \infty \right\},$$

$$Af := \left(\sum_{n=1}^\infty x_n (u_n|f) \right) u, \forall f \in D_A.$$

This mapping is a linear operator in $\mathcal{H}$. Indeed, D_A is a linear manifold in $\mathcal{H}$ because, for all $\alpha, \beta \in \mathbb{C}$ and $f, g \in \mathcal{H}$,

$$\sum_{n=1}^\infty |x_n (u_n|\alpha f + \beta g)| \leq |\alpha| \sum_{n=1}^\infty |x_n (u_n|f)| + |\beta| \sum_{n=1}^\infty |x_n (u_n|f)|$$

(cf. 5.4.2a, 5.4.6, 5.4.5). Moreover, for all $\alpha, \beta \in \mathbb{C}$ and $f, g \in D_A$,

$$\sum_{n=1}^\infty x_n (u_n|\alpha f + \beta g) = \alpha \sum_{n=1}^\infty x_n (u_n|f) + \beta \sum_{n=1}^\infty x_n (u_n|g),$$

and hence

$$A(\alpha f + \beta g) = \alpha Af + \beta Ag.$$

We notice that $u_n \in D_A$ for all $n \in \mathbb{N}$, and hence $L\{u_n\}_{n\in\mathbb{N}} \subset D_A$ (cf. 3.1.6c), and hence

$$\mathcal{H} = V\{u_n\}_{n\in\mathbb{N}} = \overline{L\{u_n\}_{n\in\mathbb{N}}} \subset \overline{D_A}$$

(cf. 4.1.13). This proves that the operator A is adjointable.

In what follows we distinguish two cases, $\{x_n\} \in \ell^2$ and $\{x_n\} \notin \ell^2$.

First, we suppose $\{x_n\} \in \ell^2$ (cf. 10.3.8d). We set $m := \left(\sum_{n=1}^\infty |x_n|^2\right)^{\frac{1}{2}}$. For all $f \in \mathcal{H}$, we have $\{(u_n|f)\} \in \ell^2$ (cf. 10.2.8b) and hence, by the Schwarz inequality in ℓ^2 and by 10.6.4d (with $M := \mathcal{H}$),

$$\sum_{n=1}^\infty |x_n (u_n|f)| = \sum_{n=1}^\infty |x_n| \, |(u_n|f)| \le m\|f\|,$$

and hence $f \in D_A$; then, by the Schwarz inequality once more,

$$\|Af\| = \left|\sum_{n=1}^\infty x_n (u_n|f)\right| \|u\| \le m\|f\|\|u\|.$$

This proves that $A \in \mathcal{B}(\mathcal{H})$, and hence that A is closed (cf. 4.4.3).

Second, we suppose $\{x_n\} \notin \ell^2$. We choose $n_0 \in \mathbb{N}$ such that $x_{n_0} \ne 0$ and define

$$f_k := -x_k u_{n_0} + x_{n_0} u_k, \forall k \in \mathbb{N};$$

clearly,

$$f_k \in D_A \text{ and } Af_k = (-x_k x_{n_0} + x_{n_0} x_k)u = 0_\mathcal{H}, \forall k \in \mathbb{N}.$$

Now, let $g \in D_{A^\dagger}$; then

$$0 = (Af_k|g) = (f_k|A^\dagger g) = -x_k (u_{n_0}|A^\dagger g) + x_{n_0} (u_k|A^\dagger g), \forall k \in \mathbb{N};$$

this implies that either

$$(u_{n_0}|A^\dagger g) \ne 0 \text{ and hence } x_k = (u_{n_0}|A^\dagger g)^{-1} x_{n_0} (u_k|A^\dagger g), \forall k \in \mathbb{N}, \tag{1}$$

or

$$(u_{n_0}|A^\dagger g) = 0 \text{ and hence } (u_k|A^\dagger g) = 0, \forall k \in \mathbb{N}; \tag{2}$$

however, alternative 1 would entail

$$\sum_{k=1}^\infty |x_k|^2 = |(u_{n_0}|A^\dagger g)|^{-2}|x_{n_0}|^2 \sum_{k=1}^\infty |(u_k|A^\dagger g)|^2 < \infty,$$

which is contrary to the assumption that $\{x_n\} \notin \ell^2$; therefore, alternative 2 is true, and hence $A^\dagger g = 0_\mathcal{H}$ (cf. 10.6.4e with $M := \mathcal{H}$). This proves that

$$A^\dagger \subset \mathbb{O}_\mathcal{H} \tag{3}$$

(for the operator $\mathbb{O}_\mathcal{H}$, cf. 3.2.9). If the operator A were closable then the operator $A^\dagger$ would be adjointable (cf. 12.1.6b) and 3 would imply $A^{\dagger\dagger} = \mathbb{O}_\mathcal{H}$ (this would follow immediately from 12.1.3B), and hence we would have $A \subset \mathbb{O}_\mathcal{H}$ (this would follow from $A \subset A^{\dagger\dagger}$), contrary to the fact that

$$u_{n_0} \in D_A \text{ and } Au_{n_0} = x_{n_0}u \ne 0_\mathcal{H}.$$

Therefore, the operator A is not closable. We point out that this implies that A is not bounded (cf. 4.4.12).

12.2 Adjoints and boundedness

12.2.1 Theorem. *Let $A \in \mathcal{O}_E(\mathcal{H})$ (for $\mathcal{O}_E(\mathcal{H})$, cf. 3.2.12). Then the operator $A^\dagger$ is bounded.*

Proof. By 12.1.3A and 10.1.7a, we have

$$|F_{A^\dagger g}f| = |(f|A^\dagger g)| = |(Af|g)| \le \|Af\|, \forall f \in D_A = \mathcal{H}, \forall g \in D_{A^\dagger} \cap \tilde{\mathcal{H}}$$

(for $F_{A^\dagger g}$, cf. 10.5.1; for $\tilde{\mathcal{H}}$, cf. 10.9.4). This proves that

$$\forall f \in \mathcal{H}, \exists m_f \in [0, \infty) \text{ such that } |F_{A^\dagger g}f| \le m_f, \forall g \in D_{A^\dagger} \cap \tilde{\mathcal{H}},$$

and hence, by 4.2.13 and 10.5.1, that

$$\exists m \in [0, \infty) \text{ such that } \|A^\dagger g\| = \|F_{A^\dagger g}\| \le m, \forall g \in D_{A^\dagger} \cap \tilde{\mathcal{H}},$$

and hence that

$$\exists m \in [0, \infty) \text{ such that } \left\| A^\dagger \left(\frac{1}{\|g\|} g \right) \right\| \le m, \text{ i.e. } \|A^\dagger g\| \le m \|g\|, \forall g \in D_{A^\dagger} - \{0_{\mathcal{H}}\},$$

and hence that the operator $A^\dagger$ is bounded. $\qquad\qquad\square$

12.2.2 Theorem. *Let A be an adjointable operator in $\mathcal{H}$. Then the following conditions are equivalent:*

(a) A is bounded;
(b) $A^\dagger \in \mathcal{B}(\mathcal{H})$ (for $\mathcal{B}(\mathcal{H})$, cf. 4.2.10);
(c) $D_{A^\dagger} = \mathcal{H}$.

If these conditions are satisfied, then

(d) $\|A^\dagger\| = \|A\|$.

Proof. $a \Rightarrow$ (*b and d*): Assuming condition a, by 4.2.6 there exists $\tilde{A} \in \mathcal{B}(\mathcal{H})$ such that $A \subset \tilde{A}$, since $\overline{D_A} = \mathcal{H}$. Then the function

$$\psi : \mathcal{H} \times \mathcal{H} \to \mathbb{C}$$
$$(f, g) \mapsto \psi(f, g) := (\tilde{A}f|g)$$

is a bounded sesquilinear form on $\mathcal{H}$ (cf. 10.5.5), and hence by 10.5.6 there exists $B \in \mathcal{B}(\mathcal{H})$ such that

$$(\tilde{A}f|g) = (f|Bg), \forall f, g \in \mathcal{H},$$

and hence such that

$$(Af|g) = (f|Bg), \forall f \in D_A, \forall g \in \mathcal{H} = D_B.$$

By 12.1.3B, this implies $B \subset A^\dagger$ and hence $B = A^\dagger$. Further, we have $\|A\| = \|\tilde{A}\|$ by 4.2.6d and $\|\tilde{A}\| = \|B\|$ by 10.1.14.

$b \Rightarrow c$: This is obvious.

$c \Rightarrow a$: Assuming condition c, by 12.2.1 we have that $A^{\dagger\dagger}$ is bounded. Since $A \subset A^{\dagger\dagger}$ (cf. 12.1.6b), by 4.2.5a we obtain that A is bounded. $\qquad\square$

12.2.3 Theorem (Closed graph theorem in Hilbert space). *If $A \in \mathcal{O}_E(\mathcal{H})$ and A is closed, then A is bounded.*

Proof. Suppose that A is a closed operator in $\mathcal{H}$ such that $D_A = \mathcal{H}$. By 12.1.6c we have $\overline{D_{A^\dagger}} = \mathcal{H}$ and $A = A^{\dagger\dagger}$. Moreover, $A^\dagger$ is bounded by 12.2.1, and hence $A^{\dagger\dagger}$ is bounded by 12.2.2. $\qquad\qquad\square$

12.2.4 Remark. Here we suppose that $\mathcal{H}$ is finite-dimensional (cf. 10.8.3B). Then, an operator A in $\mathcal{H}$ is adjointable iff $A \in \mathcal{O}_E(\mathcal{H})$. In fact, the condition $\overline{D_A} = \mathcal{H}$ is the same as the condition $D_A = \mathcal{H}$ since every linear manifold in $\mathcal{H}$ is closed (cf. 10.8.2). Moreover, for every $A \in \mathcal{O}_E(\mathcal{H})$ we have $A^\dagger \in \mathcal{O}_E(\mathcal{H})$ by 12.2.2 because A is bounded (cf. 10.8.3A).

Let N be the dimension of $\mathcal{H}$, $\{u_1, ..., u_N\}$ a c.o.n.s. in $\mathcal{H}$, and Φ_U the isomorphism from the associative algebra $\mathcal{O}_E(\mathcal{H})$ onto the associative algebra $\mathbb{M}(N)$ defined in 10.8.4. Then, for all $A \in \mathcal{O}_E(\mathcal{H})$ we have

$$\Phi_U(A^\dagger) = [(u_i|A^\dagger u_k)] = [(Au_i|u_k)] = [\overline{(u_k|Au_i)}];$$

thus, the matrix $\Phi_U(A^\dagger)$ is the complex conjugate of the transpose of the matrix $\Phi_U(A)$.

12.3   Adjoints and algebraic operations

12.3.1 Proposition. *Let $A, B \in \mathcal{O}(\mathcal{H})$ be such that $\overline{D_{A+B}} = \mathcal{H}$. Then $\overline{D_A} = \overline{D_B} = \mathcal{H}$ and:*

(a) $A^\dagger + B^\dagger \subset (A+B)^\dagger$;
(b) if $B \in \mathcal{B}(\mathcal{H})$ then $A^\dagger + B^\dagger = (A+B)^\dagger$.

Proof. We have $\overline{D_A} = \overline{D_B} = \mathcal{H}$ because $D_{A+B} = D_A \cap D_B$ (cf. 2.3.9d).
    a: By 12.1.3A we have

$$((A+B)f|g) = (Af|g) + (Bf|g) = (f|A^\dagger g) + (f|B^\dagger g) = (f|(A^\dagger + B^\dagger)g),$$
$$\forall f \in D_{A+B}, \forall g \in D_{A^\dagger} \cap D_{B^\dagger} = D_{A^\dagger + B^\dagger}.$$

By 12.1.3B, this implies $A^\dagger + B^\dagger \subset (A+B)^\dagger$.
    b: Assuming $B \in \mathcal{B}(\mathcal{H})$, we have $D_{B^\dagger} = \mathcal{H}$ by 12.2.2, and hence

$$g \in D_{(A+B)^\dagger} \Rightarrow [((A+B)f|g) = (f|(A+B)^\dagger g), \forall f \in D_{A+B}]$$
$$\Rightarrow [(Af|g) = (f| - B^\dagger g + (A+B)^\dagger g), \forall f \in D_A]$$
$$\Rightarrow g \in D_A^* = D_{A^\dagger} = D_{A^\dagger + B^\dagger},$$

where 12.1.3A, the equality $D_{A+B} = D_A$, and the definition of D_A^* have been used. This proves that $D_{(A+B)^\dagger} \subset D_{A^\dagger + B^\dagger}$, which (in view of result a) implies $A^\dagger + B^\dagger = (A+B)^\dagger$. $\qquad\qquad\square$

12.3.2 Proposition. *Let A be an adjointable operator in $\mathcal{H}$. Then the operator αA is adjointable for all $\alpha \in \mathbb{C}$ and:*

(a) $\overline{\alpha} A^\dagger = (\alpha A)^\dagger, \forall \alpha \in \mathbb{C} - \{0\}$;
(b) $0 A^\dagger \subset (0A)^\dagger = \mathbb{O}_\mathcal{H}$, *and* $0A^\dagger = (0A)^\dagger$ *iff A is bounded (for the operator $\mathbb{O}_\mathcal{H}$, cf. 3.2.9).*

Proof. For all $\alpha \in \mathbb{C}$, we have $\overline{D_{\alpha A}} = \mathcal{H}$ because $D_{\alpha A} = D_A$, and also (cf. 12.1.3A)

$$(\alpha A f | g) = \overline{\alpha}(A f | g) = (f | \overline{\alpha} A^\dagger g), \forall f \in D_{\alpha A}, \forall g \in D_{A^\dagger} = D_{\overline{\alpha} A^\dagger}.$$

By 12.1.3B, this implies $\overline{\alpha} A^\dagger \subset (\alpha A)^\dagger$.

a: If $\alpha \in \mathbb{C} - \{0\}$, we have

$$g \in D_{(\alpha A)^\dagger} \Rightarrow [(\alpha A f | g) = (f | (\alpha A)^\dagger g), \forall f \in D_{\alpha A}]$$
$$\Rightarrow [(A f | g) = \left(f | \frac{1}{\alpha} (\alpha A)^\dagger g \right), \forall f \in D_A]$$
$$\Rightarrow g \in D_A^* = D_{A^\dagger} = D_{\overline{\alpha} A^\dagger},$$

i.e. $D_{(\alpha A)^\dagger} \subset D_{\overline{\alpha} A^\dagger}$, which (in view of what was proved above) implies $\overline{\alpha} A^\dagger = (\alpha A)^\dagger$.

b: In view of what was proved above we already know that $0 A^\dagger \subset (0A)^\dagger$. Moreover,

$$(0 A f | g) = 0 = (f | \mathbb{O}_\mathcal{H} g), \forall f \in D_A, \forall g \in \mathcal{H} = D_{\mathbb{O}_\mathcal{H}},$$

proves that $\mathbb{O}_\mathcal{H} \subset (0A)^\dagger$ (cf. 12.1.3B), and hence that $\mathbb{O}_\mathcal{H} = (0A)^\dagger$. Now, $D_{0 A^\dagger} = D_{A^\dagger}$ and $D_{A^\dagger} = \mathcal{H}$ iff A is bounded (cf. 12.2.2). □

12.3.3 Remark. The equality $\mathbb{1}_\mathcal{H}^\dagger = \mathbb{1}_\mathcal{H}$ (for the operator $\mathbb{1}_\mathcal{H}$, cf. 3.2.5) follows immediately from 12.1.3B with $A := \psi := \mathbb{1}_\mathcal{H}$. Then, for every adjointable operator A in $\mathcal{H}$ and every $\alpha \in \mathbb{C}$, from 12.3.1b and 12.3.2 we obtain

$$(A + \alpha \mathbb{1}_\mathcal{H})^\dagger = A^\dagger + \overline{\alpha} \mathbb{1}_\mathcal{H}$$

(note that the operator $A + \alpha \mathbb{1}_\mathcal{H}$ is adjointable because $D_{A + \alpha \mathbb{1}_\mathcal{H}} = D_A$).

12.3.4 Proposition. *Let $A, B \in \mathcal{O}(\mathcal{H})$ be such that $\overline{D_{BA}} = \mathcal{H}$ and $\overline{D_B} = \mathcal{H}$. Then $\overline{D_A} = \mathcal{H}$ and:*

(a) $A^\dagger B^\dagger \subset (BA)^\dagger$;
(b) if $B \in \mathcal{B}(\mathcal{H})$ then $A^\dagger B^\dagger = (BA)^\dagger$.

Proof. We have $\overline{D_A} = \mathcal{H}$ because $D_{BA} := \{f \in D_A : A f \in D_B\} \subset D_A$.

a: By 12.1.3A we have

$$(B A f | g) = (A f | B^\dagger g) = (f | A^\dagger B^\dagger g), \forall f \in D_{BA}, \forall g \in D_{A^\dagger B^\dagger}.$$

By 12.1.3B, this implies $A^\dagger B^\dagger \subset (BA)^\dagger$.

b: Assuming $B \in \mathcal{B}(\mathcal{H})$, we have $D_{B^\dagger} = \mathcal{H}$ by 12.2.2, and hence

$$g \in D_{(BA)^\dagger} \Rightarrow [(BAf|g) = (f|(BA)^\dagger g), \forall f \in D_{BA}]$$
$$\Rightarrow [(Af|B^\dagger g) = (f|(BA)^\dagger g), \forall f \in D_A]$$
$$\Rightarrow B^\dagger g \in D_A^* = D_{A^\dagger} \Rightarrow g \in D_{A^\dagger B^\dagger},$$

where 12.1.3A, the equality $D_{BA} = D_A$, and the definition of D_A^* have been used. This proves that $D_{(BA)^\dagger} \subset D_{A^\dagger B^\dagger}$, which (in view of result a) implies $A^\dagger B^\dagger = (BA)^\dagger$. $\qquad\square$

12.3.5 Remark. If we have $A \in \mathcal{B}(\mathcal{H})$ in 12.3.4, then the equation $A^\dagger B^\dagger = (BA)^\dagger$ may not hold, as is proved by the following counterexample. First, we note that the equality $\mathbb{O}_\mathcal{H}^\dagger = \mathbb{O}_\mathcal{H}$ follows immediately from 12.1.3B with $A := \psi := \mathbb{O}_\mathcal{H}$. Then, let $A := \mathbb{O}_\mathcal{H}$ and let B be an adjointable operator in $\mathcal{H}$ such that $D_{B^\dagger} \neq \mathcal{H}$ (by 12.2.2, this is true iff B is not bounded). Then $BA = \mathbb{O}_\mathcal{H}$, and hence $(BA)^\dagger = \mathbb{O}_\mathcal{H}$. However, $D_{A^\dagger B^\dagger} = D_{B^\dagger} \neq \mathcal{H}$, and hence $A^\dagger B^\dagger \neq (BA)^\dagger$.

12.4   Symmetric and self-adjoint operators

12.4.1 Definition. An operator A in $\mathcal{H}$ is said to be *symmetric* if $\overline{D_A} = \mathcal{H}$ and $A \subset A^\dagger$.

12.4.2 Proposition. *If A is a symmetric operator in $\mathcal{H}$ then*

$$\overline{D_{A^\dagger}} = \mathcal{H} \text{ and } A \subset A^{\dagger\dagger} \subset A^\dagger.$$

Proof. From 12.1.4 we have $\overline{D_{A^\dagger}} = \mathcal{H}$ and $A^{\dagger\dagger} \subset A^\dagger$. From 12.1.6b we have $A \subset A^{\dagger\dagger}$. $\qquad\square$

12.4.3 Theorem. *Let A be an adjointable operator in $\mathcal{H}$. The following conditions are equivalent:*

(a) A is symmetric;
(b) $(f|Af) \in \mathbb{R}, \forall f \in D_A$;
(c) $(Af|g) = (f|Ag), \forall f, g \in D_A$.

Proof. $a \Rightarrow b$: Assuming condition a, from 12.1.3A we have

$$\overline{(f|Af)} = (Af|f) = (f|A^\dagger f) = (f|Af), \forall f \in D_A.$$

$b \Rightarrow c$: Assuming condition b, we have

$$(Af|f) = (f|Af), \forall f \in D_A.$$

In view of 10.1.10a, this implies condition c.

$c \Rightarrow a$: This follows directly from 12.1.3B. $\qquad\square$

12.4.4 Remarks.

(a) A symmetric operator A is closable, since $A^\dagger$ is densely defined (cf. 12.1.6b), and its closure is symmetric. In fact, $\overline{A} = A^{\dagger\dagger}$ (cf. 12.1.6b) and $A^{\dagger\dagger}$ is symmetric since $\overline{D_{A^{\dagger\dagger}}} = \mathcal{H}$ and $A^{\dagger\dagger} \subset A^\dagger = (A^{\dagger\dagger})^\dagger$ (cf. 12.1.6b).

(b) If A is a symmetric operator in $\mathcal{H}$ and $B \in \mathcal{O}(\mathcal{H})$ is such that $\overline{D_B} = \mathcal{H}$ and $B \subset A$, then B is symmetric since from 12.1.4 we have $B \subset A \subset A^\dagger \subset B^\dagger$.

(c) Theorem 12.4.3 is important because it provides two criteria for deciding whether a given adjointable operator is symmetric without explicitily constructing its adjoint, which might be difficult.

12.4.5 Definition. An operator A in $\mathcal{H}$ is said to be *self-adjoint* (briefly, *s.a.*) if $\overline{D_A} = \mathcal{H}$ and $A = A^\dagger$.

12.4.6 Remarks.

(a) A self-adjoint operator is closed (cf. 12.1.6a).

(b) Suppose that A and B are self-adjoint operators in $\mathcal{H}$ and that $A \subset B$. Then $B = B^\dagger \subset A^\dagger = A$ by 12.1.4, and hence $A = B$.

12.4.7 Theorem (The Hellinger–Toeplitz theorem). *Let A be a self-adjoint operator in $\mathcal{H}$. Then the following conditions are equivalent:*

(a) A is bounded;
(b) $A \in \mathcal{B}(\mathcal{H})$;
(c) $D_A = \mathcal{H}$.

Proof. The statement follows at once from 12.2.2. $\qquad\square$

12.4.8 Proposition. *Let A be a self-adjoint operator in $\mathcal{H}$ such that $N_A = \{0_{\mathcal{H}}\}$. Then, the operator A is injective and the operator A^{-1} is self-adjoint.*

Proof. The operator A is injective in view of 3.2.6a. Then, 12.1.8 implies that

$$\overline{D_{A^{-1}}} = \mathcal{H} \text{ and } (A^{-1})^\dagger = (A^\dagger)^{-1} = (A)^{-1},$$

i.e. that A^{-1} is s.a.. $\qquad\square$

12.4.9 Definition. An operator A in $\mathcal{H}$ is said to be *essentially self-adjoint* (briefly, *e.s.a.*) if it is symmetric and $A^{\dagger\dagger} = A^\dagger$.

12.4.10 Remark. For a closed symmetric operator, being essentially self-adjoint is the same as being self-adjoint (cf. 12.1.6b). Now suppose that a symmetric operator A is not closed. Its closure $\overline{A}$ is always symmetric but it is self-adjoint iff A is essentially self-adjoint (cf. 12.4.4a). If this is so then $\overline{A}$ is the only self-adjoint operator that extends A (cf. 12.4.11c). In this way an essentially self-adjoint operator leads to a unique self-adjoint operator. This accounts for its name.

12.4.11 Proposition. *Let A be an adjointable operator in $\mathcal{H}$. Then the following conditions are equivalent:*

(a) A is essentially self-adjoint;

(b) A is closable and $\overline{A}$ is self-adjoint.

If these conditions are satisfied, then

(c) A has a unique self-adjoint extension; in fact, $\overline{A}$ is the only self-adjoint operator that extends A.

Proof. $a \Rightarrow$ (b and c): Suppose that A is e.s.a.. Then it is symmetric, and hence it is closable and $\overline{A} = A^{\dagger\dagger}$ (cf. 12.4.4a). Thus, $\overline{D_{\overline{A}}} = \overline{D_{A^\dagger}} = \mathcal{H}$ and $(\overline{A})^\dagger = (A^{\dagger\dagger})^\dagger = A^\dagger = A^{\dagger\dagger} = \overline{A}$ (cf. 12.1.6b).

Now suppose that B is a s.a. operator in $\mathcal{H}$ such that $A \subset B$. From 12.1.4 we have $B^\dagger \subset A^\dagger$ and then $A^{\dagger\dagger} \subset B^{\dagger\dagger}$. Since $B^{\dagger\dagger} = B^\dagger = B$ and $A^{\dagger\dagger} = A^\dagger$, this implies $B = A^{\dagger\dagger} = \overline{A}$. This proves that condition c is true.

$b \Rightarrow a$: If A is closable, then $\overline{D_{A^\dagger}} = \mathcal{H}$ and $A^{\dagger\dagger} = \overline{A}$ by 12.1.6b. If $\overline{A}$ is s.a., we also have $A^{\dagger\dagger} = \overline{A} = (\overline{A})^\dagger = (A^{\dagger\dagger})^\dagger = A^\dagger$ (cf. 12.1.6b). Thus, $A \subset A^\dagger$ and $A^{\dagger\dagger} = A^\dagger$. $\qquad\square$

12.4.12 Proposition. *Let A be a self-adjoint operator in $\mathcal{H}$ and let M be a linear manifold in $\mathcal{H}$ such that $M \subset D_A$. Then the restriction A_M of A to M is closable by 4.4.11b since A is closed (cf. 12.4.6a). The following conditions are equivalent:*

(a) A_M is essentially self-adjoint;

(b) $\overline{A_M} = A$.

Proof. $a \Rightarrow b$: If A_M is e.s.a. then $\overline{A_M}$ is the only s.a. operator that extends A_M (cf. 12.4.11c), and hence $A = \overline{A_M}$.

$b \Rightarrow a$: If $\overline{A_M} = A$ then from the definition of $\overline{A_M}$ (cf. 4.4.10) and from 2.3.10 we obtain $D_A = D_{\overline{A_M}} \subset \overline{D_{A_M}}$, and hence $\overline{D_{A_M}} = \mathcal{H}$. Thus, A_M is adjointable and closable, and $\overline{A_M}$ is s.a.. Then, A_M is e.s.a. by 12.4.11. $\qquad\square$

12.4.13 Remark. Proposition 12.4.11 shows that an essentially self-adjoint operator has one and only one self-adjoint extension. A linear manifold M which satisfies the condition of 12.4.12 with respect to a self-adjoint operator A is called a *core* for A. Thus, to specify A uniquely one need not give the exact domain of A, but just some core for A. Usually, there are many cores for a given self-adjoint operator. This explains why, if for a given rule (cf. 1.2.1) there exists a domain on which that rule would define a self-adjoint operator, it is usually easier to guess a domain on which that rule defines an essentially self-adjoint operator than the domain of self-adjointness.

12.4.14 Theorem. *Let A be a symmetric operator in $\mathcal{H}$ and suppose that $\lambda \in \mathbb{C}$ exists so that*

$$R_{A-\lambda 1_{\mathcal{H}}} = \overline{R_{A-\overline{\lambda} 1_{\mathcal{H}}}} = \mathcal{H}.$$

Then A is self-adjoint.

Proof. The equality $R_{A-\lambda\mathbb{1}_\mathcal{H}} = \mathcal{H}$ implies that, for each $g \in D_{A^\dagger}$, there exists $f \in D_{A-\lambda\mathbb{1}_\mathcal{H}}$ so that

$$A^\dagger g - \lambda g = Af - \lambda f,$$

and hence, since $A \subset A^\dagger$, so that

$$(A^\dagger - \lambda\mathbb{1}_\mathcal{H})(g - f) = 0_\mathcal{H}.$$

Now, $A^\dagger - \lambda\mathbb{1}_\mathcal{H} = (A - \bar{\lambda}\mathbb{1}_\mathcal{H})^\dagger$ by 12.3.3, $N_{(A-\bar{\lambda}\mathbb{1}_\mathcal{H})^\dagger} = R^\perp_{A-\bar{\lambda}\mathbb{1}_\mathcal{H}}$ by 12.1.7, $R^\perp_{A-\bar{\lambda}\mathbb{1}_\mathcal{H}} = \{0_\mathcal{H}\}$ since $\overline{R_{A-\bar{\lambda}\mathbb{1}_\mathcal{H}}} = \mathcal{H}$ (cf. 10.4.4d). Thus, $g - f = 0_\mathcal{H}$ and hence $g \in D_{A-\lambda\mathbb{1}_\mathcal{H}} = D_A$. This proves that $D_{A^\dagger} \subset D_A$ and hence that A is s.a.. $\square$

12.4.15 Corollary. *If A is a symmetric operator in $\mathcal{H}$ and $R_A = \mathcal{H}$, then A is self-adjoint.*

Proof. Set $\lambda := 0$ in 12.4.14. $\square$

12.4.16 Lemma. *Let A be a symmetric operator in $\mathcal{H}$. Then $R_{A^{\dagger\dagger}+i\mathbb{1}_\mathcal{H}}$ and $R_{A^{\dagger\dagger}-i\mathbb{1}_\mathcal{H}}$ are closed subsets of $\mathcal{H}$.*

Proof. Suppose that $\{f_n\}$ is a sequence in $R_{A^{\dagger\dagger}+i\mathcal{H}}$ and that there exists $f \in \mathcal{H}$ so that $\lim_{n\to\infty} f_n = f$. Then there exists a sequence $\{g_n\}$ in $D_{A^{\dagger\dagger}+i\mathbb{1}_\mathcal{H}} = D_{A^{\dagger\dagger}}$ so that $(A^{\dagger\dagger} + i\mathbb{1}_\mathcal{H})g_n = f_n$ for all $n \in \mathbb{N}$. Now,

$$\|(A^{\dagger\dagger} + i\mathbb{1}_\mathcal{H})g\|^2 = \|A^{\dagger\dagger}g\|^2 + i\left(A^{\dagger\dagger}g|g\right) - i\left(g|A^{\dagger\dagger}g\right) + \|g\|^2$$
$$= \|A^{\dagger\dagger}g\|^2 + \|g\|^2, \forall g \in D_{A^{\dagger\dagger}},$$

since $A^{\dagger\dagger}$ is symmetric (cf. 12.4.4a and 12.4.3). Therefore, for all $n, m \in \mathbb{N}$,

$$\|g_n - g_m\| \le \|f_n - f_m\| \text{ and } \|A^{\dagger\dagger}g_n - A^{\dagger\dagger}g_m\| \le \|f_n - f_m\|.$$

This shows that $\{g_n\}$ and $\{A^{\dagger\dagger}g_n\}$ are Cauchy, and hence convergent, sequences. Since the operator $A^{\dagger\dagger}$ is closed, this implies that

$$\lim_{n\to\infty} g_n \in D_{A^{\dagger\dagger}} \text{ and } A^{\dagger\dagger}(\lim_{n\to\infty} g_n) = \lim_{n\to\infty} A^{\dagger\dagger}g_n,$$

and hence

$$\lim_{n\to\infty} g_n \in D_{A^{\dagger\dagger}} \text{ and}$$
$$f = \lim_{n\to\infty} A^{\dagger\dagger}g_n + i \lim_{n\to\infty} g_n = (A^{\dagger\dagger} + i\mathbb{1}_\mathcal{H}) \lim_{n\to\infty} g_n,$$

and hence $f \in R_{A^{\dagger\dagger}+i\mathbb{1}_\mathcal{H}}$. This proves that $R_{A^{\dagger\dagger}+i\mathbb{1}_\mathcal{H}}$ is closed (cf. 2.3.4).

The proof for $R_{A^{\dagger\dagger}-i\mathbb{1}_\mathcal{H}}$ is analogous. $\square$

12.4.17 Theorem. *Let A be a symmetric operator in $\mathcal{H}$. The following conditions are equivalent:*

(a) A is essentially self-adjoint;
(b) $N_{A^\dagger+i\mathbb{1}_\mathcal{H}} = N_{A^\dagger-i\mathbb{1}_\mathcal{H}} = \{0_\mathcal{H}\}$;
(c) $\overline{R_{A+i\mathbb{1}_\mathcal{H}}} = \overline{R_{A-i\mathbb{1}_\mathcal{H}}} = \mathcal{H}$.

Proof. $a \Rightarrow b$: If A is e.s.a. then

$$f \in N_{A^\dagger \pm i1_{\mathcal{H}}} \Rightarrow$$
$$[f \in D_{A^\dagger} = D_{A^{\dagger\dagger}} \text{ and } A^\dagger f = A^{\dagger\dagger} f = \mp if] \Rightarrow$$
$$[f \in D_{A^\dagger} = D_{A^{\dagger\dagger}} \text{ and } \pm i\,(f|f) = \left(A^\dagger f|f\right) = \left(f|A^{\dagger\dagger} f\right) = \mp i\,(f|f)] \Rightarrow$$
$$f = 0_{\mathcal{H}}.$$

This proves that $N_{A^\dagger \pm i1_{\mathcal{H}}} = \{0_{\mathcal{H}}\}$.

$b \Rightarrow c$: Assuming condition b, we have

$$R^\perp_{A \pm i1_{\mathcal{H}}} = N_{(A \pm i1_{\mathcal{H}})^\dagger} = N_{A^\dagger \mp i1_{\mathcal{H}}} = \{0_{\mathcal{H}}\},$$

by 12.1.7 and 12.3.3, and this implies condition c by 10.4.4d.

$c \Rightarrow a$: Assuming condition c, we have

$$\overline{R_{A^{\dagger\dagger} + i1_{\mathcal{H}}}} = \overline{R_{A^{\dagger\dagger} - i1_{\mathcal{H}}}} = \mathcal{H}$$

since $A \subset A^{\dagger\dagger}$, and hence

$$R_{A^{\dagger\dagger} + i1_{\mathcal{H}}} = R_{A^{\dagger\dagger} - i1_{\mathcal{H}}} = \mathcal{H}$$

by 12.4.16. Then $A^{\dagger\dagger}$ is s.a. by 12.4.14 (with A replaced by $A^{\dagger\dagger}$ and $\lambda := i$), since it is a symmetric operator (cf. 12.4.4a). Thus, $A^{\dagger\dagger} = (A^{\dagger\dagger})^\dagger = A^\dagger$ (cf. 12.1.6b). $\square$

12.4.18 Theorem. *Let A be a symmetric operator in $\mathcal{H}$. The following conditions are equivalent:*

(a) A is self-adjoint;
(b) A is closed and $N_{A^\dagger + i1_{\mathcal{H}}} = N_{A^\dagger - i1_{\mathcal{H}}} = \{0_{\mathcal{H}}\}$;
(c) $R_{A + i1_{\mathcal{H}}} = R_{A - i1_{\mathcal{H}}} = \mathcal{H}$.

Proof. $a \Rightarrow b$: Assuming condition a, A is closed (cf. 12.4.6a) and it is obviously e.s.a., and this implies

$$N_{A^\dagger + i1_{\mathcal{H}}} = N_{A^\dagger - i1_{\mathcal{H}}} = \{0_{\mathcal{H}}\}$$

by 12.4.17.

$b \Rightarrow c$: Assuming condition b, we have

$$\overline{R_{A + i1_{\mathcal{H}}}} = \overline{R_{A - i1_{\mathcal{H}}}} = \mathcal{H}$$

by 12.4.17. Since A is closed, we also have $A = A^{\dagger\dagger}$ by 12.1.6c, and hence

$$R_{A + i1_{\mathcal{H}}} = R_{A - i1_{\mathcal{H}}} = \mathcal{H}$$

by 12.4.16.

$c \Rightarrow a$: This follow directly from 12.4.14 (with $\lambda := i$). $\square$

12.4.19 Remark. Self-adjoint operators are, among symmetric operators, the important ones because the spectral theorem holds true for them. One is often given an operator A which for some reason is known to be symmetric even if its adjoint is not known (e.g., A might have been proved to be symmetric by 12.4.3), and wants to find out if A is self-adjoint, or at least essentially self-adjoint. Condition 12.4.17c is a criterion for deciding whether a symmetric operator is essentially self-adjoint in which only the operator itself appears, and condition 12.4.18c is the same for self-adjointness. If the operator A is found to be essentially self-adjoint, then it has a unique self-adjoint extension, which is $\overline{A}$ (cf. 12.4.11), and it is often possible to learn the relevant properties of the self-adjoint extension of A without explicitly constructing $\overline{A}$ or $A^{\dagger\dagger}$, but relying instead on the explicit form of A and on the abstract properties of closures and adjoints. We point out that it usually easier to find essentially self-adjoint operators then self-adjoint ones because there are usually many essentially self-adjoint operators that are restrictions of the same self-adjoint operator (cf. 12.4.13). It is worth mentioning that there exist symmetric operators that have many self-adjoint extensions and others that have no self-adjoint extension.

12.4.20 Proposition. *Let A be a symmetric operator in $\mathcal{H}$. Then:*

(A) $Ap\sigma(A) \subset \mathbb{R}$.
(B) Assuming $\sigma_p(A) \neq \emptyset$, suppose $\lambda_1, \lambda_2 \in \sigma_p(A)$ and $\lambda_1 \neq \lambda_2$. Then,

$$N_{A-\lambda_1 \mathbb{1}_{\mathcal{H}}} \subset N^{\perp}_{A-\lambda_2 \mathbb{1}_{\mathcal{H}}}.$$

Thus, eigenvectors of A corresponding to different eigenvalues are orthogonal to each other.
(C) If the Hilbert space $\mathcal{H}$ is separable then $\sigma_p(A)$ is a countable set.

Proof. A: For $\lambda \in \mathbb{C}$, set $a := \operatorname{Re} \lambda$ and $b := \operatorname{Im} \lambda$. We have

$$
\begin{aligned}
\|(A - \lambda \mathbb{1}_{\mathcal{H}})f\|^2 &= \|(A - a\mathbb{1}_{\mathcal{H}})f - ibf\|^2 \\
&= \|(A - a\mathbb{1}_{\mathcal{H}})f\|^2 - ib\left((A - a\mathbb{1}_{\mathcal{H}})f|f\right) + ib\left(f|(A - a\mathbb{1}_{\mathcal{H}})f\right) + b^2\|f\|^2 \\
&= \|(A - a\mathbb{1}_{\mathcal{H}})f\|^2 + b^2\|f\|^2 \geq b^2\|f\|^2, \forall f \in D_{A-\lambda\mathbb{1}},
\end{aligned}
$$

by 12.4.3 since $A - a\mathbb{1}_{\mathcal{H}} \subset A^{\dagger} - a\mathbb{1}_{\mathcal{H}} = (A - a\mathbb{1}_{\mathcal{H}})^{\dagger}$ (cf. 12.3.3). In view of 4.2.3, this proves that if $\operatorname{Im}\lambda \neq 0$ then $A - \lambda\mathbb{1}_{\mathcal{H}}$ is injective and $(A - \lambda\mathbb{1}_{\mathcal{H}})^{-1}$ is bounded, and hence $\lambda \notin Ap\sigma(A)$.
   B: For $f_1 \in N_{A-\lambda_1\mathbb{1}_{\mathcal{H}}}$ and $f_2 \in N_{A-\lambda_2\mathbb{1}_{\mathcal{H}}}$, we have

$$\lambda_1\left(f_1|f_2\right) = (Af_1|f_2) = (f_1|Af_2) = \lambda_2\left(f_1|f_2\right)$$

since $\lambda_1, \lambda_2 \in \mathbb{R}$ (cf. 4.5.8 and result A) and A is symmetric (cf. 12.4.3), and hence $(f_1|f_2) = 0$.
   C: This follows from B, in view of 10.7.7. Indeed, we can construct an o.n.s. in $\mathcal{H}$ by choosing an element of $N_{A-\lambda\mathbb{1}_{\mathcal{H}}} \cap \tilde{\mathcal{H}}$ for each $\lambda \in \sigma_p(A)$. $\qquad\square$

12.4.21 Theorem. *Let A be a self-adjoint operator in $\mathcal{H}$. Then:*

(a) $\sigma(A) \subset \mathbb{R}$;
(b) $Ap\sigma(A) = \sigma(A)$.

Proof. a. We prove that $\mathbb{C} - \mathbb{R} \subset \mathbb{C} - \sigma(A)$. Let $\lambda \in \mathbb{C} - \mathbb{R}$ and set $a := \operatorname{Re} \lambda$, $b := \operatorname{Im} \lambda$, $B := \frac{1}{b}(A - a\mathbb{1}_{\mathcal{H}})$. From 12.3.2a and 12.3.3 we have that the operator B is s.a.. Hence, from 12.4.18 we have

$$R_{B-i\mathbb{1}_{\mathcal{H}}} = \mathcal{H}.$$

Now, $R_{A-\lambda\mathbb{1}_{\mathcal{H}}} = R_{B-i\mathbb{1}_{\mathcal{H}}}$ because $A - \lambda\mathbb{1}_{\mathcal{H}} = b(B - i\mathbb{1}_{\mathcal{H}})$ an the range of a linear operator is a linear manifold. Thus, $R_{A-\lambda\mathbb{1}_{\mathcal{H}}} = \mathcal{H}$. Moreover, from 12.4.20A we have that $\lambda \notin Ap\sigma(A)$, and hence that $A - \lambda\mathbb{1}_{\mathcal{H}}$ is injective and $(A - \lambda\mathbb{1}_{\mathcal{H}})^{-1}$ is bounded. Therefore, $\lambda \in \rho(A) = \mathbb{C} - \sigma(A)$.

b: In view of 4.5.4 and of result a, we have $Ap\sigma(A) \subset \sigma(A) \subset \mathbb{R}$. Now we prove that $\mathbb{R} - Ap\sigma(A) \subset \mathbb{R} - \sigma(A)$. Let $\lambda \in \mathbb{R} - Ap\sigma(A)$. From 12.1.7, 12.3.3, 3.2.6a we have

$$R_{A-\lambda\mathbb{1}_{\mathcal{H}}}^{\perp} = N_{A-\lambda\mathbb{1}_{\mathcal{H}}} = \{0_{\mathcal{H}}\},$$

and hence $\overline{R_{A-\lambda\mathbb{1}_{\mathcal{H}}}} = \mathcal{H}$ by 10.4.4d. Therefore, $\lambda \in \rho(A)$, i.e. $\lambda \in \mathbb{R} - \sigma(A)$. This proves that $Ap\sigma(A) = \sigma(A)$. $\qquad\square$

12.4.22 Definition. The *continuous spectrum* of a self-adjoint operator A is the set

$$\sigma_c(A) := \sigma(A) - \sigma_p(A).$$

From 12.4.21b we have that, for $\lambda \in \mathbb{R}$,

$$\lambda \in \sigma_c(A) \text{ iff } [A - \lambda\mathbb{1}_{\mathcal{H}} \text{ is injective and } (A - \lambda\mathbb{1}_{\mathcal{H}})^{-1} \text{ is not bounded}].$$

12.4.23 Theorem. *Let A be a self-adjoint operator in $\mathcal{H}$. For $\lambda \in \mathbb{C}$, the following conditions are equivalent:*

(a) $\lambda \in \rho(A)$;
(b) $R_{A-\lambda\mathbb{1}_{\mathcal{H}}} = \mathcal{H}$.

Proof. $a \Rightarrow b$: Since the operator A is closed (cf. 12.4.6a), this follows directly from 4.5.12.

$b \Rightarrow a$: If $\lambda \notin \mathbb{R}$, then $\lambda \in \rho(A)$ by 12.4.21a. Now assume $\lambda \in \mathbb{R}$ and $R_{A-\lambda\mathbb{1}_{\mathcal{H}}} = \mathcal{H}$. Then the operator $A - \lambda\mathbb{1}_{\mathcal{H}}$ is s.a. (cf. 12.3.3) and in view of 12.1.7 we have

$$N_{A-\lambda\mathbb{1}_{\mathcal{H}}} = R_{A-\lambda\mathbb{1}_{\mathcal{H}}}^{\perp} = \{0_{\mathcal{H}}\},$$

which implies that the operator $A - \lambda\mathbb{1}_{\mathcal{H}}$ is injective and $(A - \lambda\mathbb{1}_{\mathcal{H}})^{-1}$ is s.a. (cf. 12.4.8). Then the equalities

$$D_{(A-\lambda\mathbb{1}_{\mathcal{H}})^{-1}} = R_{A-\lambda\mathbb{1}_{\mathcal{H}}} = \mathcal{H}$$

imply that the operator $(A - \lambda\mathbb{1}_{\mathcal{H}})^{-1}$ is bounded (cf. 12.4.7). Therefore, we have $\lambda \in \rho(A)$. $\qquad\square$

12.4.24 Proposition. *Suppose that A is a symmetric operator and that there exist a c.o.n.s. $\{u_n\}_{n \in \mathbb{N}}$ in $\mathcal{H}$ and a sequence $\{\lambda_n\}$ in $\mathbb{R}$ so that*

$$u_n \in D_A \text{ and } Au_n = \lambda_n u_n, \forall n \in \mathbb{N}.$$

Then:

$D_{A^\dagger} = \{g \in \mathcal{H} : \sum_{n=1}^\infty \lambda_n^2 |(u_n|g)|^2 < \infty\}$ *and*
$A^\dagger g = \sum_{n=1}^\infty \lambda_n (u_n|g) u_n, \forall g \in D_{A^\dagger}$;
the operator A is essentially self-adjoint;
$\sigma_p(A) = \sigma_p(A^\dagger) = \{\lambda_n\}_{n \in \mathbb{N}}$ *and* $\sigma(A) = \sigma(A^\dagger) = \overline{\{\lambda_n\}_{n \in \mathbb{N}}}$.

Proof. We have

$$g \in D_{A^\dagger} \Rightarrow [(u_n|A^\dagger g) = (Au_n|g) = \lambda_n (u_n|g), \forall n \in \mathbb{N}] \Rightarrow$$

$$\left[\sum_{n=1}^\infty \lambda_n^2 |(u_n|g)|^2 = \sum_{n=1}^\infty |(u_n|A^\dagger g)|^2 < \infty \text{ and}\right.$$

$$\left. A^\dagger g = \sum_{n=1}^\infty (u_n|A^\dagger g) u_n = \sum_{n=1}^\infty \lambda_n (u_n|g) u_n\right]$$

(cf. 12.1.3A, 10.2.8b, 10.6.4b) and, for $g \in \mathcal{H}$,

$$\sum_{n=1}^\infty \lambda_n^2 |(u_n|g)|^2 < \infty \Rightarrow [\text{the series } \sum_{n=1}^\infty \lambda_n (u_n|g) u_n \text{ is convergent}] \Rightarrow$$

$$\left[(Af|g) = \sum_{n=1}^\infty (Af|u_n)(u_n|g) = \sum_{n=1}^\infty (f|Au_n)(u_n|g)\right.$$

$$\left. = \sum_{n=1}^\infty \lambda_n (f|u_n)(u_n|g) = \left(f|\sum_{n=1}^\infty \lambda_n (u_n|g) u_n\right), \forall f \in D_A\right] \Rightarrow$$

$$g \in D_A^* = D_{A^\dagger}$$

(cf. 10.4.8b, 10.6.4c, 12.4.3c). This proves that

$$D_{A^\dagger} = \left\{g \in \mathcal{H} : \sum_{n=1}^\infty \lambda_n^2 |(u_n|g)|^2 < \infty\right\} \text{ and } A^\dagger g = \sum_{n=1}^\infty \lambda_n (u_n|g) u_n, \forall g \in D_{A^\dagger}.$$

. Now, it is obvious that $u_n \in D_{A^\dagger}$ for all $n \in \mathbb{N}$. Therefore, the operator $A^\dagger$ is adjointable by 10.6.5b. Moreover,

$$(g|A^\dagger g) = \sum_{n=1}^\infty \lambda_n |(u_n|g)|^2 \in \mathbb{R}, \forall g \in D_{A^\dagger}.$$

This proves that the operator $A^\dagger$ is symmetric (cf. 12.4.3), and hence that $A^\dagger = A^{\dagger\dagger}$ since we already know that $A^{\dagger\dagger} \subset A^\dagger$ (cf. 12.4.2). Thus, the operator A is e.s.a..

It is obvious that $\{\lambda_n\}_{n \in \mathbb{N}} \subset \sigma_p(A) \subset \sigma_p(A^\dagger)$. If $\lambda \in \sigma_p(A^\dagger)$ existed such that $\lambda \neq \lambda_n$ for all $n \in \mathbb{N}$, then by 12.4.20B there would exist $f \in D_{A^\dagger}$ so that $f \neq 0_\mathcal{H}$

and $(u_n|f) = 0$ for all $n \in \mathbb{N}$, and hence the o.n.s. $\{u_n\}_{n\in\mathbb{N}}$ would not be complete (cf. 10.6.4e). This proves that $\sigma_p(A^\dagger) \subset \{\lambda_n\}_{n\in\mathbb{N}}$ and hence that

$$\sigma_p(A) = \sigma_p(A^\dagger) = \{\lambda_n\}_{n\in\mathbb{N}}.$$

The inclusion $\overline{\{\lambda_n\}_{n\in\mathbb{N}}} \subset \sigma(A^\dagger)$ is true because $\{\lambda_n\}_{n\in\mathbb{N}} = \sigma_p(A^\dagger) \subset \sigma(A^\dagger)$ and $\sigma(A^\dagger)$ is a closed subset of $\mathbb{C}$ (cf. 10.4.6). Now let $\lambda \in \mathbb{C} - \overline{\{\lambda_n\}_{n\in\mathbb{N}}}$; then (cf. 2.3.10),

$$\exists \varepsilon > 0 \text{ such that } |\lambda - \lambda_n| \geq \varepsilon, \forall n \in \mathbb{N},$$

and this implies that

$$\exists \varepsilon > 0 \text{ s.t. } \|(A^\dagger - \lambda\mathbb{1}_\mathcal{H})g\|^2 = \sum_{n=1}^{\infty} |\lambda_n - \lambda|^2 |(u_n|g)|^2$$

$$\geq \varepsilon^2 \sum_{n=1}^{\infty} |(u_n|g)|^2 = \varepsilon^2 \|g\|^2, \forall g \in D_{A^\dagger}$$

(cf. 10.6.4b, 10.4.8a, 10.6.4d), and this implies that $\lambda \in \mathbb{C} - Ap\sigma(A^\dagger)$ (cf. 4.2.3), i.e. $\lambda \in \mathbb{C} - \sigma(A^\dagger)$ (cf. 12.4.21b). This proves that $\sigma(A^\dagger) \subset \overline{\{\lambda_n\}_{n\in\mathbb{N}}}$, and hence that

$$\sigma(A^\dagger) = \overline{\{\lambda_n\}_{n\in\mathbb{N}}}.$$

Finally, the equation $\sigma(A) = \sigma(A^\dagger)$ follows from 4.5.11 since the operator A is closable and $\overline{A} = A^{\dagger\dagger} = A^\dagger$ (cf. 12.4.4a). $\qquad\square$

12.4.25 Examples. The examples we examine here are operators in the Hilbert space $L^2(a,b)$. Most of the elements of $L^2(a,b)$ that we use in these examples are equivalence classes which contain an element of $\mathcal{C}(a,b)$, and we find it pointless to distinguish always between the symbol φ for an element of $\mathcal{C}(a,b)$ and the symbol $[\varphi]$ for the element of $L^2(a,b)$ that contains φ. In fact, if $\varphi \in \mathcal{C}(a,b)$ then φ is the only continuous function in the equivalence class of $[\varphi]$ (cf. 11.2.2b) and therefore it is unambiguously identified with $[\varphi]$. This is useful for avoiding some cumbersome notation. In the same spirit, we use the same symbol for a subset of $\mathcal{C}(a,b)$ and its image under the mapping ι defined in 11.2.1. For instance, in what follows we regard the set

$$\mathcal{C}_0^1(a,b) := \{\varphi \in C^1(a,b) : \varphi(a) = \varphi(b) = 0\}$$

(for $C^1(a,b)$, cf. 3.1.10f) as a subset of $L^2(a,b)$. Clearly, $\mathcal{C}_0^1(a,b)$ is a linear manifold in $L^2(a,b)$. Also, $\overline{\mathcal{C}_0^1(a,b)} = L^2(a,b)$ by 10.6.5b since $\{s_n\}_{n\in\mathbb{N}} \subset \mathcal{C}_0^1(a,b)$, where $\{s_n\}_{n\in\mathbb{N}}$ is the c.o.n.s. in $L^2(a,b)$ defined in 11.2.6.

For any $\theta \in [0, 2\pi)$ we define:

$$D_{A_\theta} := \{\varphi \in C^1(a,b) : \varphi(b) = e^{i\theta}\varphi(a)\},$$
$$A_\theta : D_{A_\theta} \to L^2(a,b)$$
$$\varphi \mapsto A_\theta\varphi := -i\varphi'.$$

Clearly, the mapping A_θ is a linear operator in $L^2(a, b)$, and it is adjointable since $\mathcal{C}_0^1(a, b) \subset D_{A_\theta}$. Moreover

$$(A_\theta \varphi | \psi) \overset{(1)}{=} i \int_a^b \overline{\varphi'}(x) \psi(x) dx \overset{(2)}{=} i \int_a^b \overline{\varphi}'(x) \psi(x) dx$$

$$\overset{(3)}{=} i(\overline{\varphi}(b) \psi(b) - \overline{\varphi}(a) \psi(a)) - i \int_a^b \overline{\varphi}(x) \psi'(x) dx$$

$$= -i \int_a^b \overline{\varphi}(x) \psi'(x) dx \overset{(4)}{=} (\varphi | A_\theta \psi), \forall \varphi, \psi \in D_{A_\theta},$$

where: 1 and 4 hold because an inner product of elements of $\mathcal{C}(a, b)$ can be written as a Riemann integral (cf. 10.1.5b); 2 holds by the definition of derivative of a complex function (cf. 1.2.21); 3 is integration by parts. By 12.4.3, this proves that the operator A_θ is symmetric. Let e_θ be the element of $\mathcal{C}(a, b)$ defined by

$$e_\theta(x) := \exp\left(i\theta \frac{x - a}{b - a}\right), \forall x \in [a, b].$$

If $\{u_n\}_{n \in \mathbb{Z}}$ is the c.o.n.s. in $L^2(a, b)$ defined in 11.2.4, it is obvious that the family $\{e_\theta u_n\}_{n \in \mathbb{Z}}$ is an o.n.s. in $L^2(a, b)$. Moreover, for $[\varphi] \in L^2(a, b)$, $\overline{e_\theta} \varphi \in \mathcal{L}^2(a, b)$ and

$$([e_\theta u_n] | [\varphi]) = ([u_n] | [\overline{e_\theta} \varphi]), \forall n \in \mathbb{Z};$$

therefore, in view of 10.6.4e,

$$([e_\theta u_n] | [\varphi]) = 0, \forall n \in \mathbb{Z} \Rightarrow \left[\exp\left(-i\theta \frac{x - a}{b - a}\right) \varphi(x) = 0 \text{ } m\text{-a.e. on } [a, b]\right]$$

$$\Rightarrow [\varphi(x) = 0 \text{ } m\text{-a.e. on } [a, b]] \Rightarrow [\varphi] = 0_{L^2(a,b)},$$

and this proves that $\{e_\theta u_n\}_{n \in \mathbb{Z}}$ is a c.o.n.s. in $L^2(a, b)$. Now,

$$e_\theta u_n \in D_{A_\theta} \text{ and } A_\theta e_\theta u_n = \frac{2\pi n + \theta}{b - a} e_\theta u_n, \forall n \in \mathbb{Z}.$$

By 12.4.24, this proves that the operator A_θ is e.s.a. and that

$$\sigma_p(A_\theta) = \sigma_p(\overline{A_\theta}) = \sigma(A_\theta) = \sigma(\overline{A_\theta}) = \left\{\frac{2\pi n + \theta}{b - a}\right\}_{n \in \mathbb{Z}}$$

(recall that $\overline{A_\theta} = A_\theta^{\dagger\dagger} = A_\theta^\dagger$). For $\theta_1, \theta_2 \in [0, 2\pi)$ such that $\theta_1 \neq \theta_2$, it is clear that $\overline{A_{\theta_1}} \neq \overline{A_{\theta_2}}$ since $\sigma(\overline{A_{\theta_1}}) \neq \sigma(\overline{A_{\theta_2}})$.

Next, we define the mapping

$$B : \mathcal{C}_0^1(a, b) \to L^2(a, b)$$

$$\varphi \mapsto B\varphi := -i\varphi'.$$

Clearly, B is a linear operator in $L^2(a, b)$, B is adjointable, and $B \subset A_\theta$ for all $\theta \in [0, 2\pi)$. Hence, B is symmetric (cf. 12.4.4b). Since $B \subset \overline{A_\theta}$ for all $\theta \in [0, 2\pi)$, B is not e.s.a. (cf. 12.4.11c).

Further, we define the mapping

$$C : C^1(a, b) \to L^2(a, b)$$
$$\varphi \mapsto C\varphi := -i\varphi'.$$

Clearly, C is a linear operator in $L^2(a, b)$ and $B \subset C$. Hence, C is adjointable. Moreover,

$$(B\varphi|\psi) = i \int_a^b \overline{\varphi'(x)}\psi(x)dx = i \int_a^b \overline{\varphi}'(x)\psi(x)dx$$
$$= i(\overline{\varphi}(b)\psi(b) - \overline{\varphi}(a)\psi(a)) - i \int_a^b \overline{\varphi}(x)\psi'(x)dx$$
$$= (\varphi|C\psi), \forall \varphi \in D_B, \forall \psi \in D_C.$$

By 12.1.3B, this proves that $B \subset C^\dagger$ and $C \subset B^\dagger$. By 12.1.6a and 4.4.11b, this proves that the operator C is closable.

Finally, we define:

$$D_D := \{\varphi \in C^1(a, b) : \varphi(a) = 0\},$$
$$D : D_D \to L^2(a, b)$$
$$\varphi \mapsto D\varphi := -i\varphi'.$$

Clearly, the mapping D is a linear operator in $L^2(a, b)$ and $B \subset D \subset C$. Therefore, D is adjointable and $D \subset B^\dagger$ (since $C \subset B^\dagger$), and hence D is closable. Letting φ_0 be the element of $C(a, b)$ defined by

$$\varphi_0(x) := x - a, \forall x \in [a, b],$$

we have $\varphi_0 \in D_D$ and

$$(D\varphi|\varphi) = i(b - a)^2 + (\varphi|D\varphi) \neq (\varphi|D\varphi);$$

in view of 12.4.3, this proves that the operator D is not symmetric, and also that the operator C is not symmetric (since $D \subset C$, cf. 12.4.4b).

Now we study the spectra of B, C, D, which are the same as the spectra of $\overline{B}, \overline{C}, \overline{D}$ (cf. 4.5.11).

For each $\lambda \in \mathbb{C}$, let $\exp i\lambda\xi$ be the element of $C(a, b)$ defined by

$$(\exp i\lambda\xi)(x) := \exp(i\lambda x), \forall x \in [a, b];$$

clearly,

$$\exp i\lambda\xi \in D_C \text{ and } C \exp i\lambda\xi = \lambda \exp i\lambda\xi.$$

Thus, $\sigma_p(C) = \mathbb{C}$ and hence $\sigma(C) = \mathbb{C}$.

For $\varphi \in C(a, b)$ and $\lambda \in \mathbb{C}$, let φ_λ be the element of $C(a, b)$ defined by

$$\varphi_\lambda(x) := i \exp(i\lambda x) \int_a^x \exp(-i\lambda s)\varphi(s)ds, \forall x \in [a, b].$$

Then, for each $\lambda \in \mathbb{C}$ we define the mapping

$$S_\lambda : \mathcal{C}(a,b) \to \mathcal{C}(a,b)$$
$$\varphi \mapsto S_\lambda \varphi := \varphi_\lambda.$$

Clearly, S_λ is a linear operator in $L^2(a,b)$. Moreover, for all $\varphi \in \mathcal{C}(a,b)$ we have

$$|\varphi_\lambda(x)| \leq M_\lambda \int_{[a,x]} |\varphi| dm \leq M_\lambda \int_{[a,b]} |\varphi| dm, \forall x \in [a,b],$$

with $M_\lambda := \sup\{|\exp(i\lambda(x-s))| : x,s \in [a,b]\}$, and hence

$$\|S_\lambda \varphi\|^2 = \int_{[a,b]} |\varphi_\lambda|^2 dm \leq (b-a) M_\lambda^2 \left(\int_{[a,b]} |\varphi| dm \right)^2$$

$$\leq (b-a) M_\lambda^2 \left(\int_{[a,b]} 1 dm \right) \left(\int_{[a,b]} |\varphi|^2 dm \right) = (b-a)^2 M_\lambda^2 \|\varphi\|^2$$

(we have used 10.1.7a for the elements $1_{[a,b]}$ and φ of $L^2(a,b)$). Thus, the operator S_λ is bounded for each $\lambda \in \mathbb{C}$.

Now, for each $\lambda \in \mathbb{C}$ we have $R_{S_\lambda} \subset D_D$ and

$$(D - \lambda \mathbb{1}_{L^2(a,b)}) S_\lambda \varphi = \varphi, \forall \varphi \in \mathcal{C}(a,b) = D_{S_\lambda}, \tag{1}$$

and also, for all $\psi \in D_D$ and $x \in [a,b]$,

$$(S_\lambda(D - \lambda \mathbb{1}_{L^2(a,b)})\psi)(x) = i \exp(i\lambda x) \int_a^x \exp(-i\lambda s)(-i\psi'(s) - \lambda\psi(s)) ds$$

$$= \exp(i\lambda x)(\exp(-i\lambda x)\psi(x) + i\lambda \int_a^x \exp(-i\lambda s)\psi(s) ds) - \lambda\psi_\lambda(x) = \psi(x),$$

which proves that

$$S_\lambda(D - \lambda \mathbb{1}_{L^2(a,b)})\psi = \psi, \forall \psi \in D_{D - \lambda \mathbb{1}_{L^2(a,b)}}. \tag{2}$$

By 1.2.16b, 1 and 2 imply that, for each $\lambda \in \mathbb{C}$,

$$D - \lambda \mathbb{1}_{L^2(a,b)} \text{ is injective and } (D - \lambda \mathbb{1}_{L^2(a,b)})^{-1} = S_\lambda.$$

Since $D_{S_\lambda} = \mathcal{C}(a,b)$ and $\overline{\mathcal{C}(a,b)} = L^2(a,b)$, this proves that

$$\rho(D) = \mathbb{C} \text{ and hence } \sigma(D) = \emptyset.$$

Furthermore, for each $\lambda \in \mathbb{C}$ we have

$$S_\lambda(B - \lambda \mathbb{1}_{L^2(a,b)})\psi = \psi, \forall \psi \in D_{B - \lambda \mathbb{1}_{L^2(a,b)}},$$

since $B \subset D$. By 1.2.16a, this implies that, for each $\lambda \in \mathbb{C}$,

$$B - \lambda \mathbb{1}_{L^2(a,b)} \text{ is injective and } (B - \lambda \mathbb{1}_{L^2(a,b)})^{-1} \subset S_\lambda,$$

and hence (cf. 4.2.5a) also that $(B - \lambda \mathbb{1}_{L^2(a,b)})^{-1}$ is bounded. This proves that

$$Ap\sigma(B) = \emptyset.$$

We also have, for each $\lambda \in \mathbb{C}$,

$$\exp i\bar{\lambda}\xi \in N_{C - \bar{\lambda}\mathbb{1}_{L^2(a,b)}} \subset N_{B^\dagger - \bar{\lambda}\mathbb{1}_{L^2(a,b)}} = N_{(B - \lambda\mathbb{1}_{L^2(a,b)})^\dagger},$$

since $C \subset B^\dagger$ (cf. also 12.3.3). Thus, for each $\lambda \in \mathbb{C}$,

$$R^\perp_{B - \lambda\mathbb{1}_{L^2(a,b)}} = N_{(B - \lambda\mathbb{1}_{L^2(a,b)})^\dagger} \neq \{0_{L^2(a,b)}\}$$

(cf. 12.1.7), and hence $\overline{R_{B - \lambda\mathbb{1}_{L^2(a,b)}}} \neq L^2(a,b)$ (cf. 10.4.4d). This proves that

$$\rho(B) = \emptyset \text{ and hence } \sigma(B) = \mathbb{C}.$$

All the operators examined above are defined by the same rule (cf. 1.2.1); actually, they are all restrictions of the operator C. It is therefore clear that their various properties depend entirely on the domains on which they are defined.

Finally, we examine some "second order" derivation operators. We have the inclusion $\{u\} \cup \{c_n\}_{n \in \mathbb{N}} \subset D_{BC}$, where $\{u\} \cup \{c_n\}_{n \in \mathbb{N}}$ is the c.o.n.s. in $L^2(a,b)$ defined in 11.2.6, and hence $\overline{D_{BC}} = L^2(a,b)$ (cf. 10.6.5b). Furthermore, we have $BC \subset C^\dagger B^\dagger \subset (BC)^\dagger$ by 12.3.4a. Thus, the operator BC is symmetric. Moreover,

$$BCu = 0_{L^2(a,b)} \text{ and } BCc_n = \left(\frac{\pi}{b - a}\right)^2 n^2 c_n, \forall n \in \mathbb{N}.$$

By 12.4.24, this proves that the operator BC is e.s.a. and that

$$\sigma_p(BC) = \sigma_p(\overline{BC}) = \sigma(BC) = \sigma(\overline{BC}) = \{0\} \cup \left\{\left(\frac{\pi}{b - a}\right)^2 n^2\right\}_{n \in \mathbb{N}}.$$

Similarly, relying on the c.o.n.s. $\{s_n\}_{n \in \mathbb{N}}$ defined in 11.2.6, we can prove that the operator CB is e.s.a., that the elements of $\{s_n\}_{n \in \mathbb{N}}$ are eigenvectors of CB, and that

$$\sigma_p(CB) = \sigma_p(\overline{CB}) = \sigma(CB) = \sigma(\overline{CB}) = \left\{\left(\frac{\pi}{b - a}\right)^2 n^2\right\}_{n \in \mathbb{N}}.$$

Similarly, relying on the c.o.n.s. $\{e_\theta u_n\}_{n \in \mathbb{Z}}$ defined above, we can prove that the operator A_θ^2 is e.s.a. for any $\theta \in [0, 2\pi)$, that the elements of $\{e_\theta u_n\}_{n \in \mathbb{Z}}$ are eigenvectors of A_θ^2, and that

$$\sigma_p(A_\theta^2) = \sigma_p(\overline{A_\theta^2}) = \sigma(A_\theta^2) = \sigma(\overline{A_\theta^2}) = \left\{\left(\frac{2\pi n + \theta}{b - a}\right)^2\right\}_{n \in \mathbb{Z}}.$$

All these "second order" derivation operators are defined by the same rule. Hence, the diversity of their spectra depends on the diversity of their domains.

12.5 Unitary operators and adjoints

12.5.1 Theorem. *For $A \in \mathcal{O}(\mathcal{H})$, the following conditions are equivalent:*

(a) $A \in \mathcal{U}(\mathcal{H})$;
(b) $D_A = \mathcal{H}$, A *is injective,* $A^{-1} = A^{\dagger}$;
(c) $A^{\dagger}A = AA^{\dagger} = \mathbb{1}_{\mathcal{H}}$;
(d) $D_A = \mathcal{H}$ *and* $A^{\dagger} \in \mathcal{U}(\mathcal{H})$;

Proof. $a \Rightarrow b$: Assuming condition a, $D_A = \mathcal{H}$ and A is injective by the definition of an automorphism of $\mathcal{H}$ (cf. 10.1.17). Further we have, for the same reason,

$$(Af|g) = \left(Af|A(A^{-1}g)\right) = \left(f|A^{-1}g\right), \forall f \in D_A, \forall g \in D_{A^{-1}},$$

and this implies $A^{-1} \subset A^{\dagger}$ by 12.1.3B, and hence $A^{-1} = A^{\dagger}$ since $D_{A^{-1}} = R_A = \mathcal{H}$ by the definition of an automorphism.

$b \Rightarrow c$: Assuming condition b, we have $A^{\dagger}A = \mathbb{1}_{\mathcal{H}}$ since $D_A = \mathcal{H}$ and $AA^{\dagger} = \mathbb{1}_{R_A}$ (cf. 3.2.6b). Now, $A^{-1} = A^{\dagger}$ implies that A^{-1} is closed (cf. 12.1.6a). Hence A is closed (cf. 4.4.7), hence A is bounded by 12.2.3, hence $D_{A^{\dagger}} = \mathcal{H}$ by 12.2.2, hence $R_A = D_{A^{-1}} = \mathcal{H}$, and hence $AA^{\dagger} = \mathbb{1}_{\mathcal{H}}$.

$c \Rightarrow d$: Assuming condition c, $A^{\dagger}A = \mathbb{1}_{\mathcal{H}}$ implies $D_A = \mathcal{H}$ and $R_{A^{\dagger}} = \mathcal{H}$, and $AA^{\dagger} = \mathbb{1}_{\mathcal{H}}$ implies $D_{A^{\dagger}} = \mathcal{H}$ (cf. 1.2.13Ab,Ac). Further we have

$$(f|g) = (A(A^{\dagger}f)|g) = (A^{\dagger}f|A^{\dagger}g), \forall f, g \in \mathcal{H},$$

by 12.1.3A. In view of 10.1.20, this proves that $A^{\dagger} \in \mathcal{U}(\mathcal{H})$.

$d \Rightarrow a$: We replace A with $A^{\dagger}$ in the implication $[a \Rightarrow d]$ already proved, and we obtain $A^{\dagger\dagger} \in \mathcal{U}(\mathcal{H})$. Now, $A^{\dagger\dagger} = A$ because $A \subset A^{\dagger\dagger}$ (cf. 12.1.6b) and $D_A = \mathcal{H}$. $\square$

12.5.2 Theorem. *Let U be a unitary operator in $\mathcal{H}$. Then:*

(A) $Ap\sigma(U) = \sigma(U)$.
(B) $\sigma(U) \subset \mathbb{T} := \{z \in \mathbb{C} : |z| = 1\}$.
(C) *Assuming $\sigma_p(U) \neq \emptyset$, suppose $\lambda_1, \lambda_2 \in \sigma_p(U)$ and $\lambda_1 \neq \lambda_2$. Then,*

$$N_{U-\lambda_1\mathbb{1}_{\mathcal{H}}} \subset N^{\perp}_{U-\lambda_2\mathbb{1}_{\mathcal{H}}}.$$

Thus, eigenvectors of U corresponding to different eigenvalues are orthogonal to each other.

Proof. A: From 12.5.1 we have $U^{\dagger}U = UU^{\dagger}$ and hence

$$\|(U - \lambda\mathbb{1}_{\mathcal{H}})f\|^2 = \left(f|(U^{\dagger} - \bar{\lambda}\mathbb{1}_{\mathcal{H}})(U - \lambda\mathbb{1}_{\mathcal{H}})f\right)$$
$$= \left(f|(U - \lambda\mathbb{1}_{\mathcal{H}})(U^{\dagger} - \bar{\lambda}\mathbb{1}_{\mathcal{H}})f\right)$$
$$= \|(U^{\dagger} - \bar{\lambda}\mathbb{1}_{\mathcal{H}})f\|^2, \forall \lambda \in \mathbb{C}, \forall f \in \mathcal{H},$$

where 12.1.3A and 12.3.3 have been used. This implies

$$N_{U-\lambda\mathbb{1}_{\mathcal{H}}} = N_{U^{\dagger}-\bar{\lambda}\mathbb{1}_{\mathcal{H}}}, \forall \lambda \in \mathbb{C}.$$

Now suppose $\lambda \in \mathbb{C} - Ap\sigma(U)$. Then,

$$N_{U-\lambda\mathbb{1}_{\mathcal{H}}} = \{0_{\mathcal{H}}\} \text{ and } (U - \lambda\mathbb{1}_{\mathcal{H}})^{-1} \text{ is bounded,}$$

and also

$$R^{\perp}_{U-\lambda\mathbb{1}_{\mathcal{H}}} = N_{U^{\dagger}-\overline{\lambda}\mathbb{1}_{\mathcal{H}}} = N_{U-\lambda\mathbb{1}_{\mathcal{H}}} = \{0_{\mathcal{H}}\}$$

by 12.1.7, and hence

$$\overline{R_{U-\lambda\mathbb{1}_{\mathcal{H}}}} = \mathcal{H}$$

by 10.4.4d. Therefore, $\lambda \in \rho(U) = \mathbb{C} - \sigma(U)$. This proves that $\sigma(U) \subset Ap\sigma(U)$, and hence that $Ap\sigma(U) = \sigma(U)$.

B: For any $\lambda \in \mathbb{C}$ we have

$$\|(U - \lambda\mathbb{1}_{\mathcal{H}})f\| = \|Uf - \lambda f\| \geq |\|Uf\| - |\lambda|\|f\|| = |1 - |\lambda||\|f\|, \forall f \in \mathcal{H}.$$

By 4.2.3 this proves that

$$U - \lambda\mathbb{1}_{\mathcal{H}} \text{ is injective and } (U - \lambda\mathbb{1}_{\mathcal{H}})^{-1} \text{ is bounded}$$

whenever $|\lambda| \neq 1$. Therefore, $\mathbb{C} - \mathbb{T} \subset \mathbb{C} - Ap\sigma(U)$, i.e. $\sigma(U) \subset \mathbb{T}$ in view of result A.

C: For $f_1 \in N_{U-\lambda_1\mathbb{1}_{\mathcal{H}}}$ and $f_2 \in N_{U-\lambda_2\mathbb{1}_{\mathcal{H}}}$ we have

$$(f_1|f_2) = (Uf_1|Uf_2) = \overline{\lambda_1}\lambda_2 (f_1|f_2).$$

Now, $\overline{\lambda_1}\lambda_2 \neq 1$ since $|\lambda_1| = 1$ (cf. result B) and $\lambda_1 \neq \lambda_2$, and hence $(f_1|f_2) = 0$. $\quad\square$

12.5.3 Theorem. *Let A be a symmetric operator in $\mathcal{H}$. Then:*
the operator $A + i\mathbb{1}_{\mathcal{H}}$ is injective;
for the operator $V := (A - i\mathbb{1}_{\mathcal{H}})(A + i\mathbb{1}_{\mathcal{H}})^{-1}$ we have

$$D_V = R_{A+i\mathbb{1}_{\mathcal{H}}}, \quad R_V = R_{A-i\mathbb{1}_{\mathcal{H}}}, \quad \|Vf\| = \|f\| \text{ for all } f \in D_V;$$

the operator $V - \mathbb{1}_{\mathcal{H}}$ is injective and $A = -i(V + \mathbb{1}_{\mathcal{H}})(V - \mathbb{1}_{\mathcal{H}})^{-1}$;
$1 \notin \sigma_p(V)$;
the operator V is unitary iff the operator A is self-adjoint.
The operator V is called the Cayley transform of A.

Proof. In view of 12.4.20A we have $-i \notin \sigma_p(A)$, and hence that the operator $A + i\mathbb{1}_{\mathcal{H}}$ is injective. We have

$$D_{(A+i\mathbb{1}_{\mathcal{H}})^{-1}} = R_{A+i\mathbb{1}_{\mathcal{H}}} \text{ and } R_{(A+i\mathbb{1}_{\mathcal{H}})^{-1}} = D_{A+i\mathbb{1}_{\mathcal{H}}} = D_A = D_{A-i\mathbb{1}_{\mathcal{H}}}$$

(cf. 1.2.11a), and from these equalities we obtain

$$D_V = R_{A+i\mathbb{1}_{\mathcal{H}}} \text{ and } R_V = R_{A-i\mathbb{1}_{\mathcal{H}}}.$$

For each $f \in D_V$, we set $g := (A + i\mathbb{1}_{\mathcal{H}})^{-1}f$ (note that $D_V = D_{(A+i\mathbb{1}_{\mathcal{H}})^{-1}}$); then,

$$g \in D_{A+i\mathbb{1}_{\mathcal{H}}} = D_A = D_{A-i\mathbb{1}_{\mathcal{H}}} \text{ and } f = (A + i\mathbb{1}_{\mathcal{H}})g, \tag{1}$$

and also

$$Vf = (A - i\mathbb{1}_{\mathcal{H}})(A + i\mathbb{1}_{\mathcal{H}})^{-1}f = (A - i\mathbb{1}_{\mathcal{H}})g; \tag{2}$$

since

$$\|(A \pm i\mathbb{1}_{\mathcal{H}})g\|^2 = \|Ag\|^2 \pm i\,(Ag|g) \mp i\,(g|Ag) + \|g\|^2 = \|Ag\|^2 + \|g\|^2$$

(cf. 12.4.3a), from 2 we have

$$\|Vf\| = \|(A - i\mathbb{1}_{\mathcal{H}})g\| = \|(A + i\mathbb{1}_{\mathcal{H}})g\| = \|f\|;$$

moreover, from 1 and 2 we have

$$(V - \mathbb{1}_{\mathcal{H}})f = (A - i\mathbb{1}_{\mathcal{H}})g - (A + i\mathbb{1}_{\mathcal{H}})g = -2ig = -2i(A + i\mathbb{1}_{\mathcal{H}})^{-1}f \tag{3}$$

and also

$$(V + \mathbb{1}_{\mathcal{H}})f = 2Ag = 2A(A + i\mathbb{1}_{\mathcal{H}})^{-1}f. \tag{4}$$

Since

$$D_{V-\mathbb{1}_{\mathcal{H}}} = D_V = R_{A+i\mathbb{1}_{\mathcal{H}}} = D_{(A+i\mathbb{1}_{\mathcal{H}})^{-1}},$$

3 implies that

$$V - \mathbb{1}_{\mathcal{H}} = -2i(A + i\mathbb{1}_{\mathcal{H}})^{-1}, \tag{5}$$

which implies (cf. 1.2.11b) that the operator $V - \mathbb{1}_{\mathcal{H}}$ is injective and hence that $1 \notin \sigma_p(V)$, and also that

$$(V - \mathbb{1}_{\mathcal{H}})^{-1} = -\frac{1}{2i}(A + i\mathbb{1}_{\mathcal{H}}).$$

From $R_{(A+i\mathbb{1}_{\mathcal{H}})^{-1}} = D_A$ we have

$$D_{A(A+i\mathbb{1}_{\mathcal{H}})^{-1}} = D_{(A+i\mathbb{1}_{\mathcal{H}})^{-1}} = D_V = D_{V+\mathbb{1}_{\mathcal{H}}}$$

(cf. 1.2.13Ad); then, 4 implies that

$$V + \mathbb{1}_{\mathcal{H}} = 2A(A + i\mathbb{1}_{\mathcal{H}})^{-1}. \tag{6}$$

Now, 5 and 6 imply that

$$-i(V + \mathbb{1}_{\mathcal{H}})(V - \mathbb{1}_{\mathcal{H}})^{-1} = A(A + i\mathbb{1}_{\mathcal{H}})^{-1}(A + i\mathbb{1}_{\mathcal{H}}) = A,$$

where the last equality holds because

$$(A + i\mathbb{1}_{\mathcal{H}})^{-1}(A + i\mathbb{1}_{\mathcal{H}}) = \mathbb{1}_{D_{A+i\mathbb{1}_{\mathcal{H}}}} = \mathbb{1}_{D_A}.$$

Finally, since

$$\|Vf\| = \|f\|, \forall f \in D_V,$$

the operator V is unitary iff $D_V = R_V = \mathcal{H}$ (cf. 10.1.20), i.e. iff

$$R_{A+i\mathbb{1}_{\mathcal{H}}} = R_{A-i\mathbb{1}_{\mathcal{H}}} = \mathcal{H},$$

i.e. iff the operator A is s.a. (cf. 12.4.18). $\qquad\square$

The next theorem must be added to what was proved in 4.6.5 (also, cf. 10.3.19) about the unitary-antiunitary equivalence of operators.

12.5.4 Theorem. *Let $\mathcal{H}_1$ and $\mathcal{H}_2$ be isomorphic Hilbert spaces, let $A \in \mathcal{O}(\mathcal{H}_1)$ and $B \in \mathcal{O}(\mathcal{H}_2)$, and suppose that there exists $U \in \mathcal{UA}(\mathcal{H}_1, \mathcal{H}_2)$ so that $B = UAU^{-1}$. Then:*

(a) if $\overline{D_A} = \mathcal{H}_1$ then $\overline{D_B} = \mathcal{H}_2$ and $B^\dagger = UA^\dagger U^{-1}$;
(b) if A is symmetric then B is symmetric;
(c) if A is self-adjoint then B is self-adjoint;
(d) if A is essentially self-adjoint then B is essentially self-adjoint.

Proof. a: Suppose $\overline{D_A} = \mathcal{H}_1$. Then $\overline{D_B} = \mathcal{H}_2$ by 4.6.4i (also, cf. 10.3.19). From 4.6.4g we have

$$G_B = T_U(G_A),$$

where T_U is the unitary or antiunitary operator from the Hilbert space $\mathcal{H}_1 \oplus \mathcal{H}_1$ onto the Hilbert space $\mathcal{H}_2 \oplus \mathcal{H}_2$ defined in 4.6.3 (also, cf. 10.3.19). We denote by W_1 and W_2 the unitary operators defined in $\mathcal{H}_1 \oplus \mathcal{H}_1$ and $\mathcal{H}_2 \oplus \mathcal{H}_2$ respectively as W was in $\mathcal{H} \oplus \mathcal{H}$ (cf. 12.1.5). We have

$$W_2 T_U(f, g) = W_2(Uf, Ug) = (Ug, -Uf)$$
$$= T_U(g, -f) = T_U W_1(f, g), \forall (f, g) \in \mathcal{H}_1 \oplus \mathcal{H}_1.$$

Then, in view of 12.1.5c,

$$G_{B^\dagger} = (W_2(G_B))^\perp = (W_2(T_U(G_A)))^\perp$$
$$= (T_U(W_1(G_A)))^\perp \overset{(1)}{=} T_U(W_1(G_A))^\perp \overset{(2)}{=} G_{UA^\dagger U^{-1}},$$

where 1 holds by 10.2.16 (also, cf. 10.3.16f) and 2 holds by 4.6.4g. This proves the equation $B^\dagger = UA^\dagger U^{-1}$.

b, c, d: These follow immediately from result a. $\qquad\square$

12.6   The C^*-algebra of bounded operators in Hilbert space

12.6.1 Definition. A C^*-*algebra* is a sextuple $(X, \sigma, \mu, \pi, \nu, \iota)$, where $(X, \sigma, \mu, \pi, \nu)$ is a Banach algebra over $\mathbb{C}$ and ι is a mapping $\iota : X \to X$ with the following properties, which we write with the shorthand notation $x^* := \iota(x)$:

(c_1^*) $(x + y)^* = x^* + y^*, \forall x, y \in X,$
(c_2^*) $(\alpha x)^* = \overline{\alpha} x^*, \forall \alpha \in \mathbb{C}, \forall x \in X,$
(c_3^*) $(xy)^* = y^* x^*, \forall x, y \in X,$
(c_4^*) $(x^*)^* = x, \forall x \in X,$
(c_5^*) $\|x^* x\| = \|x\|^2, \forall x \in X.$

The mapping ι is called an *involution*.

12.6.2 Proposition. *Let $(X, \sigma, \mu, \pi, \nu, \iota)$ be a C^*-algebra. Then,*
$$\|x^*\| = \|x\|, \forall x \in X,$$
and the mapping ι is continuous.

Proof. For every $x \in X$, by c_5^* and na (cf. 4.3.1) we have
$$\|x\|^2 = \|x^*x\| \leq \|x^*\|\|x\|,$$
and by c_5^*, c_4^* and na
$$\|x^*\|^2 = \|(x^*)^*x^*\| = \|xx^*\| \leq \|x\|\|x^*\|;$$
these inequalities imply
$$\|x\| \leq \|x^*\| \text{ and } \|x^*\| \leq \|x\|,$$
and hence $\|x\| = \|x^*\|$.

If a sequence $\{x_n\}$ in X and an element x of X are so that $x_n \to x$, then
$$\|x_n^* - x^*\| = \|(x_n - x)^*\| = \|x_n - x\| \to 0, \text{ i.e. } x_n^* \to x^*,$$
where c_1^* and c_2^* have been used. In view of 2.4.2, this proves that the mapping ι is continuous. $\qquad\square$

12.6.3 Proposition. *If x is an element of a C^*-algebra such that $x^*x = xx^*$, then*
$$\|x^n\| = \|x\|^n, \forall n \in \mathbb{N}.$$

Proof. If an element y of a C^*-algebra is such that $y^*y = yy^*$, then
$$\|y^2\|^2 = \|(y^2)^*y^2\| = \|y^*y^*yy\| = \|y^*yy^*y\| = \|(y^*y)^*(y^*y)\| = \|y^*y\|^2 = \|y\|^4,$$
where c_3^*, c_4^*, c_5^* have been used, and hence
$$\|y^2\| = \|y\|^2.$$
Now, let x be an element of a C^*-algebra such that $x^*x = xx^*$. First we prove by induction that
$$\|x^{2^n}\| = \|x\|^{2^n}, \forall n \in \mathbb{N}.$$
We already know that this equality is true for $n = 1$. If we assume that the equality is true for a given $n \in \mathbb{N}$ then we have
$$\|x^{2^{n+1}}\| = \|(x^{2^n})^2\| = \|x^{2^n}\|^2 = \|x\|^{2^{n+1}}$$
by the result proved above for y, since
$$(x^{2^n})^*x^{2^n} = (x^*)^{2^n}x^{2^n} = x^{2^n}(x^*)^{2^n} = x^{2^n}(x^{2^n})^*,$$
where c_3^* has been used. This concludes the proof by induction. Now, for any $n \in \mathbb{N}$, let $m \in \mathbb{N}$ be so that $n + m = 2^k$ for some $k \in \mathbb{N}$. Then,
$$\|x\|^n\|x\|^m = \|x\|^{n+m} = \|x^{n+m}\| = \|x^nx^m\| \leq \|x^n\|\|x^m\| \leq \|x^n\|\|x\|^m,$$
where na (cf. 4.3.1) has been used twice. If $x \neq 0_X$, this proves that
$$\|x\|^n \leq \|x^n\|,$$
which is also trivially true for $x = 0_X$. Since
$$\|x^n\| \leq \|x\|^n$$
is true by na, we have $\|x^n\| = \|x\|^n$. $\qquad\square$

12.6.4 Theorem. *The mapping*

$$\iota : \mathcal{B}(\mathcal{H}) \to \mathcal{B}(\mathcal{H})$$
$$A \mapsto \iota(A) := A^\dagger$$

is defined consistently, and $\mathcal{B}(\mathcal{H})$ is a C^*-algebra with this mapping as involution. Hence:

$$A^\dagger \in \mathcal{B}(\mathcal{H}), \forall A \in \mathcal{B}(\mathcal{H}),$$
$$(A + B)^\dagger = A^\dagger + B^\dagger, \forall A, B \in \mathcal{B}(\mathcal{H}),$$
$$(\alpha A)^\dagger = \overline{\alpha} A^\dagger, \forall \alpha \in \mathbb{C}, \forall A \in \mathcal{B}(\mathcal{H}),$$
$$(AB)^\dagger = B^\dagger A^\dagger, \forall A, B \in \mathcal{B}(\mathcal{H}),$$
$$A^{\dagger\dagger} = A, \forall A \in \mathcal{B}(\mathcal{H}),$$
$$\|A^\dagger A\| = \|A\|^2, \forall A \in \mathcal{B}(\mathcal{H}),$$
$$\|A^\dagger\| = \|A\|, \forall A \in \mathcal{B}(\mathcal{H}),$$
$$\|A^n\| = \|A\|^n, \forall n \in \mathbb{N}, \text{ if } A \in \mathcal{B}(\mathcal{H}) \text{ is such that } A^\dagger A = AA^\dagger.$$

Proof. We already know that $\mathcal{B}(\mathcal{H})$ is a Banach algebra over $\mathbb{C}$ (cf. 4.3.5). The definition of the mapping ι of the statement is consistent because

$$A^\dagger \in \mathcal{B}(\mathcal{H}), \forall A \in \mathcal{B}(\mathcal{H}),$$

by 12.2.2. Now we prove that the mapping ι of the statement has all the properties listed in 12.6.1.

c_1^*: this follows from 12.3.1b.

c_2^*: this follows from 12.3.2.

c_3^*: this follows from 12.3.4b.

c_4^*: this follows from 12.1.6b.

c_5^*: For $A \in \mathcal{B}(\mathcal{H})$ we have

$$\|Af\|^2 = (Af|Af) = (f|A^\dagger Af) \le \|f\| \|A^\dagger Af\| \le \|f\| \|A^\dagger A\| \|f\|, \forall f \in \mathcal{H},$$

by 12.1.3A, 10.1.7a, 4.2.5b, and this proves that $\|A\|^2 \le \|A^\dagger A\|$. We also have

$$|(f|A^\dagger Ag)| = |(Af|Ag)| \le \|Af\| \|Ag\| \le \|A\|^2 \|f\| \|g\|, \forall f, g \in \mathcal{H},$$

by the same reasons as above, and this proves that $\|A^\dagger A\| \le \|A\|^2$ (cf. 10.1.14).

The last two assertions of the statement follow directly from 12.6.2 and from 12.6.3. □

12.6.5 Remark. A pair A, B of self-adjoint operators in $\mathcal{H}$ is said to "satisfy" the *Heisenberg canonical commutation relation* if

$$AB - BA \subset i\mathbb{1}_{\mathcal{H}}. \tag{HCCR}$$

If this condition is satisfied, then either A or B or both A and B must be non-bounded (thus, the relation HCCR cannot be discussed without worrying about the domains of the operators, in view of 12.4.7). The proof is as follows. If A and B

were bounded, then we should have $A, B \in \mathcal{B}(\mathcal{H})$ (cf. 12.4.7) and condition HCCR would be

$$AB - BA = i\mathbb{1}_{\mathcal{H}}. \tag{1}$$

This would imply the equations

$$A^n B - BA^n = inA^{n-1}, \forall n \in \mathbb{N}. \tag{2}$$

Indeed, 1 is 2 for $n = 1$ (recall that $A^0 := \mathbb{1}_{\mathcal{H}}$, cf. 3.3.1) and, assuming that 2 is true for a given $n \in \mathbb{N}$, we have

$$A^n BA - BA^{n+1} = inA^n,$$

which in view of 1 can be written as

$$A^n(AB - i\mathbb{1}_{\mathcal{H}}) - BA^{n+1} = inA^n,$$

or

$$A^{n+1}B - BA^{n+1} = i(n+1)A^n.$$

This proves 2 by induction. From 2 we would have, by 12.6.3 (also, cf. 12.6.4),

$$n\|A\|^{n-1} = \|inA^{n-1}\| \le \|A\|^n \|B\| + \|B\| \|A\|^n, \forall n \in \mathbb{N},$$

which would imply (note that 1 implies $A \ne \mathbb{O}_{\mathcal{H}}$)

$$n \le 2\|A\| \|B\|, \forall n \in \mathbb{N},$$

which is a contradiction.

This also shows that the relation

$$AB - BA = i\mathbb{1}_{\mathcal{H}}$$

(which is clearly stronger than HCCR) is an impossible relation for two self-adjoint operators A and B, since it would imply $D_A = D_B = \mathcal{H}$, and hence for both the operators A and B to be bounded (cf. 12.4.7). We mention the fact that there are pairs of self-adjoint operators which satisfy HCCR (cf. 20.1.3b and 20.1.7).

12.6.6 Remarks.

(a) Here we make some remarks about linear operators in a one-dimensional Hilbert space which could also be deduced from 10.8.4. Thus, we suppose in what follows that $\mathcal{H}$ is a one-dimensional Hilbert space.

Since $\{0_{\mathcal{H}}\}$ and $\mathcal{H}$ are the only linear manifolds in $\mathcal{H}$, the domain of every non-trivial linear operator in $\mathcal{H}$ must be $\mathcal{H}$. Moreover, every linear operator in $\mathcal{H}$ is bounded (cf. 10.8.3). Thus, the family of non-trivial linear operators in $\mathcal{H}$ is $\mathcal{B}(\mathcal{H})$.

For $\alpha \in \mathbb{C}$, we define the mapping

$$A_\alpha : \mathcal{H} \to \mathcal{H}$$
$$f \mapsto A_\alpha f := \alpha f.$$

It is obvious that $A_\alpha \in \mathcal{B}(\mathcal{H})$. If $\alpha, \beta \in \mathbb{C}$ are such that $A_\alpha = A_\beta$ then $\alpha f = \beta f$ for all $f \in \mathcal{H}$ and hence $\alpha = \beta$. Now let $A \in \mathcal{B}(\mathcal{H})$ and fix $f_0 \in \mathcal{H} - \{0_\mathcal{H}\}$; then,

$$\exists! \alpha \in C \text{ such that } Af_0 = \alpha f_0;$$

hence, for each $f \in \mathcal{H}$, if $k_f \in \mathbb{C}$ is such that $f = k_f f_0$ then

$$Af = A(k_f f_0) = k_f \alpha f_0 = \alpha(k_f f_0) = \alpha f = A_\alpha f.$$

This proves that the mapping

$$\mathbb{C} \ni \alpha \mapsto \Phi(\alpha) := A_\alpha \in \mathcal{B}(\mathcal{H})$$

is a bijection from $\mathbb{C}$ onto $\mathcal{B}(\mathcal{H})$. It is obvious that $\mathbb{C}$ is a C^*-algebra with the modulus of complex numbers as norm and the complex conjugation as involution. Now, it is obvious that

$$A_{\alpha+\beta} = A_\alpha + A_\beta \text{ and } A_{\alpha\beta} = \alpha A_\beta = A_\alpha A_\beta, \forall \alpha, \beta \in \mathbb{C}.$$

This proves that Φ is an isomorphism from the associative algebra $\mathbb{C}$ onto the associative algebra $\mathcal{B}(\mathcal{H})$.

For each $\alpha \in \mathbb{C}$ we have

$$\|A_\alpha f\| = |\alpha| \|f\|, \forall f \in \mathcal{H};$$

by 4.2.5c, this proves that

$$\|A_\alpha\| = |\alpha|, \forall \alpha \in \mathbb{C}.$$

Moreover, for each $\alpha \in \mathbb{C}$ we have

$$(A_\alpha f | g) = \overline{\alpha}(f|g) = (f|A_{\overline{\alpha}}g), \forall f, g \in \mathcal{H};$$

by 12.1.3B, this proves that

$$A_{\overline{\alpha}} = A_\alpha^\dagger, \forall \alpha \in \mathbb{C}.$$

All this means that the bijection Φ identifies the two C^*-algebras $\mathbb{C}$ and $\mathcal{B}(\mathcal{H})$ also as to their norms and their involutions.

We also note that

$$A_\alpha \text{ is self-adjoint iff } \alpha \in \mathbb{R}.$$

Moreover we have

$$A_{\frac{1}{\alpha}} A_\alpha = \mathbb{1}_\mathcal{H}, \forall \alpha \in \mathbb{C} - \{0\};$$

in view of 1.2.16a, this proves that A_α is injective and $A_\alpha^{-1} = A_{\frac{1}{\alpha}}$ for all $\alpha \in \mathbb{C} - \{0\}$. Then,

$$\sigma(A_\alpha) = \sigma_p(A_\alpha) = \{\alpha\}, \forall \alpha \in \mathbb{C},$$

since $A_\alpha - \lambda \mathbb{1}_\mathcal{H} = A_{\alpha-\lambda}$ for all $\alpha, \lambda \in \mathbb{C}$.

Finally, the subsets $\mathbb{T} := \{z \in C : |z| = 1\}$ of $\mathbb{C}$ and $\mathcal{U}(\mathcal{H})$ of $\mathcal{B}(\mathcal{H})$ are groups (cf. 10.3.10). For $\alpha \in \mathbb{C}$ we have, in view of 10.1.20,

$$A_\alpha \in \mathcal{U}(\mathcal{H}) \text{ iff } [\|A_\alpha f\| = \|f\|, \forall f \in \mathcal{H}] \text{ iff } \alpha \in \mathbb{T}.$$

This, along with

$$A_{z_1} A_{z_2} = A_{z_1 z_2}, \forall z_1, z_2 \in \mathbb{T},$$

proves that the restriction $\Phi_\mathbb{T}$ of the mapping Φ to $\mathbb{T}$ is an isomorphism from the group $\mathbb{T}$ onto the group $\mathcal{U}(\mathcal{H})$.

(b) For a one-dimensional Hilbert space $\mathcal{H}$, theorem 12.5.3 on the Cayley transform of a self-adjoint operator can be rephrased as follows, in view of what was seen in remark a:

for all $x \in \mathbb{R}$, $\frac{x-i}{x+i} \in \mathbb{T}$ and $\frac{x-i}{x+i} \neq 1$;

if we define the function

$$\varphi : \mathbb{R} \to \mathbb{T} - \{1\}$$

$$x \mapsto \varphi(x) := \frac{x-i}{x+i},$$

then we have

$$x = -i\frac{\varphi(x)+1}{\varphi(x)-1}, \forall x \in \mathbb{R},$$

and hence the function φ is injective.

Of course, all this can be proved directly, without going through 12.5.3. The name of *Cayley transform* was originally given to the function φ.

Now, let $z \in \mathbb{T} - \{1\}$ and write $z = \exp i\theta$ with $0 < \theta < 2\pi$. Then,

$$-i\frac{z+1}{z-1} = -i\frac{\exp i\frac{\theta}{2} + \exp\left(-i\frac{\theta}{2}\right)}{\exp i\frac{\theta}{2} - \exp\left(-i\frac{\theta}{2}\right)} = -\frac{\cos\frac{\theta}{2}}{\sin\frac{\theta}{2}} \in \mathbb{R}$$

and

$$\varphi\left(-\frac{\cos\frac{\theta}{2}}{\sin\frac{\theta}{2}}\right) = \frac{\cos\frac{\theta}{2} + i\sin\frac{\theta}{2}}{\cos\frac{\theta}{2} - i\sin\frac{\theta}{2}} = \frac{\exp i\frac{\theta}{2}}{\exp\left(-i\frac{\theta}{2}\right)} = \exp i\theta = z.$$

This proves that the function φ is a bijection from $\mathbb{R}$ onto $\mathbb{T} - \{1\}$ and that its inverse is the function

$$\psi : \mathbb{T} - \{1\} \to \mathbb{R}$$

$$z \mapsto \psi(z) := -i\frac{z+1}{z-1}.$$

12.6.7 Proposition. *Let X be a non-empty set. For the Banach algebra $\mathcal{F}_B(X)$ (cf. 4.3.6a), the mapping*

$$\iota : \mathcal{F}_B(X) \to \mathcal{F}_B(X)$$

$$\varphi \mapsto \iota(\varphi) := \overline{\varphi}$$

is defined consistently, and $\mathcal{F}_B(X)$ is a C^*-algebra with this mapping as involution.

If $\mathcal{A}$ is a σ-algebra on X, the Banach algebra $\mathcal{M}_B(X, \mathcal{A})$ (cf. 6.2.29) is a C^*-algebra with the restriction $\iota_{\mathcal{M}_B(X,\mathcal{A})}$ as involution.

If a distance is defined on X, the Banach algebra $\mathcal{C}_B(X)$ (cf. 4.3.6b) is a C^*-algebra with the restriction $\iota_{\mathcal{C}_B(X)}$ as involution.

Proof. It is obvious that the mapping ι is defined consistently and that it satisfies all the conditions listed in 12.6.1. For instance, as to condition c_5^* we have, for all $\varphi \in \mathcal{F}_B(X)$,

$$[|\varphi(x)| \leq \|\varphi\|_\infty, \forall x \in X] \Rightarrow$$
$$[|\overline{\varphi}(x)\varphi(x)| = |\varphi(x)|^2 \leq \|\varphi\|_\infty^2, \forall x \in X] \Rightarrow \|\overline{\varphi}\varphi\|_\infty \leq \|\varphi\|_\infty^2$$

and

$$[|\varphi(x)|^2 = |\overline{\varphi}(x)\varphi(x)| \leq \|\overline{\varphi}\varphi\|_\infty, \forall x \in X] \Rightarrow \|\varphi\|_\infty \leq \sqrt{\|\overline{\varphi}\varphi\|_\infty} \Rightarrow \|\varphi\|_\infty^2 \leq \|\overline{\varphi}\varphi\|.$$

The restrictions $\iota_{\mathcal{M}_B(X,\mathcal{A})}$ and $\iota_{\mathcal{C}_B(X)}$ are defined because $\mathcal{M}_B(X,\mathcal{A})$ and $\mathcal{C}_B(X)$ are subsets of $\mathcal{F}_B(X)$. Moreover, $\iota(\mathcal{M}_B(X,\mathcal{A})) \subset \mathcal{M}_B(X,\mathcal{A})$ (cf. 6.2.17), $\iota(\mathcal{C}_B(X)) \subset \mathcal{C}_B(X)$, and it is obvious that $\iota_{\mathcal{M}_B(X,\mathcal{A})}$ and $\iota_{\mathcal{C}_B(X)}$ have the properties of an involution. $\qquad\square$

Chapter 13

Orthogonal Projections and Projection Valued Measures

In the first half of this chapter we study orthogonal projections, which are the building blocks of unitary and of self-adjoint operators, as the spectral theorems show. Orthogonal projections enter our formulation of the spectral theorems in the guise of projection valued measures, which we study in the second half of this chapter.

Throughout this chapter, $\mathcal{H}$ denotes an abstract Hilbert space.

13.1 Orthogonal projections

13.1.1 Definitions. Let M be a subspace of $\mathcal{H}$. In view of the orthogonal decomposition theorem (cf. 10.4.1), we can define the *orthogonal decomposition mapping*

$$\delta_M : \mathcal{H} \to M \times M^\perp$$
$$f \mapsto \delta_M(f) := (f_1, f_2) \text{ if } (f_1, f_2) \in M \times M^\perp \text{ is such that } f = f_1 + f_2.$$

We denote by π_M the projection mapping (cf. 1.2.6c)

$$\pi_M : M \times M^\perp \to M$$
$$(f, g) \mapsto \pi_M(f, g) := f,$$

and we call *orthogonal projection onto M* the composition of π_M with δ_M, i.e. the mapping P_M defined by $P_M := \pi_M \circ \delta_M$. Thus, P_M is a mapping from $\mathcal{H}$ to M. However, it is convenient to consider $\mathcal{H}$ instead of M as the final set of the mapping P_M (cf. 1.2.1). Clearly, the mapping P_M can be defined directly as follows:

$$P_M : \mathcal{H} \to \mathcal{H}$$
$$f \mapsto P_M f := f' \text{ if } f' \in M \text{ and } f - f' \in M^\perp.$$

It is expedient to denote by $\mathscr{S}(\mathcal{H})$ the family of all subspaces of $\mathcal{H}$ and to define

$$\mathscr{P}(\mathcal{H}) := \{P_M : M \in \mathscr{S}(\mathcal{H})\}.$$

The elements of $\mathscr{P}(\mathcal{H})$ are called *orthogonal projections in $\mathcal{H}$*.

13.1.2 Remark. It is obvious that

$$\delta_{\{0_{\mathcal{H}}\}}(f) = (0_{\mathcal{H}}, f) \text{ and } \delta_{\mathcal{H}}(f) = (f, 0_{\mathcal{H}}), \forall f \in \mathcal{H}.$$

Thus the orthogonal projections onto the trivial subspaces $\{0_{\mathcal{H}}\}$ and $\mathcal{H}$ are the trivial operators:

$$P_{\{0_{\mathcal{H}}\}} = \mathbb{O}_{\mathcal{H}} \text{ and } P_{\mathcal{H}} = \mathbb{1}_{\mathcal{H}}.$$

13.1.3 Theorem. *For $M \in \mathscr{S}(\mathcal{H})$, we have:*

(a) P_M is a linear operator, i.e. $P_M \in \mathcal{O}_E(\mathcal{H})$;
(b) $N_{P_M} = M^{\perp}$;
(c) $R_{P_M} = M = \{f \in \mathcal{H} : P_M f = f\} = \{f \in \mathcal{H} : \|P_M f\| = \|f\|\}$;
(d) the operator P_M is bounded, i.e. $P_M \in \mathcal{B}(\mathcal{H})$, and $\|P_M\| = 1$ if $M \neq \{0_{\mathcal{H}}\}$;
(e) $P_{M^{\perp}} = \mathbb{1}_{\mathcal{H}} - P_M$ (recall that $M^{\perp} \in \mathscr{S}(\mathcal{H})$, cf. 10.2.13), and hence $M^{\perp} = R_{\mathbb{1}_{\mathcal{H}} - P_M}$.

Proof. a: First, $D_{P_M} = \mathcal{H}$. Next, for all $\alpha, \beta \in \mathbb{C}$ and $f, g \in \mathcal{H}$, if we write

$$(f_1, f_2) := \delta_M(f) \text{ and } (g_1, g_2) := \delta_M(g)$$

then

$$\delta_M(\alpha f + \beta g) = (\alpha f_1 + \beta g_1, \alpha f_2 + \beta g_2)$$

since M and $M^{\perp}$ are linear manifolds in $\mathcal{H}$ and $\alpha f + \beta g = \alpha f_1 + \beta g_1 + \alpha f_2 + \beta g_2$, and hence

$$P_M(\alpha f + \beta g) = \alpha f_1 + \beta g_1 = \alpha P_M f + \beta P_M g.$$

This proves condition *lo* of 3.2.1 for the mapping P_M.

b: For $f \in \mathcal{H}$, if we write $(f_1, f_2) := \delta_M(f)$ then f_1 equals $0_{\mathcal{H}}$ iff f_2 equals f, in view of the equation $f = f_1 + f_2$. Then,

$$f \in N_{P_M} \Leftrightarrow P_M f = 0_{\mathcal{H}} \Leftrightarrow \delta_M(f) = (0_{\mathcal{H}}, f) \Leftrightarrow f \in M^{\perp}.$$

c: For $f \in \mathcal{H}$, if we write $(f_1, f_2) := \delta_M(f)$ then f_1 equals f iff f_2 equals $0_{\mathcal{H}}$. Then,

$$P_M f = f \Leftrightarrow \delta_M(f) = (f, 0_{\mathcal{H}}) \Leftrightarrow f \in M.$$

This proves the equation

$$M = \{f \in \mathcal{H} : P_M f = f\}.$$

Since the inclusions $R_{P_M} \subset M$ and $\{f \in \mathcal{H} : P_M f = f\} \subset R_{P_M}$ are obvious, we have the equations

$$R_{P_M} = M = \{f \in \mathcal{H} : P_M f = f\}.$$

The inclusion $\{f \in \mathcal{H} : P_M f = f\} \subset \{f \in \mathcal{H} : \|P_M f\| = \|f\|\}$ is obvious. Moreover, if for $f \in \mathcal{H}$ we write $(f_1, f_2) := \delta_M(f)$ then, in view of 10.2.3,

$$\|P_M f\| = \|f\| \Rightarrow \|f_1\|^2 = \|f\|^2 = \|f_1\|^2 + \|f_2\|^2 \Rightarrow f_2 = 0_{\mathcal{H}} \Rightarrow f = f_1 \in M.$$

This proves the inclusion $\{f \in \mathcal{H} : \|P_M f\| = \|f\|\} \subset M$. Thus, all the equations of the statement are proved.

d: For every $f \in \mathcal{H}$, if we write $(f_1, f_2) := \delta_M(f)$ then

$$\|P_M f\|^2 = \|f_1\|^2 \leq \|f_1\|^2 + \|f_2\|^2 = \|f\|^2.$$

This proves that the operator P_M is bounded and that $\|P_M\| \leq 1$ (cf. 4.2.4). If $M \neq \{0_{\mathcal{H}}\}$, then

$$\exists f \in \mathcal{H} \text{ such that } f \neq 0_{\mathcal{H}} \text{ and } \|P_M f\| = \|f\|$$

(cf. result c), and hence $\|P_M\| \geq 1$ by 4.2.5c, and hence $\|P_M\| = 1$.

e: For every $f \in \mathcal{H}$, if we write $(f_1, f_2) := \delta_M(f)$ then $\delta_{M^\perp}(f) = (f_2, f_1)$ since $M \subset M^{\perp\perp}$ (cf. 10.2.10d), and hence

$$P_{M^\perp} f = f_2 = f - f_1 = f - P_M f = (\mathbb{1}_{\mathcal{H}} - P_M)f.$$

This proves the equation $P_{M^\perp} = \mathbb{1}_{\mathcal{H}} - P_M$, which implies $M^\perp = R_{\mathbb{1}_{\mathcal{H}} - P_M}$ by result c (with M replaced by $M^\perp$). $\qquad\qquad\square$

13.1.4 Remark.

(a) For every projection A in $\mathcal{H}$ we have, in view of 13.1.3c,

$$R_A \in \mathscr{S}(\mathcal{H}) \text{ and } A = P_{R_A}.$$

We also have, in view of 13.1.3c,e,

$$P_{R_A^\perp} = \mathbb{1}_{\mathcal{H}} - A \text{ and hence } R_A^\perp = R_{\mathbb{1}_{\mathcal{H}} - A}.$$

(b) In view of 13.1.3c, the mapping

$$\mathscr{S}(\mathcal{H}) \ni M \mapsto P_M \in \mathscr{P}(\mathcal{H})$$

is injective (if $M, N \in \mathscr{S}(\mathcal{H})$ are such that $P_M = P_N$ then $M = R_{P_M} = R_{P_N} = N$) and hence it is bijective from $\mathscr{S}(\mathcal{H})$ onto $\mathscr{P}(\mathcal{H})$, and the mapping

$$\mathscr{P}(\mathcal{H}) \ni A \mapsto R_A \in \mathscr{S}(\mathcal{H})$$

is defined consistently and it is the inverse of the mapping preceding (cf. remark a).

13.1.5 Theorem. *For $A \in \mathcal{O}_E(\mathcal{H})$ the following conditions are equivalent:*

(a) $A \in \mathscr{P}(\mathcal{H})$;
(b) $A = A^\dagger$ and $A = A^2$.

Proof. $a \Rightarrow b$: Let $M \in \mathscr{S}(\mathcal{H})$ be so that $A = P_M$. For every $f \in \mathcal{H}$, if we write $(f_1, f_2) := \delta_M(f)$ then

$$(f|Af) = (f_1 + f_2|f_1) = \|f_1\|^2 \in \mathbb{R}$$

and

$$A^2 f = A f_1 = f_1 = Af,$$

in view of 13.1.3c. This proves that A is symmetric (cf. 12.4.3) and hence self-adjoint (since $D_A = \mathcal{H}$) and also that $A^2 = A$.

$b \Rightarrow a$: We assume condition b. For every $f \in \mathcal{H}$, we have obviously

$$f = Af + (f - Af)$$

and

$$Af \in R_A \text{ and hence } Af \in \overline{R_A};$$

we also have

$$(f - Af|Ag) = (A(f - Af)|g) = (Af - A^2 f|g) = 0, \forall g \in \mathcal{H},$$

and this proves that $f - Af \in R_A^\perp$ and hence $f - Af \in (\overline{R_A})^\perp$ (cf. 10.2.11); since $\overline{R_A} \in \mathscr{S}(\mathcal{H})$ (cf. 3.2.2a and 4.1.12), all this can be written as

$$\delta_{\overline{R_A}}(f) = (Af, f - Af),$$

and this implies $Af = P_{\overline{R_A}}f$. This proves that $A = P_{\overline{R_A}}$, and hence condition a. $\square$

13.1.6 Theorem. *For $A \in \mathcal{O}_E(\mathcal{H})$ the following conditions are equivalent:*

(a) $A \in \mathscr{P}(\mathcal{H})$;
(b) $A = A^2$ and $\|Af\| \leq \|f\|$ for all $f \in \mathcal{H}$.

Proof. $a \Rightarrow b$: This follows from 13.1.5 and 13.1.3d.

$b \Rightarrow a$: We assume condition b. Then $N_{\mathbb{1}_{\mathcal{H}} - A} \in \mathscr{S}(\mathcal{H})$, by 4.4.3 and 4.4.8. We write $M := N_{\mathbb{1}_{\mathcal{H}} - A}$. Suppose that there exists $f \in M^\perp$ such that $Af \neq 0_{\mathcal{H}}$. Since

$$(\mathbb{1}_{\mathcal{H}} - A)Af = Af - A^2 f = 0_{\mathcal{H}},$$

we have $Af \in M$ and hence, in view of 10.2.3,

$$\|f\|^2 + t^2\|Af\|^2 = \|f + tAf\|^2$$
$$\geq \|A(f + tAf)\|^2 = \|(1+t)Af\|^2 = (1+t)^2\|Af\|^2, \forall t \in \mathbb{R},$$

and hence

$$\|f\|^2 \geq (1 + 2t)\|Af\|^2, \forall t \in \mathbb{R},$$

which leads to a contradiction for t so that $1 + 2t > \|f\|^2\|Af\|^{-2}$. This proves that

$$Af = 0_{\mathcal{H}}, \forall f \in M^\perp.$$

Then, for every $g \in \mathcal{H}$ we have, in view of 13.1.3e and 13.1.3c (with M replaced by $M^\perp$)

$$g - P_M g = P_{M^\perp} g \in M^\perp, \text{ and hence } A(g - P_M g) = 0_{\mathcal{H}};$$

we also have, in view of 13.1.3c,

$$P_M g \in M, \text{ and hence } (\mathbb{1}_{\mathcal{H}} - A)P_M g = 0_{\mathcal{H}}, \text{ and hence } AP_M g = P_M g;$$

thus, we have

$$Ag = A((g - P_M g) + P_M g) = P_M g.$$

This proves that $A = P_M$, and hence that $A \in \mathscr{P}(\mathcal{H})$. $\square$

13.1.7 Remarks.

(a) For any normed space X, an operator $A \in \mathcal{O}_E(X)$ is called a *projection* in X if $A = A^2$. From 13.1.5 we see that a projection in $\mathcal{H}$ is an orthogonal projection iff it is self-adjoint. Besides, from 13.1.6 we see that a projection in $\mathcal{H}$ is an orthogonal projection iff it is bounded with norm not greater than one.

The only projections in $\mathcal{H}$ that we consider in this book are orthogonal projections. For this reason, we may sometimes use the word projection to mean an orthogonal projection.

(b) The plan for the proof of $b \Rightarrow a$ in 13.1.5 was suggested by the fact that if $A \in \mathscr{P}(\mathcal{H})$ then $A = P_{R_A} = P_{\overline{R_A}}$ (cf. 13.1.4.a). We point out that we could not have set out to prove the equation $A = P_{R_A}$, because we did not know yet that R_A was a subspace. However, we did know that $\overline{R_A}$ was a subspace and therefore it was sensible to set out to prove that $A = P_{\overline{R_A}}$.

The plan for the proof of $b \Rightarrow a$ in 13.1.6 was suggested by the fact that if $A \in \mathscr{P}(\mathcal{H})$ then $\mathbb{1}_\mathcal{H} - A = P_{R_A^\perp}$ (cf. 13.1.4a), and hence $N_{\mathbb{1}_\mathcal{H} - A} = R_A^{\perp\perp} = R_A$ (cf. 13.1.3b and 10.4.4a), and hence $A = P_{R_A} = P_M$ if we write $M := N_{\mathbb{1}_\mathcal{H} - A}$ (cf. 13.1.4a). We point out that the first thing we proved was that $N_{\mathbb{1}_\mathcal{H} - A}$ was a subspace.

(c) In view of 13.1.5, for each $A \in \mathscr{P}(\mathcal{H})$ we have

$$(f|Af) = (f|A^2 f) = (Af|Af) = \|Af\|^2, \forall f \in \mathcal{H}.$$

13.1.8 Theorem. *Let $\mathcal{H}_1$ and $\mathcal{H}_2$ be isomorphic Hilbert spaces, let $A \in \mathscr{P}(\mathcal{H}_1)$ and $B \in \mathcal{O}(\mathcal{H}_2)$, and suppose that there exists $U \in \mathcal{UA}(\mathcal{H}_1, \mathcal{H}_2)$ so that $B = UAU^{-1}$. Then $B \in \mathscr{P}(\mathcal{H}_2)$. In fact,*

$$U(R_A) \in \mathscr{S}(\mathcal{H}_2) \text{ and } B = P_{U(R_A)}.$$

Proof. It is obvious that $D_B = \mathcal{H}_2$. Since $A = A^\dagger$ and $A = A^2$ (cf. 13.1.5) we have $B = B^\dagger$ by 12.5.4c and

$$B^2 = UAU^{-1}UAU^{-1} = UA^2U^{-1} = UAU^{-1} = B.$$

In view of 13.1.5, this proves that $B \in \mathscr{P}(\mathcal{H}_2)$. Then, $R_B \in \mathscr{S}(\mathcal{H}_2)$ and $B = P_{R_B}$ (cf. 13.1.4a). Now, $R_B = U(R_A)$ by 4.6.4h, 10.3.12b, 10.3.19. $\qquad \square$

13.1.9 Corollary (Cor. of the closed graph theorem in Hilbert space).

Let $A \in \mathcal{O}(\mathcal{H})$. If D_A is closed and A is closed, then A is bounded.

Proof. Assume D_A closed and A closed, let P denote the orthogonal projection onto D_A, and consider the operator AP. Clearly, $D_{AP} = \mathcal{H}$. Moreover, if two vectors f, g of $\mathcal{H}$ and a sequence $\{f_n\}$ in $\mathcal{H}$ are so that

$$f_n \to f \text{ and } APf_n \to g$$

then

$$Pf_n \to Pf$$

because P is bounded (cf. 13.1.3d) and hence continuous, and hence $g = APf$ because A is closed. This proves that AP is closed and hence, by the closed graph theorem in Hilbert space (cf. 12.2.3), that AP is bounded. Since A is the same as the restriction of AP to D_A (cf. 13.1.3c), A is bounded as well. $\qquad\square$

13.1.10 Proposition. *Let $\{u_i\}_{i\in I}$ be an o.n.s. in $\mathcal{H}$. If we write $M := V\{u_i\}_{i\in I}$, then*
$$P_M f = \sum_{i\in I} (u_i|f)\, u_i, \forall f \in \mathcal{H}.$$

Proof. From 10.6.1 we have
$$\delta_M(f) = \left(\sum_{i\in I} (u_i|f)\, u_i, f - \sum_{i\in I} (u_i|f)\, u_i\right), \forall f \in \mathcal{H}.$$
$\qquad\square$

13.1.11 Corollary. *If $\mathcal{H}$ is separable then for every $A \in \mathscr{P}(\mathcal{H}) - \{\mathbb{O}_\mathcal{H}\}$ there exists a countable o.n.s. $\{u_i\}_{i\in I}$ in $\mathcal{H}$ so that*
$$Af = \sum_{i\in I} (u_i|f)\, u_i, \forall f \in \mathcal{H}.$$

Proof. This follows immediately from 10.7.2 and 13.1.10. $\qquad\square$

13.1.12 Definition. For $u \in \tilde{\mathcal{H}}$ (for $\tilde{\mathcal{H}}$, cf. 10.9.4) we write $A_u := P_{V\{u\}}$. In view of 13.1.10, we have
$$D_{A_u} = \mathcal{H} \text{ and } A_u f = (u|f)\, u, \forall f \in \mathcal{H}.$$
The operator A_u is said to be a *one-dimensional projection in $\mathcal{H}$*.

13.1.13 Remarks.

(a) If $u, v \in \tilde{\mathcal{H}}$ are such that $u \div v$ (for the relation $\div$ in $\mathcal{H}$, cf. 10.9.1) then there exists $z \in \mathbb{T}$ such that $u = zv$, and hence
$$A_u f = (zv|f)\, zv = \bar{z}z\, (v|f)\, v = (v|f)\, v = A_v f, \forall f \in \mathcal{H},$$
i.e. $A_u = A_v$. Conversely, if $u, v \in \tilde{\mathcal{H}}$ are such that $A_u = A_v$ then in particular
$$u = A_u u = A_v u = (v|u)\, v,$$
and hence $u \div v$. Therefore, the mapping
$$\hat{\mathcal{H}} \ni [u] \mapsto A_u \in \mathscr{P}(\mathcal{H})$$
(for $\hat{\mathcal{H}}$, cf. 10.9.4) can be defined consistently and it is injective; hence, it is a bjiection from $\hat{\mathcal{H}}$ onto the family of all one-dimensional projections in $\mathcal{H}$.

(b) If $\mathcal{H}_1$ and $\mathcal{H}_2$ are isomorphic Hilbert spaces, for $u \in \tilde{\mathcal{H}}_1$ and $U \in \mathcal{UA}(\mathcal{H}_1, \mathcal{H}_2)$ we have $UA_u U^{-1} = A_{Uu}$. Indeed,
$$UA_u U^{-1} f = U\left(u|U^{-1}f\right) u = U\,(Uu|f)\, u = (Uu|f)\, Uu = A_{Uu} f, \forall f \in \mathcal{H},$$
if U is unitary, and
$$UA_u U^{-1} f = U\left(u|U^{-1}f\right) u = U\,(f|Uu)\, u = (Uu|f)\, Uu = A_{Uu} f, \forall f \in \mathcal{H},$$
if U is antiunitary.

13.2 Orthogonal projections and subspaces

In this section we examine some conditions under which orthogonal projections can be constructed out of other orthogonal projections. The bijection between $\mathscr{S}(\mathcal{H})$ and $\mathscr{P}(\mathcal{H})$ examined in 13.1.4b translates relations between subspaces into relations between orthogonal projections. Examples of this can be found in 13.2.1, 13.2.4, 13.2.8, 13.2.9.

13.2.1 Theorem. *For $M, N \in \mathscr{S}(\mathcal{H})$, the following conditions are equivalent:*

(a) $P_N P_M \in \mathscr{P}(\mathcal{H})$;
(b) $P_N P_M = P_M P_N$;
(c) $(M \cap N)^\perp = (M \cap N^\perp) + (M^\perp \cap N) + (M^\perp \cap N^\perp)$;
(d) $M = (M \cap N) + (M \cap N^\perp)$.

Since condition b remains the same if M and N are interchanged, the above conditions are also equivalent to conditions a and d with M and N interchanged (condition c remains the same if M and N are interchanged in it).

Since condition b is obviously equivalent to each of the following conditions

$$P_N(\mathbb{1}_\mathcal{H} - P_M) = (\mathbb{1}_\mathcal{H} - P_M)P_N,$$
$$(\mathbb{1}_\mathcal{H} - P_N)P_M = P_M(\mathbb{1}_\mathcal{H} - P_N),$$
$$(\mathbb{1}_\mathcal{H} - P_N)(\mathbb{1}_\mathcal{H} - P_M) = (\mathbb{1}_\mathcal{H} - P_M)(\mathbb{1}_\mathcal{H} - P_N),$$

and since $\mathbb{1}_\mathcal{H} - P_M = P_{M^\perp}$ and $\mathbb{1}_\mathcal{H} - P_N = P_{N^\perp}$ (cf. 13.1.3e), conditions a, b, c, d are equivalent to the same conditions with either M or N or both M and N replaced by $M^\perp$ and $N^\perp$ respectively.

If the above conditions are satisfied, then

(e) $P_N P_M = P_{M \cap N}$*, or equivalently $R_{P_N P_M} = M \cap N$*

(recall that $M \cap N \in \mathscr{S}(\mathcal{H})$, cf. 4.1.10).

In view of the remark above, the equations also hold true that are obtained from the equations in e by replacing either M or N or both M and N with $M^\perp$ and $N^\perp$ respectively.

Proof. $a \Rightarrow b$: Assuming condition a we have

$$P_N P_M = (P_N P_M)^\dagger = P_M^\dagger P_N^\dagger = P_M P_N$$

(cf. 13.1.5 and 12.3.4b).

$b \Rightarrow c$: Assuming condition b, in view of 13.1.3c we have

$$P_M P_N f \in M \cap N, \forall f \in \mathcal{H}.$$

Assuming condition b, we also have $P_M P_{N^\perp} = P_{N^\perp} P_M$ (cf. remark in the statement) and hence

$$P_M P_{N^\perp} f \in M \cap N^\perp, \forall f \in \mathcal{H}.$$

Then,

$$f \in M \Rightarrow f = P_M f = P_M P_N f + P_M(\mathbb{1}_{\mathcal{H}} - P_N)f$$
$$= P_M P_N f + P_M P_{N^\perp} f \in (M \cap N) + (M \cap N^\perp).$$

Since the inclusion $(M \cap N) + (M \cap N^\perp) \subset M$ is obvious, this proves the equation

$$M = (M \cap N) + (M \cap N^\perp).$$

Assuming condition b, we also have $P_N P_{M^\perp} = P_{M^\perp} P_N$ (cf. the second remark in the statement), and this implies (proceeding as above) the equation

$$M^\perp = (M^\perp \cap N) + (M^\perp \cap N^\perp).$$

In view of 10.4.2a, this proves the equation

$$\mathcal{H} = (M \cap N) + (M \cap N^\perp) + (M^\perp \cap N) + (M^\perp \cap N^\perp). \tag{1}$$

Now, in view of 10.2.10b and 10.2.13, we have

$$M \cap N^\perp \subset N^\perp \subset (M \cap N)^\perp \text{ and } (M^\perp \cap N) + (M^\perp \cap N^\perp) \subset M^\perp \subset (M \cap N)^\perp,$$

and hence

$$(M \cap N^\perp) + (M^\perp \cap N) + (M^\perp \cap N^\perp) \subset (M \cap N)^\perp. \tag{2}$$

In view of 10.2.15, 1 and 2 imply condition c.

$c \Rightarrow d$: Assuming condition c, by 10.4.2a we have

$$\mathcal{H} = (M \cap N) + (M \cap N^\perp) + (M^\perp \cap N) + (M^\perp \cap N^\perp),$$

and hence

$$\mathcal{H} = (M \cap N) + (M \cap N^\perp) + M^\perp \tag{3}$$

since $(M^\perp \cap N) + (M^\perp \cap N^\perp) \subset M^\perp$. Now,

$$(M \cap N) + (M \cap N^\perp) \subset M = (M^\perp)^\perp \tag{4}$$

(cf. 10.4.4a). In view of 10.2.15, 3 and 4 imply that

$$(M \cap N) + (M \cap N^\perp) = (M^\perp)^\perp = M.$$

$d \Rightarrow (a \text{ and } e)$: Assuming condition d we have that

$$\forall g \in M, \exists!(g_1, g_2) \in (M \cap N) \times (M \cap N^\perp) \text{ so that } g = g_1 + g_2$$

(the uniqueness of (g_1, g_2) as above follows from $g_1 \in N$ and $g_2 \in N^\perp$, cf. 10.4.1). This and 10.4.1 imply that, for every $f \in \mathcal{H}$,

$$\exists!(f_1, f_2, f_3) \in (M \cap N) \times (M \cap N^\perp) \times M^\perp \text{ so that } f = f_1 + f_2 + f_3;$$

then we have:

$$\delta_{M \cap N}(f) = (f_1, f_2 + f_3)$$

since $f_2 \in M \cap N^\perp \subset N^\perp \subset (M \cap N)^\perp$ and $f_3 \in M^\perp \subset (M \cap N)^\perp$, and hence $f_2 + f_3 \in (M \cap N)^\perp$;

$$\delta_M(f) = (f_1 + f_2, f_3)$$

since $f_1, f_2 \in M$ and hence $f_1 + f_2 \in M$;

$$\delta_N(f_1 + f_2) = (f_1, f_2);$$

therefore we have

$$P_{M \cap N} f = f_1 = P_N(f_1 + f_2) = P_N P_M f.$$

This proves the equation $P_N P_M = P_{M \cap N}$, which obviously implies condition a.

Finally, the equation $P_N P_M = P_{M \cap N}$ is equivalent to the equation $R_{P_N P_M} = M \cap N$, in view of 13.1.3c and 13.1.4a. □

For any $M, N \in \mathscr{S}(\mathcal{H})$ we have $M \cap N \in \mathscr{S}(\mathcal{H})$ (cf. 4.1.10) and hence we can consider the orthogonal projection $P_{M \cap N}$. If $P_N P_M = P_M P_N$ then 13.2.1 proves that $P_{M \cap N} = P_N P_M$. The next theorem shows how $P_{M \cap N}$ can be obtained from P_M and P_N in the general case. In its proof, we follow von Neumann faithfully (Neumann, 1950).

13.2.2 Theorem. *For $M, N \in \mathscr{S}(\mathcal{H})$, let $A_1 := P_M$, and let*

$$A_{2k} := (P_N P_M)^k, \ A_{2k+1} := P_M(P_N P_M)^k, \ \forall k \in \mathbb{N}.$$

Then, for all $f \in \mathcal{H}$, the sequence $\{A_n f\}$ is convergent and

$$P_{M \cap N} f = \lim_{n \to \infty} A_n f,$$

and the sequence $\{(P_N P_M)^n f\}$ is also convergent and

$$P_{M \cap N} f = \lim_{n \to \infty} (P_N P_M)^n f.$$

Proof. For all $k, h \in \mathbb{N}$ and all $f \in \mathcal{H}$ we have, in view of 13.1.5:

$$(A_{2k} f | A_{2h} f) = ((P_M P_N)^h (P_N P_M)^k f | f) = (A_{2k+2h-1} f | f),$$
$$(A_{2k+1} f | A_{2h+1} f) = ((P_M P_N)^h P_M (P_N P_M)^k f | f) = (A_{(2k+1)+(2h+1)-1} f | f),$$
$$(A_{2k} f | A_{2h+1} f) = ((P_M P_N)^h P_M (P_N P_M)^k f | f) = (A_{2k+(2h+1)} f | f),$$
$$(A_{2k+1} f | A_{2h} f) = ((P_M P_N)^h P_M (P_N P_M)^k f | f) = (A_{(2k+1)+2h} f | f),$$

since

$$(P_M P_N)^h (P_N P_M)^k = P_M (P_N P_M)^{h-1} P_N (P_N P_M)^k$$
$$= P_M (P_N P_M)^{k+h-1} = A_{2k+2h-1}$$

and

$$(P_M P_N)^h P_M (P_N P_M)^k = P_M (P_N P_M)^h (P_N P_M)^k = A_{2k+2h+1}.$$

This proves that, for all $m, n \in \mathbb{N}$ and all $f \in \mathcal{H}$,

$$(A_m f | A_n f) = (A_{m+n-s} f | f),$$

with $s = 1$ if m and n have the same parity and $s = 0$ if m and n have different parity, and hence that

$$\|A_m f - A_n f\|^2 = (A_m f | A_m f) + (A_n f | A_n f) - (A_m f | A_n f) - (A_n f | A_m f)$$
$$= (A_{2m-1} f | f) + (A_{2n-1} f | f) - 2 (A_{m+n-s} f | f) \tag{5}$$
$$= (A_{2m-1} f | f) + (A_{2n-1} f | f) - 2 (A_{2k_{m,n}-1} f | f),$$

with $k_{m,n} \in \mathbb{N}$ such that $2k_{m,n} - 1 = m + n - s$ (note that $m + n - s$ is always odd).

Moreover, for all $i \in \mathbb{N}$ and all $f \in \mathcal{H}$, we have

$$(A_{2i-1} f | f) = (A_i f | A_i f) = \|A_i f\|^2 \text{ and}$$
$$(A_{2i+1} f | f) = (A_{i+1} f | A_{i+1} f) = \|A_{i+1} f\|^2;$$

now, $A_{i+1} f = P_M A_i f$ if i is even and $A_{i+1} f = P_N A_i f$ if i is odd; therefore (cf. 13.1.3d or 13.1.6), in any case,

$$\|A_{i+1} f\| \le \|A_i f\|.$$

This proves that, for every $f \in \mathcal{H}$, the sequence of non-negative real numbers $\{(A_{2i-1} f | f)\}$ is monotone non-increasing, and hence that it is convergent, and hence (cf. 2.6.2) that

$$\forall \varepsilon > 0, \exists N_\varepsilon \in \mathbb{N} \text{ s.t. } N_\varepsilon < i, j \Rightarrow |(A_{2i-1} f | f) - (A_{2j-1} f | f)| < \varepsilon;$$

from this and from 5 we have that

$$\forall \varepsilon > 0, \exists N_\varepsilon \in \mathbb{N} \text{ s.t. } N_\varepsilon < m, n \Rightarrow$$
$$\|A_m f - A_n f\|^2 \le |(A_{2m-1} f | f) - (A_{2k_{m,n}-1} f | f)| +$$
$$|(A_{2n-1} f | f) - (A_{2k_{m,n}-1} f | f)| < 2\varepsilon$$

(note that $N_\varepsilon < m, n$ implies $N_\varepsilon < k_{m,n}$); since $\mathcal{H}$ is a complete metric space, this proves that the sequence $\{A_n f\}$ is convergent.

Thus, we can define the mapping

$$A : \mathcal{H} \to \mathcal{H}$$
$$f \mapsto Af := \lim_{n \to \infty} A_n f.$$

It is easy to see that the mapping A is a linear operator by the continuity of vector sum and of scalar multiplication. Further, we have

$$(Af | Af) = \lim_{n \to \infty} (A_n f | A_n f) = \lim_{n \to \infty} (A_{2n-1} f | f)$$
$$= \left(\lim_{n \to \infty} A_{2n-1} f | f \right) = (Af | f), \forall f \in \mathcal{H},$$

by the continuity of inner product and by 2.1.7b (in relation to the subsequence $\{A_{2n-1} f\}$ of the sequence $\{A_n f\}$). In view of 12.4.3, this proves that the operator A is symmetric, and hence self-adjoint since $D_A = \mathcal{H}$. Then, from the equation above we also have

$$(A^2 f | f) = (Af | f), \forall f \in \mathcal{H},$$

which proves the equation $A^2 = A$, in view of 10.2.12. Thus (cf. 13.1.5), the operator A is an orthogonal projection.

Now, in view of 13.1.3c we have

$$f \in M \cap N \Rightarrow [P_M f = f \text{ and } P_N f = f] \Rightarrow$$
$$[A_n f = f, \forall n \in \mathbb{N}] \Rightarrow Af = f \Rightarrow f \in R_A,$$

and conversely

$$f \in R_A \Rightarrow f = Af = \lim_{n\to\infty} A_{2n} f \Rightarrow$$
$$P_N f = \lim_{n\to\infty} P_N A_{2n} f = \lim_{n\to\infty} A_{2n} f = f \Rightarrow f \in N$$

as well as

$$f \in R_A \Rightarrow f = Af = \lim_{n\to\infty} A_{2n+1} f \Rightarrow$$
$$P_M f = \lim_{n\to\infty} P_M A_{2n+1} f = \lim_{n\to\infty} A_{2n+1} f = f \Rightarrow f \in M$$

(we have used 2.1.7b in relation to the subsequences $\{A_{2n}f\}$ and $\{A_{2n+1}f\}$ of the sequence $\{A_n f\}$). This proves that $R_A = M \cap N$, and hence that $A = P_{M \cap N}$ (cf. 13.1.4a).

Finally we have, as already noted,

$$P_{M \cap N} f = Af = \lim_{n\to\infty} A_{2n} f = \lim_{n\to\infty} (P_N P_M)^n f, \forall f \in \mathcal{H}.$$

$\square$

13.2.3 Remark. In 13.2.2, if $P_N P_M = P_M P_N$ then $A_{2k} = A_{2k+1} = P_N P_M$ for all $k \in \mathbb{N}$ (cf. 13.1.5 or 13.1.6) and we have $P_{M \cap N} = P_N P_M$, as already proved in 13.2.1.

13.2.4 Theorem. *For $M, N \in \mathscr{S}(\mathcal{H})$, the following conditions are equivalent:*

(a) $P_M - P_N \in \mathscr{P}(\mathcal{H})$;
(b) $P_M P_N = P_N$;
(c) $P_N P_M = P_N$;
(d) $(f|P_N f) \leq (f|P_M f)$, $\forall f \in \mathcal{H}$;
(e) $N \subset M$.

If the above conditions are satisfied, then

(f) $P_M - P_N = P_{M \cap N^\perp}$, *or equivalently* $R_{P_M - P_N} = M \cap N^\perp$.

Proof. $a \Rightarrow b$: Assuming condition a, in view of 13.1.5 or 13.1.6 we have

$$P_M - P_N = (P_M - P_N)^2 = P_M - P_M P_N - P_N P_M + P_N,$$

and hence

$$P_M P_N + P_N P_M = 2P_N, \tag{6}$$

whence

$$P_M P_N + P_M P_N P_M = 2P_M P_N \text{ and } P_M P_N P_M + P_N P_M = 2P_N P_M,$$

whence

$$P_M P_N = P_N P_M.$$

Substituting $P_M P_N$ for $P_N P_M$ in 6, we obtain $P_M P_N = P_N$.

$b \Rightarrow c$: In view of 13.1.5 and 12.3.4b, condition b implies

$$P_N = P_N^\dagger = (P_M P_N)^\dagger = P_N^\dagger P_M^\dagger = P_N P_M.$$

$c \Rightarrow d$: Assuming condition c, in view of 13.1.3d or 13.1.6 we have

$$\|P_N f\| = \|P_N P_M f\| \le \|P_M f\|, \forall f \in \mathcal{H},$$

and hence condition d by 13.1.7c.

$d \Rightarrow e$: Assuming condition d, in view of 13.1.3c and 13.1.7c we have

$$f \in N \Rightarrow \|f\| = \|P_N f\| \le \|P_M f\| \Rightarrow \|f\| = \|P_M f\| \Rightarrow f \in M,$$

where the second implication is true because $\|P_M f\| \le \|f\|$ for all $f \in \mathcal{H}$ (cf. 13.1.3d or 13.1.6).

$e \Rightarrow (a \text{ and } f)$: Assuming condition e, 10.4.3 implies that

$$\forall g \in M, \exists!(g_1, g_2) \in N \times (M \cap N^\perp) \text{ so that } g = g_1 + g_2.$$

This and 10.4.1 imply that, for every $f \in \mathcal{H}$,

$$\exists!(f_1, f_2, f_3) \in N \times (M \cap N^\perp) \times M^\perp \text{ so that } f = f_1 + f_2 + f_3;$$

then we have:

$$\delta_{M \cap N^\perp}(f) = (f_2, f_1 + f_3)$$

since $f_1 \in N = N^{\perp\perp} \subset (M \cap N^\perp)^\perp$ and $f_3 \in M^\perp \subset (M \cap N^\perp)^\perp$, and hence $f_1 + f_3 \in (M \cap N^\perp)^\perp$;

$$\delta_N(f) = (f_1, f_2 + f_3)$$

since $f_2 \in M \cap N^\perp \subset N^\perp$ and $f_3 \in M^\perp \subset N^\perp$, and hence $f_2 + f_3 \in N^\perp$;

$$\delta_M(f) = (f_1 + f_2, f_3)$$

since $f_1 \in N \subset M$ and $f_2 \in M \cap N^\perp \subset M$, and hence $f_1 + f_2 \in M$; therefore we have

$$P_{M \cap N^\perp} f = f_2 = (f_1 + f_2) - f_1 = P_M f - P_N f = (P_M - P_N)f.$$

This proves the equation $P_M - P_N = P_{M \cap N^\perp}$, which obviously implies condition a.

Finally, the equation $P_M - P_N = P_{M \cap N^\perp}$ is equivalent to the equation $R_{P_M - P_N} = M \cap N^\perp$, in view of 13.1.3c and 13.1.4a. $\square$

13.2.5 Definition. The family $\mathscr{S}(\mathcal{H})$ of all subspaces of $\mathcal{H}$ is obviously a partially ordered set (cf. 1.1.5) with set inclusion as partial ordering, i.e. with the relation $\leq$ in $\mathscr{S}(\mathcal{H})$ defined as follows:

$$\text{for } N, M \in \mathscr{S}(\mathcal{H}), N \leq M \text{ if } N \subset M.$$

The l.u.b. and the g.l.b. exist for every family $\{M_i\}_{i \in I}$ of elements of $\mathscr{S}(\mathcal{H})$, and they are

$$\sup\{M_i\}_{i \in I} = V\left(\bigcup_{i \in I} M_i\right) \text{ and } \inf\{M_i\}_{i \in I} = \bigcap_{i \in I} M_i$$

(the first equation is proved by 4.1.11a,b,c). In view of the bijection existing between $\mathscr{S}(\mathcal{H})$ and $\mathscr{P}(\mathcal{H})$ (cf. 13.1.4b), we can obviously define a partial ordering in $\mathscr{P}(\mathcal{H})$ as follows:

$$\text{for } P, Q \in \mathscr{P}(\mathcal{H}), P \leq Q \text{ if } R_P \subset R_Q.$$

In view of 13.2.4, for $P, Q \in \mathscr{P}(\mathcal{H})$ we have

$$P \leq Q \Leftrightarrow PQ = P \Leftrightarrow [(f|Pf) \leq (f|Qf), \forall f \in \mathcal{H}].$$

Clearly, for every family $\{P_i\}_{i \in I}$ of elements of $\mathscr{P}(\mathcal{H})$,

$$\sup\{P_i\}_{i \in I} = P_{\hat{M}} \text{ if } \hat{M} := V\left(\bigcup_{i \in I} R_{P_i}\right)$$

and

$$\inf\{P_i\}_{i \in I} = P_{\check{M}} \text{ if } \check{M} := \bigcap_{i \in I} R_{P_i}.$$

13.2.6 Remark. Let $\{P_i\}_{i \in I}$ be an arbitrary family of elements of $\mathscr{P}(\mathcal{H})$. Since $\inf\{P_i\}_{i \in I} \leq P_k$ for all $k \in I$, we have

$$(f|(\inf\{P_i\}_{i \in I})f) \leq (f|P_k f), \forall k \in I, \forall f \in \mathcal{H}.$$

However, the equations

$$(f|(\inf\{P_i\}_{i \in I})f) = \inf\{(f|P_i f)\}_{i \in I}, \forall f \in \mathcal{H}, \tag{7}$$

need not hold true, and indeed they do not in general, not even when the elements of the family are so that $P_i P_j = P_j P_i$ for all $i, j \in I$, as is shown by the family of one-dimensional projections $\{A_{u_1}, A_{u_2}\}$ with $\{u_1, u_2\}$ an o.n.s. in $\mathcal{H}$; in fact, $\inf\{A_{u_1}, A_{u_2}\} = \mathbb{O}_{\mathcal{H}}$, while for the vector $f := u_1 + u_2$ we have $(f|A_{u_i}f) = 1$ for $i = 1, 2$. In 13.2.7 we prove that statement 7 is true if the family $\{P_i\}_{i \in I}$ is closed under multiplication, i.e. if the product $P_i P_j$ belongs to the family for all $i, j \in I$. This result is important in the theory of projection valued measures (it is used in the proof of 13.4.2).

13.2.7 Theorem. *Let $\{P_i\}_{i \in I}$ be a family of elements of $\mathscr{P}(\mathcal{H})$ and suppose that*

$$\forall i, j \in I, \exists k \in I \text{ such that } P_i P_j = P_k.$$

Then,

$$\exists! P \in \mathscr{P}(\mathcal{H}) \text{ so that } (f|Pf) = \inf\{(f|P_i f)\}_{i \in I}, \forall f \in \mathcal{H}.$$

This unique orthogonal projection P is the orthogonal projection $\inf\{P_i\}_{i \in I}$.

Proof. We write $P := \inf\{P_i\}_{i \in I}$. In what follows we prove that

$$(f|Pf) = \inf\{(f|P_i f)\}_{i \in I}, \forall f \in \mathcal{H}.$$

The uniqueness asserted in the statement then follows immediately from 10.2.12.

We define

$$Q_i := P_i - P, \forall i \in I,$$

and we observe that $Q_i \in \mathscr{P}(\mathcal{H})$ by 13.2.4. Moreover,

$$Q_i Q_j = (P_i - P)(P_j - P) = P_i P_j - P - P + P = P_i P_j - P, \forall i, j \in I,$$

and hence, by the condition assumed in the statement,

$$\forall i, j \in I, \exists k \in I \text{ such that } Q_i Q_j = Q_k$$

(we note that this implies $Q_i Q_j = Q_j Q_i$ for all $i, j \in I$, by 13.2.1). By induction, this implies that

$$\forall n \in \mathbb{N}, \forall (i_1, ..., i_n) \in I^n, \exists k \in I \text{ such that } Q_{i_1} \cdots Q_{i_n} = Q_k,$$

and hence, if we write $M_i := R_{Q_i}$ for all $i \in I$, that

$$\forall n \in \mathbb{N}, \forall (i_1, ..., i_n) \in I^n, \exists k \in I \text{ such that } M_{i_1} \cap \cdots \cap M_{i_n} = M_k$$

(we have used 13.1.4b and induction applied to 13.2.1e), and hence that

$$\forall n \in \mathbb{N}, \forall (i_1, ..., i_n) \in I^n, \exists k \in I \text{ such that } V(M_{i_1}^{\perp} \cup \cdots \cup M_{i_n}^{\perp}) = M_k^{\perp};$$

in fact, in view of 10.4.4b, 10.2.10c, 10.4.4a we have

$$V(M_{i_1}^{\perp} \cup \cdots \cup M_{i_n}^{\perp}) = (M_{i_1}^{\perp} \cup \cdots \cup M_{i_n}^{\perp})^{\perp\perp}$$
$$= (M_{i_1}^{\perp\perp} \cap \cdots \cap M_{i_n}^{\perp\perp})^{\perp} = (M_{i_1} \cap \cdots \cap M_{i_n})^{\perp}.$$

Then, in view of 3.1.7 and 4.1.13,

$$f \in L\left(\bigcup_{i \in I} M_i^{\perp}\right) \Rightarrow$$

$$[\exists n \in \mathbb{N}, \exists (i_1, ..., i_n) \in I^n \text{ such that } f \in L(M_{i_1}^{\perp} \cup \cdots \cup M_{i_n}^{\perp})] \Rightarrow$$
$$[\exists k \in I \text{ such that } f \in M_k^{\perp}].$$

This proves the inclusion

$$L\left(\bigcup_{i \in I} M_i^{\perp}\right) \subset \bigcup_{i \in I} M_i^{\perp},$$

which is obviously equivalent to the equation

$$L\left(\bigcup_{i\in I} M_i^\perp\right) = \bigcup_{i\in I} M_i^\perp. \tag{8}$$

Moreover, in view of 13.2.4f we have

$$M_i = R_{P_i} \cap R_P^\perp,$$

and hence

$$\bigcap_{i\in I} M_i = \left(\bigcap_{i\in I} R_{P_i}\right) \cap R_P^\perp = \left(\bigcap_{i\in I} R_{P_i}\right) \cap \left(\bigcap_{i\in I} R_{P_i}\right)^\perp = \{0_{\mathcal{H}}\},$$

and hence (proceeding as above)

$$V\left(\bigcup_{i\in I} M_i^\perp\right) = \left(\bigcup_{i\in I} M_i^\perp\right)^{\perp\perp} = \left(\bigcap_{i\in I} M_i^{\perp\perp}\right)^\perp$$

$$= \left(\bigcap_{i\in I} M_i\right)^\perp = \{0_{\mathcal{H}}\}^\perp = \mathcal{H}. \tag{9}$$

In view of 4.1.13, 8 and 9 imply the equation

$$\overline{\left(\bigcup_{i\in I} M_i^\perp\right)} = \mathcal{H}.$$

Since $M_i^\perp = R_{\mathbb{1}_{\mathcal{H}} - Q_i}$ (cf. 13.1.3e), in view of 2.3.12 this statement can be written as

$$\forall f \in \mathcal{H}, \forall \varepsilon > 0, \exists i_\varepsilon \in I, \exists g_\varepsilon \in \mathcal{H} \text{ such that } \|f - (\mathbb{1}_{\mathcal{H}} - Q_{i_\varepsilon})g_\varepsilon\| < \varepsilon. \tag{10}$$

Now, for every $A \in \mathscr{P}(\mathcal{H})$, in view of 10.2.3 we have

$$\|f - Af\|^2 \le \|f - Af\|^2 + \|Af - Ag\|^2 = \|f - Af + Af - Ag\|^2$$
$$= \|f - Ag\|^2, \forall f, g \in \mathcal{H};$$

in fact, in view of 13.1.5c,e we have

$$(f - Af|Af - Ag) = 0, \forall f, g \in \mathcal{H}.$$

Thus, 10 implies that

$$\forall f \in \mathcal{H}, \forall \varepsilon > 0, \exists i_\varepsilon \in I \text{ such that } \|Q_{i_\varepsilon} f\| = \|f - (\mathbb{1}_{\mathcal{H}} - Q_{i_\varepsilon})f\| < \varepsilon,$$

or (cf. 13.1.7c)

$$\forall f \in \mathcal{H}, \forall \varepsilon > 0, \exists i_\varepsilon \in I \text{ such that } (f|Q_{i_\varepsilon} f) < \varepsilon^2,$$

or

$$\forall f \in \mathcal{H}, \forall \varepsilon > 0, \exists i_\varepsilon \in I \text{ such that } (f|P_{i_\varepsilon} f) < \varepsilon^2 + (f|Pf). \tag{11}$$

On the other hand we have (as already noted in 13.2.6)

$$(f|Pf) \le (f|P_i f), \forall i \in I, \forall f \in \mathcal{H}. \tag{12}$$

Now, 11 and 12 prove that

$$(f|Pf) = \inf\{(f|P_i f)\}_{i\in I}, \forall f \in \mathcal{H}.$$

$\square$

13.2.8 Theorem. *For a sequence $\{M_n\}$ in $\mathscr{S}(\mathcal{H})$, the following conditions are equivalent (we write $P_n := P_{M_n}, \forall n \in \mathbb{N}$):*

(a) the series $\sum_{n=1}^{\infty} P_n f$ is convergent for all $f \in \mathcal{H}$ and the mapping

$$P : \mathcal{H} \to \mathcal{H}$$

$$f \mapsto Pf := \sum_{n=1}^{\infty} P_n f$$

is an orthogonal projection;
(b) $\sum_{n=1}^{\infty} \|P_n f\|^2 \le \|f\|^2, \forall f \in \mathcal{H}$;
(c) $P_i P_k = \mathbb{O}_{\mathcal{H}}$ if $i \ne k$;
(d) $M_k \subset M_i^{\perp}$ if $i \ne k$.

If the above conditions are satisfied, the subset of $\mathcal{H}$ defined by

$$\left\{ f \in \mathcal{H} : \text{there exists a sequence } \{f_n\} \in \bigcup_{n=1}^{\infty} M_n \text{ such that} \right.$$

$$\left. f_n \in M_n \text{ for all } n \in \mathbb{N}, \sum_{n=1}^{\infty} f_n \text{ is convergent}, f = \sum_{n=1}^{\infty} f_n \right\}$$

is called the orthogonal sum of the sequence of subspaces $\{M_n\}$ and is denoted by the symbol $\sum_{n=1}^{\infty \oplus} M_n$.
  If the above conditions are satisfied, then:

(e) $R_P = \sum_{n=1}^{\infty \oplus} M_n = V(\cup_{n=1}^{\infty} M_n)$, and hence $\sum_{n=1}^{\infty \oplus} M_n \in \mathscr{S}(\mathcal{H})$;
(f) if β is a bijection from $\mathbb{N}$ onto $\mathbb{N}$ then $\sum_{n=1}^{\infty} P_{\beta(n)} f = Pf, \forall f \in \mathcal{H}$.

Proof. $a \Rightarrow b$: Assuming condition a, we have

$$\sum_{n=1}^{\infty} \|P_n f\|^2 = \sum_{n=1}^{\infty} (f|P_n f) = \left(f \left| \sum_{n=1}^{\infty} P_n f \right. \right)$$

$$= (f|Pf) = \|Pf\|^2 \le \|f\|^2, \forall f \in \mathcal{H},$$

in view of 13.1.7c, of the continuity of inner product, and of 13.1.3d or 13.1.6.
  $b \Rightarrow c$: Assuming condition b, for $i, k \in \mathbb{N}$ such that $i \ne k$ we have

$$\|P_i P_k g\|^2 + \|P_k P_k g\|^2 \le \sum_{n=1}^{\infty} \|P_n (P_k g)\|^2 \le \|P_k g\|^2, \forall g \in \mathcal{H},$$

and hence, since $P_k^2 = P_k$,

$$\|P_i P_k g\|^2 \le 0, \forall g \in \mathcal{H},$$

and hence

$$P_i P_k g = 0_{\mathcal{H}}, \forall g \in \mathcal{H}.$$

  $c \Rightarrow d$: From $P_i P_k = \mathbb{O}_{\mathcal{H}}$ we obtain $R_{P_k} \subset N_{P_i}$, i.e. $M_k \subset M_i^{\perp}$ in view of 13.1.3b,c.

$d \Rightarrow (a, e, f)$: We assume condition d.

For each $f \in \mathcal{H}$, we have

$$(P_i f | P_k f) = 0 \text{ if } i \neq k$$

since $P_n f \in M_n$ for all $n \in \mathbb{N}$ (cf. 13.1.3c); then, if we define

$$I_f := \{n \in \mathbb{N} : P_n f \neq 0_{\mathcal{H}}\},$$

the family of vectors $\left\{ \frac{1}{\|P_n f\|} P_n f \right\}_{n \in I_f}$ is an o.n.s. in $\mathcal{H}$; in view of 13.1.7c we have

$$\left(\frac{1}{\|P_n f\|} P_n f \middle| f \right) = \frac{1}{\|P_n f\|} \|P_n f\|^2 = \|P_n f\|, \forall n \in I_f;$$

therefore,

$$\sum_{n=1}^{\infty} \|P_n f\|^2 = \sum_{n \in I_f} \|P_n f\|^2 = \sum_{n \in I_f} \left| \left(\frac{1}{\|P_n f\|} P_n f \middle| f \right) \right|^2 < \infty$$

by 10.2.8b; in view of 10.4.7b, this proves that the series $\sum_{n=1}^{\infty} P_n f$ is convergent. It is obvious that the mapping P is a linear operator, owing to the continuity of vector sum and of scalar multiplication. Now, for the mapping P we have

$$(f | P f) = \sum_{n=1}^{\infty} (f | P_n f) = \sum_{n=1}^{\infty} \|P_n f\|^2 \in \mathbb{R}, \forall f \in \mathbb{R}$$

(cf. 13.1.7c). By 12.4.3, this proves that the operator P is symmetric, and hence self-adjoint since $D_P = \mathcal{H}$. We also have

$$(P f | P f) = \sum_{i=1}^{\infty} \sum_{k=1}^{\infty} (P_i f | P_k f) = \sum_{i=1}^{\infty} (P_i f | P_i f)$$

$$= \sum_{i=1}^{\infty} \|P_i f\|^2 = (f | P f), \forall f \in \mathcal{H},$$

and hence

$$(f | P^2 f) = (f | P f), \forall f \in \mathcal{H},$$

which implies $P^2 = P$ by 10.2.12. Thus, P is an orthogonal projection by 13.1.5, and condition a is proved.

As to the equation in e, the inclusion $R_P \subset \sum_{n=1}^{\infty \oplus} M_n$ is obvious since $R_{P_n} = M_n$ for all $n \in \mathbb{N}$ (cf. 13.1.3c). And conversely, if $f \in \sum_{n=1}^{\infty \oplus} M_n$ then $f = \sum_{n=1}^{\infty} f_n$ with $f_n \in M_n$ for all $n \in \mathbb{N}$ and hence

$$P f = \sum_{i=1}^{\infty} P_i \left(\sum_{k=1}^{\infty} f_k \right) = \sum_{i=1}^{\infty} \sum_{k=1}^{\infty} P_i f_k = \sum_{i=1}^{\infty} P_i f_i = \sum_{i=1}^{\infty} f_i = f,$$

since $f_k \in M_k \subset M_i^{\perp} = N_{P_i}$ if $i \neq k$, and hence $P_i f_k = 0_{\mathcal{H}}$ if $i \neq k$, while $P_i f_i = f_i$ holds for all $i \in \mathbb{N}$ (cf. 13.1.3b,c). In view of 13.1.3c, this proves the inclusion $\sum_{n=1}^{\infty \oplus} M_n \subset R_P$ and hence the equation $R_P = \sum_{n=1}^{\infty \oplus} M_n$. This equation implies

$\sum_{n=1}^{\infty\oplus} M_n \in \mathscr{S}(\mathcal{H})$ (cf. 13.1.4a). Next, the inclusion $\sum_{n=1}^{\infty\oplus} M_n \subset V\left(\bigcup_{n=1}^{\infty} M_n\right)$ is obvious since $V\left(\bigcup_{n=1}^{\infty} M_n\right)$ is a subspace and it contains $\bigcup_{n=1}^{\infty} M_n$ (also, cf. 2.3.4). On the other hand, the inclusion $M_k \subset \sum_{n=1}^{\infty\oplus} M_n$ is obvious for all $k \in \mathbb{N}$; then the inclusion $V\left(\bigcup_{n=1}^{\infty} M_n\right) \subset \sum_{n=1}^{\infty\oplus} M_n$ follows from 4.1.11c since $\sum_{n=1}^{\infty\oplus} M_n$ is a subspace.

Finally, if β is a bijection from $\mathbb{N}$ onto $\mathbb{N}$, then

$$\text{the series } \sum_{n=1}^{\infty} P_{\beta(n)}f \text{ is convergent and}$$

$$\sum_{n=1}^{\infty} P_{\beta(n)}f = \sum_{n=1}^{\infty} P_n f = Pf, \forall f \in \mathcal{H}$$

by 10.4.9 since $(P_i f | P_k f) = 0$ if $i \neq k$. $\qquad\square$

13.2.9 Corollary. *For a finite family $\{M_1, ..., M_N\}$ of subspaces of $\mathcal{H}$, the following conditions are equivalent (we write $P_n := P_{M_n}, \forall n \in \{1, ..., N\}$):*

(a) $\sum_{n=1}^{N} P_n \in \mathscr{P}(\mathcal{H})$;
(b) $P_i P_k = \mathbb{O}_{\mathcal{H}}$ if $i \neq k$;
(c) $M_k \subset M_i^{\perp}$ if $i \neq k$.

If the above conditions are satisfied, the subset $\sum_{n=1}^{N\oplus} M_n$ of $\mathcal{H}$ defined by

$$\sum_{n=1}^{N\oplus} M_n := M_1 + ... + M_N$$

(cf. 3.1.8) is called the orthogonal sum of the family of subspaces $\{M_1, ..., M_N\}$ and the following equations are true (we write $P := \sum_{n=1}^{N} P_n$):

(d) $R_P = \sum_{n=1}^{N\oplus} M_n = V\left(\bigcup_{n=1}^{N} M_n\right)$, and hence $\sum_{n=1}^{N\oplus} M_n \in \mathscr{S}(\mathcal{H})$.

Proof. Define a sequence $\{M_n\}$ in $\mathscr{S}(\mathcal{H})$ by letting $M_n := \{0_{\mathcal{H}}\}$ for $n > N$. For this sequence, conditions a, b, c are equivalent to conditions a, c, d of 13.2.8 respectively, and the equations in d are the same as the equations in e of 13.2.8 since

$$\sum_{n=1}^{N\oplus} M_n = \sum_{n=1}^{\infty\oplus} M_n \text{ and } V\left(\bigcup_{n=1}^{N} M_n\right) = V\left(\bigcup_{n=1}^{\infty} M_n\right).$$

$\qquad\square$

13.2.10 Remarks.

(a) If the conditions in 13.2.8 are satisfied then

$$\sum_{n=1}^{\infty\oplus} M_n = \left\{ f \in \mathcal{H} : \quad \text{there exists a sequence } \{f_n\} \text{ in } \bigcup_{n=1}^{\infty} M_n \text{ such that} \right.$$

$$\left. f_n \in M_n \text{ for all } n \in \mathbb{N}, \sum_{n=1}^{\infty} \|f_n\|^2 < \infty, f = \sum_{n=1}^{\infty} f_n \right\}.$$

This follows immediately from 10.4.7. If $f \in \sum_{n=1}^{\infty \oplus} M_n$ then the sequence $\{f_n\}$ such that $f_n \in M_n$ and $f = \sum_{n=1}^{\infty} f_n$ is unique. In fact, suppose that $\{g_n\}$ is another sequence such that $g_n \in M_n$ and $f = \sum_{n=1}^{\infty} g_n$. Then,

$$f_k = \sum_{n=1}^{\infty} P_k f_n = P_k \sum_{n=1}^{\infty} f_n = P_k f$$

$$= P_k \sum_{n=1}^{\infty} g_n = \sum_{n=1}^{\infty} P_k g_n = g_k, \forall k \in \mathbb{N},$$

where we have used the continuity of P_k (cf. 13.1.3d) and the equations

$$P_k f_n = \delta_{k,n} f_n \text{ and } P_k g_n = \delta_{k,n} g_n, \forall k, n \in \mathbb{N},$$

which follow from 13.1.3b,c.

Similarly, if the conditions in 13.2.9 are satisfied then, for $f \in \sum_{n=1}^{N \oplus} M_n$, the N-tuple $\{f_1, ..., f_N\}$ such that $f_n \in M_n$ and $f = \sum_{n=1}^{N} f_n$ is unique.

(b) If the conditions in 13.2.8 are satisfied, the orthogonal projection P is called the *series of the sequence of projections* $\{P_n\}$ and is denoted by the symbol $\sum_{n=1}^{\infty} P_n$, i.e. one writes $\sum_{n=1}^{\infty} P_n := P$. However, unless

$$\exists k \in \mathbb{N} \text{ such that } n > k \Rightarrow P_n = \mathbb{O}_{\mathcal{H}}, \tag{13}$$

the series $\sum_{n=1}^{\infty} P_n$ is not convergent in the normed space $\mathcal{B}(\mathcal{H})$. Indeed, if condition 13 does not hold true then

$$\forall m \in \mathbb{N}, \exists n_m \in \mathbb{N} \text{ such that } n_m > m \text{ and } P_{n_m} \neq \mathbb{O}_{\mathcal{H}},$$

and hence

$$\forall m \in \mathbb{N}, \exists n_m \in \mathbb{N} \text{ such that } n_m > m \text{ and}$$

$$\left\| \sum_{k=1}^{n_m} P_k - \sum_{k=1}^{n_m-1} P_k \right\| = \|P_{n_m}\| = 1$$

(cf. 13.1.3d); this implies that the sequence $\{\sum_{k=1}^{n} P_k\}$ is not a Cauchy sequence in the normed space $\mathcal{B}(\mathcal{H})$, and hence that it is not convergent (cf. 2.6.2).

(c) From 13.2.8f we have that the projection $\sum_{i \in I} P_i$ can be defined unambiguously for any countable family $\{P_i\}_{i \in I}$ of projections such that $P_i P_j = \mathbb{O}_{\mathcal{H}}$ if $i \neq j$. Indeed, if I is denumerable, we define

$$\sum_{i \in I} P_i := \sum_{n=1}^{\infty} P_{i(n)}$$

with $\mathbb{N} \ni n \mapsto i(n) \in I$ any bijection from $\mathbb{N}$ onto I.

(d) For $M, N \in \mathscr{S}(\mathcal{H})$, if $M \subset N^{\perp}$ then $M + N \in \mathscr{S}(\mathcal{H})$ (cf. 13.2.9 with $N := 2$). A more general condition for $M + N$ to be a subspace is $P_M P_N = P_N P_M$. In fact, if this condition is true then

$$M = (M \cap N) + (M \cap N^{\perp}) \text{ and } N = (N \cap M) + (N \cap M^{\perp})$$

(cf. 13.2.1) and hence
$$M + N = (M \cap N) + (M \cap N^{\perp}) + (N \cap M^{\perp}).$$
Since the subspaces $M \cap N$, $M \cap N^{\perp}$, $N \cap M^{\perp}$ are obviously orthogonal to each other (cf. 10.2.14), 13.2.9 (with $N := 3$) proves that $M + N$ is a subspace. If $P_M P_N = P_N P_M$ then P_{M+N} can be readily obtained from P_M and P_N. In fact, in view of what was seen above, 13.2.9 implies that
$$P_{M+N} = P_{M \cap N} + P_{M \cap N^{\perp}} + P_{N \cap M^{\perp}},$$
and this equation can be written as
$$P_{M+N} = P_M P_N + P_M (\mathbb{1}_{\mathcal{H}} - P_N) + P_N (\mathbb{1}_{\mathcal{H}} - P_M) = P_M + P_N - P_M P_N,$$
in view of 13.2.1e and 13.1.3e.

(e) One can wonder whether there exist subspaces M and N in $\mathcal{H}$ so that $M + N$ is not a subspace. Now, for all $M, N \in \mathscr{S}(\mathcal{H})$ the set $M + N$ is always a linear manifold in $\mathcal{H}$ (this was already noted in 3.1.8); if M and N are finite-dimensional then so is $M + N$ and hence $M + N \in \mathscr{S}(\mathcal{H})$ in view of 10.8.1. The following example proves that there are infinite-dimensional subspaces M and N of $\mathcal{H}$ so that $M + N$ is not a subspace.

Assume that $\mathcal{H}$ is separable and that its orthogonal dimension is denumerable, let $\{u_n\}_{n \in \mathbb{N}}$ be a c.o.n.s. in $\mathcal{H}$, and define
$$v_n := \left(\sin \frac{1}{n} \right) u_{2n-1} + \left(\cos \frac{1}{n} \right) u_{2n}, \forall n \in \mathbb{N},$$
and
$$M := V\{u_{2n}\}_{n \in \mathbb{N}} \text{ and } N := V\{v_n\}_{n \in \mathbb{N}}.$$
We prove that $M + N$ is not a subspace by proving that $\overline{M + N} = \mathcal{H}$ and $M + N \neq \mathcal{H}$. First, we have
$$f \in (M + N)^{\perp} \Rightarrow [(f|u_{2n}) = (f|v_n) = 0, \forall n \in \mathbb{N}] \Rightarrow$$
$$[(f|u_{2n}) = (f|u_{2n-1}) = 0, \forall n \in \mathbb{N}] \Rightarrow f = 0_{\mathcal{H}}$$
(cf. 10.6.4e), and hence $\overline{M + N} = \mathcal{H}$ by 10.4.4d. Next, we define the vector $g := \sum_{n=1}^{\infty} \left(\sin \frac{1}{n} \right) u_{2n-1}$ (this series is convergent by 10.4.7b) and we prove that $g \notin M + N$. Indeed, it is immediate to see that $\{v_n\}_{n \in \mathbb{N}}$ is an o.n.s in $\mathcal{H}$, and hence $g \in M + N$ would imply
$$g = \sum_{n=1}^{\infty} \alpha_n u_{2n} + \sum_{n=1}^{\infty} \beta_n v_n, \text{ with } \{\alpha_n\}, \{\beta_n\} \in \ell^2$$
(cf. 10.6.6b), and hence
$$\sin \frac{1}{k} = (u_{2k-1}|g) = \left(u_{2k-1} \Big| \sum_{n=1}^{\infty} \alpha_n u_{2n} + \sum_{n=1}^{\infty} \beta_n v_n \right)$$
$$= \beta_k \sin \frac{1}{k}, \forall k \in \mathbb{N},$$
and hence $\beta_k = 1$ for all $k \in \mathbb{N}$, which is in contradiction with $\{\beta_n\} \in \ell^2$.

(f) In 13.2.8 and in 13.2.9 we introduced the symbols $\sum_{n=1}^{\infty \oplus} M_n$ and $\sum_{n=1}^{N \oplus} M_n$ to denote the orthogonal sum of a sequence $\{M_n\}$ or of a finite family $\{M_1, ..., M_N\}$ of subspaces of $\mathcal{H}$ such that $M_k \subset M_i^{\perp}$ if $i \neq k$. These symbols can be unified in the symbol $\sum_{n \in I}^{\oplus} M_n$, with $I := \mathbb{N}$ or $I := \{1, ..., N\}$. This symbol is the same as the one we used in 10.3.7 to denote the direct sum of a countable family of Hilbert spaces, and indeed the concepts of direct sum and of orthogonal sum are strictly related, as we show in what follows.

Let $\{\mathcal{H}_n\}_{n \in I}$ be a family of Hilbert spaces, with $I := \{1, ..., N\}$ or $I := \mathbb{N}$, and define

$$M_k := \left\{ \{f_n\} \in \sum_{n \in I}^{\oplus} \mathcal{H}_n : f_n = 0_{\mathcal{H}_n} \text{ if } n \neq k \right\}, \forall k \in I,$$

where $\sum_{n \in I}^{\oplus} \mathcal{H}_n$ denotes the direct sum of the family $\{\mathcal{H}_n\}_{n \in I}$. It is easy to see that M_k is a subspace of the Hilbert space $\sum_{n \in I}^{\oplus} \mathcal{H}_n$ for each $k \in I$ (one may use 2.6.6a), that $M_k \subset M_i^{\perp}$ if $i \neq k$, and that $\sum_{n \in I}^{\oplus} \mathcal{H}_n = \sum_{n \in I}^{\oplus} M_n$, where the right hand side of this equation denotes the orthogonal sum of the family of subspaces $\{M_n\}_{n \in I}$. Moreover, M_k as a Hilbert space on its own (cf. 10.3.2) is obviously isomorphic to $\mathcal{H}_k$, for each $k \in I$. Thus, a direct sum of Hilbert spaces equals an orthogonal sum of subspaces, and the terms of the two sums are pairwise isomorphic.

Conversely, let $\{M_n\}_{n \in I}$ be a finite family or a sequence of subspaces of a Hilbert space $\mathcal{H}$, and suppose that $M_k \subset M_i^{\perp}$ if $i \neq k$. For each $n \in I$, M_n can be considered as a Hilbert space on its own (cf. 10.3.2), which we denote by M_n^h. Then, it is easy to see that the mapping

$$\sum_{n \in I}^{\oplus} M_n^h \ni \{f_n\} \mapsto \sum_{n \in I}^{\oplus} f_n \in \sum_{n \in I}^{\oplus} M_n,$$

where $\sum_{n \in I}^{\oplus} M_n^h$ denotes the direct sum of the family $\{M_n^h\}_{n \in I}$ of Hilbert spaces and $\sum_{n \in I}^{\oplus} M_n$ denotes the orthogonal sum of the family $\{M_n\}_{n \in I}$ of subspaces of $\mathcal{H}$, is an isomorphism from the Hilbert space $\sum_{n \in I}^{\oplus} M_n^h$ onto the Hilbert space we obtain by considering the subspace $\sum_{n \in I}^{\oplus} M_n$ of $\mathcal{H}$ as a Hilbert space on its own (cf. 10.3.2). Thus, an orthogonal sum of subspaces is isomorphic to a direct sum of Hilbert spaces, and the terms of the two sums are pairwise equal, once considered as subspaces and once as Hilbert spaces.

13.3 Projection valued measures

Throughout this section, X stands for a non-empty set.

For any non-empty family $\mathcal{F}$ of subsets of X, any mapping $Q : \mathcal{F} \to \mathscr{P}(\mathcal{H})$, and every $f \in \mathcal{H}$, in this and in later chapters we denote by μ_f^Q the function defined by

$$\mu_f^Q : \mathcal{F} \to [0, \infty)$$

$$E \mapsto \mu_f^Q(E) := (f|Q(E)f) \quad (= \|Q(E)f\|^2)$$

(for the equality $(f|Q(E)f) = \|Q(E)f\|^2$, cf. 13.1.7c).

13.3.1 Definition. Let $\mathcal{A}_0$ be an algebra on X. A *projection valued additive mapping* (briefly, a *p.v.a.m.*) *on* $\mathcal{A}_0$ (with values in $\mathscr{P}(\mathcal{H})$, which we usually omit) is a mapping $P_0 : \mathcal{A}_0 \to \mathscr{P}(\mathcal{H})$ which has the following properties:

($pvam_1$) for every finite and disjoint family $\{E_1, ..., E_n\}$ of elements of $\mathcal{A}_0$,

$$P_0\left(\bigcup_{k=1}^n E_k\right) = \sum_{k=1}^n P_0(E_k)$$

(this property of P_0 is called *additivity*);
($pvam_2$) $P_0(X) = \mathbb{1}_{\mathcal{H}}$.

13.3.2 Proposition. *Let $\mathcal{A}_0$ be an algebra on X and P_0 a p.v.a.m. on $\mathcal{A}_0$. Then:*

(a) $P_0(\emptyset) = \mathbb{O}_{\mathcal{H}}$;
(b) if $E, F \in \mathcal{A}_0$ are such that $E \cap F = \emptyset$, then $P_0(E)P_0(F) = \mathbb{O}_{\mathcal{H}}$;
(c) $P_0(E)P_0(F) = P_0(E \cap F)$, $\forall E, F \in \mathcal{A}_0$;
(d) $P_0(E)P_0(F) = P_0(F)P_0(E)$, $\forall E, F \in \mathcal{A}_0$;
(e) if $E, F \in \mathcal{A}_0$ are such that $E \subset F$, then $P_0(E) \le P_0(F)$ (i.e. $R_{P_0(E)} \subset R_{P_0(F)}$, cf. 13.2.5);
(f) $P_0(X - E) = \mathbb{1}_{\mathcal{H}} - P_0(E)$, or equivalently $R_{P_0(X-E)} = R_{P_0(E)}^{\perp}$, $\forall E \in \mathcal{A}_0$;
(g) $\mu_f^{P_0}$ is an additive function on $\mathcal{A}_0$, $\forall f \in \mathcal{H}$;
(h) if $\{E_1, ..., E_N\}$ is a finite family of elements of $\mathcal{A}_0$ such that $P_0(E_n) = \mathbb{O}_{\mathcal{H}}$ for $n = 1, ..., N$, then $P_0\left(\bigcup_{n=1}^N E_n\right) = \mathbb{O}_{\mathcal{H}}$.

Proof. a: Set $n := 2$ and $E_1 := E_2 := \emptyset$ in $pvam_1$.

b: If $E, F \in \mathcal{A}_0$ are such that $E \cap F = \emptyset$, then $pvam_1$ implies that

$$P_0(E) + P_0(F) \in \mathscr{P}(\mathcal{H}),$$

and this implies $P_0(E)P_0(F) = \mathbb{O}_{\mathcal{H}}$ by 13.2.9.

c: For $E, F \in \mathcal{A}_0$ we have

$$E = (E \cap F) \cup (E \cap (X - F)) \ \text{ and } \ F = (F \cap E) \cup (F \cap (X - E));$$

then, in view of $pvam_1$ we have

$$P_0(E) = P_0(E \cap F) + P_0(E \cap (X - F)) \ \text{ and } \ P_0(F) = P_0(F \cap E) + P_0(F \cap (X - E)),$$

and hence, in view of result b and of 13.1.5 or 13.1.6,

$$P_0(E)P_0(F) = P_0(E \cap F)P_0(F \cap E) = P_0(E \cap F).$$

d: This follows immediately from result c.

e: This follows immediately from result c, since $R_{P_0(E)} \subset R_{P_0(F)}$ is equivalent to $P_0(E)P_0(F) = P_0(E)$ (cf. 13.2.4).

f: This follows immediately from $pvam_1$ and $pvam_2$ (cf. also 13.1.4a).

g: For every $f \in \mathcal{H}$, the function $\mu_f^{P_0}$ has property af_1 of 7.1.1 in view of result a, and it has property af_2 in view of property $pvam_1$ of P_0.

h: If $P_0(E_n) = \mathbb{O}_{\mathcal{H}}$ then $\mu_f^{P_0}(E_n) = 0$ for all $f \in \mathcal{H}$, and hence

$$\mu_f^{P_0}\left(\bigcup_{n=1}^{N} E_n\right) = 0, \forall f \in \mathcal{H},$$

by 7.1.2b, and this implies $P_0\left(\bigcup_{n=1}^{N} E_n\right) = \mathbb{O}_{\mathcal{H}}$. $\qquad\square$

13.3.3 Definition. Let $\mathcal{A}$ be a σ-algebra on X. A *projection valued measure* (briefly, a *p.v.m.*) on $\mathcal{A}$ (with values in $\mathscr{P}(\mathcal{H})$, which we usually omit) is a p.v.a.m. P on $\mathcal{A}$ which has the following property:

(pvm) for every sequence $\{E_n\}$ in $\mathcal{A}$ such that $E_i \cap E_j = \emptyset$ if $i \neq j$,

$$P\left(\bigcup_{n=1}^{\infty} E_n\right) f = \sum_{n=1}^{\infty} P(E_n)f, \forall f \in \mathcal{H}$$

(this property of P is called *σ-additivity*).

13.3.4 Remark. In condition *pvm* of 13.3.3, the series $\sum_{n=1}^{\infty} P(E_n)f$ is convergent in view of 13.3.2b and 13.2.8. With the notation introduced in 13.2.10b, condition *pvm* can be written briefly as follows:

(pvm) for every sequence $\{E_n\}$ in $\mathcal{A}$ such that $E_i \cap E_j = \emptyset$ if $i \neq j$,

$$P\left(\bigcup_{n=1}^{\infty} E_n\right) = \sum_{n=1}^{\infty} P(E_n).$$

13.3.5 Theorem. *Let A be a σ-algebra on X and P a mapping $P : \mathcal{A} \to \mathscr{P}(\mathcal{H})$. The following conditions are equivalent:*

(a) P is a p.v.m. on $\mathcal{A}$;
(b) μ_f^P is a measure on $\mathcal{A}$ and $\mu_f^P(X) = \|f\|^2, \forall f \in \mathcal{H}$;
(c) μ_u^P is a probability measure on $\mathcal{A}$ (i.e. μ_u^P is a measure on $\mathcal{A}$ and $\mu_u^P(X) = 1$), $\forall u \in \tilde{\mathcal{H}}$.

Proof. $a \Rightarrow b$: Assuming condition a, for every $f \in \mathcal{H}$ the function μ_f^P is an additive function on $\mathcal{A}$ in view of 13.3.2g and it has property *me* of 7.1.7 in view of condition *pvm* for P and of the continuity of inner product. Also, condition $pvam_2$ for P implies

$$\mu_f^P(X) = \|f\|^2, \forall f \in \mathcal{H}.$$

$b \Rightarrow c$: This is obvious.

$c \Rightarrow a$: Assuming condition c, for every finite and disjoint family $\{E_1, ..., E_n\}$ of elements of $\mathcal{A}$ we have

$$\left(u \middle| P\left(\bigcup_{k=1}^{n} E_k\right) u\right) = \mu_u^P\left(\bigcup_{k=1}^{n} E_k\right) = \sum_{k=1}^{n} \mu_u^P(E_k)$$

$$= \sum_{k=1}^{n} (u|P(E_k)u) = \left(u \middle| \sum_{k=1}^{n} P(E_k)u\right), \forall u \in \tilde{\mathcal{H}},$$

and hence $P\left(\bigcup_{k=1}^{n} E_k\right) = \sum_{k=1}^{n} P(E_k)$ by 10.2.12. We also have

$$(u|P(X)u) = \mu_u^P(X) = 1 = (u|\mathbb{1}_{\mathcal{H}} u), \forall u \in \tilde{\mathcal{H}},$$

and hence $P(X) = \mathbb{1}_{\mathcal{H}}$ by 10.2.12 (cf. also 13.2.10b). Thus, P is a p.v.a.m. on $\mathcal{A}$. Then, for a sequence $\{E_n\}$ in $\mathcal{A}$ such that $E_i \cap E_j = \emptyset$ if $i \neq j$, we have that the series $\sum_{n=1}^{\infty} P(E_n)u$ is convergent for all $u \in \tilde{\mathcal{H}}$, in view of 13.3.2b and 13.2.8, and hence

$$\left(u \middle| P\left(\bigcup_{n=1}^{\infty} E_n\right) u\right) = \mu_u^P\left(\bigcup_{n=1}^{\infty} E_n\right) = \sum_{n=1}^{\infty} \mu_u^P(E_n)$$

$$= \sum_{n=1}^{\infty} (u|P(E_n)u) = \left(u \middle| \sum_{n=1}^{\infty} P(E_n)u\right), \forall u \in \tilde{\mathcal{H}},$$

and hence $P\left(\bigcup_{n=1}^{\infty} E_n\right) = \sum_{n=1}^{\infty} P(E_n)$ by 10.2.12 (cf. also 13.2.10b). Thus, P is a p.v.m. on $\mathcal{A}$. $\qquad\square$

13.3.6 Proposition. *Let A be a σ-algebra on X and P a p.v.m. on $\mathcal{A}$. Then:*

(a) if $\{E_n\}$ is a sequence in $\mathcal{A}$ such that $E_n \subset E_{n+1}$ for all $n \in \mathbb{N}$, then

$$P\left(\bigcup_{n=1}^{\infty} E_n\right) f = \lim_{n \to \infty} P(E_n)f, \forall f \in \mathcal{H};$$

(b) if $\{E_n\}$ is a sequence in $\mathcal{A}$ such that $E_{n+1} \subset E_n$ for all $n \in \mathbb{N}$, then

$$P\left(\bigcap_{n=1}^{\infty} E_n\right) f = \lim_{n \to \infty} P(E_n)f, \forall f \in \mathcal{H};$$

(c) if $\{E_n\}$ is a sequence in $\mathcal{A}$ such that $P(E_n) = \mathbb{O}_{\mathcal{H}}$ for all $n \in \mathbb{N}$, then

$$P\left(\bigcup_{n=1}^{\infty} E_n\right) = \mathbb{O}_{\mathcal{H}}.$$

Proof. a: If $\{E_n\}$ is a sequence in $\mathcal{A}$ such that $E_n \subset E_{n+1}$ for all $n \in \mathbb{N}$, then there exists a sequence $\{F_n\}$ in $\mathcal{A}$ such that

$$\bigcup_{n=1}^{\infty} F_n = \bigcup_{n=1}^{\infty} E_n, \quad \bigcup_{k=1}^{n} F_k = E_n \text{ for all } n \in \mathbb{N}, \quad F_k \cap F_l = \emptyset \text{ if } k \neq l$$

(cf. 6.1.8). Then, in view of conditions pvm and $pvam_1$ of P, we have

$$P\left(\bigcup_{n=1}^{\infty} E_n\right) f = \sum_{n=1}^{\infty} P(F_n)f = \lim_{n\to\infty} \sum_{k=1}^{n} P(F_k)f$$

$$= \lim_{n\to\infty} P\left(\bigcup_{k=1}^{n} F_k\right) f = \lim_{n\to\infty} P(E_n)f, \forall f \in \mathcal{H}.$$

b: Let $\{E_n\}$ be a sequence in $\mathcal{A}$ such that $E_{n+1} \subset E_n$ for all $n \in \mathbb{N}$. Then, by letting $F_n := X - E_n$ for all $n \in \mathbb{N}$, we obtain a sequence $\{F_n\}$ in $\mathcal{A}$ such that $F_n \subset F_{n+1}$ for all $n \in \mathbb{N}$. Then, in view of 13.3.2f and of result a, we have

$$P\left(\bigcap_{n=1}^{\infty} E_n\right) f = P\left(X - \bigcup_{n=1}^{\infty} F_n\right) f = f - P\left(\bigcup_{n=1}^{\infty} F_n\right) f$$

$$= \lim_{n\to\infty} (f - P(F_n)f) = \lim_{n\to\infty} P(E_n)f, \forall f \in \mathcal{H}.$$

c: If $P(E_n) = \mathbb{O}_{\mathcal{H}}$ then $\mu_f^P(E_n) = 0$ for all $f \in \mathcal{H}$, and hence

$$\mu_f^P\left(\bigcup_{n=1}^{\infty} E_n\right) = 0, \forall f \in \mathcal{H},$$

by 7.1.4a, and this implies $P\left(\bigcup_{n=1}^{\infty} E_n\right) = \mathbb{O}_{\mathcal{H}}$. $\qquad\square$

13.4 Extension theorems for projection valued mappings

Throughout this section, X stands for a non-empty set.

13.4.1 Theorem. *Let $\mathcal{S}$ be a semialgebra on X and let $Q : \mathcal{S} \to \mathscr{P}(\mathcal{H})$ be a mapping from $\mathcal{S}$ to $\mathscr{P}(\mathcal{H})$ which satisfies the following conditions:*

(q_1) *for every finite and disjoint family $\{E_1, ..., E_n\}$ of elements of $\mathcal{S}$ such that $\bigcup_{k=1}^{n} E_k \in \mathcal{S}$,*

$$Q\left(\bigcup_{k=1}^{n} E_k\right) = \sum_{k=1}^{n} Q(E_k);$$

(q_2) *if $E, F \in \mathcal{S}$ are such that $E \cap F = \emptyset$, then $Q(E)Q(F) = \mathbb{O}_{\mathcal{H}}$.*

Then there exists a unique mapping $P_0 : \mathcal{A}_0(\mathcal{S}) \to \mathscr{P}(\mathcal{H})$ from $\mathcal{A}_0(\mathcal{S})$ (the algebra on X generated by $\mathcal{S}$) to $\mathscr{P}(\mathcal{H})$ which has property $pvam_1$ of 13.3.1 and is an extension of Q.

If Q satisfies the further condition

(q_3) *there exists a finite and disjoint family $\{F_1, ..., F_N\}$ of elements of $\mathcal{S}$ such that*

$$\bigcup_{k=1}^{N} F_k = X \text{ and } \sum_{k=1}^{N} Q(F_k) = \mathbb{1}_{\mathcal{H}},$$

then P_0 is a p.v.a.m. on $\mathcal{A}_0(\mathcal{S})$.

Proof. Condition q_1 implies that

$$Q(\emptyset) = Q(\emptyset \cup \emptyset) = Q(\emptyset) + Q(\emptyset),$$

and hence the following condition

(q_0) $Q(\emptyset) = \mathbb{O}_{\mathcal{H}}$.

If $\{E_1, ..., E_N\}$ is a finite and disjoint family of elements of $\mathcal{S}$, condition q_2 and 13.2.9 imply that $\sum_{n=1}^{N} Q(E_n) \in \mathscr{P}(\mathcal{H})$. Then, proceeding as in the proof of 7.3.1A with ν replaced by Q, μ_0 by P_0, the number 0 by the operator $\mathbb{O}_{\mathcal{H}}$, we can prove that conditions q_0 and q_1 imply that the mapping

$$P_0 : \mathcal{A}_0(\mathcal{S}) \to \mathscr{P}(\mathcal{H})$$

$$E \mapsto P_0(E) := \sum_{n=1}^{N} Q(E_n) \text{ if } \{E_1, ..., E_N\} \text{ is a finite and disjoint family}$$

$$\text{of elements of } \mathcal{S} \text{ s.t. } E = \cup_{n=1}^{N} E_n$$

is defined consistently and has property $pvam_1$ of 13.3.1. It is obvious that P_0 is an extension of Q. Moreover, proceeding as in the proof of 7.3.1A we can prove that P_0 is the unique mapping from $\mathcal{A}_0(\mathcal{S})$ to $\mathscr{P}(\mathcal{H})$ which has these properties.

Finally, it is obvious that condition q_3 for Q implies condition $pvam_2$ for P_0. $\square$

13.4.2 Theorem. *Let $\mathcal{A}_0$ be an algebra on X, let P_0 be a p.v.a.m. on $\mathcal{A}_0$, and suppose that $\mu_f^{P_0}$ is a premeasure on $\mathcal{A}_0$ for all $f \in \mathcal{H}$. Then there exists a unique p.v.m. on $\mathcal{A}(\mathcal{A}_0)$ (the σ-algebra on X generated by $\mathcal{A}_0$) which is an extension of P_0.*

Proof. From 7.3.2 we have that, for every $f \in \mathcal{H}$, there exists a measure μ_f on $\mathcal{A}(\mathcal{A}_0)$ which is an extension of $\mu_f^{P_0}$ and which is defined by

$$\mu_f(E) := \inf M_E, \forall E \in \mathcal{A}(\mathcal{A}_0),$$

with, for all $E \in \mathcal{A}(\mathcal{A}_0)$,

$$M_E := \left\{ \sum_{n=1}^{\infty} (f|P_0(A_n)f) : \{A_n\} \text{ a sequence in } \mathcal{A}_0 \text{ s.t.} \right.$$

$$\left. A_i \cap A_j = \emptyset \text{ if } i \neq j \text{ and } E \subset \bigcup_{n=1}^{\infty} A_n \right\}.$$

For every sequence $\{A_n\}$ in $\mathcal{A}_0$ s.t. $A_i \cap A_j = \emptyset$ if $i \neq j$, in view of 13.3.2b and 13.2.8 we can define the orthogonal projection that is denoted by $\sum_{n=1}^{\infty} P_0(A_n)$ (cf. 13.2.10b), for which we have

$$\left(f \middle| \left(\sum_{n=1}^{\infty} P_0(A_n) \right) f \right) = \sum_{n=1}^{\infty} (f|P_0(A_n)f), \forall f \in \mathcal{H}.$$

Then, for each $E \in \mathcal{A}(\mathcal{A}_0)$, we define the family of orthogonal projections

$$\mathscr{P}_E := \left\{ \sum_{n=1}^{\infty} P_0(A_n) : \{A_n\} \text{ a sequence in } \mathcal{A}_0 \text{ s.t.} \right.$$

$$\left. A_i \cap A_j = \emptyset \text{ if } i \neq j \text{ and } E \subset \bigcup_{n=1}^{\infty} A_n \right\}$$

and we have

$$\mu_f(E) := \inf\{(f|Tf) : T \in \mathscr{P}_E\}, \forall f \in \mathcal{H}.$$

Now, let $\{A_n\}$ and $\{B_n\}$ be two sequences in $\mathcal{A}_0$ such that $A_i \cap A_j = B_i \cap B_j = \emptyset$ if $i \neq j$, $E \subset \bigcup_{n=1}^{\infty} A_n$, $E \subset \bigcup_{n=1}^{\infty} B_n$; then,

$$A_n \cap B_l \in \mathcal{A}_0, \forall (n,l) \in \mathbb{N} \times \mathbb{N},$$

$$(A_m \cap B_k) \cap (A_n \cap B_l) = \emptyset \text{ if } (m,k) \neq (n,l),$$

$$E \subset \left(\bigcup_{n=1}^{\infty} A_n \right) \cap \left(\bigcup_{l=1}^{\infty} B_l \right) = \bigcup_{(n,l) \in \mathbb{N} \times \mathbb{N}} A_n \cap B_l,$$

$$\left(\sum_{n=1}^{\infty} P_0(A_n) \right) \left(\sum_{l=1}^{\infty} P_0(B_l) \right) f = \sum_{n=1}^{\infty} \sum_{l=1}^{\infty} P_0(A_n) P_0(B_l) f$$

$$= \sum_{(n,l) \in \mathbb{N} \times \mathbb{N}} P_0(A_n \cap B_l) f, \forall f \in \mathcal{H}$$

(cf. 13.1.3d and 13.3.2c). This proves that

$$T_1 T_2 \in \mathscr{P}_E, \forall T_1, T_2 \in \mathscr{P}_E.$$

Then, 13.2.7 implies that

$$\forall E \in \mathcal{A}(\mathcal{A}_0), \exists! P_E \in \mathscr{P}(\mathcal{H}) \text{ so that}$$

$$(f|P_E f) = \inf\{(f|Tf) : T \in \mathscr{P}_E\} = \mu_f(E), \forall f \in \mathcal{H}.$$

Therefore, we can define the mapping

$$P : \mathcal{A}(\mathcal{A}_0) \to \mathscr{P}(\mathcal{H})$$

$$E \mapsto P(E) := P_E.$$

Since μ_f is an extension of $\mu_f^{P_0}$ for all $f \in \mathcal{H}$, we have

$$(f|P(E)f) = \mu_f(E) = \mu_f^{P_0}(E) = (f|P_0(E)f), \forall f \in \mathcal{H}, \forall E \in \mathcal{A}_0,$$

and hence, in view of 10.2.12,

$$P(E) = P_0(E), \forall E \in \mathcal{A}_0.$$

Thus, P is an extension of P_0. Also, in view of 13.3.5, P is a p.v.m. on $\mathcal{A}(\mathcal{A}_0)$ since $\mu_f^P = \mu_f$ and $\mu_f^P(X) = \mu_f^{P_0}(X) = (f|P_0(X)f) = \|f\|^2$, for all $f \in \mathcal{H}$.

Finally, suppose that $\tilde{P}$ is a p.v.m. on $\mathcal{A}(\mathcal{A}_0)$ and that $\tilde{P}$ is an extension of P_0. Then the measure $\mu_f^{\tilde{P}}$ (cf. 13.3.5) is an extension of $\mu_f^{P_0}$ for each $f \in \mathcal{H}$, and hence $\mu_f^{\tilde{P}} = \mu_f$ for each $f \in \mathcal{H}$ by the uniqueness asserted in 7.3.2 for a σ-finite premeasure ($\mu_f^{P_0}$ is finite since $\mu_f^{P_0}(X) = \|f\|^2$). Therefore we have

$$\left(f|\tilde{P}(E)f\right) = \mu_f^{\tilde{P}}(E) = \mu_f(E) = (f|P(E)f), \forall f \in \mathcal{H}, \forall E \in \mathcal{A}(\mathcal{A}_0),$$

which implies $\tilde{P} = P$ by 10.2.12. $\square$

13.4.3 Corollary. *Let $\mathcal{S}$ be a semialgebra on X and let P and $\tilde{P}$ be projection valued measures (both with values in $\mathscr{P}(\mathcal{H})$) on $\mathcal{A}(\mathcal{S})$ (the σ-algebra on X generated by $\mathcal{S}$) such that*

$$P(E) = \tilde{P}(E), \forall E \in \mathcal{S}.$$

Then $P = \tilde{P}$.

Proof. If we denote by Q the restrictions of P and of $\tilde{P}$ to $\mathcal{S}$, then Q has properties q_1 and q_2 of 13.4.1 since P and $\tilde{P}$ satisfy conditions $pvam_1$ and 13.3.2b. Then, 13.4.1 implies that Q has a unique extension which is defined on $\mathcal{A}_0(\mathcal{S})$ and satisfies condition $pvam_1$. Hence, the restrictions of P and $\tilde{P}$ to $\mathcal{A}_0(\mathcal{S})$ must coincide, since they both extend Q and satisfy condition $pvam_1$. Now, if we denote by P_0 the restrictions of P and of $\tilde{P}$ to $\mathcal{A}_0(\mathcal{S})$, then P_0 is obviously a p.v.a.m. and $\mu_f^{P_0}$ is a premeasure on $\mathcal{A}_0(\mathcal{S})$ for all $f \in \mathcal{H}$ since it is the restriction of μ_f^P (for instance) to $\mathcal{A}_0(\mathcal{S})$ and μ_f^P is a measure on $\mathcal{A}(\mathcal{S})$ (note that $\mathcal{A}_0(\mathcal{S}) \subset \mathcal{A}(\mathcal{S})$ since $\mathcal{A}(\mathcal{S}) = \mathcal{A}(\mathcal{A}_0(\mathcal{S}))$, cf. 6.1.18). Then, 13.4.2 implies that there exists a unique p.v.m. on $\mathcal{A}(\mathcal{A}_0(\mathcal{S}))$, i.e. on $\mathcal{A}(\mathcal{S})$, which extends P_0. Therefore, $P = \tilde{P}$. $\square$

13.4.4 Theorem. *Suppose that we have a distance d on X, a semialgebra $\mathcal{S}$ on X, and a mapping $Q : \mathcal{S} \to \mathscr{P}(\mathcal{H})$ which satisfies the following conditions:*

(q_1) for every finite and disjoint family $\{E_1, ..., E_n\}$ of elements of $\mathcal{S}$ such that $\bigcup_{k=1}^n E_k \in \mathcal{S}$,

$$Q\left(\bigcup_{k=1}^n E_k\right) = \sum_{k=1}^n Q(E_k);$$

(q_2) if $E, F \in \mathcal{S}$ are such that $E \cap F = \emptyset$, then $Q(E)Q(F) = O_\mathcal{H}$;

(q_3) there exists a finite and disjoint family $\{F_1, ..., F_N\}$ of elements of $\mathcal{S}$ such that

$$\bigcup_{k=1}^N F_k = X \text{ and } \sum_{k=1}^N Q(F_k) = \mathbb{1}_\mathcal{H};$$

(q_4) $\forall f \in \mathcal{H}, \forall E \in \mathcal{S}, \forall \varepsilon > 0, \exists F \in \mathcal{S}$ such that $\overline{F} \subset E$, $\overline{F}$ is compact and $|\mu_f^Q(E) - \mu_f^Q(F)| < \varepsilon$.

Then there exists a unique p.v.m. on $\mathcal{A}(\mathcal{S})$ which is an extension of Q.

Proof. In view of 13.4.1, conditions q_1, q_2, q_3 imply that there exists a unique p.v.a.m. P_0 on $\mathcal{A}_0(\mathcal{S})$ which is an extension of Q. Then, for every $f \in \mathcal{H}$, the function $\mu_f^{P_0}$ is an additive function on $\mathcal{A}_0(\mathcal{S})$ (cf. 13.3.2g) and it is an extension of μ_f^Q; hence, condition q_4 implies that $\mu_f^{P_0}$ satisfies condition a of 7.1.6 (if $E, F \in \mathcal{S}$ and $F \subset E$ then $\mu_f^Q(F) \leq \mu_f^Q(E)$ since μ_f^Q is restriction of an additive function, cf. 7.1.2a); since $\mu_f^{P_0}(X) = \|f\|^2 < \infty$, this implies that $\mu_f^{P_0}$ is a premeasure (cf. 7.1.6). Then, 13.4.2 implies that there exists a unique p.v.m. P on $\mathcal{A}(\mathcal{A}_0(\mathcal{S}))$ which extends P_0. Since $\mathcal{A}(\mathcal{A}_0(\mathcal{S})) = \mathcal{A}(\mathcal{S})$ (cf. 6.1.18), P is a p.v.m. on $\mathcal{A}(\mathcal{S})$. The uniqueness of P follows from 13.4.3. $\qquad\square$

13.5 Product of commuting projection valued measures

13.5.1 Definition. Let $(X_1, \mathcal{A}_1)$ and $(X_2, \mathcal{A}_2)$ be measurable spaces. A p.v.m. P_1 on $\mathcal{A}_1$ and a p.v.m. P_2 on $\mathcal{A}_2$ (both P_1 and P_2 with values in $\mathscr{P}(\mathcal{H})$) are said to *commute* if

$$P_1(E_1)P_2(E_2) = P_2(E_2)P_1(E_1), \forall E_1 \in \mathcal{A}_1, \forall E_2 \in \mathcal{A}_2.$$

13.5.2 Proposition. *Let $(X_1, \mathcal{A}_1)$ and $(X_2, \mathcal{A}_2)$ be measurable spaces and let P be a p.v.m. on the σ-algebra $\mathcal{A}_1 \otimes \mathcal{A}_2$ (which is a σ-algebra on $X_1 \times X_2$, cf. 6.1.28). Then the mappings*

$$P_1 : \mathcal{A}_1 \to \mathscr{P}(\mathcal{H})$$
$$E_1 \mapsto P_1(E_1) := P(E_1 \times X_2)$$

and

$$P_2 : \mathcal{A}_2 \to \mathscr{P}(\mathcal{H})$$
$$E_2 \mapsto P_2(E_2) := P(X_1 \times E_2)$$

are projection valued measures, they commute, and

$$P_1(E_1)P_2(E_2) = P(E_1 \times E_2), \forall E_1 \in \mathcal{A}_1, \forall E_2 \in \mathcal{A}_2$$

(recall that $E_1 \times E_2 \in \mathcal{A}_1 \otimes \mathcal{A}_2$ for all $E_1 \in \mathcal{A}_1$ and $E_2 \in \mathcal{A}_2$, cf. 6.1.30a).

Proof. For every family $\{E_{1,i}\}_{i \in I}$ of elements of $\mathcal{A}_1$ we have $(\bigcup_{i \in I} E_{1,i}) \times X_2 = \bigcup_{i \in I}(E_{1,i} \times X_2)$. For $E_1, F_1 \in \mathcal{A}_1$, if $E_1 \cap F_1 = \emptyset$ then $(E_1 \times X_2) \cap (F_1 \times X_2) = \emptyset$. Then, it is obvious that P_1 has the properties of a p.v.m. on $\mathcal{A}_1$ since P is a p.v.m. on $\mathcal{A}_1 \otimes \mathcal{A}_2$. And similarly for P_2. From property 13.3.2d of P we have that P_1 and P_2 commute. Finally, for all $E_1 \in \mathcal{A}_1$ and $E_2 \in \mathcal{A}_2$,

$$P_1(E_1)P_2(E_2) = P(E_1 \times X_2)P(X_1 \times E_2)$$
$$= P((E_1 \times X_2) \cap (X_1 \times E_2)) = P(E_1 \times E_2),$$

by property 13.3.2c of P. $\qquad\square$

13.5.3 Theorem. *Let (X_1, d^1) and (X_2, d^2) be complete and separable metric spaces, let P_1 be a p.v.m. on the Borel σ-algebra $\mathcal{A}(d^1)$, let P_2 be a p.v.m. on the Borel σ-algebra $\mathcal{A}(d^2)$ (both P_1 and P_2 with values in $\mathscr{P}(\mathcal{H})$), and suppose that P_1 and P_2 commute. Then there exists a p.v.m. P on the σ-algebra $\mathcal{A}(d^1) \otimes \mathcal{A}(d^2)$ (which is the same as the Borel σ-algebra $\mathcal{A}(d^1 \times d^2)$, where $d^1 \times d^2$ denotes the product distance on $X_1 \times X_2$, cf. 6.1.31, 2.7.1, 2.7.2) such that*

$$P(E_1 \times E_2) = P_1(E_1)P_2(E_2), \forall E_1 \in \mathcal{A}(d^1), \forall E_2 \in \mathcal{A}(d^2).$$

The p.v.m. P is the unique p.v.m. on $\mathcal{A}(d^1) \otimes \mathcal{A}(d^2)$ such that

$$P(E_1 \times X_2) = P_1(E_1), \forall E_1 \in \mathcal{A}(d^1), \text{ and } P(X_1 \times E_2) = P_2(E_2), \forall E_2 \in \mathcal{A}(d^2).$$

The p.v.m. P is called the product of P_1 and P_2.

Proof. The family $\mathcal{S}$ of subsets of $X_1 \times X_2$ defined by

$$\mathcal{S} := \{E_1 \times E_2 : E_1 \in \mathcal{A}(d^1) \text{ and } E_2 \in \mathcal{A}(d^2)\}$$

is a semialgebra on $X_1 \times X_2$ (cf. 6.1.30a). Since the operator $P_1(E_1)P_2(E_2)$ is a projection for all $E_1 \in \mathcal{A}(d^1)$ and $E_2 \in \mathcal{A}(d^2)$ (cf. 13.2.1), we can define the mapping

$$Q : \mathcal{S} \to \mathscr{P}(\mathcal{H})$$
$$E_1 \times E_2 \mapsto Q(E_1 \times E_2) := P_1(E_1)P_2(E_2).$$

We prove that this mapping satisfies the conditions assumed for Q in 13.4.4.

q_1: Suppose that a finite and disjoint family $\{E_{1,1} \times E_{2,1}, ..., E_{1,n} \times E_{2,n}\}$ of elements of $\mathcal{S}$ and an element $E_1 \times E_2 \in \mathcal{S}$ are so that

$$E_1 \times E_2 = \bigcup_{k=1}^{n} (E_{1,k} \times E_{2,k}).$$

We can assume that the sets $E_{1,k}$ and $E_{2,k}$ are non-empty for all $k \in \{1, ..., n\}$. Then we have $E_p = \bigcup_{k=1}^{n} E_{p,k}$ for $p = 1, 2$. Since every σ-algebra is a semialgebra, 6.1.4 implies that, for $p = 1, 2$, there exists a finite and disjoint family $\{F_{p,j}\}_{j \in J_p}$ of elements of $\mathcal{A}(d^p)$ so that

$$\forall k \in \{1, ..., n\}, \exists J_{p,k} \subset J_p \text{ such that } E_{p,k} = \bigcup_{j \in J_{p,k}} F_{p,j}.$$

Then we have

$$E_{1,k} \times E_{2,k} = \bigcup_{(i,j) \in J_{1,k} \times J_{2,k}} (F_{1,i} \times F_{2,j}), \forall k \in \{1, ..., n\}. \tag{14}$$

Clearly, we can assume that $F_{p,j}$ is non-empty for all $j \in J_p$ and for $p = 1, 2$. Then, the condition

$$(E_{1,k} \times E_{2,k}) \cap (E_{1,h} \times E_{2,h}) = \emptyset \text{ if } k \neq h$$

and 14 imply the condition

$$(J_{1,k} \times J_{2,k}) \cap (J_{1,h} \times J_{2,h}) = \emptyset \text{ if } k \neq h. \tag{15}$$

Moreover, we can assume $E_p = \bigcup_{j \in J_p} F_{p,j}$ for $p = 1, 2$ (if this is not already true, we can replace J_p with $\bigcup_{k=1}^{n} J_{p,k}$). Then we have

$$\bigcup_{(i,j) \in J_1 \times J_2} (F_{1,i} \times F_{2,j}) = E_1 \times E_2 = \bigcup_{k=1}^{n} (E_{1,k} \times E_{2,k})$$

$$= \bigcup_{k=1}^{n} \left(\bigcup_{(i,j) \in J_{1,k} \times J_{2,k}} (F_{1,i} \times F_{2,j}) \right). \tag{16}$$

Now, the inclusion $\bigcup_{k=1}^{n} (J_{1,k} \times J_{2,k}) \subset J_1 \times J_2$ is obvious. For $(i,j) \in J_1 \times J_2$, let $(x_1, x_2) \in F_{1,i} \times F_{2,j}$; then 16 implies that there exist $k \in \{1, ..., n\}$ and $(l, m) \in J_{1,k} \times J_{2,k}$ so that $(x_1, x_2) \in F_{1,l} \times F_{2,m}$; since

$$(F_{1,i} \times F_{2,j}) \cap (F_{1,l} \times F_{2,m}) = \emptyset \text{ if } (i,j) \neq (l,m),$$

this implies $(l, m) = (i, j)$; this proves the inclusion $J_1 \times J_2 \subset \bigcup_{k=1}^{n} (J_{1,k} \times J_{2,k})$. Thus,

$$J_1 \times J_2 = \bigcup_{k=1}^{n} (J_{1,k} \times J_{2,k}).$$

This equation and 15 imply the equation

$$\sum_{(i,j) \in J_1 \times J_2} P_1(F_{1,i}) P_2(F_{2,j})$$

$$= \sum_{k=1}^{n} \left(\sum_{(i,j) \in J_{1,k} \times J_{2,k}} P_1(F_{1,i}) P_2(F_{2,j}) \right). \tag{17}$$

Now, by the additivity of P_1 and P_2 we have

$$\sum_{(i,j) \in J_1 \times J_2} P_1(F_{1,i}) P_2(F_{2,j}) = \sum_{i \in J_1} \left(\sum_{j \in J_2} P_1(F_{1,i}) P_2(F_{2,j}) \right)$$

$$= \sum_{i \in J_1} P_1(F_{1,i}) P_2 \left(\bigcup_{j \in J_2} F_{2,j} \right)$$

$$= \sum_{i \in J_1} P_1(F_{1,i}) P_2(E_2)$$

$$= P_1 \left(\bigcup_{i \in J_1} F_{1,i} \right) P_2(E_2) = P_1(E_1) P_2(E_2),$$

and similarly, for each $k \in \{1, ..., n\}$,

$$\sum_{(i,j) \in J_{1,k} \times J_{2,k}} P_1(F_{1,i}) P_2(F_{2,j}) = \sum_{i \in J_{1,k}} P_1(F_{1,i}) P_2 \left(\bigcup_{j \in J_{2,k}} F_{2,j} \right)$$

$$= \sum_{i \in J_{1,k}} P_1(F_{1,i}) P_2(E_{2,k})$$

$$= P_1 \left(\bigcup_{i \in J_{1,k}} F_{1,i} \right) P_2(E_{2,k})$$

$$= P_1(E_{1,k}) P_2(E_{2,k}).$$

These equations and 17 prove that

$$Q(E_1 \times E_2) = P_1(E_1) P_2(E_2) = \sum_{k=1}^{n} P_1(E_{1,k}) P_2(E_{2,k}) = \sum_{k=1}^{n} Q(E_{1,k} \times E_{2,k}).$$

q_2: Let $E_1 \times E_2$ and $F_1 \times F_2$ be elements of $\mathcal{S}$ such that $(E_1 \times E_2) \cap (F_1 \times F_2) = \emptyset$. Then at least one of the two conditions

$$E_1 \cap F_1 = \emptyset \text{ and } E_2 \cap F_2 = \emptyset$$

is true, and hence (cf. 13.3.2a,c) at least one of the two conditions

$$P_1(E_1) P_1(F_1) = P_1(E_1 \cap F_1) = \mathbb{O}_{\mathcal{H}} \text{ and } P_2(E_2) P_2(F_2) = P_2(E_2 \cap F_2) = \mathbb{O}_{\mathcal{H}}$$

is true, and hence

$$Q(E_1 \times E_2) Q(F_1 \times F_2) = P_1(E_1) P_2(E_2) P_1(F_1) P_2(F_2)$$

$$= P_1(E_1) P_1(F_1) P_2(E_2) P_2(F_2) = \mathbb{O}_{\mathcal{H}}.$$

q_3: We have $X_1 \times X_2 \in \mathcal{S}$ and $Q(X_1 \times X_2) = P_1(X_1) P_2(X_2) = \mathbb{1}_{\mathcal{H}} \mathbb{1}_{\mathcal{H}} = \mathbb{1}_{\mathcal{H}}$.

q_4: We fix $f \in \mathcal{H}$, $E_1 \times E_2 \in \mathcal{S}$, $\varepsilon \in (0, \infty)$. For $i = 1, 2$, in view of the fact that the measure $\mu_f^{P_i}$ is finite and the metric space (X_i, d^i) is complete and separable, 7.4.7 implies that

$$\exists F_i \in \mathcal{A}(d^i) \text{ so that } F_i \subset E_i, F_i \text{ is compact and } \mu_f^{P_i}(E_i) - \mu_f^{P_i}(F_i) < \frac{\varepsilon}{2}.$$

Now, $F_1 \times F_2 \in \mathcal{S}$, $F_1 \times F_2 \subset E_1 \times E_2$, $F_1 \times F_2$ is compact in the metric space $(X_1 \times X_2, d^1 \times d^2)$ (cf. 2.8.10), and hence $\overline{F_1 \times F_2} = F_1 \times F_2$ (cf. 2.8.6). We have

$$E_1 \times E_2 = (E_1 \times (E_2 - F_2)) \cup ((E_1 - F_1) \times F_2) \cup (F_1 \times F_2),$$

$$E_1 - F_1 \in \mathcal{A}(d^1),$$

$$E_2 - F_2 \in \mathcal{A}(d^2);$$

then, by property q_1 of Q already proved,

$$Q(E_1 \times E_2) = Q(E_1 \times (E_2 - F_2)) + Q((E_1 - F_1) \times F_2) + Q(F_1 \times F_2),$$

and hence

$$\mu_f^Q(E_1 \times E_2) - \mu_f^Q(F_1 \times F_2) = \mu_f^Q((E_1 \times (E_2 - F_2))) + \mu_f^Q(((E_1 - F_1) \times F_2));$$

this obviously implies

$$0 \leq \mu_f^Q(E_1 \times E_2) - \mu_f^Q(F_1 \times F_2);$$

moreover,

$$\begin{aligned}
\mu_f^Q(E_1 \times (E_2 - F_2)) &= \|P_1(E_1)P_2(E_2 - F_2)f\|^2 \\
&\leq \|P_2(E_2 - F_2)f\|^2 \\
&= \mu_f^{P_2}(E_2 - F_2) = \mu_f^{P_2}(E_2) - \mu_f^{P_2}(F_2) < \frac{\varepsilon}{2}
\end{aligned}$$

and

$$\begin{aligned}
\mu_f^Q((E_1 - F_1) \times F_2) &= \|P_1(E_1 - F_1)P_2(F_2)f\|^2 = \|P_2(F_2)P_1(E_1 - F_1)f\|^2 \\
&\leq \|P_1(E_1 - F_1)f\|^2 = \mu_f^{P_1}(E_1 - F_1) = \mu_f^{P_1}(E_1) - \mu_f^{P_1}(F_1) \\
&< \frac{\varepsilon}{2},
\end{aligned}$$

and therefore

$$|\mu_f^Q(E_1 \times E_2) - \mu_f^Q(F_1 \times F_2)| < \varepsilon.$$

Thus, the mapping Q satisfies all the conditions of 13.4.4, and hence there exists a unique p.v.m. P on $\mathcal{A}(\mathcal{S})$ which is an extension of Q. Now, $\mathcal{A}(\mathcal{S}) = \mathcal{A}(d^1) \otimes \mathcal{A}(d^2)$ (cf. 6.1.30a).

Finally, suppose that $\tilde{P}$ is a p.v.m. on $\mathcal{A}(d^1) \otimes \mathcal{A}(d^2)$ such that

$$\tilde{P}(E_1 \times X_2) = P_1(E_1), \forall E_1 \in \mathcal{A}(d^1), \text{ and } \tilde{P}(X_1 \times E_2) = P_2(E_2), \forall E_2 \in \mathcal{A}(d^2).$$

Then, in view of 13.3.2c,

$$\begin{aligned}
\tilde{P}(E_1 \times E_2) &= \tilde{P}((E_1 \times X_2) \cap (X_1 \times E_2)) = \tilde{P}(E_1 \times X_2)\tilde{P}(X_2 \times E_2) \\
&= P_1(E_1)P_2(E_2) = Q(E_1 \times E_2), \forall E_1 \in \mathcal{A}(d^1), \forall E_2 \in \mathcal{A}(d^2),
\end{aligned}$$

and hence $\tilde{P}$ is an extension of Q, and hence $\tilde{P} = P$. $\qquad\square$

13.6   Spectral families and projection valued measures

Our version of the spectral theorem for self-adjoint operators (cf. 15.2.1) relates self-adjoint operators to projection valued measures on the Borel σ-algebra $\mathcal{A}(d_{\mathbb{R}})$ (which is a σ-algebra on $\mathbb{R}$, cf. 6.1.22 and 2.1.4). In other books, this theorem is often phrased so that it relates self-adjoint operators to spectral families. These two versions of the spectral theorem are completely equivalent. In this section we prove the equivalence of the notions of a spectral family and of a p.v.m. on $\mathcal{A}(d_{\mathbb{R}})$. However, the results of this section are not needed in other parts of the present book.

13.6.1 Definition. A *spectral family* in $\mathcal{H}$ is a mapping $T : \mathbb{R} \to \mathscr{P}(\mathcal{H})$ which has the following properties:

(sf_1) $T(x_1) \leq T(x_2)$ if $x_1 \leq x_2$;

(sf_2) $T(x + \delta_n)f \xrightarrow[n \to \infty]{} T(x)f$, $\forall x \in \mathbb{R}$, $\forall f \in \mathcal{H}$, for every sequence $\{\delta_n\}$ in $[0, \infty)$

s.t. $\delta_n \xrightarrow[n \to \infty]{} 0$ (this property of T is called *continuity from the right*);

(sf_3) $T(x_n)f \xrightarrow[n \to \infty]{} 0_{\mathcal{H}}$, $\forall f \in \mathcal{H}$, for every sequence $\{x_n\}$ in $\mathbb{R}$ s.t. $x_n \xrightarrow[n \to \infty]{} -\infty$;

(sf_4) $T(x_n)f \xrightarrow[n \to \infty]{} f$, $\forall f \in \mathcal{H}$, for every sequence $\{x_n\}$ in $\mathbb{R}$ s.t. $x_n \xrightarrow[n \to \infty]{} \infty$.

Another name for a spectral family is *resolution of the identity*.

13.6.2 Proposition. *Let P be a p.v.m. on the σ-algebra $\mathcal{A}(d_{\mathbb{R}})$. Then the mapping*

$$T : \mathbb{R} \to \mathscr{P}(\mathcal{H})$$
$$x \mapsto T(x) := P((-\infty, x])$$

is a spectral family.

Proof. We prove that the mapping T has all the properties of a spectral family.

sf_1: This follows immediately from 13.3.2e.

sf_2: We fix $x \in \mathbb{R}$ and $f \in \mathcal{H}$. We have obviously

$$\left(-\infty, x + \frac{1}{n+1}\right] \subset \left(-\infty, x + \frac{1}{n}\right], \forall n \in \mathbb{N}, \text{ and } (-\infty, x] = \bigcap_{n=1}^{\infty} \left(-\infty, x + \frac{1}{n}\right],$$

and hence, by 13.3.6b,

$$T(x)f = P((-x, \infty])f = \lim_{n \to \infty} P\left(\left(-\infty, x + \frac{1}{n}\right]\right)f$$
$$= \lim_{n \to \infty} T\left(x + \frac{1}{n}\right)f.$$

Now, let $\{\delta_n\}$ be a sequence in $[0, \infty)$ such that $\delta_n \xrightarrow[n \to \infty]{} 0$ and fix $\varepsilon > 0$. Let $n_\varepsilon \in \mathbb{N}$ be such that

$$\left\| T(x)f - T\left(x + \frac{1}{n_\varepsilon}\right)f \right\| < \varepsilon$$

and let $N_\varepsilon \in \mathbb{N}$ be such that

$$n > N_\varepsilon \Rightarrow \delta_n < \frac{1}{n_\varepsilon}.$$

Then, for $n > N_\varepsilon$ we have $T(x + \delta_n) \leq T\left(x + \frac{1}{n_\varepsilon}\right)$ in view of property sf_1 already proved, and hence, in view of 13.2.4,

$$\|T(x + \delta_n)f - T(x)f\|^2 = (f | T(x + \delta_n)f - T(x)f)$$
$$\leq \left(f | T\left(x + \frac{1}{n_\varepsilon}\right)f - T(x)f \right)$$
$$= \left\| T\left(x + \frac{1}{n_\varepsilon}\right)f - T(x)f \right\|^2 < \varepsilon^2,$$

where the equalities hold by 13.1.7c since $T(x + \delta_n) - T(x)$ and $T\left(x + \frac{1}{n_\varepsilon}\right) - T(x)$ are orthogonal projections (cf. sf_1 and 13.2.4).

sf_3: We fix $f \in \mathcal{H}$. By 13.3.2a and 13.3.6b we have

$$0_\mathcal{H} = P(\emptyset)f = \lim_{n \to \infty} P((-\infty, n])f = \lim_{n \to \infty} T(-n)f.$$

Now, let $\{x_n\}$ be a sequence in $\mathbb{R}$ such that $x_n \xrightarrow[n \to \infty]{} -\infty$ and fix $\varepsilon > 0$. Let $n_\varepsilon \in \mathbb{N}$ be such that

$$\|T(-n_\varepsilon)f\| < \varepsilon$$

and let $N_\varepsilon \in \mathbb{N}$ be such that

$$n > N_\varepsilon \Rightarrow x_n < -n_\varepsilon.$$

Then, for $n > N_\varepsilon$ we have $T(x_n) \leq T(-n_\varepsilon)$ in view of property sf_1 and hence

$$\|T(x_n)f\|^2 = (f|T(x_n)f) \leq (f|T(-n_\varepsilon)f) = \|T(-n_\varepsilon)f\|^2 < \varepsilon^2.$$

sf_4: We fix $f \in \mathcal{H}$. By property $pvam_2$ of P and by 13.3.6a we have

$$f = P(\mathbb{R})f = \lim_{n \to \infty} P((-\infty, n])f = \lim_{n \to \infty} T(n)f.$$

Now, let $\{x_n\}$ be a sequence in $\mathbb{R}$ such that $x_n \xrightarrow[n \to \infty]{} \infty$ and fix $\varepsilon > 0$. Let $n_\varepsilon \in \mathbb{N}$ be such that

$$\|f - T(n_\varepsilon)f\| < \varepsilon$$

and let $N_\varepsilon \in \mathbb{N}$ be such that

$$n > N_\varepsilon \Rightarrow x_n > n_\varepsilon.$$

Then, for $n > N_\varepsilon$ we have $T(n_\varepsilon) \leq T(x_n)$ in view of property sf_1, and hence

$$\|f - T(x_n)f\|^2 = (f|f - T(x_n)f) \leq (f|f - T(n_\varepsilon)f) = \|f - T(n_\varepsilon)f\|^2 < \varepsilon^2,$$

where the equalities hold by 13.1.7c because $\mathbb{1}_\mathcal{H} - T(x_n)$ and $\mathbb{1}_\mathcal{H} - T(n_\varepsilon)$ are orthogonal projections (cf. 13.1.3e). $\qquad\square$

13.6.3 Theorem. *Let T be a spectral family. Then there exists a unique p.v.m. P on the σ-algebra $\mathcal{A}(d_\mathbb{R})$ such that*

$$T(x) = P((-\infty, x]), \forall x \in \mathbb{R}.$$

Proof. Here we denote by $\mathcal{S}$ the semialgebra on $\mathbb{R}$ that was denoted by $\mathcal{I}_9$ in 6.1.25. We note that

$$T(b) - T(a) \in \mathscr{P}(\mathcal{H}), \text{ for all } a, b \in \mathbb{R} \text{ so that } a < b,$$

by property sf_1 of T and by 13.2.4, and that

$$\mathbb{1}_\mathcal{H} - T(a) \in \mathscr{P}(\mathcal{H}), \forall a \in \mathbb{R},$$

by 13.1.3e. Then, we can define a mapping $Q : \mathcal{S} \to \mathscr{P}(\mathcal{H})$ by letting

$$Q(\emptyset) := O_{\mathcal{H}},$$
$$Q((a, b]) := T(b) - T(a), \text{ for all } a, b \in \mathbb{R} \text{ so that } a < b,$$
$$Q((-\infty, b]) := T(b), \forall b \in \mathbb{R},$$
$$Q((a, \infty)) := \mathbb{1}_{\mathcal{H}} - T(a), \forall a \in \mathbb{R}.$$

For every $f \in \mathcal{H}$, the function

$$F_f : \mathbb{R} \to \mathbb{R}$$
$$x \mapsto F_f(x) := (f|T(x)f)$$

has the properties that were assumed for the function F in 9.1.1, in view of properties sf_1 and sf_2 of T (also, cf. 13.2.4), and hence there exists a measure μ_f on $\mathcal{A}(d_\mathbb{R})$ such that

$$\mu_f((a, b]) = F_f(b) - F_f(a), \text{ for all } a, b \in \mathbb{R} \text{ so that } a < b.$$

For the measure μ_f we also have (cf. the proof of 9.1.1):

$$\mu_f(\emptyset) = 0 = (f|Q(\emptyset)f);$$
$$\mu_f((-\infty, b]) = F_f(b) - \lim_{n \to \infty} F_f(-n) = F_f(b)$$
$$= (f|Q((-\infty, b])f), \forall b \in \mathbb{R};$$
$$\mu_f((a, \infty)) = \lim_{n \to \infty} F_f(n) - F_f(a) = (f|f) - F_f(a)$$
$$= (f|Q((a, \infty))f), \forall a \in \mathbb{R}$$

(we have $\lim_{n \to \infty} F_f(-n) = 0$ by property sf_3 of T and $\lim_{n \to \infty} F_f(n) = (f|f)$ by property sf_4 of T). Thus, we have

$$\mu_f(E) = (f|Q(E)f), \forall E \in \mathcal{S}.$$

Now we can prove that the mapping Q satisfies conditions q_1, q_2, q_3 of 13.4.1.

q_1: For every finite and disjoint family $\{E_1, ..., E_n\}$ of elements of $\mathcal{S}$ such that $\bigcup_{k=1}^{n} E_k \in \mathcal{S}$, we have

$$\left(f\Big|Q\left(\bigcup_{k=1}^{n} E_k\right)f\right) = \mu_f\left(\bigcup_{k=1}^{n} E_k\right) = \sum_{k=1}^{n} \mu_f(E_k)$$
$$= \sum_{k=1}^{n} (f|Q(E_k)f) = \left(f\Big|\sum_{k=1}^{n} Q(E_k)f\right), \forall f \in \mathcal{H},$$

and hence

$$Q\left(\bigcup_{k=1}^{n} E_k\right) = \sum_{k=1}^{n} Q(E_k),$$

by 10.2.12.

q_2: First, fix $a, b \in \mathbb{R}$ so that $a < b$. Then:
for $c, d \in \mathbb{R}$ so that $c < d$, if $(a, b] \cap (c, d] = \emptyset$ then either $b \le c$ or $d \le a$; supposing
e.g. $b \le c$, we have

$$Q((a, b])Q((c, d]) = (T(b) - T(a))(T(d) - T(c))$$
$$= T(b) - T(a) - T(b) + T(a) = \mathbb{O}_{\mathcal{H}};$$

for $c \in \mathbb{R}$, if $(a, b] \cap (-\infty, c] = \emptyset$ then $c \le a$ and hence

$$Q((a, b])Q((-\infty, c]) = (T(b) - T(a))T(c) = T(c) - T(c) = \mathbb{O}_{\mathcal{H}};$$

for $c \in \mathbb{R}$, if $(a, b] \cap (c, \infty) = \emptyset$ then $b \le c$ and hence

$$Q((a, b])Q((c, \infty)) = (T(b) - T(a))(\mathbb{1}_{\mathcal{H}} - T(c))$$
$$= T(b) - T(a) - T(b) + T(a) = \mathbb{O}_{\mathcal{H}}.$$

Next, if $a, b \in \mathbb{R}$ are so that $(a, \infty) \cap (-\infty, b] = \emptyset$ then $b \le a$ and hence

$$Q((a, \infty))Q((-\infty, b]) = (\mathbb{1}_{\mathcal{H}} - T(a))T(b) = T(b) - T(b) = \mathbb{O}_{\mathcal{H}}.$$

We have used property sf_1 of T and 13.2.4 sistematically.

q_3: Choose $a \in \mathbb{R}$. Then $(-\infty, a]$ and (a, ∞) are elements of $\mathcal{S}$ such that

$$(-\infty, a] \cup (a, \infty) = \mathbb{R} \text{ and } Q((-\infty, a]) + Q((a, \infty)) = T(a) + \mathbb{1}_{\mathcal{H}} - T(a) = \mathbb{1}_{\mathcal{H}}.$$

Thus, the mapping Q satisfies all the conditions of 13.4.1 and hence there exists a
unique p.v.a.m. P_0 on $\mathcal{A}_0(\mathcal{S})$ which is an extension of Q. Now, for every $f \in \mathcal{H}$,
$\mu_f^{P_0}$ is an additive function on $\mathcal{A}_0(\mathcal{S})$ (cf. 13.3.2g) and it must be the same as
the restrictions of μ_f to $\mathcal{A}_0(\mathcal{S})$ (note that $\mathcal{A}_0(\mathcal{S}) \subset \mathcal{A}(d_{\mathbb{R}})$ since $\mathcal{A}(d_{\mathbb{R}}) = \mathcal{A}(\mathcal{S}) =$
$\mathcal{A}(\mathcal{A}_0(\mathcal{S}))$, cf. 6.1.25 and 6.1.18) owing to the uniqueness asserted in 7.3.1, since
the restrictions of $\mu_f^{P_0}$ and of μ_f to $\mathcal{S}$ are the same (both of them are equal to
μ_f^Q). Hence, $\mu_f^{P_0}$ is a premeasure on $\mathcal{A}_0(\mathcal{S})$. Then, 13.4.2 implies that there exists
a unique p.v.m. P on $\mathcal{A}(\mathcal{A}_0(\mathcal{S})) = \mathcal{A}(d_{\mathbb{R}})$ which is an extension of P_0. Thus, P is
also an extension of Q and we have in particular

$$P((-\infty, x]) = Q((-\infty, x]) = T(x), \forall x \in \mathbb{R}.$$

Finally, suppose that $\tilde{P}$ is a p.v.m. on $\mathcal{A}(d_{\mathbb{R}})$ such that

$$\tilde{P}((-\infty, x]) = T(x), \forall x \in \mathbb{R}.$$

Then we have

$$\tilde{P}((-\infty, a]) + \tilde{P}((a, \infty)) = \tilde{P}(\mathbb{R}) = \mathbb{1}_{\mathcal{H}}, \forall a \in \mathbb{R},$$

and hence

$$\tilde{P}((a, \infty)) = \mathbb{1}_{\mathcal{H}} - T(a), \forall a \in \mathbb{R}.$$

We also have, for all $a, b \in \mathbb{R}$ so that $a < b$,

$$\tilde{P}((a, b]) = \tilde{P}((-\infty, b] \cap (a, \infty)) = \tilde{P}((-\infty, b])\tilde{P}((a, \infty))$$
$$= T(b)(\mathbb{1}_{\mathcal{H}} - T(a)) = T(b) - T(a)$$

(cf. 13.3.2c). This proves that

$$\tilde{P}(E) = Q(E) = P(E), \forall E \in \mathcal{S},$$

and this implies $\tilde{P} = P$ by 13.4.3. $\qquad\qquad\qquad\qquad\qquad\qquad\square$

13.6.4 Remark. Some define a spectral family replacing "continuity from the right" in sf_2 with "continuity from the left" (defined in an obvious way). Clearly the two definitions are not the same, but they are equivalent in the following sense: the spectral theorem (in the formulation in which spectral families instead of projection valued measures are used) says that for any given self-adjoint operator there exists a unique spectral family for each type (i.e. either continuous from the right or continuous from the left) so that the operator "is the integral of the function ξ (cf. 11.3.2) with respect to that family". Actually, in order to prove the existence of a spectral family which does this trick one could dispose altogether of condition sf_2 and only require condition sf_1 in the definition of a spectral family. However, the spectral family (thus redefined) associated to a given self-adjoint operator would not be unique. In a way, right continuity or left continuity are "normalization conditions". Obviously, a p.v.m. P on $\mathcal{A}(d_{\mathbb{R}})$ determines and is determined uniquely by a "left continuous" spectral family T in a way similar to what was seen in 13.6.2 and in 13.6.3, and the link condition is

$$T(x) = P((-\infty, x)), \forall x \in \mathbb{R}.$$

Chapter 14

Integration with respect to a Projection Valued Measure

The spectral theorems for unitary and for self-adjoint operators will be presented in the next chapter. They consist in the representation of a unitary or a self-adjoint operator as an integral with respect to a projection valued measure. In this chapter we investigate the idea of an integral with respect to a projection valued measure and study the properties of this kind of integral.

14.1 Integrals of bounded measurable functions

In this section, $(X, \mathcal{A})$ denotes an abstract measurable space (i.e. X is a non-empty set and $\mathcal{A}$ is a σ-algebra on X), $\mathcal{H}$ denotes an abstract Hilbert space, and P denotes a projection valued measure on $\mathcal{A}$ with values in $\mathscr{P}(\mathcal{H})$.

We recall that $\mathcal{M}_B(X, \mathcal{A})$ denotes the family of all bounded $\mathcal{A}$-measurable complex functions on X (cf. 6.2.28). With the $\| \ \|_\infty$ norm (cf. 4.3.6a), $\mathcal{M}_B(X, \mathcal{A})$ is a C^*-algebra (cf. 6.2.29 and 12.6.7). The family $\mathcal{S}(X, \mathcal{A})$ of all $\mathcal{A}$-simple functions on X (cf. 6.2.22) is a subalgebra of $\mathcal{M}_B(X, \mathcal{A})$ and it is dense (with respect to the $\| \ \|_\infty$ norm) in $\mathcal{M}_B(X, \mathcal{A})$ (cf. 6.2.29). Finally, we recall that $\mathcal{B}(\mathcal{H})$ denotes the family of all bounded (i.e. continuous) linear operators on $\mathcal{H}$ (cf. 4.2.10) and that $\mathcal{B}(\mathcal{H})$ is a C^*-algebra (cf. 12.6.4).

14.1.1 Theorem. *There exists a unique mapping $\hat{J}_P : \mathcal{M}_B(X, \mathcal{A}) \to \mathcal{B}(\mathcal{H})$ such that:*

(a) $\hat{J}_P(\chi_E) = P(E)$, $\forall E \in \mathcal{A}$;
(b) $\hat{J}_P$ *is a linear operator;*
(c) $\hat{J}_P$ *is continuous.*

In addition, the following conditions are true;

(d) $\hat{J}_P(\varphi_1 \varphi_2) = \hat{J}_P(\varphi_1) \hat{J}_P(\varphi_2)$, $\forall \varphi_1, \varphi_2 \in \mathcal{M}_B(X, \mathcal{A})$;
(e) $\hat{J}_P(\overline{\varphi}) = (\hat{J}_P(\varphi))^\dagger$, $\forall \varphi \in \mathcal{M}_B(X, \mathcal{A})$;
(f) $\left(f | \hat{J}_P(\varphi) f \right) = \int_X \varphi \, d\mu_f^P$, $\forall f \in \mathcal{H}$, $\forall \varphi \in \mathcal{M}_B(X, \mathcal{A})$;
(g) $\| \hat{J}_P(\varphi) f \|^2 = \int_X |\varphi|^2 \, d\mu_f^P$, $\forall f \in \mathcal{H}$, $\forall \varphi \in \mathcal{M}_B(X, \mathcal{A})$;

425

(h) if $A \in \mathcal{B}(\mathcal{H})$ is so that $AP(E) = P(E)A$ for all $E \in \mathcal{A}$, then $A\hat{J}_P(\varphi) = \hat{J}_P(\varphi)A$ for all $\varphi \in \mathcal{M}_B(X, \mathcal{A})$.

Proof. We begin with a preliminary remark. For $n, m \in \mathbb{N}$, let $\{\alpha_1, ..., \alpha_n\}$ and $\{\beta_1, ..., \beta_m\}$ be families of elements of $\mathbb{C}$, let $\{E_1, ..., E_n\}$ and $\{F_1, ..., F_m\}$ be disjoint families of elements of $\mathcal{A}$, and suppose that

$$\sum_{k=1}^{n} \alpha_k \chi_{E_k} = \sum_{l=1}^{m} \beta_l \chi_{F_l}.$$

The same proof as the one given in 8.1.1 (with μ replaced by P) shows that

$$\sum_{k=1}^{n} \alpha_k P(E_k) = \sum_{l=1}^{m} \beta_l P(F_l).$$

Now let $\psi \in \mathcal{S}(X, \mathcal{A})$. Then there are $n \in \mathbb{N}$, a family $\{\alpha_1, ..., \alpha_n\}$ of elements of $\mathbb{C}$, and a disjoint family $\{E_1, ..., E_n\}$ of elements of $\mathcal{A}$ so that $\psi = \sum_{k=1}^{n} \alpha_k \chi_{E_k}$. We define the operator

$$A_\psi^P := \sum_{k=1}^{n} \alpha_k P(E_k),$$

which is an element of $\mathcal{B}(\mathcal{H})$ determined by ψ without ambiguity in view of the preliminary remark above. A proof similar to the one given in 8.1.4a (with μ replaced by P) shows that the mapping

$$\mathcal{S}(X, \mathcal{A}) \ni \psi \mapsto A_P(\psi) := A_\psi^P \in \mathcal{B}(\mathcal{H})$$

is a linear operator. Moreover, for every $\psi \in \mathcal{S}(X, \mathcal{A})$ we have (cf. 13.3.2b, 13.2.9, 10.2.3)

$$\|A_\psi^P f\|^2 = \sum_{k=1}^{n} |\alpha_k|^2 \|P(E_k)f\|^2$$

$$\leq \|\psi\|_\infty^2 \sum_{k=1}^{n} \|P(E_k)f\|^2 = \|\psi\|_\infty^2 \|P(\cup_{k=1}^{n} E_k)f\|^2$$

$$\leq \|\psi\|_\infty^2 \|f\|^2, \forall f \in \mathcal{H},$$

and hence $\|A_\psi^P\| \leq \|\psi\|_\infty$. This proves that

$$\|A_P(\psi)\| \leq \|\psi\|_\infty, \forall \psi \in \mathcal{S}(X, \mathcal{A}),$$

and hence that the linear operator A_P is bounded. Since $\mathcal{S}(X, \mathcal{A})$ is dense in $\mathcal{M}_B(X, \mathcal{A})$ and $\mathcal{B}(\mathcal{H})$ is a Banach space, by 4.2.6 there exists a unique bounded (and hence continuous) linear operator

$$\hat{J}_P : \mathcal{M}_B(X, \mathcal{A}) \to \mathcal{B}(\mathcal{H})$$

which is an extension of A_P, i.e. such that $\hat{J}_P(\psi) = A_\psi^P$ for all $\psi \in \mathcal{S}(X, \mathcal{A})$, and hence such that

$$\hat{J}_P(\chi_E) = A_{\chi_E}^P = P(E), \forall E \in \mathcal{A}.$$

Thus, the mapping $\hat{J}_P$ has properties a, b, c.

Next, we prove that $\hat{J}_P$ is the unique mapping from $\mathcal{M}_B(X, \mathcal{A})$ to $\mathcal{B}(\mathcal{H})$ which has properties a, b, c. Suppose that conditions a, b, c hold true for a mapping $J : \mathcal{M}_B(X, \mathcal{A}) \to \mathcal{B}(\mathcal{H})$. Then,

$$J(\psi) = A_\psi^P = A_P(\psi), \forall \psi \in \mathcal{S}(X, \mathcal{A}),$$

in view of conditions a and b, and therefore $J = \hat{J}_P$ in view of condition c and of the uniqueness asserted in 4.2.6, since $\mathcal{S}(X, \mathcal{A})$ is dense in $\mathcal{M}_B(X, \mathcal{A})$.

In what follows we prove that the mapping $\hat{J}_P$ has the additional properties of the statement.

d: For $\psi_1, \psi_2 \in \mathcal{S}(X, \mathcal{A})$, let n and m be elements of $\mathbb{N}$, $\{\alpha_1, ..., \alpha_n\}$ and $\{\beta_1, ..., \beta_m\}$ families of elements of $\mathbb{C}$, and $\{E_1, ..., E_n\}$ and $\{F_1, ..., F_m\}$ disjoint families of elements of $\mathcal{A}$ so that

$$\psi_1 = \sum_{k=1}^n \alpha_k \chi_{E_k} \text{ and } \psi_2 = \sum_{l=1}^m \beta_l \chi_{F_l}.$$

Then

$$\psi_1 \psi_2 = \sum_{k=1}^n \sum_{l=1}^m \alpha_k \beta_l \chi_{E_k \cap F_l},$$

and hence (since $\{E_k \cap F_l\}_{k=1,...,n; l=1,...,m}$ is a disjoint family of elements of $\mathcal{A}$; also, cf. 13.3.2c),

$$\hat{J}_P(\psi_1 \psi_2) = A_{\psi_1 \psi_2}^P = \sum_{k=1}^n \sum_{l=1}^m \alpha_k \beta_l P(E_k \cap F_l)$$

$$= \sum_{k=1}^n \sum_{l=1}^m \alpha_k \beta_l P(E_k) P(F_l) = \left(\sum_{k=1}^n \alpha_k P(E_k) \right) \left(\sum_{l=1}^m \beta_l P(F_l) \right)$$

$$= A_{\psi_1}^P A_{\psi_2}^P = \hat{J}_P(\psi_1) \hat{J}_P(\psi_2).$$

Now, for $\varphi_1, \varphi_2 \in \mathcal{M}_B(X, \mathcal{A})$ let $\{\psi_{1,n}\}$ and $\{\psi_{2,n}\}$ be sequences in $\mathcal{S}(X, \mathcal{A})$ such that $\varphi_1 = \lim_{n \to \infty} \psi_{1,n}$ and $\varphi_2 = \lim_{n \to \infty} \psi_{2,n}$ (in the $\| \ \|_\infty$ norm); then $\varphi_1 \varphi_2 = \lim_{n \to \infty} \psi_{1,n} \psi_{2,n}$ in view of 4.3.3, and hence

$$\hat{J}_P(\varphi_1 \varphi_2) \overset{(1)}{=} \lim_{n \to \infty} \hat{J}_P(\psi_{1,n} \psi_{2,n}) = \lim_{n \to \infty} \hat{J}_P(\psi_{1,n}) \hat{J}_P(\psi_{2,n})$$

$$\overset{(2)}{=} (\lim_{n \to \infty} \hat{J}_P(\psi_{1,n}))(\lim_{n \to \infty} \hat{J}_P(\psi_{2,n})) \overset{(3)}{=} \hat{J}_P(\varphi_1) \hat{J}_P(\varphi_2),$$

where 1 and 3 hold by continuity of $\hat{J}_P$ and 2 by 4.3.3.

e: For $\psi \in \mathcal{S}(X, \mathcal{A})$, let n be an element of $\mathbb{N}$, $\{\alpha_1, ..., \alpha_n\}$ a family of elements of $\mathbb{C}$, $\{E_1, ..., E_n\}$ a disjoint family of elements of $\mathcal{A}$ so that $\psi = \sum_{k=1}^n \alpha_k \chi_{E_k}$. Then

$$\hat{J}_P(\overline{\psi}) = A_{\overline{\psi}}^P = \sum_{k=1}^n \overline{\alpha_k} P(E_k) = \left(\sum_{k=1}^n \alpha_k P(E_k) \right)^\dagger = (A_\psi^P)^\dagger = (\hat{J}_P(\psi))^\dagger.$$

Now, for $\varphi \in \mathcal{M}_B(X, \mathcal{A})$ let $\{\psi_n\}$ be a sequence in $\mathcal{S}(X, \mathcal{A})$ such that $\varphi = \lim_{n\to\infty} \psi_n$ (in the $\| \ \|_\infty$ norm); then $\overline{\varphi} = \lim_{n\to\infty} \overline{\psi_n}$ in view of 12.6.2, and hence

$$\hat{J}_P(\overline{\varphi}) \overset{(4)}{=} \lim_{n\to\infty} \hat{J}_P(\overline{\psi_n}) = \lim_{n\to\infty} (\hat{J}_P(\psi_n))^\dagger$$

$$\overset{(5)}{=} (\lim_{n\to\infty} \hat{J}_P(\psi_n))^\dagger \overset{(6)}{=} (\hat{J}_P(\varphi))^\dagger,$$

where 4 and 6 hold by continuity of $\hat{J}_P$ and 5 by 12.6.2.

f: For $\psi \in \mathcal{S}(X, \mathcal{A})$, let $\psi = \sum_{k=1}^n \alpha_k \chi_{E_k}$ be as in the proof of e. Then

$$\left(f | \hat{J}_P(\psi) f \right) = \left(f | A_\psi^P f \right) = \sum_{k=1}^n \alpha_k \left(f | P(E_k) f \right)$$

$$= \sum_{k=1}^n \alpha_k \mu_f^P(E_k) = \int_X \psi d\mu_f^P, \forall f \in \mathcal{H}.$$

Now, for $\varphi \in \mathcal{M}_B(X, \mathcal{A})$ let $\{\psi_n\}$ be a sequence in $\mathcal{S}(X, \mathcal{A})$ such that $\varphi = \lim_{n\to\infty} \psi_n$ (in the $\| \ \|_\infty$ norm); then $\|\varphi\|_\infty = \lim_{n\to\infty} \|\psi_n\|_\infty$ in view of 4.1.6a, and hence (cf. 2.1.9)

$$\exists m \in [0, \infty) \text{ such that } |\psi_n(x)| \le \|\psi_n\|_\infty < m, \forall x \in X, \forall n \in \mathbb{N};$$

we notice that

$$\varphi(x) = \lim_{n\to\infty} \psi_n(x), \forall x \in X, \text{ and } m_X \in \mathcal{L}^1(X, \mathcal{A}, \mu_f^P), \forall f \in \mathcal{H}$$

(m_X denotes the constant function on X with value m, cf. 1.2.19; also, cf. 8.2.6); then,

$$\left(f | \hat{J}_P(\varphi) f \right) \overset{(7)}{=} \lim_{n\to\infty} \left(f | \hat{J}_P(\psi_n) f \right) = \lim_{n\to\infty} \int_X \psi_n d\mu_f^P$$

$$\overset{(8)}{=} \int_X \varphi d\mu_f^P, \forall f \in \mathcal{H},$$

where: 7 holds by continuity of $\hat{J}_P$, by 4.2.12, and by 10.1.16c; 8 holds by 8.2.11.

g: In view of conditions d, e, f, already proved, we have, for $\varphi \in \mathcal{M}_B(X, \mathcal{A})$,

$$\|\hat{J}_P(\varphi) f\|^2 = \left(f | (\hat{J}_P(\varphi))^\dagger \hat{J}_P(\varphi) f \right) = \left(f | \hat{J}_P(\overline{\varphi}) \hat{J}_P(\varphi) f \right)$$

$$= \left(f | \hat{J}_P(|\varphi|^2) f \right) = \int_X |\varphi|^2 d\mu_f^P, \forall f \in \mathcal{H}.$$

h: If $A \in \mathcal{B}(\mathcal{H})$ is so that $AP(E) = P(E)A$ for all $E \in \mathcal{A}$, then obviously

$$A\hat{J}_P(\psi) = AA_\psi^P = A_\psi^P A = \hat{J}_P(\psi)A, \forall \psi \in \mathcal{S}(X, \mathcal{A}).$$

Now, for $\varphi \in \mathcal{M}_B(X, \mathcal{A})$ let $\{\psi_n\}$ be a sequence in $\mathcal{S}(X, \mathcal{A})$ such that $\varphi = \lim_{n\to\infty} \psi_n$ (in the $\| \ \|_\infty$ norm); then, in view of 4.3.3 and of the continuity of $\hat{J}_P$,

$$A\hat{J}_P(\varphi) = A \lim_{n\to\infty} \hat{J}_P(\psi_n) = \lim_{n\to\infty} A\hat{J}_P(\psi_n)$$

$$= \lim_{n\to\infty} \hat{J}_P(\psi_n)A = (\lim_{n\to\infty} \hat{J}_P(\psi_n))A = \hat{J}_P(\psi)A.$$

$\square$

14.1.2 Remark. Let μ be a finite measure on $\mathcal{A}$. Then there exists a unique function $\hat{J}_\mu : \mathcal{M}_B(X, \mathcal{A}) \to \mathbb{C}$ such that:

(a) $\hat{J}_\mu(\chi_E) = \mu(E)$, $\forall E \in \mathcal{A}$;
(b) $\hat{J}_\mu$ is a linear operator;
(c) $\hat{J}_\mu$ is continuous.

Indeed, we can define (cf. 8.2.6)

$$\mathcal{M}_B(X, \mathcal{A}) \ni \varphi \mapsto \hat{J}_\mu(\varphi) := \int_X \varphi \, d\mu \in \mathbb{C},$$

and thus obtain a function with properties a (cf. 8.1.3a) and b (cf. 8.2.9); condition c is proved by

$$|\hat{J}_\mu(\varphi)| \leq \int_X |\varphi| d\mu \leq \mu(X) \|\varphi\|_\infty$$

(cf. 8.2.10 and 8.1.11a). If we assume conversely that conditions a, b, c hold true for a function $J : \mathcal{M}_B(X, \mathcal{A}) \to \mathbb{C}$, then

$$J(\psi) = \int_X \psi \, d\mu = \hat{J}_\mu(\psi), \forall \psi \in \mathcal{S}(X, \mathcal{A}),$$

in view of conditions a and b, and hence $J = \hat{J}_\mu$ by the uniqueness asserted in 4.2.6, since $\mathcal{S}(X, \mathcal{A})$ is dense in $\mathcal{M}_B(X, \mathcal{A})$ and both J and J_μ are continuous.

This shows that there exists a close analogy between the mappings $\hat{J}_P$ and $\hat{J}_\mu$. Owing to this analogy, for $\varphi \in \mathcal{M}_B(X, \mathcal{A})$ the operator $\hat{J}_P(\varphi)$ is called the *integral of φ with respect to P* and it is often denoted as follows

$$\int_X \varphi \, dP := \hat{J}_P(\varphi).$$

14.2   Integrals of general measurable functions

On the basis of the results of the previous section, in this section we extend the notion of an integral with respect to a projection valued measure, to measurable functions which are not necessarily bounded nor necessarily defined on the whole of X.

As before, $(X, \mathcal{A})$ denotes an abstract measurable space, $\mathcal{H}$ denotes an abstract Hilbert space, and P denotes a projection valued measure on $\mathcal{A}$ with values in $\mathscr{P}(\mathcal{H})$.

14.2.1 Definition. Let E be an element of $\mathcal{A}$ and, for each $x \in E$, let $Q(x)$ be a proposition, i.e. a statement about x which is either true or false. We write

"$Q(x)$ P-a.e. on E" or "$Q(x)$ is true P-a.e. on E"

when the following condition is satisfied

$$\exists F \in \mathcal{A} \text{ such that } P(F) = \mathbb{O}_\mathcal{H} \text{ and } Q(x) \text{ is true for all } x \in E - F.$$

It is obvious that, if

$$Q(x) \ P\text{-a.e. on } E$$

then (cf. 7.1.9)

$$Q(x) \ \mu_f^P\text{-a.e. on } E, \forall f \in \mathcal{H}.$$

14.2.2 Definition. We denote by $\mathcal{M}(X, \mathcal{A}, P)$ the family of functions from X to $\mathbb{C}$ which is defined as follows

$$\mathcal{M}(X, \mathcal{A}, P) := \{\varphi : D_\varphi \to \mathbb{C} : D_\varphi \in \mathcal{A}, P(X - D_\varphi) = \mathbb{O}_\mathcal{H}, \varphi \in \mathcal{M}(D_\varphi, \mathcal{A}^{D_\varphi})\}$$

(for $\mathcal{M}(X, \mathcal{A})$, cf. 6.2.15). It is obvious that

$$\mathcal{M}(X, \mathcal{A}, P) \subset \mathcal{M}(X, \mathcal{A}, \mu_f^P), \forall f \in \mathcal{H}$$

(for $\mathcal{M}(X, \mathcal{A}, \mu_f^P)$, cf. 8.2.1). The elements of $\mathcal{M}(X, \mathcal{A}, P)$ are called *P-measurable functions*.

In $\mathcal{M}(X, \mathcal{A}, P)$ we define a relation, denoted by $\sim$, as follows:

$$\varphi \sim \psi \text{ if } \varphi(x) = \psi(x) \ P\text{-a.e. on } D_\varphi \cap D_\psi.$$

14.2.3 Theorem. *The following statements hold true:*

(a) We have:

$$\alpha\varphi + \beta\psi \in \mathcal{M}(X, \mathcal{A}, P), \ \forall \alpha, \beta \in \mathbb{C}, \ \forall \varphi, \psi \in \mathcal{M}(X, \mathcal{A}, P);$$
$$\varphi\psi \in \mathcal{M}(X, \mathcal{A}, P), \ \forall \varphi, \psi \in \mathcal{M}(X, \mathcal{A}, P).$$

However, $\mathcal{M}(X, \mathcal{A}, P)$ is not an associative algebra nor a linear space, unless $E := \emptyset$ is the only element E of $\mathcal{A}$ such that $P(E) = \mathbb{O}_\mathcal{H}$.

(b) The relation $\sim$ in $\mathcal{M}(X, \mathcal{A}, P)$ is an equivalence relation.

(c) If $\varphi, \varphi', \psi, \psi' \in \mathcal{M}(X, \mathcal{A}, P)$ are so that $\varphi' \sim \varphi$ and $\psi' \sim \psi$, then

$$\alpha\varphi' + \beta\psi' \sim \alpha\varphi + \beta\psi, \ \forall \alpha, \beta \in \mathbb{C},$$
$$\varphi'\psi' \sim \varphi\psi.$$

Proof. The arguments used in the proofs of 8.2.2, 8.2.12, 8.2.13 can be repeated word for word, with the measure μ replaced by the projection valued measure P and references to 7.1.2b replaced by references to 13.3.2h. $\qquad\square$

14.2.4 Definitions. We denote by $\mathcal{L}^\infty(X, \mathcal{A}, P)$ the subset of $\mathcal{M}(X, \mathcal{A}, P)$ defined by

$$\mathcal{L}^\infty(X, \mathcal{A}, P) := \{\varphi \in \mathcal{M}(X, \mathcal{A}, P) : \ \exists m \in [0, \infty) \text{ such that }$$
$$|\varphi(x)| \leq m \ P\text{-a.e. on } D_\varphi\}.$$

For $\varphi \in \mathcal{L}^\infty(X, \mathcal{A}, P)$, we define the *essential supremum* of $|\varphi|$ (with respect to P) as

$$P\text{-sup}|\varphi| := \inf\{m \in [0, \infty) : |\varphi(x)| \leq m \ P\text{-a.e. on } D_\varphi\};$$

obviously, $P\text{-sup}|\varphi| \in [0, \infty)$.

Clearly,

$$\mathcal{M}_B(X, \mathcal{A}) \subset \mathcal{L}^\infty(X, \mathcal{A}, P) \text{ and } P\text{-sup}|\varphi| \leq \|\varphi\|_\infty, \forall \varphi \in \mathcal{M}_B(X, \mathcal{A}).$$

14.2.5 Theorem. *We have:*

$|\varphi(x)| \le P\text{-}sup|\varphi|$ *P-a.e. on* D_φ, $\forall \varphi \in \mathcal{L}^\infty(X, \mathcal{A}, P)$;

$[\varphi \in \mathcal{L}^\infty(X, \mathcal{A}, P)$ *and* $P\text{-}sup|\varphi| = 0] \Rightarrow [\varphi(x) = 0$ *P-a.e. on* $D_\varphi]$;

$\alpha\varphi \in \mathcal{L}^\infty(X, \mathcal{A}, P)$ *and* $P\text{-}sup|\alpha\varphi| = |\alpha|P\text{-}sup|\varphi|$, $\forall \alpha \in \mathbb{C}$, $\forall \varphi \in \mathcal{L}^\infty(X, \mathcal{A}, P)$;

$\varphi + \psi \in \mathcal{L}^\infty(X, \mathcal{A}, P)$ *and* $P\text{-}sup|\varphi + \psi| \le P\text{-}sup|\varphi| + P\text{-}sup|\psi|$,
$\forall \varphi, \psi \in \mathcal{L}^\infty(X, \mathcal{A}, P)$;

$\varphi\psi \in \mathcal{L}^\infty(X, \mathcal{A}, P)$ *and* $P\text{-}sup|\varphi\psi| \le (P\text{-}sup|\varphi|)(P\text{-}sup|\psi|)$,
$\forall \varphi, \psi \in \mathcal{L}^\infty(X, \mathcal{A}, P)$;

$\overline{\varphi} \in \mathcal{L}^\infty(X, \mathcal{A}, P)$ *and* $P\text{-}sup|\overline{\varphi}| = P\text{-}sup|\varphi|$, $\forall \varphi \in \mathcal{L}^\infty(X, \mathcal{A}, P)$.

However, $\mathcal{L}^\infty(X, \mathcal{A}, P)$ *is not an associative algebra nor a linear space, and the function*

$$\mathcal{L}^\infty(X, \mathcal{A}, P) \ni \varphi \mapsto P\text{-}sup|\varphi| \in \mathbb{R}$$

is not a norm, nor is a norm its restriction to $\mathcal{M}_B(X, \mathcal{A})$, *unless* $E := \emptyset$ *is the only element* E *of* $\mathcal{A}$ *such that* $P(E) = \mathbb{O}_\mathcal{H}$.

Proof. Let $\varphi \in \mathcal{L}^\infty(X, \mathcal{A}, P)$. For each $n \in \mathbb{N}$, there exist $m_n \in [0, \infty)$ and $E_n \in \mathcal{A}$ so that

$$m_n < (P\text{-}sup|\varphi|) + \frac{1}{n}, \quad P(E_n) = \mathbb{O}_\mathcal{H}, \quad |\varphi(x)| \le m_n \text{ for all } x \in D_\varphi - E_n;$$

letting $E := \bigcup_{n=1}^\infty E_n$, we have $E \in \mathcal{A}$, $P(E) = \mathbb{O}_\mathcal{H}$ (cf. 13.3.6c), and also

$$x \in D_\varphi - E = \bigcap_{n=1}^\infty (D_\varphi - E_n) \Rightarrow$$

$$[|\varphi(x)| < (P\text{-}sup|\varphi|) + \frac{1}{n}, \forall n \in \mathbb{N}] \Rightarrow |\varphi(x)| \le P\text{-}sup|\varphi|;$$

this proves that

$$|\varphi(x)| \le P\text{-}sup|\varphi| \text{ P-a.e. on } D_\varphi.$$

Then, it is obvious that

$$P\text{-}sup|\varphi| = 0 \Rightarrow [\varphi(x) = 0 \text{ P-a.e. on } D_\varphi].$$

We have $\alpha\varphi \in \mathcal{M}(X, \mathcal{A}, P)$ for every $\alpha \in \mathbb{C}$, by 14.2.3a. If $\alpha = 0$ then the equation

$$P\text{-}sup|\alpha\varphi| = |\alpha|P\text{-}sup|\varphi|$$

is obvious. If $\alpha \ne 0$, then

$$|\alpha\varphi(x)| = |\alpha||\varphi(x)| \le |\alpha|P\text{-}sup|\varphi| \text{ P-a.e. on } D_\varphi = D_{\alpha\varphi},$$

whence

$$\alpha\varphi \in \mathcal{L}^\infty(X, \mathcal{A}, P) \text{ and } P\text{-}sup|\alpha\varphi| \le |\alpha|P\text{-}sup|\varphi|;$$

by the same token,

$$P\text{-}sup|\varphi| = P\text{-}sup|\alpha^{-1}\alpha\varphi| \le |\alpha^{-1}|P\text{-}sup|\alpha\varphi|,$$

whence

$$|\alpha| P\text{-sup}|\varphi| \le P\text{-sup}|\alpha\varphi|;$$

therefore,

$$P\text{-sup}|\alpha\varphi| = |\alpha| P\text{-sup}|\varphi|.$$

For every $\psi \in \mathcal{L}^\infty(X,\mathcal{A},P)$ we have $\varphi + \psi \in \mathcal{M}(X,\mathcal{A},P)$ and $\varphi\psi \in \mathcal{M}(X,\mathcal{A},P)$ by 14.2.3a. Now, let E be as before and let $F \in \mathcal{A}$ be such that

$$P(F) = \mathbb{O}_\mathcal{H} \text{ and } |\psi(x)| \le P\text{-sup}|\psi|, \forall x \in D_\psi - F;$$

then $E \cup F \in \mathcal{A}$ and $P(E \cup F) = \mathbb{O}_\mathcal{H}$ (cf. 13.3.2h), and also

$$|\varphi(x) + \psi(x)| \le |\varphi(x)| + |\psi(x)| \le P\text{-sup}|\varphi| + P\text{-sup}|\psi|,$$
$$\forall x \in (D_\varphi - E) \cap (D_\psi - F) = D_{\varphi+\psi} - (E \cup F),$$

which proves that

$$\varphi + \psi \in \mathcal{L}^\infty(X,\mathcal{A},P) \text{ and } P\text{-sup}|\varphi + \psi| \le P\text{-sup}|\varphi| + P\text{-sup}|\psi|;$$

moreover

$$|\varphi(x)\psi(x)| = |\varphi(x)||\psi(x)| \le (P\text{-sup}|\varphi|)(P\text{-sup}|\psi|),$$
$$\forall x \in (D_\varphi - E) \cap (D_\psi - F) = D_{\varphi\psi} - (E \cup F),$$

which proves that

$$\varphi\psi \in \mathcal{L}^\infty(X,\mathcal{A},P) \text{ and } P\text{-sup}|\varphi\psi| \le (P\text{-sup}|\varphi|)(P\text{-sup}|\psi|).$$

It is obvious that

$$\overline{\varphi} \in \mathcal{L}^\infty(X,\mathcal{A},P) \text{ and } P\text{-sup}|\varphi| = P\text{-sup}|\overline{\varphi}|.$$

Finally, suppose that there exists $E \in \mathcal{A}$ such that $E \ne \emptyset$ and $P(E) = \mathbb{O}_\mathcal{H}$. Then the family of functions $\mathcal{L}^\infty(X,\mathcal{A},P)$ is not an associative algebra nor a linear space for the same reason why $\mathcal{M}(X,\mathcal{A},P)$ is not (cf. the proof of 8.2.2). Therefore, the function $\mathcal{L}^\infty(X,\mathcal{A},P) \ni \varphi \mapsto P\text{-sup}|\varphi| \in \mathbb{R}$ cannot be a norm. Moreover $\chi_E \in \mathcal{M}_B(X,\mathcal{A})$, $\chi_E \ne 0_X$, and $P\text{-sup}|\chi_E| = 0$; this proves that the function $\mathcal{M}_B(X,\mathcal{A}) \ni \varphi \mapsto P\text{-sup}|\varphi| \in \mathbb{R}$ is not a norm. $\square$

14.2.6 Proposition. *For every $\varphi \in \mathcal{L}^\infty(X,\mathcal{A},P)$ there exists $\varphi_e \in \mathcal{M}_B(X,\mathcal{A})$ such that $\varphi_e(x) = \varphi(x)$ P-a.e. on D_φ.*

Proof. Let $\varphi \in \mathcal{L}^\infty(X,\mathcal{A},P)$ and let $E \in \mathcal{A}$ be such that

$$P(E) = \mathbb{O}_\mathcal{H} \text{ and } |\varphi(x)| \le P\text{-sup}|\varphi|, \forall x \in D_\varphi - E.$$

Then the function

$$\varphi_e : X \to \mathbb{C}$$

$$x \mapsto \varphi_e(x) := \begin{cases} \varphi(x) & \text{if } x \in D_\varphi - E, \\ 0 & \text{if } x \in X - (D_\varphi - E) \end{cases}$$

is $\mathcal{A}$-measurable (this is true by 6.2.12, because $\varphi_{D_\varphi - E}$ is $\mathcal{A}^{D_\varphi - E}$-measurable in view of 6.2.3 and because $\text{Re}\,\varphi_e$ and $\text{Im}\,\varphi_e$ are the standard extensions of $\text{Re}\,\varphi_{D_\varphi - E}$ and of $\text{Im}\,\varphi_{D_\varphi - E}$ respectively, cf. 8.1.14) and bounded, i.e. $\varphi_e \in \mathcal{M}_B(X,\mathcal{A})$; moreover, $\varphi_e(x) = \varphi(x)$ P-a.e. on D_φ since $P(X - D_\varphi) = \mathbb{O}_\mathcal{H}$ and $P(E) = \mathbb{O}_\mathcal{H}$, and hence $P(X - (D_\varphi - E)) = P((X - D_\varphi) \cup E) = \mathbb{O}_\mathcal{H}$ (cf. 13.3.2h). $\square$

14.2.7 Proposition. *The mapping*

$$\tilde{J}_P : \mathcal{L}^\infty(X, \mathcal{A}, P) \to \mathcal{B}(\mathcal{H})$$

$$\varphi \mapsto \tilde{J}_P(\varphi) := \hat{J}_P(\varphi_e) \text{ if } \varphi_e \in \mathcal{M}_B(X, \mathcal{A}) \text{ is such that}$$

$$\varphi_e(x) = \varphi(x) \ P\text{-a.e. on } D_\varphi$$

is defined consistently and is an extension of the mapping $\hat{J}_P$.

The following conditions hold true:

(a) $\tilde{J}_P(\chi_E) = P(E), \ \forall E \in \mathcal{A}$;
(b) $\tilde{J}_P(\alpha\varphi + \beta\psi) = \alpha\tilde{J}_P(\varphi) + \beta\tilde{J}_P(\psi), \ \forall \alpha, \beta \in \mathbb{C}, \ \forall \varphi, \psi \in \mathcal{L}^\infty(X, \mathcal{A}, P)$;
(c) $\tilde{J}_P(\varphi\psi) = \tilde{J}_P(\varphi)\tilde{J}_P(\psi), \ \forall \varphi, \psi \in \mathcal{L}^\infty(X, \mathcal{A}, P)$;
(d) $\tilde{J}_P(\overline{\varphi}) = (\tilde{J}_P(\varphi))^\dagger, \ \forall \varphi \in \mathcal{L}^\infty(X, \mathcal{A}, P)$;
(e) $\left(f | \tilde{J}_P(\varphi) f\right) = \int_X \varphi \, d\mu_f^P, \ \forall f \in \mathcal{H}, \ \forall \varphi \in \mathcal{L}^\infty(X, \mathcal{A}, P)$;
(f) $\|\tilde{J}_P(\varphi)f\|^2 = \int_X |\varphi|^2 \, d\mu_f^P, \ \forall f \in \mathcal{H}, \ \forall \varphi \in \mathcal{L}^\infty(X, \mathcal{A}, P)$;
(g) *if $A \in \mathcal{B}(\mathcal{H})$ is so that $AP(E) = P(E)A$ for all $E \in \mathcal{A}$, then $A\tilde{J}_P(\varphi) = \tilde{J}_P(\varphi)A$ for all $\varphi \in \mathcal{L}^\infty(X, \mathcal{A}, P)$;*
(h) $\|\tilde{J}_P(\varphi)\| = P\text{-sup}|\varphi|, \ \forall \varphi \in \mathcal{L}^\infty(X, \mathcal{A}, P)$;
(i) *for $\varphi, \varphi' \in \mathcal{L}^\infty(X, \mathcal{A}, P)$, $\tilde{J}_P(\varphi) = \tilde{J}_P(\varphi')$ iff $\varphi(x) = \varphi'(x)$ P-a.e. on $D_\varphi \cap D_{\varphi'}$.*

Proof. Let $\varphi \in \mathcal{L}^\infty(X, \mathcal{A}, P)$. By 14.2.6, there exists $\varphi_e \in \mathcal{M}_B(X, \mathcal{A})$ such that $\varphi_e(x) = \varphi(x)$ P-a.e. on D_φ. Suppose that $\varphi_e' \in \mathcal{M}_B(X, \mathcal{A})$ also is such that $\varphi_e'(x) = \varphi(x)$ P-a.e. on D_φ. Then $\varphi_e(x) = \varphi_e'(x)$ P-a.e. on X since the relation $\sim$ in $\mathcal{M}(X, \mathcal{A}, P)$ is an equivalence relation (cf. 14.2.3b), and hence $\varphi_e(x) = \varphi_e'(x)$ μ_f^P-a.e. on X for all $f \in \mathcal{H}$, and hence (cf. 14.1.1f and 8.2.7)

$$\left(f | \hat{J}_P(\varphi_e)f\right) = \int_X \varphi_e \, d\mu_f^P = \int_X \varphi_e' \, d\mu_f^P = \left(f | \hat{J}_P(\varphi_e')f\right), \forall f \in \mathcal{H},$$

and hence $\hat{J}_P(\varphi_e) = \hat{J}_P(\varphi_e')$ (cf. 10.2.12). This proves that the mapping $\tilde{J}_P$ is defined consistently. It is obvious that $\tilde{J}_P$ is an extension of $\hat{J}_P$.

Now we prove the conditions listed in the statement. For $\varphi \in \mathcal{L}^\infty(X, \mathcal{A}, P)$, we denote by φ_e an element of $\mathcal{M}_B(X, \mathcal{A})$ such that $\varphi_e(x) = \varphi(x)$ P-a.e. on D_φ.

a: This follows at once from 14.1.1.a, since $\tilde{J}_P$ is an extension of $\hat{J}_P$.

b: Let $\alpha, \beta \in \mathbb{C}$ and $\varphi, \psi \in \mathcal{L}^\infty(X, \mathcal{A}, P)$. Then, $\alpha\varphi_e + \beta\psi_e \in \mathcal{M}_B(X, \mathcal{A})$ and

$$(\alpha\varphi_e + \beta\psi_e)(x) = (\alpha\varphi + \beta\psi)(x) \ P\text{-a.e. on } D_{\alpha\varphi + \beta\psi}$$

by 14.2.3c, and hence (cf. 14.1.1b)

$$\tilde{J}_P(\alpha\varphi + \beta\psi) = \hat{J}_P(\alpha\varphi_e + \beta\psi_e) = \alpha\hat{J}_P(\varphi_e) + \beta\hat{J}_P(\psi_e) = \alpha\tilde{J}_P(\varphi) + \beta\tilde{J}_P(\psi).$$

c: Let $\varphi, \psi \in \mathcal{L}^\infty(X, \mathcal{A}, P)$. Then $\varphi_e\psi_e \in \mathcal{M}_B(X, \mathcal{A})$ and

$$(\varphi_e\psi_e)(x) = (\varphi\psi)(x) \ P\text{-a.e. on } D_{\varphi\psi}$$

by 14.2.3c, and hence (cf. 14.1.1d)

$$\tilde{J}_P(\varphi\psi) = \hat{J}_P(\varphi_e\psi_e) = \hat{J}_P(\varphi_e)\hat{J}_P(\psi_e) = \tilde{J}_P(\varphi)\tilde{J}_P(\psi).$$

d: For every $\varphi \in \mathcal{L}^\infty(X, \mathcal{A}, P)$, it is obvious that $\overline{\varphi_e} \in \mathcal{M}_B(X, \mathcal{A})$ and that

$$\overline{\varphi_e}(x) = \overline{\varphi}(x) \ P\text{-a.e. on } D_{\overline{\varphi}};$$

then (cf. 14.1.1e)

$$\tilde{J}_P(\overline{\varphi}) = \hat{J}_P(\overline{\varphi_e}) = (\hat{J}_P(\varphi_e))^\dagger = (\tilde{J}_P(\varphi))^\dagger.$$

e: For every $\varphi \in \mathcal{L}^\infty(X, \mathcal{A}, P)$,

$$\left(f | \tilde{J}_P(\varphi) f \right) = \left(f | \hat{J}_P(\varphi_e) f \right) = \int_X \varphi_e d\mu_f^P = \int_X \varphi d\mu_f^P, \forall f \in \mathcal{H},$$

by 14.1.1f and 8.2.7 (since $\varphi_e(x) = \varphi(x) \ \mu_f^P$-a.e. on $D_\varphi, \forall f \in \mathcal{H}$).

f: For every $\varphi \in \mathcal{L}^\infty(X, \mathcal{A}, P)$,

$$\|\tilde{J}_P(\varphi) f\|^2 = \|\hat{J}_P(\varphi_e) f\|^2 = \int_X |\varphi_e|^2 d\mu_f^P = \int_X |\varphi|^2 d\mu_f^P, \forall f \in \mathcal{H},$$

by 14.1.1g and 8.2.7 (or 8.1.17c).

g: This follows at once from the definition of $\tilde{J}_P$ and 14.1.1h.

h: Let $\varphi \in \mathcal{L}^\infty(X, \mathcal{A}, P)$. Since $|\varphi(x)| \le P\text{-sup}|\varphi| \ P$-a.e. on D_φ (cf. 14.2.5), we have

$$\|\tilde{J}_P(\varphi) f\|^2 \le (P\text{-sup}|\varphi|)^2 \int_X 1_X d\mu_f^P = (P\text{-sup}|\varphi|)^2 \mu_f^P(X)$$
$$= (P\text{-sup}|\varphi|)^2 \|f\|^2, \forall f \in \mathcal{H},$$

by condition f and 8.1.17b. This proves that

$$\|\tilde{J}_P(\varphi)\| \le P\text{-sup}|\varphi|.$$

If $P\text{-sup}|\varphi| = 0$, this inequality implies $\|\tilde{J}_P(\varphi)\| = P\text{-sup}|\varphi|$. Then, suppose $P\text{-sup}|\varphi| \ne 0$. Let $n \in \mathbb{N}$ be such that $\frac{1}{n} < P\text{-sup}|\varphi|$, and define

$$E_n := |\varphi|^{-1}([(P\text{-sup}|\varphi|) - \frac{1}{n}, \infty));$$

we have $E_n \in \mathcal{A}$ (cf. 6.2.17 and 6.2.13a with $n := 7$) and $P(E_n) \ne 0_\mathcal{H}$; in fact, $P(E_n) = 0_\mathcal{H}$ would imply $P\text{-sup}|\varphi| \le (P\text{-sup}|\varphi|) - \frac{1}{n}$, which is a contradiction; therefore, there exists $f_n \in \mathcal{H}$ such that $f_n \ne 0_\mathcal{H}$ and $P(E_n) f_n = f_n$, and hence such that (cf. 13.3.2b)

$$\mu_{f_n}^P(X - E_n) = \|P(X - E_n) P(E_n) f_n\|^2 = 0,$$

and hence such that

$$\|\tilde{J}_P(\varphi) f_n\|^2 = \int_X |\varphi|^2 d\mu_{f_n}^P = \int_{E_n} |\varphi|^2 d\mu_{f_n}^P \ge \left((P\text{-sup}|\varphi|) - \frac{1}{n} \right)^2 \int_{E_n} 1_X d\mu_{f_n}^P$$
$$= \left((P\text{-sup}|\varphi|) - \frac{1}{n} \right)^2 \int_X 1_X d\mu_{f_n}^P = \left((P\text{-sup}|\varphi|) - \frac{1}{n} \right)^2 \|f_n\|^2$$

(cf. condition f, 8.3.3a, 8.1.7). In view of 4.2.5c, this proves that

$$\|\tilde{J}_P(\varphi)\| \ge (P\text{-sup}|\varphi|) - \frac{1}{n}, \forall n \in \mathbb{N},$$

and hence that

$$\|\tilde{J}_P(\varphi)\| \geq P\text{-sup}|\varphi|.$$

i: If $\varphi, \varphi' \in \mathcal{L}^\infty(X, \mathcal{A}, P)$ are such that $\varphi(x) = \varphi'(x)$ P-a.e. on $D_\varphi \cap D_{\varphi'}$, then

$$\varphi(x) = \varphi'(x) \ \mu_f^P\text{-a.e. on } D_\varphi \cap D_{\varphi'}, \forall f \in \mathcal{H},$$

and hence (cf. condition e and 8.2.7)

$$\left(f|\tilde{J}_P(\varphi)f\right) = \int_X \varphi d\mu_f^P = \int_X \varphi' d\mu_f^P = \left(f|\tilde{J}_P(\varphi')f\right), \forall f \in \mathcal{H},$$

and hence $\tilde{J}_P(\varphi) = \tilde{J}_P(\varphi')$ by 10.2.12. Conversely, if $\varphi, \varphi' \in \mathcal{L}^\infty(X, \mathcal{A}, P)$ are such that $\tilde{J}_P(\varphi) = \tilde{J}_P(\varphi')$, then (cf. conditions b and h)

$$P\text{-sup}|\varphi - \varphi'| = \|\tilde{J}_P(\varphi - \varphi')\| = \|\tilde{J}_P(\varphi) - \tilde{J}_P(\varphi')\| = 0,$$

and hence (cf. 14.2.5)

$$\varphi(x) - \varphi'(x) = 0, \text{ i.e. } \varphi(x) = \varphi'(x), \ P\text{-a.e. on } D_{\varphi - \varphi'} = D_\varphi \cap D_{\varphi'}.$$

$\square$

14.2.8 Remark. We denote by $M(X, \mathcal{A}, P)$ the quotient set defined by the equivalence relation $\sim$ in $\mathcal{M}(X, \mathcal{A}, P)$ (cf. 14.2.3.b). On the basis of 14.2.3a,c, it is easy to see that $M(X, \mathcal{A}, P)$ becomes an abelian associative algebra if we define

$$[\varphi] + [\psi] := [\varphi + \psi], \forall[\varphi], [\psi] \in M(X, \mathcal{A}, P),$$
$$\alpha[\varphi] := [\alpha\varphi], \forall \alpha \in \mathbb{C}, \forall[\varphi] \in M(X, \mathcal{A}, P),$$
$$[\varphi][\psi] := [\varphi\psi], \forall[\varphi], [\psi] \in M(X, \mathcal{A}, P)$$

(there is a close analogy between $M(X, \mathcal{A}, P)$ and $M(X, \mathcal{A}, \mu)$, cf. 8.2.13).

We can define a subset of $M(X, \mathcal{A}, P)$ by

$$L^\infty(X, \mathcal{A}, P) := \{[\varphi] \in M(X, \mathcal{A}, P) : \varphi \in \mathcal{L}^\infty(X, \mathcal{A}, P)\}.$$

Indeed, if $\varphi \in \mathcal{L}^\infty(X, \mathcal{A}, P)$ and $\varphi' \in [\varphi]$, let $E \in \mathcal{A}$ be such that

$$P(E) = \mathbb{O}_\mathcal{H} \text{ and } \exists m \in [0, \infty) \text{ s.t. } |\varphi(x)| \leq m, \forall x \in D_\varphi - E,$$

and let $F \in \mathcal{A}$ be such that

$$P(F) = \mathbb{O}_\mathcal{H} \text{ and } \varphi'(x) = \varphi(x), \forall x \in (D_{\varphi'} \cap D_\varphi) - F;$$

then,

$$|\varphi'(x)| \leq m, \forall x \in ((D_{\varphi'} \cap D_\varphi) - F) \cap (D_\varphi - E) = D_{\varphi'} - ((X - D_\varphi) \cup F \cup E),$$

and this proves that $\varphi' \in \mathcal{L}^\infty(X, \mathcal{A}, P)$, in view of 13.3.2h. Thus, the condition $\varphi \in \mathcal{L}^\infty(X, \mathcal{A}, P)$ is actually a condition for the equivalence class $[\varphi]$ even though it is expressed through a particular element of the class. On the basis of 14.2.5, it is easy to see that $L^\infty(X, \mathcal{A}, P)$ is a subalgebra (cf. 3.3.2) of $M(X, \mathcal{A}, P)$, and that it becomes a normed algebra if we define a norm by

$$\|[\varphi]\| := P\text{-sup}|\varphi|, \forall[\varphi] \in L^\infty(X, \mathcal{A}, P).$$

Proceeding as we would if P were a measure on $\mathcal{A}$, we can prove that $L^\infty(X, \mathcal{A}, P)$ is a Banach algebra (cf. e.g. Berberian, 1999, 6.6.7). Then, it is obvious that $L^\infty(X, \mathcal{A}, P)$ becomes a C^*-algebra if we define an involution by

$$[\varphi]^* := [\overline{\varphi}], \forall \varphi \in L^\infty(X, \mathcal{A}, P).$$

On the basis of 14.2.7b,c,d,h,i, we can define the mapping

$$L^\infty(X, \mathcal{A}, P) \ni [\varphi] \mapsto \Phi_P([\varphi]) := \tilde{J}_P(\varphi) \in \mathcal{B}(\mathcal{H}),$$

see that it is a homomorphism (cf. 3.3.5) from $L^\infty(X, \mathcal{A}, P)$ to $\mathcal{B}(\mathcal{H})$, and see that

$$\|\Phi_P([\varphi])\| = \|[\varphi]\|, \forall [\varphi] \in L^\infty(X, \mathcal{A}, P),$$
$$(\Phi_P([\varphi]))^\dagger = \Phi_P([\varphi]^*), \forall [\varphi] \in L^\infty(X, \mathcal{A}, P).$$

Then, it is easy to see that R_{Φ_P} is an abelian C^*-algebra (cf. 3.3.6 and 2.6.4). Thus Φ_P is an isomorphism from the C^*-algebra $L^\infty(X, \mathcal{A}, P)$ onto this C^*-algebra, and it is norm-preserving and involution-preserving.

14.2.9 Definition. Let $\varphi \in \mathcal{M}(X, \mathcal{A}, P)$. A sequence $\{\varphi_n\}$ in $\mathcal{M}(X, \mathcal{A}, P)$ is said to be φ-convergent if the following conditions are satisfied:

$\varphi_n \in \mathcal{L}^\infty(X, \mathcal{A}, P), \forall n \in \mathbb{N};$

$$\varphi(x) = \lim_{n \to \infty} \varphi_n(x) \ P\text{-a.e. on } D_\varphi \cap \left(\bigcap_{n=1}^\infty D_{\varphi_n} \right);$$

$\exists k_1, k_2 \in [0, \infty)$ such that $|\varphi_n(x)|^2 \le k_1 |\varphi(x)|^2 + k_2 \ P\text{-a.e. on } D_\varphi \cap D_{\varphi_n}, \forall n \in \mathbb{N}.$

14.2.10 Remarks.

(a) Let $\varphi \in \mathcal{M}(X, \mathcal{A}, P)$. For each $n \in \mathbb{N}$, we define the set

$$E_n := |\varphi|^{-1}([0, n]),$$

which is an element of $\mathcal{A}$ (cf. 6.2.17 and 6.2.13a with $n := 3$). It is obvious that the sequence $\{\chi_{E_n} \varphi\}$ is φ-convergent. This proves that the family of φ-convergent sequences is not empty.

(b) If $\psi \in \mathcal{L}^\infty(X, \mathcal{A}, P)$ then $|\psi(x)| \le P\text{-sup}|\psi| \ \mu_f^P\text{-a.e. on } D_\psi$ for all $f \in \mathcal{H}$ (cf. 14.2.5), and hence $\psi \in \mathcal{L}^2(X, \mathcal{A}, \mu_f^P)$ for all $f \in \mathcal{H}$ (cf. 8.2.6). Thus, if $\varphi \in \mathcal{M}(X, \mathcal{A}, P)$ and $\{\varphi_n\}$ is a φ-convergent sequence, then $\varphi_n \in \mathcal{L}^2(X, \mathcal{A}, \mu_f^P)$ for all $f \in \mathcal{H}$ and all $n \in \mathbb{N}$.

14.2.11 Proposition. *Let $\varphi \in \mathcal{M}(X, \mathcal{A}, P)$ and let $D_P(\varphi)$ be the subset of $\mathcal{H}$ defined by*

$$D_P(\varphi) := \{f \in \mathcal{H} : \varphi \in \mathcal{L}^2(X, \mathcal{A}, \mu_f^P)\} = \left\{ f \in \mathcal{H} : \int_X |\varphi|^2 d\mu_f^P < \infty \right\}.$$

Let $\{\varphi_n\}$ be a φ-convergent sequence. For $f \in \mathcal{H}$, the following conditions are equivalent:

(a) $f \in D_P(\varphi)$;
(b) the sequence $\{[\varphi_n]\}$ is convergent in the Hilbert space $L^2(X, \mathcal{A}, \mu_f^P)$;
(c) the sequence $\{\tilde{J}_P(\varphi_n)f\}$ is convergent in the Hilbert space $\mathcal{H}$.

If these conditions are satisfied, then:

(d) $[\varphi] = \lim_{n \to \infty}[\varphi_n]$ in the Hilbert space $L^2(X, \mathcal{A}, \mu_f^P)$;
(e) *if $\{\varphi_n'\}$ is any φ-convergent sequence, then*

$$\lim_{n \to \infty} \tilde{J}_P(\varphi_n')f = \lim_{n \to \infty} \tilde{J}_P(\varphi_n)f.$$

Proof. $a \Rightarrow$ (b and d): Since the sequence $\{\varphi_n\}$ is φ-convergent, for any $f \in \mathcal{H}$ we have:

$$\lim_{n \to \infty} |\varphi_n(x) - \varphi(x)|^2 = 0 \ \mu_f^P\text{-a.e. on } D_\varphi \cap \left(\bigcap_{n=1}^\infty D_{\varphi_n} \right) = \bigcap_{n=1}^\infty D_{\varphi_n - \varphi};$$

$\exists k_1, k_2 \in [0, \infty)$ such that
$$|\varphi_n(x) - \varphi(x)|^2 \le 2|\varphi_n(x)|^2 + 2|\varphi(x)|^2 \le 2(k_1 + 1)|\varphi(x)|^2 + 2k_2$$
μ_f^P-a.e. on $D_{\varphi_n - \varphi}, \forall n \in \mathbb{N}$

(cf. inequality 2 in the proof of 10.3.7). If $f \in D_P(\varphi)$ then $\varphi \in \mathcal{L}^2(X, \mathcal{A}, \mu_f^P)$ and hence

$$\lim_{n \to \infty} \|[\varphi_n] - [\varphi]\|^2_{L^2(X, \mathcal{A}, \mu_f^P)} = \lim_{n \to \infty} \int_X |\varphi_n - \varphi|^2 d\mu_f^P = 0$$

by 8.2.11 with $2(k_1 + 1)|\varphi|^2 + 2k_2$ as dominating function (recall that a constant function is μ_f^P integrable since $\mu_f^P(X) < \infty$, cf. 8.2.6).

$b \Rightarrow c$: Assuming condition b, we have

$$\int_X |\varphi_n - \varphi_m|^2 d\mu_f^P = \|[\varphi_n] - [\varphi_m]\|^2_{L^2(X, \mathcal{A}, \mu_f^P)} \to 0 \text{ as } n, m \to \infty,$$

and hence, in view of 14.2.7b,f,

$$\|\tilde{J}_P(\varphi_n)f - \tilde{J}_P(\varphi_m)f\|^2 = \|\tilde{J}_P(\varphi_n - \varphi_m)f\|^2$$
$$= \int_X |\varphi_n - \varphi_m|^2 d\mu_f^P \to 0 \text{ as } n, m \to \infty,$$

and hence condition c, since $\mathcal{H}$ is a complete metric space.

$c \Rightarrow a$: Assuming condition c, let $g_f := \lim_{n \to \infty} \tilde{J}_P(\varphi_n)f$. Then,

$$\|g_f\|^2 = \lim_{n \to \infty} \|\tilde{J}_P(\varphi_n)f\|^2 = \lim_{n \to \infty} \int_X |\varphi_n|^2 d\mu_f^P$$

(cf. 14.2.7f) and hence (cf. 2.1.9)

$$\exists M \in [0, \infty) \text{ such that } \int_X |\varphi_n|^2 d\mu_f^P \le M, \forall n \in \mathbb{N}.$$

Since

$$\varphi(x) = \lim_{n \to \infty} \varphi_n(x) \ \mu_f^P\text{-a.e. on } D_\varphi \cap \left(\bigcap_{n=1}^\infty D_{\varphi_n} \right),$$

this implies

$$\int_X |\varphi|^2 d\mu_f^P \le M$$

by 8.1.20, i.e. $f \in D_P(\varphi)$.

e: Assuming condition a, let $\{\varphi_n'\}$ be a φ-convergent sequence. From condition d (written for $\{\varphi_n\}$ and for $\{\varphi_n'\}$) we have

$$\int_X |\varphi_n' - \varphi_n|^2 d\mu_f^P = \|[\varphi_n'] - [\varphi_n]\|_{L^2(X,\mathcal{A},\mu_f^P)}^2$$

$$\le (\|[\varphi_n'] - [\varphi]\|_{L^2(X,\mathcal{A},\mu_f^P)} + \|[\varphi] - [\varphi_n]\|_{L^2(X,\mathcal{A},\mu_f^P)})^2 \xrightarrow[n\to\infty]{} 0.$$

and hence (cf. 14.2.7b,f)

$$\|\tilde{J}_P(\varphi_n')f - \tilde{J}_P(\varphi_n)f\|^2 = \|\tilde{J}_P(\varphi_n' - \varphi_n)f\|^2 = \int_X |\varphi_n' - \varphi_n|^2 d\mu_f^P \xrightarrow[n\to\infty]{} 0,$$

and hence

$$\|\lim_{k\to\infty} \tilde{J}_P(\varphi_k)f - \tilde{J}_P(\varphi_n')f\|$$

$$\le \|\lim_{k\to\infty} \tilde{J}_P(\varphi_k)f - \tilde{J}_P(\varphi_n)f\| + \|\tilde{J}_P(\varphi_n)f - \tilde{J}_P(\varphi_n')f\| \xrightarrow[n\to\infty]{} 0,$$

which is condition e. $\square$

14.2.12 Lemma. *Let* $\varphi \in \mathcal{L}^\infty(X,\mathcal{A},P)$ *and* $f \in \mathcal{H}$*. If we write* $g := \tilde{J}_P(\varphi)f$*, then*

$$\mu_g^P(E) = \int_E |\varphi|^2 d\mu_f^P, \forall E \in \mathcal{A}.$$

Proof. In view of 14.2.7a,c,f, for every $E \in \mathcal{A}$ we have

$$\mu_g^P(E) = \|P(E)\tilde{J}_P(\varphi)f\|^2 = \|\tilde{J}_P(\chi_E)\tilde{J}_P(\varphi)f\|^2$$

$$= \|\tilde{J}_P(\chi_E\varphi)f\|^2 = \int_X \chi_E|\varphi|^2 d\mu_f^P = \int_E |\varphi|^2 d\mu_f^P.$$

$\square$

14.2.13 Proposition. *For all* $\varphi \in \mathcal{M}(X,\mathcal{A},P)$*,* $\overline{D_P(\varphi)} = \mathcal{H}$*.*

Proof. Let $\varphi \in \mathcal{M}(X,\mathcal{A},P)$. For each $n \in \mathbb{N}$, we define the set

$$E_n := |\varphi|^{-1}([0,n]),$$

which is an element of $\mathcal{A}$ (cf. 14.2.10a). Clearly

$$E_n \subset E_{n+1} \text{ for all } n \in \mathbb{N} \text{ and } \bigcup_{n=1}^{\infty} E_n = D_\varphi;$$

since $P(X - D_\varphi) = \mathbb{O}_\mathcal{H}$, in view of 13.3.6a this implies that

$$f = P(X)f = P(D_\varphi)f = P\left(\bigcup_{n=1}^{\infty} E_n\right)f = \lim_{n\to\infty} P(E_n)f, \forall f \in \mathcal{H}.$$

Now, we fix $f \in \mathcal{H}$ and write $g_n := P(E_n)f$ for each $n \in \mathbb{N}$. Then $g_n = \tilde{J}_P(\chi_{E_n})f$ (cf. 14.2.7a), and hence (cf. 14.2.12)

$$\mu_{g_n}^P(E) = \int_E \chi_{E_n} d\mu_f^P, \forall E \in \mathcal{A},$$

and hence (cf. 8.3.4b and 8.1.17b)

$$\int_X |\varphi|^2 d\mu_{g_n}^P = \int_X |\varphi|^2 \chi_{E_n} d\mu_f^P \leq n^2 \int_X 1_X d\mu_f^P = n^2 \mu_f^P(X) < \infty,$$

and hence $g_n \in D_P(\varphi)$. In view of 2.3.12, this proves that $\overline{D_P(\varphi)} = \mathcal{H}$. $\qquad\square$

14.2.14 Theorem. *Let $\varphi \in \mathcal{M}(X, \mathcal{A}, P)$. Then there exists a unique linear operator J_φ^P in $\mathcal{H}$ such that:*

(a) $D_{J_\varphi^P} = D_P(\varphi)$;
(b) $(f|J_\varphi^P f) = \int_X \varphi d\mu_f^P, \forall f \in D_P(\varphi)$

(note that $\varphi \in \mathcal{L}^1(X, \mathcal{A}, \mu_f^P), \forall f \in D_P(\varphi)$; cf. 11.1.3).

In addition, the following conditions are true:

(c) for every φ-convergent sequence $\{\varphi_n\}$,

$$J_\varphi^P f = \lim_{n \to \infty} \tilde{J}_P(\varphi_n)f, \forall f \in D_P(\varphi);$$

(d) $\|J_\varphi^P f\|^2 = \int_X |\varphi|^2 d\mu_f^P, \forall f \in D_P(\varphi)$;
(e) if $A \in \mathcal{B}(\mathcal{H})$ is so that $AP(E) = P(E)A$ for all $E \in \mathcal{A}$, then $AJ_\varphi^P \subset J_\varphi^P A$.

Proof. We fix a φ-convergent sequence $\{\varphi_n\}$.

In view of 14.2.11 $(a \Rightarrow c)$, we can define the mapping

$$J_\varphi^P : D_P(\varphi) \to \mathcal{H}$$

$$f \mapsto J_\varphi^P f := \lim_{n \to \infty} \tilde{J}_P(\varphi_n)f.$$

In view of 14.2.11e, this mapping does not depend on the choice of the φ-convergent sequence $\{\varphi_n\}$. Thus, the mapping J_φ^P satisfies conditions a and c.

Let $\alpha, \beta \in \mathbb{C}$ and $f, g \in D_P(\varphi)$. The sequences $\{\tilde{J}_P(\varphi_n)f\}$ and $\{\tilde{J}_P(\varphi_n)g\}$ are convergent and hence (cf. 10.1.16a,b) the sequence $\{\alpha\tilde{J}_P(\varphi_n)f + \beta\tilde{J}_P(\varphi_n)g\}$ is convergent and

$$\lim_{n \to \infty} (\alpha\tilde{J}_P(\varphi_n)f + \beta\tilde{J}_P(\varphi_n)g) = \alpha \lim_{n \to \infty} \tilde{J}_P(\varphi_n)f + \beta \lim_{n \to \infty} \tilde{J}_P(\varphi_n)g;$$

since $\alpha\tilde{J}_P(\varphi_n)f + \beta\tilde{J}_P(\varphi_n)g = \tilde{J}_P(\varphi_n)(\alpha f + \beta g)$, this implies that $\alpha f + \beta g \in D_P(\varphi)$ (cf. 14.2.11, $c \Rightarrow a$), and that

$$J_\varphi^P(\alpha f + \beta g) = \alpha J_\varphi^P f + \beta J_\varphi^P g.$$

This proves that J_φ^P is a linear operator.

Further we have, for every $f \in D_P(\varphi)$,

$$(f|J_\varphi^P f) \overset{(1)}{=} \lim_{n \to \infty} \left(f|\tilde{J}_P(\varphi_n)f\right) \overset{(2)}{=} \lim_{n \to \infty} \int_X \varphi_n d\mu_f^P \overset{(3)}{=} \int_X \varphi d\mu_f^P,$$

where 1 holds by 10.1.16c and 2 by 14.2.7e; as to 3, from $[\varphi] = \lim_{n\to\infty}[\varphi_n]$ in the Hilbert space $L^2(X, \mathcal{A}, \mu_f^P)$ (cf. 14.2.11, $a \Rightarrow d$) we have

$$\left| \int_X \varphi_n d\mu_f^P - \int_X \varphi d\mu_f^P \right| = \left| \int_X 1_X(\varphi_n - \varphi)d\mu_f^P \right|$$

$$\leq \left(\int_X 1_X d\mu_f^P \right)^{\frac{1}{2}} \left(\int_X |\varphi_n - \varphi|^2 d\mu_f^P \right)^{\frac{1}{2}} \xrightarrow[n\to\infty]{} 0$$

by the Schwarz inequality in $L^2(X, \mathcal{A}, \mu_f^P)$. This proves that condition b is satisfied.

The uniqueness of the linear operator in $\mathcal{H}$ which satisfies conditions a and b follows from 14.2.13 and 10.2.12.

In what follows, we prove conditions d and e.

d: For every $f \in D_P(\varphi)$, we have

$$\|J_\varphi^P f\|^2 \overset{(4)}{=} \lim_{n\to\infty} \|\tilde{J}_P(\varphi_n)f\|^2 \overset{(5)}{=} \lim_{n\to\infty} \int_X |\varphi_n|^2 d\mu_f^P \overset{(6)}{=} \int_X |\varphi|^2 d\mu_f^P,$$

where 4 holds by 4.1.6a and 5 by 14.2.7f; as to 6, from $[\varphi] = \lim_{n\to\infty}[\varphi_n]$ in the Hilbert space $L^2(X, \mathcal{A}, \mu_f^P)$ we have $\|[\varphi]\|^2_{L^2(X,\mathcal{A},\mu_f^P)} = \lim_{n\to\infty} \|[\varphi_n]\|^2_{L^2(X,\mathcal{A},\mu_f^P)}$.

e: We have:

$$f \in D_P(\varphi) \overset{(7)}{\Rightarrow} \{\tilde{J}_P(\varphi_n)f\} \text{ is convergent } \overset{(8)}{\Rightarrow}$$

$$[\{A\tilde{J}_P(\varphi_n)f\} \text{ is convergent and } A\lim_{n\to\infty}\tilde{J}_P(\varphi_n)f = \lim_{n\to\infty}A\tilde{J}_P(\varphi_n)f] \overset{(9)}{\Rightarrow}$$

$$[\{\tilde{J}_P(\varphi_n)Af\} \text{ is convergent and } AJ_\varphi^P f = \lim_{n\to\infty}\tilde{J}_P(\varphi_n)Af] \overset{(10)}{\Rightarrow}$$

$$[Af \in D_P(\varphi) \text{ and } AJ_\varphi^P f = J_\varphi^P Af],$$

where: 7 holds by 14.2.11 ($a \Rightarrow c$); 8 holds because $A \in \mathcal{B}(\mathcal{H})$; 9 holds by 14.2.7g; 10 holds by 14.2.11 ($c \Rightarrow a$). Since $D_{AJ_\varphi^P} = D_P(\varphi)$, this proves condition e (cf. 3.2.3 and 3.2.4). □

14.2.15 Theorem. *For all* $\varphi \in \mathcal{M}(X, \mathcal{A}, P)$, *the operator* J_φ^P *is adjointable and*

$$(J_\varphi^P)^\dagger = J_{\overline{\varphi}}^P.$$

Proof. For every $\varphi \in \mathcal{M}(X, \mathcal{A}, P)$, 14.2.13 shows that the operator J_φ^P is adjointable.

Now, let $\varphi \in \mathcal{M}(X, \mathcal{A}, P)$ and let $\{\varphi_n\}$ be a φ-convergent sequence. The sequence $\{\overline{\varphi}_n\}$ is obviously $\overline{\varphi}$-convergent, and hence (cf. 14.2.14c and 14.2.7d)

$$\left(J_\varphi^P f | g\right) = \lim_{n\to\infty} \left(\tilde{J}_P(\varphi_n)f | g\right) = \lim_{n\to\infty} \left(f | (\tilde{J}_P(\varphi_n))^\dagger g\right)$$

$$= \lim_{n\to\infty} \left(f | \tilde{J}_P(\overline{\varphi}_n)g\right) = \left(f | J_{\overline{\varphi}}^P g\right), \forall f \in D_P(\varphi), \forall g \in D_P(\overline{\varphi})$$

In view of 12.1.3B, this proves that $J_{\overline{\varphi}}^P \subset (J_\varphi^P)^\dagger$.

Conversely, let $g \in D_{(J_\varphi^P)^\dagger}$ and write $h := (J_\varphi^P)^\dagger g$; then,

$$(h|f) = \left(g | J_\varphi^P f\right), \forall f \in D_P(\varphi).$$

For each $n \in \mathbb{N}$, define the set $E_n := |\varphi|^{-1}([0, n])$, which is an element of $\mathcal{A}$ (cf. 14.2.10a), and the vector $f_n := \tilde{J}_P(\chi_{E_n}\overline{\varphi})g$ (note that $\chi_{E_n}\overline{\varphi} \in \mathcal{L}^\infty(X, \mathcal{A}, P)$); then (cf. 14.2.12)

$$\mu_{f_n}^P(E) = \int_E \chi_{E_n}|\overline{\varphi}|^2 d\mu_g^P, \forall E \in \mathcal{A},$$

and hence (cf. 8.3.4b and 8.1.17b)

$$\int_X |\varphi|^2 d\mu_{f_n}^P = \int_X |\varphi|^2 \chi_{E_n}|\overline{\varphi}|^2 d\mu_g^P \le n^4 \int_X 1_X d\mu_g^P < \infty,$$

and hence $f_n \in D_P(\varphi)$. The sequence $\{\chi_{E_n}\varphi\}$ is φ-convergent (cf. 14.2.10a), and hence

$$J_\varphi^P f_n \overset{(1)}{=} \lim_{k \to \infty} \tilde{J}_P(\chi_{E_k}\varphi)\tilde{J}_P(\chi_{E_n}\overline{\varphi})g \overset{(2)}{=} \tilde{J}_P(\chi_{E_n}|\varphi|^2)g, \forall n \in \mathbb{N},$$

where 1 holds by 14.2.14c and 2 by 14.2.7c, since $\chi_{E_k}\chi_{E_n} = \chi_{E_n}$ if $k \ge n$. Then,

$$(h|f_n) = (g|J_\varphi^P f_n) = \left(g|\tilde{J}_P(\chi_{E_n}|\varphi|^2)g\right)$$
$$\overset{(3)}{=} \int_X \chi_{E_n}|\varphi|^2 d\mu_g^P \overset{(4)}{=} \|\tilde{J}_P(\chi_{E_n}\overline{\varphi})g\|^2 = \|f_n\|^2, \forall n \in \mathbb{N},$$

where 3 holds by 14.2.7e and 4 by 14.2.7f. Then the Schwarz inequality yields

$$\|f_n\|^2 = (h|f_n) \le \|h\|\|f_n\|, \forall n \in \mathbb{N},$$

and hence

$$\|f_n\| \le \|h\|, \forall n \in \mathbb{N},$$

and hence

$$\int_X \chi_{E_n}|\varphi|^2 d\mu_g^P = \|f_n\|^2 \le \|h\|^2, \forall n \in \mathbb{N}.$$

Since $\lim_{n \to \infty} \chi_{E_n}|\varphi(x)|^2 = |\varphi(x)|^2$, $\forall x \in D_\varphi$, by 8.1.20 we obtain

$$\int_X |\varphi|^2 d\mu_g^P \le \|h\|^2,$$

and hence $g \in D_P(\overline{\varphi})$. This proves the inclusion $D_{(J_\varphi^P)^\dagger} \subset D_P(\overline{\varphi})$, and hence that $(J_\varphi^P)^\dagger = J_{\overline{\varphi}}^P$. $\square$

14.2.16 Corollary. *For all $\varphi \in \mathcal{M}(X, \mathcal{A}, P)$, the operator J_φ^P is closed.*

Proof. For every $\varphi \in \mathcal{M}(X, \mathcal{A}, P)$, we have $J_\varphi^P = (J_{\overline{\varphi}}^P)^\dagger$ in view of 14.2.15. By 12.1.6a, this proves that the operator J_φ^P is closed. $\square$

14.2.17 Proposition. *For $\varphi \in \mathcal{M}(X, \mathcal{A}, P)$, the following conditions are equivalent:*

(a) the operator J_φ^P is bounded;
(b) $D_P(\varphi) = \mathcal{H}$;

(c) $J_\varphi^P \in \mathcal{B}(\mathcal{H})$;

(d) $\varphi \in \mathcal{L}^\infty(X, \mathcal{A}, P)$.

If these conditions are satisfied, then

(e) $J_\varphi^P = \tilde{J}_P(\varphi)$.

Proof. Equivalence of a, b, c: We know that J_φ^P is closed (cf. 14.2.16) and that $\overline{D_P(\varphi)} = \mathcal{H}$ (cf. 14.2.13). Then, the implication $a \Rightarrow b$ is true by 4.4.4 and 2.3.9c, and the implication $b \Rightarrow a$ is true by 12.2.3. In view of this, the implications $a \Rightarrow c$ and $b \Rightarrow c$ are obvious. The implications $c \Rightarrow a$ and $c \Rightarrow b$ are obvious.

$a \Rightarrow d$: We prove this by contraposition. We suppose that $\varphi \notin \mathcal{L}^\infty(X, \mathcal{A}, P)$. Then we have

$$\forall n \in \mathbb{N}, \exists k_n \in \mathbb{N} \text{ so that } k_n \geq n \text{ and } P(|\varphi|^{-1}([k_n, k_n + 1))) \neq \mathbb{O}_\mathcal{H};$$

in fact, if we had

$$\exists n \in \mathbb{N} \text{ such that } P(|\varphi|^{-1}([k, k+1))) = \mathbb{O}_\mathcal{H}, \forall k \in \mathbb{N} \text{ so that } k \geq n,$$

then we should have, by 13.3.6c,

$$\exists n \in \mathbb{N} \text{ such that } P(|\varphi|^{-1}([n, \infty))) = P\left(\bigcup_{k=n}^\infty |\varphi|^{-1}([k, k+1))\right) = \mathbb{O}_\mathcal{H},$$

and hence $\varphi \in \mathcal{L}^\infty(X, \mathcal{A}, P)$. For each $n \in \mathbb{N}$, we write $F_n := |\varphi|^{-1}([k_n, k_n + 1))$ and we choose $f_n \in \mathcal{H}$ such that $f_n \neq 0_\mathcal{H}$ and $P(F_n)f_n = f_n$; then,

$$\mu_{f_n}^P(X - F_n) = \|P(X - F_n)P(F_n)f_n\|^2 = 0$$

and hence

$$\int_X |\varphi|^2 d\mu_{f_n}^P \overset{(1)}{=} \int_{F_n} |\varphi|^2 d\mu_{f_n}^P \overset{(2)}{\leq} (k_n + 1)^2 \int_{F_n} 1_X d\mu_{f_n}^P < \infty,$$

where 1 holds by 8.3.3a and 2 by 8.1.17b, and hence $f_n \in D_P(\varphi)$; moreover,

$$\|J_\varphi^P f_n\|^2 \overset{(3)}{=} \int_X |\varphi|^2 d\mu_{f_n}^P = \int_{F_n} |\varphi|^2 d\mu_{f_n}^P$$

$$\overset{(4)}{\geq} k_n^2 \int_{F_n} 1_X d\mu_{f_n}^P = k_n^2 \mu_{f_n}^P(X) = k_n^2 \|f_n\|^2,$$

where 3 holds by 14.2.14d and 4 by 8.1.17b. Since $f_n \neq 0_\mathcal{H}$ for all $n \in \mathbb{N}$, this proves that the operator J_φ^P is not bounded.

$d \Rightarrow$ (a and e): If $\varphi \in \mathcal{L}^\infty(X, \mathcal{A}, P)$, then $\varphi_n := \varphi$ for each $n \in \mathbb{N}$ defines an obviously φ-convergent sequence. In view of 14.2.14c, this proves that $J_\varphi^P = \tilde{J}_P(\varphi)$, and therefore also that the operator J_φ^P is bounded. $\qquad\square$

14.2.18 Definitions. We define the mapping

$$J_P : \mathcal{M}(X, \mathcal{A}, P) \to \mathcal{O}(\mathcal{H})$$
$$\varphi \mapsto J_P(\varphi) := J_\varphi^P$$

(we recall that $\mathcal{O}(\mathcal{H})$ denotes the family of all linear operators in $\mathcal{H}$, cf. 3.2.1). For notational convenience, we will often write $J_P(\varphi)$ instead of J_φ^P; for the same reason, we always write $D_P(\varphi)$ instead of $D_{J_\varphi^P}$.

From 14.2.17($d \Rightarrow e$) we have that the mapping J_P is an extension of the mapping $\tilde{J}_P$, and hence of the mapping $\hat{J}_P$. The terminology adopted in 14.1.2 for $\hat{J}_P$ is extended to J_P. Thus, for $\varphi \in \mathcal{M}(X, \mathcal{A}, P)$ the operator $J_P(\varphi)$ is called the *integral of φ with respect to P* and is often denoted as follows

$$\int_X \varphi dP := J_P(\varphi).$$

A reason behind this extension of terminology will be set out in 14.3.7.

14.3 Sum, product, inverse, self-adjointness, unitarity of integrals

As before, $(X, \mathcal{A})$ denotes an abstract measurable space, $\mathcal{H}$ denotes an abstract Hilbert space, and P denotes a projection valued measure on $\mathcal{A}$ with values in $\mathscr{P}(\mathcal{H})$.

14.3.1 Proposition. *Let $\varphi \in \mathcal{M}(X, \mathcal{A}, P)$ and $\psi \in \mathcal{L}^\infty(X, \mathcal{A}, P)$. Then,*

$$J_P(\varphi) + J_P(\psi) = J_P(\varphi + \psi).$$

Proof. First we note that, for $f \in \mathcal{H}$,

$$f \in D_P(\varphi) \Leftrightarrow \varphi \in \mathcal{L}^2(X, \mathcal{A}, \mu_f^P) \overset{(1)}{\Leftrightarrow} \varphi + \psi \in \mathcal{L}^2(X, \mathcal{A}, \mu_f^P) \Leftrightarrow f \in D_P(\varphi + \psi),$$

where 1 holds because $\psi \in \mathcal{L}^2(X, \mathcal{A}, \mu_f^P)$ for all $f \in \mathcal{H}$ (cf. 14.2.10b). Hence,

$$D_{J_P(\varphi) + J_P(\psi)} \overset{(2)}{=} D_P(\varphi) = D_P(\varphi + \psi),$$

where 2 holds because $D_P(\psi) = \mathcal{H}$ (cf. 14.2.17).

Next, let $\{\varphi_n\}$ be a φ-convergent sequence. Then $\{\varphi_n + \psi\}$ is a $(\varphi + \psi)$-convergent sequence since the condition

$$\exists k_1, k_2 \in [0, \infty) \text{ such that } |\varphi_n(x)|^2 \le k_1 |\varphi(x)|^2 + k_2 \text{ } P\text{-a.e. on } D_\varphi \cap D_{\varphi_n}, \forall n \in \mathbb{N},$$

implies that there exist $k_1, k_2 \in [0, \infty)$ such that

$$|\varphi_n(x) + \psi(x)|^2 \overset{(3)}{\le} 2|\varphi_n(x)|^2 + 2|\psi(x)|^2 \le 2k_1|\varphi(x)|^2 + 2k_2 + 2|\psi(x)|^2$$
$$\overset{(4)}{\le} 4k_1|\varphi(x) + \psi(x)|^2 + 4k_1|\psi(x)|^2 + 2k_2 + 2|\psi(x)|^2$$
$$\le 4k_1|\varphi(x) + \psi(x)|^2 + (4k_1 + 2)(P\text{-sup}|\psi|)^2 + 2k_2$$
$$P\text{-a.e. on } D_\varphi \cap D_{\varphi_n} \cap D_\psi = D_{\varphi + \psi} \cap D_{\varphi_n + \psi}, \forall n \in \mathbb{N},$$

where 3 holds by inequality 2 in the proof of 10.3.7 and 4 holds by the inequality (thereby derived)

$$|\alpha|^2 = |\alpha + \beta - \beta|^2 \le 2|\alpha + \beta|^2 + 2|\beta|^2, \forall \alpha, \beta \in \mathbb{C}.$$

This yields

$$(J_P(\varphi) + J_P(\psi))f \overset{(5)}{=} \lim_{n\to\infty} \tilde{J}_P(\varphi_n)f + \tilde{J}_P(\psi)f$$

$$= \lim_{n\to\infty} (\tilde{J}_P(\varphi_n)f + \tilde{J}_P(\psi))f \overset{(6)}{=} \lim_{n\to\infty} \tilde{J}_P(\varphi_n + \psi)f$$

$$\overset{(7)}{=} J_P(\varphi + \psi)f, \forall f \in D_P(\varphi),$$

where: 5 holds by 14.2.14c and 14.2.17e; 6 holds by 14.2.7b; 7 holds by 14.2.14c. $\square$

14.3.2 Corollary. *For all $\varphi \in \mathcal{M}(X, \mathcal{A}, P)$ and $\lambda \in \mathbb{C}$,*

$$\|J_P(\varphi)f - \lambda f\|^2 = \int_X |\varphi - \lambda|^2 d\mu_f^P, \forall f \in D_P(\varphi)$$

(we recall that $\varphi - \lambda := \varphi - \lambda_X$, cf. 1.2.19).

Proof. Since $\lambda_X \in \mathcal{L}^\infty(X, \mathcal{A}, P)$,

$$J_P(-\lambda_X) = \tilde{J}_P(-\lambda_X) = -\lambda \tilde{J}_P(1_X) = -\lambda P(X) = -\lambda \mathbb{1}_{\mathcal{H}}$$

(cf. 14.2.17e and 14.2.7a,b). Then, 14.3.1 implies that

$$J_P(\varphi) - \lambda \mathbb{1}_{\mathcal{H}} = J_P(\varphi - \lambda),$$

and hence 14.2.14d implies that

$$\|J_P(\varphi)f - \lambda f\|^2 = \|J_P(\varphi - \lambda)f\|^2 = \int_X |\varphi - \lambda|^2 d\mu_f^P, \forall f \in D_P(\varphi - \lambda) = D_P(\varphi).$$

$\square$

14.3.3 Lemma. *Let $\varphi \in \mathcal{M}(X, \mathcal{A}, P)$ and $f \in D_P(\varphi)$. Let $\{\varphi_n\}$ be a sequence in $\mathcal{M}(X, \mathcal{A}, P)$ such that*

$$f \in D_P(\varphi_n), \forall n \in \mathbb{N}, \quad and \quad \lim_{n\to\infty} [\varphi_n] = [\varphi] \text{ in the Hilbert space } L^2(X, \mathcal{A}, \mu_f^P).$$

Then

$$\lim_{n\to\infty} J_P(\varphi_n)f = J_P(\varphi)f.$$

Proof. Let $\{\psi_n\}$ be a φ-convergent sequence. Then $\lim_{n\to\infty}[\psi_n] = [\varphi]$ in the Hilbert space $L^2(X, \mathcal{A}, \mu_f^P)$ (cf. 14.2.11), and hence

$$\|J_P(\varphi_n)f - J_P(\psi_n)f\|^2 \overset{(1)}{=} \|J_P(\varphi_n - \psi_n)f\|^2 \overset{(2)}{=} \int_X |\varphi_n - \psi_n|^2 d\mu_f^P$$

$$= \|[\varphi_n] - [\psi_n]\|^2_{L^2(X, \mathcal{A}, \mu_f^P)} \xrightarrow[n\to\infty]{} 0,$$

where 1 holds by 14.3.1 and 2 by 14.2.14d. Then,

$$\|J_P(\varphi_n)f - J_P(\varphi)f\| \le \|J_P(\varphi_n)f - J_P(\psi_n)f\| + \|J_P(\psi_n)f - J_P(\varphi)f\| \xrightarrow[n\to\infty]{} 0$$

since $J_P(\varphi)f = \lim_{n\to\infty} J_P(\psi_n)f$ (cf. 14.2.14c and 14.2.17e). $\square$

14.3.4 Proposition. *Let* $\varphi, \psi \in \mathcal{M}(X, \mathcal{A}, P)$. *Then,*

$$J_P(\varphi) + J_P(\psi) \subset J_P(\varphi + \psi).$$

Proof. We have

$$f \in D_P(\varphi) \cap D_P(\psi) \Rightarrow \varphi, \psi \in \mathcal{L}^2(X, \mathcal{A}, \mu_f^P) \Rightarrow$$
$$\varphi + \psi \in \mathcal{L}^2(X, \mathcal{A}, \mu_f^P) \Rightarrow f \in D_P(\varphi + \psi),$$

or $D_{J_\varphi^P + J_\psi^P} \subset D_P(\varphi + \psi)$.

Now let $f \in D_P(\varphi) \cap D_P(\psi)$, let $\{\varphi_n\}$ be a φ-convergent sequence, and let $\{\psi_n\}$ be a ψ-convergent sequence. Then (cf. 14.2.11)

$$[\varphi] = \lim_{n \to \infty} [\varphi_n] \text{ and } [\psi] = \lim_{n \to \infty} [\psi_n] \text{ in the Hilbert space } L^2(X, \mathcal{A}, \mu_f^P),$$

and hence

$$[\varphi + \psi] = \lim_{n \to \infty} [\varphi_n + \psi_n] \text{ in the Hilbert space } L^2(X, \mathcal{A}, \mu_f^P).$$

Since $f \in D_P(\varphi + \psi)$ and $D_P(\varphi_n + \psi_n) = \mathcal{H}$ for all $n \in \mathcal{H}$ (cf. 14.2.17), by 14.3.3 we have

$$J_P(\varphi + \psi)f = \lim_{n \to \infty} J_P(\varphi_n + \psi_n)f;$$

now, $J_P(\varphi_n + \psi_n) = \tilde{J}_P(\varphi_n + \psi_n) = \tilde{J}_P(\varphi_n) + \tilde{J}_P(\psi_n)$ (cf. 14.2.17e and 14.2.7b), and hence

$$J_P(\varphi + \psi)f = \lim_{n \to \infty} (\tilde{J}_P(\varphi_n) + \tilde{J}_P(\psi_n))f$$
$$= \lim_{n \to \infty} \tilde{J}_P(\varphi_n)f + \lim_{n \to \infty} \tilde{J}_P(\psi_n)f = J_P(\varphi)f + J_P(\psi)f$$

(cf. 14.2.14c). $\qquad\square$

14.3.5 Proposition. *Let* $\varphi \in \mathcal{M}(X, \mathcal{A}, P)$. *Then,*

$$\alpha J_P(\varphi) = J_P(\alpha\varphi), \forall \alpha \in \mathbb{C} - \{0\}, \text{ and } \alpha J_P(\varphi) \subset J_P(\alpha\varphi) \text{ for } \alpha := 0.$$

Proof. For every $\alpha \in \mathbb{C} - \{0\}$ we have

$$f \in D_P(\varphi) \Leftrightarrow \varphi \in \mathcal{L}^2(X, \mathcal{A}, \mu_f^P) \Leftrightarrow \alpha\varphi \in \mathcal{L}^2(X, \mathcal{A}, \mu_f^P) \Leftrightarrow f \in D_P(\alpha\varphi)$$

and hence $D_{\alpha J_\varphi^P} = D_P(\varphi) = D_P(\alpha\varphi)$.

If $\alpha = 0$ then

$$\alpha\varphi \in \mathcal{L}^2(X, \mathcal{A}, \mu_f^P), \forall f \in \mathcal{H},$$

and hence $D_{\alpha J_\varphi^P} = D_P(\varphi) \subset \mathcal{H} = D_P(\alpha\varphi)$.

For all $\alpha \in \mathbb{C}$ and all $f \in D_P(\varphi)$ we have (cf. 14.2.14b)

$$(f|J_P(\alpha\varphi)f) = \int_X \alpha\varphi d\mu_f^P = \alpha \int_X \varphi d\mu_f^P = \alpha(f|J_P(\varphi)f) = (f|\alpha J_P(\varphi)f).$$

In view of 14.2.13 and 10.2.12, this proves the statement. $\qquad\square$

14.3.6 Proposition. *For all* $\alpha, \beta \in \mathbb{C}$ *and* $\varphi, \psi \in \mathcal{M}(X, \mathcal{A}, P)$,

$$\alpha J_P(\varphi) + \beta J_P(\psi) \subset J_P(\alpha\varphi + \beta\psi).$$

Proof. This follows immediately from 14.3.5 and 14.3.4. $\qquad\square$

14.3.7 Remark. If $\varphi \in \mathcal{M}(X, \mathcal{A}, P)$ is such that $0 \le \varphi(x)$ P-a.e. on D_φ, then the sequence $\{\psi_n\}$ defined as in the proof of 8.1.15 (cf. also the proof of 6.2.26) is a φ-convergent sequence since

$$\psi_n \in \mathcal{S}^+(X, \mathcal{A}) \quad (\text{and } \mathcal{S}^+(X, \mathcal{A}) \subset \mathcal{M}_B(X, \mathcal{A}) \subset \mathcal{L}^\infty(X, \mathcal{A}, P)), \forall n \in \mathbb{N},$$

$$\psi_n(x) \le \varphi(x), \forall x \in D_\varphi, \forall n \in \mathbb{N},$$

$$\lim_{n\to\infty} \psi_n(x) = \varphi(x), \forall x \in D_\varphi.$$

Moreover,

$$\tilde{J}_P(\psi_n) = \hat{J}_P(\psi_n)$$

$$= \sum_{k=2}^{n2^n} \frac{k-1}{2^n} P\left(\varphi^{-1}\left(\left[\frac{k-1}{2^n}, \frac{k}{2^n}\right)\right)\right) + nP(\varphi^{-1}([n, \infty])), \forall n \in \mathbb{N}.$$

Then we have (cf. 14.2.14c)

$$J_P(\varphi)f$$

$$= \lim_{n\to\infty} \left[\sum_{k=2}^{n2^n} \frac{k-1}{2^n} P\left(\varphi^{-1}\left(\left[\frac{k-1}{2^n}, \frac{k}{2^n}\right)\right)\right) + nP(\varphi^{-1}([n, \infty]))\right] f,$$

$$\forall f \in D_P(\varphi).$$

This equation must be compared with 8.1.15.

Further, for every $\varphi \in \mathcal{M}(X, \mathcal{A}, P)$ we have $D_P(\varphi) \subset D_P((\operatorname{Re}\varphi)^+)$ (since $(\operatorname{Re}\varphi)^+(x) \le |\varphi(x)|$, $\forall x \in D_\varphi$) and similarly for $(\operatorname{Re}\varphi)^-$, $(\operatorname{Im}\varphi)^+$, $(\operatorname{Im}\varphi)^-$, and hence (cf. 14.3.6)

$$J_P(\varphi) = J_P((\operatorname{Re}\varphi)^+) - J_P((\operatorname{Re}\varphi)^-) + iJ_P((\operatorname{Im}\varphi)^+) - iJ_P((\operatorname{Im}\varphi)^-).$$

This equation must be compared with the definition of $\int_X \varphi d\mu$ in 8.2.3.

Thus, we see that, for $\varphi \in \mathcal{M}(X, \mathcal{A}, P)$, there is a close analogy between the vector $J_P(\varphi)f$ for any $f \in D_P(\varphi)$ and the integral $\int_X \varphi d\mu$ for any measure μ on $\mathcal{A}$ such that $\varphi \in \mathcal{L}^1(X, \mathcal{A}, \mu)$. This is the reason why the operator $J_P(\varphi)$ is called the *integral of φ with respect to P* and is often denoted as follows

$$\int_X \varphi dP := J_P(\varphi)$$

(cf. 14.2.18).

We point out that, if $\varphi \in \mathcal{L}^\infty(X, \mathcal{A}, P)$ is so that $0 \le \varphi(x)$ P-a.e. on D_φ, then for the sequence $\{\psi_n\}$ considered above we have

$$P\text{-sup}|\psi_n - \varphi| \to 0 \text{ as } n \to \infty$$

(this is true because φ satisfies condition c of 6.2.26 with $Y := D_\varphi - E$, where $E \in \mathcal{A}$ and $P(E) = \mathbb{O}_{\mathcal{H}}$), and hence (cf. 14.2.7b,h)

$$J_P(\varphi) = \tilde{J}_P(\varphi) = \lim_{n\to\infty} \left[\sum_{k=2}^{n2^n} \frac{k-1}{2^n} P\left(\varphi^{-1}\left(\left[\frac{k-1}{2^n}, \frac{k}{2^n}\right)\right)\right) + nP(\varphi^{-1}([n,\infty))) \right],$$

where the limit is with respect to the norm of $\mathcal{B}(\mathcal{H})$.

14.3.8 Lemma. *Let $\varphi \in \mathcal{M}(X, \mathcal{A}, P)$ and $f \in D_P(\varphi)$. If we write $g := J_P(\varphi)f$,* *then*

$$\mu_g^P(E) = \int_E |\varphi|^2 d\mu_f^P, \forall E \in \mathcal{A}$$

(this generalizes 14.2.12).

Proof. Let $\{\varphi_n\}$ be a φ-convergent sequence. For every $E \in \mathcal{A}$ we have (cf. 14.2.7a,c)

$$P(E)\tilde{J}_P(\varphi_n) = \tilde{J}_P(\chi_E)\tilde{J}_P(\varphi_n) = \tilde{J}_P(\chi_E\varphi_n), \forall n \in \mathbb{N},$$

and hence

$$\mu_g^P(E) = \|P(E)J_P(\varphi)f\|^2 = \|P(E)\lim_{n\to\infty}\tilde{J}_P(\varphi_n)f\|^2$$

$$= \lim_{n\to\infty}\|P(E)\tilde{J}_P(\varphi_n)f\|^2 = \lim_{n\to\infty}\|\tilde{J}_P(\chi_E\varphi_n)f\|^2 = \lim_{n\to\infty}\int_X |\chi_E\varphi_n|^2 d\mu_f^P$$

(cf. 14.2.14c and 14.2.7f). Further, we have

$$\chi_E(x)\varphi(x) = \lim_{n\to\infty}\chi_E(x)\varphi_n(x)$$

$$\mu_f^P\text{-a.e. on } D_\varphi \cap \left(\bigcap_{n=1}^\infty D_{\varphi_n}\right) = D_{\chi_E\varphi} \cap \left(\bigcap_{n=1}^\infty D_{\chi_E\varphi_n}\right),$$

and

$$\exists k_1, k_2 \in [0, \infty) \text{ such that}$$
$$|\chi_E(x)\varphi_n(x)|^2 \le k_1|\varphi(x)|^2 + k_2 \ \mu_f^P\text{-a.e. on } D_\varphi \cap D_{\chi_E\varphi_n}, \forall n \in \mathbb{N}.$$

Then, by 8.2.11 (recall that a constant function is μ_f^P-integrable) we have

$$\lim_{n\to\infty}\int_X |\chi_E\varphi_n|^2 d\mu_f^P = \int_X \chi_E|\varphi|^2 d\mu_f^P.$$

This proves the statement. $\qquad\square$

14.3.9 Proposition. *Let $\varphi, \psi \in \mathcal{M}(X, \mathcal{A}, P)$. Then*

$$D_{J_P(\psi)J_P(\varphi)} = D_P(\varphi) \cap D_P(\psi\varphi) \text{ and } J_P(\psi)J_P(\varphi) \subset J_P(\psi\varphi).$$

Proof. We prove the part of the statement about the domains.

We have

$$f \in D_{J_P(\psi)J_P(\varphi)} \Leftrightarrow [f \in D_P(\varphi) \text{ and } J_P(\varphi)f \in D_P(\psi)].$$

For $f \in D_P(\varphi)$, letting $g := J_P(\varphi)f$ we have

$$J_P(\varphi)f \in D_P(\psi) \Leftrightarrow \int_X |\psi|^2 d\mu_g^P < \infty;$$

since $\int_X |\psi|^2 d\mu_g^P = \int_X |\psi|^2 |\varphi|^2 d\mu_f^P$ (cf. 14.3.8 and 8.3.4b), we actually have

$$J_P(\varphi)f \in D_P(\psi) \Leftrightarrow \psi\varphi \in \mathcal{L}^2(X, \mathcal{A}, \mu_f^P) \Leftrightarrow f \in D_P(\psi\varphi).$$

Thus we have

$$f \in D_{J_P(\psi)J_P(\varphi)} \Leftrightarrow [f \in D_P(\varphi) \text{ and } f \in D_P(\psi\varphi)],$$

or $D_{J_P(\psi)J_P(\varphi)} = D_P(\varphi) \cap D_P(\psi\varphi)$.

Now we prove the part of the statement about the operators. We note that from the part of the statement about the domains we have $D_{J_P(\psi)J_P(\varphi)} \subset D_P(\psi\varphi)$. Thus, we need to prove that $J_P(\psi)J_P(\varphi)f = J_P(\psi\varphi)f$ for all $f \in D_{J_P(\psi)J_P(\varphi)}$.

First we assume $\psi \in \mathcal{L}^\infty(X, \mathcal{A}, P)$. If $\{\varphi_n\}$ is a φ-convergent sequence then the sequence $\{\psi\varphi_n\}$ is $(\psi\varphi)$-convergent, as can be seen easily. Hence for every $f \in D_P(\varphi)$ we have $f \in D_{J_P(\psi)J_P(\varphi)}$ (in view of 14.2.17) and

$$J_P(\psi)J_P(\varphi)f \overset{(1)}{=} \tilde{J}_P(\psi) \lim_{n\to\infty} \tilde{J}_P(\varphi_n)f \overset{(2)}{=} \lim_{n\to\infty} \tilde{J}_P(\psi)\tilde{J}_P(\varphi_n)f$$

$$\overset{(3)}{=} \lim_{n\to\infty} \tilde{J}_P(\psi\varphi_n)f \overset{(4)}{=} J_P(\psi\varphi)f;$$

where: 1 holds by 14.2.17e and 14.2.14c, since $f \in D_P(\varphi)$; 2 holds because $\tilde{J}_P(\psi)$ is continuous; 3 holds by 14.2.7c; 4 holds by 14.2.14c, since $f \in D_P(\psi\varphi)$.

Next, let ψ be any element of $\mathcal{M}(X, \mathcal{A}, P)$, let $\{\psi_n\}$ be a ψ-convergent sequence, and let $f \in D_{J_P(\psi)J_P(\varphi)}$; this implies $f \in D_P(\varphi) \cap D_P(\psi\varphi)$, or $\varphi \in \mathcal{L}^2(X, \mathcal{A}, \mu_f^P)$ and $\psi\varphi \in \mathcal{L}^2(X, \mathcal{A}, \mu_f^P)$; since $\psi_n \in \mathcal{L}^\infty(X, \mathcal{A}, P)$, we have also $\psi_n\varphi \in \mathcal{L}^2(X, \mathcal{A}, \mu_f^P)$ for all $n \in \mathbb{N}$. Since $\{\psi_n\}$ is a ψ-convergent sequence, we have

$$\lim_{n\to\infty} |\psi_n(x)\varphi(x) - \psi(x)\varphi(x)|^2 = 0 \ \mu_f^P\text{-a.e. on } D_{\psi\varphi} \cap \left(\bigcap_{n=1}^\infty D_{\psi_n\varphi} \right),$$

and also that there exist $k_1, k_2 \in [0, \infty)$ so that

$$|\psi_n(x) - \psi(x)|^2 \leq 2|\psi_n(x)|^2 + 2|\psi(x)|^2 \leq 2(k_1 + 1)|\psi(x)|^2 + 2k_2$$
$$\mu_f^P\text{-a.e on } D_\psi \cap D_{\psi_n}, \forall n \in \mathbb{N},$$

and hence so that

$$|\psi_n(x)\varphi(x) - \psi(x)\varphi(x)|^2 \leq 2(k_1 + 1)|\psi(x)\varphi(x)|^2 + 2k_2|\varphi(x)|^2$$
$$\mu_f^P\text{-a.e. on } D_{\psi\varphi} \cap D_{\psi_n\varphi}, \forall n \in \mathbb{N};$$

then we have, by 8.2.11,

$$\|[\psi_n\varphi] - [\psi\varphi]\|^2_{L^2(X,\mathcal{A},\mu^P_f)} = \int_X |\psi_n\varphi - \psi\varphi|^2 d\mu^P_f \xrightarrow[n\to\infty]{} 0.$$

In view of 14.3.3 (recall that $f \in D_P(\psi\varphi)$ and $f \in D_P(\psi_n\varphi)$ for all $n \in \mathbb{N}$), this yields

$$\lim_{n\to\infty} J_P(\psi_n\varphi)f = J_P(\psi\varphi)f;$$

moreover, $J_P(\psi_n\varphi)f = J_P(\psi_n)J_P(\varphi)f$ in view of what was proved above (since $\psi_n \in \mathcal{L}^\infty(X,\mathcal{A},P)$ and $f \in D_P(\varphi)$), and hence

$$\lim_{n\to\infty} J_P(\psi_n\varphi)f = \lim_{n\to\infty} J_P(\psi_n)J_P(\varphi)f = J_P(\psi)J_P(\varphi)f$$

in view of 14.2.14c, since $J_P(\varphi)f \in D_P(\psi)$. Thus,

$$J_P(\psi)J_P(\varphi)f = J_P(\psi\varphi)f.$$

$\square$

14.3.10 Corollary. *If $\varphi \in \mathcal{L}^\infty(X,\mathcal{A},P)$ and $\psi \in \mathcal{M}(X,\mathcal{A},P)$, then*

$$J_P(\psi)J_P(\varphi) = J_P(\psi\varphi).$$

Proof. This follows at once from 14.3.9, since $\varphi \in \mathcal{L}^\infty(X,\mathcal{A},P)$ entails $D_P(\varphi) = \mathcal{H}$ (cf. 14.2.17) and hence $D_{J_P(\psi)J_P(\varphi)} = D_P(\psi\varphi)$. $\square$

14.3.11 Proposition. *Let $\varphi, \psi \in \mathcal{M}(X,\mathcal{A},P)$. Then the operator $J_P(\varphi) + J_P(\psi)$ is closable and*

$$\overline{J_P(\varphi) + J_P(\psi)} = J_P(\varphi + \psi).$$

Proof. From 14.2.16 and 14.3.4 we have that $J_P(\varphi + \psi)$ is a closed extension of $J_P(\varphi) + J_P(\psi)$. Therefore, the operator $J_P(\varphi) + J_P(\psi)$ is closable (cf. 4.4.11b) and (cf. 4.4.10)

$$\overline{J_P(\varphi) + J_P(\psi)} \subset J_P(\varphi + \psi).$$

Now we fix $f \in D_P(\varphi + \psi)$. For each $n \in \mathbb{N}$, we define the set

$$E_n := \{x \in D_\varphi \cap D_\psi : |\varphi(x)| + |\psi(x)| \leq n\},$$

which is an element of $\mathcal{A}$, and we define the vector $g_n := P(E_n)f$. Proceeding as in the proof of 14.2.13, we see that $f = \lim_{n\to\infty} g_n$ and that $g_n \in D_P(|\varphi| + |\psi|)$ for each $n \in \mathbb{N}$; then,

$$\int_X |\varphi|^2 d\mu^P_{g_n} \leq \int_X (|\varphi| + |\psi|)^2 d\mu^P_{g_n} < \infty,$$

and this proves that $g_n \in D_P(\varphi)$; similarly, $g_n \in D_P(\psi)$. Letting $\varphi_n := \chi_{E_n}\varphi$, we have $\varphi_n \in \mathcal{L}^\infty(X,\mathcal{A},P)$ and

$$J_P(\varphi)g_n \overset{(1)}{=} J_P(\varphi)J_P(\chi_{E_n})f \overset{(2)}{=} J_P(\varphi_n)f, \forall n \in \mathbb{N},$$

where 1 holds by 14.2.7a and 14.2.17e, and 2 holds by 14.3.10; similarly, letting $\psi_n := \chi_{E_n}\psi$, we have $\psi_n \in \mathcal{L}^\infty(X, \mathcal{A}, P)$ and

$$J_P(\psi)g_n = J_P(\psi_n)f, \forall n \in \mathbb{N}.$$

Moreover, we have

$$\lim_{n \to \infty} |\varphi_n(x) + \psi_n(x) - \varphi(x) - \psi(x)|^2 = 0, \forall x \in D_{\varphi+\psi},$$

and

$$|\varphi_n(x) + \psi_n(x) - \varphi(x) - \psi(x)|^2 \le 4|\varphi(x) + \psi(x)|^2, \forall x \in D_{\varphi+\psi}, \forall n \in \mathbb{N};$$

then, by 8.2.11 (recall that $\varphi + \psi \in \mathcal{L}^2(X, \mathcal{A}, \mu_f^P)$ since $f \in D_P(\varphi + \psi)$) we have

$$\lim_{n \to \infty} \int_X |\varphi_n + \psi_n - \varphi - \psi|^2 d\mu_f^P = 0,$$

or (note that $\varphi_n + \psi_n \in \mathcal{L}^\infty(X, \mathcal{A}, P)$ implies $\varphi_n + \psi_n \in \mathcal{L}^2(X, \mathcal{A}, \mu_f^P)$, cf. 14.2.10b)

$$\lim_{n \to \infty} [\varphi_n + \psi_n] = [\varphi + \psi] \text{ in the Hilbert space } L^2(X, \mathcal{A}, \mu_f^P);$$

by 14.3.3 (note that $D_P(\varphi_n + \psi_n) = \mathcal{H}$, cf. 14.2.17), this implies

$$\lim_{n \to \infty} J_P(\varphi_n + \psi_n)f = J_P(\varphi + \psi)f.$$

Further, in view of 14.3.1 we have

$$J_P(\varphi_n + \psi_n)f = J_P(\varphi_n)f + J_P(\psi_n)f = J_P(\varphi)g_n + J_P(\psi)g_n$$
$$= (J_P(\varphi) + J_P(\psi))g_n, \forall n \in \mathbb{N}.$$

Thus, we have constructed a sequence $\{g_n\}$ in $D_P(\varphi) \cap D_P(\psi) = D_{J_P(\varphi)+J_P(\psi)}$ which is such that

$$\lim_{n \to \infty} g_n = f \text{ and the sequence } \{(J_P(\varphi) + J_P(\psi))g_n\} \text{ is convergent.}$$

This implies $f \in D_{\overline{J_P(\varphi)+J_P(\psi)}}$ (cf. 4.4.10). Since f was an arbitrary element of $D_P(\varphi + \psi)$, we have

$$D_P(\varphi + \psi) \subset D_{\overline{J_P(\varphi)+J_P(\psi)}},$$

and hence $\overline{J_P(\varphi) + J_P(\psi)} = J_P(\varphi + \psi)$. $\square$

14.3.12 Proposition. *Let $\varphi, \psi \in \mathcal{M}(X, \mathcal{A}, P)$. Then the operator $J_P(\psi)J_P(\varphi)$ is closable and*

$$\overline{J_P(\psi)J_P(\varphi)} = J_P(\varphi\psi).$$

Proof. From 14.2.16 and 14.3.9 we have that $J_P(\psi\varphi)$ is a closed extension of $J_P(\psi)J_P(\varphi)$. Therefore, the operator $J_P(\psi)J_P(\varphi)$ is closable (cf. 4.4.11b) and (cf. 4.4.10)

$$\overline{J_P(\psi)J_P(\varphi)} \subset J_P(\psi\varphi).$$

Now we fix $f \in D_P(\psi\varphi)$. For each $n \in \mathbb{N}$, we define the set

$$E_n := |\varphi|^{-1}([0, n]),$$

which is an element of $\mathcal{A}$, and we define the vector $g_n := P(E_n)f$. In the proof of 14.2.13, we saw that $f = \lim_{n\to\infty} g_n$ and that $g_n \in D_P(\varphi)$ for each $n \in \mathbb{N}$. Letting $\varphi_n := \chi_{E_n}\varphi$, we have $\varphi_n \in \mathcal{L}^\infty(X, \mathcal{A}, P)$ and

$$J_P(\varphi)g_n \overset{(1)}{=} J_P(\varphi)J_P(\chi_{E_n})f \overset{(2)}{=} J_P(\varphi_n)f, \forall n \in \mathbb{N},$$

where 1 holds by 14.2.7a and 14.2.17e, and 2 holds by 14.3.10. Moreover, we have

$$\lim_{n\to\infty} |\psi(x)\varphi_n(x) - \psi(x)\varphi(x)|^2 = 0, \forall x \in D_{\psi\varphi},$$

and

$$|\psi(x)\varphi_n(x) - \psi(x)\varphi(x)|^2 \le 4|\psi(x)\varphi(x)|^2, \forall x \in D_{\psi\varphi}, \forall n \in \mathbb{N};$$

then, by 8.2.11 (recall that $\psi\varphi \in \mathcal{L}^2(X, \mathcal{A}, \mu_f^P)$ since $f \in D_P(\psi\varphi)$) we have

$$\lim_{n\to\infty} \int_X |\psi\varphi_n - \psi\varphi|^2 d\mu_f^P = 0,$$

or $(|\psi(x)\varphi_n(x)| \le |\psi(x)\varphi(x)|, \forall x \in D_{\psi\varphi}$, implies $\psi\varphi_n \in \mathcal{L}^2(X, \mathcal{A}, \mu_f^P))$

$$\lim_{n\to\infty} [\psi\varphi_n] = [\psi\varphi] \text{ in the Hilbert space } L^2(X, \mathcal{A}, \mu_f^P);$$

by 14.3.3 (note that $f \in D_P(\psi\varphi_n)$ since $\psi\varphi_n \in \mathcal{L}^2(X, \mathcal{A}, \mu_f^P)$) this yields

$$\lim_{n\to\infty} J_P(\psi\varphi_n)f = J_P(\psi\varphi)f.$$

Further, in view of 14.3.10 we have $J_P(\psi)J_P(\varphi_n) = J_P(\psi\varphi_n)$ for each $n \in \mathbb{N}$; since $f \in D_P(\psi\varphi_n)$, this implies $J_P(\varphi_n)f \in D_P(\psi)$, i.e. $J_P(\varphi)g_n \in D_P(\psi)$, i.e. $g_n \in D_{J_P(\psi)J_P(\varphi)}$, and

$$J_P(\psi\varphi_n)f = J_P(\psi)J_P(\varphi_n)f = J_P(\psi)J_P(\varphi)g_n.$$

Thus, we have constructed a sequence $\{g_n\}$ in $D_{J_P(\psi)J_P(\varphi)}$ which is such that

$$\lim_{n\to\infty} g_n = f \text{ and the sequence } \{J_P(\psi)J_P(\varphi)g_n\} \text{ is convergent.}$$

This implies $f \in D_{\overline{J_P(\psi)J_P(\varphi)}}$ (cf. 4.4.10). Since f was an arbitrary element of $D_P(\psi\varphi)$, we have

$$D_P(\psi\varphi) \subset D_{\overline{J_P(\psi)J_P(\varphi)}},$$

and hence

$$\overline{J_P(\psi)J_P(\varphi)} = J_P(\varphi\psi).$$

$\square$

14.3.13 Remark. For every $\varphi \in \mathcal{M}(X, \mathcal{A}, P)$ and every $\alpha \in \mathbb{C} - \{0\}$, the operator $\alpha J_P(\varphi)$ is closed; this is true because $\alpha J_P(\varphi) = J_P(\alpha\varphi)$ (cf. 14.3.5 and 14.2.16), but more in general because αA is closed for any closed operator A and every $\alpha \in \mathbb{C} - \{0\}$ (as can be seen easily). If $\alpha = 0$ then $\alpha J_P(\varphi)$ is closed iff $\varphi \in \mathcal{L}^\infty(X, \mathcal{A}, P)$ (cf. 14.2.17, 14.2.13, 4.4.3, 4.4.4).

14.3.14 Proposition. *Let $\varphi \in \mathcal{M}(X, \mathcal{A}, P)$. Then,*

(a) $N_{J_P(\varphi)} = R_{P(\varphi^{-1}(\{0\}))}$.

The following conditions are equivalent:

(b) the operator $J_P(\varphi)$ is injective;
(c) $P(\varphi^{-1}(\{0\})) = \mathbb{O}_{\mathcal{H}}$;
(d) $\varphi(x) \neq 0$ P-a.e. on D_φ.

If these conditions are satisfied, then

(e) $\frac{1}{\varphi} \in \mathcal{M}(X, \mathcal{A}, P)$ and $(J_P(\varphi))^{-1} = J_P\left(\frac{1}{\varphi}\right)$ (for the function $\frac{1}{\varphi}$, cf. 1.2.19; in particular $D_{\frac{1}{\varphi}} := \{x \in D_\varphi : \varphi(x) \neq 0\}$).

Proof. a: First we point out that $\varphi^{-1}(\{0\}) \in \mathcal{A}^{D_\varphi} \subset \mathcal{A}$. Then, for $f \in \mathcal{H}$ we have

$$f \in N_{J_P(\varphi)} \Leftrightarrow [f \in D_P(\varphi) \text{ and } \|J_P(\varphi)f\| = 0] \overset{(1)}{\Leftrightarrow}$$

$$\int_X |\varphi|^2 d\mu_f^P = 0 \overset{(2)}{\Leftrightarrow} \varphi(x) = 0 \ \mu_f^P\text{-a.e. on } D_\varphi \overset{(3)}{\Leftrightarrow}$$

$$\mu_f^P(D_\varphi - \varphi^{-1}(\{0\})) = 0 \Leftrightarrow P(D_\varphi - \varphi^{-1}(\{0\}))f = 0_{\mathcal{H}} \Leftrightarrow$$

$$f = P(X)f \overset{(4)}{=} P(D_\varphi)f = P(\varphi^{-1}(\{0\}))f \overset{(5)}{\Leftrightarrow} f \in R_{P(\varphi^{-1}(\{0\}))},$$

where: 1 holds by definition of $D_P(\varphi)$ and by 14.2.14d; 2 holds by 8.1.18a; 3 holds because $\varphi^{-1}(\{0\}) \in \mathcal{A}^{D_\varphi}$ (cf. the last part of 7.1.10); 4 holds because $P(X - D_\varphi) = \mathbb{O}_{\mathcal{H}}$; 5 holds by 13.1.3c.

$b \Leftrightarrow c$: This follows from a and from 3.2.6a.

$c \Leftrightarrow d$: This is true (by an argument similar to the argument used in the last part of 7.1.10) because $\varphi^{-1}(\{0\}) \in \mathcal{A}^{D_\varphi}$.

e: We assume condition c and note that

$$D_{\frac{1}{\varphi}} = D_\varphi - \varphi^{-1}(\{0\});$$

then $D_{\frac{1}{\varphi}} \in \mathcal{A}$ and $X - D_{\frac{1}{\varphi}} = (X - D_\varphi) \cup \varphi^{-1}(\{0\})$, whence $P\left(X - D_{\frac{1}{\varphi}}\right) = \mathbb{O}_{\mathcal{H}}$ by 13.3.2h; in view of 6.2.17, this proves that $\frac{1}{\varphi} \in \mathcal{M}(X, \mathcal{A}, P)$. Moreover,

$$\varphi(x)\frac{1}{\varphi}(x) = \frac{1}{\varphi}(x)\varphi(x) = 1, \forall x \in D_{\frac{1}{\varphi}} = D_{\varphi\frac{1}{\varphi}} = D_{\frac{1}{\varphi}\varphi};$$

this implies $\varphi\frac{1}{\varphi}, \frac{1}{\varphi}\varphi \in \mathcal{L}^\infty(X, \mathcal{A}, P)$ and hence (cf. 14.2.17e and 14.2.7a,i)

$$J_P\left(\varphi\frac{1}{\varphi}\right) = J_P\left(\frac{1}{\varphi}\varphi\right) = \tilde{J}_P(1_X) = P(X) = \mathbb{1}_{\mathcal{H}};$$

then, by 14.3.9,

$$J_P(\varphi)J_P\left(\frac{1}{\varphi}\right) \subset \mathbb{1}_{\mathcal{H}}, \quad J_P\left(\frac{1}{\varphi}\right)J_P(\varphi) \subset \mathbb{1}_{\mathcal{H}},$$

$$D_{J_P(\varphi)J_P(\frac{1}{\varphi})} = D_P\left(\frac{1}{\varphi}\right), \quad D_{J_P(\frac{1}{\varphi})J_P(\varphi)} = D_P(\varphi);$$

by 1.2.16b, this implies $(J_P(\varphi))^{-1} = J_P\left(\frac{1}{\varphi}\right)$. □

14.3.15 Proposition. *Let* $\varphi, \psi \in \mathcal{M}(X, \mathcal{A}, P)$. *Then,*

$$J_P(\varphi) = J_P(\psi) \text{ iff } \varphi(x) = \psi(x) \text{ } P\text{-a.e. on } D_\varphi \cap D_\psi.$$

Proof. First we assume $\varphi(x) = \psi(x)$ P-a.e. on $D_\varphi \cap D_\psi$. Then $\varphi(x) = \psi(x)$ μ_f^P-a.e. on $D_\varphi \cap D_\psi$ for all $f \in \mathcal{H}$, and hence (cf. 8.1.17c)

$$\int_X |\varphi|^2 d\mu_f^P = \int_X |\psi|^2 d\mu_f^P, \forall f \in \mathcal{H},$$

and hence $D_P(\varphi) = D_P(\psi)$. Moreover, in view of 14.2.14b and 8.2.7,

$$(f|J_P(\varphi)f) = \int_X \varphi d\mu_f^P = \int_X \psi d\mu_f^P = (f|J_P(\psi)f), \forall f \in D_P(\varphi).$$

Then, $J_P(\varphi) = J_P(\psi)$ in view of 14.2.13 and 10.2.12.

Conversely, we assume $J_P(\varphi) = J_P(\psi)$. For each $n \in \mathbb{N}$, we define the set

$$E_n := \{x \in D_\varphi \cap D_\psi : |\varphi(x)| + |\psi(x)| \le n\},$$

which is an element of $\mathcal{A}$; then (cf. 14.3.10),

$$J_P(\varphi\chi_{E_n}) = J_P(\varphi)J_P(\chi_{E_n}) = J_P(\psi)J_P(\chi_{E_n}) = J_P(\psi\chi_{E_n});$$

since $\varphi\chi_{E_n}, \psi\chi_{E_n} \in \mathcal{L}^\infty(X, \mathcal{A}, P)$, this implies (cf. 14.2.17e and 14.2.7i) that

$$\varphi(x)\chi_{E_n}(x) = \psi(x)\chi_{E_n}(x) \text{ } P\text{-a.e. on } D_\varphi \cap D_\psi,$$

or equivalently that

$$\exists F_n \in \mathcal{A} \text{ such that } P(F_n) = \mathbb{O}_{\mathcal{H}} \text{ and } \varphi(x) = \psi(x), \forall x \in E_n \cap (X - F_n).$$

By letting $F := \bigcup_{n=1}^\infty F_n$, we have (cf. 13.3.6c)

$$F \in \mathcal{A}, P(F) = \mathbb{O}_{\mathcal{H}} \text{ and } \varphi(x) = \psi(x), \forall x \in E_n \cap (X - F), \forall n \in \mathbb{N},$$

and hence

$$F \in \mathcal{A}, P(F) = \mathbb{O}_{\mathcal{H}} \text{ and } \varphi(x) = \psi(x),$$

$$\forall x \in \left(\bigcup_{n=1}^\infty E_n\right) \cap (X - F) = D_\varphi \cap D_\psi \cap (X - F),$$

or $\varphi(x) = \psi(x)$ P-a.e. on $D_\varphi \cap D_\psi$. □

14.3.16 Remark. On the basis of 14.3.15, we can define the mapping

$$M(X, \mathcal{A}, P) \ni [\varphi] \mapsto \Phi_P([\varphi]) := J_P(\varphi) \in \mathcal{O}(\mathcal{H})$$

and see that it is injective. This mapping is an extension of the mapping denoted by the same symbol in 14.2.8. However, this mapping is not a homomorphism from the associative algebra $M(X, \mathcal{A}, P)$ to any algebra of operators. First, $\mathcal{O}(\mathcal{H})$ is not an associative algebra (it is not even a linear space, cf. 3.2.11); furthermore, and more decisively, for $[\varphi], [\psi] \in M(X, \mathcal{A}, P)$ we have

$$\Phi_P([\varphi]) + \Phi_P([\psi]) \subset \Phi_P([\varphi] + [\psi]) \text{ and } \Phi_P([\varphi])\Phi_P([\psi]) \subset \Phi_P([\varphi][\psi]),$$

and not in general the corresponding equalities. For instance, if $\varphi \in \mathcal{M}(X, \mathcal{A}, P)$ and $\varphi \notin \mathcal{L}^\infty(X, \mathcal{A}, P)$ then $D_{J_P(\varphi)} + D_{J_P(-\varphi)} = D_P(\varphi) \neq \mathcal{H}$ (cf. 14.2.17), while $D_P(\varphi - \varphi) = \mathcal{H}$; similarly, if $\varphi \in \mathcal{M}(X, \mathcal{A}, P)$ is such that $\varphi \notin \mathcal{L}^\infty(X, \mathcal{A}, P)$ and $\varphi(x) \neq 0$ P-a.e on D_φ, then $D_{J_P(\frac{1}{\varphi})J_P(\varphi)} = D_P(\varphi) \neq \mathcal{H}$ while $D_P\left(\frac{1}{\varphi}\varphi\right) = \mathcal{H}$ (cf. the proof of 14.3.14).

14.3.17 Proposition. *Let $\varphi \in \mathcal{M}(X, \mathcal{A}, P)$. Then the following conditions are equivalent:*

(a) the operator $J_P(\varphi)$ is self-adjoint;
(b) the operator $J_P(\varphi)$ symmetric;
(c) $\varphi(x) = \overline{\varphi}(x)$ P-a.e. on D_φ.

Proof. We recall that $\overline{D_P(\varphi)} = \mathcal{H}$ (cf. 14.2.13). Thus, the operator $J_P(\varphi)$ is adjointable.

$a \Rightarrow b$: This is obvious.

$b \Rightarrow c$: Assuming condition b, 14.2.15 implies $J_P(\varphi) \subset J_P(\overline{\varphi})$, and hence $J_P(\varphi) = J_P(\overline{\varphi})$ since $D_P(\varphi) = D_P(\overline{\varphi})$, and hence condition c by 14.3.15.

$c \Rightarrow a$: This follows immediately from 14.3.15 and 14.2.15. $\qquad\square$

14.3.18 Proposition. *Let $\varphi \in \mathcal{M}(X, \mathcal{A}, P)$. Then the following conditions are equivalent:*

(a) the operator $J_P(\varphi)$ is unitary;
(b) $|\varphi(x)| = 1$ P-a.e. on D_φ;
(c) $\|J_P(\varphi)f\| = \|f\|, \forall f \in D_P(\varphi)$.

Proof. $a \Rightarrow b$: Assuming condition a, $J_P(\varphi)$ is injective and $(J_P(\varphi))^{-1} = (J_P(\varphi))^\dagger$ (cf. 12.5.1b). By 14.3.14 and 14.2.15, this implies $\frac{1}{\varphi} \in \mathcal{M}(X, \mathcal{A}, P)$ and $J_P\left(\frac{1}{\varphi}\right) = J_P(\overline{\varphi})$, and this implies (cf. 14.3.15) that

$$\frac{1}{\varphi(x)} = \overline{\varphi}(x) \ P\text{-a.e. on } D_{\frac{1}{\varphi}} \cap D_\varphi = D_{\frac{1}{\varphi}},$$

or equivalently that

$$\exists F \in \mathcal{A} \text{ such that}$$
$$P(F) = \mathbb{O}_{\mathcal{H}} \text{ and } \frac{1}{\varphi}(x) = \overline{\varphi}(x), \text{ or } |\varphi(x)| = 1, \forall x \in D_{\frac{1}{\varphi}} \cap (X - F);$$

now, $D_{\frac{1}{\varphi}} \cap (X - F) = D_\varphi \cap (X - ((X - D_{\frac{1}{\varphi}}) \cup F))$ and $P((X - D_{\frac{1}{\varphi}}) \cup F) = \mathbb{O}_\mathcal{H}$ (cf. 13.3.2h); thus, condition b is proved.

$b \Rightarrow c$: Assuming condition b, 14.2.14d and 8.2.7 yield

$$\|J_P(\varphi)f\|^2 = \int_X 1_X d\mu_f^P = \mu_f^P(X) = \|f\|^2, \forall f \in D_P(\varphi).$$

$c \Rightarrow a$: We assume condition c. Then, 4.2.3 implies that $J_P(\varphi)$ is injective and $(J_P(\varphi))^{-1}$ is bounded; moreover, $(J_P(\varphi))^{-1} = J_P\left(\frac{1}{\varphi}\right)$ (cf. 14.3.14e); then, $R_{J_P(\varphi)} = D_{(J_P(\varphi))^{-1}} = \mathcal{H}$ by 14.2.17. Similarly, condition c implies that $J_P(\varphi)$ is bounded and hence that $D_P(\varphi) = \mathcal{H}$ (cf. 14.2.17). In view of 10.1.20, this proves that U is an automorphism of $\mathcal{H}$, i.e. that U is a unitary operator. $\qquad\square$

14.4 Spectral properties of integrals

As before, $(X, \mathcal{A})$ denotes an abstract measurable space, $\mathcal{H}$ denotes an abstract Hilbert space, and P denotes a projection valued measure on $\mathcal{A}$ with values in $\mathscr{P}(\mathcal{H})$.

14.4.1 Proposition. *Let* $\varphi \in \mathcal{M}(X, \mathcal{A}, P)$. *Then,*

$$\rho(J_P(\varphi)) = \{\lambda \in \mathbb{C} : J_P(\varphi) - \lambda\mathbb{1}_\mathcal{H} \text{ is injective and } (J_P(\varphi) - \lambda\mathbb{1}_\mathcal{H})^{-1} \text{ is bounded}\}$$

or equivalently

$$\sigma(J_P(\varphi)) = Ap\sigma(J_P(\varphi)).$$

Proof. We prove the statement by proving, for $\lambda \in \mathbb{C}$, the implication

$$[J_P(\varphi) - \lambda\mathbb{1}_\mathcal{H} \text{ is injective}] \Rightarrow \overline{R_{J_P(\varphi) - \lambda\mathbb{1}_\mathcal{H}}} = \mathcal{H}.$$

Now, $J_P(\varphi) - \lambda\mathbb{1}_\mathcal{H} = J_P(\varphi - \lambda)$ for all $\lambda \in \mathbb{C}$ (cf. the proof of 14.3.2); therefore, if $J_P(\varphi) - \lambda\mathbb{1}_\mathcal{H}$ is injective then (cf. 14.3.14)

$$\frac{1}{\varphi - \lambda} \in \mathcal{M}(X, \mathcal{A}, P) \text{ and } (J_P(\varphi) - \lambda\mathbb{1}_\mathcal{H})^{-1} = J_P\left(\frac{1}{\varphi - \lambda}\right),$$

and hence $R_{J_P(\varphi) - \lambda\mathbb{1}_\mathcal{H}} = D_P\left(\frac{1}{\varphi - \lambda}\right)$, and hence $\overline{R_{J_P(\varphi) - \lambda\mathbb{1}_\mathcal{H}}} = \mathcal{H}$ (cf. 14.2.13). $\qquad\square$

14.4.2 Theorem. *Let* $\varphi \in \mathcal{M}(X, \mathcal{A}, P)$ *and* $\lambda \in \mathbb{C}$. *Then the following conditions are equivalent:*

(a) $\lambda \in \sigma(J_P(\varphi))$;
(b) $P(\varphi^{-1}(B(\lambda, \varepsilon))) \neq \mathbb{O}_\mathcal{H}$, $\forall \varepsilon \in (0, \infty)$
(recall that $B(\lambda, \varepsilon) := \{z \in \mathbb{C} : |z - \lambda| < \varepsilon\}$ *and note that* $\varphi^{-1}(B(\lambda, \varepsilon)) \in \mathcal{A}$ *by 6.2.13c with* $\mathcal{G} := \mathcal{T}_{d_\mathbb{C}}$).

Proof. $a \Rightarrow b$: We prove (not b)$\Rightarrow$(not a). Assuming condition (not b), there exists $\varepsilon \in (0, \infty)$ so that $P(\varphi^{-1}(B(\lambda, \varepsilon))) = \mathbb{O}_{\mathcal{H}}$, and hence so that

$$\mu_f^P(\varphi^{-1}(B(\lambda, \varepsilon))) = \|P(\varphi^{-1}(B(\lambda, \varepsilon)))f\|^2 = 0, \forall f \in \mathcal{H},$$

and hence so that, letting $E := X - \varphi^{-1}(B(\lambda, \varepsilon))$,

$$\|(J_P(\varphi) - \lambda \mathbb{1}_{\mathcal{H}})f\|^2 = \int_X |\varphi - \lambda|^2 d\mu_f^P = \int_E |\varphi - \lambda|^2 d\mu_f^P$$

$$\geq \varepsilon^2 \int_E 1_X d\mu_f^P$$

$$= \varepsilon^2 \int_X 1_X d\mu_f^P = \varepsilon^2 \mu_f^P(X) = \varepsilon^2 \|f\|^2, \forall f \in D_P(\varphi)$$

(cf. 14.3.2, 8.3.3a, 8.1.17b). By 4.2.3 and 14.4.1, this proves that $\lambda \in \rho(J_P(\varphi))$, i.e. condition (not a).

$b \Rightarrow a$: We prove (not a)$\Rightarrow$(not b). In view of 14.4.1, condition (not a) implies that

$$J_P(\varphi) - \lambda \mathbb{1}_{\mathcal{H}} \text{ is injective and } (J_P(\varphi) - \lambda \mathbb{1}_{\mathcal{H}})^{-1} \text{ is bounded,}$$

and hence, in view of the equality $J_P(\varphi) - \lambda \mathbb{1}_{\mathcal{H}} = J_P(\varphi - \lambda)$ (cf. the proof of 14.3.2) and of 14.3.14, that

$$\frac{1}{\varphi - \lambda} \in \mathcal{M}(X, \mathcal{A}, P) \text{ and } J_P\left(\frac{1}{\varphi - \lambda}\right) \text{ is bounded,}$$

and hence, in view of 14.2.17, that

$$\exists m \in (0, \infty) \text{ such that } \left|\frac{1}{\varphi(x) - \lambda}\right| \leq m \text{ } P\text{-a.e. on } D_{\frac{1}{\varphi - \lambda}};$$

proceeding as in the proof of 14.3.18 ($a \Rightarrow b$), we see that this is equivalent to

$$\exists m \in (0, \infty) \text{ such that } |\varphi(x) - \lambda| \geq \frac{1}{m} \text{ } P\text{-a.e. on } D_\varphi;$$

proceeding as at the end of 7.1.10 (in view of $\varphi^{-1}\left(B\left(\lambda, \frac{1}{m}\right)\right) \in \mathcal{A}$), this yields

$$\exists m \in (0, \infty) \text{ such that } P\left(\varphi^{-1}\left(B\left(\lambda, \frac{1}{m}\right)\right)\right) = \mathbb{O}_{\mathcal{H}},$$

i.e. condition (not b). $\qquad \square$

14.4.3 Theorem. *Let* $\varphi \in \mathcal{M}(X, \mathcal{A}, P)$. *Then,*

$$P(\varphi^{-1}(\sigma(J_P(\varphi)))) = \mathbb{1}_{\mathcal{H}}$$

(note that $\varphi^{-1}(\sigma(J_P(\varphi))) \in \mathcal{A}$ *by 10.4.6 and by 6.2.13c with* $\mathcal{G} := \mathcal{K}_{d_C}$*), and hence*

$$\sigma(J_P(\varphi)) \neq \emptyset.$$

Proof. For each $\lambda \in \mathbb{C} - \sigma(J_P(\varphi))$, 14.4.2 implies that there exists $\varepsilon \in (0, \infty)$ such that

$$P(\varphi^{-1}(B(\lambda, \varepsilon))) = \mathbb{O}_{\mathcal{H}};$$

this condition implies

$$B(\lambda, \varepsilon) \subset \mathbb{C} - \sigma(J_P(\varphi));$$

indeed, if $z \in B(\lambda, \varepsilon)$ then there exists $\eta \in (0, \infty)$ such that $B(z, \eta) \subset B(\lambda, \varepsilon)$, and hence such that $\varphi^{-1}(B(z, \eta)) \subset \varphi^{-1}(B(\lambda, \varepsilon))$, and hence (cf. 13.3.2e) such that

$$P(\varphi^{-1}(B(z, \eta))) = \mathbb{O}_{\mathcal{H}};$$

in view of 14.4.2, this implies $z \in \mathbb{C} - \sigma(J_P(\varphi))$.

Now, for each $\lambda \in \mathbb{C} - \sigma(J_P(\varphi))$ let $\varepsilon_\lambda \in (0, \infty)$ be such that

$$P(\varphi^{-1}(B(\lambda, \varepsilon_\lambda))) = \mathbb{O}_{\mathcal{H}}.$$

Since $B(\lambda, \varepsilon_\lambda) \subset \mathbb{C} - \sigma(J_P(\varphi))$ for all $\lambda \in \mathbb{C} - \sigma(J_P(\varphi))$, we have obviously

$$\mathbb{C} - \sigma(J_P(\varphi)) = \bigcup_{\lambda \in \mathbb{C} - \sigma(J_P(\varphi))} B(\lambda, \varepsilon_\lambda).$$

Since $(\mathbb{C}, d_\mathbb{C})$ is a separable metric space (cf. 2.7.4a), by 2.3.18 there exists a countable subset $\{\lambda_n\}_{n \in I}$ of $\mathbb{C} - \sigma(J_P(\varphi))$ such that

$$\mathbb{C} - \sigma(J_P(\varphi)) = \bigcup_{n \in I} B(\lambda_n, \varepsilon_{\lambda_n}),$$

and hence such that

$$D_\varphi - \varphi^{-1}(\sigma(J_P(\varphi))) = \varphi^{-1}(\mathbb{C} - \sigma(J_P(\varphi))) = \bigcup_{n \in I} \varphi^{-1}(B(\lambda_n, \varepsilon_{\lambda_n}));$$

then (cf. 13.3.6c) $P(D_\varphi - \varphi^{-1}(\sigma(J_P(\varphi)))) = \mathbb{O}_{\mathcal{H}}$, or equivalently

$$P(\varphi^{-1}(\sigma(J_P(\varphi)))) = P(D_\varphi) = P(D_\varphi) + P(X - D_\varphi) = P(X) = \mathbb{1}_{\mathcal{H}}.$$

Obviously, this implies $\sigma(J_P(\varphi)) \neq \emptyset$ (otherwise, $\varphi^{-1}(\sigma(J_P(\varphi))) = \emptyset$ and hence $P(\varphi^{-1}(\sigma(J_P(\varphi)))) = \mathbb{O}_{\mathcal{H}}$). $\qquad \square$

14.4.4 Remark. For every $\varphi \in \mathcal{M}(X, \mathcal{A}, P)$, the equalities in 14.2.14b and in 14.3.2 can be written as follows:

$$(f|J_P(\varphi)f) = \int_{\varphi^{-1}(\sigma(J_P(\varphi)))} \varphi d\mu_f^P, \forall f \in D_P(\varphi);$$

$$\|J_P(\varphi)f - \lambda f\|^2 = \int_{\varphi^{-1}(\sigma(J_P(\varphi)))} |\varphi - \lambda|^2 d\mu_f^P, \forall f \in D_P(\varphi), \forall \lambda \in \mathbb{C}.$$

This follows from 8.3.3, since 14.4.3 implies that

$$P(X - \varphi^{-1}(\sigma(J_P(\varphi)))) = \mathbb{O}_{\mathcal{H}},$$

and hence

$$\mu_f^P(X - \varphi^{-1}(\sigma(J_P(\varphi)))) = 0, \forall f \in \mathcal{H}.$$

14.4.5 Proposition. *Let* $\varphi \in \mathcal{M}(X, \mathcal{A}, P)$. *Then the operator* $J_P(\varphi)$ *is bounded iff* $\sigma(J_P(\varphi))$ *is a bounded subset of* $\mathbb{C}$.

Proof. If the operator $J_P(\varphi)$ is bounded then $J_P(\varphi) \in \mathcal{B}(\mathcal{H})$ (cf. 14.2.17), and hence $\sigma(J_P(\varphi))$ is a bounded subset of $\mathbb{C}$ by 4.5.10.

Conversely, suppose that $\sigma(J_P(\varphi))$ is bounded and let $m \in [0, \infty)$ be such that

$$|z| \le m, \forall z \in \sigma(J_P(\varphi));$$

then (cf. 14.4.4)

$$\begin{aligned}
\|J_P(\varphi)f\|^2 &= \int_{\varphi^{-1}(\sigma(J_P(\varphi)))} |\varphi|^2 d\mu_f^P \\
&\le m^2 \int_{\varphi^{-1}(\sigma(J_P(\varphi)))} 1_X d\mu_f^P = m^2 \int_X 1_X d\mu_f^P \\
&= m^2 \mu_f^P(X) = m^2 \|f\|^2, \forall f \in D_P(\varphi),
\end{aligned}$$

and this proves that $J_P(\varphi)$ is bounded. $\qquad\square$

14.4.6 Theorem. *Let* $\varphi \in \mathcal{M}(X, \mathcal{A}, P)$. *Then,*

$$N_{J_P(\varphi) - \lambda \mathbb{1}_{\mathcal{H}}} = R_{P(\varphi^{-1}(\{\lambda\}))}.$$

For $\lambda \in \mathbb{C}$, *the following conditions are equivalent:*

(a) $\lambda \in \sigma_P(J_P(\varphi))$;
(b) $P(\varphi^{-1}(\{\lambda\})) \ne \mathbb{O}_{\mathcal{H}}$.

If $\lambda \in \sigma_P(J_P(\varphi))$ *then* $P(\varphi^{-1}(\{\lambda\}))$ *is the projection onto the corresponding eigenspace.*

Proof. For $\lambda \in \mathbb{C}$, we have $J_P(\varphi) - \lambda \mathbb{1}_{\mathcal{H}} = J_P(\varphi - \lambda)$ (cf. the proof of 14.3.2). If we define $\psi_\lambda := \varphi - \lambda$, we have $\psi_\lambda^{-1}(\{0\}) = \varphi^{-1}(\{\lambda\})$; then, from 14.3.14a we have

$$N_{J_P(\varphi) - \lambda \mathbb{1}_{\mathcal{H}}} = N_{J_P(\psi_\lambda)} = R_{P(\psi_\lambda^{-1}(\{0\}))} = R_{P(\varphi^{-1}(\{\lambda\}))}.$$

In view of this, the equivalence of conditions a and b follows directly from 4.5.7, and so does the part of the statement about eigenspaces (for which, cf. also 13.1.3c). $\quad\square$

14.5 Multiplication operators

In this section, $(X, \mathcal{A}, \mu)$ stands for an abstract measure space. At variance with what was done in Section 11.1, we denote the elements of $\mathcal{L}^2(X, \mathcal{A}, \mu)$ with the letters f, g,....

For $\varphi \in \mathcal{M}(X, \mathcal{A}, \mu)$, we define the mapping from $L^2(X, \mathcal{A}, \mu)$ to itself

$$\begin{aligned}
M_\varphi : D_{M_\varphi} &\to L^2(X, \mathcal{A}, \mu) \\
[f] &\mapsto M_\varphi[f] := [\varphi f],
\end{aligned}$$

with

$$D_{M_\varphi} := \{[f] \in L^2(X, \mathcal{A}, \mu) : \varphi f \in \mathcal{L}^2(X, \mathcal{A}, \mu)\}$$

$$= \left\{[f] \in L^2(X, \mathcal{A}, \mu) : \int_X |\varphi f|^2 d\mu < \infty\right\}$$

(note that $\varphi f \in \mathcal{M}(X, \mathcal{A}, \mu)$ for all $f \in \mathcal{L}^2(X, \mathcal{A}, \mu)$, in view of 8.2.2). It is easy to see that M_φ is a linear operator (D_{M_φ} is a linear manifold in $L^2(X, \mathcal{A}, \mu)$ by 11.1.2a).

For $E \in \mathcal{A}$, we write $P_E := M_{\chi_E}$. We have:

$$D_{P_E} = L^2(X, \mathcal{A}, \mu) \text{ since } \int_X |\chi_E f|^2 d\mu \leq \int_X |f|^2 d\mu < \infty, \forall[f] \in L^2(X, \mathcal{A}, \mu);$$

$$([f]|P_E[f]) = \int_X \chi_E |f|^2 d\mu \in \mathbb{R}, \forall[f] \in L^2(X, \mathcal{A}, \mu), \text{ hence } P_E = P_E^\dagger \text{ (cf. 12.4.3)};$$

$$P_E([f]) = [\chi_E f] = [\chi_E^2 f] = P_E^2[f], \forall[f] \in L^2(X, \mathcal{A}, \mu), \text{ hence } P_E = P_E^2.$$

This proves that P_E is a projection (cf. 13.1.5).

Now, we define the mapping

$$P : \mathcal{A} \to \mathscr{P}(L^2(X, \mathcal{A}, \mu))$$

$$E \mapsto P(E) := P_E.$$

For all $[f] \in L^2(X, \mathcal{A}, \mu)$, we have:

$$\mu_{[f]}^P(X) = ([f]|P_X[f]) = \|[f]\|^2, \forall[f] \in L^2(X, \mathcal{A}, \mu);$$

$$\mu_{[f]}^P(E) = ([f]|P_E[f]) = \int_X \chi_E |f|^2 d\mu = \int_E |f|^2 d\mu, \forall E \in \mathcal{A}.$$

In view of 8.3.4a and 13.3.5, this proves that P is a projection valued measure on $\mathcal{A}$. If $F \in \mathcal{A}$ is such that $\mu(F) = 0$, then $\mu_{[f]}^P(F) = 0$ for all $[f] \in L^2(X, \mathcal{A}, \mu)$ (cf. 8.3.4a) and hence $P(F) = \mathbb{O}_{L^2(X, \mathcal{A}, \mu)}$. Therefore, $\mathcal{M}(X, \mathcal{A}, \mu) \subset \mathcal{M}(X, \mathcal{A}, P)$.

For $\varphi \in \mathcal{M}(X, \mathcal{A}, \mu)$, we have

$$[f] \in D_{M_\varphi} \Leftrightarrow \int_X |\varphi|^2 |f|^2 d\mu < \infty \overset{(1)}{\Leftrightarrow} \int_X |\varphi|^2 d\mu_{[f]}^P < \infty \Leftrightarrow [f] \in D_{P(\varphi)},$$

where 1 holds by 8.3.4b; moreover, we have

$$([f]|M_\varphi[f]) = \int_X \varphi |f|^2 d\mu \overset{(2)}{=} \int_X \varphi d\mu_{[f]}^P, \forall[f] \in D_{M_\varphi},$$

where 2 holds by 8.3.4c. This proves that $M_\varphi = J_P(\varphi)$, by the uniqueness asserted in 14.2.14.

Now *we assume that the measure μ is σ-finite*, i.e. that there exists a countable family $\{E_n\}_{n \in I}$ of elements of $\mathcal{A}$ so that $X = \bigcup_{n \in I} E_n$ and $\mu(E_n) < \infty$ for all $n \in I$ (this implies that $\chi_{E_n} \in \mathcal{L}^2(X, \mathcal{A}, \mu)$ for all $n \in I$). If $F \in \mathcal{A}$ is such that $P(F) = \mathbb{O}_{L^2(X, \mathcal{A}, \mu)}$, then

$$\mu(F \cap E_n) = \int_X \chi_{F \cap E_n} d\mu = \int_X \chi_F |\chi_{E_n}|^2 d\mu = ([\chi_{E_n}]|P(F)[\chi_{E_n}]) = 0, \forall n \in I,$$

and hence

$$\mu(F) = \mu\left(\bigcup_{n\in I}(F\cap E_n)\right) \le \sum_{n\in I}\mu(F\cap E_n) = 0$$

(cf. 7.1.4a), and hence $\mu(F) = 0$. Now, let E be an element of $\mathcal{A}$ and, for each $x \in E$, let $Q(x)$ be a proposition. Then,

$$[Q(x) \; P\text{-a.e. on } E] \text{ is equivalent to } [Q(x) \; \mu\text{-a.e. on } E].$$

Thus, $\mathcal{M}(X,\mathcal{A},\mu) = \mathcal{M}(X,\mathcal{A},P)$ and all the statements of Sects. 14.2, 14.3, 14.4 hold true with $J_P(\varphi)$ replaced by M_φ, the projection valued measure P replaced by the measure μ, "P-a.e." replaced by "μ-a.e." ($\mathcal{L}^\infty(X,\mathcal{A},\mu)$ is defined as $\mathcal{L}^\infty(X,\mathcal{A},P)$ was, with P replaced by μ).

14.6   Change of variable. Unitary equivalence.

In some cases, there are relations between integrals constructed with respect to two different projection valued measures. In this section we examine two important cases of this kind.

14.6.1 Theorem (Change of variable theorem). *Let* $\mathcal{H}$ *be a Hilbert space,* $(X_1,\mathcal{A}_1)$ *a measurable space, and* P_1 *a projection valued measure on* $\mathcal{A}_1$ *with values in* $\mathscr{P}(\mathcal{H})$*. Let* $(X_2,\mathcal{A}_2)$ *be a measurable space, and let* $\pi : D_\pi \to X_2$ *be a mapping from* X_1 *to* X_2 *which is measurable w.r.t.* ${\mathcal{A}_1}^{D_\pi}$ *and* $\mathcal{A}_2$*, and so that* $D_\pi \in \mathcal{A}_1$ *and* $P_1(X_1 - D_\pi) = \mathbb{O}_\mathcal{H}$*.*

(a) The mapping

$$P_2 : \mathcal{A}_2 \to \mathscr{P}(\mathcal{H})$$
$$E \mapsto P_2(E) := P_1(\pi^{-1}(E))$$

is a projection valued measure on $\mathcal{A}_2$*.*
(b) For $\varphi \in \mathcal{M}(X_2,\mathcal{A}_2,P_2)$ *we have:*

$$\varphi \circ \pi \in \mathcal{M}(X_1,\mathcal{A}_1,P_1) \text{ and}$$
$$J_{P_2}(\varphi) = J_{P_1}(\varphi \circ \pi) \text{ or } \int_{X_2}\varphi dP_2 = \int_{X_1}(\varphi \circ \pi)dP_1).$$

Proof. a: For every $f \in \mathcal{H}$ we have

$$\mu_f^{P_2}(E) = \mu_f^{P_1}(\pi^{-1}(E)), \forall E \in \mathcal{A}_2;$$

then, $\mu_f^{P_2}$ is a measure on $\mathcal{A}_2$ in view of 13.3.5 ($a \Rightarrow b$) and 8.3.11a; moreover,

$$\mu_f^{P_2}(X_2) = \mu_f^{P_1}(D_\pi) \stackrel{(1)}{=} \mu_f^{P_1}(X_1) = \|f\|^2,$$

where 1 is true because $P_1(X_1 - D_\pi) = \mathbb{O}_\mathcal{H}$. In view of 13.3.5 ($b \Rightarrow a$), this proves that P_2 is a projection valued measure on $\mathcal{A}_2$.

b: For $\varphi \in (X_2, \mathcal{A}_2, P_2)$ we have:

$$D_{\varphi \circ \pi} = \pi^{-1}(D_\varphi) \in \mathcal{A}_1^{D_\pi} \subset \mathcal{A}_1;$$

$$P_1(X_1 - D_{\varphi \circ \pi}) \overset{(2)}{=} P_1(D_\pi - \pi^{-1}(D_\varphi)) = P_1(\pi^{-1}(X_2 - D_\varphi))$$
$$= P_2(X_2 - D_\varphi) = \mathbb{O}_{\mathcal{H}},$$

where 2 is true because $P_1(X_1 - D_\pi) = \mathbb{O}_{\mathcal{H}}$; further $\varphi \circ \pi$ is measurable w.r.t. $\mathcal{A}_1^{D_{\varphi \circ \pi}}$ and $\mathcal{A}(d_{\mathbb{C}})$ since π is measurable w.r.t. $\mathcal{A}_1^{D_\pi}$ and $\mathcal{A}_2$ and φ is measurable w.r.t. $\mathcal{A}_2^{D_\varphi}$ and $\mathcal{A}(d_{\mathbb{C}})$ (cf. 6.2.6). Thus, $\varphi \circ \pi \in \mathcal{M}(X_1, \mathcal{A}_1, P_1)$.

Then, in view of 8.3.11b we have

$$\int_{X_2} |\varphi|^2 d\mu_f^{P_2} = \int_{X_1} |\varphi \circ \pi|^2 d\mu_f^{P_1}, \forall f \in \mathcal{H},$$

and hence $D_{P_1}(\varphi \circ \pi) = D_{P_2}(\varphi)$. Moreover, in view of 14.2.14b and 8.3.11c we have

$$(f|J_{P_2}(\varphi)f) = \int_{X_2} \varphi \, d\mu_f^{P_2} = \int_{X_1} (\varphi \circ \pi) d\mu_f^{P_1} = (f|J_{P_1}(\varphi \circ \pi)f), \forall f \in D_{P_2}(\varphi).$$

This proves the equality $J_{P_1}(\varphi \circ \pi) = J_{P_2}(\varphi)$, in view of 14.2.13 and 10.2.12. $\quad\square$

14.6.2 Theorem. *Let $\mathcal{H}_1$ and $\mathcal{H}_2$ be isomorphic Hilbert spaces and suppose that $U \in \mathcal{UA}(\mathcal{H}_1, \mathcal{H}_2)$ (for $\mathcal{UA}(\mathcal{H}_1, \mathcal{H}_2)$, cf. 10.3.15). Let $(X, \mathcal{A})$ be a measurable space and let P_1 be a projection valued measure on $\mathcal{A}$ with values in $\mathscr{P}(\mathcal{H}_1)$. Then the mapping*

$$P_2 : \mathcal{A} \to \mathscr{P}(\mathcal{H}_2)$$
$$E \mapsto P_2(E) := U P_1(E) U^{-1}$$

is a projection valued measure on $\mathcal{A}$. We have $\mathcal{M}(X, \mathcal{A}, P_2) = \mathcal{M}(X, \mathcal{A}, P_1)$ and, for all $\varphi \in \mathcal{M}(X, \mathcal{A}, P_1)$,

$$J_{P_2}(\varphi) = U J_{P_1}(\varphi) U^{-1} \text{ if } U \in \mathcal{U}(\mathcal{H}_1, \mathcal{H}_2),$$
$$J_{P_2}(\overline{\varphi}) = U J_{P_1}(\varphi) U^{-1} \text{ if } U \in \mathcal{A}(\mathcal{H}_1, \mathcal{H}_2).$$

Proof. From 13.1.8 we have $P_2(E) \in \mathscr{P}(\mathcal{H}_2), \forall E \in \mathcal{A}$. Further, condition 13.3.5b for P_1 implies the same condition for P_2 since

$$\mu_f^{P_2}(E) = \|U P_1(E) U^{-1} f\|^2 = \|P_1(E) U^{-1} f\|^2$$
$$= \mu_{U^{-1}f}^{P_1}(E), \forall E \in \mathcal{A}, \forall f \in \mathcal{H}_2, \tag{1}$$

and since $\|U^{-1}f\| = \|f\|, \forall f \in \mathcal{H}_2$. Thus, P_2 is a projection valued measure on $\mathcal{A}$.

The equality $\mathcal{M}(X, \mathcal{A}, P_2) = \mathcal{M}(X, \mathcal{A}, P_1)$ is obvious since, for $E \in \mathcal{A}$, $P_2(E) = \mathbb{O}_{\mathcal{H}_2}$ iff $P_1(E) = \mathbb{O}_{\mathcal{H}_1}$. Now let $\varphi \in \mathcal{M}(X, \mathcal{A}, P_1)$. In view of 1 we have, for $f \in \mathcal{H}_2$,

$$\varphi \in \mathcal{L}_2(X, \mathcal{A}, \mu_f^{P_2}) \text{ iff } \varphi \in \mathcal{L}_2(X, \mathcal{A}, \mu_{U^{-1}f}^{P_1}),$$

and hence

$$D_{P_2}(\overline{\varphi}) = D_{P_2}(\varphi) = \{f \in \mathcal{H}_2 : U^{-1}f \in D_{P_1}(\varphi)\} = D_{J_{P_1}(\varphi)U^{-1}} = D_{U J_{P_1}(\varphi)U^{-1}}.$$

Moreover, for every $f \in D_{P_2}(\varphi) (= D_{P_2}(\overline{\varphi}))$, in view of 14.2.14b and of 1 we have (since $U^{-1}f \in D_{P_1}(\varphi) = D_{P_1}(\overline{\varphi})$):

$$(f|J_{P_2}(\varphi)f) = \int_X \varphi d\mu_f^{P_2} = \int_X \varphi d\mu_{U^{-1}f}^{P_1}$$
$$= (U^{-1}f|J_{P_1}(\varphi)U^{-1}f) = (f|UJ_{P_1}(\varphi)U^{-1}f),$$

if $U \in \mathcal{U}(\mathcal{H}_1, \mathcal{H}_2)$;

$$(f|J_{P_2}(\overline{\varphi})f) = \int_X \overline{\varphi} d\mu_f^{P_2} = (U^{-1}f|J_{P_1}(\overline{\varphi})U^{-1}f)$$
$$\overset{(2)}{=} (J_{P_1}(\varphi)U^{-1}f|U^{-1}f) = (f|UJ_{P_1}(\varphi)U^{-1}f),$$

if $U \in \mathcal{A}(\mathcal{H}_1, \mathcal{H}_2)$ (2 is true because $J_{P_1}(\overline{\varphi}) = (J_{P_1}(\varphi))^\dagger$, cf. 14.2.15).

This proves the equalities of the statement, in view of 14.2.13 and 10.2.12. $\square$

Chapter 15

Spectral Theorems

Unitary and self-adjoint operators can be represented in a unique way as integrals with respect to suitable projection valued measures. This is the content of the corresponding spectral theorems. Following John von Neumann, we deduce the spectral theorem for self-adjoint operators from the spectral theorem for unitary operators.

The spectral theorem for self-adjoint operators is of crucial importance in quantum mechanics. In fact, it is through this theorem that self-adjoint operators step onto the quantum mechanical stage, since quantum observables arise most naturally in the guise of projection valued measures (cf. Section 19.3).

Functions of a self-adjoint operator can be defined, on the basis of the projection valued measure associated with that operator. The mathematical idea of a function of a self-adjoint operator has its physical counterpart in the idea of a function of a quantum observable (cf. Section 19.3).

15.1 The spectral theorem for unitary operators

The proof we give of the spectral theorem for unitary operators rests on the Fejér–Riesz lemma which is proved in 15.1.2, on the Stone–Weierstrass theorem for the unit circle proved in 4.3.7, on the Riesz–Markov theorem for positive linear functionals proved in 8.5.3, and on the characterization of the family of bounded Borel functions provided in 6.3.4.

We recall that $\mathcal{P}$ denotes the family of trigonometric polynomials on the unit circle $\mathbb{T}$, that $\mathcal{P}$ is a subalgebra of the associative algebra $\mathcal{C}(\mathbb{T})$, that $\mathcal{C}(\mathbb{T}) = \mathcal{C}_B(\mathbb{T})$ since the metric subspace $(\mathbb{T}, d_{\mathbb{T}})$ of the metric space $(\mathbb{C}, d_{\mathbb{C}})$ is compact, and hence that $\mathcal{C}(\mathbb{T})$ is a normed algebra (cf. 4.3.6a,c). We note that obviously $\bar{p} \in \mathcal{P}$ for all $p \in \mathcal{P}$.

Throughout this section, $\mathcal{H}$ denotes an abstract Hilbert space. We recall that $\mathcal{B}(\mathcal{H})$ is a C^*-algebra (cf. 12.6.4).

15.1.1 Theorem. *Let U be a unitary operator in $\mathcal{H}$. Then the following definition, of the mapping $\hat{\phi}_U$, is consistent:*

$$\hat{\phi}_U : \mathcal{P} \to \mathcal{B}(\mathcal{H})$$

$$p \mapsto \hat{\phi}_U(p) := p(U),$$

where

$$p(U) := \sum_{k=-N}^{N} \alpha_k U^k$$

if $N \geq 0$ and $(\alpha_0, \alpha_1, \alpha_{-1}, ..., \alpha_N, \alpha_{-N}) \in \mathbb{C}^{2N+1}$ are so that

$$p(z) = \sum_{k=-N}^{N} \alpha_k z^k, \forall z \in \mathbb{T}$$

(we define $U^0 := \mathbb{1}_{\mathcal{H}}$, cf. 3.3.1; for $n \in \mathbb{N}$ we define $U^{-n} := (U^{-1})^n = (U^n)^{-1}$, where the second equality follows from 1.2.14B).

The mapping $\hat{\phi}_U$ has the following properties:

(a) $\hat{\phi}_U(\alpha_1 p_1 + \alpha_2 p_2) = \alpha_1 \hat{\phi}_U(p_1) + \alpha_2 \hat{\phi}_U(p_2), \forall \alpha_1, \alpha_2 \in \mathbb{C}, \forall p_1, p_2 \in \mathcal{P};$
(b) $\hat{\phi}_U(p_1 p_2) = \hat{\phi}_U(p_1)\hat{\phi}_U(p_2), \forall p_1, p_2 \in \mathcal{P};$
(c) $\hat{\phi}_U(\overline{p}) = (\hat{\phi}_U(p))^{\dagger}, \forall p \in \mathcal{P}.$

Proof. Consistency: If $p \in \mathcal{P}$ then

$$\exists N \geq 0, \exists (\alpha_0, \alpha_1, \alpha_{-1}, ..., \alpha_N, \alpha_{-N}) \in \mathbb{C}^{2N+1} \text{ so that}$$

$$p(z) = \sum_{k=-N}^{N} \alpha_k z^k, \forall z \in \mathbb{T};$$

we can assume that $|\alpha_N| + |\alpha_{-N}| \neq 0$. Suppose that $M \geq 0$ and $(\beta_0, \beta_1, \beta_{-1}, ..., \beta_M, \beta_{-M}) \in \mathbb{C}^{2M+1}$ are so that $|\beta_M| + |\beta_{-M}| \neq 0$ and

$$p(z) = \sum_{h=-M}^{M} \beta_h z^h, \forall z \in \mathbb{T}.$$

Then, supposing e.g. $M \leq N$, the equation

$$z^N \left(\sum_{k=-N}^{N} \alpha_k z^k - \sum_{h=-M}^{M} \beta_h z^h \right) = 0, \forall z \in \mathbb{T},$$

shows that $M = N$ ($M < N$ would imply $\alpha_{-N} = \alpha_N = 0$) and $\beta_k = \alpha_k$ for all $k \in \{0, \pm 1, ..., \pm N\}$. This proves that the definition of the mapping $\hat{\phi}_U$ is consistent (also, note that $p(U) \in \mathcal{B}(\mathcal{H})$ because $U, U^{-1} \in \mathcal{B}(\mathcal{H})$ and $\mathcal{B}(\mathcal{H})$ is an associative algebra, cf. 4.3.5).

a and b: These properties are obvious, since $\mathcal{B}(\mathcal{H})$ is an associative algebra.

c: This property follows from 12.3.1b, 12.3.2, 12.3.4b, 12.5.1b (also, note that $\overline{z} = z^{-1}$ for all $z \in \mathbb{T}$). □

15.1.2 Lemma (The Fejér–Riesz lemma). *Let $p \in \mathcal{P}$ be such that $0 \leq p(z)$ for all $z \in \mathbb{T}$. Then,*

$$\exists q \in \mathcal{P} \text{ such that } p = \bar{q}q.$$

Proof. Since $p \in \mathcal{P}$,

$$\exists N \geq 0, \exists (\alpha_0, \alpha_1, \alpha_{-1}, ..., \alpha_N, \alpha_{-N}) \in \mathbb{C}^{2N+1} \text{ so that}$$

$$p(z) = \sum_{k=-N}^{N} \alpha_k z^k, \forall z \in \mathbb{T}.$$

If $N = 0$, then $0 \leq \alpha_0$, and hence the element q of $\mathcal{P}$ defined by

$$q(z) := \sqrt{\alpha_0}, \forall z \in \mathbb{T},$$

is so that $p = \bar{q}q$.

In what follows we suppose $N > 0$ and $|\alpha_N| + |\alpha_{-N}| \neq 0$.

We have

$$\sum_{k=-N}^{N} \alpha_k z^k = p(z) = \overline{p(z)} = \sum_{k=-N}^{N} \overline{\alpha_k} z^{-k} = \sum_{k=-N}^{N} \overline{\alpha_{-k}} z^k, \forall z \in \mathbb{T},$$

and hence

$$\sum_{k=-N}^{N} (\alpha_k - \overline{\alpha_{-k}}) z^{k+N} = 0, \forall z \in \mathbb{T},$$

and hence

$$\alpha_k = \overline{\alpha_{-k}}, \forall k \in \{0, \pm 1, ..., \pm N\}.$$

This implies that both α_N and α_{-N} are non-zero (if one of them were zero then the other one would be zero as well, and thus we should have $|\alpha_N| + |\alpha_{-N}| = 0$). Therefore, zero cannot be a root of the polynomial P defined by

$$P(z) := \sum_{k=-N}^{N} \alpha_k z^{k+N}, \forall z \in \mathbb{C},$$

(otherwise, $\alpha_{-N} = 0$) and the degree of P is $2N$.

First we suppose $p(z) > 0$ for all $z \in \mathbb{T}$. Then the roots of P cannot be elements of the unit circle $\mathbb{T}$, since

$$p(z) = z^{-N} P(z), \forall z \in \mathbb{T}.$$

Let $\{\lambda_i\}_{i \in I}$ be the family of the roots of P inside the unit circle and $\{\mu_j\}_{j \in J}$ the family of the roots outside; let r_i be the multiplicity of the root λ_i and s_j the multiplicity of the root μ_j; thus,

$$\sum_{i \in I} r_i + \sum_{j \in J} s_j = 2N.$$

Then we have the factorization

$$P(z) = c \prod_{i \in I} (z - \lambda_i)^{r_i} \prod_{j \in J} (z - \mu_j)^{s_j}, \forall z \in \mathbb{C},$$

with $c \in \mathbb{C}$, and hence (recall that no root of P is zero)

$$\overline{p(z)} = \overline{z^{-N} P(z)} = z^N \overline{c} \prod_{i \in I} (z^{-1} - \overline{\lambda_i})^{r_i} \prod_{j \in J} (z^{-1} - \overline{\mu_j})^{s_j}$$

$$= z^N \overline{c} z^{-2N} \prod_{i \in I} \overline{\lambda_i}^{r_i} \prod_{j \in J} \overline{\mu_j}^{s_j} \prod_{i \in I} (z - \overline{\lambda_i}^{-1})^{r_i} \prod_{j \in J} (z - \overline{\mu_j}^{-1})^{s_j}$$

$$= z^{-N} c_1 \prod_{i \in I} (z - \overline{\lambda_i}^{-1})^{r_i} \prod_{j \in J} (z - \overline{\mu_j}^{-1})^{s_j}, \forall z \in \mathbb{T},$$

with $c_1 \in \mathbb{C}$, and hence

$$c \prod_{i \in I} (z - \lambda_i)^{r_i} \prod_{j \in J} (z - \mu_j)^{s_j} = z^N p(z)$$

$$= z^N \overline{p(z)} = c_1 \prod_{i \in I} (z - \overline{\lambda_i}^{-1})^{r_i} \prod_{j \in J} (z - \overline{\mu_j}^{-1})^{s_j}, \forall z \in \mathbb{T}.$$

Since the set of the roots inside the unit circle must be the same on the two sides of this equation and so must be their multiplicities (or, equivalently, since the factorization of a polynomial with respect to its roots is unique), this implies that $\{\lambda_i\}_{i \in I} = \{\overline{\mu_j}^{-1}\}_{j \in J}$, and hence the sets of indices I and J can be identified, and also that $r_i = s_i$ for all $i \in I$, and hence $\sum_{i \in I} r_i = \sum_{i \in I} s_i = N$. Thus there exists $(\nu_1, ..., \nu_N) \in \mathbb{C}^N$ (the components of this N-tuple are the roots of P outside the unit circle, each of them repeated as many times as its multiplicity) so that

$$P(z) = c \prod_{k=1}^{N} (z - \overline{\nu_k}^{-1}) \prod_{k=1}^{N} (z - \nu_k), \forall z \in \mathbb{C}.$$

Now we suppose that p is not strictly positive, i.e. that there exists $z \in \mathbb{T}$ such that $p(z) = 0$. For every $n \in \mathbb{N}$, we define the trigonometric polynomial $p_n := p + \frac{1}{n}$ and the polynomial

$$P_n(z) := \sum_{k=-N}^{N} \left(\alpha_k + \delta_{0,k} \frac{1}{n} \right) z^{k+N}, \forall z \in \mathbb{C}.$$

Since $p_n(z) > 0$ and $p_n(z) = z^{-N} P_n(z)$ for all $z \in \mathbb{T}$, proceeding as above we see that

$$P_n(z) = c(n) \prod_{k=1}^{N} (z - \overline{\nu_k(n)}^{-1}) \prod_{k=1}^{N} (z - \nu_k(n)), \forall z \in \mathbb{C},$$

where $c(n) \in \mathbb{C}$ and the components of the N-tuple $(\nu_1(n), ..., \nu_N(n))$ are the roots of P_n outside the unit circle, repeated as many times as their multiplicities. Since the roots of a polynomial depend continuously on the coefficients of the polynomial (cf. e.g. Horn and Johnson, 2013, th.D.1.), the sequence $\{\nu_k(n)\}$ converges to a root ν_k of P for each $k \in \{1, ..., N\}$. Then we have

$$P(z) = \lim_{n \to \infty} P_n(z) = c \prod_{k=1}^{N} (z - \overline{\nu}_k^{-1}) \prod_{k=1}^{N} (z - \nu_k), \forall z \in \mathbb{C},$$

where $c := \lim_{n \to \infty} c(n)$; indeed, the sequence $\{c(n)\}$ is convergent since, for $z_0 \in \mathbb{C}$ such that $P(z_0) \neq 0$, for n large enough we have

$$\prod_{k=1}^{N} (z_0 - \overline{\nu_k(n)}^{-1}) \prod_{k=1}^{N} (z_0 - \nu_k(n)) \neq 0$$

and

$$c(n) = P_n(z) \left(\prod_{k=1}^{N} (z_0 - \overline{\nu_k(n)}^{-1}) \prod_{k=1}^{N} (z_0 - \nu_k(n)) \right)^{-1}.$$

Although it is not relevant for the present proof, we note that what we have just seen proves that every root of P in the unit circle has even multiplicity.

Thus, as a consequence of the hypothesis $p(z) \geq 0$ for all $z \in T$, there exist $c \in \mathbb{C}$ and $(\nu_1, ..., \nu_N) \in \mathbb{C}^N$ so that

$$P(z) = c \prod_{k=1}^{N} (z - \overline{\nu}_k^{-1}) \prod_{k=1}^{N} (z - \nu_k), \forall z \in \mathbb{C},$$

and hence so that

$$p(z) = z^{-N} P(z) = c(-1)^N \prod_{k=1}^{N} \overline{\nu}_k^{-1} \prod_{k=1}^{N} (z^{-1} - \overline{\nu}_k) \prod_{k=1}^{N} (z - \nu_k)$$

$$= c_2 \prod_{k=1}^{N} (\overline{z} - \overline{\nu}_k) \prod_{k=1}^{N} (z - \nu_k), \forall z \in \mathbb{T},$$

with $c_2 \in \mathbb{C}$. Since there exists $z \in \mathbb{C}$ such that $p(z) > 0$, $c_2 > 0$ must be true. Then, the trigonometric polynomial q defined by

$$q(z) := \sqrt{c_2} \prod_{k=1}^{N} (z - \nu_k), \forall z \in \mathbb{T},$$

is such that $p = \overline{q}q$. □

15.1.3 Proposition. *Let U be a unitary operator in $\mathcal{H}$ and let $p \in \mathcal{P}$ be such that $0 \leq p(z)$ for all $z \in \mathbb{T}$. Then,*

$$0 \leq \left(f | \hat{\phi}_U(p) f \right), \forall f \in \mathcal{H}.$$

Proof. In view of 15.1.2, there exists $q \in \mathcal{P}$ so that $p = \overline{q}q$. Then, by 15.1.1b,c,

$$\hat{\phi}_U(p) = (\hat{\phi}_U(q))^\dagger \hat{\phi}_U(q),$$

and hence

$$0 \leq \|\hat{\phi}_U(q)f\|^2 = \left(f | (\hat{\phi}_U(q))^\dagger \hat{\phi}_U(q) f \right) = \left(f | \hat{\phi}_U(p) f \right), \forall f \in \mathcal{H}.$$

□

15.1.4 Proposition. *The mapping* $\hat{\phi}_U$ *has the following property:*

$$\|\hat{\phi}_U(p)\| \leq \|p\|_\infty, \forall p \in \mathcal{P}.$$

Proof. For $p \in \mathcal{P}$, let $\tilde{p}$ be the element of $\mathcal{P}$ defined by

$$\tilde{p}(z) := \|p\|_\infty^2 - \overline{p(z)}p(z), \forall z \in \mathbb{T};$$

obviously, we have $0 \leq \tilde{p}(z)$ for all $z \in \mathbb{T}$, and hence, in view of 15.1.3,

$$0 \leq \left(f|\hat{\phi}_U(\tilde{p})f\right), \forall f \in \mathcal{H};$$

now, 15.1.1a,b,c imply that

$$\hat{\phi}_U(\tilde{p}) = \|p\|_\infty^2 \mathbb{1}_\mathcal{H} - (\hat{\phi}_U(p))^\dagger \hat{\phi}_U(p);$$

thus,

$$0 \leq \|p\|_\infty^2 \|f\|^2 - \|\hat{\phi}_U(p)f\|^2, \forall f \in \mathcal{H},$$

which yields

$$\|\hat{\phi}_U(p)\| \leq \|p\|_\infty.$$

$\square$

15.1.5 Theorem. *Let U be a unitary operator in $\mathcal{H}$. Then there exists a unique mapping $\phi_U : \mathcal{C}(\mathbb{T}) \to \mathcal{B}(\mathcal{H})$ such that:*

(a) $\phi_U(\alpha\varphi + \beta\psi) = \alpha\phi_U(\varphi) + \beta\phi_U(\psi)$, $\forall\alpha, \beta \in \mathbb{C}$, $\forall\varphi, \psi \in \mathcal{C}(\mathbb{T})$;
(b) $\phi_U(p) = \hat{\phi}_U(p)$, $\forall p \in \mathcal{P}$;
(c) ϕ_U is continuous.

In addition, the following conditions are true:

(d) $\|\phi_U(\varphi)\| \leq \|\varphi\|_\infty$, $\forall\varphi \in \mathcal{C}(\mathbb{T})$;
(e) $\phi_U(\varphi\psi) = \phi_U(\varphi)\phi_U(\psi)$, $\forall\varphi, \psi \in \mathcal{C}(\mathbb{T})$;
(f) $\phi_U(\overline{\varphi}) = (\phi_U(\varphi))^\dagger$, $\forall\varphi \in \mathcal{C}(\mathbb{T})$;
(g) if $\varphi \in \mathcal{C}(\mathbb{T})$ is such that $0 \leq \varphi(z)$ for all $z \in \mathbb{T}$, then $0 \leq (f|\phi_U(\varphi)f)$, $\forall f \in \mathcal{H}$;
(h) if $A \in \mathcal{B}(\mathcal{H})$ is such that $AU = UA$, then $A\phi_U(\varphi) = \phi_U(\varphi)A$, $\forall\varphi \in \mathcal{C}(\mathbb{T})$.

Proof. In view of 15.1.1a and 15.1.4, the mapping $\hat{\phi}_U$ is a bounded linear operator from the normed space $\mathcal{C}(\mathbb{T})$ to the Banach space $\mathcal{B}(\mathcal{H})$. Since $\overline{\mathcal{P}} = \mathcal{C}(\mathbb{T})$ (cf. 4.3.7), 4.2.6 implies that there exists a unique linear operator $\phi_U : \mathcal{C}(\mathbb{T}) \to \mathcal{B}(\mathcal{H})$ which is an extension of $\hat{\phi}_U$ and which is bounded, i.e. continuous. This proves that there exists a unique mapping $\phi_U : \mathcal{C}(\mathbb{T}) \to \mathcal{B}(\mathcal{H})$ which has properties a, b, c.

Now we prove the additional properties of ϕ_U.

d: In view of 4.2.6d, the norm of the linear operator ϕ_U equals the norm of the linear operator $\hat{\phi}_U$. Now, 15.1.4 implies that the latter is not greater than one. Thus, we have condition d (cf. 4.2.5b).

e: For $\varphi, \psi \in \mathcal{C}(\mathbb{T})$, let $\{p_n\}$ and $\{q_n\}$ be sequences in $\mathcal{P}$ such that $\varphi = \lim_{n \to \infty} p_n$ and $\psi = \lim_{n \to \infty} q_n$. Then, $\varphi \psi = \lim_{n \to \infty} p_n q_n$ (cf. 4.3.3) and hence

$$\phi_U(\varphi\psi) = \lim_{n \to \infty} \hat{\phi}_U(p_n q_n) \stackrel{(1)}{=} \lim_{n \to \infty} \hat{\phi}_U(p_n)\hat{\phi}_U(q_n)$$

$$\stackrel{(2)}{=} \big(\lim_{n \to \infty} \hat{\phi}_U(p_n) \big)\big(\lim_{n \to \infty} \hat{\phi}_U(q_n) \big) = \phi_U(\varphi)\phi_U(\psi),$$

where 1 holds by 15.1.1b and 2 by 4.3.3.

f: For $\varphi \in \mathcal{C}(\mathbb{T})$, let $\{p_n\}$ be a sequence in $\mathcal{P}$ such that $\varphi = \lim_{n \to \infty} p_n$. Then, $\overline{\varphi} = \lim_{n \to \infty} \overline{p}_n$ (this is obvious) and hence

$$\phi_U(\overline{\varphi}) = \lim_{n \to \infty} \hat{\phi}_U(\overline{p}_n) \stackrel{(3)}{=} \lim_{n \to \infty} (\hat{\phi}_U(p_n))^\dagger$$

$$\stackrel{(4)}{=} \big(\lim_{n \to \infty} \hat{\phi}_U(p_n) \big)^\dagger = (\phi_U(\varphi))^\dagger,$$

where 3 holds by 15.1.1c and 4 by 12.6.2.

g: For $\varphi \in \mathcal{C}(\mathbb{T})$ such that $0 \leq \varphi(z)$ for all $z \in \mathbb{T}$, let ψ be the element of $\mathcal{C}(\mathbb{T})$ defined by

$$\psi(z) := \sqrt{\varphi(z)}, \forall z \in \mathbb{T}.$$

Then, $\varphi = \psi^2$ and $\overline{\psi} = \psi$ imply

$$\phi_U(\varphi) = \phi_U(\psi)\phi_U(\psi) = (\phi_U(\psi))^\dagger \phi_U(\psi)$$

(cf. conditions e and f), and hence

$$0 \leq \|\phi_U(\psi)f\|^2 = \big(f|(\phi_U(\psi))^\dagger\phi_U(\psi)f\big) = (f|\phi_U(\varphi)f), \forall f \in \mathcal{H}.$$

h: Let $A \in \mathcal{B}(\mathcal{H})$ be such that $AU = UA$. We have also

$$AU^{-1} = U^{-1}(UA)U^{-1} = U^{-1}(AU)U^{-1} = U^{-1}A.$$

These conditions imply, owing to the very definition of $\hat{\phi}_U$,

$$A\hat{\phi}_U(p) = \hat{\phi}_U(p)A, \forall p \in \mathcal{P}.$$

For $\varphi \in \mathcal{C}(\mathbb{T})$, let $\{p_n\}$ be a sequence in $\mathcal{P}$ such that $\varphi = \lim_{n \to \infty} p_n$. Then, for each $n \in \mathbb{N}$,

$$\|A\phi_U(\varphi) - \phi_U(\varphi)A\| \leq \|A\phi_U(\varphi) - A\phi_U(p_n)\| + \|A\phi_U(p_n) - \phi_U(\varphi)A\|$$

$$\stackrel{(5)}{=} \|A\phi_U(\varphi - p_n)\| + \|\phi_U(p_n - \varphi)A\|$$

$$\stackrel{(6)}{\leq} 2\|A\|\|\phi_U(\varphi - p_n)\| \stackrel{(7)}{\leq} 2\|A\|\|\varphi - p_n\|_\infty,$$

where 5 holds by condition a, 6 by 4.2.9, 7 by condition d. This proves that

$$A\phi_U(\varphi) = \phi_U(\varphi)A.$$

$\square$

15.1.6 Theorem (The spectral theorem for unitary operators). *Let U be a unitary operator in $\mathcal{H}$.*

(A) *There exists a unique projection valued measure P on the Borel σ-algebra $\mathcal{A}(d_{\mathbb{T}})$*
   on $\mathbb{T}$ (cf. 6.1.22; as before, $d_{\mathbb{T}}$ denotes the restriction of the distance $d_{\mathbb{C}}$ to
   $\mathbb{T} \times \mathbb{T}$), with values in $\mathscr{P}(\mathcal{H})$, such that $U = J_\zeta^P$, where ζ is the function
   defined by

$$\zeta : \mathbb{T} \to \mathbb{T}$$
$$z \mapsto \zeta(z) := z$$

   and J_ζ^P is the operator defined in 14.2.14.
   Equivalently, there exists a unique projection valued measure P on $\mathcal{A}(d_{\mathbb{T}})$, with
   values in $\mathscr{P}(\mathcal{H})$, such that

$$(f|Uf) = \int_{\mathbb{T}} \zeta d\mu_f^P, \forall f \in \mathcal{H}.$$

(B) *If $A \in \mathcal{B}(\mathcal{H})$ is such that $AU = UA$, then*

$$AP(E) = P(E)A, \forall E \in \mathcal{A}(d_{\mathbb{T}}).$$

Proof. A: We divide the proof into nine steps.

Step 1: For every $f \in \mathcal{H}$, the function

$$\mathcal{C}(\mathbb{T}) \ni \varphi \mapsto (f|\phi_U(\varphi)f) \in \mathbb{C}$$

is a positive linear functional, in view of 15.1.5a,g; since the metric space $(\mathbb{T}, d_{\mathbb{T}})$ is
compact, by 8.5.3 this implies that there exists a unique finite measure μ_f on $\mathcal{A}(d_{\mathbb{T}})$
so that

$$(f|\phi_U(\varphi)f) = \int_{\mathbb{T}} \varphi d\mu_f, \forall \varphi \in \mathcal{C}(\mathbb{T});$$

in particular we have, from 15.1.5b and the very definition of $\hat{\phi}_U$,

$$\int_{\mathbb{T}} 1_{\mathbb{T}} d\mu_f = \left(f | \hat{\phi}_U(1_{\mathbb{T}}) f \right) = (f|1_{\mathcal{H}}f) = \|f\|^2.$$

Step 2: For every $\varphi \in \mathcal{M}_B(\mathbb{T}, \mathcal{A}(d_{\mathbb{T}}))$, we define the function

$$\psi_\varphi : \mathcal{H} \times \mathcal{H} \to \mathbb{C}$$
$$(f, g) \mapsto \psi_\varphi(f, g) := \sum_{n=1}^{4} \frac{1}{4i^n} \int_{\mathbb{T}} \varphi d\mu_{f+i^n g}$$

(note that $\mathcal{M}_B(\mathbb{T}, \mathcal{A}(d_{\mathbb{T}})) \subset \mathcal{L}^1(\mathbb{T}, \mathcal{A}(d_{\mathbb{T}}), \mu_f)$ for all $f \in \mathcal{H}$, in view of 8.2.6).
We want to prove that:

$$\forall \varphi \in \mathcal{M}_B(\mathbb{T}, \mathcal{A}(d_{\mathbb{T}})), \exists! B_\varphi \in \mathcal{B}(\mathcal{H}) \text{ such that } (f|B_\varphi g) = \psi_\varphi(f, g), \forall f, g \in \mathcal{H}.$$

To this end, we define the family

$$\mathcal{V}_1 := \{\varphi \in \mathcal{M}_B(\mathbb{T}, \mathcal{A}(d_{\mathbb{T}})) : \psi_\varphi \text{ is a bounded sesquilinear form}\}$$

(for a bounded sesquilinear form, cf. 10.1.1 and 10.5.4).

If $\varphi \in \mathcal{C}(\mathbb{T})$ then

$$\psi_\varphi(f, g) = \sum_{n=1}^{4} \frac{1}{4i^n} \left(f + i^n g | \phi_U(\varphi)(f + i^n g) \right) \overset{(1)}{=} \left(f | \phi_U(\varphi) g \right), \forall f, g \in \mathcal{H}$$

(for 1, cf. 10.1.10a), and hence $\varphi \in \mathcal{V}_1$ by 10.5.5.

Next, suppose $\varphi \in \mathcal{F}(\mathbb{T})$ and that $\{\varphi_n\}$ is a sequence in $\mathcal{M}_B(\mathbb{T}, \mathcal{A}(d_\mathbb{T}))$ such that $\varphi_n \xrightarrow{ubp} \varphi$, i.e. such that (cf. 6.3.1):

$$\exists m \in [0, \infty) \text{ such that } |\varphi_n(z)| \leq m, \forall z \in \mathbb{T}, \forall n \in \mathbb{N};$$

$$\lim_{n \to \infty} \varphi_n(z) = \varphi(z), \forall z \in \mathbb{T}.$$

Then, $\varphi \in \mathcal{M}_B(\mathbb{T}, \mathcal{A}(d_\mathbb{T}))$ (cf. 6.3.4a). Moreover, 8.2.11 (with the constant function $m_\mathbb{T}$ as dominating function) implies that

$$\int_\mathbb{T} \varphi d\mu_f = \lim_{n \to \infty} \int_\mathbb{T} \varphi_n d\mu_f, \forall f \in \mathcal{H},$$

and hence

$$\psi_\varphi(f, g) = \lim_{n \to \infty} \sum_{k=1}^{4} \frac{1}{4i^k} \int_\mathbb{T} \varphi_n d\mu_{f + i^k g} \tag{2}$$

$$= \lim_{n \to \infty} \psi_{\varphi_n}(f, g), \forall f, g \in \mathcal{H}.$$

Now suppose also that $\varphi_n \in \mathcal{V}_1$ for all $n \in \mathbb{N}$. Then 2 implies that ψ_φ is a sesquilinear form, since so is ψ_{φ_n} for all $n \in \mathbb{N}$. Moreover, for all $u, v \in \mathcal{H}$ such that $\|u\| = \|v\| = 1$,

$$|\psi_\varphi(u, v)| \leq \frac{1}{4} \sum_{n=1}^{4} \int_\mathbb{T} |\varphi| d\mu_{u + i^n v} \leq \frac{m}{4} \sum_{n=1}^{4} \|u + i^n v\|^2 \leq 4m$$

(since $\|u + i^n v\| \leq 2$); then,

$$|\psi_\varphi(f, g)| = \|f\| \|g\| \psi_\varphi \left(\frac{1}{\|f\|} f, \frac{1}{\|g\|} g \right) \leq 4m \|f\| \|g\|, \forall f, g \in \mathcal{H} - \{0_\mathcal{H}\},$$

and this proves that the sesquilinear form ψ_φ is bounded.

Thus, $\mathcal{V}_1$ is a family of complex function on $\mathbb{T}$ which contains $\mathcal{C}(\mathbb{T})$ and which is ubp closed. Hence (cf. 6.3.4b) $\mathcal{M}_B(\mathbb{T}, \mathcal{A}(d_\mathbb{T})) \subset \mathcal{V}_1$ (actually, $\mathcal{V}_1 = \mathcal{M}_B(\mathbb{T}, \mathcal{A}(d_\mathbb{T}))$), or

$$\psi_\varphi \text{ is a bounded sesquilinear form}, \forall \varphi \in \mathcal{M}_B(\mathbb{T}, \mathcal{A}(d_\mathbb{T})).$$

Then, by 10.5.6,

$$\forall \varphi \in \mathcal{M}_B(\mathbb{T}, \mathcal{A}(d_\mathbb{T})), \exists! B_\varphi \in \mathcal{B}(\mathcal{H}) \text{ such that } (f | B_\varphi g) = \psi_\varphi(f, g), \forall f, g \in \mathcal{H}.$$

Step 3: For every $\alpha \in \mathbb{C}$ and every $f \in \mathcal{H}$, we have

$$\int_\mathbb{T} \varphi d\mu_{\alpha f} = |\alpha|^2 (f | \phi_U(\varphi) f) = |\alpha|^2 \int_\mathbb{T} \varphi d\mu_f \overset{(3)}{=} \int_\mathbb{T} \varphi d(|\alpha|^2 \mu_f), \forall \varphi \in \mathcal{C}(\mathbb{T})$$

(for 3, cf. 8.3.5a with $a_1 := |\alpha|^2$, $\mu_1 := \mu_f$, μ_k the null measure on $\mathcal{A}(d_{\mathbb{T}})$ for $k > 1$); in view of the uniqueness asserted in 8.5.3, this proves that

$$\mu_{\alpha f} = |\alpha|^2 \mu_f.$$

In particular, for every $f \in \mathcal{H}$ we have $\mu_{f+i^n f} = |1 + i^n|^2 \mu_f$ for $n = 1, 2, 3, 4$; thus,

$$\mu_{f-f} \text{ is the null measure, } \mu_{f+f} = 4\mu_f, \; \mu_{f+if} = \mu_{f-if};$$

this yields

$$\sum_{n=1}^4 \frac{1}{4i^n} \int_{\mathbb{T}} \varphi d\mu_{f+i^n f} = \frac{1}{4} 4 \int_{\mathbb{T}} \varphi d\mu_f = \int_{\mathbb{T}} \varphi d\mu_f, \forall \varphi \in \mathcal{M}_B(\mathbb{T}, \mathcal{A}(d_{\mathbb{T}})),$$

and hence

$$(f|B_\varphi f) = \psi_\varphi(f, f) = \int_{\mathbb{T}} \varphi d\mu_f, \forall \varphi \in \mathcal{M}_B(\mathbb{T}, \mathcal{A}(d_{\mathbb{T}})).$$

Step 4: Suppose $\varphi \in \mathcal{F}(\mathbb{T})$ and that $\{\varphi_n\}$ is a sequence in $\mathcal{M}_B(\mathbb{T}, \mathcal{A}(d_{\mathbb{T}}))$ such that $\varphi_n \xrightarrow{ubp} \varphi$. Then $\varphi \in \mathcal{M}_B(\mathbb{T}, \mathcal{A}(d_{\mathbb{T}}))$ and

$$(f|B_\varphi g) = \psi_\varphi(f, g) = \lim_{n \to \infty} \psi_{\varphi_n}(f, g) = \lim_{n \to \infty} (f|B_{\varphi_n} g), \forall f, g \in \mathcal{H}.$$

This follows from what we saw in step 2.

Step 5: For every $\varphi \in \mathcal{C}(\mathbb{T})$ we have (cf. step 3)

$$(f|B_\varphi f) = \int_{\mathbb{T}} \varphi d\mu_f = (f|\phi_U(\varphi) f), \forall f \in \mathcal{H},$$

whence $B_\varphi = \phi_U(\varphi)$ by 10.2.12.

Step 6: For every $\varphi \in \mathcal{M}_B(\mathbb{T}, \mathcal{A}(d_{\mathbb{T}}))$ we have (cf. step 3)

$$(f|B_\varphi^\dagger f) = \overline{(f|B_\varphi f)} = \overline{\int_{\mathbb{T}} \varphi d\mu_f} \overset{(4)}{=} \int_{\mathbb{T}} \overline{\varphi} d\mu_f = (f|B_{\overline{\varphi}} f), \forall f \in \mathcal{H}$$

(for 4, cf. 8.2.3), whence $B_\varphi^\dagger = B_{\overline{\varphi}}$ by 10.2.12.

Step 7: Here we prove that $B_\varphi B_\psi = B_{\varphi\psi}, \forall \varphi, \psi \in \mathcal{M}_B(\mathbb{T}, \mathcal{A}(d_{\mathbb{T}}))$.

We define the family

$$\mathcal{V}_2 := \{\psi \in \mathcal{M}_B(\mathbb{T}, \mathcal{A}(d_{\mathbb{T}})) : B_\psi B_\varphi = B_{\psi\varphi}, \forall \varphi \in \mathcal{C}(\mathbb{T})\}.$$

If $\psi \in \mathcal{C}(\mathbb{T})$ then

$$B_\psi B_\varphi \overset{(5)}{=} \phi_U(\psi)\phi_U(\varphi) \overset{(6)}{=} \phi_U(\psi\varphi) \overset{(7)}{=} B_{\psi\varphi}, \forall \varphi \in \mathcal{C}(\mathbb{T})$$

(for 5 and 7, cf. step 5; 6 holds by 15.1.5e), and hence $\psi \in \mathcal{V}_2$.

Next, suppose $\psi \in \mathcal{F}(\mathbb{T})$ and that $\{\psi_n\}$ is a sequence in $\mathcal{V}_2$ such that $\psi_n \xrightarrow{ubp} \psi$. Then (cf. step 4), $\psi \in \mathcal{M}_B(\mathbb{T}, \mathcal{A}(d_{\mathbb{T}}))$ and

$$(f|B_\psi g) = \lim_{n \to \infty} (f|B_{\psi_n} g), \forall f, g \in \mathcal{H},$$

and hence, for all $\varphi \in \mathcal{C}(\mathbb{T})$,

$$(f|B_\psi B_\varphi g) = \lim_{n \to \infty} (f|B_{\psi_n} B_\varphi g) = \lim_{n \to \infty} (f|B_{\psi_n \varphi} g) \overset{(8)}{=} (f|B_{\psi\varphi} g), \forall f, g \in \mathcal{H},$$

where 8 holds (in view of step 4) because $\{\psi_n\varphi\}$ is a sequence in $\mathcal{M}_B(\mathbb{T},\mathcal{A}(d_\mathbb{T}))$ such that $\psi_n\varphi \xrightarrow{ubp} \psi\varphi$; in fact, if $m \in [0,\infty)$ is such that $|\psi_n(z)| \le m$ for all $z \in \mathbb{T}$, then $|(\psi_n\varphi)(z)| \le m\|\varphi\|_\infty$ for all $z \in \mathbb{T}$. Therefore, $B_\psi B_\varphi = B_{\psi\varphi}$.

Thus, $\mathcal{V}_2 \subset \mathcal{F}(\mathbb{T})$, $\mathcal{C}(\mathbb{T}) \subset \mathcal{V}_2$, and $\mathcal{V}_2$ is ubp closed. Hence, $\mathcal{M}_B(\mathbb{T},\mathcal{A}(d_\mathbb{T})) \subset \mathcal{V}_2$, or

$$B_\psi B_\varphi = B_{\psi\varphi}, \forall \varphi \in \mathcal{C}(\mathbb{T}), \forall \psi \in \mathcal{M}_B(\mathbb{T},\mathcal{A}(d_\mathbb{T})).$$

This implies that

$$B_\varphi B_\psi \overset{(9)}{=} B_{\overline{\varphi}}^\dagger B_{\overline{\psi}}^\dagger \overset{(10)}{=} (B_{\overline{\psi}}B_{\overline{\varphi}})^\dagger = B_{\overline{\psi}\overline{\varphi}}^\dagger$$

$$\overset{(11)}{=} B_{\psi\varphi} = B_{\varphi\psi}, \forall \varphi \in \mathcal{C}(\mathbb{T}), \forall \psi \in \mathcal{M}_B(\mathbb{T},\mathcal{A}(d_\mathbb{T}))$$

(for 9 and 11, cf. step 6; 10 holds by 12.3.4b).

Now we define the family

$$\mathcal{V}_3 := \{\varphi \in \mathcal{M}_B(\mathbb{T},\mathcal{A}(d_\mathbb{T})) : B_\varphi B_\psi = B_{\varphi\psi}, \forall \psi \in \mathcal{M}_B(\mathbb{T},\mathcal{A}(d_\mathbb{T}))\}.$$

The last thing proved implies $\mathcal{C}(\mathbb{T}) \subset \mathcal{V}_3$.

Next, suppose $\varphi \in \mathcal{F}(\mathbb{T})$ and that $\{\varphi_n\}$ is a sequence in $\mathcal{V}_3$ such that $\varphi_n \xrightarrow{ubp} \varphi$. Then, proceeding exactly as above we have that, for all $\psi \in \mathcal{M}_B(\mathbb{T},\mathcal{A}(d_\mathbb{T}))$,

$$(f|B_\varphi B_\psi g) = \lim_{n\to\infty} (f|B_{\varphi_n}B_\psi g) = \lim_{n\to\infty} (f|B_{\varphi_n\psi}g) \overset{(12)}{=} (f|B_{\varphi\psi}g), \forall f,g \in \mathcal{H},$$

where 12 holds (in view of step 4) because $\{\varphi_n\psi\}$ is a sequence in $\mathcal{M}_B(\mathbb{T},\mathcal{A}(d_\mathbb{T}))$ such that $\varphi_n\psi \xrightarrow{ubp} \varphi\psi$. Therefore, $B_\varphi B_\psi = B_{\varphi\psi}$.

Thus, $\mathcal{V}_3 \subset \mathcal{F}(\mathbb{T})$, $\mathcal{C}(\mathbb{T}) \subset \mathcal{V}_3$, and $\mathcal{V}_3$ is ubp closed. Hence, $\mathcal{M}_B(\mathbb{T},\mathcal{A}(d_\mathbb{T})) \subset \mathcal{V}_3$, or

$$B_\varphi B_\psi = B_{\varphi\psi}, \forall \psi \in \mathcal{M}_B(\mathbb{T},\mathcal{A}(d_\mathbb{T})), \forall \varphi \in \mathcal{M}_B(\mathbb{T},\mathcal{A}(d_\mathbb{T})).$$

Step 8: For every $E \in \mathcal{A}(d_\mathbb{T})$, we have $\chi_E \in \mathcal{M}_B(\mathbb{T},\mathcal{A}(d_\mathbb{T}))$. Since $\overline{\chi_E} = \chi_E$, we have $B_{\chi_E}^\dagger = B_{\chi_E}$ (cf. step 6). Since $\chi_E^2 = \chi_E$, we have $B_{\chi_E}^2 = B_{\chi_E}$ (cf. step 7). Thus, $B_{\chi_E} \in \mathscr{P}(\mathcal{H})$ by 13.1.5.

Now, we define the mapping

$$P : \mathcal{A}(d_\mathbb{T}) \to \mathscr{P}(\mathcal{H})$$
$$E \mapsto P(E) := B_{\chi_E}.$$

For every $f \in \mathcal{H}$, we have (cf. step 3)

$$\mu_f^P(E) = (f|B_{\chi_E}f) = \int_\mathbb{T} \chi_E d\mu_f = \mu_f(E), \forall E \in \mathcal{A}(d_\mathbb{T});$$

this proves that μ_f^P is a measure on $\mathcal{A}(d_\mathbb{T})$, and also (cf. step 1) that

$$\mu_f^P(\mathbb{T}) = \int_\mathbb{T} 1_\mathbb{T} d\mu_f = \|f\|^2.$$

Thus, P is a projection valued measure on $\mathcal{A}(d_\mathbb{T})$, in view of 13.3.5.

Now, we have

$$D_U = \mathcal{H} = \left\{ f \in \mathcal{H} : \int_{\mathbb{T}} |\zeta|^2 d\mu_f^P < \infty \right\}$$

since the function ζ is bounded and the measure μ_f^P is finite for all $f \in \mathcal{H}$, and also

$$(f|Uf) \overset{(13)}{=} \left(f|\hat{\phi}_U(\zeta)f \right) \overset{(14)}{=} (f|\phi_U(\zeta)f) = \int_{\mathbb{T}} \zeta d\mu_f = \int_{\mathbb{T}} \zeta d\mu_f^P, \forall f \in \mathcal{H},$$

where 13 holds by the very definition of $\hat{\phi}_U$ and 14 by 15.1.5b. By the uniqueness asserted in 14.2.14, this is equivalent to $U = J_\zeta^P$.

This concludes the proof that a projection valued measure P exists as in the statement.

Step 9: Here we prove that the projection valued measure as in the statement is unique. To this end, suppose that Q is a projection valued measure on $\mathcal{A}(d_{\mathbb{T}})$, with values in $\mathscr{P}(\mathcal{H})$, such that

$$(f|Uf) = \int_{\mathbb{T}} \zeta d\mu_f^Q, \forall f \in \mathcal{H}.$$

Since the equality $D_U = \left\{ f \in \mathcal{H} : \int_{\mathbb{T}} |\zeta|^2 d\mu_f^Q < \infty \right\}$ is obvious, this is equivalent to

$$J_\zeta^Q = U.$$

In view of 14.2.17e (and of the fact that the mapping $\tilde{J}_Q$ is an extension of the mapping $\hat{J}_Q$, cf. 14.2.7), this can be written as

$$\hat{J}_Q(\zeta) = U.$$

Now, in view of 14.1.1b,d,e (and of the fact that $\zeta^{-1} = \bar{\zeta}$ and $U^\dagger = U^{-1}$, cf. 12.5.1b), this implies that

$$\hat{J}_Q(p) = \hat{\phi}_U(p), \forall p \in \mathcal{P}.$$

Moreover, the mapping $\hat{J}_Q$ is continuous (cf. 14.1.1c), and so is its restriction to $\mathcal{C}(\mathbb{T})$. Then, the uniqueness asserted in 15.1.5 implies that

$$\hat{J}_Q(\varphi) = \phi_U(\varphi), \forall \varphi \in \mathcal{C}(\mathbb{T}),$$

and hence (cf. 14.1.1f)

$$\int_{\mathbb{T}} \varphi d\mu_f^Q = \left(f|\hat{J}_Q(\varphi)f \right) = (f|\phi_U(\varphi)f) = \int_{\mathbb{T}} \varphi d\mu_f, \forall \varphi \in \mathcal{C}(\mathbb{T}), \forall f \in \mathcal{H};$$

then, by the uniqueness asserted in 8.5.3,

$$\mu_f^Q = \mu_f \overset{(15)}{=} \mu_f^P, \forall f \in \mathcal{H}$$

(for 15, cf. step 8), and hence

$$(f|Q(E)f) = \mu_f^Q(E) = \mu_f^P(E) = (f|P(E)f), \forall E \in \mathcal{A}(d_{\mathbb{T}}), \forall f \in \mathcal{H},$$

and hence, by 10.2.12,

$$Q(E) = P(E), \forall E \in \mathcal{A}(d_{\mathbb{T}}).$$

B: Let $A \in \mathcal{B}(\mathcal{H})$ be such that $AU = UA$.

We define the family

$$\mathcal{V}_4 := \{\varphi \in \mathcal{M}_B(\mathbb{T}, \mathcal{A}(d_{\mathbb{T}})) : AB_\varphi = B_\varphi A\}.$$

If $\varphi \in \mathcal{C}(\mathbb{T})$ then $B_\varphi = \phi_U(\varphi)$ (cf. step 5), and hence $\varphi \in \mathcal{V}_4$ in view of 15.1.5h.

Next, suppose $\varphi \in \mathcal{F}(\mathbb{T})$ and that $\{\varphi_n\}$ is a sequence in $\mathcal{V}_4$ such that $\varphi_n \xrightarrow{ubp} \varphi$. Then (cf. step 4), $\varphi \in \mathcal{M}_B(\mathbb{T}, \mathcal{A}(d_{\mathbb{T}}))$ and

$$(f|B_\varphi g) = \lim_{n \to \infty} (f|B_{\varphi_n} g), \forall f, g \in \mathcal{H},$$

and hence

$$(f|AB_\varphi g) = \left(A^\dagger f|B_\varphi g\right) = \lim_{n \to \infty} \left(A^\dagger f|B_{\varphi_n} g\right)$$
$$= \lim_{n \to \infty} (f|AB_{\varphi_n} g) = \lim_{n \to \infty} (f|B_{\varphi_n} Ag) = (f|B_\varphi Ag), \forall f, g \in \mathcal{H},$$

whence $AB_\varphi = B_\varphi A$.

Thus, $\mathcal{V}_4 \subset \mathcal{F}(\mathbb{T})$, $\mathcal{C}(\mathbb{T}) \subset \mathcal{V}_4$, and $\mathcal{V}_4$ is ubp closed. Hence, $\mathcal{M}_B(\mathbb{T}, \mathcal{A}(d_{\mathbb{T}})) \subset \mathcal{V}_4$, or

$$AB_\varphi = B_\varphi A, \forall \varphi \in \mathcal{M}_B(\mathbb{T}, \mathcal{A}(d_{\mathbb{T}})).$$

Then, in particular

$$AP(E) = AB_{\chi_E} = B_{\chi_E} A = P(E)A, \forall E \in \mathcal{A}(d_{\mathbb{T}}).$$

$\square$

15.2 The spectral theorem for self-adjoint operators

The spectral theorem for self-adjoint operators is deduced from the spectral theorem for unitary operators, by means of the Cayley transform (in both its incarnations, as an operator and as a function).

Throughout this section, $\mathcal{H}$ denotes an abstract Hilbert space.

15.2.1 Theorem (The spectral theorem for self-adjoint operators). *Let A be a self-adjoint operator in $\mathcal{H}$.*

(A) There exists a unique projection valued measure P on the Borel σ-algebra $\mathcal{A}(d_{\mathbb{R}})$ on $\mathbb{R}$, with values in $\mathscr{P}(\mathcal{H})$, such that $A = J_\xi^P$, where ξ is the function defined by

$$\xi : \mathbb{R} \to \mathbb{C}$$
$$x \mapsto \xi(x) := x$$

and J_ξ^P is the operator defined in 14.2.14.

Equivalently, there exists a unique projection valued measure P on $\mathcal{A}(d_{\mathbb{R}})$, with values in $\mathcal{P}(\mathcal{H})$, such that

$$D_A = \left\{ f \in \mathcal{H} : \int_{\mathbb{R}} \xi^2 d\mu_f^P < \infty \right\},$$

$$(f|Af) = \int_{\mathbb{R}} \xi d\mu_f^P, \forall f \in D_A.$$

(B) If $B \in \mathcal{B}(\mathcal{H})$ is such that $BA \subset AB$, then

$$BP(E) = P(E)B, \forall E \in \mathcal{A}(d_{\mathbb{R}}).$$

Proof. *A* existence: Let V be the Cayley transform of A (cf. 12.5.3). Since V is a unitary operator in $\mathcal{H}$, 15.1.6A grants that there exists a unique projection valued measure P^V on the Borel σ-algebra $\mathcal{A}(d_{\mathbb{T}})$, with values in $\mathcal{P}(\mathcal{H})$, such that

$$V = J_\zeta^{P^V} = J_{P^V}(\zeta)$$

(J_{P^V} is the mapping defined in 14.2.18). In view of 14.2.17e and 14.2.7a,b, this implies that

$$J_{P^V}(-i(\zeta + 1)) = -i(J_{P^V}(\zeta) + J_{P^V}(1_{\mathbb{T}})) = -i(V + \mathbb{1}_{\mathcal{H}})$$

and

$$J_{P^V}(\zeta - 1) = V - \mathbb{1}_{\mathcal{H}}.$$

Now, $1 \notin \sigma_p(V)$ (cf. 12.5.3) and hence, in view of 14.4.6,

$$P^V(\{1\}) = P^V(\zeta^{-1}(\{1\})) = \mathbb{O}_{\mathcal{H}};$$

thus, $\zeta - 1 \neq 0$ P^V-a.e. on $\mathbb{T}$, and hence 14.3.14 implies that $\frac{1}{\zeta-1} \in \mathcal{M}(\mathbb{T}, \mathcal{A}(d_{\mathbb{T}}), P^V)$ (for $\frac{1}{\zeta-1}$, cf. 1.2.19) and that

$$J_{P^V}\left(\frac{1}{\zeta - 1}\right) = (J_{P^V}(\zeta - 1))^{-1} = (V - \mathbb{1}_{\mathcal{H}})^{-1}$$

(it was already known that the operator $V - \mathbb{1}_{\mathcal{H}}$ was injective, cf. 12.5.3). Now the function ψ defined in 12.6.6b can be written as

$$\psi = -i(\zeta + 1)\frac{1}{\zeta - 1}.$$

Therefore, in view of 14.3.9,

$$-i(V + \mathbb{1}_{\mathcal{H}})(V - \mathbb{1}_{\mathcal{H}})^{-1} = J_{P^V}(-i(\zeta + 1))J_{P^V}\left(\frac{1}{\zeta - 1}\right) \subset J_{P^V}(\psi).$$

Now, $-i(V + \mathbb{1}_{\mathcal{H}})(V - \mathbb{1}_{\mathcal{H}})^{-1} = A$ (cf. 12.5.3) and $J_{P^V}(\psi)$ is self-adjoint since $\psi(z) \in \mathbb{R}$ for all $z \in D_\psi$ (cf. 14.3.17). Hence (cf. 12.4.6b),

$$J_{P^V}(\psi) = A.$$

Now a change of variable is made, by 14.6.1 with

$$X_1 := \mathbb{T}, \quad \mathcal{A}_1 := \mathcal{A}(d_{\mathbb{T}}), \quad P_1 := P^V, \quad X_2 := \mathbb{R}, \quad \mathcal{A}_2 := \mathcal{A}(d_{\mathbb{R}}), \quad \pi := \psi$$

(note that $X_1 - D_\pi = \{1\}$ and hence $P_1(X_1 - D_\pi) = \mathbb{O}_\mathcal{H}$). Then the mapping

$$P : \mathcal{A}(d_\mathbb{R}) \to \mathscr{P}(\mathcal{H})$$
$$E \mapsto P(E) := P^V(\psi^{-1}(E))$$

is a projection valued measure on $\mathcal{A}(d_\mathbb{R})$ (cf. 14.6.1a) and

$$J_P(\xi) = J_{PV}(\xi \circ \psi) = J_{PV}(\psi) = A$$

(cf. 14.6.1b with $\varphi := \xi$).

Finally, for a linear operator A, the equality $A = J_P(\xi)$ is equivalent to

$$D_A = \left\{ f \in \mathcal{H} : \int_\mathbb{R} \xi^2 d\mu_f^P < \infty \right\} \text{ and } (f|Af) = \int_\mathbb{R} \xi d\mu_f^P, \forall f \in D_A,$$

owing to the uniqueness asserted in 14.2.14.

A uniqueness: Suppose that Q is a projection valued measure on $\mathcal{A}(d_\mathbb{R})$, with values in $\mathscr{P}(\mathcal{H})$, such that

$$A = J_\xi^Q = J_Q(\xi).$$

In view of 14.3.1 and 14.3.5, this implies

$$J_Q(\xi - i) = J_Q(\xi) - iJ_Q(1_\mathbb{R}) = A - i\mathbb{1}_\mathcal{H}$$

and

$$J_Q(\xi + i) = A + i\mathbb{1}_\mathcal{H}.$$

Since $x + i \neq 0$ for all $x \in \mathbb{R}$, 14.3.14 implies that

$$J_Q\left(\frac{1}{\xi + i}\right) = (J_Q(\xi + i))^{-1} = (A + i\mathbb{1}_\mathcal{H})^{-1}$$

(it was already known that the operator $A + i\mathbb{1}_\mathcal{H}$ was injective, cf. 12.5.3). Now the Cayley transform φ (cf. 12.6.6b) can be written as

$$\varphi = (\xi - i)\frac{1}{\xi + i}.$$

Therefore, in view of 14.3.10 (note that $\frac{1}{\xi+i} \in \mathcal{M}_B(\mathbb{R}, \mathcal{A}(d_\mathbb{R}))$),

$$(A - i\mathbb{1}_\mathcal{H})(A + i\mathbb{1}_\mathcal{H})^{-1} = J_Q(\xi - i)J_Q\left(\frac{1}{\xi + i}\right) = J_Q(\varphi).$$

Since $V = (A - i\mathbb{1}_\mathcal{H})(A + i\mathbb{1}_\mathcal{H})^{-1}$ (cf. 12.5.3), this can be written as

$$V = J_Q(\varphi).$$

Now a change of variable is made, by 14.6.1 with

$$X_1 := \mathbb{R}, \quad \mathcal{A}_1 := \mathcal{A}(d_\mathbb{R}), \quad P_1 := Q, \quad X_2 := \mathbb{T}, \quad \mathcal{A}_2 := \mathcal{A}(d_\mathbb{T}), \quad \pi := \varphi$$

(then, $X_1 - D_\pi = \emptyset$ and hence $P_1(X_1 - D_\pi) = \mathbb{O}_\mathcal{H}$). Then the mapping

$$Q^V : \mathcal{A}(d_\mathbb{T}) \to \mathscr{P}(\mathcal{H})$$
$$F \mapsto Q^V(F) := Q(\varphi^{-1}(F))$$

is a projection valued measure on $\mathcal{A}(d_{\mathbb{T}})$ (cf. 14.6.1a) and
$$J_{Q^V}(\zeta) = J_Q(\zeta \circ \varphi) = J_Q(\varphi) = V$$
(cf. 14.6.1b with $\varphi := \zeta$; naturally, here φ stands for the symbol used in 14.6.1.b and not for the Cayley transform). By the uniqueness asserted in 15.1.6A, this implies $Q^V = P^V$. Then,
$$Q(E) \overset{(1)}{=} Q(\varphi^{-1}(\varphi(E))) \overset{(2)}{=} Q^V(\varphi(E)) = P^V(\varphi(E))$$
$$\overset{(3)}{=} P^V(\psi^{-1}(E)) \overset{(4)}{=} P(E), \forall E \in \mathcal{A}(d_{\mathbb{R}}),$$
where: 1 holds because φ is injective; φ is a bijection from $\mathbb{R}$ onto $\mathbb{T} - \{1\}$ (cf. 12.6.6b) and hence, for every subset S of $\mathbb{T} - \{1\}$, $\varphi^{-1}(S)$ can be interpreted as the image of S under the function φ^{-1} or as the counterimage of S under the function φ, and it is the latter interpretation the one that upholds 2; finally, $\varphi = \psi^{-1}$ (cf. 12.6.6b) and, for every subset T of $\mathbb{R}$, $\psi^{-1}(T)$ can be interpreted as the image of T under the function ψ^{-1} (this interpretation upholds 3) or as the counterimage of T under the function ψ (this interpretation upholds 4), as ψ is a bijection from $\mathbb{T} - \{1\}$ onto $\mathbb{R}$.

B: First note that 12.4.21 and 12.4.18 imply that the operator $A + i\mathbb{1}_{\mathcal{H}}$ is injective and $D_{(A+i\mathbb{1}_{\mathcal{H}})^{-1}} = \mathcal{H}$.

Now, let $B \in \mathcal{B}(\mathcal{H})$ be such that $BA \subset AB$, i.e. such that
$$f \in D_A \Rightarrow [Bf \in D_A \text{ and } BAf = ABf]. \tag{5}$$
Fix $g \in \mathcal{H}$. Then $(A + i\mathbb{1}_{\mathcal{H}})^{-1}g \in D_A$ and hence, in view of 5, $B(A + i\mathbb{1}_{\mathcal{H}})^{-1}g \in D_A$ and
$$\begin{aligned}(A - i\mathbb{1}_{\mathcal{H}})B(A + i\mathbb{1}_{\mathcal{H}})^{-1}g &= AB(A + i\mathbb{1}_{\mathcal{H}})^{-1}g - iB(A + i\mathbb{1}_{\mathcal{H}})^{-1}g \\ &= BA(A + i\mathbb{1}_{\mathcal{H}})^{-1}g - iB(A + i\mathbb{1}_{\mathcal{H}})^{-1}g \quad (6) \\ &= B(A - i\mathbb{1}_{\mathcal{H}})(A + i\mathbb{1}_{\mathcal{H}})^{-1}g,\end{aligned}$$
and also
$$(A + i\mathbb{1}_{\mathcal{H}})B(A + i\mathbb{1}_{\mathcal{H}})^{-1}g = B(A + i\mathbb{1}_{\mathcal{H}})(A + i\mathbb{1}_{\mathcal{H}})^{-1}g = Bg,$$
and hence
$$B(A + i\mathbb{1}_{\mathcal{H}})^{-1}g = (A + i\mathbb{1}_{\mathcal{H}})^{-1}Bg. \tag{7}$$
Now, 6 and 7 yield
$$\begin{aligned}B(A - i\mathbb{1}_{\mathcal{H}})(A + i\mathbb{1}_{\mathcal{H}})^{-1}g &= (A - i\mathbb{1}_{\mathcal{H}})B(A + i\mathbb{1}_{\mathcal{H}})^{-1}g \\ &= (A - i\mathbb{1}_{\mathcal{H}})(A + i\mathbb{1}_{\mathcal{H}})^{-1}Bg.\end{aligned}$$
Since $V = (A - i\mathbb{1}_{\mathcal{H}})(A + i\mathbb{1}_{\mathcal{H}})^{-1}$ (cf. 12.5.3) and since g was arbitrary, this proves that
$$BV = VB.$$
This, in view of 15.1.6 B, implies that
$$BP^V(F) = P^V(F)B, \forall F \in \mathcal{A}(d_{\mathbb{T}}),$$
and hence, owing to the way P was defined, that
$$BP(E) = P(E)B, \forall E \in \mathcal{A}(d_{\mathbb{R}}).$$

$\square$

15.2.2 Remarks. The mapping

$$P \mapsto J_\xi^P$$

is a bijection from the family of all projection valued measures on the Borel σ-algebra $\mathcal{A}(d_\mathbb{R})$, with values in $\mathscr{P}(\mathcal{H})$, onto the family of all self-adjoint operators in $\mathcal{H}$. Indeed, 14.3.17 proves that the operator J_ξ^P is self-adjoint for every projection valued measure P on $\mathcal{A}(d_\mathbb{R})$. Further, 15.2.1A proves that, for every self-adjoint operator A, there exists one and only one projection valued measure P on $\mathcal{A}(d_\mathbb{R})$ such that $A = J_\xi^P$.

For a given projection valued measure P on $\mathcal{A}(d_\mathbb{R})$, we sometimes denote by A^P the operator J_ξ^P. Conversely, for a given self-adjoint operator we always denote by P^A the unique projection valued measure P on $\mathcal{A}(d_\mathbb{R})$ which is so that $A = J_\xi^P$, and we call it *the projection valued measure of A*. Thus, the bijection discussed above is defined by

$$P \mapsto A^P,$$

while its inverse is the bijection from the family of all self-adjoint operators onto the family of all projection valued measures on $\mathcal{A}(d_\mathbb{R})$ which is defined by

$$A \mapsto P^A.$$

Explicitly, for a given projection valued measure P on $\mathcal{A}(d_\mathbb{R})$, A^P is the linear operator in $\mathcal{H}$ characterized by the conditions (cf. 14.2.11 and 14.2.14):

(a) $D_{A^P} = \{f \in \mathcal{H} : \int_\mathbb{R} \xi^2 d\mu_f^P < \infty\}$;
(b) $(f|A^P f) = \int_\mathbb{R} \xi d\mu_f^P, \forall f \in D_{A^P}$.

From 14.3.2 we also have

(c) $\|A^P f - \lambda f\|^2 = \int_\mathbb{R} |\xi - \lambda|^2 d\mu_f^P, \forall f \in D_{A^P}, \forall \lambda \in \mathbb{C}$.

For every self-adjoint operator A, from the more general results of Chapter 14 we have the following results, since $A = J_\xi^{P^A}$:

(d) $P^A(\sigma(A)) = \mathbb{1}_\mathcal{H}$ (equivalently, $P^A(\mathbb{R} - \sigma(A)) = \mathbb{O}_\mathcal{H}$; we recall that $\sigma(A)$ is a closed subset of $\mathbb{R}$, cf. 10.4.6 and 12.4.21a) and hence $\sigma(A) \neq \emptyset$ (cf. 14.4.3);
(e) $D_A = \{f \in \mathcal{H} : \int_\mathbb{R} \xi^2 d\mu_f^{P^A} < \infty\} = \{f \in \mathcal{H} : \int_{\sigma(A)} \xi^2 d\mu_f^{P^A} < \infty\}$,
$(f|Af) = \int_\mathbb{R} \xi d\mu_f^{P^A} = \int_{\sigma(A)} \xi d\mu_f^{P^A}, \forall f \in D_A$,
$\|Af - \lambda f\|^2 = \int_\mathbb{R} |\xi - \lambda|^2 d\mu_f^{P^A} = \int_{\sigma(A)} |\xi - \lambda|^2 d\mu_f^{P^A}, \forall f \in D_A$
(cf. 14.4.4);
(f) A is a bounded operator iff $\sigma(A)$ is a bounded subset of $\mathbb{R}$ (cf. 14.4.5).

For convenience, we collect the results (also the ones already known before this chapter) for the spectrum and the point spectrum of a self-adjoint operator in the next two theorems, after defining two numbers which are of great importance in quantum mechanics.

15.2.3 Definitions. Let A be a self-adjoint operator in $\mathcal{H}$ and let $u \in D_A \cap \tilde{\mathcal{H}}$ (for $\tilde{\mathcal{H}}$, cf. 10.9.4). We define the numbers

$$\langle A \rangle_u := (u|Au) \text{ and } \Delta_u A := \|Au - \langle A \rangle_u u\|.$$

Thus, if P is a projection valued measure on $\mathcal{A}(d_{\mathbb{R}})$, for every $u \in D_{A^P} \cap \tilde{\mathcal{H}}$ we have (cf. 15.2.2b,c)

$$\langle A^P \rangle_u = \int_{\mathbb{R}} \xi d\mu_u^P \text{ and } \Delta_u A^P = \left(\int_{\mathbb{R}} (\xi - \langle A^P \rangle_u)^2 d\mu_u^P \right)^{\frac{1}{2}}.$$

15.2.4 Theorem. *Let A be a self-adjoint operator in $\mathcal{H}$. For $\lambda \in \mathbb{R}$, the following conditions are equivalent:*

(a) $\lambda \in \sigma(A)$;
(b) $\forall \varepsilon > 0, \exists f_\varepsilon \in D_A$ such that $\|Af_\varepsilon - \lambda f_\varepsilon\| < \varepsilon \|f_\varepsilon\|$ (hence, $f_\varepsilon \neq 0_{\mathcal{H}}$);
(c) $P^A((\lambda - \varepsilon, \lambda + \varepsilon)) \neq \mathbb{O}_{\mathcal{H}}, \forall \varepsilon > 0$;
(d) $\forall \varepsilon > 0, \exists u_\varepsilon \in D_A \cap \tilde{\mathcal{H}}$ such that $|\langle A \rangle_{u_\varepsilon} - \lambda| < \varepsilon$ and $\Delta_{u_\varepsilon} A < 2\varepsilon$.

Proof. $a \Leftrightarrow b$: Cf. 12.4.21b, 4.5.2, 4.5.3.

$a \Leftrightarrow c$: Cf. 14.4.2.

$b \Rightarrow d$: Fix $\varepsilon \in (0, \infty)$. Condition b implies that there exists $u_\varepsilon \in D_A \cap \tilde{\mathcal{H}}$ such that

$$\|Au_\varepsilon - \lambda u_\varepsilon\| < \varepsilon.$$

Then, by the Schwarz inequality we have

$$|\langle A \rangle_{u_\varepsilon} - \lambda| = |(u_\varepsilon|Au_\varepsilon - \lambda u_\varepsilon)| \leq \|Au_\varepsilon - \lambda u_\varepsilon\| < \varepsilon$$

and then also

$$\Delta_{u_\varepsilon} A = \|Au_\varepsilon - \langle A \rangle_{u_\varepsilon} u_\varepsilon\| \leq \|Au_\varepsilon - \lambda u_\varepsilon\| + \|\lambda u_\varepsilon - \langle A \rangle_{u_\varepsilon} u_\varepsilon\| < 2\varepsilon.$$

$d \Rightarrow b$: Fix $\varepsilon \in (0, \infty)$. Condition d implies that there exists $u_\varepsilon \in D_A \cap \tilde{\mathcal{H}}$ such that

$$\|Au_\varepsilon - \lambda u_\varepsilon\| \leq \|Au_\varepsilon - \langle A \rangle_{u_\varepsilon} u_\varepsilon\| + \|\langle A \rangle_{u_\varepsilon} u_\varepsilon - \lambda u_\varepsilon\| = \Delta_{u_\varepsilon} A + |\langle A \rangle_{u_\varepsilon} - \lambda| < 3\varepsilon.$$

This proves condition b (take $f_\varepsilon := u_{\frac{\varepsilon}{3}}$). $\square$

15.2.5 Theorem. *Let A be a self-adjoint operator in $\mathcal{H}$. For $\lambda \in \mathbb{R}$, the following conditions are equivalent:*

(a) $\lambda \in \sigma_p(A)$;
(b) $\exists f \in D_A$ such that $f \neq 0_{\mathcal{H}}$ and $Af = \lambda f$;
(c) $P^A(\{\lambda\}) \neq \mathbb{O}_{\mathcal{H}}$;
(d) $\exists u \in D_A \cap \tilde{\mathcal{H}}$ such that $\langle A \rangle_u = \lambda$ and $\Delta_u A = 0$.

Moreover,

(e) $N_{A-\lambda 1_{\mathcal{H}}} = R_{P^A(\{\lambda\})}, \forall \lambda \in \mathbb{R}$;

thus, if $\lambda \in \sigma_p(A)$ *then* $P^A(\{\lambda\})$ *is the projection onto the corresponding eigenspace.*

Proof. $a \Leftrightarrow b$: Cf. 4.5.6 and 4.5.7.

$a \Leftrightarrow c$, and e: Cf. 14.4.6.

$b \Rightarrow d$: Condition b implies that there exists $u \in D_A \cap \tilde{\mathcal{H}}$ such that $Au = \lambda u$; then

$$\langle A \rangle_u = (u|Au) = \lambda \text{ and } \Delta_u A = \|Au - \lambda u\| = 0.$$

$d \Rightarrow b$: If $u \in D_A \cap \tilde{\mathcal{H}}$ is such that $\langle A \rangle_u = \lambda$ and $\Delta_u A = 0$, then

$$\|Au - \lambda u\| = \|Au - \langle A \rangle_u u\| = \Delta_u A = 0,$$

whence $Au = \lambda u$. $\qquad\square$

15.2.6 Remark. Let A be a self-adjoint operator in $\mathcal{H}$ and let λ be an isolated point of $\sigma(A)$, i.e.

$$\lambda \in \mathbb{R} \text{ and } \exists \delta \in (0, \infty) \text{ such that } (\lambda - \delta, \lambda + \delta) \cap \sigma(A) = \{\lambda\}.$$

Then $\lambda \in \sigma_p(A)$. Indeed, $(\lambda - \delta, \lambda) \subset \mathbb{R} - \sigma(A)$ implies $P^A((\lambda - \delta, \lambda)) = \mathbb{O}_{\mathcal{H}}$ (cf. 15.2.2d and 13.3.2e); similarly, $P^A((\lambda, \lambda + \delta)) = \mathbb{O}_{\mathcal{H}}$. Therefore

$$P^A(\{\lambda\}) = P^A((\lambda - \delta, \lambda)) + P^A(\{\lambda\}) + P^A((\lambda, \lambda + \delta)) = P^A((\lambda - \delta, \lambda + \delta)) \neq \mathbb{O}_{\mathcal{H}}$$

by 15.2.4, and hence $\lambda \in \sigma_p(A)$ by 15.2.5.

15.2.7 Theorem. *Let* $(X, \mathcal{A})$ *be a measurable space and* P *a projection valued measure on* $\mathcal{A}$ *with values in* $\mathscr{P}(\mathcal{H})$. *Let* $\psi \in \mathcal{M}(X, \mathcal{A}, P)$ *be so that* $\overline{\psi} = \psi$. *Then the operator* $A := J_\psi^P$ *is self-adjoint and*

$$P^A(E) = P(\psi^{-1}(E)), \forall E \in \mathcal{A}(d_{\mathbb{R}}).$$

Proof. The operator A is self-adjoint by 14.3.17.

Now we resort to 14.6.1 with

$$X_1 := X, \quad \mathcal{A}_1 := \mathcal{A}, \quad P_1 := P, \quad X_2 := \mathbb{R}, \quad \mathcal{A}_2 := \mathcal{A}(d_{\mathbb{R}}), \quad \pi := \psi$$

(we note that $X_1 - D_\pi = X - D_\psi$ and hence $P_1(X_1 - D_\pi) = \mathbb{O}_{\mathcal{H}}$). Then the mapping

$$Q : \mathcal{A}(d_{\mathbb{R}}) \to \mathscr{P}(\mathcal{H})$$
$$E \mapsto Q(E) := P(\psi^{-1}(E))$$

is a projection valued measure on $\mathcal{A}(d_{\mathbb{R}})$ (cf. 14.6.1a) and

$$J_Q(\xi) = J_P(\xi \circ \psi) = J_P(\psi) = A$$

(cf. 14.6.1b with $\varphi := \xi$). Then $P^A = Q$ by definition of P^A. $\qquad\square$

The next theorem can be proved directly (cf. e.g. Simmons, 1963, Chapter 11). Instead, we deduce it from the results proved in this section.

15.2.8 Theorem (Finite-dimensional spectral th. for s.a. operators).
Suppose that the Hilbert space $\mathcal{H}$ is finite-dimensional, and let A be a self-adjoint operator in $\mathcal{H}$. Then $\sigma_p(A)$ is a non-empty finite set. If N is the number of the eigenvalues of A, letting
$$\{\lambda_1, ..., \lambda_N\} := \sigma_p(A) \text{ and } P_n := P^A(\{\lambda_n\}), \forall n \in \{1, ..., N\},$$
we have:
$$P_n \neq O_{\mathcal{H}}, \forall n \in \{1, ..., N\};$$
$$P_i P_j = O_{\mathcal{H}} \text{ if } i \neq j;$$
$$\sum_{n=1}^{N} P_n = \mathbb{1}_{\mathcal{H}};$$
$$A = \sum_{n=1}^{N} \lambda_n P_n.$$

Proof. We know that $\sigma(A) \neq \emptyset$ (cf. 15.2.2d). Now let $\lambda \in \sigma(A)$. Since $\sigma(A) = Ap\sigma(A)$ (cf. 12.4.21b) and since every linear operator in $\mathcal{H}$ is bounded (cf. 10.8.3B), we have that the operator $A - \lambda \mathbb{1}_{\mathcal{H}}$ is not injective, i.e. that $\lambda \in \sigma_p(A)$. This proves that $\sigma_p(A)$ is a non-empty set, and also (in view of 4.5.8) that
$$\sigma_p(A) = \sigma(A).$$
In view of 12.4.20B, $\sigma_p(A)$ must be a finite set: if it were not, then by choosing an element of $N_{A-\lambda \mathbb{1}_{\mathcal{H}}} \cap \tilde{\mathcal{H}}$ for each $\lambda \in \sigma_p(A)$ we could construct a non-finite o.n.s. in $\mathcal{H}$ and hence (cf. 10.7.3) there would exist a non-finite c.o.n.s. in $\mathcal{H}$, contrary to the hypothesis that $\mathcal{H}$ is finite-dimensional. Thus, we can write
$$\{\lambda_1, ..., \lambda_N\} := \sigma_p(A).$$
In view of 15.2.5 we have
$$P_n := P^A(\{\lambda_n\}) \neq O_{\mathcal{H}}, \forall n \in \{1, ..., N\}.$$
Moreover, we have
$$P_i P_j = P^A(\{\lambda_i\}) P^A(\{\lambda_j\}) = O_{\mathcal{H}} \text{ if } i \neq j$$
(cf. 13.3.2b) and also
$$\sum_{n=1}^{N} P_n = \sum_{n=1}^{N} P^A(\{\lambda_n\}) = P^A(\sigma_p(A)) = P^A(\sigma(A)) = \mathbb{1}_{\mathcal{H}}$$
(cf. 15.2.2d). Finally, we note that $D_A = \mathcal{H}$ in view of 10.8.3B and 12.4.7, and that
$$P_n f = P^A(\{\lambda_n\}) f \in N_{A-\lambda_n \mathbb{1}_{\mathcal{H}}} \text{ and hence } AP_n f = \lambda_n P_n f, \forall f \in \mathcal{H}$$
(cf. 15.2.5e). This yields
$$Af = \sum_{n=1}^{N} AP_n f = \sum_{n=1}^{N} \lambda_n P_n f, \forall f \in \mathcal{H},$$
or $A = \sum_{n=1}^{N} \lambda_n P_n$. $\square$

15.3 Functions of a self-adjoint operator

Throughout this section, $\mathcal{H}$ stands for an abstract Hilbert space.

15.3.1 Definition. Let A be a self-adjoint operator in $\mathcal{H}$.

For a function $\varphi \in \mathcal{M}(\mathbb{R}, \mathcal{A}(d_{\mathbb{R}}), P^A)$, we write

$$\varphi(A) := J_{\varphi}^{P^A}.$$

This operator is said to be a *function of A*. This name is justified by the fact that a function of A as defined here is often nothing else that the function of A as defined in an obvious direct way. An important instance of this is the subject of 15.3.5.

We note that the equality $\xi(A) = A$ is obvious, by the very definition of P^A.

15.3.2 Remark. Let A be a self-adjoint operator in $\mathcal{H}$.

For a function $\varphi \in \mathcal{M}(\mathbb{R}, \mathcal{A}(d_{\mathbb{R}}), P^A)$, we obtain immediately a great number of results for $\varphi(A)$ from the corresponding more general results of Chapter 14; for quick reference, we list here some of them (cf. 14.2.14a,b, 14.2.15, 14.2.17, 14.3.2, 14.3.6, 14.3.9, 14.3.15):

(a) $D_{\varphi(A)} = \{f \in \mathcal{H} : \varphi \in \mathcal{L}^2(\mathbb{R}, \mathcal{A}(d_{\mathbb{R}}), \mu_f^{P^A})\}$;

(b) $(f|\varphi(A)f) = \int_{\mathbb{R}} \varphi d\mu_f^{P^A}, \forall f \in D_{\varphi(A)}$;

(c) $(\varphi(A))^{\dagger} = \overline{\varphi}(A)$;

(d) $\varphi(A) \in \mathcal{B}(\mathcal{H})$ iff $\varphi \in \mathcal{L}^{\infty}(\mathbb{R}, \mathcal{A}(d_{\mathbb{R}}), P^A)$;

(e) $\|\varphi(A)f - \lambda f\|^2 = \int_{\mathbb{R}} |\varphi - \lambda|^2 d\mu_f^{P^A}, \forall f \in D_{\varphi(A)}, \forall \lambda \in \mathbb{C}$;

(f) $\alpha\varphi(A) + \beta\psi(A) \subset (\alpha\varphi + \beta\psi)(A), \forall \alpha, \beta \in \mathbb{C}, \forall \psi \in \mathcal{M}(\mathbb{R}, \mathcal{A}(d_{\mathbb{R}}), P^A)$;

(g) $D_{\psi(A)\varphi(A)} = D_{\varphi(A)} \cap D_{(\psi\varphi)(A)}$ and $\psi(A)\varphi(A) \subset (\psi\varphi)(A)$, $\forall \psi \in \mathcal{M}(\mathbb{R}, \mathcal{A}(d_{\mathbb{R}}), P^A)$;

(h) for $\psi \in \mathcal{M}(\mathbb{R}, \mathcal{A}(d_{\mathbb{R}}), P^A)$, $\varphi(A) = \psi(A)$ iff $\varphi(x) = \psi(x)$ P^A-a.e. on $D_{\varphi} \cap D_{\psi}$.

15.3.3 Remark. For every self-adjoint operator A in $\mathcal{H}$ and every $E \in \mathcal{A}(d_{\mathbb{R}})$, we have $D_{\chi_E(A)} = \mathcal{H}$ and

$$(f|\chi_E(A)f) = \int_{\mathbb{R}} \chi_E d\mu_f^{P^A} = \mu_f^{P^A}(E) = (f|P^A(E)f), \forall f \in \mathcal{H}$$

(cf. 15.3.2a,b), and hence $\chi_E(A) = P^A(E)$ (cf. 10.2.12). Obviously, this equation cannot be used for the construction of the projection valued measure P^A by means of A, since it is actually based on the previous existence of P^A.

15.3.4 Examples.

(A) We set $(X, \mathcal{A}, \mu) := (\mathbb{R}, \mathcal{A}(d_{\mathbb{R}}), m)$ in the discussion of Section 14.5; we recall that m denotes the Lebesgue measure on $\mathbb{R}$. Thus, $L^2(\mathbb{R}, \mathcal{A}, \mu) = L^2(\mathbb{R})$. The projection valued measure P of Section 14.5 is now defined on $\mathcal{A}(d_{\mathbb{R}})$ and we define the operator $Q := J_{\xi}^P$, which is a self-adjoint operator in $L^2(\mathbb{R})$. This operator is denoted by Q since in non-relativistic quantum mechanics it represents the observable "position of a quantum particle in one dimension" (cf. 20.3.6c; the operator that represents the observable "linear momentum of a quantum particle in one dimension" is denoted by P, and the symbols Q and P are chosen on analogy of the symbols q and p used in classical mechanics; however, in the present discussion P denotes the projection valued measure of Section 14.5). We have $P^Q = P$ by definition of P^Q.

It is easy to see that

$$P^Q((\lambda - \varepsilon, \lambda + \varepsilon)) \neq \mathbb{O}_{L^2(\mathbb{R})}, \forall \lambda \in \mathbb{R}, \forall \varepsilon > 0;$$

thus, $\sigma(Q) = \mathbb{R}$ (cf. 15.2.4). It is obvious that

$$P^Q(\{\lambda\}) = \mathbb{O}_{L^2(\mathbb{R})}, \forall \lambda \in \mathbb{R};$$

thus, $\sigma_p(Q) = \emptyset$ (cf. 15.2.5).

Since $J_\xi^P = M_\xi$, we have:

$$D_Q = \left\{ [f] \in L^2(\mathbb{R}) : \int_{\mathbb{R}} |\xi f|^2 dm < \infty \right\};$$
$$Q[f] = [\xi f], \forall [f] \in D_Q.$$

Since the measure m is σ-finite, $\mathcal{M}(\mathbb{R}, \mathcal{A}(d_\mathbb{R}), m) = \mathcal{M}(\mathbb{R}, \mathcal{A}(d_\mathbb{R}), P^Q)$ and hence

$$M_\varphi = J_{P^Q}(\varphi) = \varphi(Q), \forall \varphi \in \mathcal{M}(\mathbb{R}, \mathcal{A}(d_\mathbb{R}), m).$$

(B) Suppose that $\mathcal{H}$ is a separable Hilbert space and let A be a self-adjoint operator in $\mathcal{H}$. In view of 12.4.20C, $\sigma_p(A)$ is a countable set and hence $\sigma_p(A) \in \mathcal{A}(d_\mathbb{R})$. The following conditions are equivalent:

(a) $P^A(\mathbb{R} - \sigma_p(A)) = \mathbb{O}_\mathcal{H}$, or equivalently $P^A(\sigma_p(A)) = \mathbb{1}_\mathcal{H}$;

(b) there exists a family $\{(\lambda_n, P_n)\}_{n \in I}$, with $I = \{1, ..., N\}$ or $I = \mathbb{N}$, so that

   $\lambda_n \in \mathbb{R}, P_n \in \mathscr{P}(\mathcal{H}), P_n \neq \mathbb{O}_\mathcal{H}, \forall n \in I,$
   $\lambda_i \neq \lambda_j$ and $P_i P_j = \mathbb{O}_\mathcal{H}$ if $i \neq j,$
   $\sum_{n \in I} P_n f = f, \forall f \in \mathcal{H},$
   $D_A = \{f \in \mathcal{H} : \sum_{n \in I} \lambda_n^2 \|P_n f\|^2 < \infty\},$
   $Af = \sum_{n \in I} \lambda_n P_n f, \forall f \in D_A$
   (we note that, if $I = \mathbb{N}$, the series $\sum_{n \in I} \lambda_n P_n f$ is convergent for all $f \in D_A$, in view of 13.2.8d and 10.4.7b).

Indeed, if condition a is true then we define $\{\lambda_n\}_{n \in I} := \sigma_p(A)$, with the condition $\lambda_i \neq \lambda_j$ if $i \neq j$ and with $I := \{1, ..., N\}$ or $I := \mathbb{N}$ as the case may be, and $P_n := P^A(\{\lambda_n\})$ for each $n \in I$. Then we have:

   $\lambda_n \in \mathbb{R}, P_n \in \mathscr{P}(\mathcal{H}), P_n \neq \mathbb{O}_\mathcal{H}$ (cf. 15.2.5c), $\forall n \in I;$
   $P_i P_j = \mathbb{O}_\mathcal{H}$ if $i \neq j$ and $\sum_{n \in I} P_n f = P^A(\sigma_p(A))f = f, \forall f \in \mathcal{H}$
   (since P^A is a projection valued measure).

Moreover, if we define $I_E := \{n \in I : \lambda_n \in E\}$ for all $E \in \mathcal{A}(d_\mathbb{R})$, then we have

$$P^A(E)f \overset{(1)}{=} P^A(E \cap \sigma_p(A))f = P^A\left(\bigcup_{n \in I_E} \{\lambda_n\}\right)f$$

$$= \sum_{n \in I_E} P_n f, \forall f \in \mathcal{H}, \forall E \in \mathcal{A}(d_\mathbb{R})$$

(1 holds by 13.3.2c), and hence

$$\mu_f^{P^A}(E) = \sum_{n \in I_E} \|P_n f\|^2, \forall E \in \mathcal{A}(d_\mathbb{R}), \forall f \in \mathcal{H},$$

and hence

$$D_A = \left\{f \in \mathcal{H} : \int_\mathbb{R} \xi^2 d\mu_f^{P^A} < \infty\right\} = \left\{f \in \mathcal{H} : \sum_{n \in I} \lambda_n^2 \|P_n f\|^2 < \infty\right\},$$

by 15.2.2a and 8.3.8. Thus, if $I = \mathbb{N}$, the series $\sum_{n \in I} \lambda_n P_n f$ is convergent for all $f \in D_A$, and, for either $I = \{1, ..., N\}$ or $I = \mathbb{N}$, we can define the mapping

$$B : D_A \to \mathcal{H}$$

$$f \mapsto Bf := \sum_{n \in I} \lambda_n P_n f$$

which is obviously a linear operator. Now we have

$$(f|Af) = \int_\mathbb{R} \xi d\mu_f^{P^A} = \sum_{n \in I} \lambda_n (f|P_n f) = (f|Bf), \forall f \in D_A,$$

by 15.2.2b and 8.3.8, and hence $A = B$ by 10.2.12, or

$$Af = Bf = \sum_{n \in I} \lambda_n P_n f, \forall f \in D_A.$$

This proves that condition a implies condition b.
Conversely, if condition b is true then

$$\|Af - \lambda f\|^2 = \left\|\sum_{n \in I} \lambda_n P_n f - \sum_{n \in I} \lambda P_n f\right\|^2$$

$$= \sum_{n \in I} |\lambda_n - \lambda|^2 \|P_n f\|^2, \forall \lambda \in \mathbb{C}, \forall f \in D_A,$$

by 13.2.9c and 10.2.3 or by 13.2.8d and 10.4.7a; thus, for $f \in D_A$,

$$[\lambda \in \mathbb{C} - \{\lambda_n\}_{n \in I} \text{ and } Af = \lambda f] \Rightarrow$$

$$[P_n f = 0_\mathcal{H}, \forall n \in I] \Rightarrow f = \sum_{n \in I} P_n f = 0_\mathcal{H}$$

and, for every $k \in I$,

$$f \in N_{A - \lambda_k \mathbf{1}_\mathcal{H}} \Leftrightarrow [P_n f = 0_\mathcal{H}, \forall n \in I - \{k\}] \Leftrightarrow f = P_k f,$$

i.e. $N_{A-\lambda_k \mathbb{1}_{\mathcal{H}}} = R_{P_k}$ (cf. 13.1.3c). Since $P_n \neq \mathbb{O}_{\mathcal{H}}$ for all $n \in I$, this proves that

$$\sigma_p(A) = \{\lambda_n\}_{n\in I} \text{ and (cf. 15.2.5e and 13.1.4b) } P^A(\{\lambda_n\}) = P_n, \forall n \in I.$$

Therefore,

$$f = \sum_{n\in I} P^A(\{\lambda_n\})f = P^A(\sigma_p(A))f, \quad \forall f \in \mathcal{H},$$

i.e. $P^A(\sigma_p(A)) = \mathbb{1}_{\mathcal{H}}$, which is condition a.

Since every operator determines uniquely its point spectrum and its eigenspaces, what we saw above proves the uniqueness of the family $\{(\lambda_n, P_n)\}_{n\in I}$ of condition b.

The following condition

(c) there exists a c.o.n.s. $\{v_j\}_{j\in J}$ in $\mathcal{H}$ whose elements are eigenvectors of A, i.e. so that

$$\forall j \in J, v_j \in D_A \text{ and } \exists \mu_j \in \mathbb{R} \text{ such that } Av_j = \mu_j v_j$$

is a further condition which is equivalent to condition b (and hence to condition a).

Indeed, suppose that condition b is true. Then, for each $n \in I$, we fix a countable o.n.s. $\{u_{n,s}\}_{s\in I_n}$ which is complete in the subspace R_{P_n} (cf. 10.7.2). Then the set $\bigcup_{n\in I}\{u_{n,s}\}_{s\in I_n}$ in an o.n.s. in $\mathcal{H}$ (cf. 13.2.8d or 13.2.9c) and it is complete in $\mathcal{H}$ by 10.6.4 (with $M := \mathcal{H}$) since

$$f = \sum_{n\in I} P_n f \overset{(2)}{=} \sum_{n\in I}\sum_{s\in I_n} (u_{n,s}|f)\, u_{n,s}, \quad \forall f \in \mathcal{H},$$

where 2 holds by 13.1.10 (note that $\sum_{n\in I}\sum_{s\in I_n}(u_{n,s}|f)\,u_{n,s}$ can be construed as a single series with the set of indices $\bigcup_{n\in I}\{(n,s)\}_{s\in I_n}$, and that there is no need to specify what ordering is used to define this series, in view of 10.4.10). Since $u_{n,s}$ is an eigenvector of A (cf. the proof of b $\Rightarrow$ a above), this proves that condition b implies condition c.

Conversely, if condition c is true, then from 12.4.24 we have:

$D_A = \{f \in \mathcal{H} : \sum_{j\in J}\mu_j^2| (v_j|f)|^2 < \infty\}$;
$Af = \sum_{j\in J}\mu_j (v_j|f)\, v_j, \forall f \in D_A$;
$\sigma_p(A) = \{\mu_j\}_{j\in J}$

(actually, 12.4.24 is written on the assumption that the orthogonal dimension of $\mathcal{H}$ is denumerable; if the orthogonal dimension of $\mathcal{H}$ is finite, a simplified version of the proof of 12.4.24 leads to the equations written above).

Now, we define $\{\lambda_n\}_{n\in I} := \{\mu_j\}_{j\in J}$ with the condition $\lambda_i \neq \lambda_j$ if $i \neq j$ and with $I := \{1, ..., N\}$ or $I := \mathbb{N}$; moreover, for each $n \in I$, we define

$$J_n := \{j \in J : \mu_j = \lambda_n\}$$

and P_n as the projection such that

$$P_n f = \sum_{j \in J_n} (v_j | f) v_j, \forall f \in \mathcal{H}$$

(cf. 13.1.10). Then,

$$P_n \neq \mathbb{O}_\mathcal{H}, \forall n \in I, \text{ and } P_i P_j = \mathbb{O}_\mathcal{H} \text{ if } i \neq j,$$

$$\sum_{n \in I} P_n f = \sum_{n \in I} \sum_{j \in J_n} (v_j | f) v_j = \sum_{j \in J} (v_j | f) v_j = f, \forall f \in \mathcal{H}$$

(cf. 10.4.10 and 10.6.4b). Moreover,

$$D_A = \left\{ f \in \mathcal{H} : \sum_{n \in I} \lambda_n^2 \sum_{j \in J_n} |(v_j | f)|^2 < \infty \right\} = \left\{ f \in \mathcal{H} : \sum_{n \in I} \lambda_n^2 \| P_n f \|^2 < \infty \right\}$$

(cf. 5.4.7 and 10.2.3 or 10.4.8a), and

$$Af = \sum_{n \in I} \lambda_n \sum_{j \in J_n} (v_j | f) v_j = \sum_{n \in I} \lambda_n P_n f, \forall f \in D_A$$

(cf. 10.4.10). This proves that condition c implies condition b.

Now suppose that conditions a, b, c hold true.

Then every function $\varphi : \{\lambda_n\}_{n \in I} \to \mathbb{C}$ is an element of $\mathcal{M}(\mathbb{R}, \mathcal{A}(d_\mathbb{R}), P^A)$, because it is obviously $\mathcal{A}(d_\mathbb{R})^{D_\varphi}$-measurable and because $\{\lambda_n\}_{n \in I} = \sigma_p(A)$ (see above) and hence $P^A(\mathbb{R} - \{\lambda_n\}_{n \in I}) = \mathbb{O}_\mathcal{H}$. From

$$\mu_f^{P^A}(E) = \sum_{n \in I_E} \| P_n f \|^2, \forall E \in \mathcal{A}(d_\mathbb{R}), \forall f \in \mathcal{H}$$

(see above), we have (in view of 15.3.2a,b and of 8.3.8)

$$D_{\varphi(A)} = \left\{ f \in \mathcal{H} : \int_\mathbb{R} |\varphi|^2 d\mu_f^{P^A} < \infty \right\}$$

$$= \left\{ f \in \mathcal{H} : \sum_{n \in I} |\varphi(\lambda_n)|^2 \| P_n f \|^2 < \infty \right\},$$

$$(f | \varphi(A) f) = \int_\mathbb{R} \varphi \, d\mu_f^{P^A} = \sum_{n \in I} \varphi(\lambda_n) \| P_n f \|^2$$

$$= \left(f | \sum_{n \in I} \varphi(\lambda_n) P_n f \right), \forall f \in D_{\varphi(A)};$$

since the mapping $D_{\varphi(A)} \ni f \mapsto \sum_{n \in I} \varphi(\lambda_n) P_n f \in \mathcal{H}$ is obviously a linear operator (its definition is consistent by 10.4.7b), in view of 10.2.12 this implies that

$$\varphi(A) f = \sum_{n \in I} \varphi(\lambda_n) P_n f, \forall f \in D_{\varphi(A)}.$$

If $\{u_{n,s}\}_{s\in I_n}$ is as before for each $n \in I$, we have

$$P_n f = \sum_{s\in I_n} (u_{n,s}|f)\, u_{n,s} \text{ and}$$

$$\|P_n f\|^2 = \sum_{s\in I_n} |(u_{n,s}|f)|^2, \forall f \in \mathcal{H}, \forall n \in I$$

(cf. 13.1.10, and 10.2.3 or 10.4.8a), and hence

$$D_{\varphi(A)} = \left\{ f \in \mathcal{H} : \sum_{n\in I} |\varphi(\lambda_n)|^2 \sum_{s\in I_n} |(u_{n,s}|f)|^2 < \infty \right\},$$

$$\varphi(A)f = \sum_{n\in I} \varphi(\lambda_n) \sum_{s\in I_n} (u_{n,s}|f) u_{n,s}, \quad \forall f \in D_{\varphi(A)}.$$

(C) If the Hilbert space $\mathcal{H}$ is finite-dimensional then 15.2.8 proves that condition b of example B holds true for every self-adjoint operator A in $\mathcal{H}$. Then condition a holds true as well (this was also seen directly in the proof of 15.2.8), and so does condition c. Thus, for every self-adjoint operator A in a finite-dimensional Hilbert space $\mathcal{H}$ there exists a c.o.n.s. in $\mathcal{H}$ whose elements are eigenvectors of A.

(D) Let M be a subspace of $\mathcal{H}$. The mapping

$$P : \mathcal{A}(d_{\mathbb{R}}) \to \mathscr{P}(\mathcal{H})$$

$$E \mapsto P(E) := \chi_E(0)P_{M^\perp} + \chi_E(1)P_M$$

is a projection valued measure in view of 13.3.5. Indeed, for every $f \in \mathcal{H}$, μ_f^P is the measure μ defined in 8.3.8 with

$$I := \{1,2\}, \; x_1 := 0, \; x_2 := 1, \; a_1 := (f|P_{M^\perp}f), \; a_2 := (f|P_M f);$$

moreover, this entails $\mu_f^P(\mathbb{R}) = a_1 + a_2 = \|f\|^2 - \|P_M f\|^2 + \|P_M f\|^2 = \|f\|^2$. The operator A^P is the projection P_M since (cf. 8.3.8 and 15.2.2a,b)

$$\int_{\mathbb{R}} \xi^2 d\mu_f^P = \int_{\mathbb{R}} \xi d\mu_f^P = 0a_1 + 1a_2 = (f|P_M f),$$

and hence

$$\left\{ f \in \mathcal{H} : \int_{\mathbb{R}} \xi^2 d\mu_f^P < \infty \right\} = \mathcal{H} = D_{P_M}$$

and

$$\int_{\mathbb{R}} \xi d\mu_f^P = (f|P_M f), \forall f \in \mathcal{H}.$$

15.3.5 Proposition. *Let A be a self-adjoint operator in $\mathcal{H}$. Let p be a polynomial, i.e. there exist $N \geq 0$ and $(\alpha_0, \alpha_1, ..., \alpha_N) \in \mathbb{C}^{N+1}$ (we assume $\alpha_N \neq 0$) so that*

$$p = \sum_{k=0}^N \alpha_k \xi^k \text{ (we define } \xi^0 := 1_{\mathbb{R}}).$$

Then,

$$p(A) = \sum_{k=0}^{N} \alpha_k A^k \ (we \ define \ A^0 := \mathbb{1}_{\mathcal{H}}).$$

Let q be a non-trivial polynomial, i.e. there exist $M \geq 1$ and $(\beta_0, \beta_1, ..., \beta_M) \in \mathbb{C}^{M+1}$ with $\beta_M \neq 0$ so that

$$q = \sum_{i=0}^{M} \beta_i \xi^i.$$

If the roots of q are not elements of $\sigma_p(A)$, then $\frac{1}{q} \in \mathcal{M}(\mathbb{R}, \mathcal{A}(d_{\mathbb{R}}), P^A)$ (where $\frac{1}{q}$ is defined as in 1.2.19), the operator $\sum_{i=0}^{M} \beta_i A^i$ is injective and

$$\frac{1}{q}(A) = \left(\sum_{i=0}^{M} \beta_i A^i \right)^{-1}.$$

If, further, the roots of q are not elements of $\sigma(A)$, then (letting $\frac{p}{q} := p\frac{1}{q}$)

$$\frac{p}{q}(A) = \left(\sum_{k=0}^{N} \alpha_k A^k \right) \left(\sum_{i=0}^{M} \beta_i A^i \right)^{-1}.$$

Proof. First we note that $A^0 = \mathbb{1}_{\mathcal{H}} = 1_{\mathbb{R}}(A) = \xi^0(A)$. Now we prove by induction the proposition

$$A^n = \xi^n(A), \forall n \in \mathbb{N}.$$

For $n = 1$, $A = \xi(A)$ means $A = J_{\xi}^{P^A}$, which is obvious by definition of P^A. Then we assume, for a fixed positive integer n, that

$$A^n = \xi^n(A).$$

From the inequality

$$|x|^n \leq |x|^{n+1} + 1, \forall x \in \mathbb{R}$$

(for $x \neq 0$, $1 \leq |x| + |x|^{-n}$ as $1 \leq |x|^{-n}$ if $|x| \leq 1$) we have (cf. 15.3.2a)

$$f \in D_{\xi^{n+1}(A)} \Rightarrow \xi^{n+1} \in \mathcal{L}^2(\mathbb{R}, \mathcal{A}(d_{\mathbb{R}}), \mu_f^{P^A}) \Rightarrow$$
$$\xi^{n+1} + 1 \in \mathcal{L}^2(\mathbb{R}, \mathcal{A}(d_{\mathbb{R}}), \mu_f^{P^A}) \Rightarrow \xi^n \in \mathcal{L}^2(\mathbb{R}, \mathcal{A}(d_{\mathbb{R}}), \mu_f^{P^A}) \Rightarrow$$
$$f \in D_{\xi^n(A)},$$

i.e. $D_{\xi^{n+1}(A)} \subset D_{\xi^n(A)}$; on the other hand, we also have (cf. 15.3.2g)

$$D_{\xi(A)\xi^n(A)} = D_{\xi^n(A)} \cap D_{\xi^{n+1}(A)} \ and \ \xi(A)\xi^n(A) \subset \xi^{n+1}(A);$$

therefore,

$$D_{\xi(A)\xi^n(A)} = D_{\xi^{n+1}(A)} \ and \ \xi(A)\xi^n(A) = \xi^{n+1}(A),$$

and hence (by the assumption made)

$$A^{n+1} = AA^n = \xi(A)\xi^n(A) = \xi^{n+1}(A).$$

Thus, the proposition is proved.

Next, 15.3.2f implies that

$$\sum_{k=0}^{N} \alpha_k A^k = \sum_{k=0}^{N} \alpha_k \xi^k(A) \subset p(A).$$

If we define $B := \sum_{k=0}^{N} \alpha_k A^k$, we have

$$D_B = \bigcap_{k=0}^{N} D_{A^k} = D_{A^N}$$

since obviously $D_{A^{k+1}} \subset D_{A^k}$ for all $k \in \mathbb{N}$. Now, it is easy to prove that there exists a bounded interval I so that

$$\frac{1}{2}|\alpha_N||x|^N \le |p(x)|, \forall x \in \mathbb{R} - I;$$

therefore,

$$f \in D_{p(A)} \Rightarrow p \in \mathcal{L}^2(\mathbb{R}, \mathcal{A}(d_{\mathbb{R}}), \mu_f^{P^A}) \Rightarrow$$
$$\xi^N \in \mathcal{L}^2(\mathbb{R}, \mathcal{A}(d_{\mathbb{R}}), \mu_f^{P^A}) \Rightarrow f \in D_{\xi^N(A)} = D_{A^N}.$$

This proves that $D_{p(A)} \subset D_B$, and hence that

$$p(A) = \sum_{k=0}^{N} \alpha_k A^k.$$

This proves the first part of the statement. In what follows, we prove the second part.

If the roots of q are not elements of $\sigma_p(A)$, then (cf. 15.2.5c)

$$P^A(q^{-1}(\{0\})) = \mathbb{O}_{\mathcal{H}}$$

and hence (cf. 14.3.14)

the operator $q(A)$ is injective, $\quad \dfrac{1}{q} \in \mathcal{M}(\mathbb{R}, \mathcal{A}(d_{\mathbb{R}}), P^A), \quad \dfrac{1}{q}(A) = (q(A))^{-1};$

now, $q(A) = \sum_{i=0}^{M} \beta_i A^i$ in view of the first part of the statement.

To prove the last part of the statement, let $\{\lambda_1, ..., \lambda_M\}$ be the roots of q (each value is repeated as many times as its multiplicity); then

$$\frac{1}{q} = \beta_M^{-1} \frac{1}{\xi - \lambda_1} \cdots \frac{1}{\xi - \lambda_M}.$$

Now, suppose $\lambda_i \notin \sigma(A)$ for all $i \in \{1, ..., M\}$. Then, for each $i \in \{1, ..., M\}$, the operator $A - \lambda_i \mathbb{1}_{\mathcal{H}}$ is injective and the operator $(A - \lambda_i \mathbb{1}_{\mathcal{H}})^{-1}$ is bounded (cf.

12.4.21b, 4.5.2, 4.5.3); moreover, $A - \lambda_i \mathbb{1}_\mathcal{H} = (\xi - \lambda_i)(A)$ (cf. the first part of the statement); then, 14.3.14 and 14.2.17 imply that

$$\frac{1}{\xi - \lambda_i} \in \mathcal{L}^\infty(\mathbb{R}, \mathcal{A}(d_\mathbb{R}), P^A).$$

Thus, $\frac{1}{q} \in \mathcal{L}^\infty(\mathbb{R}, \mathcal{A}(d_\mathbb{R}), P^A)$ (cf. 14.2.5) and hence, in view of 14.3.10,

$$\frac{p}{q}(A) = J_{P^A}\left(p\frac{1}{q}\right) = J_{P^A}(p) J_{P^A}\left(\frac{1}{q}\right) = p(A)\frac{1}{q}(A),$$

or

$$\frac{p}{q}(A) = \left(\sum_{k=0}^N \alpha_k A^k\right)\left(\sum_{i=0}^M \beta_i A^i\right)^{-1},$$

in view of what has already been proved. □

15.3.6 Remark. If A is a self-adjoint operator in $\mathcal{H}$ then the operator $\sum_{k=0}^N \alpha_k A^k$ is self-adjoint for every $N \geq 0$ and every $(\alpha_0, \alpha_1, ..., \alpha_N) \in \mathbb{R}^{N+1}$. This follows at once from 15.3.5 and 15.3.2c. Hence, in particular, the operator A^n is self-adjoint for all $n \in \mathbb{N}$.

15.3.7 Proposition. *Let A be a self-adjoint operator in $\mathcal{H}$ and let $B \in \mathcal{B}(\mathcal{H})$ be such that $BA \subset AB$. Then*

$$B\varphi(A) \subset \varphi(A)B, \forall \varphi \in \mathcal{M}(\mathbb{R}, \mathcal{A}(d_\mathbb{R}), P^A).$$

Proof. From 15.2.1B we have

$$BP^A(E) = P^A(E)B, \forall E \in \mathcal{A}(d_\mathbb{R}).$$

Then the statement is proved by 14.2.14e. □

15.3.8 Proposition. *Let A be a self-adjoint operator in $\mathcal{H}$ and suppose that a function $\varphi \in \mathcal{M}(\mathbb{R}, \mathcal{A}(d_\mathbb{R}), P^A)$ is such that $\overline{\varphi} = \varphi$. Then the operator $\varphi(A)$ is self-adjoint and*

$$P^{\varphi(A)}(E) = P^A(\varphi^{-1}(E)), \forall E \in \mathcal{A}(d_\mathbb{R}).$$

Proof. The statement follows at once from 15.2.7. □

15.3.9 Theorem. *Let A be a self-adjoint operator in $\mathcal{H}$ such that $0 \leq (f|Af)$ for all $f \in D_A$. Then $\sigma(A) \subset [0, \infty)$ and there exists a unique self-adjoint operator B in $\mathcal{H}$ such that*

$$0 \leq (f|Bf), \forall f \in D_B, \quad and \quad A = B^2.$$

If $T \in \mathcal{B}(\mathcal{H})$ is such that $TA \subset AT$ then $TB \subset BT$. If A is bounded then $B \in \mathcal{B}(\mathcal{H})$.

Proof. For every $\lambda \in (-\infty, 0)$ we have, by the Schwarz inequality,

$$\|f\|\|(A - \lambda \mathbb{1}_{\mathcal{H}})f\| \geq (f|(A - \lambda \mathbb{1}_{\mathcal{H}})f) \geq -\lambda \|f\|^2, \forall f \in D_A$$

(recall that $(f|Af) \in \mathbb{R}, \forall f \in D_A$), and hence $\lambda \notin Ap\sigma(A)$ (cf. 4.5.2 and 4.5.3), and hence $\lambda \notin \sigma(A)$ (cf. 12.4.21b). This proves that $\sigma(A) \subset [0, \infty)$.

Since $\sigma(A) \subset [0, \infty)$, from 15.2.2d and 13.3.2e we have $P^A((-\infty, 0)) = \mathbb{O}_{\mathcal{H}}$. Then the function

$$\varphi : [0, \infty) \to \mathbb{C}$$
$$x \mapsto \varphi(x) := \sqrt{x}$$

(we remind the reader that we always take the square root of a positive real number to be positive) is an element of $\mathcal{M}(\mathbb{R}, \mathcal{A}(d_{\mathbb{R}}), P^A)$ and we can define the operator $B := \varphi(A)$. Since $\bar{\varphi} = \varphi$, the operator B is self-adjoint (cf. 15.3.2c). Further (cf. 15.3.2b)

$$(f|Bf) = \int_{\mathbb{R}} \varphi d\mu_f^{P^A} \geq 0, \forall f \in D_B,$$

since $\varphi \in \mathcal{L}^+(\mathbb{R}, \mathcal{A}(d_{\mathbb{R}}), \mu_f^{P^A})$ for all $f \in \mathcal{H}$. Moreover,

$$D_{\varphi(A)\varphi(A)} = D_{\varphi(A)} \cap D_{\varphi^2(A)} \text{ and } \varphi(A)\varphi(A) \subset \varphi^2(A)$$

(cf. 15.3.2g); now, the inequality

$$x \leq \frac{1}{2}(1 + x^2), \forall x \in [0, \infty)$$

shows that $D_{\varphi^2(A)} \subset D_{\varphi(A)}$, and hence that $D_{\varphi(A)\varphi(A)} = D_{\varphi^2(A)}$, and hence that

$$\varphi(A)\varphi(A) = \varphi^2(A).$$

Now, $\varphi^2(x) = \xi(x)$ for all $x \in D_{\varphi^2} \cap D_\xi (= D_\varphi)$, and hence (cf. 15.3.2h)

$$B^2 = \varphi(A)\varphi(A) = \varphi^2(A) = \xi(A) = A.$$

This proves the existence of B. Before proving its uniqueness, we will prove the last two assertions of the statement.

Let $T \in \mathcal{B}(\mathcal{H})$ be such that $TA \subset AT$. Then, $TB \subset BT$ by 15.3.7.

If A is bounded then $\sigma(A)$ is bounded (cf. 15.2.2f); then, $\varphi \in \mathcal{L}^\infty(\mathbb{R}, \mathcal{A}(d_{\mathbb{R}}), P^A)$ since $P^A(\mathbb{R} - \sigma(A)) = \mathbb{O}_{\mathcal{H}}$ (cf. 15.2.2d), and hence $B \in \mathcal{B}(\mathcal{H})$ by 15.3.2d.

To prove the uniqueness of B, suppose that C is a self-adjoint operator in $\mathcal{H}$ such that

$$0 \leq (f|Cf), \forall f \in D_C, \text{ and } A = C^2.$$

As before for A, we have $P^C((-\infty, 0)) = \mathbb{O}_{\mathcal{H}}$; moreover, from 15.3.5 we have $A = \xi^2(C)$. Then the function

$$\psi : [0, \infty) \to \mathbb{C}$$
$$x \mapsto \psi(x) := x^2$$

is an element of $\mathcal{M}(\mathbb{R}, \mathcal{A}(d_\mathbb{R}), P^C)$, $\psi(x) = \xi^2(x)$ for all $x \in D_\psi \cap D_{\xi^2} (= D_\psi)$, and hence (cf. 15.3.2h)

$$\psi(C) = \xi^2(C) = A.$$

Now we resort to 14.6.1 with

$$X_1 := \mathbb{R}, \quad \mathcal{A}_1 := \mathcal{A}(d_\mathbb{R}), \quad P_1 := P^C, \quad X_2 := \mathbb{R}, \quad \mathcal{A}_2 := \mathcal{A}(d_\mathbb{R}), \quad \pi := \psi$$

(then, $X_1 - D_\pi = (-\infty, 0)$ and hence $P_1(X_1 - D_\pi) = \mathbb{O}_\mathcal{H}$). Then the mapping

$$Q : \mathcal{A}(d_\mathbb{R}) \to \mathscr{P}(\mathcal{H})$$
$$E \mapsto Q(E) := P^C(\psi^{-1}(E))$$

is a projection valued measure on $\mathcal{A}(d_\mathbb{R})$ and

$$J_Q(\xi) = J_{P^C}(\xi \circ \psi) = J_{P^C}(\psi) = \psi(C) = A.$$

By definition of P^A, this implies $Q = P^A$, i.e.

$$P^C(\psi^{-1}(E)) = P^A(E), \forall E \in \mathcal{A}(d_\mathbb{R});$$

now,

$$P^B(E) = P^A(\varphi^{-1}(E)), \forall E \in \mathcal{A}(d_\mathbb{R})$$

(cf. 15.3.8), and hence

$$P^B(E) = P^C(\psi^{-1}(\varphi^{-1}(E))), \forall E \in \mathcal{A}(d_\mathbb{R});$$

then we note that

$$\psi^{-1}(\varphi^{-1}(E)) = E \cap [0, \infty), \forall E \in \mathcal{A}(d_\mathbb{R}),$$

and hence

$$P^C(\psi^{-1}(\varphi^{-1}(E))) = P^C(E)P^C([0, \infty)) = P^C(E), \forall E \in \mathcal{A}(d_\mathbb{R})$$

by 13.3.2c and the equality $P^C([0, \infty)) = \mathbb{1}_\mathcal{H}$. Thus, $P^B = P^C$ and hence (cf. 15.2.2)

$$B = J_\xi^{P^B} = J_\xi^{P^C} = C.$$

$\square$

15.3.10 Theorem. *Let A be a self-adjoint operator in $\mathcal{H}$ and let $\varphi : \sigma(A) \to \mathbb{C}$ be a continuous function (with respect to the metric subspace $(\sigma(A), d_{\sigma(A)})$ of the metric space $(\mathbb{R}, d_\mathbb{R})$). Then $\sigma(\varphi(A))$ is the closure of $\varphi(\sigma(A))$, i.e.*

$$\sigma(\varphi(A)) = \overline{\varphi(\sigma(A))}.$$

If A is bounded, then $\sigma(\varphi(A)) = \varphi(\sigma(A))$.

Proof. Let $z \in \mathbb{C} - \overline{\varphi(\sigma(A))}$; then there exists $\eta \in (0, \infty)$ so that

$$B(z, \eta) \subset \mathbb{C} - \overline{\varphi(\sigma(A))} \subset \mathbb{C} - \varphi(\sigma(A)),$$

and hence so that $\varphi^{-1}(B(z, \eta)) = \emptyset$, and hence so that $P^A(\varphi^{-1}(B(z, \eta))) = \mathbb{O}_{\mathcal{H}}$; in view of 14.4.2, this implies $z \notin \sigma(\varphi(A))$. Thus,

$$\sigma(\varphi(A)) \subset \overline{\varphi(\sigma(A))}.$$

Conversely, let $z \in \varphi(\sigma(A))$. Then there exists $\lambda \in \sigma(A)$ such that $z = \varphi(\lambda)$, and (since φ is continuous) $\forall \varepsilon \in (0, \infty)$, $\exists \delta \in (0, \infty)$ such that

$$[x \in \sigma(A) \text{ and } |x - \lambda| < \delta] \Rightarrow |z - \varphi(x)| < \varepsilon,$$

or $\sigma(A) \cap (\lambda - \delta, \lambda + \delta) \subset \varphi^{-1}(B(z, \varepsilon))$, and hence such that

$$P^A((\lambda - \delta, \lambda + \delta)) \overset{(1)}{=} P^A(\sigma(A) \cap (\lambda - \delta, \lambda + \delta)) \overset{(2)}{\le} P^A(\varphi^{-1}(B(z, \varepsilon))),$$

where 1 holds by 13.3.2c (since $P^A(\sigma(A)) = \mathbb{1}_{\mathcal{H}}$, cf. 15.2.2d) and 2 holds by 13.3.2e; now, $\lambda \in \sigma(A)$ implies $P^A((\lambda - \delta, \lambda + \delta)) \ne \mathbb{O}_{\mathcal{H}}$ (cf. 15.2.4). Therefore

$$P^A(\varphi^{-1}(B(z, \varepsilon))) \ne \mathbb{O}_{\mathcal{H}}, \forall \varepsilon \in (0, \infty),$$

and hence $z \in \sigma(\varphi(A))$ by 14.4.2. Thus,

$$\varphi(\sigma(A)) \subset \sigma(\varphi(A)),$$

and hence

$$\overline{\varphi(\sigma(A))} \subset \sigma(\varphi(A))$$

since $\sigma(\varphi(A))$ is a closed subset of $\mathbb{C}$ (cf. 10.4.6). This concludes the proof of the first equation of the statement.

Finally, suppose that A is bounded. Then $\sigma(A)$ is bounded (cf. 15.2.2f) and hence it is a compact subset of $\mathbb{R}$ (cf. 10.4.6 and 2.8.7). Then $\varphi(\sigma(A))$ is a compact subset of $\mathbb{C}$ (cf. 2.8.12) and hence it is closed (cf. 2.8.6), i.e. such that $\varphi(\sigma(A)) = \overline{\varphi(\sigma(A))}$. $\qquad\square$

15.4 Unitary equivalence

15.4.1 Theorem. *Let $\mathcal{H}_1$ and $\mathcal{H}_2$ be isomorphic Hilbert spaces and suppose that $U \in \mathcal{UA}(\mathcal{H}_1, \mathcal{H}_2)$. Let A_1 and A_2 be self-adjoint operators in $\mathcal{H}_1$ and in $\mathcal{H}_2$ respectively. Then the following conditions are equivalent:*

(a) $P^{A_2}(E) = U P^{A_1}(E) U^{-1}, \forall E \in \mathcal{A}(d_{\mathbb{R}})$;
(b) $A_2 = U A_1 U^{-1}$.

Proof. $a \Rightarrow b$: This follows immediately from 14.6.2.

$b \Rightarrow a$: We define the mapping

$$Q : \mathcal{A}(d_{\mathbb{R}}) \to \mathscr{P}(\mathcal{H}_2)$$
$$E \mapsto Q(E) := U P^{A_1}(E) U^{-1}.$$

Then (cf. 14.6.2) Q is a projection valued measure on $\mathcal{A}(d_{\mathbb{R}})$ and

$$J_\xi^Q = U J_\xi^{P^{A_1}} U^{-1} = U A_1 U^{-1} = A_2.$$

Thus, $Q = P^{A_2}$ by the definition of P^{A_2}, and hence condition a. $\qquad\square$

Chapter 16

One-Parameter Unitary Groups and Stone's Theorem

The subject of this chapter is fundamental for quantum mechanics. Indeed, continuous one-parameter unitary groups and Stone's theorem are the mathematical basis for the description of time evolution of conservative and reversible quantum systems (cf. Section 19.6).

Moreover, if G is a Lie group which is considered to be a symmetry group for a quantum system, then a continuous one-parameter unitary group is found to be associated with each element of the Lie algebra of G, and the generators of these one-parameter groups are self-adjoint operators which are interpreted as observables representing the elements of the Lie algebra. However, this topic is outside the scope of this book (cf. e.g. Thaller, 1992, 2.3.1).

16.1 Continuous one-parameter unitary groups

Throughout this section, $\mathcal{H}$ denotes an abstract Hilbert space. We recall that $\mathcal{U}(\mathcal{H})$ denotes the group of unitary operators in $\mathcal{H}$ (cf. 10.3.9 and 10.3.10).

16.1.1 Definition. A *continuous one-parameter unitary group* (briefly, a *c.o.p.u.g.*) in $\mathcal{H}$ is a mapping

$$U : \mathbb{R} \to \mathcal{U}(\mathcal{H})$$

such that:

(ug_1) U is homomorphism from the additive group $\mathbb{R}$ to the group $\mathcal{U}(\mathcal{H})$, i.e.

$$U(t_1)U(t_2) = U(t_1 + t_2), \forall t_1, t_2 \in \mathbb{R};$$

(ug_2) the mapping

$$\mathbb{R} \ni t \mapsto U_f(t) := U(t)f \in \mathcal{H}$$

is continuous, for all $f \in \mathcal{H}$.

From condition ug_1 we have (cf. 1.3.3 and 12.5.1b)

$$U(0) = \mathbb{1}_{\mathcal{H}} \text{ and } U(-t) = U(t)^{-1} = U(t)^{\dagger}, \forall t \in \mathbb{R};$$

we also have (cf. 1.3.5b)

$$U(t_1)U(t_2) = U(t_2)U(t_1), \forall t_1, t_2 \in \mathbb{R}.$$

Obviously, $U_f(0) = f$ for all $f \in \mathcal{H}$.

16.1.2 Proposition. *Let $U : \mathbb{R} \to \mathcal{U}(\mathcal{H})$ be a homomorphism from $\mathbb{R}$ to $\mathcal{U}(\mathcal{H})$ (i.e., condition ug_1 holds true). Then the following conditions are equivalent:*

(a) U is a c.o.p.u.g. (i.e., condition ug_2 holds true);
(b) the function

$$\mathbb{R} \ni t \mapsto (U(t)f|g) \in \mathbb{C}$$

is continuous, for all $f, g \in \mathcal{H}$;
(c) there exists $t_0 \in \mathbb{R}$ such that the function

$$\mathbb{R} \ni t \mapsto (U(t)f|g) \in \mathbb{C}$$

is continuous at t_0, for all $f, g \in \mathcal{H}$;
(d) there exists $t_0 \in \mathbb{R}$ such that the mapping U_f is continuous at t_0, for all $f \in \mathcal{H}$.

Proof. $a \Rightarrow b$: This is obvious, in view of 10.1.16c.

$b \Rightarrow c$: This is obvious, by definition of a continuous function (cf. 2.4.1).

$c \Rightarrow d$: Assuming condition c, fix $f \in \mathcal{H}$ and let $\{t_n\}$ be a sequence in $\mathbb{R}$ such that $t_n \to t_0$; then

$$\begin{aligned}
\|U_f(t_n) - U_f(t_0)\|^2 &= \|U(t_n)f - U(t_0)f\|^2 \\
&= 2\|f\|^2 - 2\operatorname{Re}(U(t_n)f|U(t_0)f) \\
&\to 2\|f\|^2 - 2\operatorname{Re}(U(t_0)f|U(t_0)f) = 0.
\end{aligned}$$

This proves that condition d is true (cf. 2.4.2).

$d \Rightarrow a$: Assuming condition d, fix $f \in \mathcal{H}$ and $t \in \mathbb{R}$, and let $\{t_n\}$ be a sequence in $\mathbb{R}$ such that $t_n \to t$; then $t_0 - t + t_n \to t_0$ and hence (cf. 2.4.2)

$$\|U_f(t_n) - U_f(t)\| = \|U(t_0 - t)(U(t_n)f - U(t)f)\| = \|U_f(t_0 - t + t_n) - U_f(t_0)\| \to 0.$$

This proves that condition a is true (cf. 2.4.2), since f and t were arbitrary. $\qquad \square$

16.1.3 Definitions. Let X be a normed space.

For a mapping $\psi : \mathbb{R} - \{0\} \to X$, we say that $\lim_{t \to 0} \psi(t)$ *exists* if there exists $f \in X$ so that

$$[\{t_n\} \text{ is a sequence in } \mathbb{R} - \{0\}, t_n \to 0] \Rightarrow \|\psi(t_n) - f\| \to 0;$$

if this condition is true then we define

$$\lim_{t \to 0} \psi(t) := f$$

(it is immediate to see that, if f as above exists, then it is unique).

For a mapping $\varphi : \mathbb{R} \to X$ and $t_0 \in \mathbb{R}$, we say that φ *is differentiable at* t_0 if

$$\lim_{t \to 0} \frac{1}{t}(\varphi(t_0 + t) - \varphi(t_0)) \text{ exists;}$$

if this condition is true then we define the *derivative of* φ *at* t_0 as

$$\frac{d\varphi}{dt}\bigg|_{t_0} := \lim_{t \to 0} \frac{1}{t}(\varphi(t_0 + t) - \varphi(t_0)).$$

More formally, we define the function

$$\mathbb{R} - \{0\} \ni t \mapsto \psi_{\varphi,t_0}(t) := \frac{1}{t}(\varphi(t_0 + t) - \varphi(t_0)) \in X$$

and we say that φ is differentiable at t_0 if $\lim_{t\to 0} \psi_{\varphi,t_0}(t)$ exists, in which case we write

$$\frac{d\varphi}{dt}\bigg|_{t_0} := \lim_{t \to 0} \psi_{\varphi,t_0}(t).$$

Obviously, if $X = \mathbb{C}$ then these definitions agree with the ones given in 1.2.21 (cf. 2.7.6 and 2.4.2).

16.1.4 Proposition. *Let* $U : \mathbb{R} \to \mathcal{U}(\mathcal{H})$ *be a homomorphism from* $\mathbb{R}$ *to* $\mathcal{U}(\mathcal{H})$ *(i.e., condition ug$_1$ holds true). Then, for* $f \in \mathcal{H}$*, the following conditions are equivalent:*

(a) the mapping U_f *is differentiable at* t_0*, for all* $t_0 \in \mathbb{R}$*;*
(b) the mapping U_f *is differentiable at* 0*.*

If these conditions hold true, then the following condition also holds true:

(c) for each $t_0 \in \mathbb{R}$*, if* $g := U(t_0)f$ *then the mapping* U_g *is differentiable at* 0 *and*

$$\frac{dU_g}{dt}\bigg|_0 = \frac{dU_f}{dt}\bigg|_{t_0} = U(t_0)\left(\frac{dU_f}{dt}\bigg|_0\right).$$

Proof. $a \Rightarrow b$: This is obvious.

$b \Rightarrow (a$ and $c)$: We assume condition b and fix $t_0 \in \mathbb{R}$. Then, if $\{t_n\}$ is a sequence in $\mathbb{R} - \{0\}$ such that $t_n \to 0$, we have

$$\left\|\frac{1}{t_n}(U_f(t_0 + t_n) - U_f(t_0)) - U(t_0)\left(\frac{dU_f}{dt}\bigg|_0\right)\right\|$$

$$= \left\|\frac{1}{t_n}(U_f(t_n) - U_f(0)) - \frac{dU_f}{dt}\bigg|_0\right\| \to 0.$$

This shows that the mapping U_f is differentiable at t_0 and also that

$$\frac{dU_f}{dt}\bigg|_{t_0} = U(t_0)\left(\frac{dU_f}{dt}\bigg|_0\right).$$

Next, we note that

$$U_f(t_0 + t) - U_f(t_0) = U(t)U(t_0)f - U(t_0)f = U_g(t) - U_g(0), \forall t \in \mathbb{R};$$

then, from above we have

$$\left\| \frac{1}{t_n}(U_g(t_n) - U_g(0)) - U(t_0)\left(\left.\frac{dU_f}{dt}\right|_0\right) \right\| \to 0,$$

and this shows that the mapping U_g is differentiable at 0 and also that

$$\left.\frac{dU_g}{dt}\right|_0 = U(t_0)\left(\left.\frac{dU_f}{dt}\right|_0\right).$$

Since t_0 was arbitrary, this proves that condition b implies conditions a and c. $\square$

16.1.5 Proposition. *Suppose that, for a self-adjoint operator A in $\mathcal{H}$ and a c.o.p.u.g. U in $\mathcal{H}$, the following condition holds true:*

(sa-ug) the mapping U_f is differentiable at 0 and $\left.\frac{dU_f}{dt}\right|_0 = iAf$, $\forall f \in D_A$.

Then the following conditions also hold true:

(a) $D_A = \{f \in \mathcal{H} : U_f \text{ is differentiable at } 0\}$;

(b) U_f is differentiable at t_0 , $U_f(t_0) \in D_A$, $\left.\frac{dU_f}{dt}\right|_{t_0} = iAU_f(t_0)$,

   $\forall f \in D_A, \forall t_0 \in \mathbb{R}$;

(c) $U(t_0)A \subset AU(t_0)$, $\forall t_0 \in \mathbb{R}$.

Moreover:

(d) A is the only self-adjoint operator in $\mathcal{H}$ which satisfies condition sa-ug with U;

(e) U is the only c.o.p.u.g. in $\mathcal{H}$ which satisfies condition sa-ug with A.

Proof. a: We define the set

$$D := \{g \in \mathcal{H} : U_g \text{ is differentiable at } 0\}.$$

Condition sa-ug implies $D_A \subset D$. Now let $g \in D$ and let $\{t_n\}$ be a sequence in $\mathbb{R} - \{0\}$ such that $t_n \to 0$; then, by 10.1.16c and by condition sa-ug, we have

$$-i\left(f \left|\left.\frac{dU_g}{dt}\right|_0\right.\right) = -i\left(f\left|\lim_{n\to\infty}\frac{1}{t_n}(U_g(t_n) - U_g(0))\right.\right)$$

$$= -i\lim_{n\to\infty}\left(\frac{1}{t_n}(U(t_n)^\dagger - \mathbb{1}_\mathcal{H})f|g\right)$$

$$= i\left(\lim_{n\to\infty}\frac{1}{-t_n}(U(-t_n)f - f)|g\right)$$

$$= \left(-i\left.\frac{dU_f}{dt}\right|_0 |g\right) = (Af|g), \forall f \in D_A;$$

by the very definition of $D_{A^\dagger}$ (cf. 12.1.1), this shows that $g \in D_{A^\dagger}$. Thus we have $D \subset D_{A^\dagger}$, i.e. $D \subset D_A$ since A is self-adjoint, and hence $D_A = D$, which is condition a.

b and c: We fix $f \in D_A$ and $t_0 \in \mathbb{R}$. From condition sa-ug we have, by 16.1.4:

$$U_f \text{ is differentiable at } t_0 \quad \text{and} \quad \left.\frac{dU_f}{dt}\right|_{t_0} = U(t_0)\left(\left.\frac{dU_f}{dt}\right|_0\right) = U(t_0)(iAf),$$

and also, letting $g := U(t_0)f$,

$$U_g \text{ is differentiable at } 0 \quad \text{and} \quad \frac{dU_g}{dt}\bigg|_0 = U(t_0)(iAf).$$

Since U_g is differentiable at 0, in view of condition a already proved we have

$$U_f(t_0) = U(t_0)f = g \in D_A,$$

and hence, in view of condition sa-ug,

$$\frac{dU_g}{dt}\bigg|_0 = iAg = iAU(t_0)f.$$

Then we have

$$\frac{dU_f}{dt}\bigg|_{t_0} = U(t_0)(iAf) = \frac{dU_g}{dt}\bigg|_0 = iAU(t_0)f = iAU_f(t_0).$$

Since $f \in D_A$ and $t_0 \in \mathbb{R}$ were arbitrary, this proves condition b and also condition c, since condition c can be written as follows:

$$U(t_0)f \in D_A \text{ and } U(t_0)Af = AU(t_0)f, \forall f \in D_A, \forall t_0 \in \mathbb{R}.$$

d: Suppose that B is a self-adjoint operator in $\mathcal{H}$ which satisfies condition sa-ug with U. Then $D_B \subset D_A$ in view of condition a already proved, and hence $B \subset A$ since

$$Bf := -i\,\frac{dU_f}{dt}\bigg|_0 = Af, \forall f \in D_B.$$

Then, $B = A$ by 12.4.6b.

e: Suppose that V is a c.o.p.u.g. in $\mathcal{H}$ which satisfies condition sa-ug with A. Then A and V satisfy condition b as well. Then fix $f \in D_A$, let t_0 be an arbitrary element of $\mathbb{R}$, and let $\{t_n\}$ be a sequence in $\mathbb{R} - \{0\}$ such that $t_n \to 0$. The equation

$$\|U_f(t_0 + t_n) - V_f(t_0 + t_n)\|^2 - \|U_f(t_0) - V_f(t_0)\|^2$$
$$= (U_f(t_0 + t_n) - V_f(t_0 + t_n) - U_f(t_0) + V_f(t_0)|U_f(t_0 + t_n) - V_f(t_0 + t_n))$$
$$+ (U_f(t_0) - V_f(t_0)|U_f(t_0 + t_n) - V_f(t_0 + t_n) - U_f(t_0) + V_f(t_0)), \forall n \in \mathbb{N},$$

implies, in view of condition b for U and for V (also, cf. 10.1.16),

$$\lim_{n \to \infty} \frac{1}{t_n}(\|U_f(t_0 + t_n) - V_f(t_0 + t_n)\|^2 - \|U_f(t_0) - V_f(t_0)\|^2)$$
$$= (iAU_f(t_0) - iAV_f(t_0)|U_f(t_0) - V_f(t_0))$$
$$+ (U_f(t_0) - V_f(t_0)|iAU_f(t_0) - iAV_f(t_0)) = 0.$$

This shows that the function

$$\mathbb{R} \ni t \mapsto \|U_f(t) - V_f(t)\|^2 \in \mathbb{R}$$

is differentiable at every point of $\mathbb{R}$ and that its derivative is zero at every point of $\mathbb{R}$. Therefore, this function is a constant function and hence

$$\|U_f(t) - V_f(t)\| = \|U_f(0) - V_f(0)\| = \|f - f\| = 0, \forall t \in \mathbb{R}.$$

Since f was an arbitrary element of D_A, this proves that

$$U(t)f = V(t)f, \forall f \in D_A, \forall t \in \mathbb{R}.$$

Now, for any $t \in \mathbb{R}$, $U(t)$ and $V(t)$ are elements of $\mathcal{B}(\mathcal{H})$ and $\overline{D_A} = \mathcal{H}$. Then,

$$U(t) = V(t), \forall t \in \mathbb{R},$$

by the uniqueness asserted in 4.2.6. $\qquad\square$

16.1.6 Theorem. *Let A be a self-adjoint operator in $\mathcal{H}$ and let U^A be the mapping defined by*

$$U^A : \mathbb{R} \to \mathcal{U}(\mathcal{H})$$
$$t \mapsto U(t) := \varphi_t(A),$$

where φ_t is the function

$$\mathbb{R} \ni x \mapsto \varphi_t(x) := e^{itx} \in \mathbb{C},$$

for all $t \in \mathbb{R}$ (we recall that $\varphi_t(A) = J_{\varphi_t}^{P^A} = J_{P^A}(\varphi_t)$, cf. 15.3.1 and 14.2.18).
   Then, U^A is a c.o.p.u.g. and condition sa-ug holds true for A and U^A, i.e.

(sa-ug) the mapping U_f^A is differentiable at 0 and $\left.\dfrac{dU_f^A}{dt}\right|_0 = iAf$, $\forall f \in D_A$

(we set $U_f^A := (U^A)_f$).

Proof. First we prove that the mapping U^A is a c.o.p.u.g.

We have $U^A(t) \in \mathcal{U}(\mathcal{H})$ for all $t \in \mathbb{R}$, in view of 14.3.18. We also have, in view of 14.3.10,

$$U^A(t_1)U^A(t_2) = J_{P^A}(\varphi_{t_1})J_{P^A}(\varphi_{t_2}) = J_{P^A}(\varphi_{t_1}\varphi_{t_2})$$
$$= J_{P^A}(\varphi_{t_1+t_2}) = U^A(t_1 + t_2), \forall t_1, t_2 \in \mathbb{R}.$$

Thus, U^A has property gu_1. Finally, for each $t \in \mathbb{R}$ and every sequence $\{t_n\} \in \mathbb{R}$ such that $t_n \to t$, we have

$$\lim_{n\to\infty} |\varphi_{t_n}(x) - \varphi_t(x)| = 0, \forall x \in \mathbb{R},$$
$$|\varphi_{t_n}(x) - \varphi_t(x)| \le 2, \forall x \in \mathbb{R}, \forall n \in \mathbb{N};$$

then we have, for all $f \in \mathcal{H}$,

$$\|U_f^A(t_n) - U_f^A(t)\|^2 = \|J_{P^A}(\varphi_{t_n} - \varphi_t)f\|^2 = \int_{\mathbb{R}} |\varphi_{t_n} - \varphi_t|^2 d\mu_f^{P^A} \to 0,$$

by 14.3.1, 14.2.14d, 8.2.11 (we recall that a constant function is $\mu_f^{P^A}$-integrable, since the measure $\mu_f^{P^A}$ is finite). This proves that U^A has property ug_2.

Next we prove that A and U^A satisfy condition sa-ug.

Let $\{t_n\}$ be any sequence in $\mathbb{R} - \{0\}$ such that $t_n \to 0$. Then we have, by 14.3.5, 14.3.1, 14.2.14d,

$$\left\| \frac{1}{t_n}(U_f^A(t_n) - U_f^A(0)) - iAf \right\|^2$$

$$= \left\| \frac{1}{t_n}(J_{P^A}(\varphi_{t_n}) - J_{P^A}(1_{\mathbb{R}}))f - iJ_{P^A}(\xi)f \right\|^2$$

$$= \left\| J_{P^A}\left(\frac{1}{t_n}(\varphi_{t_n} - 1_{\mathbb{R}}) - i\xi \right)f \right\|^2$$

$$= \int_{\mathbb{R}} \left| \frac{1}{t_n}(\varphi_{t_n} - 1_{\mathbb{R}}) - i\xi \right|^2 d\mu_f^{P^A}, \forall f \in D_A, \forall n \in \mathbb{N}.$$

We also have

$$\lim_{n\to\infty} \left| \frac{1}{t_n}(\varphi_{t_n}(x) - 1) - ix \right| = 0, \forall x \in \mathbb{R}$$

(since $\frac{de^{itx}}{dt}\big|_0 = ix, \forall x \in \mathbb{R}$), and

$$\left| \frac{1}{t_n}(\varphi_{t_n}(x) - 1) - ix \right| \le \left| \frac{1}{t_n}(e^{it_n x} - 1) \right| + |x| \le 2|x|, \forall x \in \mathbb{R}, \forall n \in \mathbb{N}$$

(we have used the inequality $|e^{i\alpha} - 1| \le |\alpha|, \forall \alpha \in \mathbb{R}$). Since $\xi \in \mathcal{L}^2(\mathbb{R}, \mathcal{A}(d_\mathbb{R}), \mu_f^{P^A})$ for all $f \in D_A$, by 8.2.11 (with $4\xi^2$ as dominating function) we have

$$\lim_{n\to\infty} \left\| \frac{1}{t_n}(U_f^A(t_n) - U_f^A(0)) - iAf \right\| = 0, \forall f \in D_A.$$

This proves condition sa-ug. □

16.1.7 Remark. For every self-adjoint operator A in $\mathcal{H}$, 16.1.6 and 16.1.5e show that the mapping U^A defined by

$$\mathbb{R} \ni t \mapsto U^A(t) := \varphi_t(A) \in \mathcal{U}(\mathcal{H})$$

is a c.o.p.u.g. and that it is the only c.o.p.u.g. U in $\mathcal{H}$ which satisfies with A the condition

(sa-ug) the mapping U_f is differentiable at 0 and $\frac{dU_f}{dt}\big|_0 = iAf$, $\forall f \in D_A$.

We point out that, for each $t \in \mathbb{R}$, $U^A(t)$ is the unique linear operator in $\mathcal{H}$ such that $D_{U^A(t)} = \mathcal{H}$ and

$$(f|U^A(t)f) = \int_\mathbb{R} \varphi_t d\mu_f^{P^A}, \forall f \in \mathcal{H}$$

(cf. 14.2.14). The operator $U^A(t)$ is often denoted as e^{itA}.

Finally, 16.1.5d shows that the mapping

$$A \mapsto U^A,$$

from the family of all self-adjoint operators in $\mathcal{H}$ to the family of all c.o.p.u.g.'s in $\mathcal{H}$, is injective.

16.1.8 Proposition. *Let A be a self-adjoint operator in $\mathcal{H}$.*

(a) Let $\lambda \in \mathbb{R}$. Then, the operator $B := A + \lambda 1_\mathcal{H}$ is self-adjoint and the following conditions are true:

$$P^B(E) = P^A(E - \lambda), \forall E \in \mathcal{A}(d_\mathbb{R})$$

(we recall that $E - \lambda := \{x - \lambda : x \in E\}$, cf. 9.2.1a);

$$U^B(t) = e^{i\lambda t}U^A(t), \forall t \in \mathbb{R}.$$

(b) Let $\mu \in \mathbb{R} - \{0\}$. Then, the operator $C := \mu A$ is self-adjoint and the following conditions are true:

$$P^C(E) = P^A(\mu^{-1}E), \forall E \in \mathcal{A}(d_\mathbb{R})$$

(we recall that $\mu^{-1}E := \{\mu^{-1}x : x \in E\}$, cf. 9.2.2a);

$$U^C(t) = U^A(\mu t), \forall t \in \mathbb{R}.$$

Proof. a: If we write $\psi_\lambda := \xi + \lambda$, then (cf. 15.3.5)

$$B = \psi_\lambda(A).$$

Hence, in view of 15.3.8, the operator B is self-adjoint and

$$P^B(E) = P^A(\psi_\lambda^{-1}(E)) = P^A(E - \lambda), \forall E \in \mathcal{A}(d_\mathbb{R}).$$

For each $t \in \mathbb{R}$, we can use 14.6.1, with

$$X_1 := \mathbb{R}, \quad \mathcal{A}_1 := \mathcal{A}(d_\mathbb{R}), \quad P_1 := P^A, \quad X_2 := \mathbb{R}, \quad \mathcal{A}_2 := \mathcal{A}(d_\mathbb{R}), \quad \pi := \psi_\lambda,$$

and hence $P_2 = P^B$,

to obtain

$$U^B(t) = J_{P^B}(\varphi_t) = J_{P^A}(\varphi_t \circ \psi_\lambda) = J_{P^A}\left(e^{it\lambda}\varphi_t\right) = e^{i\lambda t}J_{P^A}(\varphi_t) = e^{i\lambda t}U^A(t)$$

(we have used 16.1.6 and 14.3.5).

b: If we write $\gamma_\mu := \mu\xi$, then (cf. 15.3.5)

$$C = \gamma_\mu(A).$$

Hence, in view of 15.3.8, the operator C is self-adjoint and

$$P^C(E) = P^A(\gamma_\mu^{-1}(E)) = P^A(\mu^{-1}E), \forall E \in \mathcal{A}(d_\mathbb{R}).$$

For each $t \in \mathbb{R}$, we can use 14.6.1, with

$$X_1 := \mathbb{R}, \quad \mathcal{A}_1 := \mathcal{A}(d_\mathbb{R}), \quad P_1 := P^A, \quad X_2 := \mathbb{R}, \quad \mathcal{A}_2 := \mathcal{A}(d_\mathbb{R}), \quad \pi := \gamma_\mu,$$

and hence $P_2 = P^C$,

to obtain

$$U^C(t) = J_{P^C}(\varphi_t) = J_{P^A}(\varphi_t \circ \gamma_\mu) = J_{P^A}(\varphi_{\mu t}) = U^A(\mu t)$$

(we have used 16.1.6). $\qquad\square$

16.1.9 Theorem. *Let U be a c.o.p.u.g. in $\mathcal{H}$ and let D be a linear manifold in $\mathcal{H}$ such that:*

(a) $\overline{D} = \mathcal{H}$;
(b) $U(t)f \in D$, $\forall f \in D$, $\forall t \in \mathbb{R}$;
(c) the mapping U_f is differentiable at 0, $\forall f \in D$.

Then the mapping

$$A_0 : D \to \mathcal{H}$$

$$f \mapsto A_0 f := -i \left. \frac{dU_f}{dt} \right|_0$$

is an essentially self-adjoint operator (hence, A_0 is closable and the operator $\overline{A_0}$ is self-adjoint) and, letting $A := \overline{A_0}$, $U = U^A$.

Proof. In view of 10.1.16a,b, it is easy to see that the mapping A_0 is a linear operator.

Let $\{t_n\}$ be a sequence in $\mathbb{R} - \{0\}$ such that $t_n \to 0$; then, by 10.1.16c,

$$(A_0 f | g) = i \left(\lim_{n \to \infty} \frac{1}{t_n} (U_f(t_n) - U_f(0)) | g \right) = i \lim_{n \to \infty} \left(\frac{1}{t_n} (U(t_n) f - f) | g \right)$$

$$= i \lim_{n \to \infty} \left(f | \frac{1}{t_n} (U(t_n)^\dagger - \mathbb{1}_{\mathcal{H}}) g \right) = -i \lim_{n \to \infty} \left(f | \frac{1}{-t_n} (U_g(-t_n) - U_g(0)) \right)$$

$$= (f | A_0 g), \forall f, g \in D;$$

Since the operator A_0 is adjointable (cf. condition a), this proves that A_0 is symmetric (cf. 12.4.3).

Now let $h \in N_{A_0^\dagger - i \mathbb{1}_{\mathcal{H}}}$, i.e. $h \in D_{A_0^\dagger}$ and $A_0^\dagger h = ih$, and fix $f \in D$. For any $t_0 \in \mathbb{R}$, letting $g := U(t_0) f$ we have $g \in D$ by condition b; hence, U_g is differentiable at 0 by condition c and

$$\left. \frac{dU_g}{dt} \right|_0 = i A_0 g$$

by definition of A_0; moreover, in view of 16.1.4, we have that U_f is differentiable at t_0 and

$$\left. \frac{dU_f}{dt} \right|_{t_0} = \left. \frac{dU_g}{dt} \right|_0 ;$$

then, for every sequence $\{t_n\}$ in $\mathbb{R} - \{0\}$ such that $t_n \to 0$, we have

$$\lim_{n \to \infty} \frac{1}{t_n} ((U(t_0 + t_n) f | h) - (U(t_0) f | h)) = \left(\left. \frac{dU_f}{dt} \right|_{t_0} | h \right)$$

$$= (i A_0 U(t_0) f | h) = -i \left(U(t_0) f | A_0^\dagger h \right) = (U(t_0) f | h) .$$

This shows that the function

$$\mathbb{R} \ni t \mapsto \psi_{h,f}(t) := (U(t) f | h) \in \mathbb{C}$$

is differentiable at t_0 for all $t_0 \in \mathbb{R}$, and that it satisfies the differential equation

$$\psi'_{h,f}(t) = \psi_{h,f}(t), \forall t \in \mathbb{R}.$$

Then, there exists $k \in \mathbb{C}$ so that

$$\psi_{h,f}(t) = k e^t, \forall t \in \mathbb{R};$$

now, by the Schwarz inequality,

$$|\psi_{h,f}(t)| \le \|f\|\|h\|, \forall t \in \mathbb{R},$$

and hence $k = 0$, and hence $\psi_{h,f}(0) = 0$, i.e.

$$(f|h) = 0.$$

Since f was an arbitrary element of D, this proves that $h \in D^\perp$, and hence (cf. condition a and 10.4.4d) that $h = 0_{\mathcal{H}}$. This proves that

$$N_{A_0^\dagger - i\mathbb{1}_{\mathcal{H}}} = \{0_{\mathcal{H}}\},$$

The equation $N_{A_0^\dagger + i\mathbb{1}_{\mathcal{H}}} = \{0_{\mathcal{H}}\}$ can be proved in a similar way.

Thus, the operator A_0 is essentially self-adjoint by 12.4.17, and hence (cf. 12.4.11) A_0 is closable and the operator $A := \overline{A_0}$ is self-adjoint.

For every $f \in D$ and every $t_0 \in \mathbb{R}$, we have already seen that U_f is differentiable at t_0 and

$$U(t_0)f \in D \quad \text{and} \quad \left.\frac{dU_f}{dt}\right|_{t_0} = iA_0 U(t_0)f;$$

since $A_0 \subset A$, this yields

$$U(t_0)f \in D_A \quad \text{and} \quad \left.\frac{dU_f}{dt}\right|_{t_0} = iAU(t_0)f;$$

since $f \in D_A$, we also have (cf. 16.1.6 and 16.1.5b) that U_f^A is differentiable at t_0 and

$$U^A(t_0)f \in D_A \quad \text{and} \quad \left.\frac{dU_f^A}{dt}\right|_{t_0} = iAU^A(t_0)f.$$

Then, proceeding exactly as in the proof of 16.1.5e (with D_A replaced by D and V replaced by U^A), we have

$$U(t)f = U^A(t)f, \forall f \in D, \forall t \in \mathbb{R},$$

and hence

$$U(t) = U^A(t), \forall t \in \mathbb{R},$$

by the uniqueness asserted in 4.2.6, since $U(t)$ and $V(t)$ are elements of $\mathcal{B}(\mathcal{H})$ for all $t \in \mathbb{R}$ and since $\overline{D} = \mathcal{H}$. $\square$

16.1.10 Theorem (Stone's theorem). *Let U be a c.o.p.u.g. in $\mathcal{H}$. Then there exists a self-adjoint operator A in $\mathcal{H}$ such that $U = U^A$.*

Proof. First we define a symbol. For $\varphi \in C(\mathbb{R})$ and $a, b \in \mathbb{R}$, if $a < b$ then $\int_a^b \varphi(x)dx$ denotes the Riemann integral (cf. 9.3.2) of the restriction of φ to the interval $[a, b]$; otherwise, we define:

$$\int_a^b \varphi(x)dx := -\int_b^a \varphi(x)dx \quad \text{if } b < a;$$
$$\int_a^b \varphi(x)dx := 0 \qquad \text{if } a = b.$$

Now we come to the proof of the theorem.

For all $f, g \in \mathcal{H}$, the function

$$\mathbb{R} \ni x \mapsto (U(x)f|g) \in \mathbb{C}$$

is continuous (cf. 16.1.2). Thus, for each $t \in \mathbb{R} - \{0\}$, we can define the function

$$\mathcal{H} \times \mathcal{H} \ni (f, g) \mapsto \psi_t(f, g) := \frac{1}{t} \int_0^t (U(x)f|g)\, dx \in \mathbb{C},$$

which is clearly a sesquilinear form; moreover, by the Schwarz inequality, we have

$$|\psi_t(f, g)| \leq \frac{1}{t} \int_0^t |(U(x)f|g)|\, dx \leq \|f\|\|g\|, \forall f, g \in \mathcal{H}, \tag{1}$$

and this proves that the sesquilinear form is bounded; then, by 10.5.6,

$$\exists! B_t \in \mathcal{B}(\mathcal{H}) \text{ such that } (B_t f|g) = \frac{1}{t} \int_0^t (U(x)f|g)\, dx, \forall f, g \in \mathcal{H};$$

in view of 1 and of 10.1.14, we have $\|B_t\| \leq 1$, and hence (cf. 4.2.5b)

$$\|B_t f\| \leq \|f\|, \forall f \in \mathcal{H}. \tag{2}$$

For each $f \in \mathcal{H}$, by Riemann's fundamental theorem of calculus the function

$$\mathbb{R} \ni t \mapsto \int_0^t (U(x)f|f)\, dx \in \mathbb{C}$$

is differentiable at 0 and its derivative at 0 is the number $(U(0)f|f)$; hence, for any sequence $\{t_n\}$ in $\mathbb{R} - \{0\}$ such that $t_n \to 0$, we have

$$\|f\|^2 = (U(0)f|f) = \lim_{n \to \infty} \frac{1}{t_n} \int_0^{t_n} (U(x)f|f)\, dx = \lim_{n \to \infty} (B_{t_n} f|f), \tag{3}$$

and hence also

$$\|f\|^2 = \lim_{n \to \infty} \overline{(B_{t_n} f|f)} = \lim_{n \to \infty} (f|B_{t_n} f); \tag{4}$$

now, in view of 2 we have

$$\|B_{t_n} f - f\|^2 = \|B_{t_n} f\|^2 + \|f\|^2 - (B_{t_n} f|f) - (f|B_{t_n} f)$$
$$\leq 2\|f\|^2 - (B_{t_n} f|f) - (f|B_{t_n} f), \forall n \in \mathbb{N},$$

and hence, in view of 3 and 4,

$$\lim_{n \to \infty} \|B_{t_n} f - f\| = 0.$$

This proves that

$$\lim_{t \to 0} B_t f \text{ exists} \quad \text{and} \quad \lim_{t \to 0} B_t f = f, \forall f \in \mathcal{H}. \tag{5}$$

Now we define the set

$$S := \bigcup_{t \in \mathbb{R} - \{0\}} R_{B_t}.$$

If $f \in S^{\perp}$ then, for a sequence $\{t_n\}$ in $\mathbb{R} - \{0\}$ such that $t_n \to 0$, in view of 5 we have

$$(f|f) = \left(f \middle| \lim_{n \to \infty} B_{t_n} f\right) = \lim_{n \to \infty} (f|B_{t_n} f) = 0,$$

and hence $f = 0_{\mathcal{H}}$. This proves that

$$S^{\perp} = \{0_{\mathcal{H}}\}. \tag{6}$$

For each $t \in \mathbb{R} - \{0\}$, we define the operator

$$C_t := \frac{1}{t}(U(t) - \mathbb{1}_{\mathcal{H}}).$$

For all $f, g \in \mathcal{H}$ and all $s, t \in \mathbb{R} - \{0\}$, we have

$$
\begin{aligned}
(C_t B_s f | g) &= \left(B_s f | C_t^{\dagger} g\right) = \frac{1}{s} \int_0^s \left(U(x) f | C_t^{\dagger} g\right) dx = \frac{1}{s} \int_0^s (C_t U(x) f | g) \, dx \\
&= \frac{1}{st} \int_0^s (U(t+x) f | g) \, dx - \frac{1}{st} \int_0^s (U(x) f | g) \, dx \\
&\overset{(7)}{=} \frac{1}{st} \int_t^{s+t} (U(x) f | g) \, dx - \frac{1}{st} \int_0^t (U(x) f | g) \, dx + \frac{1}{st} \int_s^t (U(x) f | g) \, dx \\
&= \frac{1}{st} \int_s^{s+t} (U(x) f | g) \, dx - \frac{1}{st} \int_0^t (U(x) f | g) \, dx \\
&\overset{(8)}{=} \frac{1}{st} \int_0^t (U(x+s) f | g) \, dx - \frac{1}{st} \int_0^t (U(x) f | g) \, dx \\
&= \frac{1}{t} \int_0^t \left(U(x) \frac{1}{s}(U(s) - \mathbb{1}_{\mathcal{H}}) f | g\right) dx = (B_t C_s f | g),
\end{aligned}
$$

where 7 and 8 are changes of variables for the Riemann integrals; they can be justified on the basis of 9.3.3, 8.3.2 and 9.2.1b; for 7 we use the translation $x \mapsto x - t$ (note that $\chi_{[0,s]}(x - t) = \chi_{[t,s+t]}(x)$ if $0 < s$ and $\chi_{[s,0]}(x - t) = \chi_{[s+t,t]}(x)$ if $s < 0$) and for 8 we use the translation $x \mapsto x + s$ (note that $\chi_{[s,s+t]}(x + s) = \chi_{[0,t]}(x)$ if $0 < t$ and $\chi_{[s+t,s]}(x + s) = \chi_{[t,0]}(x)$ if $t < 0$). This proves that

$$C_t B_s = B_t C_s, \forall s, t \in \mathbb{R} - \{0\}. \tag{9}$$

In view of 5, we have that

$$\lim_{t \to 0} B_t C_s f \text{ exists}, \forall f \in \mathcal{H}, \forall s \in \mathbb{R} - \{0\};$$

hence, in view of 9, we have that

$$\lim_{t \to 0} C_t B_s f \text{ exists}, \forall f \in \mathcal{H}, \forall s \in \mathbb{R} - \{0\}. \tag{10}$$

Now, we define the mapping $A_0 := D_{A_0} \to \mathcal{H}$ by

$$D_{A_0} := \{f \in \mathcal{H} : \lim_{t \to 0} C_t f \text{ exists}\} \text{ and } A_0 f := -i \lim_{t \to 0} C_t f, \forall f \in D_{A_0}.$$

Since C_t is a linear operator for all $t \in \mathbb{R} - \{0\}$, D_{A_0} is a linear manifold in view of 10.1.16a,b. Moreover, 10 shows that $S \subset D_{A_0}$; from this and from 6 we have (cf. 10.2.10b)

$$D_{A_0}^{\perp} \subset S^{\perp} = \{0_{\mathcal{H}}\},$$

and hence (cf. 10.4.4d) $\overline{D_{A_0}} = \mathcal{H}$. Furthermore we have, for each $t_0 \in \mathbb{R}$,

$$f \in D_{A_0} \Rightarrow \lim_{t \to 0} C_t f \text{ exists} \overset{(11)}{\Rightarrow} \lim_{t \to 0} U(t_0) C_t f \text{ exists} \overset{(12)}{\Rightarrow}$$
$$\lim_{t \to 0} C_t U(t_0) f \text{ exists} \Rightarrow U(t_0) f \in D_{A_0},$$

where 11 holds because $U(t_0) \in \mathcal{B}(\mathcal{H})$ and 12 holds because $U(t_0) C_t = C_t U(t_0)$ for all $t \in \mathbb{R} - \{0\}$. Finally we note that, for $f \in \mathcal{H}$,

$$\lim_{t \to 0} C_t f \text{ exists iff } U_f \text{ is differentiable at } 0$$

and, if $\lim_{t \to 0} C_t f$ exists, then

$$\lim_{t \to 0} C_t f = \frac{dU_f}{dt} \bigg|_0 .$$

Thus, U and A_0 satisfy all the conditions that held for U and A_0 in 16.1.9. Therefore, A_0 is an essentially self-adjoint operator, the operator $A := \overline{A_0}$ is self-adjoint, and $U = U^A$. This completes the proof. However, we note that as a matter of fact $A_0 = A$. Indeed, 16.1.5a implies that

$$D_A = \{f \in \mathcal{H} : U_f \text{ is differentiable at } 0\};$$

therefore, $D_{A_0} = D_A$ and hence $A_0 = A$ (since $A_0 \subset \overline{A_0}$). Note that, in 16.1.9, D_{A_0} was not assumed to be the family of *all* the vectors f for which U_f was differentiable; for this reason, the operator A_0 of 16.1.9 did not need to be self-adjoint. $\square$

16.1.11 Remarks.

(a) Stone's theorem proves that the mapping

$$A \mapsto U^A$$

(cf. 16.1.7) is a surjection, and hence a bijection from the family of all self-adjoint operators in $\mathcal{H}$ onto the family of all c.o.p.u.g.'s in $\mathcal{H}$.

For a c.o.p.u.g. U, the self-adjoint operator A such that $U = U^A$ is called the *generator* of U.

(b) For every self-adjoint operator A, it is obvious that the mapping

$$\mathbb{R} \ni t \mapsto V(t) := U^A(-t) \in \mathcal{U}(\mathcal{H})$$

is a c.o.p.u.g.. Moreover, it is obvious that

the mapping V_f is differentiable at 0 and
$$\frac{dV_f}{dt} \bigg|_0 = -iAf, \forall f \in D_{-A} \quad (= D_A).$$

Therefore, $V = U^{-A}$ (cf. 16.1.7).

16.2 Norm-continuous one-parameter unitary groups

The theorem we present in this section determines when the results of the previous section can be expressed entirely within the Banach algebra structure of $\mathcal{B}(\mathcal{H})$.

16.2.1 Theorem. *Let A be a self-adjoint operator in a Hilbert space $\mathcal{H}$. The following conditions are equivalent:*

(a) the mapping U^A is norm-continuous, i.e.

$$\forall t_0 \in \mathbb{R}, \forall \varepsilon > 0, \exists \delta > 0 \text{ such that } |t - t_0| < \delta \Rightarrow \|U(t) - U(t_0)\| < \varepsilon$$

(this condition is stronger than condition ug_1, cf. 2.4.2 and 4.2.12);
(b) $A \in \mathcal{B}(\mathcal{H})$;

If these conditions hold true then the following conditions also hold true:

(c) the series $\sum_{n=0}^{\infty} \frac{1}{n!}(it)^n A^n$ is convergent in the normed space $\mathcal{B}(\mathcal{H})$ and

$$U^A(t) = \sum_{n=0}^{\infty} \frac{1}{n!}(it)^n A^n, \forall t \in \mathbb{R};$$

(d) the mapping U^A is differentiable at t_0 and $\left.\frac{dU^A}{dt}\right|_{t_0} = iAU(t_0)$, $\forall t_0 \in \mathbb{R}$, in the normed space $\mathcal{B}(\mathcal{H})$, i.e.

$$\lim_{n \to \infty} \left\| \frac{1}{t_n}(U^A(t_0 + t_n) - U^A(t_0)) - iAU^A(t_0) \right\| = 0,$$

for every $t_0 \in \mathbb{R}$ and every sequence $\{t_n\}$ in $\mathbb{R} - \{0\}$ such that $t_n \to 0$.

Proof. $a \Rightarrow b$: With reference to the proof of 16.1.10, we have

$$R_{B_t} \subset D_A, \forall t \in \mathbb{R} - \{0\}. \tag{1}$$

Moreover, for each $t \in \mathbb{R} - \{0\}$, by the Schwarz inequality we have

$$|\left((B_t - \mathbb{1}_{\mathcal{H}})f|g\right)| = \left| \frac{1}{t} \int_0^t \left((U(x) - \mathbb{1}_{\mathcal{H}})f|g\right) dx \right|$$

$$\leq (\sup\{\|U(x) - \mathbb{1}_{\mathcal{H}}\| : x \in [0, t]\})\|f\|\|g\|, \forall f, g \in \mathcal{H},$$

since

$$\|(U(x) - \mathbb{1}_{\mathcal{H}})f\| \leq \|U(x) - \mathbb{1}_{\mathcal{H}}\|\|f\|, \forall f \in \mathcal{H}, \forall x \in \mathbb{R}$$

(cf. 4.2.5b); by 10.1.14, this implies that

$$\|B_t - \mathbb{1}_{\mathcal{H}}\| \leq \sup\{\|U(x) - \mathbb{1}_{\mathcal{H}}\| : x \in [0, t]\}. \tag{2}$$

Now we assume condition a. Then there exists $\delta \in (0, \infty)$ such that

$$|x| < \delta \Rightarrow \|U(x) - \mathbb{1}_{\mathcal{H}}\| < \frac{1}{2};$$

then, for $s := \frac{\delta}{2}$, 2 implies that

$$\|B_s - \mathbb{1}_{\mathcal{H}}\| < 1;$$

then (cf. 4.5.10)

$$-1 \in \rho(B_s - \mathbb{1}_{\mathcal{H}});$$

since the operator $B_s - \mathbb{1}_{\mathcal{H}}$ is closed (cf. 4.4.3), this is equivalent to (cf. 4.5.13)

$$B_s \text{ is injective} \quad \text{and} \quad B_s^{-1} \in \mathcal{B}(\mathcal{H});$$

then

$$R_{B_s} = D_{B_s^{-1}} = \mathcal{H}.$$

In view of 1, this implies $D_A = \mathcal{H}$ and hence $A \in \mathcal{B}(\mathcal{H})$ (cf. 12.4.7).

$b \Rightarrow a$: In this and in the ensuing parts of the proof we need the equation

$$U^A(t) = \varphi_t(A) = J_{P^A}(\varphi_t), \forall t \in \mathbb{R},$$

with

$$\mathbb{R} \ni t \mapsto \varphi_t(x) := e^{itx} \in \mathbb{C}$$

(cf. 16.1.6).

Now we assume condition b. Then $\sigma(A)$ is a bounded subset of $\mathbb{R}$ (cf. 15.2.2f) and we can fix $m \in (0, \infty)$ such that $\sigma(A) \subset [-m, m]$. Then,

$$P^A(\mathbb{R} - [-m, m]) = \mathbb{O}_{\mathcal{H}} \tag{3}$$

(cf. 15.2.2d and 13.3.2e). For every $t \in \mathbb{R}$ and every sequence $\{t_n\}$ in $\mathbb{R}$ such that $t_n \to t$, we have

$$|\varphi_{t_n}(x) - \varphi_t(x)| = |e^{i(t_n - t)x} - 1| \le |t_n - t||x|, \forall x \in \mathbb{R}, \forall n \in \mathbb{N} \tag{4}$$

(we have used the inequality $|e^{i\alpha} - 1| \le |\alpha|, \forall \alpha \in \mathbb{R}$); then,

$$\|U^A(t_n) - U^A(t)\| = \|J_{P^A}(\varphi_{t_n}) - J_{P^A}(\varphi_t)\| \overset{(5)}{=} \|\tilde{J}_{P^A}(\varphi_{t_n} - \varphi_t)\|$$

$$\overset{(6)}{=} P^A\text{-sup}|\varphi_{t_n} - \varphi_t| \overset{(7)}{\le} |t_n - t|P^A\text{-sup}|\xi|$$

$$\overset{(8)}{\le} |t_n - t|m, \forall n \in \mathbb{N},$$

where 5 holds by 14.2.17e and 14.2.7b, 6 holds by 14.2.7h, 7 holds in view of 4, 8 holds in view of 3. This proves that

$$\|U^A(t_n) - U^A(t)\| \to 0.$$

Thus, in view of 2.4.2, condition a is proved.

$b \Rightarrow c$: We assume condition b and fix $m \in (0, \infty)$ as above. For each $t \in \mathbb{R}$, it is a known fact that

$$s_n(t) := \sup \left\{ \left| e^{itx} - \sum_{k=0}^n \frac{1}{k!}(it)^k x^k \right| : x \in [-m, m] \right\} \to 0 \text{ as } n \to \infty; \tag{9}$$

moreover, we have

$$\left\| U^A(t) - \sum_{k=0}^{n} \frac{1}{k!}(it)^k A^k \right\| \overset{(10)}{=} \left\| J_{P^A}(\varphi_t) - \sum_{k=0}^{n} \frac{1}{k!}(it)^k J_{P^A}(\xi^k) \right\|$$

$$\overset{(11)}{=} \left\| \tilde{J}_{P^A} \left(\varphi_t - \sum_{k=0}^{n} \frac{1}{k!}(it)^k \xi^k \right) \right\|$$

$$\overset{(12)}{=} P^A\text{-sup} \left| \varphi_t - \sum_{k=0}^{n} \frac{1}{k!}(it)^k \xi^k \right|$$

$$\overset{(13)}{\leq} s_n(t), \forall n \in \mathbb{N},$$

where 10 holds by 15.3.5, 11 holds by 14.2.17e and 14.2.7b since 3 that implies that $\xi \in \mathcal{L}^\infty(\mathbb{R}, \mathcal{A}(d_\mathbb{R}), P^A)$, 12 holds by 14.2.7h, 13 holds in view of 3. In view of 9, this proves that

$$\left\| U^A(t) - \sum_{k=0}^{n} \frac{1}{k!}(it)^k A^k \right\| \to 0 \text{ as } n \to \infty, \forall t \in \mathbb{R},$$

i.e. condition c.

$b \Rightarrow d$: We assume condition b and fix $m \in (0,\infty)$ as above. Let $t_0 \in \mathbb{R}$ and let $\{t_n\}$ be any sequence in $\mathbb{R} - \{0\}$ such that $t_n \to 0$. Proceeding as before (using 16.1.5c, which now reads $U^A(t_0)A = AU^A(t_0)$ since $A \in \mathcal{B}(\mathcal{H})$, and also 4.2.9 and the equation $\|U(t_0)\| = 1$) we see that

$$\left\| \frac{1}{t_n}(U^A(t_0 + t_n) - U^A(t_0)) - iAU(t_0) \right\|$$

$$\leq \left\| \frac{1}{t_n}(U^A(t_n) - \mathbb{1}_\mathcal{H}) - iA \right\| = \left\| \tilde{J}_{P^A} \left(\frac{1}{t_n}(\varphi_{t_n} - \mathbb{1}_\mathbb{R}) - i\xi \right) \right\| \quad (14)$$

$$= P^A\text{-sup} \left| \frac{1}{t_n}(\varphi_{t_n} - \mathbb{1}_\mathbb{R}) - i\xi \right|, \forall n \in \mathbb{N}.$$

Next we fix $\varepsilon > 0$; since $\left. \frac{de^{is}}{ds} \right|_0 = i$, there exists $\delta_\varepsilon > 0$ such that

$$0 < s < \delta_\varepsilon \Rightarrow \left| \frac{1}{s}(e^{is} - 1) - i \right| < \frac{\varepsilon}{m};$$

now let $N_\varepsilon \in \mathbb{N}$ be such that

$$n > N_\varepsilon \Rightarrow |t_n| < \frac{\delta_\varepsilon}{m};$$

then, for every $x \in [-m, m]$ such that $x \neq 0$, we have

$$n > N_\varepsilon \Rightarrow |t_n x| \leq |t_n|m < \delta_\varepsilon \Rightarrow$$

$$\left| \frac{1}{t_n}(e^{it_n x} - 1) - ix \right| = |x| \left| \frac{1}{t_n x}(e^{it_n x} - 1) - i \right| < m\frac{\varepsilon}{m} = \varepsilon;$$

hence, in view of 3 (also, note that $\frac{1}{t_n}(e^{it_n x} - 1) - ix = 0$ for $x = 0$), we have

$$n > N_\varepsilon \Rightarrow P^A\text{-sup} \left| \frac{1}{t_n}(\varphi_{t_n} - 1_\mathbb{R}) - i\xi \right| < \varepsilon,$$

and hence, in view of 14,

$$n > N_\varepsilon \Rightarrow \left\| \frac{1}{t_n}(U^A(t_0 + t_n) - U^A(t_0)) - iAU^A(t_0) \right\| < \varepsilon.$$

Thus, condition d is proved. $\qquad\square$

16.2.2 Remark. From 12.6.1 we have that the restriction of the mapping $A \mapsto U^A$ (cf. 16.1.7 and 16.1.11a) to the family of all bounded self-adjoint operators is a bijection from this family onto the family of all norm-continuous c.o.p.u.g.'s.

16.2.3 Theorem (Stone's theorem in one dimension). *Let γ be a continuous homomorphism from the additive group $\mathbb{R}$ to the multiplicative group $\mathbb{T}$, i.e.*

$$\text{the function } \gamma : \mathbb{R} \to \mathbb{T} \text{ is continuous,}$$

$$\gamma(t_1)\gamma(t_2) = \gamma(t_1 + t_2), \forall t_1, t_2 \in \mathbb{R}.$$

Then,

$$\exists! a \in \mathbb{R} \text{ so that } \gamma(t) = e^{iat}, \forall t \in \mathbb{R}.$$

Proof. Let $\mathcal{H}$ be a one-dimensional Hilbert space. We recall (cf. 12.6.6a) that there exists an isomorphism

$$\mathbb{C} \ni \alpha \mapsto A_\alpha \in \mathcal{B}(\mathcal{H})$$

from the associative algebra $\mathbb{C}$ onto the associative algebra $\mathcal{B}(\mathcal{H})$ such that:

$$\|A_\alpha\| = |\alpha|, \forall \alpha \in \mathbb{C};$$

$$A_\alpha \text{ is self-adjoint iff } \alpha \in \mathbb{R};$$

$$A_\alpha \in \mathcal{U}(\mathcal{H}) \text{ iff } \alpha \in \mathbb{T}.$$

Then, the mapping

$$\mathbb{R} \ni t \mapsto A_{\gamma(t)} \in \mathcal{U}(\mathcal{H})$$

is a norm-continuous c.o.p.u.g. in $\mathcal{H}$. Therefore (cf. 16.1.10 and 16.2.1c), there exists $a \in \mathbb{R}$ so that

$$A_{\gamma(t)} = \sum_{n=0}^{\infty} \frac{1}{n!}(it)^n A_a^n, \forall t \in \mathbb{R},$$

in the normed space $\mathcal{B}(\mathcal{H})$, and hence so that

$$\gamma(t) = \sum_{n=0}^{\infty} \frac{1}{n!}(it)^n a^n = e^{iat}, \forall t \in \mathbb{R},$$

in the normed space $\mathbb{C}$. The uniqueness of a is shown e.g. by 16.2.1d, which in the normed space $\mathbb{C}$ gives

$$\left. \frac{d\gamma}{dt} \right|_0 = ia.$$

$\qquad\square$

16.3 Unitary equivalence

16.3.1 Theorem. *Let $\mathcal{H}_1$ and $\mathcal{H}_2$ be isomorphic Hilbert spaces and suppose that $V \in \mathcal{UA}(\mathcal{H}_1, \mathcal{H}_2)$. Let A_1 and A_2 be self-adjoint operators in $\mathcal{H}_1$ and in $\mathcal{H}_2$ respectively. Then the following conditions are equivalent:*

(a) $A_2 = V A_1 V^{-1}$;
(b) $U^{A_2}(t) = V U^{A_1}(t) V^{-1}, \forall t \in \mathbb{R}$, if $V \in \mathcal{U}(\mathcal{H}_1, \mathcal{H}_2)$, or
   $U^{A_2}(-t) = V U^{A_1}(t) V^{-1}, \forall t \in \mathbb{R}$, if $V \in \mathcal{A}(\mathcal{H}_1, \mathcal{H}_2)$.

Proof. $a \Rightarrow b$: Condition a implies (cf. 15.4.1)

$$P^{A_2}(E) = V P^{A_1}(E) V^{-1}, \forall E \in \mathcal{A}(d_\mathbb{R}),$$

and this implies (cf. 14.6.2)

$$\varphi_t(A_2) = V \varphi_t(A_1) V^{-1}, \forall t \in \mathbb{R}, \text{ if } V \in \mathcal{U}(\mathcal{H}_1, \mathcal{H}_2), \text{ or}$$
$$\overline{\varphi}_t(A_2) = V \varphi_t(A_1) V^{-1}, \forall t \in \mathbb{R}, \text{ if } V \in \mathcal{A}(\mathcal{H}_1, \mathcal{H}_2).$$

Now,

$$\varphi_t(A_2) = U^{A_2}(t) \text{ and } \overline{\varphi}_t(A_2) = \varphi_{-t}(A_2) = U^{A_2}(-t)$$

(cf. 16.1.6). Thus, condition b is proved.

$b \Rightarrow a$: We assume condition b and define the operator B in $\mathcal{H}_2$ by

$$B := V A_1 V^{-1};$$

the operator B is self-adjoint (cf. 12.5.4c). Then, in view of $a \Rightarrow b$ already proved, we have

$$U^B(t) = V U^{A_1}(t) V^{-1}, \forall t \in \mathbb{R}, \text{ if } V \in \mathcal{U}(\mathcal{H}_1, \mathcal{H}_2), \text{ or}$$
$$U^B(-t) = V U^{A_1}(t) V^{-1}, \forall t \in \mathbb{R}, \text{ if } V \in \mathcal{A}(\mathcal{H}_1, \mathcal{H}_2).$$

In either case we have $U^B = U^{A_2}$ and hence $B = A_2$ since the mapping $A \mapsto U^A$ is injective, and this is condition a. $\qquad\square$

16.4 One-parameter groups of automorphisms

The main theorem of this section (cf. 16.4.11) is a special case of a much more general theorem proved by Valentine Bargmann (Bargmann, 1954). In the analysis of time evolution of conservative and reversible quantum systems (cf. Section 19.6) and of symmetries (for an example, cf. Section 20.3), one is led to consider what we call continuous one-parameter groups of automorphisms. The special case of Bargmann's theorem we consider here is the essential link between these and c.o.p.u.g.'s.

Throughout this section, $\mathcal{H}$ denotes an abstract Hilbert space of dimension greater than one. We recall that $\operatorname{Aut}\hat{\mathcal{H}}$ denotes the group of automorphisms of the projective Hilbert space $(\hat{\mathcal{H}}, \tau)$ (cf. 10.9.4).

16.4.1 Definition. A *continuous one-parameter group of automorphisms* of $\hat{\mathcal{H}}$ is a mapping

$$\mathbb{R} \ni t \mapsto \omega_t \in \operatorname{Aut}\hat{\mathcal{H}}$$

such that:

(ag_1) the mapping is a homomorphism from the additive group $\mathbb{R}$ to the group $\operatorname{Aut}\hat{\mathcal{H}}$, i.e.

$$\omega_{t_1} \circ \omega_{t_2} = \omega_{t_1+t_2}, \forall t_1, t_2 \in \mathbb{R};$$

(ag_2) the function

$$\mathbb{R} \ni t \mapsto \tau([u], \omega_t([v])) \in [0,1]$$

is continuous, $\forall u, v \in \tilde{\mathcal{H}}$ (we note that $\tau([u],[v]) \leq 1$ for all $u, v \in \tilde{\mathcal{H}}$, by the Schwarz inequality).

16.4.2 Remarks.

(a) For *any* mapping $\mathbb{R} \ni t \mapsto \omega_t \in \operatorname{Aut}\hat{\mathcal{H}}$, by Wigner's theorem (cf. 10.9.6) we have that, for each $t \in \mathbb{R}$, there exists a family of operators $U_t \in \mathcal{UA}(\mathcal{H})$ which are such that

$$\omega_{U_t} = \omega_t, \text{ i.e. } [U_t u] = \omega_t([u]), \forall u \in \tilde{\mathcal{H}},$$

and that, given an operator of this family, all the others are multiplies of this one by a factor in $\mathbb{T}$. Hence, for each $t \in \mathbb{R}$, either all the operators $U_t \in \mathcal{UA}(\mathcal{H})$ which are such that $\omega_{U_t} = \omega_t$ are unitary or all of them are antiunitary.

(b) Let $\mathbb{R} \ni t \mapsto \omega_t \in \operatorname{Aut}\hat{\mathcal{H}}$ be a homomorphism from the additive group $\mathbb{R}$ to $\operatorname{Aut}\hat{\mathcal{H}}$.

First, we have (cf. 1.3.3 and 1.3.5b):

$$\omega_0 = id_{\hat{\mathcal{H}}};$$
$$\omega_{-t} = \omega_t^{-1}, \forall t \in \mathbb{R};$$
$$\omega_{t_1} \circ \omega_{t_2} = \omega_{t_2} \circ \omega_{t_1}, \forall t_1, t_2 \in \mathbb{R}.$$

Next, for each $t \in \mathbb{R}$ and any choice of U_t and of $U_{\frac{t}{2}}$ in $\mathcal{UA}(\mathcal{H})$ such that $\omega_{U_t} = \omega_t$ and $\omega_{U_{\frac{t}{2}}} = \omega_{\frac{t}{2}}$, we have

$$[U_t u] = \omega_t([u]) = \omega_{\frac{t}{2}} \circ \omega_{\frac{t}{2}}([u]) = [U_{\frac{t}{2}}^2 u], \forall u \in \tilde{\mathcal{H}};$$

then (cf. a) there exists $z \in \mathbb{T}$ so that $U_t = zU_{\frac{t}{2}}^2$, and hence $U_t \in \mathcal{U}(\mathcal{H})$ (cf. 10.3.16c). Thus, the operators $U_t \in \mathcal{UA}(\mathcal{H})$ such that $\omega_{U_t} = \omega_t$ are unitary, for each $t \in \mathbb{R}$.

16.4.3 Proposition. *Let U be a c.o.p.u.g. in $\mathcal{H}$. Then:*

(a) the mapping

$$\mathbb{R} \ni t \mapsto \omega_{U(t)} \in \operatorname{Aut} \hat{\mathcal{H}}$$

is a continuous one-parameter group of automorphisms;

(b) if V is a c.o.p.u.g. in $\mathcal{H}$ such that $\omega_{V(t)} = \omega_{U(t)}$ for all $t \in \mathbb{R}$, then

$$\exists! a \in \mathbb{R} \text{ so that } V(t) = e^{iat} U(t), \forall t \in \mathbb{R}.$$

Proof. a: For all $t_1, t_2 \in \mathbb{R}$, we have

$$\omega_{U(t_1)} \circ \omega_{U(t_2)}([u]) = [U(t_1)U(t_2)u]$$
$$= [U(t_1 + t_2)u] = \omega_{U(t_1+t_2)}([u]), \forall u \in \tilde{\mathcal{H}},$$

and hence

$$\omega_{U(t_1)} \circ \omega_{U(t_2)} = \omega_{U(t_1+t_2)}.$$

For all $u, v \in \tilde{\mathcal{H}}$, the function

$$\mathbb{R} \ni t \mapsto (u|U(t)v) \in \mathbb{C}$$

is continuous (cf. 16.1.2b), and hence the function

$$\mathbb{R} \ni t \mapsto \tau([u], \omega_{U(t)}([v])) = |(u|U(t)v)| \in [0,1]$$

is continuous.

b: If V is a c.o.p.u.g. in $\mathcal{H}$ such that $\omega_{V(t)} = \omega_{U(t)}$ for all $t \in \mathbb{R}$, then (cf. 10.9.6) there exists a function

$$\mathbb{R} \ni t \mapsto \gamma(t) \in \mathbb{T}$$

so that

$$V(t) = \gamma(t)U(t) \text{ or } U(-t)V(t) = \gamma(t)\mathbb{1}_{\mathcal{H}}, \forall t \in \mathbb{R}.$$

Now we fix $u \in \tilde{\mathcal{H}}$. For every $t \in \mathbb{R}$ and every sequence $\{t_n\}$ in $\mathbb{R}$ such that $t_n \to t$, we have

$$\gamma(t_n) = (u|U(-t_n)V(t_n)u) = (U(t_n)u|V(t_n)u) \xrightarrow[n \to \infty]{} (U(t)u|V(t)u) = \gamma(t),$$

in view of condition ug_2 and of 10.1.16c. Thus, the function γ is continuous (cf. 2.4.2). Moreover, for all $t_1, t_2 \in \mathbb{R}$ we have

$$\gamma(t_1)\gamma(t_2) = \gamma(t_1)(u|U(-t_2)V(t_2)u) = \gamma(t_1)(U(t_2)u|V(t_2)u)$$
$$= (U(t_2)u|U(-t_1)V(t_1)V(t_2)u) = (u|U(-t_1 - t_2)V(t_1 + t_2)u)$$
$$= \gamma(t_1 + t_2).$$

Then the result follows from 16.2.3. $\qquad\square$

16.4.4 Lemma. *We define*

$$d([u], [v]) := 2^{\frac{1}{2}}(1 - \tau([u], [v]))^{\frac{1}{2}}, \forall u, v \in \tilde{\mathcal{H}}.$$

Then:

(a) $1 - \tau([u], [v])^2 \le d([u], [v])^2$, $\forall u, v \in \tilde{\mathcal{H}}$;

(b) $|\tau([u], [v]) - \tau([u], [w])| \le d([v], [w])$, $\forall u, v, w \in \tilde{\mathcal{H}}$.

Proof. a: For all $u, v \in \tilde{\mathcal{H}}$, we have

$$1 - \tau([u], [v]) = \frac{1}{2}d([u], [v])^2,$$

and hence

$$1 + \tau([u], [v]) = 2 - \frac{1}{2}d([u], [v])^2,$$

and hence

$$1 - \tau([u], [v])^2 = \left(1 - \frac{1}{4}d([u], [v])^2\right)d([u], [v])^2 \le d([u], [v])^2.$$

b: Fix $u, v, w \in \tilde{\mathcal{H}}$, and let $z \in \mathbb{T}$ be such that

$$(v|w) = z|\,(v|w)\,| = z\tau([v], [w]).$$

If we put $v_0 := zv$ then we have

$$(v_0|w) = \bar{z}\,(v|w) = \bar{z}z\tau([v], [w]) = \tau([v], [w]),$$

and hence

$$2(1 - \tau([v], [w])) = 2(1 - (v_0|w)) = (v_0|v_0) + (w|w) - (v_0|w) - (w|v_0) = \|v_0 - w\|^2,$$

and hence, by the Schwarz inequality,

$$\begin{aligned}|\tau([u], [v]) - \tau([u], [w])| &= |\,|\,(u|v_0)\,| - |\,(u|w)\,|\,| \\ &\le |\,(u|v_0) - (u|w)\,| \le \|u\|\|v_0 - w\| \\ &= 2^{\frac{1}{2}}(1 - \tau([v], [w]))^{\frac{1}{2}} = d([v], [w]).\end{aligned}$$

$\square$

16.4.5 Proposition. *Let* $\mathbb{R} \ni t \mapsto \omega_t \in \operatorname{Aut}\hat{\mathcal{H}}$ *be a homomorphism from the additive group* $\mathbb{R}$ *to* $\operatorname{Aut}\hat{\mathcal{H}}$ *(i.e., condition* ag_1 *holds true). Then the following conditions are equivalent:*

(a) $\mathbb{R} \ni t \mapsto \omega_t \in \operatorname{Aut}\hat{\mathcal{H}}$ *is a continuous one-parameter group of automorphisms (i.e., condition* ag_2 *holds true);*

(b) *the function* $\mathbb{R} \ni t \mapsto \tau([u], \omega_t([u])) \in [0, 1]$ *is continuous at 0,* $\forall u \in \tilde{\mathcal{H}}$.

Proof. $a \Rightarrow b$: This is obvious.

$b \Rightarrow a$: We assume condition b. Then, for every $v \in \tilde{\mathcal{H}}$ and for every sequence $\{s_n\}$ in $\mathbb{R}$ such that $s_n \to 0$, we have (cf. 2.4.2)

$$\tau([v], \omega_{s_n}([v])) \xrightarrow[n\to\infty]{} \tau([v], \omega_0([v])) = \tau([v], [v]) = 1,$$

and hence, for every $w \in \tilde{\mathcal{H}}$,

$$|\tau([w], [v]) - \tau([w], \omega_{s_n}([v]))| \overset{(1)}{\le} d([v], \omega_{s_n}([v])) = 2^{\frac{1}{2}}(1 - \tau([v], \omega_{s_n}([v])))^{\frac{1}{2}} \to 0$$

(1 holds by 16.4.4b), or

$$\tau([w], \omega_{s_n}([v])) \xrightarrow[n \to \infty]{} \tau([w], [v]).$$

Then, for all $u, v \in \tilde{\mathcal{H}}$, for every $t \in \mathbb{R}$, and for every sequence $\{t_n\}$ in $\mathbb{R}$ such that $t_n \to t$, we have

$$\tau([u], \omega_{t_n}([v])) = \tau(\omega_{-t}([u]), \omega_{t_n-t}([v])) \xrightarrow[n \to \infty]{} \tau(\omega_{-t}([u]), [v]) = \tau([u], \omega_t([v]))$$

since $t_n - t \to 0$. Thus, condition ag_2 holds true (cf. 2.4.2). $\square$

The next theorem is instrumental in proving the special case of Bargmann's theorem we mentioned at the beginning of this section. It was proved (for a case more general than the one of interest here) by Valentine Bargmann, who followed arguments which had been put forward by Eugene P. Wigner before. In the proof of 16.4.6 we reproduce Bargmann's proof (cf. Bargmann, 1954, th.1.1).

16.4.6 Theorem. *Let $\mathbb{R} \ni t \mapsto \omega_t \in \text{Aut}\,\hat{\mathcal{H}}$ be a mapping such that:*

(a) $\omega_0 = id_{\hat{\mathcal{H}}}$;
(b) the operators $U_t \in \mathcal{UA}(\mathcal{H})$ such that $\omega_{U_t} = \omega_t$ are unitary, $\forall t \in \mathbb{R}$;
(c) the function $\mathbb{R} \ni t \mapsto \tau([u], \omega_t([v])) \in [0,1]$ is continuous, $\forall u, v \in \tilde{\mathcal{H}}$.

Then there exists $a \in (0, \infty)$ and a mapping

$$(-a, a) \ni t \mapsto V_t \in \mathcal{U}(\mathcal{H})$$

so that:

$V_0 = \mathbb{1}_{\mathcal{H}}$;
$\omega_t = \omega_{V_t}, \forall t \in (-a, a)$;
the mapping $(-a, a) \ni t \mapsto V_t f \in \mathcal{H}$ is continuous, $\forall f \in \mathcal{H}$.

Proof. We fix $h \in \tilde{\mathcal{H}}$ and $\delta \in (0, 1)$ throughout the proof. We divide the proof into five steps.

Step 1: Here we define $a \in (0, \infty)$ and $V(t) \in \mathcal{U}(\mathcal{H})$ for all $t \in (-a, a)$ so that $V_0 = \mathbb{1}_{\mathcal{H}}$ and $\omega_t = \omega_{V_t}, \forall t \in (-a, a)$.

The function

$$\mathbb{R} \ni r \mapsto \rho_r := \tau([h], \omega_r([h])) \in [0,1]$$

is continuous and $\rho_0 = 1$, in view of conditions c and a. Hence, we can choose $a \in (0, \infty)$ so that

$$r \in (-a, a) \Rightarrow \delta < \rho_r \leq 1.$$

For each $r \in (-a, a)$, there exists a unique $V_r \in \mathcal{U}(\mathcal{H})$ so that $\omega_{V_r} = \omega_r$ and

$$(h|V_r h) = |(h|V_r h)| = \rho_r \tag{1}$$

(in view of condition b, there exists $U_r \in \mathcal{U}(\mathcal{H})$ such that $\omega_{U_r} = \omega_r$; then, we define $V_r := z_r U_r$, with $z_r := |(h|U_r h)|\,(h|U_r h)^{-1}$; the uniqueness of V_r is obvious, since for any other $V'_r \in \mathcal{U}(\mathcal{H})$ such that $\omega_{V'_r} = \omega_r$ we would have $(h|V'_r h) = z\rho_r$ with $z \neq 1$, cf. 16.4.2a). Clearly,

$$V_0 = \mathbb{1}_{\mathcal{H}}.$$

Step 2: Here we prove two auxiliary relations.

For all $u \in \tilde{\mathcal{H}}$ and $r, s \in (-a, a)$, we define

$$d_{r,s}(u) := d(\omega_r([u]), \omega_s([u]));$$
$$\sigma_{r,s}(u) := (V_r u | V_s u);$$
$$z_{r,s}(u) := V_s u - \sigma_{r,s}(u) V_r u.$$

We have

$$(V_r u | z_{r,s}(u)) = 0,$$

and hence

$$1 = \|V_s u\|^2 = \|z_{r,s}(u) + \sigma_{r,s}(u) V_r u\|^2 = \|z_{r,s}(u)\|^2 + |\sigma_{r,s}(u)|^2,$$

and hence

$$\|z_{r,s}(u)\|^2 = 1 - |\sigma_{r,s}(u)|^2 = 1 - \tau(\omega_r([u]), \omega_s([u]))^2 \overset{(2)}{\leq} d_{r,s}(u)^2, \qquad (3)$$

where 2 holds by 16.4.4a. Moreover, we have

$$\|V_s u - V_r u\|^2 = 2 - 2 \operatorname{Re}(V_r u | V_s u) \leq 2|1 - (V_r u | V_s u)| = 2|1 - \sigma_{r,s}(u)|. \qquad (4)$$

Step 3: Here we prove that, for every $t \in (-a, a)$ and every sequence $\{t_n\}$ in $(-a, a)$ such that $t_n \to t$, we have $d_{t,t_n}(u) \xrightarrow[n \to \infty]{} 0$ for all $u \in \tilde{\mathcal{H}}$.

Indeed, for each $u \in \tilde{\mathcal{H}}$, by condition c we have

$$\tau(\omega_t([u]), \omega_{t_n}([u])) \xrightarrow[n \to \infty]{} \tau(\omega_t([u]), \omega_t([u])) = 1$$

(we have used the continuity at t of the function $s \mapsto \tau(\omega_t([u]), \omega_s([u]))$); therefore,

$$d_{t,t_n}(u) = d(\omega_t([u]), \omega_{t_n}([u])) = 2^{\frac{1}{2}}(1 - \tau(\omega_t([u]), \omega_{t_n}([u])))^{\frac{1}{2}} \xrightarrow[n \to \infty]{} 0.$$

Step 4: Here we prove that, for every $t \in (-a, a)$ and every sequence $\{t_n\}$ in $(-a, a)$ such that $t_n \to t$, we have $\|V_{t_n} h - V_t h\| \xrightarrow[n \to \infty]{} 0$.

For all $r, s \in (-a, a)$, we have

$$(h|z_{r,s}(h)) = (h|V_s h) - \sigma_{r,s}(h)(h|V_r h) \overset{(5)}{=} \rho_s - \sigma_{r,s}(h)\rho_r$$

(5 holds in view of 1), and hence

$$1 - \sigma_{r,s}(h) = \rho_r^{-1}(\rho_r - \rho_s + (h|z_{r,s}(h))),$$

and hence

$$\|V_s h - V_r h\|^2 \overset{(6)}{\leq} 2|1 - \sigma_{r,s}(h)| \leq 2\rho_r^{-1}(|\rho_r - \rho_s| + |(h|z_{r,s}(h))|)$$

$$\overset{(7)}{\leq} 2\rho_r^{-1}(|\tau([h], \omega_r([h])) - \tau([h], \omega_s([h]))| + \|h\|\|z_{r,s}(h)\|)$$

$$\overset{(8)}{\leq} 2\rho_r^{-1}(d(\omega_r([h]), \omega_s([h])) + d_{r,s}(h)) = 4\rho_r^{-1} d_{r,s}(h) < 4\delta^{-1} d_{r,s}(h),$$

where 6 holds by 4, 7 by the Schwarz inequality, 8 by 16.4.4b and by 3. Thus, for every $t \in (-a, a)$ and every sequence $\{t_n\}$ in $\mathbb{R}$ such that $t_n \to t$, we have

$$d_{t,t_n}(h) \xrightarrow[n \to \infty]{} 0$$

(cf. step 3) and hence, in view of the inequality just proved,

$$\|V_{t_n} h - V_t h\| \xrightarrow[n \to \infty]{} 0.$$

Step 5: Here we prove the continuity of $(-a, a) \ni t \mapsto V_t f \in \mathcal{H}$ for all $f \in \mathcal{H}$ (this concludes the proof of the theorem).

Let g be any element of $\tilde{\mathcal{H}}$ such that $(g|h) = 0$, and set

$$k := 2^{-\frac{1}{2}}(h + g);$$

clearly, $k \in \tilde{\mathcal{H}}$. For all $r, s \in (-a, a)$, we have

$$(V_r h | z_{r,s}(k)) = (V_r h | V_s k - \sigma_{r,s}(k) V_r k)$$
$$= (V_r h - V_s h | V_s k) + (V_s h | V_s k) - \sigma_{r,s}(k)(V_r h | V_r k),$$

and hence (since $(V_s h | V_s k) = (V_r h | V_r k) = (h|k) = 2^{-\frac{1}{2}}$)

$$2^{-\frac{1}{2}}(1 - \sigma_{r,s}(k)) = (V_r h | z_{r,s}(k)) + (V_s h - V_r h | V_s k),$$

and hence

$$\|V_s k - V_r k\|^2 \overset{(9)}{\leq} 2|1 - \sigma_{r,s}(k)| \leq 2^{\frac{3}{2}}(|(V_r h | z_{r,s}(k))| + |(V_s h - V_r h | V_s k)|)$$
$$\overset{(10)}{\leq} 2^{\frac{3}{2}}(\|z_{r,s}(k)\| + \|V_s h - V_r h\|) \overset{(11)}{\leq} 2^{\frac{3}{2}}(d_{r,s}(k) + \|V_s h - V_r h\|),$$

where 9 holds by 4, 10 by the Schwarz inequality, 11 by 3. Thus, for every $t \in (-a, a)$ and every sequence $\{t_n\}$ in $(-a, a)$ such that $t_n \to t$, we have

$$d_{t,t_n}(k) \xrightarrow[n \to \infty]{} 0$$

(cf. step 3) and

$$\|V_{t_n} h - V_t h\| \to 0, \quad \text{or} \quad V_{t_n} h \xrightarrow[n \to \infty]{} V_t h,$$

(cf. step 4), and hence, in view of the inequality just proved,

$$\|V_{t_n} k - V_t k\| \to 0, \quad \text{or} \quad V_{t_n} k \xrightarrow[n \to \infty]{} V_t k,$$

and hence, in view of 10.1.16a,b,

$$V_{t_n} g = \sqrt{2} V_{t_n} k - V_{t_n} h \xrightarrow[n \to \infty]{} \sqrt{2} V_t k - V_t h = V_t g.$$

Now, for each $f \in \mathcal{H}$, there exist $\lambda_1, \lambda_2 \in \mathbb{C}$ and $g \in \tilde{\mathcal{H}}$ so that $(g|h) = 0$ and

$$f = \lambda_1 h + \lambda_2 g$$

(cf. 10.4.1 with $M := V\{h\}$; also, cf. 4.1.15 and 10.2.11); then, for every $t \in (-a, a)$ and every sequence $\{t_n\}$ in $(-a, a)$ such that $t_n \to t$, we have

$$V_{t_n} f = \lambda_1 V_{t_n} h + \lambda_2 V_{t_n} g \xrightarrow[n \to \infty]{} \lambda_1 V_t h + \lambda_2 V_t g = V_t f.$$

This proves that the mapping $(-a, a) \ni t \mapsto V_t f \in \mathcal{H}$ is continuous (cf. 2.4.2). $\square$

We need the next four results in the proof of 16.4.11, which is the above-mentioned special case of Bargmann's general theorem.

16.4.7 Lemma. *Let $U, V \in \mathcal{U}(\mathcal{H})$ and let $\{U_n\}, \{V_n\}$ be sequences in $\mathcal{U}(\mathcal{H})$ such that:*

$$U_n f \xrightarrow[n \to \infty]{} Uf, \forall f \in \mathcal{H};$$
$$V_n f \xrightarrow[n \to \infty]{} Vf, \forall f \in \mathcal{H}.$$

Then:

$$U_n V_n^{-1} f \xrightarrow[n \to \infty]{} UV^{-1}f, \forall f \in \mathcal{H};$$
$$V_n^{-1} f \xrightarrow[n \to \infty]{} V^{-1}f, \forall f \in \mathcal{H};$$
$$U_n V_n f \xrightarrow[n \to \infty]{} UVf, \forall f \in \mathcal{H}.$$

Proof. For every $f \in \mathcal{H}$, we have

$$\|U_n V_n^{-1} f - U_n V^{-1} f\| = \|V_n^{-1} f - V^{-1} f\| = \|f - V_n V^{-1} f\|$$
$$= \|VV^{-1}f - V_n V^{-1}f\|, \forall n \in \mathbb{N},$$

and hence

$$\begin{aligned}
\|U_n V_n^{-1} f - UV^{-1} f\| &\leq \|U_n V_n^{-1} f - U_n V^{-1} f\| + \|U_n V^{-1} f - UV^{-1} f\| \\
&= \|V(V^{-1}f) - V_n(V^{-1}f)\| + \|U_n(V^{-1}f) - U(V^{-1}f)\| \\
&\xrightarrow[n \to \infty]{} 0.
\end{aligned}$$

Thus,

$$U_n V_n^{-1} f \xrightarrow[n \to \infty]{} UV^{-1}f, \forall f \in \mathcal{H}. \tag{1}$$

If we set $U := U_n := \mathbb{1}_{\mathcal{H}}$ in the statement, from 1 we have

$$V_n^{-1} f \xrightarrow[n \to \infty]{} V^{-1}f, \forall f \in \mathcal{H}. \tag{2}$$

Since 2 is true, we can substitute V^{-1} for V and V_n^{-1} for V in the statement, and thus obtain from 1

$$U_n V_n f \xrightarrow[n \to \infty]{} UVf, \forall f \in \mathcal{H}.$$

$\square$

16.4.8 Lemma. *Let $\mu : \mathbb{R}^2 \to \mathbb{T}$ be a continuous function such that $\mu(0,0) = 1$. Then there exists a continuous function $\xi : \mathbb{R}^2 \to \mathbb{R}$ such that*

$$\xi(0,0) = 0 \text{ and } \mu(r,s) = e^{i\xi(r,s)}, \forall (r,s) \in \mathbb{R}^2.$$

Proof. It is outside the scope of this book to prove this result, which is a special case of a theorem of topology about liftings (cf. e.g. Greenberg and Harper, 1981, 6.1 and 6.6). $\square$

16.4.9 Lemma. *Let $\xi : \mathbb{R}^2 \to \mathbb{C}$ be a continuous function and let $\varphi \in C_c(\mathbb{R})$ (cf. 3.1.10g). Then the function*

$$\mathbb{R} \ni x \mapsto \lambda(x) := \int_{\mathbb{R}} \xi(x,t)\varphi(t)dm(t) \in \mathbb{C}$$

(m denotes the Lebesgue measure on $\mathbb{R}$) is continuous.

Proof. First we note that the integral which defines the function λ exists by 2.8.14 and 8.2.6.

Let $a, b \in \mathbb{R}$ be so that $a < b$ and $\varphi(t) = 0$ for all $t \in \mathbb{R} - [a, b]$. Then,

$$\lambda(x) = \int_{[a,b]} \xi(x,t)\varphi(t)dm(t), \forall x \in \mathbb{R}.$$

Now we fix $x_0 \in \mathbb{R}$ and $d \in (0, \infty)$. The function

$$[x_0 - d, x_0 + d] \times [a, b] \ni (x,t) \mapsto \xi(x,t)\varphi(t) \in \mathbb{C}$$

is continuous, and hence it is uniformly continuous (cf. 2.8.7 and 2.8.15). Hence, for every $\varepsilon \in (0, \infty)$ there exists $\delta_\varepsilon \in (0, \infty)$ such that $\delta_\varepsilon < d$ and

$$|x_0 - x| < \delta_\varepsilon \Rightarrow [d_2((x_0,t),(x,t)) < \delta_\varepsilon, \forall t \in [a,b]] \Rightarrow$$
$$[|\xi(x_0,t)\varphi(t) - \xi(x,t)\varphi(t)| < \varepsilon, \forall t \in [a,b]] \Rightarrow$$
$$|\lambda(x_0) - \lambda(x)| \leq \int_{[a,b]} |\xi(x_0,t)\varphi(t) - \xi(x,t)\varphi(t)|dm(t) \leq \varepsilon(b - a).$$

This proves that the function λ is continuous at x_0, and hence that it is continuous since x_0 was arbitrary. $\qquad\qquad\square$

16.4.10 Lemma. *Let $\xi : \mathbb{R}^2 \to \mathbb{C}$ be a continuous function and let $\varphi \in C_c(\mathbb{R})$ be such that φ is differentiable at all points of $\mathbb{R}$ and the function φ' (the derivative of φ) is continuous. Let ψ be the function defined by*

$$\mathbb{R}^2 \ni (x,y) \mapsto \psi(x,y) := \int_{\mathbb{R}} \xi(x,t)\varphi(t-y)dm(t) \in \mathbb{C}.$$

Then:

the partial derivative $\frac{\partial \psi}{\partial y}(x,y)$ exists and $\frac{\partial \psi}{\partial y}(x,y) = -\int_{\mathbb{R}} \xi(x,t)\varphi'(t-y)dm(t)$,

$\forall (x,y) \in \mathbb{R}^2$;

the function $\mathbb{R} \ni x \mapsto \frac{\partial \psi}{\partial y}(x,0) \in \mathbb{C}$ is continuous.

Proof. We fix $x_0 \in \mathbb{R}$ and define the function

$$\mathbb{R}^2 \ni (t,y) \mapsto \chi(t,y) := \xi(x_0,t)\varphi(t-y) \in \mathbb{C}.$$

For each $(t,y) \in \mathbb{R}^2$, the partial derivative $\frac{\partial \chi}{\partial y}(t,y)$ exists and

$$\frac{\partial \chi}{\partial y}(t,y) = -\xi(x_0,t)\varphi'(t-y). \tag{1}$$

Let $a, b \in \mathbb{R}$ be so that $a < b$ and $\varphi(s) = 0$ for all $s \in \mathbb{R} - [a, b]$. Then, for each $y \in \mathbb{R}$,

$$\varphi(t - y) = 0, \forall t \in \mathbb{R} - [a + y, b + y].$$

We fix $y_0 \in \mathbb{R}$ and $d \in (0, \infty)$, and define the interval

$$I(y_0, d) := [a + y_0 - d, b + y_0 + d];$$

then, for each $y \in [y_0 - d, y_0 + d]$,

$$\varphi(t - y) = 0, \forall t \in \mathbb{R} - I(y_0, d).$$

Since $\varphi'(s) = 0$ for all $s \notin [a, b]$, the same reasoning as above proves that, for each $y \in [y_0 - d, y_0 + d]$,

$$\varphi'(t - y) = 0, \forall t \in \mathbb{R} - I(y_0, d).$$

Hence, for each $h \in \mathbb{R} - \{0\}$ such that $|h| \le d$, we have (cf. 1)

$$\frac{1}{h}(\psi(x_0, y_0 + h) - \psi(x_0, y_0)) + \int_{\mathbb{R}} \xi(x_0, t)\varphi'(t - y_0)dm(t)$$
$$= \int_{I(y_0, d)} \left(\frac{1}{h}(\chi(t, y_0 + h) - \chi(t, y_0)) - \frac{\partial \chi}{\partial y}(t, y_0) \right) dm(t) \tag{2}$$

(note that the function χ depends on x_0, which however is fixed).

Now, the function

$$I(y_0, d) \times [y_0 - d, y_0 + d] \ni (t, y) \mapsto \frac{\partial \chi}{\partial y}(t, y) \in \mathbb{C}$$

is continuous (cf. 1); hence it is uniformly continuous (cf. 2.8.7 and 2.8.15), and hence for each $\varepsilon \in (0, \infty)$ there exists $\delta_\varepsilon \in (0, \infty)$ such that $\delta_\varepsilon < d$ and

$$|y - y_0| < \delta_\varepsilon \Rightarrow [d_2((t, y), (t, y_0)) < \delta_\varepsilon, \forall t \in I(y_0, d)] \Rightarrow$$
$$\left[\left| \frac{\partial \chi}{\partial y}(t, y) - \frac{\partial \chi}{\partial y}(t, y_0) \right| < \varepsilon, \forall t \in I(y_0, d) \right]. \tag{3}$$

For each $h \in \mathbb{R} - \{0\}$, the mean value theorem implies that

$$\forall t \in \mathbb{R}, \exists y_t \in \mathbb{R} \text{ s.t. } |y_t - y_0| \le |h| \text{ and } \frac{1}{h}(\chi(t, y_0 + h) - \chi(t, y_0)) = \frac{\partial \chi}{\partial y}(t, y_t);$$

this and 3 imply that, if $0 < |h| < \delta_\varepsilon$, then

$$\left| \frac{1}{h}(\chi(t, y_0 + h) - \chi(t, y_0)) - \frac{\partial \chi}{\partial y}(t, y_0) \right| < \varepsilon, \forall t \in I(y_0, d);$$

hence, in view of 2, if $0 < |h| < \delta_\varepsilon$ then

$$\left| \frac{1}{h}(\psi(x_0, y_0 + h) - \psi(x_0, y_0)) + \int_{\mathbb{R}} \xi(x_0, t)\varphi'(t - y_0)dm(t) \right|$$
$$\le \int_{I(y_0, d)} \left| \frac{1}{h}(\chi(t, y_0 + h) - \chi(t, y_0)) - \frac{\partial \chi}{\partial y}(t, y_0) \right| dm(t) \le \varepsilon(b - a + 2d).$$

This proves that the partial derivative $\frac{\partial \psi}{\partial y}$ exists at (x_0, y_0) and that

$$\frac{\partial \psi}{\partial y}(x_0, y_0) = -\int_{\mathbb{R}} \xi(x_0, t)\varphi'(t - y_0)dm(t).$$

Since x_0 and y_0 were arbitrary, the first conclusion of the statement is proved.

Since

$$\frac{\partial \psi}{\partial y}(x, 0) = -\int_{\mathbb{R}} \xi(x, t)\varphi'(t)dm(t), \forall x \in \mathbb{R},$$

the second conclusion of the statement follows from 16.4.9.   □

The following theorem is the special case of Bargmann's theorem we mentioned before. In the first part of the proof (steps 1, 2 and 3) we exploit the special nature of $\mathbb{R}$ in order to extend the result of 16.4.6 to the whole of $\mathbb{R}$ and to obtain what Bargmann calls an exponent. In the second part we follow closely Bargmann's exposition (cf. Bargmann, 1954, p.19–21), which however deals with a more general situation than the one of interest for us here.

16.4.11 Theorem. *Let $\mathbb{R} \ni t \mapsto \omega_t \in \operatorname{Aut}\hat{\mathcal{H}}$ be a continuous one-parameter group of automorphisms. Then there exists a c.o.p.u.g. U in $\mathcal{H}$ such that*

$$\omega_t = \omega_{U(t)}, \forall t \in \mathbb{R}.$$

Proof. The mapping $\mathbb{R} \ni t \mapsto \omega_t \in \operatorname{Aut}\hat{\mathcal{H}}$ we are now considering has all the properties assumed for the mapping considered in 16.4.6 (cf. 16.4.2b and ag_2). Then there exists $a \in (0, \infty)$ and a mapping

$$(-a, a) \ni t \mapsto V_t \in \mathcal{U}(\mathcal{H})$$

with the properties listed in 16.4.6.

We divide the proof of the theorem into five steps.

Step 1: Here we define a mapping $\mathbb{R} \ni t \mapsto T(t) \in \mathcal{U}(\mathcal{H})$ such that

$$T(0) = \mathbb{1}_{\mathcal{H}} \quad \text{and} \quad \omega_t = \omega_{T(t)}, \forall t \in \mathbb{R}.$$

We set $b := \frac{1}{2}a$. Then,

$$\forall t \in \mathbb{R}, \exists!(k(t), r(t)) \in \mathbb{Z} \times [0, b) \text{ such that } t = k(t)b + r(t).$$

Thus we can define the mapping

$$T : \mathbb{R} \to \mathcal{U}(\mathcal{H})$$

$$t \mapsto T(t) := V_b^{k(t)}V_{r(t)}$$

(we recall that $V_b^0 := \mathbb{1}_{\mathcal{H}}$). We see that $T(0) = \mathbb{1}_{\mathcal{H}}$. Moreover, in view of ag_1,

$$\omega_{T(t)}([u]) = [V_b^{k(t)}V_{r(t)}u] = (\omega_b \circ \cdots k(t) \text{ times} \cdots \circ \omega_b \circ \omega_{r(t)})([u])$$

$$= \omega_{k(t)b+r(t)}([u]) = \omega_t([u]), \forall u \in \tilde{\mathcal{H}}, \forall t \in \mathbb{R},$$

and hence

$$\omega_{T(t)} = \omega_t, \forall t \in \mathbb{R}.$$

Step 2: Here we prove that the mapping $\mathbb{R} \ni t \mapsto T(t)f \in \mathcal{H}$ is continuous, for all $f \in \mathcal{H}$.

In what follows, we fix $f \in \mathcal{H}$.

In the first place we suppose that $t \in \mathbb{R}$ is such that $t \neq kb$ for all $k \in \mathbb{Z}$; then,

$$k(t)b < t < (k(t)+1)b;$$

if $\{t_n\}$ is a sequence in $\mathbb{R}$ such that $t_n \to t$, then there exists $n_0 \in \mathbb{N}$ so that

$$n > n_0 \Rightarrow k(t)b < t_n < (k(t)+1)b \Rightarrow t_n = k(t)b + r(t_n) \Rightarrow T(t_n) = V_b^{k(t)} V_{r(t_n)},$$

and hence

$$\|T(t_n)f - T(t)f\| \overset{(1)}{=} \|V_b^{k(t)} V_{r(t_n)} f - V_b^{k(t)} V_{r(t)} f\| = \|V_{r(t_n)} f - V_{r(t)} f\| \xrightarrow[n \to \infty]{} 0$$

(1 holds for $n > n_0$) since $r(t_n) - r(t) = t_n - t$ for $n > n_0$. This proves that the mapping $t \mapsto T(t)f$ is continuous at t.

In the second place we suppose that $t \in \mathbb{R}$ is such that $t = k(t)b$; hence, $T(t) = V_b^{k(t)}$. First let $\{t_n\}$ be a sequence in $\mathbb{R}$ such that $t_n \to t$ and such that there exists $n_1 \in \mathbb{N}$ so that

$$n > n_1 \Rightarrow k(t)b = t \leq t_n < (k(t)+1)b \Rightarrow T(t_n) = V_b^{k(t)} V_{r(t_n)};$$

then,

$$\|T(t_n)f - T(t)f\| \overset{(2)}{=} \|V_b^{k(t)} V_{r(t_n)} f - V_b^{k(t)} f\| = \|V_{r(t_n)} f - f\| \xrightarrow[n \to \infty]{} 0$$

(2 holds for $n > n_1$) since $r(t_n) = t_n - t$ for $n > n_1$; by the argument used in the proof of 2.4.2 ($b \Rightarrow a$), this implies that

$$\forall \varepsilon > 0, \exists \delta_\varepsilon^+ > 0 \text{ such that } t \leq s < t + \delta_\varepsilon^+ \Rightarrow \|T(t)f - T(s)f\| < \varepsilon.$$

Next let $\{t_n\}$ be a sequence in $\mathbb{R}$ such that $t_n \to t$ and such that there exists $n_2 \in \mathbb{N}$ so that

$$n > n_2 \Rightarrow (k(t)-1)b \leq t_n < t = k(t)b \Rightarrow T(t_n) = V_b^{k(t)-1} V_{r(t_n)};$$

then $t - t_n = b - r(t_n)$ for $n > n_2$ and hence

$$[t_n \to t] \Rightarrow [r(t_n) \to b] \Rightarrow [V_{r(t_n)} f \xrightarrow[n \to \infty]{} V_b f],$$

and hence, since $V_b^{k(t)-1} \in \mathcal{B}(\mathcal{H})$,

$$T(t_n)f \overset{(3)}{=} V_b^{k(t)-1} V_{r(t_n)} f \xrightarrow[n \to \infty]{} V_b^{k(t)-1} V_b t = T(t)f$$

(3 holds for $n > n_2$); by the argument used in the proof of 2.4.2 ($b \Rightarrow a$), this implies that

$$\forall \varepsilon > 0, \exists \delta_\varepsilon^- \text{ such that } t - \delta_\varepsilon^- < s < t \Rightarrow \|T(t)f - T(s)f\| < \varepsilon.$$

Letting $\delta_\varepsilon := \min\{\delta_\varepsilon^+, \delta_\varepsilon^-\}$, we have proved that

$$\forall \varepsilon > 0, \exists \delta_\varepsilon > 0 \text{ such that } |s - t| < \delta_\varepsilon \Rightarrow \|T(t)f - T(s)f\| < \varepsilon,$$

and hence that the mapping $t \mapsto T(t)f$ is continuous at t.

Since f was arbitrary, step 2 is concluded.

Step 3: The exponent ξ.

We have (cf. step 1 and ag_1)

$$[T(r)T(s)u] = \omega_r \circ \omega_s([u]) = \omega_{r+s}([u]) = [T(r+s)u], \forall u \in \tilde{\mathcal{H}}, \forall (r,s) \in \mathbb{R}^2,$$

and hence (cf. 10.9.6)

$$\forall (r,s) \in \mathbb{R}^2, \exists! \mu(r,s) \in \mathbb{T} \text{ such that } T(r)T(s) = \mu(r,s)T(r+s); \qquad (4)$$

this defines a function $\mathbb{R}^2 \ni (r,s) \mapsto \mu(r,s) \in \mathbb{T}$, for which we have

$$\begin{aligned}
\mu(r,s)\mu(r+s,t)T(r+s+t) &= \mu(r,s)T(r+s)T(t) = T(r)T(s)T(t) \\
&= T(r)\mu(s,t)T(s+t) \\
&= \mu(s,t)\mu(r,s+t)T(r+s+t), \forall (r,s,t) \in \mathbb{R}^3,
\end{aligned}$$

and hence

$$\mu(r,s)\mu(r+s,t) = \mu(s,t)\mu(r,s+t), \forall (r,s,t) \in \mathbb{R}^3. \qquad (5)$$

Moreover, from 4 we have

$$\mu(r,s)\mathbb{1}_{\mathcal{H}} = T(r)T(s)T(r+s)^{-1}, \forall (r,s) \in \mathbb{R}^2.$$

Since $T(0) = \mathbb{1}_{\mathcal{H}}$ (cf. step 1), this gives $\mu(0,0) = 1$. Moreover, for each $(r,s) \in \mathbb{R}^2$, if $\{(r_n, s_n)\}$ is a sequence in $\mathbb{R}^2$ such that $(r_n, s_n) \to (r,s)$ then $r_n \to r$, $s_n \to s$, $r_n + s_n \to r + s$, and hence (cf. step 2)

$$T(r_n)f \xrightarrow[n\to\infty]{} T(r)f, \; T(s_n)f \xrightarrow[n\to\infty]{} T(s)f, \; T(r_n + s_n)f \xrightarrow[n\to\infty]{} T(r+s)f, \forall f \in \mathcal{H},$$

and hence (cf. 16.4.7)

$$T(s_n)T(r_n + s_n)^{-1}f \xrightarrow[n\to\infty]{} T(s)T(r+s)^{-1}f, \forall f \in \mathcal{H},$$

and hence (cf. 16.4.7)

$$T(r_n)T(s_n)T(r_n + s_n)^{-1}f \xrightarrow[n\to\infty]{} T(r)T(s)T(r+s)^{-1}f, \forall f \in \mathcal{H},$$

and hence, for $u \in \tilde{\mathcal{H}}$,

$$\mu(r_n, s_n) = \left(u|T(r_n)T(s_n)T(r_n + s_n)^{-1}u\right) \xrightarrow[n\to\infty]{} \left(u|T(r)T(s)T(r+s)^{-1}u\right) = \mu(r,s);$$

thus, the function μ is continuous (cf. 2.4.2). Therefore, by 16.4.8 there exists a continuous function $\xi : \mathbb{R}^2 \to \mathbb{R}$ such that

$$\xi(0,0) = 0 \text{ and } \mu(r,s) = e^{i\xi(r,s)}, \forall (r,s) \in \mathbb{R}^2. \qquad (6)$$

The function

$$\mathbb{R}^3 \ni (r,s,t) \mapsto \xi(r,s) + \xi(r+s,t) - \xi(s,t) - \xi(r,s+t) \in \mathbb{R}$$

is obviously continuous; then, since $(\mathbb{R}^3, d_3)$ is a connected metric space (cf. 2.9.10), its range can only be either $\mathbb{R}$ or an interval or a singleton set (cf. 2.9.6); now, 5 implies that

$$\forall (r, s, t) \in \mathbb{R}^3, \exists n_{r,s,t} \in \mathbb{Z} \text{ such that}$$

$$\xi(r, s) + \xi(r + s, t) - \xi(s, t) - \xi(r, s + t) = 2n_{r,s,t}\pi;$$

hence,

$$\exists n \in \mathbb{Z} \text{ such that } \xi(r, s) + \xi(r + s, t) - \xi(s, t) - \xi(r, s + t) = 2n\pi, \forall (r, s, t) \in \mathbb{R}^3;$$

if we set $r = s = t = 0$ in this, we see that $n = 0$ since $\xi(0, 0) = 0$; thus,

$$\xi(r, s) + \xi(r + s, t) = \xi(s, t) + \xi(r, s + t), \forall (r, s, t) \in \mathbb{R}^3. \tag{7}$$

If we set $r = s = 0$ in this, we obtain

$$\xi(0, t) = 0, \forall t \in \mathbb{R}. \tag{8}$$

Step 4: The exponent ξ_0.

Throughout this step we fix a real function $\varphi \in \mathcal{C}_c(\mathbb{R})$ which is differentiable at all points of $\mathbb{R}$, and also such that the derivative φ' is continuous and $\int_{\mathbb{R}} \varphi dm = 1$. A possible choice is

$$\varphi(x) := \begin{cases} \frac{1}{2\pi}(\cos x + 1) & \text{if } x \in [-\pi, \pi] \\ 0 & \text{if } x \notin [-\pi, \pi]. \end{cases}$$

We define the function

$$\mathbb{R} \ni r \mapsto \lambda(r) := -\int_{\mathbb{R}} \xi(r, t)\varphi(t)dm(t) \in \mathbb{R}; \tag{9}$$

this function is continuous, in view of 16.4.9; moreover, $\lambda(0) = 0$ in view of 8.

Next we define the mapping

$$\mathbb{R} \ni t \mapsto W(t) := e^{i\lambda(t)}T(t) \in \mathcal{U}(\mathcal{H}).$$

In view of step 1 we have

$$W(0) = \mathbb{1}_{\mathcal{H}} \text{ and } \omega_t = \omega_{W(t)}, \forall t \in \mathbb{R}; \tag{10}$$

in view of step 2 and of 10.1.16b we have that

$$\text{the mapping } \mathbb{R} \ni t \mapsto W(t)f \in \mathcal{H} \text{ is continuous, } \forall f \in \mathcal{H}. \tag{11}$$

In view of 4 and 6 we see that

$$W(r)W(s) = e^{i(\lambda(r)+\lambda(s)-\lambda(r+s)+\xi(r,s))}W(r + s), \forall (r, s) \in \mathbb{R}^2,$$

or

$$W(r)W(s) = e^{i\xi_0(r,s)}W(r + s), \forall (r, s) \in \mathbb{R}^2, \tag{12}$$

where ξ_0 is the function defined by

$$\mathbb{R}^2 \ni (r, s) \mapsto \xi_0(r, s) := \xi(r, s) + \lambda(r) + \lambda(s) - \lambda(r + s) \in \mathbb{R}; \tag{13}$$

the function ξ_0 is obviously continuous; moreover, $\xi_0(0,0) = 0$ in view of 8. Then we can repeat the reasoning which led from 4 and 6 to 7, to obtain

$$\xi_0(r,s) + \xi_0(r+s,t) = \xi_0(s,t) + \xi_0(r,s+t), \forall(r,s,t) \in \mathbb{R}^3; \tag{14}$$

in we set $s = t = 0$ in this, we obtain

$$\xi_0(r,0) = 0, \forall r \in \mathbb{R}. \tag{15}$$

For each $(r,s) \in \mathbb{R}^2$, since obviously

$$\xi(r,s) = \int_{\mathbb{R}} \xi(r,s)\varphi(t)dm(t),$$

we have

$$\xi_0(r,s) \overset{(16)}{=} \int_{\mathbb{R}} (\xi(r,s) - \xi(r,t) - \xi(s,t) + \xi(r+s,t))\varphi(t)dm(t)$$

$$\overset{(17)}{=} \int_{\mathbb{R}} \xi(r,s+t)\varphi(t)dm(t) - \int_{\mathbb{R}} \xi(r,t)\varphi(t)dm(t)$$

$$\overset{(18)}{=} \int_{\mathbb{R}} \xi(r,t)\varphi(t-s)dm(t) - \int_{\mathbb{R}} \xi(r,t)\varphi(t)dm(t),$$

where 16 holds in view of 9 and 13, 17 holds in view of 7, 18 holds by 9.2.1b; by 16.4.10, this shows that the partial derivative $\frac{\partial\xi_0}{\partial s}(r,s)$ exists and that

$$\text{the function } \mathbb{R} \ni r \mapsto \frac{\partial\xi_0}{\partial s}(r,0) \in \mathbb{R} \text{ is continuous.} \tag{19}$$

Step 5: Here we prove the statement of the theorem.

We define the function

$$\mathbb{R}^2 \ni (r,s) \mapsto \psi(r,s) := \frac{\partial\xi_0}{\partial s}(r,s) \in \mathbb{R}.$$

If we derive 14 with respect to t at $t = 0$ we obtain

$$\psi(r+s,0) = \psi(s,0) + \psi(r,s), \forall(r,s) \in \mathbb{R}^2 \tag{20}$$

From 19 and 20 we have that the function ψ is continuous.

Next we define the function

$$\mathbb{R} \ni t \mapsto \lambda_0(t) := \int_0^t \psi(r,0)dr \in \mathbb{R}$$

(the symbol $\int_a^b$ has in the present proof the same meaning it has in the proof of 16.1.10), which is continuous.

For all $(t_1,t_2) \in \mathbb{R}^2$, we have

$$\lambda_0(t_1 + t_2) - \lambda_0(t_1) - \lambda_0(t_2)$$

$$= \int_0^{t_1+t_2} \psi(r,0)dr - \int_0^{t_1} \psi(r,0)dr - \int_0^{t_2} \psi(r,0)dr$$

$$= \int_{t_1}^{t_1+t_2} \psi(r,0)dr - \int_0^{t_2} \psi(r,0)dr \overset{(21)}{=} \int_0^{t_2} \psi(t_1+r,0)dr - \int_0^{t_2} \psi(r,0)dr$$

$$\overset{(22)}{=} \int_0^{t_2} \psi(t_1,r)dr \overset{(23)}{=} \xi_0(t_1,t_2) - \xi_0(t_1,0) \overset{(24)}{=} \xi_0(t_1,t_2),$$

where 21 holds by a change of variable (cf. the explanation of 7 and 8 in the proof of 16.1.10), 22 holds in view of 20, 23 holds by Riemann's fundamental theorem of calculus, 24 holds in view of 15.

Finally, we define the mapping

$$\mathbb{R} \ni t \mapsto U(t) := e^{i\lambda_0(t)} W(t) \in \mathcal{U}(\mathcal{H}).$$

In view of 10 we have

$$\omega_t = \omega_{U(t)}, \forall t \in \mathbb{R}.$$

In view of 11 and of 10.1.16b we have that

the mapping $\mathbb{R} \ni t \mapsto U(t)f \in \mathcal{H}$ is continuous, $\forall f \in \mathcal{H}$;

moreover, in view of 12 and of the equation proved above we have

$$U(t_1)U(t_2) = e^{i(\lambda_0(t_1) + \lambda_0(t_2) - \lambda_0(t_1 + t_2) + \xi_0(t_1, t_2))} U(t_1 + t_2)$$
$$= U(t_1 + t_2), \forall t_1, t_2 \in \mathbb{R};$$

thus, the mapping U is a c.o.p.u.g. $\qquad\square$

Chapter 17

Commuting Operators and Reducing Subspaces

As usual, $\mathcal{H}$ denotes an abstract Hilbert space throughout this chapter.

Outside $\mathcal{B}(\mathcal{H})$, the idea of commutativity of operators needs careful examination even from its definition, and the idea of a reducing subspace for an operator is subtler than the idea of an invariant subspace. The subject of the present chapter is the study of these ideas, which are closely connected with each other.

17.1 Commuting operators

The results of operations performed on elements of $\mathcal{O}(\mathcal{H})$ (cf. 3.2.1) can be misunderstood if they are interpreted as if $\mathcal{O}(\mathcal{H})$ were an algebra, since $\mathcal{O}(\mathcal{H})$ is not an algebra and not even a linear space (cf. 3.2.11). This is true in particular for the commutator of two elements of $\mathcal{O}(\mathcal{H})$.

17.1.1 Definitions.

(a) Let A, B be elements of $\mathcal{O}(\mathcal{H})$, i.e. linear operators in $\mathcal{H}$. The *commutator* of A and B is the linear operator denoted by the symbol $[A, B]$ and defined by

$$[A, B] := AB - BA.$$

We note that

$$D_{[A,B]} = \{f \in D_A \cap D_B : Af \in D_B \text{ and } Bf \in D_A\}.$$

(b) Two elements A and B of $\mathcal{B}(\mathcal{H})$ are said to *commute* if $AB = BA$, i.e. if

$$[A, B] = \mathbb{O}_{\mathcal{H}}.$$

The definition given in 17.1.1b is the natural one for elements of $\mathcal{B}(\mathcal{H})$, since $\mathcal{B}(\mathcal{H})$ is an algebra (cf. 4.3.5). It might be thought that this definition could be generalized meaningfully to arbitrary elements of $\mathcal{O}(\mathcal{H})$ in a direct way, by saying that two elements A and B of $\mathcal{O}(\mathcal{H})$ commute if $[A, B] \subset \mathbb{O}_{\mathcal{H}}$. However, this definition would not be very useful. Firstly, it is clear that the content of the relation $[A, B] \subset \mathbb{O}_{\mathcal{H}}$ depends on the size of $D_{[A,B]}$ (it is even void if $D_{[A,B]} = \{0_{\mathcal{H}}\}$). Moreover, if A and B are self-adjoint elements of $\mathcal{B}(\mathcal{H})$ then the relation $[A, B] = \mathbb{O}_{\mathcal{H}}$

has consequences (cf. 17.1.4 and 17.1.7) which are not granted in general by the relation $[A, B] \subset \mathbb{O}_{\mathcal{H}}$ when A and B are self-adjoint elements of $\mathcal{O}(\mathcal{H})$ which are not defined on the whole of $\mathcal{H}$ (we recall that, for a self-adjoint operator A, $D_A = \mathcal{H}$ is equivalent to $A \in \mathcal{B}(\mathcal{H})$, cf. 12.4.7); this is proved by examples (cf. 17.1.8). In general, the condition $[A, B] \subset \mathbb{O}_{\mathcal{H}}$ does not seem to lead to interesting results for self-adjoint operators A and B which are not in $\mathcal{B}(\mathcal{H})$. The main task of this section is to find a commutativity condition for self-adjoint operators which has the same consequences whether or not the operators are in $\mathcal{B}(\mathcal{H})$.

We start off by noting that there is a condition of commutativity which has already played a role in previous chapters.

17.1.2 Definition. An element B of $\mathcal{B}(\mathcal{H})$ is said to *commute* with an element A of $\mathcal{O}(\mathcal{H})$ if

$$BA \subset AB,$$

i.e. if the following implication holds true (notice that $D_{BA} = D_A$)

$$f \in D_A \Rightarrow [Bf \in D_A \text{ and } BAf = ABf].$$

It is obvious that, if both A and B are elements of $\mathcal{B}(\mathcal{H})$, then B commutes with A iff $[A, B] = \mathbb{O}_{\mathcal{H}}$. In this sense, the definition now given generalizes the one given in 17.1.1b.

17.1.3 Remarks.

(a) Let $(X, \mathcal{A})$ be a measurable space and let $P : \mathcal{A} \to \mathscr{P}(\mathcal{H})$ be a projection valued measure. If $B \in \mathcal{B}(\mathcal{H})$ is such that $[B, P(E)] = \mathbb{O}_{\mathcal{H}}$ for all $E \in \mathcal{A}$, then B commutes with J_{φ}^{P} for all $\varphi \in \mathcal{M}(X, \mathcal{A}, P)$ (cf. 14.2.14e).

(b) The previous remark implies that, if A is a self-adjoint operator in $\mathcal{H}$ and $B \in \mathcal{B}(\mathcal{H})$ is such that $[B, P^A(E)] = \mathbb{O}_{\mathcal{H}}$ for all $E \in \mathcal{A}(d_{\mathbb{R}})$, then B commutes with $\varphi(A)$ for all $\varphi \in \mathcal{M}(\mathbb{R}, \mathcal{A}(d_{\mathbb{R}}), P^A)$; in particular, B commutes with A since $A = \xi(A)$.

(c) If A is a self-adjoint operator in $\mathcal{H}$ and $B \in \mathcal{B}(\mathcal{H})$ commutes with A, then $[B, P^A(E)] = \mathbb{O}_{\mathcal{H}}$ for all $E \in \mathcal{A}(d_{\mathbb{R}})$ (cf. 15.2.1B).

(d) Remarks b and c imply that, if A is a self-adjoint operator in $\mathcal{H}$ and $B \in \mathcal{B}(\mathcal{H})$ commutes with A, then B commutes with $\varphi(A)$ for all $\varphi \in \mathcal{M}(\mathbb{R}, \mathcal{A}(d_{\mathbb{R}}), P^A)$.

We note that, while the definition provided in 17.1.1b sets up a relation in $\mathcal{B}(\mathcal{H})$ which is obviously symmetric, for $A \in \mathcal{O}(\mathcal{H})$ and $B \in \mathcal{B}(\mathcal{H})$ the condition $BA \subset AB$ is asymmetric if $D_A \neq \mathcal{H}$. The implication equivalent to $BA \subset AB$ that is written in 17.1.2 makes this asymmetry immediately clear. If neither of two linear operators A and B is an element of $\mathcal{B}(\mathcal{H})$ then we do not try at all to define anything like commutativity for A and B, unless both operators are self-adjoint. Indeed, 17.1.4 proves that if both $A \in \mathcal{O}(\mathcal{H})$ and $B \in \mathcal{B}(\mathcal{H})$ are self-adjoint then the condition $BA \subset AB$ is in fact equivalent to a relation in which A and B have equal roles, and suggests how this condition can be generalized to a symmetric relation between any kind of self-adjoint operators (cf. 17.1.5). After that, 17.1.7 shows that this generalization is a meaningful condition of commutativity for self-adjoint operators.

17.1.4 Proposition. *Let A and B be self-adjoint operators in $\mathcal{H}$ and suppose that $B \in \mathcal{B}(\mathcal{H})$. Then the following conditions are equivalent:*

(a) B commutes with A, i.e. $BA \subset AB$;
(b) the projection valued measures P^A and P^B commute, i.e. (cf. 13.5.1)

$$[P^A(E), P^B(F)] = \mathbb{O}_{\mathcal{H}}, \forall E, F \in \mathcal{A}(d_{\mathbb{R}}).$$

Proof. $a \Rightarrow b$: We assume condition a and fix $E \in \mathcal{A}(d_{\mathbb{R}})$. Then we have $[B, P^A(E)] = \mathbb{O}_{\mathcal{H}}$ (cf. 17.1.3c). Thus, $P^A(E)$ commutes with B and hence $[P^A(E), P^B(F)] = \mathbb{O}_{\mathcal{H}}$ for all $F \in \mathcal{A}(d_{\mathbb{R}})$ (cf. 17.1.3c with $A := B$ and $B := P^A(E)$). Since E was an arbitrary element of $\mathcal{A}(d_{\mathbb{R}})$, this proves condition b.

$b \Rightarrow a$: We assume condition b. Then, for every $E \in \mathcal{A}(d_{\mathbb{R}})$, $P^A(E)$ commutes with B (cf. 17.1.3b with $A := B$ and $B := P^A(E)$), i.e. $[B, P^A(E)] = \mathbb{O}_{\mathcal{H}}$. By 17.1.3b, this implies condition a. $\square$

In view of 17.1.4, the following definition is consistent with the definition given in 17.1.2.

17.1.5 Definition. Two self-adjoint operators A and B in $\mathcal{H}$ are said to *commute* if the projection valued measures P^A and P^B commute, i.e. if

$$[P^A(E), P^B(F)] = \mathbb{O}_{\mathcal{H}}, \forall E, F \in \mathcal{A}(d_{\mathbb{R}}).$$

17.1.6 Remarks.

(a) Let A and B be bounded self-adjoint operators in $\mathcal{H}$. Then, A and B commute (in the sense of 17.1.5) iff $[A, B] = \mathbb{O}_{\mathcal{H}}$. Indeed, A and B are elements of $\mathcal{B}(\mathcal{H})$ (cf. 12.4.7) and hence condition a of 17.1.4 reads $[A, B] = \mathbb{O}_{\mathcal{H}}$.
(b) Every self-adjoint operator A in $\mathcal{H}$ commutes with itself, i.e.

$$[P^A(E), P^A(F)] = \mathbb{O}_{\mathcal{H}}, \forall E, F \in \mathcal{A}(d_{\mathbb{R}})$$

(cf. 13.3.2d).
(c) Let A and B be commuting (in the sense of 17.1.5) self-adjoint operators, let φ be a real element of $\mathcal{M}(\mathbb{R}, \mathcal{A}(d_{\mathbb{R}}), P^A)$, let ψ be a real element of $\mathcal{M}(\mathbb{R}, \mathcal{A}(d_{\mathbb{R}}), P^B)$. Then the operators $\varphi(A)$ and $\psi(B)$ are self-adjoint and they commute (in the sense of 17.1.5). This follows immediately from 15.3.8.

Every self-adjoint operator defines the family of operators that contains the operator itself and the ranges of the projection valued measure and of the continuous one-parameter unitary group determined by the operator. The next theorem proves that two self-adjoint operators commute (in the sense of 17.1.5) iff any bounded element of the family defined as above by one of them commutes (in the sense of 17.1.2) with any element of the family defined as above by the other one.

17.1.7 Theorem. *Let A and B be self-adjoint operators in $\mathcal{H}$. Then the following conditions are equivalent:*

(a) A and B commute (in the sense defined in 17.1.5);
(b) $P^A(E)$ commutes with B, i.e. $P^A(E)B \subset BP^A(E)$, $\forall E \in \mathcal{A}(d_{\mathbb{R}})$;
(c) $[P^A(E), U^B(t)] = \mathbb{O}_{\mathcal{H}}$, $\forall E \in \mathcal{A}(d_{\mathbb{R}})$, $\forall t \in \mathbb{R}$;
(d) $[U^A(s), U^B(t)] = \mathbb{O}_{\mathcal{H}}$, $\forall s, t \in \mathbb{R}$;
(e) $U^A(s)B = BU^A(s)$, $\forall s \in \mathbb{R}$;
(f) $U^A(s)$ commutes with B, i.e. $U^A(s)B \subset BU^A(s)$, $\forall s \in \mathbb{R}$.

Obviously, in view of the symmetry between A and B in condition a, the above conditions are also equivalent to conditions b, c, e, f with A and B interchanged, and hence in particular to condition

(g) $[U^A(s), P^B(F)] = \mathbb{O}_{\mathcal{H}}$, $\forall s \in \mathbb{R}$, $\forall F \in \mathcal{A}(d_{\mathbb{R}})$.

If the above conditions are satisfied, then the following condition holds true:

(h) $[A, B] \subset \mathbb{O}_{\mathcal{H}}$.

Proof. $a \Rightarrow b$: We assume condition a. Then, for every $E \in \mathcal{A}(d_{\mathbb{R}})$, $P^A(E)$ commutes with B (cf. 17.1.3b with $A := B$ and $B := P^A(E)$).

$b \Rightarrow c$: This follows from 17.1.3d (with $A := B$, $B := P^A(E)$, $\varphi := \varphi_t$, cf. 16.1.7).

$c \Rightarrow d$: This follows from 17.1.3b (with $B := U^B(t)$ and $\varphi := \varphi_s$).

$d \Rightarrow e$: We assume condition d and fix $s \in \mathbb{R}$. Then,

$$U^B(t) = U^A(s)U^B(t)(U^A(s))^{-1}, \forall t \in \mathbb{R}.$$

By 16.3.1, this yields

$$B = U^A(s)B(U^A(s))^{-1},$$

whence (cf. 3.2.10b_1)

$$BU^A(s) = U^A(s)B.$$

$e \Rightarrow f$: This is obvious.

$f \Rightarrow a$: We assume condition f. Then we have, for every $s \in \mathbb{R}$,

$$[U^A(s), P^B(F)] = \mathbb{O}_{\mathcal{H}}, \forall F \in \mathcal{A}(d_{\mathbb{R}})$$

(cf. 17.1.3c with $A := B$ and $B := U^A(s)$). Now we fix $F \in \mathcal{A}(d_{\mathbb{R}})$. We have

$$P^B(F)U^A(s) = U^A(s)P^B(F), \forall s \in \mathbb{R}. \tag{1}$$

Let $f \in D_A$; for any sequence $\{t_n\}$ in $\mathbb{R} - \{0\}$ such that $t_n \to 0$, if we define

$$g_n := \frac{1}{t_n}(U^A(t_n) - \mathbb{1}_{\mathcal{H}})f, \forall n \in \mathbb{N},$$

then we have

$$Af = -i \lim_{n\to\infty} g_n$$

(cf. 16.1.6); since $P^B(F) \in \mathcal{B}(\mathcal{H})$, this implies that

the sequence $\{P^B(F)g_n\}$ is convergent and $P^B(F)Af = -i \lim_{n\to\infty} P^B(F)g_n$;

now, 1 implies that

$$P^B(F)g_n = \frac{1}{t_n}(U^A(t_n) - \mathbb{1}_{\mathcal{H}})P^B(F)f;$$

this proves that

$$P^B(F)f \in D_A \text{ and } P^B(F)Af = AP^B(F)f$$

(cf. 16.1.6 and 16.1.5a). Thus, $P^B(F)$ commutes with A (in the sense of 17.1.2).
Then,

$$[P^B(F), P^A(E)] = \mathbb{O}_{\mathcal{H}}, \forall E \in \mathcal{A}(d_{\mathbb{R}}),$$

in view of 17.1.3c (with $B := P^B(F)$). Since F was an arbitrary element of $\mathcal{A}(d_{\mathbb{R}})$,
condition a is proved.

h: We assume condition f. For every $f \in D_{[A,B]}$ and every $g \in D_B$, we have

$$(g|U^A(s)Bf) = (g|BU^A(s)f) = (Bg|U^A(s)f), \forall s \in \mathbb{R},$$

and hence, for a sequence $\{t_n\}$ in $\mathbb{R} - \{0\}$ such that $t_n \to 0$,

$$(g|iABf) = \lim_{n\to\infty} \frac{1}{t_n}\left(g|(U^A(t_n) - \mathbb{1})Bf\right)$$

$$= \lim_{n\to\infty} \frac{1}{t_n}\left(Bg|(U^A(t_n) - \mathbb{1}_{\mathcal{H}})f\right) = (Bg|iAf)$$

(cf. 16.1.6; note that $Bf \in D_A$ and $f \in D_A$), and hence

$$(g|ABf) = (g|BAf)$$

(note that $Af \in D_B$). Since $\overline{D_B} = \mathcal{H}$, this implies (cf. 10.2.11 and 10.2.10a)

$$ABf = BAf, \forall f \in D_{[A,B]},$$

which is condition h. $\qquad\square$

From 17.1.7 we see that if two self-adjoint operators A and B commute (in the sense defined in 17.1.5) then $[A, B] \subset \mathbb{O}_{\mathcal{H}}$. It is almost obvious that the converse cannot be true in general because two self-adjoint operators A and B can be such that $[A, B] \subset \mathbb{O}_{\mathcal{H}}$, but with $D_{[A,B]}$ so little that the relation $[A, B] \subset \mathbb{O}_{\mathcal{H}}$ is of no consequence. Now, one might conjecture that, if $[A, B] \subset \mathbb{O}_{\mathcal{H}}$ and $D_{[A,B]}$ is dense in $\mathcal{H}$, then A and B should commute. Example a in 17.1.8 proves that this conjecture is false. Then one might go one step further and conjecture that if $D_{[A,B]}$ is not only dense in $\mathcal{H}$ but also large enough so that two self-adjoint operators A and B are uniquely determined by their restrictions to $D_{[A,B]}$, then $[A, B] \subset \mathbb{O}_{\mathcal{H}}$ could be a sufficient condition to guarantee commutativity of A and B. In particular, one might conjecture that this should be the case when the restrictions of A and B to $D_{[A,B]}$ are essentially self-adjoint, since an essentially self-adjoint operator has a unique self-adjoint extension, its closure (cf. 12.4.11). However this conjecture is proved false by example b in 17.1.8.

17.1.8 Examples.

(a) The Hilbert space of this example is $L^2(a,b)$. As in 12.4.25, here we do not distinguish between a symbol φ for an element of $C(a,b)$ and the symbol $[\varphi]$ for the element of $L^2(a,b)$ that contains φ. Accordingly, the family of functions $C_0^\infty(a,b)$ defined in 11.4.17 is identified with the subset $\iota(C_0^\infty(a,b))$ of $L^2(a,b)$. We consider the operators A_0 and A_1 defined as A_θ in 12.4.25, with $\theta := 0$ and $\theta := 1$. It is obvious that

$$[A_0, A_1] \subset \mathbb{O}_{L^2(a,b)}.$$

Moreover, it is obvious that $C_0^\infty(a,b) \subset D_{[A_0,A_1]}$. Now, $C_0^\infty(a,b)$ is dense in $L^2(a,b)$ (cf. 11.4.21) and hence so is $D_{[A_0,A_1]}$. Then we have

$$\mathbb{O}_{L^2(a,b)} = \mathbb{O}_{L^2(a,b)}^\dagger \subset [A_0, A_1]^\dagger$$

(cf. 12.1.4; the equation $\mathbb{O}_{L^2(a,b)} = \mathbb{O}_{L^2(a,b)}^\dagger$ follows e.g. from 12.1.3B), and hence

$$[A_0, A_1]^\dagger = \mathbb{O}_{L^2(a,b)},$$

and hence

$$A_1^\dagger A_0^\dagger - A_0^\dagger A_1^\dagger \subset (A_0 A_1)^\dagger - (A_1 A_0)^\dagger \subset (A_0 A_1 - A_1 A_0)^\dagger = \mathbb{O}_{L^2(a,b)}$$

(cf. 12.3.4a and 12.3.1a). Thus, $A_0^\dagger$ and $A_1^\dagger$ are self-adjoint operators (since A_0 and A_1 are essentially self-adjoint, cf. 12.4.25) such that

$$[A_0^\dagger, A_1^\dagger] \subset \mathbb{O}_{L^2(a,b)}$$

and

$$D_{[A_0^\dagger, A_1^\dagger]} \text{ is dense in } L^2(a,b)$$

(note that $D_{[A_0,A_1]} \subset D_{[A_0^\dagger, A_1^\dagger]}$ since $A_0 \subset A_0^\dagger$ and $A_1 \subset A_1^\dagger$).

Now, the conditions of 15.3.4B hold true for both the self-adjoint operators $A_0^\dagger$ and $A_1^\dagger$ (cf. 12.4.25). The number 0 is eigenvalue of $A_0^\dagger$ and its eigenspace is the one-dimensional subspace generated by the element u of $L^2(a,b)$ defined by

$$u(x) := \left(\frac{1}{b-a}\right)^{\frac{1}{2}}, \forall x \in [a,b];$$

therefore we have

$$P^{A_0^\dagger}(\{0\})\varphi = (u|\varphi)\,u, \forall \varphi \in L^2(a,b)$$

(we identify the symbols φ and $[\varphi]$ also for an element φ of $\mathcal{L}^2(a,b)$). The number $\frac{1}{b-a}$ is eigenvalue for $A_1^\dagger$ and its eigenspace is the one-dimensional subspace generated by the element v of $L^2(a,b)$ defined by

$$v(x) := \left(\frac{1}{b-a}\right)^{\frac{1}{2}} \exp\left(i\frac{x-a}{b-a}\right), \forall x \in [a,b];$$

therefore we have

$$P^{A_1^\dagger}\left(\left\{\frac{1}{b-a}\right\}\right)\varphi = (v|\varphi)\,v, \forall\varphi \in L^2(a,b).$$

Thus, we have

$$P^{A_0^\dagger}(\{0\})P^{A_1^\dagger}\left(\left\{\frac{1}{b-a}\right\}\right)v = (u|v)\,u$$

and

$$P^{A_1^\dagger}\left(\left\{\frac{1}{b-a}\right\}\right)P^{A_0^\dagger}(\{0\})v = (u|v)\,(v|u)\,v.$$

Now,

$$(u|v) = \frac{1}{b-a}\int_a^b \exp\left(i\frac{x-a}{b-a}\right)dx = \int_0^1 e^{is}ds \neq 0,$$

and hence

$$\left[P^{A_0^\dagger}(\{0\}), P^{A_1^\dagger}\left(\left\{\frac{1}{b-a}\right\}\right)\right] \neq \mathbb{O}_{L^2(a,b)}.$$

This proves that the self-adjoint operators $A_0^\dagger$ and $A_1^\dagger$ do not commute (in the sense defined in 17.1.5).

We point out that $D_{[A_0^\dagger, A_1^\dagger]}$, though dense in $L^2(a,b)$, cannot be so that the restrictions of $A_0^\dagger$ and $A_1^\dagger$ to $D_{[A_0^\dagger, A_1^\dagger]}$ are essentially self-adjoint, since these restrictions are equal but the self-adjoint operators $A_0^\dagger$ and $A_1^\dagger$ are not (cf. 12.4.11c and 12.4.13).

(b) This example is due to Edward Nelson (cf. Reed and Simon, 1980, 1972, p.306), and its key-point is the proof of the following proposition.

There exists a Hilbert space $\mathcal{K}$, a linear manifold D dense in $\mathcal{K}$, and two linear operators A and B in $\mathcal{K}$ so that:

(a) $D_A = D_B = D$, $A(D) \subset D$, $B(D) \subset D$;
(b) $ABf - BAf = 0_\mathcal{K}, \forall f \in D$;
(c) A and B are essentially self-adjoint;
(d) $\exists f \in D$ such that $U^{\bar{A}}(1)U^{\bar{B}}(1)f \neq U^{\bar{B}}(1)U^{\bar{A}}(1)f$.

We do not prove this proposition. A scheme of its proof can be found at p.273–274 of (Reed and Simon, 1980, 1972).

We note that from condition a and b we have

$$AB - BA \subset \mathbb{O}_\mathcal{K},$$

and hence (in view of 12.1.4), since condition a implies $D_{BA-AB} = D$ and hence $\overline{D_{BA-AB}} = \mathcal{K}$,

$$\mathbb{O}_{\mathcal{K}} = \mathbb{O}_{\mathcal{K}}^{\dagger} \subset (BA - AB)^{\dagger},$$

and hence

$$(BA - AB)^{\dagger} = \mathbb{O}_{\mathcal{K}}.$$

Then from condition c we have (in view of 12.1.6b, 12.3.4a, 12.3.1a)

$$\overline{A}\,\overline{B} - \overline{B}\,\overline{A} = A^{\dagger}B^{\dagger} - B^{\dagger}A^{\dagger} \subset (BA)^{\dagger} - (AB)^{\dagger} \subset (BA - AB)^{\dagger} = \mathbb{O}_{\mathcal{K}}.$$

However, condition d proves (in view of 17.1.7) that the self-adjoint operators $\overline{A}$ and $\overline{B}$ do not commute (in the sense defined in 17.1.5).

Two self-adjoint operators commute (in the sense of 17.1.5) iff they are functions of a third self-adjoint operator (cf. 17.1.10 a $\Leftrightarrow$ c). The difficult part of this equivalence is proved by the next theorem. The main idea for the proof we provide is drawn from Section 130 of (Riesz and Sz.-Nagy, 1972). We will write this proof in full detail even at the risk of belabouring the obvious.

17.1.9 Theorem. *Let A_1 and A_2 be self-adjoint operators in $\mathcal{H}$, and suppose that they commute (in the sense defined in 17.1.5). Then there exist a self-adjoint operator B in $\mathcal{H}$ and two real functions $\varphi_i \in \mathcal{M}(\mathbb{R}, \mathcal{A}(d_{\mathbb{R}}), P^B)$ so that $A_i = \varphi_i(B)$, for $i = 1, 2$.*

Proof. We divide the proof into four steps.

Step 1: The projection valued measure P on $\mathcal{A}(d_2)$ and the operators $J_{\pi_i}^P$.

Let $\varphi : \mathbb{R} \to \mathbb{R}$ be a bijection from $\mathbb{R}$ onto an interval (α, β) such that $0 \le \alpha < \beta \le 1$ and such that both φ and φ^{-1} are continuous (e.g., the function defined by $\varphi(x) := \frac{1}{2}(1 + e^x)^{-1}, \forall x \in \mathbb{R}$, has these properties).

The operators $\varphi(A_1)$ and $\varphi(A_2)$ are self-adjoint and they commute (cf. 17.1.6c). Therefore (cf. 13.5.3) there exists a unique projection valued measure P on the Borel σ-algebra $\mathcal{A}(d_2)$ such that

$$P(E_1 \times E_2) = P^{\varphi(A_1)}(E_1)P^{\varphi(A_2)}(E_2), \forall E_1, E_2 \in \mathcal{A}(d_{\mathbb{R}}).$$

We note that (cf. 15.3.8)

$$\begin{aligned} P([0,1) \times [0,1)) &= P^{A_1}(\varphi^{-1}([0,1)))P^{A_2}(\varphi^{-1}([0,1))) \\ &= P^{A_1}(\mathbb{R})P^{A_2}(\mathbb{R}) = \mathbb{1}_{\mathcal{H}}. \end{aligned} \tag{1}$$

For $i = 1, 2$, we define the function

$$\mathbb{R}^2 \ni (x_1, x_2) \mapsto \pi_i(x_1, x_2) := x_i \in \mathbb{R},$$

and the operator $\tilde{A}_i := J_{\pi_i}^P$. Both the operators $\tilde{A}_i$ are elements of $\mathcal{B}(\mathcal{H})$ because $\pi_i \in \mathcal{L}^{\infty}(\mathbb{R}^2, \mathcal{A}(d_2), P)$ (cf. 14.2.17); moreover, they are self-adjoint and

$$P^{\tilde{A}_1}(E) = P(\pi_1^{-1}(E)) = P(E \times \mathbb{R}) = P^{\varphi(A_1)}(E), \forall E \in \mathcal{A}(d_{\mathbb{R}}),$$
$$P^{\tilde{A}_2}(E) = P(\pi_2^{-1}(E)) = P(\mathbb{R} \times E) = P^{\varphi(A_2)}(E), \forall E \in \mathcal{A}(d_{\mathbb{R}})$$

(cf. 15.2.7); therefore,

$$J_{\pi_i}^P = \varphi(A_i) \text{ for } i = 1, 2. \tag{2}$$

Step 2: The partitions $\mathcal{F}^m$ of $[0,1)$ and $\mathcal{G}^m$ of $[0,1) \times [0,1)$.

For all $m \in \mathbb{N}$, we define a partition $\mathcal{F}^m$ of the interval $[0,1)$ by letting

$$\mathcal{F}^m := \{\iota_n^m\}_{n=1,...,4^m},$$

with $\iota_n^m := [(n-1)4^{-m}, n4^{-m})$ for $n = 1, ..., 4^m$.

For all $m \in \mathbb{N}$, we define a partition $\mathcal{G}^m$ of the square $[0,1) \times [0,1)$ by letting

$$\mathcal{G}^m := \{\sigma_n^m\}_{n=1,...,4^m},$$

where $\{\sigma_n^m\}_{n=1,...,4^m}$ is the family of half-open squares, of the $[w, x) \times [y, z)$ type, that is defined inductively as follows: for $m = 1$ we define $\{\sigma_n^1\}_{n=1,...,4}$ by

$$[0,1) \times [0,1) = \begin{array}{|c|c|} \hline \sigma_1^1 & \sigma_4^1 \\ \hline \sigma_2^1 & \sigma_3^1 \\ \hline \end{array} \quad ;$$

for $m > 1$, supposing that $\{\sigma_n^m\}_{n=1,...,4^m}$ has already been defined, we define $\{\sigma_n^{m+1}\}_{n=1,...,4^{m+1}}$ by

$$\sigma_n^m = \begin{array}{|c|c|} \hline \sigma_{4n-3}^{m+1} & \sigma_{4n}^{m+1} \\ \hline \sigma_{4n-2}^{m+1} & \sigma_{4n-1}^{m+1} \\ \hline \end{array} \quad , \text{ for all } n \in \{1, ..., 4^m\}. \tag{3}$$

It can be easily proved by induction that, for all $m \in \mathbb{N}$, for all $n \in \{1, ..., 4^m\}$, for all $l \in \mathbb{N}$ such that $m < l$,

$$\iota_n^m = \bigcup_{s \in I(m,l)} \iota_s^l \text{ and } \sigma_n^m = \bigcup_{s \in I(m,l)} \sigma_s^l, \tag{4}$$

with $I(m,l) := \{n4^{l-m} - r : r = 0, 1, ..., 4^{l-m} - 1\}$.

For all $m \in \mathbb{N}$ and for all $l \in \mathbb{N}$ such that $m \le l$, in view of 4 we have, for $n = 1, ..., 4^m$ and $s = 1, ..., 4^l$:

$$\text{either } \iota_s^l \subset \iota_n^m \text{ or } \iota_s^l \cap \iota_n^m = \emptyset; \tag{5}$$

$$\text{either } \sigma_s^l \subset \sigma_n^m \text{ or } \sigma_s^l \cap \sigma_n^m = \emptyset; \tag{6}$$

$$\iota_s^l \subset \iota_n^m \text{ iff } \sigma_s^l \subset \sigma_n^m; \tag{7}$$

$$\iota_s^l \cap \iota_n^m = \emptyset \text{ iff } \sigma_s^l \cap \sigma_n^m = \emptyset; \tag{8}$$

For $l, m \in \mathbb{N}$, let I_1 and I_2 be subfamilies of $\{1, ..., 4^l\}$ and of $\{1, ..., 4^m\}$ respectively. First we note that if $l = m$ then (in view of 5)

$$\left[\bigcup_{s \in I_1} \iota_s^m = \bigcup_{n \in I_2} \iota_n^m \right] \Rightarrow \left[I_1 = I_2 \text{ and hence } \bigcup_{s \in I_1} \sigma_s^m = \bigcup_{n \in I_2} \sigma_n^m \right]. \tag{9}$$

Next we suppose $m < l$ and

$$\bigcup_{s \in I_1} \iota_s^l = \bigcup_{n \in I_2} \iota_n^m; \tag{10}$$

then (in view of 5)

$$\forall s \in I_1, \exists n \in I_2 \text{ s.t. } \iota_s^l \subset \iota_n^m \text{ and hence (in view of 7) s.t. } \sigma_s^l \subset \sigma_n^m;$$

this proves that

$$\bigcup_{s \in I_1} \sigma_s^l \subset \bigcup_{n \in I_2} \sigma_n^m;$$

therefore, if $\bigcup_{s \in I_1} \sigma_s^l \neq \bigcup_{n \in I_2} \sigma_n^m$ were true, we should have

$$\bigcup_{n \in I_2} \left(\sigma_n^m - \bigcup_{s \in I_1} \sigma_s^l \right) = \left(\bigcup_{n \in I_2} \sigma_n^m \right) - \bigcup_{s \in I_1} \sigma_s^l \neq \emptyset,$$

and hence there would exists $\overline{n} \in I_2$ such that

$$\sigma_{\overline{n}}^m - \bigcup_{s \in I_1} \sigma_s^l \neq \emptyset,$$

and hence (in view of 4) there would exist $t \in \{1, ..., 4^l\}$ such that

$$\sigma_t^l \subset \sigma_{\overline{n}}^m - \bigcup_{s \in I_1} \sigma_s^l,$$

i.e. such that

$$\sigma_t^l \subset \sigma_{\overline{n}}^m \text{ and } \sigma_t^l \cap \sigma_s^l = \emptyset, \forall l \in I_1,$$

i.e. (in view of 7 and 8) such that

$$\iota_t^l \subset \iota_{\overline{n}}^m \text{ and } \iota_t^l \cap \iota_s^l = \emptyset, \forall l \in I_1,$$

but this would be in contradiction with 10. This proves that

$$\left[m < l \text{ and } \bigcup_{s \in I_1} \iota_s^l = \bigcup_{n \in I_2} \iota_n^m \right] \Rightarrow \left[\bigcup_{s \in I_1} \sigma_s^l = \bigcup_{n \in I_2} \sigma_n^m \right]. \tag{11}$$

Step 3: We prove that there exists a projection valued measure T on $\mathcal{A}(d_{\mathbb{R}})$ such that $T(\iota_n^m) = P(\sigma_n^m), \forall m \in \mathbb{N}, \forall n \in \{1, ..., 4^m\}$.

Let $\mathcal{S}$ be the collection of subsets of $[0, 1)$ whose elements are the empty set and all the intervals $[a, b)$ such that $0 \leq a < b \leq 1$ and

$$a = 0 \text{ or } a = n_a 4^{-m_a}, \ b = 1 \text{ or } b = n_b 4^{-m_b} \tag{12}$$

with $m_a, m_b \in \mathbb{N}$ and n_a, n_b elements of $\mathbb{N}$ which are not multiples of 4 (equivalently, if $a \neq 0$ then m_a is the least positive integer so that $a = n_a 4^{-m_a}$ with $n_a \in \mathbb{N}$, and similarly for m_b if $b \neq 1$). It is obvious that $\mathcal{S}$ is a semialgebra on $[0, 1)$.

For $E := [a, b) \in \mathcal{S}$, we define $m(E)$ as follows:

$$m(E) := 1 \text{ if } a = 0 \text{ and } b = 1;$$
$$m(E) := m_b \text{ if } a = 0 \text{ and } b \neq 1;$$
$$m(E) := m_a \text{ if } a \neq 0 \text{ and } b = 1;$$
$$m(E) := \max\{m_a, m_b\} \text{ if } a \neq 0 \text{ and } b \neq 1;$$

then there exists a subfamily $I(E)$ of $\{1, ..., 4^{m(E)}\}$ such that

$$E = \bigcup_{n \in I(E)} \iota_n^{m(E)},$$

and we define

$$Q_E := \sum_{n \in I(E)} P(\sigma_n^{m(E)}) = P\left(\bigcup_{n \in I(E)} \sigma_n^{m(E)}\right)$$

(the second equality is true because P is a projection valued measure). Moreover, we define

$$Q_\emptyset := 0_{\mathcal{H}}.$$

We prove below that the mapping

$$Q : \mathcal{S} \to \mathscr{P}(\mathcal{H})$$
$$E \mapsto Q(E) := Q_E,$$

satisfies all the conditions of 13.4.4. Then there exists a unique projection valued measure Q_e on $\mathcal{A}(\mathcal{S})$ which is an extension of Q.

Now we note that $\mathcal{A}(\mathcal{S}) = \mathcal{A}(d_{\mathbb{R}})^{[0,1)}$. On the one hand, $\mathcal{A}(\mathcal{S}) \subset \mathcal{A}(d_{\mathbb{R}})^{[0,1)}$ follows immediately from $\mathcal{S} \subset \mathcal{A}(d_{\mathbb{R}})^{[0,1)}$ (cf. 6.1.25 with $n = 2$). On the other hand, let $a, b \in \mathbb{R}$ be so that $0 \leq a < b \leq 1$; for every $n \in \mathbb{N}$ there are multiples a_n and b_n of 4^{-n-1} so that $a < a_n \leq a + 4^{-n}$ and $b - 4^{-n} \leq b_n \leq b$, and hence so that $(a, b) = \bigcup_{n=1}^{\infty} [a_n, b_n)$ (if $b_n \leq a_n$, then $[a_n, b_n) := \emptyset$); this proves that $(a, b) \in \mathcal{A}(\mathcal{S})$. Thus,

$$\{(a, b) : 0 \leq a < b \leq 1\} \subset \mathcal{A}(\mathcal{S});$$

since $[b, 1) = (0, 1) - (0, b)$, this implies that

$$\{[b, 1) : 0 \leq b < 1\} \subset \mathcal{A}(\mathcal{S}),$$

and hence that

$$\{[a, b) : 0 \leq a < b \leq 1\} \subset \mathcal{A}(\mathcal{S}).$$

Since

$$\mathcal{A}(\{[a, b) : 0 \leq a < b \leq 1\}) = \mathcal{A}(d_{\mathbb{R}})^{[0,1)}$$

(cf. 6.1.20 and 6.1.25 with $n = 2$), we have $\mathcal{A}(d_{\mathbb{R}})^{[0,1)} \subset \mathcal{A}(\mathcal{S})$ and hence $\mathcal{A}(\mathcal{S}) = \mathcal{A}(d_{\mathbb{R}})^{[0,1)}$.

Thus, Q_e is projection valued measure on $\mathcal{A}(d_{\mathbb{R}})^{[0,1)}$. Then, it is obvious that the mapping

$$T : \mathcal{A}(d_{\mathbb{R}}) \to \mathscr{P}(\mathcal{H})$$
$$E \mapsto T(E) := Q_e(E \cap [0, 1))$$

is a projection valued measure and that

$$T(\iota_n^m) = Q(\iota_n^m) = P(\sigma_n^m), \forall m \in \mathbb{N}, \forall n \in \{1, ..., 4^m\}.$$

Now it is time to prove that the mapping Q satisfies all the conditions of 13.4.4.
q_1: Let $\{E_1, ..., E_N\}$ be a disjoint family of elements of $\mathcal{S}$ such that

$$E := \bigcup_{k=1}^N E_k \in \mathcal{S}.$$

We define $m := \max\{m(E), m(E_1), ..., m(E_N)\}$. In view of 4 there are subsets of $\{1, ..., 4^m\}$, I and I_k for $k = 1, ..., N$, so that

$$E = \bigcup_{s \in I} \iota_s^m \text{ and } E_k = \bigcup_{s \in I_k} \iota_s^m \text{ for } k = 1, ..., N,$$

and hence so that

$$\bigcup_{n \in I(E)} \iota_n^{m(E)} = \bigcup_{s \in I} \iota_s^m \text{ and } \bigcup_{r \in I(E_k)} \iota_r^{m(E_k)} = \bigcup_{s \in I_k} \iota_s^m \text{ for } k = 1, ..., N.$$

In view of 9 and 11, this yields

$$\bigcup_{n \in I(E)} \sigma_n^{m(E)} = \bigcup_{s \in I} \sigma_s^m \text{ and } \bigcup_{r \in I(E_k)} \sigma_r^{m(E_k)} = \bigcup_{s \in I_k} \sigma_s^m \text{ for } k = 1, ..., N;$$

moreover,

$$\bigcup_{s \in I} \iota_s^m = E = \bigcup_{k=1}^N E_k = \bigcup_{k=1}^N \left(\bigcup_{s \in I_k} \iota_s^m \right)$$

implies (in view of 9) $I = \bigcup_{k=1}^N I_k$, and hence

$$\bigcup_{n \in I(E)} \sigma_n^{m(E)} = \bigcup_{s \in I} \sigma_s^m = \bigcup_{k=1}^N \left(\bigcup_{s \in I_k} \sigma_s^m \right) = \bigcup_{k=1}^N \left(\bigcup_{r \in I(E_k)} \sigma_r^{m(E_k)} \right),$$

and hence

$$Q(E) = P\left(\bigcup_{n \in I(E)} \sigma_n^{m(E)} \right) = \sum_{k=1}^N P\left(\bigcup_{r \in I(E_k)} \sigma_r^{m(E_k)} \right) = \sum_{k=1}^N Q(E_k),$$

where the second equality is true because P is a projection valued measure (if $k \neq h$ then $\sigma_r^{m(E_k)} \cap \sigma_t^{m(E_h)} = \emptyset$ in view of 8 since $\iota_r^{m(E_k)} \cap \iota_t^{m(E_h)} = \emptyset$, for all $r \in I(E_k)$ and $t \in I(E_h)$).

q_2: Let $E, F \in \mathcal{S}$ be such that $E \cap F = \emptyset$. Then

$$\iota_n^{m(E)} \cap \iota_s^{m(F)} = \emptyset, \forall n \in I(E), \forall s \in I(F),$$

and hence (in view of 8)

$$\sigma_n^{m(E)} \cap \sigma_s^{m(F)} = \emptyset, \forall n \in I(E), \forall s \in I(F),$$

and hence (cf. 13.3.2b)

$$P(\sigma_n^{m(E)})P(\sigma_s^{m(F)}) = \mathbb{O}_{\mathcal{H}}, \forall n \in I(E), \forall s \in I(F),$$

and hence

$$Q(E)Q(F) = \sum_{n \in I(E)} \sum_{s \in I(F)} P(\sigma_n^{m(E)})P(\sigma_s^{m(F)}) = \mathbb{O}_{\mathcal{H}}.$$

q_3: We have $[0, 1) \in \mathcal{S}$ and (cf.1)

$$Q([0, 1)) = P\left(\bigcup_{s=1}^{4} \sigma_s^1\right) = P([0, 1) \times [0, 1)) = \mathbb{1}_{\mathcal{H}}.$$

q_4: We fix $f \in \mathcal{H}$ and $E \in \mathcal{S}$. We write $E = [a, b)$ with a and b as in 12. Thus,

$$\exists k \in \mathbb{N}, \exists h \in \mathbb{N} \text{ so that } b = k4^{-h} \tag{13}$$

(if $b = 1$, we choose h in whatever way and set $k = 4^h$). Let $m_0 \in \mathbb{N}$ be such that $4^{-m_0} < b - a$ and $h < m_0$. For each $m \in \mathbb{N}$ such that $m_0 \leq m$, we have

$$[a, b) = [a, b - 4^{-m}) \cup [b - 4^{-m}, b),$$

and it is obvious that $[a, b - 4^{-m})$ and $[b - 4^{-m}, b)$ are elements of $\mathcal{S}$. In view of condition q_1, already proved, we have

$$Q([a, b) = Q([a, b - 4^{-m})) + Q([b - 4^{-m}, b)),$$

and hence

$$\mu_f^Q([a, b)) - \mu_f^Q([a, b - 4^{-m}))| = \mu_f^Q([b - 4^{-m}, b)).$$

From 13 we have

$$[b - 4^{-m}, b) = [(k4^{m-h} - 1)4^{-m}, (k4^{m-h})4^{-m}) = \iota_{n_m}^m,$$

with $n_m := k4^{m-h}$, and hence

$$\mu_f^Q([b - 4^{-m}, b)) = \mu_f^P(\sigma_{n_m}^m).$$

Now we note that, for each $m \geq m_0$,

$$n_{m+1} = 4n_m;$$

thus, $\sigma_{n_{m+1}}^{m+1}$ is the top-right square of the four squares into which $\sigma_{n_m}^m$ is divided (cf. 3 with $n = n_m$). Since these squares are top-right open, we have

$$\bigcap_{m=m_0}^{\infty} \sigma_{n_m}^m = \emptyset,$$

and hence (cf. 13.3.6b)

$$\lim_{m\to\infty} P(\sigma_{n_m}^m)f = 0_{\mathcal{H}},$$

and hence

$$\lim_{m\to\infty} \mu_f^Q([b - 4^{-m}, b)) = 0$$

To conclude the proof that condition q_4 holds true, we note that obviously

$$\overline{[a, b - 4^{-m})} = [a, b - 4^{-m}] \subset [a, b), \forall m > m_0,$$

and recall 2.8.7.

Step 4: We prove the statement of the theorem.

We define the self-adjoint operator $B := J_\xi^T$. Clearly,

$$T = P^B. \tag{14}$$

For all $m \in \mathbb{N}$ and $n \in \{1, ..., 4^m\}$, we denote by (x_n^m, y_n^m) the bottom-left corner of the square σ_n^m. For each $m \in \mathbb{N}$, we define the function

$$\rho_m := \sum_{n=1}^{4^m} x_n^m \chi_{\iota_n^m},$$

which is obviously an element of $\mathcal{L}^\infty(\mathbb{R}, \mathcal{A}(d_\mathbb{R}), T)$. We note that

$$\rho_m(x) \le \rho_{m+1}(x), \forall x \in \mathbb{R}, \forall m \in \mathbb{N};$$

indeed, fix $x \in [0, 1)$ and for each $m \in \mathbb{N}$ let $n_m(x) \in \{1, ..., 4^m\}$ be such that $x \in \iota_{n_m(x)}^m$; then,

$$\iota_{n_{m+1}(x)}^{m+1} \subset \iota_{n_m(x)}^m$$

and hence (cf. 7)

$$\sigma_{n_{m+1}(x)}^{m+1} \subset \sigma_{n_m(x)}^m,$$

and hence $x_{n_m(x)}^m \le x_{n_{m+1}(x)}^{m+1}$; now, $x_{n_m(x)}^m = \rho_m(x)$. Thus, we can define the function

$$\mathbb{R} \ni x \mapsto \psi_1(x) := \lim_{m\to\infty} \rho_m(x) \in \mathbb{R}.$$

We have $\psi_1 \in \mathcal{L}^\infty(\mathbb{R}, \mathcal{A}(d_\mathbb{R}), T)$ (cf. 6.2.20c) and it is obvious that the sequence $\{\rho_m\}$ is ψ_1-convergent. Then,

$$D_T(\psi_1) = \mathcal{H} \text{ and } J_{\psi_1}^T f = \lim_{m\to\infty} J_{\rho_m}^T f, \forall f \in \mathcal{H} \tag{15}$$

(cf. 14.2.17e and 14.2.14c).

Now, for each $m \in \mathbb{N}$, we define the function

$$\tau_m := \sum_{n=1}^{4^m} x_n^m \chi_{\sigma_n^m},$$

which is obviously an element of $\mathcal{L}^\infty(\mathbb{R}^2, \mathcal{A}(d_2), P)$. We note that

$$|\pi_1(x,y) - \tau_m(x,y)| < 2^{-m}, \forall (x,y) \in [0,1) \times [0,1), \forall m \in \mathbb{N};$$

indeed, fix $(x,y) \in [0,1) \times [0,1)$ and for each $m \in \mathbb{N}$ let $n_m(x,y) \in \{1, ..., 4^m\}$ be such that $(x,y) \in \sigma^m_{n_m(x,y)}$; then,

$$|\pi_1(x,y) - \tau_m(x,y)| = |x - x^m_{n_m(x,y)}| < 2^{-m}$$

since $(x^m_{n_m(x,y)}, y^m_{n_m(x,y)})$ is a corner of the square $\sigma^m_{n_m(x,y)}$ whose sides are 2^{-m} long; moreover,

$$\tau_m(x,y) \leq \pi_1(x,y), \forall (x,y) \in [0,1) \times [0,1), \forall m \in \mathbb{N},$$

since (x^m_n, y^m_n) is the bottom-left corner of the square σ^m_n. This proves that the sequence $\{\tau_m\}$ is π_1-convergent (recall that $P(\mathbb{R} - [0,1) \times [0,1)) = \mathbb{O}_\mathcal{H}$, cf. 1), and hence that

$$J^P_{\pi_1} f = \lim_{m \to \infty} J^P_{\tau_m} f, \forall f \in \mathcal{H}. \tag{16}$$

Now we note that

$$J^T_{\rho_m} = \sum_{n=1}^{4^m} x^m_n T(\iota^m_n) = \sum_{n=1}^{4^m} x^m_n P(\sigma^m_n) = J^P_{\tau_m}, \forall m \in \mathbb{N}.$$

In view of 15 and 16, this implies that

$$J^T_{\psi_1} = J^P_{\pi_1},$$

and hence, in view of 2, that

$$J^T_{\psi_1} = \varphi(A_1).$$

In view of 15.2.7 and 15.3.8, and of 14, this yields

$$P^B(\psi_1^{-1}(E)) = P^{A_1}(\varphi^{-1}(E)), \forall E \in \mathcal{A}(d_\mathbb{R}). \tag{17}$$

Now, for each $F \in \mathcal{A}(d_\mathbb{R})$, $\varphi(F) = (\varphi^{-1})^{-1}(F)$ (where φ^{-1} is the inverse function, and $(\varphi^{-1})^{-1}(F)$ can be understood either as the counterimage of F under φ^{-1} or as the image of F under the inverse of φ^{-1}, i.e. under φ, since φ^{-1} is injective), and hence $\varphi(F) \in \mathcal{A}(d_\mathbb{R})$ (for the counterimage $(\varphi^{-1})^{-1}(F)$ under φ^{-1}, we have $(\varphi^{-1})^{-1}(F) \in \mathcal{A}(d_\mathbb{R})^{(\alpha,\beta)}$ since $D_{\varphi^{-1}} = (\alpha, \beta)$, and hence $(\varphi^{-1})^{-1}(F) \in \mathcal{A}(d_\mathbb{R})$ in view of 6.1.19a). Then, 17 implies that

$$P^B(\psi_1^{-1}(\varphi(F))) = P^{A_1}(\varphi^{-1}(\varphi(F))) = P^{A_1}(F), \forall F \in \mathcal{A}(d_\mathbb{R}); \tag{18}$$

we have in particular

$$P^B(\psi_1^{-1}(\varphi(\mathbb{R}))) = P^{A_1}(\mathbb{R}) = \mathbb{1}_\mathcal{H}, \tag{19}$$

and this proves that $\psi_1^{-1}(D_{\varphi^{-1}}) = \psi_1^{-1}(\varphi(\mathbb{R})) \neq \emptyset$; thus, we can define the function $\varphi_1 := \varphi^{-1} \circ \psi_1$; moreover, we have $D_{\varphi_1} \in \mathcal{A}(d_\mathbb{R})$ and

$$P^B(D_{\varphi_1}) = \mathbb{1}_\mathcal{H}, \text{ or } P^B(\mathbb{R} - D_{\varphi_1}) = \mathbb{O}_\mathcal{H},$$

in view of 19 and of the equalities

$$D_{\varphi_1} = \psi_1^{-1}(D_{\varphi^{-1}}) = \psi_1^{-1}(\varphi(\mathbb{R}));$$

thus, we have $\varphi_1 \in \mathcal{M}(\mathbb{R}, \mathcal{A}(d_{\mathbb{R}}), P^B)$ and we can define the self-adjoint operator

$$C_1 := \varphi_1(B).$$

Then we have (cf. 15.3.8 and 18)

$$P^{C_1}(F) = P^B(\varphi_1^{-1}(F)) = P^B(\psi_1^{-1}(\varphi(F))) = P^{A_1}(F), \forall F \in \mathcal{A}(d_{\mathbb{R}}),$$

and hence $A_1 = C_1 = \varphi_1(B)$.

The proof for A_2 would be similar (the functions ρ_m and τ_m would be defined with y_n^m in lieu of x_n^m, and we should use the function π_2 in lieu of π_1). □

For two self-adjoint operators, 17.1.7 lists a number of conditions equivalent to their commuting. The result of the last theorem can be collected in a similar way with other results already proved, to obtain the following corollary.

17.1.10 Corollary. *Let A_1 and A_2 be self-adjoint operators in $\mathcal{H}$. The following conditions are equivalent:*

(a) A_1 and A_2 commute (in the sense defined in 17.1.5);
(b) there exists a projection valued measure P on the Borel σ-algebra $\mathcal{A}(d_2)$ so that

$$P^{A_1}(E) = P(E \times \mathbb{R}) \quad and \quad P^{A_2}(E) = P(\mathbb{R} \times E), \forall E \in \mathcal{A}(d_{\mathbb{R}});$$

(c) there exist a self-adjoint operator B in $\mathcal{H}$ and two real functions $\varphi_i \in \mathcal{M}(\mathbb{R}, \mathcal{A}(d_{\mathbb{R}}), P^B)$ so that $A_i = \varphi_i(B)$, for $i = 1, 2$.

If these conditions hold true, then the projection valued measure P is unique.

Proof. $a \Rightarrow (b$ and uniqueness of P): Cf. 13.5.3.
$b \Rightarrow a$: This follows from 13.3.2d.
$a \Rightarrow c$: Cf. 17.1.9.
$c \Rightarrow a$: Cf. 17.1.6b,c. □

17.1.11 Definition. Let A_1 and A_2 be commuting (in the sense of 17.1.5) self-adjoint operators in $\mathcal{H}$, and let P be the projection valued measure of 17.1.10b. For a function $\varphi \in \mathcal{M}(\mathbb{R}^2, \mathcal{A}(d_2), P)$, we write

$$\varphi(A_1, A_2) := J_\varphi^P$$

and we say that this operator is a *function of A_1 and A_2*. This name is justified by the fact that $\varphi(A_1, A_2)$ is often the closure of the function φ of A_1 and A_2 defined in an obvious direct way. In 17.1.12 we examine two instances of this.

17.1.12 Proposition. *Let A_1 and A_2 be commuting (in the sense of 17.1.5) self-adjoint operators in $\mathcal{H}$.*

(a) The operator $A_1 + A_2$ is essentially self-adjoint.
   For the function

$$\mathbb{R}^2 \ni (x_1, x_2) \mapsto \varphi(x_1, x_2) := x_1 + x_2 \in \mathbb{R},$$

we have $\varphi(A_1, A_2) = \overline{(A_1 + A_2)} = (A_1 + A_2)^\dagger$.
If $A_1 \in \mathcal{B}(\mathcal{H})$ then the operator $A_1 + A_2$ is self-adjoint and $\varphi(A_1, A_2) = A_1 + A_2$.
(b) The operators $A_1 A_2$ and $A_2 A_1$ are essentially self-adjoint.
   For the function

$$\mathbb{R}^2 \ni (x_1, x_2) \mapsto \psi(x_1, x_2) := x_1 x_2 \in \mathbb{R},$$

we have $\psi(A_1, A_2) = \overline{A_1 A_2} = \overline{A_2 A_1} = (A_1 A_2)^\dagger = (A_2 A_1)^\dagger$.
If $A_1 \in \mathcal{B}(\mathcal{H})$ then the operator $A_2 A_1$ is self-adjoint and $\psi(A_1, A_2) = A_2 A_1$.
If $A_2 \in \mathcal{B}(\mathcal{H})$ then the operator $A_1 A_2$ is self-adjoint and $\psi(A_1, A_2) = A_1 A_2$.

Proof. Preliminary remark: Let P be the projection valued measure of 17.1.10b. For $i = 1, 2$, we define the function

$$\mathbb{R}^2 \ni (x_1, x_2) \mapsto \pi_i(x_1, x_2) := x_i \in \mathbb{R},$$

and the operator $B_i := J_{\pi_i}^P$. The operators B_1 and B_2 are self-adjoint and

$$P^{B_1}(E) = P(\pi_1^{-1}(E)) = P(E \times \mathbb{R}) = P^{A_1}(E), \forall E \in \mathcal{A}(d_{\mathbb{R}}),$$
$$P^{B_2}(E) = P(\pi_2^{-1}(E)) = P(\mathbb{R} \times E) = P^{A_2}(E), \forall E \in \mathcal{A}(d_{\mathbb{R}})$$

(cf. 15.2.7). Therefore,

$$J_{\pi_i}^P = A_i \text{ for } i = 1, 2. \tag{1}$$

a: We note that $\pi_1 + \pi_2 = \varphi$. Then, 1 and 14.3.11 imply that

the operator $A_1 + A_2$ is closable and $\overline{A_1 + A_2} = J_\varphi^P = \varphi(A_1, A_2)$.

The operator $A_1 + A_2$ is adjointable since $D_{\overline{A_1 + A_2}} \subset \overline{D_{A_1 + A_2}}$ (cf. 4.4.10) and $\overline{D_P(\varphi)} = \mathcal{H}$ (cf. 14.2.13). Moreover, the operator $\varphi(A_1, A_2)$ is self-adjoint (cf. 14.3.17). Therefore, $A_1 + A_2$ is essentially self-adjoint by 12.4.11. Then we have $\overline{A_1 + A_2} = (A_1 + A_2)^\dagger$ (cf. 12.1.6b).
   If $A_1 \in \mathcal{B}(\mathcal{H})$ then (cf. 12.3.1b) $(A_1 + A_2)^\dagger = A_1^\dagger + A_2^\dagger = A_1 + A_2$.
   b: We note that $\pi_1 \pi_2 = \pi_2 \pi_1 = \psi$. Then, 1 and 14.3.12 imply that

the operators $A_1 A_2$ and $A_2 A_1$ are closable and $\overline{A_1 A_2} = \overline{A_2 A_1} = J_\psi^P = \psi(A_1, A_2)$.

Proceeding as in part a, we see that the operators $A_1 A_2$ and $A_2 A_1$ are essentially self-adjoint, that the operator $\psi(A_1 A_2)$ is self-adjoint, and that $\overline{A_1 A_2} = (A_1 A_2)^\dagger$ and $\overline{A_2 A_1} = (A_2 A_1)^\dagger$.
   If $A_1 \in \mathcal{B}(\mathcal{H})$ then (cf. 12.3.4b) $(A_1 A_2)^\dagger = A_2^\dagger A_1^\dagger = A_2 A_1$.
   Similarly, if $A_2 \in \mathcal{B}(\mathcal{H})$ then $(A_2 A_1)^\dagger = A_1^\dagger A_2^\dagger = A_1 A_2$. $\square$

The next result has an important role in the discussion of compatible quantum observables (cf. 19.5.23 and 19.5.24f). It may be interesting to note that in a way it extends to a pair of commuting self-adjoint operators what happens for a single self-adjoint operator (cf. 15.2.4 $a \Rightarrow d$).

17.1.13 Theorem. *Let A_1 and A_2 be commuting (in the sense of 17.1.5) self-adjoint operators in $\mathcal{H}$, and let $\lambda_1 \in \sigma(A_1)$. Then, for every $\varepsilon > 0$, there exist $\lambda_2 \in \sigma(A_2)$ and $u_\varepsilon \in D_{A_1} \cap D_{A_2} \cap \tilde{\mathcal{H}}$ so that*

$$|\langle A_i \rangle_{u_\varepsilon} - \lambda_i| \leq \varepsilon \ \text{ and } \ \Delta_{u_\varepsilon} A_i \leq 2\varepsilon, \ \text{ for } i = 1, 2$$

(for $\langle A \rangle_u$ and $\Delta_u A$, cf. 15.2.3).

Proof. We fix $\varepsilon \in (0, \infty)$. Then $P^{A_1}((\lambda_1 - \varepsilon, \lambda_1 + \varepsilon)) \neq O_{\mathcal{H}}$ (cf. 15.2.4). We define the mapping

$$Q : \mathcal{A}(d_{\mathbb{R}}) \to \mathscr{P}(\mathcal{H})$$
$$E \mapsto Q(E) := P^{A_1}((\lambda_1 - \varepsilon, \lambda_1 + \varepsilon))P^{A_2}(E);$$

we point out that this definition is consistent (in view of 13.2.1) because A_1 and A_2 commute. We note that, for each $f \in \mathcal{H}$ and all $E \in \mathcal{A}(d_{\mathbb{R}})$,

$$\mu_f^Q(E) = \left(P^{A_1}((\lambda_1 - \varepsilon, \lambda_1 + \varepsilon))f | P^{A_2}(E)P^{A_1}((\lambda_1 - \varepsilon, \lambda_1 + \varepsilon))f \right) = \mu_g^{P^{A_2}}(E)$$

if $g := P^{A_1}((\lambda_1 - \varepsilon, \lambda_1 + \varepsilon))f$; thus μ_f^Q is a measure on $\mathcal{A}(d_{\mathbb{R}})$ for all $f \in \mathcal{H}$ (cf. 13.3.5). This implies that, if $E, F \in \mathcal{A}(d_{\mathbb{R}})$ are such that $E \subset F$ and $Q(F) = O_{\mathcal{H}}$, then (cf. 7.1.2a)

$$(f|Q(E)f) = \mu_f^Q(E) \leq \mu_f^Q(F) = (f|Q(F)f) = 0, \forall f \in \mathcal{H},$$

and hence $Q(E) = O_{\mathcal{H}}$.

We define the set

$$G := \{\mu \in \mathbb{R} : \exists \varepsilon_\mu > 0 \text{ such that } Q((\mu - \varepsilon_\mu, \mu + \varepsilon_\mu)) = O_{\mathcal{H}}\}.$$

For each $\mu \in G$ we choose $\varepsilon_\mu > 0$ as above, and we see that

$$\mu' \in (\mu - \varepsilon_\mu, \mu + \varepsilon_\mu) \Rightarrow [\exists \varepsilon_{\mu'} > 0 \text{ s.t. } (\mu' - \varepsilon_{\mu'}, \mu' + \varepsilon_{\mu'}) \subset (\mu - \varepsilon_\mu, \mu + \varepsilon_\mu)] \Rightarrow$$
$$[\exists \varepsilon_{\mu'} > 0 \text{ such that } Q((\mu' - \varepsilon_{\mu'}, \mu' + \varepsilon_{\mu'})) = O_{\mathcal{H}}] \Rightarrow \mu' \in G.$$

This proves that

$$(\mu - \varepsilon_\mu, \mu + \varepsilon_\mu) \subset G, \forall \mu \in G,$$

and hence that

$$G = \bigcup_{\mu \in G} (\mu - \varepsilon_\mu, \mu + \varepsilon_\mu).$$

Then, by 2.3.16 and 2.3.18 there is a countable subset $\{\mu_n\}_{n \in J}$ of G so that

$$G = \bigcup_{n \in J} (\mu_n - \varepsilon_{\mu_n}, \mu_n + \varepsilon_{\mu_n}).$$

By 7.1.4a, this yields

$$\mu_f^Q(G) \le \sum_{n \in J} \mu_f^Q((\mu_n - \varepsilon_{\mu_n}, \mu_n + \varepsilon_{\mu_n})) = 0, \forall f \in \mathcal{H},$$

and hence $Q(G) = \mathbb{O}_{\mathcal{H}}$. Since

$$Q(\mathbb{R}) = P^{A_1}((\lambda_1 - \varepsilon, \lambda_1 + \varepsilon)) \neq \mathbb{O}_{\mathcal{H}},$$

this proves that $G \neq \mathbb{R}$ and hence that there exists $\lambda_2 \in \mathbb{R}$ such that

$$Q((\lambda_2 - \eta, \lambda_2 + \eta)) \neq \mathbb{O}_{\mathcal{H}}, \forall \eta > 0.$$

First, this implies obviously that

$$P^{A_2}((\lambda_2 - \eta, \lambda_2 + \eta)) \neq \mathbb{O}_{\mathcal{H}}, \forall \eta > 0,$$

and hence that $\lambda_2 \in \sigma(A_2)$ (cf. 15.2.4). Second, we can choose $u_\varepsilon \in \tilde{\mathcal{H}}$ such that

$$Q((\lambda_2 - \varepsilon, \lambda_2 + \varepsilon))u_\varepsilon = u_\varepsilon$$

(cf. 13.1.3c). Then we have (cf. 13.3.2b)

$$\mu_{u_\varepsilon}^{P^{A_1}}(\mathbb{R} - (\lambda_1 - \varepsilon, \lambda_1 + \varepsilon))$$
$$= \|P^{A_1}(\mathbb{R} - (\lambda_1 - \varepsilon, \lambda_1 + \varepsilon))P^{A_1}((\lambda_1 - \varepsilon, \lambda_1 + \varepsilon))P^{A_2}((\lambda_2 - \varepsilon, \lambda_2 + \varepsilon))u_\varepsilon\|^2 = 0$$

and

$$\mu_{u_\varepsilon}^{P^{A_2}}(\mathbb{R} - (\lambda_2 - \varepsilon, \lambda_2 + \varepsilon))$$
$$= \|P^{A_2}(\mathbb{R} - (\lambda_2 - \varepsilon, \lambda_2 + \varepsilon))P^{A_1}((\lambda_1 - \varepsilon, \lambda_1 + \varepsilon))P^{A_2}((\lambda_2 - \varepsilon, \lambda_2 + \varepsilon))u_\varepsilon\|^2 = 0,$$

and hence (cf. 8.3.3 and 15.2.2e), for $i = 1, 2$,

$$\int_{\mathbb{R}} \xi^2 d\mu_{u_\varepsilon}^{P^{A_i}} = \int_{(\lambda_i - \varepsilon, \lambda_i + \varepsilon)} \xi^2 d\mu_{u_\varepsilon}^{P^{A_i}} < \infty, \text{ i.e. } u_\varepsilon \in D_{A_i},$$

and

$$\|A_i u_\varepsilon - \lambda_i u_\varepsilon\|^2 = \int_{(\lambda_i - \varepsilon, \lambda_i + \varepsilon)} |\xi - \lambda_i|^2 d\mu_{u_\varepsilon}^{P^{A_i}} \le \varepsilon^2 \mu_{u_\varepsilon}^{P^{A_i}}((\lambda_i - \varepsilon, \lambda_i + \varepsilon)) = \varepsilon^2.$$

Then, by the Schwarz inequality we obtain, for $i = 1, 2$,

$$|\langle A_i \rangle_{u_\varepsilon} - \lambda_i| = |(u_\varepsilon | A_i u_\varepsilon - \lambda_i u_\varepsilon)| \le \|A_i u_\varepsilon - \lambda_i u_\varepsilon\| \le \varepsilon,$$

and hence also

$$\Delta_{u_\varepsilon} A_i = \|A_i u_\varepsilon - \langle A_i \rangle_{u_\varepsilon} u_\varepsilon\| \le \|A_i u_\varepsilon - \lambda_i u_\varepsilon\| + \|\lambda_i u_\varepsilon - \langle A_i \rangle_{u_\varepsilon} u_\varepsilon\| \le 2\varepsilon.$$

$\square$

17.1.14 Example. Suppose that $\mathcal{H}$ is a separable Hilbert space and let A_1 and A_2 be self-adjoint operators in $\mathcal{H}$ such that conditions a, b, c of 15.3.4B hold true for both of them. Then the following conditions are equivalent:

(α) A_1 and A_2 commute (in the sense defined in 17.1.5);

(β) if, for $k = 1, 2$, $\{(\lambda_n^k, P_n^k)\}_{n \in I_k}$ is the family associated with A_k as $\{(\lambda_n, P_n)\}_{n \in I}$ was associated with A in 15.3.4B, then $[P_n^1, P_l^2] = \mathbb{O}_{\mathcal{H}}$ for all $(n, l) \in I_1 \times I_2$;

(γ) there exists a c.o.n.s. $\{v_j\}_{j \in J}$ in $\mathcal{H}$ whose elements are eigenvectors of both A_1 and A_2.

Indeed, the equivalence of conditions α and β follows at once from

$$P^{A_k}(E)f = \sum_{n \in I_E^k} P_n^k f, \forall f \in \mathcal{H}, \forall E \in \mathcal{A}(d_{\mathbb{R}}),$$

with $I_E^k := \{n \in I_k : \lambda_n^k \in E\}$, for $k = 1, 2$ (cf. 15.3.4B).

Moreover, if condition β is true then $\{P_n^1 P_l^2\}_{(n,l) \in I_1 \times I_2}$ is a family of projections (cf. 13.2.1) which is so that

$$(P_n^1 P_l^2)(P_m^1 P_j^2) = \mathbb{O}_{\mathcal{H}} \text{ if } (n, l) \neq (m, j); \tag{1}$$

if we set $I_0 := \{(n, l) \in I_1 \times I_2 : P_n^1 P_l^2 \neq \mathbb{O}_{\mathcal{H}}\}$, we have

$$\begin{aligned}
\sum_{(n,l) \in I_0} P_n^1 P_l^2 f &= \sum_{n \in I_1} P_n^1 \left(\sum_{l \in I_2} P_l^2 f \right) \\
&= \sum_{n \in I_1} P_n^1 f = f, \forall f \in \mathcal{H};
\end{aligned} \tag{2}$$

for each $(n, l) \in I_0$, we fix an o.n.s. $\{u_{n,l,s}\}_{s \in I_{n,l}}$ which is complete in the subspace $R_{P_n^1 P_l^2}$ (cf. 10.7.2); then the set $\bigcup_{(n,l) \in I_0} \{u_{n,l,s}\}_{s \in I_{n,l}}$ is an o.n.s. in $\mathcal{H}$ in view of 1 (cf. 13.2.8d and 13.2.9c) and it is complete in $\mathcal{H}$ by 10.6.4 (with $M := \mathcal{H}$) since (cf. 2 and 13.1.10)

$$f = \sum_{(n,l) \in I_0} P_n^1 P_l^2 f = \sum_{(n,l) \in I_0} \sum_{s \in I_{n,l}} (u_{n,l,s}|f) \, u_{n,l,s}, \forall f \in \mathcal{H};$$

moreover, all the elements of this c.o.n.s. are eigenvectors of both A_1 and A_2, since

$$R_{P_n^1 P_l^2} = R_{P_n^1} \cap R_{P_l^2}, \forall (n, l) \in I_0$$

(cf. 13.2.1e) and since all the non-null elements of $R_{P_n^k}$ are eigenvectors of A_k, for all $n \in I_k$ and for $k = 1, 2$ (cf. 15.3.4B). This proves that condition β implies condition γ.

Conversely, assume that condition γ is true. Then (cf. 15.3.4B):

$$\forall n \in I_1, \exists J_n^1 \subset J \text{ s.t. } P_n^1 f = \sum_{j \in J_n^1} (v_j|f) \, v_j, \forall f \in \mathcal{H};$$

$$\forall l \in I_2, \exists J_l^2 \subset J \text{ s.t. } P_l^2 f = \sum_{j \in J_l^2} (v_j|f) \, v_j, \forall f \in \mathcal{H}.$$

This implies that

$$P_n^1 P_l^2 f = \sum_{j \in J_n^1 \cap J_l^2} (v_j|f) \, v_j = P_l^2 P_n^1 f, \forall f \in \mathcal{H}, \forall (n, l) \in I_1 \times I_2$$

(if $J_n^1 \cap J_l^2 = \emptyset$ then the sum of the series is defined to be $0_{\mathcal{H}}$), and this proves that condition β is true.

Now suppose that A_1 and A_2 commute and let $\{(\lambda_n^k, P_n^k)\}_{n \in I_k}$ be as in condition β, for $k = 1, 2$. For each $G \in \mathcal{A}(d_2)$, let $J_G := \{(n, l) \in I_0 : (\lambda_n^1, \lambda_l^2) \in G\}$ and let P_G be the projection defined by

$$P_G f := \sum_{(n,l) \in J_G} P_n^1 P_l^2 f, \forall f \in \mathcal{H}$$

(if $J_G = \emptyset$ then $P_G := 0_{\mathcal{H}}$; if $J_G \neq \emptyset$ then 13.2.8 or 13.2.9 proves that P_G is indeed a projection). Then define the mapping

$$\mathcal{A}(d_2) \ni G \mapsto P(G) := P_G \in \mathscr{P}(\mathcal{H}).$$

For every $f \in \mathcal{H}$,

$$\mu_f^P(G) = \sum_{(n,l) \in J_G} \|P_n^1 P_l^2 f\|^2, \forall G \in \mathcal{A}(d_2);$$

hence, μ_f^P is a measure (cf. 8.3.8 with $(X, \mathcal{A}) := (\mathbb{R}^2, \mathcal{A}(d_2))$, $I := I_0$, $x_{(n,l)} := (\lambda_n^1, \lambda_l^2)$, $a_{(n,l)} := \|P_n^1 P_l^2 f\|^2$) and

$$\mu_f^P(\mathbb{R}^2) = \sum_{(n,l) \in I_0} \|P_n^1 P_l^2 f\|^2 = \left\| \sum_{(n,l) \in I_0} P_n^1 P_l^2 f \right\|^2 = \|f\|^2.$$

Therefore, μ_f^P is a projection valued measure (cf. 13.3.5). Furthermore, for every $E \in \mathcal{A}(d_{\mathbb{R}})$,

$$(f | P(E \times \mathbb{R}) f) = \sum_{n \in I_E^1} \sum_{l \in I_2} \|P_n^1 P_l^2 f\|^2 = \sum_{n \in I_E^1} \sum_{l \in I_2} \|P_l^2 P_n^1 f\|^2$$

$$= \sum_{n \in I_E^1} \|P_n^1 f\|^2 = \mu_f^{P^{A_1}}(E) = (f | P^{A_1}(E) f), \forall f \in \mathcal{H}$$

(cf. 10.2.3 or 10.4.7a, and 15.3.4B), and hence

$$P(E \times \mathbb{R}) = P^{A_1}(E)$$

(cf. 10.2.12); similarly,

$$P(\mathbb{R} \times E) = P^{A_2}(E), \forall E \in \mathcal{A}(d_{\mathbb{R}}).$$

This proves that P is the projection valued measure through which functions of A_1 and A_2 are defined (cf. 17.1.10 and 17.1.11).

Since $P(\mathbb{R}^2 - \{(\lambda_n^1, \lambda_l^2)\}_{(n,l) \in I_0}) = 0_{\mathcal{H}}$, each function $\varphi : \{(\lambda_n^1, \lambda_l^2)\}_{(n,l) \in I_0} \to \mathbb{C}$ is an element of $\mathcal{M}(\mathbb{R}^2, \mathcal{A}(d_2), P)$ (such a function is always $\mathcal{A}(d_2)^{D_\varphi}$-measurable), and (cf. 14.2.14a,b and 8.3.8)

$$D_{\varphi(A_1, A_2)} = \left\{ f \in \mathcal{H} : \int_{\mathbb{R}^2} |\varphi|^2 d\mu_f^P < \infty \right\}$$

$$= \left\{ f \in \mathcal{H} : \sum_{(n,l) \in I_0} |\varphi(\lambda_n^1, \lambda_l^2)|^2 \|P_n^1 P_l^2 f\|^2 < \infty \right\},$$

$$(f|\varphi(A_1, A_2)f) = \int_{\mathbb{R}^2} \varphi d\mu_f^P = \sum_{(n,l)\in I_0} \varphi(\lambda_n^1, \lambda_l^2)\|P_n^1 P_l^2 f\|^2$$

$$= \left(f\Big| \sum_{(n,l)\in I_0} \varphi(\lambda_n^1, \lambda_l^2) P_n^1 P_l^2 f\right), \forall f \in D_{\varphi(A_1, A_2)};$$

since the mapping $D_{\varphi(A_1, A_2)} \ni f \mapsto \sum_{(n,l)\in I_0} \varphi(\lambda_n^1, \lambda_l^2) P_n^1 P_l^2 f$ is obviously a linear operator (its definition is consistent by 10.4.7b), in view of 10.2.12 this implies that

$$\varphi(A_1, A_2)f = \sum_{(n,l)\in I_0} \varphi(\lambda_n^1, \lambda_l^2) P_n^1 P_l^2 f, \forall f \in D_{\varphi(A_1, A_2)}.$$

For each $(n, l) \in I_0$, if $\{u_{n,l,s}\}_{s\in I_{n,l}}$ is as before then

$$P_n^1 P_l^2 f = \sum_{s\in I_{n,l}} (u_{n,l,s}|f)\, u_{n,l,s} \text{ and } \|P_n^1 P_l^2 f\|^2 = \sum_{s\in I_{n,l}} |(u_{n,l,s}|f)|^2, \forall f \in \mathcal{H}$$

(cf. 13.1.10, and 10.2.3 or 10.4.8a). Therefore,

$$D_{\varphi(A_1, A_2)} = \left\{ f \in \mathcal{H} : \sum_{(n,l)\in I_0} |\varphi(\lambda_n^1, \lambda_l^2)|^2 \sum_{s\in I_{n,l}} |(u_{n,l,s}|f)|^2 < \infty \right\},$$

$$\varphi(A_1, A_2)f = \sum_{(n,l)\in I_0} \varphi(\lambda_n^1, \lambda_l^2) \sum_{s\in I_{n,l}} (u_{n,l,s}|f)\, u_{n,l,s}, \forall f \in D_{\varphi(A_1, A_2)}.$$

17.2   Invariant and reducing subspaces

If $\lambda \in \mathbb{C}$ is an eigenvalue of a linear operator A in $\mathcal{H}$ then the corresponding eigenspace is the set M_λ of the vectors f in D_A such that $Af = \lambda f$ (cf. 4.5.7), and we have trivially the inclusion $A(M_\lambda) \subset M_\lambda$ since M_λ is a linear manifold. If the operator A is closed then M_λ is a subspace of $\mathcal{H}$ (cf. 4.5.9). More general than the concept of a closed eigenspace is the concept of an invariant subspace.

17.2.1 Definition. Let A be a linear operator in $\mathcal{H}$, i.e. $A \in \mathcal{O}(\mathcal{H})$, and let M be a subspace of $\mathcal{H}$, i.e. $M \in \mathscr{S}(\mathcal{H})$. We say that M is an *invariant subspace* for A if

$$Af \in M, \forall f \in D_A \cap M.$$

If this condition is true, we denoted by A^M the restriction $A_{D_A \cap M}$ of A to $D_A \cap M$ (cf. 1.2.5 and 3.2.3) when M is regarded as the final set of the mapping $A_{D_A \cap M}$ (cf. 1.2.1). Then it is obvious that A^M is a linear operator in the Hilbert space M (cf. 10.3.2), i.e. that $A^M \in \mathcal{O}(M)$ (in particular, $D_A \cap M$ is a linear manifold in M, cf. 3.1.5 and 3.1.4b).

If $M \in \mathscr{S}(\mathcal{H})$ is an invariant subspace for $A \in \mathcal{O}(\mathcal{H})$, then $M^\perp$ may or may not be an invariant subspace for A. Let us suppose that both M and $M^\perp$ are invariant subspaces for the operator A. Does the study of the operator A reduce in this case to the study of the two operators A^M and $A^{M^\perp}$? The answer is clearly in the affirmative if $D_A = \mathcal{H}$. Indeed, if $D_A = \mathcal{H}$ then for each $f \in D_A$ we have the unique representation

$$f = f_1 + f_2,$$

where $f_1 \in M$ and $f_2 \in M^\perp$ (cf. 10.4.1), from which it follows that

$$Af = A^M f_1 + A^{M^\perp} f_2.$$

However, if $D_A \neq \mathcal{H}$ then we do not have in general

$$D_A = (D_A \cap M) + (D_A \cap M^\perp)$$

(cf. 3.1.8 for the sum of two subsets of a linear space). In fact, we have the following proposition, which is preliminary to the idea of a reducing subspace.

17.2.2 Proposition. *Let D be a linear manifold in $\mathcal{H}$ and let M be a subspace of $\mathcal{H}$. The following conditions are equivalent:*

(a) $D = (D \cap M) + (D \cap M^\perp)$;
(b) $P_M(D) = D \cap M$ and $P_{M^\perp}(D) = D \cap M^\perp$;
(c) $P_M(D) \subset D$;
(d) $P_{M^\perp}(D) \subset D$.

Proof. $a \Rightarrow b$: We assume condition a. Then, for every $f \in D$,

$$f = f_1 + f_2, \text{ with } f_1 \in D \cap M \text{ and } f_2 \in D \cap M^\perp,$$

and hence $P_M f = f_1 \in D \cap M$. This proves the inclusion

$$P_M(D) \subset D \cap M.$$

Moreover, the implications

$$f \in D \cap M \Rightarrow [f \in D \text{ and } P_M f = f] \Rightarrow f \in P_M(D)$$

(cf. 13.3.3c) prove the inclusion

$$D \cap M \subset P_M(D).$$

This reasoning can be repeated with M replaced by $M^\perp$ (since $M = M^{\perp\perp}$, cf. 10.4.4a).

$b \Rightarrow c$: This is obvious.

$c \Rightarrow d$: We assume condition c. Then,

$$f \in D \Rightarrow [f \in D \text{ and } P_M f \in D] \Rightarrow P_{M^\perp} f = f - P_M f \in D$$

(for $P_{M^\perp} = \mathbb{1}_{\mathcal{H}} - P_M$, cf. 13.1.3e).

$d \Rightarrow a$: We assume condition d. Then,

$$f \in D \Rightarrow [f \in D \text{ and } P_{M^\perp} f \in D] \Rightarrow P_M f = f - P_{M^\perp} f \in D,$$

and hence

$$f \in D \Rightarrow [P_M f \in D \text{ and } P_{M^\perp} f \in D] \Rightarrow$$
$$[P_M f \in D \cap M \text{ and } P_{M^\perp} f \in D \cap M^\perp] \Rightarrow$$
$$[\exists (f_1, f_2) \in (D \cap M) \times (D \cap M^\perp) \text{ so that } f = f_1 + f_2] \Rightarrow$$
$$f \in (D \cap M) + (D \cap M^\perp).$$

This proves the inclusion

$$D \subset (D \cap M) + (D \cap M^\perp).$$

On the other hand, the inclusion

$$(D \cap M) + (D \cap M^\perp) \subset D$$

is obvious.   □

We point out that, if D is a subspace N of $\mathcal{H}$, then condition a of 17.2.2 is condition d of 13.2.1 with M and N interchanged. Thus, if D is a subspace N of $\mathcal{H}$, all the conditions of 17.2.2 are equivalent to conditions a, b, c, d of 13.2.1.

The next theorem selects conditions which are equivalent to each other, and which embody the conditions of 17.2.2 with the domain of a linear operator A as D. It proves that, if a subspace M and its orthogonal complement $M^\perp$ are invariant subspaces for an operator A, then the study of A can be reduced to the study of the operators A^M and $A^{M^\perp}$ *provided that* the additional condition $P_M(D_A) \subset D_A$ is satisfied (this condition can obviously be replaced by any of the conditions to which it is equivalent in view of 17.2.2).

17.2.3 Theorem. *Let $A \in \mathcal{O}(\mathcal{H})$ and $M \in \mathscr{S}(\mathcal{H})$. The following conditions are equivalent:*

(a) M and $M^\perp$ are invariant subspaces for A (i.e. $Af \in M$, $\forall f \in D_A \cap M$, and $Ag \in M^\perp$, $\forall g \in D_A \cap M^\perp$) and $P_M(D_A) \subset D_A$;

(b) $P_M(D_A) \subset D_A$ and there exist $A_1 \in \mathcal{O}(M)$, $A_2 \in \mathcal{O}(M^\perp)$ so that
$$D_{A_1} = P_M(D_A), \quad D_{A_2} = P_{M^\perp}(D_A),$$
$$Af = A_1 P_M f + A_2 P_{M^\perp} f, \forall f \in D_A;$$

(c) $P_M A \subset A P_M$ (i.e. P_M commutes with A, in the sense defined in 17.1.2).

If these conditions are satisfied, then

(d) the operators A_1 and A_2 are uniquely determined by condition b; in fact, $A_1 = A^M$ and $A_2 = A^{M^\perp}$.

Proof. $a \Rightarrow b$: We assume condition a. Then, in view of 17.2.2,

$$D_{A^M} = D_A \cap M = P_M(D_A) \text{ and } D_{A^{M^\perp}} = D_A \cap M^\perp = P_{M^\perp}(D_A).$$

Moreover,

$$Af = A(P_M f + P_{M^\perp} f) = A^M P_M f + A^{M^\perp} P_{M^\perp} f, \forall f \in D_A.$$

This proves condition b, with $A_1 := A^M$ and $A_2 := A^{M^\perp}$.

$b \Rightarrow c$: We assume condition b. Then we have

$$f \in D_{P_M A} \Rightarrow f \in D_A \Rightarrow P_M f \in D_A,$$

i.e. $D_{P_M A} \subset D_{AP_M}$. Moreover, for every $f \in D_{P_M A} (= D_A)$ we have

$$P_M A f = P_M (A_1 P_M f + A_2 P_{M^\perp} f) = A_1 P_M f$$

(since $A_1 P_M f \in M$ and $A_2 P_{M^\perp} f \in M^\perp$, cf. 13.1.3b,c), and also

$$A P_M f = A_1 P_M^2 f + A_2 P_{M^\perp} P_M f = A_1 P_M f,$$

and hence

$$P_M A f = A P_M f.$$

$c \Rightarrow a$: We assume condition c. Then we have $D_{P_M A} \subset D_{AP_M}$ and hence

$$f \in D_A \Rightarrow f \in D_{P_M A} \Rightarrow f \in D_{AP_M} \Rightarrow P_M f \in D_A,$$

i.e. $P_M(D_A) \subset D_A$. Moreover we have

$$f \in D_A \cap M \Rightarrow A f \overset{(1)}{=} A P_M f \overset{(2)}{=} P_M A f \in M$$

(1 holds true because $f \in M$ and 2 because $f \in D_A = D_{P_M A}$), and also

$$f \in D_A \cap M^\perp \Rightarrow A f \overset{(3)}{=} A P_{M^\perp} f \overset{(4)}{=} A f - A P_M f \overset{(5)}{=} A f - P_M A f = P_{M^\perp} A f \in M^\perp$$

(3 holds true because $f \in M^\perp$, 4 because $f \in D_A = D_{P_M A}$ and hence $f \in D_{AP_M}$, 5 because $f \in D_{P_M A}$).

d: We suppose that $P_M(D_A) \subset D_A$ and that $A_1 \in \mathcal{O}(M)$, $A_2 \in \mathcal{O}(M^\perp)$ are so that condition b holds true. Then condition a holds true as well and we have

$$D_{A_1} = P_M(D_A) = D_A \cap M = D_{A^M},$$

in view of 17.2.2, and

$$A_1 f = A_1 P_M f = A_1 P_M^2 f + A_2 P_{M^\perp} P_M f = A f = A^M f, \forall f \in D_{A_1}.$$

This proves that $A_1 = A^M$. The proof of the equation $A_2 = A^{M^\perp}$ is similar. $\qquad \square$

17.2.4 Definition. Let A be an operator in $\mathcal{H}$, i.e. $A \in \mathcal{O}(\mathcal{H})$, and let M be a subspace in $\mathcal{H}$, i.e. $M \in \mathscr{S}(\mathcal{H})$. We say that M is a *reducing subspace* for A, or that A *is reduced* by M, if the conditions of 17.2.3 hold true for A and M, e.g. if

$$P_M A \subset A P_M.$$

We see that A is reduced by M iff A is reduced by $M^\perp$, in view of condition a of 17.2.3, of the equivalence between conditions c and d of 17.2.2, and of the equality $M = M^{\perp\perp}$ (cf. 10.4.4a). We note that if A is reduced by M then M is an invariant subspace for A and hence the operator A^M is defined.

Obviously, all operators in $\mathcal{H}$ are reduced by the trivial subspaces $\{0_\mathcal{H}\}$ and $\mathcal{H}$.

If an operator A is reduced by a subspace M, many properties of A are inherited by the operator A^M in the Hilbert space M, as is shown by the following propositions.

17.2.5 Proposition. *Let A be a closable operator in $\mathcal{H}$, let M be a subspace of $\mathcal{H}$, and suppose that A is reduced by M. Then A^M is a closable operator in the Hilbert space M, the operator $\overline{A}$ is reduced by M, and*

$$\overline{(A^M)} = (\overline{A})^M,$$

where $\overline{(A^M)}$ denotes the closure of the operator A^M in the Hilbert space M (hence, $\overline{(A^M)} \in \mathcal{O}(M)$).

Proof. Let $(0_M, g) \in \overline{G_{A^M}}$; then there exists a sequence $\{(f_n, g_n)\}$ in G_{A^M} so that

$$f_n \to 0_M \text{ and } g_n \to g;$$

now, $(f_n, g_n) \in G_A$ for all $n \in \mathbb{N}$ and hence (since $0_M = 0_{\mathcal{H}}$) $(0_{\mathcal{H}}, g) \in \overline{G_A}$, and hence (since A is closable) $g = 0_{\mathcal{H}} = 0_M$. By 4.4.11a, this proves that A^M is closable.

We have $P_M A \subset A P_M$ by hypothesis. Let $f \in D_{\overline{A}} (= D_{P_M \overline{A}})$; then there exists a sequence $\{f_n\}$ in D_A so that

$$f_n \to f, \{A f_n\} \text{ is convergent, } \overline{A} f = \lim_{n\to\infty} A f_n$$

(cf. 4.4.10); now, the sequence $\{P_M f_n\}$ is in D_A and (since P_M is continuous)

$$P_M f_n \to P_M f, \{P_M A f_n\} \text{ is convergent i.e. } \{A P_M f_n\} \text{ is convergent;}$$

therefore,

$$P_M f \in D_{\overline{A}} \text{ and}$$
$$\overline{A} P_M f = \lim_{n\to\infty} A P_M f_n = \lim_{n\to\infty} P_M A f_n = P_M \lim_{n\to\infty} A f_n = P_M \overline{A} f.$$

This proves that $P_M \overline{A} \subset \overline{A} P_M$, i.e. that the operator $\overline{A}$ is reduced by M.

Moreover, the following implications are true:

$$f \in D_{\overline{(A^M)}} \Rightarrow$$

[there exists $\{f_n\}$ in D_{A^M} s.t. $f_n \to f, \{A^M f_n\}$ is convergent,

$$\overline{(A^M)} f = \lim_{n\to\infty} A^M f_n] \overset{(1)}{\Rightarrow}$$

$[f \in D_{\overline{A}} \cap M = D_{(\overline{A})^M}$ and $\overline{(A^M)} f = \overline{A} f = (\overline{A})^M f]$

(1 holds because $D_{A^M} \subset D_A$ and $A^M \subset A$, and because M is closed). This proves the inclusion $\overline{(A^M)} \subset (\overline{A})^M$. Conversely, let $f \in D_{(\overline{A})^M}$; then (since $D_{(\overline{A})^M} \subset D_{\overline{A}}$) there exists a sequence $\{f_n\}$ in D_A so that

$$f_n \to f \text{ and } \{A f_n\} \text{ in convergent;}$$

now, the sequence $\{P_M f_n\}$ is in D_{A^M} (since $P_M(D_A) = D_{A^M}$, cf. 17.2.3b,d) and

$$P_M f_n \to P_M f = f, \{P_M A f_n\} \text{ is convergent i.e. } \{A^M P_M f_n\} \text{ is convergent;}$$

therefore, $f \in D_{\overline{(A^M)}}$. This proves the inclusion $D_{(\overline{A})^M} \subset D_{\overline{(A^M)}}$. $\qquad\square$

17.2.6 Corollary. *Let A be a closed operator in $\mathcal{H}$, let M be a subspace of $\mathcal{H}$, and suppose that A is reduced by M. Then the operator A^M is closed.*

Proof. In view of 17.2.5 we have $\overline{(A^M)} = (\overline{A})^M = A^M$. Hence, A^M is closed (cf. 4.4.10).   □

17.2.7 Proposition. *Let A be an adjointable operator in $\mathcal{H}$ (i.e. $\overline{D_A} = \mathcal{H}$), let M be a subspace of $\mathcal{H}$, and suppose that A is reduced by M. Then A^M is an adjointable operator in the Hilbert space M, the operator $A^\dagger$ is reduced by M, and*

$$(A^M)^\dagger = (A^\dagger)^M,$$

where $(A^M)^\dagger$ denotes the adjoint of the operator A^M in the Hilbert space M (hence, $(A^M)^\dagger \in \mathcal{O}(M)$).

Proof. We note that $(D_A \cap M)^\perp \cap M$ is the orthogonal complement of $D_{A^M} = D_A \cap M$ in the Hilbert space M. Let $g \in D_A$; we have

$$g = g_1 + g_2 \text{ with } g_1 \in D_A \cap M \text{ and } g_2 \in D_A \cap M^\perp$$

(cf. 17.2.2a); then, for every $f \in (D_A \cap M)^\perp \cap M$, we have

$$(f|g_1) = (f|g_2) = 0$$

(because $f \in (D_A \cap M)^\perp$ and $g_1 \in D_A \cap M$, and because $f \in M$ and $g_2 \in M^\perp$), and hence

$$(f|g) = 0.$$

Since g was an arbitrary element of D_A and $D_A^\perp = \{0_\mathcal{H}\}$ (cf. 10.4.4d), this proves that

$$(D_A \cap M^\perp) \cap M = \{0_\mathcal{H}\} = \{0_M\},$$

and hence that the operator A^M in the Hilbert space M is adjointable.

By hypothesis we have $P_M A \subset A P_M$. In view of 13.1.5, 12.3.4a,b, 12.1.4, this implies that

$$P_M A^\dagger \subset (A P_M)^\dagger \subset (P_M A)^\dagger = A^\dagger P_M,$$

and hence that $A^\dagger$ is reduced by M.

In what follows, we denote by a subscript whether a given inner product is to be regarded as pertaining to the Hilbert space $\mathcal{H}$ or to the Hilbert space M. We have

$$\left(A^M f | g\right)_M = (Af|g)_\mathcal{H} \overset{(1)}{=} \left(f|A^\dagger g\right)_\mathcal{H} = \left(f|(A^\dagger)^M g\right)_M, \forall f \in D_{A^M}, \forall g \in D_{(A^\dagger)^M}$$

(1 holds because $D_{(A^\dagger)^M} \subset D_{A^\dagger}$). By 12.1.3B, this proves that

$$(A^\dagger)^M \subset (A^M)^\dagger. \tag{2}$$

Now let $g \in D_{(A^M)^\dagger}$; then we have, for all $f \in D_A$,

$$
(Af|g)_{\mathcal{H}} \overset{(3)}{=} \left(A^M P_M f + A^{M^\perp} P_{M^\perp} f | g \right)_{\mathcal{H}}
$$

$$
\overset{(4)}{=} \left(A^M P_M f | g \right)_M = \left(P_M f | (A^M)^\dagger g \right)_M
$$

$$
\overset{(5)}{=} \left(P_M f + P_{M^\perp} f | (A^M)^\dagger g \right)_{\mathcal{H}} = \left(f | (A^M)^\dagger g \right)_{\mathcal{H}}
$$

(3 holds in view of 17.2.3b,d; 4 holds because $A^{M^\perp} P_{M^\perp} f \in M^\perp$ and $g \in M$ since $(A^M)^\dagger$ denotes the adjoint of the operator A^M in the Hilbert space M; 5 holds because $P_{M^\perp} f \in M^\perp$ and $(A^M)^\dagger g \in M$); therefore $g \in D_{A^\dagger}$ and hence $g \in D_{A^\dagger} \cap M = D_{(A^\dagger)M}$. This proves that

$$
D_{(A^M)^\dagger} \subset D_{(A^\dagger)M},
$$

and hence, in view of 2, that $(A^\dagger)^M = (A^M)^\dagger$. $\square$

17.2.8 Corollary. *Let A be a symmetric, or a self-adjoint, or an essentially self-adjoint operator in $\mathcal{H}$, and suppose that A is reduced by a subspace M of $\mathcal{H}$. Then A^M is a symmetric, or a self-adjoint, or an essentially self-adjoint operator in the Hilbert space M.*

Proof. This follows immediately from 17.2.7. $\square$

17.2.9 Proposition. *Let $A \in \mathcal{B}(\mathcal{H})$ and $M \in \mathscr{S}(\mathcal{H})$. The following conditions are equivalent:*

(a) A is reduced by M;
(b) M is an invariant subspace for both A and $A^\dagger$ (i.e. $Af \in M$ and $A^\dagger f \in M$, $\forall f \in M$; recall that $D_{A^\dagger} = \mathcal{H}$, cf. 12.2.2).

If these conditions are satisfied, then:

(c) $A^M \in \mathcal{B}(\mathcal{H})$ and $\|A^M\| \leq \|A\|$.

Proof. $a \Rightarrow b$: If A is reduced by M then M is an invariant subspace for A (cf. 17.2.3a) and $A^\dagger$ is reduced by M (cf. 17.2.7), and hence M is an invariant subspace for $A^\dagger$ as well.

$b \Rightarrow a$: We assume condition b. Then, in view of 13.1.3c we have

$$
P_M A P_M f = A P_M f \text{ and } P_M A^\dagger P_M f = A^\dagger P_M f, \forall f \in \mathcal{H},
$$

and hence (cf. 12.1.6b, 12.3.4b, 13.1.5, and recall that $A^\dagger \in \mathcal{B}(\mathcal{H})$, cf. 12.2.2)

$$
A P_M = P_M A P_M = (P_M A^\dagger P_M)^\dagger = (A^\dagger P_M)^\dagger = P_M A,
$$

i.e. condition a.

c: If M is an invariant subspace for A then it is obvious that condition c is true, since A^M is a restriction of A. $\square$

17.2.10 Proposition. *Let* $A \in \mathcal{U}(\mathcal{H})$ *and* $M \in \mathcal{S}(\mathcal{H})$, *and suppose that* A *is reduced by* M. *Then* $A^M \in \mathcal{U}(M)$.

Proof. We have $D_{A^M} = M$ and

$$\|A^M f\|_M = \|Af\|_\mathcal{H} = \|f\|_\mathcal{H} = \|f\|_M, \forall f \in M.$$

Moreover, for every $f \in M$ we have

$$A^{-1}f = A^\dagger f \in M$$

(cf. 12.5.1 and 17.2.9) and hence

$$f = A(A^{-1}f) = A^M(A^{-1}f).$$

This proves that $R_{A^M} = M$, and hence that $A^M \in \mathcal{U}(M)$ by 10.1.20. $\qquad\square$

17.2.11 Proposition. *Let* $U : G \to \mathcal{U}(\mathcal{H})$ *be a homomorphism from a group* G *to the group* $\mathcal{U}(\mathcal{H})$ *(cf. 10.3.10; U is then called a unitary representation of G) and let M be a subspace of $\mathcal{H}$. The following conditions are equivalent:*

(a) M is a reducing subspace for $U(g)$, $\forall g \in G$;
(b) M is an invariant subspace for $U(g)$, $\forall g \in G$.

If these conditions hold true, then

(c) The mappings

$$G \ni g \mapsto (U(g))^M \in \mathcal{U}(M) \text{ and } G \ni g \mapsto (U(g))^{M^\perp} \in \mathcal{U}(M^\perp)$$

are homomorphisms from G to the groups $\mathcal{U}(M)$ and $\mathcal{U}(M^\perp)$ respectively.

Proof. $a \Rightarrow b$: This is obvious (cf. 17.2.3a).

$b \Rightarrow a$: If we assume condition b then we have, for every $g \in G$,

$$U(g)f \in M \text{ and } (U(g))^\dagger f = (U(g))^{-1}f = U(g^{-1})f \in M, \forall f \in M$$

(cf. 12.5.1 and 1.3.3), and this implies condition a by 17.2.9.

c: If condition a holds true, then

$$(U(g))^M \in \mathcal{U}(M) \text{ and } (U(g))^{M^\perp} \in \mathcal{U}(M^\perp), \forall g \in G,$$

by 17.2.10. Moreover, for all $g, g' \in G$,

$$(U(g))^M (U(g'))^M f = U(g)U(g')f = U(gg')f = (U(gg'))^M f, \forall f \in M,$$

and similarly for $M^\perp$. $\qquad\square$

17.2.12 Proposition. *Let* $A \in \mathcal{P}(\mathcal{H})$ *and* $M \in \mathcal{S}(\mathcal{H})$, *and suppose that* A *is reduced by* M. *Then* $A^M \in \mathcal{P}(M)$. *More precisely, A^M is the orthogonal projection onto the subspace $R_A \cap M$ of the Hilbert space M.*

Proof. In view of 13.1.5, we have $A^M = (A^M)^\dagger$ (cf. 17.2.8) and

$$A^M f = Af = A^2 f = (A^M)^2 f, \forall f \in M,$$

and hence $A \in \mathscr{P}(M)$.

Moreover, $R_A \cap M$ is a subspace of the Hilbert space M (cf. 3.1.5, 3.1.4b, 13.1.4a, 2.3.3) and we have, for $f \in M$,

$$A^M f = f \text{ iff } Af = f \text{ iff } f \in R_A \text{ iff } f \in R_A \cap M$$

(cf. 13.1.3c). This proves that A^M is the orthogonal projection onto $R_A \cap M$ (cf. 13.1.3c once again). $\qquad\square$

17.2.13 Proposition. *Let A be a self-adjoint operator in $\mathcal{H}$ and M a subspace of $\mathcal{H}$. The following conditions are equivalent:*

(a) A is reduced by M;
(b) $P^A(E)$ is reduced by M (or equivalently $P^A(E)f \in M$, $\forall f \in M$), $\forall E \in \mathcal{A}(d_\mathbb{R})$;
(c) $U^A(t)$ is reduced by M (or equivalently $U^A(t)f \in M$, $\forall f \in M$), $\forall t \in \mathbb{R}$.

If these conditions hold true then A^M is a self-adjoint operator in the Hilbert space M (cf. 17.2.8) and:

(d) the mapping

$$\mathcal{A}(d_\mathbb{R}) \ni E \mapsto P^{A,M}(E) := (P^A(E))^M \in \mathscr{P}(M)$$

is the projection valued measure of A^M;
(e) the mapping

$$\mathbb{R} \ni t \mapsto (U^A(t))^M \in \mathcal{U}(M)$$

is the continuous one-parameter unitary group whose generator is A^M.

Proof. The parenthetic equivalence in condition b follows from 13.1.5 and 17.2.9. The parenthetic equivalence in condition c follows from 17.2.11.

Condition a is $P_M A \subset A P_M$ and hence, in view of 17.1.4, it is equivalent to condition a of 17.1.7 with $B := P_M$. Condition b is $P_M P^A(E) = P^A(E) P_M$ for all $E \in \mathcal{A}(d_\mathbb{R})$, and hence it is condition b of 17.1.7 with $B := P_M$. Condition c is $P_M U^A(t) = U^A(t) P_M$ for all $t \in \mathbb{R}$, and hence it is condition e of 17.1.7 with $B := P_M$. This proves that conditions a, b, c are equivalent.

d: We assume conditions a and b. Then we have

$$P^{A,M}(E) \in \mathscr{P}(M), \forall E \in \mathcal{A}(d_\mathbb{R}),$$

by 17.2.12, and also

$$\mu_f^{P^{A,M}} = \mu_f^{P^A} \text{ and } \mu_f^{P^{A,M}}(\mathbb{R}) = \left(f|\mathbb{1}_\mathcal{H}^M f\right)_M = (f|\mathbb{1}_M f)_M = \|f\|_M^2, \ \forall f \in M.$$

In view of 13.3.5, this proves that $P^{A,M}$ is a projection valued measure with values in $\mathscr{P}(M)$. In view of 15.2.2e, we have, for $f \in M$,

$$\int_\mathbb{R} \xi^2 d\mu_f^{P^{A,M}} < \infty \text{ iff } \int_\mathbb{R} \xi^2 d\mu_f^{P^A} < \infty \text{ iff } f \in D_A \text{ iff } f \in D_A \cap M = D_{A^M},$$

and also

$$\left(f|A^M f\right)_M = (f|Af)_{\mathcal{H}} = \int_{\mathbb{R}} \xi d\mu_f^{P^A} = \int_{\mathbb{R}} \xi d\mu_f^{P^{A,M}}, \forall f \in D_{A^M}.$$

This proves that $P^{A,M}$ is the projection valued measure of the self-adjoint operator A^M (cf. 15.2.1).

e: We assume conditions a and c. Then, for every $t \in \mathbb{R}$, $(U^A(t))^M$ is a linear operator in M and (cf. 16.1.7)

$$\left(f|(U^A(t))^M f\right)_M = \left(f|U^A(t)f\right)_{\mathcal{H}} = \int_{\mathbb{R}} \varphi_t d\mu_f^{P^A} = \int_{\mathbb{R}} \varphi_t d\mu_f^{P^{A,M}}, \forall f \in M.$$

By 16.1.7, this proves that $(U^A(t))^M = U^{A^M}(t)$ for all $t \in \mathbb{R}$. □

In the next two theorems, the first statements generalize the content of 17.2.3b,d. On the basis of these theorems, the study of the structure of a (closed) operator can be carried out through the investigation of its reducing subspaces and its restrictions to the intersections of its domain with them.

17.2.14 Proposition. *Let $N \in \mathbb{N}$, let $\{M_1, ..., M_N\}$ be a family of subspaces of $\mathcal{H}$ such that, writing $P_n := P_{M_n}$ for all $n \in \{1, ..., N\}$, the following conditions are true*

$$M_k \subset M_i^{\perp} \ \text{if} \ i \neq k \ \text{and} \ \sum_{n=1}^{N} P_n = \mathbb{1}_{\mathcal{H}},$$

and suppose that an operator A in $\mathcal{H}$ is reduced by M_n for all $n \in \{1, ..., N\}$. Writing $A_n := A^{M_n}$ for all $n \in \{1, ..., N\}$, we have:

(a) *$D_A = \{f \in \mathcal{H} : P_n f \in D_{A_n}, \forall n \in \{1, ..., N\}\}$,*
 $Af = \sum_{n=1}^{N} A_n P_n f, \ \forall f \in D_A$;
(b) *A is closed iff A_n is a closed operator in the Hilbert space M_n, $\forall n \in \{1, ..., N\}$;*
(c) *A is adjointable iff A_n is an adjointable operator in the Hilbert space M_n, $\forall n \in \{1, ..., N\}$;*
 if these conditions hold true, then
 $D_{A^{\dagger}} = \{f \in \mathcal{H} : P_n f \in D_{A_n^{\dagger}}, \forall n \in \{1, ..., N\}\}$,
 $A^{\dagger} f = \sum_{n=1}^{N} A_n^{\dagger} P_n f, \ \forall f \in D_{A^{\dagger}}$,
 where $A_n^{\dagger}$ denotes the adjoint of A_n in the Hilbert space M_n;
(d) *A is symmetric iff A_n is symmetric, $\forall n \in \{1, ..., N\}$;*
 A is self-adjoint iff A_n is self-adjoint, $\forall n \in \{1, ..., N\}$;
 A is essentially self-adjoint iff A_n is essentially self-adjoint, $\forall n \in \{1, ..., N\}$;
(e) *$A \in \mathcal{B}(\mathcal{H})$ iff $A_n \in \mathcal{B}(M_n)$, $\forall n \in \{1, ..., N\}$;*
(f) *$A \in \mathcal{U}(\mathcal{H})$ iff $A_n \in \mathcal{U}(M_n)$, $\forall n \in \{1, ..., N\}$;*
(g) *$A \in \mathscr{P}(\mathcal{H})$ iff $A_n \in \mathscr{P}(M_n)$, $\forall n \in \{1, ..., N\}$.*

Moreover, if A is self-adjoint then:

(h) *$P^A(E) = \sum_{n=1}^{N} P^{A_n}(E) P_n, \ \forall E \in \mathcal{A}(d_{\mathbb{R}})$;*

(i) $U^A(t) = \sum_{n=1}^{N} U^{A_n}(t) P_n$, $\forall t \in \mathbb{R}$.

Proof. a: For all $n \in \{1, ..., N\}$, we have

$$P_n f \in D_A \cap M_n = D_{A_n}, \forall f \in D_A, \tag{1}$$

and

$$P_n A f = A P_n f = A_n P_n f, \forall f \in D_A. \tag{2}$$

From 1 we have

$$D_A \subset \{f \in \mathcal{H} : P_n f \in D_{A_n}, \forall n \in \{1, ..., N\}\};$$

on the other hand, for $f \in \mathcal{H}$ we have

$$[P_n f \in D_{A_n}, \forall n \in \{1, ..., N\}] \Rightarrow f = \sum_{n=1}^{N} P_n f \in D_A,$$

since $D_{A_n} \subset D_A$ and D_A is a linear manifold. This proves the part of statement a about D_A. Moreover, from 2 we have

$$Af = \sum_{n=1}^{N} P_n A f = \sum_{n=1}^{N} A_n P_n f, \forall f \in D_A.$$

b: The "only if" part of statement b follows from 17.2.6.

Now we assume that A_n is a closed operator in the Hilbert space M_n, for all $n \in \{1, ..., N\}$. Suppose that a sequence $\{f_k\}$ in D_A and two vectors $f, g \in \mathcal{H}$ are so that

$$f_k \xrightarrow[k \to \infty]{} f \text{ and } A f_k \xrightarrow[k \to \infty]{} g.$$

Then, for all $n \in \{1, ..., N\}$,

$$P_n f_k \in D_{A_n}, \forall k \in \mathbb{N}, \quad P_n f_k \xrightarrow[k \to \infty]{} P_n f, \quad P_n A f_k \xrightarrow[k \to \infty]{} P_n g;$$

we have $P_n A f_k = A_n P_n f_k$, $\forall k \in \mathbb{N}$ (cf. 2); hence, since A_n is closed, we have

$$P_n f \in D_{A_n} \text{ and } P_n g = A_n P_n f.$$

In view of result a, this implies that

$$f \in D_A \text{ and } g = \sum_{n=1}^{N} P_n g = \sum_{n=1}^{N} A_n P_n f = Af.$$

This proves that the operator A is closed.

c: The "only if" part of statement c follows from 17.2.7.

Now we assume that A_n is an adjointable operator in the Hilbert space M_n, for all $n \in \{1, ..., N\}$. Let $g \in D_A^\perp$; then,

$$(P_n g | P_n f)_{\mathcal{H}} = (g | P_n f)_{\mathcal{H}} = 0, \forall f \in D_A, \forall n \in \{1, ..., N\}$$

(since $P_n(D_A) \subset D_A$), and hence

$$(P_n g | h)_{M_n} = 0, \forall h \in D_{A_n}, \forall n \in \{1, ..., N\}$$

(since $P_n(D_A) = D_{A_n}$), and hence

$$P_n g = 0_{M_n} = 0_{\mathcal{H}}, \forall n \in \{1, ..., N\}$$

(since the orthogonal complement of D_{A_n} in the Hilbert space M_n is $\{0_{M_n}\}$, by 10.4.4d), and hence

$$g = \sum_{n=1}^{N} P_n g = 0_{\mathcal{H}}.$$

By 10.4.4d, this proves that A is adjointable.

Now we assume that A is adjointable. Then for all $n \in \{1, ..., N\}$, $A^\dagger$ is reduced by M_n and $(A^\dagger)^{M_n} = A_n^\dagger$ (cf. 17.2.7), and hence

$$P_n f \in D_{A^\dagger} \cap M_n = D_{(A^\dagger)^{M_n}} = D_{A_n^\dagger}, \forall f \in D_{A^\dagger}, \tag{3}$$

and

$$P_n A^\dagger f = A^\dagger P_n f = (A^\dagger)^{M_n} P_n f = A_n^\dagger P_n f, \forall f \in D_{A^\dagger}. \tag{4}$$

From 3 we have

$$D_{A^\dagger} \subset \{f \in \mathcal{H} : P_n f \in D_{A_n^\dagger}, \forall n \in \{1, ..., N\}\};$$

on the other hand, for $f \in \mathcal{H}$ we have

$$[P_n f \in D_{A_n^\dagger}, \forall n \in \{1, ..., N\}] \Rightarrow f = \sum_{n=1}^{N} P_n f \in D_{A^\dagger},$$

since $D_{A_n^\dagger} = D_{(A^\dagger)^{M_n}} \subset D_{A^\dagger}$ and $D_{A^\dagger}$ is a linear manifold. This proves the part of the statement about $D_{A^\dagger}$. Moreover, from 4 we have

$$A^\dagger f = \sum_{n=1}^{N} P_n A^\dagger f = \sum_{n=1}^{N} A_n^\dagger P_n f, \forall f \in D_{A^\dagger}.$$

d: The "only if" parts of statement d follow from 17.2.8. The "if" parts follow from results a and c.

e: The "only if" part of statement e follows from 17.2.9c.

Now we assume $A_n \in \mathcal{B}(M_n)$, for all $n \in \{1, ..., N\}$. From result a we have $D_A = \mathcal{H}$ and also (cf. 4.2.5b)

$$\|Af\|^2 = \sum_{n=1}^{N} \|A_n P_n f\|^2 \leq \max\{\|A_n\|^2 : n \in \{1, ..., N\}\} \sum_{n=1}^{N} \|P_n f\|^2$$
$$= \max\{\|A_n\|^2 : n \in \{1, ..., N\}\} \|f\|^2, \forall f \in \mathcal{H}.$$

f: The "only if" part of statement f follows from 17.2.10.

Now we assume $A_n \in \mathcal{U}(M_n)$, for all $n \in \{1, ..., N\}$. From results a and c we have $D_A = D_{A^\dagger} = \mathcal{H}$, and also (in view of 12.5.1)

$$
AA^\dagger f = \sum_{n=1}^{N} A_n P_n \left(\sum_{k=1}^{N} A_k^\dagger P_k f \right) = \sum_{n=1}^{N} A_n A_n^\dagger P_n f
$$

$$
= \sum_{n=1}^{N} P_n f = f, \forall f \in \mathcal{H},
$$

i.e. $AA^\dagger = \mathbb{1}_{\mathcal{H}}$, and similarly $A^\dagger A = \mathbb{1}_{\mathcal{H}}$. By 12.5.1, this proves that $A \in \mathcal{U}(\mathcal{H})$.

g: The "only if" part of statement g follows from 17.2.12.

Now we assume $A_n \in \mathscr{P}(M_n)$, for all $n \in \{1, ..., N\}$. From 13.1.5 and result d we have $A = A^\dagger$; from result a and 13.1.5 we have $D_A = \mathcal{H}$ and

$$
A^2 f = \sum_{n=1}^{N} A_n P_n \left(\sum_{k=1}^{N} A_k P_k f \right) = \sum_{n=1}^{N} A_n^2 P_n f
$$

$$
= \sum_{n=1}^{N} A_n P_n f = Af, \forall f \in \mathcal{H}.
$$

By 13.1.5, this proves that $A \in \mathscr{P}(\mathcal{H})$.

h: We suppose that A is self-adjoint. Then (cf. 17.2.13), for every $E \in \mathcal{A}(d_{\mathbb{R}})$, $P^A(E)$ is reduced by M_n for all $n \in \{1, ..., N\}$, and we have (cf. result a written with $P^A(E)$ in place of A, and 17.2.13d)

$$
P^A(E) = \sum_{n=1}^{N} (P^A(E))^{M_n} P_n = \sum_{n=1}^{N} P^{A_n}(E) P_n.
$$

i: We suppose that A is self-adjoint. Then (cf. 17.2.13), for every $t \in \mathbb{R}$, $U^A(t)$ is reduced by M_n for all $n \in \{1, ..., N\}$, and we have (cf. result a written with $U^A(t)$ in place of A, and 17.2.13e)

$$
U^A(t) = \sum_{n=1}^{N} (U^A(t))^{M_n} P_n = \sum_{n=1}^{N} U^{A_n}(t) P_n.
$$

$\square$

17.2.15 Proposition. *Let $\{M_n\}$ be a sequence of subspaces of $\mathcal{H}$ such that*

$$
M_k \subset M_i^{\perp} \text{ if } i \neq k \text{ and } \sum_{n=1}^{\infty} P_n f = f, \forall f \in \mathcal{H},
$$

where $P_n := P_{M_n}$ for all $n \in \mathbb{N}$. Suppose that an operator A in $\mathcal{H}$ is reduced by M_n for all $n \in \mathbb{N}$. Then, writing $A_n := A^{M_n}$ for all $n \in \mathbb{N}$, we have:

(a) $D_A \subset \{f \in \mathcal{H} : P_n f \in D_{A_n}, \forall n \in \mathbb{N}, \text{ and } \sum_{n=1}^{\infty} A_n P_n f \text{ is convergent}\}$,
  $Af = \sum_{n=1}^{\infty} A_n P_n f, \forall f \in D_A$;

(b) A is closed iff
$[D_A = \{f \in \mathcal{H} : P_n f \in D_{A_n}, \forall n \in \mathbb{N},$ *and* $\sum_{n=1}^{\infty} \|A_n P_n f\|^2 < \infty\}$
and A_n is closed, $\forall n \in \mathbb{N}]$;

(c) A is adjointable iff A_n is adjointable, $\forall n \in \mathbb{N}$; if these conditions hold true, then
$D_{A^\dagger} = \{f \in \mathcal{H} : P_n f \in D_{A_n^\dagger}, \forall n \in \mathbb{N},$ *and* $\sum_{n=1}^{\infty} \|A_n^\dagger P_n f\|^2 < \infty\},$
$A^\dagger f = \sum_{n=1}^{\infty} A_n^\dagger P_n f, \forall f \in D_{A^\dagger}.$

(d) A is symmetric iff A_n is symmetric, $\forall n \in \mathbb{N}$;
A is self-adjoint iff $[A_n$ is self-adjoint, $\forall n \in \mathbb{N}$, and A is closed];

(e) $A \in \mathcal{B}(\mathcal{H})$ iff
$[A_n \in \mathcal{B}(M_n), \forall n \in \mathbb{N},$ *and* $\sup\{\|A_n\| : n \in \mathbb{N}\} < \infty$, *and A is closed];*

(f) $A \in \mathcal{U}(\mathcal{H})$ iff $[A_n \in \mathcal{U}(M_n), \forall n \in \mathbb{N}$, and A is closed];

(g) $A \in \mathscr{P}(\mathcal{H})$ iff $[A_n \in \mathscr{P}(M_n), \forall n \in \mathbb{N}$, and A is closed].

Moreover, if A is self-adjoint then:

(h) $P^A(E)f = \sum_{n=1}^{\infty} P^{A_n}(E)P_n f, \forall f \in \mathcal{H}, \forall E \in \mathcal{A}(d_\mathbb{R})$;
(i) $U^A(t)f = \sum_{n=1}^{\infty} U^{A_n}(t)P_n f, \forall f \in \mathcal{H}, \forall t \in \mathbb{R}.$

Proof. a: As in the proof of 17.2.14a, we see that, for all $n \in \mathbb{N}$,

$$P_n f \in D_{A_n} \text{ and } P_n A f = A_n P_n f, \forall f \in D_A. \tag{1}$$

Then, for all $f \in D_A$, the series $\sum_{n=1}^{\infty} A_n P_n f$ is convergent because so is the series $\sum_{n=1}^{\infty} P_n A f$ (cf. 13.2.8), and

$$Af = \sum_{n=1}^{\infty} P_n A f = \sum_{n=1}^{\infty} A_n P_n f.$$

b: First we suppose that A is closed. Then A_n is closed for all $n \in \mathbb{N}$, by 17.2.6. Moreover, let $f \in \mathcal{H}$ be such that

$$P_n f \in D_{A_n}, \forall n \in \mathbb{N}, \text{ and } \sum_{n=1}^{\infty} \|A_n P_n f\|^2 < \infty; \tag{2}$$

since $D_{A_n} \subset D_A$ and D_A is a linear manifold, the first condition in 2 implies that

$$\sum_{k=1}^{n} P_k f \in D_A \text{ and } A \sum_{k=1}^{n} P_k f = \sum_{k=1}^{n} A_k P_k f, \forall n \in \mathbb{N};$$

now, the sequence $\{\sum_{k=1}^{n} P_k f\}$ is convergent (cf. 13.2.8); moreover, the second condition in 2 implies that the series $\sum_{n=1}^{\infty} A P_n f$ is convergent (cf. 10.4.7b), and hence that the sequence $\{A \sum_{k=1}^{n} P_k f\}$ is convergent; since A is supposed to be closed, this implies that

$$f = \lim_{n \to \infty} \sum_{k=1}^{n} P_k f \in D_A.$$

This proves that

$$\left\{ f \in \mathcal{H} : P_n f \in D_{A_n}, \forall n \in \mathbb{N}, \text{ and } \sum_{n=1}^{\infty} \|A_n P_n f\|^2 < \infty \right\} \subset D_A.$$

The opposite inclusion follows from result a and 10.4.7a. This concludes the proof of the "only if" part of statement b.

Now we suppose that A_n is closed, for all $n \in \mathbb{N}$, and that

$$D_A = \left\{ f \in \mathcal{H} : P_n f \in D_{A_n}, \forall n \in \mathbb{N}, \text{ and } \sum_{n=1}^{\infty} \|A_n P_n f\|^2 < \infty \right\}.$$

Let a sequence $\{f_k\}$ in D_A and two vectors $f, g \in \mathcal{H}$ be so that

$$f_k \xrightarrow[k \to \infty]{} f \text{ and } A f_k \xrightarrow[k \to \infty]{} g.$$

As in the proof of 17.2.14b, we see that

$$P_n f \in D_{A_n} \text{ and } P_n g = A_n P_n f, \ \forall n \in \mathbb{N}.$$

Since $\sum_{n=1}^{\infty} \|P_n g\|^2 < \infty$ (cf. 13.2.8), we have $\sum_{n=1}^{\infty} \|A_n P_n f\|^2 < \infty$, and hence $f \in D_A$. Moreover, in view of result a, we have

$$Af = \sum_{n=1}^{\infty} A_n P_n f = \sum_{n=1}^{\infty} P_n g = g.$$

This proves that the operator A is closed.

c: If A is adjointable then A_n is adjointable for all $n \in \mathbb{N}$, by 17.2.7.

If A_n is adjointable for all $n \in \mathbb{N}$, then as in the proof of 17.2.14c we see that

$$g \in D_A^{\perp} \Rightarrow P_n g = 0_{\mathcal{H}}, \forall n \in \mathbb{N},$$

and hence

$$g \in D_A^{\perp} \Rightarrow g = \sum_{n=1}^{\infty} P_n g = 0_{\mathcal{H}}.$$

By 10.4.4d, this proves that A is adjointable.

Now we suppose that A is adjointable. Then $A^{\dagger}$ is reduced by M_n and $(A^{\dagger})^{M_n} = A_n^{\dagger}$, for all $n \in \mathbb{N}$ (cf. 17.2.7), and $A^{\dagger}$ is closed (cf. 12.1.6a). Then, we use result a and the "only if" part of result b, with A replaced by $A^{\dagger}$, to obtain the second part of statement c.

d: The "only if" parts of statement d follow from 17.2.8 and 12.1.6a.

If A_n is symmetric for all $n \in \mathbb{N}$, then from result c we see that A is adjointable, and from results a and c that

$$f \in D_A \Rightarrow [f \in D_{A^{\dagger}} \text{ and } Af = A^{\dagger} f].$$

This proves that A is symmetric.

If A is closed and A_n is self-adjoint for all $n \in \mathbb{N}$, then A is self-adjoint in view of results a, b, c.

e: The "only if" part of statement e follows from 17.2.9c and 4.4.3.

Now we assume that A is closed, $A_n \in \mathcal{B}(M_n)$ for all $n \in \mathbb{N}$, and

$$m := \sup\{\|A_n\| : n \in \mathbb{N}\} < \infty.$$

Then we have

$$\sum_{n=1}^{\infty} \|A_n P_n f\|^2 \le m^2 \sum_{n=1}^{\infty} \|P_n f\|^2 = m^2 \|f\|^2, \forall f \in \mathcal{H},$$

and hence $D_A = \mathcal{H}$, in view of result b. Moreover, in view of result a, we have (cf. 10.4.7a)

$$\|Af\|^2 = \left\|\sum_{n=1}^{\infty} A_n P_n f\right\|^2 = \sum_{n=1}^{\infty} \|A_n P_n f\|^2 \le m^2 \|f\|^2, \forall f \in \mathcal{H}.$$

Thus, $A \in \mathcal{B}(\mathcal{H})$.

f, g, h, i: The proofs of these statements are analogous to those of statements f, g, h, i of 17.2.14, on the basis of results a, b, c. $\qquad\square$

17.2.16 Remark. The condition that A be closed cannot be disposed of in the "if" parts of statements d (second part), e, f, g of 17.2.15. This is shown by the following example.

Let $\mathcal{H}$ be a separable Hilbert space which is not finite-dimensional, and let $\{u_n\}_{n\in\mathbb{N}}$ be a c.o.n.s. in $\mathcal{H}$. For each $n \in \mathbb{N}$, we define the one-dimensional subspace $M_n := V\{u_n\}$; then, $P_n := P_{M_n}$ is the projection defined by

$$P_n f := (u_n|f)\, u_n, \forall f \in \mathcal{H}$$

(cf. 13.1.12), and we have

$$P_k P_i = \mathbb{O}_{\mathcal{H}} \text{ if } i \ne k \text{ and } \sum_{n=1}^{\infty} P_n f = \sum_{n=1}^{\infty} (u_n|f)\, u_n = f, \forall f \in \mathcal{H}.$$

For any function $\varphi : \mathbb{N} \to \mathbb{C}$, we define a linear operator A by letting

$$D_A := L\{u_n\}_{n\in\mathbb{N}} \text{ and } Af := \sum_{n=1}^{\infty} \varphi(n)\,(u_n|f)\, u_n, \forall f \in D_A$$

(we note that $L\{u_n\}_{n\in\mathbb{N}} = \{f \in \mathcal{H} : \exists n_f \in \mathbb{N} \text{ s.t. } n > n_f \Rightarrow (u_n|f) = 0\}$). We see that, for all $n \in \mathbb{N}$,

$$P_n f \in D_A \text{ and } AP_n f = \varphi(n)\,(u_n|f)\, u_n = P_n Af, \forall f \in D_A,$$

i.e. A is reduced by M_n. Thus, $\{M_n\}_{n\in\mathbb{N}}$ and A are as in 17.2.15.

We consider the following cases:

(a) $\varphi(n) \in \mathbb{R}, \forall n \in \mathbb{N}$.

In this case, A_n is self-adjoint for all $n \in \mathbb{N}$, but A is not self-adjoint; indeed,

$$D_{A^\dagger} = \left\{f \in \mathcal{H} : \sum_{n=1}^{\infty} |\varphi(n)|^2\,|(u_n|f)|^2 < \infty\right\};$$

hence, if we define

$$f := \sum_{n=1}^{\infty} \frac{1}{\varphi(n) + i}\frac{1}{n} u_n$$

(this series is convergent by 10.4.7b), we have $f \in D_{A^\dagger}$ but $f \notin D_A$.

(b) $\exists m \in [0, \infty)$ so that $|\varphi(n)| \leq m, \forall n \in \mathbb{N}$.

In this case, $A_n \in \mathcal{B}(M_n)$ for all $n \in \mathbb{N}$ and A is bounded, but $A \notin \mathcal{B}(\mathcal{H})$ since $D_A \neq \mathcal{H}$.

(c) $|\varphi(n)| = 1, \forall n \in \mathbb{N}$.

In this case, $A_n \in \mathcal{U}(M_n)$ for all $n \in \mathbb{N}$ and $\|Af\| = \|f\|$ for all $f \in D_A$, but $A \notin \mathcal{U}(\mathcal{H})$ since $D_A \neq \mathcal{H}$.

(d) $\varphi(n) \in \{0, 1\}, \forall n \in \mathbb{N}$.

In this case, $A_n \in \mathscr{P}(M_n)$ for all $n \in \mathbb{N}$ and $A = A^2$, but $A \notin \mathscr{P}(\mathcal{H})$ since A is not self-adjoint.

We point out that, if we had defined A by letting

$$D_A := \left\{ f \in \mathcal{H} : \sum_{n=1}^{\infty} |\varphi(n)|^2 \, |(u_n|f)|^2 < \infty \right\},$$

$$Af := \sum_{n=1}^{\infty} \varphi(n) \, (u_n|\varphi) \, u_n, \forall f \in D_A,$$

we would have had: A self-adjoint in case a, $A \in \mathcal{B}(\mathcal{H})$ in case b, $A \in \mathcal{U}(\mathcal{H})$ in case c, $A \in \mathscr{P}(\mathcal{H})$ in case d. In fact, A would have been a closed operator.

To conclude this section, we present an example which shows that for an invariant subspace it is possible not to be a reducing subspace even when its orthogonal complement is invariant as well.

17.2.17 Example. Let $a, b, c \in \mathbb{R}$ be so that $a < c < b$. We define two subsets of the Hilbert space $L^2(a, b)$ by letting

$$M_1 := \{[\varphi] \in L^2(a, b) : \varphi(x) = 0 \text{ } m\text{-a.e. on } [c, b]\},$$
$$M_2 := \{[\varphi] \in L^2(a, b) : \varphi(x) = 0 \text{ } m\text{-a.e. on } [a, c]\}.$$

It is obvious that $M_1 \subset M_2^{\perp}$ and that

$$[\varphi] = [\chi_{[a,c]}\varphi] + [\chi_{[c,b]}\varphi], \forall [\varphi] \in L^2(a, b). \tag{1}$$

By 10.2.15, this proves that $M_1 = M_2^{\perp}$ and $M_2 = M_1^{\perp}$. Therefore, M_1 is a subspace of $L^2(a, b)$ (cf. 10.2.13) and M_2 is its orthogonal complement. From 1 we have that the projection P_{M_1} is defined by

$$P_{M_1}[\varphi] := [\chi_{[a,c]}\varphi], \forall [\varphi] \in L^2(a, b).$$

Now we consider the operator A_θ defined in 12.4.25, with $\theta := 0$ (actually, the results we obtain are true for every $\theta \in [0, 2\pi)$; we fix $\theta := 0$ for simplicity). It is obvious that M_1 and M_2 are invariant subspaces for A_0, i.e. that

$$A_0[\varphi] = -i[\varphi'] \in M_k, \forall [\varphi] \in D_{A_0} \cap M_k, \text{ for } k = 1, 2.$$

However, M_1 is not a reducing subspace for A_0 since the inclusion $P_{M_1}(D_{A_0}) \subset D_{A_0}$ is false. Indeed, for the element u_0 of $C(a, b)$ (cf. 11.2.4 and 12.4.25) we have

$$[u_0] \in D_{A_0} \text{ but } P_{M_1}[u_0] \notin D_{A_0},$$

since the equivalence class $P_{M_1}[u_0]$ does not contain any continuous function (cf. 11.2.2c).

We note that not even the self-adjoint operator $\overline{A_0}$ (cf. 12.4.25) is reduced by the subspace M_1. Indeed, if $\overline{A_0}$ were reduced by M_1 then we should have

$$P_{M_1}\overline{A_0} \subset \overline{A_0}P_{M_1},$$

and this would imply

$$[P_{M_1}, P^{\overline{A_0}}(E)] = \mathbb{O}_{\mathcal{H}}, \forall E \in \mathcal{A}(d_{\mathbb{R}})$$

(cf. 17.1.3c), and hence in particular

$$P_{M_1}P^{\overline{A_0}}(\{0\}) = P^{\overline{A_0}}(\{0\})P_{M_1}; \tag{2}$$

now, from 12.4.25 and 15.3.4B we have

$$P^{\overline{A_0}}(\{0\})[\varphi] = ([u_0]|[\varphi])\,[u_0], \forall[\varphi] \in L^2(a,b);$$

then, 2 would imply

$$P_{M_1}[u_0] = P_{M_1}P^{\overline{A_0}}(\{0\})[u_0] = P^{\overline{A_0}}(\{0\})P_{M_1}[u_0] = ([u_0]|P_{M_1}[u_0])\,[u_0],$$

which cannot be true since $P_{M_1}[u_0]$ does not contain any continuous function.

17.3 Irreducibility

In Section 17.2 we studied what happened when an operator was reduced by a subspace (and hence by its orthogonal complement as well), or more generally by the subspaces of a countable family $\{M_n\}_{n\in I}$ such that $\sum_{n\in I}^{\oplus} M_n = \mathcal{H}$ (i.e. such that $\sum_{n\in I} P_{M_n}f = f$ for all $f \in \mathcal{H}$, cf. 13.2.8 and 13.2.9). However, when a set of operators is taken to represent a mathematical or a physical structure, this set is often required not to be reduced simultaneously by any non-trivial subspace; if this is true, the set is said to be irreducible. In this section we study the condition of irreducibility for a set of self-adjoint operators, and some of its consequences.

17.3.1 Definition. A set $\{A_i\}_{i\in I}$ of operators in $\mathcal{H}$ is said to be *irreducible* if there does not exist any non-trivial subspace of $\mathcal{H}$ which is reducing for A_i for all $i \in I$, i.e. if

$$[P \in \mathscr{P}(\mathcal{H}) \text{ and } PA_i \subset A_iP, \forall i \in I] \Rightarrow P \in \{\mathbb{O}_{\mathcal{H}}, \mathbb{1}_{\mathcal{H}}\}.$$

17.3.2 Lemma. *Let A be a self-adjoint operator in $\mathcal{H}$. The following conditions are equivalent:*

(a) $P^A(E) \in \{\mathbb{O}_{\mathcal{H}}, \mathbb{1}_{\mathcal{H}}\}, \forall E \in \mathcal{A}(d_{\mathbb{R}})$;
(b) $\exists\lambda \in \mathbb{R}$ so that $A = \lambda\mathbb{1}_{\mathcal{H}}$.

Proof. $a \Rightarrow b$: We assume condition a. Since $\sigma(A) \neq \emptyset$ (cf. 15.2.2d), there exists $\lambda \in \mathbb{R}$ such that

$$P^A\left(\left(\lambda - \frac{1}{n}, \lambda + \frac{1}{n}\right)\right) \neq O_{\mathcal{H}}, \text{ and hence } P^A\left(\left(\lambda - \frac{1}{n}, \lambda + \frac{1}{n}\right)\right) = \mathbb{1}_{\mathcal{H}}, \forall n \in \mathbb{N}$$

(cf. 15.2.4), and hence such that

$$P^A(\{\lambda\})f = P^A\left(\bigcap_{n=1}^{\infty}\left(\lambda - \frac{1}{n}, \lambda + \frac{1}{n}\right)\right)f$$

$$= \lim_{n \to \infty} P^A\left(\left(\lambda - \frac{1}{n}, \lambda + \frac{1}{n}\right)\right)f = f, \forall f \in \mathcal{H}$$

(cf. 13.3.6b), and hence such that

$$P^A(\{\lambda\}) = \mathbb{1}_{\mathcal{H}} \text{ and hence } P^A(\mathbb{R} - \{\lambda\}) = O_{\mathcal{H}}.$$

Then,

$$\mu_f^{P^A} = \|f\|^2 \mu_\lambda, \forall f \in \mathcal{H},$$

where μ_λ is the Dirac measure in λ (cf. 8.3.6 and 8.3.5b with $\mu := \mu_\lambda$ and ν the null measure on $\mathcal{A}(d_{\mathbb{R}})$). In view of 15.2.2e, this implies that

$$D_A = \left\{f \in \mathcal{H} : \int_{\mathbb{R}} \xi^2 d\mu_f^{P^A} < \infty\right\} = \mathcal{H},$$

since $\int_{\mathbb{R}} \xi^2 d\mu_f^{P^A} = \lambda^2 \|f\|^2$ for all $f \in \mathcal{H}$, and that

$$(f|Af) = \int_{\mathbb{R}} \xi d\mu_f^{P^A} = \lambda \|f\|^2 = (f|\lambda \mathbb{1}_{\mathcal{H}} f), \forall f \in \mathcal{H}.$$

Then, $A = \lambda \mathbb{1}_{\mathcal{H}}$ by 10.2.12.

$b \Rightarrow a$: We assume condition b. By 13.3.5, the mapping

$$\mathcal{A}(d_{\mathbb{R}}) \ni E \mapsto P(E) := \chi_E(\lambda)\mathbb{1}_{\mathcal{H}} \in \mathscr{P}(\mathcal{H})$$

is a projection valued measure since

$$\mu_f^P = \|f\|^2 \mu_\lambda.$$

We see that:

$$D_A = \mathcal{H} = \left\{f \in \mathcal{H} : \int_{\mathbb{R}} \xi^2 d\mu_f^P < \infty\right\},$$

$$(f|Af) = (f|\lambda \mathbb{1}_{\mathcal{H}} f) = \lambda \|f\|^2 = \int_{\mathbb{R}} \xi d\mu_f^P, \forall f \in \mathcal{H}.$$

This proves that $P^A = P$ (cf. 15.2.1), and hence that condition a is true. $\quad\square$

17.3.3 Theorem. *Let $\{A_i\}_{i \in I}$ be a set of self-adjoint operators in $\mathcal{H}$. The following conditions are equivalent:*

(a) the set $\{A_i\}_{i \in I}$ is irreducible;

(b) if A is a self-adjoint operator in $\mathcal{H}$ which commutes with A_i (in the sense defined in 17.1.5) for all $i \in I$, then there exists $\lambda \in \mathbb{R}$ so that $A = \lambda \mathbb{1}_{\mathcal{H}}$.

Proof. $a \Rightarrow b$: Let A be a self-adjoint operator in $\mathcal{H}$ which commutes with A_i for all $i \in I$. Then,

$$P^A(E)A_i \subset A_i P^A(E), \forall i \in I, \forall E \in \mathcal{A}(d_{\mathbb{R}})$$

(cf. 17.1.7b with $B := A_i$), and hence, if we assume condition a,

$$P^A(E) \in \{\mathbb{O}_{\mathcal{H}}, \mathbb{1}_{\mathcal{H}}\}, \forall E \in \mathcal{A}(d_{\mathbb{R}}).$$

By 17.3.2, this implies that there exists $\lambda \in \mathbb{R}$ so that $A = \lambda \mathbb{1}_{\mathcal{H}}$.

$b \Rightarrow a$: Let $P \in \mathscr{P}(\mathcal{H})$ be such that

$$PA_i \subset A_i P, \forall i \in I.$$

Then P commutes with A_i (in the sense defined in 17.1.5) for all $i \in I$, by 17.1.4, and hence, if we assume condition b,

$$\exists \lambda \in \mathbb{R} \text{ so that } P = \lambda \mathbb{1}_{\mathcal{H}}.$$

Since P is a projection, from $P = P^2$ (cf. 13.1.5) we have $\lambda = \lambda^2$, and hence

$$P \in \{\mathbb{O}_{\mathcal{H}}, \mathbb{1}_{\mathcal{H}}\}.$$

$\square$

17.3.4 Corollary. *Let $\{A_i\}_{i \in I}$ be an irreducible set of self-adjoint operators in $\mathcal{H}$. Then,*

$$[B \in \mathcal{B}(\mathcal{H}) \text{ and } BA_i \subset A_i B, \forall i \in I] \Rightarrow [\exists \alpha \in \mathbb{C} \text{ so that } B = \alpha \mathbb{1}_{\mathcal{H}}].$$

Proof. Let $B \in \mathcal{B}(\mathcal{H})$ be such that

$$BA_i \subset A_i B, \forall i \in I. \tag{1}$$

In view of 12.3.4a,b and 12.1.4, we have

$$B^\dagger A_i \subset (A_i B)^\dagger \subset (BA_i)^\dagger = A_i B^\dagger, \forall i \in I. \tag{2}$$

We define the self-adjoint operators

$$B_1 := \frac{1}{2}(B + B^\dagger) \text{ and } B_2 := \frac{1}{2i}(B - B^\dagger),$$

From 1 and 2 we have

$$B_1 A_i = \frac{1}{2}(BA_i + B^\dagger A_i) \subset \frac{1}{2}(A_i B + A_i B^\dagger) \overset{(3)}{\subset} A_i B_1, \forall i \in I,$$

where 3 holds because, for all $i \in I$,

$$f \in D_{A_i B} \cap D_{A_i B^\dagger} \Rightarrow Bf, B^\dagger f \in D_{A_i} \Rightarrow B_1 f \in D_{A_i}.$$

Similarly, we have

$$B_2 A_i \subset A_i B_2, \forall i \in I.$$

By 17.1.4, this implies that B_k (for $k = 1, 2$) commutes (in the sense of 17.1.5) with A_i for all $i \in I$, and hence (in view of 17.3.3) that there exists $\lambda_k \in \mathbb{R}$ so that $B_k = \lambda_k \mathbb{1}_{\mathcal{H}}$. Then, if we set $\alpha := \lambda_1 + i\lambda_2$, we have

$$B = B_1 + iB_2 = \alpha \mathbb{1}_{\mathcal{H}}.$$

$\square$

17.3.5 Corollary (Shur's lemma). *Let* $\{A_i\}_{i \in I}$ *be an irreducible set of elements of* $\mathcal{B}(\mathcal{H})$ *such that*

$$\forall i \in I, \exists j \in I \text{ so that } A_i^\dagger = A_j.$$

Then,

$$[B \in \mathcal{B}(\mathcal{H}) \text{ and } BA_i = A_iB, \forall i \in I] \Rightarrow [\exists \alpha \in \mathbb{C} \text{ so that } B = \alpha \mathbb{1}_{\mathcal{H}}].$$

Proof. Let $B \in \mathcal{B}(\mathcal{H})$ be such that

$$BA_i = A_iB, \forall i \in I. \tag{1}$$

Since for all $i \in I$ there exists $j \in I$ so that $A_i^\dagger = A_j$, this implies that

$$BA_i^\dagger = A_i^\dagger B, \forall i \in I. \tag{2}$$

We define the self-adjoint operators

$$A_i' := \frac{1}{2}(A_i + A_i^\dagger) \text{ and } A_i'' := \frac{1}{2i}(A_i - A_i^\dagger), \forall i \in I.$$

We note that the set $\{A_i', A_i''\}_{i \in I}$ is irreducible; indeed, for $P \in \mathscr{P}(\mathcal{H})$,

$$[PA_i' = A_i'P \text{ and } PA_i'' = A_i''P, \forall i \in I] \overset{(3)}{\Rightarrow} [PA_i = A_iP, \forall i \in I] \Rightarrow P \in \{\mathbb{O}_{\mathcal{H}}, \mathbb{1}_{\mathcal{H}}\},$$

where 3 holds because $A_i = A_i' + iA_i''$ for all $i \in I$. From 1 and 2 we have

$$BA_i' = A_i'B \text{ and } BA_i'' = A_i''B, \forall i \in I.$$

By 17.3.4, this implies that there exists $\alpha \in \mathbb{C}$ so that $B = \alpha \mathbb{1}_{\mathcal{H}}$. $\square$

17.3.6 Remark. If $U : G \to \mathcal{U}(\mathcal{H})$ is a homomorphism from a group G to the group $\mathcal{U}(\mathcal{H})$, then R_U is a set of elements of $\mathcal{B}(\mathcal{H})$ such that

$$U(g)^\dagger = (U(g))^{-1} = U(g^{-1}), \forall g \in G$$

(cf. 12.5.1 and 1.3.3). Therefore, if the set R_U is irreducible (U is then called a *unitary irreducible representation* of G) then the multiples of the identity operator $\mathbb{1}_{\mathcal{H}}$ are the only elements of $\mathcal{B}(\mathcal{H})$ which commute with $U(g)$ for all $g \in G$.

Chapter 18

Trace Class and Statistical Operators

Statistical operators were devised by John von Neumann in order to represent the most general statistical ensembles of a given quantum system (cf. Neumann, 1932, Chapter IV). In this representation, those particular ensembles which von Neumann denoted as homogeneous (and which we call pure states in Chapter 19) are represented by one-dimensional projections, which are a special case of statistical operators.

In this chapter we study statistical operators. Before that, we need to study the polar decomposition for elements of $\mathcal{B}(\mathcal{H})$ and a subset of $\mathcal{B}(\mathcal{H})$ which is called the trace class. As usual, $\mathcal{H}$ denotes an abstract Hilbert space throughout the chapter.

18.1 Positive operators and polar decomposition

In this section we find a decomposition for elements of $\mathcal{B}(\mathcal{H})$ which is the generalization of the decomposition $z = |z| \exp(i \arg z)$ for a complex number z. First, we must find the right analogous of a positive number and of the absolute value $|z|$ of a complex number z.

18.1.1 Definition. An operator $A \in \mathcal{B}(\mathcal{H})$ is said to be *positive* if $(f|Af) \geq 0$ for all $f \in \mathcal{H}$.

18.1.2 Remarks.

(a) If an operator $A \in \mathcal{B}(\mathcal{H})$ is positive then $(f|Af) \in \mathbb{R}$ for all $f \in \mathcal{H}$, and hence A is self-adjoint (cf. 12.4.3).

(b) If an operator $A \in \mathcal{B}(\mathcal{H})$ is positive then there exists a unique positive operator $B \in \mathcal{B}(\mathcal{H})$ such that $A = B^2$ (cf. 15.3.9). The operator B will be denoted by the symbol $A^{\frac{1}{2}}$.

If $A \in \mathcal{B}(\mathcal{H})$ is positive and $T \in \mathcal{B}(\mathcal{H})$ is such that $[T, A] = \mathbb{O}_{\mathcal{H}}$, then $[T, A^{\frac{1}{2}}] = \mathbb{O}_{\mathcal{H}}$ (cf. 15.3.9).

(c) If an operator $A \in \mathcal{B}(\mathcal{H})$ is positive then the operator UAU^{-1} is a positive element of $\mathcal{B}(\mathcal{H})$ for all $U \in \mathcal{UA}(\mathcal{H})$, since

$$(f|UAU^{-1}f) = (U^{-1}f|AU^{-1}f) \geq 0, \forall f \in \mathcal{H},$$

and $(UAU^{-1})^{\frac{1}{2}} = UA^{\frac{1}{2}}U^{-1}$, as can be seen easily.

18.1.3 Definition. Let $A \in \mathcal{B}(\mathcal{H})$. Then the operator $A^{\dagger}A$ is obviously a positive element of $\mathcal{B}(\mathcal{H})$, and we define $|A| := (A^{\dagger}A)^{\frac{1}{2}}$. Thus, $|A|$ is the unique positive element of $\mathcal{B}(\mathcal{H})$ such that $A^{\dagger}A = |A|^2$.

If A is a positive element of $\mathcal{B}(\mathcal{H})$ then obviously $|A| = A$.

18.1.4 Proposition. *Let $A \in \mathcal{B}(\mathcal{H})$. Then:*

(a) $\||A|f\| = \|Af\|, \forall f \in \mathcal{H}$;
(b) $\||A|\| = \|A\|$;
(c) $N_{|A|} = N_A$;
(d) $|UAU^{-1}| = U|A|U^{-1}, \forall U \in \mathcal{UA}(\mathcal{H})$;
(e) $|\alpha A| = |\alpha||A|, \forall \alpha \in \mathbb{C}$.

Proof. a: We have

$$\||A|f\|^2 = (f||A|^2f) = (f|A^{\dagger}Af) = \|Af\|^2, \forall f \in \mathcal{H}.$$

b and c: These follow immediately from result a.

d: Let $U \in \mathcal{UA}(\mathcal{H})$. The operator $U|A|U^{-1}$ is a positive element of $\mathcal{B}(\mathcal{H})$ (cf. 18.1.2c) and

$$(UAU^{-1})^{\dagger}(UAU^{-1}) = (UA^{\dagger}U^{-1})(UAU^{-1}) = UA^{\dagger}AU^{-1}$$
$$= U|A|^2U^{-1} = (U|A|U^{-1})^2$$

(cf. 12.5.4a). Therefore, $|UAU^{-1}| = U|A|U^{-1}$.

e: Let $\alpha \in \mathbb{C}$. The operator $|\alpha||A|$ is obviously a positive element of $\mathcal{B}(\mathcal{H})$ and

$$(\alpha A)^{\dagger}(\alpha A) = \bar{\alpha}\alpha A^{\dagger}A = |\alpha|^2|A|^2 = (|\alpha||A|)^2$$

(cf. 12.3.2). Therefore, $|\alpha A| = |\alpha||A|$. $\qquad\qquad\qquad\qquad\qquad\square$

The definitions given in 18.1.1 and in 18.1.3 are generalizations from $\mathbb{C}$ to $\mathcal{B}(\mathcal{H})$. Indeed, if $\mathcal{H}$ is a one-dimensional Hilbert space then every complex number can be identified with an element of $\mathcal{B}(\mathcal{H})$ (cf. 12.6.6a). In this identification, positive numbers are identified with positive operators; moreover, if A_{α} is the operator that corresponds to the complex number α, then $|A_{\alpha}|$ corresponds to $|\alpha|$. In the decomposition $z = |z| \exp(i \arg z)$ for a complex number z, the number $\exp(i \arg z)$ is an element of $\mathbb{T}$ and hence it can be identified with an element of $\mathcal{U}(\mathcal{H})$ (cf. 12.6.6a). However, in order to obtain the decomposition for elements of $\mathcal{B}(\mathcal{H})$ we are after, the right generalization of $\mathbb{T}$ is wider than $\mathcal{U}(\mathcal{H})$ when $\mathcal{H}$ is not a one-dimensional Hilbert space.

18.1.5 Definitions. An operator $U \in \mathcal{B}(\mathcal{H})$ is called an *isometry* if $\|Uf\| = \|f\|$ for all $f \in \mathcal{H}$, while it is more generally called a *partial isometry*, or it is said to be *partially isometric*, if $\|Uf\| = \|f\|$ for all $f \in N_U^{\perp}$.

If $U \in \mathcal{B}(\mathcal{H})$ is partially isometric then $I(U) := N_U^\perp$ is called the *initial subspace* of U ($I(U)$ is actually a subspace of $\mathcal{H}$ by 10.2.13).

Each element of $\mathcal{U}(\mathcal{H})$ is obviously an isometry. Each element of $\mathscr{P}(\mathcal{H})$ is a partial isometry in view of 13.1.3b,c and 10.4.4a.

18.1.6 Proposition. *Let $U \in \mathcal{B}(\mathcal{H})$ be partially isometric. Then:*

(a) if $U \neq \mathbb{O}_\mathcal{H}$ then $\|U\| = 1$;
(b) R_U is a closed subset of $\mathcal{H}$; $F(U) := R_U$ is called the final subspace of U;
(c) the mapping $U_{I(U)}$ (the restriction of U to $N_U^\perp$) is a unitary operator from the Hilbert space $I(U)$ onto the Hilbert space $F(U)$;
(d) $U^\dagger U = P_{I(U)}$ and $U U^\dagger = P_{F(U)}$;
(e) the operator $U^\dagger$ is partially isometric;
(f) the mapping $U_{F(U)}^\dagger$ (the restriction of $U^\dagger$ to R_U) is a unitary operator from the Hilbert space $F(U)$ onto the Hilbert space $I(U)$.

Proof. Preliminary remark: Since N_U is a subspace of $\mathcal{H}$ (cf. 4.4.3 and 4.4.8), we can write

$$f = P_{N_U} f + P_{I(U)} f, \forall f \in \mathcal{H}$$

(cf. 13.1.3e). Then,

$$U f = U P_{I(U)} f, \forall f \in \mathcal{H},$$

i.e. $U = U P_{I(U)}$, since $P_{N_U} f \in N_U$ for all $f \in \mathcal{H}$ (clearly, this is true for every $U \in \mathcal{B}(\mathcal{H})$ if we define $I(U) := N_U^\perp$ for every $U \in \mathcal{B}(\mathcal{H})$).

a: We have

$$\|U f\| = \|U P_{I(U)} f\| = \|P_{I(U)} f\| \leq \|f\|, \forall f \in \mathcal{H},$$

in view of the preliminary remark and 13.1.3d. This proves that $\|U\| \leq 1$. If $U \neq \mathbb{O}_\mathcal{H}$ then there exists $f \neq 0_\mathcal{H}$ such that $f \in N_U^\perp$ (cf. 10.4.4d), and hence such that $\|U f\| = \|f\|$. In view of 4.2.5c, this proves that $\|U\| = 1$ if $U \neq \mathbb{O}_\mathcal{H}$.

b: Let $\{g_n\}$ be a Cauchy sequence in R_U. For each $n \in \mathbb{N}$, we choose $f_n \in \mathcal{H}$ so that $g_n = U f_n$ and we set $f_n' := P_{I(U)} f_n$; then, $g_n = U f_n'$ (cf. the preliminary remark). Thus, $\{f_n'\}$ is a sequence in $N_U^\perp$ and it is a Cauchy sequence since

$$\|f_n' - f_m'\| = \|U(f_n' - f_m')\| = \|g_n - g_m\|, \forall n, m \in \mathbb{N}.$$

Then there exists $f \in \mathcal{H}$ so that

$$f_n' \to f$$

and hence (since U is continuous) so that

$$g_n = U f_n' \to U f.$$

This proves that R_U is a complete metric subspace of the metric space $\mathcal{H}$, and hence that R_U is a closed subset of $\mathcal{H}$ (cf. 2.6.6a).

c: First we point out that $I(U)$ and $F(U)$ can be considered Hilbert spaces since they are subspaces of $\mathcal{H}$ (cf. 10.3.2). Next we notice that the linear operator $U_{I(U)}$ is surjective onto R_U since $Uf = UP_{I(U)}f$ for all $f \in \mathcal{H}$ (cf. the preliminary remark). Then, statement c holds true in view of 10.1.20.

d: We have

$$\left(f|U^\dagger U f\right) = \|Uf\|^2 = \|UP_{I(U)}f\|^2 = \|P_{I(U)}f\|^2 = \left(f|P_{I(U)}f\right), \forall f \in \mathcal{H}.$$

By 10.2.12, this proves that $U^\dagger U = P_{I(U)}$.

Moreover, we have

$$(UU^\dagger)^\dagger = U^{\dagger\dagger}U^\dagger = UU^\dagger$$

(cf. 12.3.4b and 12.1.6b), and also

$$(UU^\dagger)^2 = U(U^\dagger U)U^\dagger = UP_{I(U)}U^\dagger = UU^\dagger.$$

By 13.1.5, this proves that $UU^\dagger \in \mathscr{P}(\mathcal{H})$. Further, we have

$$\left(g|U^\dagger f\right) = (Ug|f) = 0, \forall g \in N_U, \forall f \in \mathcal{H},$$

and hence $R_{U^\dagger} \subset N_U^\perp$; therefore, we have

$$\|UU^\dagger f\| = \|U^\dagger f\|, \forall f \in \mathcal{H},$$

and hence

$$N_{UU^\dagger} = N_{U^\dagger} = R_U^\perp$$

(cf. 12.1.7), and hence (in view of 13.1.3b,c, 10.4.4a, and statement b)

$$R_{UU^\dagger} = N_{UU^\dagger}^\perp = R_U = F(U),$$

which is equivalent to $UU^\dagger = P_{F(U)}$.

e: From $N_{U^\dagger} = R_U^\perp$ we have $N_{U^\dagger}^\perp = R_U = F(U)$, and hence (in view of statement d and 13.1.3c)

$$\|U^\dagger f\|^2 = \left(f|UU^\dagger f\right) = \left(f|P_{F(U)}f\right) = \|P_{F(U)}f\|^2 = \|f\|^2, \forall f \in N_{U^\dagger}^\perp.$$

Thus, $U^\dagger$ is partially isometric.

f: From statements e and c (written with $U^\dagger$ in place of U) we have that $U^\dagger_{I(U^\dagger)}$ is a unitary operator from the Hilbert space $I(U^\dagger)$ onto the Hilbert space $F(U^\dagger)$. Now,

$$I(U^\dagger) = N_{U^\dagger}^\perp = R_U = F(U).$$

Moreover, from

$$N_U = N_{U^{\dagger\dagger}} = R_{U^\dagger}^\perp$$

(cf. 12.1.6b and 12.1.7) we have

$$F(U^\dagger) = R_{U^\dagger} = N_U^\perp = I(U)$$

by 10.4.4a since $R_{U^\dagger}$ is a subspace of $\mathcal{H}$, in view of statements e and b (written with $U^\dagger$ in place of U). □

18.1.7 Theorem. *Let $A \in \mathcal{B}(\mathcal{H})$. Then there exists a unique partially isometric operator $U \in \mathcal{B}(\mathcal{H})$ such that*

$$A = U|A| \quad and \quad N_U = N_A.$$

Moreover,

$$R_U = \overline{R_A} \quad and \quad |A| = U^\dagger A.$$

The equality $A = U|A|$ is called the polar decomposition of A.

Proof. Existence: We define the mapping

$$V : R_{|A|} \to \mathcal{H}$$
$$f \mapsto Vf := Ag \text{ if } g \in \mathcal{H} \text{ is so that } f = |A|g.$$

This definition is consistent because, for $g_1, g_2 \in \mathcal{H}$,

$$|A|g_1 = |A|g_2 \Rightarrow g_1 - g_2 \in N_{|A|} \Rightarrow g_1 - g_2 \in N_A \Rightarrow Ag_1 = Ag_2$$

(cf. 18.1.4c). Moreover, let $f_1, f_2 \in R_{|A|}$ and let $g_1, g_2 \in \mathcal{H}$ be so that $f_1 = |A|g_1$, $f_2 = |A|g_2$; then, for all $\alpha, \beta \in \mathbb{C}$, $\alpha f_1 + \beta f_2 = |A|(\alpha g_1 + \beta g_2)$ and hence

$$V(\alpha f_1 + \beta f_2) = A(\alpha g_1 + \beta g_2) = \alpha Ag_1 + \beta Ag_2 = \alpha Vf_1 + \beta Vf_2;$$

this proves that the mapping V is a linear operator.

Let $f \in D_V (= R_{|A|})$ and let $g \in \mathcal{H}$ be so that $f = |A|g$; then,

$$\|Vf\| = \|Ag\| = \||A|g\| = \|f\| \tag{1}$$

(cf. 18.1.4a). We denote by $\tilde{V}$ the bounded linear operator such that

$$D_{\tilde{V}} = \overline{D_V} \quad and \quad V \subset \tilde{V}$$

(cf. 4.2.6). Let $f \in D_{\tilde{V}}$ and let $\{f_n\}$ be a sequence in D_V such that $f_n \to f$ (cf. 2.3.10); then $\tilde{V}f = \lim_{n\to\infty} Vf_n$, and hence (in view of 1)

$$\|\tilde{V}f\| = \lim_{n\to\infty} \|Vf_n\| = \lim_{n\to\infty} \|f_n\| = \|f\|. \tag{2}$$

Since the inclusion $R_V \subset R_A$ is obvious, the implications

$$h \in R_A \Rightarrow$$
$$[\exists g \in \mathcal{H} \text{ s.t. } h = Ag, \text{ and hence s.t. } V(|A|g) = Ag = h] \Rightarrow$$
$$h \in R_V$$

prove the equality

$$R_V = R_A. \tag{3}$$

Let $f \in D_{\tilde{V}}$ and let $\{f_n\}$ be a sequence in D_V such that $f_n \to f$; then $\tilde{V}f = \lim_{n\to\infty} Vf_n$, and hence $\tilde{V}f \in \overline{R_V}$ (cf. 2.3.10). This proves the inclusion $R_{\tilde{V}} \subset \overline{R_V}$. Conversely, let $h \in \overline{R_V}$ and let $\{h_n\}$ be a sequence in R_V such that $h_n \to h$; for each $n \in \mathbb{N}$, we choose $g_n \in D_V$ so that $h_n = Vg_n$; in view of 1, the sequence $\{g_n\}$ is a Cauchy sequence; therefore, there exists $g \in \overline{D_V}$ such that $g_n \to g$, and hence such that $h = \lim_{n \to \infty} V g_n = \tilde{V} g$; thus, $h \in R_{\tilde{V}}$. Thus proves the inclusion $\overline{R_V} \subset R_{\tilde{V}}$, and hence (in view of 3) the equalities

$$R_{\tilde{V}} = \overline{R_V} = \overline{R_A}. \tag{4}$$

Now we set $M := \overline{R_{|A|}}$ and define the operator

$$U := \tilde{V} P_M,$$

which is an element of $\mathcal{B}(\mathcal{H})$ (note that $D_{\tilde{V}} = M$). In what follows we prove that U satisfies the conditions of the statement.

From the definition of U and from 4 we have

$$R_U = R_{\tilde{V}} = \overline{R_A}.$$

From the definitions of $U, \tilde{V}, V$, from 4 and from 13.1.3c we have

$$U|A|g = \tilde{V}|A|g = V|A|g = Ag, \forall g \in \mathcal{H},$$

i.e. $A = U|A|$.

Moreover, we have

$$N_U \overset{(5)}{=} N_{P_M} \overset{(6)}{=} (\overline{R_{|A|}})^\perp \overset{(7)}{=} R_{|A|}{}^\perp \overset{(8)}{=} N_{|A|} \overset{(9)}{=} N_A, \tag{10}$$

where 5 follows from 2, 6 from 13.1.3b, 7 from 10.2.11, 8 from 12.1.7 (since $|A|$ is self-adjoint), 9 from 18.1.4c.

Furthermore, from 10 and from 10.4.4c we have

$$N_U^\perp = R_{|A|}^{\perp\perp} = \overline{R_{|A|}} = M, \tag{11}$$

and hence, in view of the definition of U and of 13.1.3c,

$$Uf = \tilde{V}f, \forall f \in N_U^\perp,$$

and hence, in view of 2,

$$\|Uf\| = \|f\|, \forall f \in N_U^\perp.$$

Thus, the operator U is partially isometric.

Finally, from $U|A| = A$ we have

$$U^\dagger U|A| = U^\dagger A;$$

now, from 18.1.6d and 11, we have $U^\dagger U = P_M$ and hence (in view of 13.1.3c)

$$|A| = U^\dagger A.$$

Uniqueness: Suppose that T is a partially isometric element of $\mathcal{B}(\mathcal{H})$ such that

$$A = T|A| \text{ and } N_T = N_A.$$

Let $f \in R_{|A|}$ and let $g \in \mathcal{H}$ be so that $f = |A|g$; then,

$$Uf = U|A|g = Ag = T|A|g = Tf.$$

Since both U and T are continuous, this implies that

$$Uf = Tf, \forall f \in \overline{R_{|A|}}. \tag{12}$$

Now we notice that

$$N_T = N_A = N_U$$

and set $P := P_{N_A}$; then (in view of 13.1.3e and of 11),

$$\mathbb{1}_{\mathcal{H}} - P = P_M,$$

where $M := \overline{R_{|A|}}$. Then,

$$Tf = TPf + T(\mathbb{1}_{\mathcal{H}} - P)f = TP_M f$$
$$\overset{(13)}{=} UP_M f = UPf + U(\mathbb{1}_{\mathcal{H}} - P)f = Uf, \forall f \in \mathcal{H},$$

where 13 follows from 12. Thus, $T = U$. $\qquad\square$

18.1.8 Remark. If $\mathcal{H}$ is a one-dimensional Hilbert space, then $\mathbb{C}$ can be identified with $\mathcal{B}(\mathcal{H})$ and $\mathbb{T}$ with $\mathcal{U}(\mathcal{H})$ (cf. 12.6.6a and the discussion before 18.1.5). In this identification, the decomposition $z = |z| \exp(i \arg z)$ is actually the polar decomposition of a complex number z.

18.1.9 Remark. The analogy between the symbols $|A|$ for $A \in \mathcal{B}(\mathcal{H})$ and $|z|$ for $z \in \mathbb{C}$ must not induce the reader to expect other properties for $|A|$ than the ones discussed above. To see this, for any $u, v \in \tilde{\mathcal{H}}$ (cf. 10.9.4) we define the mapping

$$A_{u,v} : \mathcal{H} \to \mathcal{H}$$
$$f \mapsto A_{u,v}f := (u|f)\, v.$$

We notice that, if $u = v$, then $A_{u,v} = A_u$ (the one-dimension projection defined in 13.1.12). It is obvious that $A_{u,v} \in \mathcal{B}(\mathcal{H})$ (use the Schwarz inequality, cf. 10.1.9). Moreover, the equation

$$(A_{u,v}f|g) = \overline{(u|f)}\,(v|g) = (f|A_{v,u}g)\,, \forall f,g \in \mathcal{H},$$

proves that $A^{\dagger}_{u,v} = A_{v,u}$ (cf. 12.1.3B). Then, the equation

$$A^{\dagger}_{u,v} A_{u,v} f = (u|f)\, u = A_u f, \forall f \in \mathcal{H},$$

proves that $A^{\dagger}_{u,v} A_{u,v} = A_u$, and hence (since A_u is positive and $A_u^2 = A_u$) that

$$|A_{u,v}| = A_u.$$

By the same token, we also have

$$|A^{\dagger}_{u,v}| = |A_{v,u}| = A_v.$$

Moreover, from

$$A_{v,u} A_{u,v} = A_u$$

we have

$$|A_{v,u} A_{u,v}| = A_u,$$

while

$$|A_{v,u}||A_{u,v}| = A_v A_u.$$

This proves that the relations

$$|\bar{z}| = |z|, \forall z \in \mathbb{C},$$
$$|zw| = |z||w|, \forall z, w \in \mathbb{C}$$

cannot be extended to $\mathcal{B}(\mathcal{H})$ through the symbol $|A|$ (if $\mathcal{H}$ is a one-dimensional Hilbert space and if complex numbers are identified with elements of $\mathcal{B}(\mathcal{H})$, the adjoint of a complex number z as a linear operator is identified with $\bar{z}$, and the product of two complex numbers z and w as linear operators is identified with zw; cf. 12.6.6a).

18.2   The trace class

In this and in the next section, the Hilbert space $\mathcal{H}$ is assumed to be separable. All definitions, statements and proofs are written on the hypothesis that the orthogonal dimension of $\mathcal{H}$ is denumerable. If the orthogonal dimension of $\mathcal{H}$ was finite then all the arguments presented would get simplified in an obvious way and some conditions would become trivial.

18.2.1 Theorem. *Let A be a positive element of $\mathcal{B}(\mathcal{H})$ and let $\{u_n\}_{n \in \mathbb{N}}$ be a c.o.n.s. in $\mathcal{H}$.*

If $\{v_n\}_{n \in \mathbb{N}}$ is another c.o.n.s. in $\mathcal{H}$ then

$$\sum_{n=1}^{\infty} (u_n | A u_n) = \sum_{n=1}^{\infty} (v_n | A v_n)$$

(these sums of series are defined as in 5.4.1).

The sum of the series $\sum_{n=1}^{\infty} (u_n | A u_n)$ is an element of $[0, \infty]$, which is called the trace of A and denoted by $\operatorname{tr} A$; in view of the proposition above, it is independent of the c.o.n.s. in $\mathcal{H}$ chosen to compute it. Thus,

$$\operatorname{tr} A = \sum_{n=1}^{\infty} (v_n | A v_n)$$

for whichever c.o.n.s. $\{v_n\}_{n \in \mathbb{N}}$ in $\mathcal{H}$.

Proof. If $\{v_n\}_{n \in \mathbb{N}}$ is a c.o.n.s. in $\mathcal{H}$, then

$$\sum_{n=1}^{\infty} (u_n | A u_n) = \sum_{n=1}^{\infty} \|A^{\frac{1}{2}} u_n\|^2 \overset{(1)}{=} \sum_{n=1}^{\infty} \left(\sum_{m=1}^{\infty} \left| \left(v_m | A^{\frac{1}{2}} u_n \right) \right|^2 \right)$$

$$\overset{(2)}{=} \sum_{m=1}^{\infty} \left(\sum_{n=1}^{\infty} \left| \left(u_n | A^{\frac{1}{2}} v_m \right) \right|^2 \right) \overset{(3)}{=} \sum_{m=1}^{\infty} \|A^{\frac{1}{2}} v_m\|^2$$

$$= \sum_{m=1}^{\infty} (v_m | A v_m),$$

where 1 and 3 hold true by 10.6.4d and 2 by 5.4.7. $\qquad\qquad\qquad\square$

18.2.2 Proposition. *Let A be a positive element of $\mathcal{B}(\mathcal{H})$. Then:*

(a) if B is a positive element of $\mathcal{B}(\mathcal{H})$ then the operator $A + B$ is positive and

$$\operatorname{tr}(A + B) = \operatorname{tr} A + \operatorname{tr} B;$$

(b) if $a \in [0, \infty)$ then the operator aA is positive and

$$\operatorname{tr}(aA) = a \operatorname{tr} A;$$

(c) if $B \in \mathcal{B}(\mathcal{H})$ is such that $(f|Af) \leq (f|Bf)$ for all $f \in \mathcal{H}$, then B is positive and

$$\operatorname{tr} A \leq \operatorname{tr} B;$$

(d) if $U \in \mathcal{UA}(\mathcal{H})$ then the operator $U A U^{-1}$ is positive and

$$\operatorname{tr}(U A U^{-1}) = \operatorname{tr} A;$$

(e) if V is a partially isometric element of $\mathcal{B}(\mathcal{H})$, then the operator $V A V^{\dagger}$ is positive and

$$\operatorname{tr}(V A V^{\dagger}) \leq \operatorname{tr} A;$$

if, in particular, V is an isometry then

$$\operatorname{tr}(V A V^{\dagger}) = \operatorname{tr} A;$$

(f) if $\{v_i\}_{i \in I}$ is an o.n.s. in $\mathcal{H}$ (note that every o.n.s. in $\mathcal{H}$ is countable by 10.7.7) then

$$\sum_{i \in I} \|A v_i\|^2 \leq (\operatorname{tr} A)^2 := (\operatorname{tr} A)(\operatorname{tr} A).$$

The sum $\operatorname{tr} A + \operatorname{tr} B$ in statement a is defined as in 5.3.1d; the products $a \operatorname{tr} A$ in statement b and $(\operatorname{tr} A)(\operatorname{tr} A)$ in statement f are defined as in 5.3.1c; the total ordering $\leq$ in statements c, e and f is the one defined in 5.1.1.

Proof. a: The positivity of $A + B$ is obvious. Then use 5.4.6.

b: The positivity of aA is obvious. Then use 5.4.5.

c: The positivity of B is obvious. Then use 5.4.2a.

d: We already know that the operator UAU^{-1} is positive (cf. 18.1.2c). If $\{u_n\}_{n \in \mathbb{N}}$ is a c.o.n.s. in $\mathcal{H}$ then $\{U^{-1}u_n\}_{n \in \mathbb{N}}$ too is a c.o.n.s. in $\mathcal{H}$ (cf. 10.6.8b), and hence

$$\text{tr}(UAU^{-1}) = \sum_{n=1}^{\infty} \left(u_n | UAU^{-1}u_n \right) = \sum_{n=1}^{\infty} \left(U^{-1}u_n | AU^{-1}u_n \right) = \text{tr } A.$$

e: Let $V \in \mathcal{B}(\mathcal{H})$ be partially isometric. The operator $VAV^\dagger$ is positive since

$$\left(f | VAV^\dagger f \right) = \left(V^\dagger f | AV^\dagger f \right) \geq 0, \forall f \in \mathcal{H}.$$

If $V = \mathbb{O}_\mathcal{H}$ then the equality of statement e is obvious. In what follows, we suppose $V \neq \mathbb{O}_\mathcal{H}$. Then let $\{u_i\}_{i \in I}$ be a o.n.s. which is complete in the subspace R_V (cf. 18.1.6b and 10.7.2) and let $\{v_j\}_{j \in J}$ be a o.n.s. in $\mathcal{H}$ which is complete in the subspace $R_V^\perp$ (provided $R_V^\perp \neq \{0_\mathcal{H}\}$; otherwise, the sum over J below is void). The set $\{u_i\}_{i \in I} \cup \{v_j\}_{j \in J}$ is a c.o.n.s. in $\mathcal{H}$ (for this, cf. the proof of 10.7.3 since $R_V^\perp = (V\{u_i\}_{i \in I})^\perp = (\{u_i\}_{i \in I})^\perp$ by 10.2.11) and hence

$$\text{tr}(VAV^\dagger) \overset{(1)}{=} \sum_{i \in I} \left(u_i | VAV^\dagger u_i \right) + \sum_{j \in J} \left(v_j | VAV^\dagger v_j \right)$$

$$= \sum_{i \in I} \left(u_i | VAV^\dagger u_i \right), \tag{2}$$

since $R_V^\perp = N_{V^\dagger}$ (cf. 12.1.7). We point out that 1 holds true by an easy corollary to 5.4.7 (in 5.4.7, take $a_{n,m} := 0$ for $n > 2$). Now, the restriction of $V^\dagger$ to R_V is a unitary operator from the Hilbert space R_V onto the Hilbert space $N_V^\perp$ (cf. 18.1.6f), and hence $\{V^\dagger u_i\}_{i \in I}$ is a c.o.n.s. in the Hilbert space $N_V^\perp$ (cf. 10.6.5c and 10.6.8b), and hence it is an o.n.s. in $\mathcal{H}$ which is complete in the subspace $N_V^\perp$ (cf. 10.6.5c). In V is an isometry then $N_V = \{0_\mathcal{H}\}$, and hence $\{V^\dagger u_i\}_{i \in I}$ is a c.o.n.s. in $\mathcal{H}$, and hence (in view of 2)

$$\text{tr } A = \sum_{i \in I} \left(V^\dagger u_i | AV^\dagger u_i \right) = \sum_{i \in I} \left(u_i | VAV^\dagger u_i \right) = \text{tr}(VAV^\dagger).$$

If $N_V \neq \{0_\mathcal{H}\}$, let $\{w_k\}_{k \in K}$ be an o.n.s. in $\mathcal{H}$ which is complete in the subspace N_V (cf. 4.4.3 and 4.4.8); then the set $\{V^\dagger u_i\}_{i \in I} \cup \{w_k\}_{k \in K}$ is a c.o.n.s. in $\mathcal{H}$ and hence

$$\text{tr } A = \sum_{i \in I} \left(V^\dagger u_i | AV^\dagger u_i \right) + \sum_{k \in K} \left(w_k | Aw_k \right)$$

$$\geq \sum_{i \in I} \left(u_i | VAV^\dagger u_i \right) = \text{tr}(VAV^\dagger).$$

f: Let $\{v_i\}_{i \in I}$ be an o.n.s. in $\mathcal{H}$ and let $\{u_n\}_{n \in \mathbb{N}}$ be a c.o.n.s. in $\mathcal{H}$ which contains $\{v_i\}_{i \in I}$ (cf. 10.7.3.). Then,

$$\sum_{i \in I} \|Av_i\|^2 \leq \sum_{n=1}^{\infty} \|Au_n\|^2 = \sum_{n=1}^{\infty} \left(\sum_{k=1}^{\infty} |(u_k|Au_n)|^2 \right)$$

$$\overset{(3)}{\leq} \sum_{n=1}^{\infty} \left[\sum_{k=1}^{\infty} (u_k|Au_k)(u_n|Au_n) \right]$$

$$\overset{(4)}{=} \sum_{n=1}^{\infty} \left[\left(\sum_{k=1}^{\infty} (u_k|Au_k) \right) (u_n|Au_n) \right]$$

$$\overset{(5)}{=} \left(\sum_{k=1}^{\infty} (u_k|Au_k) \right) \left(\sum_{n=1}^{\infty} (u_n|Au_n) \right) = (\operatorname{tr} A)(\operatorname{tr} A),$$

where 4 and 5 hold true by 5.4.5 and 3 by 5.4.2a since, for all $k, n \in \mathbb{N}$,

$$|(u_k|Au_n)|^2 = \left| \left(A^{\frac{1}{2}} u_k | A^{\frac{1}{2}} u_n \right) \right|^2 \leq \|A^{\frac{1}{2}} u_k\|^2 \|A^{\frac{1}{2}} u_n\|^2 = (u_k|Au_k)(u_n|Au_n).$$

$\square$

18.2.3 Definition. The subset of $\mathcal{B}(\mathcal{H})$ defined by

$$\mathcal{T}(\mathcal{H}) := \{A \in \mathcal{B}(\mathcal{H}) : \operatorname{tr} |A| < \infty\}$$

is called the *trace class*. The elements of $\mathcal{T}(\mathcal{H})$ are called *trace class operators*.

18.2.4 Theorem. *The following properties of $\mathcal{T}(\mathcal{H})$ are true:*

(a) *if $A, B \in \mathcal{T}(\mathcal{H})$ then $A + B \in \mathcal{T}(\mathcal{H})$ and $\operatorname{tr}|A+B| \leq \operatorname{tr}|A| + \operatorname{tr}|B|$;*
(b) *if $\alpha \in \mathbb{C}$ and $A \in \mathcal{T}(\mathcal{H})$ then $\alpha A \in \mathcal{T}(\mathcal{H})$ and $\operatorname{tr}|\alpha A| = |\alpha| \operatorname{tr}|A|$;*
(c) *if $A \in \mathcal{T}(\mathcal{H})$ and $\operatorname{tr}|A| = 0$ then $A = \mathbb{O}_{\mathcal{H}}$;*
(d) *if $A \in \mathcal{T}(\mathcal{H})$ then $\|A\| \leq \operatorname{tr}|A|$;*
(e) *if $A \in \mathcal{T}(\mathcal{H})$ then $A^{\dagger} \in \mathcal{T}(\mathcal{H})$ and $\operatorname{tr}|A^{\dagger}| = \operatorname{tr}|A|$;*
(f) *if $U \in \mathcal{UA}(\mathcal{H})$ and $A \in \mathcal{T}(\mathcal{H})$ then $UAU^{-1} \in \mathcal{T}(\mathcal{H})$ and $\operatorname{tr}|UAU^{-1}| = \operatorname{tr}|A|$.*

Proof. a: Let $A, B \in \mathcal{T}(\mathcal{H})$ and let U, V, W be partially isometric elements of $\mathcal{B}(\mathcal{H})$ such that

$$|A + B| = U^{\dagger}(A + B), \quad A = V|A|, \quad B = W|B|$$

(cf. 18.1.7). Let $\{u_n\}_{n \in \mathbb{N}}$ be a c.o.n.s. in $\mathcal{H}$. Then the operator $U^{\dagger}V|A|V^{\dagger}U$ is positive and

$$\sum_{n=1}^{\infty} \||A|^{\frac{1}{2}} V^{\dagger} U u_n\|^2 = \sum_{n=1}^{\infty} (u_n | U^{\dagger} V |A| V^{\dagger} U u_n)$$

$$= \operatorname{tr}(U^{\dagger} V |A| V^{\dagger} U) \leq \operatorname{tr}(V|A|V^{\dagger}) \leq \operatorname{tr}|A|,$$

in view of 18.2.2e (once for V and once for $U^\dagger$, which is partially isometric by 18.1.6e). Moreover,

$$\sum_{n=1}^{\infty} \||A|^{\frac{1}{2}} u_n\|^2 = \sum_{n=1}^{\infty} (u_n||A|u_n) = \operatorname{tr}|A|.$$

Then we have, by the Schwarz inequality in $\mathcal{H}$, by 5.4.2a, and by the Schwarz inequality in ℓ^2 (cf. 10.3.8d),

$$\sum_{n=1}^{\infty} |\left(u_n|U^\dagger V|A|u_n\right)| = \sum_{n=1}^{\infty} \left|\left(|A|^{\frac{1}{2}} V^\dagger U u_n||A|^{\frac{1}{2}} u_n\right)\right|$$

$$\leq \sum_{n=1}^{\infty} \||A|^{\frac{1}{2}} V^\dagger U u_n\|\,\||A|^{\frac{1}{2}} u_n\|$$

$$\leq \left(\sum_{n=1}^{\infty} \||A|^{\frac{1}{2}} V^\dagger U u_n\|^2\right)^{\frac{1}{2}} \left(\sum_{n=1}^{\infty} \||A|^{\frac{1}{2}} u_n\|^2\right)^{\frac{1}{2}} \leq \operatorname{tr}|A|.$$

We can prove in the same way that

$$\sum_{n=1}^{\infty} |\left(u_n|U^\dagger W|B|u_n\right)| \leq \operatorname{tr}|B|.$$

Then we have, by 5.4.2a and 5.4.6,

$$\sum_{n=1}^{\infty} (u_n||A+B|u_n) = \sum_{n=1}^{\infty} (u_n|U^\dagger(A+B)u_n)$$

$$\leq \sum_{n=1}^{\infty} \left(|\left(u_n|U^\dagger A u_n\right)| + |\left(u_n|U^\dagger B u_n\right)|\right)$$

$$= \sum_{n=1}^{\infty} |\left(u_n|U^\dagger V|A|u_n\right)| + \sum_{n=1}^{\infty} |\left(u_n|U^\dagger W|B|u_n\right)|$$

$$\leq \operatorname{tr}|A| + \operatorname{tr}|B|.$$

This proves statement a.

b: Let $\alpha \in \mathbb{C}$ and $A \in \mathcal{T}(\mathcal{H})$. Then $|\alpha A| = |\alpha||A|$ (cf. 18.1.4e) and hence (cf. 18.2.2b)

$$\operatorname{tr}|\alpha A| = |\alpha|\operatorname{tr}|A|.$$

This proves statement b.

c and d: Let $A \in \mathcal{T}(\mathcal{H})$. For each $f \in \mathcal{H} - \{0_{\mathcal{H}}\}$, the set $\{\|f\|^{-1}f\}$ is an o.n.s. in $\mathcal{H}$, and hence

$$\|f\|^{-2}\||A|f\|^2 \leq (\operatorname{tr}|A|)^2$$

(cf. 18.2.2f). This proves that

$$\|Af\| = \||A|f\| \leq (\operatorname{tr}|A|)\|f\|, \forall f \in \mathcal{H}$$

(cf. 18.1.4a), and hence that $\|A\| \le \operatorname{tr}|A|$, which is statement d. From this we have obviously

$$A = \mathbb{O}_{\mathcal{H}} \text{ if } \operatorname{tr}|A| = 0,$$

which is statement c.

e: Let $A \in \mathcal{T}(\mathcal{H})$ and let U be a partially isometric element of $\mathcal{B}(\mathcal{H})$ such that

$$A = U|A| \text{ and } N_U = N_A$$

(cf. 18.1.7). Then $A^\dagger = |A|U^\dagger$ (cf. 12.6.4) and hence

$$A^\dagger = U^\dagger U|A|U^\dagger$$

by 13.1.3c, since $U^\dagger U$ is the orthogonal projection onto the subspace

$$N_U^\perp = N_A^\perp = N_{|A|}^\perp$$

(cf. 18.1.6d and 18.1.4c) and since the equality $N_{|A|} = R_{|A|}^\perp$ (cf. 12.1.7) implies the inclusion

$$N_{|A|}^\perp \supset R_{|A|}$$

(cf. 10.2.10d). Thus,

$$(A^\dagger)^\dagger A^\dagger = AA^\dagger = U|A|U^\dagger U|A|U^\dagger.$$

In view of 18.2.2e, the operator $U|A|U^\dagger$ is positive. Therefore,

$$U|A|U^\dagger = |A^\dagger|.$$

Then, by 18.2.2e once more,

$$\operatorname{tr}|A^\dagger| = \operatorname{tr}(U|A|U^\dagger) \le \operatorname{tr}|A|.$$

This proves that $A^\dagger \in \mathcal{T}(\mathcal{H})$.

Now, since in the reasoning above A was an arbitrary element of $\mathcal{T}(\mathcal{H})$, we can replace A with $A^\dagger$ and obtain

$$\operatorname{tr}|A| = \operatorname{tr}|(A^\dagger)^\dagger| \le \operatorname{tr}|A^\dagger|,$$

and hence

$$\operatorname{tr}|A^\dagger| = \operatorname{tr}|A|.$$

f: If $U \in \mathcal{UA}(\mathcal{H})$ and $A \in \mathcal{B}(\mathcal{H})$, then $|UAU^{-1}| = U|A|U^{-1}$ (cf. 18.1.4d), and hence

$$\operatorname{tr}|UAU^{-1}| = \operatorname{tr}(U|A|U^{-1}) = \operatorname{tr}|A|$$

(cf. 18.2.2d). Therefore, if $A \in \mathcal{T}(\mathcal{H})$ then $UAU^{-1} \in \mathcal{T}(\mathcal{H})$. $\qquad\square$

18.2.5 Definition. We define the function

$$\nu_1 : \mathcal{T}(\mathcal{H}) \to \mathbb{R}$$
$$A \mapsto \nu_1(A) := \operatorname{tr}|A|.$$

In view of 18.2.4a,b,c, $\mathcal{T}(\mathcal{H})$ is a linear manifold in the linear space $\mathcal{B}(\mathcal{H})$ and hence $\mathcal{T}(\mathcal{H})$ itself is a linear space (cf. 3.1.3), and moreover ν_1 is a norm for $\mathcal{T}(\mathcal{H})$.

(a) In view of 18.2.4d, if a sequence in $\mathcal{T}(\mathcal{H})$ is convergent to an element of $\mathcal{T}(\mathcal{H})$ with respect to the norm ν_1 then it is convergent to the same operator also with respect to the norm defined in 4.2.11a.

(b) In view of 18.2.4e, the mapping $\mathcal{T}(\mathcal{H}) \ni A \mapsto A^\dagger \in \mathcal{T}(\mathcal{H})$ is continuous with respect to the norm ν_1; in fact, if $A \in \mathcal{T}(\mathcal{H})$ and $\{A_n\}$ is a sequence in $\mathcal{T}(\mathcal{H})$ such that $\operatorname{tr}|A_n - A| \to 0$, then

$$\operatorname{tr}|A_n^\dagger - A^\dagger| = \operatorname{tr}|(A_n - A)^\dagger| = \operatorname{tr}|A_n - A| \to 0.$$

(c) In view of 18.2.4f, for $U \in \mathcal{UA}(\mathcal{H})$ the mapping $\mathcal{T}(\mathcal{H}) \ni A \mapsto UAU^{-1} \in \mathcal{T}(\mathcal{H})$ is an automorphism of the normed space $(\mathcal{T}(\mathcal{H}), \nu_1)$.

18.2.6 Lemma. *Let $A \in \mathcal{B}(\mathcal{H})$. Then there exist $U_1, U_2, U_3, U_4 \in \mathcal{U}(\mathcal{H})$ so that*

$$A = \frac{1}{2}\|A\|(U_1 + U_2 - iU_3 - iU_4).$$

Proof. First let $B \in \mathcal{B}(\mathcal{H})$ be self-adjoint and such that $\|B\| \leq 1$. Then the operator $\mathbb{1}_\mathcal{H} - B^2$ is positive, since

$$(f|(\mathbb{1}_\mathcal{H} - B^2)f) = \|f\|^2 - \|Bf\|^2 \geq 0, \forall f \in \mathcal{H}$$

(cf. 4.2.5b), and hence we can define the operator $(\mathbb{1}_\mathcal{H} - B^2)^{\frac{1}{2}}$. We have

$$(B \pm i(\mathbb{1}_\mathcal{H} - B^2)^{\frac{1}{2}})^\dagger = B \mp i(\mathbb{1}_\mathcal{H} - B^2)^{\frac{1}{2}}$$

and hence

$$\begin{aligned}
(B \pm i(\mathbb{1}_\mathcal{H} - B^2)^{\frac{1}{2}})^\dagger(B \pm i(\mathbb{1}_\mathcal{H} - B^2)^{\frac{1}{2}}) \\
= B^2 \mp i(\mathbb{1}_\mathcal{H} - B^2)^{\frac{1}{2}}B \pm iB(\mathbb{1}_\mathcal{H} - B^2)^{\frac{1}{2}} + \mathbb{1}_\mathcal{H} - B^2 = \mathbb{1}_\mathcal{H},
\end{aligned}$$

since $[B, (\mathbb{1}_\mathcal{H} - B^2)^{\frac{1}{2}}] = \mathbb{O}_\mathcal{H}$ (cf. 18.1.2b). Similarly, we have

$$(B \pm i(\mathbb{1}_\mathcal{H} - B^2)^{\frac{1}{2}})(B \pm i(\mathbb{1}_\mathcal{H} - B^2)^{\frac{1}{2}})^\dagger = \mathbb{1}_\mathcal{H}.$$

In view of 12.5.1, this proves that $B + i(\mathbb{1}_\mathcal{H} - B^2)^{\frac{1}{2}}$ and $B - i(\mathbb{1}_\mathcal{H} - B^2)^{\frac{1}{2}}$ are unitary operators. Moreover,

$$\frac{1}{2}(B + i(\mathbb{1}_\mathcal{H} - B^2)^{\frac{1}{2}}) + \frac{1}{2}(B - i(\mathbb{1}_\mathcal{H} - B^2)^{\frac{1}{2}}) = B.$$

Thus, there exist $V_1, V_2 \in \mathcal{U}(\mathcal{H})$ so that $B = \frac{1}{2}(V_1 + V_2)$.

Next we notice that, for all $A \in \mathcal{B}(\mathcal{H}) - \{\mathbb{O}_\mathcal{H}\}$:

$$A = \|A\| \left[\frac{1}{2}\|A\|^{-1}(A + A^\dagger) - i\frac{1}{2}\|A\|^{-1}i(A - A^\dagger) \right];$$

$$\frac{1}{2}\|A\|^{-1}(A + A^\dagger) \text{ and } \frac{1}{2}\|A\|^{-1}i(A - A^\dagger) \text{ are self-adjoint;}$$

$$\left\| \frac{1}{2}\|A\|^{-1}(A + A^\dagger) \right\| \leq 1.$$

The two things proved above prove the statement. $\qquad\square$

18.2.7 Theorem. *Suppose that $A \in \mathcal{T}(\mathcal{H})$ and $B \in \mathcal{B}(\mathcal{H})$. Then $BA \in \mathcal{T}(\mathcal{H})$ and $AB \in \mathcal{T}(\mathcal{H})$.*

Proof. Let $U \in \mathcal{U}(\mathcal{H})$. The operator $U^{-1}|A|U$ is positive (cf. 18.1.2c) and

$$(U^{-1}|A|U)^2 = U^{-1}|A|^2 U = U^{-1}A^\dagger A U = (AU)^\dagger (AU)$$

(cf. 12.6.4 and 12.5.1b). Therefore,

$$U^{-1}|A|U = |AU| \text{ and hence } \operatorname{tr}|AU| = \operatorname{tr}|A| < \infty$$

(cf. 18.2.2d). Moreover,

$$|A|^2 = A^\dagger A = A^\dagger U^\dagger U A = (UA)^\dagger (UA)$$

(cf. 12.5.1c) proves that

$$|A| = |UA| \text{ and hence } \operatorname{tr}|UA| = \operatorname{tr}|A| < \infty.$$

Since U was an arbitrary element of $\mathcal{U}(\mathcal{H})$, this proves that

$$AU, UA \in \mathcal{T}(\mathcal{H}), \forall U \in \mathcal{U}(\mathcal{H}),$$

and this proves the statement, in view of 18.2.6 and 18.2.4a,b. $\qquad\square$

18.2.8 Theorem. *Let A be a positive element of $\mathcal{T}(\mathcal{H})$ and suppose that $A \neq \mathbb{O}_{\mathcal{H}}$. Then there exist an o.n.s. $\{u_n\}_{n \in I}$ (with $I := \{1, ..., N\}$ or $I := \mathbb{N}$) in $\mathcal{H}$ and a family $\{\lambda_n\}_{n \in I}$ of elements of $(0, \infty)$ (not necessarily different from each other) so that (denoting by $\sum_{n \in I}$ either $\sum_{n=1}^{N}$ or $\sum_{n=1}^{\infty}$)*

$$A = \sum_{n \in I} \lambda_n A_{u_n} \text{ and } \operatorname{tr} A = \sum_{n \in I} \lambda_n.$$

If $I = \mathbb{N}$, the first series is convergent with respect to the norm for $\mathcal{B}(\mathcal{H})$ defined in 4.2.11a. The one-dimensional projection A_{u_n} is defined as in 13.1.12.

Proof. We set

$$E_k := \left(\frac{1}{k+1}\|A\|, \frac{1}{k}\|A\| \right] \text{ and } P_k := P^A(E_k), \forall k \in \mathbb{N}.$$

Since $\sigma(A) \subset [0, \|A\|]$ (cf. 15.3.9 and 4.5.10), we have

$$P^A(\{0\})f + \sum_{k=1}^{\infty} P_k f = P^A(\sigma(A))f = f, \forall f \in \mathcal{H} \tag{1}$$

(cf. 15.2.2d).

For all $k \in \mathbb{N}$, the subspace $M_k := R_{P_k}$ is finite-dimensional. Indeed, if we fix $k \in \mathbb{N}$ then for all $f \in M_k$ we have

$$\mu_f^{P^A}(\mathbb{R} - E_k) = \|P^A(\mathbb{R} - E_k)f\|^2 = \|P^A(\mathbb{R} - E_k)P^A(E_k)f\|^2 = 0$$

(cf. 13.1.3c and 13.3.2b), and hence

$$\|Af\|^2 = \int_{\mathbb{R}} \xi^2 d\mu_f^{P^A} \geq \frac{1}{(k+1)^2}\|A\|^2 \int_{\mathbb{R}} 1_{\mathbb{R}} d\mu_f^{P^A} = \frac{1}{(k+1)^2}\|A\|^2\|f\|^2$$

(cf. 15.2.2e and 8.1.11a). In view of 18.2.2f, this proves that each o.n.s. contained in M_k must be finite, and hence that the orthogonal dimension of M_k is finite.

For all $k \in \mathbb{N}$, we have

$$[A, P_k] = \mathbb{O}_{\mathcal{H}}$$

(cf. 15.2.1B). Hence, the operator A is reduced by the subspace M_k (cf. 17.2.4) and $A_k := A^{M_k}$ is a self-adjoint operator in the Hilbert space M_k (cf. 17.2.8). Therefore, if $M_k \neq \{0_{\mathcal{H}}\}$ then there exists an o.n.s. $\{v_{k,i}\}_{i \in I_k}$ which is complete in the subspace M_k and whose elements are eigenvectors of A_k (cf. 15.3.4C and 10.6.5c), i.e. so that

$$\forall i \in I_k, \exists \mu_{k,i} \in \mathbb{R} \text{ such that } A_k v_{k,i} = \mu_{k,i} v_{k,i}.$$

For each $i \in I_k$, it is obvious that $\mu_{k,i}$ is an eigenvalue of A; then, $\mu_{k,i} \in [0, \infty)$; moreover, $P_k P^A(\{0\}) = \mathbb{O}_{\mathcal{H}}$ (cf. 13.3.2b) implies $v_{k,i} \in N_A^\perp$ by 13.2.9 (since $P^A(\{0\})$ is the orthogonal projection onto N_A, cf. 15.2.5e), and hence $\mu_{k,i} \in (0, \infty)$. Further, we have

$$AP_k f = A_k P_k f = \sum_{i \in I_k} (v_{k,i} | A_k P_k f)_{M_k} v_{k,i}$$

$$= \sum_{i \in I_k} (P_k A v_{k,i} | f)_{\mathcal{H}} v_{k,i} = \sum_{i \in I_k} \mu_{k,i} (v_{k,i} | f) v_{k,i}, \forall f \in \mathcal{H} \tag{2}$$

(cf. 10.6.4b). Letting $J := \{k \in \mathbb{N} : M_k \neq \{0_{\mathcal{H}}\}\}$, from 1 and 2 and from the continuity of A we infer that

$$Af = AP^A(\{0\})f + \sum_{k=1}^{\infty} AP_k f$$

$$= \sum_{k \in J} \left(\sum_{i \in I_k} \mu_{k,i} (v_{k,i} | f) v_{k,i} \right), \forall f \in \mathcal{H}. \tag{3}$$

Now let $I := \{1, ..., N\}$ or $I := \mathbb{N}$ be so that there is a bijection from I onto the set $\bigcup_{k \in J} I_k$, and for each $n \in I$ let

$$u_n := v_{k,i} \text{ and } \lambda_n := \mu_{k,i} \text{ if } n \text{ corresponds to the pair } (k, i).$$

Then $\{u_n\}_{n \in I}$ is obviously an o.n.s. in $\mathcal{H}$ (since $M_k \subset M_h^\perp$ if $k \neq h$), $\{\lambda_n\}_{n \in I}$ is a family of elements of $(0, \infty)$, and 3 can be written as

$$Af = \sum_{n \in I} \lambda_n (u_n | f) u_n, \forall f \in \mathcal{H}, \tag{4}$$

in view of 10.4.10 (note that every series which may appear in 3 is convergent in view of 13.2.8 and 10.6.1).

Now let $\{w_j\}_{j \in \mathbb{N}}$ be a c.o.n.s. in $\mathcal{H}$ which contains $\{u_n\}_{n \in I}$ (cf. 10.7.3). Then,

$$Aw_j = 0_{\mathcal{H}} \text{ if } j \in \mathbb{N} \text{ is such that } w_j \notin \{u_n\}_{n \in I}$$

(this is clear from 4), and hence

$$\operatorname{tr} A = \sum_{j=1}^{\infty} (w_j | A w_j) = \sum_{n \in I} (u_n | A u_n) = \sum_{n \in I} \lambda_n.$$

Thus, if $I = \mathbb{N}$, the series $\sum_{n=1}^{\infty} \lambda_n$ is convergent since $A \in \mathcal{T}(\mathcal{H})$, and hence the series $\sum_{n=1}^{\infty} \lambda_n A_{u_n}$ is absolutely convergent in the Banach space $\mathcal{B}(\mathcal{H})$ (cf. 4.2.11b) since $\|A_{u_n}\| = 1$ (cf. 13.1.3d) for all $n \in \mathbb{N}$, and hence the series $\sum_{n=1}^{\infty} \lambda_n A_{u_n}$ is convergent (cf. 4.1.8b).

Finally, from 4 (and from 4.2.12, if $I = \mathbb{N}$) we have

$$\left(\sum_{n \in I} \lambda_n A_{u_n} \right) f = \sum_{n \in I} \lambda_n A_{u_n} f = Af, \forall f \in \mathcal{H}.$$

$\square$

18.2.9 Corollary. *Let $A \in \mathcal{T}(\mathcal{H})$ and suppose that $A \neq \mathbb{O}_{\mathcal{H}}$. Then there exist two orthonormal systems $\{u_n\}_{n \in I}$ and $\{v_n\}_{n \in I}$ (with $I := \{1, ..., N\}$ or $I := \mathbb{N}$) in $\mathcal{H}$ and a family $\{\lambda_n\}_{n \in I}$ of elements of $(0, \infty)$ (not necessarily different from each other) so that (denoting by $\sum_{n \in I}$ either $\sum_{n=1}^{N}$ or $\sum_{n=1}^{\infty}$)*

$$A = \sum_{n \in I} \lambda_n A_{u_n, v_n}, \ |A| = \sum_{n \in I} \lambda_n A_{u_n}, \ \operatorname{tr}|A| = \sum_{n \in I} \lambda_n.$$

If $I = \mathbb{N}$, the first two series are convergent with respect to the norm for $\mathcal{B}(\mathcal{H})$ defined in 4.2.11a. The operator A_{u_n, v_n} is defined as in 18.1.9.

Proof. Let U be a partially isometric element of $\mathcal{B}(\mathcal{H})$ such that

$$A = U|A| \text{ and } N_U = N_A$$

(cf. 18.1.7). Moreover, let $\{u_n\}_{n \in I}$ (with $I := \{1, ..., N\}$ or $I := \mathbb{N}$) be an o.n.s. in $\mathcal{H}$ and $\{\lambda_n\}_{n \in I}$ a family of elements of $(0, \infty)$ so that

$$|A| = \sum_{n \in I} \lambda_n A_{u_n} \text{ and } \operatorname{tr}|A| = \sum_{n \in I} \lambda_n$$

(cf. 18.2.8 with $|A|$ in place of A). If $I = \mathbb{N}$, the first series is convergent with respect to the norm for $\mathcal{B}(\mathcal{H})$ defined in 4.2.11a. Since $N_{|A|} = N_A$ (cf.18.1.4c) and since $u_n \in N_{|A|}^{\perp}$ (note that $|A|u_n = \lambda_n u_n$ and then use 12.4.20B), we have

$$u_n \in N_U^{\perp}, \forall n \in I.$$

Now, the restriction of the operator U to the subspace $N_U^{\perp}$ is a unitary operator from the Hilbert space $N_U^{\perp}$ onto the Hilbert space R_U (cf. 18.1.6c). Thus, if we set $v_n := U u_n$ for all $n \in I$, $\{v_n\}_{n \in I}$ is an o.n.s. in $\mathcal{H}$ (cf. 10.6.5c and 10.6.8a). Moreover,

$$U A_{u_n} = A_{u_n, v_n}, \forall n \in I,$$

as can be seen immediately. Therefore we have

$$A = U \left(\sum_{n \in I} \lambda_n A_{u_n} \right) = \sum_{n \in I} \lambda_n U A_{u_n} = \sum_{n \in I} \lambda_n A_{u_n, v_n}.$$

If $I = \mathbb{N}$, we have used the continuity of the operator product in the Banach algebra $\mathcal{B}(\mathcal{H})$ (cf. 4.3.5 and 4.3.3). Thus, if $I = \mathbb{N}$, all the series written above are convergent with respect to the norm for $\mathcal{B}(\mathcal{H})$ defined in 4.2.11a. $\qquad \square$

18.2.10 Theorem. *Let $A \in \mathcal{T}(\mathcal{H})$ and let $\{v_n\}_{n \in \mathbb{N}}$ be a c.o.n.s. in $\mathcal{H}$. Then the series $\sum_{n=1}^{\infty} (v_n | A v_n)$ is absolutely convergent and hence it is convergent. The sum of this series is independent of the c.o.n.s. $\{v_n\}_{n \in \mathbb{N}}$ in $\mathcal{H}$ chosen to compute it, and it is called the trace of A and denoted by $\operatorname{tr} A$. Thus,*

$$\operatorname{tr} A := \sum_{n=1}^{\infty} (w_n | A w_n)$$

for whichever c.o.n.s. $\{w_n\}_{n \in \mathbb{N}}$ in $\mathcal{H}$. It is obvious that, if A is positive, this definition agrees with the one given in 18.2.1.

The following inequalities hold true:

(a) $|\operatorname{tr} BA| \le \|B\| \operatorname{tr} |A|, \ \forall B \in \mathcal{B}(\mathcal{H})$;
(b) $\operatorname{tr} |BA| \le \|B\| \operatorname{tr} |A|, \ \forall B \in \mathcal{B}(\mathcal{H})$;
(c) $|\operatorname{tr} A| \le \operatorname{tr} |A|$.

Proof. For $A = \mathbb{O}_{\mathcal{H}}$ the whole statement is trivially true. Thus, we suppose $A \ne \mathbb{O}_{\mathcal{H}}$. Then, in view of 18.2.8, there are an o.n.s. $\{u_n\}_{n \in I}$ (with $I := \{1, ..., N\}$ or $I := \mathbb{N}$) in $\mathcal{H}$ and a family $\{\lambda_n\}_{n \in I}$ of elements of $(0, \infty)$ so that

$$|A| f = \sum_{n \in I} \lambda_n (u_n | f) u_n, \forall f \in \mathcal{H}, \quad \text{and} \quad \sum_{n \in I} \lambda_n = \operatorname{tr} |A|.$$

We notice that, if P is the projection defined by

$$P f := \sum_{n \in I} (u_n | f) u_n, \forall f \in \mathcal{H}$$

(cf. 13.1.10), we have $P|A| = |A|$.

Let U be a partially isometric element of $\mathcal{B}(\mathcal{H})$ such that

$$A = U|A|$$

(cf. 18.1.7) and let $\{v_n\}_{n \in \mathbb{N}}$ be a c.o.n.s. in $\mathcal{H}$. For each $n \in I$, we have

$$\sum_{k=1}^{\infty} |(|A| u_n | v_k) (v_k | U u_n)| = \lambda_n \sum_{k=1}^{\infty} |(u_n | v_k)| |(v_k | U u_n)|$$

$$\overset{(1)}{\le} \lambda_n \left(\sum_{k=1}^{\infty} |(u_n | v_k)|^2 \right)^{\frac{1}{2}} \left(\sum_{k=1}^{\infty} |(v_k | U u_n)|^2 \right)^{\frac{1}{2}}$$

$$\overset{(2)}{=} \lambda_n \|u_n\| \|U u_n\| \overset{(3)}{\le} \lambda_n$$

(1 hold true by the Schwarz inequality in ℓ^2, cf. 10.2.8b and 10.3.8d; 2 holds true by 10.6.4d with $M := \mathcal{H}$; 3 holds true by 18.1.6a and 4.2.5b). Then we have

$$\sum_{n \in I} \left(\sum_{k=1}^{\infty} | (U^{\dagger} v_k | u_n) (u_n || A | v_k) | \right) \leq \sum_{n \in I} \lambda_n = \mathrm{tr} \, |A| < \infty, \tag{4}$$

and hence

$$\sum_{k=1}^{\infty} | (v_k | A v_k) | = \sum_{k=1}^{\infty} | (U^{\dagger} v_k || A | v_k) | = \sum_{k=1}^{\infty} | (U^{\dagger} v_k | P | A | v_k) |$$

$$= \sum_{k=1}^{\infty} \left| \sum_{n \in I} (U^{\dagger} v_k | u_n) (u_n || A | v_k) \right|$$

$$\overset{(5)}{\leq} \sum_{k=1}^{\infty} \sum_{n \in I} | (U^{\dagger} v_k | u_n) (u_n || A | v_k) |$$

$$\overset{(6)}{=} \sum_{n \in I} \left(\sum_{k=1}^{\infty} | (U^{\dagger} v_k | u_n) (u_n || A | v_k) | \right) < \infty$$

(5 holds true by 5.4.2a, and by 5.4.10 if $I = \mathbb{N}$; 6 holds true by 5.4.6 if $I = \{1, ..., N\}$ or by 5.4.7 if $I = \mathbb{N}$). Thus, the series $\sum_{k=1}^{\infty} (v_k | A v_k)$ is absolutely convergent and hence it is convergent by 4.1.8b. Moreover, we have

$$\sum_{k=1}^{\infty} (v_k | A v_k) = \sum_{k=1}^{\infty} (U^{\dagger} v_k || A | v_k) = \sum_{k=1}^{\infty} (U^{\dagger} v_k | P | A | v_k)$$

$$= \sum_{k=1}^{\infty} \left(\sum_{n \in I} (U^{\dagger} v_k | u_n) (u_n || A | v_k) \right)$$

$$\overset{(7)}{=} \sum_{n \in I} \left(\sum_{k=1}^{\infty} (U^{\dagger} v_k | u_n) (u_n || A | v_k) \right)$$

$$= \sum_{n \in I} \left(\sum_{k=1}^{\infty} (|A| u_n | v_k) (v_k | U u_n) \right) \overset{(8)}{=} \sum_{n \in I} (|A| u_n | U u_n)$$

(7 holds true by 8.4.14b, since 4 proves that the conditions in 8.4.14a are satisfied; 8 holds true by 10.6.4c with $M := \mathcal{H}$). Since the last term of this equation is independent of the choice of the c.o.n.s. $\{v_n\}_{n \in \mathbb{N}}$, this proves that the sum of the series $\sum_{k=1}^{\infty} (v_k | A v_k)$ is independent as well (note that, if $I = \mathbb{N}$, the series $\sum_{n=1}^{\infty} (|A| u_n | U u_n)$ is shown to be convergent by the way the last term of the equation above has been obtained).

Now we prove the inequalities of the statement.

a: Let $\{w_j\}_{j \in \mathbb{N}}$ be a c.o.n.s. in $\mathcal{H}$ which contains the o.n.s. $\{u_n\}_{n \in I}$ (cf. 10.7.3). If $j \in \mathbb{N}$ is such that $w_j \notin \{u_n\}_{n \in I}$, then

$$(w_j | u_n) = 0, \forall n \in I,$$

and hence

$$w_j \in N_{|A|} = N_A$$

(cf. 18.1.4c). Therefore we have, for all $B \in \mathcal{B}(\mathcal{H})$,

$$
\begin{aligned}
|\operatorname{tr} BA| &= \left| \sum_{j=1}^{\infty} (w_j | BA w_j) \right| = \left| \sum_{n \in I} (u_n | BA u_n) \right| \\
&\overset{(9)}{\leq} \sum_{n \in I} |(u_n | BA u_n)| \overset{(10)}{\leq} \sum_{n \in I} \|BA u_n\| \\
&\overset{(11)}{\leq} \sum_{n \in I} \|B\| \|A u_n\| \overset{(12)}{=} \|B\| \sum_{n \in I} \||A| u_n\| = \|B\| \sum_{n \in I} \lambda_n = \|B\| \operatorname{tr} |A|,
\end{aligned}
$$

where 9 holds by 5.4.10 if $I = \mathbb{N}$, 10 by the Schwarz inequality and by 5.4.2a if $I = \mathbb{N}$, 11 by 4.2.5b and by 5.4.2a if $I = \mathbb{N}$, 12 by 5.4.5.

b: Let $B \in \mathcal{B}(\mathcal{H})$ and let V be a partially isometric element of $\mathcal{B}(\mathcal{H})$ such that

$$|BA| = V^\dagger BA$$

(cf. 18.1.7). Then, in view of inequality a (with $V^\dagger B$ in place of B), we have

$$
\begin{aligned}
\operatorname{tr} |BA| &= \operatorname{tr}(V^\dagger BA) = |\operatorname{tr}(V^\dagger BA)| \leq \|V^\dagger B\| \operatorname{tr} |A| \\
&\leq \|V^\dagger\| \|B\| \operatorname{tr} |A| \leq \|B\| \operatorname{tr} |A|
\end{aligned}
$$

(cf. 4.2.9), since $\|V^\dagger\| = 1$ if $V^\dagger \neq \mathbb{O}_\mathcal{H}$ (cf. 18.1.6a,e).

c: This follows immediately from inequality a with $B := \mathbb{1}_\mathcal{H}$. $\qquad\square$

18.2.11 Theorem. *The function defined by*

$$
\operatorname{tr} : \mathcal{T}(\mathcal{H}) \to \mathbb{C}
$$
$$
A \mapsto \operatorname{tr} A
$$

has the following properties:

(a) tr *is a linear functional, continuous with respect to the norm ν_1 defined in 18.2.5;*
(b) $\operatorname{tr} A^\dagger = \overline{\operatorname{tr} A}$, $\forall A \in \mathcal{T}(\mathcal{H})$;
(c) $\operatorname{tr}(AB) = \operatorname{tr}(BA)$, $\forall A \in \mathcal{T}(\mathcal{H})$, $\forall B \in \mathcal{B}(\mathcal{H})$;
(d) $\operatorname{tr}(UAU^{-1}) = \operatorname{tr} A$, $\forall A \in \mathcal{T}(\mathcal{H})$, $\forall U \in \mathcal{U}(\mathcal{H})$;
(e) $\operatorname{tr}(VAV^{-1}) = \overline{\operatorname{tr} A}$, $\forall A \in \mathcal{T}(\mathcal{H})$, $\forall V \in \mathcal{A}(\mathcal{H})$.

Proof. a: For the function tr, property lo_1 of 3.2.1 is obvious and properties lo_2, lo_3 follow directly from the property ip_1 of an inner product and from the continuity of sum and product in $\mathbb{C}$. The continuity of tr follows from 18.2.10c and from 4.2.2.

b: This follows directly from 12.1.3A, from property ip_2 of an inner product, and from the continuity of complex conjugation.

c: Let $A \in \mathcal{T}(\mathcal{H})$. For all $U \in \mathcal{U}(\mathcal{H})$ we have, if $\{v_n\}_{n \in \mathbb{N}}$ is a c.o.n.s. in $\mathcal{H}$,

$$\mathrm{tr}(AU) = \sum_{n=1}^{\infty} (v_n | AUv_n) \overset{(1)}{=} \sum_{n=1}^{\infty} (U^\dagger Uv_n | AUv_n)$$

$$= \sum_{n=1}^{\infty} (Uv_n | UAUv_n) \overset{(2)}{=} \mathrm{tr}(UA),$$

where 1 holds true by 12.5.1c and 2 because $\{Uv_n\}_{n \in \mathbb{N}}$ is a c.o.n.s in $\mathcal{H}$ (cf. 10.6.8b). In view of 18.2.6 and property a, this proves that

$$\mathrm{tr}(AB) = \mathrm{tr}(BA), \forall B \in \mathcal{B}(\mathcal{H}).$$

d: This follows immediately from property c.

e: Let $A \in \mathcal{T}(\mathcal{H})$ and $V \in \mathcal{A}(\mathcal{H})$. If $\{v_n\}_{n \in \mathbb{N}}$ is a c.o.n.s. in $\mathcal{H}$ then

$$\mathrm{tr}(VAV^{-1}) = \sum_{n=1}^{\infty} (v_n | VAV^{-1}v_n) = \sum_{n=1}^{\infty} (AV^{-1}v_n | V^{-1}v_n) = \overline{\mathrm{tr}\,A},$$

since $\{V^{-1}v_n\}_{n \in \mathbb{N}}$ is a c.o.n.s. in $\mathcal{H}$ (cf. 10.6.8b). $\qquad \square$

18.2.12 Lemma. *Let $P \in \mathscr{P}(\mathcal{H})$. Then:*

(a) $P \in \mathcal{T}(\mathcal{H})$ iff the orthogonal dimension of the subspace R_P is finite;

(b) if $P \in \mathcal{T}(\mathcal{H})$ then $\mathrm{tr}\,P$ equals the orthogonal dimension of R_P;

(c) if $P \in \mathcal{T}(\mathcal{H})$ then

$$\mathrm{tr}(PA) = \sum_{n \in I} (u_n | Au_n),$$

for each $A \in \mathcal{B}(\mathcal{H})$ and for each o.n.s. $\{u_n\}_{n \in I}$ in $\mathcal{H}$ which is complete in the subspace R_P;

(d) if $P \in \mathcal{T}(\mathcal{H})$ then

$$0 \le \mathrm{tr}(PA) \le \mathrm{tr}\,A$$

for each positive element A of $\mathcal{B}(\mathcal{H})$.

Proof. In what follows, let $\{u_n\}_{n \in I}$ be a c.o.n.s. in $\mathcal{H}$ which is complete in the subspace R_P (cf. 10.7.2) and let $\{w_j\}_{j \in \mathbb{N}}$ be a c.o.n.s. in $\mathcal{H}$ which contains $\{u_n\}_{n \in I}$ (cf. 10.7.3). Then,

$$w_j \in N_P \text{ if } j \in \mathbb{N} \text{ is such that } w_j \notin \{u_n\}_{n \in I}$$

(cf. 13.1.10).

a and b: We notice that P is positive, in view of 13.1.7c. Then we have

$$\mathrm{tr}\,|P| = \mathrm{tr}\,P = \sum_{j=1}^{\infty} (w_j | Pw_j) = \sum_{n \in I} (u_n | Pu_n) = \sum_{n \in I} (u_n | u_n)$$

(cf. 13.1.3c). This proves both a and b.

c: Suppose $P \in \mathcal{T}(\mathcal{H})$. Then, for all $A \in \mathcal{B}(\mathcal{H})$, we have

$$\text{tr}(PA) = \sum_{j=1}^{\infty} (w_j|PAw_j) = \sum_{j=1}^{\infty} (Pw_j|Aw_j) = \sum_{n \in I} (u_n|Au_n).$$

d: Suppose $P \in \mathcal{T}(\mathcal{H})$. Then, for all positive element A of $\mathcal{B}(\mathcal{H})$, in view of c we have

$$0 \leq \text{tr}(PA) \leq \sum_{j=1}^{\infty} (w_j|Aw_j) = \text{tr } A.$$

$\square$

18.2.13 Lemma. *Let A be a positive element of $\mathcal{B}(\mathcal{H})$. Then*

$$\text{tr } A = \sup\{\text{tr}(PA) : P \in \mathscr{P}(\mathcal{H}) \cap \mathcal{T}(\mathcal{H})\}$$

(this l.u.b. is meant with respect to the total ordering defined in 5.1.1).

Proof. In view of 18.2.12d, we have

$$\sup\{\text{tr}(PA) : P \in \mathscr{P}(\mathcal{H}) \cap \mathcal{T}(\mathcal{H})\} \leq \text{tr } A.$$

Now let $\{u_n\}_{n \in \mathbb{N}}$ be a c.o.n.s. in $\mathcal{H}$ and, for each $N \in \mathbb{N}$, let P_N be the orthogonal projection onto the subspace $V\{u_1, ..., u_N\}$. Then

$$\text{tr}(P_N A) = \sum_{n=1}^{N} (u_n|Au_n)$$

(cf. 18.2.12c), and hence

$$\text{tr } A = \sum_{n=1}^{\infty} (u_n|Au_n) = \sup_{N \geq 1} \sum_{n=1}^{N} (u_n|Au_n) = \sup_{N \geq 1} \text{tr}(P_N A)$$

(cf. 5.4.1). This proves the inequality

$$\text{tr } A \leq \sup\{\text{tr}(PA) : P \in \mathscr{P}(\mathcal{H}) \cap \mathcal{T}(\mathcal{H})\},$$

and hence the equality of the statement. $\square$

18.2.14 Theorem. *The normed space $(\mathcal{T}(\mathcal{H}), \nu_1)$ (i.e. the linear space $\mathcal{T}(\mathcal{H})$ with the norm ν_1, cf. 18.2.5) is a Banach space.*

Proof. Let $\{A_n\}$ be a sequence in $\mathcal{T}(\mathcal{H})$ such that

$$\forall \varepsilon > 0, \exists N_\varepsilon \in \mathbb{N} \text{ so that } N_\varepsilon < n, m \Rightarrow \nu_1(A_n - A_m) < \varepsilon.$$

We need to prove that there exists $A \in \mathcal{T}(\mathcal{H})$ such that $\nu_1(A_n - A) \to 0$. We note that, if such A exists, then it must be so that $\|A_n - A\| \to 0$ (cf. 18.2.4d).

Since

$$\|A_n - A_m\| \leq \text{tr } |A_n - A_m| = \nu_1(A_n - A_m)$$

(cf. 18.2.4d), by 4.2.11b there exists $A \in \mathcal{B}(\mathcal{H})$ so that

$$\|A_n - A\| \to 0.$$

We fix $\varepsilon \in (0, \infty)$. Let $n > N_\varepsilon$ and let U_n be a partially isometric element of $\mathcal{B}(\mathcal{H})$ such that

$$|A_n - A| = U_n^\dagger(A_n - A)$$

(cf. 18.1.7). Now let $P \in \mathscr{P}(\mathcal{H}) \cap \mathcal{T}(\mathcal{H})$ and let $\{u_i\}_{i \in I}$ be an o.n.s. in $\mathcal{H}$ which is complete in the subspace R_P (hence the set I is finite, cf. 18.2.12a); then we have

$$0 \overset{(1)}{\leq} \operatorname{tr}(P|A_n - A|) \overset{(2)}{=} \sum_{i \in I} \left(u_i | U_n^\dagger(A_n - A)u_i\right)$$

$$\overset{(3)}{=} \lim_{m \to \infty} \sum_{i \in I} \left(u_i | U_n^\dagger(A_n - A_m)u_i\right)$$

(1 holds true by 18.2.12d, 2 by 18.2.12c, 3 by 4.2.12 and the continuity of $\tilde{U}_n^\dagger$); moreover we have, for all $m > N_\varepsilon$,

$$\left| \sum_{i \in I} \left(u_i | U_n^\dagger(A_n - A_m)u_i\right) \right| \overset{(4)}{=} |\operatorname{tr}(P U_n^\dagger(A_n - A_m))|$$

$$\overset{(5)}{\leq} \|P U_n^\dagger\| \operatorname{tr}|A_n - A_m| \overset{(6)}{\leq} \nu_1(A_n - A_m) < \varepsilon$$

(4 holds by 18.2.12c, 5 by 18.2.10a, 6 by 4.2.9). Therefore we have

$$0 \leq \operatorname{tr}(P|A_n - A|) = \left| \lim_{m \to \infty} \sum_{i \in I} \left(u_i | U_n^\dagger(A_n - A_m)u_i\right) \right|$$

$$= \lim_{m \to \infty} \left| \sum_{i \in I} \left(u_i | U_n^\dagger(A_n - A_m)u_i\right) \right| \leq \varepsilon.$$

Since P was an arbitrary element of $\mathscr{P}(\mathcal{H}) \cap \mathcal{T}(\mathcal{H})$, by 18.2.13 we have

$$\operatorname{tr}|A_n - A| \leq \varepsilon.$$

This proves in the first place that $A_n - A \in \mathcal{T}(\mathcal{H})$ and hence that $A \in \mathcal{T}(\mathcal{H})$ (by 18.2.4a,b since $A = A_n - (A_n - A)$), and in the second place that

$$\nu_1(A_n - A) = \operatorname{tr}|A_n - A| \to 0 \text{ as } n \to \infty$$

(since ε was an arbitrary element of $(0, \infty)$). $\qquad\qquad\square$

18.2.15 Theorem. *Let $\{u_n\}_{n \in I}$ and $\{v_n\}_{n \in I}$ be families of elements of $\tilde{\mathcal{H}}$ (cf. 10.9.4) and let $\{\lambda_n\}_{n \in I}$ be a family of elements of $\mathbb{C}$, with $I := \{1, ..., N\}$ or $I := \mathbb{N}$. If $I = \mathbb{N}$, suppose that $\sum_{n=1}^\infty |\lambda_n| < \infty$.*

If $I = \{1, ..., N\}$ then the operator defined by

$$A := \sum_{n=1}^N \lambda_n A_{u_n, v_n}$$

is an element of $\mathcal{T}(\mathcal{H})$.

If $I = \mathbb{N}$ then the series $\sum_{n=1}^{\infty} \lambda_n A_{u_n,v_n}$ is convergent in the normed space $(\mathcal{T}(\mathcal{H}), \nu_1)$ (and hence also in the normed space $\mathcal{B}(\mathcal{H})$ with respect to the norm for $\mathcal{B}(\mathcal{H})$ defined in 4.2.11a) and therefore the operator defined by

$$A := \sum_{n=1}^{\infty} \lambda_n A_{u_n,v_n}$$

is an element of $\mathcal{T}(\mathcal{H})$.

In both cases we have (denoting by $\sum_{n\in I}$ either $\sum_{n=1}^{N}$ or $\sum_{n=1}^{\infty}$)

$$\text{tr}(AB) = \text{tr}(BA) = \sum_{n\in I} \lambda_n (u_n|Bv_n), \forall B \in \mathcal{B}(\mathcal{H}),$$

and hence in particular (for $B := \mathbb{1}_{\mathcal{H}}$)

$$\text{tr}\, A = \sum_{n\in I} \lambda_n (u_n|v_n).$$

Proof. First we recall that, for $u, v \in \tilde{\mathcal{H}}$, we have $|A_{u,v}| = A_u$ (cf. 18.1.9), and hence $\text{tr}\, |A_{u,v}| = 1$ (cf. 18.2.12b), and hence $A_{u,v} \in \mathcal{T}(\mathcal{H})$. Moreover, if $\{w_n\}_{n\in\mathbb{N}}$ is a c.o.n.s. in $\mathcal{H}$ which contains $\{u\}$ (cf. 10.7.3), then we have

$$\text{tr}(BA_{u,v}) = \sum_{n=1}^{\infty} (w_n|BA_{u,v}w_n) = (u|Bv), \forall B \in \mathcal{B}(\mathcal{H}).$$

Since $\mathcal{T}(\mathcal{H})$ is a linear manifold in $\mathcal{B}(\mathcal{H})$ and since the function tr is a linear functional (cf. 18.2.11a), this proves the whole statement for $I = \{1, ..., N\}$.

Now we suppose $I = \mathbb{N}$. We notice that, in the normed space $(\mathcal{T}(\mathcal{H}), \nu_1)$, the series $\sum_{n=1}^{\infty} \lambda_n A_{u_n,v_n}$ is absolutely convergent since

$$\nu_1(\lambda_n A_{u_n,v_n}) = |\lambda_n|\, \text{tr}\, |A_{u_n,v_n}| = |\lambda_n|, \forall n \in \mathbb{N}.$$

Then, in view of 18.2.14 and 4.1.8b, the series $\sum_{n=1}^{\infty} \lambda_n A_{u_n,v_n}$ is convergent in the normed space $(\mathcal{T}(\mathcal{H}), \nu_1)$, and hence also in the normed space $\mathcal{B}(\mathcal{H})$ with respect to the norm for $\mathcal{B}(\mathcal{H})$ defined in 4.2.11a (cf. 18.2.5a). For all $B \in \mathcal{B}(\mathcal{H})$, we have (cf. 18.2.10b)

$$\text{tr}\left| BA - B\sum_{k=1}^{n} \lambda_k A_{u_k,v_k}\right| \le \|B\|\, \text{tr}\left| A - \sum_{k=1}^{n} \lambda_k A_{u_k,v_k}\right| \xrightarrow[n\to\infty]{} 0;$$

therefore, in view of the continuity of the linear functional tr (cf. 18.2.11a), we have

$$\text{tr}(BA) = \lim_{n\to\infty} \text{tr}\left(B\sum_{k=1}^{n} \lambda_k A_{u_k,v_k}\right)$$

$$= \lim_{n\to\infty} \sum_{k=1}^{n} \lambda_k (u_k|Bv_k) = \sum_{n=1}^{\infty} \lambda_n (u_n|Bv_n);$$

finally, the equality $\text{tr}(AB) = \text{tr}(BA)$ holds true by 18.2.11c. $\qquad\square$

18.2.16 Remark. In view of 18.2.15, the series of operators which appear in 18.2.8 and in 18.2.9 (if $I = \mathbb{N}$) are convergent not only with respect to the norm defined in 4.2.11a but also with respect to the norm ν_1.

18.2.17 Proposition. *Let M and N be subspaces of $\mathcal{H}$, let $T_1 := P_M$, and let*
$$T_{2h} := (P_N P_M)^h, \quad T_{2h+1} := P_M(P_N P_M)^h, \forall h \in \mathbb{N}.$$
Then
$$\mathrm{tr}(BP_{M\cap N} A P_{M\cap N}) = \lim_{k\to\infty} \mathrm{tr}(BT_k A T_k^\dagger), \forall A \in \mathcal{T}(\mathcal{H}), \forall B \in \mathcal{B}(\mathcal{H}),$$
and hence (for $B := \mathbb{1}_\mathcal{H}$)
$$\mathrm{tr}(P_{M\cap N} A P_{M\cap N}) = \lim_{k\to\infty} \mathrm{tr}(T_k A T_k^\dagger), \forall A \in \mathcal{T}(\mathcal{H}).$$
If A is a positive element of $\mathcal{T}(\mathcal{H})$ and $\mathrm{tr}(P_{M\cap N} A P_{M\cap N}) \neq 0$, then
$$\mathrm{tr}(T_k A T_k^\dagger) \neq 0, \forall k \in \mathbb{N}.$$

Proof. If $A = \mathbb{O}_\mathcal{H}$ then the statement is trivially true. In what follows, we assume $A \in \mathcal{T}(\mathcal{H}) - \{\mathbb{O}_\mathcal{H}\}$, we fix $B \in \mathcal{B}(\mathcal{H})$, and we set $P := P_{M\cap N}$. Let $\{u_n\}_{n\in I}, \{v_n\}_{n\in I}$, $\{\lambda_n\}_{n\in I}$ be with respect to A as in 18.2.9. In view of 18.2.11c and 18.2.15, we have
$$\mathrm{tr}(BPAP) = \mathrm{tr}(PBPA) = \sum_{n\in I} \lambda_n (u_n|PBPv_n) = \sum_{n\in I} \lambda_n (Pu_n|BPv_n),$$
and
$$\mathrm{tr}(BT_k A T_k^\dagger) = \mathrm{tr}(T_k^\dagger B T_k A) = \sum_{n\in I} \lambda_n (T_k u_n | B T_k v_n), \forall k \in \mathbb{N}.$$
Moreover, by 13.2.2 (and by the continuity of B) we have
$$(Pu_n | BPv_n) = \lim_{k\to\infty} (T_k u_n | B T_k v_n), \forall n \in I.$$
If $I = \{1, ..., N\}$, this proves that
$$\mathrm{tr}(BPAP) = \lim_{k\to\infty} \mathrm{tr}(BT_k A T_k^\dagger).$$
Now we suppose $I = \mathbb{N}$. We notice that
$$|(T_k u_n | B T_k v_n)| \leq \|T_k u_n\| \|B T_k v_n\| \leq \|B\|, \forall n \in \mathbb{N}, \forall k \in \mathbb{N}$$
(cf. 10.1.9, 4.2.5b, 4.2.9), and that
$$\sum_{n=1}^\infty |\lambda_n| \|B\| = \|B\| \sum_{n=1}^\infty |\lambda_n| < \infty.$$
Then, by 8.3.10a and 8.2.11 (with the sequence $\{|\lambda_n| \|B\|\}$ as dominating function) we have
$$\mathrm{tr}(BPAP) = \sum_{n=1}^\infty \lambda_n (Pu_n | BPv_n)$$
$$= \lim_{k\to\infty} \sum_{n=1}^\infty \lambda_n (T_k u_n | B T_k v_n) = \lim_{k\to\infty} \mathrm{tr}(BT_k A T_k^\dagger).$$
Finally, we suppose that A is positive. Then the operator $T_k A T_k^\dagger$ is positive for all $k \in \mathbb{N}$, as can be seen easily. Therefore, if $k \in \mathbb{N}$ exists so that $\mathrm{tr}(T_k A T_k^\dagger) = 0$ then $T_k A T_k^\dagger = \mathbb{O}_\mathcal{H}$ (cf. 18.2.4c), and hence $T_m A T_m^\dagger = \mathbb{O}_\mathcal{H}$ for all $m > k$ since
$$\forall m > k, \exists S_{m,k} \in \mathcal{B}(\mathcal{H}) \text{ s.t. } T_m A T_m^\dagger = S_{m,k} T_k A T_k^\dagger S_{m,k}^\dagger,$$
and hence $\lim_{k\to\infty} \mathrm{tr}(T_k A T_k^\dagger) = 0$. $\qquad\qquad\square$

18.3   Statistical operators

Statistical operators are nothing else than positive trace class operators which are normalized with respect to the norm ν_1 for $\mathcal{T}(\mathcal{H})$ (i.e., their trace is one). Thus, the results we prove in this section are essentially exercises about positive trace class operators and they are of interest especially in view of the role played by statistical operators in quantum mechanics.

     Throughout this section, $\mathcal{H}$ denotes a separable Hilbert space whose orthogonal dimension is denumerable. For a finite-dimensional Hilbert space, everything holds in an obviously simplified fashion.

18.3.1 Definition. An operator $W \in \mathcal{B}(\mathcal{H})$ is said to be a *statistical operator* if it is positive and $\operatorname{tr} W = 1$. The family of all statistical operators in $\mathcal{H}$ is denoted by the symbol $\mathcal{W}(\mathcal{H})$.

     Clearly, $\mathcal{W}(\mathcal{H}) \subset \mathcal{T}(\mathcal{H})$.

     Another name for a statistical operator is *density matrix*.

18.3.2 Remarks.

(a) If $W \in \mathcal{W}(\mathcal{H})$ and $U \in \mathcal{UA}(\mathcal{H})$, then $UWU^{-1} \in \mathcal{W}(\mathcal{H})$. This follows from 18.2.2d.

(b) For each $u \in \tilde{\mathcal{H}}$, the one-dimensional projection A_u is a statistical operator. In fact A_u is positive (so are all orthogonal projections, in view of 13.1.7c) and $\operatorname{tr} A_u = 1$ (cf. 18.2.12b). From 18.2.12c we have

$$\operatorname{tr}(BA_u) = (u|Bu), \forall B \in \mathcal{B}(\mathcal{H}).$$

In view of 18.2.12a,b, the one-dimensional projections are the only orthogonal projections which are statistical operators.

(c) If $W \in \mathcal{W}(\mathcal{H})$ then, in view of 18.2.8, there exist an o.n.s. $\{u_n\}_{n \in I}$ (with $I := \{1, ..., N\}$ or $I := \mathbb{N}$) and a family $\{\lambda_n\}_{n \in I}$ of elements of $(0, \infty)$ so that

$$W = \sum_{n \in I} \lambda_n A_{u_n} \quad \text{and} \quad \sum_{n \in I} \lambda_n = \operatorname{tr} W = 1; \tag{1}$$

thus $\lambda_n \in (0, 1]$ for all $n \in I$. If $I = \mathbb{N}$ then the first of these series is convergent with respect to the norm for $\mathcal{B}(\mathcal{H})$ defined in 4.2.11a and also with respect to the norm ν_1 for $\mathcal{T}(\mathcal{H})$ (cf. 18.2.16), and we have

$$Wf = \sum_{n=1}^{\infty} \lambda_n (u_n|f) u_n, \forall f \in \mathcal{H},$$

by 4.2.12. Moreover, in view of 18.2.15, we have

$$\operatorname{tr}(BW) = \sum_{n \in I} \lambda_n (u_n|Bu_n), \forall B \in \mathcal{B}(\mathcal{H}).$$

In view of 15.3.4B, $\{\lambda_n\}_{n \in I}$ is the family of all non-zero eigenvalues of W (recall that $\{\lambda_n\}_{n \in I}$ stands for the range of the mapping $I \ni n \mapsto \lambda_n \in (0, 1]$, cf. 1.2.1); therefore, this family is uniquely determined (if $\{u_n\}_{n\in I}$ is required, as above, to be an o.n.s.). The family $\{A_{u_n}\}_{n\in I}$ is uniquely determined iff the eigenspaces of all non-zero eigenvalues of W are one-dimensional (if this is true then A_{u_n} is the orthogonal projection on the eigenspace corresponding to λ_n). However, even in this case, a decomposition of W as in 1 is not unique if the family $\{u_n\}_{n\in I}$ is not required to be an o.n.s. but only to consist of elements of $\tilde{\mathcal{H}}$, unless W is a one-dimensional projection. This will be proved in 18.3.7.

18.3.3 Proposition. *Let $W \in \mathcal{W}(\mathcal{H})$ be such that* $\operatorname{tr} W^2 = 1$. *Then W is a one-dimensional projection.*

Proof. Let $\{u_n\}_{n\in I}$ be an o.n.s. in $\mathcal{H}$ (with $I := \{1, ..., N\}$ or $I := \mathbb{N}$) and $\{\lambda_n\}_{n\in I}$ a family of elements of $(0, 1]$ so that

$$W = \sum_{n\in I} \lambda_n A_{u_n} \text{ and } \sum_{n\in I} \lambda_n = 1,$$

as in 18.3.2c. We have

$$W^2 = \sum_{n\in I} \lambda_n^2 A_{u_n}$$

since $A_{u_n} A_{u_m} = \delta_{n,m} A_{u_n}$ for all $n, m \in I$ (if $I = \mathbb{N}$, we have used also the continuity of the operator product in $\mathcal{B}(\mathcal{H})$, cf. 4.3.5 and 4.3.3). We notice that $\lambda_n^2 \leq \lambda_n$ and hence $\sum_{n\in I} \lambda_n^2 < \infty$. Then, in view of 18.2.15, $W^2 \in \mathcal{T}(\mathcal{H})$ and

$$1 = \operatorname{tr} W^2 = \sum_{n\in I} \lambda_n^2 (u_n|u_n) = \sum_{n\in I} \lambda_n^2.$$

Therefore,

$$\sum_{n\in I} (\lambda_n - \lambda_n^2) = 0$$

and hence

$$\lambda_n \in \{0, 1\}, \forall n \in I.$$

This implies $I = \{1\}$ and hence $W = A_{u_1}$. $\qquad\square$

18.3.4 Proposition. *Let $\{W_n\}_{n\in I}$ be a family (with $I := \{1, ..., N\}$ or $I := \mathbb{N}$) of elements of $\mathcal{W}(\mathcal{H})$ and let $\{w_n\}_{n\in I}$ be a family of elements of $(0, 1]$ so that $\sum_{n\in I} w_n = 1$ (in the whole section, $\sum_{n\in I}$ stands for either $\sum_{n=1}^{N}$ or $\sum_{n=1}^{\infty}$; the reader must be warned that, while in previous parts of this chapter the symbol w_n represented a vector, in the present section it represents an element of $(0, 1]$; the reason for the use of this symbol is that the elements of a family $\{w_n\}_{n\in I}$ as in the present proposition are called "weights" in quantum mechanics and w is the first letter of the word "weight", cf. 19.3.5b). Then:*

(a) if $I = \mathbb{N}$, the series $\sum_{n=1}^{\infty} w_n W_n$ is convergent in the normed space $(\mathcal{T}(\mathcal{H}), \nu_1)$ and also with respect to the norm for $\mathcal{B}(\mathcal{H})$ defined in 4.2.11a;

(b) the operator

$$W := \sum_{n \in I} w_n W_n$$

is an element of $\mathcal{W}(\mathcal{H})$;
(c) for all $B \in \mathcal{B}(\mathcal{H})$,

$$\text{tr}(BW) = \sum_{n \in I} w_n \, \text{tr}(BW_n).$$

Proof. a: If $I = \mathbb{N}$ then the series $\sum_{n=1}^{\infty} w_n W_n$ is absolutely convergent in the normed space $(\mathcal{T}(\mathcal{H}), \nu_1)$ since

$$\nu_1(w_n W_n) = w_n \, \text{tr} \, W_n = w_n, \forall n \in \mathbb{N},$$

and hence it is convergent in this normed space (cf. 18.2.14 and 4.1.8b). Then, this series is convergent also with respect to the norm for $\mathcal{B}(\mathcal{H})$ defined in 4.2.11a (cf. 18.2.5a).

b: From 18.2.4a,b (if $I = \{1, ..., N\}$) or from result a (if $I = \mathbb{N}$) we have $W \in \mathcal{T}(\mathcal{H})$. Moreover,

$$(f|Wf) = \sum_{n \in I} w_n \, (f|W_n f) \geq 0, \forall f \in \mathcal{H}$$

(if $I = \mathbb{N}$, we have used 4.2.12), shows that W is positive. Finally, we have

$$\text{tr} \, W = \sum_{n \in I} w_n \, \text{tr} \, W_n = \sum_{n \in I} w_n = 1$$

by 18.2.11a.

c: If $I = \{1, ..., N\}$, this follows from the linearity of the function tr (cf. 18.2.11a). Now we suppose that $I = \mathbb{N}$ and fix $B \in \mathcal{B}(\mathcal{H})$. Then the series $\sum_{n=1}^{\infty} w_n B W_n$ is absolutely convergent in the normed space $(\mathcal{T}(\mathcal{H}), \nu_1)$ since

$$\nu_1(w_n B W_n) = w_n \, \text{tr} \, |BW_n| \leq w_n \|B\| \, \text{tr} \, W_n = w_n \|B\|, \forall n \in \mathbb{N},$$

(cf. 18.2.10b), and hence it is convergent in this normed space, and hence it is convergent with respect to the norm defined in 4.2.11a. Then for its sum we have

$$\sum_{n=1}^{\infty} w_n B W_n = B \left(\sum_{n=1}^{\infty} w_n W_n \right) = BW,$$

in view of the continuity of the operator product with respect to the norm defined in 4.2.11a. Hence we have

$$\text{tr}(BW) = \text{tr} \left(\sum_{n=1}^{\infty} w_n B W_n \right) = \sum_{n=1}^{\infty} w_n \, \text{tr}(BW_n),$$

in view of the continuity of the function tr with respect to the norm ν_1 (cf. 18.2.11a).

$\square$

18.3.5 Corollary. *Let $I := \{1, ..., N\}$ or $I := \mathbb{N}$, let $\{u_n\}_{n \in I}$ be a family of elements of $\tilde{\mathcal{H}}$, let $\{w_n\}_{n \in I}$ be a family of elements of $(0, 1]$ such that $\sum_{n \in I} w_n = 1$. Then:*

(a) *if $I = \mathbb{N}$, the series $\sum_{n=1}^{\infty} w_n A_{u_n}$ is convergent in the normed space $(\mathcal{T}(\mathcal{H}), \nu_1)$ and also with respect to the norm for $\mathcal{B}(\mathcal{H})$ defined in 4.2.11a;*
(b) *the operator*

$$W := \sum_{n \in I} w_n A_{u_n}$$

is an element of $\mathcal{W}(\mathcal{H})$;
(c) *for all $B \in \mathcal{B}(\mathcal{H})$,*

$$\operatorname{tr}(BW) = \sum_{n \in I} w_n \operatorname{tr}(BA_{u_n}) = \sum_{n \in I} w_n (u_n|Bu_n).$$

Proof. Everything follows from 18.3.2b and 18.3.4. □

18.3.6 Corollary. *Let $W \in \mathcal{O}_E(\mathcal{H})$ (i.e., W is a linear operator in $\mathcal{H}$ and $D_W = \mathcal{H}$). Then the following conditions are equivalent:*

(a) *$W \in \mathcal{W}(\mathcal{H})$;*
(b) *there exist a family $\{u_n\}_{n \in I}$ (with $I := \{1, ..., N\}$ or $I := \mathbb{N}$) of elements of $\tilde{\mathcal{H}}$ and a family $\{w_n\}_{n \in I}$ of elements of $(0, 1]$ so that*

$$A_{u_i} \neq A_{u_k} \text{ if } i \neq k,$$
$$\sum_{n \in I} w_n = 1,$$
$$Wf = \sum_{n \in I} w_n A_{u_n} f, \forall f \in \mathcal{H}.$$

Proof. $a \Rightarrow b$: Cf. 18.3.2c.
$b \Rightarrow a$: This follows from 18.3.5 and 4.2.12. □

18.3.7 Proposition. *Let $W \in \mathcal{W}(\mathcal{H})$. The following conditions are equivalent:*

(a) *the representation of W as in 18.3.6b is unique (i.e. the families $\{A_{u_n}\}_{n \in I}$ and $\{w_n\}_{n \in I}$ as in 18.3.6b are uniquely determined);*
(b) *W is a one-dimensional projection.*

Thus, if $W = A_u$ with $u \in \tilde{\mathcal{H}}$ then A_u is the only representation of W as in 18.3.6b.

Proof. $a \Rightarrow b$: We prove (not b) $\Rightarrow$ (not a). We consider a decomposition of W as in 18.3.2c, i.e.

$$W = \sum_{n \in I} w_n A_{u_n}, \tag{1}$$

with $\{u_n\}_{n\in I}$ an o.n.s. in $\mathcal{H}$ and $\{w_n\}_{n\in I}$ a family of elements of $(0,1]$ such that $\sum_{n\in I} w_n = 1$. We suppose that W is not a one-dimensional projection. Then the index set I must contain more than one element, and we define the vectors

$$v_1 := 2^{-\frac{1}{2}}(u_1 + u_2) \text{ and } v_2 := 2^{-\frac{1}{2}}(u_1 - u_2),$$

which are elements of $\tilde{\mathcal{H}}$. It is easy to see that

$$A_{u_1} + A_{u_2} = A_{v_1} + A_{v_2}.$$

We set $J := I - \{1,2\}$. If $w_1 = w_2$, we have

$$W = w_1 A_{v_1} + w_2 A_{v_2} + \sum_{n\in J} w_n A_{u_n} \tag{2}$$

(if $I = \{1,2\}$ then $\sum_{n\in J} w_n A_{u_n} := \mathbb{O}_{\mathcal{H}}$). If $w_1 \neq w_2$ and (for instance) $w_1 < w_2$, we have

$$W = w_1 A_{v_1} + w_1 A_{v_2} + (w_2 - w_1) A_{u_2} + \sum_{n\in J} w_n A_{u_n}. \tag{3}$$

Now, the decompositions of W in 2 and in 3 are different than the decomposition in 1, and both comply with the conditions set down in 18.3.6b.

$b \Rightarrow a$: We suppose that there exists $u \in \tilde{\mathcal{H}}$ so that $W = A_u$. Let $\{u_n\}_{n\in I}$ be a family (with $I := \{1,...,N\}$ and $N > 1$, or $I := \mathbb{N}$) of elements of $\tilde{\mathcal{H}}$ and $\{w_n\}_{n\in I}$ a family of elements of $(0,1]$ so that

$$\sum_{n\in I} w_n = 1,$$

$$W f = \sum_{n\in I} w_n A_{u_n} f, \forall f \in \mathcal{H}.$$

We fix $k \in I$ and note that

$$w_k < 1 \tag{4}$$

because either $I = \{1,...,N\}$ with $N > 1$ or $I = \mathbb{N}$. We define the set of indices

$$J := I - \{k\}$$

and the operator

$$\tilde{W} := \sum_{n\in J} (1 - w_k)^{-1} w_n A_{u_n}.$$

We have $\tilde{W} \in \mathcal{W}(\mathcal{H})$ by 18.3.5, and also

$$A_u = w_k A_{u_k} + (1 - w_k)\tilde{W}$$

(if $I = \mathbb{N}$, we have used the continuity of scalar multiplication and of vector sum in $\mathcal{B}(\mathcal{H})$), and hence

$$A_u = A_u^2 = w_k^2 A_{u_k} + (1 - w_k)w_k \tilde{W} A_{u_k} + w_k(1 - w_k) A_{u_k} \tilde{W} + (1 - w_k)^2 \tilde{W}^2,$$

and hence (cf. 18.2.11a,c and 18.3.2b)

$$1 = \operatorname{tr} A_u = w_k^2 + 2w_k(1 - w_k)\operatorname{tr}(\tilde{W} A_{u_k}) + (1 - w_k)^2 \operatorname{tr} \tilde{W}^2. \tag{5}$$

Moreover, we have

$$\operatorname{tr}(\tilde{W} A_{u_k}) \leq \operatorname{tr} \tilde{W} = 1 \tag{6}$$

(cf. 18.2.11c and 18.2.12d), and also

$$\operatorname{tr} \tilde{W}^2 \leq \|\tilde{W}\| \operatorname{tr} \tilde{W} \leq \operatorname{tr} \tilde{W} = 1,$$

in view of 18.2.10a and of the inequality $\|\tilde{W}\| \leq \operatorname{tr} \tilde{W} = 1$ (cf. 18.2.4d). Now, $\operatorname{tr} \tilde{W}^2 < 1$ would imply (in view of 4, 5, 6)

$$1 < w_k^2 + 2w_k(1 - w_k) + (1 - w_k)^2 = 1,$$

which is a contradiction. Therefore, $\operatorname{tr} \tilde{W}^2 = 1$ and hence (cf. 18.3.3) there exists $v \in \tilde{\mathcal{H}}$ so that $\tilde{W} = A_v$. Then (cf. 18.3.2b)

$$\operatorname{tr}(\tilde{W} A_{u_k}) = \operatorname{tr}(A_v A_{u_k}) = (u_k | A_v u_k) = |(u_k | v)|^2,$$

and 5 reads

$$1 = w_k^2 + 2w_k(1 - w_k)|(u_k | v)|^2 + (1 - w_k)^2.$$

Hence, $|(u_k | v)| < 1$ would imply

$$1 < w_k^2 + 2w_k(1 - w_k) + (1 - w_k)^2 = 1.$$

Therefore,

$$|(u_k | v)| = 1$$

and hence (cf. 10.1.7b) there exists $z \in \mathbb{T}$ such that $v = z u_k$, and hence (cf. 13.1.13a)

$$\tilde{W} = A_{u_k},$$

and hence

$$A_u = w_k A_{u_k} + (1 - w_k) A_{u_k} = A_{u_k}.$$

Since k was an arbitrary element of I, this proves that

$$A_{u_n} = A_u, \forall n \in I.$$

Thus, the condition

$$A_{u_i} \neq A_{u_k} \text{ if } i \neq k$$

(cf. 18.3.6b) cannot be true, and the only representation of W as in 18.3.6b is the one given by

$$I := \{1\}, \ w_1 := 1 \ A_{u_1} := A_u,$$

i.e. a tautology. $\qquad\qquad\qquad\qquad\qquad\qquad\qquad\qquad\qquad\qquad\qquad\qquad\qquad\quad \square$

18.3.8 Proposition. *Let $W \in \mathcal{W}(\mathcal{H})$ and $P \in \mathscr{P}(\mathcal{H})$. Then:*

(a) $\mathrm{tr}(PW) = \sum_{n \in I} (u_n | W u_n)$
   for each o.n.s. $\{u_n\}_{n \in I}$ in $\mathcal{H}$ which is complete in the subspace R_P;
(b) $0 \le \mathrm{tr}(PW) \le 1$.

Proof. Let $\{u_n\}_{n \in I}$ be an o.n.s. in $\mathcal{H}$ which is complete in the subspace R_P and let $\{v_n\}_{n \in \mathbb{N}}$ be a c.o.n.s in $\mathcal{H}$ which contains $\{u_n\}_{n \in I}$ (cf. 10.7.3).
   a: We have

$$\mathrm{tr}(PW) = \mathrm{tr}(WP) = \sum_{n=1}^{\infty} (v_n | W P v_n) = \sum_{n \in I} (u_n | W u_n)$$

(cf. 18.2.11c, 13.1.3b,c, 10.2.11).
   b: We have

$$0 \le \sum_{n \in I} (u_n | W u_n) \le \sum_{n=1}^{\infty} (v_n | W v_n) = \mathrm{tr}\, W = 1.$$

In view of statement a, this proves statement b. $\qquad\square$

18.3.9 Proposition. *Let $W \in \mathcal{W}(\mathcal{H})$ and $P \in \mathscr{P}(\mathcal{H})$. Then the following conditions are equivalent:*

(a) $\mathrm{tr}(PW) = 1$;
(b) $R_W \subset R_P$;
(c) $PW = W$;
(d) $PWP = W$;

Proof. Let $\{u_n\}_{n \in I}$ and $\{\lambda_n\}_{n \in I}$ be as in 18.3.2c, so that

$$W = \sum_{n \in I} \lambda_n A_{u_n}.$$

   $a \Rightarrow b$: Condition a implies

$$\sum_{n \in I} \lambda_n (u_n | P u_n) = \mathrm{tr}(PW) = 1,$$

and hence (since $\lambda_n > 0$ for each $n \in I$ and $\sum_{n \in I} \lambda_n = 1$)

$$\|P u_n\|^2 = (u_n | P u_n) = 1, \forall n \in I,$$

and hence (cf. 13.1.3c)

$$u_n \in R_P, \forall n \in I.$$

Since $R_W \subset V\{u_n\}_{n \in I}$, this proves that $R_W \subset R_P$.
   $b \Rightarrow c$: We assume condition b. We have

$$u_n = \lambda_n^{-1} W u_n, \forall n \in I,$$

and hence

$$u_n \in R_P, \forall n \in I,$$

and hence

$$PA_{u_n} = A_{u_n}, \forall n \in I$$

(cf. 13.1.3c). This implies $PW = W$ (if $I = \mathbb{N}$, use e.g. the continuity of the operator product in $\mathcal{B}(\mathcal{H})$).

$c \Rightarrow d$: We have

$$PW = W \Rightarrow WP = (PW)^\dagger = W$$

(cf. 12.3.4b). Therefore,

$$PW = W \Rightarrow PWP = WP = W.$$

$d \Rightarrow a$: Condition d implies

$$\operatorname{tr}(PW) = \operatorname{tr}(P^2 W) = \operatorname{tr}(PWP) = \operatorname{tr}(W) = 1$$

(cf. 13.1.5 and 18.2.11c). $\qquad\square$

18.3.10 Corollary. *Let $W \in \mathcal{W}(\mathcal{H})$ and $u \in \tilde{\mathcal{H}}$. Then the following conditions are equivalent:*

(a) $\operatorname{tr}(A_u W) = 1$;
(b) $W = A_u$.

Proof. $a \Rightarrow b$: We assume condition a. Then we have $R_W \subset V\{u\}$ by 18.3.9 $(a \Rightarrow b)$, since $R_{A_u} = V\{u\}$ (cf. 13.1.12). Now, if $\{u_n\}_{n\in I}$ and $\{\lambda_n\}_{n\in I}$ are as in 18.3.2c, this implies

$$u_n \in V\{u\}, \forall n \in I,$$

and hence

$$I = \{1\}, \lambda_1 = 1, A_{u_1} = A_u,$$

and hence $W = A_u$.
$b \Rightarrow a$: If $W = A_u$ then

$$\operatorname{tr}(A_u W) = \operatorname{tr} A_u = 1.$$

$\qquad\square$

18.3.11 Corollary. *Let $W \in \mathcal{W}(\mathcal{H})$ and $P \in \mathscr{P}(\mathcal{H})$. Then the following conditions are equivalent:*

(a) $\operatorname{tr}(PW) = 0$;
(b) $R_W \subset N_P$;
(c) $PW = \mathbb{O}_\mathcal{H}$;
(d) $PWP = \mathbb{O}_\mathcal{H}$.

Proof. We notice that:

$$\mathrm{tr}(PW) = 0 \Leftrightarrow \mathrm{tr}((\mathbb{1}_\mathcal{H} - P)W) = 1$$

(cf. 18.2.11a);

$$N_P = R_{\mathbb{1}_\mathcal{H}-P}$$

(cf. 13.1.3b,e);

$$PW = \mathbb{O}_\mathcal{H} \Leftrightarrow (\mathbb{1}_\mathcal{H} - P)W = W.$$

Thus, conditions a, b, c are in fact conditions a, b, c of 18.3.9 written with the projection $\mathbb{1}_\mathcal{H} - P$ in place of P, and therefore they are equivalent.

It is obvious that condition c implies condition d. Condition d implies condition a because $\mathrm{tr}(PW) = \mathrm{tr}(P^2W) = \mathrm{tr}(PWP)$. $\qquad\square$

18.3.12 Proposition. *Let $W \in \mathcal{W}(\mathcal{H})$ and let $\{P_n\}$ be a sequence in $\mathscr{P}(\mathcal{H})$ such that $P_iP_k = \mathbb{O}_\mathcal{H}$ if $i \neq k$. Then*

$$\mathrm{tr}\left(\left(\sum_{n=1}^{\infty} P_n\right) W\right) = \sum_{n=1}^{\infty} \mathrm{tr}(P_nW)$$

(the orthogonal projection $\sum_{n=1}^{\infty} P_n$ is defined as in 13.2.10b).

Proof. We define the set $I := \{n \in \mathbb{N} : P_n \neq \mathbb{O}_\mathcal{H}\}$. For each $n \in I$, let $\{u_{n,s}\}_{(n,s)\in I_n}$ be an o.n.s. in $\mathcal{H}$ which is complete in the subspace R_{P_n} (cf. 10.7.2). We define the set $J := \bigcup_{n\in I} I_n$. Then, $\{u_{n,s}\}_{(n,s)\in J}$ in an o.n.s. in $\mathcal{H}$ (cf. 13.2.8d or 13.2.9c). Let $\{v_k\}_{k\in\mathbb{N}}$ be a c.o.n.s. in $\mathcal{H}$ which contains $\{u_{n,s}\}_{(n,s)\in J}$ (cf. 10.7.3). For each $k \in \mathbb{N}$, we have

$$\left(\sum_{n=1}^{\infty} P_n\right) v_k = \sum_{n=1}^{\infty} P_n v_k = \begin{cases} 0_\mathcal{H} & \text{if } v_k \notin \{u_{n,s}\}_{(n,s)\in J} \\ u_{m,s} & \text{if } (m,s) \in J \text{ is s.t. } v_k = u_{m,s} \end{cases}$$

(cf. 13.1.3b,c and 10.2.11).

Since $W\left(\sum_{n=1}^{\infty} P_n\right) \in \mathcal{T}(\mathcal{H})$, the series $\sum_{k=1}^{\infty} \left(v_k|W\left(\sum_{n=1}^{\infty} P_n\right) v_k\right)$ is absolutely convergent (cf. 18.2.10), and hence

$$\sum_{(n,s)\in J} |(u_{n,s}|Wu_{n,s})| = \sum_{k=1}^{\infty} \left|\left(v_k|W\left(\sum_{n=1}^{\infty} P_n\right) v_k\right)\right| < \infty. \qquad (1)$$

Then,

$$\mathrm{tr}\left(\left(\sum_{n=1}^{\infty} P_n\right) W\right) \overset{(2)}{=} \mathrm{tr}\left(W\left(\sum_{n=1}^{\infty} P_n\right)\right) = \sum_{k=1}^{\infty} \left(v_k|W\left(\sum_{n=1}^{\infty} P_n\right) v_k\right)$$

$$= \sum_{(n,s)\in J} (u_{n,s}|Wu_{n,s}) \overset{(3)}{=} \sum_{n\in I} \left(\sum_{(n,s)\in I_n} (u_{n,s}|Wu_{n,s})\right)$$

$$\overset{(4)}{=} \sum_{n\in I} \mathrm{tr}(P_nW) = \sum_{n=1}^{\infty} \mathrm{tr}(P_nW),$$

where: 2 holds true by 18.2.11c; 3 by 8.4.15b since 1 proves that the conditions in 8.4.15a are satisfied; 4 by 18.3.8a. $\qquad\square$

18.3.13 Proposition. *Let $(X, \mathcal{A})$ be a measurable space, let P be a projection valued measure on $\mathcal{A}$ with values in $\mathscr{P}(\mathcal{H})$, and let $W \in \mathcal{W}(\mathcal{H})$. Then the function*

$$\mu_W^P : \mathcal{A} \to [0, 1]$$
$$E \mapsto \mu_W^P(E) := \mathrm{tr}(P(E)W)$$

is a probability measure on $\mathcal{A}$.

Proof. The range of the function μ_W^P is indeed a subset of $[0, 1]$, in view of 18.3.8b. The rest of the statement follows immediately from the definition of a projection valued measure and from 18.3.12. $\qquad\square$

18.3.14 Definitions. Let A be a self-adjoint operator in $\mathcal{H}$ and let $W \in \mathcal{W}(\mathcal{H})$.
We say that A is *computable* in W if

$$\int_{\mathbb{R}} \xi^2 d\mu_W^{P^A} < \infty,$$

where ξ is the function defined in 15.2.1A.
If A is computable in W then the function ξ is $\mu_W^{P^A}$-integrable since the measure $\mu_W^{P^A}$ is finite (cf. 11.1.3) and we can define the real number

$$\langle A \rangle_W := \int_{\mathbb{R}} \xi d\mu_W^{P^A},$$

and the function $(\xi - \langle A \rangle_W)^2$ is also $\mu_W^{P^A}$-integrable (cf. 11.1.2a) and we can define the real number

$$\Delta_W A := \left(\int_{\mathbb{R}} (\xi - \langle A \rangle_W)^2 d\mu_W^{P^A} \right)^{\frac{1}{2}}.$$

18.3.15 Proposition. *Let A be a self-adjoint operator in $\mathcal{H}$ and let $u \in \tilde{\mathcal{H}}$. The following conditions are equivalent:*

(a) $u \in D_A$;
(b) A is computable in A_u;
(c) $AA_u \in \mathcal{T}(\mathcal{H})$.

If the above conditions are satisfied, then:

(d) $\langle A \rangle_{A_u} = \mathrm{tr}(AA_u) = (u|Au) = \langle A \rangle_u$;
(e) $\Delta_{A_u} A = \|Au - \langle A \rangle_u u\| = \Delta_u A$

(for $\langle A \rangle_u$ and $\Delta_u A$, cf. 15.2.3).

Proof. Preliminary remark: In view of 18.3.2b, we have

$$\mu_{A_u}^{P^A}(E) = \mathrm{tr}(P^A(E)A_u) = (u|P^A(E)u) = \mu_u^{P^A}(E), \forall E \in \mathcal{A}(d_{\mathbb{R}}). \tag{1}$$

$a \Leftrightarrow b$: In view of 15.2.2e and 1, we have

$$u \in D_A \Leftrightarrow \int_{\mathbb{R}} \xi^2 d\mu_u^{P^A} < \infty \Leftrightarrow \int_{\mathbb{R}} \xi^2 d\mu_{A_u}^{P^A} < \infty.$$

$c \Rightarrow a$: We assume $AA_u \in \mathcal{T}(\mathcal{H})$. Then $D_{AA_u} = \mathcal{H}$ and hence $u = A_u u \in D_A$.
$a \Rightarrow [c, d, e]$: We assume $u \in D_A$. Then $D_{AA_u} = \mathcal{H}$ and

$$AA_u f = (u|f) \, Au, \forall f \in \mathcal{H}.$$

If $Au = 0_{\mathcal{H}}$ then $AA_u = \mathbb{O}_{\mathcal{H}}$ and hence $AA_u \in \mathcal{T}(\mathcal{H})$, and also

$$\mathrm{tr}(AA_u) = 0 = (u|Au) = \langle A \rangle_u.$$

If $Au \neq 0_{\mathcal{H}}$ then, letting $v := \|Au\|^{-1} Au$, we have

$$AA_u = \|Au\| A_{u,v},$$

and hence (cf. 18.2.15) $AA_u \in \mathcal{T}(\mathcal{H})$ and also

$$\mathrm{tr}(AA_u) = \|Au\| \, (u|v) = (u|Au) = \langle A \rangle_u.$$

Finally, for both $Au = 0_{\mathcal{H}}$ and $Au \neq 0_{\mathcal{H}}$, A is computable in A_u in view of the implication $a \Rightarrow b$ proved above, and the equalities

$$\langle A \rangle_{A_u} = \langle A \rangle_u \text{ and } \Delta_{A_u} = \Delta_u A$$

follow from 15.2.2e and 1. □

18.3.16 Proposition. *Let A be a self-adjoint operator in $\mathcal{H}$ and let $W \in \mathcal{W}(\mathcal{H})$. Let a family $\{u_n\}_{n \in I}$ of elements of $\tilde{\mathcal{H}}$ and a family $\{w_n\}_{n \in I}$ of elements of $(0, 1]$ (with $I := \{1, ..., N\}$ or $I := \mathbb{N}$) be so that*

$$W = \sum_{n \in I} w_n A_{u_n} \text{ and } \sum_{n \in I} w_n = 1$$

(if $I = \mathbb{N}$, cf. 18.3.5a for the convergence of the series $\sum_{n=1}^{\infty} w_n A_{u_n}$; families $\{u_n\}_{n \in I}$ and $\{w_n\}_{n \in I}$ as above exist in view of 18.3.2c).
  The following conditions are equivalent:

(a) A is computable in W;
(b) $u_n \in D_A$ for all $n \in I$ and $\sum_{n \in I} w_n \|Au_n\|^2 < \infty$.

If conditions a and b are satisfied, then:

  $AW \in \mathcal{T}(\mathcal{H})$;
  $\langle A \rangle_W = \mathrm{tr}(AW) = \sum_{n \in I} w_n \langle A \rangle_{u_n}$;
  $\Delta_W A = \left(\sum_{n \in I} w_n \|Au_n - \langle A \rangle_W u_n\|^2 \right)^{\frac{1}{2}}$.

Proof. Preliminary remark: In view of 18.3.5c, we have

$$\mu_W^{P^A}(E) = \sum_{n \in I} w_n \left(u_n | P^A(E) u_n \right) = \sum_{n \in I} w_n \mu_{u_n}^{P^A}(E), \forall E \in \mathcal{A}(d_{\mathbb{R}}). \tag{1}$$

$a \Leftrightarrow b$: In view of 8.3.5 and 1, we have

$$\sum_{n \in I} w_n \int_{\mathbb{R}} \xi^2 d\mu_{u_n}^{P^A} = \int_{\mathbb{R}} \xi^2 d\mu_W^{P^A}. \tag{2}$$

If condition a holds true then we have (in view of 2, and since $w_n > 0$ for all $n \in I$)

$$\int_{\mathbb{R}} \xi^2 d\mu_{u_n}^{P^A} < \infty, \forall n \in I,$$

and hence (cf. 15.2.2e)

$$u_n \in D_A, \forall n \in I,$$

and also

$$\|Au_n\|^2 = \int_{\mathbb{R}} \xi^2 d\mu_{u_n}^{P^A}, \forall n \in I,$$

and hence (cf. 2) also

$$\sum_{n \in I} w_n \|Au_n\|^2 < \infty.$$

Thus, condition b holds true.

If condition b holds true then we have (in view of 15.2.2e and 2)

$$\int_{\mathbb{R}} \xi^2 d\mu_W^{P^A} = \sum_{n \in I} w_n \|Au_n\|^2 < \infty,$$

and this proves that condition a holds true.

In what follows we assume that conditions a and b are satisfied.

If $I = \{1, ..., N\}$ then $D_{AW} = \mathcal{H}$ since

$$R_W \subset L\{u_1, ..., u_N\} \subset D_A,$$

and also

$$AWf = \sum_{n=1}^{N} w_n (u_n|f) Au_n, \forall f \in \mathcal{H}. \tag{3}$$

Now we suppose $I = \mathbb{N}$. We fix $f \in \mathcal{H}$. Then,

$$\sum_{n=1}^{N} w_n (u_n|f) u_n \in D_A \text{ and}$$

$$A \left(\sum_{n=1}^{N} w_n (u_n|f) u_n \right) = \sum_{n=1}^{N} w_n (u_n|f) Au_n, \forall N \in \mathbb{N}.$$

Moreover, the inequality

$$\sum_{n=1}^{\infty} \|w_n (u_n|f) u_n\| \overset{(4)}{\leq} \sum_{n=1}^{\infty} w_n \|f\| = \|f\| \sum_{n=1}^{\infty} w_n < \infty$$

(4 holds true by the Schwarz inequality) proves that the series $\sum_{n=1}^{\infty} w_n (u_n|f) u_n$ is convergent (cf. 4.1.8b). Similarly, the inequalities

$$\sum_{n=1}^{\infty} \|w_n (u_n|f) Au_n\| \overset{(5)}{\leq} \sum_{n=1}^{\infty} w_n \|f\| \|Au_n\|$$

$$\overset{(6)}{\leq} \|f\| \left(\sum_{n=1}^{\infty} w_n \right)^{\frac{1}{2}} \left(\sum_{n=1}^{\infty} w_n \|Au_n\|^2 \right)^{\frac{1}{2}} < \infty$$

(5 holds true by the Schwarz inequality in $\mathcal{H}$; 6 holds true by the Schwarz inequality in ℓ^2 for the two sequences $\{w_n^{\frac{1}{2}}\}$ and $\{w_n^{\frac{1}{2}}\|Au_n\|\}$, cf. 10.3.8d) prove that the series $\sum_{n=1}^{\infty} w_n\,(u_n|f)\,Au_n$ is convergent. Since the operator A is closed (cf. 12.4.6a), this implies that

$$\sum_{n=1}^{\infty} w_n\,(u_n|f)\,u_n \in D_A \text{ and } A\left(\sum_{n=1}^{\infty} w_n\,(u_n|f)\,u_n\right) = \sum_{n=1}^{\infty} w_n\,(u_n|f)\,Au_n.$$

Since f was an arbitrary element of $\mathcal{H}$, this proves that

$$Wf \in D_A \text{ for all } f \in \mathcal{H}, \text{ or } D_{AW} = \mathcal{H},$$

and

$$AWf = \sum_{n=1}^{\infty} w_n\,(u_n|f)\,Au_n, \forall f \in \mathcal{H}. \tag{7}$$

In what follows, I can be either $\{1,...,N\}$ or $\mathbb{N}$. We define the set of indices

$$J := \{n \in I : Au_n \neq 0_{\mathcal{H}}\}.$$

If $J = \emptyset$ then from either 3 or 7 we have $AW = \mathbb{O}_{\mathcal{H}}$ and hence $AW \in \mathcal{T}(\mathcal{H})$, and also

$$\mathrm{tr}(AW) = 0 = \sum_{n \in I} w_n\,(u_n|Au_n) = \sum_{n \in I} w_n\langle A\rangle_{u_n}.$$

If $J \neq \emptyset$, we define

$$v_n := \|Au_n\|^{-1} Au_n, \forall n \in J;$$

then from either 3 or 7 we have

$$AWf = \sum_{n \in J} w_n\|Au_n\|A_{u_n,v_n}f, \forall f \in \mathcal{H},$$

and hence (cf. 18.2.15; note that, if $I = \mathbb{N}$, 6 proves that $\sum_{n=1}^{\infty} w_n\|Au_n\| < \infty$) $AW \in \mathcal{T}(\mathcal{H})$ and

$$\mathrm{tr}(AW) = \sum_{n \in J} w_n\|Au_n\|\,(u_n|v_n) = \sum_{n \in I} w_n\,(u_n|Au_n) = \sum_{n \in I} w_n\langle A\rangle_{u_n}.$$

Finally, for both $J = \emptyset$ and $J \neq \emptyset$, we have

$$\langle A\rangle_W = \int_{\mathbb{R}} \xi d\mu_W^{P^A} \overset{(8)}{=} \sum_{n \in I} w_n \int_{\mathbb{R}} \xi d\mu_{u_n}^{P^A} \overset{(9)}{=} \sum_{n \in I} w_n\langle A\rangle_{u_n}$$

and

$$\Delta_W A = \left(\int_{\mathbb{R}} (\xi - \langle A\rangle_W)^2 d\mu_W^{P^A}\right)^{\frac{1}{2}} \overset{(10)}{=} \left(\sum_{n \in I} w_n \int_{\mathbb{R}} (\xi - \langle A\rangle_W)^2 d\mu_{u_n}^{P^A}\right)^{\frac{1}{2}}$$

$$\overset{(11)}{=} \left(\sum_{n \in I} w_n\|Au_n - \langle A\rangle_W u_n\|^2\right)^{\frac{1}{2}}$$

(8 and 10 hold true by 8.3.5, in view of 1; 9 and 11 hold true by 15.2.2e). $\square$

18.3.17 Corollary. *Let A be a self-adjoint operator in $\mathcal{H}$, let $W \in \mathcal{W}(\mathcal{H})$, and suppose that the operator A^2 (which is self-adjoint, cf. 15.3.6) is computable in W. Then A is computable in W and*

$$\Delta_W A = \left(\langle A^2 \rangle_W - \langle A \rangle_W^2\right)^{\frac{1}{2}}.$$

Proof. For notational reasons we denote by P the projection valued measure of the self adjoint operator A^2. From 15.3.5 and 15.3.8 we have

$$P(E) = P^A(\pi^{-1}(E)), \forall E \in \mathcal{A}(d_\mathbb{R}),$$

with $\pi := \xi^2$, and hence

$$\mu_W^P(E) = \mu_W^{P^A}(\pi^{-1}(E)), \forall E \in \mathcal{A}(d_\mathbb{R}),$$

and hence, by 8.3.11b,

$$\int_\mathbb{R} \xi^4 d\mu_W^{P^A} = \int_\mathbb{R} \xi^2 d\mu_W^P < \infty,$$

since A^2 is computable in W. This implies

$$\int_\mathbb{R} \xi^2 d\mu_W^{P^A} < \infty,$$

since the measure $\mu_W^{P^A}$ is finite (cf. 11.1.3 and 8.2.4). Thus, A is computable in W (cf. 11.1.3 and 8.2.4).

Now let $\{u_n\}_{n \in I}$ and $\{w_n\}_{n \in I}$ be as in 18.3.16. In view of 18.3.16, written with A^2 in place of A, we have

$$u_n \in D_{A^2}, \forall n \in I,$$

and

$$\langle A^2 \rangle_W = \sum_{n \in I} w_n \langle A^2 \rangle_{u_n}.$$

Therefore we have, in view of 18.3.16,

$$(\Delta_W A)^2 = \sum_{n \in I} w_n \| A u_n - \langle A \rangle_W u_n \|^2$$

$$= \sum_{n \in I} w_n \left((u_n | A^2 u_n) - 2 \langle A \rangle_W (u_n | A u_n) + \langle A \rangle_W^2 \right)$$

$$= \sum_{n \in I} w_n \langle A^2 \rangle_{u_n} - 2 \langle A \rangle_W \sum_{n \in I} w_n \langle A \rangle_{u_n} + \langle A \rangle_W^2 \sum_{n \in I} w_n$$

$$= \langle A^2 \rangle_W - 2 \langle A \rangle_W^2 + \langle A \rangle_W^2 = \langle A^2 \rangle_W - \langle A \rangle_W^2.$$

□

18.3.18 Corollary. *If a self-adjoint operator is bounded, then it is computable in all statistical operators.*

Proof. Let $W \in \mathcal{W}(\mathcal{H})$, let $\{u_n\}_{n \in I}$ and $\{w_n\}_{n \in I}$ be as in 18.3.16, and let A be a self-adjoint operator in $\mathcal{H}$. If A is bounded then $D_A = \mathcal{H}$ (cf. 12.4.7) and hence

$$u_n \in D_A, \forall n \in I;$$

moreover,

$$\sum_{n \in I} w_n \|Au_n\|^2 \leq \sum_{n \in I} w_n \|A\|^2 \|u_n\|^2 = \|A\|^2 \sum_{n \in I} w_n < \infty$$

(cf. 4.2.5b). By 18.3.16 ($b \Rightarrow a$), this proves that A is computable in W. $\square$

18.3.19 Remarks.

(a) Let A be a self-adjoint operator in $\mathcal{H}$ and $W \in \mathcal{W}(\mathcal{H})$. If conditions a and b in 18.3.16 hold true, then

$$AW \in \mathcal{T}(\mathcal{H}) \text{ and } \langle A \rangle_W = \text{tr}(AW).$$

If A is bounded then $A \in \mathcal{B}(\mathcal{H})$ (cf. 12.4.7) and hence we have

$$WA \in \mathcal{T}(\mathcal{H}) \text{ and } \langle A \rangle_W = \text{tr}(WA),$$

by 18.2.7 and by 18.2.11c respectively.

If A is not bounded then $D_A \neq \mathcal{H}$ (cf. 12.4.7), and hence $D_{WA} \neq \mathcal{H}$, and hence the operator WA cannot be trace class and the formula $\text{tr}(WA)$ is meaningless.

(b) We recall that, for a statistical operator W, the decomposition $W = \sum_{n \in I} w_n A_{u_n}$ as in 18.3.6b is not unique unless W is a one-dimensional projection (cf. 18.3.7). From 18.3.16 we have that, if a self-adjoint operator A is computable in W, then

$$u_n \in D_A \text{ for all } n \in I \text{ and } \sum_{n \in I} w_n \|Au_n\|^2 < \infty$$

for whichever representation of W as in 18.3.6.b.

Chapter 19

Quantum Mechanics in Hilbert Space

In this chapter we examine how the theory of Hilbert space operators is used in quantum mechanics. This chapter is not meant to be a short treatise on quantum mechanics, since only the basic mathematical structure of the quantum theories is discussed and no applications are provided.

The predictions that are provided by quantum mechanics are in general statistical ones. And indeed, in what follows, quantum mechanics is presented as a theoretical scheme which can account for the probabilistic distributions of measurements, in experiments where measurements are repeatedly carried out on a large number of suitably-prepared copies of a physical system. The probabilities are interpreted as the theoretical predictions of the relative frequencies with which results are obtained when measurements are made on a large number of identically-prepared copies of the physical system under consideration. Quantum mechanics shares a good deal of its theoretical framework with other statistical theories, e.g. classical statistical mechanics and theories of games of chance (we will refer to all these theories as "classical statistical theories"). In the first section of this chapter we give an outline of this shared framework, which we call a "general statistical theory". For the abstract concepts we introduce, we use the names that are commonly used for them in quantum mechanics. In the second section we examine how this general statistical theory is implemented in the classical theories, and in the third section how it is implemented in the quantum theories. In the fourth and fifth sections, other topics are examined which are specific to quantum mechanics: state reduction, compatibility of observables, uncertainty relations.

Up to Section 19.5 we think of time as standing still: the time intervals between operational procedures are always supposed to be sufficiently small that there is no need to consider the internal time evolution of the system. At times, this is indicated by the use of the locution "immediately after". In Section 19.6 we examine time evolution in non-relativistic quantum mechanics.

19.1 Elements of a general statistical theory

Since we are mainly concerned with the mathematical aspects of the foundations of quantum mechanics, we could set off in an axiomatic way by simply saying that we are given two abstract sets Π and Σ and a function $p : \Pi \times \Sigma \to [0, 1]$, specifying that Π represents the family of all propositions pertaining to a physical system and Σ the family of all states of the system, and that $p(\pi, \sigma)$ is the probability that the proposition π is true when the system is in the state σ. However, we prefer to explain by what kind of reasoning these abstract objects are brought about in what we call a general statistical theory.

19.1.1 Definitions. A *state preparation* (or, simply, a *state*) is a collection of instructions for a set of physical operations to be performed on an array of objects, so that:

> the operations can be repeated, at least in principle, an indefinite number of times;

> the objects are macroscopic bodies, in the sense that the instructions are governed by standard classical logic.

A *proposition* is an event so that:

> the occurrence or non-occurrence of the event is to be decided immediately after a state preparation has been performed;

> when the event occurs, it takes place in a macroscopic device, in the sense that the procedure for ascertaining whether the event has occurred is governed by standard classical logic;

> when the event occurs, it leaves a long-lasting record in the device and its occurrence is ascertained by verifying this record, which has an objective meaning (the record can be read by any number of scientific observers and all of them agree about its meaning).

When the procedures which define a state σ and a proposition π are implemented, they have obviously space and time positions. However, it is assumed that these "absolute" positions are immaterial (and therefore they are not specified in the definitions of σ and π) and that, if π is to be decided immediately after σ, then the relative space positions of π and σ are suitable for an "interaction" between σ and π to take place (suitable, that is, according to the picture one has of a possible "interaction" between σ and π).

For a given proposition π and a given state σ, we define the following course of action: we perform the operations prescribed by σ and immediately after we ascertain whether the event π has occurred, and we do this a large number N of times. If we implement this course of action twice, and if N'_π (respectively, N''_π) denotes the number of times when π has occurred in the first (respectively, second) implementation, we cannot expect N'_π and N''_π to be equal in general. We say that π and σ are *correlated* when, as N grows, the difference between the *relative frequencies* $\frac{N'_\pi}{N}$ and $\frac{N''_\pi}{N}$ approaches zero and the relative frequencies approach a limit (clearly, the term "approach" has here an informal meaning and so has the term "limit"). When π and σ are correlated, the limit is called the probability of the occurrence of the proposition π immediately after the state preparation σ or, simply, the *probability of π in σ*.

We say that we have a *physical system* (or, simply, a *system*) if we have a family Σ of states and a family Π of propositions so that π and σ are correlated for each pair $(\pi, \sigma) \in \Pi \times \Sigma$ and if we think that the resulting probabilities are liable to be organized in a consistent theory. If $(\pi, \sigma) \in \Pi \times \Sigma$, a single implementation of the operations prescribed by σ is said to be a *copy of the system prepared in the state σ* (or, simply, a *copy in σ*), and the ascertainment whether π has occurred immediately after an implementation of the operations prescribed by σ is said to be the *determination of π for a copy in σ*; if π has occurred, then π is said to be *true in that copy*. We say that we have a *statistical theory* of the system if we have a theoretical scheme whereby a function $p : \Pi \times \Sigma \to [0, 1]$ can be obtained so that $p(\pi, \sigma)$ is, for all $(\pi, \sigma) \in \Pi \times \Sigma$, the probability of the occurrence of π immediately after the state preparation σ. The function p is called a *probability function*. If the theory supplies such a function, then $p(\pi, \sigma)$ is the *theoretical prediction* of the relative frequency $\frac{N_\pi}{N}$, where N_π is the number of copies in which a proposition π has turned out to be true out of N copies of the system prepared in a state σ, provided that N is large enough. It is important to note that, although a state σ and a proposition π are procedures to be operated on a single copy of the system, the number $p(\pi, \sigma)$ that the theory assigns to the pair (π, σ) can be compared with the experimental results only if we have a large (hypothetically infinite) collection of copies of the system, for each of which the determination of π is carried out immediately after the copy has been prepared in the state σ. A large collection of copies, all prepared in the state σ, is sometimes called an *ensemble* representing σ.

19.1.2 Remarks.

(a) While in the classical statistical theories it is often obvious what is to be considered the physical system under consideration (e.g. the gas contained in a vessel, a coin, a pair of dice, a roulette table), this is not so in the quantum theories, and in our opinion it is convenient to consider a quantum system as an "interaction channel" between a definite set of states and a definite set of propositions, according to the definition given in 19.1.1. For instance, if the state is to switch on a "source" to the left of a Stern–Gerlach apparatus (the source and the Stern–Gerlach apparatus are thus the objects which appear in the abstract definition of state) and the proposition is the event defined by the reaction of a detector to the right of the Stern–Gerlach apparatus, then the physical system is called a spin-half (for instance) particle. As another example, if the state is to operate an accelerator in a given mode and to arrange magnetic analysers and collimating slits in a given way, and the proposition is once again the reaction of a detector, then the case may be that the physical system is called a meson. In any case, while in a classical theory it is non-controversial that a physical system is defined also independently of the operational framework outlined above, in a general statistical theory which aims to include quantum mechanics it is expedient to consider a physical system as defined by a collection of statistical experiments.

(b) The definition we have given of a state may seem not suitable for describing states in classical physics, where the state of a system is usually considered to be the result of its previous history. However, also in the case of classical physics, in order to define a state one often has to make operations (if nothing else, a selection) on macroscopic bodies. We must acknowledge that the condition we have assumed of indefinite reproducibility of a state preparation clearly limits the cases which can be described by our theory; for instance, in astrophysics or in geophysics the states one deals with are most of the times the result of a previous history of the system which cannot be reproduced. On the other hand, in games of chance, the definition we have given of a state is well suited for the description of what happens in obvious cases: a state may be a definite way of tossing a coin or a definite way of rolling the dice; in the game of roulette it is a definite way of spinning the wheel and throwing the ball. It may be convenient to keep in mind these examples of classical statistical theories throughout this section.

(c) In the classical theories, the macroscopic device in which the event that defines a proposition is produced (or is not produced) is often (but not always) what we identify with the physical system itself: e.g., propositions are "the coin has landed in such a way that heads shows", "the dice have landed in such a way that the sum of the values on top is five", "the colour of the roulette wheel compartment in which the ball has fallen is red". On the contrary, in the case of quantum mechanics, the event that defines a proposition can never be produced in what we perhaps picture as the physical system, since it must be produced in a macroscopic device; actually, it should be considered as occurring (or not occurring) in a macroscopic device which is in an unstable condition so that a suitable state preparation can trigger a lasting macroscopic effect, i.e. a lasting effect which can be detected by classical means: although the device is obviously made up of quantum particles, it must be possible to handle it in a such way that only its classical properties are involved in the detection of the event. For instance, a proposition in quantum physics can be "a certain detector has reacted", and the assessment that the detector has reacted can be based on hearing or otherwise recording a sound in a counter in which we imagine that a "particle" has triggered the macroscopic effect known as electrical discharge, or the assessment can be based on seeing or otherwise registering a blackening in a definite region of a photographic plate, where we imagine that a "particle" has initiated a chemical reaction in a grain of the emulsion and so has caused the macroscopic effect known as latent image.

(d) Experimental evidence supports the assumption that only the *relative* positions in space (as specified in 19.1.1) and time (a proposition immediately after a state) are important. Actually, this is perhaps the first invariance law discovered in the history of physics. This fact has allowed physics to be tackled as an experimental science.

19.1.3 Definition. The event that defines a proposition π can be used to define another event, which defines another proposition; this new proposition is called the *negation* of π and denoted by the symbol $\neg\pi$, and the event that defines $\neg\pi$ is the non-occurrence of the event that defines π.

19.1.4 Remarks.

(a) If there is a statistical theory for the physical system defined by a collection Π of propositions and a collection Σ of states, then the theory is consistent only if, for the probability function p, we have

$$p(\neg\pi, \sigma) = 1 - p(\pi, \sigma), \quad \forall\pi \in \Pi, \forall\sigma \in \Sigma. \tag{1}$$

(b) We point out that nothing has been said about the possibility, for two propositions π and π', of defining the event that is said to have occurred if and only if both the events that define π and π' have occurred (such new event would define the proposition "π and π'") or of defining the event that is said to have occurred if and only if at least one of the events that define π and π' has occurred (such new event would define the proposition "π or π'"). Indeed, this requires that it is feasible to determine both π and π' for a single copy prepared in any state. This feasibility is actually assumed in all classical statistical theories. We will see that this is one of the aspects in which the quantum theories differ from the classical statistical ones.

19.1.5 Definitions. Let Π and Σ be families of propositions and states so that $\Pi \times \Sigma$ defines a physical system for which a probability function p is given. We define an equivalence relation $\mathcal{R}_\Pi$ in Π by letting

$$\mathcal{R}_\Pi := \{(\pi', \pi'') \in \Pi \times \Pi : p(\pi', \sigma) = p(\pi'', \sigma), \forall\sigma \in \Sigma\},$$

and similarly we define an equivalence relation $\mathcal{R}_\Sigma$ in Σ by letting

$$\mathcal{R}_\Sigma := \{(\sigma', \sigma'') \in \Sigma \times \Sigma : p(\pi, \sigma') = p(\pi, \sigma''), \forall\pi \in \Pi\}.$$

We denote by $\hat\Pi$ and $\hat\Sigma$ the quotient sets which are thus defined, we denote by $\hat\pi$ and $\hat\sigma$, and still call *proposition* and *states*, the equivalence classes containing $\pi \in \Pi$ and $\sigma \in \Sigma$, and we define the function

$$\hat{p} : \hat\Pi \times \hat\Sigma \to [0, 1]$$

$$(\hat\pi, \hat\sigma) \mapsto \hat{p}(\hat\pi, \hat\sigma) := p(\pi, \sigma),$$

which we still call a *probability function*. Obviously, we have

$$\hat{p}(\hat\pi', \hat\sigma) = \hat{p}(\hat\pi'', \hat\sigma), \forall\hat\sigma \in \hat\Sigma \Rightarrow \hat\pi' = \hat\pi'' \text{ and}$$

$$\hat{p}(\hat\pi, \hat\sigma') = \hat{p}(\hat\pi, \hat\sigma''), \forall\hat\pi \in \hat\Pi \Rightarrow \hat\sigma' = \hat\sigma''.$$

19.1.6 Remarks.

(a) It is clear that, for a physical system defined by a family Π of propositions and a family Σ of states, there may be state preparations $\sigma', \sigma'' \in \Sigma$ which are different (i.e. they are different collections of instructions which refer to different arrays of objects), but which nonetheless lead to the same experimental statistical results in the sense that, for every $\pi \in \Pi$, the probability of π in σ' equals the probability of π in σ''. This means that the differences between σ' and σ'' are immaterial as far as the statistical study of the physical system under consideration goes. And a similar remark can be made for the elements of Π. Thus, a statistical theory of the system must contain mathematical representations of the quotient sets $\hat{\Pi}$, $\hat{\Sigma}$ defined above, through which a formula which defines the function $\hat{p}$ must then be written.

(b) From condition 1 of 19.1.4 we see that, for each $\hat{\pi} \in \hat{\Pi}$, the family of all negations of the representatives of $\hat{\pi}$ constitute an equivalence class, which we still call the *negation* of $\hat{\pi}$ and denote by the symbol $\neg\hat{\pi}$.

19.1.7 Definitions. For every physical system, two (trivial) propositions always exist, which we denote by the symbols π_0 and π_1. The proposition π_0 is defined by the event which is said to occur if and only if no copy of the system has been prepared: this event never occurs if π_0 is determined immediately after a state preparation has been performed. The proposition π_1 is defined by the event which is said to occur if and only if a copy of the system has been prepared: this event always occurs if π_1 is determined immediately after a state preparation has been performed. Clearly, the equivalence classes $\hat{\pi}_0$ and $\hat{\pi}_1$ are characterized by the following conditions

$$\hat{p}(\hat{\pi}_0, \hat{\sigma}) = 0, \forall \hat{\sigma} \in \hat{\Sigma}, \text{ and } \hat{p}(\hat{\pi}_1, \hat{\sigma}) = 1, \forall \hat{\sigma} \in \hat{\Sigma},$$

where Σ denotes as usual the family of states that defines the system.

19.1.8 Definition. Consider a physical system for which a statistical theory is given (i.e., a probability function p is defined), and let Σ and Π be the families of states and propositions that define the system. Let $(X, \mathcal{A})$ be a measurable space (cf. 6.1.13). An *X-valued observable* pertaining to the system is a mapping $\alpha : \mathcal{A} \to \Pi$ which is so that, for every state $\sigma \in \Sigma$, the function

$$\mu_\sigma^\alpha : \mathcal{A} \to [0, 1]$$
$$E \mapsto \mu_\sigma^\alpha(E) := p(\alpha(E), \sigma)$$

is a probability measure on $\mathcal{A}$, and also so that it can be associated with a measuring instrument (as explained in 19.1.9a).

If X is a Borel subset of $\mathbb{R}^n$, $\mathcal{A}$ is always assumed to be the Borel σ-algebra $\mathcal{A}(d_n)^X$ (cf. 2.2.2, 2.7.4b, 6.1.19, 6.1.22). An $\mathbb{R}$-valued observable is simply called an *observable*.

19.1.9 Remarks.

(a) In 19.1.8, the measurable space $(X, \mathcal{A})$ provides a representation of the events into which the measurement of a physical quantity can be analysed. In our view, an X-valued physical quantity is defined by an ideal apparatus which comprises a dial, which is represented by X (i.e. there is a mapping, not necessarily injective, from the dial to X), and a pointer. Immediately after a copy of the system has been prepared in a state in such a way that an "interaction" between the copy and the apparatus can happen (i.e., the state preparation must be implemented in a suitable spatial position with respect to the apparatus), the apparatus gives an X-result by the position of the pointer on the dial, which identifies a point of the dial (since the apparatus is an ideal one) and hence a point of X. *The pointer and the dial are assumed to be macroscopic objects, to wit the ascertainment of the possible positions of the pointer is governed by standard classical logic.* An element E of $\mathcal{A}$ is a subset of X to which it is deemed sensible to assign the probability that, in a given state, the X-result is a point of E (a probability is always considered in this book to be a normalized measure on a σ-algebra, cf. 7.1.7); for instance, if X is endowed with a distance then a natural choice for $\mathcal{A}$ is the Borel σ-algebra on X (cf. 6.1.22).
All this leads to an X-valued observable α if, for each $E \in \mathcal{A}$, we define $\alpha(E)$ to be the event which is said to have occurred if and only if the X-result given by the apparatus has been an element of E and if we assume that this event defines a proposition of the system. In fact, the condition that μ_σ^α be a probability measure on $\mathcal{A}$ for each $\sigma \in \Sigma$ can be accounted for as follows. We assume that all propositions $\alpha(E)$, for $E \in \mathcal{A}$, can be determined simultaneously for any single copy prepared in any state. Moreover, we assume that, if $\{E_n\}$ is a sequence in $\mathcal{A}$ such that $E_i \cap E_k = \emptyset$ for $i \neq k$, then the proposition $\alpha(\cup_{n=1}^\infty E_n)$ is true in a copy prepared in a state if and only if there exists exactly one $E_{\overline{n}}$ such that $\alpha(E_{\overline{n}})$ is true in that copy. Actually, the basis for these two assumptions is the macroscopic nature assumed before for the pointer and the dial. Now let $\{E_n\}$ be a sequence in $\mathcal{A}$ such that $E_i \cap E_k = \emptyset$ for $i \neq k$; if we determine all propositions $\alpha(E_n)$, for $n \in \mathbb{N}$, as well as the proposition $\alpha(\cup_{n=1}^\infty E_n)$ (which is possible on account of the first assumption above) for a large number N of copies of the system prepared in a state σ, and if we denote by N_n the number of copies in which the proposition $\alpha(E_n)$ is true and by N_U the number of copies in which the proposition $\alpha(\cup_{n=1}^\infty E_n)$ is true, then we have (on account of the second assumption above)

$$\sum_{n=1}^\infty \frac{N_n}{N} = \frac{N_U}{N}$$

(the series on the left hand side is actually a sum); since $p(\pi, \sigma)$ is the theoretical prediction of the relative frequency of a proposition $\pi \in \Pi$ being true in a large number of copies all prepared in the state σ, the consistency of the theory leads to the equation

$$\sum_{n=1}^{\infty} \mu_\sigma^\alpha(E_n) = \sum_{n=1}^{\infty} p(\alpha(E_n), \sigma) = p\left(\alpha\left(\cup_{n=1}^\infty E_n\right), \sigma\right) = \mu_\sigma^\alpha\left(\cup_{n=1}^\infty E_n\right).$$

Thus, the function μ_σ^α is σ-additive. Moreover, since we have assumed that the apparatus (which is an ideal one) always gives an X-result immediately after a copy has been prepared in some suitable state, the proposition $\alpha(X)$ is always true in every state, and therefore the consistency of the theory leads to the condition $\mu_\sigma^\alpha(X) = 1$ for every state σ. Thus, μ_σ^α is a probability measure on $\mathcal{A}$ for every state σ.

As to the assumption that $\mathcal{A} = \mathcal{A}(d_{\mathbb{R}})^X$ when X is a Borel subset of $\mathbb{R}$, we note that intervals are most naturally associated with a dial which is represented by a subset of $\mathbb{R}$, and that $\mathcal{A}(d_{\mathbb{R}})$ is the σ-algebra on $\mathbb{R}$ generated by the family of intervals (cf. 6.1.25). And similarly for the general case $\mathcal{A}(d_n)^X$.

The determination of all propositions $\alpha(E)$ (i.e. of $\alpha(E)$ for all $E \in \mathcal{A}$) for a copy, prepared in some state, can be performed in actual fact by determining only some propositions, on account of the assumption that the ascertainment of the position of the pointer is governed by classical logic (e.g., if X is so that $\{x\} \in \mathcal{A}$ and if $\alpha(\{x\})$ is true in a copy, then the proposition $\alpha(E)$ is true in that copy if and only if $x \in E$). The determination of all propositions $\alpha(E)$ for a copy is said to be a *measurement* of the observable α in that copy.

(b) The "position of the pointer on the dial" may be actually implemented by an apparatus which does not comprise a needle-like object over a graduated scale: it may be the blackening of a grain in a photographic plate (and then X is a subset of $\mathbb{R}^2$), or the formation of a bubble in a bubble-chamber (and then X is a subset of $\mathbb{R}^3$), or the digital reading of an instrument (and then X is a subset of $\mathbb{R}$). In any case, the apparatus that was considered above was clearly an ideal one inasmuch as the position of the pointer was supposed to identify a point of the dial. But reference to ideal instruments is a common feature of all mathematical physics (however, we shall see that in quantum mechanics we do not need an ideal apparatus in order to get exact measurements, if the observable is quantized; nor do we need an ideal apparatus in the game of dice or in the game of roulette).

(c) The analysis which was carried out in remark a was aimed at showing that it is reasonable to represent an instrument, which can measure a physical quantity, by a mapping $\alpha : \mathcal{A} \to \Pi$ which is so that the function μ_σ^α is a probability measure on $\mathcal{A}$ for every state σ, where $\mathcal{A}$ is a σ-algebra which represents the sensible parts of the dial of the instrument. However, even when $(X, \mathcal{A})$ is chosen in a conservative way (e.g. $(X, \mathcal{A}) = (\mathbb{R}, \mathcal{A}(d_{\mathbb{R}}))$, since the dial of most measuring instruments can be identified with some part of $\mathbb{R}$), it would be hard to justify in general the assumption that *any* mapping $\alpha : \mathcal{A} \to \Pi$ for which μ_σ^α is a probability measure for all $\sigma \in \Sigma$ should be taken to represent a measuring instrument, and therefore should be considered a bona-fide X-valued observable. And indeed, whenever we say that a mapping $\alpha : \mathcal{A} \to \Pi$ is an X-valued observable, we will assume that μ_σ^α is a probability measure for all $\sigma \in \Sigma$ *and* that α is related to a measuring apparatus in the way described in remark a.

(d) It is a well-known fact that technology often provides instruments which are equivalent, in the sense that they give equal results under equal conditions. Of this, care will be taken in 19.1.10.

19.1.10 Definition. Let $\Sigma, \Pi, (X, \mathcal{A})$ be as in 19.1.8. Two X-valued observables α_1 and α_2 are said to be equivalent if $\mu_\sigma^{\alpha_1} = \mu_\sigma^{\alpha_2}$ for each $\sigma \in \Sigma$, and this indeed defines an equivalence relation in the family of all X-valued observables of the system. It is easy to see that the quotient set which is thus defined can be identified with a family of mappings $\hat{\alpha} : \mathcal{A} \to \hat{\Pi}$ which are so that the function

$$\mu_{\hat{\sigma}}^{\hat{\alpha}} : \mathcal{A} \to [0, 1]$$
$$E \mapsto \mu_{\hat{\sigma}}^{\hat{\alpha}}(E) := \hat{p}(\hat{\alpha}(E), \hat{\sigma})$$

is a probability measure for each $\hat{\sigma} \in \hat{\Sigma}$. We still call X-valued observables such mappings.

19.1.11 Proposition. *Let* $\Sigma, \Pi, (X, \mathcal{A})$ *be as in 19.1.8 and let* $\hat{\alpha}$ *be an X-valued observable. We have:*

$$\hat{\alpha}(\emptyset) = \hat{\pi}_0,$$
$$\hat{\alpha}(X) = \hat{\pi}_1,$$
$$\hat{\alpha}(X - E) = \neg\hat{\alpha}(E), \ \forall E \in \mathcal{A}.$$

Proof. The equalities of the statement follow from the following facts, which are true because $\mu_{\hat{\sigma}}^{\hat{\alpha}}$ is a probability measure: for each $\hat{\sigma} \in \hat{\Sigma}$,

$$p(\hat{\alpha}(\emptyset), \hat{\sigma}) = \mu_{\hat{\sigma}}^{\hat{\alpha}}(\emptyset) = 0 = \hat{p}(\hat{\pi}_0, \hat{\sigma}),$$
$$p(\hat{\alpha}(X), \hat{\sigma}) = \mu_{\hat{\sigma}}^{\hat{\alpha}}(X) = 1 = \hat{p}(\hat{\pi}_1, \hat{\sigma}),$$
$$p(\hat{\alpha}(X - E), \hat{\sigma}) = \mu_{\hat{\sigma}}^{\hat{\alpha}}(X - E) = 1 - \mu_{\hat{\sigma}}^{\hat{\alpha}}(E) = 1 - \hat{p}(\hat{a}(E), \hat{\sigma}) = \hat{p}(\neg\hat{\alpha}(E), \hat{\sigma}).$$

$\square$

19.1.12 Remark. Throughout the rest of this chapter, we always assume that we are dealing with a fixed, although general, physical system for which we assume that a probability function is defined. The symbols Σ and Π always denote the families of states and propositions which define the system.

As a rule we drop the carets in the symbols $\hat{\Sigma}, \hat{\Pi}, \hat{\sigma}, \hat{\pi}, \hat{\alpha}$ *and we leave it to the* reader to understand whether we refer to an equivalence class or to a representative of it. If σ is an equivalence class of states, a representative of σ is sometimes called an implementation of σ; and similarly for propositions.

19.1.13 Proposition. *Let* $(X_1, \mathcal{A}_1)$ *and* $(X_2, \mathcal{A}_2)$ *be measurable spaces, let* α *be an X_1-valued observable, let* $\varphi : D_\varphi \to X_2$ *be a mapping from X_1 to X_2 which is* *measurable w.r.t.* $\mathcal{A}_1{}^{D_\varphi}$ *and* $\mathcal{A}_2$ *and so that* $D_\varphi \in \mathcal{A}_1$ *and* $\mu_\sigma^\alpha(X_1 - D_\varphi) = 0$ *for each* $\sigma \in \Sigma$. *Then, the mapping* $\varphi(\alpha)$ *defined by*

$$\varphi(\alpha) : \mathcal{A}_2 \to \Pi \quad E \mapsto \varphi(\alpha)(E) := \alpha(\varphi^{-1}(E))$$

is so that the function $\mu_\sigma^{\varphi(\alpha)}$ *is a probability measure on* $\mathcal{A}_2$ *for each state* $\sigma \in \Sigma$.

Proof. For each $\sigma \in \Sigma$, the function $\mu_\sigma^{\varphi(\alpha)}$ is a measure on $\mathcal{A}_2$, since μ_σ^α is a measure on $\mathcal{A}_1$ and

$$\mu_\sigma^{\varphi(\alpha)}(E) = \mu_\sigma^\alpha(\varphi^{-1}(E)), \forall E \in \mathcal{A}_2$$

(cf. 8.3.11a). Moreover,

$$\mu_\sigma^{\varphi(\alpha)}(X_2) = \mu_\sigma^\alpha(\varphi^{-1}(X_2)) = \mu_\sigma^\alpha(D_\varphi) = \mu_\sigma^\alpha(X_1) = 1$$

since μ_σ^α is a probability measure. Thus, $\mu_\sigma^{\varphi(\alpha)}$ is a probability measure on $\mathcal{A}_2$. $\square$

19.1.14 Remark. If α and φ are as in the statement of 19.1.13, then $\varphi(\alpha)$ can be considered an X_2-valued observable, which is called the *function of* α *according to* φ. Indeed, there exists a measuring instrument which is represented by $\varphi(\alpha)$ (cf. 19.1.9c), since $\varphi(\alpha)$ can be interpreted as the X_2-valued observable that is defined operationally by the same apparatus that defines α (cf. 19.1.9a), in which however a change of scale has been made: while the dial of the apparatus is represented by X_1 when the apparatus is related to α, the dial is represented by X_2 when the apparatus is related to $\varphi(\alpha)$. Assuming first $D_\varphi = X_1$, if a point of the dial is represented by $x \in X_1$ when the scale that defines α is used, then that same point of the dial is represented by $\varphi(x) \in X_2$ when the scale that defines $\varphi(\alpha)$ is used; thus, in a copy prepared in some state and for any $E \in \mathcal{A}_2$, the operational meaning of the proposition $\varphi(\alpha)(E)$ is so that the proposition $\varphi(\alpha)(E)$ is true if and only if the X_2-value given by the apparatus that defines $\varphi(\alpha)$ is in E, and this is true if and only if the X_1-value given by the apparatus that defines α is in $\varphi^{-1}(E)$, and this is true (by the operational meaning of the proposition $\alpha(\varphi^{-1}(E))$) if and only if the proposition $\alpha(\varphi^{-1}(E))$ is true; thus, the proposition $\varphi(\alpha)(E)$ must coincide with the proposition $\alpha(\varphi^{-1}(E))$. We notice that, in the reasoning just made, there was no need for φ to be injective (e.g., in the game of roulette all observables are functions of the observable α which assigns a number from 0 to 36 to any copy of the system, and for instance the observable which assigns the colour-values "rouge", "noir" or "nul" is defined by a non-injective function). If $D_\varphi \neq X_1$ we can extend φ to a function $\tilde{\varphi}$ defined on the whole X_1 in any way that makes it $\mathcal{A}_1$-measurable, and repeat the reasoning for this extension $\tilde{\varphi}$. For each $E \in \mathcal{A}_2$ we have

$$p(\alpha(\tilde{\varphi}^{-1}(E)), \sigma) = \mu_\sigma^\alpha(\tilde{\varphi}^{-1}(E))$$
$$= \mu_\sigma^\alpha(\tilde{\varphi}^{-1}(E) \cap D_\varphi) + \mu_\sigma^\alpha(\tilde{\varphi}^{-1}(E) \cap (X_1 - D_\varphi))$$
$$= \mu_\sigma^\alpha(\tilde{\varphi}^{-1}(E) \cap D_\varphi) = \mu_\sigma^\alpha(\varphi^{-1}(E)) = p(\alpha(\varphi^{-1}(E)), \sigma), \forall \sigma \in \Sigma,$$

where the monotonicity of μ_σ^α and the condition $\mu_\sigma^\alpha(X_1 - D_\varphi) = 0$ have been used; this proves that $\alpha(\tilde{\varphi}^{-1}(E)) = \alpha(\varphi^{-1}(E))$. Thus, the reasoning we made before can indeed be referred to the extension $\tilde{\varphi}$, but the observable $\tilde{\varphi}(\alpha)$ we obtain does not depend on the extension we use: it depends only on φ, and therefore can be denoted by the symbol $\varphi(\alpha)$. Furthermore, it can be defined directly trough φ as in the statement of 19.1.13.

In what follows, we are concerned mainly with $\mathbb{R}$-valued observables, which are simply called observables.

19.1.15 Definitions. Let α be an observable. A number $\lambda \in \mathbb{R}$ is said to be a *possible result for* α if the following condition is satisfied

$$\forall \varepsilon > 0, \exists \sigma \in \Sigma \text{ so that } \mu_\sigma^\alpha((\lambda - \varepsilon, \lambda + \varepsilon)) \neq 0.$$

A number $\lambda \in \mathbb{R}$ is said to be an *impossible result for* α if it is not a possible result, i.e. if the following condition is satisfied

$$\exists \varepsilon > 0 \text{ so that } \mu_\sigma^\alpha((\lambda - \varepsilon, \lambda + \varepsilon)) = 0, \forall \sigma \in \Sigma.$$

The set of all possible results for α, i.e. the set sp_α defined by

$$sp_\alpha := \{\lambda \in \mathbb{R} : \forall \varepsilon > 0, \exists \sigma \in \Sigma \text{ so that } \mu_\sigma^\alpha((\lambda - \varepsilon, \lambda + \varepsilon)) \neq 0\},$$

is called the *spectrum* of α.

The observable α is said to be *bounded* if sp_α is a bounded set.

19.1.16 Remarks.

(a) If a number $\lambda \in \mathbb{R}$ happens to be so that, for an observable α, there exists $\sigma \in \Sigma$ such that

$$\mu_\sigma^\alpha(\{\lambda\}) \neq 0, \tag{2}$$

then obviously λ must be considered a possible result for α from an operational point of view: in N repetitions of the measurement of α in copies of the system prepared in the state σ, the result λ occurs so often that its relative frequency will approach a non-null number as N grows. In this case, it is obvious (owing to the monotonicity of μ_σ^α) that λ fulfills the condition that we have given in 19.1.15 to characterize a possible result for α, and λ is said to be an *exact result* for α.

However, condition 2 need not be fulfilled by every number which can occur as the result obtained in the measurement of α in some copy: in N repetitions of the measurement of α in a copy prepared in a state, a number λ can indeed occur, but so seldom that its relative frequency will approach zero as N grows. This is indeed what we expect to happen if λ belongs to what our theoretical image of the system depicts as a continuum of possible results (unless state preparations are assumed to exist that are so "precise" as to pinpoint a value of an observable amid a continuum of possible values; such state preparations are not realistic; however, classical mechanics is indeed based on such state preparations, while quantum mechanics is not, as we shall see). If such is the case, for λ to be considered a possible result we rather naturally require that we obtain in a substantial way (i.e. with non-null probability) values around λ with any margin of error, in suitable states. If, on the contrary, there is a whole interval centered in λ so that no value in it ever occurs, then we are led to consider λ an impossible result for α.

These are the ideas that are formalized in 19.1.15.

(b) Since by "proposition" and "observable" we actually mean equivalence classes, a real number λ is a possible result for an observable α if and only if

$$\alpha((\lambda - \varepsilon, \lambda + \varepsilon)) \neq \pi_0 \text{ for all } \varepsilon > 0$$

(cf. 19.1.7).

19.1.17 Proposition. *For every observable α, the spectrum sp_α is a closed subset of $\mathbb{R}$ and we have $\mu_\sigma^\alpha(\mathbb{R} - sp_\alpha) = 0$, or equivalently $\mu_\sigma^\alpha(sp_\alpha) = 1$, for all $\sigma \in \Sigma$.*

Proof. For each $\lambda \in \mathbb{R} - sp_\alpha$, let $\varepsilon_\lambda > 0$ be such that $\mu_\sigma^\alpha((\lambda - \varepsilon_\lambda, \lambda + \varepsilon_\lambda)) = 0$, $\forall \sigma \in \Sigma$. Then we have

$$\mathbb{R} - sp_\alpha = \cup_{\lambda \in \mathbb{R} - sp_\alpha}(\lambda - \varepsilon_\lambda, \lambda + \varepsilon_\lambda);$$

indeed, if $\lambda \in \mathbb{R} - sp_\alpha$ and $\mu \in (\lambda - \varepsilon_\lambda, \lambda + \varepsilon_\lambda)$, then there exists $\eta > 0$ such that

$$(\mu - \eta, \mu + \eta) \subset (\lambda - \varepsilon_\lambda, \lambda + \varepsilon_\lambda),$$

and therefore (owing to the monotonicity of μ_σ^α) such that

$$\mu_\sigma^\alpha((\mu - \eta, \mu + \eta)) = 0, \forall \sigma \in \Sigma,$$

and this proves that $\mu \in \mathbb{R} - sp_\alpha$ and therefore that $(\lambda - \varepsilon_\lambda, \lambda + \varepsilon_\lambda) \subset \mathbb{R} - sp_\alpha$. Thus, $\mathbb{R} - sp_\alpha$ is an open set and sp_α is closed set.

Furthermore, by Lindelöf's theorem (cf. 2.3.16 and 2.3.18) there exists a countable subset $\{\lambda_n\}_{n \in I}$ of $\mathbb{R} - sp_\alpha$ so that $\mathbb{R} - sp_\alpha = \cup_{n \in I}(\lambda_n - \varepsilon_{\lambda_n}, \lambda_n + \varepsilon_{\lambda_n})$, and this implies (by the σ-subadditivity of μ_σ^α) that

$$\mu_\sigma^\alpha(\mathbb{R} - sp_\alpha) = 0, \forall \sigma \in \Sigma,$$

which is equivalent to $\mu_\sigma^\alpha(sp_\alpha) = 1$ for all $\sigma \in \Sigma$, since μ_σ^α is a probability measure. $\square$

19.1.18 Definition. An observable α is said to be *discrete* if there exists a countable family $\{\lambda_n\}_{n \in I}$ of real numbers so that $\mu_\sigma^\alpha(\{\lambda_n\}_{n \in I}) = 1$ for all $\sigma \in \Sigma$.

19.1.19 Remarks. In 19.1.18 we may assume that

$$\forall n \in I, \exists \sigma \in \Sigma \text{ so that } \mu_\sigma^\alpha(\{\lambda_n\}) \neq 0. \tag{3}$$

Indeed, if this was not the case, we could eliminate from the family $\{\lambda_n\}_{n \in I}$ each element λ_n which was such that $\mu_\sigma^\alpha(\{\lambda_n\}) = 0$ for all $\sigma \in \Sigma$, without altering the property the family is required to have in 19.1.18. Then, if 3 is true, all the elements of the family $\{\lambda_n\}_{n\in I}$ are exact results for α (cf. 19.1.16a).

We cannot say that in general every possible result for α is an element of $\{\lambda_n\}_{n\in I}$, since $sp_\alpha = \overline{\{\lambda_n\}_{n\in I}}$. Indeed, while $\overline{\{\lambda_n\}_{n\in I}} \subset sp_\alpha$ is obvious since sp_α is closed, for $\lambda \in \mathbb{R}$ we have (by the monotonicity of μ_σ^α)

$$\lambda \notin \overline{\{\lambda_n\}_{n\in I}} \Rightarrow$$
$$[\exists \varepsilon > 0 \text{ s.t. } (\lambda - \varepsilon, \lambda + \varepsilon) \subset \mathbb{R} - \{\lambda_n\}_{n\in I} \text{ and hence s.t.}$$
$$\mu_\sigma^\alpha((\lambda - \varepsilon, \lambda + \varepsilon)) = 0, \forall \sigma \in \Sigma],$$

which proves that $\mathbb{R} - \overline{\{\lambda_n\}_{n\in I}} \subset \mathbb{R} - sp_\alpha$.

At any rate, if $\{\lambda_n\}_{n\in I}$ is a finite family then actually $sp_\alpha = \{\lambda_n\}_{n\in I}$ and we can say that $\{\lambda_n\}_{n\in I}$ is the family of all possible results for α. Note that this is indeed the case for all observables in most games of chance (e.g. in tossing a coin, dice, roulette).

Finally we note that, if the spectrum of an observable is finite (i.e. the family of all its possible results is finite), then obviously the observable is discrete and all its possible results are exact.

19.1.20 Definitions. Let α be an observable and $\sigma \in \Sigma$. Let ξ be the function defined in 15.2.1A. If $\xi \in \mathcal{L}^1(\mathbb{R}, \mathcal{A}(d_\mathbb{R}), \mu_\sigma^\alpha)$, the real number

$$\langle \alpha \rangle_\sigma := \int_\mathbb{R} \xi d\mu_\sigma^\alpha$$

is called the *expected result* of α in σ.

If $\xi \in \mathcal{L}^1(\mathbb{R}, \mathcal{A}(d_\mathbb{R}), \mu_\sigma^\alpha)$ and $\xi - \langle \alpha \rangle_\sigma \in \mathcal{L}^2(\mathbb{R}, \mathcal{A}(d_\mathbb{R}), \mu_\sigma^\alpha)$, the finite positive number

$$\Delta_\sigma \alpha := \left(\int_\mathbb{R} (\xi - \langle \alpha \rangle_\sigma)^2 d\mu_\sigma^\alpha \right)^{\frac{1}{2}}$$

is called the *uncertainty* of α in σ.

19.1.21 Proposition. *Let α be an observable and $\sigma \in \Sigma$. The following conditions are equivalent:*

(a) $\xi \in \mathcal{L}^1(\mathbb{R}, \mathcal{A}(d_\mathbb{R}), \mu_\sigma^\alpha)$ *and* $\xi - \langle \alpha \rangle_\sigma \in \mathcal{L}^2(\mathbb{R}, \mathcal{A}(d_\mathbb{R}), \mu_\sigma^\alpha)$;
(b) $\int_\mathbb{R} \xi^2 d\mu_\sigma^\alpha < \infty$.

The observable α is said to be evaluable in the state σ if these conditions are satisfied.

Proof. $a \Rightarrow b$: From condition a we have $\xi \in \mathcal{L}^2(\mathbb{R}, \mathcal{A}(d_\mathbb{R}), \mu_\sigma^\alpha)$ because a constant function is an element of $\mathcal{L}^2(\mathbb{R}, \mathcal{A}(d_\mathbb{R}), \mu_\sigma^\alpha)$ since the measure μ_σ^α is finite (cf. 8.2.6) and $\xi = (\xi - \langle \alpha \rangle_\sigma) + \langle \alpha \rangle_\sigma$.

$b \Rightarrow a$: From $\xi \in \mathcal{L}^2(\mathbb{R}, \mathcal{A}(d_\mathbb{R}), \mu_\sigma^\alpha)$ we have both conditions $\xi \in \mathcal{L}^1(\mathbb{R}, \mathcal{A}(d_\mathbb{R}), \mu_\sigma^\alpha)$ and $\xi - \langle \alpha \rangle_\sigma \in \mathcal{L}^2(\mathbb{R}, \mathcal{A}(d_\mathbb{R}), \mu_\sigma^\alpha)$ since the measure μ_σ^α is finite (cf. 11.1.3). $\qquad \square$

19.1.22 Remarks.

(a) Suppose that α is a discrete observable with a finite family $\{\lambda_k\}_{k\in I}$ of possible results. Then $\mu_\sigma^\alpha(\{\lambda_k\}_{k\in I}) = 1$ for all $\sigma \in \Sigma$ and the results obtained in any collection of measurements of α are bound to be elements of the family $\{\lambda_k\}_{k\in I}$. Now suppose that measurements of α are performed in N copies of the system, all prepared in the same state $\sigma \in \Sigma$. Two important quantities connected with these measurements are the *average of the results* and the *standard deviation of the results*, which are defined respectively by

$$\mathbf{A}_{\sigma,N}(\alpha) := \sum_{k\in I} \lambda_k \frac{N_k}{N} \text{ and } \mathbf{D}_{\sigma,N}(\alpha) := \left(\sum_{k\in I} (\lambda_k - \mathbf{A}_{\sigma,N}(\alpha))^2 \frac{N_k}{N} \right)^{\frac{1}{2}},$$

if N_k denotes the number of copies for which the result λ_k has been obtained. For N large enough, the theoretical predictions of $\mathbf{A}_{\sigma,N}$ and $\mathbf{D}_{\sigma,N}$ are respectively

$$\mathbf{A}_{\sigma,th}(\alpha) := \sum_{k\in I} \lambda_k p(\alpha(\{\lambda_k\}),\sigma) \text{ and}$$

$$\mathbf{D}_{\sigma,th}(\alpha) := \left(\sum_{k\in I} (\lambda_k - \mathbf{A}_{\sigma,th}(\alpha))^2 p(\alpha(\{\lambda_k\}),\sigma) \right)^{\frac{1}{2}},$$

which can be written as

$$\mathbf{A}_{\sigma,th}(\alpha) = \int_{\mathbb{R}} \xi d\mu_\sigma^\alpha \text{ and } \mathbf{D}_{\sigma,th}(\alpha) = \left(\int_{\mathbb{R}} (\xi - \mathbf{A}_{\sigma,th}(\alpha))^2 d\mu_\sigma^\alpha \right)^{\frac{1}{2}}$$

since

$$\mu_\sigma^\alpha(\mathbb{R} - \{\lambda_k\}_{k\in I}) = 0 \text{ and } p(\alpha(\{\lambda_k\}),\sigma) = \mu_\sigma^\alpha(\{\lambda_k\}), \forall k \in I$$

(cf. 8.3.9 and 8.3.8). Thus, for a discrete observable with a finite number of possible results, the expected result and the uncertainty defined in 19.1.20 are the theoretical predictions of the average and of the standard deviation of the results obtained in a large number of measurements.

The analysis above cannot be carried out for an observable with an infinite number of possible results, since for an observable α of this kind there might be possible results λ such that $p(\alpha(\{\lambda\}),\sigma) = 0$ for all $\sigma \in \Sigma$ (this would represent the existence of a continuum of possible results). One could argue that for every actual measuring instrument there exists a finite set which contains all the results that the instrument can produce, and therefore every actual measuring instrument must be represented by an observable with only a finite number of possible results. Thus, one could be tempted into discarding observables with an infinite number of possible results on the grounds that they are not realistic. However, physical theories can hardly ever be formulated in terms of actual measuring instruments and the use of idealized observables is common practice in physics (for instance, the position and the velocity of a particle are observables which in both the classical and the quantum mechanics are not discrete, even though no actual measuring instrument can pinpoint their alleged values better than assigning them to intervals related to the resolution of the instrument; moreover, as to velocity in the classical mechanics, no actual instrument can really compute a derivative). Hence, idealized observables must be taken into consideration. The idealistic import of this is lessened by the fact that every observable α can be considered as the limit of a sequence of realistic observables in the sense explained below.

Let α be an observable. For each $n \in \mathbb{N}$, let E_n be a bounded interval, let $\{F_{n,k}\}_{k \in I_n}$ be a finite partition of E_n such that $F_{n,k}$ is an interval for all $k \in I_n$, let $\lambda_{n,k}$ be a non-null element of $F_{n,k}$ for all $k \in I_n$; further, assume that $E_n \subset E_{n+1}$ for each $n \in \mathbb{N}$, that $\bigcup_{n=1}^\infty E_n = \mathbb{R}$, and that $\lim_{n \to \infty} \ell_n = 0$ if ℓ_n denotes the maximum length of the intervals of the family $\{F_{n,k}\}_{k \in I_n}$. For instance, we could have $E_n := \left[-n, n + \frac{1}{2^n}\right)$, $I_n := \{0, \pm 1, \pm 2, ..., \pm n2^n\}$, $F_{n,k} := \left[\frac{k}{2^n}, \frac{k+1}{2^n}\right)$, $\lambda_{n,k} := \frac{k+\frac{1}{2}}{2^n}$. We define the function $\xi_n := \sum_{k \in I_n} \lambda_{n,k} \chi_{F_{n,k}}$ and the observable $\alpha_n := \xi_n(\alpha)$. The observable α_n is discrete and it has a finite number of possible results since

$$\mu_\sigma^{\alpha_n}(\{\lambda_{n,k}\}_{k \in I_n} \cup \{0\}) = \mu_\sigma^\alpha(\xi_n^{-1}(\{\lambda_{n,k}\}_{k \in I_n} \cup \{0\})) = \mu_\sigma^\alpha(\mathbb{R}) = 1, \forall \sigma \in \Sigma.$$

The observable α_n can be considered a realistic approximation of the observable α. In fact, α_n is obtained by replacing the scale that defines α with another scale (cf. 19.1.14) which yields the same conventional results $\lambda_{n,k}$ for any ideal result (i.e. a result according to α) that belongs to the interval $F_{n,k}$, and which gives an approximation of just a limited part of the ideal scale that defines α since a conventional value (chosen here to be zero, but what follows would be the same if we chose any other number) is assigned to the event $\alpha(\mathbb{R} - E_n)$ which occurs when the ideal result is not in E_n. Thus, the observable α_n (which is interpreted here as an equivalence class, cf. 19.1.10) might at least in principle be implemented by a realistic measuring instrument, which would register only a limited range of results and would not distinguish between values that lie in the same interval $F_{n,k}$ (the maximum length ℓ_n would be related to the resolving power of the instrument).

Inasmuch as the sequence $\{\alpha_n\}$ of realistic observables is considered an approximation of the observable α, the theoretical prediction of the average of the results obtained when α is measured in a large number of copies prepared in a state $\sigma \in \Sigma$ must be given by $\lim_{n \to \infty} \mathbf{A}_{\sigma,th}(\alpha_n)$, provided this limit exists and is independent of the particular choice of E_n, $\{F_{n,k}\}_{k \in I_n}$, $\{\lambda_{n,k}\}_{k \in I_n}$. And similarly for the theoretical prediction of the standard deviation of the results and $\lim_{n \to \infty} \mathbf{D}_{\sigma,th}(\alpha_n)$. Now, the following proposition can be proved (see below):

Proposition. *Let $\sigma \in \Sigma$. Then the sequences $\{\mathbf{A}_{\sigma,th}(\alpha_n)\}$ and $\{\mathbf{D}_{\sigma,th}(\alpha_n)\}$ are convergent iff $\int_{\mathbb{R}} \xi^2 d\mu_\sigma^\alpha < \infty$; if these sequences are convergent then*

$$\lim_{n \to \infty} \mathbf{A}_{\sigma,th}(\alpha_n) = \int_{\mathbb{R}} \xi d\mu_\sigma^\alpha \ and$$

$$\lim_{n \to \infty} \mathbf{D}_{\sigma,th}(\alpha_n) = \left(\int_{\mathbb{R}} (\xi - \left(\int_{\mathbb{R}} \xi d\mu_\sigma^\alpha \right))^2 d\mu_\sigma^\alpha \right)^{\frac{1}{2}}.$$

This gives an operational interpretation to the definitions given in 19.1.20 and in 19.1.21.

We end these remarks by proving the proposition above.

Preliminarily we note that, for each $n \in \mathbb{N}$,

$$\mathbf{A}_{\sigma,th}(\alpha_n) := \sum_{k \in I_n} \lambda_{n,k} p(\alpha_n(\{\lambda_{n,k}\}), \sigma) = \int_{\mathbb{R}} \xi_n d\mu_\sigma^\alpha$$

and

$$\mathbf{D}_{\sigma,th}(\alpha_n) := \left(\sum_{k \in I_n} (\lambda_{n,k} - \mathbf{A}_{\sigma,th}(\alpha_n))^2 p(\alpha_n(\{\lambda_{n,k}\}), \sigma) \right)^{\frac{1}{2}}$$

$$= \left(\int_{\mathbb{R}} (\xi_n - \mathbf{A}_{\sigma,th}(\alpha_n))^2 d\mu_\sigma^\alpha \right)^{\frac{1}{2}} = \left(\int_{\mathbb{R}} \xi_n^2 d\mu_\sigma^\alpha - (\mathbf{A}_{\sigma,th}(\alpha_n))^2 \right)^{\frac{1}{2}},$$

since $p(\alpha_n(\{\lambda_{n,k}\}), \sigma) = p(\alpha(\xi_n^{-1}(\{\lambda_{n,k}\})), \sigma) = \mu_\sigma^\alpha(F_{n,k})$ for all $k \in I_n$ (the equalities above are in agreement with what is proved more in general in 19.1.23).

First we assume $\int_{\mathbb{R}} \xi^2 d\mu_\sigma^\alpha < \infty$. Because the measure μ_σ^α is finite, we have $\xi \in \mathcal{L}^1(\mathbb{R}, \mathcal{A}(d_{\mathbb{R}}), \mu_\sigma^\alpha)$ (cf. 11.1.3) and hence $|\xi| + \ell_n \in \mathcal{L}^1(\mathbb{R}, \mathcal{A}(d_{\mathbb{R}}), \mu_\sigma^\alpha)$ since $1_{\mathbb{R}} \in \mathcal{L}^1(\mathbb{R}, \mathcal{A}(d_{\mathbb{R}}), \mu_\sigma^\alpha)$ (cf. 8.2.6). Moreover,

$$|\xi_n(x)| \leq |x| + \ell_n \ and \ \xi_n(x) \xrightarrow[n \to \infty]{} x, \forall x \in \mathbb{R}.$$

Then, by Lebesgue's dominated convergence theorem (cf. 8.2.11) we have

$$\int_{\mathbb{R}} \xi_n d\mu_\sigma^\alpha \xrightarrow[n \to \infty]{} \int_{\mathbb{R}} \xi d\mu_\sigma^\alpha.$$

Also, we have $(|\xi| + \ell_n)^2 \in \mathcal{L}^1(\mathbb{R}, \mathcal{A}(d_{\mathbb{R}}), \mu_\sigma^\alpha)$ since $1_{\mathbb{R}} \in \mathcal{L}^2(\mathbb{R}, \mathcal{A}(d_{\mathbb{R}}), \mu_\sigma^\alpha)$ (cf. 11.1.2a). Moreover,

$$|\xi_n^2(x)| \leq (|x| + \ell_n)^2 \ and \ \xi_n^2(x) \xrightarrow[n \to \infty]{} x^2, \forall x \in \mathbb{R}.$$

Then, by Lebesgue's dominated convergence theorem we have

$$\int_{\mathbb{R}} \xi_n^2 d\mu_\sigma^\alpha \xrightarrow[n \to \infty]{} \int_{\mathbb{R}} \xi^2 d\mu_\sigma^\alpha$$

and hence

$$\left(\int_{\mathbb{R}} \xi_n^2 d\mu_\sigma^\alpha - \left(\int_{\mathbb{R}} \xi_n d\mu_\sigma^\alpha \right)^2 \right)^{\frac{1}{2}} \xrightarrow[n \to \infty]{} \left(\int_{\mathbb{R}} \xi^2 d\mu_\sigma^\alpha - \left(\int_{\mathbb{R}} \xi d\mu_\sigma^\alpha \right)^2 \right)^{\frac{1}{2}}$$

$$= \left(\int_{\mathbb{R}} \left(\xi - \left(\int_{\mathbb{R}} \xi d\mu_\sigma^\alpha \right) \right)^2 d\mu_\sigma^\alpha \right)^{\frac{1}{2}}.$$

Next and conversely we assume that the sequences $\{\mathbf{A}_{\sigma,th}(\alpha_n)\}$ and $\{\mathbf{D}_{\sigma,th}(\alpha_n)\}$ are convergent. Then the sequence $\{\int_{\mathbb{R}} \xi_n^2 d\mu_\sigma^\alpha\}$ is convergent since

$$(\mathbf{D}_{\sigma,th}(\alpha_n))^2 - (\mathbf{A}_{\sigma,th}(\alpha_n))^2 = \int_{\mathbb{R}} \xi_n^2 d\mu_\sigma^\alpha,$$

and hence (cf. 2.1.9)

$$\exists M \in [0,\infty) \text{ such that } \int_{\mathbb{R}} \xi_n^2 d\mu_\sigma^\alpha \leq M, \forall n \in \mathbb{N}.$$

By Fatou's lemma (cf. 8.1.20), this implies that $\int_{\mathbb{R}} \xi^2 d\mu_\sigma^\alpha \leq M$.

(b) Suppose that an observable α and a state $\sigma \in \Sigma$ are so that α is evaluable in σ and $\Delta_\sigma \alpha = 0$. Then $\int_{\mathbb{R}} (\xi - \langle\alpha\rangle_\sigma)^2 d\mu_\sigma^\alpha = 0$, and hence (cf. 8.1.12a) $x - \langle\alpha\rangle_\sigma = 0$ μ_σ^α-a.e. on $\mathbb{R}$, and hence $\mu_\sigma^\alpha(\mathbb{R} - \{\langle\alpha\rangle_\sigma\}) = 0$, and hence $\mu_\sigma^\alpha(\{\langle\alpha\rangle_\sigma\}) = 1$. This means that there is a result λ which is obtained with certainty for any number of copies prepared in the state σ (then, obviously, $\langle\alpha\rangle_\sigma = \lambda$).

Conversely, suppose that for $\sigma \in \Sigma$ there is $\lambda \in \mathbb{R}$ such that $\mu_\sigma^\alpha(\{\lambda\}) = 1$. Then $\mu_\sigma^\alpha(\mathbb{R} - \{\lambda\}) = 0$. Thus, μ_σ^α is the Dirac measure in λ and we have (cf. 8.3.6) $\int_{\mathbb{R}} \xi^2 d\mu_\sigma^\alpha = \lambda^2 < \infty$ (hence, α is evaluable in σ), $\langle\alpha\rangle_\sigma = \lambda$, $\Delta_\sigma \alpha = 0$.

19.1.23 Proposition. *Let $(X, \mathcal{A})$ be a measurable space and let α be an X-valued observable. Let $\varphi : D_\varphi \to \mathbb{R}$ be a function from X to $\mathbb{R}$ which is $\mathcal{A}^{D_\varphi}$-measurable and which is so that $D_\varphi \in \mathcal{A}$ and $\mu_\sigma^\alpha(X - D_\varphi) = 0$ for each $\sigma \in \Sigma$. For the observable $\varphi(\alpha)$ we have:*

$\varphi(\alpha)$ is evaluable in a state $\sigma \in \Sigma$ iff $\int_X \varphi^2 d\mu_\sigma^\alpha < \infty$;
if $\varphi(\alpha)$ is evaluable in $\sigma \in \Sigma$, then

$$\langle\varphi(\alpha)\rangle_\sigma = \int_X \varphi d\mu_\sigma^\alpha \text{ and } \Delta_\sigma \varphi(\alpha) = \left(\int_X (\varphi - \langle\varphi(\alpha)\rangle_\sigma)^2 d\mu_\sigma^\alpha \right)^{\frac{1}{2}}.$$

Proof. Since $\mu_\sigma^{\varphi(\alpha)}(E) = \mu_\sigma^\alpha(\varphi^{-1}(E))$ for all $E \in \mathcal{A}(d_{\mathbb{R}})$, we obtain the statement from 8.3.11 (π is there what φ is here). $\qquad\square$

19.1.24 Corollary. *Let α be an observable and let $\alpha^2 := \xi^2(\alpha)$. If α^2 is evaluable in a state $\sigma \in \Sigma$, then α is evaluable in σ and*

$$\Delta_\sigma \alpha = \left(\langle\alpha^2\rangle_\sigma - \langle\alpha\rangle_\sigma^2 \right)^{\frac{1}{2}}.$$

Proof. If α^2 is evaluable in $\sigma \in \Sigma$ then $\xi^2 \in \mathcal{L}^2(\mathbb{R}, \mathcal{A}(d_{\mathbb{R}}), \mu_\sigma^\alpha)$ (cf. 19.1.23), and hence $\xi^2 \in \mathcal{L}^1(\mathbb{R}, \mathcal{A}(d_{\mathbb{R}}), \mu_\sigma^\alpha)$ (since the measure μ_σ^α is finite, cf. 11.1.3), i.e. α is evaluable in σ. Moreover,

$$\Delta_\sigma \alpha = \left(\int_{\mathbb{R}} (\xi - \langle\alpha\rangle_\sigma)^2 d\mu_\sigma^\alpha \right)^{\frac{1}{2}} = \left(\int_{\mathbb{R}} \xi^2 d\mu_\sigma^\alpha - \langle\alpha\rangle_\sigma^2 \right)^{\frac{1}{2}} = \left(\langle\alpha^2\rangle_\sigma - \langle\alpha\rangle_\sigma^2 \right)^{\frac{1}{2}},$$

since $\int_{\mathbb{R}} \xi^2 d\mu_\sigma^\alpha = \langle\alpha^2\rangle_\sigma$ (cf. 19.1.23). $\qquad\square$

19.1.25 Proposition. *A bounded observable is evaluable in every state $\sigma \in \Sigma$.*

Proof. If α is bounded then there exists $k \in [0, \infty)$ such that $|\xi^2(x)| = x^2 \leq k$ for all $x \in sp_\alpha$, and hence μ_σ^α-a.e. on $\mathbb{R}$ for every $\sigma \in \Sigma$, since $\mu_\sigma^\alpha(\mathbb{R} - sp_\alpha) = 0$ (cf. 19.1.17). The result then follows from 8.2.6. $\qquad\square$

19.1.26 Definition. For each proposition $\pi \in \Pi$, we define the observable α_π by letting

$$\alpha_\pi : \mathcal{A}(d_\mathbb{R}) \to \Pi$$

$$E \mapsto \alpha_\pi(E) := \begin{cases} \pi_0 \text{ if } 0 \notin E \text{ and } 1 \notin E, \\ \pi \text{ if } 0 \notin E \text{ and } 1 \in E, \\ \neg\pi \text{ if } 0 \in E \text{ and } 1 \notin E, \\ \pi_1 \text{ if } 0 \in E \text{ and } 1 \in E. \end{cases}$$

For every state $\sigma \in \Sigma$, the function $\mu_\sigma^{\alpha_\pi}$ is indeed a probability measure on $\mathcal{A}(d_\mathbb{R})$, since $p(\pi_0, \sigma) = 0$, $p(\neg\pi, \sigma) = 1 - p(\pi, \sigma)$, $p(\pi_1, \sigma) = 1$; it is the measure defined in 8.3.8, with $I := \{1, 2\}$, $x_1 := 1$, $x_2 := 0$, $a_1 := p(\pi, \sigma)$, $a_2 := p(\neg\pi, \sigma)$. Moreover, it is obvious that there exists a measuring apparatus which supports α_π: it is the same piece of equipment that defines the proposition π (cf. 19.1.9c).

The observable α_π is said to be a *yes-no observable* or a *two-valued observable*, since the apparatus which defines α_π gives the result 1 when the event π occurs and the result 0 when the event π does not occur (i.e., when the event $\neg\pi$ occurs). Thus, the possible result for α_π are 1 and 0, provided $\pi \neq \pi_1$ (if $\pi = \pi_1$ then the only possible result is 1) and $\pi \neq \pi_0$ (if $\pi = \pi_0$ then then only possible result is 0).

19.1.27 Proposition. *For each proposition $\pi \in \Pi$, the observable α_π is evaluable in every state $\sigma \in \Sigma$ and we have*

$$\langle \alpha_\pi \rangle_\sigma = p(\pi, \sigma) \text{ and } \Delta_\sigma \alpha_\pi = (p(\pi, \sigma)(1 - p(\pi, \sigma)))^{\frac{1}{2}}.$$

Proof. Since α_π is such that $\mu_\sigma^{\alpha_\pi}(\mathbb{R} - \{0, 1\}) = 0$, $\mu_\sigma^{\alpha_\pi}(\{1\}) = p(\pi, \sigma)$, $\mu_\sigma^{\alpha_\pi}(\{0\}) = p(\neg\pi, \sigma) = 1 - p(\pi, \sigma)$ for every $\sigma \in \Sigma$, by 8.3.9 and 8.3.8 we have that α_π is evaluable in every state $\sigma \in \Sigma$ and

$$\langle \alpha_\pi \rangle_\sigma = 0\mu_\sigma^{\alpha_\pi}(\{0\}) + 1\mu_\sigma^{\alpha_\pi}(\{1\}) = p(\pi, \sigma),$$
$$(\Delta_\sigma \alpha_\pi)^2 = (0 - \langle \alpha_\pi \rangle_\sigma)^2 \mu_\sigma^{\alpha_\pi}(\{0\}) + (1 - \langle \alpha_\pi \rangle_\sigma)^2 \mu_\sigma^{\alpha_\pi}(\{1\})$$
$$= (0 - p(\pi, \sigma))^2(1 - p(\pi, \sigma)) + (1 - p(\pi, \sigma))^2 p(\pi, \sigma)$$
$$= p(\pi, \sigma)(1 - p(\pi, \sigma)).$$

$\qquad\square$

19.2 States, propositions, observables in classical statistical theories

Classical statistical theories, although very diverse, have some common features, some of which we set out here axiomatically. As before, we denote by Σ and Π the families of states and of propositions that define a fixed physical system, which is assumed in this section to be described by a classical statistical theory. By Σ, Π and p we denote what was denoted by $\hat{\Sigma}, \hat{\Pi}, \hat{p}$ in 19.1.5 (cf. 19.1.12).

19.2.1 Axiom (Axiom C1). In a classical statistical theory it is assumed that, for every pair of propositions $\pi, \pi' \in \Pi$, both π and π' can be determined simultaneously in any copy of the system prepared in any state $\sigma \in \Sigma$, and that it is possible to define a proposition of the system by the event which is considered to have occurred if and only if both the events that define π and π' have occurred, and also another proposition by the event which is considered to have occurred if and only if at least one of the events that define π and π' has occurred. The proposition defined by the first event will be denoted by $\pi \wedge \pi'$ (π and π') and the proposition defined by the second event by $\pi \vee \pi'$ (π or π').

19.2.2 Remark. The reason behind axiom C1 is that, in a classical theory, the determination of any proposition π for any copy prepared in any state σ is held to be implementable in such an "unobtrusive" way that, immediately after the proposition has been determined, the copy can still be considered as if it had just been prepared in the state σ. That is to say, recalling that π stands for an equivalence class, there is an event which belongs to the class π and which requires an interaction, between a copy prepared in σ and the apparatus in which the event possibly occurs, which involves so little transfers of e.g. energy, momentum, angular momentum that they can be considered negligible, so that it is as if nothing had happened to the copy, which therefore can be considered still in the state σ. This makes it possible to determine two propositions, one immediately after the other, in the same copy and assume that the determination of the first of them has had no influence on the outcome of the determination of the second. Moreover, it makes it possible to consider immaterial the order in which the two propositions are determined.

19.2.3 Axiom (Axiom C2). In a classical statistical theory it is assumed that there is a subfamily S of the family Σ of all states which is so that

$$p(\pi, s) \in \{0, 1\}, \forall \pi \in \Pi, \forall s \in S$$

(it is a matter of convenience to denote the elements of S by the letter s).

For each $\pi \in \Pi$ we define $S_\pi := \{s \in S : p(\pi, s) = 1\}$, and we denote by $\mathcal{A}$ the σ-algebra on S which is generated by the family $\{S_\pi : \pi \in \Pi\}$. Then, it is assumed that $\{s\} \in \mathcal{A}$ for all $s \in S$ and that for every $\sigma \in \Sigma$ there is a probability measure μ_σ on $\mathcal{A}$ such that

$$p(\pi, \sigma) = \mu_\sigma(S_\pi), \forall \pi \in \Pi$$

(μ_σ is uniquely defined by this condition, as will be seen in 19.2.5b).

19.2.4 Remark. The assumptions of axiom C2 can be derived from a picture of the essentials of a classical theory which can be summarized as follows.

In a classical theory, any copy of a physical system is assumed to be, at any given time, in a "real condition" so that every proposition is either certainly true or certainly false. If a state preparation is completely accurate, then we must know the "real condition" of the copy it has produced (or it has selected, as it is more often thought to have done in classical theories); the elements of S are these completely accurate state preparations (they are sometimes called *microstates*) and therefore we must have either $p(\pi, s) = 1$ or $p(\pi, s) = 0$ for all $\pi \in \Pi$ and $s \in S$. Moreover, for every possible "real condition" it is assumed that there exists a state preparation which produces (or selects) with certainty a copy in that condition; this corresponds to the notion that every "real condition" must have an operational counterpart, at least in principle. In this way, the family of "real conditions" can be identified with the family S of completely accurate states (for this reason, S is sometimes called the *phase space* of the system). If a state preparation $\sigma \in \Sigma$ is not completely accurate, then the copies it produces are not all in the same "real condition"; however, σ is assumed to determine the probability that a copy it produces is in any real condition: σ determines a probability measure μ_σ on $\mathcal{A}$ so that, for every measurable subset E of S, $\mu_\sigma(E)$ is the probability that a copy produced by σ is in the "real condition" in which the copy would be if one of the microstates of E had been performed. Now, if N copies of the system are produced by the preparation procedure σ and N_π denotes the number of times when a proposition $\pi \in \Pi$ has been found to be true, then N_π is also the number of times when the copy is in the "real condition" associated with a microstate of S_π (this is so by the definition of S_π, recalling that $p(\pi, s) \in \{0, 1\}$ for all $s \in S$); thus, $\frac{N_\pi}{N}$ approaches both $p(\pi, \sigma)$ and $\mu_\sigma(S_\pi)$ as N grows, and we are led to the condition

$$p(\pi, \sigma) = \mu_\sigma(S_\pi).$$

We point out that, since a copy of the system is considered to be always in a "real condition" corresponding to an element of S, the probability measure μ_σ which represents a state σ is a measure of our lack of knowledge of what that "real condition" actually happens to be. For this reason, the probabilities that arise in a classical statistical theory are said to be of an *epistemic* nature. Otherwise stated, statistical aspects emerge in a classical theory only when we consider state preparations which are not completely precise, while the theory does not contain any statistical aspect if we restrict ourselves to considering only the absolutely precise states of the family S (e.g., this is the case in classical mechanics).

As to the assumption that $\{s\} \in \mathcal{A}$ for all $s \in S$, we note that a stronger assumption would be to suppose that every state $s \in S$ defines the proposition π_s that is determined to be true if and only if the copy of the system is found to be in the real condition which would have been produced by s. Then we would have $\{s\} = S_{\pi_s}$. However, this hypothetical proposition π_s can be difficult to implement unless one can find a simple event, or a collection of such, that identifies the real condition corresponding to s. The assumption $\{s\} \in \mathcal{A}$ is a rather milder request than the request of having the hypothetical proposition π_s as an element of Π. In any case, the main role of the condition $\{s\} \in \mathcal{A}$ is to make it possible to claim that there exists a probability measure μ_s on $\mathcal{A}$ such that

$$p(\pi, s) = \mu_s(S_\pi), \forall \pi \in \Pi$$

(and indeed in 19.2.5c we will see that this condition implies that μ_s is the Dirac measure in s, which requires the condition $\{s\} \in \mathcal{A}$ in order to be defined).

19.2.5 Proposition. *The following statements are true:*

(a) the family $\{S_\pi : \pi \in \Pi\}$ is an algebra on S;
(b) for $\sigma \in \Sigma$, if μ is a measure on $\mathcal{A}$ such that

$$p(\pi, \sigma) = \mu(S_\pi), \forall \pi \in \Pi,$$

then $\mu = \mu_\sigma$;
(c) for every $s \in S$, μ_s is the Dirac measure in s (the Dirac measure in s should be denoted by μ_s according to the notation introduced in 8.3.6; however, μ_s denotes here the measure μ_σ defined as in 19.2.3 for $\sigma := s$, and therefore in the proof we shall temporarily denote the Dirac measure in s by a different symbol).
(d) if $\sigma, \sigma' \in \Sigma$ are so that $\mu_\sigma = \mu_{\sigma'}$, then $\sigma = \sigma'$;
(e) if $\pi, \pi' \in \Pi$ are so that $S_\pi = S_{\pi'}$, then $\pi = \pi'$.
(f) for $\pi \in \Pi$, $\pi \neq \pi_0$ iff there exists $s \in S$ such that $p(\pi, s) = 1$.

Proof. a: From the definition of $\pi \vee \pi'$ that was given in 19.2.1, we have

$$S_\pi \cup S_{\pi'} = S_{\pi \vee \pi'}, \forall \pi, \pi' \in \Pi.$$

From the definition of $\neg \pi$ that was given in 19.1.3, we have (cf. 19.1.4a)

$$S - S_\pi = S_{\neg \pi}, \forall \pi \in \Pi.$$

Thus, $\{S_\pi : \pi \in \Pi\}$ is an algebra on S.

b: For μ as in the statement, we have $\mu = \mu_\sigma$ because $\{S_\pi : \pi \in \Pi\}$ is an algebra and μ_σ is a finite measure (cf. 7.3.2).

c: If $s \in S$ and μ^s is the Dirac measure in s, directly from the definition of S_π and of Dirac measure, we have

$$p(\pi, s) = \mu^s(S_\pi), \forall \pi \in \Pi;$$

in view of statement b, this proves that $\mu^s = \mu_s$.

d: For $\sigma, \sigma' \in \Sigma$, we have (cf. 19.1.5; recall that we are "dropping the carets")

$$\mu_\sigma = \mu_{\sigma'} \Rightarrow [p(\pi, \sigma) = \mu_\sigma(S_\pi) = \mu_{\sigma'}(S_\pi) = p(\pi, \sigma'), \forall \pi \in \Pi] \Rightarrow \sigma = \sigma'.$$

e: For $\pi, \pi' \in \Pi$, we have (cf. 19.1.5)

$$S_\pi = S_{\pi'} \Rightarrow [p(\pi, \sigma) = \mu_\sigma(S_\pi) = \mu_\sigma(S_{\pi'}) = p(\pi', \sigma), \forall \sigma \in \Sigma] \Rightarrow \pi = \pi'.$$

f: The "if" part is obvious (cf. 19.1.7). As to the "only if" part, we have

$$\pi \neq \pi_0 \Rightarrow [\exists \sigma \in \Sigma \text{ s.t. } \mu_\sigma(S_\pi) = p(\pi, \sigma) \neq 0] \Rightarrow S_\pi \neq \emptyset \Rightarrow$$
$$[\exists s \in S \text{ s.t. } s \in S_\pi, \text{ and hence s.t. } p(\pi, s) = \mu_s(S_\pi) = 1].$$

$\square$

19.2.6 Proposition. *Let α be an observable. Then:*

(a) *for every $s \in S$ there exists $\alpha_s \in \mathbb{R}$ so that μ_s^α is the Dirac measure in α_s; then α is evaluable in s, $\langle \alpha \rangle_s = \alpha_s$, $\Delta_s \alpha = 0$;*

(b) *there exists a unique function $\varphi_\alpha : S \to \mathbb{R}$ such that*

$$S_{\alpha(E)} = \varphi_\alpha^{-1}(E), \forall E \in \mathcal{A}(d_\mathbb{R});$$

the function φ_α is $\mathcal{A}$-measurable and it is defined by $\varphi_\alpha(s) := \langle \alpha \rangle_s$ for all $s \in S$; we have $sp_\alpha = \overline{R_{\varphi_\alpha}}$; then, α is a bounded observable iff φ_α is a bounded function;

for a state $\sigma \in \Sigma$, α is evaluable in σ iff $\int_S \varphi_\alpha^2 d\mu_\sigma < \infty$; if α is evaluable in σ then

$$\langle \alpha \rangle_\sigma = \int_S \varphi_\alpha d\mu_\sigma \text{ and } \Delta_\sigma \alpha = \left(\int_S (\varphi_\alpha - \langle \alpha \rangle_\sigma)^2 d\mu_\sigma \right)^{\frac{1}{2}};$$

if β is an observable such that $\varphi_\alpha = \varphi_\beta$, then $\alpha = \beta$.

(c) *for a function $\psi : D_\psi \to \mathbb{R}$ from $\mathbb{R}$ to $\mathbb{R}$ which is $\mathcal{A}(d_\mathbb{R})^{D_\psi}$-measurable and such that $D_\psi \in \mathcal{A}(d_\mathbb{R})$, the condition $\mu_\sigma^\alpha(\mathbb{R} - D_\psi) = 0$ for all $\sigma \in \Sigma$ (which is the condition for the definition of the observable $\psi(\alpha)$) is equivalent to the condition $R_{\varphi_\alpha} \subset D_\psi$; if these conditions are satisfied then $\varphi_{\psi(\alpha)} = \psi \circ \varphi_\alpha$.*

Proof. a: For each $s \in S$, the measure μ_s^α is defined by

$$\mu_s^\alpha(E) = p(\alpha(E), s), \forall E \in \mathcal{A}(d_\mathbb{R})$$

(cf. 19.1.8). Hence, in view of 19.2.3 we have $\mu_s^\alpha(E) \in \{0, 1\}$ for all $E \in \mathcal{A}(d_\mathbb{R})$. By 8.3.7, this implies that there exists $\alpha_s \in \mathbb{R}$ so that μ_s^α is the Dirac measure in α_s, and this implies (cf. 8.3.6) that

$$\int_\mathbb{R} \xi^2 d\mu_s^\alpha = \alpha_s^2 < \infty, \langle \alpha \rangle_s = \int_\mathbb{R} \xi d\mu_s^\alpha = \alpha_s, \Delta_s \alpha = \left(\int_\mathbb{R} (x - \langle \alpha \rangle_s)^2 d\mu_s^\alpha(x) \right)^{\frac{1}{2}} = 0.$$

b: We define the function

$$\varphi_\alpha : S \to \mathbb{R}$$
$$s \mapsto \varphi_\alpha(s) := \langle \alpha \rangle_s.$$

For every $E \in \mathcal{A}(d_\mathbb{R})$ we have (directly from the definitions and from the fact that μ_s^α is the Dirac measure in $\langle \alpha \rangle_s$), for $s \in S$,

$$s \in S_{\alpha(E)} \Leftrightarrow p(\alpha(E), s) = 1 \Leftrightarrow \mu_s^\alpha(E) = 1 \Leftrightarrow \langle \alpha \rangle_s \in E \Leftrightarrow s \in \varphi_\alpha^{-1}(E),$$

and hence $S_{\alpha(E)} = \varphi_\alpha^{-1}(E)$. This also proves that φ_α is $\mathcal{A}$-measurable.

Now suppose that a function $\varphi : S \to \mathbb{R}$ is such that

$$S_{\alpha(E)} = \varphi^{-1}(E), \forall E \in \mathcal{A}(d_\mathbb{R});$$

then φ is $\mathcal{A}$-measurable and

$$\mu_s^\alpha(E) = p(\alpha(E), s) = \mu_s(S_{\alpha(E)}) = \mu_s(\varphi^{-1}(E)), \forall E \in \mathcal{A}(d_\mathbb{R}), \forall s \in S.$$

Then, by 8.3.11 and 19.2.5c,

$$\varphi(s) = \int_S \varphi d\mu_s = \int_{\mathbb{R}} \xi d\mu_s^\alpha = \langle \alpha \rangle_s = \varphi_\alpha(s), \forall s \in S.$$

Now we prove that $sp_\alpha = \overline{R_{\varphi_\alpha}}$. If $\lambda \in R_{\varphi_\alpha}$ then there exists $s \in S$ such that $s \in \varphi_\alpha^{-1}(\{\lambda\})$ and hence

$$\mu_s^\alpha(\{\lambda\}) = \mu_s(S_{\alpha(\{\lambda\})}) = \mu_s(\varphi_\alpha^{-1}(\{\lambda\})) = 1;$$

this proves that $R_{\varphi_\alpha} \subset sp_\alpha$ and hence $\overline{R_{\varphi_\alpha}} \subset sp_\alpha$ since sp_α is closed (cf. 19.1.17). If conversely $\lambda \notin \overline{R_{\varphi_\alpha}}$ then there exists $\varepsilon > 0$ such that $\varphi_\alpha(s) \notin (\lambda - \varepsilon, \lambda + \varepsilon)$, and hence $\mu_s(\varphi_\alpha^{-1}((\lambda - \varepsilon, \lambda + \varepsilon))) = 0$, for all $s \in S$ (since $\varphi_\alpha(s) \notin (\lambda - \varepsilon, \lambda + \varepsilon)$ is equivalent to $s \notin \varphi_\alpha^{-1}((\lambda - \varepsilon, \lambda + \varepsilon))$); then we have, for every $\sigma \in \Sigma$,

$$\mu_\sigma^\alpha((\lambda - \varepsilon, \lambda + \varepsilon)) = \mu_\sigma(\varphi_\alpha^{-1}((\lambda - \varepsilon, \lambda + \varepsilon)))$$
$$= \int_S \chi_{\varphi_\alpha^{-1}((\lambda - \varepsilon, \lambda + \varepsilon))}(s) d\mu_\sigma(s)$$
$$= \int_S \mu_s(\varphi_\alpha^{-1}((\lambda - \varepsilon, \lambda + \varepsilon))) d\mu_\sigma(s) = 0,$$

and hence $\lambda \notin sp_\alpha$. This proves that $sp_\alpha \subset \overline{R_{\varphi_\alpha}}$.

Then we also have: α is bounded iff sp_α is bounded iff $\overline{R_{\varphi_\alpha}}$ is bounded iff R_{φ_α} is bounded iff φ_α is bounded.

For $\sigma \in \Sigma$ we have

$$\mu_\sigma^\alpha(E) = p(\alpha(E), \sigma) = \mu_\sigma(S_{\alpha(E)}) = \mu_\sigma(\varphi_\alpha^{-1}(E)), \forall E \in \mathcal{A}(d_{\mathbb{R}}),$$

and this implies, by 8.3.11, that

$$\int_{\mathbb{R}} \xi^2 d\mu_\sigma^\alpha = \int_S \varphi_\alpha^2 d\mu_\sigma$$

and, if α is evaluable in σ, that

$$\langle \alpha \rangle_\sigma = \int_{\mathbb{R}} \xi d\mu_\sigma^\alpha = \int_S \varphi_\alpha d\mu_\sigma \text{ and}$$
$$\Delta_\sigma \alpha = \left(\int_{\mathbb{R}} (\xi - \langle \alpha \rangle_\sigma)^2 d\mu_\sigma^\alpha \right)^{\frac{1}{2}} = \left(\int_S (\varphi_\alpha - \langle \alpha \rangle_\sigma)^2 d\mu_\sigma \right)^{\frac{1}{2}}.$$

Finally, if β is an observable such that $\varphi_\alpha = \varphi_\beta$, then

$$S_{\alpha(E)} = \varphi_\alpha^{-1}(E) = \varphi_\beta^{-1}(E) = S_{\beta(E)}, \forall E \in \mathcal{A}(d_{\mathbb{R}}),$$

and hence, by 19.2.5e,

$$\alpha(E) = \beta(E), \forall E \in \mathcal{A}(d_{\mathbb{R}}),$$

and hence $\alpha = \beta$.

c: Let $\psi : D_\psi \to \mathbb{R}$ be a function from $\mathbb{R}$ to $\mathbb{R}$ which is $\mathcal{A}(d_{\mathbb{R}})^{D_\psi}$-measurable and such that $D_\psi \in \mathcal{A}(d_{\mathbb{R}})$. First, we note that $\mu_s^\alpha(\mathbb{R} - D_\psi) = 0$ for all $s \in S$ is equivalent (in view of statements a and b) to $\varphi_\alpha(s) \notin \mathbb{R} - D_\psi$ for all $s \in S$, which is equivalent to $\varphi_\alpha(s) \in D_\psi$ for all $s \in S$, which is equivalent to $R_{\varphi_\alpha} \subset D_\psi$. Next, we note that if $\mu_\sigma^\alpha(\mathbb{R} - D_\psi) = 0$ for all $\sigma \in \Sigma$ then obviously $\mu_s^\alpha(\mathbb{R} - D_\psi) = 0$ for all $s \in S$. Finally, we note that if $\mu_s^\alpha(\mathbb{R} - D_\psi) = 0$ for all $s \in S$ then $S_{\alpha(\mathbb{R} - D_\psi)} = \emptyset$, and hence $\mu_\sigma^\alpha(\mathbb{R} - D_\psi) = \mu_\sigma(S_{\alpha(\mathbb{R} - D_\psi)}) = 0$ for all $\sigma \in \Sigma$.

Thus, the condition $R_{\varphi_\alpha} \subset D_\psi$ holds true if and only if the observable $\psi(\alpha)$ can be defined (cf. 19.1.23). In this case, $\psi \circ \varphi_\alpha$ is an $\mathcal{A}$-measurable function from S to $\mathbb{R}$ such that $D_{\psi \circ \varphi_\alpha} = S$, and we have

$$\varphi_{\psi(\alpha)}(s) = \langle \psi(\alpha) \rangle_s = \int_\mathbb{R} \psi \, d\mu_s^\alpha = \psi(\langle \alpha \rangle_s) = \psi(\varphi_\alpha(s)) = (\psi \circ \varphi_\alpha)(s), \forall s \in S,$$

where the first equation holds by statement b, the second by 19.1.23, and the third because μ_s^α is the Dirac measure in $\langle \alpha \rangle_s$ (cf. statement a). $\qquad \square$

19.2.7 Proposition. *A partial ordering is defined in* Π *by letting, for* $\pi, \pi' \in \Pi$,

$$\pi \le \pi' \text{ if } S_\pi \subset S_{\pi'}.$$

For each pair $\{\pi, \pi'\}$ *of elements of* Π, *the g.l.b. exists and we have* $\inf\{\pi, \pi'\} = \pi \wedge \pi'$, *and the l.u.b. exists and we have* $\sup\{\pi, \pi'\} = \pi \vee \pi'$. *Further, we have:*

$$\pi \wedge (\pi' \vee \pi'') = (\pi \wedge \pi') \vee (\pi \wedge \pi'') \text{ and}$$
$$\pi \vee (\pi' \wedge \pi'') = (\pi \vee \pi') \wedge (\pi \vee \pi''), \forall \pi, \pi', \pi'' \in \Pi$$

(Π *is thus what is called a distributive lattice).*

We also have:

$\pi_0 \le \pi$ *and* $\pi \le \pi_1$, $\forall \pi \in \Pi$;
$\pi \wedge (\neg\pi) = \pi_0$ *and* $\pi \vee (\neg\pi) = \pi_1$, $\forall \pi \in \Pi$;
$\neg(\neg\pi) = \pi$, $\forall \pi \in \Pi$;
$\pi \le \pi' \Rightarrow \neg\pi' \le \neg\pi$, $\forall \pi, \pi' \in \Pi$

(Π *is thus what is called a Boolean algebra).*

Proof. The mapping $\pi \mapsto S_\pi$ is a bijection from Π onto $\{S_\pi : \pi \in \Pi\}$ (cf. 19.2.5e). All the facts of the statement follow from this and from the equivalent facts for the family $\{S_\pi : \pi \in \Pi\}$ of subsets of S, which hold true trivially since we have, directly from the definitions:

$S_\pi \cap S_{\pi'} = S_{\pi \wedge \pi'}$ and $S_\pi \cup S_{\pi'} = S_{\pi \vee \pi'}$, $\forall \pi, \pi' \in \Pi$;
$S_{\pi_0} = \emptyset$ and $S_{\pi_1} = S$;
$S_{\neg\pi} = S - S_\pi$, $\forall \pi \in \Pi$.

$\qquad \square$

19.2.8 Remark. We can summarize the basic mathematical structure of a classical statistical theory as follows.

There are a measurable space $(S, \mathcal{A})$, an injective mapping $\pi \mapsto S_\pi$ from the family Π of all propositions to the σ-algebra $\mathcal{A}$, an injective mapping $\sigma \mapsto \mu_\sigma$ from the family Σ of all states to the family of all probability measures on $\mathcal{A}$, so that $\mathcal{A}$ is the σ-algebra generated by the family $\{S_\pi : \pi \in \Pi\}$ and

$$p(\pi, \sigma) = \mu_\sigma(S_\pi), \forall \pi \in \Pi, \forall \sigma \in \Sigma$$

(in this subsection, we denote by $(S, \mathcal{A})$ an abstract measurable space and therefore we must denote the family of microstates by a different symbol than the symbol S used before; in what follows the family of microstates is denoted by the symbol Σ_0).

Further, there is an injective mapping $\alpha \mapsto \varphi_\alpha$ from the family of all observables to the family of all $\mathcal{A}$-measurable real functions so that for each observable α we have:

- $S_{\alpha(E)} = \varphi_\alpha^{-1}(E), \forall E \in \mathcal{A}(d_{\mathbb{R}})$;
- α is a bounded observable iff φ_α is a bounded function;
- for a state σ, α is evaluable in σ iff $\int_S \varphi_\alpha^2 d\mu_\sigma < \infty$;
- for a state σ, if α is evaluable in σ then

$$\langle \alpha \rangle_\sigma = \int_S \varphi_\alpha d\mu_\sigma \text{ and } \Delta_\sigma \alpha = \left(\int_S (\varphi_\alpha(s) - \langle \alpha \rangle_\sigma)^2 d\mu_\sigma(s) \right)^{\frac{1}{2}};$$

- if $\psi : D_\psi \to \mathbb{R}$, with $D_\psi \in \mathcal{A}(d_{\mathbb{R}})$, is such that $\psi(\alpha)$ can be defined, then $\varphi_{\psi(\alpha)} = \psi \circ \varphi_\alpha$.

Moreover, there is a subfamily Σ_0 of Σ so that $p(\pi, \sigma) \in \{0, 1\}$ for all $\pi \in \Pi$ and $\sigma \in \Sigma_0$. For $\pi \in \Pi$, $\pi \neq \pi_0$ if and only if there exists $\sigma \in \Sigma_0$ so that $p(\pi, \sigma) = 1$. Every observable α is evaluable in every state $\sigma \in \Sigma_0$ and $\Delta_\sigma \alpha = 0$; thus, every observable α can be said to have a definite value (equal to $\langle \alpha \rangle_\sigma$) in every state $\sigma \in \Sigma_0$ (cf. 19.1.22b). It is possible to identify Σ_0 with S: for $\sigma \in \Sigma$, we have $\sigma \in \Sigma_0$ iff μ_σ is the Dirac measure in a point s_σ of S; also, if $\sigma \in \Sigma_0$, then $\langle \alpha \rangle_\sigma = \varphi_\alpha(s_\sigma)$.

Finally, the family π of all propositions has the structure of a Boolean algebra.

We shall see that the mathematical structure of quantum mechanics is altogether different.

19.2.9 Remark. If we consider *only one* observable in the general statistical theory of Section 19.1 (and therefore in a quantum theory as a special case), we can note a similarity between the nature of the probabilities that played a role in that situation (cf. 19.1.8 and 19.1.9) and the nature of probabilities in a classical statistical theory. In fact, for a state σ, while the nature of the probability $p(\pi, \sigma)$ for a general proposition π is completely unspecified in the general statistical theory (and indeed $p(\pi, \sigma)$ will be obtained in a quantum theory by an algorithm altogether different from the one used in a classical theory, cf. 19.3.1 and 19.2.3), if a fixed observable α is considered then there is a σ-algebra $\mathcal{A}$ so that an element E of $\mathcal{A}$ represents the proposition "the position of the pointer is in the section of the dial identified with E" and the probability of this proposition is $\mu_\sigma^\alpha(E)$, where μ_σ^α is a probability measure on $\mathcal{A}$, and this is similar to what happens in a classical statistical theory. Actually, this is due to the classical nature we assumed for the dial and the pointer of any measuring instrument. However, at variance with what we have in a classical statistical theory, it is not true that, for any observable α, there exists a family Σ_0 of states so that $\mu_\sigma^\alpha(E) \in \{0, 1\}$ for all $E \in \mathcal{A}$ and all $\sigma \in \Sigma_0$ (these states would be the absolutely precise preparation procedures which we will mention later, in 19.3.12b).

19.3 States, propositions, observables in quantum mechanics

Quantum mechanics is a family of statistical theories, called quantum theories, which are structured in accordance with the axioms that we set out in this and in later sections of this chapter.

In a classical theory, if one considers only microstates then one has a theory which is not really statistical, since all probabilities are trivial (i.e., they are either zero or one); in a classical theory, microstates represent possible "real conditions" of the system, and one can often relate the mathematical representation of a simple classical system to one's common experience. In a quantum theory there are no states for which all probabilities are trivial, there are no states which can be related to a "real condition" of the system (this concept has no place in a quantum theory), and the mathematical representation of the physical system allows no intuitive imagery; in fact, this mathematical representation is utterly abstract and the only parts of the mathematical machinery that can be directly linked to common experience are the probabilities which can be computed through it. Moreover, in a quantum theory, states, propositions and observables cannot be handled with the same kind of logic that can be used in a classical theory. This impossibility is encoded in the structure of their mathematical representations.

In this section we examine how states, propositions, observables are represented in a quantum theory, and how probabilities, expected results and uncertainties can be computed. As always, we denote by Σ and Π the families of states and propositions that define a fixed physical system, which is assumed in this and in the next sections of this chapter to be described by a quantum theory. We remind the reader that we denote by Σ, Π, p what was denoted by $\hat{\Sigma}, \hat{\Pi}, \hat{p}$ in 19.1.5 (cf. 19.1.12).

19.3.1 Axiom (Axiom Q1). A quantum theory is a statistical theory for which a separable Hilbert space $\mathcal{H}$ is assumed to exists so that:

(a) there is a bijective mapping $\Sigma \ni \sigma \mapsto W_\sigma \in \mathcal{W}(\mathcal{H})$ from the family Σ of all states onto the family $\mathcal{W}(\mathcal{H})$ of all statistical operators in $\mathcal{H}$;

(b) there is a bijective mapping $\Pi \ni \pi \mapsto P_\pi \in \mathscr{P}(\mathcal{H})$ from the family Π of all propositions onto the family $\mathscr{P}(\mathcal{H})$ of all orthogonal projections in $\mathcal{H}$;

(c) $p(\pi, \sigma) = \mathrm{tr}(P_\pi W_\sigma)$, $\forall \pi \in \Pi$, $\forall \sigma \in \Sigma$.

19.3.2 Remarks.

(a) For all $P \in \mathscr{P}(\mathcal{H})$ and $W \in \mathcal{W}(\mathcal{H})$ we have $0 \leq \mathrm{tr}(PW) \leq 1$ (cf. 18.3.8b). Thus, condition c in 19.3.1 is consistent with the fact that p is a probability function.

(b) The structure which emerges from 19.3.1 is a truly statistical one. In a statistical theory, the probabilistic aspects become trivial only when there is a pair proposition-state (π, σ) such that the probability $p(\pi, \sigma)$ is either 0 or 1: the proposition π is then either never true or always true in all copies of the system prepared in the state σ. Consider then a proposition π such that $P_\pi \neq \mathbb{O}_\mathcal{H}$ and $P_\pi \neq \mathbb{1}$ (this is possible if the dimension of $\mathcal{H}$ is greater than one, which we assume), and a state σ such that $W_\sigma = A_u$, with $u \in \tilde{\mathcal{H}}$ (cf. 18.3.2b). Then we have

$$p(\pi, \sigma) = (u|P_\pi u) = \|P_\pi u\|^2,$$

and hence $p(\pi, \sigma) \neq 0$ and $p(\pi, \sigma) \neq 1$ whenever $u \notin N_{P_\pi} \cup R_{P_\pi}$ (cf. 13.1.3c). Now, there are infinitely many operators A_u such that $u \notin N_{P_\pi} \cup R_{P_\pi}$.

(c) There are quantum theories, which are said to be "with superselection rules", for which the mappings of conditions a and b in 19.3.1 are not surjective. These theories are outside the scope of this book. Thus, all quantum theories we discuss are "without superselection rules".

19.3.3 Proposition. *Condition c in 19.3.1 is consistent with the implications (cf. 19.1.5 and 19.1.12):*

$$[\pi', \pi'' \in \Pi \text{ and } p(\pi', \sigma) = p(\pi'', \sigma), \forall \sigma \in \Sigma] \Rightarrow \pi' = \pi'',$$
$$[\sigma', \sigma'' \in \Sigma \text{ and } p(\pi, \sigma') = p(\pi, \sigma''), \forall \pi \in \Pi] \Rightarrow \sigma' = \sigma''.$$

Proof. For $P', P'' \in \mathscr{P}(\mathcal{H})$, if

$$\mathrm{tr}(P'W) = \mathrm{tr}(P''W), \forall W \in \mathcal{W}(\mathcal{H}),$$

then in particular (cf. 18.3.2b)

$$(u|P'u) = \mathrm{tr}(P'A_u) = \mathrm{tr}(P''A_u) = (u|P''u), \forall u \in \tilde{\mathcal{H}},$$

and hence $P' = P''$ by 10.2.12. This proves the first implication of the statement. The proof of the second implication is similar. $\qquad\square$

19.3.4 Proposition. *We have $P_{\pi_0} = \mathbb{O}_\mathcal{H}$, $P_{\pi_1} = \mathbb{1}_\mathcal{H}$, and $P_{\neg\pi} = \mathbb{1}_\mathcal{H} - P_\pi$ for every proposition π.*

Proof. We have

$$\mathrm{tr}(P_{\pi_0}W_\sigma) = p(\pi_0, \sigma) = 0 \text{ and } \mathrm{tr}(P_{\pi_1}W_\sigma) = p(\pi_1, \sigma) = 1, \forall \sigma \in \Sigma.$$

Since the mapping $\Sigma \ni \sigma \mapsto W_\sigma \in \mathcal{W}(\mathcal{H})$ is surjective, this implies (cf. 18.3.2b)

$$(u|P_{\pi_0}u) = 0 = (u|\mathbb{O}_\mathcal{H}u) \text{ and } (u|P_{\pi_1}u) = 1 = (u|\mathbb{1}_\mathcal{H}u), \forall u \in \tilde{\mathcal{H}},$$

and hence $P_{\pi_0} = \mathbb{O}_{\mathcal{H}}$ and $P_{\pi_1} = \mathbb{1}_{\mathcal{H}}$ (cf. 10.2.12).

For every $\pi \in \Pi$ we have

$$\mathrm{tr}(P_{\neg\pi} W_\sigma) = p(\neg\pi, \sigma) = 1 - p(\pi, \sigma) = 1 - \mathrm{tr}(P_\pi W_\sigma), \forall \sigma \in \Sigma,$$

and hence

$$(u|P_{\neg\pi}u) = (u|(\mathbb{1}_{\mathcal{H}} - P_\pi)u), \forall u \in \tilde{\mathcal{H}},$$

and hence $P_{\neg\pi} = \mathbb{1}_{\mathcal{H}} - P_\pi$. $\square$

19.3.5 Remarks.

(a) We always assume that the dimension of the Hilbert space $\mathcal{H}$ in 19.3.1 is greater than one, for otherwise the only projections in $\mathcal{H}$ would be $\mathbb{O}_{\mathcal{H}}$ and $\mathbb{1}_{\mathcal{H}}$ and hence the only propositions of the system would be the trivial propositions π_0 and π_1.

(b) Let $\sigma \in \Sigma$ be a state such that W_σ is not a one dimensional projection. Then (cf. 18.3.6) there exist countable families $\{u_n\}_{n\in I}$ of elements of $\tilde{\mathcal{H}}$ and $\{w_n\}_{n\in I}$ of elements of $(0, 1]$, so that I contains more than one index, $A_{u_i} \neq A_{u_k}$ if $i \neq k$, $\sum_{n\in I} w_n = 1$, and

$$W_\sigma f = \sum_{n\in I} w_n A_{u_n} f, \forall f \in \mathcal{H}. \tag{1}$$

If we denote by σ_n the element of Σ such that $W_{\sigma_n} = A_{u_n}$, then we have (cf. 18.3.5c)

$$p(\pi, \sigma) = \mathrm{tr}(P_\pi W_\sigma) = \sum_{n\in I} w_n \, \mathrm{tr}(P_\pi A_{u_n}) = \sum_{n\in I} w_n p(\pi, \sigma_n), \forall \pi \in \Pi.$$

Consider now the state preparation procedure σ_0 which is defined as follows. When a copy of the system is prepared in σ_0, then it is actually as if it had been prepared in one of the states σ_n; however, it is not known in which σ_n the copy actually is, but only the probability w_n is known that the copy is in σ_n. This lack of knowledge, of which σ_n does take effect when σ_0 is implemented, could arise from technological fluctuations of the equipment that defines σ_0 (and then w_n would be a classical probability, since that equipment is made of macroscopic bodies) or else from the procedure σ_n being triggered by a previous quantum event, pertaining to a perhaps different system. Now, suppose we have N copies of the system prepared according to the procedure σ_0. If N_n is the number of the copies which are as if they had been prepared according to σ_n, and if $N_{n,\pi}$ is the number of these copies in which a proposition π is true, then we obviously have

$$\frac{\sum_{n\in I} N_{n,\pi}}{N} = \sum_{n\in I} \frac{N_n}{N} \frac{N_{n,\pi}}{N_n}.$$

Since probabilities are theoretical predictions of frequencies, we expect that, as N grows, N_n grows as well (since $w_n \neq 0$), and that

$$\frac{\sum_{n \in I} N_{n,\pi}}{N} \text{ approaches } p(\pi, \sigma_0),$$

$$\frac{N_n}{N} \text{ approaches } w_n,$$

$$\frac{N_{n,\pi}}{N_n} \text{ approaches } p(\pi, \sigma_n).$$

Thus, we are led to conclude that

$$p(\pi, \sigma_0) = \sum_{n \in I} w_n p(\pi, \sigma_n) = p(\pi, \sigma), \forall \pi \in \Pi,$$

and hence that $\sigma_0 = \sigma$ (as equivalence classes).

Therefore, the probability $p(\pi, \sigma)$ can be interpreted, within the procedure σ_0, as a mixture of different probabilities: the probabilities $p(\pi, \sigma_n)$ are part of the quantum statistical theory we are discussing, while the probabilities w_n are part of a different statistical theory, whose role is here to quantify to what extent the preparation procedures σ_n can be controlled. For this reason, a state $\sigma \in \Sigma$ such that W_σ is not a one dimensional projection is said to be a *mixed state*.

It must be pointed out that decomposition 1 for a mixed state is never unique (cf. 18.3.7). Thus, the analysis carried out above cannot be considered as *the* interpretation of the mixed state σ, but it must be regarded as a description of how *one of the many equivalent* procedures that are contained in this equivalence class σ can be implemented on the basis of procedures which implement the states σ_n. Failure to acknowledge the non-uniqueness of these many equivalent procedures may well lead to some of the so-called "paradoxes" of quantum mechanics.

All that was said above can be generalized to the case of a state $\sigma \in \Sigma$ such that there are countable families $\{W_n\}_{n \in I}$ of elements of $\mathcal{W}(\mathcal{H})$ and $\{w_n\}_{n \in I}$ of elements of $(0, 1]$ so that $W_i \neq W_k$ for $i \neq k$, $\sum_{n \in I} w_n = 1$, and

$$W_\sigma f = \sum_{n \in I} w_n W_n f, \forall f \in \mathcal{H}$$

(cf. 18.3.4). In this case, if σ_n denotes the state which is such that $W_{\sigma_n} = W_n$, the state σ is said to be *a mixture of the family* $\{\sigma_n\}_{n \in I}$ *of states*, and the elements of the family $\{w_n\}_{n \in I}$ are said to be the *weights* of the decomposition.

(c) A state σ such that W_σ is a one-dimensional projection cannot be decomposed into a mixture of other states (cf. 18.3.7). Thus, the probabilities $p(\pi, \sigma)$ that arise in connection with σ are not mixtures of probabilities intrinsic to the quantum theory that is being discussed and probabilities of a different kind; they are, that is, purely quantum probabilities. For this reason, a state $\sigma \in \Sigma$ such that W_σ is a one-dimensional projection is said to be a *purely quantum state*, or simply a *pure state*. Since the mapping $\hat{\mathcal{H}} \ni [u] \mapsto A_u \in \mathscr{P}(\mathcal{H})$

is a bijection from the family $\hat{\mathcal{H}}$ of all rays of $\mathcal{H}$ onto the family of all one-dimensional projections in $\mathcal{H}$ (cf. 13.1.13a), if we denote by Σ_0 the family of all pure states we have a bijection $\Sigma_0 \ni \sigma \mapsto [u_\sigma] \in \hat{\mathcal{H}}$, where $[u_\sigma]$ denotes, for any $\sigma \in \Sigma_0$, the ray such that $W_\sigma = A_{u_\sigma}$, i.e. such that (cf. 18.3.2b)

$$p(\pi, \sigma) = \text{tr}(P_\pi A_{u_\sigma}) = (u_\sigma | P_\pi u_\sigma), \forall \pi \in \Pi.$$

(d) Suppose we are given a countable family $\{\sigma_n\}_{n \in I}$ of pure states, and for each $n \in I$ let u_n be an element of $\hat{\mathcal{H}}$ such that $W_{\sigma_n} = A_{u_n}$. Moreover, suppose we are given a family $\{\alpha_n\}_{n \in I}$ of complex numbers so that $\sum_{n \in I} \alpha_n u_n$ converges (if it is a series) and $\left\| \sum_{n \in I} \alpha_n u_n \right\| = 1$. Then, the bijectivity of the mapping $\Sigma_0 \ni \sigma \mapsto [u_\sigma] \in \hat{\mathcal{H}}$ allows considering the pure state σ_p which is such that $[u_{\sigma_p}] = \left[\sum_{n \in I} \alpha_n u_n \right]$, i.e. such that $W_{\sigma_p} = A_u$ with $u := \sum_{n \in I} \alpha_n u_n$. This state is said to be a *coherent superposition of the family* $\{\sigma_n\}$ *of pure states*. Note that, in spite of its name, the state σ_p actually depends not only on the family $\{\sigma_n\}$ but also on the choice of the representative u_n in each equivalence class $[u_n]$.

The bijectivity of the mapping $\Sigma_0 \ni \sigma \mapsto u_\sigma \in \hat{\mathcal{H}}$ is called *superposition principle*.

We point out that, on the basis of the family $\{\sigma_n\}_{n \in I}$ of pure states considered above, we can obtain a mixed state for any family $\{w_n\}_{n \in I}$ of elements of $(0, 1]$ such that $\sum_{n \in I} w_n = 1$, defined as the state σ_m such that

$$W_{\sigma_m} f = \sum_{n \in I} w_n W_{\sigma_n} f = \sum_{n \in I} w_n A_{u_n} f, \forall f \in \mathcal{H}.$$

Clearly, this state depends only on the equivalence classes $[u_n]$ and not on their representatives.

(e) Suppose we are given an o.n.s. $\{u_n\}_{n \in I}$ in $\mathcal{H}$ and a family $\{\alpha_n\}_{n \in I}$ of complex numbers so that $\sum_{n \in I} |\alpha_n|^2 = 1$. Then we can consider the pure state $\sigma_p \in \Sigma_0$ which is such that $W_{\sigma_p} = A_u$, with $u := \sum_{n \in I} \alpha_n u_n$, or else we can consider the mixed state σ_m which is such that $W_{\sigma_m} f = \sum_{n \in I} |\alpha_n|^2 A_{u_n} f$ for each $f \in \mathcal{H}$. For each $\pi \in \Pi$ we have:

$$p(\pi, \sigma_p) = (u | P_\pi u) = \sum_{n \in I} |\alpha_n|^2 (u_n | P_\pi u_n) + \sum_{\substack{n,m \in I \\ n \neq m}} \overline{\alpha_n} \alpha_m (u_n | P_\pi u_m);$$

$$p(\pi, \sigma_m) = \sum_{n \in I} |\alpha_n|^2 \text{tr}(P_\pi A_{u_n}) = \sum_{n \in I} |\alpha_n|^2 (u_n | P_\pi u_n).$$

This shows in which way a coherent superposition of a family of pure states is different from a mixture of the same family, in the particular case we have when the family of pure states corresponds to an o.n.s. in $\mathcal{H}$. The real number $\sum_{\substack{n,m \in I \\ n \neq m}} \overline{\alpha_n} \alpha_m (u_n | P_\pi u_m)$, which makes the difference, is said to be an *interference term*.

(f) For a proposition $\pi \in \Pi$ we have (cf. 19.3.4 and 19.1.7, recalling that we "drop the carets" in conformity with 19.1.12)

$$P_\pi \neq \mathbb{O}_\mathcal{H} \Leftrightarrow \pi \neq \pi_0 \Leftrightarrow [\exists \sigma \in \Sigma \text{ s.t. } p(\pi, \sigma) \neq 0].$$

Moreover, for $\sigma \in \Sigma_0$ we have $p(\pi, \sigma) = (u_\sigma | P_\pi u_\sigma) = \| P_\pi u_\sigma \|^2$ (cf. remark c), and therefore (cf. 13.1.3c)

$$p(\pi, \sigma) = 1 \Leftrightarrow u_\sigma \in R_{P_\pi} \text{ and } p(\pi, \sigma) = 0 \Leftrightarrow u_\sigma \in N_{P_\pi};$$

these equivalences show that, for each $\sigma \in \Sigma_0$, there are propositions $\pi \in \Pi$ such that $p(\pi, \sigma) \notin \{0, 1\}$ (e.g., assume π such that $P_\pi = A_u$, with $u \in \tilde{\mathcal{H}}$ and $(u | u_\sigma) \notin \{0, 1\}$); from the first equivalence we also have

$$P_\pi \neq \mathbb{O}_\mathcal{H} \Leftrightarrow R_{P_\pi} \neq \{0_\mathcal{H}\} \Leftrightarrow [\exists \sigma \in \Sigma_0 \text{ s.t. } p(\pi, \sigma) = 1].$$

For a proposition $\pi \in \Pi$ and a state $\sigma \in \Sigma$ we have (cf. 18.3.9 and 18.3.11)

$$p(\pi, \sigma) = 1 \Leftrightarrow R_{W_\sigma} \subset R_{P_\pi} \Leftrightarrow P_\pi W_\sigma = W_\sigma \Leftrightarrow P_\pi W_\sigma P_\pi = W_\sigma$$

and

$$p(\pi, \sigma) = 0 \Leftrightarrow R_{W_\sigma} \subset N_{P_\pi} \Leftrightarrow P_\pi W_\sigma = \mathbb{O}_\mathcal{H} \Leftrightarrow P_\pi W_\sigma P_\pi = \mathbb{O}_\mathcal{H}.$$

If P_π is a one-dimensional projection, i.e. $P_\pi = A_u$ with $u \in \tilde{\mathcal{H}}$, then for a state $\sigma \in \Sigma$ we have $p(\pi, \sigma) = 1$ if and only if σ is a pure state and $[u_\sigma] = [u]$ (cf. 18.3.10).

(g) If, for two propositions $\pi, \pi' \in \Pi$, we have

$$\{\sigma \in \Sigma_0 : p(\pi, \sigma) = 1\} = \{\sigma \in \Sigma_0 : p(\pi', \sigma) = 1\},$$

then $\pi = \pi'$. In fact, the above condition can be written as

$$\{u \in \tilde{\mathcal{H}} : (u | P_\pi u) = 1\} = \{u \in \tilde{\mathcal{H}} : (u | P_{\pi'} u) = 1\},$$

and this can be written as

$$\{u \in \tilde{\mathcal{H}} : \| P_\pi u \| = \| u \|\} = \{u \in \tilde{\mathcal{H}} : \| P_{\pi'} u \| = \| u \|\},$$

and this is equivalent to $R_{P_\pi} = R_{P_{\pi'}}$, in view of 13.1.3c. Then, $P_\pi = P_{\pi'}$ and hence $\pi = \pi'$.

19.3.6 Definitions. Let $(X, \mathcal{A})$ be a measurable space and α an X-valued observable. We define the projection valued mapping

$$P_\alpha : \mathcal{A} \to \mathscr{P}(\mathcal{H})$$

$$E \mapsto P_\alpha(E) := P_{\alpha(E)}.$$

For every $u \in \tilde{\mathcal{H}}$, the function $\mu_u^{P_\alpha}$ (cf. Section 13.3 for the definition of $\mu_u^{P_\alpha}$) is a probability measure on $\mathcal{A}$ since $\mu_u^{P_\alpha} = \mu_{\sigma_u}^\alpha$ if σ_u is the pure state such that $W_{\sigma_u} = A_u$:

$$\mu_u^{P_\alpha}(E) = (u | P_{\alpha(E)} u) = \text{tr}(P_{\alpha(E)} A_u) = p(\alpha(E), \sigma_u) = \mu_{\sigma_u}^\alpha(E), \forall E \in \mathcal{A}.$$

Thus (cf. 13.3.5) P_α is a projection valued measure on $\mathcal{A}$.

If α is an observable (i.e. an $\mathbb{R}$-valued observable), then P_α defines a unique self-adjoint operator A^{P_α} (cf. 15.2.2), and we write $A_\alpha := A^{P_\alpha}$. Thus, A_α is the unique self-adjoint operator in $\mathcal{H}$ such that (cf. 15.2.2)

$$P^{A_\alpha}(E) = P_{\alpha(E)}, \forall E \in \mathcal{A}(d_\mathbb{R}).$$

19.3.7 Proposition. *If $(X, \mathcal{A})$ is a measurable space and α and β are X-valued observables such that $P_\alpha = P_\beta$, then $\alpha = \beta$.*

The mapping $\alpha \mapsto A_\alpha$, from the family of all observables to the family of all self-adjoint operators in $\mathcal{H}$, is injective.

Proof. If α and β are X-valued observables and $P_\alpha = P_\beta$, then $P_{\alpha(E)} = P_{\beta(E)}$ and hence (by the injectivity of the mapping of 19.3.1b) $\alpha(E) = \beta(E)$ for all $E \in \mathcal{A}$, i.e. $\alpha = \beta$.

If α and β are observables and $A_\alpha = A_\beta$, then $P_\alpha = P_\beta$ (cf. 15.2.2) and hence $\alpha = \beta$. $\qquad\qquad\square$

19.3.8 Remark. If the assumption is made that every mapping $\alpha : \mathcal{A}(d_\mathbb{R}) \to \Pi$, for which μ_σ^α is a probability measure for all $\sigma \in \Sigma$, must be considered an observable, then every self-adjoint operator in $\mathcal{H}$ represents an observable. Indeed, if A is a self-adjoint operator in $\mathcal{H}$, we can define the mapping $\alpha_A : \mathcal{A}(d_\mathbb{R}) \to \Pi$ by letting $\alpha_A(E)$ be the proposition such that $P_{\alpha_A(E)} = P^A(E)$, for all $E \in \mathcal{A}(d_\mathbb{R})$. Then we have, for each $\sigma \in \Sigma$,

$$\mu_\sigma^{\alpha_A}(E) = p(\alpha_A(E), \sigma) = \mathrm{tr}(P^A(E)W_\sigma), \forall E \in \mathcal{A}(d_\mathbb{R}),$$

which shows (cf. 18.3.13) that $\mu_\sigma^{\alpha_A}$ is a probability measure on $\mathcal{A}(d_\mathbb{R})$. Thus, α_A could be an observable. If α_A actually is an observable, then we obviously have $A_{\alpha_A} = A$. This shows that the hypothetical assumption above is equivalent to the assumption that the mapping $\alpha \mapsto A_\alpha$, from the family of all observables to the family of all self-adjoint operators (cf. 19.3.7), is bijective.

We do not make the assumption above, but merely claim that every self-adjoint operator A in $\mathcal{H}$ is capable of representing an observable (this claim is actually equivalent to the spectral theorem for self-adjoint operators, cf. 15.2.1A), and that it does represent an observable whenever it can be properly justified that the mapping α_A represents a measuring instrument (cf. 19.1.9c). Some actually assume that, in a quantum theory without superselection rules, every self-adjoint operator represents an observable. However, for most operators it is very difficult to imagine measuring instruments that could support the corresponding observables.

In any case, in every quantum theory without superselection rules, at least all projections are self-adjoint operators which represent observables, owing to the surjectivity of the mapping $\Pi \ni \pi \mapsto P_\pi \in \mathscr{P}(\mathcal{H})$ of 19.3.1b. Indeed, for each proposition $\pi \in \Pi$ we can define the observable α_π (cf. 19.1.26), and we have, for $E \in \mathcal{A}(d_\mathbb{R})$,

$$P^{A_{\alpha_\pi}}(E) = P_{\alpha_\pi(E)} = \begin{cases} P_{\pi_0} = \mathbb{O}_\mathcal{H} & \text{if } 0 \notin E \text{ and } 1 \notin E, \\ P_\pi & \text{if } 0 \notin E \text{ and } 1 \in E, \\ P_{\neg\pi} = \mathbb{1}_\mathcal{H} - P_\pi & \text{if } 0 \in E \text{ and } 1 \notin E, \\ P_{\pi_1} = \mathbb{1}_\mathcal{H} & \text{if } 0 \in E \text{ and } 1 \in E; \end{cases}$$

now, this is the projection valued measure of the self-adjoint operator P_π (cf. 15.3.4D and 13.1.3e); thus $A_{\alpha_\pi} = P_\pi$. Furthermore, for every projection $P \in \mathscr{P}(\mathcal{H})$

there exists a proposition π such that $P = P_\pi$, owing to the surjectivity of the mapping $\Pi \ni \pi \mapsto P_\pi \in \mathscr{P}(\mathcal{H})$.

19.3.9 Proposition. *Let $(X, \mathcal{A})$ be a measurable space, let α be an X-valued observable, let $\varphi : D_\varphi \to \mathbb{R}$ be a function from X to $\mathbb{R}$ which is $\mathcal{A}^{D_\varphi}$-measurable and so that $D_\varphi \in \mathcal{A}$. Then the observable $\varphi(\alpha)$ can be defined (i.e. $\mu_\sigma^\alpha(X - D_\varphi) = 0$ for each $\sigma \in \Sigma$, cf. 19.1.13) if and only if the operator $J_\varphi^{P_\alpha}$ can be defined (i.e. $P_\alpha(X - D_\varphi) = \mathbb{O}_\mathcal{H}$, or $\varphi \in \mathcal{M}(X, \mathcal{A}, P_\alpha)$, cf. 14.2.14). If these conditions are true then $A_{\varphi(\alpha)} = J_\varphi^{P_\alpha}$. If, further, $(X, \mathcal{A}) = (\mathbb{R}, \mathcal{A}(d_\mathbb{R}))$, then $A_{\varphi(\alpha)} = \varphi(A_\alpha)$. Owing to this, the mapping $\alpha \mapsto A_\alpha$, from the family of all observables to the family of all self-adjoint operators, is said to be function preserving.*

Proof. The condition $\mu_\sigma^\alpha(X - D_\varphi) = 0$ for each $\sigma \in \Sigma$ is equivalent to $\alpha(X - D_\varphi) = \pi_0$, and hence to $P_\alpha(X - D_\varphi) = \mathbb{O}_\mathcal{H}$. Assuming that these conditions are true, we have (cf. 15.2.7 and 19.1.13)

$$P_\varphi^{J_\varphi^{P_\alpha}}(E) = P_\alpha(\varphi^{-1}(E)) = P_{\alpha(\varphi^{-1}(E))} = P_{\varphi(\alpha)(E)} = P^{A_{\varphi(\alpha)}}(E), \forall E \in \mathcal{A}(d_\mathbb{R}),$$

and hence (cf. 15.2.2) $J_\varphi^{P_\alpha} = A_{\varphi(\alpha)}$. If, further, $(X, \mathcal{A}) = (\mathbb{R}, \mathcal{A}(d_\mathbb{R}))$, then (cf. 15.3.1)

$$\varphi(A_\alpha) = J_\varphi^{P^{A_\alpha}} = J_\varphi^{P_\alpha} = A_{\varphi(\alpha)}.$$

$\square$

19.3.10 Proposition. *For an observable α, the following statements hold true:*

(a) $sp_\alpha = \sigma(A_\alpha)$.
(b) The following conditions are equivalent:

> *α is a bounded observable;*
> *A_α is a bounded operator;*
> *$D_{A_\alpha} = \mathcal{H}$.*

(c) The following conditions are equivalent:

> *α is a discrete observable;*
> *there exists a c.o.n.s. in $\mathcal{H}$ the elements of which are eigenvectors of A_α.*

Proof. a: For $\lambda \in \mathbb{R}$ we have

$$\lambda \in sp_\alpha \Leftrightarrow [\forall \varepsilon > 0, \exists \sigma \in \Sigma \text{ s.t. } \mu_\sigma^\alpha((\lambda - \varepsilon, \lambda + \varepsilon)) \neq 0] \Leftrightarrow$$
$$[\forall \varepsilon > 0, P^{A_\alpha}((\lambda - \varepsilon, \lambda + \varepsilon)) \neq \mathbb{O}_\mathcal{H}] \Leftrightarrow \lambda \in \sigma(A_\alpha),$$

where we have used 19.1.15, 19.3.5f, 15.2.4.

b: From statement a and the fact that $\sigma(A_\alpha)$ is a bounded set iff A_α is a bounded operator (cf. 15.2.2f) we have that α is a bounded observable iff A_α is a bounded operator. From 12.4.7 we have that A_α is a bounded operator iff $D_{A_\alpha} = \mathcal{H}$.

c: Assume first that α is a discrete observable, and let $\{\lambda_n\}_{n \in I}$ be a countable family of real numbers so that $\mu_\sigma^\alpha(\{\lambda_n\}_{n \in I}) = 1$ for all $\sigma \in \Sigma$ and so that (cf. 19.1.19)

$$\forall n \in I, \exists \sigma \in \Sigma \text{ so that } \mu_\sigma^\alpha(\{\lambda_n\}) \neq 0.$$

Then $\mu_\sigma^\alpha(\mathbb{R} - \{\lambda_n\}_{n \in I}) = 0$ for all $\sigma \in \Sigma$, and hence (by the monotonicity of μ_σ^α) $\mu_\sigma^\alpha(\{\lambda\}) = 0$ for all $\sigma \in \Sigma$ and for each $\lambda \in \mathbb{R} - \{\lambda_n\}_{n \in I}$, and hence (cf. 19.3.5f)

$$P^{A_\alpha}(\{\lambda\}) = \mathbb{O}_\mathcal{H}, \forall \lambda \in \mathbb{R} - \{\lambda_n\}_{n \in I};$$

moreover, by 19.3.5f,

$$P^{A_\alpha}(\{\lambda_n\}) \neq \mathbb{O}_\mathcal{H}, \forall n \in I.$$

Thus, $\{\lambda_n\}_{n \in I} = \sigma_p(A_\alpha)$ by 15.2.5, and hence $\mu_\sigma^\alpha(\mathbb{R} - \sigma_p(A_\alpha)) = 0$ for all $\sigma \in \Sigma$, and hence (cf. 19.3.5f) $P^{A_\alpha}(\mathbb{R} - \sigma_p(A_\alpha)) = \mathbb{O}_\mathcal{H}$, and hence (cf. 15.3.4B) there exists a c.o.n.s in $\mathcal{H}$ whose elements are eigenvectors of A_α.

Assume, next and conversely, that there exists a c.o.n.s. in $\mathcal{H}$ whose elements are eigenvectors of A_α. Then (cf. 15.3.4B) $P^{A_\alpha}(\mathbb{R} - \sigma_p(A_\alpha)) = \mathbb{O}_\mathcal{H}$, and hence (cf. 19.3.5f) $\mu_\sigma^\alpha(\mathbb{R} - \sigma_p(A_\alpha)) = 0$ for all $\sigma \in \Sigma$. Since $\sigma_p(A_\alpha)$ is countable (cf. 12.4.20C), this proves that the observable α is discrete. $\square$

19.3.11 Remark. There are quantum theories in which it is unavoidable to have observables which are not bounded (e.g., the observable which is interpreted as the energy of the system). From 19.3.10b it follows that it is then unavoidable to have self-adjoint operators which are not bounded (or, equivalently, which are not defined on the whole space and which are touchy about their domains, cf. for example 12.4.25).

19.3.12 Remarks. Let $(X, \mathcal{A})$ be a measurable space and α an X-valued observable. We recall that, for every $E \in \mathcal{A}$ and every state $\sigma \in \Sigma$, the number $\mu_\sigma^\alpha(E)$ is the probability $p(\alpha(E), \sigma)$ of the proposition $\alpha(E)$ in the state σ, i.e. the probability that the apparatus underlying α gives a result which is an element of E, for a collection of copies of the system prepared in the state σ (cf. 19.1.9a). We also recall that $p(\alpha(E), \sigma) = \text{tr}(P_\alpha(E)W_\sigma)$.

(a) For an observable α, a real number λ, and a state σ, the number $\mu_\sigma^\alpha(\{\lambda\})$ is the probability of obtaining λ as result for α in the state σ. For α and λ we have (cf. 15.2.5 and 19.3.5f and recall that $P^{A_\alpha} = P_\alpha$)

$$\lambda \in \sigma_p(A_\alpha) \Leftrightarrow P^{A_\alpha}(\{\lambda\}) \neq \mathbb{O}_\mathcal{H} \Leftrightarrow [\exists \sigma \in \Sigma \text{ s.t. } \mu_\sigma^\alpha(\{\lambda\}) \neq 0] \Leftrightarrow$$
$$[\exists \sigma \in \Sigma_0 \text{ s.t. } \mu_\sigma^\alpha(\{\lambda\}) = 1].$$

Thus, λ is as an exact result for α (cf. 19.1.16a) if and only if λ is an eigenvalue of A_α, and this is true if and only if there exists a pure state in which the result λ is certain.

If α is an observable and a real number λ is an isolated point of $\sigma(A_\alpha)$, i.e. if

$$\exists \delta > 0 \text{ such that } (\lambda - \delta, \lambda + \delta) \cap \sigma(A_\alpha) = \{\lambda\},$$

then $\lambda \in \sigma_p(A_\alpha)$ (cf. 15.2.6) and, to produce copies of the system in which the result for α is λ with certainty, we only need a preparation procedure σ so that $\mu_\sigma^\alpha((\lambda - \delta, \lambda + \delta)) = 1$, i.e. a state which produces with certainty the result λ with a margin of error not greater than $\frac{\delta}{2}$. "For example, if we know of a hydrogen atom that it contains less energy than is necessary for the second lowest energy level, then we know its energy content with absolute precision: it is the lowest energy value" (Neumann, 1932, p.222). We call an isolated point of $\sigma(A_\alpha)$ a *quantized result* for α.

Suppose now that α is a discrete observable. This is equivalent to the assumption that $P^{A_\alpha}(\mathbb{R} - \sigma_p(A_\alpha)) = \mathbb{O}_\mathcal{H}$ (cf. 19.3.10c and 15.3.4B). Now, this does not entail that the elements of $\sigma_p(A_\alpha)$ are isolated points of $\sigma(A_\alpha)$. Whether the elements of $\sigma_p(A_\alpha)$ are isolated points of $\sigma(A_\alpha)$ actually depends on the scale which is used in the instrument that defines α. In fact, since $\sigma_p(A_\alpha)$ is countable (cf. 12.4.20C) we can write $\sigma_p(A_\alpha) = \{\lambda_n\}_{n \in I}$ with $I := \{1, ..., N\}$ or $I := \mathbb{N}$; then we fix $\delta > 0$ and define the function

$$\varphi : \{\lambda_n\}_{n \in I} \to \mathbb{R}$$
$$\lambda_n \mapsto \varphi(\lambda_n) := n\delta;$$

this function is trivially $\mathcal{A}(d_\mathbb{R})^{D_\varphi}$-measurable and we have

$$\mu_\sigma^\alpha(\mathbb{R} - D_\varphi) = \operatorname{tr}\left(P^{A_\alpha}(\mathbb{R} - \sigma_p(A_\alpha))W_\sigma\right) = 0, \forall \sigma \in \Sigma;$$

thus, the observable $\varphi(\alpha)$ can be defined and we have $A_{\varphi(\alpha)} = \varphi(A_\alpha)$ (cf. 19.3.9); now, we have the equations $\sigma_p(\varphi(A_\alpha)) = \sigma(\varphi(A_\alpha)) = \{n\delta : n \in I\}$ and $P^{\varphi(A_\alpha)}(\mathbb{R} - \sigma_p(\varphi(A_\alpha))) = \mathbb{O}_\mathcal{H}$ (cf. 15.3.4B), and therefore $\varphi(\alpha)$ is a discrete observable, the entire spectrum of $A_{\varphi(\alpha)}$ is made up of isolated points, and to pinpoint a result for $\varphi(\alpha)$ we only need a preparation procedure which produces with certainty a result with a margin of error not greater than $\frac{\delta}{2}$. We may assume that such a procedure exists at least in principle since it is easy to see that, for each $n \in I$, statistical operators W exist so that $\operatorname{tr}\left(P_{\varphi(\alpha)}\left((n\delta - \frac{\delta}{2}, n\delta + \frac{\delta}{2})\right)W\right) = 1$, and hence states σ so that $\mu_\sigma^{\varphi(\alpha)}\left((n\delta - \frac{\delta}{2}, n\delta + \frac{\delta}{2})\right) = 1$ (however, it may be difficult to attain procedures which define such states in practice; among these states there are the states σ for which $\mu_\sigma^{\varphi(\alpha)}(\{n\delta\}) = 1$; states σ for which only the milder condition $\mu_\sigma^{\varphi(\alpha)}\left((n\delta - \frac{\delta}{2}, n\delta + \frac{\delta}{2})\right) = 1$ is requested may be easier to implement). Now, an apparatus which underlies $\varphi(\alpha)$ is obtained by a change of scale in an apparatus which underlies α (cf. 19.1.14); and indeed the two observables α and $\varphi(\alpha)$ are operationally equivalent; in fact we also have $\alpha = \varphi^{-1}(\varphi(\alpha))$, as can be easily proved, and therefore if the result $n\delta$ has been obtained for $\varphi(\alpha)$ then the result λ_n can be said to have been obtained for α. We also note that no change of scale, defined by a function $\psi : D_\psi \to \mathbb{R}$ which meets the conditions of 19.3.9 with $D_\psi \in \mathcal{A}(d_\mathbb{R})$, can give an observable $\psi(\alpha)$ which is not discrete. This can be seen from

$$\mu_\sigma^{\psi(\alpha)}(\psi(\{\lambda_n\}_{n\in I})) = \mu_\sigma^\alpha(\psi^{-1}(\psi(\{\lambda_n\}_{n\in I}))) \geq \mu_\sigma^\alpha(\{\lambda_n\}_{n\in I}) = 1, \forall\sigma \in \Sigma$$

(we have used the monotonicity of μ_σ^α and $\{\lambda_n\}_{n\in I} \subset \psi^{-1}(\psi(\{\lambda_n\}_{n\in I}))$). Thus, the discreteness of the observable α is a property which is shared by all functions of α.

What we have just seen shows that a discrete observable is an observable so that at least in principle there are realistic states (i.e. preparation procedures which do not demand absolute precision for their implementation) in which an exact result is obtainable with certainty. An observable is said to be *quantized* if it is discrete. This idea was expressed by John von Neumann as follows: "In the method of observation of classical mechanics ... we assign to each quantity α in each state [what is meant here is 'in each microstate'] a completely determined value. At the same time, however, we recognize that each conceivable measuring apparatus, as a consequence of the imperfections of human means of observations (which result in the reading of the position of a pointer or in locating the blackening of a photographic plate with only limited accuracy), can furnish this value only with a certain (never vanishing) margin of error. This margin of error can, by sufficient refinement of the method of measurement, be made arbitrarily close to zero but it is never exactly zero. One expects that this will also be true in quantum theory for those quantities which ... are not quantized; for example, for the cartesian coordinates of an electron (which can take on every value between $-\infty$ and $+\infty$, and whose operators have continuous spectra [what is meant here is that their point spectra are empty]). On the other hand, for those quantities which ... are 'quantized', the converse is true: since these are capable of assuming only discrete values, it suffices to observe them with just sufficient precision that no doubt can exist as to which one of these 'quantized' values is occurring. That value is then as good as 'observed' with absolute precision. ... This division into quantized and unquantized quantities corresponds ... to the division into quantities α with an operator A_α that has a pure discrete spectrum [what is mean here is that $P^{A_\alpha}(\mathbb{R} - \sigma_p(A_\alpha)) = \mathbb{O}_\mathcal{H}$], and into such quantities for which this is not the case. And it was for the former, and only for these, that we found a possibility of an absolutely precise measurement — while the latter could be observed only with arbitrarily good (but never absolute) precision" (Neumann, 1932, p.221–222).

(b) Let α be an observable, and suppose that $\lambda \in \sigma_c(A_\alpha)$ (cf. 12.4.22). Then the result λ can never be obtained exactly with certainty, because $\lambda \notin \sigma_p(A_\alpha)$. However, from 19.3.10a and 19.3.5f we have that

$$\forall\varepsilon > 0, \exists\sigma \in \Sigma_0 \text{ such that } \mu_\sigma^\alpha((\lambda - \varepsilon, \lambda + \varepsilon)) = 1.$$

This means that the result λ can be obtained with certainty with arbitrarily good precision. Thus, $\sigma_c(A_\alpha)$ can be interpreted as representing a continuum of possible results for α. To obtain one of these results with absolute precision would require an absolutely precise preparation procedure (the situation is in a certain sense opposite to the one discussed in remark a). The treatment of quantum mechanics based on Hilbert space does not allow these rather idealistic procedures, which are instead part of the treatments of quantum mechanics that use the notion of "improper eigenfunction" to represent them. Now let von Neumann speak. "It should be observed that the introduction of an eingenfunction which is 'improper', i.e. which does not belong to Hilbert space, gives a less good approach to reality than our treatment here. For such a method pretends the existence of such states in which quantities with continuous spectra take on certain values exactly, although this never occurs. Although such idealizations have often been advanced, we believe that it is necessary to discard them on these grounds, in addition to their mathematical untenability" (Neumann, 1932, p.223). We point out that, in this respect, quantum mechanics in Hilbert space is a construction which requires a smaller amount of idealization than classical mechanics, which has at its core states (the microstates) in which all quantities take on exact values with certainty.

What was considered as "mathematically untenable" by von Neumann in 1932 was Dirac's notion of bras and kets (Dirac, 1958, 1947, 1935, 1930), which was actually systematized later by the mathematical theory of "rigged Hibert spaces". However, this theory relies heavily on von Neumann's spectral theorem and "we must emphasize that we regard the spectral theorem as sufficient for any argument where a nonrigorous approach might rely on Dirac notation; thus, we only recommend the abstract rigged space approach to readers with a strong emotional attachment to the Dirac formalism" (Reed and Simon, 1980, 1972, p.244).

19.3.13 Proposition. *For an observable α and a state $\sigma \in \Sigma$, the following facts are true.*

(a) α is evaluable in σ iff A_α is computable in W_σ; if α is evaluable in σ, then

$$A_\alpha W_\sigma \in \mathcal{T}(\mathcal{H}),$$
$$\langle \alpha \rangle_\sigma = \langle A_\alpha \rangle_{W_\sigma} = \mathrm{tr}(A_\alpha W_\sigma),$$
$$\Delta_\sigma \alpha = \Delta_{W_\sigma} A_\alpha.$$

(b) If α^2 is evaluable in σ, then α is evaluable in σ and

$$A_\alpha^2 W_\sigma \in \mathcal{T}(\mathcal{H}),$$
$$A_\alpha W_\sigma \in \mathcal{T}(\mathcal{H}),$$
$$\Delta_\sigma \alpha = \left(\mathrm{tr}(A_\alpha^2 W_\sigma) - (\mathrm{tr}(A_\alpha W_\sigma))^2\right)^{\frac{1}{2}} = \left(\langle \alpha^2 \rangle_\sigma - \langle \alpha \rangle_\sigma^2\right)^{\frac{1}{2}}.$$

(c) If $\{u_n\}_{n \in I}$ is a countable family of elements of $\tilde{\mathcal{H}}$ and $\{w_n\}_{n \in I}$ is a family of elements of $(0, 1]$ so that $\sum_{n \in I} w_n = 1$ and $W_\sigma f = \sum_{n \in I} w_n A_{u_n} f$ for all $f \in \mathcal{H}$ (cf. 18.3.6 and 19.3.5b), then:

α is evaluable in σ iff $[u_n \in D_{A_\alpha}$ for all $n \in I$ and $\sum_{n \in I} w_n \|A_\alpha u_n\|^2 < \infty]$;
if α is evaluable in σ, then

$$\langle \alpha \rangle_\sigma = \sum_{n \in I} w_n \, (u_n | A_\alpha u_n),$$

$$\Delta_\sigma \alpha = \left(\sum_{n \in I} w_n \|A_\alpha u_n - \langle \alpha \rangle_\sigma u_n\|^2 \right)^{\frac{1}{2}}.$$

(d) *If σ is a pure state, then:*

α is evaluable in σ iff $u_\sigma \in D_{A_\alpha}$;
if α is evaluable in σ, then

$$\langle \alpha \rangle_\sigma = (u_\sigma | A_\alpha u_\sigma) = \langle A_\alpha \rangle_{u_\sigma},$$

$$\Delta_\sigma \alpha = \|A_\alpha u_\sigma - \langle \alpha \rangle_\sigma u_\sigma\| = \Delta_{u_\sigma} A_\alpha.$$

Proof. a: We have (cf. 19.3.6 and 18.3.13 for the definition of $\mu_{W_\sigma}^{P^{A_\alpha}}$)

$$\mu_\sigma^\alpha(E) = p(\alpha(E), \sigma) = \text{tr}(P^{A_\alpha}(E)W_\sigma) = \mu_{W_\sigma}^{P^{A_\alpha}}(E), \forall E \in \mathcal{A}(d_{\mathbb{R}});$$

the results then follow from the definitions given in 19.1.20, 19.1.21, 18.3.14, and from 18.3.16.

b: Since $\alpha^2 := \xi^2(\alpha)$, from 19.3.9 we have $A_{\alpha^2} = \xi^2(A_\alpha)$; since $\xi^2(A_\alpha) = A_\alpha^2$ (cf. 15.3.5), we have $A_{\alpha^2} = A_\alpha^2$. Then the results follow from the results in part a and from 18.3.17.

c: The results follow from the results in part a and from 18.3.16.

d: If σ is a pure state, then $W_\sigma = A_{u_\sigma}$ (cf. 19.3.5c). Hence the results are the particularization of the results of part c to the case of I containing just one index (cf. also the definitions of $\langle A \rangle_u$ and $\Delta_u A$ in 15.2.3). $\qquad \square$

19.3.14 Remarks.

(a) If α is not a bounded observable then $D_{A_\alpha} \neq \mathcal{H}$ (cf. 19.3.10b) and therefore $W_\sigma A_\alpha$ is not an element of $\mathcal{T}(\mathcal{H})$ and we cannot write $\langle \alpha \rangle_\sigma = \text{tr}(W_\sigma A_\alpha)$ even if α is evaluable in σ (cf. also 18.3.19a).

(b) The results of 19.3.13c,d show that, if a mixed state $\sigma \in \Sigma$ is the mixture of a countable family $\{\sigma_n\}_{n \in I}$ of pure states with weights $\{w_n\}_{n \in I}$, then for an observable α which is evaluable in σ we have that α is evaluable in every pure state σ_n and

$$\langle \alpha \rangle_\sigma = \sum_{n \in I} w_n \langle \alpha \rangle_{\sigma_n}.$$

This supports the idea (cf. 19.3.5b) that σ can be implemented using implementations of the states σ_n, by the procedure which is put into effect by carrying out with probability w_n the plan of action σ_n (this procedure is not precise, because each time it is put into effect we do not know which plan of action σ_n is actually going into effect, but it is not utterly at random, because the probabilities w_n are defined). However, we remind the reader that the decomposition of a mixed state into a mixture is never unique, and thus σ cannot be interpreted as being necessarily implemented by this procedure: in fact, as an equivalence class, σ contains all the procedures that can be constructed as above, on the basis of any decomposition of σ into a mixture of other states.

19.3.15 Remarks. The results we have obtained for a quantum theory are consistent with the results we obtained for a general statistical theory in Section 19.1. This could be checked systematically. We examine here five instances of this consistency.

(a) For an observable α we have $sp_\alpha = \sigma(A_\alpha)$ (cf. 19.3.10a). Then sp_α is closed because such is the spectrum of every operator in $\mathcal{H}$ (cf. 10.4.6), and this is consistent with 19.1.17.

(b) For an observable α and a function φ as in 19.3.9 we have $A_{\varphi(\alpha)} = \varphi(A_\alpha)$. Then, for a pure state σ, 19.3.13d, 15.3.2 and $\mu_{u_\sigma}^{P^{A_\alpha}} = \mu_\sigma^\alpha$ (cf. 19.3.6) imply 19.1.23.

(c) If an observable α is bounded then the operator A_α is bounded (cf. 19.3.10b), and hence A_α is computable in W_σ for every $\sigma \in \Sigma$ (cf. 18.3.18), and hence α is evaluable in every $\sigma \in \Sigma$ (cf. 19.3.13a). This is consistent with 19.1.25. In a quantum theory we can also prove the converse of 19.1.25: if an observable α is evaluable in every state, then α is evaluable in every pure state, and hence $D_{A_\alpha} = \mathcal{H}$ (cf. 19.3.13d), and hence α is bounded (cf. 19.3.10b).

(d) For each $\pi \in \Pi$ we have $A_{\alpha_\pi} = P_\pi$ (cf. 19.3.8). Then, since P_π is bounded (cf. 13.1.3d), A_{α_π} is computable in W_σ for every $\sigma \in \Sigma$ (cf. 18.3.18), and hence α_π is evaluable in every $\sigma \in \Sigma$ (cf. 19.3.13a). Moreover, for each $\sigma \in \Sigma$, 19.3.13a implies that

$$\langle \alpha_\pi \rangle_\sigma = \mathrm{tr}(A_{\alpha_\pi} W_\sigma) = \mathrm{tr}(P_\pi W_\sigma) = p(\pi, \sigma)$$

and, since $P_\pi^2 = P_\pi$ (cf. 13.1.5), 19.3.13b implies that

$$\Delta_\sigma \alpha_\pi = \left(\mathrm{tr}(A_{\alpha_\pi}^2 W_\sigma) - (\mathrm{tr}(A_{\alpha_\pi} W_\sigma))^2 \right)^{\frac{1}{2}}$$
$$= (\mathrm{tr}(P_\pi W_\sigma)(1 - \mathrm{tr}(P_\pi W_\sigma)))^{\frac{1}{2}} = (p(\pi, \sigma)(1 - p(\pi, \sigma)))^{\frac{1}{2}} .$$

Now, these results are consistent with 19.1.27.

(e) The results obtained in 19.3.13b are consistent with 19.1.24.

19.3.16 Proposition. *For an observable α and a real number λ, the following conditions are equivalent:*

(a) $\lambda \in \sigma(A_\alpha)$;
(b) $\forall \varepsilon > 0$, $\exists \sigma_\varepsilon \in \Sigma_0$ such that α is evaluable in σ_ε, $|\langle \alpha \rangle_{\sigma_\varepsilon} - \lambda| < \varepsilon$, $\Delta_{\sigma_\varepsilon} \alpha < 2\varepsilon$.

Proof. The asserted equivalence follows from 15.2.4 and 19.3.13d. $\qquad\square$

19.3.17 Remark. On account of the equality $\sigma(A_\alpha) = sp_\alpha$ (cf. 19.3.10a), we already have a physical interpretation of the spectrum of the self-adjoint operator A_α that represents an observable α: $\sigma(A_\alpha)$ coincides with the spectrum of α, i.e. with the set of possible results for α (cf. 19.1.15 and 19.1.16a). Thus, on the grounds of 19.3.16 we can further analyse the idea of a possible result in quantum mechanics: a real number λ is a possible result for an observable α if and only if, for each given $\varepsilon > 0$, there exists a pure state σ_ε so that the average of the results obtained measuring α in a large number of copies prepared in σ_ε is predicted to differ from λ by less than ε, and this with a standard deviation which is predicted to be smaller than 2ε (cf. 19.1.22a).

19.3.18 Proposition. *For an observable α and a real number λ, the following conditions are equivalent:*

(a) $\lambda \in \sigma_p(A_\alpha)$;
(b) $\exists \sigma \in \Sigma_0$ such that α is evaluable in σ, $\langle \alpha \rangle_\sigma = \lambda$, $\Delta_\sigma \alpha = 0$.

Proof. The result follows from 15.2.5 and 19.3.13d. $\square$

19.3.19 Remarks.

(a) The result in 19.3.18 confirms the interpretation that was made in 19.3.12a of $\sigma_p(A_\alpha)$, for an observable α in quantum mechanics: a real number λ is an eigenvalue of A_α if and only if there exists a pure state σ so that λ is the result that is always obtained when α is measured for any number of copies prepared in σ; in fact (cf. 19.1.22b) the meaning of $\Delta_\sigma \alpha = 0$ is that the same result is always obtained for any number of measurements (then, of course, this result is also the mean result). It is also clear from 19.3.13d that, for a pure state σ in which α is evaluable, we have $\langle \alpha \rangle_\sigma = \lambda$ and $\Delta_\sigma \alpha = 0$ if and only if λ is an eigenvalue of A_α and u_σ is an eigenvector of A_α corresponding to λ; and indeed this is true if and only if (cf. 15.2.5e and 13.1.3c) $\mu_\sigma^\alpha(\{\lambda\}) = \|P^{A_\alpha}(\{\lambda\})u_\sigma\|^2 = 1$, in agreement with what was seen in 19.3.12a.

(b) For an observable α, a pure state σ, a real number λ, in remark a we saw that $\mu_\sigma^\alpha(\{\lambda\}) = 1$ if and only if λ is an eigenvalue of A_α and u_σ is an eigenvector of A_α corresponding to λ. More in general we have (cf. 19.3.5c)

$$\mu_\sigma^\alpha(\{\lambda\}) = \left(u_\sigma | P^{A_\alpha}(\{\lambda\})u_\sigma \right).$$

Thus, if $\lambda \in \sigma_p(A_\alpha)$ and $\{u_{\lambda,d}\}_{d \in I_\lambda}$ is an o.n.s. in $\mathcal{H}$ which is complete in $N_{A_\alpha - \lambda \mathbb{1}_\mathcal{H}}$, i.e. so that $V\{u_{\lambda,d}\}_{d \in I_\lambda} = N_{A_\alpha - \lambda \mathbb{1}_\mathcal{H}}$, we have (cf. 15.2.5e and 13.1.10)

$$\mu_\sigma^\alpha(\{\lambda\}) = \sum_{d \in I_\lambda} |(u_{\lambda,d}|u_\sigma)|^2.$$

If the dimension of $N_{A_\alpha - \lambda \mathbb{1}_\mathcal{H}}$ is one, i.e. if λ is a non-degenerate eigenvalue of A_α, we have

$$\mu_\sigma^\alpha(\{\lambda\}) = |(u_\lambda|u_\sigma)|^2,$$

where u_λ is any element of $\tilde{\mathcal{H}} \cap N_{A_\alpha - \lambda \mathbb{1}_{\mathcal{H}}}$.

19.3.20 Proposition. *For an observable α and a state $\sigma \in \Sigma$, the following conditions are equivalent:*

(a) α is evaluable in σ and $\Delta_\sigma \alpha = 0$;
(b) $\exists \lambda \in \mathbb{R}$ such that $\mu_\sigma^\alpha(\{\lambda\}) = 1$;
(c) $\exists \lambda \in \mathbb{R}$ such that $R_{W_\sigma} \subset R_{P^{A_\alpha}(\{\lambda\})}$.

If these conditions are satisfied then there is only one real number λ such that condition b, or condition c, is satisfied; λ is the same number for both conditions and we have:

(d) $\lambda \in \sigma_p(A)$ and $\langle \alpha \rangle_\sigma = \lambda$.

Proof. $a \Leftrightarrow b$: Cf. 19.1.22b.

$b \Leftrightarrow c$, and uniqueness of λ: In condition b, λ is clearly unique since $\mu_\sigma^\alpha(\{\lambda\}) = 1$ implies $\mu_\sigma^\alpha(\mathbb{R} - \{\lambda\}) = 0$; moreover, for $\lambda \in \mathbb{R}$ we have

$$\mu_\sigma^\alpha(\{\lambda\}) = 1 \Leftrightarrow R_{W_\sigma} \subset R_{P^{A_\alpha}(\{\lambda\})}$$

by 19.3.5f, since $\mu_\sigma^\alpha(\{\lambda\}) = p(\alpha(\{\lambda\}), \sigma)$ and $P_{\alpha(\{\lambda\})} = P^{A_\alpha}(\{\lambda\})$.

d: Condition c implies obviously $P^{A_\alpha}(\{\lambda\}) \neq O_{\mathcal{H}}$ and hence $\lambda \in \sigma_p(A_\alpha)$ (cf. 15.2.5). In 19.1.22b it was proved that condition b implies $\langle \alpha \rangle_\sigma = \lambda$. $\qquad \square$

19.3.21 Remark. From 19.3.20 we have that, for an observable α and a state σ in which α is evaluable, $\Delta_\sigma \alpha = 0$ is possible if and only if $\sigma_p(A_\alpha) \neq \emptyset$; moreover, if $\sigma_p(A_\alpha) \neq \emptyset$ then $\Delta_\sigma \alpha = 0$ is true if and only if there exists an eigenvalue λ of A_α so that $\mu_\sigma^\alpha(\{\lambda\}) = 1$, namely an eigenvalue of A_α which is the result that is always obtained when α is measured in any number of copies prepared in σ.

From 19.3.13c we also have that an observable α is evaluable in a state σ and $\Delta_\sigma \alpha = 0$ if and only if any collection of pure states, into a mixture of which σ can be decomposed, is comprised of states represented by eigenvectors of A_α corresponding to $\langle \alpha \rangle_\sigma$, which is then the eigenvalue λ of A_α such that $\mu_\sigma^\alpha(\{\lambda\}) = 1$, or equivalently such that $R_{W_\sigma} \subset R_{P^{A_\alpha}(\{\lambda\})}$. If the state σ is pure, we have $\Delta_\sigma \alpha = 0$ if and only if u_σ is an eigenvector of A_α; if this holds true, then $\langle \alpha \rangle_\sigma$ is the eigenvalue of A_α to which u_σ corresponds. Thus, we have derived the results of 19.3.19a as a special case of the results obtained in the present remark.

19.3.22 Remark. Here we make a summary of the basic mathematical structure of a quantum theory.

There are a separable Hilbert space $\mathcal{H}$ of dimension greater than one, a bijective mapping $\sigma \mapsto W_\sigma$ from the family Σ of all states onto the family $\mathcal{W}(\mathcal{H})$ of all statistical operators in $\mathcal{H}$, a bijective mapping $\pi \mapsto P_\pi$ from the family Π of all propositions onto the family $\mathscr{P}(\mathcal{H})$ of all orthogonal projections in $\mathcal{H}$, so that

$$p(\pi, \sigma) = \mathrm{tr}(P_\pi W_\sigma), \forall \pi \in \Pi, \forall \sigma \in \Sigma.$$

Further, there is an injective mapping $\alpha \mapsto A_\alpha$ from the family of all observables to the family of all self-adjoint operators in $\mathcal{H}$, which is defined by the condition

$$P^{A_\alpha}(E) = P_{\alpha(E)}, \forall E \in \mathcal{A}(d_\mathbb{R}), \text{ for every observable } \alpha.$$

For every observable α, the following facts hold true:

- the spectrum sp_α of α is equal to the spectrum $\sigma(A_\alpha)$ of A_α;
- a real number is an exact result for α iff it is an eigenvalue of A_α;
- α is a bounded observable iff A_α is a bounded operator;
- α is a discrete observable iff there exists a c.o.n.s. in $\mathcal{H}$ the elements of which are eigenvectors of A_α;
- α is evaluable in a state σ iff A_α is computable in W_σ;
- if α is evaluable in a state σ then $\langle\alpha\rangle_\sigma = \text{tr}(A_\alpha W_\sigma)$ and, if α^2 is also evaluable in σ, $\Delta_\sigma\alpha = \left(\text{tr}(A_\alpha^2 W_\sigma) - (\text{tr}(A_\alpha W_\sigma))^2\right)^{\frac{1}{2}}$;
- for an $\mathcal{A}(d_\mathbb{R})^{D_\varphi}$-measurable function $\varphi : D_\varphi \to \mathbb{R}$ with $D_\varphi \in \mathcal{A}(d_\mathbb{R})$, the observable $\varphi(\alpha)$ can be defined iff the operator $\varphi(A_\alpha)$ can be defined; if they can be defined, then $A_{\varphi(\alpha)} = \varphi(A_\alpha)$.

For the subfamily Σ_0 of Σ defined by $\Sigma_0 := \{\sigma \in \Sigma : \exists[u_\sigma] \in \hat{\mathcal{H}} \text{ s.t. } W_\sigma = A_{u_\sigma}\}$, the following facts hold true:

- for a proposition $\pi \in \Pi$, $\pi \neq \pi_0$ iff there exists $\sigma \in \Sigma_0$ such that $p(\pi, \sigma) = 1$;
- for all $\sigma \in \Sigma_0$, there exists $\pi \in \Pi$ such that $p(\pi, \sigma) \notin \{0, 1\}$;
- the mapping $\Sigma_0 \ni \sigma \mapsto [u_\sigma] \in \hat{\mathcal{H}}$ is a bijection from Σ_0 onto the family of all rays of $\mathcal{H}$;
- an observable α is evaluable in $\sigma \in \Sigma_0$ iff $u_\sigma \in D_{A_\alpha}$;
- if an observable α is evaluable in $\sigma \in \Sigma_0$, then $\langle\alpha\rangle_\sigma = (u_\sigma|A_\alpha u_\sigma)$ and $\Delta_\sigma\alpha = \|A_\alpha u_\sigma - \langle\alpha\rangle_\sigma u_\sigma\|$;
- an observable α is evaluable in $\sigma \in \Sigma_0$ and $\Delta_\sigma\alpha = 0$ iff u_σ is an eigenvector of A_α;
- every element of Σ which is not an element of Σ_0 can be decomposed into a mixture of elements of Σ_0;
- Σ_0 is the family of all the states that cannot be decomposed into mixtures of other states.

This summary should be compared with the one given in 19.2.8.

19.3.23 Remark. In a quantum theory, states, propositions, observables are represented by operators which are defined in a separable Hilbert space $\mathcal{H}$; indeed we have the mappings $\sigma \mapsto W_\sigma$, $\pi \mapsto P_\pi$, $\alpha \mapsto A_\alpha$ (cf. 19.3.22). Suppose now that $\mathcal{H}'$ is a separable Hilbert space so that $\mathcal{H}$ and $\mathcal{H}'$ have the same orthogonal dimension ($\mathcal{H}'$ could be the same as $\mathcal{H}$), and that U is a unitary or antiunitary operator from $\mathcal{H}$ onto $\mathcal{H}'$ (cf. 10.7.14). Then, if we define

$$W_\sigma' := UW_\sigma U^{-1}, \forall\sigma \in \Sigma,$$

$$P'_\pi := UP_\pi U^{-1}, \forall \pi \in \Pi,$$

we have a "new" representation of states and propositions ($UWU^{-1} \in \mathcal{W}(\mathcal{H})$ if $W \in \mathcal{W}(\mathcal{H})$, cf. 18.3.2a, and $UPU^{-1} \in \mathscr{P}(\mathcal{H})$ if $P \in \mathscr{P}(\mathcal{H})$, cf. 13.1.8), which is equivalent to the "old" one because

$$\mathrm{tr}(P'_\pi W'_\sigma) = \mathrm{tr}(UP_\pi W_\sigma U^{-1}) = \mathrm{tr}(P_\pi W_\sigma) = p(\pi, \sigma), \forall \pi \in \Pi, \forall \sigma \in \Sigma$$

(cf. 18.2.11d,e and 18.3.8b). In the "new" representation, an observable α is represented by the projection valued measure $\mathcal{A}(d_\mathbb{R}) \ni E \mapsto UP^{A_\alpha}U^{-1} \in \mathscr{P}(\mathcal{H}')$, and hence by the operator $A'_\alpha := UA_\alpha U^{-1}$ (cf. 15.4.1), and a pure state $\sigma \in \Sigma_0$ by the ray $[u'_\sigma] := [Uu_\sigma]$ (cf. 13.1.13b). The "new" and the "old" representations are then easily checked to be wholly equivalent as to everything that has been examined in this section (in particular, they give the same numbers for every expected result $\langle\alpha\rangle_\sigma$ and every uncertainty $\Delta_\sigma\alpha$). The "new" and the "old" representations are said to be *unitarily-antiunitarily equivalent*. If the operator U is unitary (or antiunitary), they are said to be *unitarily* (or *antiunitarily*) equivalent.

Thus, the same quantum theory can be formulated in many unitarily-antiunitarily equivalent ways.

19.4 State reduction in quantum mechanics

The subject of this section is sometimes known as von Neumann's and Lüders' reduction postulates. We start by examining in 19.4.1 two experiments which we consider to be paradigmatic of what we later analyse in the abstract.

As before, in this and in the following sections Σ and Π denote the families of equivalence classes (cf. 19.1.12) of states and propositions of a given quantum system (i.e., a system described by a quantum theory), and $\mathcal{H}$ denotes the Hilbert space in which they are represented as summarized in 19.3.22.

19.4.1 Remarks.

(a) The determination of a proposition for a copy of a system prepared in a state is a procedure which is frequently performed by actually determining a possibly different proposition which can be related to the spatial position of the copy. This is indeed the case in the two examples we examine below. The determination of the position of a copy of a system which is considered to be of atomic or subatomic size (we shall call such a system a *microparticle*) can be carried out by means of a variety of detectors: Wilson chambers, bubble chambers, Geiger counters, etc.; the simplest of them all is a photographic plate. And this procedure often destroys the copy (for example, when a photon activates a Geiger counter, it is absorbed in the process) or else renders it of no further interest so far as the study of the system in question is concerned (for example, when an electron hits a photographic plate, it is lost among the electrons of the emulsion). Even when these catastrophic events do not occur, for other determination techniques, the analysis of the process of determination of a proposition in the physics of microparticles (initiated by Werner Heisenberg) leads to the conclusion that the determination of a proposition is a process which is bound to alter in a substantial way the copy for which the determination is carried out. As a matter of fact, an alteration takes place in classical physics too, but in classical physics it is assumed that the determination of any proposition in any state can always be implemented by probing the copy is such a way that the alteration of the copy is negligible (cf. 19.2.2). Since this is not the case for microparticles, in quantum mechanics (which deals mostly with microparticles) we must acknowledge that a proposition is true in a copy, or it is not true, only upon its determination, and not in general also immediately after that. However the case may be that the experimental set-up which is used for the determination of a proposition π can be modified so that it selects copies for which π is certainly true: if π is determined for any number of copies "emerging" from the modified set-up, then π will be found to be true in all of them. In what follows, we provide two examples of this sort.

(b) As a first example, we consider the method depicted in fig. 1 (all figures are on page 656) for determining the magnitude of the linear momentum (in what follows, briefly, "momentum") of a charged particle. To the left of the screen S_1 a particle of known charge e is produced which, after passing through the narrow openings O_1 and O_2 in the screens S_1 and S_2, is deflected by a uniform magnetic field $\overrightarrow{H}$, which is present to the right of the screen S_2 and orthogonal to the plane of the drawing. In D there is a detector. If the particle is detected in the region D, the magnitude of the momentum of the particle is determined to be $p_D = eHr_D$ (in suitable units), where r_D is as in fig. 1. In fact, if the particle is classical (i.e., a charged particle which is not a microparticle) then its trajectory is a circle with a radius depending on the momentum as in the formula just used; thus, from the region of localization of the particle we can deduce the magnitude of its momentum, and indeed the fact that detection of a charged classical particle in D corresponds to the magnitude $p_D = eHr_D$ is uncontentious. If the particle is not a classical particle, but it is a microparticle instead, the whole description given above, which is based on the idea of a trajectory, is meaningless (for a microparticle the concept of a trajectory loses its meaning, as first pointed out by Werner Heisenberg); the observable "magnitude of momentum" is then *defined* as the observable to which the result $p_D := eHr_D$ is ascribed if detection in the region D occurs (and other results in other similar experiments); indeed, if the particle is a microparticle, the experimental arrangement described above is one of those which give an empirical meaning to the concept of momentum of a microparticle. One could ask the question: "how can I know that the macroscopic event that happened in a detector located in D (as for instance the blackening of a spot of a photographic plate or the click of a detector) is due to a microparticle, and that the magnitude of the momentum of that particle is really $p_D = eHr_D$?".

But this question would be empirically meaningless because the only knowledge that we can have about microparticles is the one we obtain from events which happen in macroscopic objects (cf. e.g. Heisenberg, 1925). For both a classical particle and a microparticle, the detection at D can be performed by means of a device which absorbs the particle (e.g., a photographic plate for a microparticle), in which case the result p_D clearly refers only to the momentum of the particle upon its detection. Or else, the particle can be localized at D by shining a beam of light on the region D and registering whether light is reflected from it. Now, light has a certain momentum, which is imparted to the objects on which it impinges; however, this momentum transfer can be considered completely negligible if the particle is not a microparticle, and therefore we can assume that if a classical particle is detected at D by this method then its momentum has magnitude p_D also immediately after detection; on the contrary, if we are dealing with a microparticle we cannot reach this conclusion because the uncontrollable momentum transfer can no longer be considered irrelevant; therefore, even with this method of detection at D we must regard the result p_D as referring to the momentum of the microparticle upon detection and not also immediately after that.

The analysis carried out above is actually oversimplified, and we need to further distinguish between a classical particle and a microparticle. We note that the procedure discussed above cannot actually lead to an exact result p_D. In fact, owing to the non-null size of the detection region D and the non-null width of the openings O_1 and O_2 in the screens S_1 and S_2, in both the classical and the quantum cases we can only conclude that the magnitude of the momentum lies in a range E_D of possible values. However, in the classical case we can assume that this range can be made arbitrarily little by reducing the size of the detection region D and the width of the openings. In the case of a microparticle this assumption is untenable because reducing the width produces "diffraction effects", which are revealed in another experiment as follows: if a detector (e.g., a photographic plate) is placed behind two parallel screens each having a narrow enough aperture, a microparticle can be detected not only along any straight line passing through the two apertures, but elsewhere as well (in this experiment, no magnetic field is present). Thus, in the case of a microparticle, to reach the conclusion that detection in D corresponds to a fairly limited range E_D of possible values for the magnitude of the momentum, we must use openings O_1 and O_2 wide enough so that diffraction effects can be neglected (and this also requires a detector in D wide enough) and narrow enough so that E_D is still a limited range; how to reach this compromise is discussed e.g. in (Wichmann, 1971, Chapter 6). In this case, we can say that the experimental procedure just discussed implements the proposition "the magnitude of the

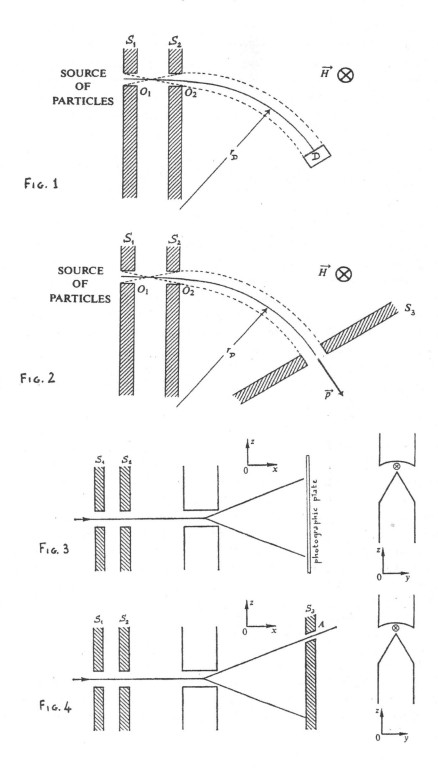

FIG. 1

FIG. 2

FIG. 3

FIG. 4

momentum of the microparticle belongs to the subset E_D of $\mathbb{R}$", which we will denote by the symbol π_D: the proposition π_D is true in a copy of the system (i.e., of the microparticle) if and only if a detection in the region D occurs. As already remarked, the procedure just discussed says nothing, in the case of a microparticle, as to the magnitude of the momentum immediately after detection, and it is designed not to produces copies of the system with some definite property but to analyse a given state (which is, in this case, a preparation procedure which takes place to the left of the screen S_1) by comparing the relative frequency of the different results for the magnitude of the momentum which are experimentally obtained (this would be done by varying the position of the region D) with the probability distribution that is predicted by the quantum-theory of the system.

The experimental set-up described above can be converted into a contrivance that selects copies for which the proposition π_D is certainly true. To obtain this, we replace detection in D with a filtering procedure, as shown in fig. 2: there is a third screen S_3 with an aperture which corresponds to the region D. Then each microparticle which is not absorbed by the screen S_3 has a momentum of magnitude that lies in E_D, and in the direction indicated in fig. 2. This can be proved in the following way: to the right of the screen S_3 we set up a slightly modified replica of the experimental arrangement of fig. 2, with the replica screens S_1' and S_2' parallel to S_3, with the replica openings O_1' and O_2' aligned with the aperture in S_3, and with an array of detectors (or with just one large detector, e.g. a photographic plate) which cover the space that would be occupied by a screen that was placed with respect to S_2' as S_3 is to S_2; then, in many repetitions of the experiment, we see that the only detector that reacts (if some detector does react), is the one placed in the position that is to S_2' as D was to S_2.

On these grounds, the experimental arrangement of fig. 2 (without, of course, its replica) can be considered a filtering device in the following sense: the copies that emerge from the arrangement are so that the proposition π_D "is true" in all of them, i.e. in all of them π_D would certainly be determined to be true if a determination of it was carried out. We point out that this experimental arrangement does not produce copies of the system, but it *selects* (among the copies produced in some state to the left of the screen S_1) copies to which a definite property can be attributed (in all of them, if the proposition π_D was determined then π_D would turn out to be true). Thus, a state preparation σ which takes place to the left of the screen S_1 plus the filtering device of fig. 2 must be considered to be a new state preparation procedure σ', provided $p(\pi_D, \sigma) \neq 0$; note in fact that, if $p(\pi_D, \sigma) = 0$, then no copies are ever revealed in the region D and accordingly no copies can be produced by σ supplemented with the filtering device with the aperture in the region D; if conversely $p(\pi_D, \sigma) \neq 0$, then out of a large number N of copies produced in the state σ the new procedure σ' selects a number of copies which will be approximately $p(\pi_D, \sigma)N$. We note that, since $p(\pi_D, \sigma') = 1$, we must have $R_{W_{\sigma'}} \subset R_{P_{\pi_D}}$ (cf. 19.3.5f). The study of the momentum of a microparticle in other experiments leads to the conclusion that P_{π_D} is not a one-dimensional projection; therefore, there are many possible states σ' which are so that $R_{W_{\sigma'}} \subset R_{P_{\pi_D}}$. However, even if the state σ was completely unknown, we now have a state σ' for which something is known, and this knowledge may be useful in other experiments.

(c) We consider a second example of a procedure for determining propositions which can be converted into a filtering device linked with a proposition. Its schematic experimental layout is sketched in fig. 3. There are two screens S_1 and S_2, each with a narrow opening in it. To the left of the screen S_1 copies are produced of a microparticle which has a magnetic (dipole) moment. To the right of the screen S_2 an inhomogeneous magnetic field is established by a pair of shaped magnets (magnetic poles), and to the right of the magnets there is a photographic plate. This experimental set-up is called a Stern–Gerlach device. If the particle were classical we should expect that a great number of copies of the particle, produced to the left of S_1 with random orientations of their magnetic moments, left random marks on the photographic plate. It is found instead that the experiment produces marks which are grouped in n well separated regions along the z axis; in fig. 3, $n = 2$. By definition, the spin of the microparticle is taken to be $\frac{n-1}{2}$; thus, fig. 3 shows the two possibilities for copies of a spin one-half microparticle. For a spin one-half microparticle, the Stern–Gerlach device can be used to determine two proposition π_{z+} and π_{z-}: π_{z+} (respectively, π_{z-}) is true in a copy of the microparticle if the mark "left by that copy" is in the upper (respectively lower) region. The observable "z-component of the spin" is then the mapping s_z defined by

$$s_z : \mathcal{A}(d_{\mathbb{R}}) \to \Pi$$

$$E \mapsto s_z(E) := \begin{cases} \pi_0 & \text{if } -\frac{1}{2} \notin E \text{ and } \frac{1}{2} \notin E, \\ \pi_{z+} & \text{if } -\frac{1}{2} \notin E \text{ and } \frac{1}{2} \in E, \\ \pi_{z-} & \text{if } -\frac{1}{2} \in E \text{ and } \frac{1}{2} \notin E, \\ \pi_1 & \text{if } -\frac{1}{2} \in E \text{ and } \frac{1}{2} \in E \end{cases}$$

(this mapping is an observable since $\pi_{z-} = \neg\pi_{z+}$; this is due to the fact that all marks are left in either the upper or the lower region, and nowhere else). Thus, when a copy leaves a mark in the upper (respectively lower) region of the plate we can say that $\frac{1}{2}$ (respectively $-\frac{1}{2}$) is the exact result for s_z, since in that case the proposition $s_z\left(\left\{\frac{1}{2}\right\}\right)$ (respectively $s_z\left(\left\{-\frac{1}{2}\right\}\right)$) is true.

If we replace, in the experimental set-up examined above, the photographic plate with a screen S_3 in which an aperture A is opened in the same position where the upper blackening region was, as in fig. 4, then we have a filtering procedure which selects copies in which the proposition π_{z+} is certainly true. Indeed, if we arrange a second Stern–Gerlach device to the right of the screen

S_3, with the apertures of the "new" screens S_1' and S_2' on the line of the hypothetical beam coming out of A (we are now using, as "new" source of copies, the source to the left of the screen S_1 plus the first Stern–Gerlach device with the photographic plate replaced by the screen S_3) and with a "new" photographic plate, we see that all the copies that are detected by the "new" photographic plate leave marks in the upper region of the plate. As in remark b, if the state σ in which the copies are prepared to the left of the screen S_1 is so that $p(\pi_{z+}, \sigma) \neq 0$, then σ plus the modified Stern–Gerlach device of fig. 4 amounts to a new state preparation procedure σ' which is so that $R_{W_{\sigma'}} \subset R_{P_{\pi_{z+}}}$ (cf. 19.3.5f). Now, a spin one-half microparticle is wholly described by a quantum theory the Hilbert space of which is not two-dimensional. However, if one is interested in studying spin (beside s_z, there are other spin observables, one for each direction in three-dimensional space) and nothing else, then one can give a partial description of a spin one-half microparticle in a two-dimensional Hilbert space, e.g. $\mathbb{C}^2$ (at the opposite end, if spin is disregarded one can give a partial description of the same microparticle in $L^2(\mathbb{R}^3)$). In that case, the projection $P_{\pi_{z+}}$ is one-dimensional and one can conclude that, whatever the state σ to the left of S_1, if $p(\pi_{z+}, \sigma) \neq 0$ then the copies that are selected by the procedure described above are in the pure state σ' represented by the ray $[u_{\sigma'}]$ of $\mathbb{C}^2$ (cf. 19.3.5c) which is so that $A_{u_{\sigma'}} = P_{\pi_{z+}}$; indeed, $p(\pi_{z+}, \sigma') = 1$ implies now $W_{\sigma'} = P_{\pi_{z+}}$ since $P_{\pi_{z+}}$ is now one-dimensional (cf. 19.3.5f). Thus, in the partial description in which just the spin observables are represented, the procedure described above can be interpreted, provided $p(\pi_{z+}, \sigma) \neq 0$, as an implementation of the pure state represented by the ray which contains the normalized eigenvectors of the self-adjoint operator A_{s_z} corresponding to the eigenvalue $\frac{1}{2}$. If a large number N of copies are prepared in the state σ to the left of S_1, this procedure selects approximately $p(\pi_{z+}, \sigma)N$ copies which are in this pure state.

19.4.2 Definition. We say that we have a *filter for a proposition* $\pi \in \Pi$ if we have, for every state preparation $\sigma \in \Sigma$ such that $p(\pi, \sigma) \neq 0$, an experimental set-up which can be added to a definite experimental implementation of σ and which affects a collection of copies prepared in σ as follows:

some copies are "absorbed" or "destroyed" by the set-up (i.e., "after" the setup, no effect can be observed that can be related to those copies);

there is the probability $p(\pi, \sigma)$ that a copy will not be absorbed;

there is a state (as an equivalence class) σ' which depends on σ and for which $p(\pi, \sigma') = 1$, so that if a copy has not been absorbed then it is in σ' (a copy which has not been absorbed by the set-up is said to *have gone through the filter*).

Thus, if $p(\pi, \sigma) \neq 0$ then there is a state σ' so that $p(\pi, \sigma') = 1$ and so that the experimental implementation of σ combined with the experimental set-up of the filter amounts to an experimental implementation of σ'.

19.4.3 Remarks.

(a) In 19.4.2 it is not asserted that a filter exists for every proposition π. Indeed, such claim would be an axiom. However, an even stronger assumption will actually be made in 19.4.6.

(b) In 19.4.2 it is not maintained that a filter *produces* copies of the system. Rather, we can say that a filter *selects and modifies* copies of the system. In fact, the definition of a filter implies that, *if* a state preparation procedure $\sigma \in \Sigma$ is activated *then* the filter affects the copy so that the copy is either absorbed or modified into a new copy (i.e., a copy in a new state). If $p(\pi, \sigma) \neq 0$, we can say that the filter transforms the state σ into a new state σ'. This transformation is called a *state reduction*. Note that, in a given experimental situation, σ and σ' are represented by different ensembles: if we have an ensemble consisting of a large number N of copies prepared in σ, then "after" the filter we have a *new* ensemble consisting of approximately $p(\pi, \sigma)N$ copies prepared in σ'.

(c) For a proposition $\pi \in \Pi$ there may exist essentially different filters. In fact, if $p(\pi, \sigma) \neq 0$, the state σ' is only subject to the condition $R_{W_{\sigma'}} \subset R_{P_\pi}$ (cf. 19.3.5f). Thus, there may exist different experimental set-ups which act as filters for the same proposition but lead to different state-reductions.

(d) It is expedient to define an equivalence relation in the family of filters for a proposition $\pi \in \Pi$, by defining two filters equivalent if they transform in the same way any state $\sigma \in \Sigma$ such that $p(\pi, \sigma) \neq 0$ (it is obvious that this defines an equivalence relation). An equivalence class is still called a filter. A representative of an equivalence class is sometimes called an implementation of the filter.

(e) If $\pi \in \Pi$ is such that P_π is a one-dimensional projection, that is to say $P_\pi = A_u$ with $u \in \tilde{\mathcal{H}}$, then just one filter (as an equivalence class) can exist, because $p(\pi, \sigma') = 1$ then implies $W_{\sigma'} = A_u$ (cf. 19.3.5f). This can be rephrased as follows: if π is represented by a one dimensional projection A_u, then the only state that can be obtained by supplementing any state with a filter for π is the pure state represented by the ray $[u]$ (cf. 19.3.5c).

(f) Suppose we have, for a proposition $\pi \in \Pi$, an experimental implementation of π which includes a detector so that the event which defines π is declared to have occurred when the detector "clicks". Then it is often possible to convert this apparatus into a filter for π by replacing the detector with a suitably oriented screen in which an aperture is opened in the shape of the detector. This is in fact what was done in the two examples of 19.4.1, which are examples of how filters can be obtained by modifying pieces of equipment originally designed for determining propositions.

19.4.4 Definition. A filter for a proposition $\pi \in \Pi$ is said to be an *ideal filter* if it transforms a state $\sigma \in \Sigma$ such that $p(\pi, \sigma) \neq 0$ into the state σ' represented by the statistical operator

$$W_{\sigma'} = W_{\sigma,\pi} := \frac{1}{\text{tr}(P_\pi W_\sigma P_\pi)} P_\pi W_\sigma P_\pi \left(= \frac{1}{\text{tr}(P_\pi W_\sigma)} P_\pi W_\sigma P_\pi \right).$$

19.4.5 Remarks.

(a) The condition that defines an ideal filter in 19.4.4 is consistent. Indeed, for every projection $P \in \mathscr{P}(\mathcal{H})$ and every statistical operator $W \in \mathcal{W}(\mathcal{H})$ we have:

$PWP \in \mathcal{T}(\mathcal{H})$ by 18.2.7;

$0 \leq (Pf|WPf) = (f|PWPf), \forall f \in \mathcal{H}$, since $P = P^\dagger$;

$\text{tr}(PWP) = \text{tr}(P^2W) = \text{tr}(PW)$ by 18.2.11c, since $P = P^2$;

this shows that, if $\text{tr}(PW) \neq 0$, then $\frac{1}{\text{tr}(PWP)}PWP \in \mathcal{W}(\mathcal{H})$. Also, recall that $\text{tr}(P_\pi W_\sigma) = p(\pi,\sigma)$. Furthermore, it is clear that $p(\pi,\sigma') = 1$ since $P_\pi^2 = P_\pi$ implies $P_\pi W_{\sigma,\pi} = W_{\sigma,\pi}$ (cf. 19.3.5f).

An ideal filter can be regarded as a filter which alters any "incoming" state σ as little as possible. In fact, for the "outgoing" state σ' we must have $R_{W_{\sigma'}} \subset R_{P_\pi}$ (cf. 19.4.3c), and the operator $P_\pi W_\sigma P_\pi$ is so to speak just the operator W_σ "reduced" to the subspace R_{P_π}.

(b) If an ideal filter for a proposition π exists then it is clearly unique (as an equivalence class), owing to the injectivity of the mapping $\sigma \mapsto W_\sigma$.

(c) For a proposition π which is represented by a one-dimensional projection, only one filter can exist (cf. 19.4.3e). Actually, if a filter exists then it is the ideal filter. Indeed, if a filter for π exists then it transforms every state $\sigma \in \Sigma$ such that $p(\pi,\sigma) \neq 0$ into the state σ' which is represented by the statistical operator $W_{\sigma'} = A_u$, if $P_\pi = A_u$ with $u \in \tilde{\mathcal{H}}$ (cf. 19.4.3e). Now, for each $W \in \mathcal{W}(\mathcal{H})$ and each $u \in \tilde{\mathcal{H}}$ so that $\text{tr}(A_u W) \neq 0$ we have $R_{A_u W A_u} = V\{u\}$ (notice that $A_u W A_u \neq \mathbb{O}$ since $(u|A_u W A_u u) = (u|Wu) = \text{tr}(A_u W)$) and hence $\frac{1}{\text{tr}(A_u W)} A_u W A_u = A_u$ (this follows easily from 18.3.2c). Thus $W_{\sigma'} = W_{\sigma,\pi}$.

(d) If the ideal filter exists for a proposition π, then it transforms each pure state σ such that $p(\pi,\sigma) \neq 0$ into the pure state represented by the ray $\left[\frac{1}{\|P_\pi u_\sigma\|} P_\pi u_\sigma\right]$ (cf. 19.3.5c). Indeed, for $P \in \mathscr{P}(\mathcal{H})$ and $u \in \tilde{\mathcal{H}}$ we have

$PA_u Pf = (u|Pf) Pu = (Pu|f) Pu, \forall f \in \mathcal{H}$,

$\text{tr}(PA_u) = (u|Pu) = \|Pu\|^2$ (cf. 18.3.2b),

and hence, if $\text{tr}(PA_u) \neq 0$, $\frac{1}{\text{tr}(PA_u)}PA_u P = A_{u'}$ with $u' := \frac{1}{\|Pu\|}Pu$.

(e) For a proposition π and a state σ, the ideal filter for π (if it exists) transforms σ into itself , i.e. we have $\sigma' = \sigma$ in 19.4.4, if and only if $p(\pi,\sigma) = 1$. This follows at once from the equivalence between $p(\pi,\sigma) = 1$ and $P_\pi W_\sigma P_\pi = W_\sigma$ (cf. 19.3.5f). Indeed, $p(\pi,\sigma) = 1$ implies $P_\pi W_\sigma P_\pi = W_\sigma$ and $\text{tr}(P_\pi W_\sigma) = 1$, and hence $W_{\sigma,\pi} = P_\pi W_\sigma P_\pi = W_\sigma$. Conversely, since obviously $P_\pi W_{\sigma,\pi} P_\pi = W_{\sigma,\pi}$, $W_{\sigma,\pi} = W_\sigma$ implies $P_\pi W_\sigma P_\pi = W_\sigma$ and hence $p(\pi,\sigma) = 1$.

19.4.6 Axiom (Axiom Q2). The ideal filter exists for every proposition $\pi \in \Pi$.

19.4.7 Remarks.

(a) Axiom Q2 is a version of what is sometimes called Lüder's reduction axiom. A milder version of the axiom would be to assume that a filter exists for every proposition represented by a one-dimensional projection. This milder version would be a version of what is sometimes called von Neumann's reduction axiom, or projection postulate.

We point out that in our approach to quantum mechanics, in which states correspond to ensembles of copies prepared in a definite way, the transformation of a state σ into a pure state σ' such that $[u_{\sigma'}] = [u]$, upon action of a filter for the proposition represented by a one-dimensional projection A_u, is an immediate consequence of the definition of filter (cf. 19.4.3e) and it does not need to be assumed. However, it is not obvious that a filter does exist for every one-dimensional proposition (even less, that a filter exists for every proposition).

(b) For all $u, v \in \tilde{\mathcal{H}}$, axiom Q2 implies that there exists an experimental set-up which can be used in conjunction with an apparatus which implements the pure state σ represented by the ray $[v]$ (cf. 19.3.5c) so that, when the set-up is used, there is the probability $|(u|v)|^2$ that a copy prepared in σ is modified into a copy in the pure state σ' represented by the ray $[u]$. Indeed, any implementation of the filter for the proposition π represented by the one-dimensional projection A_u is such an experimental set-up, since

$$|(u|v)|^2 = (v|A_u v) = p(\pi, \sigma)$$

(cf. 19.4.2 and 19.4.3e). For this reason, the number $|(u|v)|^2$ is called the *transition probability* from the pure state represented by v to the pure state represented by u. We point out that the transition probability from one pure state to another is one if and only if the two states coincide (cf. 10.1.7b and 13.1.13a; also, this is a special case of 19.4.5e).

19.4.8 Definitions. We say that we have a *first kind implementation* of a proposition $\pi \in \Pi$ if we have experimental procedures for determining π in any state which are so that, immediately after π has been determined by these procedures in a copy prepared in a state $\sigma \in \Sigma$, we have a copy in a new state σ' which depends on σ and is such that $p(\pi, \sigma') = 1$ if π has been determined to be true, or else such that $p(\neg\pi, \sigma') = 1$ if $\neg\pi$ has been determined to be true. A first kind implementation is called an *ideal-implementation* if the state σ' that we have immediately after the determination of π is the state represented by the statistical operator $W_{\sigma,\pi}$ (defined in 19.4.4) if π has been determined to be true, or else by the statistical operator $W_{\sigma,\neg\pi}$ if $\neg\pi$ has been determined to be true.

We say that we have a *first kind* (respectively an *ideal*) *determination* of π if a first kind (respectively an ideal) implementation of π is carried out. A *second kind implementation*, or *determination*, of a proposition is one which is not first kind.

19.4.9 Remarks.

(a) If we have a first kind implementation of a proposition, this happens notwithstanding the cautionary remarks of 19.4.1a. Some assume that there are first kind implementations for all propositions, but we do not make this assumption.

(b) Clearly, a first kind (respectively an ideal) implementation of a proposition π is a collection of procedures which amounts to a filter (respectively an ideal filter) for π if they are supplemented with devices which absorb all the copies in which $\neg\pi$ has been found to be true (i.e., in which π has not been found to be true).

(c) If a proposition π is represented by a one-dimensional projection then a first kind implementation of π is necessarily an ideal one (cf. 19.4.5c).

19.4.10 Definitions. Let $(X, \mathcal{A})$ be a measurable space and α an X-valued observable. A *first kind* (respectively an *ideal*) *measurement* of α is a measurement of α (cf. 19.1.9a) which is performed by means of first kind (respectively ideal) determinations of all propositions $\alpha(E)$ for $E \in \mathcal{A}$.

A *second kind measurement* of α is one which is not first kind.

19.4.11 Remarks. Wolfgang Pauli introduced the distinction between first and second kind measurements (Pauli, 1933), when he distinguished between two types of measurements. The first type of measurement brings (or leaves) the copy of the system into a state in which the observable that has been measured surely gives the result that has been the outcome of the measurement if it is measured a second time. The second type of measurement either destroys the copy or else changes its state arbitrarily. For an example of each type, we quote from Josef M. Jauch (note that Jauch calls "value" what we call "result"). "First we consider the measurement of the position of some elementary particle by a counter with a finite sensitive volume. After the measurement has been performed and the counter has recorded the presence of a particle inside its sensitive volume, we know for certain that the particle, at the instant of the triggering, is actually inside the sensitive volume. By this we mean the following: Suppose we repeated the measurement immediately after it has occurred (this is of course an idealization, since counters are notorious for having a dead time after they are triggered), then we would with *certainty* observe the particle inside the volume of the counter. In the second example, we consider a momentum measurement with a counter which analyzes the pulse height of a recoil particle. Here the situation is quite different. The experiment will permit us to determine the value of the momentum only *before* the collision occurred. If we repeat the measurement immediately after it has occurred, then we find that the momentum of the particle will have a quite different value from its measured value. The very act of measurement has changed the momentum, and it is this change which produced the observable effect. We shall call a measurement which will give the same value when immediately repeated a measurement of the *first kind*. The second example is then a measurement of the *second kind*" (Jauch, 1968, p.165).

19.4.12 Remarks.

(a) In what follows we assume that α is a discrete observable. Then the self-adjoint operator A_α that represents α is the operator determined by a family $\{(\lambda_n, P_n)\}_{n \in I}$ as A was in 15.3.4B (cf. 19.3.10c). Since $\{\lambda_n\}_{n \in I} = \sigma_p(A_\alpha)$, $\{\lambda_n\}_{n \in I}$ is the family of all exact results for α (cf. 19.3.12a); moreover, $P_n = P^{A_\alpha}(\{\lambda_n\}) = P_{\alpha(\{\lambda_n\})}$ for each $n \in I$. In what follows we consider a definite state $\sigma \in \Sigma$.

First, suppose that we have an ideal measurement of α in an ensemble representing σ, i.e. in a large number N of copies of the system all prepared in the state σ, and that, for a definite $n \in I$, a device is installed which absorbs all the copies in which the result λ_n has not been found, i.e. in which the proposition $\alpha(\{\lambda_n\})$ is not true. Then we have an ideal filter for $\alpha(\{\lambda_n\})$ (cf. 19.4.9b). This *selects*, from the original ensemble of copies, a *subensemble* containing approximately $p(\alpha(\{\lambda_n\}), \sigma)N$ copies which are in the state σ_n' represented by the statistical operator

$$W_{\sigma_n'} = W_{\sigma, \alpha(\{\lambda_n\})} = \frac{1}{\mathrm{tr}(P_{\alpha(\{\lambda_n\})} W_\sigma)} P_{\alpha(\{\lambda_n\})} W_\sigma P_{\alpha(\{\lambda_n\})}$$

$$= \frac{1}{\mathrm{tr}(P_n W_\sigma)} P_n W_\sigma P_n,$$

provided this subensemble is not empty, i.e. provided $p(\alpha(\{\lambda_n\}), \sigma) = \mathrm{tr}(P_n W_\sigma) \neq 0$ (cf. 19.4.3b and 19.4.4).

Next suppose that we are in a different situation, and that we have just one copy which had been previously prepared in σ and in which an ideal measurement of α has given the exact result λ_n. Then immediately after the measurement the copy is in the state σ_n'. Indeed, since we are considering the copy after the proposition $\alpha(\{\lambda_n\})$ has been determined to be true in it, there is no need to select the copy since everything is as if the copy had gone through an ideal filter for $\alpha(\{\lambda_n\})$ (if we had provided a device that would absorb the copies in which $\alpha(\{\lambda_n\})$ was not true, our copy would not have been absorbed).

Suppose once again that we have an ideal measurement of α in a copy prepared in σ, but this time the result obtained has not been recorded; i.e., there has been a result which was necessarily one of the numbers in $\{\lambda_n\}_{n \in I}$ (since a measurement of α means that all propositions $\alpha(\{\lambda_n\})$ have been determined, and hence one of them has been found to be true because the elements of $\{\lambda_n\}_{n \in I}$ are the only numbers that can be obtained as results in view of the fact that $P^{A_\alpha}(\mathbb{R} - \{\lambda_n\}_{n \in I}) = O_\mathcal{H}$ and this implies that $p(\alpha(\mathbb{R} - \{\lambda_n\}_{n \in I}), \sigma) = \mathrm{tr}(P^{A_\alpha}(\mathbb{R} - \{\lambda_n\}_{n \in I})W_\sigma) = 0)$, but the measuring apparatus has failed to keep record of the result (if we include ourselves as observers in the measuring apparatus, this could mean that we have not registered the result in our memories or elsewhere). Then we only know that immediately after the measurement the copy has probability

$$\mathrm{tr}(P_n W_\sigma) = \mathrm{tr}(P_{\alpha(\{\lambda_n\})} W_\sigma) = p(\alpha(\{\lambda_n\}), \sigma)$$

of being in the state σ'_n, and thus we must conclude (cf. 19.3.5b) that the state of the copy after the measurement is the mixed state σ'' represented by the statistical operator $W_{\sigma''}$ defined by

$$W_{\sigma''} f := \sum_{n \in I_0} \left(\mathrm{tr}(P_n W_\sigma) \right) W_{\sigma'_n} f = \sum_{n \in I} P_n W_\sigma P_n f, \forall f \in \mathcal{H},$$

where $I_0 := \{n \in I : \mathrm{tr}(P_n W_\sigma) \neq 0\}$ and the second equality follows from the fact that if $\mathrm{tr}(P_n W_\sigma) = 0$ then $P_n W_\sigma P_n = \mathbb{O}_{\mathcal{H}}$ (cf. 19.3.5f). We point out that, in what we have just done, the probabilities $p(\alpha(\{\lambda_n\}), \sigma)$ have not been used as theoretical predictions of frequencies, but rather to quantify our ignorance of which exact result has actually been obtained (but not recorded) by the measuring apparatus. Thus, they are of an epistemic nature, like classical probabilities (cf. 19.2.4).

Suppose for the third time that we have an ideal measurement of α in a copy prepared in σ, and that we only know that the result obtained belongs to a definite subset E of $\mathbb{R}$. This implies that the proposition $\alpha(\{\lambda_n\}_{n \in I_E})$ has been determined to be true, with $I_E := \{n \in I : \lambda_n \in E\}$. Then we know that the probability for the copy to be, immediately after the measurement, in the state σ'_n is:

$$
\begin{array}{ll}
0 & \text{if } n \notin I_E, \\[2mm]
\dfrac{\mathrm{tr}(P_n W_\sigma)}{\mathrm{tr}(P_{\alpha(\{\lambda_k\}_{k \in I_E})} W_\sigma)} = \dfrac{p(\alpha(\{\lambda_n\}), \sigma)}{p(\alpha(\{\lambda_k\}_{k \in I_E}), \sigma)} & \text{if } n \in I_E;
\end{array}
$$

in fact, our ignorance is smaller than it was in the previous case, and we modify the probabilities of the previous case as we should do if they were classical probabilities. Proceeding as before and observing that

$$P_{\alpha(\{\lambda_k\}_{k \in I_E})} = P^{A_\alpha}(\{\lambda_k\}_{k \in I_E}) = \sum_{k \in I_E} P_k$$

(cf. 15.3.4B), we see that immediately after the measurement the copy is in the state σ''_E represented by the statistical operator $W_{\sigma''_E}$ defined by

$$W_{\sigma''_E} f := \frac{1}{\sum_{k \in I_E} \mathrm{tr}(P_k W_\sigma)} \sum_{n \in I_E} P_n W_\sigma P_n f, \forall f \in \mathcal{H}$$

(note that $\mathrm{tr}\left((\sum_{k \in I_E} P_k) W_\sigma \right) = \sum_{k \in I_E} \mathrm{tr}(P_k W_\sigma)$ by 18.3.12 and that $\mathrm{tr}\, W_{\sigma''_E} = 1$ since $\mathrm{tr}(P_n W_\sigma P_n) = \mathrm{tr}(P_n W_\sigma)$).

We must underline the fact that, in the last two cases considered above (when σ was transformed into σ'' or σ''_E), there is a measuring apparatus which "interacts" with a copy of the system in such a way as to turn out an exact result, and that only the recording section of the apparatus is defective. Indeed, if in the last case considered above the apparatus was only capable of determining whether the proposition $\alpha(\{\lambda_n\}_{n \in I_E})$ was true, then we would only have an ideal determination of this proposition (and not an ideal measurement of α) and, after an "interaction" with the apparatus in which this proposition was determined to be true, the copy would be in the state σ'_E represented by the statistical operator $W_{\sigma'_E}$ defined by

$$
\begin{aligned}
W_{\sigma'_E} f &= W_{\sigma,\alpha(\{\lambda_k\}_{k \in I_E})} f \\
&= \frac{1}{\mathrm{tr}(P_{\alpha(\{\lambda_k\}_{k \in I_E})} W_\sigma)} P_{\alpha(\{\lambda_k\}_{k \in I_E})} W_\sigma P_{\alpha(\{\lambda_k\}_{k \in I_E})} f \\
&= \frac{1}{\sum_{k \in I_E} \mathrm{tr}(P_k W_\sigma)} \sum_{n,m \in I_E} P_n W_\sigma P_m f, \forall f \in \mathcal{H},
\end{aligned}
$$

which is clearly not the same as $W_{\sigma''_E}$ (and hence σ'_E is not the same as σ''_E). Finally, suppose that we have ideal measurements of α in an ensemble of N copies all prepared in σ. We have already seen what happens if we make a selection by keeping just those copies in which a particular exact result has been obtained. If instead no selection is made, then after the measurements we have an ensemble which still contains N copies, all of them in the state σ''. If only a coarse selection is made by keeping just those copies for which a result has been obtained that belongs to a definite subset E of $\mathbb{R}$, then after the measurements and the selection we have an ensemble which contains approximately $\left(\sum_{k \in I_E} \mathrm{tr}(P_k W_\sigma)\right) N$ copies, all of them in the state σ''_E.
All transformations considered above of σ into another state (σ'_n, σ'',σ''_E,σ'_E) are called *state reductions* (for the transformations of σ into σ'_n or into σ'_E, this name was already known from 19.4.3b).

(b) We suppose here that α is an observable which is not discrete and, for the sake of simplicity, we also suppose that A_α has no eigenvalues, i.e. (cf. 19.3.12a) that there are no real numbers which are exact results for α. What happens then if an ideal measurement of α is carried out? Naturally, a result is obtained which is identified with a real number λ, but there is no state in which this result has non zero probability of being obtained, since $\alpha(\{\lambda\}) = \pi_0$ for each $\lambda \in \mathbb{R}$. Indeed, in N repetitions of the measurement of α we will obtain N results, but each of them so seldom that its relative frequency approaches zero as N grows (cf. 19.1.16a). However, an observable with no exact results (or, more generally, a non discrete observable) is an idealization which is useful (under some respects, even essential) on the theoretical level but which on the operational level actually stands for a sequence of more realistic discrete observables which correspond to more realistic measuring instruments and which can be assumed to be functions of α, as for instance the observables α_n defined in 19.1.22a. In order to perform a non-fictional measurement of α, we must actually measure one of these more realistic discrete observables, for instance one of the observables α_n, and hence the analysis of remark a applies.
As already observed in 19.1.22a, the relation between the observable α and the more realistic discrete observables which approximate α is conceptually similar to the one that exists, in classical mechanics, between derivatives used to represent values of speed and the way speed is actually measured. When

speed is measured, only difference quotients are actually measured; however, the notion of speed as derivative is essential for the laws of classical mechanics.

19.5 Compatible observables and uncertainty relations in quantum mechanics

In discussions about quantum mechanics the issue is often addressed of whether two observables are compatible with each other, something which is often regarded as being equivalent to the condition that they can be measured simultaneously. However, it is not always clear what is meant by a "simultaneous measurement". And indeed the idea of an interaction of a copy of a quantum system with two measuring instruments at the same time does not seem experimentally very sound. A perhaps more promising idea might be that two observables α and β are simultaneously measurable if a measurement of α followed immediately by a measurement of β yields the same results as when the order of the α and β measurements is reversed. In the first part of this section we endeavour to deal with this topic on mainly statistical grounds.

In the second part of this section we discuss uncertainty relations, an issue which in the early days of quantum mechanics seemed to involve deep epistemological and even philosophical questions. However, a strict statistical interpretation of uncertainty relations as presented here is quite unproblematic.

As usual, states, propositions, observables are referred to a given quantum system (cf. also 19.1.12) and they are represented as summarized in 19.3.22.

19.5.1 Remarks.

(a) First, for a proposition $\pi' \in \Pi$ and a state $\sigma \in \Sigma$, we consider the occurrence

o_1: *a copy of the system, prepared in σ, goes through the ideal filter for π'.*

The probability of o_1 is $p(\pi', \sigma)$, by the definition of a filter for π' (cf. 19.4.2). Next, for one more proposition $\pi'' \in \Pi$, we consider the occurrence

o_2: *π'' is ascertained to be true in a copy of the system which, immediately after being prepared in σ, has just gone through the ideal filter for π'.*

The occurrence o_2 can actually happen only if some copies can exist which, after being prepared in σ, go through the ideal filter for π', i.e. only if $p(\pi', \sigma) \neq 0$. If such is the case, the probability of o_2 is $p(\pi'', \sigma')$ with σ' the state that is represented by the statistical operator $W_{\sigma, \pi'}$, by the definition of the ideal filter for π' (cf. 19.4.4).

Finally, we consider the occurrence

o: *a copy of the system, prepared in σ, goes through the ideal filter for π' and, immediately after that, π'' is ascertained to be true in that copy.*

We denote by $p(\pi'', \pi', \sigma)$ the probability of o. Now, if $p(\pi', \sigma) \neq 0$, the occurrence o is the joint happening of the occurrences o_1 and o_2, and therefore we have

$$p(\pi'', \pi', \sigma) = p(\pi'', \sigma')p(\pi', \sigma) = \mathrm{tr}(P_{\pi''} W_{\sigma, \pi'}) \, \mathrm{tr}(P_{\pi'} W_\sigma)$$
$$= \mathrm{tr}(P_{\pi''} P_{\pi'} W_\sigma P_{\pi'}).$$

Moreover, if $p(\pi', \sigma) = 0$ then the occurrence o can never happen and hence $p(\pi'', \pi', \sigma) = 0$, and $p(\pi', \sigma) = 0$ also implies $P_{\pi'} W_\sigma = \mathbb{O}_{\mathcal{H}}$ (cf. 19.3.5f) and hence $\mathrm{tr}(P_{\pi''} P_{\pi'} W_\sigma P_{\pi'}) = 0$. Thus we have

$$p(\pi'', \pi', \sigma) = \mathrm{tr}(P_{\pi''} P_{\pi'} W_\sigma P_{\pi'})$$

whatever the value of $p(\pi', \sigma)$, and this equation can be written as

$$p(\pi'', \pi', \sigma) = \mathrm{tr}(P_{\pi'} P_{\pi''} P_{\pi'} W_\sigma) = \mathrm{tr}(P_{\pi''} P_{\pi'} W_\sigma P_{\pi'} P_{\pi''})$$

(cf. 18.2.11c and, for the second equality, $P_{\pi''} = P_{\pi''}^2$).
If σ is a pure state, i.e. $W_\sigma = A_u$ with $u \in \tilde{\mathcal{H}}$, then (cf. 18.3.2b and 13.1.5)

$$p(\pi'', \pi', \sigma) = \mathrm{tr}(P_{\pi'} P_{\pi''} P_{\pi'} A_u) = (u|P_{\pi'} P_{\pi''} P_{\pi'} u) = \|P_{\pi''} P_{\pi'} u\|^2.$$

(b) Suppose that there exists an ideal implementation of a proposition $\pi' \in \Pi$. Then, for every proposition $\pi'' \in \Pi$ and every state $\sigma \in \Sigma$, $p(\pi'', \pi', \sigma)$ is also the probability that both π' and π'' turn out to be true if an ideal determination of π' is carried out in a copy prepared in σ and a determination of π'' is carried out immediately after that; this follows from the definition of an ideal determination given in in 19.4.8. Moreover, $p(\pi'', \pi', \sigma)$ is also the probability for a copy prepared in σ to go through the ideal filter for π' and, immediately after that, through a filter for π'' as well; this too follows from the definitions.

19.5.2 Definition. Two propositions $\pi', \pi'' \in \Pi$ are said to be *compatible* if

$$p(\pi'', \pi', \sigma) = p(\pi', \pi'', \sigma), \forall \sigma \in \Sigma.$$

19.5.3 Proposition. *For two propositions $\pi', \pi'' \in \Pi$, the following conditions are equivalent:*

(a) π' and π'' are compatible;
(b) $p(\pi'', \pi', \sigma) = p(\pi', \pi'', \sigma), \forall \sigma \in \Sigma_0$;
(c) $[P_{\pi'}, P_{\pi''}] = \mathbb{O}_{\mathcal{H}}$.

Proof. $a \Rightarrow b$: This is obvious.

$b \Rightarrow c$: Assume condition b. Owing to the bijection that exists from Σ_0 onto the family of all rays of $\mathcal{H}$ (cf. 19.3.5c), condition b implies (cf. 19.5.1a)

$$(u|P_{\pi'} P_{\pi''} P_{\pi'} u) = (u|P_{\pi''} P_{\pi'} P_{\pi''} u), \forall u \in \tilde{\mathcal{H}},$$

and hence (cf. 10.2.12)

$$P_{\pi'} P_{\pi''} P_{\pi'} = P_{\pi''} P_{\pi'} P_{\pi''},$$

and hence (since $P_{\pi'}^2 = P_{\pi'}$ and $P_{\pi''}^2 = P_{\pi''}$)

$$P_{\pi'}P_{\pi''}P_{\pi'} = P_{\pi'}P_{\pi''}P_{\pi'}P_{\pi''} \text{ and } P_{\pi''}P_{\pi'}P_{\pi''}P_{\pi'} = P_{\pi''}P_{\pi'}P_{\pi''}.$$

Then, if we define $A := [P_{\pi'}, P_{\pi''}]$, by 13.1.5 we have

$$\|Af\|^2 = (Af|Af) = (f|P_{\pi''}P_{\pi'}P_{\pi''}f) - (f|P_{\pi'}P_{\pi''}P_{\pi'}P_{\pi''}f)$$
$$- (f|P_{\pi''}P_{\pi'}P_{\pi''}P_{\pi'}f) + (f|P_{\pi'}P_{\pi''}P_{\pi'}f) = 0, \forall f \in \mathcal{H},$$

and hence $A = \mathbb{O}_{\mathcal{H}}$.

$c \Rightarrow a$: If condition c is true then we have (cf. 19.5.1a)

$$p(\pi'', \pi', \sigma) = \operatorname{tr}(P_{\pi''}P_{\pi'}W_\sigma P_{\pi'}P_{\pi''})$$
$$= \operatorname{tr}(P_{\pi'}P_{\pi''}W_\sigma P_{\pi''}P_{\pi'}) = p(\pi', \pi'', \sigma), \forall \sigma \in \Sigma.$$

$\square$

19.5.4 Proposition. *Let $\pi', \pi'' \in \Pi$. For a state $\sigma \in \Sigma$ such that $p(\pi'', \pi', \sigma) \neq 0$, consider the copy of the system that results from a copy which, after being prepared in σ, has gone through the ideal filter for π' and, immediately after that, through the ideal filter for π''. This copy is in the state $\tilde{\sigma}$ represented by the statistical operator*

$$W_{\tilde{\sigma}} := \frac{1}{\operatorname{tr}(P_{\pi''}P_{\pi'}W_\sigma P_{\pi'}P_{\pi''})} P_{\pi''}P_{\pi'}W_\sigma P_{\pi'}P_{\pi''}.$$

While π'' is certainly true in this copy, i.e. $p(\pi'', \tilde{\sigma}) = 1$, for π' the following conditions are equivalent:

(a) $p(\pi', \tilde{\sigma}) = 1$ for each $\sigma \in \Sigma$ such that $p(\pi'', \pi', \sigma) \neq 0$;
(b) π' and π'' are compatible.

Proof. First we notice that the denominator in the statement is non-zero since $\operatorname{tr}(P_{\pi''}P_{\pi'}W_\sigma P_{\pi'}P_{\pi''}) = p(\pi'', \pi', \sigma)$ (cf. 19.5.1a). Next, from 19.4.4 we have that the state $\tilde{\sigma}$ is represented by the statistical operator $W_{\sigma', \pi''}$ with $W_{\sigma'} = W_{\sigma, \pi'}$, and hence by the statistical operator

$$\frac{1}{\operatorname{tr}(P_{\pi''}W_{\sigma, \pi'}P_{\pi''})} P_{\pi''}W_{\sigma, \pi'}P_{\pi''} = \frac{1}{\operatorname{tr}(P_{\pi''}P_{\pi'}W_\sigma P_{\pi'}P_{\pi''})} P_{\pi''}P_{\pi'}W_\sigma P_{\pi'}P_{\pi''}.$$

From $P_{\pi''}^2 = P_{\pi''}$ we have $P_{\pi''}W_{\tilde{\sigma}} = W_{\tilde{\sigma}}$, and hence $p(\pi'', \tilde{\sigma}) = 1$ by 19.3.5f.

We prove now the equivalence between conditions a and b.

$a \Rightarrow b$: Assume condition a. Then, for each $\sigma \in \Sigma$ such that $p(\pi'', \pi', \sigma) \neq 0$, we have

$$W_{\tilde{\sigma}} = W_{\tilde{\sigma}, \pi'}$$

(cf. 19.4.5e). This equality is true in particular for each pure state $\sigma \in \Sigma_0$ such that $\|P_{\pi''}P_{\pi'}u_\sigma\|^2 = p(\pi'', \pi', \sigma) \neq 0$ (cf. 19.5.1a), for which it can be written as (cf. 19.4.5d)

$$A_{\tilde{u}_\sigma} = A_{\tilde{u}'_\sigma}$$

with

$$\tilde{u}_\sigma := \frac{1}{\|P_{\pi''}P_{\pi'}u_\sigma\|}P_{\pi''}P_{\pi'}u_\sigma \text{ and } \tilde{u}'_\sigma := \frac{1}{\|P_{\pi'}P_{\pi''}P_{\pi'}u_\sigma\|}P_{\pi'}P_{\pi''}P_{\pi'}u_\sigma,$$

and this implies (cf. 13.1.13a) that there exists $\alpha \in \mathbb{C}$ so that

$$P_{\pi''}P_{\pi'}u_\sigma = \alpha P_{\pi'}P_{\pi''}P_{\pi'}u_\sigma;$$

applying $P_{\pi'}$ to the left of both sides of this equality we get

$$P_{\pi'}P_{\pi''}P_{\pi'}u_\sigma = \alpha P_{\pi'}P_{\pi''}P_{\pi'}u_\sigma,$$

and hence $\alpha = 1$ since $P_{\pi'}P_{\pi''}P_{\pi'}u_\sigma \neq 0_\mathcal{H}$. Owing to the bijection that exists from Σ_0 onto the family of all rays of $\mathcal{H}$ (cf. 19.3.5c), this proves that

$$P_{\pi''}P_{\pi'}u = P_{\pi'}P_{\pi''}P_{\pi'}u \text{ for each } u \in \tilde{\mathcal{H}} \text{ such that } P_{\pi''}P_{\pi'}u \neq 0_\mathcal{H};$$

since the same is trivially true for each $u \in \tilde{\mathcal{H}}$ such that $P_{\pi''}P_{\pi'}u = 0_\mathcal{H}$, we have

$$P_{\pi''}P_{\pi'} = P_{\pi'}P_{\pi''}P_{\pi'}.$$

By taking the adjoints of both sides we get

$$P_{\pi'}P_{\pi''} = P_{\pi'}P_{\pi''}P_{\pi'}$$

(cf. 12.3.4b), and hence $[P_{\pi'},P_{\pi''}] = 0_\mathcal{H}$, and hence condition b by 19.5.3.

$b \Rightarrow a$: If π' and π'' are compatible, then $[P_{\pi'},P_{\pi''}] = 0_\mathcal{H}$ by 19.5.3, and hence $P_{\pi'}W_{\tilde{\sigma}} = W_{\tilde{\sigma}}$, and hence $p(\pi',\tilde{\sigma}) = 1$ by 19.3.5f. $\qquad\square$

19.5.5 Proposition. *Suppose that we have ideal implementations of two propositions $\pi', \pi'' \in \Pi$, and consider for a state $\sigma \in \Sigma$ the two occurrences*

$o_{\pi',\pi''}$: *π' is ideally determined to be true (i.e., it is ascertained to be true by means of its ideal determination) in a copy prepared in σ and, immediately after that, π'' is determined to be true;*

$o_{\pi'',\pi'}$: *π'' is ideally determined to be true in a copy prepared in σ and, immediately after that, π' is determined to be true.*

The probability of the occurrence $o_{\pi',\pi''}$ is $p(\pi'',\pi',\sigma)$, and the probability of the occurrence $o_{\pi'',\pi'}$ is $p(\pi',\pi'',\sigma)$ (cf. 19.5.1b).

The following conditions are equivalent:

(a) there exists a proposition $\pi \in \Pi$ such that $p(\pi'',\pi',\sigma) = p(\pi,\sigma)$ for each $\sigma \in \Sigma$;
(b) π' and π'' are compatible.

If these conditions are satisfied, then the proposition π (as an equivalence class) is unique and we have:

(c) $p(\pi',\pi'',\sigma) = p(\pi,\sigma)$ for each $\sigma \in \Sigma$;
(d) $P_\pi = P_{\pi''}P_{\pi'}$.

Proof. First, we observe that if condition a is satisfied then the proposition π, as an equivalence class (cf. 19.1.12), is unique by the very definition of the equivalence relation in Π (cf. 19.1.5).

$a \Rightarrow (b$ and $d)$: We assume condition a. Then we have (cf. 19.5.1a)

$$\text{tr}(P_{\pi'} P_{\pi''} P_{\pi'} W_\sigma) = \text{tr}(P_\pi W_\sigma), \forall \sigma \in \Sigma,$$

and hence in particular (cf. 18.3.2b)

$$(u|P_{\pi'} P_{\pi''} P_{\pi'} u) = (u|P_\pi u), \forall u \in \tilde{\mathcal{H}},$$

and hence, by 10.2.12,

$$P_{\pi'} P_{\pi''} P_{\pi'} = P_\pi.$$

Then we have

$$\|P_{\pi'} P_{\pi''} P_{\pi'} f\|^2 = \|P_\pi f\|^2 = (f|P_\pi f) = (f|P_{\pi'} P_{\pi''} P_{\pi'} f) = \|P_{\pi''} P_{\pi'} f\|^2, \forall f \in \mathcal{H},$$

and hence (cf. 13.1.3c, with $P_M := P_{\pi'}$)

$$P_{\pi'} P_{\pi''} P_{\pi'} f = P_{\pi''} P_{\pi'} f, \forall f \in \mathcal{H},$$

and hence

$$P_\pi = P_{\pi''} P_{\pi'},$$

which is condition d. Moreover, this implies $[P_{\pi'}, P_{\pi''}] = \mathbb{O}_{\mathcal{H}}$ by 13.2.1, and hence condition b by 19.5.3.

$b \Rightarrow (a$ and $c)$: We assume condition b. Then we have $[P_{\pi'}, P_{\pi''}] = \mathbb{O}_{\mathcal{H}}$ by 19.5.3, and hence $P_{\pi''} P_{\pi'} \in \mathscr{P}(\mathcal{H})$ by 13.2.1. Letting π be the proposition such that $P_\pi = P_{\pi''} P_{\pi'}$, we have (cf. 19.5.1a)

$$p(\pi', \pi'', \sigma) = p(\pi'', \pi', \sigma) = \text{tr}(P_{\pi''} P_{\pi'} W_\sigma P_{\pi'} P_{\pi''})$$
$$= \text{tr}(P_\pi W_\sigma P_\pi) = \text{tr}(P_\pi W_\sigma) = p(\pi, \sigma), \forall \sigma \in \Sigma.$$

Thus, both conditions a and c are proved. $\qquad\square$

19.5.6 Remarks.

(a) The equivalence between conditions a and b in 19.5.5 shows that *we cannot accept all occurrences related to a quantum system as bonafide events which define propositions*. Indeed, the meaning of condition a is that the occurrence $o_{\pi',\pi''}$ is actually a quantum event which defines a proposition, and the equivalence between conditions a and b shows that this is true if and only if π' and π'' are compatible. Condition c shows that if π' and π'' are compatible then both the occurrences $o_{\pi',\pi''}$ and $o_{\pi'',\pi'}$ are implementations of the same proposition π. Thus, if π' and π'' are compatible, we can say that an event in the equivalence class of π is the "simultaneous occurrence" of the events that define π' and π''; actually, the experimental determinations of π' and π'' will require to determine first one of them and then, immediately afterwards, the other one; however, the order is immaterial since $o_{\pi',\pi''}$ and $o_{\pi'',\pi'}$ define propositions which are in the same equivalence class. This equivalence class, which we have denoted by π up to now, will be denoted by the symbol $\pi' \wedge \pi''$ henceforth (thus, this symbol implies that π' and π'' are compatible and that there exist ideal implementations of them).

(b) If π' and π'' are compatible propositions and ideal implementations of them are available, then the proposition we have denoted by $\pi' \wedge \pi''$ is represented by the orthogonal projection $P_{\pi' \wedge \pi''} = P_{\pi''} P_{\pi'}$ (cf. 19.5.5d), i.e. by the orthogonal projection defined by the subspace $R_{P_{\pi'}} \cap R_{P_{\pi''}}$ (cf. 13.2.1e).

(c) We remark that, for two propositions π' and π'', the operator $P_{\pi''} P_{\pi'}$ is an orthogonal projection if and only if π' and π'' are compatible (cf. 19.5.3 and 13.2.1). However, for any pair of propositions π', π'' there is always (i.e., with no conditions on π', π'') an orthogonal projection which is defined by the subspace $R_{P_{\pi'}} \cap R_{P_{\pi''}}$ (cf. 4.1.10), and hence there is always a proposition, which we still denote by π, such that $R_{P_\pi} = R_{P_{\pi'}} \cap R_{P_{\pi''}}$, since the mapping of 19.3.1b is bijective. For a state σ we have

$$p(\pi, \sigma) = 1 \Leftrightarrow R_{W_\sigma} \subset R_{P_\pi} = R_{P_{\pi'}} \cap R_{P_{\pi''}} \Leftrightarrow p(\pi', \sigma) = p(\pi'', \sigma) = 1$$

(cf. 19.3.5f). Thus, π is certainly true in a state if and only if both π' and π'' are certainly true in that state. We note that, if π' and π'' were proposition in a classical theory, then the classical proposition $\pi' \wedge \pi''$ (defined in 19.2.1) would be certainly true in a state if and only if both π' and π'' were certainly true in that state. Indeed, for a state σ, in a classical theory we would have (cf. 19.2.8 and the proof of 19.2.7)

$$p(\pi' \wedge \pi'', \sigma) = \mu_\sigma(S_{\pi' \wedge \pi''}) = \mu_\sigma(S_{\pi'} \cap S_{\pi''}) = 1 \Leftrightarrow$$
$$[p(\pi', \sigma) = \mu_\sigma(S_{\pi'}) = 1 \text{ and } p(\pi'', \sigma) = \mu_\sigma(S_{\pi''}) = 1];$$

in fact, one implication follows immediately from the monotonicity of μ_σ and for the other one we have

$$\mu_\sigma(S_{\pi'}) = \mu_\sigma(S_{\pi''}) = 1 \Rightarrow \mu_\sigma(S - S_{\pi'}) = \mu_\sigma(S - S_{\pi''}) = 0 \Rightarrow$$
$$\mu_\sigma(S - (S_{\pi'} \cap S_{\pi''})) = \mu_\sigma((S - S_{\pi'}) \cup (S - S_{\pi''})) = 0 \Rightarrow \mu_\sigma(S_{\pi'} \cap S_{\pi''}) = 1.$$

This could suggest interpreting π as the proposition "π' and π''" also in the quantum theory. However, if pursued in the quantum theory, this interpretation *must not* lead to thinking that in general π' and π'' can be determined in the same copies (as instead they could in a classical theory); actually, $p(\pi', \sigma) = 1$ means that π' is found to be true in all copies of an ensemble representing σ and $p(\pi'', \sigma) = 1$ means that π'' is found to be true in all copies of a *different* ensemble representing σ. Moreover, determining π' in a copy and then π'' in the resulting copy is a procedure which is not in general equivalent to determining first π'' and then π', as 19.5.3 shows. However, if π' and π'' are compatible and if ideal implementations are available for both of them, then we saw in 19.5.6a that an ideal determination of one of them in a copy immediately followed by a determination of the other one in the resulting copy defines an event which lies in the equivalence class of π. Thus, when π' and π'' are compatible there are experimentally reasonable grounds for interpreting the proposition π as the proposition "π' and π''". In any case, we will reserve the symbol $\pi' \wedge \pi''$ for the case of compatible propositions π', π'' for which ideal implementations are available.

(d) Suppose that two propositions π' and π'' are compatible and that ideal implementations are available for both of them. Then the pairs π' and $\neg\pi''$, $\neg\pi'$ and
π'', $\neg\pi'$ and $\neg\pi''$ are all compatible; this follows at once from 19.5.3 and 19.3.4.
Thus, for every state σ, the probabilities for the joint results of π' and π'' are
independent from the order in which the determinations are made, i.e.

$$p(\pi^*, \pi^{**}, \sigma) = p(\pi^{**}, \pi^*, \sigma) \text{ for } \pi^* = \pi', \neg\pi' \text{ and } \pi^{**} = \pi'', \neg\pi''.$$

19.5.7 Remark. Let $\pi', \pi'' \in \Pi$ and consider, as in 19.5.6b, the proposition $\pi \in \Pi$
which is so that $R_{P_\pi} = R_{P_{\pi'}} \cap R_{P_{\pi''}}$, i.e. so that, for a state $\sigma \in \Sigma$,

$$p(\pi, \sigma) = 1 \Leftrightarrow p(\pi', \sigma) = p(\pi'', \sigma) = 1.$$

If π' and π'' are compatible, then $P_\pi = P_{\pi''} P_{\pi'}$ (cf. 19.5.3 and 13.2.1) even when no
ideal implementations of π' and π'' are available. If π' and π'' are compatible, then
it is clear from the form of the statistical operator $W_{\tilde\sigma}$ in 19.5.4 that $W_{\tilde\sigma} = W_{\sigma,\pi}$,
and therefore that implementations of the ideal filters for π' and π'', applied the
one after the other in either order, amount to an implementation of the ideal filter
for π.

Even when π' and π'' are not compatible it is possible to design, at least in
principle, an implementation of the ideal filter for the proposition π which is so
that $R_{P_\pi} = R_{P_{\pi'}} \cap R_{P_{\pi''}}$, by means of implementations of the ideal filters for π' and
π''. The procedure is as follows. We have a copy, prepared in a state σ, go through
a pack of n ideal filters for π' and π'', arranged in an alternate sequence (first a
filter for π', second a filter for π'', third a filter for π', and so on n times). If a copy
does go through this pack of filters then, proceeding as in the proof of 19.5.4 (cf.
also 12.3.4b), we see that afterwards this copy is in the state σ_n represented by the
statistical operator

$$W_{\sigma_n} = \frac{1}{\text{tr}(T_n W_\sigma T_n^\dagger)} T_n W_\sigma T_n^\dagger$$

with the operator T_n defined as in 18.2.17 for $P_M := P_{\pi'}$ and $P_N := P_{\pi''}$. If we
admit that it is experimentally meaningful to pursue this course of action for any
number n of filters, then we have a statistical approximation as good as we want of
the ideal filter for π, since for every projection $P \in \mathscr{P}(\mathcal{H})$ we have

$$\text{tr}(PW_{\sigma_n}) = \frac{1}{\text{tr}(T_n W_\sigma T_n^\dagger)} \text{tr}(PT_n W_\sigma T_n^\dagger)$$

$$\xrightarrow[n\to\infty]{} \frac{1}{\text{tr}(P_\pi W_\sigma P_\pi)} \text{tr}(PP_\pi W_\sigma P_\pi) = \text{tr}(PW_{\sigma,\pi})$$

whenever $p(\pi, \sigma) \neq 0$ (cf. 18.2.17). In fact this shows that all the probabilities that
are determined (according to 19.3.1c) by the state into which σ is transformed by
the ideal filter for π can be approximated as well as we want by the probabilities
that are determined by the state into which σ is transformed by the ideal filters for
π' and π'', used n times alternatingly. We point out that what was written above is

consistent because if $\mathrm{tr}(P_\pi W_\sigma P_\pi) = p(\pi, \sigma) \neq 0$ then $\mathrm{tr}(T_n W_\sigma T_n^\dagger) \neq 0$ for all $n \in \mathbb{N}$ (cf. 18.2.17). We also point out the obvious fact that, if π' and π'' are compatible, then this procedure is equivalent to the one in which only two filters are used, one for π' and the other for π''. In fact, if π' and π'' are compatible then $T_n = P_{\pi''} P_{\pi'}$ for all $n > 1$ (cf. 19.5.3).

19.5.8 Remark. To understand better the meaning of the results obtained so far in this section, it is useful to examine what we should have if, in the situations discussed, we were considering a classical statistical theory (for which we refer to Section 19.2).

In a classical statistical theory, the action of an ideal filter for a proposition π would be to transform any state σ such that $\mu_\sigma(S_\pi) = p(\pi, \sigma) \neq 0$ into the state σ' represented by the probability measure $\mu_{\sigma,\pi}$ on $\mathcal{A}$ defined by

$$\mu_{\sigma,\pi}(E) := \frac{1}{\mu_\sigma(S_\pi)} \mu_\sigma(E \cap S_\pi), \forall E \in \mathcal{A}.$$

Note that this obviously defines a probability measure and that $p(\pi, \sigma') = \mu_{\sigma,\pi}(S_\pi) = 1$; thus, the reduction from μ_σ to $\mu_{\sigma,\pi}$ would indeed represent the action of a filter for π; moreover, $\mu_{\sigma,\pi}$ is obtained from the original measure μ_σ by altering it to the least degree consistent with the condition $\mu_{\sigma,\pi}(S_\pi) = 1$, as an ideal filter should do.

Then, for two propositions π', π'' and a state σ in a classical statistical theory, if $p(\pi', \sigma) \neq 0$ we should have, reasoning as in 19.5.1,

$$p(\pi'', \pi', \sigma) = p(\pi'', \sigma')p(\pi', \sigma),$$

where σ' would be the state represented by the probability measure $\mu_{\sigma,\pi'}$, and hence

$$p(\pi'', \pi', \sigma) = \mu_{\sigma,\pi'}(S_{\pi''})\mu_\sigma(S_{\pi'}) = \mu_\sigma(S_{\pi''} \cap S_{\pi'});$$

since $p(\pi', \sigma) = 0$ implies that the occurrence o defined in 19.5.1 can never happen and hence $p(\pi'', \pi', \sigma) = 0$, and also implies $\mu_\sigma(S_{\pi''} \cap S_{\pi'}) = 0$ (by the monotonicity of μ_σ), we should have

$$p(\pi'', \pi', \sigma) = \mu_\sigma(S_{\pi''} \cap S_{\pi'})$$

whatever the value of $p(\pi', \sigma)$. And similarly we should have

$$p(\pi', \pi'', \sigma) = \mu_\sigma(S_{\pi'} \cap S_{\pi''}).$$

Thus, in a classical statistical theory we should have

$$p(\pi'', \pi', \sigma) = p(\pi', \pi'', \sigma)$$

for every pair of propositions and every state, in contrast with the result of 19.5.3.

As to 19.5.4, in a classical statistical theory a copy, initially prepared in a state σ, after going through an ideal filter for a proposition π' and through an ideal filter for a proposition π'' would be in the state $\tilde{\sigma}$ represented by the probability measure $\mu_{\tilde{\sigma}}$ on $\mathcal{A}$ defined by

$$\mu_{\tilde{\sigma}}(E) := \frac{1}{\mu_\sigma(S_{\pi''} \cap S_{\pi'})} \mu_\sigma(E \cap S_{\pi''} \cap S_{\pi'}), \forall E \in \mathcal{A},$$

and hence we should have

$$p(\pi', \tilde{\sigma}) = \mu_{\tilde{\sigma}}(S_{\pi'}) = 1$$

with no conditions on the pair π', π'' except the obvious one $p(\pi'', \pi', \sigma) \neq 0$ (if $p(\pi'', \pi', \sigma) = 0$ then no copy can go through the two filters), in contrast with the result of 19.5.4.

Finally, and in contrast with the result of 19.5.5., the result obtained above for $p(\pi'', \pi', \sigma)$ shows that in a classical theory we should have, for every pair of propositions π', π'' and every state σ,

$$p(\pi', \pi'', \sigma) = p(\pi'', \pi', \sigma) = p(\pi' \wedge \pi'', \sigma),$$

since the classical proposition $\pi' \wedge \pi''$ is the proposition such that $S_{\pi' \wedge \pi''} = S_{\pi'} \cap S_{\pi''}$.

Thus, if and only if a pair of quantum propositions are compatible do they behave with respect to each other as any pair of classical propositions would.

We point out that the results obtained here for the classical case derive from the fact that, in a classical theory, each copy of the system is in a "real condition" so that each proposition is certainly true or certainly false in that copy (cf. 19.2.4), and an ideal filter for a proposition π only selects the copies in which π is true while leaving unaltered their "real condition", so that the properties that were true in a copy before the selection are true also after it. That this is not the case in a quantum theory is proved by the results of this section. The result of 19.5.4 is particularly clear-cut in this respect.

19.5.9 Definition. Two observables α_1, α_2 are said to be *compatible* if the propositions $\alpha_1(E_1)$ and $\alpha_2(E_2)$ are compatible for all $E_1, E_2 \in \mathcal{A}(d_{\mathbb{R}})$.

19.5.10 Proposition. *Two observables α_1 and α_2 are compatible if and only if the operators A_{α_1} and A_{α_2} commute.*

Proof. This result follows from 19.5.3, from the definitions of the operators A_{α_1} and A_{α_2} (cf. 19.3.6), and from the definition of commutativity for two self-adjoint operators (cf. 17.1.5). $\qquad\square$

19.5.11 Remarks.

(a) Suppose that we have an $\mathbb{R}^2$-valued observable α. Then α represents a measuring instrument which yields a result by the position of a pointer in a dial which is represented by $\mathbb{R}^2$ (cf. 19.1.9a).

We can define the mapping

$$\alpha_1 : \mathcal{A}(d_{\mathbb{R}}) \to \Pi$$

$$E \mapsto \alpha_1(E) := \alpha(E \times \mathbb{R}),$$

and we see that α_1 is an observable since $\alpha_1 = \varphi_1(\alpha)$, with

$$\varphi_1 : \mathbb{R}^2 \to \mathbb{R}$$

$$(x_1, x_2) \mapsto \varphi_1(x_1, x_2) := x_1$$

(cf. 19.1.13 and 19.1.14); indeed,

$$\varphi_1(\alpha)(E) = \alpha(\varphi_1^{-1}(E)) = \alpha(E \times \mathbb{R}) = \alpha_1(E), \forall E \in \mathcal{A}(d_{\mathbb{R}}).$$

The observable α_1 is supported by the same measuring instrument that is represented by α, in which however only a partial recording of the results obtained is made: if the instrument brings forth the result which is represented by the element (x_1, x_2) of $\mathbb{R}^2$, then just the number x_1 is recorded. And similarly we can define α_2 by letting $\alpha_2(E) := \alpha(\mathbb{R} \times E)$ for each $E \in \mathcal{A}(d_{\mathbb{R}})$.
The two observables α_1 and α_2 are compatible by 19.5.10, since

$$P^{A_{\alpha_1}}(E) = P_\alpha(E \times \mathbb{R}) \text{ and } P^{A_{\alpha_2}}(E) = P_\alpha(\mathbb{R} \times E), \forall E \in \mathcal{A}(d_{\mathbb{R}}),$$

implies that A_{α_1} and A_{α_2} commute (cf. $b \Rightarrow a$ in 17.1.10). We note that it is consistent to say that the proposition $\alpha_1(E_1)$ can be determined simultaneously with the proposition $\alpha_2(E_2)$ for any $E_1, E_2 \in \mathcal{A}(d_{\mathbb{R}})$, since $\alpha_1(E_1)$ and $\alpha_2(E_2)$ are propositions in the range of α and we assumed that all the propositions in the range of an observable can be determined simultaneously for any single copy prepared in any state (cf. 19.1.9a). We recall that the basis for that assumption was the macroscopic, to wit classical, nature of pointer and dial in a measuring instrument that underlies an observable.

(b) We examine here a situation in a sense opposite to the one discussed in remark a. Suppose that we have two compatible observables α_1 and α_2. Then, by 19.5.10 and $a \Rightarrow b$ in 17.1.10, there exists a unique projection valued measure P on $\mathcal{A}(d_2)$ such that

$$P_{\alpha_1(E)} = P(E \times \mathbb{R}) \text{ and } P_{\alpha_2(E)} = P(\mathbb{R} \times E), \forall E \in \mathcal{A}(d_{\mathbb{R}}),$$

and hence, owing to the bijectivity of the mapping of 19.3.1b and to 13.3.5, there is a unique mapping $\alpha : \mathcal{A}(d_2) \to \Pi$ which is so that

$$\alpha_1(E) = \alpha(E \times \mathbb{R}) \text{ and } \alpha_2(E) = \alpha(\mathbb{R} \times E), \forall E \in \mathcal{A}(d_{\mathbb{R}}),$$

and so that μ_σ^α is a probability measure for all $\sigma \in \Sigma_0$. Actually, for each $E \in \mathcal{A}(d_2)$, $\alpha(E)$ is the proposition such that $P_{\alpha(E)} = P(E)$. Then, μ_σ^α is a probability measure for all $\sigma \in \Sigma$ (cf. 18.3.13).
To what extent can the mapping α be considered to be an $\mathbb{R}^2$-valued observable? That is (cf. 19.1.9c), to what extent can α be taken to represent a measuring apparatus (whose dial would then be represented by $\mathbb{R}^2$)? We note that, for each $(E_1, E_2) \in \mathcal{A}(d_{\mathbb{R}}) \times \mathcal{A}(d_{\mathbb{R}})$,

$$
\begin{aligned}
p(\alpha(E_1 \times E_2), \sigma) &= \text{tr}(P_{\alpha(E_1 \times E_2)} W_\sigma) = \text{tr}(P(E_1 \times E_2) W_\sigma) \\
&= \text{tr}(P(\mathbb{R} \times E_2) P(E_1 \times \mathbb{R}) W_\sigma) \\
&= \text{tr}(P_{\alpha_2(E_2)} P_{\alpha_1(E_1)} W_\sigma) = p(\alpha_2(E_2), \alpha_1(E_1), \sigma), \forall \sigma \in \Sigma
\end{aligned}
$$

(cf. 13.3.2c). Assume then that we have ideal implementations of all the propositions in the range of α_1 and α_2. Then, for each $(E_1, E_2) \in \mathcal{A}(d_{\mathbb{R}}) \times \mathcal{A}(d_{\mathbb{R}})$, we have the proposition $\alpha_1(E_1) \wedge \alpha_2(E_2)$ (cf. 19.5.6a), which is defined by the pieces of equipment that define $\alpha_1(E_1)$ and $\alpha_2(E_2)$, and hence by the measuring instruments represented by α_1 and α_2, and for which we have (cf. 19.5.5)

$$p(\alpha_2(E_2), \alpha_1(E_1), \sigma) = p(\alpha_1(E_1) \wedge \alpha_2(E_2), \sigma), \forall \sigma \in \Sigma,$$

and hence

$$\alpha(E_1 \times E_2) = \alpha_1(E_1) \wedge \alpha_2(E_2).$$

This gives an operational interpretation to the proposition $\alpha(E)$ on the basis of the measuring instruments represented by α_1 and α_2, for each $E \in S :=$ $\{E_1 \times E_2 : (E_1, E_2) \in \mathcal{A}(d_{\mathbb{R}}) \times \mathcal{A}(d_{\mathbb{R}})\}$. In particular, for each $(x_1, x_2) \in \mathbb{R}^2$ we can say that the determination of the proposition $\alpha(\{(x_1, x_2)\})$ is, in any state, "the simultaneous determination" of the propositions $\alpha_1(\{x_1\})$ and $\alpha_2(\{x_2\})$, in the sense specified in 19.5.6a.

The reason why we define the $\mathbb{R}^2$-valued observable α on $\mathcal{A}(d_2)$ and not just on S is that we want the probability functions μ_σ^α to be bona fide measures and hence to be defined on a σ-algebra (S is just a semialgebra and $\mathcal{A}(d_2)$ is the σ-algebra generated by S, cf. 6.1.30a and 6.1.32). However, an operational meaning for the proposition $\alpha(E)$ for each $E \in \mathcal{A}(d_2)$ cannot be inferred from the operational interpretation given above to all propositions $\alpha(E)$ with $E \in S$, because there is no constructive procedure for obtaining each element of $\mathcal{A}(d_2)$ starting from elements of S. Still, we know that, for every $\sigma \in \Sigma$, the measure μ_σ^α is uniquely determined by its values on S (this follows from 6.1.18, from the uniqueness asserted in 7.3.1A, and from the uniqueness asserted in 7.3.2 for a σ-finite premeasure); in this respect, the operational grounds found above for the propositions $\alpha(E)$ with $E \in S$ provide operational grounds for the probability measures μ_σ^α.

(c) Suppose that we have an $\mathbb{R}^2$-valued observable α and a function $\varphi : D_\varphi \to \mathbb{R}$ such that $D_\varphi \in \mathcal{A}(d_2)$, $P_\alpha(\mathbb{R}^2 - D_\varphi) = \mathbb{O}_\mathcal{H}$, φ is $\mathcal{A}(d_2)^{D_\varphi}$-measurable. We can define the observable $\varphi(\alpha)$ (cf. 19.1.13, 19.1.14, 19.3.9), which is supported by the same measuring instrument that defines α: if a measurement of α yields the result $(x_1, x_2) \in \mathbb{R}^2$ then we attribute the result $\varphi(x_1, x_2)$ to $\varphi(\alpha)$. Consider now the two compatible observables α_1 and α_2 that are related to α as above: either α_1 and α_2 are obtained from α as in remark a, or α is obtained from α_1 and α_2 as in remark b. Then the observable $\varphi(\alpha)$ can be considered a function of α_1 and α_2: if a "simultaneous measurement" of α_1 and α_2 brings out the pair of results x_1, x_2 then the result (x_1, x_2), as an element of $\mathbb{R}^2$, is assigned to α and hence the result $\varphi(x_1, x_2)$ is assigned to $\varphi(\alpha)$. For this reason, the observable $\varphi(\alpha)$ is also called the *function of α_1, α_2 according to φ* and denoted by the symbol $\varphi(\alpha_1, \alpha_2)$. Thus $\varphi(\alpha_1, \alpha_2) := \varphi(\alpha)$ and we have

$$P^{A_{\varphi(\alpha_1, \alpha_2)}}(E) = P^{A_{\varphi(\alpha)}}(E) = P_\alpha(\varphi^{-1}(E)) = P^{\varphi(A_{\alpha_1}, A_{\alpha_2})}(E), \forall E \in \mathcal{A}(d_{\mathbb{R}})$$

(cf. the proof of 19.3.9, 17.1.11, 15.2.7, noticing that the relation between the pairs of commuting self-adjoint operators $A_{\alpha_1}, A_{\alpha_2}$ and the projection valued measure P_α is the same as the one between the pair A_1, A_2 and P in 17.1.10b), and hence (cf. 15.2.2) $A_{\varphi(\alpha_1,\alpha_2)} = \varphi(A_{\alpha_1}, A_{\alpha_2})$. This extends the function preserving property of the representation of observables by self-adjoint operators that was noted in 19.3.9.

Suppose in particular that we have two compatible observables α_1 and α_2, that we have ideal implementations of all the propositions in the ranges of α_1 and α_2, and that we wish to define, using the measuring instruments that are represented by α_1 and α_2, a new observable to which the result $x_1 + x_2$ (or $x_1 x_2$) is assigned when the "simultaneous" results x_1 and x_2 are obtained for α_1 and α_2 respectively. Then, from what we saw above and from 17.1.12 it follows that this new observable is represented by the self-adjoint extension of the essentially self-adjoint operator $A_1 + A_2$ (or $A_1 A_2$), which actually coincides with $A_1 + A_2$ (or $A_1 A_2$) whenever A_2 is bounded.

19.5.12 Remark. The results of 19.5.11a,b are based on the equivalence between conditions a and b in 17.1.10, and can be summarised as follows: two observables α_1 and α_2 are compatible if and only if there exists an $\mathbb{R}^2$-valued observable α such that

$$\alpha_1(E) = \alpha(E \times \mathbb{R}) \text{ and } \alpha_2(E) = \alpha(\mathbb{R} \times E), \forall E \in \mathcal{A}(d_\mathbb{R})$$

(actually, for the "only if" part we have to assume that there are ideal implementations of all the propositions in the ranges of α_1 and α_2). This gives, in our opinion, a nice characterization of the compatibility of two observables.

However, in standard quantum mechanics textbooks, the only X-valued observables that are considered are observables. Now, it is possible to give a characterization of the compatibility of two observables in which only observables are used. This is accomplished on the basis of the equivalence between conditions a and c in 17.1.10. Indeed, if two observables α_1 and α_2 are functions of an observable β, then by 19.3.9 the self-adjoint operators A_{α_1} and A_{α_2} are functions of the self-adjoint operator A_β, and hence A_{α_1} and A_{α_2} commute by c $\Rightarrow$ a in 17.1.10, and hence α_1 and α_2 are compatible by 19.5.10. If conversely two observable α_1 and α_2 are compatible, then the self-adjoint operators A_{α_1} and A_{α_2} commute by 19.5.10, and hence there are a self-adjoint operator B and two functions φ_i so that $A_{\alpha_i} = \varphi_i(B)$ for $i = 1, 2$, by a $\Rightarrow$ c in 17.1.10; now, it would be hard to give in general an operational meaning (as instead we did for the mapping α in 19.5.11b) to the mapping $\beta : \mathcal{A}(d_\mathbb{R}) \rightarrow \Pi$ which is defined by letting $\beta(E)$ be the proposition such that $P_{\beta(E)} = P^B(E)$, for all $E \in \mathcal{A}(d_\mathbb{R})$; this is due to the fact that the construction of the projection valued measure P^B out of the projection valued measures P^{A_1} and P^{A_2}, in the proof of 17.1.9, is utterly abstract (whereas condition b in 17.1.10 relates directly the projection valued measure P to the projection valued measures P^{A_1} and P^{A_2}); however, every self-adjoint operator is taken to represent an observable in standard quantum mechanics textbooks, and hence according to their rules we can say that there exists an observable β which is represented by the self-adjoint operator B, and hence such that $\alpha_i = \varphi_i(\beta)$ since

$$P_{\alpha_i(E)} = P^{A_{\alpha_i}}(E) = P^{\varphi_i(B)}(E) = P^B(\varphi_i^{-1}(E))$$
$$= P_{\beta(\varphi_i^{-1}(E))} = P_{\varphi_i(\beta)(E)}, \forall E \in \mathcal{A}(d_{\mathbb{R}}),$$

for $i = 1, 2$ (cf. 15.3.8 and 19.1.13). Thus, within the rules of standard quantum mechanics textbooks, two observables α_1 and α_2 are compatible if and only if there exists an observable β of which both α_1 and α_2 are functions.

19.5.13 Proposition. *For a proposition $\pi \in \Pi$, a discrete observable α, a state $\sigma \in \Sigma$, we denote by $p(\pi, \alpha, \sigma)$ the probability that π is true in a copy which is produced by an ideal measurement of α with any result, carried out in a copy initially prepared in the state σ. Thus, $p(\pi, \alpha, \sigma)$ is the theoretical prediction of the relative frequency of π being found true in an ensemble of copies which, after being prepared in σ, have gone through an ideal measurement of α without being selected according to any particular set of results for α.*

The following conditions are equivalent:

(a) $p(\pi, \alpha, \sigma) = p(\pi, \sigma), \forall \sigma \in \Sigma$;
(b) π and $\alpha(E)$ are compatible, $\forall E \in \mathcal{A}(d_{\mathbb{R}})$.

Proof. Let $\{(\lambda_n, P_n)\}_{n \in I}$ be the family related to the self-adjoint operator A_α as in 15.3.4B with $A := A_\alpha$ (cf. 19.3.10c). From 19.4.12a we see that, for every $\sigma \in \Sigma$,

$$p(\pi, \alpha, \sigma) = p(\pi, \sigma'') = \operatorname{tr}(P_\pi W_{\sigma''}) = \sum_{n \in I} \operatorname{tr}(P_\pi P_n W_\sigma P_n)$$

(the third equality follows from 18.3.4c).

We prove now the equivalence between conditions a and b.

$a \Rightarrow b$: Assuming condition a, we have in particular

$$p(\pi, \alpha, \sigma) = p(\pi, \sigma), \forall \sigma \in \Sigma_0,$$

which is equivalent to

$$\sum_{n \in I} \operatorname{tr}(P_\pi P_n A_u P_n) = \operatorname{tr}(P_\pi A_u), \forall u \in \tilde{\mathcal{H}}.$$

We note that, if I is infinite, the series $\sum_{n \in I} P_n P_\pi P_n f$ is convergent for each $f \in \mathcal{H}$ by 10.4.7b, since

$$(P_i P_\pi P_i f | P_j P_\pi P_j f) = (P_\pi P_i f | P_i P_j P_\pi P_j f) = 0 \text{ if } i \neq j,$$
$$\|P_n P_\pi P_n f\| \leq \|P_n f\| \text{ (cf. 13.1.3d)},$$
$$\sum_{n \in I} \|P_n f\|^2 < \infty \text{ (cf. 13.2.8)};$$

thus, we can define the operator

$$\sum_{n \in I} P_n P_\pi P_n : \mathcal{H} \to \mathcal{H}$$

$$f \mapsto \left(\sum_{n \in I} P_n P_\pi P_n \right) f := \sum_{n \in I} P_n P_\pi P_n f.$$

Then we have

$$\left(u \middle| \left(\sum_{n \in I} P_n P_\pi P_n\right) u\right) = \sum_{n \in I} (u|P_n P_\pi P_n u) = \sum_{n \in I} \mathrm{tr}(P_n P_\pi P_n A_u)$$

$$= \sum_{n \in I} \mathrm{tr}(P_\pi P_n A_u P_n) = \mathrm{tr}(P_\pi A_u) = (u|P_\pi u), \forall u \in \tilde{\mathcal{H}},$$

and hence, by 10.2.12,

$$\sum_{n \in I} P_n P_\pi P_n = P_\pi.$$

From this we obtain, for each $k \in I$,

$$P_k P_\pi f = \sum_{n \in I} P_k P_n P_\pi P_n f = P_k P_\pi P_k f$$

$$= \sum_{n \in I} P_n P_\pi P_n P_k f = P_\pi P_k f, \forall f \in \mathcal{H},$$

and hence, for every $E \in \mathcal{A}(d_\mathbb{R})$,

$$[P_{\alpha(E)}, P_\pi]f = [P^{A_\alpha}(E), P_\pi]f = \sum_{n \in I_E} [P_n, P_\pi]f = 0_\mathcal{H}, \forall f \in \mathcal{H}$$

(where I_E is defined as in 15.3.4B), which is equivalent to condition b by 19.5.3.

$b \Rightarrow a$: Assuming condition b, by 19.5.3 we have in particular

$$[P_\pi, P_n] = [P_\pi, P^{A_\alpha}(\{\lambda_n\})] = [P_\pi, P_{\alpha(\{\lambda_n\})}] = 0_\mathcal{H}, \forall n \in I,$$

and hence, for every $\sigma \in \Sigma$,

$$p(\pi, \alpha, \sigma) = \sum_{n \in I} \mathrm{tr}(P_n P_\pi W_\sigma P_n) = \sum_{n \in I} \mathrm{tr}(P_n P_\pi W_\sigma)$$

$$= \mathrm{tr}\left(\left(\sum_{n \in I} P_n P_\pi\right) W_\sigma\right) = \mathrm{tr}(P_\pi W_\sigma) = p(\pi, \sigma),$$

where we have used 18.2.11c and 18.3.12 (note that $P_n P_\pi \in \mathscr{P}(\mathcal{H})$ for each $n \in I$ by 13.2.1, and that $(P_i P_\pi)(P_k P_\pi) = P_i P_k P_\pi = 0_\mathcal{H}$ if $i \neq k$) and the equality $\sum_{n \in I} P_n = \mathbb{1}_\mathcal{H}$ (cf. 15.3.4B). $\square$

19.5.14 Corollary. *For a discrete observable α and any observable β, the following conditions are equivalent:*

(a) $p(\beta(E), \alpha, \sigma) = p(\beta(E), \sigma)$, $\forall E \in \mathcal{A}(d_\mathbb{R})$, $\forall \sigma \in \Sigma$;
(b) α and β are compatible.

Proof. The result follows immediately from 19.5.13 and the definition of compatibility for two observables. $\square$

19.5.15 Remark. For two observables α and β in a classical statistical theory, we presume that in any state it is possible to measure α in such an "undisturbing" way that the results we obtain when we measure β immediately after measuring α (and having kept all the copies in which the measurements of α have been made) are the same as the ones we should obtain if α had not been measured (cf. 19.2.2).

In a quantum situation, the most "undisturbing" method for measuring an observable is to use an ideal measurement (cf. 19.4.5a). Now, the result of 19.5.14 says that, in the quantum case, if α is a discrete observable then, for any observable β, if and only if α and β are compatible is it statistically inconsequential whether α has been ideally measured before β or not. If the observable α is not discrete, when β and α are compatible then so are β and a realistic, and therefore discrete, approximation of α (in the sense discussed in 19.1.22a and in 19.4.12b), since a realistic approximation of α is assumed to be a function of α and therefore all the propositions in its range are in the range of α as well. Thus, if we maintain the idea that a measurement of α is at the operational level actually a measurement of one of its realistic approximations, we can still say that the compatibility of α and β ensures that it is immaterial, for the statistics of the results we obtain in a long series of measurements of β in a state σ, whether we have used directly copies prepared in σ or copies which, after being prepared in σ, have gone through an ideal measurement of α in which no selection was made according to any particular set of results for α.

19.5.16 Remark. Let α be a discrete observable and let $\{(\lambda_n, P_n)\}_{n\in I}$ be the family related to the self-adjoint operator A_α as in 15.3.4B with $A := A_\alpha$ (cf. 19.3.10c). Suppose that the projection P_n is one-dimensional, i.e. that there exists $u_n \in \tilde{\mathcal{H}}$ such that $P_n = A_{u_n}$, for each $n \in I$, and that we have a procedure for carrying out a first kind measurement of α. If a first kind measurement of α is made in a copy of the system prepared in a state $\sigma \in \Sigma$ and if the result λ_n is obtained (the elements of $\{\lambda_n\}_{n\in I}$ are the only numbers that can be obtained as results, since $P^{A_\alpha}(\mathbb{R} - \{\lambda_n\}_{n\in I}) = \mathbb{O}_\mathcal{H}$ and this implies $p(\alpha(\mathbb{R} - \{\lambda_n\}_{n\in I}), \sigma) = \mathrm{tr}(P^{A_\alpha}(\mathbb{R} - \{\lambda_n\}_{n\in I})W_\sigma) = 0$), then immediately after the measurement we have a copy in the pure state represented by the ray $[u_n]$, whatever the state σ was; this follows from 19.4.3e, since $P_{\alpha(\{\lambda_n\})} = P^{A_\alpha}(\{\lambda_n\}) = P_n$ (cf. 15.3.4B). We also note that, for $i \neq j$, $P_i P_j = \mathbb{O}_\mathcal{H}$ implies $(u_i|u_j) = 0$, and also that

$$f = P^{A_\alpha}(\mathbb{R})f = \sum_{n\in I} P_n f = \sum_{n\in I} (u_n|f)\, u_n, \forall f \in \mathcal{H}$$

(cf. 15.3.4B). This proves that the family $\{u_n\}_{n\in I}$ is a c.o.n.s. in $\mathcal{H}$ (cf. 10.6.4).

Thus, if we have a discrete observable α such that the self-adjoint operator A_α has one-dimensional eigenspaces and a procedure for a first kind measurement of α, we actually have a procedure for preparing pure states, and a great deal of them (one for each element of a c.o.n.s. in $\mathcal{H}$). However, observables with these characteristics are seldom available. More often, their function in preparing pure states is fulfilled by a set of observables with the features specified in 19.5.17, as is explained in 19.5.18.

19.5.17 Definition. Let $\{\alpha_1, \alpha_2, ..., \alpha_\ell\}$ be a finite family of discrete observables and, for $k = 1, 2, ..., \ell$, let $\{(\lambda_n^k, P_n^k)\}_{n \in I_k}$ be the family associated with the self-adjoint operator A_{α_k} as the family $\{(\lambda_n, P_n)\}_{n \in I}$ was associated with the self-adjoint operator A in 15.3.4B. The family $\{\alpha_1, \alpha_2, ..., \alpha_\ell\}$ is said to be a *complete set of compatible observables* if the observables of the family are pairwise compatible and if the projection $P_{n_1}^1 P_{n_2}^2 \cdots P_{n_\ell}^\ell$ is either one-dimensional or the operator $\mathbb{O}_\mathcal{H}$, for all $(n_1, n_2, ..., n_\ell) \in I_1 \times I_2 \times \cdots \times I_\ell$ (the operator $P_{n_1}^1 P_{n_2}^2 \cdots P_{n_\ell}^\ell$ is a projection by 19.5.10, 17.1.14, 13.2.1).

19.5.18 Remark. Let the family $\{\alpha_1, \alpha_2, ..., \alpha_\ell\}$ be as in 19.5.17, and suppose that it is a complete set of compatible observables. Suppose further that procedures are available for performing ideal measurements of all observables α_k. If ideal measurements are made for all observables α_k, one immediately after the other in whichever order, in a copy of the system initially prepared in whatever state σ, and if $\lambda_{n_1}^1, \lambda_{n_2}^2, ..., \lambda_{n_\ell}^\ell$ are the results obtained, then immediately after the ℓ measurements we have a copy which is in the pure state represented by the ray $[u_{n_1, n_2, ..., n_\ell}]$ if $P_{n_1}^1 P_{n_2}^2 \cdots P_{n_\ell}^\ell = A_{u_{n_1, n_2, ..., n_\ell}}$. Indeed, reasoning as in 19.5.1a we see that the probability of obtaining the results $\lambda_{n_1}^1, \lambda_{n_2}^2, ..., \lambda_{n_\ell}^\ell$ was, before the measurements, $\mathrm{tr}(P_{n_1}^1 P_{n_2}^2 \cdots P_{n_\ell}^\ell W_\sigma)$; thus, if the results $\lambda_{n_1}^1, \lambda_{n_2}^2, ..., \lambda_{n_\ell}^\ell$ have actually been obtained then $P_{n_1}^1 P_{n_2}^2 \cdots P_{n_\ell}^\ell \neq \mathbb{O}_\mathcal{H}$ and hence the projection $P_{n_1}^1 P_{n_2}^2 \cdots P_{n_\ell}^\ell$ is one-dimensional; then, reasoning as in the proof of 19.5.4 we see that after the ℓ measurements we have a copy which is in the state represented by the statistical operator

$$\frac{1}{\mathrm{tr}(P_{n_1}^1 P_{n_2}^2 \cdots P_{n_\ell}^\ell W_\sigma)} P_{n_1}^1 P_{n_2}^2 \cdots P_{n_\ell}^\ell W_\sigma P_{n_1}^1 P_{n_2}^2 \cdots P_{n_\ell}^\ell,$$

which is the same as $A_{u_{n_1, n_2, ..., n_\ell}}$, for whatever state σ such that $\mathrm{tr}(P_{n_1}^1 P_{n_2}^2 \cdots P_{n_\ell}^\ell W_\sigma) \neq 0$ (cf. 19.4.5c). This gives us a method for preparing pure states, one for each element of a c.o.n.s. in $\mathcal{H}$. To see this, define

$$J := \{(n_1, n_2, ..., n_\ell) \in I_1 \times I_2 \times \cdots \times I_\ell : P_{n_1}^1 P_{n_2}^2 \cdots P_{n_\ell}^\ell \neq \mathbb{O}_\mathcal{H}\}$$

and let $u_{n_1, n_2, ..., n_\ell} \in \tilde{\mathcal{H}}$ be such that $P_{n_1}^1 P_{n_2}^2 \cdots P_{n_\ell}^\ell = A_{u_{n_1, n_2, ..., n_\ell}}$ for $(n_1, n_2, ..., n_\ell) \in J$. The condition $P_{n_k}^k P_{n_k'}^k = \mathbb{O}_\mathcal{H}$ if $n_k \neq n_k'$ (cf. 15.3.4B) implies that

$$\left(u_{n_1, n_2, ..., n_\ell} | u_{n_1', n_2', ..., n_\ell'}\right) = 0 \text{ if } (n_1, n_2, ..., n_\ell) \neq (n_1', n_2', ..., n_\ell');$$

moreover, the condition $\mathbb{1} = P^{A_{\alpha_k}}(\mathbb{R}) = \sum_{n_k \in I_k} P_{n_k}^k$ (cf. 15.3.4B) implies that

$$f = \sum_{n_1 \in I_1} \sum_{n_2 \in I_2} \cdots \sum_{n_\ell \in I_\ell} P_{n_1}^1 P_{n_2}^2 \cdots P_{n_\ell}^\ell f$$

$$= \sum_{(n_1, n_2, ..., n_\ell) \in J} (u_{n_1, n_2, ..., n_\ell} | f) \, u_{n_1, n_2, ..., n_\ell}, \forall f \in \mathcal{H}.$$

This proves that the family $\{u_{n_1,n_2,...,n_\ell}\}_{(n_1,n_2,...,n_\ell)\in J}$ is a c.o.n.s. in $\mathcal{H}$ (cf. 10.6.4).

19.5.19 Proposition. *Let α and β be two observables and σ a state in which both α and β are evaluable, and let $\{u_n\}_{n\in I}$ and $\{w_n\}_{n\in I}$ be as in 19.3.13c so that $W_\sigma f = \sum_{n\in I} w_n A_{u_n} f$ for all $f \in \mathcal{H}$. Then:*

$$u_n \in D_{A_\alpha} \cap D_{A_\beta}, \forall n \in I;$$

$$\Delta_\sigma \alpha \Delta_\sigma \beta \geq \frac{1}{2} \sum_{n\in I} w_n |(A_\alpha u_n | A_\beta u_n) - (A_\beta u_n | A_\alpha u_n)|.$$

If in particular σ is a pure state, then:

$$u_\sigma \in D_{A_\alpha} \cap D_{A_\beta} \text{ and } \Delta_\sigma \alpha \Delta_\sigma \beta \geq \frac{1}{2} |(A_\alpha u_\sigma | A_\beta u_\sigma) - (A_\beta u_\sigma | A_\alpha u_\sigma)|.$$

Proof. From 19.3.13c we have $u_n \in D_{A_\alpha} \cap D_{A_\beta}$ for all $n \in I$. For the product $\Delta_\sigma \alpha \Delta_\sigma \beta$ we have

$$\Delta_\sigma \alpha \Delta_\sigma \beta = \sqrt{\sum_{n\in I} w_n \|A_\alpha u_n - \langle\alpha\rangle_\sigma u_n\|^2} \sqrt{\sum_{n\in I} w_n \|A_\beta u_n - \langle\beta\rangle_\sigma u_n\|^2}$$

$$\geq \sum_{n\in I} w_n \|A_\alpha u_n - \langle\alpha\rangle_\sigma u_n\| \|A_\beta u_n - \langle\beta\rangle_\sigma u_n\|;$$

the equality follows from 19.3.13c and the inequality is the Schwarz inequality in $\mathbb{C}^N$ if I contains N elements or in ℓ^2 if I is denumerable (cf. 10.3.8c,d; if I is denumerable, the sequences $\{\sqrt{w_n}\|A_\alpha u_n - \langle\alpha\rangle_\sigma u_n\|\}$ and $\{\sqrt{w_n}\|A_\beta u_n - \langle\beta\rangle_\sigma u_n\|\}$ are elements of ℓ^2, cf. 19.3.13c). Further, for each $n \in I$ we have (using the fact that the operators A_α and A_β are symmetric, cf. 12.4.3c)

$$\|A_\alpha u_n - \langle\alpha\rangle_\sigma u_n\| \|A_\beta u_n - \langle\beta\rangle_\sigma u_n\|$$

$$\geq |(A_\alpha u_n - \langle\alpha\rangle_\sigma u_n | A_\beta u_n - \langle\beta\rangle_\sigma u_n)|$$

$$\geq |\text{Im}\,(A_\alpha u_n - \langle\alpha\rangle_\sigma u_n | A_\beta u_n - \langle\beta\rangle_\sigma u_n)|$$

$$= \frac{1}{2} |(A_\alpha u_n - \langle\alpha\rangle_\sigma u_n | A_\beta u_n - \langle\beta\rangle_\sigma u_n) - (A_\beta u_n - \langle\beta\rangle_\sigma u_n | A_\alpha u_n - \langle\alpha\rangle_\sigma u_n)|$$

$$= \frac{1}{2} |(A_\alpha u_n | A_\beta u_n) - (A_\beta u_n | A_\alpha u_n)|.$$

Thus, the first part of the statement is proved. The second part follows immediately from the first since $W_\sigma = A_{u_\sigma}$ if σ is a pure state. $\square$

19.5.20 Corollary. *Let α and β be two observables and σ a state in which both α and β are evaluable, and also such that $A_\alpha A_\beta W_\sigma \in \mathcal{T}(\mathcal{H})$ and $A_\beta A_\alpha W_\sigma \in \mathcal{T}(\mathcal{H})$. Then:*

$$[A_\alpha, A_\beta] W_\sigma \in \mathcal{T}(\mathcal{H}) \text{ and } \Delta_\sigma \alpha \Delta_\sigma \beta \geq \frac{1}{2} |\text{tr}([A_\alpha, A_\beta] W_\sigma)|.$$

If in particular σ is a pure state, the above conditions for σ are equivalent to the one condition $u_\sigma \in D_{[A_\alpha, A_\beta]}$ and, if they are fulfilled, the following inequality holds:

$$\Delta_\sigma \alpha \Delta_\sigma \beta \geq \frac{1}{2} |(u_\sigma | [A_\alpha, A_\beta] u_\sigma)|.$$

Proof. Let $\{u_n\}_{n\in I}$ and $\{w_n\}_{n\in I}$ be as in 19.5.19, with $\{u_n\}_{n\in I}$ an o.n.s. in $\mathcal{H}$ (cf. 18.3.2c); then $u_n \in R_{W_\sigma}$ for each $n \in I$. Since $D_{A_\alpha A_\beta W_\sigma} = D_{A_\beta A_\alpha W_\sigma} = \mathcal{H}$, we have $A_\beta u_n \in D_{A_\alpha}$ and $A_\alpha u_n \in D_{A_\beta}$, and hence $u_n \in D_{[A_\alpha, A_\beta]}$ for each $n \in I$. Then, since the operators A_α and A_β are symmetric, from 19.5.19 we obtain

$$\Delta_\sigma \alpha \Delta_\sigma \beta \geq \frac{1}{2} \sum_{n\in I} w_n \left| (A_\alpha u_n | A_\beta u_n) - (A_\beta u_n | A_\alpha u_n) \right|$$

$$\geq \frac{1}{2} \left| \sum_{n\in I} w_n (u_n | [A_\alpha, A_\beta] u_n) \right|.$$

If in particular σ is a pure state, then $u_\sigma \in D_{[A_\alpha, A_\beta]}$ and

$$\Delta_\sigma \alpha \Delta_\sigma \beta \geq \frac{1}{2} \left| (u_\sigma | [A_\alpha, A_\beta] u_\sigma) \right|.$$

In the general case, from 18.2.4a,b we have $[A_\alpha, A_\beta] W_\sigma \in \mathcal{T}(\mathcal{H})$, and we can compute $\mathrm{tr}([A_\alpha, A_\beta] W_\sigma)$ by means of a c.o.n.s. in $\mathcal{H}$ which contains $\{u_n\}_{n\in I}$ (cf. 10.7.3); then we have

$$\mathrm{tr}([A_\alpha, A_\beta] W_\sigma) = \sum_{n\in I} (u_n | [A_\alpha, A_\beta] W_\sigma u_n) = \sum_{n\in I} w_n (u_n | [A_\alpha, A_\beta] u_n).$$

Finally, if σ is a pure state and $u_\sigma \in D_{[A_\alpha, A_\beta]}$, then $u_\sigma \in D_{A_\alpha} \cap D_{A_\beta}$ and hence both α and β are evaluable in σ (cf. 19.3.13d); moreover,

$$A_\alpha A_\beta W_\sigma f = (u_\sigma | f) A_\alpha A_\beta u_\sigma \text{ and } A_\beta A_\alpha W_\sigma f = (u_\sigma | f) A_\beta A_\alpha u_\sigma, \forall f \in \mathcal{H},$$

and this proves that $A_\alpha A_\beta W_\sigma \in \mathcal{T}(\mathcal{H})$ and $A_\beta A_\alpha W_\sigma \in \mathcal{T}(\mathcal{H})$. Indeed, if $A_\alpha A_\beta u_\sigma \neq 0_\mathcal{H}$ then $A_\alpha A_\beta W_\sigma = \lambda A_{u,v}$ with $\lambda := \|A_\alpha A_\beta u_\sigma\|$, $u := u_\sigma$, $v := \lambda^{-1} A_\alpha A_\beta u_\sigma$, and hence $A_\alpha A_\beta W_\sigma \in \mathcal{T}(\mathcal{H})$ in view of 18.2.15; and similarly for $A_\beta A_\alpha W_\sigma$. $\qquad \square$

19.5.21 Proposition. *Let α and β be two observables, and suppose that β is bounded. Then β is evaluable in every state and*

$$\forall \varepsilon > 0, \exists \sigma_\varepsilon \in \Sigma_0 \text{ so that } \alpha \text{ is evaluable in } \sigma_\varepsilon \text{ and } \Delta_{\sigma_\varepsilon} \alpha \Delta_{\sigma_\varepsilon} \beta < \varepsilon.$$

Proof. Since β is bounded, β is evaluable in every state (cf. 19.3.15c), the operator A_β is bounded, and $D_{A_\beta} = \mathcal{H}$ (cf. 19.3.10b). For each pure state $\sigma \in \Sigma_0$, in view of 19.3.13d we have

$$|\langle \beta \rangle_\sigma| = |(u_\sigma | A_\beta u_\sigma)| \leq \|A_\beta u_\sigma\|$$

by the Schwarz inequality, and hence (cf. 4.2.5b)

$$\Delta_\sigma \beta = \|A_\beta u_\sigma - \langle \beta \rangle_\sigma u_\sigma\| \leq 2\|A_\beta u_\sigma\| \leq 2\|A_\beta\|.$$

If $\|A_\beta\| = 0$ then we have $\Delta_\sigma \beta = 0$ and hence $\Delta_\sigma \alpha \Delta_\sigma \beta = 0$ for each state $\sigma \in \Sigma_0$ in which α is evaluable. Assuming $\|A_\beta\| \neq 0$, 19.3.16 implies that for every $\varepsilon > 0$ there exists a pure state $\sigma_\varepsilon \in \Sigma_0$ such that α is evaluable in σ_ε and $\Delta_{\sigma_\varepsilon} \alpha < \frac{\varepsilon}{2\|A_\beta\|}$, and hence such that

$$\Delta_{\sigma_\varepsilon} \alpha \Delta_{\sigma_\varepsilon} \beta < \frac{\varepsilon}{2\|A_\beta\|} 2\|A_\beta\| = \varepsilon.$$

$\qquad \square$

19.5.22 Proposition. *Let α and β be two compatible observables. Then:*

$$(A_\alpha f|A_\beta f) - (A_\beta f|A_\alpha f) = 0, \forall f \in D_{A_\alpha} \cap D_{A_\beta}.$$

Proof. First we notice that, for every self-adjoint operator A in $\mathcal{H}$, condition sa-ug in 16.1.6 and the continuity of the inner product imply that, for $g \in \mathcal{H}$ and $f \in D_A$, the function $\mathbb{R} \ni t \mapsto \left(g|U_f^A(t)\right)$ is differentiable at 0 and $\frac{d}{dt}\left(g|U_f^A(t)\right)\Big|_0 = (g|iAf)$. And similarly $\frac{d}{dt}\left(U_f^A(t)|g\right)\Big|_0 = (iAf|g)$.

For $f \in D_{A_\alpha} \cap D_{A_\beta}$, from 19.5.10 and 17.1.7 we have

$$\left(U_f^{A_\alpha}(-t)|A_\beta f\right) = \left(U^{A_\alpha}(-t)f|A_\beta f\right)$$

$$= (f|U^{A_\alpha}(t)A_\beta f) = (f|A_\beta U^{A_\alpha}(t)f) = \left(A_\beta f|U_f^{A_\alpha}(t)\right)$$

(recall that $U^{A_\alpha}(-t) = U^{A_\alpha}(t)^{-1} = U^{A_\alpha}(t)^\dagger$, cf. 16.1.1), and hence

$$(-iA_\alpha f|A_\beta f) = \frac{d}{dt}\left(U_f^{A_\alpha}(-t)|A_\beta f\right)\Big|_0 = \frac{d}{dt}\left(A_\beta f|U_f^{A_\alpha}(t)\right)\Big|_0 = (A_\beta f|iA_\alpha f).$$

$\square$

19.5.23 Proposition. *Let α_1 and α_2 be two compatible observables. Then for each possible result λ_1 for α_1 and each $\varepsilon > 0$ there exist a possible result λ_2 for α_2 and a pure state $\sigma_\varepsilon \in \Sigma_0$ so that*

$$\alpha_k \text{ is evaluable in } \sigma_\varepsilon, \quad |\langle \alpha_k \rangle_{\sigma_\varepsilon} - \lambda_k| < \varepsilon, \quad \Delta_{\sigma_\varepsilon}\alpha_k < 2\varepsilon, \quad \text{for } k = 1, 2.$$

Proof. Everything follows from 17.1.13 and 19.5.10 since, for each observable α, $\sigma(A_\alpha)$ is the set sp_α of all possible results for α (cf. 19.3.10a), α is evaluable in a pure state $\sigma \in \Sigma_0$ if and only if $u_\sigma \in D_{A_\alpha}$ (cf. 19.3.13d), if α is evaluable in a pure state $\sigma \in \Sigma_0$ then $\langle \alpha \rangle_\sigma = \langle A_\alpha \rangle_{u_\sigma}$ and $\Delta_\sigma \alpha = \Delta_{u_\sigma} A_\alpha$ (cf. 19.3.13d). $\square$

19.5.24 Remarks.

(a) We saw in 19.3.16 that, for each observable α, the uncertainty $\Delta_\sigma \alpha$ can be made arbitrarily small by a suitable choice of the state σ. One can wonder if a similar possibility exists for two observables α and β, i.e. if the following proposition is true

$$\text{P} : \forall \varepsilon > 0, \exists \sigma_\varepsilon \in \Sigma \text{ so that } \alpha \text{ and } \beta \text{ are evaluable in } \sigma_\varepsilon \text{ and } \Delta_{\sigma_\varepsilon}\alpha\Delta_{\sigma_\varepsilon}\beta < \varepsilon.$$

We must emphasize the fact that, whether proposition P is true or not, for any state σ the product $\Delta_\sigma \alpha \Delta_\sigma \beta$ has for us only the statistical meaning that is based on the interpretation of $\Delta_\sigma \alpha$ as the theoretical prediction of the standard deviation of the results obtained when measuring an observable α in a large number of copies all prepared in σ (cf. 19.1.22a). In particular, considering the product $\Delta_\sigma \alpha \Delta_\sigma \beta$ does not imply for us any idea of carrying out measurements of α and of β in the same copies of the quantum system. In fact, an experimental test for the value of $\Delta_\sigma \alpha \Delta_\sigma \beta$ rests on measuring α in a large collection of copies prepared in σ and, independently of that, on measuring β in a *different* large collection of copies prepared in σ. Thus, if proposition P is not true, i.e. if there exists $\mu > 0$ so that $\Delta_\sigma \alpha \Delta_\sigma \beta \geq \mu$ for all σ, then μ sets a limit to the joint precision with which the results for α and β can be predicted for any state preparation used to prepare *two different ensembles*, one for the measurements of α and the other for the measurements of β.

(b) In many quantum mechanics textbooks, the discussion of proposition P revolves around the inequalities proved in 19.5.20, which are called *uncertainty relations*. However, it would be better if they referred to the inequalities proved in 19.5.19 because, while in 19.5.19 the state σ is only required the physically meaningful condition that α and β be evaluable in σ, in 19.5.20 σ is also required to be such that $A_\alpha A_\beta W_\sigma \in \mathcal{T}(\mathcal{H})$ and $A_\beta A_\alpha W_\sigma \in \mathcal{T}(\mathcal{H})$ (or such that $u_\sigma \in D_{[A_\alpha, A_\beta]}$ if σ is a pure state) and these additional conditions have no physical meaning.

(c) Clearly, from 19.5.19 we obtain the falsification of proposition P if

$$\inf\{|\,(A_\alpha u|A_\beta u) - (A_\beta u|A_\alpha u)\,| : u \in \tilde{\mathcal{H}} \cap D_{A_\alpha} \cap D_{A_\beta}\} > 0.$$

This happens in a drastic way when α and β are the observables position and linear momentum (in a given direction) of a non-relativistic quantum particle, in which case

$$|\,(A_\alpha u|A_\beta u) - (A_\beta u|A_\alpha u)\,| = (2\pi)^{-1}h, \forall u \in \tilde{\mathcal{H}} \cap D_{A_\alpha} \cap D_{A_\beta},$$

where h is Planck's constant (cf. Section 20.3).

(d) The result of 19.5.21 shows that proposition P is true whenever at least one of the two observables α and β is bounded.

(e) For a state σ in which two observables α and β are evaluable, $\Delta_\sigma \alpha \Delta_\sigma \beta = 0$ is true if and only if for at least one out of α and β, suppose for α, $\sigma_p(A_\alpha) \neq 0$ and there is $\lambda \in \sigma_p(A_\alpha)$ such that $R_{W_\sigma} \subset R_{P^{A_\alpha}(\{\lambda\})}$ (cf. 19.3.21); if such is the case, for $\{u_n\}_{n \in I}$ as in 19.5.19 we have

$$A_\alpha u_n = \langle\alpha\rangle_\sigma u_n, \forall n \in I,$$

and this explains why the right hand sides of the inequalities in 19.5.19 and 19.5.20 vanish. In particular, for a pure state σ in which α and β are evaluable, i.e. such that $u_\sigma \in D_{A_\alpha} \cap D_{A_\beta}$, $\Delta_\sigma \alpha \Delta_\sigma \beta = 0$ is true if and only if u_σ is eigenvector of A_α or A_β. Thus, $\Delta_\sigma \alpha \Delta_\sigma \beta$ can be zero even when the operator $[A_\alpha, A_\beta]$ is not (a restriction of) the operator $\mathbb{O}_\mathcal{H}$. As an example, if α and β are two components of the orbital angular momentum for the system of a single quantum particle, then $[A_\alpha, A_\beta]$ is not a restriction of $\mathbb{O}_\mathcal{H}$; however, A_α and A_β have one common eigenvector and therefore α and β are evaluable in the pure state σ represented by this vector and we have $\Delta_\sigma \alpha \Delta_\sigma \beta = 0$ since $\Delta_\sigma \alpha = \Delta_\sigma \beta = 0$.

(f) For two observables α and β, in many quantum mechanics textbooks the condition $[A_\alpha, A_\beta] \subset \mathbb{O}_\mathcal{H}$ is considered equivalent to the condition that α and β be compatible. However, this is wrong because α and β can fail to be compatible, and hence (cf. 19.5.10) the self-adjoint operators A_α and A_β can fail to commute (in the sense defined in 17.1.5), but nonetheless be such that $[A_\alpha, A_\beta] \subset \mathbb{O}_\mathcal{H}$, with a mathematically very meaningful domain $D_{[A_\alpha, A_\beta]}$ to boot (cf. 17.1.8). It must be granted that, *if α and β are bounded*, then α and β are compatible if and only if $[A_\alpha, A_\beta] = \mathbb{O}_\mathcal{H}$ (cf. 19.3.10b, 19.5.10, 17.1.6a); but in this case 19.5.20 is of no real use since in this case the truthfulness of proposition P is assured by 19.5.21. What is true in general is that if α and β are compatible then $[A_\alpha, A_\beta] \subset \mathbb{O}_\mathcal{H}$ (cf. 19.5.10 and 17.1.7h), but it would be sensible to use this fact together with 19.5.20 only if we did not have the stronger result of 19.5.22, which shows that for compatible α and β the result of 19.5.19 does not exert any constraint on $\Delta_\sigma \alpha \Delta_\sigma \beta$ for any state σ in which α and β are evaluable (without the additional condition on σ that we should need if we were to use 19.5.20). Actually, 19.5.23 shows that if α and β are compatible then an even stronger proposition than proposition P is true. We point out that, *while for the results previously obtained about the compatibility of two observables we had to assume that an ideal measurement was available for at least one of them* (cf. 19.5.13 and 19.5.14), *this assumption is not required in* 19.5.23. We notice that the result of 19.5.23 holds trivially for every pair of classical observables; indeed, in the classical case, for each microstate $s \in S$ we have $\Delta_s \alpha = 0$ for each observable α (cf. 19.2.6a). Thus, two compatible quantum observables exhibit once again a behaviour similar to the one they would display if they were any pair of classical observables. The behaviour of two compatible quantum observables is not in general equal, but only similar to the one of two classical observables because we do not assume that for every quantum observable α and for every possible result λ for α there is a state σ such that $\langle \alpha \rangle_\sigma = \lambda$ and $\Delta_\sigma \alpha = 0$ (in our treatment of quantum mechanics, there is such a state if and only if $\sigma_p(A_\alpha) \neq \emptyset$ and $\lambda \in \sigma_p(A_\alpha)$, cf. 19.3.21; there would be such a state for every observable α and every $\lambda \in \sigma(A_\alpha)$ if we admitted in our treatment the absolute precision state preparations represented by elements which do not belong to the Hilbert space that we mentioned in 19.3.12b).

(g) An observable α is discrete if and only if there exists a c.o.n.s. $\{v_j\}_{j \in J}$ in $\mathcal{H}$ so that, letting σ_j be the pure state such that $u_{\sigma_j} = v_j$, α is evaluable in σ_j and $\Delta_{\sigma_j} \alpha = 0$, for all $j \in J$; this follows from 19.3.10c and from the fact that an observable α is evaluable in a pure state σ and $\Delta_\sigma \alpha = 0$ if and only if u_σ is an eigenvector of A_α (cf. 19.3.21). Thus, for a discrete observable there are many pure states (one for each element of a c.o.n.s. in $\mathcal{H}$) in which α behaves as a classical observable does in a microstate.

Let α and β be discrete observables. Then α and β are compatible if and only if there exists a c.o.n.s. $\{v_j\}_{j \in J}$ in $\mathcal{H}$ so that, letting σ_j be the pure state such that $u_{\sigma_j} = v_j$, α and β are evaluable in σ_j and $\Delta_{\sigma_j} \alpha = \Delta_{\sigma_j} \beta = 0$ (which is a stronger result than $\Delta_{\sigma_j} \alpha \Delta_{\sigma_j} \beta = 0$) for all $j \in J$ (cf. 19.5.10, 17.1.14, 19.3.21). Thus, if two discrete observables are compatible then there are many pure states (one for every element of a c.o.n.s. in $\mathcal{H}$) in which both of them behave as a pair of classical observables do in a microstate.

19.6   Time evolution in non-relativistic quantum mechanics

Up to now we have assumed that any procedure discussed in connection with a physical system could be carried out at a single instant of time; of course, this is an idealization which can never have an exact counterpart in real experiments. Moreover, when we considered more than one procedure, we always assumed that they were executed one immediately after the other; this is obviously a further idealization, and what we really meant was that the copy of the system did not change appreciably between successive procedures. Thus, time has played no actual role until now. In this section we examine how the flow of time enters the scheme of quantum mechanics. We restrict our discussion to the non-relativistic case, where time is an objective real parameter. This section is centred around axiom Q3, and it begins by showing how one can arrive at this axiom.

As usual, $\mathcal{H}$ denotes the Hilbert space in which a quantum system is represented, as summarized in 19.3.22.

19.6.1 Remark. For a given quantum system, let σ be a state preparation procedure and suppose that it is carried out at a definite instant of time t_0. In all the sections preceding, a copy prepared in σ was used in a second procedure (the determination of a proposition, the measurement of an observable, the passage through a filter) which took place immediately after time t_0. However, at least in principle it is possible to wait for a positive time interval t before activating the second procedure, and carry out this second procedure at time $t_0 + t$; if this is done, the second procedure takes place immediately after the *new* first procedure that can be described as follows: perform procedure σ and wait for the time interval t. Now, this new first procedure is not in general equivalent to the procedure σ. We *assume* that this new first procedure is still a state preparation procedure, which we denote by σ_t, and we say that the state σ at time t_0 evolves into the state σ_t at time $t_0 + t$. We also *assume* that σ_t is a pure state whenever σ is a pure state; thus, we have the mapping Γ_t defined by

$$\Sigma_0 \ni \sigma \mapsto \Gamma_t(\sigma) := \sigma_t \in \Sigma_0.$$

In what follows *we confine our attention* to quantum systems for which Γ_t does not depend on t_0 but only on the time interval t (this was already anticipated by the symbol Γ_t, where t_0 does not appear); these systems are called *conservative*. Also, *we confine our attention* to quantum systems for which the mapping Γ_t is a bijection from Σ_0 onto itself, for every positive t; these systems are called *reversible*. Completely isolated quantum systems are experimentally seen to be conservative and reversible. We denote the identity mapping of Σ_0 by Γ_0 and write $\Gamma_{-t} := (\Gamma_t)^{-1}$

for every positive t; for every pure state σ and any time t_0, if we prepare the state $\Gamma_{-t}(\sigma)$ at time $t_0 - t$ and we wait until time t_0, then at time t_0 we have a copy of the system in the state σ. For every pair of positive t_1, t_2 we have $\Gamma_{t_1} \circ \Gamma_{t_2} = \Gamma_{t_1+t_2}$; this is simply due to the fact that waiting for the time interval t_2 and then for the time interval t_1 is the same as waiting for the time interval $t_1 + t_2$ (and to the fact that Γ_t depends only on the time interval t). Then, it is easy to prove that we have $\Gamma_{t_1} \circ \Gamma_{t_2} = \Gamma_{t_1+t_2}$ for all $t_1, t_2 \in \mathbb{R}$. Further, we *assume* that, for every pair of pure states σ_1, σ_2 and every positive t, the transition probability (cf. 19.4.7b) from $\Gamma_t(\sigma_1)$ to $\Gamma_t(\sigma_2)$ is the same as the transition probability from σ_1 to σ_2; indeed, the probability that a copy prepared at time t_0 in a pure state σ_1 gets modified (by the action of a suitable filter) to become as if it had been prepared at time t_0 in another pure state σ_2 is experimentally seen to be the same immediately after t_0 as at any later time. Since $\Gamma_{-t} = (\Gamma_t)^{-1}$, this entails that the same is true for negative t. Next, we *assume* that for every pure state σ the transition probability from the pure state $\Gamma_t(\sigma)$ to the pure state σ approaches one as t approaches zero; the meaning of this continuity condition is obvious.

Now, since Γ_t is a bijection from the family Σ_0 of pure states onto itself, in view of the bijection between Σ_0 and the projective Hilbert space $\hat{\mathcal{H}}$ defined in 19.3.5c there exists, for all $t \in \mathbb{R}$, a unique mapping $\omega_t : \hat{\mathcal{H}} \to \hat{\mathcal{H}}$ which is a bijection from $\hat{\mathcal{H}}$ onto itself and also such that

$$[u_{\Gamma_t(\sigma)}] = \omega_t([u_\sigma]), \forall \sigma \in \Sigma_0.$$

Since Γ_t preserves the transition probability between pure states, we have

$$\tau(\omega_t([u_{\sigma_1}]), \omega_t([u_{\sigma_2}])) = |\left(u_{\Gamma_t(\sigma_1)} | u_{\Gamma_t(\sigma_2)} \right)| = |\left(u_{\sigma_1} | u_{\sigma_2} \right)|$$
$$= \tau([u_{\sigma_1}], [u_{\sigma_2}]), \forall \sigma_1, \sigma_2 \in \Sigma_0, \forall t \in \mathbb{R},$$

where τ is the function defined in 10.9.1; thus, ω_t is an automorphism of the projective Hilbert space $(\hat{\mathcal{H}}, \tau)$, for all $t \in \mathbb{R}$ (cf. 10.9.4). Moreover, the condition

$$\Gamma_{t_1} \circ \Gamma_{t_2} = \Gamma_{t_1+t_2}, \forall t_1, t_2 \in \mathbb{R},$$

is obviously equivalent to the condition

$$\omega_{t_1} \circ \omega_{t_2} = \omega_{t_1+t_2}, \forall t_1, t_2 \in \mathbb{R}.$$

Furthermore, the continuity condition assumed above is obviously equivalent to the condition that

the function $\mathbb{R} \ni t \mapsto \tau([u], \omega_t([u])) \in [0, 1]$ is continuous at $0, \forall u \in \tilde{\mathcal{H}}$.

Therefore, in view of 16.4.5, the mapping

$$\mathbb{R} \ni t \mapsto \omega_t \in \text{Aut } \hat{\mathcal{H}}$$

is a continuous one-parameter group of automorphisms. Consequently, in view of 16.4.11, there exists a continuous one-parameter unitary group U in $\mathcal{H}$ so that

$$\omega_t([u]) = [U(t)u], \forall u \in \tilde{\mathcal{H}}, \forall t \in \mathbb{R},$$

and hence, in view of 16.1.10, there exists a self-adjoint operator H in $\mathcal{H}$ so that

$$[u_{\sigma_t}] = [u_{\Gamma_t(\sigma)}] = \omega_t([u_\sigma]) = [U^H(t)u_\sigma], \forall \sigma \in \Sigma_0, \forall t \in \mathbb{R},$$

or equivalently (cf. 13.1.13b) so that

$$W_{\sigma_t} = W_{\Gamma_t(\sigma)} = A_{U^H(t)u_\sigma} = U^H(t)A_{u_\sigma}U^H(t)^{-1}$$
$$= U^H(t)W_\sigma U^H(t)^{-1}, \forall \sigma \in \Sigma_0, \forall t \in \mathbb{R}.$$

From 16.4.3b, 16.1.8a, 16.1.5d we see that the operator H is unique up to an additive multiple of the identity operator $\mathbb{1}_\mathcal{H}$.

Finally, we *assume* that the way in which pure states change over time determines the way in which every state changes, as follows: if a state $\sigma \in \Sigma$ prepared at time t_0 is the mixture of a family $\{\sigma_n\}_{n \in I}$ of pure states with weights $\{w_n\}_{n \in I}$ as in 19.3.5b, then for every positive time interval t the state σ_t is the mixture of the family $\{\Gamma_t(\sigma_n)\}_{n \in I}$ of pure states with the same weights, i.e.

$$W_{\sigma_t}f = \sum_{n \in I} w_n W_{\Gamma_t(\sigma_n)}f, \forall f \in \mathcal{H},$$

and hence

$$W_{\sigma_t}f = U^H(t) \sum_{n \in I} w_n W_{\sigma_n} U^H(t)^{-1}f = U^H(t)W_\sigma U^H(t)^{-1}f, \forall f \in \mathcal{H},$$

i.e. $W_{\sigma_t} = U^H(t)W_\sigma U^H(t)^{-1}$. We point out that, although this equality has been obtained on the basis of a particular decomposition of σ into a mixture of pure states, there is no trace of that particular decomposition in the final result, as must be since that decomposition in not unique unless σ is a pure state (cf. 19.3.5b,c). We also note that $U^H(t)W_\sigma U^H(t)^{-1}$ is indeed a statistical operator by 18.3.2a. Now we notice that, for every positive t, the mapping $\Sigma \ni \sigma \mapsto \sigma_t \in \Sigma$ results to be a bijection from Σ onto itself because the mapping of 19.3.1a is a bijection and the mapping $\mathcal{W}(\mathcal{H}) \ni W \mapsto U^H(t)WU^H(t)^{-1} \in \mathcal{W}(\mathcal{H})$ is a bijection from $\mathcal{W}(\mathcal{H})$ onto itself, as can be easily seen. Thus, for every state $\sigma \in \Sigma$ and every positive t we can define σ_{-t} as the state that evolves into the state σ at any time t_0 if it is prepared at time $t_0 - t$; clearly, we have

$$W_{\sigma_{-t}} = U^H(t)^{-1}W_\sigma U^H(t) = U^H(-t)W_\sigma U^H(-t)^{-1}.$$

Thus, we have

$$W_{\sigma_t} = U^H(t)W_\sigma U^H(t)^{-1}, \forall \sigma \in \Sigma, \forall t \in \mathbb{R}.$$

This outcome of the assumptions above can be stated as the axiom below.

19.6.2 Axiom (Axiom Q3). There are quantum systems for which there exists a self-adjoint operator H (in the Hilbert space in which the system is represented) so that for every $t_0 \in \mathbb{R}$, every positive t, and every state $\sigma \in \Sigma$, a copy prepared at time t_0 in σ becomes after the time interval t the same as a copy prepared at time $t_0 + t$ in the state σ_t represented by the statistical operator

$$W_{\sigma_t} := U^H(t)W_\sigma U^H(t)^{-1},$$

and a copy prepared at time $t_0 - t$ in the state σ_{-t} represented by the statistical operator

$$W_{\sigma_{-t}} := U^H(-t)W_\sigma U^H(-t)^{-1}$$

becomes after the time interval t the same as a copy prepared at time t_0 in σ.

19.6.3 Remarks.

(a) In 19.6.1 we proved that the assumptions made there implied axiom Q3, and it is easy to see that axiom Q3 implies those assumptions (13.1.13b and 16.4.3a must be used). Then, we see in particular that *the quantum systems for which axiom Q3 holds are the conservative and reversible quantum systems. In what follows we consider only conservative and reversible quantum systems.*

(b) The self-adjoint operator H of axiom Q3 is unique up to an additive multiple of the unit operator. This is already clear from 19.6.1. In any case, to see it directly, assume that H' is a self-adjoint operator which plays the same role as H in axiom Q3. Then we have

$$U^H(t)WU^H(t)^{-1} = U^{H'}(t)WU^{H'}(t)^{-1}, \forall W \in \mathcal{W}(\mathcal{H}), \forall t \in \mathbb{R},$$

and hence in particular (cf. 18.3.2b)

$$U^H(t)A_u U^H(t)^{-1} = U^{H'}(t)A_u U^{H'}(t)^{-1}, \forall u \in \tilde{\mathcal{H}}, \forall t \in \mathbb{R},$$

which is equivalent to (cf. 13.1.13)

$$[U^H(t)u] = [U^{H'}(t)u], \forall [u] \in \hat{\mathcal{H}}, \forall t \in \mathbb{R},$$

which can be written as

$$\omega_{U^H(t)} = \omega_{U^{H'}(t)}, \forall t \in \mathbb{R}.$$

From 16.4.3b we see that this implies that there exists $k \in \mathbb{R}$ so that

$$U^{H'}(t) = e^{ikt}U^H(t), \forall t \in \mathbb{R},$$

and hence (cf. 16.1.8a and 16.1.5d) so that

$$H' = H + k\mathbb{1}_{\mathcal{H}}.$$

(c) The self-adjoint operator $-H$ is called the *Hamiltonian* of the system, and it is interpreted as the self-adjoint operator which represents the observable "energy" of the system. This is consistent with its being unique only up to an additive multiple of the unit operator, since physically the observable energy of any system is only defined up to an additive constant (note that, for $k \in \mathbb{R}$, $\sigma(H + k\mathbb{1}_{\mathcal{H}}) = \sigma(H) + k$ and $\sigma_p(H + k\mathbb{1}_{\mathcal{H}}) = \sigma_p(H) + k$, as is obvious from 15.2.4b and 15.2.5b; then, cf. 19.3.10a and 19.3.12a).

19.6.4 Remark. The relationship between a state and the state which at any time evolves from it or has evolved into it, as implied by axiom Q3, is strictly causal, in spite of the acausal character of quantum mechanics when it is referred to a single copy of a system, as reflected in the impossibility of making more than statistical statements about the results to be expected from determinations of propositions or from measurements of observables. Thus, when referred to ensembles and not to single copies, quantum mechanics is as deterministic as classical mechanics if the quantum system is conservative and reversible, hence in particular if it is a completely isolated system. An altogether different change of state happens when there is state reduction (cf. 19.4.3b), produced by the interaction of copies of the system with a filter or with a measuring instrument in a first kind measurement (cf. 19.4.8 and 19.4.10). We point out that the number of copies in an ensemble representing a state does not change in the time evolution of axiom Q3, while it does in a state reduction.

19.6.5 Remark. For a quantum system whose time evolution is determined by a self-adjoint operator H as in axiom Q3, for each state $\sigma \in \Sigma$ activated at any time t_0 we can define the mapping $\mathbb{R} \ni t \mapsto \sigma_t \in \Sigma$, which is called the *trajectory* of the state σ. For a pure state σ, the trajectory of σ corresponds to the mapping

$$\mathbb{R} \ni t \mapsto u_\sigma(t) := U_{u_\sigma}^H(t) = U^H(t)u_\sigma \in \tilde{\mathcal{H}}$$

(cf. 19.6.1; for $U_{u_\sigma}^H$, cf. 16.1.1). Now, for $u_\sigma \in D_H$ we have (cf. 16.1.5b)

$$u_\sigma(t) \in D_H \text{ and } \frac{du_\sigma(t)}{dt} = iHu_\sigma(t), \forall t \in \mathbb{R}.$$

Thus, this is the condition that is obeyed by the pure states whose representatives (as in 19.3.5c) are rays which lie in D_H, and this is the abstract form of what is known as the *Schrödinger equation*. In many specific cases, the Hilbert space $\mathcal{H}$ is a space of equivalence classes of functions on $\mathbb{R}^n$ and H is a differential operator; then u_σ becomes a function (actually, an equivalence class of functions) and the Schrödinger equation is often written as

$$\frac{\partial u_\sigma}{\partial t}(x_1, ..., x_n, t) = iHu_\sigma(x_1, ..., x_n, t);$$

however, the use of the symbol $\frac{\partial u_\sigma}{\partial t}$ is misleading, since $\frac{du_\sigma(t)}{dt}$ has the meaning that is defined in 16.1.3 (with the limit taken with respect to the distance defined in 10.1.15). Finding the continuous one-parameter unitary group U^H is sometimes dubbed "solving the Schrödinger equation"; however, it must be noted that the trajectories of all states are known if U^H is known, while only the trajectories of the pure states represented by vectors in D_H appear in the Schrödinger equation, and it is physically impossible to have $D_H = \mathcal{H}$ (cf. 19.3.11).

19.6.6 Proposition. *If the time evolution of a quantum system is determined by a self-adjoint operator H as in axiom Q3 and the energy of the system is a discrete observable then for every pure state $\sigma \in \Sigma_0$ we have*

$$u_\sigma(t) := U^H(t)u_\sigma = \sum_{n \in I} e^{it\lambda_n} P_n u_\sigma, \forall t \in \mathbb{R},$$

where $\{\lambda_n\}_{n \in I} = \sigma_p(H)$ and $P_n = P_{N_{H-\lambda_n 1_{\mathcal{H}}}}$ for each $n \in I$, or equivalently

$$u_\sigma(t) = \sum_{n \in I} \sum_{s \in I_n} e^{it\lambda_n} (u_{n,s}|u_\sigma) u_{n,s}, \forall t \in \mathbb{R},$$

where $\{u_{n,s}\}_{s \in I_n}$ is an o.n.s. in $\mathcal{H}$ which is complete in the eigenspace of H corresponding to λ_n, for each $n \in I$.

Proof. From 19.3.10c we have that the conditions of 15.3.4B hold true for the self-adjoint operator H, since $-H$ represents the observable energy (cf. 19.6.3c). The result then follows from 16.1.6 and from the explicit forms of the operator $\varphi(A)$ in 15.3.4B. $\qquad\square$

19.6.7 Remark. The result of 19.6.6 shows why, if the energy of a quantum system is a discrete observable, knowing the eigenvalues of H and a c.o.n.s. comprised of eigenvectors of H allows one to "solve the Schrödinger equation".

19.6.8 Definition. A state $\sigma \in \Sigma$ such that $\sigma_t = \sigma$ for each $t \in \mathbb{R}$ is said to be a *stationary state*.

19.6.9 Proposition. *For a pure state $\sigma \in \Sigma_0$ of a quantum system whose time evolution is determined by a self-adjoint operator H as in axiom Q3, the following conditions are equivalent:*

(a) σ is a stationary state;
(b) $\sigma_p(H) \neq \emptyset$ and u_σ is an eigenvector of H.

Proof. $a \Rightarrow b$: Assume that $\sigma \in \Sigma_0$ is a stationary state. Then (cf. 19.6.1)

$$A_{U^H(t)u_\sigma} = W_{\sigma_t} = W_\sigma = A_{u_\sigma}, \forall t \in \mathbb{R},$$

and hence (cf. 13.1.13a) there exists a function $\rho : \mathbb{R} \to \mathbb{C}$ so that $U^H(t)u_\sigma = \rho(t)u_\sigma$ for each $t \in \mathbb{R}$. We have:

$$\rho(t_1 + t_2)u_\sigma = U^H(t_1 + t_2)u_\sigma = U^H(t_1)U^H(t_2)u_\sigma$$
$$= U^H(t_1)\rho(t_2)u_\sigma = \rho(t_1)\rho(t_2)u_\sigma,$$

and hence $\rho(t_1 + t_2) = \rho(t_1)\rho(t_2)$, $\forall t \in \mathbb{R}$;

$$\rho(t) = (u_\sigma|U^H(t)u_\sigma), \forall t \in \mathbb{R},$$

and hence the function ρ is continuous in view of 16.1.2;

$$|\rho(t)| = \|\rho(t)u_\sigma\| = \|U^H(t)u_\sigma\| = \|u_\sigma\| = 1, \forall t \in \mathbb{R}.$$

Then, by 16.2.3 there exists $\lambda \in \mathbb{R}$ so that

$$\rho(t) = e^{i\lambda t}, \forall t \in \mathbb{R},$$

and hence so that

$$U^H(t)u_\sigma = e^{i\lambda t}u_\sigma, \forall t \in \mathbb{R}.$$

Obviously, this implies that

$$U_{u_\sigma}^H \text{ is differentiable at } 0 \text{ and } \left.\frac{dU_{u_\sigma}^H}{dt}\right|_0 = i\lambda u_\sigma,$$

and this implies (cf. 16.1.5a and 16.1.6) that

$$u_\sigma \in D_H \text{ and } i\lambda u_\sigma = iHu_\sigma.$$

This proves that condition b holds true.

$b \Rightarrow a$: Assume condition b, and let $\lambda \in \mathbb{R}$ be such that $P^H(\{\lambda\})u_\sigma = u_\sigma$ (cf. 15.2.5 and 13.1.3c); then, for each $E \in \mathcal{A}(d_\mathbb{R})$,

$$\mu_{u_\sigma}^{P^H}(E) = \left(u_\sigma|P^H(E)u_\sigma\right) = \left(u_\sigma|P^H(E)P^H(\{\lambda\})u_\sigma\right)$$

$$= \left(u_\sigma|P^H(E \cap \{\lambda\})u_\sigma\right) = \begin{cases} 1 & \text{if } \lambda \in E, \\ 0 & \text{if } \lambda \notin E \end{cases}$$

(cf. 13.3.2b,c); this shows that $\mu_{u_\sigma}^{P^H}$ is the Dirac measure in λ (cf. 8.3.6). Then from 16.1.6 and 15.3.2e we have

$$\|U_t^H u_\sigma - e^{it\lambda}u_\sigma\|^2 = \int_\mathbb{R} |e^{itx} - e^{it\lambda}|^2 d\mu_{u_\sigma}^{P^H}(x) = 0, \forall t \in \mathbb{R},$$

and hence

$$U_t^H u_\sigma = e^{it\lambda}u_\sigma, \forall t \in \mathbb{R},$$

and hence (cf. 19.6.1 and 13.1.13a)

$$W_{\sigma_t} = A_{U^H(t)u_\sigma} = A_{u_\sigma} = W_\sigma, \forall t \in \mathbb{R},$$

which is equivalent to $\sigma_t = \sigma$ for each $t \in \mathbb{R}$. $\square$

19.6.10 Remark. The result of 19.6.9 shows why the point spectrum of the Hamiltonian of a quantum system is of interest: the eigenvectors represent stationary states of the system. The typical reaction of an atom to outside stimuli is to transform its state from one stationary state to another emitting light whose frequency is proportional to the difference between the corresponding eigenvalues.

19.6.11 Definition. Let $(X, \mathcal{A})$ be a measurable space. An X-valued observable α is said to be a *constant of motion* if

$$p(\alpha(E), \sigma_t) = p(\alpha(E), \sigma), \forall E \in \mathcal{A}, \forall t \in \mathbb{R}, \forall \sigma \in \Sigma.$$

19.6.12 Proposition. *Let α be an observable of a quantum system whose time evolution is determined by a self-adjoint operator H as in axiom Q3. The following conditions are equivalent:*

(a) α is a constant of motion;
(b) for each state $\sigma \in \Sigma$ in which α is evaluable,

$$\alpha \text{ is evaluable in } \sigma_t, \quad \langle\alpha\rangle_{\sigma_t} = \langle\alpha\rangle_\sigma, \quad \Delta_{\sigma_t}\alpha = \Delta_\sigma\alpha, \quad \forall t \in \mathbb{R};$$

(c) for each pure state $\sigma \in \Sigma_0$ in which α is evaluable,

$$\alpha \text{ is evaluable in } \sigma_t, \quad \langle \alpha \rangle_{\sigma_t} = \langle \alpha \rangle_\sigma, \quad \Delta_{\sigma_t} \alpha = \Delta_\sigma \alpha, \quad \forall t \in \mathbb{R};$$

(d) for each pure state $\sigma \in \Sigma_0$ in which α is evaluable,

$$\alpha \text{ is evaluable in } \sigma_t \text{ and } \langle \alpha \rangle_{\sigma_t} = \langle \alpha \rangle_\sigma, \forall t \in \mathbb{R};$$

(e) the self-adjoint operators A_α and H commute;
(f) $p(\alpha(E), \sigma_t) = p(\alpha(E), \sigma), \forall E \in \mathcal{A}(d_\mathbb{R}), \forall t \in \mathbb{R}, \forall \sigma \in \Sigma_0$.

Proof. $a \Rightarrow b$: From the definition of μ_σ^α (cf. 19.1.8) we see that condition a is the same as

$$\mu_{\sigma_t}^\alpha(E) = \mu_\sigma^\alpha(E), \forall E \in \mathcal{A}(d_\mathbb{R}), \forall t \in \mathbb{R}, \forall \sigma \in \Sigma.$$

Now, this implies condition b by the very definitions given in 19.1.21 and in 19.1.20.

$b \Rightarrow c$: This is obvious.

$c \Rightarrow d$: This is obvious.

$d \Rightarrow e$: Assume condition d. Recalling that for a pure state $\sigma \in \Sigma_0$ we have $u_{\sigma_t} = U^H(t)u_\sigma$ (cf. 19.6.1), from 19.3.13d we have

$$U^H(t)u \in D_{A_\alpha} \text{ and } \left(U^H(t)u | A_\alpha U^H(t)u\right) = (u|A_\alpha u), \forall t \in \mathbb{R}, \forall u \in \tilde{\mathcal{H}} \cap D_{A_\alpha},$$

which is equivalent to

$$D_{A_\alpha} \subset D_{U^H(t)^{-1}A_\alpha U^H(t)} \text{ and}$$
$$\left(u | U^H(t)^{-1}A_\alpha U^H(t)u\right) = (u|A_\alpha u), \forall t \in \mathbb{R}, \forall u \in \tilde{\mathcal{H}} \cap D_{A_\alpha}.$$

This implies (cf. 10.2.12)

$$A_\alpha \subset U^H(t)^{-1}A_\alpha U^H(t), \forall t \in \mathbb{R},$$

and this implies (cf. $3.2.10b_1$)

$$U^H(t)A_\alpha \subset A_\alpha U^H(t), \forall t \in \mathbb{R}.$$

Then, A_α and H commute by 17.1.7.

$e \Rightarrow a$: Assume condition e. Then by 17.1.7 we have

$$U^H(t)^{-1}P^{A_\alpha}(E)U^H(t) = P^{A_\alpha}(E), \forall E \in \mathcal{A}(d_\mathbb{R}), \forall t \in \mathbb{R},$$

and hence (cf. 18.2.11c), for each state $\sigma \in \Sigma$,

$$p(\alpha(E), \sigma_t) = \text{tr}(P^{A_\alpha}(E)W_{\sigma_t}) = \text{tr}(P^{A_\alpha}(E)U^H(t)W_\sigma U^H(t)^{-1})$$
$$= \text{tr}(U^H(t)^{-1}P^{A_\alpha}(E)U^H(t)W_\sigma) = \text{tr}(P^{A_\alpha}(E)W_\sigma)$$
$$= p(\alpha(E), \sigma), \forall t \in \mathbb{R}.$$

$a \Rightarrow f$: This is obvious.

$f \Rightarrow c$: We proceed as in the proof of $a \Rightarrow b$, since condition f can be rephrased as follows:

$$\mu_{\sigma_t}^\alpha(E) = \mu_\sigma^\alpha(E), \forall E \in \mathcal{A}(d_\mathbb{R}), \forall t \in \mathbb{R}, \forall \sigma \in \Sigma_0.$$

$\square$

19.6.13 Remark. The way of describing the time evolution of a conservative and reversible quantum system that has been discussed in this section is called the *Schrödinger picture*. There is a mathematically equivalent way of doing the same, which can at times be useful for practical calculations.

For each proposition $\pi \in \Pi$ and each state $\sigma \in \Sigma$, we see that (cf. 18.2.11c)

$$p(\pi, \sigma_t) = \text{tr}(P_\pi U^H(t) W_\sigma U^H(t)^{-1})$$
$$= \text{tr}(U^H(t)^{-1} P_\pi U^H(t) W_\sigma) = p(\pi_t, \sigma), \forall t \in \mathbb{R},$$

if we define π_t as the proposition such that $P_{\pi_t} = U^H(t)^{-1} P_\pi U^H(t)$ (this operator is an orthogonal projection in view of 13.1.8). Similarly, if α is an observable, σ is a state, and α is evaluable in σ_t for some $t \in \mathbb{R}$, we see that (cf. 19.3.13a)

$$\langle \alpha \rangle_{\sigma_t} = \text{tr}(A_\alpha U^H(t) W_\sigma U^H(t)^{-1}) = \text{tr}(A_{\alpha,t} W_\sigma),$$

if we define $A_{\alpha,t} := U^H(t)^{-1} A_\alpha U^H(t)$ (this operator is self-adjoint in view of 12.5.4c). This mathematical way of dealing with time evolution is called the *Heisenberg picture*.

Chapter 20

Position and Momentum in Non-Relativistic Quantum Mechanics

Oddly enough, Galilean-relativistic physics is called non-relativistic, while the name of relativistic physics is reserved for the physical theories which are in accord with Einstein's special relativity. In this chapter we deal with Galilean-relativistic quantum mechanics, as we already did in Chapter 19, where we expounded on the principles of the theory. The rules we discussed there are very general, and need to be supplemented with additional assumptions when a particular system is under discussion. This applies in particular to the choice of an actual Hilbert space for a given physical system, and the identification of specific operators as the representatives of specific observables. Quite often, this task is carried out on the basis of symmetry principles. In section 20.3 we examine how this can be done for the observables position and linear momentum of a quantum particle, assuming the Galilei group as symmetry group (this is equivalent to the assumption that the theory of the system is in accord with Galilei's relativity).

The first two sections of this chapter discuss mathematical ideas which play an essential role in the more physical discussion of Section 20.3.

The subject of this chapter can be discussed with a higher level of sophistication than here, within the framework of the theories of C^*-algebras (for Sections 20.1 and 20.2) and of induced representations (for Section 20.3). However, our way of dealing with the topics of this chapter has the advantage of offering a good example of the theory of Hilbert space operators directly at work.

20.1 The Weyl commutation relation

In this section we study a commutation relation which is related to the Heisenberg canonical commutation relation (cf. 12.6.5). This commutation relation was introduced by Hermann Weyl in order to rid the discussion about the Heisenberg relation of problems caused by the presence of non-bounded operators. Actually, Weyl's relation is strictly stronger than Heisenberg's (cf. 20.1.3b and 20.1.4), but we will see in Section 20.3 that Weyl's relation is exaclty the one that characterizes the self-adjoint operators which represent the observables position and linear momentum of a non-relativistic quantum particle.

In this section, $\mathcal{H}$ stands for an abstract Hilbert space. We recall that, for a self-adjoint operator A in $\mathcal{H}$, P^A denotes the projection valued measure of A (cf. 15.2.2) and U^A denotes the continuous one-parameter unitary group whose generator is A (cf. 16.1.6 and 16.1.11a).

20.1.1 Theorem. *Let A and B be self-adjoint operators in $\mathcal{H}$. Then the following conditions are equivalent:*

(a) $U^A(t)U^B(s) = e^{-its}U^B(s)U^A(t)$, $\forall (s,t) \in \mathbb{R}^2$;
(b) $U^A(t)P^B(E)U^A(-t) = P^B(E+t)$, $\forall E \in \mathcal{A}(d_\mathbb{R})$, $\forall t \in \mathbb{R}$;
(c) $U^A(t)BU^A(-t) = B - t\mathbb{1}_\mathcal{H}$, $\forall t \in \mathbb{R}$;
(d) $U^B(s)P^A(E)U^B(-s) = P^A(E-s)$, $\forall E \in \mathcal{A}(d_\mathbb{R})$, $\forall s \in \mathbb{R}$;
(e) $U^B(s)AU^B(-s) = A + s\mathbb{1}_\mathcal{H}$, $\forall s \in \mathbb{R}$.

The sets $E+t$ and $E-s$ are defined as in 9.2.1a.

Proof. In view of 16.1.8a, condition a can be written as

(a') $U^A(t)U^B(s)U^A(-t) = U^{B-t\mathbb{1}_\mathcal{H}}(s)$, $\forall s \in \mathbb{R}$, $\forall t \in \mathbb{R}$,

and condition b can be written as

(b') $U^A(t)P^B(E)U^A(-t) = P^{B-t\mathbb{1}_\mathcal{H}}(E)$, $\forall E \in \mathcal{A}(d_\mathbb{R})$, $\forall t \in \mathbb{R}$.

Now, conditions a' and c are equivalent in view of 16.3.1, and conditions b' and c are equivalent in view of 15.4.1. Thus, conditions a, b, c are equivalent. It can be proved in a similar way that conditions a, d, e are equivalent. $\qquad\square$

20.1.2 Definition. A pair of self-adjoint operators A, B in $\mathcal{H}$ is said to be a *representation of the Weyl commutation relation* (briefly, a *representation of WCR*) if the conditions of 20.1.1 hold true for A and B and if the Hilbert space $\mathcal{H}$ is non-zero.

20.1.3 Theorem. *Let a pair of self-adjoint operators A, B in $\mathcal{H}$ be a representation of WCR. Then:*

(a) $|(Au|Bu) - (Bu|Au)| = 1$, $\forall u \in D_A \cap D_B \cap \tilde{\mathcal{H}}$, *and hence* $\Delta_W A \Delta_W B \geq \frac{1}{2}$ *for each* $W \in \mathcal{W}(\mathcal{H})$ *in which both A and B are computable ($\Delta_W A$ and $\Delta_W B$ are defined as in 18.3.14);*
(b) *A and B "satisfy" the Heisenberg canonical commutation relation, i.e. (cf. 12.6.5) $[A,B] \subset i\mathbb{1}_\mathcal{H}$.*

Proof. Preliminary remark: As already noticed in the proof of 19.5.22, condition sa-ug in 16.1.6 and the continuity of the inner product imply that, for any self-adjoint operator T in $\mathcal{H}$, for all $f \in D_T$ and all $g \in \mathcal{H}$, the function

$$\mathbb{R} \ni t \mapsto (g|U^T(t)f) \in \mathbb{C}$$

is differentiable at 0 and

$$\frac{d}{dt}\left(g|U^T(t)f\right)\big|_0 = i\left(g|Tf\right),$$

and similarly

$$\frac{d}{dt}\left(U^T(-t)f|g\right)\big|_0 = i\left(Tf|g\right).$$

After this preliminary remark, now we give the proofs of statements a and b.

a: Condition a in 20.1.1 (cf. also 16.1.1) implies that

$$\left(U^A(-t)f|U^B(s)g\right) = e^{-its}\left(U^B(-s)f|U^A(t)g\right),\forall f,g\in\mathcal{H},\forall(s,t)\in\mathbb{R}^2.$$

Then for each $f\in D_A\cap D_B$ we have, by the preliminary remark,

$$i\left(U^A(-t)f|Bf\right) = \frac{\partial}{\partial s}\left(U^A(-t)f|U^B(s)f\right)\big|_0$$

$$= \frac{\partial}{\partial s}\left(e^{-its}\left(U^B(-s)f|U^A(t)f\right)\right)\big|_0$$

$$= -it\left(f|U^A(t)f\right) + i\left(Bf|U^A(t)f\right),\forall t\in\mathbb{R},$$

and hence

$$(Af|Bf) = -i\frac{d}{dt}\left(U^A(-t)f|Bf\right)\big|_0$$

$$= -i\frac{d}{dt}\left(-t\left(f|U^A(t)f\right) + \left(Bf|U^A(t)f\right)\right)\big|_0$$

$$= i\left(f|f\right) + (Bf|Af),$$

and hence

$$(Af|Bf) - (Bf|Af) = i\|f\|^2.$$

In view of 19.5.19 (the proof of 19.5.19 is actually effective for every pair of self-adjoint operators and every statistical operator in which they are both computable; cf. also 19.3.13a), this proves condition a.

b: Condition c in 20.1.1 is equivalent to

$$U^A(t)B = (B - t\mathbb{1}_\mathcal{H})U^A(t),\forall t\in\mathbb{R}$$

(cf. $3.2.10b_1$). For all $g\in D_B$ and $f\in D_{AB-BA}$, this implies that

$$\left(g|U^A(t)Bf\right) = \left(Bg|U^A(t)f\right) - t\left(g|U^A(t)f\right),\forall t\in\mathbb{R},$$

and hence, by the preliminary remark,

$$i\left(g|ABf\right) = \frac{d}{dt}\left(g|U^A(t)Bf\right)\big|_0$$

$$= \frac{d}{dt}\left(\left(Bg|U^A(t)f\right) - t\left(g|U^A(t)f\right)\right)\big|_0$$

$$= i\left(Bf|Af\right) - \left(g|f\right),$$

and hence

$$(g|ABf - BAf - if) = 0.$$

Since $\overline{D_B} = \mathcal{H}$, in view of 10.2.11 this yields

$$ABf - BAf - if = 0_\mathcal{H}, \forall f\in D_{AB-BA},$$

which is equivalent to statement b. $\qquad\square$

20.1.4 Remark. From 20.1.3b we see that, if a pair of self-adjoint operators A, B is a representation of WCR, then A and B "satisfy" the Heisenberg canonical commutation relation. Conversely, one might conjecture that, if a pair of self-adjoint operators A, B "satisfy" the Heisenberg canonical commutation relation, then A and B are a representation of WCR. A slight modification of Nelson's example (cf. 17.1.8b) proves that this is not necessarily true, even when the domain $D_{[A,B]}$ is so large that the restrictions of A and B to this linear manifold are essentially self-adjoint. This example is constructed on purely mathematical grounds, but other examples occur so to speak spontaneously in some quantum mechanical systems, e.g. in connection with the Aharonov–Bohm effect (cf. Reeh, 1988).

20.1.5 Definitions. A pair of self-adjoint operators A, B in $\mathcal{H}$ is said to be *jointly irreducible* if the set of operators $\{A, B\}$ is irreducible (cf. 17.3.1).

A pair of continuous one-parameter unitary groups U, V in $\mathcal{H}$ is said to be *jointly irreducible* if the set of operators $\{U(t) : t \in \mathbb{R}\} \cup \{V(t) : t \in \mathbb{R}\}$ is irreducible.

We see from 17.2.13 that a pair of self-adjoint operators A, B in $\mathcal{H}$ is jointly irreducible iff the pair of continuous one-parameter groups U^A, U^B is jointly irreducible.

A pair of self-adjoint operators A, B in $\mathcal{H}$ is said to be an *irreducible representation of WCR* if it is a representation of WCR and it is jointly irreducible.

20.1.6 Proposition. *Let A, B be a representation of WCR in $\mathcal{H}$ and suppose that there exists a non-trivial subspace M of $\mathcal{H}$ which is reducing for A and for B. Then the pair A^M, B^M is a representation of WCR in the Hilbert space M.*

Proof. The operators A^M and B^M are self-adjoint operators in the Hilbert space M (cf. 17.2.8). Moreover, the operators $U^A(t)$ and $U^B(t)$ are reduced by M for all $t \in \mathbb{R}$ and the mappings

$$\mathbb{R} \ni t \mapsto (U^A(t))^M \in \mathcal{U}(M) \text{ and } \mathbb{R} \ni t \mapsto (U^B(t))^M \in \mathcal{U}(M)$$

are continuous one-parameter unitary groups whose generators are A^M and B^M (cf. 17.2.13). Now, condition a in 20.1.1 implies obviously that

$$(U^A(t))^M (U^B(s))^M = e^{-its}(U^B(s))^M (U^A(t))^M, \forall (s,t) \in \mathbb{R}^2.$$

Thus, the pair A^M, B^M is a representation of WCR in the Hilbert space M. $\square$

20.1.7 Theorem (The Schrödinger representation of WCR).

(a) *The operator Q defined in 15.3.4A is a self-adjoint operator in the Hilbert space $L^2(\mathbb{R})$. The continuous one-parameter unitary group U^Q is so that*

$$U^Q(t)[f] = [f^t], \text{ with } f^t(x) := e^{itx}f(x), \forall x \in D_f,$$

for all $[f] \in L^2(\mathbb{R})$ and all $t \in \mathbb{R}$.

(b) The mapping

$$P_0 : D \to L^2(\mathbb{R})$$
$$[\varphi] \mapsto P_0[\varphi] := -i[\varphi'],$$

with

$$D := \{[\varphi] \in L^2(\mathbb{R}) : \varphi \in \mathcal{S}(\mathbb{R})\}$$

(cf. 11.3.6b), is an essentially self-adjoint operator in $L^2(\mathbb{R})$. The unique self-adjoint extension of P_0 (cf. 12.4.11c) is the generator P of the continuous one-parameter unitary group U^P defined by

$$U^P(t)[f] := [f_{-t}], \quad with \ \ f_{-t}(x) := f(x+t), \forall x \in D_f - t,$$

for all $[f] \in L^2(\mathbb{R})$ and all $t \in \mathbb{R}$.
The operators Q and P are unitarily equivalent through the Fourier transform F on $L^2(\mathbb{R})$, since

$$Q = FPF^{-1}.$$

(c) The pair Q, P is a representation of WCR, called the Schrödinger representation.

Proof. a: The operator Q is self-adjoint in view of 14.3.17. In view of 16.1.6,

$$U^Q(t) = \varphi_t(Q), \forall t \in \mathbb{R},$$

where φ_t is the function

$$\mathbb{R} \ni x \mapsto \varphi_t(x) := e^{itx} \in \mathbb{C}.$$

Now, in view of 15.3.4A and Section 14.5,

$$\varphi_t(Q) = M_{\varphi_t} = U_t, \forall t \in \mathbb{R},$$

where U_t is the operator defined in 11.4.15. This proves statement a.

b: The mapping

$$\mathbb{R} \ni t \mapsto V_t \in \mathcal{B}(L^2(\mathbb{R}))$$

defined in 11.4.15 is a continuous one-parameter unitary group. Indeed, the equation

$$V_t = F^{-1}U_t F, \forall t \in \mathbb{R}$$

(cf. 11.4.16) implies

$$V_t \in \mathcal{U}(L^2(\mathbb{R})), \forall t \in \mathbb{R}$$

(cf. 4.6.2b) and

$$V_{t_1} V_{t_2} = F^{-1}U_{t_1}U_{t_2}F = F^{-1}U_{t_1+t_2}F = V_{t_1+t_2}, \forall t_1, t_2 \in \mathbb{R};$$

furthermore, the mapping

$$\mathbb{R} \ni t \mapsto V_t[f] \in L^2(\mathbb{R})$$

is continuous for all $[f] \in L^2(\mathbb{R})$ (cf. 11.4.16). Then, if we denote by P the generator of this continuous one-parameter group, we have

$$P = F^{-1}QF$$

by 16.3.1.

The set D is obviously a linear manifold in $L^2(\mathbb{R})$ and $D \subset D_Q$; moreover $\overline{D} = L^2(\mathbb{R})$ in view of 11.3.3 and 10.6.5b (or in view of 11.4.19). The restriction Q_D of Q to D is a symmetric operator (cf. e.g. 12.4.3); moreover, for each $\varphi \in \mathcal{S}(\mathbb{R})$, the two functions

$$\mathbb{R} \ni x \mapsto \varphi_{\pm}(x) := (x \pm i)^{-1}\varphi(x) \in \mathbb{C}$$

are obviously elements of $\mathcal{S}(\mathbb{R})$ and

$$(Q_D \pm i \mathbb{1}_{L^2(\mathbb{R})})[\varphi_{\pm}] = [\varphi];$$

in view of 12.4.17, this proves that the operator Q_D is essentially self-adjoint. Therefore, the operator $F^{-1}Q_D F$ is also essentially self-adjoint (cf. 12.5.4d). We see that

$$D_{F^{-1}Q_D F} = \{[f] \in L^2(\mathbb{R}) : F[f] \in D\}.$$

Now, for $[f] \in L^2(\mathbb{R})$, we have

$$F[f] \in D \Rightarrow [\exists [g] \in D \text{ s.t. } F[f] = [g]] \Rightarrow [\exists [g] \in D \text{ s.t. } [f] = F^{-1}[g]] \Rightarrow [f] \in D$$

and

$$[f] \in D \Rightarrow F[f] \in D$$

(cf. 11.4.6). Therefore,

$$D_{F^{-1}Q_D F} = D.$$

Moreover we have

$$F^{-1}Q_D F[\varphi] = [(\xi \hat{\varphi})^{\vee}] = -i[((\hat{\varphi})^{\vee})^{(1)}] = -i[\varphi'] = P_0[\varphi], \forall \varphi \in \mathcal{S}(\mathbb{R})$$

(cf. 11.4.2 and 11.4.9). This proves that

$$F^{-1}Q_D F = P_0.$$

Then P_0 is essentially self-adjoint (cf. 12.5.4d) and P is its unique self-adjoint extension, since P is self-adjoint and $Q_D \subset Q$ implies $P_0 \subset P$ (cf. 12.4.11c). This concludes the proof of statement b.

c: For all $(s,t) \in \mathbb{R}^2$ and all $[f] \in L^2(\mathbb{R})$, we have

$$(f_{-s})^t(x) = e^{itx} f(x+s) = e^{-its} e^{it(x+s)} f(x+s) = e^{-its} (f^t)_{-s}(x), \forall x \in D_f - s,$$

and hence

$$U^Q(t)U^P(s)[f] = e^{-its} U^P(s)U^Q(t)[f].$$

This proves statement c. $\square$

20.1.8 Remark. In view of 20.1.7c, 20.1.3b, 12.6.5, we know that either operator P or Q or both operators P and Q must be non-bounded. Then both P and Q are non-bounded, in view of their unitary equivalence (cf. 20.1.7b) and of 4.6.5b.

This can be proved more directly in the following way. The operator Q is not bounded in view of 14.2.17 (cf. also Section 14.5 and 15.3.4A). Then the operator P is not bounded because Q and P are unitarily equivalent.

20.2   The Stone–von Neumann uniqueness theorem

In 20.1.7 we saw a special representation of the Weyl commutation relation, called the Schrödinger representation. The content of the Stone–von Neumann uniqueness theorem, which is the subject of the present section, is that the Schrödinger representation is irreducible and all irreducible representations of the Weyl commutation relation are unitarily equivalent. Thus, the Schrödinger representation is the unique irreducible representation of the Weyl commutation relation, up to unitary equivalence. But what was the main motivation behind the Stone–von Neumann theorem? To cut a long story short to the extreme, in the mid 1920s there were two competing formalisms for the emerging theory of quantum mechanics: the matrix mechanics of Werner Heisenberg and the wave mechanics of Erwin Schrödinger. Heisenberg's formalism was based on "infinite size matrices" q and p which satisfied the Heisenberg canonical commutation relation

$$qp - pq = i(2\pi)^{-1}h,$$

where h is Planck's constant, while Schrödinger's formalism used transformations Q and P in the space of wave functions which satisfied the same relation

$$QP - PQ = i(2\pi)^{-1}h.$$

There was the serious problem of the equivalence of these two approaches to quantum mechanics, which were initially mutually antagonistic. In 1926 Schrödinger found out a way to obtain the matrix elements of Heisenberg's q and p by using his Q and P together with what would be recognized later as a c.o.n.s. in $L^2(\mathbb{R})$ (Schrödinger, 1926). In the same year, Pascual Jordan provided a heuristic argument for the equivalence of the two formalisms (Jordan, 1926). However, these and other similar observations made by Paul A. M. Dirac and Wolfgang Pauli fell far short of an equivalence proof of matrix mechanics and wave mechanics (as is sometimes claimed), let alone of an actual mathematical understanding of quantum mechanics. Much work remained to be done before the problem of equivalence could even be described in a form suitable for real mathematical treatment. First, quantum theory had to be formulated in Hilbert space, a crucial step begun by David Hilbert himself and made explicit in 1927 by John von Neumann. Heisenberg's "infinite size matrices" were recognized as operators in the Hilbert space ℓ^2 (cf. 10.3.8d) and Schrödinger's transformations Q and P became, after some mathematical reconditioning, the operators discussed in 20.1.7 (up to the multiplicative factor $(2\pi)^{-1}h$ for P). Then the fact had to be addressed that the Heisenberg canonical commutation relation could not be understood as an operator equation on all of a Hilbert space, because there exists no implementation of this commutation relation by bounded self-adjoint operators (cf. 12.6.5). In 1927 Hermann Weyl realized that a way out was to replace the Heisenberg canonical commutation relation with what is now called the Weyl commutation relation, i.e. to replace self-adjoint operators with the continuous one-parameter unitary groups they generate (Weyl, 1927). In 1930 Marshall H. Stone stated (Stone, 1930) and in 1931 von

Neumann proved (Neumann, 1931) what is now called the Stone–von Neumann uniqueness theorem. This theorem was the real final proof of the equivalence of Heisenberg's and Schrödinger's formulations of quantum mechanics.

In section 20.3 we use the Stone–von Neumann theorem in our discussion of the position and linear momentum observables for a non-relativistic quantum particle.

There are various proofs of the Stone–von Neumann uniqueness theorem. The proof we present here is von Neumann's original one, mainly because this proof is a nice opportunity to put into action several theorems we saw in previous chapters. Some facts we use in the proof are collected as preliminary remarks in 20.2.2, part of the proof is set forth as a lemma in 20.2.3, and the theorem is stated and proved in 20.2.4. Before all that, in 20.2.1 we compute an integral which has an important role in the proof of 20.2.3.

20.2.1 Lemma. *For all $x, y \in \mathbb{R}$, the function*

$$\mathbb{R} \ni t \mapsto e^{-\frac{1}{2}(t+x+iy)^2} \in \mathbb{C}$$

is an element of $\mathcal{L}^1(\mathbb{R})$ and

$$\int_{\mathbb{R}} e^{-\frac{1}{2}(t+x+iy)^2} \, dm(t) = \sqrt{2\pi}.$$

Proof. For all $y \in \mathbb{R}$, the function

$$\mathbb{R} \ni t \mapsto e^{-\frac{1}{2}(t+iy)^2} \in \mathbb{C}$$

is an element of $\mathcal{L}^1(\mathbb{R})$ in view of 11.4.7, because

$$\left| e^{-\frac{1}{2}(t+iy)^2} \right| = e^{\frac{1}{2}y^2} e^{-\frac{1}{2}t^2}, \forall t \in \mathbb{R}.$$

Hence so is the function of the statement for all $x, y \in \mathbb{R}$, in view of 9.2.1b.

Now we have (cf. 11.4.8 with $a := 1$)

$$\int_{\mathbb{R}} e^{-\frac{1}{2}(t+iy)^2} \, dm(t) = e^{\frac{1}{2}y^2} \int_{\mathbb{R}} e^{-iyt} e^{-\frac{1}{2}t^2} \, dm(t)$$

$$= e^{\frac{1}{2}y^2} \sqrt{2\pi} \hat{\gamma}_1(y) = e^{\frac{1}{2}y^2} \sqrt{2\pi} e^{-\frac{1}{2}y^2} = \sqrt{2\pi}, \forall y \in \mathbb{R},$$

and hence (cf. 9.2.1b)

$$\int_{\mathbb{R}} e^{-\frac{1}{2}(t+x+iy)^2} \, dm(t) = \sqrt{2\pi}, \forall x, y \in \mathbb{R}.$$

$\square$

20.2.2 Remarks. Let A, B be a representation of WCR in a Hilbert space $\mathcal{H}$. We define

$$W(s,t) := e^{-i\frac{1}{2}st} U^B(s) U^A(t), \forall (s,t) \in \mathbb{R}^2.$$

(a) In view of 20.1.1a, a direct computation proves that

$$W(s_1, t_1) W(s_2, t_2) = e^{i\frac{1}{2}(s_1 t_2 - t_1 s_2)} W(s_1 + s_2, t_1 + t_2),$$

$$\forall (s_1, t_1), (s_2, t_2) \in \mathbb{R}^2.$$

(b) In view of remark a, we have

$$W(s,t)W(-s,t) = W(0,0) = \mathbb{1}_{\mathcal{H}}, \forall (s,t) \in \mathbb{R}^2,$$

and hence

$$W(-s,-t) = W(s,t)^{-1} = W(s,t)^{\dagger}, \forall (s,t) \in \mathbb{R}^2$$

(the second equality follows from 12.5.1b since $W(s,t) \in \mathcal{U}(\mathcal{H})$).

(c) For a subspace M of $\mathcal{H}$, the following conditions are equivalent:

the operators A and B are reduced by M;
$W(s,t)f \in M, \forall f \in M, \forall (s,t) \in \mathbb{R}^2$;
the operator $W(s,t)$ is reduced by M, $\forall (s,t) \in \mathbb{R}^2$.

Indeed, the first condition implies the second in view of 17.2.13; the second implies the third in view of remark b and 17.2.9; the third implies the first in view of 17.2.13.

(d) For each $f \in \mathcal{H}$, the operators A and B are reduced by the subspace

$$M_f := V\{W(s,t)f : (s,t) \in \mathbb{R}^2\}.$$

Indeed, if we fix $f \in \mathcal{H}$ then we have, for all $g \in M_f^{\perp}$ and all $(s',t') \in \mathbb{R}^2$,

$$(W(s',t')g|W(s,t)f) = (g|W(-s',-t')W(s,t)f) = 0, \forall (s,t) \in \mathbb{R}^2$$

(cf. remarks a and b), and hence

$$W(s',t')g \in \{W(s,t)f : (s,t) \in \mathbb{R}^2\}^{\perp} = M_f^{\perp}$$

(cf. 10.2.11). In view of remark c, this proves that A and B are reduced by the subspace $M_f^{\perp}$ and hence by the subspace M_f as well (cf. 17.2.4).

20.2.3 Lemma. *Let A, B be a representation of WCR in a Hilbert space $\mathcal{H}$ (hence, $\mathcal{H}$ is not a zero Hilbert space). Then:*

(a) There exists a unique operator $T \in \mathcal{B}(\mathcal{H})$ such that

$$(f|Tg) = \int_{\mathbb{R}^2} \gamma(s,t) \, (f|W(s,t)g) \, dm_2(s,t), \forall f, g \in \mathcal{H},$$

where $W(s,t)$ is defined by A, B as in 20.2.2 and the function $\gamma : \mathbb{R}^2 \to \mathbb{C}$ is defined by

$$\gamma(s,t) := (2\pi)^{-1} e^{-\frac{1}{4}(s^2+t^2)}, \forall (s,t) \in \mathbb{R}^2;$$

the operator T is self-adjoint and not $\mathbb{O}_{\mathcal{H}}$.

(b) We have

$$TW(s,t)T = 2\pi\gamma(s,t)T, \forall (s,t) \in \mathbb{R}^2.$$

(c) The operator T is an orthogonal projection and we have, for all $f, g \in R_T$,

$$(W(s_1,t_1)f|W(s_2,t_2)g) = e^{-\frac{1}{4}(s_2-s_1)^2-\frac{1}{4}(t_2-t_1)^2-i\frac{1}{2}(s_1 t_2 - t_1 s_2)} (f|g),$$
$$\forall (s_1,t_1), (s_2,t_2) \in \mathbb{R}^2.$$

(d) *Suppose that the operators A and B are reduced by a non-trivial subspace M of $\mathcal{H}$. Then the operator T is reduced by M and the restriction T^M of T to M (cf. 17.2.1) is the same as the element $T^{(M)}$ of $\mathcal{B}(M)$ which is defined by A^M and B^M as T is defined by A and B in statement a (recall that the pair A^M, B^M is a representation of WCR in the Hilbert space M, cf. 20.1.6).*

(e) *If the orthogonal dimension of the subspace R_T is one, then the pair of self-adjoint operators A, B is jointly irreducible, and hence the pair A, B is an irreducible representation of WCR.*

Proof. a: In view of the Schwarz inequality (cf. 10.1.9) we have

$$|\gamma(s,t)\,(f|W(s,t)g)\,| \leq \gamma(s,t)\|f\|\|g\|, \forall(s,t) \in \mathbb{R}^2, \forall f,g \in \mathcal{H}. \tag{1}$$

Now, $\gamma \in \mathcal{L}^1(\mathbb{R}^2, \mathcal{A}(d_2), m_2)$ in view of 8.4.9 and 11.4.7. Therefore the function

$$\mathbb{R}^2 \ni (s,t) \mapsto \gamma(s,t)\,(f|W(s,t)g) \in \mathbb{C}$$

is an element of $\mathcal{L}^1(\mathbb{R}^2, \mathcal{A}(d_2), m_2)$ for all $f, g \in \mathcal{H}$, and

$$\left| \int_{\mathbb{R}^2} \gamma(s,t)\,(f|W(s,t)g)\,dm_2(s,t) \right| \leq \left(\int_{\mathbb{R}^2} \gamma(s,t)dm_2(s,t) \right) \|f\|\|g\|, \forall f,g \in \mathcal{H}.$$

Then 10.5.6 implies that there exists a unique operator $T \in \mathcal{B}(\mathcal{H})$ such that

$$(f|Tg) = \int_{\mathbb{R}^2} \gamma(s,t)\,(f|W(s,t)g)\,dm_2(s,t), \forall f,g \in \mathcal{H}.$$

We have, for all $f, g \in \mathcal{H}$,

$$(Tg|f) = \overline{(f|Tg)} \overset{(2)}{=} \int_{\mathbb{R}^2} \gamma(s,t)\overline{(f|W(s,t)g)}dm_2(s,t)$$

$$\overset{(3)}{=} \int_{\mathbb{R}^2} \gamma(s,t)\,(g|W(-s,-t)f)\,dm_2(s,t)$$

$$\overset{(4)}{=} \int_{\mathbb{R}^2} \gamma(-s,-t)\,(g|W(s,t)f)\,dm_2(s,t) \overset{(5)}{=} (g|Tf),$$

where 2 holds true because complex conjugation commutes with integration (cf. 8.2.3), 3 holds true by 20.2.2b, 4 by 9.2.4b (with $A(s,t) := (-s,-t)$), 5 by the equality $\gamma(-s,-t) = \gamma(s,t)$. This proves that the operator T is self-adjoint (cf. 12.4.3).

Now we want to prove that $T \neq 0_\mathcal{H}$. We assume to the contrary that $T = 0_\mathcal{H}$ and we fix $f, g \in \mathcal{H}$. Then we have, for all $(s', t') \in \mathbb{R}^2$,

$$0 \overset{(6)}{=} (f|W(-s',-t')TW(s',t')g) = (W(s',t')f|TW(s',t')g)$$

$$= \int_{\mathbb{R}^2} \gamma(s,t)\,(W(s',t')f|W(s,t)W(s',t')g)\,dm_2(s,t)$$

$$\overset{(7)}{=} \int_{\mathbb{R}^2} \gamma(s,t)e^{i(st'-ts')}\,(f|W(s,t)g)\,dm_2(s,t)$$

$$\overset{(8)}{=} \int_{\mathbb{R}} e^{it's} \left(\int_{\mathbb{R}} e^{-is't}\gamma(s,t)\,(f|W(s,t)g)\,dm(t) \right) dm(s),$$

where: 6 is obvious; 7 follows from 20.2.2a,b; 8 holds true by 8.4.10c since 1 implies that the function

$$\mathbb{R}^2 \ni (s,t) \mapsto \gamma(s,t) e^{i(st'-ts')} (f|W(s,t)g) \in \mathbb{C}$$

is an element of $\mathcal{L}^1(\mathbb{R}^2, \mathcal{A}(d_2), m_2)$. For all $s' \in \mathbb{R}$, 1 and 11.4.7 imply that the function

$$\mathbb{R} \ni t \mapsto e^{-is't} \gamma(s,t) (f|W(s,t)g) \in \mathbb{C}$$

is an element of $\mathcal{L}^1(\mathbb{R})$ for all $s \in \mathbb{R}$; then we can define the function

$$\mathbb{R} \ni s \mapsto \varphi_{s'}(s) := \int_{\mathbb{R}} e^{-is't} \gamma(s,t) (f|W(s,t)g) \, dm(t) \in \mathbb{C},$$

which is an element of $\mathcal{L}^1(\mathbb{R})$ in view of 8.4.10b; thus, the result obtained above by the equalities from 6 to 8 can be written as

$$\check{\varphi}_{s'}(t') = 0, \forall t' \in \mathbb{R}, \forall s' \in \mathbb{R}. \tag{9}$$

Moreover, for all $s' \in \mathbb{R}$, 1 implies also that

$$|\varphi_{s'}(s)| \le (2\pi)^{-1} e^{-\frac{1}{4}s^2} \|f\| \|g\| \int_{\mathbb{R}} e^{-\frac{1}{4}t^2} \, dm(t), \forall s \in \mathbb{R},$$

and hence that $\varphi_{s'} \in \mathcal{L}^2(\mathbb{R})$ (cf. also 11.4.7). Then, in view of 11.4.22 we can write 9 as

$$F^{-1}[\varphi_{s'}] = 0_{L^2(\mathbb{R})}, \forall s' \in \mathbb{R},$$

and this implies that

$$[\varphi_{s'}] = 0_{L^2(\mathbb{R})}, \forall s' \in \mathbb{R}. \tag{10}$$

For all $s' \in \mathbb{R}$, the function $\varphi_{s'}$ is continuous, as can be proved by 8.2.11 with

$$\mathbb{R} \ni t \mapsto (2\pi)^{-1} \|f\| \|g\| e^{-\frac{1}{4}t^2} \in [0, \infty)$$

as dominating function. Then (cf. 11.3.6d) 10 implies that

$$\varphi_{s'}(s) = 0, \forall s \in \mathbb{R}, \forall s' \in \mathbb{R},$$

or

$$\int_{\mathbb{R}} e^{-is't} \gamma(s,t) (f|W(s,t)g) \, dm(t) = 0, \forall s' \in \mathbb{R}, \forall s \in \mathbb{R}. \tag{11}$$

Now we fix $s \in \mathbb{R}$; 1 and 11.4.7 imply that the function

$$\mathbb{R} \ni t \mapsto \beta_s(t) := \gamma(s,t) (f|W(s,t)g) \in \mathbb{C}$$

is an element of $\mathcal{L}^2(\mathbb{R}) \cap \mathcal{L}^1(\mathbb{R})$; then, in view of 11.4.22, 11 implies that

$$F[\beta_s] = [\hat{\beta}_s] = 0_{L^2(\mathbb{R})},$$

and this implies that

$$[\beta_s] = 0_{L^2(\mathbb{R})};$$

since the function β_s is continuous, this implies that

$$\beta_s(t) = 0, \forall t \in \mathbb{R}.$$

Since s was an arbitrary element of $\mathbb{R}$, we have

$$\gamma(s,t)\,(f|W(s,t)g) = 0, \forall(s,t) \in \mathbb{R}^2.$$

Since f and g were arbitrary elements of $\mathcal{H}$, this yields

$$W(s,t) = \mathbb{O}_{\mathcal{H}}, \forall(s,t) \in \mathbb{R}^2.$$

This has been obtained as a consequence of the assumption $T = \mathbb{O}_{\mathcal{H}}$. However, we have $W(0,0) = \mathbb{1}_{\mathcal{H}}$. Therefore, $T \neq \mathbb{O}_{\mathcal{H}}$.

b: We fix $(s,t) \in \mathbb{R}^2$ and $f,g \in \mathbb{R}^2$. We have

$(f|TW(s,t)Tg)$

$$\stackrel{(12)}{=} \int_{\mathbb{R}^2} \gamma(s',t') \left(\int_{\mathbb{R}^2} \gamma(s'',t'')e^{\frac{1}{2}i(st''-ts'')}e^{\frac{1}{2}i[s'(t+t'')-t'(s+s'')]} \right.$$

$$\left. (f|W(s'+s+s'',t'+t+t'')g)\,dm_2(s'',t'') \right) dm_2(s',t')$$

$$\stackrel{(13)}{=} \int_{\mathbb{R}^2} \gamma(s',t') \left(\int_{\mathbb{R}^2} \gamma(\tilde{s}-s'-s,\tilde{t}-t'-t)e^{\frac{1}{2}i[s(\tilde{t}-t'-t)-t(\tilde{s}-s'-s)+s'(\tilde{t}-t')-t'(\tilde{s}-s')]} \right.$$

$$\left. (f|W(\tilde{s},\tilde{t})g)\,dm_2(\tilde{s},\tilde{t}) \right) dm_2(s',t')$$

$$\stackrel{(14)}{=} \int_{\mathbb{R}^2} \left(\int_{\mathbb{R}^2} \gamma(s',t')\gamma(\tilde{s}-s'-s,\tilde{t}-t'-t)e^{\frac{1}{2}i[s(\tilde{t}-t'-t)-t(\tilde{s}-s'-s)+s'\tilde{t}-t'\tilde{s}]} \right.$$

$$\left. dm_2(s',t') \right) (f|W(\tilde{s},\tilde{t})g)\,dm_2(\tilde{s},\tilde{t})$$

where:

12 follows from a direct computation on the basis of the definition of T and of 20.2.2a;

13 follows from the change of variable $(\tilde{s},\tilde{t}) := (s'+s+s'',t'+t+t'')$, in view of 9.2.1b;

14 holds true by 8.4.8 and 8.4.10c, since

$$\int_{\mathbb{R}^2} \gamma(s',t') \left(\int_{\mathbb{R}^2} \gamma(\tilde{s}-s'-s,\tilde{t}-t'-t)dm_2(\tilde{s},\tilde{t}) \right) dm_2(s',t')$$

$$\stackrel{(15)}{=} \left(\int_{\mathbb{R}^2} \gamma(u,v)dm_2(u,v) \right)^2 < \infty$$

(15 follows from the change of variable $(u,v) := (\tilde{s}-s'-s,\tilde{t}-t'-t)$).

Moreover we have, for all $(\tilde{s},\tilde{t}) \in \mathbb{R}^2$,

$$\int_{\mathbb{R}^2} \gamma(s',t')\gamma(\tilde{s}-s'-s,\tilde{t}-t'-t)e^{\frac{1}{2}i[s(\tilde{t}-t'-t)-t(\tilde{s}-s'-t)+s'\tilde{t}-t'\tilde{s}]}dm_2(s',t')$$

$$\stackrel{(16)}{=} \int_{\mathbb{R}^2} \gamma(\hat{s}-s,\hat{t}-t)\gamma(\tilde{s}-\hat{s},\tilde{t}-\hat{t})e^{\frac{1}{2}i[s(\tilde{t}-\hat{t})-t(\tilde{s}-\hat{s})+(\hat{s}-s)\tilde{t}-(\hat{t}-t)\tilde{s}]}dm_2(\hat{s},\hat{t})$$

$$= (2\pi)^{-2}e^{-\frac{1}{4}(s^2+t^2)}e^{-\frac{1}{4}(\tilde{s}^2+\tilde{t}^2)} \int_{\mathbb{R}^2} e^{\frac{1}{2}[-\hat{s}^2+((s+\tilde{s})+i(t+\tilde{t}))\hat{s}-\hat{t}^2+((t+\tilde{t})-i(s+\tilde{s}))\hat{t}]}dm_2(\hat{s},\hat{t})$$

$$\stackrel{(17)}{=} 2\pi\gamma(s,t)\gamma(\tilde{s},\tilde{t}),$$

where:

16 follows from the change of variable $(\hat{s}, \hat{t}) := (s' + s, t' + t)$;

17 holds true because

$$\int_{\mathbb{R}^2} e^{\frac{1}{2}\left[-\hat{s}^2 + ((s+\tilde{s}) + i(t+\tilde{t}))\hat{s} - \hat{t}^2 + ((t+\tilde{t}) - i(s+\tilde{s}))\hat{t}\right]} dm_2(\hat{s}, \hat{t})$$

$$\stackrel{(18)}{=} \int_{\mathbb{R}^2} e^{\frac{1}{2}\left[-\left(\hat{s} - \frac{1}{2}((s+\tilde{s}) + i(t+\tilde{t}))\right)^2 - \left(\hat{t} - \frac{1}{2}((t+\tilde{t}) - i(s+\tilde{s}))\right)^2\right]} dm_2(\hat{s}, \hat{t})$$

$$\stackrel{(19)}{=} \left(\int_{\mathbb{R}} e^{-\frac{1}{2}\left(\hat{s} - \frac{1}{2}((s+\tilde{s}) + i(t+\tilde{t}))\right)^2} dm(\hat{s})\right) \left(\int_{\mathbb{R}} e^{-\frac{1}{2}\left(\hat{t} - \frac{1}{2}((t+\tilde{t}) - i(s+\tilde{s}))\right)^2} dm(\hat{t})\right)$$

$$\stackrel{(20)}{=} 2\pi$$

(18 holds true because $(a + ib)^2 + (b - ia)^2 = 0$ for all $a, b \in \mathbb{R}$; 19 holds true by 20.2.1, 8.4.9, 8.4.10c; 20 follows from 20.2.1).

Therefore we have

$$(f|TW(s,t)Tg) = 2\pi\gamma(s,t) \int_{\mathbb{R}^2} \gamma(\tilde{s}, \tilde{t}) \left(f|W(\tilde{s}, \tilde{t})g\right) dm_2(\tilde{s}, \tilde{t}) = 2\pi\gamma(s,t) (f|Tg).$$

Since f and g were arbitrary elements of $\mathcal{H}$ and (s, t) was an arbitrary element of $\mathbb{R}^2$, this proves that

$$TW(s,t)T = 2\pi\gamma(s,t)T, \forall(s,t) \in \mathbb{R}^2.$$

c: If we set $s := t := 0$ in statement b, we obtain

$$T^2 = T.$$

Since T is a self-adjoint element of $\mathcal{B}(\mathcal{H})$ (cf. statement a), this proves that T is an orthogonal projection (cf. 13.1.5).

For all $(s_1, t_1), (s_2, t_2) \in \mathbb{R}^2$ and all $f, g \in R_T$, we have

$$(W(s_1, t_1)f|W(s_2, t_2)g) \stackrel{(21)}{=} (W(s_1, t_1)Tf|W(s_2, t_2)Tg)$$

$$\stackrel{(22)}{=} (f|TW(-s_1, -t_1)W(s_2, t_2)Tg)$$

$$\stackrel{(23)}{=} e^{\frac{1}{2}i(-s_1 t_2 + t_1 s_2)} (f|TW(-s_1 + s_2, -t_1 + t_2)Tg)$$

$$\stackrel{(24)}{=} e^{-\frac{1}{2}i(s_1 t_2 - t_1 s_2) - \frac{1}{4}(s_2 - s_1)^2 - \frac{1}{4}(t_2 - t_1)^2} (f|g),$$

where 21 holds true by 13.1.3c, 22 by 20.2.2b, 23 by 20.2.2a, 24 by statement b and 13.1.3c.

d: For all $g \in M$ we have, in view of 20.2.2.c,

$$(f|Tg) = \int_{\mathbb{R}^2} \gamma(s,t) (f|W(s,t)g) dm_2(s,t) = 0, \forall f \in M^{\perp},$$

and hence $Tg \in M^{\perp\perp}$. Since $M = M^{\perp\perp}$ (cf. 10.4.4a), this proves that M is an invariant subspace for T, and hence that T is reduced by M (cf. 17.2.9).

The operator $W(s,t)$ is reduced by M, for all $(s,t) \in \mathbb{R}^2$ (cf. 20.2.2c). Now we recall that

$$U^{A^M}(t) = (U^A(t))^M \text{ and } U^{B^M}(t) = (U^B(t))^M, \forall t \in \mathbb{R}$$

(cf. 17.2.13). Then,

$$(W(s,t))^M = e^{-i\frac{1}{2}st} U^{B^M}(s) U^{A^M}(t), \forall (s,t) \in \mathbb{R}^2,$$

and hence

$$\left(h|T^M g\right)_M = (h|Tg)_{\mathcal{H}} = \int_{\mathbb{R}^2} \gamma(s,t) \, (h|W(s,t)g)_{\mathcal{H}} \, dm_2(s,t)$$

$$= \int_{\mathbb{R}^2} \gamma(s,t) \, \left(h|(W(s,t))^M g\right)_M dm_2(s,t) = \left(h|T^{(M)} g\right)_M, \forall h,g \in M,$$

and hence $T^M = T^{(M)}$ (we have denoted by a subscript whether a given inner product is to be regarded as pertaining to the Hilbert space $\mathcal{H}$ or to the Hilbert space M).

e: We prove statement e by contraposition. We assume that there exists a non trivial subspace M of $\mathcal{H}$ so that A and B are reduced by M. Then A and B are reduced by the non-trivial subspace $M^\perp$ as well. Then T^M and $T^{M^\perp}$ are non-null orthogonal projections in the Hilbert spaces M and $M^\perp$ respectively (cf. statements a, c, d). In view of 13.1.3c, there exist two normalized vectors u_1, u_2 so that:

$$u_1 \in R_{T^M}, \text{ and hence } Tu_1 = T^M u_1 = u_1;$$
$$u_2 \in R_{T^{M^\perp}}, \text{ and hence } Tu_2 = T^{M^\perp} u_2 = u_2.$$

Therefore, $\{u_1, u_2\}$ is an o.n.s. contained in R_T (cf. 13.1.3c). Hence R_T cannot be a one dimensional subspace (cf. e.g. 10.7.3).   □

20.2.4 Theorem (The Stone–von Neumann uniqueness theorem).

(a) *Two irreducible representations of WCR are unitarily equivalent:*
   Let A, B be an irreducible representation of WCR in a Hilbert space $\mathcal{H}$, and let $\tilde{A}, \tilde{B}$ be an irreducible representation of WCR in a Hilbert space $\tilde{\mathcal{H}}$. Then there exists $V \in \mathcal{U}(\mathcal{H}, \tilde{\mathcal{H}})$ so that

$$U^{\tilde{A}}(t) = V U^A(t) V^{-1} \text{ and } U^{\tilde{B}}(t) = V U^B(t) V^{-1}, \quad \forall t \in \mathbb{R},$$

   or equivalently

$$\tilde{A} = V A V^{-1} \text{ and } \tilde{B} = V B V^{-1}.$$

(b) *The Schrödinger representation of WCR is irreducible.*

(c) *Let A, B be a representation of WCR in a separable Hilbert space $\mathcal{K}$. Then there exists a family $\{M_n\}_{n\in I}$, with $I := \{1, ..., N\}$ or $I := \mathbb{N}$, of subspaces of $\mathcal{K}$ so that:*

   $M_k \subset M_i^\perp$ if $i \neq k$;
   $\sum_{n\in I}^{\oplus} M_n = \mathcal{K}$ (for $\sum_{n\in I}^{\oplus} M_n$, cf. 13.2.10f);
   for each $n \in I$, A and B are reduced by M_n and the pair of self-adjoint operators A^{M_n} and B^{M_n} is jointly irreducible (hence, the pair A^{M_n}, B^{M_n} is an irreducible representation of WCR in the Hilbert space M_n, cf. 20.1.6).

Proof. a: Let $W(s,t)$ and $\tilde{W}(s,t)$ be defined as in 20.2.2 for all $(s,t) \in \mathbb{R}^2$, with respect to the pair A, B and the pair $\tilde{A}, \tilde{B}$ respectively. Moreover, let T and $\tilde{T}$ be defined as in 20.2.3, with respect to A, B and $\tilde{A}, \tilde{B}$ respectively. Since the projections T and $\tilde{T}$ are non-zero (cf. 20.2.3a,c), we can fix two normalized vectors $u \in R_T$ and $\tilde{u} \in R_{\tilde{T}}$. Since the operators A and B are reduced by the subspace

$$M_u := V\{W(s,t)u : (s,t) \in \mathbb{R}^2\}$$

(cf. 20.2.2d) and since M_u cannot be $\{0_{\mathcal{H}}\}$ (because $W(0,0) = \mathbb{1}_{\mathcal{H}}$), the equality $M_u = \mathcal{H}$ must be true. Similarly, $M_{\tilde{u}} = \tilde{\mathcal{H}}$.

For all $L \in \mathbb{N}$, all $(\alpha_1, ..., \alpha_L) \in \mathbb{C}^L$, all $(s_1, t_1, ..., s_L, t_L) \in \mathbb{R}^{2L}$, we have

$$\left\| \sum_{l=1}^{L} \alpha_l W(s_l, t_l) u \right\|_{\mathcal{H}}^2 = \sum_{h,l=1}^{L} \overline{\alpha}_h \alpha_l \left(W(s_h, t_h)u | W(s_l, t_l)u \right)_{\mathcal{H}}$$

$$\overset{(1)}{=} \sum_{h,l=1}^{L} \overline{\alpha}_h \alpha_l \left(\tilde{W}(s_h, t_h)\tilde{u} | \tilde{W}(s_l, t_l)\tilde{u} \right)_{\tilde{\mathcal{H}}}$$

$$= \left\| \sum_{l=1}^{L} \alpha_l \tilde{W}(s_l, t_l)\tilde{u} \right\|_{\tilde{\mathcal{H}}}^2,$$

where 1 holds true in view of 20.2.3c. Then we have, for all $N, M \in \mathbb{N}$, all $(\beta_1, ..., \beta_N) \in \mathbb{C}^N$, all $(\gamma_1, ..., \gamma_M) \in \mathbb{C}^M$, all $(s_1, t_1, ..., s_N, t_N) \in \mathbb{R}^{2N}$, all $(x_1, y_1, ..., x_M, y_M) \in \mathbb{R}^{2M}$,

$$\sum_{n=1}^{N} \beta_n W(s_n, t_n)u = \sum_{m=1}^{M} \gamma_m W(x_m, y_m)u \Rightarrow$$

$$\left\| \sum_{n=1}^{N} \beta_n W(s_n, t_n)u - \sum_{m=1}^{M} \gamma_m W(x_m, y_m)u \right\|_{\mathcal{H}} = 0 \Rightarrow$$

$$\left\| \sum_{n=1}^{N} \beta_n \tilde{W}(s_n, t_n)\tilde{u} - \sum_{m=1}^{M} \gamma_m \tilde{W}(x_m, y_m)\tilde{u} \right\|_{\tilde{\mathcal{H}}} = 0 \Rightarrow$$

$$\sum_{n=1}^{N} \beta_n \tilde{W}(s_n, t_n)\tilde{u} = \sum_{m=1}^{M} \gamma_m \tilde{W}(x_m, y_m)\tilde{u}.$$

Therefore we can define a mapping

$$V_0 : L\{W(s,t)u : (s,t) \in \mathbb{R}^2\} \to L\{\tilde{W}(s,t)\tilde{u} : (s,t) \in \mathbb{R}^2\}$$

by letting

$$V_0 \left(\sum_{n=1}^{N} \alpha_n W(s_n, t_n)u \right) := \sum_{n=1}^{N} \alpha_n \tilde{W}(s_n, t_n)\tilde{u},$$

$$\forall N \in \mathbb{N}, \forall (\alpha_1, ..., \alpha_N) \in \mathbb{C}^N, \forall (s_1, t_1, ..., s_N, t_N) \in \mathbb{R}^{2N}$$

(cf. 3.1.7). It is obvious that V_0 is a linear operator from $\mathcal{H}$ to $\tilde{\mathcal{H}}$ and that

$$R_{V_0} = L\{\tilde{W}(s,t)\tilde{u} : (s,t) \in \mathbb{R}^2\}.$$

In view of 4.1.13 and 4.6.6, there exists $V \in \mathcal{U}(\mathcal{H}, \tilde{\mathcal{H}})$ such that
$$V_0 \subset V \text{ and } V^{-1}(\tilde{W}(s,t)\tilde{u}) = W(s,t)u, \forall (s,t) \in \mathbb{R}^2.$$

Then we have, for all $(s', t') \in \mathbb{R}^2$,
$$
\begin{aligned}
VW(s',t')V^{-1}(\tilde{W}(s,t)\tilde{u}) &= VW(s',t')W(s,t)u \\
&\overset{(2)}{=} V\left(e^{i\frac{1}{2}(s't-t's)}W(s'+s,t'+t)u\right) \\
&= e^{i\frac{1}{2}(s't-t's)}\tilde{W}(s'+s,t'+t)\tilde{u} \\
&\overset{(3)}{=} \tilde{W}(s',t')(\tilde{W}(s,t)\tilde{u}), \forall (s,t) \in \mathbb{R}^2
\end{aligned}
$$

(2 and 3 hold true in view of 20.2.2a), and hence by linearity
$$VW(s',t')V^{-1}\tilde{f} = \tilde{W}(s',t')\tilde{f}, \forall \tilde{f} \in L\{\tilde{W}(s,t)\tilde{u} : (s,t) \in \mathbb{R}^2\},$$

and hence
$$VW(s',t')V^{-1} = \tilde{W}(s',t')$$

in view of 4.2.6. Therefore we have
$$VU^A(t)V^{-1} = VW(0,t)V^{-1} = \tilde{W}(0,t) = U^{\tilde{A}}(t), \forall t \in \mathbb{R},$$

and
$$VU^B(s)V^{-1} = VW(s,0)V^{-1} = \tilde{W}(s,0) = U^{\tilde{B}}(s), \forall s \in \mathbb{R}.$$

By 16.3.1, these conditions are equivalent to
$$VAV^{-1} = \tilde{A} \text{ and } VBV^{-1} = \tilde{B}.$$

b: In view of 20.2.3e, we prove statement b by proving that the orthogonal projection T defined as in 20.2.3 with $A := Q$ and $B := P$ (where Q and P are the operators discussed in 20.1.7) is so that the orthogonal dimension of the subspace R_T is one. In what follows, for simplicity we do not distinguish between the symbol f for an element of $\mathcal{L}^2(\mathbb{R})$ and the symbol $[f]$ for the element of $L^2(\mathbb{R})$ that contains f.

For all $(s, t) \in \mathbb{R}^2$ and all $g \in \mathcal{L}^2(\mathbb{R})$ we have
$$(W(s,t)g)(x) = e^{-\frac{1}{2}ist}(g^t)_{-s}(x) = e^{it(x+\frac{1}{2}s)}g(x+s), \forall x \in D_g - s.$$

Now we fix $f, g \in \mathcal{L}^2(\mathbb{R})$, and suppose that the representative $g \in \mathcal{L}^2(\mathbb{R})$ is such that $D_g = \mathbb{R}$ (cf. 8.2.12). We have
$$\int_{\mathbb{R}} |\bar{f}(x)(W(s,t)g)(x)|dm(x) \le \|f\|\|W(s,t)g\| = \|f\|\|g\|, \forall (s,t) \in \mathbb{R}^2,$$

by the Schwarz inequality (cf. 10.1.9) for the elements $|f|$ and $|W(s,t)g|$ of $L^2(\mathbb{R})$, and hence
$$
\int_{\mathbb{R}^2} \gamma(s,t)\left(\int_{\mathbb{R}} |\bar{f}(x)(W(s,t)g)(x)|dm(x)\right)dm_2(s,t)
$$
$$
\le \|f\|\|g\|\int_{\mathbb{R}^2}\gamma(s,t)dm_2(s,t) < \infty, \forall (s,t) \in \mathbb{R}^2.
$$

Then, by Tonelli's theorem (cf. 8.4.8) followed by Fubini's theorem (cf. 8.4.10c with $\mu_1 := m_2$ and $\mu_2 := m$) we have

$$(f|Tg) = \int_{\mathbb{R}^2} \gamma(s,t) \left(\int_{\mathbb{R}} \overline{f}(x)(W(s,t)g)(x)dm(x) \right) dm_2(s,t)$$

$$= \int_{\mathbb{R}} \overline{f}(x) \left(\int_{\mathbb{R}^2} \gamma(s,t)(W(s,t)g)(x)dm_2(s,t) \right) dm(x).$$

Moreover we have, for all $(s,t) \in \mathbb{R}^2$ and all $x \in \mathbb{R}$,

$$\int_{\mathbb{R}^2} \gamma(s,t)(W(s,t)g)(x)dm_2(s,t)$$

$$= (2\pi)^{-1} \int_{\mathbb{R}^2} e^{-\frac{1}{4}(s^2+t^2)} e^{it\left(x+\frac{1}{2}s\right)} g(x+s)dm_2(s,t)$$

$$\overset{(4)}{=} (2\pi)^{-1} \int_{\mathbb{R}} e^{-\frac{1}{4}s^2} g(x+s) \left(\int_{\mathbb{R}} e^{-\frac{1}{4}t^2} e^{it\left(x+\frac{1}{2}s\right)} dm(t) \right) dm(s)$$

$$\overset{(5)}{=} \pi^{-\frac{1}{2}} \int_{\mathbb{R}} e^{-\frac{1}{4}s^2} g(x+s) e^{-\left(x+\frac{1}{2}s\right)^2} dm(s)$$

$$\overset{(6)}{=} \pi^{-\frac{1}{2}} e^{-\frac{1}{2}x^2} \int_{\mathbb{R}} e^{-\frac{1}{2}y^2} g(y)dm(y),$$

where:

4 holds true by 8.4.9 and 8.4.10c since, for all $x \in \mathbb{R}$,

$$\left| e^{-\frac{1}{4}(s^2+t^2)} e^{it\left(x+\frac{1}{2}s\right)} g(x+s) \right| = e^{-\frac{1}{4}t^2} e^{-\frac{1}{4}s^2} g_{-x}(s), \forall(s,t) \in \mathbb{R}^2,$$

and the function $t \mapsto e^{-\frac{1}{4}t^2}$ is an element of $\mathcal{L}^1(\mathbb{R})$ (cf. 11.4.7), and so is the function $s \mapsto e^{-\frac{1}{4}s^2} g_{-x}(s)$ (cf. 11.1.2b);

5 holds true by 11.4.8 with $a := \frac{1}{2}$;

6 follows from the change of variable $y := s + x$, by 9.2.1.b.

Now we define an element u of $L^2(\mathbb{R})$ by letting

$$u(x) := \pi^{-\frac{1}{4}} e^{-\frac{1}{2}x^2}, \forall x \in \mathbb{R};$$

we have $\|u\| = 1$ from 11.4.7. From the results obtained above, we have

$$(f|Tg) = \int_{\mathbb{R}} \overline{f}(x)u(x)\,(u|g)\,dm(x) = (u|g)\,(f|u) = (f|A_u g)\,,$$

where A_u is the one-dimensional projection defined in 13.1.12. Since f and g were arbitrary elements of $L^2(\mathbb{R})$, this proves that $T = A_u$ and hence that the orthogonal dimension of the subspace R_T is one.

c: Let T be defined by A, B as in 20.2.3. Since $T \neq \mathbb{O}_\mathcal{K}$ (cf. 20.2.3a) and $\mathcal{K}$ is separable, there exists an o.n.s. $\{u_n\}_{n \in I}$, with $I := \{1, ..., N\}$ or $I := \mathbb{N}$, which is complete in the subspace R_T (cf. 10.7.2). For each $n \in I$, we define the subspace

$$M_n := M_{u_n}$$

(cf. 20.2.2d). If $i \neq k$, we have (in view of 20.2.3c)

$$(W(s_1,t_1)u_i|W(s_2,t_2)u_k) = 0, \forall(s_1,t_1), (s_2,t_2) \in \mathbb{R}^2,$$

or

$$\{W(s,t)u_i : (s,t) \in \mathbb{R}^2\} \subset \{W(s,t)u_k : (s,t) \in \mathbb{R}^2\}^\perp,$$

and hence

$$M_k = \{W(s,t)u_k : (s,t) \in \mathbb{R}^2\}^{\perp\perp} \subset \{W(s,t)u_i : (s,t) \in \mathbb{R}^2\}^\perp = M_i^\perp$$

(cf. 10.4.4b, 10.2.10b, 10.2.11). For each $n \in I$, M_n is a reducing subspace for A and B (cf. 20.2.2d) and hence for the operator $W(s,t)$ as well, for all $(s,t) \in \mathbb{R}^2$ (cf. 20.2.2c; the operator $W(s,t)$ is defined by A, B as in 20.2.2). Then it is obvious that the subspace $\sum_{n\in I}^{\oplus} M_n$ is an invariant subspace for the operator $W(s,t)$, for all $(s,t) \in \mathbb{R}^2$. In view of 17.2.9 and 20.2.2b, this implies that $\sum_{n\in I}^{\oplus} M_n$ is a reducing subspace for $W(s,t)$ for all $(s,t) \in \mathbb{R}^2$, and hence also for A and B (cf. 20.2.2c). Therefore, the subspace

$$M_0 := \left(\sum_{n\in I}^{\oplus} M_n\right)^\perp$$

is a reducing subspace for A and B (cf. 17.2.4). Now we prove by contradiction that $M_0 = \{0_\mathcal{K}\}$. Suppose to the contrary $M_0 \neq \{0_\mathcal{K}\}$. Then the restrictions of A and B to M_0 are a representation of WCR (cf. 20.1.6) and they define a non-zero orthogonal projection $T^{(M_0)}$ in the Hilbert space M_0 which is the same as the restriction of T to M_0 (cf. 20.2.3d). Then there exists a non-zero vector $u_0 \in R_{T^{(M_0)}}$. In view of 13.1.3c, we have

$$u_0 = T^{(M_0)}u_0 = Tu_0,$$

and hence $u_0 \in R_T$. Since $u_0 \in \left(\sum_{n\in I}^{\oplus} M_n\right)^\perp$ implies that

$$(u_0|u_n) = 0, \forall n \in I$$

(note that $u_n = W(0,0)u_n \in M_n$ for all $n \in I$), we have a contradiction with the fact that $\{u_n\}_{n\in I}$ is a c.o.n.s. in R_T (cf. 10.6.4). This proves that $M_0 = \{0_\mathcal{K}\}$ and hence that

$$\sum_{n\in I}^{\oplus} M_n = \mathcal{K}.$$

Now we prove that, for each $n \in I$, the pair of self-adjoint operators A^{M_n}, B^{M_n} is jointly irreducible by proving that the projection $T^{(M_n)}$ is one-dimensional (cf. 20.2.3e). Indeed, $T^{(M_n)} = T^{M_n}$ (cf. 20.2.3d) and hence

$$T^{(M_n)}f = Tf = \sum_{k\in I}(u_k|f)_\mathcal{H}\, u_k = (u_n|f)_\mathcal{H}\, u_n = (u_n|f)_{M_n}\, u_n, \forall f \in M_n$$

(the second equality follows from 13.1.10), and this proves that $T^{(M_n)}$ is a one-dimensional projection in the Hilbert space M_n (cf. 13.1.12). $\qquad\square$

20.2.5 Remarks.

(a) We stated and proved part c of 20.2.4 even if we do not use it in this book, because we thought it better to reproduce the whole content of von Neumann's article. The other parts of the Stone–von Neumann theorem play an essential role in Section 20.3.

(b) For any irreducible representation of WCR, the Hilbert space $\mathcal{H}$ in which it is defined is separable and of denumerable orthogonal dimension. Indeed statements a and b in 20.2.4 imply that $\mathcal{H}$ and $L^2(\mathbb{R})$ are isomorphic; moreover the Hilbert space $L^2(\mathbb{R})$ is separable (cf. 11.3.4) and of denumerable orthogonal dimension (cf. 11.3.3); then so is $\mathcal{H}$, in view of 10.7.14.

20.3 Position and momentum as Galilei-covariant observables

In the first part of this section we try to explain briefly how relativistic ideas can be implemented in quantum mechanics. Even though we refer our discussion directly to the Galilei group, the experienced reader will notice that what we say can be easily adapted to the discussion of any group of space-time transformations which is considered a symmetry group (in the so-called passive approach). In the second part of the section, a mathematical model for a non-relativistic quantum particle is discussed. Some of the ideas we use are drawn from (Holevo, 1982), (Jauch, 1968) and (Mackey, 1978).

The nature of the discussion in this section makes it unsuitable to mark every bit of it with three numbers, as customary in the rest of the book. However, the purely mathematical propositions are still marked in that way.

According to both Galilean relativity and Einstein's special relativity, each observer who describes physical reality must use a frame of reference which consists of a spatial coordinate system and a method for measuring time. In both these relativity theories, there is a special class of observers, who are called inertial observers (we assume that the reader is already familiar with these ideas). In both theories, *the principle of relativity is assumed* which says that the laws of physics are the same for all inertial observers. For any pair of inertial observers O and O' there exists one and only one element g of a group of transformations of $\mathbb{R}^4$ (which is the Galilei group in Galilean relativity and the inhomogeneous Lorentz group in Einstein's special relativity) so that, if $(\boldsymbol{x}, x_0)$ are the space-time coordinates of a space-time point according to the frame of reference used by O, then $g(\boldsymbol{x}, x_0)$ are the coordinates of that point according to the frame of reference used by O'. *It is assumed that*, given the observer O, there is just one observer O' for whom this is true, and we denote this observer by $g(O)$. Each inertial observer has his (or her, this alternative is understood in all that follows) own representation of physical reality, which depends at least partially on the representation of space-time points that is given by his frame of reference. In this book we consider only Galilean relativity (we know that we can use Galilean relativity when all relevant speeds are very small compared with the speed of light). Since in what follows we only want to relate the representations of physical reality given by different inertial observer, without discussing time evolution (in fact we will only study the kinematic aspects of a quantum particle, without discussing the dynamic ones), we need only consider the subgroup G of the Galilei group which contains the transformations of $\mathbb{R}^4$ of the form

$$\mathbb{R}^4 \ni (\boldsymbol{x}, x_0) \mapsto g_{(\boldsymbol{R}, \boldsymbol{s}, \boldsymbol{v})} := (\boldsymbol{R}\boldsymbol{x} + \boldsymbol{s} + x_0\boldsymbol{v}, x_0) \in \mathbb{R}^4,$$

where $\boldsymbol{R}$ is a rotation in $\mathbb{R}^3$ and $\boldsymbol{s}, \boldsymbol{v}$ two elements of $\mathbb{R}^3$, and where the group product is the composition of mappings defined in 1.2.12. The subgroup G is called the *kinematic Galilei group* (the full Galilei group contains time translations in addition).

We *assume* that, for every inertial observer O and every $g \in G$, there exists some kind of dictionary which makes it possible to translate the representation of physical reality given by O into the representation given by $g(O)$. Moreover, we *assume* that this dictionary depends only on g and not on the particular inertial observer O whose representation we want to translate. In what follows we suppose that, for all inertial observers, the representation of a quantum system is along the lines of Section 19.3. In particular, the Hilbert space by which a quantum system is represented is never one-dimensional (cf. 19.3.5a). However, *we do not assume that the Hilbert space by which we represent a non-relativistic quantum particle is separable (in fact, this will be deduced)*. Indeed, the separability of the Hilbert space by which a quantum system was represented in Section 19.3 was necessary only in the mathematical discussion of mixed states, since statistical operators had been studied in Chapter 18 in the context of separable Hilbert spaces; had the Hilbert space been non-separable, mixed states should have been discussed in a mathematically different way. In what follows we use only pure states, which we still assume represented by rays as in Section 19.3; moreover, we still assume that there is a bijective mapping from the family of all pure states onto the family of all the rays of the Hilbert space by which we represent a non-relativistic quantum particle.

In what follows we consider a fixed quantum system. We *assume* that all inertial observers represent the system by the same Hilbert space $\mathcal{H}$ (the principle of relativity requires that all inertial observers give equivalent representations of physical reality, and hence it would lead only to assume that any two inertial observers use isomorphic Hilbert spaces; to avoid cumbersome notation we assume that all these Hilbert spaces are actually the same).

We recall that a state preparation is a collection of instructions for operations to be performed on macroscopic bodies. Since every macroscopic body is related directly to each frame of reference, each inertial observer describes the instructions for any given state in his own way, with respect to his own frame of reference. We *assume* that, for an inertial observer O and each $g \in G$, if O represents a given pure state σ by the ray $[u_\sigma]$ then $g(O)$ will represent the same pure state σ by a ray $[u_\sigma^g]$ which will not be in general the same as $[u_\sigma]$, and that there exists a bijective mapping ω_g from the projective Hilbert space $\hat{\mathcal{H}}$ onto itself so that

$$[u_\sigma^g] = \omega_g([u_\sigma])$$

for each pure state σ (this is consistent with the principle of relativity, which requires that all inertial observers give equivalent descriptions of physical reality; the totality of pure states for one observer must be the totality of pure states for another observer). We point out that ω_g must depend on g but not on O, in accordance with the assumption made above that the dictionary from O to $g(O)$ depends on g but not on O. Moreover, for every pair of pure states σ_1 and σ_2, the transition probability from σ_1 to σ_2 must be the same whether it is computed by O or by $g(O)$ (the principle of relativity implies that the statistics of all experiments must be the same for all inertial observers). Therefore, we *assume* that

$$\tau(\omega_g([u_{\sigma_1}]), \omega_g([u_{\sigma_2}])) = \tau([u_{\sigma_1}], [u_{\sigma_2}])$$

for every pair of pure states σ_1, σ_2 (τ is the function defined in 10.9.1). In view of the bijection existing from the family of all pure states onto the projective Hilbert space $\hat{\mathcal{H}}$ (cf. 19.3.5c), ω_g turns out to be an automorphism of the projective Hilbert space $(\hat{\mathcal{H}}, \tau)$. Then, by Wigner's theorem (cf. 10.9.6), there exists an implementation of ω_g, i.e. $U_g \in \mathcal{UA}(\mathcal{H})$ so that

$$\omega_g([u]) = [U_g u], \forall u \in \tilde{\mathcal{H}},$$

and U_g is unique up to a multiplicative factor in $\mathbb{T}$. After this, we consider an inertial observer O and two elements g_1, g_2 of G. The observer $(g_1 g_2)(O)$ is the same as the observer $g_1(g_2(O))$, by the definition we gave above of the inertial observer $g(O)$ and by the definition of the group product in G. Then consistency requires that the translation of the description of physical reality given by O into the description given by $(g_1 g_2)(O)$ is the same as the translation of the description given by O into the description given by $g_2(O)$ followed by the translation of the description given by $g_2(O)$ into the description given by $g_1(g_2(O))$. Since ω_g does not depend on O, this implies that

$$\omega_{g_1} \circ \omega_{g_2} = \omega_{g_1 g_2}.$$

Thus, the mapping

$$G \ni g \mapsto \omega_g \in \operatorname{Aut} \hat{\mathcal{H}}$$

is a homomorphism from the group G to the group $\operatorname{Aut} \hat{\mathcal{H}}$ (cf. 10.9.4). Finally, the group G can be given the structure of a metric space in an obvious way through an identification of G with a subset of $\mathbb{R}^n$ for a suitable integer n, in such a way that the group operations "product" and "inverse" are continuous. Then we *assume* the following *continuity condition*:

$$\lim_{n \to \infty} \tau([u_\sigma], \omega_{g_n}([u_\sigma])) = \tau([u_\sigma], [u_\sigma]) \, (= 1),$$

for each pure state σ and for each sequence $\{g_n\}$ in G which converges to the identity of G. When interpreted in terms of transitions probabilities, this assumption follows from the idea that the difference between the description of physical reality given by an inertial observer O and the description given by the observer $g(O)$ becomes negligible when g is close enough to the identity of G.

We recall that a proposition is an event which does or does not occur in a macroscopic device. Therefore each inertial observer describes this event in his own way, with respect to his own frame of reference. We *assume* that, for an inertial observer O and each $g \in G$, if O represents a given proposition π by an orthogonal projection P_π in $\mathcal{H}$ then the inertial observer $g(O)$ will represent the same proposition π by an orthogonal projection P_π^g which will not be in general the same as P_π, while $g(O)$ will represent by the same projection P_π the proposition that he describes (with respect to his own frame of reference) in the same way as O describes π (with respect to O's own frame of reference). The relation between P_π and P_π^g follows from the relation obtained above between the representations of pure states given by O and by $g(O)$. In fact, the principle of relativity implies that the probability $p(\pi, \sigma)$ is the same for O and $g(O)$, for all pure states σ. This implies that

$$(u|P_\pi u) = (U_g u|P_\pi^g U_g u) = \left(u|U_g^{-1} P_\pi^g U_g u\right), \forall u \in \tilde{\mathcal{H}},$$

and hence (cf. 10.2.12) $P_\pi = U_g^{-1} P_\pi^g U_g$, or

$$P_\pi^g = U_g P_\pi U_g^{-1},$$

where U_g is an element of $\mathcal{UA}(\mathcal{H})$ which is an implementation of ω_g as implied by Wigner's theorem. Thus, the mapping

$$\tilde{\omega}_g : \mathscr{P}(\mathcal{H}) \to \mathscr{P}(\mathcal{H})$$
$$P \mapsto \tilde{\omega}_g(P) := U_g P U_g^{-1}$$

(this definition is consistent in view of 13.1.8) is so that

$$P_\pi^g = \tilde{\omega}_g(P_\pi)$$

for each proposition π. We point out that $\tilde{\omega}_g$ depends on ω_g and not on the particular element U_g of $\mathcal{UA}(\mathcal{H})$ (among those which implement ω_g) that has been used to define $\tilde{\omega}_g$, because in $U_g P U_g^{-1}$ an arbitrary multiplicative factor in front of U_g is immaterial.

We consider an X-valued observable α, where $(X, \mathcal{A})$ is a measurable space. The equivalence of the descriptions of physical reality given by all inertial observers, embodied in the principle of relativity, accounts for the *assumption* that all inertial observers represent the dial of the measuring instrument that underlies α by the same measurable space $(X, \mathcal{A})$ (cf. 19.1.9a,b). Since the pointer and the dial are macroscopic objects, the position of the pointer on the dial is described by each inertial observer by means of his own frame of reference. Therefore, if an inertial observer O represents a position of the pointer on the dial by a point x of X then the inertial observer $g(O)$ (for any $g \in G$) will represent the same position by a point x_g of X which will not be in general the same as x. We *assume* that, for each $g \in G$, there exists a bijective measurable mapping t_g^α from X onto itself so that

$$x_g = t_g^\alpha(x), \forall x \in X.$$

As before for ω_g, we can establish that t_g^α does not depend on O and that

$$t_{g_1}^\alpha \circ t_{g_2}^\alpha = t_{g_1 g_2}^\alpha, \forall g_1, g_2 \in G.$$

We observe that, for $E \in \mathcal{A}$, the symbol $\alpha(E)$ denotes different propositions when it is used by different inertial observers, since it denotes the proposition determined by the event "the pointer of the measuring instrument is in the section of the dial represented by E", but which section of the dial is represented by E depends on the observer. The proposition denoted as $\alpha(E)$ by an inertial observer O is in fact the proposition denoted as $\alpha(t_g^\alpha(E))$ by the observer $g(O)$, for all $g \in G$. However, O and $g(O)$ represent the X-valued observable α by the same projection valued measure

$$P_\alpha : \mathcal{A} \to \mathscr{P}(\mathcal{H})$$
$$E \mapsto P_\alpha(E) := P_{\alpha(E)}.$$

Indeed we assumed above that, if O represents a proposition π by a projection P_π, then $g(O)$ represents by the same projection P_π the proposition (in general different from π) that is described by $g(O)$ as π is described by O. Now we fix $E \in \mathcal{A}$ and consider the proposition that is denoted as $\alpha(E)$ by O. The representation of this proposition given by O is $P_{\alpha(E)}$. According to what we saw above, the representation of this proposition given by $g(O)$ must be

$$\tilde{\omega}_g(P_{\alpha(E)}),$$

and it must be

$$P_{\alpha(t_g^\alpha(E))}$$

as well. Thus, consistency requires that

$$U_g P_{\alpha(E)} U_g^{-1} = \tilde{\omega}_g(P_{\alpha(E)}) = P_{\alpha(t_g^\alpha(E))}, \forall E \in \mathcal{A}, \forall g \in G,$$

where U_g is an implementation of g. This condition is called a *Galilei-covariance* relation and the X-valued observable α is said to be *Galilei-covariant*. If the relation above is written for a subgroup G_0 of the kinematic Galilei group G, then it is called a covariance relation with respect to G_0.

The case may be that, for a Galilei-covariant X-valued observable α, the mapping t_g^α is the identity mapping of X for all $g \in G$. This means that the representations of the positions of the pointer in the dial are the same for all inertial observers. Then the X-valued observable α is said to be *Galilei-invariant* and this case of covariance condition is called a *Galilei-invariance* condition.

We remark that, if α is an observable (i.e., an $\mathbb{R}$-observable), then the covariance condition can be written as

$$U_g P^{A_\alpha}(E) U_g^{-1} = P^{A_\alpha}(t_g^\alpha(E)), \forall E \in \mathcal{A}(d_\mathbb{R}), \forall g \in G,$$

and the invariance condition as

$$[U_g, P^{A_\alpha}(E)] = \mathbb{O}_{\mathcal{H}}, \forall E \in \mathcal{A}(d_{\mathbb{R}}), \forall g \in G.$$

An observable α is said to be *trivial* if the self-adjoint operator A_α which represents α is a multiple of the identity operator. The reason for this name is clear from 17.3.2: if an observable is trivial then in its range there are only the trivial propositions π_0 and π_1 (cf. 19.1.7 and 19.3.4). If an observable is trivial then it is Galilei-invariant; this follows immediately from 17.3.2.

Now we want to study the observables position and linear momentum for a non-relativistic quantum particle, i.e. for a quantum particle in the framework of Galilean relativity. To keep the discussion as simple as possible, we limit our analysis to one-dimensional space.

For all $(s, v) \in \mathbb{R}^2$, we define the mapping

$$g_{(s,v)} : \mathbb{R}^2 \to \mathbb{R}^2$$

$$(x_1, x_0) \mapsto g_{(s,v)}(x_1, x_0) := (x_1 + s + x_0 v, x_0).$$

This mapping is interpreted as the transformation of the space-time coordinates (x_1, x_0) of a space-time point according to an inertial observer O, into the coordinates of the same point according to the inertial observer $g_{(s,v)}(O)$, assuming that: the two observers use the same units for measuring space and time, their clocks are synchronized, at time $x_0 = 0$ the space-origin of the frame of reference of $g_{(s,v)}(O)$ has space-coordinate $-s$ according to O, the frame of reference of $g_{(s,v)}$ moves with velocity $-v$ according to O. The set

$$G := \{g_{(s,v)} : (s, v) \in \mathbb{R}^2\},$$

with the composition of mappings as product, is the kinematic Galilei group we need to consider for Galilei-covariance or Galilei-invariance of observables. The mapping

$$\mathbb{R}^2 \ni (s, v) \mapsto g_{(s,v)} \in G$$

is an isomorphism from the additive group $\mathbb{R}^2$ onto the group G, and it is an isomorphism of metric spaces too (actually, the distance on G is defined exactly through this mapping). In what follows, the group $\mathbb{R}^2$ will often be substituted for the group G. The subsets

$$S := \{g_{(s,0)} : s \in \mathbb{R}\} \quad \text{and} \quad V := \{g_{(0,v)} : v \in \mathbb{R}\}$$

are subgroups of G and each of them is isomorphic to the additive group $\mathbb{R}$. The elements of S are called *space translations* and those of V are called *velocity transformations*. For a Galilei-covariant observable α it is enough to determine its covariance relations with respect to S and V since

$$g_{(s,v)} = g_{(s,0)}g_{(0,v)}, \forall(s, v) \in \mathbb{R}^2,$$

and hence

$$\omega_{(s,v)} = \omega_{(s,0)} \circ \omega_{(0,v)} \quad \text{and} \quad t^\alpha_{(s,v)} = t^\alpha_{(s,0)} \circ t^\alpha_{(0,v)}, \forall(s, v) \in \mathbb{R}^2.$$

We have set here, as we do in what follows,

$$\omega_{(s,v)} := \omega_{g_{(s,v)}} \text{ and } t^\alpha_{(s,v)} := t^\alpha_{g_{(s,v)}}, \forall (s,v) \in \mathbb{R}^2.$$

Any mathematical discussion requires a clear-cut definition of what is being discussed. We say that a *mathematical model for a non-relativistic quantum particle in one-dimension and without internal degrees of freedom (briefly, a quantum particle model)*, is a mathematical representation of a quantum system which has the following requisites:

(qp_1) the analysis carried out above for Galilean-relativistic quantum mechanics holds true for the system;

(qp_2) the only observables of the system which are Galilei-invariant are the trivial ones represented by multiples of the identity operator;

(qp_3) there are two observables of the system which can be interpreted as position and linear momentum of a particle.

In what follows we discuss the meaning of these requisites and unfold the mathematical structure they contain. An essential task will be to prove that a quantum particle model as above does exist. After this, its uniqueness will have to be investigated.

In 20.3.2 we examine the part of our quantum particle model that concerns the homomorphism

$$\mathbb{R}^2 \ni (s,v) \mapsto \omega_{(s,v)} \in \text{Aut } \hat{\mathcal{H}},$$

endowed with the continuity property discussed above, which is included in requisite qp_1, where $\mathcal{H}$ denotes the Hilbert space in which the system is represented (however, in the numbered propositions below, $\mathcal{H}$ denotes any Hilbert space). Before that, we need to prove a lemma in 20.3.1.

20.3.1 Lemma. *Suppose that a function $\varphi : \mathbb{R}^2 \to \mathbb{T}$ is such that:*

(a) $\varphi(x_1 + x_2, y) = \varphi(x_1, y)\varphi(x_2, y), \forall x_1, x_2 \in \mathbb{R}, \forall y \in \mathbb{R}$;
(b) the function $\mathbb{R} \ni x \mapsto \varphi(x, y) \in \mathbb{T}$ is continuous, $\forall y \in \mathbb{R}$;
(c) $\varphi(x, y_1 + y_2) = \varphi(x, y_1)\varphi(x, y_2), \forall y_1, y_2 \in \mathbb{R}, \forall x \in \mathbb{R}$;
(d) the function $\mathbb{R} \ni y \mapsto \varphi(x, y) \in \mathbb{T}$ is continuous, $\forall x \in \mathbb{R}$.

Then

$$\exists! \mu \in \mathbb{R} \text{ such that } \varphi(x,y) = e^{i\mu xy}, \forall (x,y) \in \mathbb{R}^2.$$

Proof. In view of 16.2.3, condition a and b imply that there exists a function

$$\mathbb{R} \ni y \mapsto a(y) \in \mathbb{R}$$

so that

$$\varphi(x,y) = e^{ia(y)x}, \forall x \in \mathbb{R}, \forall y \in \mathbb{R}. \tag{1}$$

Moreover, in view of 16.2.3, condition c and d imply that

$$\exists! \mu \in \mathbb{R} \text{ so that } \varphi(1, y) = e^{i\mu y}, \forall y \in \mathbb{R}. \tag{2}$$

From 1 and 2 we have

$$e^{ia(y)} = e^{i\mu y}, \forall y \in \mathbb{R},$$

and hence that there exists a function $\lambda : \mathbb{R} \to \mathbb{Z}$ so that

$$a(y) = \mu y + \lambda(y)2\pi, \forall y \in \mathbb{R}. \tag{3}$$

From 1 and 3 we have

$$e^{i\lambda(y)2\pi x} = e^{-i\mu yx}\varphi(x, y), \forall(x, y) \in \mathbb{R}^2. \tag{4}$$

From this and condition d we have that, for each $p \in \mathbb{N}$, the function

$$\mathbb{R} \ni y \mapsto \psi_p(y) := e^{i\lambda(y)2\pi p^{-1}} \in \mathbb{T}$$

is continuous. For each $p \in \mathbb{N}$, the range of ψ_p is a finite set since

$$R_{\psi_p} \subset \{e^{ir2\pi p^{-1}} : r = 0, ..., p - 1\};$$

therefore R_{ψ_p} contains just one number; indeed, if R_{ψ_p} contained more than one number then there would exist two non-empty closed subsets F_1 and F_2 in the metric subspace R_{ψ_p} of $\mathbb{C}$ such that

$$F_1 \cap F_2 = \emptyset \text{ and } F_1 \cup F_2 = R_{\psi_p}$$

(cf. 2.3.5), and hence R_{ψ_p} would not be connected (cf. 2.9.2), in contradiction with the continuity of ψ_p (cf. 2.9.10 and 2.9.4). Now we prove by contradiction that the function λ has just one value. Let $y_1, y_2 \in \mathbb{R}$ be such that $y_1 \neq y_2$. The equalities

$$\psi_p(y_1) = \psi_p(y_2), \forall p \in \mathbb{N},$$

imply that

$$\forall p \in \mathbb{N}, \exists n_p \in \mathbb{Z} \text{ so that } \lambda(y_1)2\pi p^{-1} - \lambda(y_2)2\pi p^{-1} = n_p 2\pi,$$

whence

$$\forall p \in \mathbb{N}, \exists n_p \in \mathbb{Z} \text{ so that } |\lambda(y_1) - \lambda(y_2)| = n_p p.$$

Therefore, $\lambda(y_1) \neq \lambda(y_2)$ would imply $n_p \neq 0$ for all $p \in \mathbb{N}$, and hence

$$|\lambda(y_1) - \lambda(y_2)| \geq p, \forall p \in \mathbb{N},$$

which is a contradiction. Thus, there exists $k \in \mathbb{Z}$ so that

$$\lambda(y) = k, \forall y \in \mathbb{R}, \tag{5}$$

and hence, in view of 3, so that

$$a(y) = \mu y + k2\pi, \forall y \in \mathbb{R}. \tag{6}$$

Now, condition c implies obviously that

$$\varphi(x, 0) = 1, \forall x \in \mathbb{R}.$$

This, together with 1 and 6, implies that

$$1 = e^{ia(0)x} = e^{ik2\pi x}, \forall x \in \mathbb{R},$$

and this implies $k = 0$ (e.g. by the uniqueness asserted in 16.2.3). This, together with 4 and 5, implies that

$$\varphi(x, y) = e^{i\mu yx}, \forall(x, y) \in \mathbb{R}^2.$$

$$\square$$

20.3.2 Proposition. *Let $\mathcal{H}$ be a Hilbert space which is neither a zero nor a one-dimensional linear space.*

For a mapping $\mathbb{R}^2 \ni (s,v) \mapsto \omega_{(s,v)} \in \operatorname{Aut} \hat{\mathcal{H}}$, the following conditions are equivalent:

(a) *the mapping $\mathbb{R}^2 \ni (s,v) \mapsto \omega_{(s,v)} \in \operatorname{Aut} \hat{\mathcal{H}}$ is a homomorphism from the additive group $\mathbb{R}^2$ to the group $\operatorname{Aut} \hat{\mathcal{H}}$ and the following implication holds true*

$$[(s,v) \in \mathbb{R}^2, \{(s_n, v_n)\} \text{ a sequence in } \mathbb{R}^2, \lim_{n \to \infty} (s_n, v_n) = (0,0)] \Rightarrow$$

$$[\lim_{n \to \infty} \tau([u], \omega_{(s_n, v_n)}([u])) = 1, \forall u \in \tilde{\mathcal{H}}];$$

(b) *there exist $\mu \in \mathbb{R}$ and two continuous one parameter unitary groups U^1_μ, U^2_μ in $\mathcal{H}$ such that*

$$U^2_\mu(v)U^1_\mu(s) = e^{i\mu sv} U^1_\mu(s)U^2_\mu(v), \forall (s,v) \in \mathbb{R}^2,$$

and

$$\omega_{(s,v)}([u]) = [U^1_\mu(s)U^2_\mu(v)u], \forall u \in \tilde{\mathcal{H}}, \forall (s,v) \in \mathbb{R}^2.$$

If these conditions are satisfied, then

(c) *the real number μ as in condition b is unique for a given homomorphism as in condition a.*

Proof. $a \Rightarrow (b$ and c): We assume condition a. Then the mapping

$$\mathbb{R} \ni s \mapsto \omega_{(s,0)} \in \operatorname{Aut} \hat{\mathcal{H}}$$

is a continuous one-parameter group of automorphisms (cf. 16.4.5) and hence there exists a continuous one-parameter unitary group U^1 in $\mathcal{H}$ such that

$$\omega_{(s,0)}([u]) = [U^1(s)u], \forall u \in \tilde{\mathcal{H}}, \forall s \in \mathbb{R}$$

(cf. 16.4.11). Similarly, there exists a continuous one-parameter unitary group U^2 in $\mathcal{H}$ such that

$$\omega_{(0,v)}([u]) = [U^2(v)u], \forall u \in \tilde{\mathcal{H}}, \forall v \in \mathbb{R}.$$

Now we fix two continuous one-parameter unitary groups U^1, U^2 in $\mathcal{H}$ which satisfy the above conditions. For all $(s,v) \in \mathbb{R}^2$, we have

$$\omega_{(s,0)}\omega_{(0,v)} = \omega_{(s,v)} = \omega_{(0,v)}\omega_{(s,0)},$$

and hence

$$[U^1(s)U^2(v)u] = \omega_{(s,v)}([u]) = [U^2(v)U^1(s)u], \forall u \in \tilde{\mathcal{H}},$$

and hence there exist $z_{s,v} \in \mathbb{T}$ so that

$$U^2(v)U^1(s) = z_{s,v}U^1(s)U^2(v)$$

(cf. 10.9.6). Since $z_{s,v}$ is uniquely determined by this condition (recall that U^1 and U^2 have been fixed), we have the function

$$\mathbb{R}^2 \ni (s,v) \mapsto \varphi(s,v) := z_{s,v} \in \mathbb{T},$$

which is such that

$$U^2(v)U^1(s) = \varphi(s,v)U^1(s)U^2(v), \forall (s,v) \in \mathbb{R}^2.$$

We see that, for all $s, s' \in \mathbb{R}$ and all $v \in \mathbb{R}$,

$$\begin{aligned}
&\varphi(s,v)^{-1}\varphi(s+s',v)U^1(s+s')U^2(v)\\
&= \varphi(s,v)^{-1}U^2(v)U^1(s+s')\\
&= \varphi(s,v)^{-1}U^2(v)U^1(s)U^1(s') = U^1(s)U^2(v)U^1(s')\\
&= \varphi(s',v)U^1(s)U^1(s')U^2(v) = \varphi(s',v)U^1(s+s')U^2(v),
\end{aligned}$$

and hence

$$\varphi(s+s',v) = \varphi(s,v)\varphi(s',v).$$

Similarly we can prove that

$$\varphi(s,v+v') = \varphi(s,v)\varphi(s,v'), \forall v, v' \in \mathbb{R}, \forall s \in \mathbb{R}.$$

Moreover, let $(s,v) \in \mathbb{R}^2$ and a sequence $\{(s_n, v_n)\}$ in $\mathbb{R}^2$ be so that $(s,v) = \lim_{n\to\infty}(s_n, v_n)$, and fix $u \in \tilde{\mathcal{H}}$. By condition ug_2 in 16.1.1 and by 16.4.7, we have

$$\lim_{n\to\infty} U^1(s_n)U^2(v_n)u = U^1(s)U^2(v)u \text{ and}$$

$$\lim_{n\to\infty} U^2(v_n)U^1(s_n)u = U^2(v)U^1(s)u,$$

and hence, by the continuity of the inner product,

$$\begin{aligned}
\varphi(s,v) &= \left(U^1(s)U^2(v)u|\varphi(s,v)U^1(s)U^2(v)u\right) = \left(U^1(s)U^2(v)u|U^2(v)U^1(s)u\right)\\
&= \lim_{n\to\infty}\left(U^1(s_n)U^2(v_n)u|U^2(v_n)U^1(s_n)u\right) = \lim_{n\to\infty}\varphi(s_n, v_n).
\end{aligned}$$

Thus, the function φ satisfies condition a, b, c, d in 20.3.1. Therefore there exists a unique $\mu \in \mathbb{R}$ so that

$$U^2(v)U^1(s) = e^{i\mu sv}U^1(s)U^2(v), \forall (s,v) \in \mathbb{R}^2.$$

Letting $U^1_\mu := U^1$ and $U^2_\mu := U^2$, we have two continuous one-parameter unitary groups as in condition b.

Now let $\tilde{U}^1$ and $\tilde{U}^2$ be two continuous one-parameter unitary groups such that

$$\omega_{(s,v)}([u]) = [\tilde{U}^1(s)\tilde{U}^2(v)u], \forall u \in \tilde{\mathcal{H}}, \forall (s,v) \in \mathbb{R}^2.$$

Then

$$[\tilde{U}^1(s)u] = [U^1(s)u], \forall u \in \tilde{\mathcal{H}}, \forall s \in \mathbb{R},$$

and hence there exists a function $\gamma_1 : \mathbb{R} \to \mathbb{T}$ such that

$$\tilde{U}^1(s) = \gamma_1(s)U^1(s), \forall s \in \mathbb{R}.$$

Similarly there exists a function $\gamma_2 : \mathbb{R} \to \mathbb{T}$ such that

$$\tilde{U}^2(v) = \gamma_2(v)U^2(v), \forall v \in \mathbb{R}.$$

Then we have, for all $(s, v) \in \mathbb{R}^2$,

$$\tilde{U}^2(v)\tilde{U}^1(s) = \gamma_2(v)\gamma_1(s)U^2(v)U^1(s)$$
$$= \gamma_2(v)\gamma_1(s)e^{i\mu sv}U^1(s)U^2(v) = e^{i\mu sv}\tilde{U}^1(s)\tilde{U}^2(v).$$

This proves statement c.

$b \Rightarrow a$: We assume condition b. Then, for all $(s_1, v_1), (s_2, v_2) \in \mathbb{R}^2$, we have

$$\omega_{(s_1,v_1)}(\omega_{(s_2,v_2)}([u])) = [U^1_\mu(s_1)U^2_\mu(v_1)U^1_\mu(s_2)U^2_\mu(v_2)u]$$
$$= [e^{i\mu s_2 v_1}U^1_\mu(s_1 + s_2)U^2_\mu(v_1 + v_2)u]$$
$$= [U^1_\mu(s_1 + s_2)U^2_\mu(v_1 + v_2)u]$$
$$= \omega_{(s_1+s_2,v_1+v_2)}([u]), \forall u \in \tilde{\mathcal{H}},$$

and hence

$$\omega_{(s_1,v_1)} \circ \omega_{(s_2,v_2)} = \omega_{(s_1+s_2,v_1+v_2)}.$$

Moreover, let $\{(s_n, v_n)\}$ be a sequence in $\mathbb{R}^2$ so that $\lim_{n\to\infty}(s_n, v_n) = (0, 0)$. Then, for all $u \in \tilde{\mathcal{H}}$, we have

$$\lim_{n\to\infty} U^1_\mu(s_n)U^2_\mu(v_n)u = u$$

(cf. condition ug_2 in 16.1.1 and 16.4.7), and hence

$$\lim_{n\to\infty} \tau([u], \omega_{(s_n,v_n)}([u])) = \lim_{n\to\infty} |(u|U^1_\mu(s_n)U^2_\mu(v_n)u)| = 1.$$

This completes the proof. $\square$

As a consequence of 20.3.2, having a homomorphism

$$\mathbb{R}^2 \ni (s, v) \mapsto \omega_{(s,v)} \in \operatorname{Aut} \hat{\mathcal{H}}$$

(where $\mathcal{H}$ is the Hilbert space in which the system is represented) endowed with the required continuity property, as implied by requisite qp_1, is the same as having a real number μ and a pair of continuous one-parameter unitary groups U^1_μ, U^2_μ in $\mathcal{H}$ such that

$$U^2_\mu(v)U^1_\mu(s) = e^{i\mu sv}U^1_\mu(s)U^2_\mu(v), \forall(s, v) \in \mathbb{R}^2. \tag{7}$$

The link is

$$\omega_{(s,v)}([u]) = [U^1_\mu(s)U^2_\mu(v)u], \forall u \in \tilde{\mathcal{H}}, \forall(s, v) \in \mathbb{R}^2. \tag{8}$$

Furthermore, the real number μ is uniquely determined by the homomorphism $\mathbb{R}^2 \ni (s, v) \mapsto \omega_{(s,v)} \in \operatorname{Aut} \hat{\mathcal{H}}$.

Then we suppose we have all this and we examine requisite qp_2. This requisite corresponds to the idea that there are no "internal degrees of freedom" for the quantum particle we want to represent, or equivalently that each non-trivial observable must exhibit some connection with "external space". Now we prove that this requisite is equivalent to the joint irreducibility of the pair of continuous one-parameter unitary groups U_μ^1, U_μ^2. Indeed, suppose that requisite qp_2 is fulfilled. For each orthogonal projection P in $\mathcal{H}$ there exists a proposition π such that $P_\pi = P$ (this is the surjectivity of the mapping in 19.3.1b), and hence there exists the yes-no observable α_π, for which $A_{\alpha_\pi} = P$ (cf. 19.3.8). Moreover, in the range of the projection valued measure of the self-adjoint operator P there are only the projections $\mathbb{O}_\mathcal{H}, \mathbb{1}_\mathcal{H}, P, \mathbb{1}_\mathcal{H} - P$ (cf. 19.3.8), and the only projections which are multiples of the identity operator are $\mathbb{O}_\mathcal{H}$ and $\mathbb{1}_\mathcal{H}$. In view of all this, requisite qp_2 entails that, for $P \in \mathscr{P}(\mathcal{H})$, the following implications are true

$$[U_\mu^1(s)PU_\mu^1(-s) = U_\mu^2(v)PU_\mu^2(-v) = P, \forall(s,v) \in \mathbb{R}^2] \Rightarrow$$
$$[U_\mu^1(s)U_\mu^2(v)PU_\mu^2(-v)U_\mu^1(-s) = P, \forall(s,v) \in \mathbb{R}^2] \Rightarrow P \in \{\mathbb{O}_\mathcal{H}, \mathbb{1}_\mathcal{H}\},$$

and this is the condition that the pair U_μ^1, U_μ^2 is jointly irreducible (cf. 17.3.1). Conversely, suppose that the pair U_μ^1, U_μ^2 is jointly irreducible and that an observable α is Galilei-invariant. Then we have

$$U_\mu^1(s)P_{\alpha(E)}U_\mu^1(-s) = U_\mu^2(v)P_{\alpha(E)}U_\mu^2(-v) = P_{\alpha(E)}, \forall E \in \mathcal{A}(d_\mathbb{R}), \forall(s,v) \in \mathbb{R}^2,$$

and hence, by the irreducibility of the pair U_μ^1, U_μ^2,

$$P^{A_\alpha}(E) = P_{\alpha(E)} \in \{\mathbb{O}_\mathcal{H}, \mathbb{1}_\mathcal{H}\}, \forall E \in \mathcal{A}(d_\mathbb{R}),$$

and hence, by 17.3.2,

$$\exists \lambda \in \mathbb{R} \text{ so that } A_\alpha = \lambda \mathbb{1}_\mathcal{H}.$$

Thus, requisite qp_2 is fulfilled.

In view of the discussion above we assume that, if the homomorphism from $\mathbb{R}^2$ to $\mathrm{Aut}\,\hat{\mathcal{H}}$ of the quantum-particle model is implemented by a real number μ and a pair of continuous one-parameter groups U_μ^1, U_μ^2 as in 7 and 8, then this pair is jointly irreducible. The next proposition proves that this implies $\mu \neq 0$.

20.3.3 Proposition. *Let $\mu \in \mathbb{R}$ and let an irreducible pair U_μ^1, U_μ^2 of continuous one-parameter unitary groups in a Hilbert space $\mathcal{H}$ be so that*

$$U_\mu^2(v)U_\mu^1(s) = e^{i\mu sv}U_\mu^1(s)U_\mu^2(v), \forall(s,v) \in \mathbb{R}^2.$$

If $\mathcal{H}$ is neither a zero nor a one-dimensional linear space then $\mu \neq 0$.

Proof. The proof is by contraposition. Since the pair U_μ^1, U_μ^2 is jointly irreducible, the following implication holds true:

$$[B \in \mathcal{B}(\mathcal{H}) \text{ and } [B, U_\mu^1(s)] = [B, U_\mu^2(v)] = \mathbb{O}_\mathcal{H}, \forall(s,v) \in \mathbb{R}^2] \Rightarrow$$
$$[\exists \alpha \in \mathbb{C} \text{ so that } B = \alpha \mathbb{1}_\mathcal{H}]$$

(cf. 17.3.5). Now suppose $\mu = 0$. Then, for all $(s,v) \in \mathbb{R}^2$, $U_\mu^1(s)$ and $U_\mu^2(v)$ satisfy the first condition for B above, and hence $U_\mu^1(s)$ and $U_\mu^2(v)$ are multiples of $\mathbb{1}_\mathcal{H}$. Since the pair U_μ^1, U_μ^2 is jointly irreducible, this implies that $\mathcal{H}$ is either a zero or a one-dimensional linear space. $\qquad\square$

Thus, requisites qp_1 and qp_2 are fulfilled if we have a non-zero real number μ and an irreducible pair of continuous one-parameter unitary groups U^1_μ, U^2_μ with property 7. The next proposition proves that an irreducible pair of continuous one-parameter groups with this property does exist, for each $\mu \neq 0$.

20.3.4 Proposition. *Let $\mu \in \mathbb{R} - \{0\}$ and let $\mathcal{H}$ be a non-zero Hilbert space.*
For two mappings $U^1 : \mathbb{R} \to \mathcal{U}(\mathcal{H})$ and $U^2 : \mathbb{R} \to \mathcal{U}(\mathcal{H})$, the following conditions are equivalent:

(a) U^1 and U^2 are continuous one-parameter unitary groups and
$$U^2(v)U^1(s) = e^{i\mu sv}U^1(s)U^2(v), \forall(s, v) \in \mathbb{R}^2;$$

(b) there exists a pair of self-adjoint operators A, B in $\mathcal{H}$ which is a representation of WCR and also so that
$$U^1(s) = U^B(-s), \forall s \in \mathbb{R}, \quad and \quad U^2(v) = U^A(\mu v), \forall v \in \mathbb{R}.$$

If these conditions are satisfied, then

(c) the pair U^1, U^2 is irreducible iff the pair A, B is irreducible.

Proof. $a \Rightarrow b$: We assume condition a and define the mappings
$$V^1 : \mathbb{R} \to \mathcal{U}(\mathcal{H})$$
$$x \mapsto V^1(x) := U^1(-x)$$
and
$$V^2 : \mathbb{R} \to \mathcal{U}(\mathcal{H})$$
$$y \mapsto V^2(y) := U^2(\mu^{-1}y).$$
It is obvious that V^1 and V^2 are continuous one-parameter unitary groups. Moreover,
$$V^2(y)V^1(x) = e^{-ixy}V^1(x)V^2(y), \forall(x, y) \in \mathbb{R}^2.$$
Then, the generators A, B of V^2, V^1 respectively satisfy condition a in 20.1.1, and hence the pair of self-adjoint operators A, B is a representation of WCR, and also
$$U^1(s) = V^1(-s) = U^B(-s), \forall s \in \mathbb{R},$$
and
$$U^2(v) = V^2(\mu v) = U^A(\mu v), \forall v \in \mathbb{R}.$$
$b \Rightarrow a$: We assume condition b. Then it is obvious that U^1 and U^2 are continuous one-parameter unitary groups. Moreover,
$$U^2(v)U^1(s) = U^A(\mu v)U^B(-s) = e^{i\mu vs}U^B(-s)U^A(\mu v)$$
$$= e^{i\mu vs}U^1(s)U^2(v), \forall(s, v) \in \mathbb{R}^2.$$
c: We assume conditions a and b. Obviously, the ranges of the mappings U^1 and U^B are equal and so are the ranges of U^2 and U^A. Then condition c follows from the equivalence between the irreducibility of the pair U^A, U^B and the irreducibility of the pair A, B (cf. 20.1.5). $\square$

Thus, having a representation of a quantum system which matches requisites qp_1 and qp_2 is equivalent to having a non-zero real number μ and a pair of self-adjoint operators A, B which are an irreducible representation of WCR in the Hilbert space $\mathcal{H}$ of the system. If μ, A, B are given, then the homomorphism

$$\mathbb{R}^2 \ni (s, v) \mapsto \omega_{(s,v)} \in \operatorname{Aut} \hat{\mathcal{H}}$$

which is included in requisite qp_1 is given by

$$\omega_{(s,v)}([u]) := [U^B(-s)U^A(\mu v)u], \forall u \in \tilde{\mathcal{H}}, \forall(s, v) \in \mathbb{R}^2. \tag{9}$$

Since we know that irreducible representations of WCR do exist (one of them was constructed in 20.1.7), now we also know that the quantum particle model can be implemented as far as requisites qp_1 and qp_2 are concerned. Moreover, now we know that *the Hilbert space required by the quantum particle model is necessarily separable and of denumerable orthogonal dimension* (cf. 20.2.5b).

Now we discuss requisite qp_3, assuming that requisites qp_1 and qp_2 are fulfilled as above by means of a non-zero real number μ and a pair of self-adjoint operators A, B which are an irreducible representation of WCR. We claim that a non-relativistic quantum particle "has mass m", where m is a positive real number which is fixed for that particle, and that its "position" and its "linear momentum" (or, briefly, "momentum") are two observables which satisfy the following conditions, where q denotes the observable position and p denotes the observable momentum (A_q, A_p are the self-adjoint operators that represent q, p respectively):

$$U^B(-s)P^{A_q}(E)U^B(s) = P^{A_q}(E + s), \forall E \in \mathcal{A}(d_{\mathbb{R}}), \forall s \in \mathbb{R}, \tag{10}$$

$$U^A(\mu v)P^{A_q}(E)U^A(-\mu v) = P^{A_q}(E), \forall E \in \mathcal{A}(d_{\mathbb{R}}), \forall v \in \mathbb{R}, \tag{11}$$

$$U^B(-s)P^{A_p}(E)U^B(s) = P^{A_p}(E), \forall E \in \mathcal{A}(d_{\mathbb{R}}), \forall s \in \mathbb{R}, \tag{12}$$

$$U^A(\mu v)P^{A_p}(E)U^A(-\mu v) = P^{A_p}(E + mv), \forall E \in \mathcal{A}(d_{\mathbb{R}}), \forall v \in \mathbb{R}. \tag{13}$$

Of course we need to explain the reasoning behind the claim we have just made.

We imagine the observable "position" of a quantum particle in one dimension as the abstract representation of an array of detectors which are ideally infinitely small and cover the whole of one-dimensional space, and which are so that one and only of them reacts immediately after a copy of the system has been prepared (this is exactly one of the particle-like aspects of a quantum particle). Now suppose that a copy has been prepared, that this happens at time zero for an inertial observer O (time zero is chosen for simplicity), and that O assigns the result x to "his" observable "position" if x is the space coordinate, according to his own frame of reference, of the detector that has reacted (the detectors are classical objects and therefore each of them has a position at all times, in the classical sense). Then, on the basis of the same procedure and of the same reaction, the inertial observer $g_{(s,0)}(O)$ (for any $s \in \mathbb{R}$) will assign the result $x + s$ to "his" observable "position", while the inertial observer $g_{(0,v)}(O)$ (for any $v \in \mathbb{R}$) will assign the result x to "his" observable "position" (at time zero, the space-origins of the frames of reference of O

and of $g_{(0,v)}(O)$ are in the same place). Well, this is exactly what would happen if the system were a classical particle and its position were being measured (by means of detectors suitable for classical particles). By this analogy with the classical case, the quantum observable described above is given the name of "position" and is denoted by q.

We imagine the observable "momentum" of a quantum particle in one dimension as the abstract representation of a pair of detectors which are placed, each time a measurement is made, on either side of the apparatus that prepares a copy of the system; no forces act on these detectors and therefore they move with constant velocities with respect to all inertial observers before a copy has been prepared; moreover, they are so that one and only one of them reacts by changing its velocity after a copy of the system has been prepared (this too is one of the particle-like aspects of a quantum particle). Now suppose that a copy has been prepared. Then each inertial observer assigns, as result to "his" observable "momentum", the difference between the values of the momentum of the detector that has reacted, measured by him (with respect to his own frame of reference) before and after the reaction (the detectors are classical objects and therefore each of them has a momentum at all times, in the classical sense). If an inertial observer O assigns the result y to "his" observable "momentum", then on the basis of the same reaction the inertial observer $g_{(s,0)}(O)$ (for any $s \in \mathbb{R}$) will assign the same result to "his" observable "momentum" (the frames of reference of O and of $g_{(s,0)}(O)$ are stationary with respect to each other), while the inertial observer $g_{(0,v)}(O)$ (for any $v \in \mathbb{R}$) will assign a different result. The idea that the particle "has mass m" is supported by the experimental evidence that the result assigned by $g_{(0,v)}(O)$ is $y + mv$, where m is a positive number independent of v. Well, this is exactly what would happen if the system were a classical particle of mass m and its momentum were being measured (by techniques suitable for classical particles). By this analogy with the classical case, the quantum particle is said to "have mass m" and the quantum observable described above is given the name of "momentum" and is denoted by p.

These observations give the transformations $t^q_{(s,0)}$, $t^q_{(0,v)}$, $t^p_{(s,0)}$, $t^p_{(0,v)}$ (for all $s, v \in \mathbb{R}$) to be used in the covariance conditions for the observables q and p with respect to the subgroups S and V of the kinematic Galilei group G (and hence, with respect to any other element of G). They are:

$$t^q_{(s,0)}(x) = x + s, \forall x \in \mathbb{R}, \forall s \in \mathbb{R};$$
$$t^q_{(0,v)}(x) = x, \forall x \in \mathbb{R}, \forall v \in \mathbb{R};$$
$$t^p_{(s,0)}(y) = y, \forall y \in \mathbb{R}, \forall s \in \mathbb{R};$$
$$t^p_{(0,v)}(y) = y + mv, \forall y \in \mathbb{R}, \forall v \in \mathbb{R}.$$

Then we see that conditions 10, 11, 12, 13 are nothing else than the covariance conditions for the observables q and p with respect to S and V, since $U^B(-s)$ and $U^A(\mu v)$ are implementations of $\omega_{(s,0)}$ and $\omega_{(0,v)}$ respectively (cf. 9).

The outcome of the discussion above is that the structure of the quantum particle model for a definite mass m is equivalent to the structure made up by a pair of self-adjoint operators A, B which are an irreducible representation of WCR, together with a non-zero real number μ and a pair of self-adjoint operators A_q and A_p which satisfy conditions 10, 11, 12, 13 with the pair A, B. The operators A_q and A_p represent the observables position and momentum of the quantum particle, while the pair A, B and the number μ are related as in 9 to the homomorphism from $\mathbb{R}^2$ to $\mathrm{Aut}\,\hat{\mathcal{H}}$ that represents the action of the kinematic Galilei group in the quantum particle model.

The question of existence and uniqueness of implementations of these structures will be addressed on the basis of the next proposition.

20.3.5 Proposition. *Let A, B be an irreducible representation of WCR in a Hilbert space $\mathcal{H}$, and let m be a fixed positive number.*

(A) Let $\mu \in \mathbb{R}-\{0\}$. For two self-adjoint operators T_1, T_2 in $\mathcal{H}$, the set of conditions listed in a_1 is equivalent to the set listed in a_2:

(a_1)

$$U^B(-s)P^{T_1}(E)U^B(s) = P^{T_1}(E+s), \forall E \in \mathcal{A}(d_\mathbb{R}), \forall s \in \mathbb{R}, \tag{14}$$
$$U^A(\mu v)P^{T_1}(E)U^A(-\mu v) = P^{T_1}(E), \forall E \in \mathcal{A}(d_\mathbb{R}), \forall v \in \mathbb{R}, \tag{15}$$
$$U^B(-s)P^{T_2}(E)U^B(s) = P^{T_2}(E), \forall E \in \mathcal{A}(d_\mathbb{R}), \forall s \in \mathbb{R}, \tag{16}$$
$$U^A(\mu v)P^{T_2}(E)U^A(-\mu v) = P^{T_2}(E+mv), \forall E \in \mathcal{A}(d_\mathbb{R}), \forall v \in \mathbb{R}; \tag{17}$$

(a_2)

$$\exists k_1 \in \mathbb{R} \text{ so that } T_1 = A + k_1 \mathbb{1}_\mathcal{H},$$
$$\exists k_2 \in \mathbb{R} \text{ so that } T_2 = \mu^{-1}mB + k_2 \mathbb{1}_\mathcal{H}.$$

(B) Let $\mu \in \mathbb{R} - \{0\}$ and $k_1, k_2 \in \mathbb{R}$. Then:

(b_1) there exists $U \in \mathcal{U}(\mathcal{H})$ so that

$$U(A + k_1 \mathbb{1}_\mathcal{H})U^{-1} = A \text{ and } U(\mu^{-1}mB + k_2 \mathbb{1}_\mathcal{H})U^{-1} = \mu^{-1}mB;$$

(b_2) for every $U \in \mathcal{U}(\mathcal{H})$, the equalities in b_1 are equivalent to the equations

$$UU^A(\mu v)U^{-1} = e^{-ik_1\mu v}U^A(\mu v), \forall v \in \mathbb{R},$$
$$UU^B(-s)U^{-1} = e^{im^{-1}\mu k_2 s}U^B(-s), \forall s \in \mathbb{R}.$$

(C) Let $\mu \in \mathbb{R} - \{0\}$. Then:

(c_1) there exists $W \in \mathcal{A}(\mathcal{H})$ so that

$$WAW^{-1} = A \text{ and } W(\mu^{-1}mB)W^{-1} = -\mu^{-1}mB;$$

(c_2) for every $W \in \mathcal{A}(\mathcal{H})$, the equalities in c_1 are equivalent to the equations

$$WU^A(\mu v)W^{-1} = U^A(-\mu v), \forall v \in \mathbb{R},$$
$$WU^B(-s)W^{-1} = U^B(-s), \forall s \in \mathbb{R}.$$

(D) *Let* $\mu_1, \mu_2 \in \mathbb{R} - \{0\}$ *and suppose* $\mu_1 \neq \pm\mu_2$. *Then, there does not exist any unitary or antiunitary operator* V *in* $\mathcal{H}$ *so that*

$$VAV^{-1} = A \quad \text{and} \quad V(\mu_2^{-1}mB)V^{-1} = \mu_1^{-1}mB.$$

Proof. Preliminarily we recall that, since the pair A, B is a representation of WCR,

$$U^A(t)U^B(s) = e^{-its}U^B(s)U^A(t), \forall (s,t) \in \mathbb{R}^2. \tag{18}$$

A: In view of 15.4.1, 16.1.8a, 16.3.1 (or, more directly, in view of 20.1.1), the conditions in a_1 can be written equivalently as

$$U^B(-s)U^{T_1}(x)U^B(s) = e^{-isx}U^{T_1}(x), \forall x \in \mathbb{R}, \forall s \in \mathbb{R}, \tag{14'}$$
$$U^A(\mu v)U^{T_1}(x)U^A(-\mu v) = U^{T_1}(x), \forall x \in \mathbb{R}, \forall v \in \mathbb{R}, \tag{15'}$$
$$U^B(-s)U^{T_2}(y)U^B(s) = U^{T_2}(y), \forall y \in \mathbb{R}, \forall s \in \mathbb{R}, \tag{16'}$$
$$U^A(\mu v)U^{T_2}(y)U^A(-\mu v) = e^{-imvy}U^{T_2}(y), \forall y \in \mathbb{R}, \forall v \in \mathbb{R}. \tag{17'}$$

Now we prove the equivalence between a_1 and a_2.

$a_1 \Rightarrow a_2$: We assume that the equations in a_1 hold true. From 14' and 18 we have

$$U^B(s)(U^{T_1}(x)U^A(-x)) = e^{isx}U^{T_1}(x)U^B(s)U^A(-x)$$
$$= (U^{T_1}(x)U^A(-x))U^B(s), \forall s \in \mathbb{R}, \forall x \in \mathbb{R},$$

and from 15' (with the change of variable $z := \mu v$) we have

$$U^A(z)(U^{T_1}(x)U^A(-x)) = (U^{T_1}(x)U^A(-x))U^A(z), \forall z \in \mathbb{R}, \forall x \in \mathbb{R}.$$

Since the pair U^A, U^B is jointly irreducible (cf. 20.1.5), by 17.3.5 this implies that there exists a function $\alpha : \mathbb{R} \to \mathbb{C}$ so that

$$U^{T_1}(x)U^A(-x) = \alpha(x)\mathbb{1}_{\mathcal{H}}, \forall x \in \mathbb{R}.$$

It is easy to see that α is a continuous homomorphism from the additive group $\mathbb{R}$ to the multiplicative group $\mathbb{T}$. Hence, by 16.2.3, there exists $k_1 \in \mathbb{R}$ so that

$$\alpha(x) = e^{ik_1x}, \forall x \in \mathbb{R},$$

and hence so that

$$U^{T_1}(x) = e^{ik_1x}U^A(x), \forall x \in \mathbb{R},$$

and hence, in view of 16.1.8a and 16.1.5d (cf. also 16.1.7), so that

$$T_1 = A + k_1\mathbb{1}_{\mathcal{H}}.$$

Similarly, on the basis of 16', 17', 18 and of the joint irreducibility of the pair U^A, U^B, we can prove that there exists $k_2 \in \mathbb{R}$ so that

$$U^{T_2}(y) = e^{ik_2y}U^B(\mu^{-1}my), \forall y \in \mathbb{R},$$

and hence, in view of 16.1.8a,b, so that

$$U^{T_2}(y) = U^{\tilde{B}}(y), \forall y \in \mathbb{R},$$

if $\tilde{B} := \mu^{-1}mB + k_2\mathbb{1}_{\mathcal{H}}$. In view of 16.1.5d (cf. also 16.1.7), this implies

$$T_2 = \mu^{-1}mB + k_2\mathbb{1}_{\mathcal{H}}.$$

$a_2 \Rightarrow a_1$: We assume that the equalities in a_2 hold true. Then, in view of 16.1.8a,b, we have

$$U^{T_1}(x) = e^{ik_1x}U^A(x), \forall x \in \mathbb{R}, \tag{19}$$

and

$$U^{T_2}(y) = e^{ik_2y}U^B(\mu^{-1}my), \forall y \in \mathbb{R}. \tag{20}$$

From 18 and 19 we have

$$U^B(-s)U^{T_1}(x) = e^{ik_1x}e^{-isx}U^A(x)U^B(-s)$$
$$= e^{-isx}U^{T_1}(x)U^B(-s), \forall x \in \mathbb{R}, \forall s \in \mathbb{R},$$

and this is condition 14'. Conditions 15' follows immediately from 19, and so does condition 16' from 20. Finally, from 18 and 20 we have

$$U^A(\mu v)U^{T_2}(y) = e^{ik_2y}e^{-imvy}U^B(\mu^{-1}my)U^A(\mu v)$$
$$= e^{-imvy}U^{T_2}(y)U^A(\mu v), \forall y \in \mathbb{R}, \forall v \in \mathbb{R},$$

and this is condition 17'.

B: It is actually more convenient to prove first that $U \in \mathcal{U}(\mathcal{H})$ exists so that the equations in b_2 are true, and second that these equations are equivalent to the equalities in b_1.

We define the unitary operator $U := U^B(-k_1)U^A(m^{-1}\mu k_2)$. From 18 we have

$$UU^A(\mu v)U^{-1} = U^B(-k_1)U^A(\mu v)U^B(k_1) = e^{-ik_1\mu v}U^A(\mu v), \forall v \in \mathbb{R},$$

and

$$U^{-1}U^B(-s)U = U^A(-m^{-1}\mu k_2)U^B(-s)U^A(m^{-1}\mu k_2)$$
$$= e^{-im^{-1}\mu k_2 s}U^B(-s), \forall s \in \mathbb{R},$$

or equivalently

$$UU^B(-s)U^{-1} = e^{im^{-1}\mu k_2 s}U^B(-s), \forall s \in \mathbb{R}.$$

Thus, the equations in b_2 are proved.

For *any* $U \in \mathcal{U}(\mathcal{H})$, in view of 16.1.8a,b, the equations in b_2 (with the changes of variables $x := \mu v$ and $y := -s$) are equivalent to the equations

$$UU^A(x)U^{-1} = U^{A'}(x), \forall x \in \mathbb{R},$$
$$UU^B(y)U^{-1} = U^{B'}(y), \forall y \in \mathbb{R},$$

if $A' := A - k_1\mathbb{1}_{\mathcal{H}}$ and $B' := B - m^{-1}\mu k_2\mathbb{1}_{\mathcal{H}}$. In view of 16.3.1, this equations are equivalent to the equalities

$$UAU^{-1} = A - k_1\mathbb{1}_{\mathcal{H}},$$
$$UBU^{-1} = B - m^{-1}\mu k_2\mathbb{1}_{\mathcal{H}},$$

and hence to the equalities in b_1.

C: We consider the Schrödinger representation of WCR discussed in 20.1.7 and define the mapping

$$C : L^2(\mathbb{R}) \to L^2(\mathbb{R})$$

$$[f] \mapsto C[f] := [\overline{f}],$$

which is obviously an antiunitary operator in $L^2(\mathbb{R})$. Moreover, it is obvious that

$$C^{-1} = C,$$

$$CQC^{-1} = Q,$$

$$CP_0 C^{-1} = -P_0.$$

Now, the operator CPC^{-1} is self-adjoint (cf. 12.5.4) and so is the operator $-P$. Moreover, both the operators CPC^{-1} and $-P$ extend the essentially self-adjoint operator $-P_0$. Since the self-adjoint extension of an essentially self-adjoint operator is unique (cf. 12.4.11c), this proves the equality

$$CPC^{-1} = -P.$$

In view of 20.2.4a there exists $V \in \mathcal{U}(\mathcal{H}, L^2(\mathbb{R}))$ so that

$$VAV^{-1} = Q \text{ and } VBV^{-1} = P.$$

Then the operator $W := V^{-1}CV$ is an antiunitary operator in $\mathcal{H}$ (cf. 10.3.16c), and we have

$$WAW^{-1} = V^{-1}CQC^{-1}V = V^{-1}QV = A,$$

$$WBW^{-1} = V^{-1}CPC^{-1}V = -V^{-1}PV = -B,$$

and hence

$$W(\mu^{-1}mB)W^{-1} = -\mu^{-1}mB.$$

Thus, the equalities in c_1 are proved.

For *any* $W \in \mathcal{A}(\mathcal{H})$, the equalities in c_1 are equivalent to the equations in c_2, in view of 16.3.1 (note that $U^{-B}(s) = U^B(-s)$, in view of 16.1.8b).

D: We notice that $D_{[A,B]} \neq \{0_\mathcal{H}\}$. Indeed, for the Schrödinger representation, $D_{[Q,P]}$ is dense in $L^2(\mathbb{R})$ (it contains $[\varphi]$ for all $\varphi \in \mathcal{S}(\mathbb{R})$; then use 11.3.3 and 10.6.5b) and hence $D_{[A,B]}$ is dense in $\mathcal{H}$, by 20.2.4a. Then let $f \in D_{[A,B]}$ be such that $f \neq 0_\mathcal{H}$ and suppose that, for $\mu_1, \mu_2 \in \mathbb{R} - \{0\}$, there exists $V \in \mathcal{U}\mathcal{A}(\mathcal{H})$ (this operator has nothing to do with the operator denoted by the same symbol in the proof of statement C) so that

$$VAV^{-1} = A \text{ and } V(\mu_2^{-1}mB)V^{-1} = \mu_1^{-1}mB.$$

Then (cf. 3.2.10b_1,b_2',b_3)

$$\mu_1^{-1}m[A, B] = \mu_2^{-1}mV[A, B]V^{-1}.$$

Therefore, $V^{-1}f \in D_{[A,B]}$ and, in view of 20.1.3b,

$$i\mu_1^{-1}mf = \mu_1^{-1}m[A, B]f = \mu_2^{-1}mV[A, B]V^{-1}f = \pm i\mu_2^{-1}mf,$$

where the plus or minus sign depends on whether V is unitary or antiunitary. Hence, $\mu_1 = \pm\mu_2$. This proves statement D by contraposition. $\qquad\square$

In what follows, m is a fixed positive number.

Since we know that irreducible representations of WCR exist, propositions 20.3.2, 20.3.3, 20.3.4, 20.3.5A ($a_2 \Rightarrow a_1$) prove that our model for a quantum particle of mass m can be implemented. In fact they prove that, for each $\mu \in \mathbb{R} - \{0\}$, there exist a homomorphism from $\mathbb{R}^2$ to Aut $\hat{\mathcal{H}}$ (where $\mathcal{H}$ is the Hilbert space of the representation) which is implemented by an irreducible representation A, B of WCR as in 9, and self-adjoint operators A_q, A_p which satisfy the covariance conditions 10, 11, 12, 13. For a given irreducible representation A, B of WCR, 20.3.5A actually determines all the pairs of self-adjoint operators that can be used as representatives A_q and A_p of the observables position q and momentum p: they are the pairs

$$(A + k_1 \mathbb{1}_{\mathcal{H}}, \mu^{-1} m B + k_2 \mathbb{1}_{\mathcal{H}}), \forall \mu \in \mathbb{R} - \{0\}, \forall k_1, k_2 \in \mathbb{R}.$$

Now, it seems that not only do we have pairs which fit our scheme, but we have too many of them: what value of μ and which pair $(A + k_1 \mathbb{1}_{\mathcal{H}}, \mu^{-1} m B + k_2 \mathbb{1}_{\mathcal{H}})$ should be used to represent a quantum particle of mass m?

For a fixed value of $\mu \in \mathbb{R} - \{0\}$, 20.3.5$b_1$ shows that all the pairs related to that value of μ are unitarily equivalent to each other. If we transform, by means of a unitary operator, a pair related to a value of μ to another related to the same value, perhaps we want to transform the operators $U^B(-s)$ and $U^A(\mu v)$ as well, since they are implementations of the automorphisms $\omega_{(s,0)}$ and $\omega_{(0,v)}$ respectively. Then 20.3.5b_2 shows that these operators get just multiplied by factors in $\mathbb{T}$, and hence in the new representation the same automorphisms $\omega_{(s,0)}$ and $\omega_{(0,v)}$ are implemented as in the old one. In view of all this and of 19.3.23, we consider two pairs with the same value of μ to be equivalent for the description of position and momentum of a quantum particle of mass m.

For a fixed value of $\mu \in \mathbb{R} - \{0\}$, 20.3.5$b_1, c_1$ show that all the pairs defined by a value of μ are antiunitarily equivalent to all the pairs defined by the opposite value. If we transform, by means of an antiunitary operator, a pair defined by a value of μ into another defined by the opposite value, perhaps also in this case we want to transform the operators $U^B(-s)$ and $U^A(\mu v)$. Then 20.3.5b_2, c_2 show that these operators, besides being multiplied by inessential multiplicative factors in $\mathbb{T}$, get changed into $U^B(-s)$ and $U^A(-\mu v)$; now, these operators implement the automorphism $\omega_{(s,0)}$ and $\omega_{(0,-v)}$. Thus it appears that, in the new representation, the direction of the flow of time has been reversed. However, since we do not want to study time evolution, in view of 19.3.23 we consider pairs defined by opposite values of μ to be equivalent.

Finally, 20.3.5D (together with 20.3.5b_1, c_1) shows that, if $\mu_1, \mu_2 \in \mathbb{R} - \{0\}$ are such that $\mu_1 \neq \pm \mu_2$, then no pair defined by μ_2 is either unitarily of antiunitarily equivalent to any pair defined by μ_1.

In view of all this, for a given irreducible representation A, B of WCR, we need only consider the pairs

$$(A, \mu^{-1} m B), \text{ for all } \mu > 0,$$

but we must consider all of them. For each $\mu > 0$, they implement in inequivalent ways our quantum particle model of mass m, with the assignements

$$A_q := A \text{ and } A_p := \mu^{-1}mB,$$

and with the kinematic Galilei group represented by the automorphism of $\hat{\mathcal{H}}$ defined by

$$\omega_{(s,v)}([u]) := [U^B(-s)U^A(\mu v)u], \forall u \in \tilde{\mathcal{H}}, \forall (s, v) \in \mathbb{R}^2.$$

In addition, we recall that the Stone–von Neumann uniqueness theorem (cf. 20.2.4a) implies that, if a pair $\tilde{A}, \tilde{B}$ is a different irreducible representation of WCR, then for each $\mu \in \mathbb{R}-\{0\}$ the pair $(\tilde{A}, \mu^{-1}m\tilde{B})$ is unitarily equivalent to the pair $(A, \mu^{-1}mB)$, and so is the pair $U^{\tilde{A}}, U^{\tilde{B}}$ to the pair U^A, U^B. Thus, nothing is gained by considering irreducible representations of WCR different from A, B.

Since the quantum models defined by different positive values of μ are not unitarily or antiunitarily equivalent, the question is now what value of μ should be used to represent a quantum particle of mass m. Mathematical reasoning cannot help us here, and in fact we must turn to experimental outcomes. Indeed suppose that, for a definite positive value of μ, we have the representation

$$A_q := A \text{ and } A_p := \mu^{-1}mB.$$

This representation yields statistical estimates that do depend on μ. For instance, from 20.1.3a and 19.3.13a we have

$$\Delta_\sigma q \Delta_\sigma p \geq \frac{1}{2}\mu^{-1}m,$$

for each state σ in which both q and p are evaluable. The above representation of q and p is in accordance with experimental evidence for the value

$$\mu := \hbar^{-1}m,$$

where $\hbar := (2\pi)^{-1}h$ and h is Planck's constant. Thus, also on the basis of experimental physics, the quantum particle model of mass m is given by

$$A_q := A,$$
$$A_p := \hbar B,$$
$$\omega_{(s,v)}([u]) := [U^B(-s)U^A(\hbar^{-1}mv)u], \forall u \in \tilde{\mathcal{H}}, \forall (s, v) \in \mathbb{R}^2.$$

20.3.6 Remarks.

(a) The discussion above shows that *the Hilbert space, in which a non-relativistic quantum particle without internal degrees of freedom is represented, is necessarily separable and of denumerable dimension.*

(b) In the representation of a quantum particle of mass m obtained above, the value m of the mass does not have a role in the operators A_q and A_p which represent the observables position and momentum. However it does in the implementation of the homomorphism from $\mathbb{R}^2$ to Aut $\hat{\mathcal{H}}$ which represents the kinematic Galilei group. On the basis of 20.3.5D it is easy to see that implementations related to different values of m are not unitarily or antiunitarily equivalent.

(c) Historically, the first mathematical representation of a quantum particle of mass m was obtained in what is now called the Schrödinger representation of WCR. In this representation we have

$$\mathcal{H} := L^2(\mathbb{R}), \quad A := Q, \quad B := P,$$

where Q and P are the operators discussed in 20.1.7, and hence

$$A_q := Q,$$
$$A_p := \hbar P,$$
$$\omega_{(s,v)}([f]) := [U^P(-s)U^Q(\hbar^{-1}mv)f],$$
$$\text{for each ray } [f] \text{ in } L^2(\mathbb{R}) \text{ and each } (s,v) \in \mathbb{R}^2$$

(here, for $f \in \mathcal{L}^2(\mathbb{R})$, the element $[f]$ of $L^2(\mathbb{R})$ is denoted by the same symbol f; here, for a unit vector f of $L^2(\mathbb{R})$, $[f]$ denotes the ray that contains f). More explicitly, for all $f \in L^2(\mathbb{R})$ and all $(s,v) \in \mathbb{R}^2$, we have (assuming for simplicity $D_f = \mathbb{R}$, cf. 8.2.12)

$$(U^P(-s)U^Q(\hbar^{-1}mv)f)(x) = e^{i\hbar^{-1}mv(x-s)}f(x-s), \forall x \in \mathbb{R}$$

(cf. 20.1.7).

If a pure state σ is represented by a ray $[f_\sigma]$ in $L^2(\mathbb{R})$, it is possible to put a direct statistical interpretation on the function $|f_\sigma|^2$. In fact, from 15.3.4A and from Section 14.5 we see that

$$P_{q(E)}f_\sigma = P^Q(E)f_\sigma = \chi_E f_\sigma, \forall E \in \mathcal{A}(d_\mathbb{R}),$$

and hence

$$p(q(E),\sigma) = (f_\sigma|P_{q(E)}f_\sigma) = \int_\mathbb{R} \chi_E |f_\sigma|^2 dm, \forall E \in \mathcal{A}(d_\mathbb{R}).$$

We recall that $p(q(E),\sigma)$ is the probability that a measurement of the position of the particle yields a result in E (for any $E \in \mathcal{A}(d_\mathbb{R})$) when the particle is prepared in the state σ (crf. 19.1.9a), or the probability of "finding the particle in E" when the particle "is in the state σ".

If a pure state σ is represented by a ray $[f_\sigma]$ in $L^2(\mathbb{R})$ and if $\tilde{f}_\sigma$ denotes the vector Ff_σ (F is the Fourier transform on $L^2(\mathbb{R})$), it is possible to put a direct statistical interpretation on the function $|\tilde{f}_\sigma|^2$ too. In fact, in view of 16.1.8b and 15.4.1, and of the equality $P = F^{-1}QF$ (cf. 20.1.7), we have

$$P_{p(E)} = P^{\hbar P}(E) = P^P(\hbar^{-1}E) = F^{-1}P^Q(\hbar^{-1}E)F, \forall E \in \mathcal{A}(d_\mathbb{R}),$$

and hence

$$p(p(E), \sigma) = \left(f_\sigma | P_{p(E)} f_\sigma \right) = \left(\tilde{f}_\sigma | P^Q (\hbar^{-1} E) \tilde{f}_\sigma \right)$$

$$= \int_{\mathbb{R}} \chi_{\hbar^{-1} E} |\tilde{f}_\sigma|^2 dm, \forall E \in \mathcal{A}(d_{\mathbb{R}}),$$

and $p(p(E), \sigma)$ is the probability that a measurement of the momentum of the particle yields a result in E when the particle is prepared in the state σ.

Bibliography

Apostol, T. M. (1974). *Mathematical Analysis*, 2nd edn. (Addison Wesley Publishing Company, Reading).

Bargmann, V. (1954). *On the Unitary Ray Representations of Continuous Groups* (Annals of Mathematics 59), p.1-46.

Bargmann, V. (1964). *Note on Wigner's Theorem on Symmetry Operations* (Journal of Mathematical Physics 5), p.862-868.

Berberian, S. K. (1999). *Fundamentals of Real Analysis* (Springer, New York).

Dirac, P. A. M. (1958, 1947, 1935, 1930). *The Principles of Quantum Mechanics* (Clarendon Press, Oxford).

Greenberg, M. J. and Harper, J. R. (1981). *Algebraic Topology: a First Course* (Addison-Wesley Publishing Company, Redwood City, California).

Heisenberg, W. (1925). *Über Quantentheoretische Umdeutung Kinematischer und Mechanischer Beziehungen* (Zeitschr. f. Phys. 33), p.879-893.

Hewitt, E. and Stromberg, K. (1965). *Real and Abstract Analysis* (Springer-Verlag, New York).

Hilbert, D., Neumann, J. v., and Nordheim, L. (1927). *Über die Grundlagen der Quantenmechanik* (Mathematische Annalen 98(1)), p.1-30.

Holevo, A. S. (1982). *Probabilistic and Statistical Aspects of Quantum Theory.* (North-Holland Publishing Company, Amsterdam), second English edition published by Scuola Normale Superiore, Pisa, 2011.

Horn, R. A. and Johnson, C. R. (2013). *Matrix Analysis*, 2nd edn. (Cambridge University Press).

Jauch, J. M. (1968). *Foundations of Quantum Mechanics* (Addison-Wesley Publishing Company, Reading, Massachusetts).

Jordan, P. (1926). *Über Kanonische Transformationen in der Quantenmechanik* (Zeitschr. f. Phys. 37), p.383-386.

Mackey, G. W. (1978). *Unitary Group Representations in Physics, Probability, and Number Theory* (The Benjamin/Cummings Publishing Company, Reading, Massachusetts).

Munkres, J. R. (1991). *Analysis on Manifolds* (Addison-Wesley Publishing Company, Redwood City, California).

Parthasarathy, K. R. (2005). *Introduction to Probability and Measure* (Hindustan Book Agency (India), New Delhi).

Pauli, W. (1933). *Die Allgemeinen Prinzipien der Wellenmechanik* (Handbuch der Physik 24), p.83-272.

Reed, M. and Simon, B. (1980, 1972). *Methods of Modern Mathematical Physics I: Functional Analysis* (Academic Press, New York).

Reeh, H. (1988). *A Remark Concerning Canonical Commutation Relation* (Journal of Mathematical Physics 29), p.1535-1536.

Riesz, F. and Sz.-Nagy, B. (1972). *Leçons d'Analyse Fonctionnelle*, 6th edn. (Akadémiai Kiadó, Budapest), English translation of the 2nd edition: *Functional Analysis*, Dover Publications, New York, 1990.

Royden, H. L. (1988). *Real Analysis* (Macmillan Publishing Company, New York).

Rudin, W. (1976). *Principles of Mathematical Analysis*, 3rd edn. (McGraw-Hill Book Company, New York).

Rudin, W. (1987). *Real and Complex Analysis*, 3rd edn. (McGraw-Hill Book Company, New York).

Schrödinger, E. (1926). *Über das Verhältnis der Heisenberg-Born-Jordanschen Quantenmechanik zu der Meinen* (Annalen der Physik 79), p.734-756.

Shilov, G. E. (1973). *Mathematical Analysis, Vol. 1* (MIT Press, Cambridge), (re-issued as *Elementary Real and Complex Analysis* by Dover Publications, Mineola, 1996).

Shilov, G. E. (1974). *Mathematical Analysis, Vol. 2* (MIT Press, Cambridge), (re-issued as *Elementary Functional Analysis* by Dover Publications, Mineola, 1996).

Shilov, G. E. and Gurevich, B. L. (1966). *Integral, Measure, and Derivative: a Unified Approach* (Prentice Hall, Englewood Cliffs), (re-issued by Dover Publications, Mineola, 1977).

Simmons, G. F. (1963). *Introduction to Topology and Modern Analysis* (McGraw-Hill Book Company, New York).

Stone, M. H. (1930). *Linear Transformations in Hilbert Space III: Operational Methods and Group Theory* (Proc. Nat. Acad. Sci. U.S.A. 16), p. 172-175.

Thaller, B. (1992). *The Dirac Equation* (Springer-Verlag, Berlin).

von Neumann, J. (1931). *Die Eindeutigkeit der Schrödingerschen Operatoren* (Math. Ann. 104), p.570-578.

von Neumann, J. (1932). *Mathematische Grundlagen der Quantenmechanik* (Springer-Verlag, Berlin), pages are quoted from the English translation, *Mathematical Foundations of Quantum Mechanics*, Princeton University Press, Princeton, 1955.

von Neumann, J. (1950). *Functional Operators*, Vol. 2 (Princeton University Press, Princeton).

Weidmann, J. (1980). *Linear Operators in Hilbert Spaces* (Springer-Verlag, New York).

Weyl, H. (1927). *Quantenmechanik und Gruppentheorie* (Zeitschr. f. Phys. 46), p.1-46.

Wichmann, E. H. (1971). *Quantum Physics: Berkeley Physics Course, Vol. 4* (McGraw-Hill, New York).

Index

Printed in the United States
By Bookmasters